Lecture Notes in Computer Science 9472

Advanced Research in Computing and Software Science

Subline of Lecture Notes in Computer Science

More information about this series at http://www.springer.com/series/7407

Khaled Elbassioni · Kazuhisa Makino (Eds.)

Algorithms and Computation

26th International Symposium, ISAAC 2015 Nagoya, Japan, December 9–11, 2015 Proceedings

Editors
Khaled Elbassioni
Masdar Institute
Abu Dhabi
United Arab Emirates

Kazuhisa Makino
Kyoto University
Kyoto
Japan

ISSN 0302-9743 ISSN 1611-3349 (electronic)
Lecture Notes in Computer Science
ISBN 978-3-662-48970-3 ISBN 978-3-662-48971-0 (eBook)
DOI 10.1007/978-3-662-48971-0

Library of Congress Control Number: 2015955363

LNCS Sublibrary: SL1 – Theoretical Computer Science and General Issues

Springer Heidelberg New York Dordrecht London

Printed on acid-free paper

Springer-Verlag GmbH Berlin Heidelberg is part of Springer Science+Business Media
(www.springer.com)

Preface

The 26th International Symposium on Algorithms and Computation (ISAAC 2015) was held during December 9–11, 2015, in Nagoya, Japan. ISAAC is a well-established annual international symposium that covers a wide range of topics in algorithms and theory of computation, and provides a forum for researchers where they can exchange ideas in this active research community.

The technical program of the symposium included 65 contributed papers selected by the Program Committee from 180 submissions received in response to the call for papers. Each submission was reviewed by at least three Program Committee members, possibly with the assistance of external reviewers. Two special issues of *Algorithmica* and *International Journal of Computational Geometry and Applications* will publish selected papers among the contributed ones. The best paper award was given to "Trading Off Worst and Expected Cost in Decision Tree Problems" by Aline Saettler, Eduardo Laber, and Ferdinando Cicalese. In addition to selected papers, the program also included three invited talks by Constantinos Daskalakis, Ravindran Kannan, and Thomas Rothvoss.

We thank all the people who made this meeting possible: the authors for submitting papers, the Program Committee members and external reviewers for volunteering their time to review the submissions. We would like to extend special thanks to Conference Co-chairs Tomio Hirata and Ken-ichi Kawarabayashi, Organizing Committee members, and all conference volunteers for their dedication that made ISAAC 2015 a successful event.

We would like also to acknowledge the sponsors of ISAAC 2015 for their generous support: Kayamori Foundation of Information Science Advancement, Support Center for Advanced Telecommunications Technology Research (SCAT), The Telecommunications Advancement Foundation, Nagoya University, Special Interest Group on Algorithms (SIGAL) of IPSJ, and the Technical Committee of Theoretical Foundation of Computing (COMP) of IEICE. The symposium was partially supported by the following grants: Grant-in-Aid for Scientific Research on Innovative Areas, Exploring the Limits of Computation (ELC), JST CREST Foundations of Innovative Algorithms for Big Data, and the JST ERATO Kawarabayashi Large Graph Project.

Last but not least, we would like to acknowledge the excellent environment provided by EasyChair, without which an enormous amount of very time consuming work would have been necessary to finish this task.

December 2015

Khaled Elbassioni
Kazuhisa Makino

Organization

Program Committee

Mark de Berg	Eindhoven University of Technology, The Netherlands
Hubert Chan	University of Hong Kong, Hong Kong, SAR China
Kai-Min Chung	Academia Sinica, Taiwan
Adrian Dumitrescu	University of Wisconsin-Milwaukee, USA
Khaled Elbassioni	Masdar Institute, UAE
Amr Elmasry	Alexandria University, Egypt
Xin Han	Dalian University of Technology, China
Seok-Hee Hong	University of Sydney, Australia
Tsan-Sheng Hsu	Academia Sinica, Taiwan
Zhiyi Huang	University of Hong Kong, Hong Kong, SAR China
Naonori Kakimura	University of Tokyo, Japan
Shuji Kijima	Kyushu University, Japan
Tamas Kiraly	Eotvos University, Hungary
Yusuke Kobayashi	University of Tsukuba, Japan
Michal Koucky	Academy of Sciences of the Czech Republic, Czech Republic
Michael Lampis	Université Paris Dauphine, France
Meena Mahajan	The Institute of Mathematical Sciences, Chennai, India
Kazuhisa Makino	Kyoto University, Japan
Julian Mestre	University of Sydney, Australia
Danupon Nanongkai	KTH Royal Institute of Technology, Sweden
Konstantinos Panagiotou	Ludwig Maximilian University of Munich, Germany
Periklis Papakonstantinou	Rutgers University, USA
Seth Pettie	University of Michigan, USA
Saurabh Ray	NYU Abu Dhabi, UAE
Thomas Sauerwald	University of Cambridge, UK
Saket Saurabh	Institute of Mathematical Sciences, India
Jeong Seop Sim	Inha University, Republic of Korea
Rahul Shah	Louisiana State University, USA
Rene Sitters	Vrije Universiteit, The Netherlands
Rob van Stee	University of Leicester, UK
Chaitanya Swamy	University of Waterloo, Canada
Hing-Fung Ting	University of Hong Kong, Hong Kong, SAR China
Sang Won Bae	Kyonggi University, Republic of Korea
Yuichi Yoshida	National Institute of Informatics, Japan
Guochuan Zhang	Zhejiang University, China

Additional Reviewers

Agarwal, Pankaj
Ahmadian, Sara
Ahn, Hee-Kap
Aldecoa, Rodrigo
Allender, Eric
Angelini, Patrizio
Austrin, Per
Balko, Martin
Balogh, János
Bei, Xiaohui
Bein, Wolfgang
Belmonte, Rémy
Benchetrit, Yohann
Bender, Michael
Bhaskar, Umang
Biswas, Sudip
Bohler, Cecilia
Bohmova, Katerina
Bonichon, Nicolas
Bonnet, Edouard
Bose, Prosenjit
Brandes, Ulrik
Bringmann, Karl
Buchbinder, Niv
Bérczi, Kristóf
Cabello, Sergio
Cannon, John
Caskurlu, Bugra
Chan, Timothy M.
Chen, Ke
Chen, Lin
Chen, Yi-Hsiu
Cheng, Christine
Cheng, Siu-Wing
Choi, Joonsoo
Chrobak, Marek
Chun, Jinhee
Columbus, Tobias
Cseh, Ágnes
Dadush, Daniel
Das, Gautam K.
Datta, Samir
De Boer, Frank
de Wolf, Ronald
Dominguez-Sal, David
Doty, David
Drange, Pål Grønås
Dregi, Markus Sortland
Driemel, Anne
Dutta, Kunal
Dvorak, Zdenek
Elbassioni, Khaled
Erlebach, Thomas
Ernst, Matthias
Etscheid, Michael
Evans, William
Fagerberg, Rolf
Farshi, Mohammad
Feijao, Pedro
Fernau, Henning
Francis, Mathew
Fu, Norie
Fuhs, Carsten
Fukunaga, Takuro
Gajarský, Jakub
Ganguly, Arnab
Gaur, Daya
Gavinsky, Dmitry
Ghaffari, Mohsen
Ghosh, Anirban
Gibson, Matt
Golovach, Petr
Grabowski, Szymon
Green, Oded
Groß, Martin
Gudmundsson, Joachim
Gupta, Sushmita
Han, Li
Harks, Tobias
Hatano, Kohei
Haviv, Ishay
Heeringa, Brent
Herskovics, Dávid
Hirai, Hiroshi
Hu, Xiaocheng
Ilcinkas, David
Ito, Takehiro
Jacob, Riko
Jain, Rahul
Jankó, Zsuzsanna
Jiang, Tao
Johnson, Matthew
Jones, Mitchell
Jørgensen, Allan Grønlund
Kamali, Shahin
Kamiyama, Naoyuki
Kamiński, Marcin
Kanade And Andrew Wan, Varun
Kang, Ning
Kao, Mong-Jen
Kaplan, Haim
Katajainen, Jyrki
Kayal, Neeraj
Khalafallah, Ayman
Khramtcova, Elena
Kim, Eun Jung
Kim, Jin Wook
Kim, Sung-Ryul
Kimelfeld, Benny
Kirkpatrick, David
Kis, Tamas
Kisfaludi-Bak, Sándor
Kiyomi, Masashi
Klein, Philip
Klein, Rolf
Kociumaka, Tomasz
Kolay, Sudeshna
Korman, Matias
Kothari, Robin
Kratsch, Dieter
Krebs, Andreas
Krizanc, Danny
Krohmer, Anton
Kudahl, Christian
Kumar, Nirman
Kwon, O-Joung
Lagerqvist, Victor
Langerman, Stefan
Laudahn, Moritz
Lee, Inbok
Lee, Mun-Kyu
Li, Minming
Lin, Cedric Yen-Yu

Linhares, Andre
Liu, Sixue
Loff, Bruno
Loiseau, Patrick
Lu, Pinyan
Lu, Zaixin
Luo, Yuping
Löffler, Maarten
Mandal, Ritankar
Manea, Florin
Manlove, David
Matuschke, Jannik
Mehrabi, Saeed
Mercas, Robert
Meunier, Pierre-Étienne
Meyerhenke, Henning
Miele, Andrea
Miltzow, Tillmann
Mitsou, Valia
Miyazaki, Shuichi
Mouawad, Amer
Mozes, Shay
Mustafa, Nabil
Müller-Hannemann, Matthias
Na, Joong Chae
Nabeshima, Hidetomo
Nakagawa, Kotaro
Nandy, Subhas
Narayanaswamy, N.S.
Nielsen, Jesper Sindahl
Ning, Li
O'Rourke, Joseph
Okamoto, Yoshio
Ollinger, Nicolas
Ordyniak, Sebastian
Otachi, Yota
Pajak, Dominik
Pajor, Thomas
Pap, Gyula
Park, Heejin
Patil, Manish
Philip, Geevarghese
Pilipczuk, Marcin
Pontecorvi, Matteo
Popa, Alexandru
Pralat, Pawel
Provençal, Xavier
Pálvölgyi, Dömötör
Raichel, Benjamin
Rajgopal, Ninad
Raman, Rajiv
Raman, Venkatesh
Rao B.V., Raghavendra
Rescigno, Adele
Riondato, Matteo
Rosenbaum, David
Roy, Sasanka
Rytter, Wojciech
Räcke, Harald
Sabharwal, Yogish
Samal, Robert
Santhanam, Rahul
Sarma, Jayalal
Satti, Srinivasa Rao
Saurabh, Nitin
Scarpa, Giannicola
Schaefer, Marcus
Schmidt, Jens M.
Schmidt, Melanie
Schneider, Stefan
Scozzari, Francesca
Sgall, Jiří
Sharma, Roohani
Shin, Chan-Su
Shurbevski, Aleksandar
Sidiropoulos, Anastasios
Sikora, Florian
Son, Wanbin
Song, Fang
Speidel, Leo
Sreekumaran, Harikrishnan
Staals, Frank
Stephens-Davidowitz, Noah
Sudholt, Dirk
Syed Mohammad, Meesum
Takamatsu, Mizuyo
Tamaki, Suguru
Tamura, Takeyuki
Tan, Guang
Tan, Li-Yang
Teruyama, Junichi
Tewari, Raghunath
Thankachan, Sharma V.
Thierauf, Thomas
Toth, Csaba
Toussaint, Godfried
Tsin, Yung
Tulsiani, Madhur
Uchizawa, Kei
Ueckerdt, Torsten
Uehara, Ryuhei
Uno, Takeaki
Upfal, Eli
Vadhan, Salil
Valicov, Petru
van Stee, Rob
Variyam, Vinodchandran
Varré, Jean-Stéphane
Vialette, Stéphane
Vigneron, Antoine
Wang, Haitao
Wang, Mingqiang
Watrigant, Rémi
Watson, Thomas
Wieder, Udi
Wiese, Andreas
Witt, Carsten
Wu, Xiaodi
Wu, Xiaowei
Wu, Zhiwei Steven
Wulff-Nilsen, Christian
Xia, Ge
Xiao, Mingyu
Yamanaka, Katsuhisa
Yang, Guang
Ye, Deshi
Zehavi, Meirav
Zenklusen, Rico
Zhang, Qiang
Zhang, Qin
Zhang, Shengyu
Zhang, Yong
Zhou, Gelin
Zhu, Binhai
Zhu, Shenglong
Żyliński, Paweł

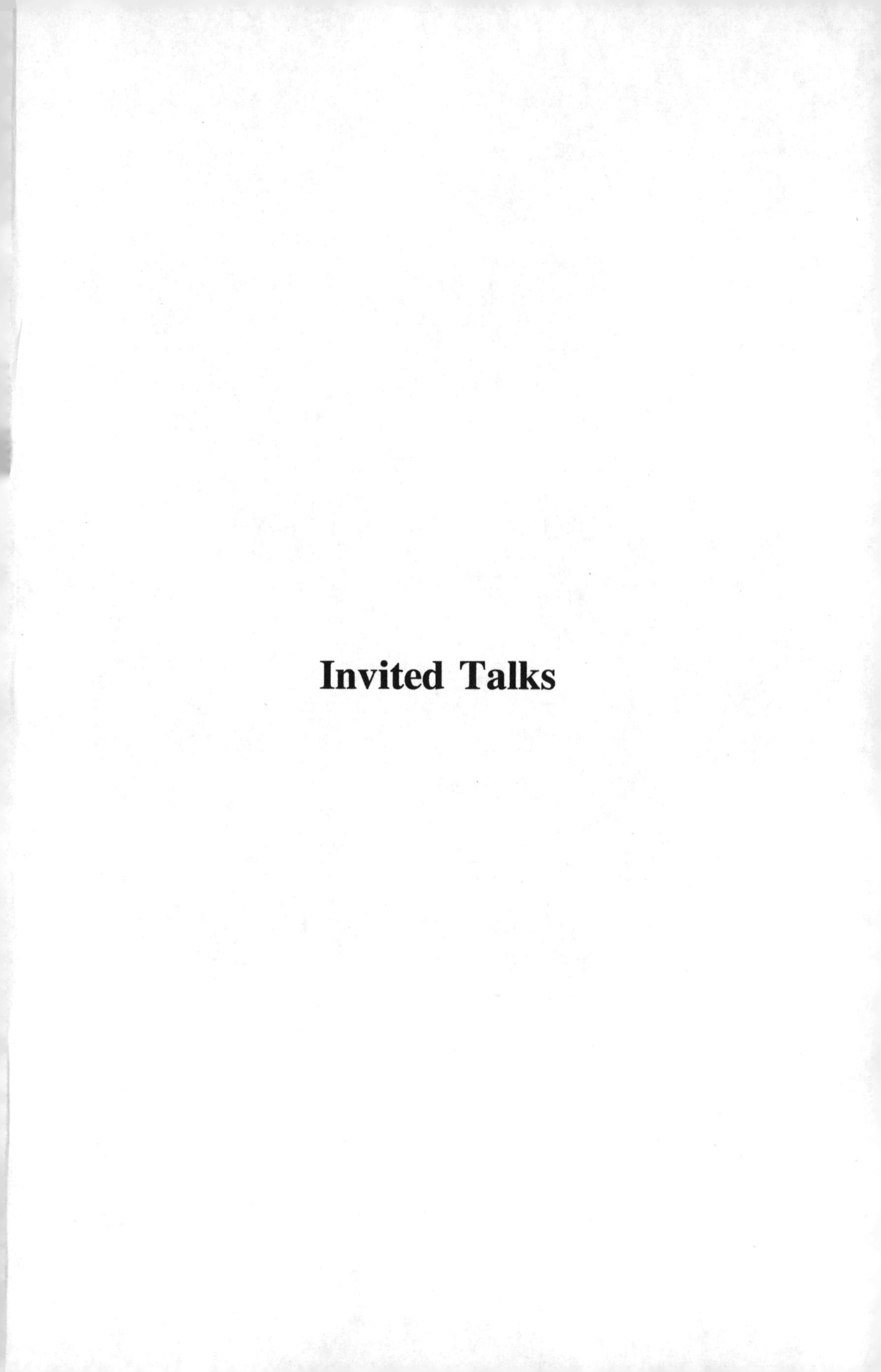

Invited Talks

Soft Clustering: Models and Algorithms

Ravi Kannan

Microsoft Research, India
kannan@microsoft.com

Abstract. Traditional Clustering partitions a set of data points into clusters. In a number of problems, each data point does not belong just to one cluster, but is best described as belonging fractionally to many clusters. One way of formalizing a Soft Clustering problem is via Non-negative Matrix Factorization (NMF). In NMF, we hypothesize that the given non-negative m by n data matrix A (with each column a data point) is approximately equal to the product of two non-negative matrices B and C, where B is m by k and C is k by n for some k much smaller than m and n. The columns of B may be thought of centers of the k clusters, each column of C is the weight a data point puts on each cluster. In general, A may or may not be stochastic. Topic Modeling is a special case of NMF, where matrix C is stochastic (often with Dirichlet distribution for the weights) with independent columns. Overlapping Community Detection is another area with a soft clustering core: given whether each pair among a population knowns each other or not, one is to soft cluster them into communities. Each of these problems and others has considerable literature and different models.

The purpose of this talk is to first discuss the problems and models and argue that there are two essential common elements:

(i) Each data point puts much higher weight on one cluster than the others. [We call this the Dominant Cluster hypothesis.]

(ii) Each cluster has a set of dominant features. Features are coordinates. Dominant features have high values. [We call this the Dominant Features hypothesis.]

We formalize these assumptions plus a few technical ones. We then give an algorithm to find the dominant cluster of each data point. The main difficulty is that traditional mixture models (Gaussian mixtures, Stochastic Block Models etc.), are hard clustering models and so the expected value of all data points in each cluster is the same. Here, the two hypotheses imply that expected value in the dominant part of a cluster is higher, but certainly not the same. This can make the spread inside each cluster larger and can also reduce the inter-cluster separation. Thus traditional hard clustering methods do not work.

We solve this problem with a crucial thresholding step at the outset which finds a suitable threshold for each feature. We show that after thresholding, we can do Singular Value Decomposition to find a decent starting clustering and then run Lloyd's algorithm which we will prove will yield a good clustering. We will also present empirical evidence that the assumptions do hold as well as results on the effectiveness of the algorithm.

Parts Joint with subsets of T. Bansal, C. Bhattacharyya, N. Goyal, J. Pani.

Computing on Strategic Inputs

Constantinos Daskalakis

EECS and CSAIL, MIT
costis@csail.mit.edu

Abstract. Algorithmic mechanism design centers around the following question: How much harder is optimizing an objective over inputs that are furnished by strategic agents compared to when the inputs to the optimization are known? The challenge is that, when agents controlling the inputs care about the output of the optimization, they may misreport them to influence the output. How does one take into account strategic behavior in optimization?

We present computationally efficient, approximation-preserving reductions from mechanism design (i.e.optimizing over strategic inputs) to algorithm design (i.e. optimizing over known inputs) in general Bayesian settings. We also explore whether structural properties about optimal mechanisms can be inferred from these reductions. As an application, we present extensions of Myerson's celebrated single-item auction to multi-item settings.

Lower Bounds on the Size of Linear Programs

Thomas Rothvoß

University of Washington, Seattle
rothvoss@uw.edu

Abstract. Linear programs are at the heart of combinatorial optimization as they allow to model a large class of polynomial time solvable problems such as flows, matchings and matroids. The concept of LP duality lead in many cases to structural insights that in turn lead to specialized polynomial time algorithms. In practice, general LP solvers turn out to be very competitive for many problems, even in cases in which specialized algorithms have the better theoretical running time. Hence it is particularly interesting to model problems with as few linear constraints as possible. For example, it is possible to model the convex hull of all spanning trees in a graph using $O(n^3)$ many linear constraints and variables.

A natural question that emerges is which polytopes do *not* admit a compact formulation. The first progress was made by [Yannakakis 1991] who showed that any *symmetric* extended formulation for the matching polytope and the TSP polytope must have exponential size. Conveniently, this allowed to reject a sequence of flawed $\mathbf{P} = \mathbf{NP}$ proofs, which claimed to have (complicated) polynomial size LPs for TSP.

The major breakthrough by [Fiorini, Massar, Pokutta, Tiwary and de Wolf 2012] showed that several well studied polytopes, including the correlation polytope and the TSP polytope, have exponential extension complexity (without relying on the symmetry assumption). More precisely, they show that the *rectangle covering lower bound* for the correlation polytope is exponential, for which they use known tools from communication complexity such as Razborov's *rectangle corruption lemma* [Razborov 1990].

A completely independent line of research was given by [Chan, Lee, Raghavendra and Steurer 2013] who use techniques from Fourier analysis to show that for constraint satisfaction problems, known integrality gaps for the Sherali-Adams LP translate to lower bounds for any LPs of a certain size. For example they show that no LP of size $n^{O(\log n/\log\log n)}$ can approximate MaxCut better than $2 - \varepsilon$. This is particularly interesting as in contrast the gap of the SDP relaxation is around 1.13 [Goemans, Williamson 1995].

A very prominent polytope in combinatorial optimization is the *perfect matching polytope*, which is the convex hull of all characteristic vectors of perfect matchings in a complete n-node graph $G = (V, E)$. A seminal work of [Edmonds 1965] gives an exact description of this polytope with $2^{\Theta(n)}$ many inequalities. Finally, [Rothvoss 2013] proved that the extension complexity is $2^{\Theta(n)}$ as well, while only a quadratic lower bound was known before.

Contents

Computational Geometry I

Data Structures

Combinatorial Optimization and Approximation Algorithms I

Computational Complexity I

Graph Drawing and Planar Graphs

Computational Complexity II

Computational Geometry I

An Optimal Algorithm for Tiling the Plane with a Translated Polyomino

Andrew Winslow(✉)

Université Libre de Bruxelles, 1050 Brussels, Belgium
andrew.winslow@ulb.ac.be

Abstract. We give a $O(n)$-time algorithm for determining whether translations of a polyomino with n edges can tile the plane. The algorithm is also a $O(n)$-time algorithm for enumerating all regular tilings, and we prove that at most $\Theta(n)$ such tilings exist.

1 Introduction

A *plane tiling* is a partition of the plane into shapes each congruent to a fixed set of *tiles*. As the works of M.C. Escher attest, plane tilings are both artistically beautiful and mathematically interesting (see [20] for a survey of both aspects). In the 1960s, Golomb [8] initiated the study of *polyomino* tiles: polygons whose edges are axis-aligned and unit-length.

Building on work of Berger [2], Golomb [9] proved that no algorithm exists for determining whether a set of polyomino tiles has a plane tiling. Ollinger [17] proved that this remains true even for sets of at most 5 tiles. It is a long-standing conjecture that there exists an algorithm for deciding whether a single tile admits a plane tiling (see [10,11]).

Motivated by applications in parallel computing, Shapiro [21] studied tilings of polyomino tiles on a common integer lattice using translated copies of a polyomino. For the remainder of the paper, only these tilings are considered. Ollinger [17] proved that no algorithm exists for determining whether sets of at most 11 tiles admit a tiling, while Wijshoff and van Leeuwen [22] obtained a polynomial-time-testable criterion for a single tile to admit a tiling. Beauquier and Nivat [1,7] improved on the result of Wijshoff and van Leeuwen by giving a simpler criterion called the *Beauquier-Nivat criterion.*

Informally, a tile satisfies the Beauquier-Nivat criterion if it can be surrounded by copies of itself (see Fig. 1). Such a surrounding must correspond to a *regular* tiling (also called *isohedral*) in which all tiles share an identical neighborhood. Using a naive algorithm, the Beauquier-Nivat criterion can be applied to a polyomino with n vertices in $O(n^4)$ time.

The $O(n^4)$ algorithm of [1] is implicit; the main achievement of [1] is a concise characterization of exact tiles, akin to Conway's criterion (see [19]). Gambini and Vuillon [6] gave an improved $O(n^2)$-time algorithm utilizing structural and algorithmic results on words describing boundaries of polyominoes. Around the

K. Elbassioni and K. Makino (Eds.): ISAAC 2015, LNCS 9472, pp. 3–13, 2015.
DOI: 10.1007/978-3-662-48971-0_1

Fig. 1. A polyomino tile (dark gray), a surrounding of the tile (gray), and the induced regular tiling (white).

same time, Brlek et al. [3,4] also used a word-based approach to achieve $O(n)$-time algorithms for two special cases: (1) the boundary contains no consecutive repeated sections larger than $O(\sqrt{n})$, and (2) testing a restricted version of the Beauquier-Nivat criterion (surroundable by just four copies). Provençal [18] further improved on the algorithm of Gambini and Vuillon for the general case, obtaining $O(n \log^3(n))$ running time. In a recent survey of the combinatorics of Escher's tilings, Massé et al. [16] conjecture that a $O(n)$-time algorithm exists. In this work, we confirm their conjecture by giving such an algorithm (Theorem 2).

The algorithm doubles as an algorithm for enumerating all surroundings (regular tilings) of the polyomino. As part of the proof of the algorithm's running time, we prove a claim of Provençal [18] that the number of surroundings of a tile with itself is $O(n)$ (Corollary 1). This complements the tight bounds on a special class of surroundings by Blondin Massé et al. [14,15], and proves that our $O(n+k)$-time algorithm for enumerating all k surroundings (Lemma 10) is also a $O(n)$-time algorithm.

2 Definitions

Here we give precise formulations of terms used throughout the paper. The definitions are similar to those of Beauquier and Nivat [1] and Brlek et al. [4].

2.1 Words

A *letter* is a symbol $x \in \Sigma = \{\mathbf{u}, \mathbf{d}, \mathbf{l}, \mathbf{r}\}$. The *complement* of a letter x, written $\bar{x}$, is defined by the following bijection on Σ: $\bar{\mathbf{u}} = \mathbf{d}$, $\bar{\mathbf{r}} = \mathbf{l}$, $\bar{\mathbf{d}} = \mathbf{u}$, and $\bar{\mathbf{l}} = \mathbf{r}$.

A *word* is a sequence of letters and the *length* of a word W, denoted $|W|$, is the number of letters in W. For an integer $i \in \{1, 2, \ldots, |W|\}$, $W[i]$ refers to the ith letter of W and $W[-i]$ refers to the ith from the last letter of W.

The notation l^k or W^k denotes the word consisting of k repeats of a letter l or word W, respectively.

There are several functions mapping a word W to another word of the same length. The *complement* of W, written $\overline{W}$, is the word obtained by replacing each letter of W with its complement. The *reverse* of W, written $\widetilde{W}$, are the letters of W in reverse order. The *backtrack* of W, written $\widehat{W}$, is defined as $\widehat{W} = \overline{\widetilde{W}}$. Note that for any two words X and Y, $\widehat{AB} = \widehat{B}\widehat{A}$.

2.2 Factors

A *factor of* W is an occurrence of a word in W, written $X \preceq W$. For integers $1 \leq i, j \leq |W|$ with $i \leq j$, $W[i..j]$ denotes the factor of W from $W[i]$ to $W[j]$, inclusive. A factor X *starts* or *ends* at $W[i]$ if $W[i]$ is the first or last letter of X, respectively.

Two factors $X, Y \preceq W$ may refer the same letters of W or merely have the same letters in common. In the former case, X and Y are *equal*, written $X = Y$, while in the latter, X and Y are *congruent*, written $X \equiv Y$. For instance, if $W =$ **uuulruuu** then $W[1..3] \equiv W[6..8]$. A *factorization* of W is a partition of W into consecutive factors F_1 through F_k, written $W = F_1 F_2 \ldots F_k$.

2.3 Special Words and Factors

A word X is a *prefix* or *suffix* of a word W provided $W = XU$ or $W = UX$, respectively. A word X *is a period of* W provided $|X| \leq |W|$ and W is a prefix of X^k for some $k \geq 1$ (introduced in [13]). Alternatively, X is a prefix of W and $W[i] = W[i + |X|]$ for all $1 \leq i \leq |W| - |X|$.

A factor $X \preceq W$ is a *prefix* if X starts at $W[1]$, written $X \preceq_{\text{pre}} W$. Similarly, $X \preceq W$ is a *suffix* if X ends at $W[-1]$, written $X \preceq_{\text{suff}} W$. A factor $X \preceq W$ that is either a prefix or suffix is an *affix*, written $X \preceq_{\text{aff}} W$. A factor $X \preceq W$ that is not an affix is a *middle*, written $X \preceq_{\text{mid}} W$.

The factor $X \preceq W$ such that $W = UXV$, $|U| = |V|$, and $|X| \in \{1, 2\}$ is the *center of* W. A factor $X \preceq W$ is a *mirror*, written $X \preceq_{\text{mir}} W$, provided $W = XUYV$ with $Y \equiv \widehat{X}$ and $|U| = |V|$. For any $X \preceq_{\text{mir}} W$, $\widehat{X}$ refers to the factor Y in the definition.

A mirror factor is *admissible* provided $U[1] \neq \overline{U[-1]}$, $V[1] \neq \overline{V[-1]}$. Observe that each admissible factor is the maximum-length mirror factor with its center. Thus any two admissible factors have distinct centers.

2.4 Polyominoes and Boundary Words

A *cell* is a unit square with lower-leftmost vertex $(x, y) \in \mathbb{Z}^2$ and remaining vertices $(x + 1, y)$, $(x, y + 1)$, $(x + 1, y + 1)$. A *polyomino* is a simply connected union of cells whose boundary is a simple closed curve.

The boundary of a polyomino consists of cell edges. The *boundary word* of a polyomino P, denoted $\mathcal{B}(P)$, is the circular word of letters corresponding to the sequence of directions traveled along cell edges during a clockwise traversal of the polyomino's boundary (see Fig. 2).

Boundary words are *circular*: the last and first letters are defined to be consecutive. Thus for any indices $i, j \in \mathbb{Z} \setminus \{0\}$, $W[i]$ and $W[i..j]$ are defined. For the boundary word $W = \mathbf{urrdll}$, $W[10] = W[-9] = \mathbf{d}$ and $W[6..2] = \mathbf{lur}$.

Fig. 2. A regular tiling (left) and non-regular tiling (right) of a polyomino with boundary word $\mathbf{ururdrurd}^3\mathbf{luldlul}$. The copies in the regular tiling have a common neighborhood factorization $ABC\widehat{A}\widehat{B}\widehat{C}$, with $A = \mathbf{u}$, $B = \mathbf{ru}$, $C = \mathbf{rdrurd}$.

2.5 Tilings

For a polyomino P, a *tiling* of P is an infinite set $\mathscr{T}$ of translations of P, called *copies*, such that every cell in the plane is in exactly one copy. A tiling is *regular* (e.g. *isohedral*) provided there exist vectors $\boldsymbol{o}, \boldsymbol{u}, \boldsymbol{v}$ such that the set of lower-leftmost vertices of copies in the tiling is $\boldsymbol{o} + \{i\boldsymbol{u} + j\boldsymbol{v} : i, j \in \mathbb{Z}\}$. Two tilings $\mathcal{T}$ and $\mathcal{T}'$ are equal provided there exists a vector $\boldsymbol{v}$ such that $\mathcal{T}' = \boldsymbol{v} + \mathcal{T}$.

Copies of a tiling intersect only along boundaries, and copies with non-empty boundary intersection are *neighbors*. Lemma 3.5 of [22] implies that the intersection between a pair of neighbors corresponds to a *neighbor factor* of each neighbor's boundary word and these factors form a *neighborhood factorization*. Every regular tiling has a neighbor factorization common to all copies in the tiling.

3 The Beauquier-Nivat Criterion

Recall that $\widehat{X}$ is the reverse complement of X. Thus $\widehat{X}$ is the same path as X but traversed in the opposite direction. So any pair of factors X and $\widehat{X}$ appearing on the boundary of a polyomino are translations of each other with the interior of the boundary on opposite sites. Beauquier and Nivat [1] gave the following criterion for determining whether a polyomino tile admits a tiling:

Definition 1. *A factorization* $W = ABC\widehat{A}\widehat{B}\widehat{C}$ *of a boundary word* W *is a* BN factorization.

Lemma 1 (Theorem 3.2 of [1]). *A polyomino* P *has a tiling if and only if* $\mathcal{B}(P)$ *has a BN factorization.*

As seen in Fig. 3, a BN factorization corresponds to the neighborhood factorization of a regular tiling. We prove this formally by reusing results from the proof of Lemma 1.

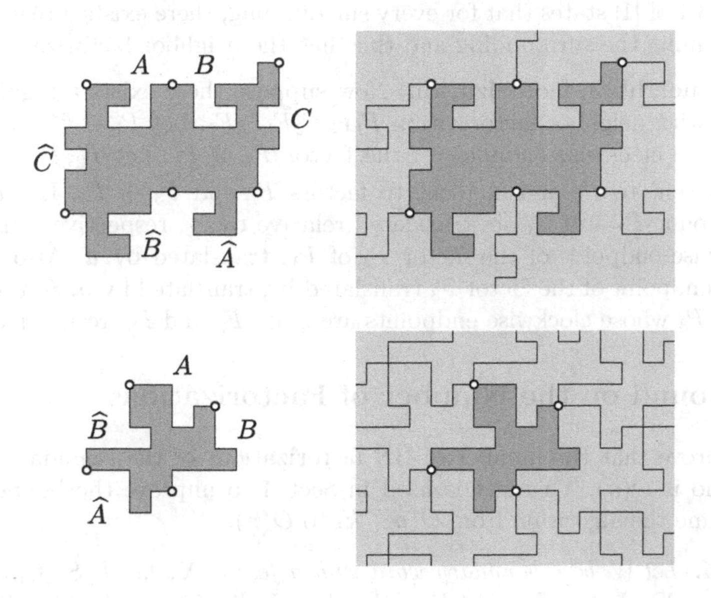

Fig. 3. BN factorizations (left) and the regular tilings induced by these factorizations (right). For one polyomino (bottom), two of the factors are zero length. However, no BN factorization can have more than two length-0 factors.

Lemma 2 (Corollary 3.2 of [1]). *Let* P *be a polyomino. There exists a factorization* $\mathcal{B}(P) = F_1\widehat{F_3}F_2\widehat{F_1}F_3\widehat{F_2}$ *if and only if there exists a tiling* $\mathcal{T}$ *of* P *with three copies* P_1, P_2, P_3 *such that:*

- P_1, P_2, P_3 *appear clockwise consecutively around a common point* q.
- F_i *is the last neighbor factor of* P_i *whose clockwise endpoint is incident to* q.

Lemma 3. *Let* P *be a polyomino. A factorization of* $\mathcal{B}(P)$ *is a BN factorization if and only if a regular tiling of* P *has this neighbor factorization.*

Proof. The factorization $\mathcal{B}(P) = F_1\widehat{F_3}F_2\widehat{F_1}F_3\widehat{F_2}$ is a generic BN factorization. So it suffices to prove that there exists a tiling $\mathcal{T}$ of P satisfying the conditions of Lemma 2 if and only if there exists a regular tiling $\mathcal{T}_{\text{reg}}$ with neighbor factorization $\mathcal{B}(P) = F_1\widehat{F_3}F_2\widehat{F_1}F_3\widehat{F_2}$.

Tiling ⇒ neighbor factorization. Let $\mathcal{T}$ be a tiling and $P_1, P_2, P_3 \in \mathcal{T}$ be copies as defined in the statement of Lemma 2. Let $\boldsymbol{u}$ and $\boldsymbol{v}$ be the amount P_2 and P_3 are translated relative to P_1, respectively. Lemma 3.2 of [1] states that the copies obtained by translating P_1 by $\boldsymbol{u}, \boldsymbol{v}, \boldsymbol{v} - \boldsymbol{u}$, $-\boldsymbol{u}$, $-\boldsymbol{v}$, and $\boldsymbol{u} - \boldsymbol{v}$ is a *surrounding* of P_1: a set of interior-disjoint copies such that every edge of C_1 is shared by a copy. Since P_3 is a copy of P_2 translated by $\boldsymbol{v} - \boldsymbol{u}$, the neighbor factor of P_1 incident to the copy translated by $\boldsymbol{v} - \boldsymbol{u}$ is F_2. By similar reasoning, P_1 has neighbor factors $\widehat{F_1}$, F_3, and $\widehat{F_2}$ incident to the copies translated by $-\boldsymbol{u}$, $-\boldsymbol{v}$, and $\boldsymbol{u} - \boldsymbol{v}$, respectively. So P_1 has neighbor factorization $\mathcal{B}(P) = F_1\widehat{F_3}F_2\widehat{F_1}F_3\widehat{F_2}$. Corollary 3.1 of [1] states that for every surrounding, there exists a regular tiling of P containing the surrounding and thus has the neighbor factorization of P_1.

Tiling ⇐ neighbor factorization. Now suppose there exists a regular tiling $\mathcal{T}_{\text{reg}}$ of P with neighbor factorization $F_1\widehat{F_3}F_2\widehat{F_1}F_3\widehat{F_2}$. Let $P_1 \in \mathcal{T}_{\text{reg}}$ be a copy and q be the clockwise endpoint of the factor F_1 of P_1. Let $P_2, P_3 \in \mathcal{T}_{\text{reg}}$ be copies adjacent to P_1 and incident to factors F_1 and $\widehat{F_3}$ of P_1. Let $\boldsymbol{u}$ and $\boldsymbol{v}$ be the amount P_2 and P_3 are translated relative to P_1, respectively. Then q is the clockwise endpoint of the factor F_2 of P_1, translated by $\boldsymbol{u}$. Also, q is the clockwise endpoint of the factor F_3 translated by, translated by $\boldsymbol{v}$. So the factors of P_2 and P_3 whose clockwise endpoints are q are F_2 and F_3, respectively. □

4 A Bound on the Number of Factorizations

Here we prove that the number of BN factorizations of the boundary word of an n-omino is $O(n)$. This fact is used in Sect. 4 to improve the bound on the running time the algorithm from $O(n + k)$ to $O(n)$.

Lemma 4. *Let W be a boundary word with a factor X. Let $P, S \preceq_{\text{mir}} W$ such that $P \preceq_{\text{pre}} X$, $S \preceq_{\text{suff}} X$, and $P \neq S$. Then X has a period of length $2|X| - (|P| + |S|)$.*

Proof. Since P and S are mirror, there exists $X' \preceq W$ with $|X'| = |X|$, $\widehat{P} \preceq_{\text{pre}} X'$, and $\widehat{S} \preceq_{\text{suff}} X'$. Observe that X has a period of length $r \geq 1$ if and only if $X[i] = X[i + r]$ for all $1 \leq i \leq |X| - r$. Let $1 \leq i \leq |P| + |S| - |X|$. Then $1 \leq |P| + 1 - i \leq |X|$ and $1 \leq |P| + 1 + |\widehat{S}| - |X'| - i \leq |\widehat{S}|$. So:

$$\begin{aligned}
X[i] &= P[i] \\
&= \overline{\widehat{P}}[|P| + 1 - i] \\
&= \overline{X'}[|P| + 1 - i] \\
&= \overline{\widehat{S}}[|P| + 1 + |\widehat{S}| - |X'| - i] \\
&= \overline{\widehat{S}}[|\widehat{S}| + 1 - (i + |X'| - |P|)] \\
&= S[i + |X'| - |P|] \\
&= X[i + |X'| - |P| + (|X| - |S|)] \\
&= X[i + 2|X| - (|P| + |S|)]
\end{aligned}$$

Since $P \neq S$, $2|X| - (|P| + |S|) \geq 2|X| - (2|X| - 1) = 1$. So X has a period of length $2|X| - (|P| + |S|)$. □

Lemma 5. *Let W be a boundary word with $X \preceq W$. Let $P, S \preceq_{\text{mir}} W$ such that $P \preceq_{\text{pre}} X$, $S \preceq_{\text{suff}} X$, and $P \neq S$. Any factor $Y \preceq_{\text{mid}} X$ with $|Y| > 2|X| - (|P| + |S|)$ is not an admissible factor of W.*

Proof. By Lemma 4, X has a period of length $r = 2|X| - (|P| + |S|)$. Let $Y \preceq_{\text{mid}} X$ and $|Y| > r$.

Let $X' \preceq W$ with $|X'| = |X|$ and the center of X' exactly $|W|/2$ letters from the center of X. Then $\widehat{P} \preceq_{\text{pre}} X'$, $\widehat{S} \preceq_{\text{suff}} X'$, and $\widehat{Y} \preceq_{\text{mid}} X'$. Again by Lemma 4, X' has a period of length r.

Let $U, V \preceq W$ such that $W = YU\widehat{Y}V$. Since Y is a middle factor of X, the letter $U[1]$ is in X. Since X has a period of length r and $|Y| > r$, $U[1] = Y[|Y| + 1 - r] = \overline{\widehat{Y}[r]}$. Since $\widehat{Y}$ is a middle factor of X' and X' has a period of length r, $U[-1] = \widehat{Y}[r]$. So $U[1] = \overline{U[-1]}$ and Y is not admissible. □

Lemma 6. *Let W be a boundary word. There exists a set $\mathscr{F}$ of $O(1)$ factors of W such that every $F \preceq_{\text{adm}} W$ with $|F| \geq |W|/6$ is an affix factor of an element of $\mathscr{F}$.*

Proof. **A special case on three factors.** Let $P_1, P_2, P_3 \preceq_{\text{adm}} W$ with $|P_1|, |P_2|, |P_3| \geq |W|/6$ and centers contained in a factor of W with length at most $|W|/14$. Let $X \preceq W$ be the shortest factor such that $P_1, P_2, P_3 \preceq X$, and so $P_i \preceq_{\text{pre}} X$ and $P_j \preceq_{\text{suff}} X$ for some $i, j \in \{1, 2, 3\}$. We prove that if $i \neq j$, then $P_1, P_2, P_3 \preceq_{\text{aff}} X$.

Without loss of generality, suppose $i = 1$, $j = 2$ and so $P_3 \preceq_{\text{mid}} X$. By Lemma 5, since $P_3 \preceq_{\text{adm}} W$, $|P_3| \leq 2|X| - (|P_1| + |P_2|) \leq |P_1| + |W|/7 + |P_2| - (|P_1| + |P_2|) = |W|/7 < |W|/6$, a contradiction. So $P_3 \preceq_{\text{aff}} X$.

All nearby factors. Consider a set $\mathscr{I} = \{F_1, F_2, \ldots, F_m\}$ of at least three admissible factors of W of length at least $|W|/6$ such that the centers of the factors are contained in a common factor of W of length $|W|/14$. We will prove that every element of $\mathscr{I}$ is an affix factor of one of two factors of W.

Let $G \preceq W$ be the shortest factor such that $F_i \preceq G$ for every $F_i \in \mathscr{I}$. It is either the case that there exist distinct $F_l, F_r \in \mathscr{I}$ with $F_l \preceq_{\text{pre}} G$, $F_r \preceq_{\text{suff}} G$, or that $G \in \mathscr{I}$ and every $F_i \in \mathscr{I}$ besides G has $F_i \preceq_{\text{mid}} G$.

In the first case, $F_i \preceq_{\text{aff}} G$ for any $i \neq l, r$ by the previous claim regarding three factors. Also $F_l, F_r \preceq_{\text{aff}} G$. So every factor in $\mathscr{I}$ is an affix factor of G.

In the second case, let $G' \preceq G$ be the shortest factor with the same center as G such that every factor in $\mathscr{I}$ excluding G is a factor of G'. Clearly $G' \preceq_{\text{mir}} W$ and $G' \not\preceq_{\text{adm}} W$. Without loss of generality, there exists $F_p \in \mathscr{I}$ such that $F_p \preceq_{\text{pre}} G'$. Since $F_p \preceq_{\text{adm}} W$ and $G' \not\preceq_{\text{adm}} W$, $F_p \neq G'$.

Applying Lemma 5 with $X = G'$, $P = F_p$, $S = G'$, every middle factor of G' in $\mathscr{I}$ has length at most $2|G'| - (|G'| + |F_p|) \leq |G'| - |F_p| \leq |W|/7 < |W|/6$. So every factor of G' in $\mathscr{I}$ is an affix factor of G'. Thus every factor in $\mathscr{I}$ is either G or an affix factor of G'.

All factors. Partition W into 15 factors $I_1, I_2, \ldots, I_{15}$ each of length at most $|W|/14$. Let $\mathscr{I}_i$ be the set of admissible factors with centers containing letters in I_i. Then by the previous claim regarding more than three factors, there exists a set $\mathscr{F}_i$ (G and possibly G') such that every element of $\mathscr{I}_i$ is an affix factor of an element of $\mathscr{F}_i$ and $|\mathscr{F}_i| \leq 2$. So every $F \preceq_{\text{adm}} W$ with $|F| \geq |W|/6$ is an affix factor of an element of $\mathscr{F} = \bigcup_{i=1}^{15} \mathscr{F}_i$ and $|\mathscr{F}| \leq 2 \cdot 15$. □

Theorem 1. *A boundary word W has $O(|W|)$ BN factorizations.*

Proof. Consider the choices for the three factors A, B, C of BN factorization $W = ABC\widehat{A}\widehat{B}\widehat{C}$. In any factorization, some factor has size at least $|W|/6$. By Lemma 6, there exists a $O(1)$-sized set of factors $\mathscr{F}$ such that any factor with length at least $|W|/6$ is an affix factor of an element of $\mathscr{F}$. Without loss of generality, either $|A| \geq |W|/6$ and A is a prefix of a factor in $\mathscr{F}$ or $|C| \geq |W|/6$ and C is a suffix of a factor in $\mathscr{F}$.

Let $H = ABC$ be the factor formed by consecutive factors A, B, C of a BN factorization. Then since $|H| = |W|/2$ and shares either the first or last letter with a factor in $\mathscr{F}$, there are $O(1)$ total factors H. For a fixed H, choosing the center of B determines B (since B is admissible) and thus A and C. So there are at most $2(|W|/2)$ factorizations for a fixed factor H. □

Since Lemma 3 proves that factorizations and tilings are equivalent, the previous theorem implies a linear upper bound on the number of regular tilings of a polyomino:

Corollary 1. *An n-omino has $O(n)$ regular tilings.*

As pointed out by Provençal [18], it is easy to construct polyominoes with $\Omega(n)$ such tilings. For instance, the polyomino with boundary word $W = \mathbf{u}\mathbf{r}^i\mathbf{d}\mathbf{l}^i$ with $i \geq 1$ has $|W|/2 - 1$ regular tilings.

5 An Algorithm for Enumerating Factorizations

The bulk of this section describes a $O(|W|)$-time algorithm for enumerating the factorizations of a polyomino boundary word W. The algorithm combines algorithmic ideas of Brlek et al. [4] and a structural result based on a well-known lemma of Galil and Seirferas [5].

Lemma 7 (Corollary 5 of [4]). *Every factor of a BN factorization is admissible.*

Lemma 8 is a variation of Lemma C4 of Galil and Seirferas [5]. We reproduce their proof with minor modifications.

Lemma 8. *Let A and B be two words of the same length. Moreover, let $A = X_1X_2 = Y_1Y_2 = Z_1Z_2$ and $B = X_Q\widehat{X_2} = \widehat{Y_1}\widehat{Y_2} = \widehat{Z_1}Z_Q$ with $|X_1| < |Y_1| < |Z_1|$. Then $X_Q = \widehat{X_1}$ and $Z_Q = \widehat{Z_2}$.*

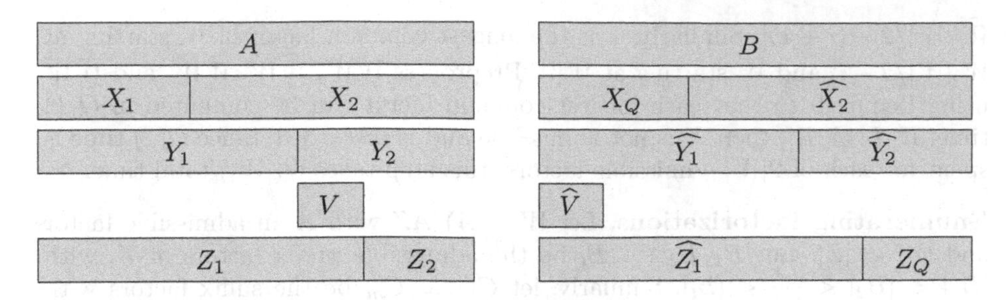

Fig. 4. The words used in the proof of Lemma 8.

Proof. Let V be the word such that $Y_1V = Z_1$ (see Fig. 4).

Claim (1): $\widehat{V}$ is a period of $\widehat{Z_1}$. Since $Y_1V = Z_1$, then $\widehat{Z_1} = \widehat{Y_1V} = \widehat{V}\widehat{Y_1}$ is a prefix of B. So $\widehat{Y_1}$ is a prefix of $\widehat{Z_1} = \widehat{V}\widehat{Y_1}$ and thus $\widehat{V}$ is a period of $\widehat{Y_1}$. So $\widehat{V}$ is a period of $\widehat{V}\widehat{Y_1} = \widehat{Z_1}$.

Claim (2): V is a prefix of X_2. Since V is a prefix of Y_2, $\widehat{V}$ is a suffix of $\widehat{Y_2}$. So $\widehat{V}$ is a suffix of $\widehat{X_2}$ and V is a prefix of X_2.

Claim (3): X_1V is a prefix of Z_1. Since V is a prefix of X_2, X_1V is a prefix of Y_1V. Since $|X_1V| < |Y_1V| = |Z_1|$, X_1V is also a prefix of Z_1.

Claim (4): $\widehat{V}$ is a period of $\widehat{X_1}$. By claim (1), $\widehat{V}$ is a period of $\widehat{Z_1}$, so Z_1 has a period of length $|\widehat{V}| = |V|$. By claim (3), X_1V is a prefix of Z_1 and so also has a period of length $|V|$. Then $\widehat{X_1V} = \widehat{V}\widehat{X_1}$ has a period of length $|V|$, namely $\widehat{V}$. So $\widehat{V}$ is also a period of $\widehat{X_1}$.

Finally, combining claims (1) and (4), since $\widehat{V}$ is a period of both X_Q and $\widehat{X_1}$, $X_Q = \widehat{X_1}$. By symmetry, the same proof also implies $Z_Q = \widehat{Z_2}$. □

Lemma 9 (Theorem 9.1.1 of [12]). *Two non-circular words X, Y can be preprocessed in $O(|X| + |Y|)$ time to support the following queries in $O(1)$-time: what is the longest common factor of X and Y starting at $X[i]$ and $Y[j]$?*

Lemma 10. *Let W be a polyomino boundary word. Then the BN factorizations of W can be enumerated in $O(|W|)$ time.*

Proof. Lemma 7 states that BN factorizations consist entirely of admissible factors. The algorithm first computes all admissible factors, then searches for factorizations consisting of them.

Computing admissible factors. Lemma 7 implies that there are at most $2|W|$ admissible factors, since admissible factor has a distinct center. For each center $W[i..i]$ or $W[i..i+1]$, the admissible factor with this center is LR, where R is the longest common factor of W starting at $W[i+1]$ and $\widehat{W}$ starting at

$\widehat{W}[|W|/2-(i+1)]$. Similarly, L is the longest common factor of $\widetilde{W}$ starting at $\widetilde{W}[|W|/2-i]$ and $\overline{W}$ starting at $\overline{W}[i]$. Preprocess WW, $\widehat{W}\widehat{W}$, $\widetilde{W}\widetilde{W}$, and $\overline{W}\overline{W}$ using Lemma 9 so that each longest common factor can be computed in $O(1)$ time. If $|L| \neq |R|$, then X is not admissible and is discarded. Since $O(1)$ time is spent for each of $2|W|$ admissible factors, this step takes $O(|W|)$ total time.

Enumerating factorizations. Let $W = AY\widehat{A}Z$ with A an admissible factor and $|Y| = |Z|$. Let $B_1, B_2, \ldots, B_l$ be the admissible prefix factors of Y, with $|B_1| < |B_2| < \cdots < |B_l|$. Similarly, let $C_1, \ldots, C_m$ be the suffix factors with $|C_1| < \cdots < |C_m|$. Lemma 8 implies that for fixed A, there exist intervals $[b, l]$, $[c, m]$ such that the BN factorizations $AB_iC_j\widehat{A}\widehat{B_i}\widehat{C_j}$ are exactly those with $i \in [b, l]$ or $j \in [c, m]$.

First, construct a length-sorted list of the admissible factors starting at each $W[k]$ in $O(|W|)$ time using counting sort. Do the same for all factors ending at each $W[k]$.

Next, use a two-finger scan to find, for each factor A that ends at $W[k]$, the longest factor B_l starting at $W[k+1]$ such that $|A| + |B_l| \leq |W|/2$. Then check whether C_j, the factor following B_l such that $|AB_lC_j| = |W|/2$, is admissible and report the factorization $AB_lC_j\widehat{A}\widehat{B_l}\widehat{C_j}$ if so. Checking whether C_j is admissible takes $O(1)$ time using an array mapping each center to the unique admissible factor with this center.

Additional BN factorizations containing A are enumerated by checking factors B_i with $i = l-1, l-2, \ldots$ for an admissible following factor C_j. Either C_j is admissible and the factorization is reported, or $i = b-1$ and the iteration stops.

Finally, use a similar two-finger scan to find, for each factor A that starts at $W[k]$, the longest factor C_m that ends at $W[k + |W|/2 - 1]$ such that $|A| + |C_m| \leq |W|/2$, check whether B_i preceeding C_m such that $|AB_iC_m| = |W|/2$ is admissible, and report the possible BN factorization. Then check and report similar factorizations with C_j for $j = m-1, m-2, \ldots$ until $j = c-1$.

In total, the two-finger scans take $O(|W|)$ time plus $O(1)$ time to report each factorization. Reporting duplicate factorizations can be avoided by only reporting a factorization if $A[1]$ appears before $B[1]$, $C[1]$, $\widehat{A}[1]$, $\widehat{B}[1]$, and $\widehat{C}[1]$ in W. Then by Theorem 1, reporting factorizations also takes $O(|W|)$ time. □

Combining this algorithm with Lemmas 1 and 3 yields the desired algorithmic result:

Theorem 2. *Let P be a polyomino with n edges. In $O(n)$ time, it can be determined if P admits a tiling and the regular tilings of P can be enumerated.*

Acknowledgments. The author thanks Stefan Langerman for fruitful discussions and comments that greatly improved the paper, and anonymous reviewers for pointing out an error in an earlier version of the paper.

References

1. Beauquier, D., Nivat, M.: On translating one polyomino to tile the plane. Discrete Comput. Geom. **6**, 575–592 (1991)
2. Berger, R.: The undecidability of the domino problem. In: Memoirs of the American Mathematical Society, vol. 66 (1966)
3. Brlek, S., Provençal, X.: An optimal algorithm for detecting pseudo-squares. In: Kuba, A., Nyúl, L.G., Palágyi, K. (eds.) DGCI 2006. LNCS, vol. 4245, pp. 403–412. Springer, Heidelberg (2006)
4. Brlek, S., Provençal, X., Fédou, J.-M.: On the tiling by translation problem. Discrete Appl. Math. 157, 464–475 (2009)
5. Galil, Z., Seiferas, J.: A linear-time on-line recognition algorithm for "Palstar". J. ACM **25**(1), 102–111 (1978)
6. Gambini, L., Vuillon, L.: An algorithm for deciding if a polyomino tiles the plane by translations. RAIRO - Theor. Inf. Appl. **41**(2), 147–155 (2007)
7. Girault-Beauquier, D., Nivat., M.: Tiling the plane with one tile. In: 6th Annual Symposium on Computational Geometry, pp. 128–138 (1990)
8. Golomb, S.W.: Polyominoes. Scribner's, New York (1965)
9. Golomb, S.W.: Tiling with sets of polyominoes. J. Comb. Theory **9**(1), 60–71 (1970)
10. Goodman-Strauss, C.: Open questions in tilings. preprint (2000). http://comp.uark.edu/~strauss/papers/survey.pdf
11. Goodman-Strauss, C.: Can't decide? Undecide!. Not. Am. Math. Soc. **57**, 343–356 (2010)
12. Gusfield, D.: Algorithms on Strings, Trees, and Sequences: Computer Science and Computational Biology. Cambridge University Press, Cambridge (1997)
13. Knuth, D.E., Morris, J.H., Pratt, V.R.: Fast pattern matching in strings. SIAM J. Comput. **6**(2), 323–350 (1977)
14. Blondin-Massé, A., Brlek, S., Garon, A., Labbé, S.: Christoffel and Fibonacci tiles. In: Brlek, S., Reutenauer, C., Provençal, X. (eds.) DGCI 2009. LNCS, vol. 5810, pp. 67–78. Springer, Heidelberg (2009)
15. Massé, A.B., Brlek, S., Garon, A., Labbé, S.: Every polyomino yields at most two square tilings. In: 7th International Conference on Lattice Paths and Applications (Lattice Paths 2010), pp. 57–61 (2010)
16. Masseé, A.B., Brlek, S., Labbé, S.: Combinatorial aspects of Escher tilings. In: 22nd International Conference on Formal Power Series and Algebraic Combinatorics (FPSAC 2010), pp. 533–544 (2010)
17. Ollinger, N.: Tiling the plane with a fixed number of polyominoes. In: Dediu, A.H., Ionescu, A.M., Martín-Vide, C. (eds.) LATA 2009. LNCS, vol. 5457, pp. 638–647. Springer, Heidelberg (2009)
18. Provençal, X.: Combinatoire des mots, géométrie discrète et pavages. Ph. D. thesis, Université du Québec à Montréal (2008)
19. Schattschneider, D.: Will it tile? Try the Conway criterion!. Math. Monthly **53**(4), 224–233 (1980)
20. Schattschneider, D.: Visions of Symmetry: Notebooks, Periodic Drawings, and Related Work of M.C. Escher. W. H. Freeman and Company, New York (1990)
21. Shapiro, H.D.: Theoretical limitations on the efficient use of parallel memories. IEEE Trans. Comput. **27**(5), 421–428 (1978)
22. Wijshoff, H.A.G., van Leeuwen, J.: Arbitrary versus periodic storage schemes and tessellations of the plane using one type of polyomino. Inf. Control **62**, 1–25 (1984)

Adaptive Point Location in Planar Convex Subdivisions

Siu-Wing Cheng(✉) and Man-Kit Lau

Department of Computer Science and Engineering, HKUST,
Hong Kong, China
scheng@cse.ust.hk

Abstract. We present a planar point location structure for a convex subdivision S. Given a query sequence of length m, the total running time is $O(\text{OPT} + m \log\log n + n)$, where n is the number of vertices in S and OPT is the minimum running time to process the same query sequence by any linear decision tree for answering planar point location queries in S. The running time includes the preprocessing time. Therefore, for $m \geq n$, our running time is only worse than the best possible bound by $O(\log\log n)$ per query, which is much smaller than the $O(\log n)$ query time offered by an worst-case optimal planar point location structure.

Keywords: Point location · Convex subdivision · Adaptive data structure

1 Introduction

There has been extensive research on planar point location—a fundamental problem in computational geometry—to obtain worst-case optimal query time, preprocessing time, and space complexity [2,16,20–25]. Some of them are now standard results in textbooks in computational geometry [8,11]. Planar point location can be seen as a generalization of the one-dimensional dictionary problem to two dimensions. In any dimension, the information theoretic lower bound in processing a sequence of m queries follows from Shannon's work [26] and the *entropy-based* lower bound is $\sum_z f(z) \cdot \log \frac{m}{f(z)}$, where $f(z)$ denotes the access frequency of an item z in the sequence of length m. The splay tree [27] has been designed such that, given an initially empty structure and a sequence of m insertions, deletions, and queries, the total running time for manipulating the data structure to process these operations is $O\left(\sum_z f(z) \cdot \log \frac{m}{f(z)}\right)$, where every insertion and deletion of z also contributes one to the access frequency of z. Notice that the access frequencies of items are unknown beforehand. As a result, $o(\log n)$ amortized query time is possible in one dimension if the access frequencies of the items are substantially unequal.

Supported by FSGRF14EG26, HKUST.

K. Elbassioni and K. Makino (Eds.): ISAAC 2015, LNCS 9472, pp. 14–22, 2015.
DOI: 10.1007/978-3-662-48971-0_2

For point location in a planar subdivision S, there are also previous works on making the performance adaptive to the access frequencies. When the regions in S have constant complexities, and the query distribution is fixed and available as part of the input, there are several works by Arya et al. [4–7] and Iacono [17] to construct a data structure such that the expected query time is $O\left(\sum_z p_z \log \frac{1}{p_z}\right)$, where p_z is the probability of a query point falling into the region z. The algorithm of Iacono [17] uses $O(n)$ space and $O(n)$ preprocessing time. The algorithm of Arya et al. [7] uses $O(n)$ space and $O(n \log n)$ preprocessing time, and its expected running time per query is optimal up to the leading constant factor modulo some additive lower-order terms. Subsequently, analogous results have been obtained for connected subdivisions [13] and disconnected subdivisions [1,9,10] in which the regions may have arbitrary complexities. In the aforementioned results, the query distribution is fixed and available as part of the input. A natural question is whether we can obtain a self-adjusting planar point location structure that can adapt to a query sequence without knowing the access frequencies of the regions beforehand. There has been only one such result in the case that S is a triangulation by Iacono and Mulzer [19]. They present a method that achieves a total running time of $O\left(n + \sum_z f(z) \cdot \log \frac{m}{f(z)}\right)$, including the preprocessing time to construct the initial structure before processing the query sequence.

In this paper, we study the adaptive point location problem for a convex subdivision S. That is, every region in S is a convex polygon (except the outer unbounded region). We do not require the regions in S to have constant complexities. One cannot just triangulate S, apply the result for triangulation by Iacono and Mulzer [19], and hope to achieve the entropy-based lower bound. Suppose that we encode the names of the regions using bit vectors of possibly different lengths. Then, the entropy-based lower bound is the minimum number of bits needed to encode the sequence of output region names corresponding to the m queries under the prescribed access frequencies. Each output bit requires at least one unit of processing time, and therefore, the entropy-based lower bound is also a lower bound for the total running time. Consequently, geometry is not taken into consideration at all. Arya et al. [7] show that one can design a convex polygon of n sides and a query distribution so that a query point lies in the polygon with probability $1/2$ and the expected number of point-line comparisons needed to decide whether a query point lies in the polygon is $\Omega(\log n)$. However, the entropy-based lower bound for a single query is only a constant in this case. This shows that the entropy-based lower bound is too weak for a convex subdivision. As in [13], we compare our result with the best linear decision tree for answering point location queries in S. This is reasonable because the linear decision tree models the process for answering a query by point-line comparisons, and many existing point location structures are based on point-line comparisons.[1]

[1] Methods that employ indexing (e.g. [15]) and bit tricks (e.g. [12]) do not fall under the linear decision tree model.

Given a sequence of m queries, our method runs in $O(\mathrm{OPT} + m \log\log n + n)$ total time, where OPT is the minimum time to process the same query sequence by any linear decision tree for answering point location queries in S. Our time bound includes the preprocessing time before processing the query sequence. Therefore, for $m \geq n$, our running time is only worse than the best possible bound by $O(\log\log n)$ time per query, which is much smaller than the $O(\log n)$ query time offered by an worst-case optimal planar point location structure.

One can build another auxiliary planar point location structure so that a query can be executed on our adaptive structure and this auxiliary point location structures simultaneously until one of the two structures returns an answer. The advantage is that this auxiliary point location structure can offer additional properties. For example, if one uses the distance-sensitive planar point location structure [3], it means that queries far away from any region boundary can be answered fast too. Alternatively, if one uses the proximate planar point location structure [18] as the auxiliary structure, then a query can be answered faster if the query point is close to the previous one.

2 Triangulation of a Convex Polygon

Let P be a convex region in S with n_P vertices in counterclockwise order $(v_0, v_1, ..., v_{n_P-1})$. We triangulate P as follows. Select every other vertex of P. (When n_P is odd, the last vertex selected is adjacent to the first vertex selected.) Let P_1 be the convex hull of these selected vertices. Clearly, $P_1 \subset P$, $P \setminus P_1$ is a collection of triangles, and the number of vertices of P_1 is at most $\lceil n_P/2 \rceil$. Then, we recurse on P_1 to construct P_2 and so on until we produce a convex hull P_j that is a single triangle or a single line segment. The triangulation of P is the collection of triangles in $P \setminus P_1$, $P_1 \setminus P_2$, etc. We denote this triangulation of P by T_P. Figure 1 shows an example. This hierarchical triangulation was first introduced by Dobkin and Kirkpatrick [14] in the context of detecting intersection between two convex polygons and polyhedra. Note that $O(\log n)$ P_i's are constructed because the size of the P_i's decreases repeatedly by a constant factor. The time to produce each P_i is $\lceil n_P/2^i \rceil$. Therefore, the total time to compute T_P is $O\left(\sum_{i=0}^{\infty} n_P/2^i\right) = O(n_P)$. A line segment ℓ in P intersects the boundary of each P_i's in at most two points. It follows that ℓ intersects at most two triangles in $P_{i-1} \setminus P_i$, and therefore, ℓ intersects $O(\log n)$ triangles in T_P. Interestingly, this simple hierarchical triangulation T_P leads to a query performance that is adaptive and only slightly worse than the best possible bound. In the following, we prove an upper bound on the entropy of T_P that is closely related to the performance of any linear decision tree.

Lemma 1. *Let P be a convex polygon in $\mathbb{R}^2$. Let $H(T_P)$ denote the entropy of T_P. Let $\mathcal{D}$ be an arbitrary linear decision tree for determining whether a query point in $\mathbb{R}^2$ lies in P. Let $L_\mathcal{D}$ be the set of leaves of $\mathcal{D}$ and for every leaf $\nu \in L_\mathcal{D}$, let r_ν denote the convex region represented by ν. Consider an arbitrary query sequence of length m. For any region $r \subseteq \mathbb{R}^2$, let $f(r)$ denote the number of queries that fall inside r. Then, the following inequality is satisfied.*

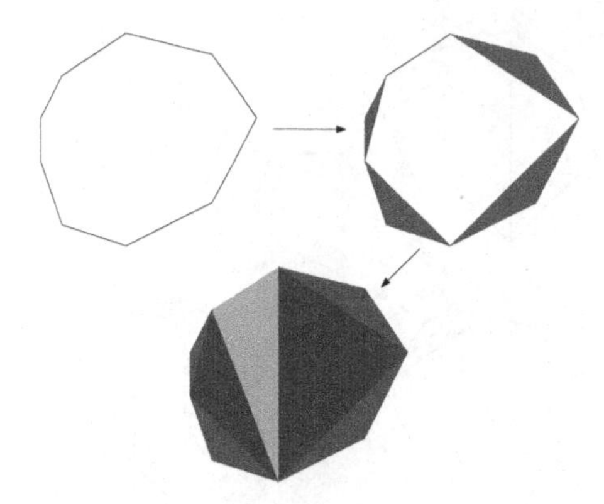

Fig. 1. Triangulation by convex hull. The red, blue and green triangles are obtained from the first, second and third convex hulls respectively (Color figure online)

$$
\begin{aligned}
H(T_P) &= \sum_{t \in T_P} f(t) \cdot \log \frac{m}{f(t)} \\
&\leq \sum_{\nu \in L_{\mathcal{D}}} f(r_\nu) \cdot (\text{depth}(\nu) + O(\log(\text{depth}(\nu))) + O(\log \log n))
\end{aligned}
$$

Proof. For any line segment ℓ inside P, ℓ intersects at most two triangles in $P_i \setminus P_{i+1}$ in each level of the hierarchical triangulation T_P. Therefore, ℓ intersects at most $O(\log n)$ triangles in T_P. Let q be a query point that falls in the convex polygon r_ν for some leaf $\nu \in L_{\mathcal{D}}$. Let k be the number of sides of r_ν. We have $\text{depth}(\nu) \geq k$ because each internal node on the path from the root of $\mathcal{D}$ to ν corresponds to a cut along a line.

We can expand the linear decision tree $\mathcal{D}$ to another linear decision tree $\mathcal{D}''$ that allows us to identify the triangle $t \in T_P$ containing q. The construction of $\mathcal{D}''$ works in two steps as follows. For each leaf $\nu \in L_{\mathcal{D}}$, we recursively add a chord to split r_ν into two convex polygons, each having at most $(\lceil \frac{k}{2} \rceil + 1)$ sides. At the same time, we attach two child nodes of ν to represent these smaller convex polygons. The recursion stops when r_ν is triangulated. Figure 2 gives an example of the recursive triangulation of r_ν. The recursive triangulation of the leaves in $L_{\mathcal{D}}$ produces a subtree rooted at ν of height $O(\log k) = O(\log(\text{depth}(\nu)))$. Let $\mathcal{D}'$ denote this intermediate linear decision tree obtained. Each leaf of $\mathcal{D}'$ represents a triangle t' that lies in r_ν for some $\nu \in L_{\mathcal{D}}$. The boundary of t' intersects $O(\log n)$ triangles in T_P. Therefore, for any query point that lies in t', we can determine which triangle $t \in T_P$ contains that query point in $O(\log \log n)$ time by applying binary search on the $O(\log n)$ triangles that intersect t'. This motivates us to expand $\mathcal{D}'$ further as follows. For every leaf ν' of $\mathcal{D}'$, replace ν' by a linear decision tree that corresponds to a binary search on the triangles in T_P that intersects the triangle corresponding to ν'. The resulting linear decision tree is $\mathcal{D}''$. The height of $\mathcal{D}''$ is $O(\log \log n)$ more than the height of $\mathcal{D}'$.

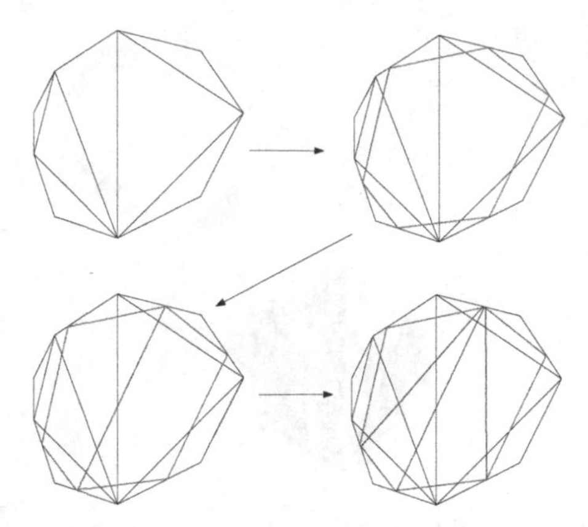

Fig. 2. The red lines represent the boundary of the convex k-gon of a leaf node of $\mathcal{D}$ and the blue lines represent the split of the convex k-gon (Color figure online).

If q is a query point inside r_ν for some $\nu \in L_\mathcal{D}$, then we can follow the path from the root of $\mathcal{D}$ to ν and then from ν to a leaf ν'' of $\mathcal{D}''$. The length of the path traversed is $\mathrm{depth}(\nu'') \le \mathrm{depth}(\nu) + O(\log(\mathrm{depth}(\nu))) + O(\log\log n)$. The entropy of T_P is an information-theoretic lower bound to answering point location queries in T_P. In particular, this lower bound applies to the linear decision tree $\mathcal{D}''$. Therefore,

$$\begin{aligned} H(T_P) &= \sum_{t \in T_P} f(t) \cdot \log \frac{m}{f(t)} \\ &\le \sum_{\text{leaf } \nu'' \text{ of } \mathcal{D}''} f(r_{\nu''}) \cdot \mathrm{depth}(\nu'') \\ &\le \sum_{\nu \in L_\mathcal{D}} f(r_\nu) \cdot (\mathrm{depth}(\nu) + O(\log(\mathrm{depth}(\nu))) + O(\log\log n)) \end{aligned}$$

□

3 Point Location in a Convex Subdivision

Let S be an input convex subdivision. For each convex region P in S, we triangulate P hierarchically as described in Sect. 2. The collection of all triangles in all convex regions in S form a triangulation T of S. Clearly, $\sum_{P \in S} n_P = O(n)$, and therefore, T has $O(n)$ triangles and T can be constructed in $O(n)$ time. Next, we invoke the previous work of Iacono and Mulzer [19] for building an adaptive point location structure for planar triangulations. This gives us a point location data structure for T. We will prove that this point location data structure guarantees that any query sequence of length m can be answered in $O(\mathrm{OPT}{+}m \log\log n{+}n)$ time, where OPT is the minimum time needed by any linear decision tree to process that query sequence.

The method of Iacono and Mulzer [19] is based on rebuilding from time to time. Initially, an optimal worst-case data structure W_0 is built on all triangles in T, and we start answering queries using W_0 until $\Theta(n^\alpha)$ queries have been answered for some $\alpha \in (0,1)$. Then we identify the n^β most frequently queried triangles for some $\beta \in (0,1)$ such that $\alpha \in (\beta, 1-\beta)$, triangulate their exterior, and then build a point location structure W_1 that is distribution-sensitive with respect to frequency counts in these n^β triangles [17]. These frequency counts are fixed when the rebuilding starts. The counts and this distribution-sensitive structure will not be updated as more queries are processed. Until the next rebuilding after another $\Theta(n^\alpha)$ queries, we first submit every query to W_1, and if W_1 does not report a triangle in the input triangulation, we resort to W_0 to answer the query. The challenge in [19] lies in proving that the total time to answer any query sequence of length m matches the entropy bound.

We prove below that by constructing Iacono and Mulzer's data structure on the triangulation T of S, we can obtain a query performance that is adaptive to the query sequence.

Theorem 1. *Let S be a convex subdivision of n vertices in $\mathbb{R}^2$. Our algorithm is a point-line comparison based algorithm that answers any point location query sequence of length m in $O(\mathrm{OPT} + m \log\log n + n)$ time, where* OPT *is the minimum time to process the same query sequence by any linear decision tree for answering point location queries in S. The preprocessing time is included in our running time bound.*

Proof. Let T be the triangulation of S obtained by triangulating every convex region in S as described in Sect. 2. We apply Theorem 2 in [19] to construct a point location structure on T. This total time spent by this structure on any query sequence of length m is

$$O\left(n + \sum_{t \in T} f(t) \cdot \log \frac{m}{f(t)}\right).$$

By manipulating the terms, we obtain

$$O\left(n + \sum_{t \in T} f(t) \cdot \log \frac{m}{f(t)}\right) = O\left(n + \sum_{P \in S} \sum_{t \in T_P} f(t) \cdot \log \frac{m}{f(t)}\right).$$

Then Lemma 1 implies that

$$\begin{aligned} & O\left(n + \sum_{t \in T} f(t) \cdot \log \frac{m}{f(t)}\right) \\ &= O\left(n + \sum_{P \in S} \sum_{\nu \in L_{\mathcal{D}}|_P} f(r_\nu) \cdot (\mathrm{depth}(\nu) + O(\log(\mathrm{depth}(\nu))) + O(\log\log n))\right), \end{aligned}$$

where $\mathcal{D}$ is an arbitrary linear decision tree for answering point location queries in S and we use $L_{\mathcal{D}}|_P$ to denote the subset of leaves of $\mathcal{D}$ that correspond to subset of points in P. Some explanation is in order why Lemma 1 is applicable. Clearly, a linear decision tree for answering point location queries in S is also a linear decision tree for answering point location queries in P, so Lemma 1 is applicable.

Since a leaf of $\mathcal{D}$ must correspond to a subset of points in at most one convex region P in S, the total running time for answering any query sequence of length m is

$$O\left(n + \sum_{\nu \in L_{\mathcal{D}}} f(r_\nu) \cdot (\text{depth}(\nu) + O(\log(\text{depth}(\nu))) + O(\log\log n))\right)$$

$$= O\left(\sum_{\nu \in L_{\mathcal{D}}} f(r_\nu) \cdot \text{depth}(\nu)\right) + O(m \log\log n + n).$$

The first term is $O(\text{OPT})$ because we can choose $\mathcal{D}$ to be the optimal linear decision tree. □

4 Conclusion

One can build another auxiliary planar point location structure so that a query can be executed on our adaptive structure and this auxiliary point location structures simultaneously until one of the two structures returns an answer. The advantage is that this auxiliary point location structure can offer additional properties. For example, if one uses the distance-sensitive planar point location structure [3], it means that queries far away from any region boundary can be answered fast too. Alternatively, if one uses the proximate planar point location structure [18] as the auxiliary structure, then a query can be answered faster if the query point is close to the previous one. Notice that the performance of these auxiliary structures are independent from the access frequencies. Therefore, such an auxiliary structure is constructed only once at the beginning, and it does not need to be rebuilt periodically as our point location structure.

Acknowledgment. We thank the anonymous referees for their helpful comments.

References

1. Afshani, P., Barbay, J., Chan, T.: Instance optimal geometric algorithms. In: Proceedings of the 50th Annual IEEE Symposium on Foundations of Computer Science, pp. 129–138 (2009)
2. Adamy, U., Seidel, R.: On the exact worst case query complexity of planar point location. In: Proceedings of the 9th Annual ACM-SIAM Symposium on Discrete Algorithms, pp. 609–618 (1998)

3. Aronov, B., de Berg, M., Roeloffzen, M., Speckmann, B.: Distance-sensitive planar point location. In: Dehne, F., Solis-Oba, R., Sack, J.-R. (eds.) WADS 2013. LNCS, vol. 8037, pp. 49–60. Springer, Heidelberg (2013)
4. Arya, S., Cheng, S.W., Mount, D.M., Ramesh, H.: Efficient expected-case algorithms for planar point location. In: Proceedings of the 7th Scandinavian Workshop on Algorithm Theory, pp. 353–366 (2000)
5. Arya, S., Malamatos, T., Mount, D.M.: Nearly optimal expected-case planar point location. In: Proceedings of the 41st Annual Symposium on Foundations of Computer Science, pp. 208–218 (2000)
6. Arya, S., Malamatos, T., Mount, D.M.: A simple entropy-based algorithm for planar point location. ACM Trans. Algorithms **3**(2), article 17 (2007)
7. Arya, S., Malamatos, T., Mount, D., Wong, K.: Optimal expected-case planar point location. SIAM J. Comput. **37**(2), 584–610 (2007)
8. Boissonnat, J.D., Yvinec, M.: Algorithmic Geometry. Cambridge University Press, Cambridge (1998)
9. Bose, P., Devroye, L., Douïeb, K., Dujmovic, V., King, J., Morin, P.: Point location in disconnected planar subdivisions. arXiv:1001.2763v1 [cs.CG], 15 January 2010
10. Bose, P., Devroye, L., Douïeb, K., Dujmovic, V., King, J., Morin, P.: Odds-On Trees, arXiv:1002.1092v1 [cs.CG], 5 February 2010
11. de Berg, M., Cheong, O., van Kreveld, M., Overmars, M.: Computational Geometry: Algorithms and Applications. Springer, New York (2008)
12. Chan, T.M., Pătraşcu, M.: Transdichotomous results in computational geometry, I: point location in sublogarithmic time. SIAM J. Comput. **39**(2), 703–729 (2009)
13. Collette, S., Dujmović, V., Iacono, J., Langerman, S., Morin, P.: Entropy, triangulation, and point location in planar subdivisions. ACM Trans. Algorithms **8**(3), article 29 (2012)
14. Dobkin, D.P., Kirkpatrick, D.G.: Determining the separation of preprocessed polyhedra–a unified approach. In: Proceedings of the 17th International Colloquium on Automata, Languages and Programming, pp. 400–413 (1990)
15. Edahiro, M., Kokubo, I., Asano, T.: A new point-location algorithm and its practical efficiency–comparison with existing algorithms. ACM Trans. Graph. **3**(2), 86–109 (1984)
16. Edelsbrunner, H., Guibas, L.J., Stolfi, J.: Optimal point location in a monotone subdivision. SIAM J. Comput. **15**(2), 317–340 (1986)
17. Iacono, J.: Expected asymptotically optimal planar point location. Comput. Geom. Theory Appl. **29**(1), 19–22 (2004)
18. Iacono, J., Langerman, S.: Proximate planar point location. In: Proceedings of the 19th Annual Symposium on Computational Geometry, pp. 220–226 (2003)
19. Iacono, J., Mulzer, W.: A static optimality transformation with applications to planar point location. Int. J. Comput. Geom. Appl. **22**(4), 327–340 (2012)
20. Kirkpatrick, D.G.: Optimal search in planar subdivisions. SIAM J. Comput. **12**(1), 28–35 (1983)
21. Lee, D.T., Preparata, F.P.: Location of a point in a planar subdivision and its applications. SIAM J. Comput. **6**(3), 594–606 (1977)
22. Mulmuley, K.: A fast planar partition algorithm, I. J. Symbolic Comput. **10**(3–4), 253–280 (1990)
23. Preparata, F.P.: A new approach to planar point location. SIAM J. Comput. **10**(3), 473–483 (1981)
24. Sarnak, N., Tarjan, R.E.: Planar point location using persistent search trees. Commun. ACM **29**(7), 669–679 (1986)

25. Seidel, R.: A simple and fast incremental randomized algorithm for computing trapezoidal decompositions and for triangulating polygons. Comput. Geom. Theory Appl. **1**(1), 51–64 (1991)
26. Shannon, C.E.: A mathematical theory of communication. ACM SIGMOBILE Mob. Comput. Commun. Rev. **5**(1), 3–55 (2001)
27. Sleator, D.D., Tarjan, R.E.: Self-adjusting binary search trees. J. ACM **32**(3), 652–686 (1985)

Competitive Local Routing with Constraints

Prosenjit Bose[1], Rolf Fagerberg[2], André van Renssen[3,4](✉), and Sander Verdonschot[1]

[1] School of Computer Science, Carleton University, Ottawa, Canada
jit@scs.carleton.ca, sander@cg.scs.carleton.ca

[2] Department of Mathematics and Computer Science, University of Southern Denmark, Odense, Denmark
rolf@imada.sdu.dk

[3] National Institute of Informatics (NII), Tokyo, Japan
andre@nii.ac.jp

[4] JST, ERATO, Kawarabayashi Large Graph Project, Tokyo, Japan

Abstract. Let P be a set of n vertices in the plane and S a set of non-crossing line segments between vertices in P, called constraints. Two vertices are visible if the straight line segment connecting them does not properly intersect any constraints. The constrained θ_m-graph is constructed by partitioning the plane around each vertex into m disjoint cones with aperture $\theta = 2\pi/m$, and adding an edge to the 'closest' visible vertex in each cone. We consider how to route on the constrained θ_6-graph. We first show that no deterministic 1-local routing algorithm is $o(\sqrt{n})$-competitive on all pairs of vertices of the constrained θ_6-graph. After that, we show how to route between any two visible vertices using only 1-local information, while guaranteeing that the returned path has length at most 2 times the Euclidean distance between the source and destination. To the best of our knowledge, this is the first local routing algorithm in the constrained setting with guarantees on the path length.

1 Introduction

A fundamental problem in any graph is the question of how to route a message from one vertex to another. What makes this more challenging is that often this must be done *locally*, i.e. it can only use knowledge of the source and destination vertex, the current vertex and all vertices directly connected to the current vertex. Routing algorithms are considered *geometric* when the graph that is routed on is embedded in the plane, with edges being straight line segments connecting pairs of vertices and weighted by the Euclidean distance between their endpoints. Geometric routing algorithms are important in wireless sensor networks (see [10,11] for surveys of the area) since they offer routing strategies that use the coordinates of the vertices to guide the search, instead of the more traditional routing tables.

Research supported in part by NSERC, Carleton University's President's 2010 Doctoral Fellowship, and the Danish Council for Independent Research, Natural Sciences.

K. Elbassioni and K. Makino (Eds.): ISAAC 2015, LNCS 9472, pp. 23–34, 2015.
DOI: 10.1007/978-3-662-48971-0_3

We study this problem in the presence of line segment *constraints*. Specifically, let P be a set of vertices in the plane and let S be a set of line segments between vertices in P, with no two line segments intersecting properly. The line segments of S are called *constraints*. Two vertices u and v can *see each other* if and only if either the line segment uv does not properly intersect any constraint or uv is itself a constraint. If two vertices u and v can see each other, the line segment uv is a *visibility edge*. The *visibility graph* of P with respect to a set of constraints S, denoted $\text{Vis}(P, S)$, has P as vertex set and all visibility edges as edge set. In other words, it is the complete graph on P minus all non-constraint edges that properly intersect one or more constraints in S.

This setting has been studied extensively within the context of motion planning amid obstacles. Clarkson [8] was one of the first to study this problem and showed how to construct a $(1 + \epsilon)$-spanner of $\text{Vis}(P, S)$ with a linear number of edges. A subgraph H of G is called a t-spanner of G (for $t \geq 1$) if for each pair of vertices u and v, the shortest path in H between u and v has length at most t times the shortest path in G between u and v. The smallest value t for which H is a t-spanner is the *spanning ratio* of H. Following Clarkson's result, Das [9] showed how to construct a spanner of $\text{Vis}(P, S)$ with constant spanning ratio and constant degree. Bose and Keil [6] showed that the Constrained Delaunay Triangulation is a 2.42-spanner of $\text{Vis}(P, S)$. Recently, the constrained half-θ_6-graph (which is identical to the constrained Delaunay graph whose empty visible region is an equilateral triangle) was shown to be a plane 2-spanner of $\text{Vis}(P, S)$ [4] and all constrained θ-graphs with at least 6 cones were shown to be spanners as well [7].

However, though it is known that these graphs contain short paths, it is not known how to route in a local fashion. To address this issue, we look at k-local routing algorithms in the constrained setting, i.e. routing algorithms that must decide which vertex to forward a message to based solely on knowledge of the source and destination vertex, the current vertex and all vertices that can be reached from the current vertex by following at most k edges. Furthermore, we require our algorithms to be *competitive*, i.e. the length of the returned path needs to be related to the length of the shortest path in the graph.

In the unconstrained setting, there exists a 1-local 0-memory routing algorithm that is 2-competitive on the θ_6-graph and $5/\sqrt{3}$-competitive on the half-θ_6-graph (the θ_6-graph consists of the union of two half-θ_6-graphs) [3]. In the same paper, the authors also show that these ratios are the best possible, i.e. there are matching lower bounds.

In this paper, we show that the situation in the constrained setting is quite different: no deterministic 1-local routing algorithm is $o(\sqrt{n})$-competitive on all pairs of vertices of the constrained θ_6-graph, regardless of the amount of memory it is allowed to use. Despite our lower bound, we describe a 1-local 0-memory routing algorithm between any two *visible* vertices of the constrained θ_6-graph that guarantees that the length of the path traveled is at most 2 times the Euclidean distance between the source and destination. Additionally, we provide a 1-local $O(1)$-memory 18-competitive routing algorithm between any two visible vertices in the constrained half-θ_6-graph. To the best of our knowledge, these are the first local routing algorithms in the constrained setting with guarantees on the path length.

2 Preliminaries

We define a *cone* C to be the region in the plane between two rays originating from a single vertex, the apex of the cone. We let six rays originate from each vertex, with angles to the positive x-axis being multiples of $\pi/3$ (see Fig. 1). Each pair of consecutive rays defines a cone. We write C_i^u to indicate the i-th cone of a vertex u, or C_i if the apex is clear from the context. For ease of exposition, we only consider point sets in general position: no two vertices define a line parallel to one of the rays that define the cones and no three vertices are collinear.

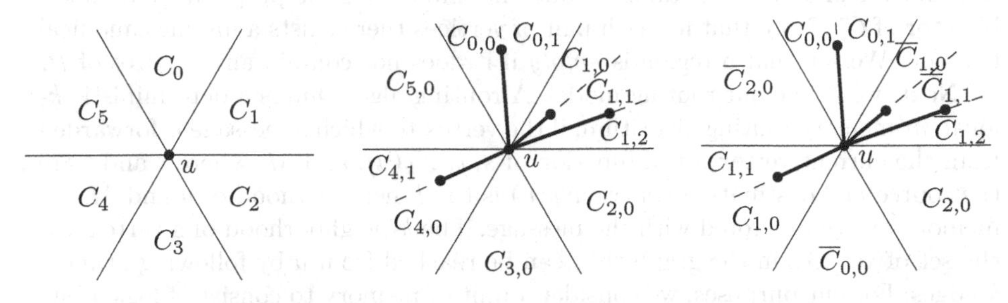

Fig. 1. The cones having apex u in the θ_6-graph

Fig. 2. The subcones having apex u in the constrained θ_6-graph.

Fig. 3. The subcones having apex u in the constrained half-θ_6-graph.

Let vertex u be an endpoint of a constraint and let the other endpoint lie in cone C_i^u. The lines through all such constraints split C_i^u into several *subcones* (see Fig. 2). We use $C_{i,j}^u$ to denote the j-th subcone of C_i^u. When a constraint $c = (u, v)$ splits a cone of u into two subcones, we define v to lie in both of these subcones. We consider a cone that is not split to be a single subcone.

The constrained θ_6-graph is constructed as follows: for each subcone $C_{i,j}$ of each vertex u, add an edge from u to the closest visible vertex in that subcone, where distance is measured along the bisector of the original cone, not the subcone. More formally, we add an edge between two vertices u and v if v can see u, $v \in C_{i,j}$, and for all vertices $w \in C_{i,j}$ that can see u, $|uv'| \leq |uw'|$, where v' and w' denote the orthogonal projection of v and w on the bisector of C_i. Note that our general position assumptions imply that each vertex adds at most one edge per subcone to the graph.

Next, we define the constrained half-θ_6-graph. This is a generalized version of the half-θ_6-graph as described by Bonichon *et al.* [1]. The constrained half-θ_6-graph is similar to the constrained θ_6-graph with one major difference: edges are only added in every second cone. More formally, its cones are categorized as positive and negative. Let $(C_0, \overline{C}_2, C_1, \overline{C}_0, C_2, \overline{C}_1)$ be the sequence of cones in counterclockwise order starting from the positive y-axis. The cones C_0, C_1, and C_2 are called *positive* cones and $\overline{C}_0$, $\overline{C}_1$, and $\overline{C}_2$ are called *negative* cones. Note that the positive cones coincide with the even cones of the constrained θ_6-graph and the negative cones coincide with the odd ones. We add edges only in the positive cones (and their subcones). We use C_i^u and $\overline{C}_i^u$ to denote cones C_i and $\overline{C}_i$ with apex u. For any two vertices u and v, $v \in C_i^u$ if and only if $u \in \overline{C}_i^v$ (see Fig. 3). Analogous to the subcones defined for the θ_6-graph, constraints can split cones into subcones. We call a subcone of a positive cone a positive subcone and a subcone of a negative cone a negative subcone (see Fig. 3). We look at the undirected version of these graphs, i.e. when an edge is added, both vertices are allowed to use it. This is consistent with previous work on θ-graphs.

Given a vertex w in a positive cone C_i^u of vertex u, we define the *canonical triangle* T_{uw} to be the triangle defined by the borders of C_i^u (not the borders of the subcone of u that constains w) and the line through w perpendicular to the bisector of C_i^u. Note that for each pair of vertices there exists a unique canonical triangle. We say that a region is *empty* if it does not contain any vertices of P.

Next, we define our routing model. A routing algorithm is a deterministic k-local, m-memory routing algorithm, if the vertex to which a message is forwarded from the current vertex u is a function of s, t, $N_k(u)$, and M, where s and t are the source and destination vertex, $N_k(u)$ is the k-neighborhood of u and M is a memory of size m, stored with the message. The k-neighborhood of a vertex u is the set of vertices in the graph that can be reached from u by following at most k edges. For our purposes, we consider a unit of memory to consist of $\log_2 n$ bits or a point in $\mathbb{R}^2$. Our model also assumes that the only information stored at each vertex of the graph is $N_k(u)$. Since our graphs are geometric, we identify each vertex by its coordinates in the plane. Unless otherwise noted, all routing algorithms we consider in this paper are deterministic 0-memory algorithms.

There are essentially two notions of *competitiveness* of a routing algorithm. One is to look at the Euclidean shortest path between the two vertices, i.e. the shortest path in the visibility graph, and the other is to compare the routing path to the shortest path in the graph. A routing algorithm is *c-competitive with respect to the Euclidean shortest path (resp. shortest path in the graph)* provided that the total distance traveled by the message is not more than c times the Euclidean shortest path length (resp. shortest path length) between source and destination. The *routing ratio* of an algorithm is the smallest c for which it is c-competitive.

Since the shortest path in the graph between two vertices is at least as long as the Euclidean shortest path between them, an algorithm that is c-competitive with respect to the Euclidean shortest path is also c-competitive with respect to the shortest path in the graph. We use competitiveness with respect to the Euclidean shortest path when proving upper bounds and with respect to the shortest path in the graph when proving lower bounds.

To be able to talk about points at intersections of lines, we distinguish between *vertices* and *points*. A *point* is any point in $\mathbb{R}^2$, while a *vertex* is part of the input.

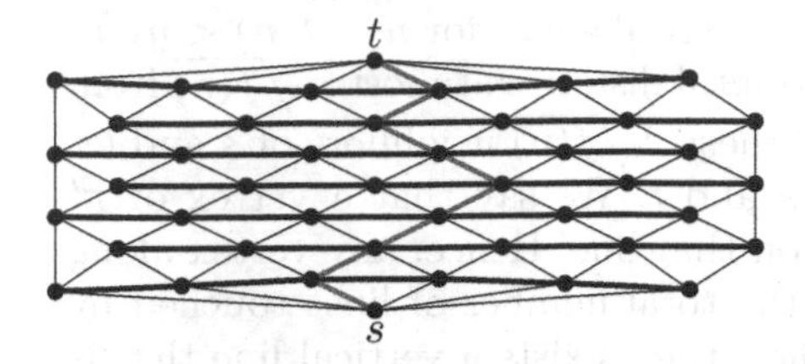

Fig. 4. The constrained θ_6-graph starting from a grid, using horizontal constraints to block vertical edges, and the red path of the routing algorithm(Color figure online)

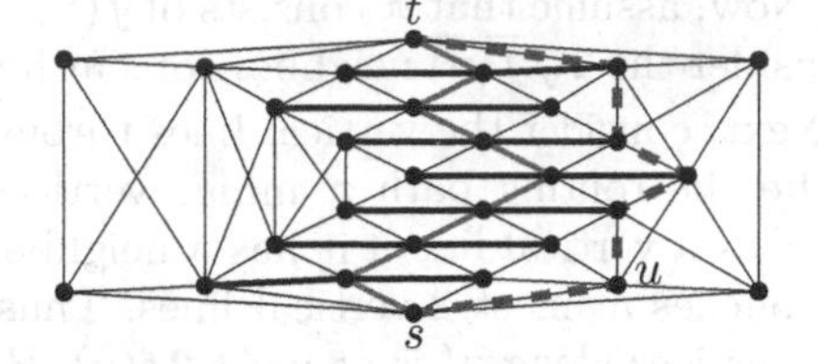

Fig. 5. The constrained θ_6-graph that looks the same from the red path of the routing algorithm, but has an almost vertical dashed blue path(Color figure online)

3 Lower Bound on Local Routing

We modify the proof by Bose *et al.* [2] (that shows that no deterministic routing algorithm is $o(\sqrt{n})$-competitive for all triangulations) to show the following lower bound.

Theorem 3.1. *No deterministic 1-local routing algorithm is* $o(\sqrt{n})$*-competitive with respect to the shortest path on all pairs of vertices of the* θ_6*-graph, regardless of the amount of memory it is allowed to use.*

Due to space constraints, we present a shortened version of the proof of this theorem. The full proof can be found in the arXiv version [5].

Proof. Consider an $n \times n$ grid and shift every second row to the right by half a unit. We stretch the grid, such that each horizontal edge has length n (see Fig. 4). Next, we replace each horizontal edge by a constraint to prevent vertical visibility edges. Finally, we add two additional vertices, origin s and destination t, centered horizontally at one unit below the bottom row and one unit above the top row, respectively.

We move all vertices by at most some arbitrarily small amount ϵ, such that no two vertices define a line parallel to one of the rays that define the cones and no three vertices are collinear. In particular, we ensure that all vertices on the bottom row have s as the closest vertex in one of their subcones and all vertices on the top row have t as the closest vertex in one of their subcones. On this point set and these constraints, we build the constrained θ_6-graph G.

Consider any deterministic 1-local ∞-memory routing algorithm and let π be the path this algorithm takes when routing from s to t. If π consists of at least $n\sqrt{n}$ non-vertical steps, the total length of the path is $\Omega(n^2\sqrt{n})$. However, G contains a path of length $O(n^2)$ between s and t: the path that follows a diagonal edge to the left of line st, followed by a diagonal edge to the right, until it reaches t. Hence, in this case, the local routing algorithm is not $o(\sqrt{n})$-competitive.

Now, assume that π consists of $f(n)$ non-vertical steps, for $n < f(n) < n\sqrt{n}$. Consider the $2\sqrt{f(n)}$ neighbors of s at horizontal distance at most $n\sqrt{f(n)}$ from s. Next, consider the vertical lines through these $2\sqrt{f(n)}$ neighbors of s and let π' be the routing path π minus vertices s and t. We say that a vertex of π' *touches* a vertical line if it has a neighbor on that line. Hence, any vertex along π' touches at most 2 vertical lines. Thus, the total number of lines touched by the vertices along π' is at most $2f(n)$. Hence, there exists a vertical line that is touched at most $\sqrt{f(n)}$ times. Let u be the neighbor of s on the vertical line that is touched the fewest number of times.

We now create a new constrained θ_6-graph G' such that the deterministic 1-local routing algorithm follows the same path, but G' contains a short 'almost vertical' path via u. We start with s, t, and all vertices of π. Next, we add all vertices and constraints connected to these vertices in G. On this point set and these constraints, we build the constrained θ_6-graph G' (see Fig. 5).

Since the horizontal distance between vertices is far larger than their vertical distance, an 'almost vertical' path from u to the top row of G' is formed. This almost vertical path is a path that is vertical whenever possible and uses detours to avoid path π (see Fig. 6): If π arrives at a vertex v that has a neighbor on the vertical line through u, we avoid π by following one edge away from π, followed by an edge back to the vertical line through u (see Fig. 6a). If π arrives at a vertex on the vertical line through u, we avoid the vertex before and after v on π as before, and meet π at v (see Fig. 6b). Since no edge along the left and right boundary of G touches the vertical line through u, this vertical line is touched by at most $\sqrt{f(n)}$ vertices of π and only $O\left(\sqrt{f(n)}\right)$ of these detour edges are required. Hence, G' contains a path from s to t of length $O\left(n\sqrt{f(n)}\right)$.

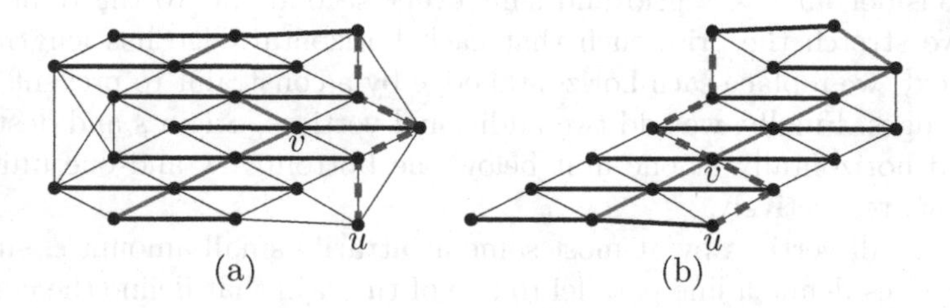

Fig. 6. The two types of detour: (a) when π does not visit the vertical line through u, (b) when π visits the vertical line through u

Since the 1-local routing algorithm is deterministic and the 1-local information of the vertices of π in G' is the same as in G, the algorithm follows the same path. The remainder of the proof uses a case distinction in order to compare the length of the routing path with the length of the shortest path. The general idea is that since most edges in G' have length at least n, π has length $\Omega(nf(n))$, which implies that π is not $o(\sqrt{n})$-competitive, as $f(n) \geq n+1$. Hence, since G' can be constructed for any deterministic 1-local routing algorithm, we have shown that no deterministic 1-local routing algorithm is $o(\sqrt{n})$-competitive on all pairs of vertices. □

4 Routing on the Constrained θ_6-Graph

In this section, we provide a 1-local routing algorithm on the constrained θ_6-graph for any pair of visible vertices. Since the constrained θ_6-graph is the union of two constrained half-θ_6-graphs, we start by describing a routing algorithm for the constrained half-θ_6-graph for the case where the destination t lies in a positive subcone of the origin s. Throughout this section, we use the following auxiliary lemma proven by Bose *et al.* [4].

Lemma 4.1. *Let u, v, and w be three arbitrary points in the plane such that uw and vw are visibility edges and w is not the endpoint of a constraint intersecting the interior of triangle uvw. Then there exists a convex chain of visibility edges from u to v in triangle uvw, such that the polygon defined by uw, wv and the convex chain is empty and does not contain any constraints.*

4.1 Positive Routing on the Constrained Half-θ_6-Graph

Before describing how to route when t lies in a positive subcone of s, we first show that there exists a path in canonical triangle T_{st}.

Lemma 4.2. *Given two vertices u and w such that u and w see each other and w lies in a positive subcone $C^u_{i,j}$, there exists a path between u and w in the triangle T_{uw} in the constrained half-θ_6-graph.*

The proof of this lemma is a straightforward modification of Theorem 1 in [4].

Positive Routing Algorithm for the Constrained Halfθ_6-Graph. Next, we describe how to route from s to t, when s can see t and t lies in a positive subcone $C^s_{i,j}$ (see Fig. 7): When we are at s, we follow the edge to the closest vertex in the subcone that contains t. When we are at any other vertex u, we look at all edges in the subcones of C^u_i and all edges in the subcones of the adjacent negative cone $\overline{C}^u$ that is intersected by st. An edge in a subcone of $\overline{C}^u$ is considered only if it does not cross st. For example, in Fig. 7, we do not consider the edge to v_1 since it lies in $\overline{C}^u$ and crosses st. It follows that we can cross st only when we follow an edge in C^u_i.

Let z be the intersection of st and the boundary of $\overline{C}^u$ that is not a boundary of C^u_i. We follow the edge uv that minimizes the unsigned angle $\angle zuv$. For example, in Fig. 7, when we are at vertex u we follow the edge to v_2 since, out of the two remaining edges uv_2 and uv_3, $\angle zuv_2$ is smaller than $\angle zuv_3$. We also note that during the routing process, t does not necessarily lie in C^u_i. Finally, since the algorithm uses only information about the location of s and t and the neighbors of the current vertex, it is a 1-local routing algorithm.

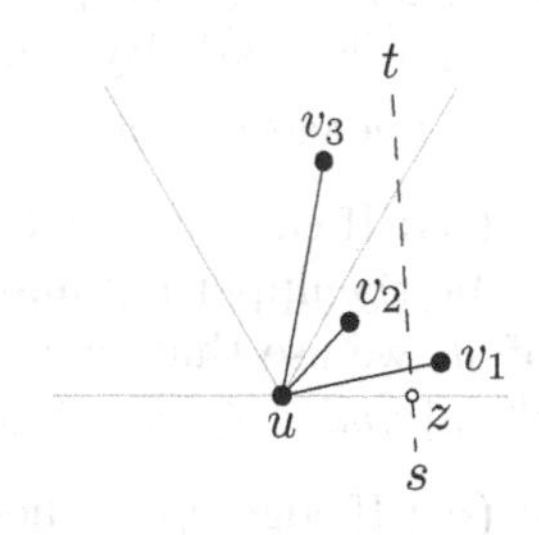

Fig. 7. An example of routing from s to $t \in C^s_0$. The dashed line represents the visibility line between s and t

We proceed by proving that the above routing algorithm can always perform a step, i.e. at every vertex reached there exists an edge that is considered by the algorithm. Due to space constraints, we only state the lemma and refer the reader to the arXiv version [5] for the proof.

Lemma 4.3. *The routing algorithm can always perform a step in the constrained half-θ_6-graph.*

Theorem 4.4. *Given two vertices s and t in the half-θ_6-graph such that s and t can see each other and t lies in a positive subcone of s, there exists a 1-local routing algorithm that routes from s to t and is 2-competitive with respect to the Euclidean distance.*

Proof. We assume without loss of generality that $t \in C_0^s$. The routing algorithm will thus only take steps in $C_0^{v_i}$, $\overline{C}_1^{v_i}$, and $\overline{C}_2^{v_i}$, where v_i is an arbitrary vertex along the routing path. Let a and b be the upper left and right corner of T_{st}. To bound the length of the routing path, we first bound the length of each edge. We consider three cases: (a) edges in subcones of $\overline{C}_1^{v_i}$ or $\overline{C}_2^{v_i}$, (b) edges in subcones of $C_0^{v_i}$ that do not cross st, (c) edges in subcones of $C_0^{v_i}$ that cross st. For ease of notation we use v_0 and v_k to denote s and t.

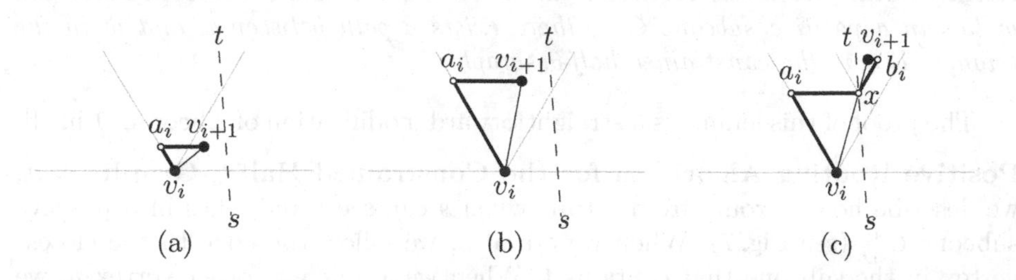

Fig. 8. Bounding the edge lengths: (a) an edge in a subcone of $\overline{C}_1^u$, (b) an edge in a subcone of C_0^u that does not cross st, and (c) an edge in a subcone of C_0^u that crosses st

Case (a): If edge $v_i v_{i+1}$ lies in a subcone of $\overline{C}_1^{v_i}$, let a_i be the upper corner of $T_{v_{i+1} v_i}$ (see Fig. 8a). By the triangle inequality, we have that $|v_i v_{i+1}| \leq |v_i a_i| + |a_i v_{i+1}|$. The case where $v_i v_{i+1}$ lies in $\overline{C}_2^{v_i}$ is analogous.

Case (b): If edge $v_i v_{i+1}$ lies in a subcone of $C_0^{v_i}$ and does not cross st, let a_i and b_i be the upper left and right corner of $T_{v_i v_{i+1}}$ (see Fig. 8b). If v_i lies to the left of st, we use that $|v_i v_{i+1}| \leq |v_i a_i| + |a_i v_{i+1}|$. If v_i lies to the right of st, we use that $|v_i v_{i+1}| \leq |v_i b_i| + |b_i v_{i+1}|$.

Case (c): If edge $v_i v_{i+1}$ lies in a subcone of $C_0^{v_i}$ and crosses st, we split it into two parts, one for each side of st (see Fig. 8c). Let x be the intersection of st and $v_i v_{i+1}$. If u lies to the left of st, let a_i be the upper left corner of $T_{v_i x}$ and let b_i be the upper right corner of $T_{x v_{i+1}}$. By the triangle inequality, we have that

$|v_i v_{i+1}| \leq |v_i a_i| + |a_i x| + |x b_i| + |b_i v_{i+1}|$. If u lies to the right of st, let a_i be the upper left corner of $T_{x v_{i+1}}$ and let b_i be the upper right corner of $T_{v_i x}$. By triangle inequality, we have that $|v_i v_{i+1}| \leq |v_i b_i| + |b_i x| + |x a_i| + |a_i v_{i+1}|$.

To bound the length of the full path, let x and x' be two consecutive points where the routing path crosses st and let $v_i v_{i+1}$ be the edge that crosses st at x and let $v_{i'} v_{i'+1}$ be the edge that crosses st at x'. Let a_x and b_x be the upper left and right corner of $T_{xx'}$. If the path between x and x' lies to the left of st, this part of the path is bounded by $|x a_i| + \sum_{j=i}^{i'-1} |a_j v_{j+1}| + \sum_{j=i+1}^{i'} |v_j a_j| + |a_{i'} x'|$. Since $x a_i$ and all $v_j a_j$ are parallel to $x a_x$ and all $a_x v_{j+1}$ are horizontal, we have that $|x a_i| + \sum_{j=i+1}^{i'} |v_j a_j| = |x a_x|$. Similarly, since $a_{i'} x'$ and all $a_j v_{j+1}$ are parallel and have disjoint projections onto $a_x x'$, we have that $\sum_{j=i}^{i'-1} |a_j v_{j+1}| + |a_{i'} x'| = |a_x x'|$. Thus, the length of a path to the left of st is at most $|x a_x| + |a_x x'|$. If the path between x and x' lies to the right of st, this part of the path is bounded by $|x b_i| + \sum_{j=i}^{i'-1} |b_j v_{j+1}| + \sum_{j=i+1}^{i'} |v_j b_j| + |b_{i'} x'| = |x b_x| + |b_x x'|$ (see Fig. 9a).

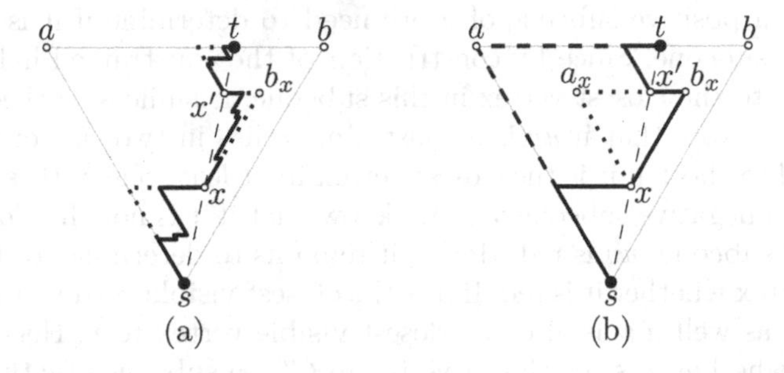

Fig. 9. Bounding the total length: (a) the bounds (solid lines) are unfolded (dotted lines) and (b) the unfolded bounds (solid lines) are flipped to the longer of the two sides (dotted lines) and unfolded again (dashed lines)

Next, we flip all unfolded bounds to the longer of the two sides at and bt: if $|at| \geq |bt|$, we replace all bounds of the form $|x b_x| + |b_x x'|$ by $|x a_x| + |a_x x'|$ and if $|at| < |bt|$, we replace all bounds of the form $|x a_x| + |a_x x'|$ by $|x b_x| + |b_x x'|$ (see Fig. 9b). Note that this can only increase the length of the bounds. Finally, we sum these bounds and get $\max\{|sa| + |at|, |sa| + |bt|\}$, which is at most $2 \cdot |st|$. □

4.2 Routing on the Constrained θ_6-Graph

To route on the constrained θ_6-graph, we split it into two constrained half-θ_6-graphs: the constrained half-θ_6-graph oriented as in Fig. 3 and the constrained half-θ_6-graph where positive and negative cones are inverted. When we route from s to t, we pick the constrained half-θ_6-graph in which t lies in a positive subcone of s, referred to as G^+ in the remainder of this section, and apply the routing algorithm described in the previous section. Since this routing algorithm is 1-local and 2-competitive, we obtain a 1-local and 2-competitive routing algorithm for the constrained θ_6-graph, provided that we can determine locally, while routing, whether an edge is part of G^+. When at a vertex u, we consider the edges in order of increasing angle with the horizontal halfline through u that intersects st.

Lemma 4.5. *While executing the positive routing algorithm for two visible vertices s and t, we can determine locally at a vertex u for any edge uv in the constrained θ_6-graph whether it is part of G^+.*

Proof. Suppose we color the edges of the constrained θ-graph red and blue such that red edges form G^+ and blue edges form the constrained half-θ_6-graph, where t lies in a negative subcone of s. At u, we need to determine locally whether uv is red. Since an edge can be part of both constrained half-θ_6-graphs, it can be red and blue at the same time. This makes it harder to determine whether an edge is red, since determining that it is blue does not imply that it is not red.

If v lies in a positive subcone of u, we need to determine if it is the closest vertex in that subcone. Since by construction of the constrained half-θ_6-graph, u is connected to the closest vertex in this subcone, it suffices to check whether this vertex is v. Note that if uv is a constraint, v lies in two subcones of u and hence we need to check if it is the closest vertex in at least one of these subcones.

If v lies in a negative subcone of u, we know that if it is not the closest visible vertex in that subcone, uv is red. Hence, it remains to determine for the edge to the closest vertex whether it is red: If it is the closest visible vertex, it is blue, but it may be red as well if u is also the closest visible vertex to v. Hence, we need to determine whether u is the closest vertex in $C^v_{i,j}$, a subcone of v that contains u. We consider two cases: (a) uv is a constraint, (b) uv is not a constraint.

Case (a): Since uv is a constraint, it cannot cross st. Since we are considering uv, all edges that make a smaller angle with the horizontal halfline through u that intersects st are not red. Hence, uv is either part of the boundary of the routing path or the constraint is contained in the interior of the region bounded by the routing path and st. However, by the invariant of Lemma 4.3, the region bounded by the routing path and st does not contain any constraints in its interior. Thus, uv is part of the boundary of the routing path and uv is red.

Case (b): If uv is not a constraint, let regions A and B be the intersection of C^v_i and the two subcones of u adjacent to $\overline{C}^u_i$ and let C be the intersection of $C^v_{i,j}$ and the negative subcone of u that contains v (see Fig. 10). We first note that since uv lies in a negative subcone of u, the invariant of Lemma 4.3 implies that B is empty. Furthermore, since v is the closest visible vertex to u, C does not contain any vertices that can see u or v.

Since C does not contain any vertices that can see u or v, any constraint in $\overline{C}_i^u$ that has u as an endpoint and lies above uv, ensures that v cannot see A, i.e. it cannot block visibility of this region only partially. Hence, if such a constraint exists, u is the closest visible vertex to v in $C_{i,j}^v$, since neither B nor C contain any vertices visible to v. Therefore, uv is red.

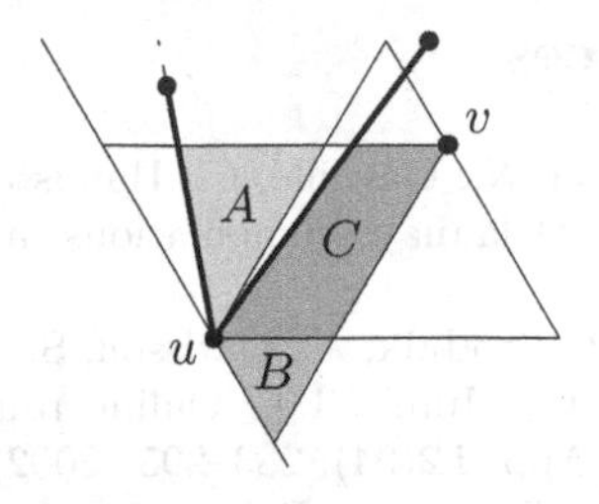

Fig. 10. Determining whether an edge is part of the constrained half-θ_6-graph

If v can see A, we show that uv is red, if and only if the closest visible vertex in the subcone of u that contains A does not lie in A. We first show that if uv is red, then the closest visible vertex in the subcone of u that contains A does not lie in A. We prove the contrapositive of this statement. Since A is visible to v, u is not the endpoint of a constraint in $\overline{C}_i^u$ above uv. Hence, we have two visibility edges uv and ux and u is not the endpoint of a constraint intersecting the interior of triangle uxv. Thus, by Lemma 4.1, we have a convex chain between x and v. Let y be the vertex adjacent to v along this chain. Since the polygon defined by ux, uv, and the convex chain is empty and does not contain any constraints, y lies in $C_{i,j}^v$. Thus, u is not the closest visible vertex in $C_{i,j}^v$ and uv is not red.

Next, we show that if the closest visible vertex x in the subcone of u that contains A does not lie in A, then uv is red. We prove this by contradiction, so assume that uv is not red. This implies that there exists a vertex $y \in C_{i,j}^v$ that is visible to v and closer than u. Since B is empty and C does not contain any vertices that can see v, y lies in A. Since uv and vy are visibility edges and v is not the endpoint of a constraint intersecting the interior of triangle uyv, by Lemma 4.1 there exists a convex chain of visibility edges between u and y. Furthermore, since C does not contain any vertices that can see u, the vertex adjacent to u along this chain lies in A. Since any vertex in A is closer to u than x, this leads to a contradiction, completing the proof. □

4.3 Negative Routing on the Constrained Half-θ_6-Graph

To complement the positive routing algorithm on the constrained half-θ_6-graph, we also provide a negative routing algorithm on this graph. Due to space constraints, we refer the reader to [5] for details on the routing algorithm. We note that negative routing is harder than positive routing, since there need not be an edge to a vertex in the cone of s that contains t. This also caused the separation between spanning ratio and routing ratio in the unconstrained setting [3].

Theorem 4.6. *There exists an $O(1)$-memory 1-local 18-competitive routing algorithm for negative routing in the constrained half-θ_6-graph.*

References

1. Bonichon, N., Gavoille, C., Hanusse, N., Ilcinkas, D.: Connections between theta-graphs, Delaunay triangulations, and orthogonal surfaces. In: WG, pp. 266–278 (2010)
2. Bose, P., Brodnik, A., Carlsson, S., Demaine, E.D., Fleischer, R., López-Ortiz, A., Morin, P., Munro, I.J.: Online routing in convex subdivisions. Int. J. Comput. Geom. App. **12**(04), 283–295 (2002)
3. Bose, P., Fagerberg, R., van Renssen, A., Verdonschot, S.: Competitive routing in the half-θ_6-graph. In: SODA, pp. 1319–1328 (2012). To appear in SIAM J. Comput
4. Bose, P., Fagerberg, R., van Renssen, A., Verdonschot, S.: On plane constrained bounded-degree spanners. In: Fernández-Baca, D. (ed.) LATIN 2012. LNCS, vol. 7256, pp. 85–96. Springer, Heidelberg (2012)
5. Bose, P., Fagerberg, R., van Renssen, A.,Verdonschot, S.: Competitive local routing with constraints. ArXiv e-prints (2014). arXiv:1412.0760 [cs.CG]
6. Bose, P., Keil, J.M.: On the stretch factor of the constrained Delaunay triangulation. In: ISVD, pp. 25–31 (2006)
7. Bose, P., van Renssen, A.: Upper bounds on the spanning ratio of constrained theta-graphs. In: Pardo, A., Viola, A. (eds.) LATIN 2014. LNCS, vol. 8392, pp. 108–119. Springer, Heidelberg (2014)
8. Clarkson, K.: Approximation algorithms for shortest path motion planning. In: STOC, pp. 56–65 (1987)
9. Das, G.: The visibility graph contains a bounded-degree spanner. In: CCCG, pp. 70–75 (1997)
10. Misra, S.C., Woungang, I., Misra, S. (eds.): Guide to Wireless Sensor Networks. Springer, London (2009)
11. Räcke, H.: Survey on oblivious routing strategies. In: Ambos-Spies, K., Löwe, B., Merkle, W. (eds.) CiE 2009. LNCS, vol. 5635, pp. 419–429. Springer, Heidelberg (2009)

Navigating Weighted Regions with Scattered Skinny Tetrahedra

Siu-Wing Cheng[1](✉), Man-Kwun Chiu[2,3], Jiongxin Jin[4], and Antoine Vigneron[5]

[1] Department of Computer Science and Engineering, HKUST, Hong Kong, Hong Kong
scheng@cse.ust.hk
[2] National Institute of Informatics (NII), Tokyo, Japan
[3] JST, ERATO, Kawarabayashi Large Graph Project, Tokyo, Japan
[4] Google Inc., Seattle, USA
[5] Visual Computing Center, King Abdullah University of Science and Technology (KAUST), Thuwal, Saudi Arabia

Abstract. We propose an algorithm for finding a $(1+\varepsilon)$-approximate shortest path through a weighted 3D simplicial complex $\mathcal{T}$. The weights are integers from the range $[1, W]$ and the vertices have integral coordinates. Let N be the largest vertex coordinate magnitude, and let n be the number of tetrahedra in $\mathcal{T}$. Let ρ be some arbitrary constant. Let κ be the size of the largest connected component of tetrahedra whose aspect ratios exceed ρ. There exists a constant C dependent on ρ but independent of $\mathcal{T}$ such that if $\kappa \leq \frac{1}{C}\log\log n + O(1)$, the running time of our algorithm is polynomial in n, $1/\varepsilon$ and $\log(NW)$. If $\kappa = O(1)$, the running time reduces to $O(n\varepsilon^{-O(1)}(\log(NW))^{O(1)})$.

Keywords: Weighted region · Shortest path · Approximation algorithm

1 Introduction

Finding shortest paths are classical geometric optimization problems (e.g. [4,10–12,15]). In 2D, researchers have also studied cost models in applications that are non-L_p metrics and anisotropic (e.g. [1,2,5–9,14,17]). In 3D, other than motion planning, shortest path is a popular tool for simulating seismic raytracing in ray-based tomography schemes for studying some geological properties (e.g. [13]).

The weighted region problem is a way to model the unequal difficulties in traversing different regions [14]. In 3D, we are given a simplicial complex $\mathcal{T}$ of n tetrahedra. These tetrahedra and their vertices, edges and triangles are called the *simplices* of $\mathcal{T}$. Given two simplices in $\mathcal{T}$, either they are disjoint or their intersection is another simplex in $\mathcal{T}$. Every vertex has integral coordinates and

S.-W. Cheng—Supported by Research Grants Council, Hong Kong, China (project no. 611812).

K. Elbassioni and K. Makino (Eds.): ISAAC 2015, LNCS 9472, pp. 35–45, 2015.
DOI: 10.1007/978-3-662-48971-0_4

let N denote the largest vertex coordinate magnitude. Each tetrahedron τ is associated with an integral weight $\omega_\tau \in [1, W]$. For every edge or triangle, its weight is equal to the minimum weight among the tetrahedra incident to that edge or triangle. The cost of a path that lies in a simplex σ is equal to the path length multiplied by ω_σ. Given a path P in $\mathcal{T}$, we denote its length by $\|P\|$ and its cost by $\text{cost}(P) = \sum_{\text{simplex } \sigma} \omega_\sigma \|P \cap \sigma\|$. The weighted region problem is to find the least-cost path from a given source vertex to a given destination vertex.

The weighted region problem in 2D has been studied extensively. Fully polynomial time approximation schemes are known [7,14]. There are also successful discretization schemes whose running time is linear in the input size and dependent on some geometric parameter of the polygonal domain [2,17]. In contrast, only one algorithm for the weighted region problem in 3D has been proposed (Aleksandrov et al. [3]). The authors [3] present a $(1+\varepsilon)$-approximation algorithm whose running time is $O\left(Kn\varepsilon^{-2.5} \log \frac{n}{\varepsilon} \log^3 \frac{1}{\varepsilon}\right)$, where K is asymptotically at least the cubic power of the maximum aspect ratio of the tetrahedra in the worst case. (Aspect ratio is defined in Sect. 2.) It is an open problem whether an FPTAS exists for the 3D weighted region problem.

Let ρ be an arbitrary constant independent of $\mathcal{T}$. We call a tetrahedron *skinny* if its aspect ratio exceeds ρ. Two skinny tetrahedra are *connected* if their boundaries touch, and the transitive closure of this relation gives the connected components of skinny tetrahedra. Let κ be the number of tetrahedra in the largest connected component of skinny tetrahedra.

We present a $(1+\varepsilon)$-approximation algorithm for the 3D weighted region problem. It runs in $O\big(2^{2^{O(\kappa)}} n\varepsilon^{-7} \log^2 \frac{W}{\varepsilon} \log^2 \frac{NW}{\varepsilon}\big)$ time. The hidden constant in the exponent $O(\kappa)$ is dependent on ρ but independent of $\mathcal{T}$. Thus, there exists a constant C dependent on ρ but independent of $\mathcal{T}$ such that if $\kappa \le \frac{1}{C} \log\log n + O(1)$, the running time is polynomial in n, $1/\epsilon$ and $\log(NW)$. If $\kappa = O(1)$, the running time is linear in n. In comparison, the running time in [3] has the advantage of being independent from N and W, but K can be arbitrarily large even if there are only $O(1)$ skinny tetrahedra. Putting the result in [3] in our model, K is a function of N and n in the worst case, and K can be $\Omega(\frac{1}{n}N^3 + 1)$.

2 Preliminaries

A path P in $\mathcal{T}$ consists of *links* and *nodes*. A link is a maximal segment that lies in a simplex of $\mathcal{T}$. Nodes are link endpoints. We assume that P does not bend in the interior of any simplex because such a bend can be shortcut. So the nodes of P lie at vertices, edges and triangles. Given two points x and y in this order in P, we use $P[x, y]$ to denote the subpath between them.

The *simplex sequence* of a path P is the ordered sequence Σ of vertices, edges and triangles that intersect the interior of P from u to v. If P has the minimum cost among all paths from u to v with simplex sequence Σ, we call P a *locally shortest path* (with respect to Σ). The shortest path from u to v is the locally shortest path with the minimum cost among all possible simplex sequences.

Let $B(x, r)$ denote a closed ball centered at a point x with radius r.

The aspect ratio of a tetrahedron τ is the ratio of the radius of the smallest sphere that encloses τ to the radius of the largest sphere inscribed in τ. If the aspect ratio is bounded by a constant, all angles of τ are bounded from below and above by some constants. A tetrahedron is *skinny* if its aspect ratio exceeds some arbitrary constant ρ fixed *a priori*. If a tetrahedron is not skinny, it is *fat*.

Two tetrahedra are *connected* if their boundaries touch. The equivalence classes of the transitive closure of this relation are called *connected components* of tetrahedra. Two tetrahedra are *edge-connected* if they share at least one edge. The equivalence classes of the transitive closure of this relation are called *edge-connected components* of tetrahedra. A *cluster* is a connected component of skinny tetrahedra. Recall that every cluster has at most κ tetrahedra.

For every simplex σ in $\mathcal{T}$, star(σ) denotes the set of tetrahedra that have σ as a boundary simplex. Given a set $\mathcal{K}$ of simplices, $|\mathcal{K}|$ denotes the union of all simplices in $\mathcal{K}$ and bd($\mathcal{K}$) denotes the set of simplices in the boundary of $|\mathcal{K}|$.

For simplicity, we will show a $1 + O(\varepsilon)$ approximation ratio, which can be reduced to $1 + \varepsilon$ by tuning some constants. Our algorithm discretizes $\mathcal{T}$ and builds an edge-weighted graph $\mathcal{G}$ so that the shortest path in $\mathcal{G}$ is a $1 + O(\varepsilon)$ approximation. This approach is also taken in [3]. However, in order to allow for skinny tetrahedra, we discretize the fat tetrahedra only, and the edges in $\mathcal{G}$ represent approximate shortest paths that may not lie within a single tetrahedron.

Let $\{u, v\}$ be a pair of vertices of $\mathcal{G}$. If u and v lie in a cluster, we would ideally connect them by an edge with weight equal to the shortest path cost between u and v within the cluster. However, even if a simplex sequence is given, finding the locally shortest path requires solving a nonlinear system derived using Snell's law. It is unclear how to do this exactly. Instead, we switch to convex distance functions induced by convex polytopes with $O(1/\varepsilon)$ vertices, so that the modified metrics give $1+O(\varepsilon)$ approximations of the original metrics. Under the modified metrics, the locally shortest path with respect to Σ can be obtained by linear programming. We enumerate all possible simplex sequences to find the shortest path cost within the cluster under the modified metrics.

3 Placement of Steiner Points

For every vertex v in $\mathcal{T}$, the fat tetrahedra in star(v) may form multiple edge-connected components and we call each a *fat substar*. For an edge or triangle σ, there is at most one fat substar in star(σ).

Definition 1. *Let x be a point in the union of vertices, edges and triangles of $\mathcal{T}$. Let σ be the simplex of lowest dimension containing x. For every fat substar F of σ, define $\delta_F(x)$ to be the minimum distance from x to a simplex in* bd(F) *that does not contain x. When σ is an edge or triangle, there is at most one fat substar of σ and so we simplify the notation to $\delta(x)$.*

Remark 1: For a vertex v of $\mathcal{T}$, $\delta_F(v)$ is the distance between v and a triangle opposite v in some tetrahedron $\tau \in F$. Since the tetrahedra in F have bounded aspect ratio and there are $O(1)$ of them, $\delta_F(v) = \Theta(\|e\|)$ for every edge $e \in F$.

Remark 2: For a point x in the interior of an edge e, $\delta(x)$ is the distance between x and an edge or triangle σ that bounds a fat tetrahedron incident to e and shares only a vertex v with e. Also, $\delta(x) = \Theta(\|vx\|)$.

For every vertex v of $\mathcal{T}$ and every fat substar F of v, define a *vertex-ball* $B_{v,F} = B(v, \frac{\varepsilon}{3W}\delta_F(v))$. Let N_v be the union of $B_{v,F} \cap F$ over all fat substars F.

Let uv be an edge of a fat tetrahedron in $\mathcal{T}$. We place Steiner points in uv outside N_u and N_v as follows. Initialize $\mathcal{B}$ to be the union of the interiors of N_u and N_v. Find the point $p \in uv \setminus \mathcal{B}$ such that $\delta(p)$ is maximum. Make p a Steiner point. Define an *edge-ball* $B_p = B(p, \frac{\varepsilon}{3}\delta(p))$. Add the interior of B_p to $\mathcal{B}$. Repeat until $uv \setminus \mathcal{B}$ is empty. Finally, make the intersection point q between uv and the boundary of N_u a Steiner point and introduce an edge-ball $B_q = B(q, \frac{\varepsilon}{3}\delta(q))$. Repeat the same for the intersection point between uv and the boundary of N_v.

As we will see below, the edge-balls centered at two consecutive Steiner points strictly outside N_u and N_v overlap significantly. After placing Steiner points strictly outside N_u and N_v, an extreme edge-ball may have a tiny overlap with N_u or N_v. In this case, if x is a point on some triangle incident to uv such that x lies close to this tiny overlap, then $\delta(x)$ can be arbitrarily small. This will cause a problem in discretizing triangles. Thus, we place two more edge-balls at the intersection points between uv and the boundaries of N_u and N_v.

Lemma 1. *Let uv be an edge of a fat tetrahedron. The edge uv is covered by the union of N_u, N_v, and the edge-balls centered at the Steiner points in uv. For every consecutive pair of Steiner points $p, q \in uv$ strictly outside N_u and N_v, $\|pq\| \geq \frac{\varepsilon}{3} \cdot \max\{\delta(p), \delta(q)\}$, and either p lies on the boundary of B_q or q lies on the boundary of B_p. There are $O\left(\frac{1}{\varepsilon}\log\frac{W}{\varepsilon}\right)$ Steiner points in uv.*

Proof. The construction ensures the coverage of uv. Assume that q was placed after p. By construction, q is not inside $B(p, \frac{\varepsilon}{3}\delta(p))$ and so $\|pq\| \geq \frac{\varepsilon}{3}\delta(p)$. As q is placed after p, $\delta(q) \leq \delta(p)$ and so $\|pq\| \geq \frac{\varepsilon}{3}\delta(p) \geq \frac{\varepsilon}{3}\delta(q)$.

In the interior of uv, $\delta(x)$ increases linearly from a limit of zero at u and then decreases linearly to a limit of zero at v. The placement of Steiner points strictly outside N_u and N_v begins with the point $p \in uv$ that maximizes $\delta(p)$. Therefore, the point $q \in uv$ that maximizes $\delta(q)$ outside the interiors of N_u, N_v, and B_p must lie on the boundary of B_p. Repeating this argument establishes the third property in the lemma.

Let F be the fat substar of u that contains uv. At the intersection point x between uv and the boundary of N_u, $\delta(x) \leq \|ux\| = \frac{\varepsilon}{3W}\delta_F(u) = \Theta(\frac{\varepsilon}{W}\|uv\|)$ by Remark 1. By Remark 2, $\delta(x) = \Omega(\min\{\|ux\|, \|vx\|\}) = \Omega(\min\{\frac{\varepsilon}{W}\|uv\|, (1 - \frac{\varepsilon}{W})\|uv\|\}$. So $\delta(x) = \Theta(\frac{\varepsilon}{W}\|uv\|)$. Similarly, at the intersection point x between uv and the boundary of N_v, $\delta(x) = \Theta\left(\frac{\varepsilon}{W}\|uv\|\right)$. The maximum value of $\delta(x)$ in the interior of uv is at most $\frac{1}{2}\|uv\|$. Let $p, q \in uv$ be two consecutive Steiner points strictly outside N_u and N_v such that $\delta(x)$ increases linearly from a limit of zero from u to p and then to q. By Remark 2, $\delta(p) = \Theta(\|pu\|)$. We have shown that $\|pq\| \geq \frac{\varepsilon}{3}\delta(p)$. By the linear increase in $\delta(\cdot)$, we get $\delta(q) = (1 + \|pq\|/\|pu\|)\delta(p) \geq (1+\Theta(\varepsilon))\delta(p)$. The next Steiner point after q is thus at distance at least $\frac{\varepsilon}{3}\delta(q) \geq \frac{\varepsilon}{3}\left(1 + \Theta(\varepsilon)\right)\delta(p)$ from q. In other words, the distance between consecutive Steiner points strictly outside N_u and N_v increases repeatedly by at least a factor $1+\Theta(\varepsilon)$ from $\Omega(\frac{\varepsilon^2}{W}\|uv\|)$ at the boundary of N_u to $O(\varepsilon\|uv\|)$ in the interior of uv. The same holds for the sequence of Steiner points from N_v. Hence, there are $O(\log_{1+\Theta(\varepsilon)} \frac{W}{\varepsilon}) = O\left(\frac{1}{\varepsilon}\log\frac{W}{\varepsilon}\right)$ Steiner points. □

Lemma 2. *Placing Steiner points on an edge takes* $O\left(\frac{1}{\varepsilon}\log\frac{W}{\varepsilon}\right)$ *time.*

The placement of Steiner points in a triangle uvw of a fat tetrahedron is slightly more involved. In the interior of uvw, the value of $\delta(x)$ is determined by the triangles of at most two fat tetrahedra incident to uvw. Consider one triangle t out of these candidates. Orient space so that uvw is horizontal. The graph of the distance function from x to t is a plane that makes an angle $\arctan(\sin\theta)$ with the horizontal, where θ is the dihedral angle between t and uvw (which is bounded from below and above by some constants). The graph of $\delta(x)$ is thus a lower envelope of planes. Moreover, this lower envelope H is supported by exactly three planes induced by three triangles that share with uvw the edges uv, vw and uw. Let ℓ denote the longest edge length of uvw. The maximum height of H is $h_{\max} = \Theta(\ell)$ as the tetrahedra defining $\delta(x)$ have bounded aspect ratios. For each point x in the interior of uvw that are close to and outside the vertex-balls and edge-balls at the boundary of uvw, $\delta(x) \geq c\varepsilon^2\ell/W$ for some constant $c > 0$.[1] Let H^+ denote the portion of H at height $h_{\min} = c\varepsilon^2\ell/W^2$ or above. We will place Steiner points in the projection of H^+ in uvw. By the geometry of H, a cross-section of H bounds a triangle that has the same angles as uvw and projects to the interior of uvw.

Define $h_0 = h_{\max}$ and for $i \geq 1$, $h_i = h_{i-1}/(1+\varepsilon)$. Let $A_i \subset uvw$ be the triangular annulus that the portion of H between heights h_i and h_{i+1} projects to. Both the inner and outer boundaries of this annulus are similar to uvw. The area of A_i is $\Theta((h_i - h_{i+1})(h_i + h_{i+1})) = \Theta(\varepsilon h_i^2)$. We place Steiner points in each A_i as follows. Initialize $\mathcal{B} = \emptyset$. Make an arbitrary point $p \in A_i \setminus \mathcal{B}$ a Steiner point. Define a *triangle-ball* $B_p = B(p, \frac{\varepsilon}{3}\delta(p))$. Add the interior of B_p to $\mathcal{B}$. Repeat until $A_i \setminus \mathcal{B}$ is empty.

Lemma 3. *Let uvw be a triangle of a fat tetrahedron. The triangle uvw is covered by the union of N_u, N_v, N_w, and edge-balls and triangle-balls with centers in uvw. There are* $O\left(\frac{1}{\varepsilon^2}\log\frac{W}{\varepsilon}\right)$ *Steiner points in uvw.*

Proof. The construction ensures the coverage of uvw. We can show as in the proof of Lemma 1 that $\|pq\| \geq \frac{\varepsilon}{3}\max\{\delta(p), \delta(q)\}$ for every pair of Steiner points p and q placed in A_i. The value of $\delta(x)$ in A_i is between h_i and h_{i+1}. Therefore, if we place disks of radii $\frac{\varepsilon}{6}h_{i+1}$ centered at the Steiner points in A_i, the disks are disjoint. At least a constant fraction of each such disk lies inside A_i. Therefore, there are $O(\varepsilon h_i^2/(\varepsilon^2 h_{i+1}^2)) = O(1/\varepsilon)$ Steiner points in A_i. As i increases, h_i decreases and approaches $h_{\min} = \Theta(\varepsilon^2 h_{\max}/W^2)$. Observe that $h_i = (1+\varepsilon)^{-i}h_{\max}$. Hence, $(1+\varepsilon)^{-i}h_{\max} \geq h_{\min}$, which implies that $i = O\left(\log_{1+\varepsilon}\frac{W}{\varepsilon}\right) = O\left(\frac{1}{\varepsilon}\log\frac{W}{\varepsilon}\right)$. It follows that there are $O\left(\frac{1}{\varepsilon^2}\log\frac{W}{\varepsilon}\right)$ Steiner points in uvw. □

[1] The smallest value of $\delta(x)$ occurs near the edge-ball centered at the intersection point between uv and the boundary of N_u or the boundary of N_v.

Lemma 4. *Placing Steiner points in uvw takes $O(\frac{1}{\varepsilon^4}\log\frac{W}{\varepsilon})$ time.*

4 Steiner Graph and Snapping

The vertices of $\mathcal{T}$ and the Steiner points form the vertices of $\mathcal{G}$. Before defining the edges of $\mathcal{G}$, we first define *extended clusters.* An extended cluster C^* consists of the skinny tetrahedra in a cluster C and the tetrahedra in contact with C. The tetrahedra in $C^* \setminus C$ are fat, and therefore, there are $O(\kappa)$ tetrahedra in C^*. If a boundary simplex σ of C^* is in contact with the boundary of C, then σ must also be a boundary simplex of $\mathcal{T}$.

There are two kinds of edges in $\mathcal{G}$. Each edge of the first kind connects two graph vertices x and y in the same extended cluster C^*. The edge weight is $1+O(\varepsilon)$ times the shortest path cost in C^* from x to y. We will show in Sect. 5 how to compute such an edge weight. Each edge of the second kind connects two graph vertices in a vertex star free of skinny tetrahedra. The edge weight is $1 + O(\varepsilon)$ times the shortest path cost in that vertex star, which can also be computed by the method in Sect. 5. Notice that $\mathcal{T}$ is covered by the extended clusters and vertex stars free of skinny tetrahedra. Due to the overlap among extended clusters and vertex stars, we may construct multiple edges between two graph vertices, and if so, we keep the edge between them with the lowest weight.

Assuming that $\mathcal{G}$ has been computed, we prove below that a shortest path in $\mathcal{G}$ is a $(1 + O(\varepsilon))$-approximate shortest path in $\mathcal{T}$. We need three technical lemmas (Lemmas 5, 6, and 7) that snap a path to vertices and Steiner points.

Lemma 5. *Let v be a vertex of a fat tetrahedron. Let F be a fat substar of v. Let x be a point in $|F|$ such that $\|vx\| \geq \delta_F(v)/2$. Let P be a path such that a subpath of P in $|F|$ connects x to a point $y \in B_{v,F}$. We can convert $P[x,y]$ to a path Q from x to y so that $Q \subset |F|$, Q passes through v, and cost$(Q) \leq (1+O(\varepsilon)) \cdot$ cost$(P[x,y])$.*

Proof. Let x' be the first entry point of $P[x,y]$ into $B_{v,F}$. We replace $P[x,y]$ by $P[x,x'] \cup x'v \cup vy$. We have $\text{cost}(x'v) \leq W\|x'v\| = \frac{\varepsilon}{3}\delta_F(v) \leq \frac{2\varepsilon}{3-2\varepsilon}\|xx'\| \leq O(\varepsilon) \cdot \text{cost}(P[x,x']) \leq O(\varepsilon) \cdot \text{cost}(P[x,y])$. Similarly, $\text{cost}(vy) \leq O(\varepsilon) \cdot \text{cost}(P[x,y])$. □

Lemma 6. *Let t be a triangle of a fat tetrahedron τ. Let p be a Steiner point in the interior of t, and let B_p denote the triangle-ball centered at p. Let P be a path such that a subpath of P in τ connects a point x in a boundary simplex of τ other than t to a point $y \in B_p \cap t$. We can convert $P[x,y]$ to a path Q from x to y so that $Q \subset \tau$, Q passes through p, and cost$(Q) \leq (1+O(\varepsilon)) \cdot$ cost$(P[x,y])$.*

Proof. $P[x,y] \subset \tau$ by assumption. Let x' be the last entry point of $P[x,y]$ into B_p. Retrace $P[x,x']$ from x' towards x until we hit a boundary simplex of τ other than t for the first time at a point $\hat{x}$. Note that $\delta(p) \leq \|p\hat{x}\|$. We replace $P[x,y]$ by $P[x,x'] \cup x'p \cup py$. Figure 1 illustrates the three cases below.

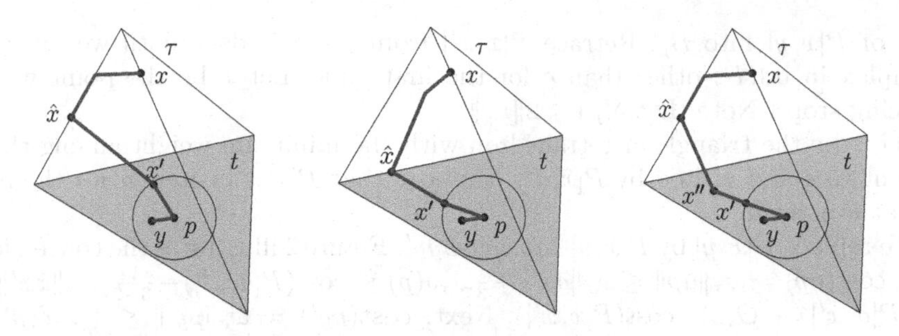

Fig. 1. The path Q in cases 1, 2 and 3 in the proof of Lemma 6 from left to right.

Case 1: $P[\hat{x}, x']$ is a segment whose interior lies in the interior of τ. We have $\text{cost}(x'p) = \frac{\varepsilon}{3}\omega_\tau\delta(p) \leq \frac{\varepsilon}{3-\varepsilon}\omega_\tau\|\hat{x}x'\| \leq O(\varepsilon)\cdot\text{cost}(P[x,x']) \leq O(\varepsilon)\cdot\text{cost}(P[x,y])$. Similarly, $\text{cost}(py) \leq \frac{\varepsilon}{3}\omega_t\delta(p) \leq \frac{\varepsilon}{3}\omega_\tau\delta(p) \leq O(\varepsilon)\cdot\text{cost}(P[x,y])$.

Case 2: $P[\hat{x}, x']$ is a segment whose interior lies in the interior of t. Then the interior of $P[\hat{x}, y]$ lies in the interior of t. We analyze the extra cost as in Case 1 with ω_τ replaced by ω_t.

Case 3: $P[\hat{x}, x']$ consists of two segments $\hat{x}x''$ and $x''x'$ whose interiors lie in the interiors of τ and t, respectively. Then the interior of $P[x'', y]$ lies in the interior of t. If $\|\hat{x}x''\| \geq \frac{1}{2}\|\hat{x}x'\|$, then we adapt the analysis in Case 1 using the relation $\delta(p) \leq \frac{6}{3-\varepsilon}\|\hat{x}x''\|$. Otherwise, $\|x''x'\| \geq \frac{1}{2}\|\hat{x}x'\|$ and we adapt the analysis in Case 2 using the relation $\delta(p) \leq \frac{6}{3-\varepsilon}\|x''x'\|$. □

Lemma 7. *Let e be an edge of a fat tetrahedron. Let F denote the fat substar of e. Let p be a Steiner point in the interior of e, and let B_p denote the edge-ball centered at p. Let x be a point in $|F|$ such that $\|px\| \geq \delta(p)/2$. Let P be a path such that a subpath of P in $|F|$ connects x to a point $y \in B_p \cap t$, where t is a triangle in F incident to e. Suppose that y lies outside every triangle-ball B_q where $q \in t$. Then, we can convert $P[x, y]$ to a path Q from x to y so that $Q \subset |F|$, Q passes through p, and cost$(Q) \leq (1 + O(\varepsilon))\cdot$ cost$(P[x, y])$.*

Proof. Since y lies outside every triangle-ball B_q where $q \in t$, y is at distance $O(\frac{\varepsilon^2}{W^2}\|e\|)$ from e. Let y' be the closest point in e to y. Let x' be the first entry

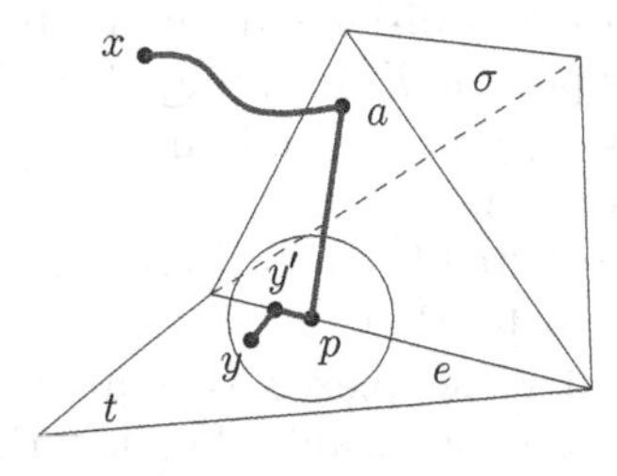

Fig. 2. The resulting path Q in the proof of Lemma 7.

point of $P[x, y]$ into B_p. Retrace $P[x, x']$ from x' towards x until we hit x or a simplex in $\mathrm{bd}(F)$ other than e for the first time. Let $\hat{x}$ be the point where retracing stops. Note that $\delta(p) \leq 2\|p\hat{x}\|$.

Let σ be the triangle or tetrahedron with the minimum weight among those incident to e and visited by $P[\hat{x}, x']$. Suppose that $P[\hat{x}, x']$ enters σ for the first time at a point a.

We replace $P[x, y]$ by $P[x, a] \cup ap \cup py' \cup yy'$. Figure 2 illustrates the conversion. First, $\mathrm{cost}(ap) = \omega_\sigma \|ap\| \leq \omega_\sigma \|ax'\| + \frac{\varepsilon}{3}\omega_\sigma \delta(p) \leq \mathrm{cost}(P[a, x']) + \frac{2\varepsilon}{3-2\varepsilon}\omega_\sigma \|\hat{x}x'\| \leq \mathrm{cost}(P[a, x']) + O(\varepsilon) \cdot \mathrm{cost}(P[x, x'])$. Next, $\mathrm{cost}(py') = \omega_e \|py'\| \leq \frac{\varepsilon}{3}\omega_\sigma \delta(p) \leq \frac{2\varepsilon}{3-2\varepsilon}\omega_\sigma \|\hat{x}x'\| \leq O(\varepsilon) \cdot \mathrm{cost}(P[x, x'])$. Also, $\mathrm{cost}(yy') \leq W\|yy'\| \leq O(W \cdot \frac{\varepsilon^2}{W^2}\|e\|)$. Recall that p is not inside the vertex-balls at the endpoints of e, and these vertex-balls have radius $\Omega(\varepsilon\|e\|/W)$. Therefore, $\delta(p) = \Omega(\varepsilon\|e\|/W)$ by Remark 2. Hence, $\mathrm{cost}(yy') \leq O(\varepsilon) \cdot \delta(p) \leq O(\varepsilon) \cdot \|\hat{x}x'\| \leq O(\varepsilon) \cdot \mathrm{cost}(P[x, x'])$. □

Next, we convert a path P from v_s to v_d to a path Q such that the nodes $v_s = u_{i_1}, u_{i_2}, u_{i_3}, \cdots, u_{i_m} = v_d$ in Q are vertices of $\mathcal{G}$, and for all $j \geq 1$, $Q[u_{i_j}, u_{i_{j+1}}]$ is contained in an extended cluster or a vertex star free of skinny tetrahedra. Moreover, $\mathrm{cost}(Q) \leq (1 + O(\varepsilon)) \cdot \mathrm{cost}(P)$. Therefore, $\mathcal{G}$ gives a $1 + O(\varepsilon)$ approximation because $\mathcal{G}$ contains the edges $\{u_{i_j}, u_{i_{j+1}}\}$ with weight $(1 + O(\varepsilon)) \cdot \mathrm{cost}(Q[u_{i_j}, u_{i_{j+1}}])$.

Lemma 8. *Let P be a path in $\mathcal{T}$ from v_s to v_d. We can convert P to a path Q in $\mathcal{T}$ from v_s to v_d such that the nodes $v_s = u_{i_1}, u_{i_2}, u_{i_3}, \cdots, u_{i_m} = v_d$ in Q are vertices of $\mathcal{G}$, and for all $j \geq 1$, $Q[u_{i_j}, u_{i_{j+1}}]$ is contained in an extended cluster or a vertex star free of skinny tetrahedra. Moreover, cost(Q) $\leq (1 + O(\varepsilon)) \cdot$ cost(P).*

Proof. Let P_0 denote a path from v_s to v_d in $\mathcal{T}$.

Suppose that v_s is disjoint from all clusters. If P_0 does not leave $\mathrm{star}(v_s)$, then v_d is a vertex in $\mathrm{star}(v_s)$ and the lemma is trivially true. Assume that P_0 leaves $\mathrm{star}(v_s)$ for the first time at a point y. Then y lies in a boundary simplex σ of $\mathrm{star}(v_s)$ disjoint from v_s. We modify $P_0[v_s, y]$ by applying Lemmas 5, 6, or 7 to make a detour to a vertex or Steiner point $p \in \sigma$.

Suppose that v_s is contained in a cluster C. Recall that C^* denotes the extended cluster corresponding to C. If P_0 does not leave C^*, then v_d is a vertex in C^* and there is nothing to prove. Assume that P_0 leaves C^* for the first time at a point y. Let x be the point in the boundary of C that P_0 leaves C for the last time before reaching y. Let σ be the simplex of lowest dimension in $\mathrm{bd}(C^*)$ that contains y. The simplex σ is disjoint from the boundary of C; otherwise, σ would be a boundary simplex of $\mathcal{T}$, meaning that P_0 cannot leave C^* at y, a contradiction. We modify $P_0[x, y]$ by applying Lemmas 5, 6, or 7 to make a detour to a vertex or Steiner point $p \in \sigma$.

Let P_1 denote the path resulted from modifying P_0. The extra cost of $O(\varepsilon) \cdot \mathrm{cost}(P_0[v_s, y])$ can be charged to $P_0[v_s, y]$. Then we work on $P_1[y, v_d]$. Recall that y belongs to the boundary simplex σ of $\mathrm{star}(v_s)$ or an extended cluster containing v_s, whichever case is applicable. We identify a vertex v as follows. If σ is a vertex, let $v = \sigma$. If σ is an edge, let v be the closest endpoint of σ to y. If σ is a triangle, let e be the closest edge of σ to y and then let v be the closest endpoint of e to y. Then, we repeat the previous analysis on v and $P_1[y, v_d]$. That is, we check the exit of $P_1[y, v_d]$ from star(v) or an extended cluster containing v, whichever case is applicable. The vertex or Steiner point p to which P_0 was snapped belongs to σ and p is already a vertex of $\mathcal{G}$. The next application of Lemmas 5, 6, or 7 will convert $P_1[y, v_d]$ to a path $P_2[y, v_d]$ that passes through a vertex or Steiner point q such that $P_2[p, q]$ lies in an extended cluster or a vertex star free of skinny tetrahedra. The extra charge in converting P_1 to P_2 can be charged to a subpath of $P_1[y, v_d]$. Repeating the argument proves the lemma. □

5 Processing Extended Clusters and Vertex Stars

Let Γ be a connected set of $O(\kappa)$ tetrahedra. Let p and q be two points in the union of vertices, edges, and triangles in Γ. We present an algorithm to compute a $(1 + O(\varepsilon))$-approximate shortest path in Γ from p to q.

5.1 Locally Shortest Path

For every triangle $t \in \Gamma$, its *unit disk* is the Euclidean disk D_t that is centered at the origin, lies on a plane parallel to t, and has radius $1/\omega_t$. The travel cost from a point x to a point y in t is λ if changing the radius of $D_t + x$ to λ/ω_t puts y on the boundary of the shrunk or expanded disk. To approximate D_t, we place $\Theta(1/\sqrt{\varepsilon})$ points roughly uniformly on the boundary of D_t as follows. Enclose D_t by a concentric unit square. Place points on the square boundary at distance $\sqrt{\varepsilon}$ apart. Project these points radially onto the boundary of D_t. Let D_t^* denote the convex hull of the points on the boundary of D_t. One can measure the travel cost from x to y by shrinking or expanding $D_t^* + x$ instead. It is easy to check that D_t^* ensures a $1 + O(\varepsilon)$ approximation of the cost under D_t. For every tetrahedron $\tau \in \Gamma$, its *unit ball* D_τ is the Euclidean ball centered at the origin with radius $1/\omega_\tau$. Similar to the 2D case, D_τ can be approximated by a convex hull D_τ^* with $O(1/\varepsilon)$ vertices. Computing D_t^* and D_τ^* for all triangles and tetrahedra takes $O\left(\frac{n}{\varepsilon}\log\frac{1}{\varepsilon}\right)$ time.

Let $\Sigma = (\sigma_1, \sigma_2, \cdots, \sigma_m)$ be a given simplex sequence. Let p and q be two points in some tetrahedra incident to σ_1 and σ_m, respectively. We show how to compute the locally shortest path from p to q with respect to Σ by linear programming. Consider the case that every σ_i is a triangle denoted by $v_{i,1}v_{i,2}v_{i,3}$. The case of some σ_i being vertices or edges can be handled similarly.

Let $x_i x'_{i+1}$ be a possible path link where $x_i \in \sigma_i$ and $x'_{i+1} \in \sigma_{i+1}$. Let τ_i denote the tetrahedron bounded by σ_i and σ_{i+1}. Using barycentric coordinates, the variable $x_i \in \mathbb{R}^3$ satisfies the constraint $x_i = \sum_{j=1}^{3} \alpha_{i,j} v_{i,j}$ for some non-negative variables $\alpha_{i,j} \in \mathbb{R}$ such that $\sum_{j=1}^{3} \alpha_{i,j} = 1$. Similarly, the variable $x'_{i+1} \in \mathbb{R}^3$ satisfies $x'_{i+1} = \sum_{j=1}^{3} \alpha'_{i+1,j} v_{i+1,j}$ for some non-negative variables $\alpha'_{i+1,j} \in \mathbb{R}$ such that $\sum_{j=1}^{3} \alpha'_{i+1,j} = 1$. For convenience, assume that $v_{0,j} = p$ and $v_{m+1,j} = q$ for $j \in [1, 3]$. We need the facet g of $D_{\tau_i}^*$ that contains the direction of the vector $x'_{i+1} - x_i$ because the cost of $x_i x'_{i+1}$ is equal to $\langle x'_{i+1} - x_i, n_g\rangle / \langle n_g, n_g\rangle$, where $\langle\cdot,\cdot\rangle$ denotes the inner product operator and n_g denotes the vector that goes from the origin to a point in the support plane of g such that $n_g \perp g$. By the convexity of $D^*_{\tau_i}$, the facet f of $D^*_{\tau_i}$ that gives the largest $\langle x'_{i+1}-x_i, n_f\rangle/\langle n_f, n_f\rangle$ is the correct facet g. Therefore, we introduce a variable $z_i \in \mathbb{R}$ and require $z_i \geq \langle x'_{i+1}-x_i, n_f\rangle/\langle n_f, n_f\rangle$ for every facet f of $D^*_{\tau_i}$. Part of the total path cost is $\sum_{i=0}^{m} z_i$. The minimization ensures that $z_i = \langle x'_{i+1}-x_i, n_g\rangle/\langle n_g, n_g\rangle$ at the end. We also allow for potential critical refraction at σ_{i+1}, i.e., allow for the link $x'_{i+1}x_{i+1} \subset \sigma_{i+1}$. To capture the cost of $x'_{i+1}x_{i+1}$, we introduce another variable z'_{i+1} and require $z'_{i+1} \geq \langle x_{i+1}-x'_{i+1}, n_f\rangle/\langle n_f, n_f\rangle$ for every edge f of $D^*_{\sigma_{i+1}}$. The objective is to minimize $\sum_{i=0}^{m} z_i + \sum_{i=1}^{m} z'_i$.

There are $\Theta(m\varepsilon^{-1})$ constraints and $\Theta(m)$ variables. The coefficients in the constraints $x_i = \sum_{j=1}^{3} \alpha_{i,j} v_{i,j}$ and $x'_i = \sum_{j=1}^{3} \alpha'_{i,j} v_{i,j}$ have magnitudes N or less because every coordinate of $v_{i,j}$ has magnitude at most N. Roughly speaking, the vertex coordinates in $D^*_{\tau_i}$ result from multiplying $1/\omega_{\tau_i}$ with the coordinates of the grid vertices on the unit cube. The grid box side length is $\sqrt{\varepsilon}$. Therefore, $O\left(\log\frac{W}{\varepsilon}\right)$ bits suffice for a vertex coordinate in $D^*_{\tau_i}$. For every facet f of $D^*_{\tau_i}$, we first compute an outward normal ν_f of f by taking cross-product using the vertices of f. The coordinates of ν_f thus require $O(\log\frac{W}{\varepsilon})$ bits. Let u be a vertex of f. We solve the linear equation $\langle\frac{1}{\alpha}\nu_f, \frac{1}{\alpha}\nu_f - u\rangle = 0$ for $\alpha \in \mathbb{R}$ such that $\frac{1}{\alpha}\nu_f$ lies on the support plane of f, i.e., $n_f = \frac{1}{\alpha}\nu_f$. Thus, α requires $O(\log\frac{W}{\varepsilon})$ bits and so does n_f. The same conclusion applies to the constraints $\langle x_i - x'_i, n_f\rangle/\langle n_f, n_f\rangle$ for every edge f of $D^*_{\sigma_i}$. In summary, the total number of bits to encode the linear program is $O\left(m\varepsilon^{-1}\log\frac{NW}{\varepsilon}\right)$. The ellipsoid method [16] solves the above linear program in $O(m^7\varepsilon^{-3}\log^2\frac{NW}{\varepsilon} + m^8\varepsilon^{-2}\log^2\frac{NW}{\varepsilon})$ arithmetic operations.

5.2 Approximate Shortest Path

To compute the approximate shortest path in Γ from p to q, our strategy is to enumerate all possible simplex sequences from p to q, use the method in Sect. 5.1 to compute a $1 + O(\varepsilon)$ approximation of the locally shortest path with respect to each simplex sequence, and finally select the shortest one among these locally shortest paths. The remaining questions are how long a simplex sequence and how many simplex sequences we need to consider.

Consider a shortest path P in Γ from p to q. Let $\sigma_1, \sigma_2, \cdots$ be the simplices in Γ in non-decreasing order of weights. We can assume that $P \cap \sigma_1$ is connected. Otherwise, we can shortcut P by joining the two connected components in $P \cap \sigma_1$ by a line segment in σ_1 without increasing the path cost. For a similar reason, we can assume that $P \cap \sigma_2$ has at most two connected components. In general, $P \cap \sigma_i$ has at most 2^{i-1} connected components. This argument is best visualized as arranging the connected components in a full binary tree with $P \cap \sigma_1$ at the root, two nodes of $P \cap \sigma_2$ at the next level, and so on. It follows that the simplex sequence is at most $2^{O(\kappa)}$ long. Consequently, there are at most $2^{2^{O(\kappa)}}$ simplex sequences. There are $O(\frac{\kappa^2}{\varepsilon^4}\log^2\frac{W}{\varepsilon})$ pairs of vertices and Steiner points in an extended cluster or vertex star free of skinny tetrahedra. We repeat the approximate shortest path computation $O(n \cdot \frac{\kappa^2}{\varepsilon^4}\log^2\frac{W}{\varepsilon})$ times, invoking the result in Sect. 5.1 at most $2^{2^{O(\kappa)}}$ times with $m = 2^{O(\kappa)}$ for each approximate shortest path computation.

Theorem 1. *Let ρ be an arbitrary constant. Let $\mathcal{T}$ be a simplicial complex of n tetrahedra such that vertices have integral coordinates with magnitude at most N and tetrahedra have integral weights in the range $[1, W]$. Let κ be the number of tetrahedra in the largest connected component of tetrahedra whose aspect ratios exceed ρ. For all $\varepsilon \in (0,1)$ and for every pair of source and destination vertices v_s and v_d in $\mathcal{T}$, we can find a $(1+\varepsilon)$-approximate shortest path in $\mathcal{T}$ from v_s to v_d in $O\big(2^{2^{O(\kappa)}} n\varepsilon^{-7} \log^2 \frac{W}{\varepsilon} \log^2 \frac{NW}{\varepsilon}\big)$ time.*

References

1. Ahmed, M.: Constrained Shortest Paths in Terrains and Graphs. Ph.D. Thesis, University of Waterloo, Canada (2009)
2. Aleksandrov, L., Maheshwari, A., Sack, J.-R.: Determining approximate shortest paths on weighted polyhedral surfaces. J. ACM **52**, 25–53 (2005)
3. Aleksandrov, L., Djidjev, H., Maheshwari, A., Sack, J.-R.: An approximation algorithm for computing shortest paths in weighted 3-d domains. Discrete. Comput. Geom. **50**, 124–184 (2013)
4. Chen, J., Han, Y.: Shortest paths on a polyhedron. Int. J. Comput. Geom. Appl. **6**, 127–144 (1996)
5. Cheng, S.-W., Jin, J.: Approximate shortest descending paths. SIAM J. Comput. **43**, 410–428 (2014)
6. Cheng, S.-W., Jin, J.: Shortest paths on polyhedral surfaces and terrains. In: Proceedings of ACM Sympoisum on Theory of Computing, pp. 373–382 (2014)
7. Cheng, S.-W., Jin, J., Vigneron, A.: Triangulation refinement and approximate shortest paths in weighted regions. In: Proceedings of the ACM-SIAM Symposium on Discrete Algorithms, pp. 1626–1640 (2015)
8. Cheng, S.-W., Na, H.-S., Vigneron, A., Wang, Y.: Approximate shortest paths in anisotropic regions. SIAM J. Comput. **38**, 802–824 (2008)
9. Cheng, S.-W., Na, H.-S., Vigneron, A., Wang, Y.: Querying approximate shortest paths in anisotropic regions. SIAM J. Comput. **39**, 1888–1918 (2010)
10. Choi, J., Sellen, J., Yap, C.-K.: Approximate Euclidean shortest path in 3-space. In: Proceedings of the Annual Symposium on Computational Geometry, pp. 41–48 (1994)
11. Clarkson, K.L.: Approximation algorithms for shortest path motion planning. In: Proceedings of the ACM Symposium on Theory Computing, pp. 56–65 (1987)
12. Hershberger, J., Subhash, S.: An optimal algorithm for Euclidean shortest paths in the plane. SIAM J. Comput. **28**, 2215–2256 (1999)
13. Menke, W.: Geophysical Data Analysis: Discrete Inverse Theory. Academic Press, New York (2012)
14. Mitchell, J.S.B., Papadimitrou, C.H.: The weighted region problem: finding shortest paths through a weighted planar subdivision. J. ACM **8**, 18–73 (1991)
15. Papadimitriou, C.H.: An algorithm for shortest-path motion in three dimensions. Inf. Process. Lett. **20**, 259–263 (1985)
16. Papadimitriou, C.H., Steiglitz, K.: Combinatorial Optimization: Algorithms and Complexity. Dover, New York (1998)
17. Sun, Z., Reif, J.: On finding approximate optimal paths in weighted regions. J. Alg. **58**, 1–32 (2006)

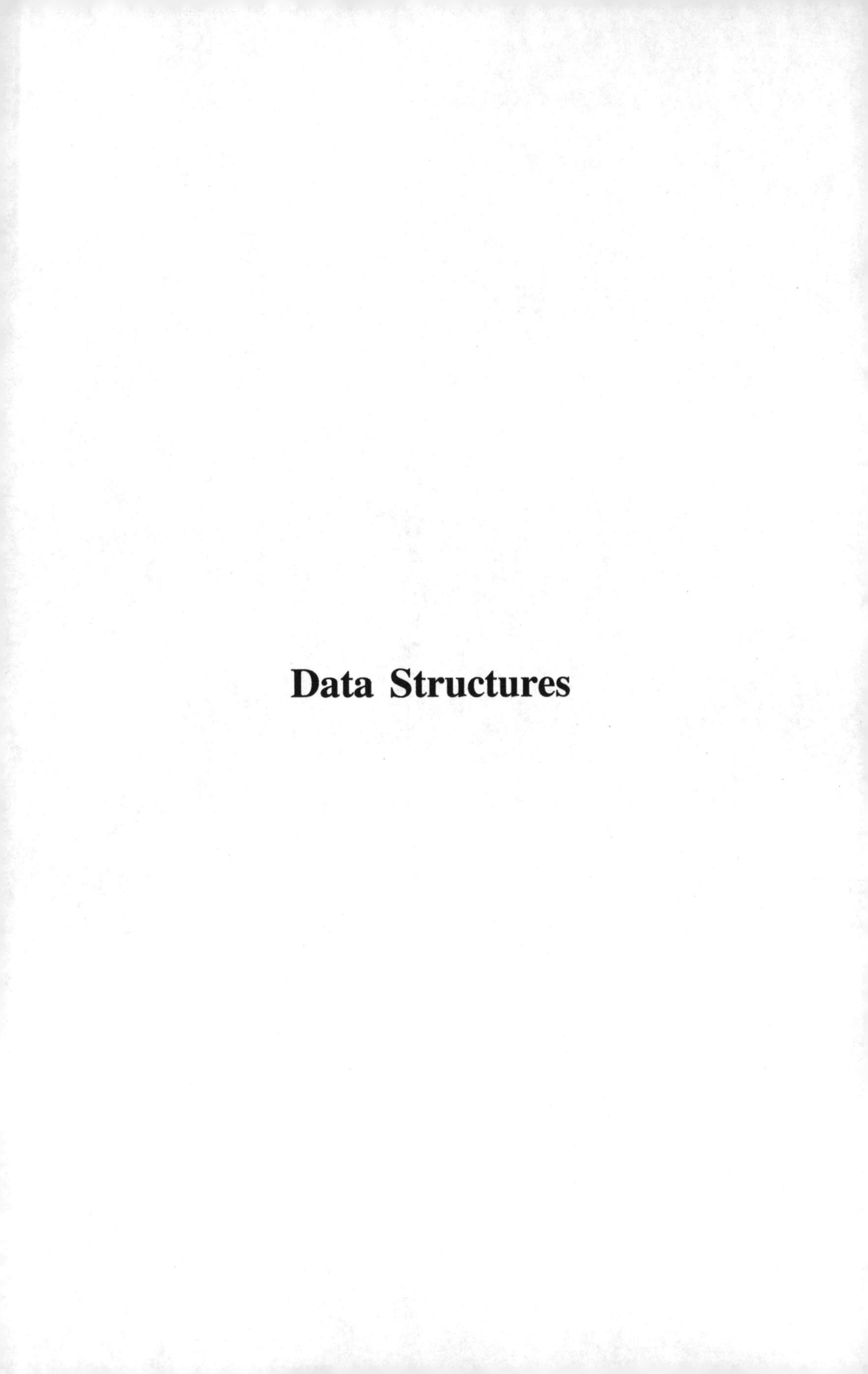

Data Structures

On the Succinct Representation of Unlabeled Permutations

Hicham El-Zein(✉), J. Ian Munro, and Siwei Yang

Cheriton School of Computer Science, University of Waterloo,
Waterloo, ON N2L 3G1, Canada
{helzein,imunro,siwei.yang}@uwaterloo.ca

Abstract. We investigate the problem of succinctly representing an arbitrary unlabeled permutation π, so that $\pi^k(i)$ can be computed quickly for any i and any integer power k. We consider the problem in several scenarios:

- Labeling schemes where we assign labels to elements and the query is to be answered by just examining the labels of the queried elements: we show that a label space of $\sum_{i=1}^{n} \lfloor \frac{n}{i} \rfloor \cdot i$ is necessary and sufficient. In other words, $2 \lg n$ bits of space are necessary and sufficient for representing each of the labels.
- Succinct data structures for the problem where we assign labels to the n elements from the label set $\{1, \ldots, cn\}$ where $c \geq 1$: we show that $\Theta(\sqrt{n})$ bits are necessary and sufficient to represent the permutation. Moreover, we support queries in such a structure in $O(1)$ time in the standard word-RAM model.
- Succinct data structures for the problem where we assign labels to the n elements from the label set $\{1, \ldots, cn^{1+\epsilon}\}$ where c is a constant and $0 < \epsilon < 1$: we show that $\Theta(n^{(1-\epsilon)/2})$ bits are necessary and sufficient to represent the permutation. We can also support queries in such a structure in $O(1)$ time in the standard word-RAM model.

1 Introduction and Motivation

A permutation π is a bijection from the set $\{1, \ldots, n\}$ to itself. Given a permutation π on an n element set, our problem is to preprocess the set, assigning a unique label to each element, to obtain a data structure with minimum space to support the following query: given a label i, determine $\pi^k(i)$ quickly. We denote such queries by $\pi^k()$. Moreover, we assume that k is bounded by some polynomial function in n.

We are interested in *succinct*, or highly-space efficient data structures. Our aim is to develop data structures whose size is within a constant factor of the information theoretic lower bound. Designing succinct data structures is an area of interest in theory and practice motivated by the need of storing large amount

This work was sponsored by the NSERC of Canada and the Canada Research Chairs Program.

K. Elbassioni and K. Makino (Eds.): ISAAC 2015, LNCS 9472, pp. 49–59, 2015.
DOI: 10.1007/978-3-662-48971-0_5

of data using the smallest space possible. For succinct representations of dictionaries, trees, arbitrary graphs, partially ordered sets and equivalence relations see [1,3,5,6,11,12,14].

Permutations are fundamental in computer science and are studied extensively. Several papers have looked into problems related to permutation generation [15], permuting in place [7] etc. Others have dealt with the problem of space-efficient representation of restricted classes of permutations, like the permutations representing the lexicographic order of the suffixes of a string [8,10], or the so-called approximately min-wise independent permutations [2], which are used for document similarity estimation. Since there are exactly $n!$ permutations, the number of bits required to represent a permutation of length n is $\lceil \lg(n!) \rceil \sim n \lg n - n \lg e + O(\lg n)$[1] bits. Munro et al. [13] studied the space efficient representation of general permutations where general powers can be computed quickly. They gave a representation taking the optimal $\lceil \lg(n!) \rceil + o(n)$ bits, and a representation taking $((1+\epsilon)n \lg n)$ bits where $\pi^k()$ can be computed in constant time.

Our paper is the first to study the space-efficient representation of permutations where labels can be freely reassigned. This problem is similar to the problem of representing unlabeled equivalence relations [5,11]. However, our problem differs from representing equivalence relations when the label space exceeds n. In our case we must know the size of each cycle, while for equivalence relations it is not necessary to know the exact size of the equivalence classes. Thus, as we increase the label space we will not witness a drastic decrease in auxiliary storage size. We study this problem in several scenarios; thus, showing the tradeoffs between label space and auxiliary storage size for the stated problem. In Sect. 3, we cover the scenario where queries are to be answered by just examining the labels of the queried elements. We show that a label space of $\sum_{i=1}^{n} \lfloor \frac{n}{i} \rfloor \cdot i$ is necessary and sufficient. Then, we show that with a label space of n^2 queries can be answered in constant time. In Sect. 4, we cover the scenario where labels can be assigned from the set $\{1, \ldots, n\}$. We show that $\Theta(\sqrt{n})$ bits are necessary and sufficient to represent the permutation. We use the same data structure as the main structure in [11]. However, we optimize it to achieve constant query time while using only $O(\sqrt{n})$ bits; thus, solving an open problem from [11]. Note that the details of this improvement are also found in the first author's thesis [4]. Section 5 contains the main result of this paper. We cover the scenario where labels can be assigned from the set $\{1, \ldots, cn^{1+\epsilon}\}$ where c is a constant and $0 < \epsilon < 1$. We show that $\Theta(n^{(1-\epsilon)/2})$ bits are necessary and sufficient to represent the permutation, and we support queries in such a structure in $O(1)$ time in the standard word-RAM model.

Finally as an application to our new data structures, we give a representation of a labeled permutation that takes $s(n) + O(\sqrt{n})$ bits and can answer $\pi^k()$ in $O(t_f + t_i)$ time, where $s(n)$ denotes the number of bits required for a representation R to store a labeled permutation, and t_f and t_i are the time needed for R to support $\pi()$ and $\pi^{-1}()$. This result improves Theorem 3.3 in [13].

[1] We use $\lg n$ to denote $\log_2 n$.

2 Definitions and Preliminaries

A *permutation* π is a bijection from the set $\{1,\ldots,n\}$ to itself, and we denote its inverse bijection as π^{-1}. We also extend the definition to arbitrary integer power of π as follows:

$$\pi^k(i) = \begin{cases} \pi^{k+1}(\pi^{-1}(i)) & k < 0 \\ i & k = 0 \\ \pi^{k-1}(\pi(i)) & k > 0 \end{cases}$$

A permutation can be viewed as a set of disjoint cycles. Since we are working with unlabeled permutations, we have the freedom to assign the labels in any way. In all our labeling schemes, we give elements within the same cycle and cycles of the same length consecutive labels. For example the elements of the first cycle of length l will get labels from the interval $[s, s+l-1]$, such that $\pi(i) = i+1$ for $i \in [s, s+l-2]$ and $\pi(s+l-1) = s$. The elements of the second cycle of length l will get labels in the range $[s+l, s+2l-1]$, and so on. Thus given a label i and an integer k, to answer $\pi^k(i)$ it is sufficient to compute l the length of the cycle that i belongs to, and s the smallest index of an element that belongs to a cycle of length l. Now, it is not hard to verify that $\pi^k(i) = s + rl + ((p+k)\%l)$[2] where $r = \lfloor (i-s)/l \rfloor$ and $p = i - (s+rl)$.

Notice that the multiset formed by the cycles lengths of a given permutation π over an n-element set will form an integer partition of the integer n. An *integer partition* p of n is a multiset of positive integers that sum to n. We call these positive integers the *elements* of p, and we denote by $|p|$ this number of elements. We say that an integer partition p of n *dominates* an integer partition q of m where $n > m$ if q is a subset of p. For example, the integer partition $\{5,5,10\}$ of 20 dominates the integer partition $\{5,5\}$ of 10, but not the integer partition $\{4,6\}$ of 10. Given an integer partition p of n, we define a *part* q of size k to be a collection of elements in p that sum to k. We say that an integer s *fills* q if q contains $\lfloor k/s \rfloor$ integers s and one integer $k \bmod s$. Furthermore, we say that two parts *intersect* if they share at least one common element; otherwise, they are *non-intersecting*. For example the integer partition $\{1,4,5\}$ of 10 contains the following parts: part $\{1\}$ of size 1, part $\{4\}$ of size 4, part $\{5\}$ of size 5, part $\{1,4\}$ of size 5, part $\{1,5\}$ of size 6, part $\{4,5\}$ of size 9 and part $\{1,4,5\}$ of size 10. We say that 5 fills the parts $\{5\}$ and $\{4,5\}$ but not the part $\{1,4,5\}$. The parts $\{4,5\}$ and $\{4\}$ are intersecting, while the parts $\{4,5\}$ and $\{1\}$ are non-intersecting.

Finally, we give two observations that we will use repeatedly.

Observation 1. *M not necessarily distinct integers $m_0,\ldots,m_{M-1}$ ordered such that $m_i \leq m_{i+1}$ in the range $[0, N-1]$, can be represented in $O(N+M)$ bits such that the ith integer m_i can be accessed in $O(1)$ time.*

[2] We use % to denote the modulo operation.

Observation 2. *M positive integers $m_0, \ldots, m_{M-1}$ that sum to N can be represented in $O(N+M)$ bits such that the ith integer m_i can be accessed in $O(1)$ time, the partial sum $\sum_{j=1}^{i} m_j$ can be computed in $O(1)$ time, and given an integer x we can compute the biggest index i such that $\sum_{j=1}^{i} m_j \leq x$ in $O(1)$ time.*

The proof of both observations is found in the appendix. Note that if we are allowed to reorder the numbers in Observation 2, we can reduce the size of the representation to $O(\sqrt{N})$ bits without compromising the constant runtime of the stated operations.

3 Direct Labeling Scheme

In this section we cover the problem where queries are answered by computing directly from the labels without using any auxiliary storage except for the value of n. We show that a label space of $\sum_{i=1}^{n} \lfloor \frac{n}{i} \rfloor \cdot i$ is necessary and sufficient to represent the permutation. Moreover, we show that with a label space of n^2 $\pi^k()$ can be computed in constant time.

Theorem 3. *Given a permutation π, a label space of $\sum_{i=1}^{n} \lfloor \frac{n}{i} \rfloor \cdot i$ is necessary and sufficient to represent the permutation.*

For the proof of Theorem 3 check the appendix. To answer queries in constant time we extend the label space to n^2. Then we assign labels from the set of integers in the range $[0, n-1]$ for all the elements in cycles of length 1, and labels from the set of integers in the range $[n(i-1)+(r-1)i, n(i-1)+ri-1]$ for the elements in the rth cycle of length i, where $1 \leq r \leq \lfloor n/i \rfloor$. Given a label x, to answer a query $\pi^k(x)$ find $l = \lfloor x/n \rfloor + 1$. Next, compute $s = (l-1)n$, $r = \lfloor (x-s)/l \rfloor$ and $p = x - (s + rl)$, then return $s + rl + ((p+k)\%l)$.

Theorem 4. *Given a permutation π, we can assign to each of the elements a label in the range of $\{1, \ldots, n^2\}$ such that $\pi^k()$ can be computed in constant time by looking only at the labels.*

4 Succinct Data Structures with Label Space n

In this section we consider the scenario where the n elements are to be assigned labels in the range 1 to n. The queries can be answered by looking at an auxiliary data structure. Moreover, we have the freedom to assign the labels in any way.

Following [11], the information theoretic lower bound for the representation of a permutation is the number of partitions of n, which by the Hardy-Ramanujan formula [9] is asymptotically equivalent to $\frac{1}{4n\sqrt{3}} e^{\pi\sqrt{\frac{2n}{3}}}$. Thus the information theoretic lower bound for representing a permutation is $\Theta(\sqrt{n})$ bits of space.

We will use the same data structure as the main structure in [11], however we will optimize it to achieve constant query time while using only $O(\sqrt{n})$ bits. Given π let k be the number of distinct cycle sizes in π. For $i = 1$ to k, let s_i

be the distinct sizes of the cycles, and let n_i be the number of cycles of size s_i. Order the cycles in non-decreasing order by $\gamma_i = s_i n_i$ so that for $i = 1$ to $k-1$, $s_i n_i \leq s_{i+1} n_{i+1}$. Notice that since

$$\sum_{i=1}^{k} s_i n_i = n \text{ and } s_i n_i \geq i \text{ for } i = 1, \ldots, k, \tag{1}$$

k is at most $\sqrt{2n}$. The primary data structure is made up of two sequences:

- the sequence $\boldsymbol{\delta}$ that consists of $\delta_1 = s_1 n_1$ and $\delta_i = s_i n_i - s_{i-1} n_{i-1}$, for $i = 2, \ldots, k$ and
- the sequence $\boldsymbol{n}$ that consists of n_i, for $i = 1, \ldots, k$.

Elements of the two sequences are represented in binary. Since the length of each element may vary, we store two other sequences that shadow the primary sequences. The shadow sequences have a 1 at the starting point of each element in the shadowed sequence and a 0 elsewhere. Also store a select structure on the two shadow sequences in order to identify the 1s quickly. It is proved in [11] these sequences can be stored in $O(\sqrt{n})$ bits.

The sequence $\boldsymbol{\delta}$ gives an implicit ordering of the elements. Assign the first $s_1 n_1$ labels to the elements of the cycles with length s_1, the elements of the next n_2 cycles are assigned the next $s_2 n_2$ labels and so on.

Denote by the predecessor of an element x to be $\max\{j \mid \sum_{i=1}^{j} s_i n_i < x\}$. Store an array A, where $A[i] = \max\{j \mid \sum_{t=1}^{j} s_t n_t \leq i(i+1)/2\}$, for $i = 1$ to $\sqrt{2n}$. Next, we prove a modified version of Lemma 2 in [11].

Lemma 1. *The predecessor $p(x)$ of an integer x in the sequence $\sum_{t=1}^{i} s_t n_t$, $i = 1$ to k is in the range $[A[\lfloor\sqrt{2x}\rfloor - 1], A[\lfloor\sqrt{2x}\rfloor - 1] + 5]$.*

Proof. Let $i = \lfloor\sqrt{2x}\rfloor - 1$. Without loss of generality assume that $i \geq 6$, since for $x < 25$ we can store $p(x)$ explicitly in $O(\lg n)$ bits. Notice that

$$i(i+1)/2 \leq (\sqrt{2x} - 1)\sqrt{2x}/2 \leq x$$

and

$$x \leq \sqrt{2x}(\sqrt{2x} + 1)/2 \leq (i+2)(i+3)/2$$

For $j = A[i] + 1$, $\sum_{t=1}^{j-1} s_t n_t \leq i(i+1)/2$, so $j - 1 \leq i$ and $j \leq i + 1$. Since $\sum_{t=1}^{j} s_t n_t > i(i+1)/2$, $s_j n_j \geq i(i+1)/(2j) \geq i/2$. Hence, $\sum_{t=1}^{j+5} s_t n_t \geq (i+2)(i+3)/2 \geq x$. $\square$

The actual value of $p(x)$ can be obtained by checking at most six numbers. Moreover, A can be stored using $O(\sqrt{n})$ bits using the method described in Observation 1.

In the standard word-RAM model, computing $\sqrt{x}$ is not a constant time operation. The standard Newton's iterative method uses $O(\lg \lg n)$ operations. Following [11], we can use a look-up to precomputed tables and finds $\sqrt{x}$ in constant time. We use two tables, one when the number of bits up to the most significant bit of x is odd, denoted by O, and one when the number of bits is even, denoted by E. For $i = 1, \ldots, \lceil\sqrt{2n}\rceil$, we store in $E[i]$ the value of $\lfloor\sqrt{i2^{\lceil \lg i\rceil}}\rfloor$, and in $O[i]$ the value of $\lfloor\sqrt{i2^{\lceil \lg i\rceil -1}}\rfloor$. E and O can be stored in $O(\sqrt{n})$ bits by storing them using the method described in Observation 1.

Lemma 2. *For $i \le n$, $\lfloor\sqrt{i}\rfloor$ can be computed in constant time using a precomputed table of $O(\sqrt{n})$ bits.*

For each i, where at least one of δ_i's bits locations in $\boldsymbol{\delta}$ is a multiple of $(\epsilon \lg n)$, store the partial sum value $\sum_{j=1}^{i}(s_j n_j)$ and the value of $s_i n_i$. Moreover, for every possible sequence of δ values $\delta_1, \delta_2, \ldots, \delta_i$ of length $(\epsilon \lg n)$ and its corresponding shadow sequence, store in a table T the values i and $\sum_{j=1}^{i}(\sum_{k=1}^{j}\delta_k)$. To compute $\sum_{j=1}^{i}(s_j n_j)$ for an arbitrary index i, find the biggest index $k \le i$ that has it's partial sum value stored. Notice that $\sum_{j=1}^{i}(s_j n_j) = \sum_{j=1}^{k}(s_j n_j) + (i-k)s_k n_k + \sum_{j=k+1}^{i}(\sum_{l=k+1}^{i}\delta_l)$. Since these values can be obtained using table lookup on T, we can compute the partial sum at an arbitrary index in constant time. Moreover, we can compute the value of $s_i n_i$ for an arbitrary index i by computing the partial sum at $i-1$ and subtracting it from the partial sum at i. Finally, we can compute s_i by computing $s_i n_i$ and dividing it by n_i. By choosing $\epsilon < 1/4$, the size of T becomes $o(\sqrt{n})$ bits.

Answering Queries: Given a label x, to compute $\pi^k(x)$ we first find the predecessor $p(x)$ of x by querying A and checking at most 6 different values. Next we compute the partial sum value $s = \sum_{i=1}^{p(x)-1}(n_i s_i)$. Then, compute $r = \lfloor(x-s)/s_{p(x)}\rfloor$ and $p = x - (s + r s_{p(x)})$, then return $s + r s_{p(x)} + ((p+k)\%l)$.

Theorem 5. *Given an unlabeled permutation of n elements, $\Theta(\sqrt{n})$ bits are necessary and sufficient for storing the permutation if each element is to be given a unique label in the range $\{1, 2, \ldots, n\}$. Moreover, $\pi^k()$ can be computed in $O(1)$ time in such a structure.*

5 Succinct Data Structures with Label Space $cn^{1+\epsilon}$

In this section we consider the scenario where the n elements are to be assigned labels in the range 1 to $cn^{1+\epsilon}$ where c is a constant and $0 < \epsilon < 1$. As in Sect. 4 we assign an implicit ordering of the elements, and queries can be answered by looking at an auxiliary data structure.

Given π, we divide the cycles in π into four different groups and handle each group appropriately. For $i = 1$ to k_3, let s_i be the distinct sizes of the cycles of size $\le n^{(1+\epsilon)/2}$, and let n_i be the number of cycles of size s_i. Without loss of generality, assume that:

- $\gamma_i = s_i n_i \le (\sqrt{c}n^{(1+\epsilon)/2})/2 = \eta$, for $1 \le i \le k_1$.
- $s_i \le n^{(1-\epsilon)/2}$ and $\gamma_i > \eta$, for $k_1 < i \le k_2$.

– $n^{(1-\epsilon)/2} < s_i \leq n^{(1+\epsilon)/2}$ and $\gamma_i > \eta$, for $k_2 < i \leq k_3$.

Let $l_{k_3+1}, \ldots, l_{k_4}$ be the size of the cycles that are bigger than $n^{(1+\epsilon)/2}$. Note that the l_i ($i = k_3 + 1$ to k_4) values are not necessarily unique.

Case 1: Reserve the first $(cn^{1+\epsilon})/4$ labels to handle all possible cycle sizes when $\gamma_i \leq \eta$. Assign labels to the elements in the cycles that satisfy this criteria in a similar method to the labeling scheme described in Theorem 4. To be more specific, we assign labels from the set of integers in the range $[0, \eta - 1]$ for all the elements in cycles of length 1, and assign labels from the set of integers in the range $[\eta(j-1), \eta j - 1]$ for all the elements in cycles of length j, where $2 \leq j \leq \eta$. This covers all the elements of the cycles of sizes $s_1, \ldots, s_{k_1}$, and increases the label space by at most $\eta^2 = (cn^{1+\epsilon})/4$. Let $B_1 = (cn^{1+\epsilon})/4$.

Case 2 ($k_1 + 1 \leq i \leq k_2$): Order the s_i values in increasing order. Make sure that all cycles of size s_i, fill a part whose length is $c_i\eta$ a multiple of η. Notice that $(k_2 - k_1) < n/\eta$ since $\gamma_i > \eta$, so the label space will increase by at most n. Since $\sum_{i=k_1+1}^{k_2}(c_i) \leq (2n)/\eta = O(n^{(1-\epsilon)/2})$, we can store the c_i values in $O(n^{(1-\epsilon)/2})$ bits using the method described in Observation 2. Moreover, we store a bit vector ψ of size $n^{(1-\epsilon)/2}$ to identify the s_i values, and we store a select structure on ψ to identify the 1s quickly. Assign labels in the range $[B_1, B_1 + c_{(k_1+1)}\eta - 1]$ to the elements in cycles of size $s_{(k_1+1)}$, then assign the next $c_{(k_1+2)}\eta$ labels to elements in cycles of size $s_{(k_1+2)}$, and so on. Let $B_2 = B_1 + \sum_{j=k_1+1}^{k_2} c_j\eta$.

Case 3 ($k_2 + 1 \leq i \leq k_3$): Make sure that all cycles of size s_i, fill a part whose length is $c_i\eta$ a multiple of η. As in case 2, store the c_i values in $O(n^{(1-\epsilon)/2})$ bits using the method described in Observation 2. To identify the s_i values: order them in increasing order of $r_i = s_i\%(16n^{(1-\epsilon)/2}/c)$ and store the r_i values in $O(n^{(1-\epsilon)/2})$ bits using the method described in Observation 1, then store the value of $q_i = s_i/(16n^{(1-\epsilon)/2}/c) \leq (cn^\epsilon/16)$ in the label of each element that is in a cycle of size s_i. Now $s_i = q_i(16n^{(1-\epsilon)/2}/c) + r_i$. Let β_1 be equal to $\sum_{i=k_2+1}^{k_3} c_i\eta$. Assign labels in the range

$$\left[B_2 + q_i 2^{\lceil \lg(\beta_1) \rceil} + \sum_{j=k_2+1}^{i-1} c_j\eta, B_2 + q_i 2^{\lceil \lg(\beta_1) \rceil} + \sum_{j=k_2+1}^{i} c_j\eta - 1\right]$$

to the elements in the cycles of size s_i. The label space will increase by at most $(cn^\epsilon/16)2^{\lceil \lg(\beta_1) \rceil} + \beta_1 \leq (cn^{1+\epsilon})/4 + O(n)$. Let $B_3 = B_2 + (cn^\epsilon/16)2^{\lceil \lg(\beta_1) \rceil} + \beta_1$.

Case 4 ($k_3 + 1 \leq i \leq k_4$): For the cycles of length l_i, make sure that each cycle fills a part whose length is $c_i\eta$ a multiple of η. As in the previous cases, store the c_i values in $O(n^{(1-\epsilon)/2})$ bits using the method described in Observation 2. To identify the l_i values: order them by $r_i = (l_i\%\eta)\%(8n^{(1-\epsilon)/2}/\sqrt{c})$ and store the r_i values in $O(n^{(1-\epsilon)/2})$ bits using the method described in Observation 1, then store the value of $q_i = (l_i\%\eta)/(8n^{(1-\epsilon)/2}/\sqrt{c}) \leq (cn^\epsilon/16)$ in the label of each element that is in a cycle of size l_i. Now $l_i = q_i(8n^{(1-\epsilon)/2}/\sqrt{c}) + r_i + (c_i - 1)\eta$. Let β_2 be equal to $\sum_{i=k_3+1}^{k_4} c_i\eta$. Assign labels in the range

$$\left[B_3 + q_i 2^{\lceil \lg(\beta_2) \rceil} + \sum_{j=k_3+1}^{i-1} c_j \eta, B_3 + q_i 2^{\lceil \lg(\beta_2) \rceil} + \sum_{j=k_3+1}^{i} c_j \eta - 1\right]$$

to the elements in the cycle of size l_i.

The total size of the structures used is $O(n^{(1-\epsilon)/2})$ bits, and the total address space increased to at most $(3cn^{1+\epsilon})/4 + O(n) \leq cn^{1+\epsilon}$ as required.

Answering Queries: Given a label x, to compute $\pi^k(x)$ we distinguish between four different cases:

Case 1 $x < B_1$**:** Compute the value of $l = \lfloor x/\eta \rfloor + 1$, $s = (l-1)\eta$, $r = \lfloor (x-s)/l \rfloor$, and $p = x - (s + rl)$. Then, return $s + rl + ((p+k)\%l)$.

Case 2 $B_1 \leq x < B_2$**:** Compute the value $m = (x - B_1)/\eta$. Then get the biggest index i such that $\sum_{j=k_1+1}^{i} c_j \leq m$. This operation can be done in $O(1)$ time using the structure from Observation 2. Next, find l the index of the ith one in ψ; l is the size of the cycle that x belongs to. Compute $s = B_1 + \sum_{j=k_1+1}^{i-1} c_j \eta$, $r = \lfloor (x-s)/l \rfloor$, and $p = x - (s + rl)$. Then, return $s + rl + ((p+k)\%l)$.

Case 3 $B_2 \leq x < B_3$**:** Compute the value $m = ((x - B_2)\%\beta_1)/\eta$. Then get the biggest index i such that $\sum_{j=k_2+1}^{i} c_j \leq m$. Next calculate $q_i = \lfloor (x - B_2)/2^{\lceil \lg(\beta_1) \rceil} \rfloor$ and $l = q_i(16n^{(1-\epsilon)/2}/c) + r_i$; l is the size of the cycle that x belongs to. Compute $s = B_2 + q_i 2^{\lceil \lg(\beta_1) \rceil} + \sum_{j=k_2+1}^{i-1} c_j \eta$, $r = \lfloor (x-s)/l \rfloor$, and $p = x - (s + rl)$. Then, return $s + rl + ((p+k)\%l)$.

Case 4 $B_3 \leq x$**:** Compute the value $m = ((x - B_3)\%\beta_2)/\eta$. Then get the biggest index i such that $\sum_{j=k_3+1}^{i} c_j \leq m$. Next calculate $q_i = \lfloor (x - B_3)/2^{\lceil \lg(\beta_2) \rceil} \rfloor$ and $l = q_i(8n^{(1-\epsilon)/2}/\sqrt{c}) + r_i + (c_i - 1)\eta$; l is the size of the cycle that x belongs to. Compute $s = B_3 + q_i 2^{\lceil \lg(\beta_2) \rceil} + \sum_{j=k_3+1}^{i-1} c_j \eta$, $r = \lfloor (x-s)/l \rfloor$, and $p = x - (s + rl)$. Then, return $s + rl + ((p+k)\%l)$.

All operations used take constant time, so $\pi^k(x)$ can be computed in $O(1)$ time.

Theorem 6. *Given an unlabeled permutation of n elements, $\Theta(n^{(1-\epsilon)/2})$ bits are sufficient for storing the permutation if each element is to be given a unique label in the range $\{1, \ldots, cn^{1+\epsilon}\}$ for any constant $c > 1$ and $\epsilon < 1$. Moreover, $\pi^k()$ can be computed in $O(1)$ time in such a structure.*

Note that ϵ doesn't need to be a constant. By setting $\epsilon = \alpha + \beta \lg\lg n/\lg n$ where α and β are constants, and $0 < \alpha < 1$ we get the following theorem:

Theorem 7. *Given an unlabeled permutation of n elements, $\Theta(n^{(1-\alpha)/2}/\lg^{\beta/2} n)$ bits are sufficient for storing the permutation if each element is to be given a unique label in the range $\{1, \ldots, cn^{1+\alpha} \lg^{\beta} n\}$ for any constant c, α, β where $0 < \alpha < 1$. Moreover, $\pi^k()$ can be computed in $O(1)$ time in such a structure.*

6 Lower Bounds

In this section we provide lower bounds on the auxiliary data size as the label space increases.

6.1 Lower Bound for Auxiliary Data with Label Space cn

In [5] El-Zein et al. showed that for the problem of representing unlabeled equivalence relations, increasing the label space by a constant factor causes the size of the auxiliary data structure to decrease from $O(\sqrt{n})$ to $O(\lg n)$ bits.

In contrast to the problem of representing unlabeled equivalence relations, in this section we show that for the problem of representing unlabeled permutations increasing the label space by a constant factor will not affect the size of the auxiliary data structure asymptotically.

For any integer $c > 1$, let S_{cn} be the set of all partitions of $\lfloor cn \rfloor$ and S_n the set of all partitions of n. Without loss of generality assume that $\sqrt{n}$ is an integer that is divisible by c. While one partition of cn can dominate many partitions of n, we argue that at least $\binom{c\sqrt{n}}{\sqrt{n}/c}/\binom{\sqrt{n}}{\sqrt{n}/c}$ partitions of cn are necessary to dominate all partitions of n. Let $\mathcal{S}$ be the smallest set of partitions of cn that dominates all the partitions of n. We claim that:

Lemma 3. $|\mathcal{S}| \geq \binom{c\sqrt{n}}{\sqrt{n}/c}/\binom{\sqrt{n}}{\sqrt{n}/c}$. *The proof of Lemma 3 is found in the appendix. The information theoretic lower bound for the space needed to represent a permutation of size n once labels are assigned from the set* $\{1, \ldots, cn\}$ *is*

$$\begin{aligned}\lg(|\mathcal{S}|) &\geq \lg(\binom{c\sqrt{n}}{\sqrt{n}/c}/\binom{\sqrt{n}}{\sqrt{n}/c}) \\ &\in \Omega(\sqrt{n}).\end{aligned}$$

Theorem 8. *Given an unlabeled permutation of* n *elements,* $\Theta(\sqrt{n})$ *bits are necessary and sufficient for storing the permutation if each element is to be given a unique label in the range* $\{1, \ldots, cn\}$ *for any constant* $c > 1$*. Moreover,* $\pi^k()$ *can be computed in* $O(1)$ *time in such a structure.*

6.2 Lower Bound for Auxiliary Data with Label Space $cn^{1+\epsilon}$

Using techniques that are similar to the techniques presented in the previous subsection, we show that for the problem of representing unlabeled permutations an auxiliary data structure of size $O(n^{(1-\epsilon)/2})$ bits is necessary when the label space is $cn^{1+\epsilon}$, where c is any constant and $0 < \epsilon < 1$.

Denote by $S_{cn^{1+\epsilon}}$ the set of all partitions of $cn^{1+\epsilon}$ and by S_n the set of all partitions of n. We argue that at least $\binom{(c+1)n^{(1+\epsilon)/2}}{n^{(1-\epsilon)/2}/(c+1)}/\binom{cn^{(1+\epsilon)/2}/(c+1)}{n^{(1-\epsilon)/2}/(c+1)}$ are necessary to dominate all partitions of n. Let $\mathcal{S}$ be the smallest set of partitions of $cn^{1+\epsilon}$ that dominates all partitions of n. We claim that:

Lemma 4. $|\mathcal{S}| \geq \binom{(c+1)n^{(1+\epsilon)/2}}{n^{(1-\epsilon)/2}/(c+1)} / \binom{cn^{(1+\epsilon)/2}/(c+1)}{n^{(1-\epsilon)/2}/(c+1)}$. *The proof of Lemma 4 is found in the appendix. The information theoretic lower bound for space to represent a permutation of size n once labels are assigned from the set* $\{1, \ldots, cn^{1+\epsilon}\}$ *is*

$$\lg(|\mathcal{S}|) \geq \lg(\binom{(c+1)n^{(1+\epsilon)/2}}{n^{(1-\epsilon)/2}/(c+1)} / \binom{cn^{(1+\epsilon)/2}/(c+1)}{n^{(1-\epsilon)/2}/(c+1)})$$
$$\in \Omega(n^{(1-\epsilon)/2}).$$

Theorem 9. *Given an unlabeled permutation of n elements,* $\Theta(n^{(1-\epsilon)/2})$ *bits are necessary and sufficient for storing the permutation if each element is to be given a unique label in the range* $\{1, \ldots, cn^{1+\epsilon}\}$ *for any constant* $c > 1$ *and* $\epsilon < 1$. *Moreover,* $\pi^k()$ *can be computed in* $O(1)$ *time in such a structure.*

7 Applications

As an application to our data structures, we give a representation of a labeled permutation that takes $s(n) + O(\sqrt{n})$ bits and can answer $\pi^k()$ in $O(t_f + t_i)$ time, where $s(n)$ denotes the number of bits required for a representation R to store a labeled permutation, and t_f and t_i are the time needed for R to support $\pi()$ and $\pi^{-1}()$.

This result improves Theorem 3.3 in [13]: Suppose there is a representation R taking $s(n)$ bits to store an arbitrary permutation π on $\{1, \ldots, n\}$, that supports $\pi()$ in time t_f, and $\pi^{-1}()$ in time t_i. Then there is a representation for an arbitrary permutation on $\{1, \ldots, n\}$ taking $s(n) + O(n \lg n / \lg \lg n)$ bits in which $\pi^k()$ can be supported in $t_f + t_i + O(1)$ time, and one taking $s(n) + O(\sqrt{n} \lg n)$ bits in which $\pi^k()$ can be supported in $t_f + t_i + O(\lg \lg n)$ time.

Theorem 10. *Suppose there is a representation R taking* $s(n)$ *bits to store an arbitrary permutation* π *on* $\{1, \ldots, n\}$, *that supports* $\pi()$ *and* $\pi^{-1}()$ *in time* t_f *and* t_i. *Then there is a representation for an arbitrary permutation on* $\{1, \ldots, n\}$ *taking* $s(n) + O(\sqrt{n})$ *bits in which* $\pi^k()$ *can be supported in* $t_f + t_i + O(1)$ *time.*

Proof. Given π, treat it as an unlabeled permutation and build the data structure from Theorem 5 on it. Call this structure P. Notice that the bijection between the labels generated by P and the real labels of π form a permutation. Store this permutation using the given scheme in a structure P'. Now $\pi^k(i) = \pi_{P'}^{-1}(\pi_P^k(\pi_{P'}^1(i)))$ can be computed in $t_f + t_i + O(1)$ time, and the total space used is $s(n) + O(\sqrt{n})$ bits. □

8 Conclusion

We have provided a complete breakdown for the label space-auxiliary storage size tradeoff for the problem of representing unlabeled permutations. As there is a huge body of research in 'labeling schemes', investigation into such a tradeoff for other problems maybe interesting. Moreover as an application to our new data structures, we showed how to improve the general representation of permutations. Given that permutations are fundamental in computer science, we feel that our structures will find applications in many other scenarios.

References

1. Barbay, J., Aleardi, L.C., He, M., Munro, J.I.: Succinct representation of labeled graphs. Algorithmica **62**(1–2), 224–257 (2012)
2. Broder, A.Z., Charikar, M., Frieze, A.M., Mitzenmacher, M.: Min-wise independent permutations. J. Comput. Syst. Sci. **60**(3), 630–659 (2000)
3. Brodnik, A., Munro, J.I.: Membership in constant time and almost-minimum space. SIAM J. Comput. **28**(5), 1627–1640 (1999)
4. El-Zein, H.: On the succinct representation of equivalence classes (2014)
5. El-Zein, H., Munro, J.I., Raman, V.: Tradeoff between label space and auxiliary space for representation of equivalence classes. In: Ahn, H.-K., Shin, C.-S. (eds.) ISAAC 2014. LNCS, vol. 8889, pp. 543–552. Springer, Heidelberg (2014)
6. Farzan, A., Munro, J.I.: Succinct representations of arbitrary graphs. In: Halperin, D., Mehlhorn, K. (eds.) ESA 2008. LNCS, vol. 5193, pp. 393–404. Springer, Heidelberg (2008)
7. Fich, F.E., Munro, J.I., Poblete, P.V.: Permuting in place. SIAM J. Comput. **24**(2), 266–278 (1995)
8. Grossi, R., Vitter, J.S.: Compressed suffix arrays and suffix trees with applications to text indexing and string matching. SIAM J. Comput. **35**(2), 378–407 (2005)
9. Hardy, G.H., Ramanujan, S.: Asymptotic formulae in combinatory analysis. Proc. London Math. Soc. **2**(1), 75–115 (1918)
10. He, M., Munro, J.I., Rao, S.S.: A categorization theorem on suffix arrays with applications to space efficient text indexes. In: Proceedings of ACM-SIAM Symposium on Discrete Algorithms (SODA), pp. 23–32. SIAM (2005)
11. Lewenstein, M., Munro, J.I., Raman, V.: Succinct data structures for representing equivalence classes. In: Cai, L., Cheng, S.-W., Lam, T.-W. (eds.) ISAAC 2013. LNCS, vol. 8283, pp. 502–512. Springer, Heidelberg (2013)
12. Munro, J.I., Nicholson, P.K.: Succinct posets. In: Epstein, L., Ferragina, P. (eds.) ESA 2012. LNCS, vol. 7501, pp. 743–754. Springer, Heidelberg (2012)
13. Munro, J.I., Raman, R., Raman, V., Rao, S.S.: Succinct representations of permutations and functions. Theoret. Comput. Sci. **438**, 74–88 (2012)
14. Raman, R., Raman, V., Satti, S.R.: Succinct indexable dictionaries with applications to encoding k-ary trees, prefix sums and multisets. ACM Trans. Algorithms **3**(4) (2007). Article no 43
15. Sedgewick, R.: Permutation generation methods. ACM Comput. Surv. **9**(2), 137–164 (1977)

How to Select the Top k Elements from Evolving Data?

Qin Huang[1], Xingwu Liu[1,2](✉), Xiaoming Sun[1], and Jialin Zhang[1]

[1] Institute of Computing Technology, Chinese Academy of Sciences, Beijing, China
{huangqin,liuxingwu,sunxiaoming,zhangjialin}@ict.ac.cn

[2] State Key Laboratory of Software Development Environment, Beihang University, Beijing, China

Abstract. In this paper we investigate the top-k-selection problem, i.e. to determine and sort the top k elements, in the dynamic data model. Here dynamic means that the underlying total order evolves over time, and that the order can only be probed by pair-wise comparisons. It is assumed that at each time step, only one pair of elements can be compared. This assumption of restricted access is reasonable in the dynamic model, especially for massive data set where it is impossible to access all the data before the next change occurs. Previously only two special cases were studied [1] in this model: selecting the element of a given rank, and sorting all elements. This paper systematically deals with $k \in [n]$. Specifically, we identify the critical point k^* such that the top-k-selection problem can be solved error-free with probability $1 - o(1)$ if and only if $k = o(k^*)$. A lower bound of the error when $k = \Omega(k^*)$ is also determined, which actually is tight under some conditions. In contrast, we show that the top-k-set problem, which means finding the top k elements without sorting them, can be solved error-free with probability $1 - o(1)$ for all $1 \le k \le n$. Additionally, we consider some extensions of the dynamic data model and show that most of these results still hold.

1 Introduction

Sorting, a fundamental primitive in algorithms, has been an active research topic in computer science for decades. In the era of big data, it is the cornerstone of numerous vital applications – Web search, online ads, and recommendation systems to name but a few. While sorting has been extensively studied, little is known when the data is dynamic. Actually, dynamic data is common in practical applications: the linking topology of Web pages, the friendship network of Facebook, the daily sales of Amazon, and so on, all keep changing. The basic challenge in dealing with dynamic, massive data is that the access to the data is too restricted to catch the changes.

The work is partially supported by National Natural Science Foundation of China (61173009, 61170062, 61222202, 61433014, 61502449), State Key Laboratory of Software Development Environment Open Fund (SKLSDE-2014KF-01), and the China National Program for support of Top-notch Young Professionals.

K. Elbassioni and K. Makino (Eds.): ISAAC 2015, LNCS 9472, pp. 60–70, 2015.
DOI: 10.1007/978-3-662-48971-0_6

For example, it is impossible to get an exact snapshot of Web, and a third-party vendor can query the Facebook network only via a rate-limited API. As a result, this paper is devoted to studying the sorting problem on dynamic, access-restricted data.

In the seminal paper [1], Anagnostopoulos et al. formulated a model for dynamic data as follows. Given a set U of n elements, at every discrete time t, there is an underlying total order π^t on U. For every $t \geq 1$, π^t is obtained from π^{t-1} by sequentially swapping α random pairs of consecutive elements, where α is a constant number. The only way to probe π^t is querying the relative rank of ONE pair of elements in U at every time step. The goal is to learn about the true order π^t. Obviously, it is impossible to always exactly find out the orders, so our objective is that at any time t, the algorithm estimates the correct answer (or an approximate answer) with high probability. In this paper, "with high probability" and "with probability $1 - o(1)$" are used interchangeably.

Anagnostopoulos et al. [1] proved that the Kendall tau distance between π^t and $\tilde{\pi}^t$, defined in Sect. 2 and denoted by $\mathrm{KT}(\pi^t, \tilde{\pi}^t)$, is lower-bounded by $\Omega(n)$ with high probability at every t, where $\tilde{\pi}^t$ is the order estimated by any algorithm. This lower bound is nearly tight, since they proposed an algorithm with $\mathrm{KT}(\pi^t, \tilde{\pi}^t) = O(n \ln \ln n)$. Furthermore, they designed an algorithm that with high probability, exactly identifies the element of a given rank.

Though elegant, this model is too restricted: the evolution is extremely slow since α is constant, and is extremely local since only consecutive elements are swapped. Hence, it is extended in this paper by allowing α to be a function of n, and is called the consecutive-swapping model. We further generalize it to the Gaussian-swapping model by relaxing the locality condition.

Inspired by [1], we study the general top-k-selection problem: at every time t, figure out the top k elements and sort them, where $k \in \{1, 2, ...n\}$. Its two extreme cases where $k = n$ and $k = 1$ correspond to the sorting problem and the selection problem in [1], respectively. The error-free solvability of the selection problem suggests that the error in solving the top-k-selection problem may vanish as k decreases, so it is natural to investigate the critical point where the error vanishes and to find the optimal solution beyond the critical point. Another motivation lies in the wide application of top-k-selection, also known as partial sorting. It has been used in a variety of areas such as Web and multimedia search systems and distributed systems, where massive data has to be dealt with efficiently [2].

Additionally, we consider a closely related top-k-set problem: at every time t, identify the set of the top k elements. The top-k-set problem is weaker in that it does not require to sort the elements. In the static data setting, when a selection algorithm identifies the kth element, it automatically determines the set of the top k elements (see for example Knuth's book [3]). However, this is not apparent in the dynamic data model.

Our Contributions. The main results of this paper lie in two aspects in the consecutive-swapping model. First, it is shown that the top-k-set problem can be solved error-free with high probability for any $1 \leq k \leq n$. Second and more important, $k^* = \Theta(\sqrt{\frac{n}{\alpha}})$ is proven to be the critical point of k for the top-k-selection problem, which means that this problem can be solved error-free with high probability if and only if $k = o(k^*)$.

In addition, for k beyond k^*, we obtain tight lower bounds of $\mathrm{KT}(\tilde{\pi}_k^t, \pi_k^t)$, the Kendall tau distance between the true order π_k^t and the algorithmically estimated order $\tilde{\pi}_k^t$ of the top k elements. Specifically, if $k = \Omega(\sqrt{\frac{n}{\alpha}})$, then for any algorithm, $\mathrm{KT}(\tilde{\pi}_k^t, \pi_k^t) \neq 0$ with constant probability. When $k = \omega(\sqrt{n})$ and $\alpha = O(1)$, for any algorithm, $\mathrm{KT}(\tilde{\pi}_k^t, \pi_k^t) = \Omega(\frac{k^2}{n})$ with high probability at every t. These lower bounds can be reached by ONE algorithm with parameter k, (see Algorithm 2), hence being tight.

The results of the top-k-selection problem in the consecutive-swapping model are summarized in Table 1. Most of the results are also generalized to the Gaussian-swapping model with constant α, as summarized in Table 2.

Table 1. Results in the consecutive-swapping model

k	$X \triangleq \mathrm{KT}(\tilde{\pi}_k^t, \pi_k^t)$
$o(\sqrt{\frac{n}{\alpha}})$	$\Pr(X = 0) = 1 - o(1)$
$\Theta(\sqrt{\frac{n}{\alpha}})$	$\Pr(X = 0) = \Theta(1) = \Pr(X > 0)$
$\omega(\sqrt{\frac{n}{\alpha}})$	$\Pr(X = O(\frac{k^2\alpha}{n})) = 1 - o(1)^a$

In [a] case, this upper bound of X is tight for constant α. See Sect. 3

Table 2. Results in the Gaussian-swapping model

k	$X \triangleq \mathrm{KT}(\tilde{\pi}_k^t, \pi_k^t)$
$o(\frac{\sqrt{n}}{\ln^{0.25} n})$	$\Pr(X = 0) = 1 - o(1)$
$\Theta(\frac{\sqrt{n}}{\ln^{0.25} n})$	$\Pr(X = 0) = \Theta(1)$
$\omega(\frac{\sqrt{n}}{\ln^{0.25} n})$	$\Pr(X = O(\frac{k^2 \ln n}{n})) = 1 - o(1)$

Related Work. The sorting/selection problem has been actively investigated for decades [2,4–6], but the study of this problem in dynamic data setting was initiated very recently [1]. In [1], Anagnostopoulos et al. considered two special cases of the top-k-selection problem, namely $k = n$ and $k = 1$, in the consecutive-swapping model with constant α. Their work has inspired the problem and the data model in this paper. The theoretical results in [1] were experimentally verified by Moreland [7] in 2014.

Dynamic data is also studied in the graph setting. [8] considered two classical graph connectivity problems (path connectivity and minimum spanning trees) where the graph keeps changing over time and the algorithm, unaware of the changes, probes the graph to maintain a path or spanning tree. Bahmani et al. [9] designed an algorithm to approximately compute the PageRank of evolving graphs, and Zhuang et al. [10] considered the influence maximization problem in dynamic social networks. On the other hand, Labouseur et al. [11] and Ren [12] dealt with the data structure and management issues, respectively, enabling efficient query processing for dynamic graphs.

It is worth noting that our dynamic data model is essentially different from noisy information model [13,14]. In computing with noisy information, the main difficulty is brought about by misleading information. On the contrary, in our model, the query results are correct, while the difficulty comes from the restricted access to the dynamic data. The ground truth can be probed only by local observation, so it is impossible to capture all changes in the data. The key issue is to choose query strategies in order to approximate the real data with high probability.

In the algorithm community, there are many other models dealing with dynamic and uncertain data, from various points of view. However, none of them captures the two crucial aspects of our dynamic data model: the underlying data keeps changing, and the data exposes limited information to the algorithm by probing. For example, data stream algorithms [15] deal with a stream of data, typically with limited space, but the algorithms can observe the entire data that has arrived; local algorithms on graphs [16,17] probe the underlying graphs by a limited number of query, but typically the graphs are static; in online algorithms [18], though the data comes over time and is processed without knowledge of the future data, the algorithms know all the data up to now; the multi-armed-bandit model [19] tends to optimize the total gain in a finite exploration-exploitation process, while our framework concerns the performance of the algorithm at every time step in an infinite process.

The rest of the paper is organized as follows. In Sect. 2, we provide the formal definition of the models and formulate the problems. Section 3 is devoted to solving the top-k-set problem and the top-k-selection problem in the consecutive-swapping model. In Sect. 4, the problems are studied in the Gaussian-swapping model. Section 5 concludes the paper. Due to the limitation of space, all proofs of the theorems will be omitted.

2 Preliminaries

We now formalize our dynamic data model.

Let $U = \{u_1, ..., u_n\}$ be a set with n elements, and $\mathcal{U}$ be the set of all total orders over U, that is, $\mathcal{U} = \{\pi : U \rightarrow [n] \mid \forall i \neq j, \pi(u_i) \neq \pi(u_j)\}$, where $[n] \triangleq \{1, 2, ...n\}$. For any $\pi \in \mathcal{U}$ and $k \in [n]$, we define $\pi^{-1}(k)$ to be the kth element and $\pi(u)$ to be the rank of u relative to π. If $\pi(u) < \pi(v)$, we say $u >_\pi v$ or simply by $u > v$ when π can be inferred from context.

In this paper, we consider the process where the order on U gradually changes over time. Time is discretized into steps sequentially numbered by nonnegative integers. At every time step t, there is an underlying total order π^t on U. For every $t \geq 1$, π^t is obtained from π^{t-1} by sequentially swapping α random pairs of consecutive elements, where α is an integer function of n. This is our consecutive-swapping model.

Now we introduce the Gaussian-swapping model whose defining feature is that non-consecutive pairs can be swapped in the evolution. Specifically, for every $t \geq 1$, π^t is still obtained from π^{t-1} by sequentially swapping α pairs of elements. However, each pair (not necessarily consecutive) is selected as follows, rather than uniformly randomly. First, d is sampled from a truncated Gaussian distribution $\Pr(D = d) = \beta e^{\frac{-d^2}{2}}$ where β is the normalizing factor. Then, a pair of elements whose ranks differ by d is chosen uniformly randomly from all such pairs. Thus, the overall probability that a pair (u, v) gets swapped is $\frac{\beta e^{\frac{-d^2}{2}}}{n-d}$, where d is the difference between the ranks of u and v, related to π^{t-1}.

In either model, at any time step t, the changes of π^t are unknown by the algorithms running on the data. The only way to probe the underlying order is by comparative queries. At any time t, given an arbitrary pair of elements $u, v \in U$, an algorithm can query whether $\pi^t(u) > \pi^t(v)$ or not. At most *one* pair of elements can be queried at each time step.

Now we define $\mathcal{I}$-sorting problem for any index set $\mathcal{I} \subseteq [n]$: at each time step t, find out all the elements whose ranks belong to $\mathcal{I}$, and sort them according to π^t. The concept of $\mathcal{I}$-sorting problem unifies both the sorting problem ($|\mathcal{I}| = n$) and the selection problem ($|\mathcal{I}| = 1$). This paper mainly studies the top-k-selection problem, a special case of the $\mathcal{I}$-sorting problem with $\mathcal{I} = [k]$ for $k \in [n]$. For convenience, in this paper we use notation π_k^t to represent the true order on the top k elements at time t. A closely-related problem, called the top-k-set problem, is also studied. It requires to find out $(\pi^t)^{-1}([k])$ at each time t, without sorting them.

We then define the performance metrics of the algorithms. In the top-k-set problem, we want to maximize the probability that the output set is exactly the same as the true set for sufficiently large t. In the top-k-selection problem, we try to minimize the Kendall tau distance between the output order and the true order on the top k elements, for sufficiently large t. Since an algorithm solving the top-k-selection problem may output an order on a wrong set, we extend the definition of Kendall tau distance to orders on different sets. Specifically, given total orders σ on set V and δ on set W with $|V| = |W|$, their Kendall tau distance is defined to be $\mathrm{KT}(\sigma, \delta) = |\{(x, y) \in V^2 : \sigma(x) < \sigma(y) \text{ and } (x \notin W \text{ or } y \notin W \text{ or } \delta(x) > \delta(y))\}|$. Intuitively, it is the number of pairs that either are not shared by W and V or are ordered inconsistently by the two total orders.

Throughout this paper, one building block of the algorithms is the randomized quick-sort algorithm. We describe the randomized quick-sort algorithm briefly. Given an array, it works as follows: (1) Uniformly randomly pick an element, called a pivot, from the array. (2) Compare all elements with the pivot, resulting in two sub-arrays: one consisting of all the elements smaller than the pivot, and the other consisting of the other elements except the pivot. (3) Recursively apply steps 1 and 2 to the two sub-arrays until all the sub-arrays are singletons.

3 Consecutive-Swapping Model

In this section, we consider the top-k-set problem and the top-k-selection problem in the consecutive-swapping model. For the top-k-set problem, Sect. 3.1 shows an algorithm which is error-free with probability $1 - o(1)$ for arbitrary k. Section 3.2 is devoted to the top-k-selection problem. It presents an algorithm that is optimal when α is constant or k is small.

3.1 An Algorithm for the Top-k-set Problem

The basic idea is to repeatedly run quick-sort over the data U, extract the set of the top k elements from the resulting order, and output this set during the next run. But an issue should be addressed: since the running time of quick-sort is $\Omega(n \ln n)$ with high probability, the set of the top k elements will change with high probability during the next run, leading to out-of-date outputs. Because the rank of every element does not change too much during the next run of quick-sort, a solution is to parallel sort a small subset of U that contains the top k elements with high probability.

Algorithm 1. Top-k-set

Input: A set U of n elements
Output: $\widetilde{T}$
1: Initialize $\tilde{\pi}, L, C, \tilde{\pi}_C$, and $\widetilde{T}$ arbitrarily
2: **while** (true) **do**
3: **Execute in odd steps:** /*QS_1*/
4: $\tilde{\pi} \leftarrow$ quick_sort(U)
5: $L \leftarrow \tilde{\pi}^{-1}([k - c\alpha \ln n])$ and $C \leftarrow \tilde{\pi}^{-1}([k + c\alpha \ln n]) \setminus L$ /*The constant c will be determined in the proof of Theorem 1*/
6: **Execute in even steps:** /*QS_2*/
7: $\tilde{\pi}_C \leftarrow$ quick_sort(C)
8: $\widetilde{T} \leftarrow L \bigcup \tilde{\pi}_C^{-1}([c\alpha \ln n])$
9: **end while**

Specifically, the algorithm Top-k-set consists of two interleaving procedures (denoted by QS_1 and QS_2, respectively), each of which restarts once it terminates. In the odd steps, QS_1 calls quick-sort to sort U, preparing two sets L and C. The set L consists of the elements that will remain among top k during the next run of QS_1 with high probability, while C contains the uncertain elements that might be among top k in this period. Then, QS_2 will sort the set C computed by the last run of QS_1 to produce the estimated set of top k elements. At any time t, the output $\widetilde{T}_t$ of the algorithm is the set $\widetilde{T}$ computed by the previous run of QS_2.

Theorem 1 shows that Algorithm 1 is error-free with high probability.

Theorem 1. *Assume that* $\alpha = o(\frac{\sqrt{n}}{\ln n})$. *For any* $k \in [n]$, $\Pr(\widetilde{T}_t = (\pi^t)^{-1}([k])) = 1 - o(1)$, *where* $\widetilde{T}_t$ *is the output of Algorithm 1 at time* t, π^t *is the true order on* U *at time* t, *and* t *is sufficiently large.*

The basic idea of the proof lies in two aspects. First, with high probability, the estimated rank of every element with respect to $\tilde{\pi}$ is at most $O(\alpha \ln n)$ away from the true rank, implying that all the elements in L are among top k and all top k elements are in $L \bigcup C$. Second, with high probability, the kth element of U does not swap throughout sorting C, so the set of top k elements remains unchanged and is exactly contained in $\widetilde{T}$. The detailed proof will be omitted.

3.2 An Algorithm for the Top-*k*-selection Problem

Now we present an algorithm to solve the top-k-selection problem. The basic idea is to repeatedly run quick-sort over the data U, extracting a small subset that includes all the elements that can be among top k during the next run. To exactly identify the top k elements in order, the small set is sorted and the order of the top k elements is produced accordingly. Like in designing the top-k-set algorithm, there is also an issue to address: since sorting the small set takes time $\Omega(k \ln k)$, the order of the top k elements will soon become out of date. Again note that with high probability the rank of each element does not change too much during sorting the small set, so the order of the top k elements can be regulated locally and keeps updated.

Specifically, Algorithm 2 consists of four interleaving procedures (QS_1, QS_2, QS_3, and Local-sort), each of which restarts once it terminates. At the $(4t+1)$-th time steps, QS_1 invokes a quick-sort on U, preparing a set C of size $k + O(\alpha \ln n)$ which with high probability, contains all the elements among top k during the next run of QS_1. At the $(4t+2)$-th time steps, QS_2 calls another quick-sort on the latest C computed by QS_1, producing a set P of size k. With high probability, the set P exactly consists of the top k elements of U during the next run of QS_2. At the $(4t+3)$-th time steps, the other quick-sort is invoked by QS_3 on the latest P computed by QS_2, periodically updating the estimated order over P. The resulting order is actually close to the true order over P during the next run of QS_3. Finally, at the $(4t)$-th time steps, an algorithm Local-sort is executed on the total order over P that is produced by the last run of QS_3, so as to locally regulate the order. At any time t, the output $\tilde{\pi}_k^t$ of Algorithm 2 is the last $\tilde{\pi}_k$ computed by Local-sort.

The main idea of Algorithm 3 (Local-sort) is to regulate the order over P block by block. Since block-by-block processing takes linear time, the errors can be corrected in time and few new errors will emerge during one run of Algorithm 3. Considering that the elements may move across blocks, it is necessary to make the blocks overlap. Actually, for each j, the element of the lowest rank in the j-th block is found, regarded as the j-th element of the final order, and removed from the block. The rest elements of the j-th block, together with the lowest-ranked element in P (according to the latest order produced by QS_3) that has not yet been processed, forms the $(j+1)$-th block. The element of the

Algorithm 2. Top-k-selection

Input: A set U of n elements
Output: $\tilde{\pi}_k$

1: Let t be the time
2: Initialize $\tilde{\pi}, C, \tilde{\pi}_C, P, \tilde{\pi}_P,$ and $\tilde{\pi}_k$ arbitrarily
3: **while** (true) **do**
4: **Execute in** $t \equiv 1$(**mod** 4) **steps** /*QS_1*/
5: $\tilde{\pi} \leftarrow$ quick_sort(U)
6: $C \leftarrow \tilde{\pi}^{-1}([k + c'\alpha \ln n])$ /*The constant c' will be determined in the proof of Theorem 2*/
7: **Execute in** $t \equiv 2$(**mod** 4) **steps** /*QS_2*/
8: $\tilde{\pi}_C \leftarrow$ quick_sort(C)
9: $P \leftarrow \tilde{\pi}_C^{-1}([k])$
10: **Execute in** $t \equiv 3$(**mod** 4) **steps** /*QS_3*/
11: $\tilde{\pi}_P \leftarrow$ quick_sort(P)
12: **Execute in** $t \equiv 0$(**mod** 4) **steps** /*Local-sort*/
13: $\tilde{\pi}_k \leftarrow$ Local-sort($P, \tilde{\pi}_P, 4c+1$) /*The constant c will be determined in the proof of Theorem 2*/
14: **end while**

Algorithm 3. Local-sort

Input: A set P; an order π over P; an integer c
Output: $\tilde{\pi}$

1: $m \leftarrow |P|$
2: $B_1 \leftarrow \pi^{-1}([c])$ /* Define the first block */
3: $\tilde{\pi}^{-1}(1) \leftarrow$ Maximum-Find(B_1)
4: $j = 2$
5: **while** $(c + j - 1 \leq m)$ **do**
6: $B_j \leftarrow (B_{j-1} \backslash \tilde{\pi}^{-1}(j-1)) \bigcup \pi^{-1}(c + j - 1)$ /* Define the j-th block */
7: $\tilde{\pi}^{-1}(j) \leftarrow$ Maximum-Find(B_j)
8: $j = j + 1$
9: **end while**
10: $B_e \leftarrow B_{j-1}$ /*Deal with the final block*/
11: **while** $|B_e| \geq 1$ **do**
12: $\tilde{\pi}^{-1}(j) \leftarrow$ Maximum-Find(B_e)
13: $B_e \leftarrow B_e \backslash \tilde{\pi}^{-1}(j)$
14: $j = j + 1$
15: **end while**

lowest rank in each block is found by calling Algorithm 4, which repeatedly runs sequential comparison. Both Algorithms 3 and 4 are self-explained, so detailed explanation is omitted here.

Theorem 2. *Assume* $\alpha = o(\frac{\sqrt{n}}{\ln n})$ *and* $k = O((\frac{n}{\alpha \ln n})^{1-\epsilon})$, *where* $\epsilon > 0$. *Let* $\tilde{\pi}_k^t$ *be the output of Algorithm 2 and* π_k^t *be the true order over the top k elements at time t. For sufficiently large t, we have that:*

Algorithm 4. Maximum-Find

Input: B
Output: u_{max}
1: $u_{max} \leftarrow B(1)$
2: $j = 2$
3: **while** $(j \leq |B|)$ **do**
4: **if** $u_{max} < B(j)$ **then**
5: $u_{max} \leftarrow B(j)$
6: **end if**
7: $j = j + 1$
8: **end while**

1. *If* $k^2\alpha = o(n)$, $\Pr(\mathrm{KT}(\tilde{\pi}_k^t, \pi_k^t) = 0) = 1 - o(1)$,
2. *If* $k^2\alpha = \Theta(n)$, $\Pr(\mathrm{KT}(\tilde{\pi}_k^t, \pi_k^t) = 0) = \Theta(1)$, *and*
3. *If* $k^2\alpha = \omega(n)$, $\Pr(\mathrm{KT}(\tilde{\pi}_k^t, \pi_k^t) = O(\frac{k^2\alpha}{n})) = 1 - o(1)$.

We sketch the basic idea of the proof. First, with high probability, the rank of every element with respect to $\tilde{\pi}$ is at most $O(\alpha \ln n)$ away from the true rank, implying that all the top k elements are contained in C. Second, with high probability, the kth element of U does not swap throughout sorting C, so P is exactly the set of top k elements and the resulting rank of every element deviates from the true rank by at most a constant. Third, due to the small rank deviation of every element, the ordering can be corrected locally by sorting blocks of constant length. The detailed proof will be omitted.

3.3 Lower Bounds for the Top-k-selection Problem

Now we analyze the lower bounds of the performance of any top-k-selection algorithm. The lower bounds hold for both randomized and deterministic algorithms.

Let A be an arbitrary algorithm which takes our dynamic data as input and outputs a total order $\tilde{\pi}_k^t$ on a subset of size k at every time step t. Let π_k^t be the true order on the top k elements. The following theorems characterize the difference between $\tilde{\pi}_k^t$ and π_k^t when k is large.

Theorem 3. *Given* $k = \Omega(\sqrt{\frac{n}{\alpha}})$ *and* $\alpha = o(n)$, $\Pr(\mathrm{KT}(\tilde{\pi}_k^t, \pi_k^t) > 0) = \Theta(1)$ *for every* $t > k$.

The main idea of the proof is that with a constant probability, in any period of $\Theta(\sqrt{\frac{n}{\alpha}})$, exactly one swap occurs among the top k elements and the swap is not observed. The detailed proof will be omitted.

Theorem 4. *Given* $k = \omega(\sqrt{n})$ *and* $\alpha = O(1)$, $\mathrm{KT}(\tilde{\pi}_k^t, \pi_k^t) = \Omega(\frac{k^2}{n})$ *in expectation and with probability* $1 - o(1)$ *for every* $t > k/8$.

The basic idea of the proof is that with high probability, in any period of $\Theta(k)$, $\Omega(\frac{k^2}{n})$ swaps occur among the top k elements and a majority of the swaps are not observed. The detailed proof will be omitted.

From Theorems 2 and 3, we know that $\Theta(\sqrt{n/\alpha})$ is the critical point of k, and it is impossible to generally improve Algorithm 2 even if $k = \omega(\sqrt{n/\alpha})$. The term *critical point* means the least upper bound of k such that top-k-selection problem can be solved error-free with probability $1 - o(1)$.

4 Gaussian-Swapping Model

This section is devoted to extending the algorithms for the consecutive-swapping model to the Gaussian-swapping model. We focus on the special case where α is a constant, and still assume that at each time step only one pair of elements can be compared.

Algorithms 1 and 2 can be slightly adapted to solve the top-k-set problem and the top-k-selection problem in this model, respectively. Specifically, replacing α in lines 5 and 8 of Algorithm 1 with $\ln^{0.5} n$, one gets Algorithm 5; likewise, in Algorithm 2, replacing α in line 6 with $\ln^{0.5} n$ and $4c + 1$ in lines 13 with $4c\ln^{0.5} n + 1$, we get Algorithm 6. The following theorems state the performance of these algorithms, and the proofs are omitted.

Theorem 5. *For any $k \in [n]$, we have $\Pr(\widetilde{T}_t = (\pi^t)^{-1}([k])) = 1 - o(1)$, where $\widetilde{T}_t$ is the output of Algorithm 5 at time t, π^t is the true order at time t, and t is sufficiently large.*

Theorem 6. *Assume that $k = O((\frac{n}{\ln n})^{1-\epsilon})$, where $\epsilon > 0$. Let $\tilde{\pi}_k^t$ be the output of Algorithm 6 and π_k^t be the true order over the top k elements at time t. For sufficiently large t, we have:*

1. *If $k = o(\frac{\sqrt{n}}{\ln^{0.25} n})$, $\Pr(\mathrm{KT}(\tilde{\pi}_k^t, \pi_k^t) = 0) = 1 - o(1)$,*
2. *If $k = \Theta(\frac{\sqrt{n}}{\ln^{0.25} n})$, $\Pr(\mathrm{KT}(\tilde{\pi}_k^t, \pi_k^t) = 0) = \Theta(1)$, and*
3. *If $k = \omega(\frac{\sqrt{n}}{\ln^{0.25} n})$, $\Pr(\mathrm{KT}(\tilde{\pi}_k^t, \pi_k^t) = O(\frac{k^2 \ln n}{n})) = 1 - o(1)$.*

Except for the Gaussian distribution, d can also be determined by other discrete distributions, for example, $p(d) = \frac{\beta}{d^\gamma}$, where γ is a constant and β is a normalizing factor. When γ is large enough (say, $\gamma > 10$), the results similar to those in the Gaussian-swapping model can be obtained.

5 Conclusions

In this paper we identify the critical point k^* such that the top-k-selection problem can be solved error-free with high probability if and only if $k = o(k^*)$. A lower bound of the error when $k = \Omega(k^*)$ is also determined, which actually is tight under some condition. On the contrary, it is shown that the top-k-set problem can be solved error-free with probability $1 - o(1)$, for all $k \in [n]$. These results hold in the consecutive-swapping model and most of them can be extended to the Gaussian-swapping model.

A number of problems remain open for the top-k-selection problem in the consecutive-swapping model. For $\alpha = \omega(1)$, we have not shown whether the upper bound $O(\frac{k^2\alpha}{n})$ of error is tight when $k = \omega(\sqrt{\frac{n}{\alpha}})$. For $\alpha = O(1)$, there exists a gap between $k = n$ and $k = O((\frac{n}{\ln n})^{1-\epsilon})$, where the lower bound $\Omega(\frac{k^2}{n})$ of error has not yet shown to be tight. We conjecture that these bounds are tight.

References

1. Anagnostopoulos, A., Kumar, R., Mahdian, M., Upfal, E.: Sort me if you can: how to sort dynamic data. In: Albers, S., Marchetti-Spaccamela, A., Matias, Y., Nikoletseas, S., Thomas, W. (eds.) ICALP 2009, Part II. LNCS, vol. 5556, pp. 339–350. Springer, Heidelberg (2009)
2. Ilyas, I., Beskales, G., Soliman, M.: A survey of top-k query processing techniques in relational database systems. ACM Comput. Surv. 40(4) (2008). Article 11
3. Knuth, D.E.: The Art of Computer Programming, vol. 3. Addison-Wesley, Boston (1973)
4. Kislitsyn, S.S.: On the selection of the kth element of an ordered set by pairwise comparison. Sibirskii Mat. Zhurnal **5**, 557–564 (1964)
5. Blum, M., Floyd, R., Pratt, V., Rivest, R., Tarjan, R.: Time bounds for selection. J. Comput. Syst. Sci. **7**(4), 448–461 (1973)
6. Dor, D., Zwick, U.: Selecting the median. In: SODA 1995, pp. 28–37 (1995)
7. Moreland, A.: Dynamic Data: Model, Sorting, Selection. Technical report (2014)
8. Anagnostopoulos, A., Kumar, R., Mahdian, M., Upfal, E., Vandin, F.:Algorithms on evolving graphs. In: 3rd Innovations in Theoretical Computer Science Conference (ITCS), pp. 149–160. ACM, New York (2012)
9. Bahmani, B., Kumar, R., Mahdian, M., Upfal, E.: Pagerank on an evolving graph. In: 18th ACM International Conference on Knowledge Discovery and Data Mining (SIGKDD), pp. 24–32. ACM (2012)
10. Zhuang, H., Sun, Y., Tang, J., Zhang J., Sun, X.: Influence maximization in dynamic social networks. In: 13th IEEE International Conference on Data Mining (ICDM), pp. 1313–1318. IEEE (2013)
11. Labouseur, A.G., Olsen, P.W., Hwang, J.H.: Scalable and robust management of dynamic graph data. In: 1st International Workshop on Big Dynamic Distributed Data (BD3@VLDB), pp. 43–48 (2013)
12. Ren, C.: Algorithms for evolving graph analysis. Doctoral dissertation. The University of Hong Kong (2014)
13. Ajtai, M., Feldman, V., Hassidim, A., Nelson, J.: Sorting and selection with imprecise comparisons. In: Albers, S., Marchetti-Spaccamela, A., Matias, Y., Nikoletseas, S., Thomas, W. (eds.) ICALP 2009, Part I. LNCS, vol. 5555, pp. 37–48. Springer, Heidelberg (2009)
14. Feige, U., Raghavan, P., Peleg, D., Upfal, E.: Computing with noisy information. SIAM J. Comput. **23**(5), 1001–1018 (1994)
15. Babcock, B., Babu, S., Datar, M., Motwani, R., Widom, J.: Models and issues in data stream systems. In: 21st ACM SIGMOD-SIGACT-SIGART Symposium on Principles of Database Systems (PODS), pp. 1–16. ACM (2002)
16. Bressan, M., Peserico, E., Pretto, L.: Approximating PageRank locally with sublinear query complexity. ArXiv preprint (2014). arXiv:1404.1864
17. Fujiwara, Y., Nakatsuji, M., Shiokawa, H., Mishima, T., Onizuka, M.: Fast and exact top-k algorithm for pagerank. In: 27th AAAI Conference on Artificial Intelligence, pp. 1106–1112 (2013)
18. Albers, S.: Online algorithms: a survey. Math. Prog. **97**(1–2), 3–26 (2003)
19. Kuleshov, V., Precup, D.: Algorithms for multi-armed bandit problems. ArXiv preprint (2014). arXiv:1402.6028

Optimal Search Trees with 2-Way Comparisons

Marek Chrobak[1], Mordecai Golin[2], J. Ian Munro[3], and Neal E. Young[1](✉)

[1] University of California – Riverside, Riverside, CA, USA
neal.young@ucr.edu
[2] Hong Kong University of Science and Technology, Hong Kong, China
[3] University of Waterloo, Waterloo, Canada

Abstract. In 1971, Knuth gave an $O(n^2)$-time algorithm for the classic problem of finding an optimal binary search tree. Knuth's algorithm works only for search trees based on 3-way comparisons, but most modern computers support only 2-way comparisons ($<$, $\leq$, $=$, $\geq$, and $>$). Until this paper, the problem of finding an optimal search tree using 2-way comparisons remained open — poly-time algorithms were known only for restricted variants. We solve the general case, giving (i) an $O(n^4)$-time algorithm and (ii) an $O(n \log n)$-time additive-3 approximation algorithm. For finding optimal *binary split trees*, we (iii) obtain a linear speedup and (iv) prove some previous work incorrect.

1 Background and Statement of Results

In 1971, Knuth [10] gave an $O(n^2)$-time dynamic-programming algorithm for a classic problem: *given a set $\mathcal{K}$ of keys and a probability distribution on queries, find an optimal binary-search tree T*. As shown in Fig. 1, a search in such a tree for a given value v compares v to the root key, then (i) recurses left if v is smaller, (ii) stops if v equals the key, or (iii) recurses right if v is larger, halting at a leaf. The comparisons made in the search must suffice to determine the relation of v to all keys in $\mathcal{K}$. (Hence, T must have $2|\mathcal{K}| + 1$ leaves.) T is optimal if it has minimum *cost*, defined as the expected number of comparisons assuming the query v is chosen randomly from the specified probability distribution.

Knuth assumed *three-way* comparisons at each node. With the rise of higher-level programming languages, most computers began supporting only two-way comparisons ($<, \leq, =, \geq, >$). In the 2nd edition of Volume 3 of *The Art of Computer Programming* [11, Sect. 6.2.2 ex. 33], Knuth commented

> *... machines that cannot make three-way comparisons at once... will have to make two comparisons... it may well be best to have a binary tree whose internal nodes specify either an equality test* **or** *a less-than test but not both.*

This is an extended abstract; a full version is available here: [2].
M. Chrobak—Research funded by NSF grants CCF-1217314 and CCF-1536026.
M. Golin—Research funded by HKUST/RGC grant FSGRF14EG28.
J.I. Munro—Research funded by NSERC and the Canada Research Chairs Programme.

K. Elbassioni and K. Makino (Eds.): ISAAC 2015, LNCS 9472, pp. 71–82, 2015.
DOI: 10.1007/978-3-662-48971-0_7

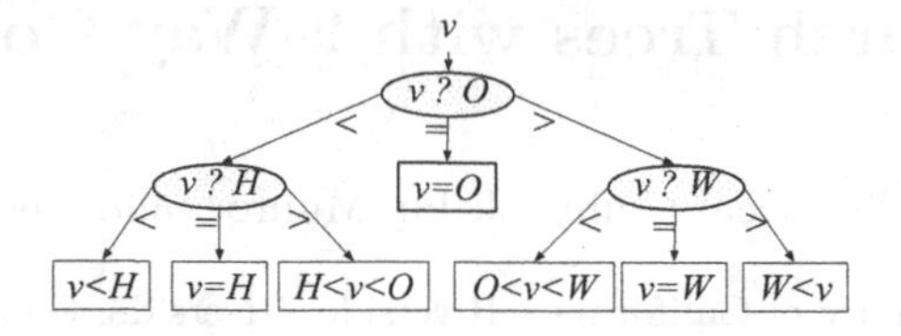

Fig. 1. A binary search tree T using 3-way comparisons, for $\mathcal{K} = \{H, O, W\}$.

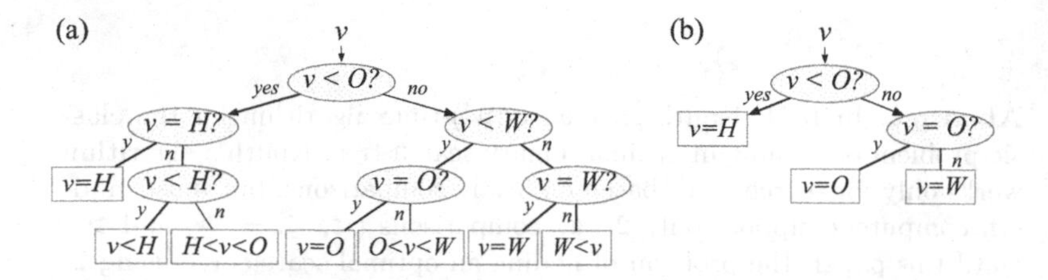

Fig. 2. Two 2WCSTs for $\mathcal{K} = \{H, O, W\}$; tree (b) only handles *successful* queries.

But Knuth gave no algorithm to find a tree built from *two-way* comparisons (a 2WCST, as in Fig. 2(a)), and, prior to the current paper, poly-time algorithms were known only for restricted variants. Most notably, in 2002 Anderson et al. [1] gave an $O(n^4)$-time algorithm for the *successful-queries* variant of 2WCST, in which each query v must be a key in $\mathcal{K}$, so only $|\mathcal{K}|$ leaves are needed (Fig. 2(b)). The *standard* problem allows arbitrary queries, so $2|\mathcal{K}| + 1$ leaves are needed (Fig. 2(a)). For the standard problem, no polynomial-time algorithm was previously known. We give one for a more general problem that we call 2WCST:

Theorem 1. 2WCST *has an* $O(n^4)$*-time algorithm.*

We specify an instance $\mathcal{I}$ of 2WCST as a tuple $\mathcal{I} = (\mathcal{K} = \{K_1, \ldots, K_n\}, \mathcal{Q}, \mathcal{C}, \alpha, \beta)$. The set $\mathcal{C}$ of allowed comparison operators can be any subset of $\{<, \leq, =, \geq, >\}$. The set $\mathcal{Q}$ specifies the queries. A solution is an optimal 2WCST T among those using operators in $\mathcal{C}$ and handling all queries in $\mathcal{Q}$. This definition generalizes both standard 2WCST (let $\mathcal{Q}$ contain each key and a value between each pair of keys), and the successful-queries variant (take $\mathcal{Q} = \mathcal{K}$ and $\alpha \equiv 0$). It further allows any query set $\mathcal{Q}$ between these two extremes, even allowing $\mathcal{K} \not\subseteq \mathcal{Q}$. As usual, β_i is the probability that v equals K_i; α_i is the probability that v falls between keys K_i and K_{i+1} (except $\alpha_0 = \Pr[v < K_1]$ and $\alpha_n = \Pr[v > K_n]$).[1]

[1] As defined here, a 2WCST T must determine the relation of the query v to every key in $\mathcal{K}$. More generally, one could specify any partition $\mathcal{P}$ of $\mathcal{Q}$, and only require T to determine, if at all possible using keys in $\mathcal{K}$, which set $S \in \mathcal{P}$ contains v. For example, if $\mathcal{P} = \{\mathcal{K}, \mathcal{Q} \setminus \mathcal{K}\}$, then T would only need to determine whether $v \in \mathcal{K}$. We note without proof that Theorem 1 extends to this more general formulation.

To prove Theorem 1, we prove Spuler's 1994 "maximum-likelihood" conjecture: *in any optimal* 2WCST *tree, each equality comparison is to a key in* $\mathcal{K}$ *of maximum likelihood, given the comparisons so far* [14, Sect. 6.4 Conj. 1]. As Spuler observed, the conjecture implies an $O(n^5)$-time algorithm; we reduce this to $O(n^4)$ using standard techniques and a new perturbation argument. Anderson et al. proved the conjecture for their special case [1, Cor. 3]. We were unable to extend their proof directly; our proof uses a different local-exchange argument.

We also give a fast additive-3 approximation algorithm:

Theorem 2. *Given any instance* $\mathcal{I} = (\mathcal{K}, \mathcal{Q}, \mathcal{C}, \alpha, \beta)$ *of* 2WCST, *one can compute a tree of cost at most the optimum plus 3, in* $O(n \log n)$ *time.*

Comparable results were known for the successful-queries variant ($\mathcal{Q} = \mathcal{K}$) [1,16]. We approximately reduce the general case to that case.

Binary split trees "split" each 3-way comparison in Knuth's 3-way-comparison model into two 2-way comparisons within the same node: an equality comparison (which, by definition, must be to the maximum-likelihood key) and a "<" comparison (to any key) [3,6,8,12,13]. The fastest algorithms to find an optimal binary split tree take $O(n^5)$-time: from 1984 for the successful-queries-only variant ($Q = K$) [8]; from 1986 for the standard problem (Q contains queries in all possible relations to the keys in K) [6]. We obtain a linear speedup:

Theorem 3. *Given any instance* $\mathcal{I} = (\mathcal{K} = \{K_1, \ldots, K_n\}, \alpha, \beta)$ *of the standard binary-split-tree problem, an optimal tree can be computed in* $O(n^4)$ *time.*

The proof uses our new perturbation argument (Sect. 3.1) to reduce to the case when all β_i's are distinct, then applies a known algorithm [6]. The perturbation argument can also be used to simplify Anderson et al.'s algorithm [1].

Generalized binary split trees (GBSTs) are binary split trees without the maximum-likelihood constraint. Huang and Wong [9] (1984) observe that relaxing this constraint allows cheaper trees — the maximum-likelihood conjecture fails here — and propose an algorithm to find optimal GBSTs. We prove it incorrect!

Theorem 4. *Lemma 4 of [9] is incorrect: there exists an instance — a query distribution* β *— for which it does not hold, and on which their algorithm fails.*

This flaw also invalidates two algorithms, proposed in Spuler's thesis [15], that are based on Huang and Wong's algorithm. We know of no poly-time algorithm to find optimal GBSTs. Of course, optimal 2WCSTs are at least as good.

2WCST*without equality tests.* Finding an optimal *alphabetical encoding* has several poly-time algorithms: by Gilbert and Moore — $O(n^3)$ time, 1959 [5]; by Hu and Tucker — $O(n \log n)$ time, 1971 [7]; and by Garsia and Wachs — $O(n \log n)$ time but simpler, 1979 [4]. The problem is equivalent to finding an optimal 3-way-comparison search tree when the probability of querying any key is zero ($\beta \equiv 0$) [11, Sect. 6.2.2]. It is also equivalent to finding an optimal 2WCST in the successful-queries variant with only "<" comparisons allowed ($\mathcal{C} = \{<\}, \mathcal{Q} = \mathcal{K}$) [1, Sect. 5.2]. We generalize this observation to prove Theorem 5:

Theorem 5. *Any* 2WCST *instance* $\mathcal{I} = (\mathcal{K} = \{K_1, \ldots, K_n\}, \mathcal{Q}, \mathcal{C}, \alpha, \beta)$ *where* $=$ *is not in* $\mathcal{C}$ *(equality tests are not allowed), can be solved in* $O(n \log n)$ *time.*

Definition 1. *Fix any* 2WCST *instance* $\mathcal{I} = (\mathcal{K}, \mathcal{Q}, \mathcal{C}, \alpha, \beta)$.

For any node N *in any* 2WCST T *for* $\mathcal{I}$, N*'s* query subset, $\mathcal{Q}_N$, *contains queries* $v \in \mathcal{Q}$ *such that the search for* v *reaches* N. *The* weight $\omega(N)$ *of* N *is the probability that a random query* v *(from distribution* (α, β)*) is in* $\mathcal{Q}_N$. *The weight* $\omega(T')$ *of any subtree* T' *of* T *is* $\omega(N)$ *where* N *is the root of* T'.

Let $\langle v < K_i \rangle$ *denote an internal node having key* K_i *and comparison operator* $<$ *(define* $\langle v \leq K_i \rangle$ *and* $\langle v = K_i \rangle$ *similarly). Let* $\langle K_i \rangle$ *denote the leaf* N *such that* $\mathcal{Q}_N = \{K_i\}$. *Abusing notation,* $\omega(K_i)$ *is a synonym for* $\omega(\langle K_i \rangle)$, *that is,* β_i.

Say T *is* irreducible *if, for every node* N *with parent* N', $\mathcal{Q}_N \neq \mathcal{Q}_{N'}$.

In the remainder of the paper, we assume that only comparisons in $\{<, \leq, =\}$ are allowed (i.e., $\mathcal{C} \subseteq \{<, \leq, =\}$). This is without loss of generality, as "$v > K_i$" and "$v \geq K_i$" can be replaced, respectively, by "$v \leq K_i$" and "$v < K_i$."

2 Proof of Spuler's Conjecture

Fix any irreducible, optimal 2WCST T for any instance $\mathcal{I} = (\mathcal{K}, \mathcal{Q}, \mathcal{C}, \alpha, \beta)$.

Theorem 6 (Spuler's conjecture). *The key* K_a *in any equality-comparison node* $N = \langle v = K_a \rangle$ *is a maximum-likelihood key:* $\beta_a = \max_i \{\beta_i : K_i \in \mathcal{Q}_N\}$.

The theorem will follow easily from Lemma 1:

Lemma 1. *Let internal node* $\langle v = K_a \rangle$ *be the ancestor of internal node* $\langle v = K_z \rangle$. *Then* $\omega(K_a) \geq \omega(K_z)$. *That is,* $\beta_a \geq \beta_z$.

Proof (Lemma 1). Throughout, "$\langle v \prec K_i \rangle$" denotes a node in T that does an inequality comparison ($\leq$ or $<$, not $=$) to key K_i. Abusing notation, in that context, "$x \prec K_i$" (or "$x \not\prec K_i$") denotes that x passes (or fails) that comparison.

Assumption 1. (i) *All nodes on the path from* $\langle v = K_a \rangle$ *to* $\langle v = K_z \rangle$ *do inequality comparisons.* **(ii)** *Along the path, some other node* $\langle v \prec K_s \rangle$ separates *key* K_a *from* K_z*: either* $K_a \prec K_s$ *but* $K_z \not\prec K_s$, *or* $K_z \prec K_s$ *but* $K_a \not\prec K_s$.

It suffices to prove the lemma assuming (i) and (ii) above. (Indeed, if the lemma holds given (i), then, by transitivity, the lemma holds in general. Given (i), if (ii) doesn't hold, then exchanging the two nodes preserves correctness, changing the cost by $(\omega(K_a) - \omega(K_z)) \times d$ for $d \geq 1$, so $\omega(K_a) \geq \omega(K_z)$ and we are done.)

By Assumption 1, the subtree rooted at $\langle v = K_a \rangle$, call it T', is as in Fig. 3(a): Let child $\langle v \prec K_b \rangle$, with subtrees T_0 and T_1, be as in Fig. 3.

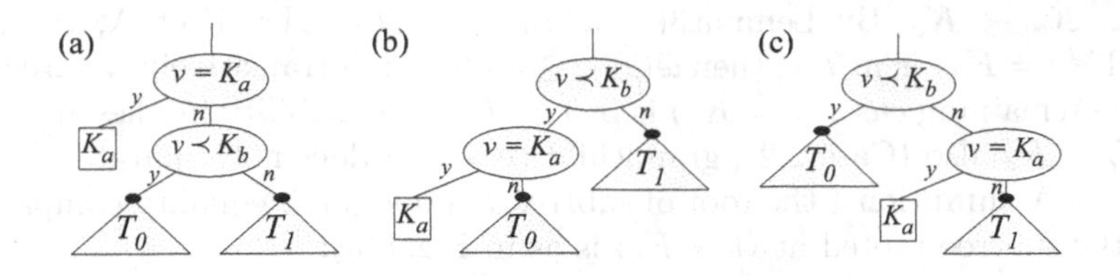

Fig. 3. (a) The subtree T' rooted at $\langle v = K_a \rangle$ and possible replacements (b), (c).

Lemma 2. *If* $K_a \prec K_b$*, then* $\omega(K_a) \geq \omega(T_1)$*, else* $\omega(K_a) \geq \omega(T_0)$.

(This and subsequent lemmas in this section are proved in [2, Sect. 7.2]. The idea behind this one is that correctness is preserved by replacing T' by subtree (b) if $K_a \prec K_b$ or (c) otherwise, implying the lemma by the optimality of T.)

Case 1: *Child* $\langle v \prec K_b \rangle$ *separates* K_a *from* K_z. If $K_a \prec K_b$, then $K_z \not\prec K_b$, so descendant $\langle v = K_z \rangle$ is in T_1, and, by this and Lemma 2, $\omega(K_a) \geq \omega(T_1) \geq \omega(K_z)$, and we're done. Otherwise $K_a \not\prec K_b$, so $K_z \prec K_b$, so descendant $\langle v = K_z \rangle$ is in T_0, and, by this and Lemma 2, $\omega(K_a) \geq \omega(T_0) \geq \omega(K_z)$, and we're done.

Case 2: *Child* $\langle v \prec K_b \rangle$ *does not separate* K_a *from* K_z. Assume also that descendant $\langle v = K_z \rangle$ is in T_1. (If descendant $\langle v = K_z \rangle$ is in T_0, the proof is symmetric, exchanging the roles of T_0 and T_1.) Since descendant $\langle v = K_z \rangle$ is in T_1, and child $\langle v \prec K_b \rangle$ does not separate K_a from K_z, we have $K_a \not\prec K_b$ and two facts:

Fact A: $\omega(K_a) \geq \omega(T_0)$ (by Lemma 2), and

Fact B: the root of T_1 does an inequality comparison (by Assumption 1).

By Fact B, subtree T' rooted at $\langle v = K_a \rangle$ is as in Fig. 4(a):
As in Fig. 4(a), let the root of T_1 be $\langle v \prec K_c \rangle$, with subtrees T_{10} and T_{11}.

Lemma 3. *(i)* $\omega(T_0) \geq \omega(T_{11})$*. (ii) If* $K_a \not\prec K_c$*, then* $\omega(K_a) \geq \omega(T_1)$.

(As replacing T' by (b) or (c) preserves correctness; proof in [2, Sect. 7.2].)
Case 2.1: $K_a \not\prec K_c$. By Lemma 3(ii), $\omega(K_a) \geq \omega(T_1)$. Descendant $\langle v = K_z \rangle$ is in T_1, so $\omega(T_1) \geq \omega(K_z)$. Transitively, $\omega(K_a) \geq \omega(K_z)$, and we are done.

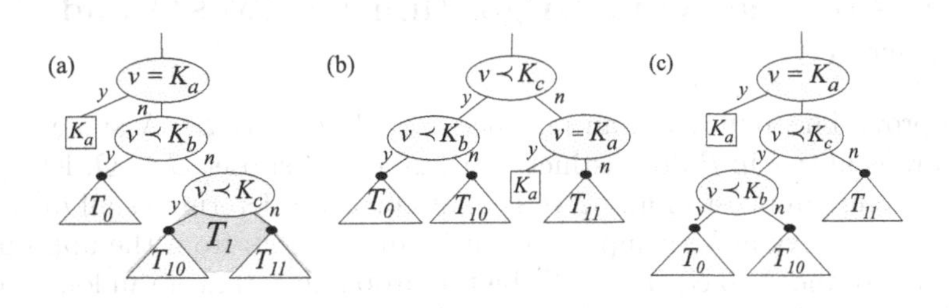

Fig. 4. (a) The subtree T' in Case 2, two possible replacements (b), (c).

Case 2.2: $K_a \prec K_c$. By Lemma 3(i), $\omega(T_0) \geq \omega(T_{11})$. By Fact A, $\omega(K_a) \geq \omega(T_{11})$. If $\langle v = K_z \rangle$ is in T_{11}, then $\omega(T_{11}) \geq \omega(K_z)$ and transitively we are done.

In the remaining case, $\langle v = K_z \rangle$ is in T_{10}. T's irreducibility implies $K_z \prec K_c$. Since $K_a \prec K_c$ also (Case 2.2), grandchild $\langle v \prec K_c \rangle$ does not separate K_a from K_z, and by Assumption 1 the root of subtree T_{10} does an inequality comparison. Hence, the subtree rooted at $\langle v \prec K_b \rangle$ is as in Fig. 5(a):

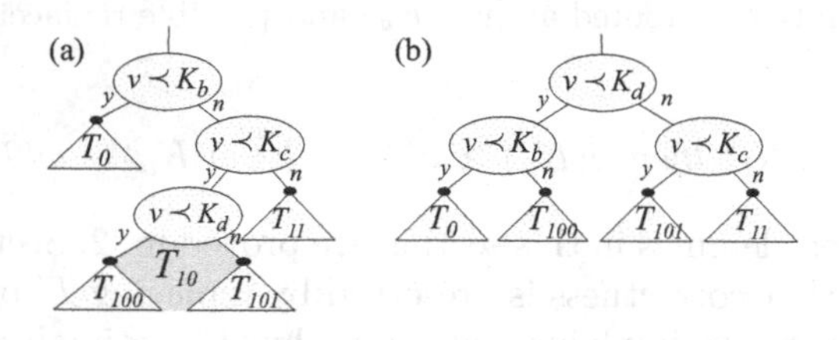

Fig. 5. (a) The subtree rooted at $\langle v \prec K_b \rangle$ in Case 2.2. (b) A possible replacement.

Lemma 4. $\omega(T_0) \geq \omega(T_{10})$.

(Because replacing (a) by (b) preserves correctness; proof in [2, Sect. 7.2].)

Since descendant $\langle v = K_z \rangle$ is in T_{10}, Lemma 4 implies $\omega(T_0) \geq \omega(T_{10}) \geq \omega(K_z)$. This and Fact A imply $\omega(K_a) \geq \omega(K_z)$. This proves Lemma 1. □

Proposition 1. *If any leaf node* $\langle K_\ell \rangle$*'s parent* P *does not do an equality comparison against key* K_ℓ*, then changing* P *so that it does so gives an irreducible* 2WCST T' *of the same cost.*

Proof. Since $\mathcal{Q}_{\langle K_\ell \rangle} = \{K_\ell\}$ and P's comparison operator is in $\mathcal{C} \subseteq \{<, \leq, =\}$, it must be that $K_\ell = \max \mathcal{Q}_P$ or $K_\ell = \min \mathcal{Q}_P$. So changing P to $\langle v = K_\ell \rangle$ (with $\langle K_\ell \rangle$ as the"yes" child and the other child the "no" child) maintains correctness, cost, and irreducibility. □

Proof (Theorem 6). Consider any equality-testing node $N = \langle v = K_a \rangle$ and any key $K_z \in \mathcal{Q}_N$. Since $K_z \in \mathcal{Q}_N$, node N has descendant leaf $\langle K_z \rangle$. Without loss of generality (by Proposition 1, leaf $\langle K_z \rangle$'s parent is $\langle v = K_z \rangle$. That parent is a descendant of $\langle v = K_a \rangle$, so $\omega(K_a) \geq \omega(K_z)$ by Lemma 1. □

3 Proofs of Theorem 1 (Algorithm for 2WCST) and Theorem 3

First we prove Theorem 1. Fix an instance $\mathcal{I} = (\mathcal{K}, \mathcal{Q}, \mathcal{C}, \alpha, \beta)$. Assume for now that all probabilities in β are distinct. For any query subset $\mathcal{S} \subseteq \mathcal{Q}$, let $\mathsf{opt}(\mathcal{S})$ denote the minimum cost of any 2WCST that correctly determines all queries in subset $\mathcal{S}$ (using keys in $\mathcal{K}$, comparisons in $\mathcal{C}$, and weights from the appropriate restriction of α and β to $\mathcal{S}$). Let $\omega(\mathcal{S})$ be the probability that a random query v is in $\mathcal{S}$. The cost of any tree for $\mathcal{S}$ is the weight of the root ($= \omega(\mathcal{S})$) plus the cost of its two subtrees, yielding the following dynamic-programming recurrence:

Lemma 5. *For any query set $\mathcal{S} \subseteq \mathcal{Q}$ not handled by a single-node tree,*

$$\mathsf{opt}(\mathcal{S}) = \omega(\mathcal{S}) + \min \begin{cases} \min_k \mathsf{opt}(\mathcal{S} \setminus \{k\}) & \textit{(if "=" is in } \mathcal{C}\textit{, else } \infty\textit{)} \quad (i) \\ \min_{k,\prec} \mathsf{opt}(\mathcal{S}_k^{\prec}) + \mathsf{opt}(\mathcal{S} \setminus \mathcal{S}_k^{\prec}), & (ii) \end{cases}$$

where k ranges over $\mathcal{K}$, and $\prec$ ranges over the allowed inequality operators (if any), and $\mathcal{S}_k^{\prec} = \{v \in \mathcal{S} : v \prec k\}$.

Using the recurrence naively to compute $\mathsf{opt}(\mathcal{Q})$ yields exponentially many query subsets $\mathcal{S}$, because of line (i). But, by Theorem 6, we can restrict k in line (i) to be the maximum-likelihood key in $\mathcal{S}$. With this restriction, the only subsets $\mathcal{S}$ that arise are intervals within $\mathcal{Q}$, minus some most-likely keys. Formally, for each of $O(n^2)$ key pairs $\{k_1, k_2\} \subseteq \mathcal{K} \cup \{-\infty, \infty\}$ with $k_1 < k_2$, define four *key intervals*

$$(k_1, k_2) = \{v \in \mathcal{Q} : k_1 < v < k_2\}, \quad [k_1, k_2] = \{v \in \mathcal{Q} : k_1 \le v \le k_2\},$$
$$(k_1, k_2] = \{v \in \mathcal{Q} : k_1 < v \le k_2\}, \quad [k_1, k_2) = \{v \in \mathcal{Q} : k_1 \le v < k_2\}.$$

For each of these $O(n^2)$ key intervals I, and each integer $h \le n$, define $\mathsf{top}(I, h)$ to contain the h keys in I with the h largest β_i's. Define $\mathcal{S}(I, h) = I \setminus \mathsf{top}(I, h)$. Applying the restricted recurrence to $\mathcal{S}(I, h)$ gives a simpler recurrence:

Lemma 6. *If $\mathcal{S}(I, h)$ is not handled by a one-node tree, then $\mathsf{opt}(\mathcal{S}(I, h))$ equals*

$$\omega(\mathcal{S}(I,h)) + \min \begin{cases} \mathsf{opt}(\mathcal{S}(I, h+1)) & \textit{(if equality is in } \mathcal{C}\textit{, else } \infty\textit{)} \quad (i) \\ \min_{k,\prec} \mathsf{opt}(\mathcal{S}(I_k^{\prec}, h_k^{\prec})) + \mathsf{opt}(\mathcal{S}(I \setminus I_k^{\prec}, h - h_k^{\prec})), & (ii) \end{cases}$$

where key interval $I_k^{\prec} = \{v \in I : v \prec k\}$, and $h_k^{\prec} = |\mathsf{top}(I, h) \cap I_k^{\prec}|$.

Now, to compute $\mathsf{opt}(\mathcal{Q})$, each query subset that arises is of the form $\mathcal{S}(I, h)$ where I is a key interval and $0 \le h \le n$. With care, each of these $O(n^3)$ subproblems can be solved in $O(n)$ time, giving an $O(n^4)$-time algorithm. In particular, represent each key-interval I by its two endpoints. For each key-interval I and integer $h \le n$, precompute $\omega(\mathcal{S}(I, h))$, and $\mathsf{top}(I, h)$, and the h'th largest key in I. Given these $O(n^3)$ values (computed in $O(n^3 \log n)$ time), the recurrence for $\mathsf{opt}(\mathcal{S}(I, h))$ can be evaluated in $O(n)$ time. In particular, for line (ii), one can enumerate all $O(n)$ pairs $(k, h_k^{\prec})$ in $O(n)$ time total, and, for each, compute $I_k^{\prec}$ and $I \setminus I_k^{\prec}$ in $O(1)$ time. Each base case can be recognized and handled (by a cost-0 leaf) in $O(1)$ time, giving total time $O(n^4)$. This proves Theorem 1 when all probabilities in β are distinct; Sect. 3.1 finishes the proof.

3.1 Perturbation Argument; Proofs of Theorems 1 and 3

Here we show that, without loss of generality, in looking for an optimal search tree, one can assume that the key probabilities (the β_i's) are all distinct. Given any instance $\mathcal{I} = (\mathcal{K}, \mathcal{Q}, \mathcal{C}, \alpha, \beta)$, construct instance $\mathcal{I}' = (\mathcal{K}, \mathcal{Q}, \mathcal{C}, \alpha, \beta')$, where

$\beta'_j = \beta_j + j\varepsilon$ and ε is a positive infinitesimal (or ε can be understood as a sufficiently small positive rational). To compute (and compare) costs of trees with respect to $\mathcal{I}'$, maintain the infinitesimal part of each value separately and extend linear arithmetic component-wise in the natural way:

1. Compute $z \times (x_1 + x_2\,\varepsilon)$ as $(zx_1) + (zx_2)\varepsilon$, where z, x_1, x_2 are any rationals,
2. compute $(x_1 + \varepsilon x_2) + (y_1 + \varepsilon y_2)$ as $(x_1 + x_2) + (y_1 + y_2)\varepsilon$,
3. and say $x_1 + \varepsilon x_2 < y_1 + \varepsilon y_2$ iff $x_1 < y_1$, or $x_1 = y_1 \wedge x_2 < y_2$.

Lemma 7. *In the instance $\mathcal{I}'$, all key probabilities β'_i are distinct. If a tree T is optimal w.r.t. $\mathcal{I}'$, then it is also optimal with respect to $\mathcal{I}$.*

Proof. Let A be a tree that is optimal w.r.t. $\mathcal{I}'$. Let B be any other tree, and let the costs of A and B under $\mathcal{I}'$ be, respectively, $a_1 + a_2\varepsilon$ and $b_1 + b_2\varepsilon$. Then their respective costs under $\mathcal{I}$ are a_1 and b_1. Since A has minimum cost under $\mathcal{I}'$, $a_1 + a_2\varepsilon \leq b_1 + b_2\varepsilon$. That is, either $a_1 < b_1$, or $a_1 = b_1$ (and $a_2 \leq b_2$). Hence $a_1 \leq b_1$: that is, A costs no more than B w.r.t. $\mathcal{I}$. Hence A is optimal w.r.t. $\mathcal{I}$. □

Doing arithmetic this way increases running time by a constant factor.[2] This completes the proof of Theorem 1. The reduction can also be used to avoid the significant effort that Anderson et al. [1] devote to non-distinct key probabilities.

For computing optimal *binary split trees* for unrestricted queries, the fastest known time is $O(n^5)$, due to [6]. But [6] also gives an $O(n^4)$-time algorithm for the case of distinct key probabilities. With the above reduction, the latter algorithm gives $O(n^4)$ time for the general case, proving Theorem 3.

4 Proof of Theorem 2 (Additive-3 Approximation Algorithm)

Fix any instance $\mathcal{I} = (\mathcal{K}, \mathcal{Q}, \mathcal{C}, \alpha, \beta)$. If $\mathcal{C}$ is $\{=\}$ then the optimal tree can be found in $O(n \log n)$ time, so assume otherwise. In particular, $<$ and/or $\leq$ are in $\mathcal{C}$. Assume that $<$ is in $\mathcal{C}$ (the other case is symmetric).

The entropy $H_{\mathcal{I}} = -\sum_i \beta_i \log_2 \beta_i - \sum_i \alpha_i \log_2 \alpha_i$ is a lower bound on $\mathsf{opt}(\mathcal{I})$. For the case $\mathcal{K} = \mathcal{Q}$ and $\mathcal{C} = \{<\}$, Yeung's $O(n)$-time algorithm [16] constructs a 2WCST that uses only $<$-comparisons whose cost is at most $H_{\mathcal{I}} + 2 - \beta_1 - \beta_n$. We reduce the general case to that one, adding roughly one extra comparison.

Construct $\mathcal{I}' = (\mathcal{K}' = \mathcal{K}, \mathcal{Q}' = \mathcal{K}, \mathcal{C}' = \{<\}, \alpha', \beta')$ where each $\alpha'_i = 0$ and each $\beta'_i = \beta_i + \alpha_i$ (except $\beta'_1 = \alpha_0 + \beta_1 + \alpha_1$). Use Yeung's algorithm [16] to construct tree T' for $\mathcal{I}'$. Tree T' uses only the $<$ operator, so any query $v \in \mathcal{Q}$ that reaches a leaf $\langle K_i \rangle$ in T' must satisfy $K_i \leq v < K_{i+1}$ (or $v < K_2$ if $i = 1$). To distinguish $K_i = v$ from $K_i < v < K_{i+1}$, we need only add one additional comparison at each leaf (except, if $i = 1$, we need two).[3] By Yeung's guarantee, T' costs at most $H_{\mathcal{I}'} + 2 - \beta'_1 - \beta'_n$. The modifications can be done so as to increase the cost by at most $1 + \alpha_0 + \alpha_1$, so the final tree costs at most $H_{\mathcal{I}'} + 3$. By standard properties of entropy, $H_{\mathcal{I}'} \leq H_{\mathcal{I}} \leq \mathsf{opt}(\mathcal{I})$, proving Theorem 2.

[2] For an algorithm that works with linear (or $O(1)$-degree polynomial) functions of β.

[3] If it is possible to distinguish $v = K_i$ from $K_i < v < K_{i+1}$, then $\mathcal{C}$ must have at least one operator other than $<$, so we can add either $\langle v = K_i \rangle$ or $\langle v \leq K_i \rangle$.

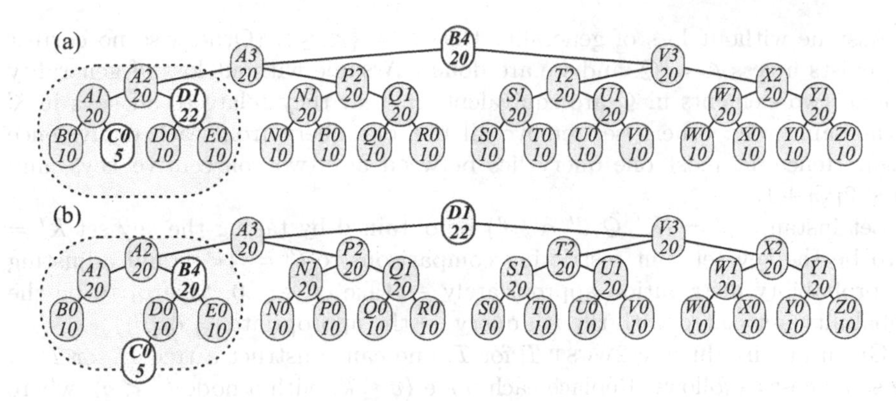

Fig. 6. Two GBSTs for an instance. Keys are ordered alphabetically ($A0 < A1 < A2 < A3 < B0 < \cdots$). Each node shows its equality key and the frequency of that key; split keys are not shown. The algorithm of [9] gives (a), of cost 1763, but (b) costs 1762.

5 Proof of Theorem 4 (Errors in Work on Binary Split Trees)

A *generalized binary split tree* (GBST) is a rooted binary tree where each node N has an *equality* key e_N and a *split* key s_N. A search for query $v \in \mathcal{Q}$ starts at the root r. If $v = e_r$, the search halts. Otherwise, the search recurses on the left subtree (if $v < s_r$) or the right subtree (if $v \geq s_r$). The *cost* of the tree is the expected number of nodes (including, by convention, leaves) visited for a random query v. Figure 6 shows two GBSTs for a single instance.

To prove Theorem 4, we observe that [9]'s Lemma 4 and algorithm fail on the instance in Fig. 6. There is a solution of cost only 1762 (in Fig. 6(b)), but the algorithm gives cost 1763 for the instance (as in Fig. 6(a)), as can be verified by executing the Python code for the algorithm in Appendix A.1. The intuition is that the optimal substructure property fails for the subproblems defined by [9]: the circled subtree in (a) (with root $A2$) is cheaper than the corresponding subtree in (b), but leads to larger global cost. For more intuition and the full proof, see the full paper [2, Sect. 7.3].

6 Proof of Theorem 5 ($O(n \log n)$ Time Without Equality)

Fix any 2WCST instance $\mathcal{I} = (\mathcal{K}, \mathcal{Q}, \mathcal{C}, \alpha, \beta)$ with $\mathcal{C} \subseteq \{<, \leq\}$. Let $n = |\mathcal{K}|$. We show that, in $O(n \log n)$ time, one can compute an equivalent instance $\mathcal{I}' = (\mathcal{K}', \mathcal{Q}', \mathcal{C}', \alpha', \beta')$ with $\mathcal{K}' = \mathcal{Q}'$, $\mathcal{C}' = \{<\}$, and $|\mathcal{K}'| \leq 2n+1$. (*Equivalent* means that, given an optimal 2WCST T' for $\mathcal{I}'$, one can compute in $O(n \log n)$ time an optimal 2WCST T for $\mathcal{I}$.) The idea is that, when $\mathcal{C} \subseteq \{<, \leq\}$, open intervals are functionally equivalent to keys.

Assume without loss of generality that $\mathcal{C} = \{<, \leq\}$. (Otherwise no correct tree exists unless $\mathcal{K} = \mathcal{Q}$, and we are done.) Assume without loss of generality that no two elements in $\mathcal{Q}$ are equivalent (in that they relate to all keys in $\mathcal{K}$ in the same way; otherwise, remove all but one query from each equivalence class). Hence, at most one query lies between any two consecutive keys, and $|\mathcal{Q}| \leq 2|\mathcal{K}| + 1$.

Let instance $\mathcal{I}' = (\mathcal{K}', \mathcal{Q}, \mathcal{C}', \alpha', \beta')$ be obtained by taking the key set $\mathcal{K}' = \mathcal{Q}$ to be the key set, but restricting comparisons to $\mathcal{C}' = \{<\}$ (and adjusting the probability distribution appropriately — take $\alpha' \equiv= 0$, take β_i to be the probability associated with the ith query — the appropriate α_j or β_j).

Given any irreducible 2WCST T for $\mathcal{I}$, one can construct a tree T' for $\mathcal{I}'$ of the same cost as follows. Replace each node $\langle v \leq k \rangle$ with a node $\langle v < q \rangle$, where q is the least query value larger than k (there must be one, since $\langle v \leq k \rangle$ is in T and T is irreducible). Likewise, replace each node $\langle v < k \rangle$ with a node $\langle v < q \rangle$, where q is the least query value greater than or equal to k (there must be one, since $\langle v < k \rangle$ is in T and T is irreducible). T' is correct because T is.

Conversely, given any irreducible 2WCST T' for $\mathcal{I}'$, one can construct an equivalent 2WCST T for $\mathcal{I}$ as follows. Replace each node $N' = \langle v < q \rangle$ as follows. If $q \in \mathcal{K}$, replace N' by $\langle v < k \rangle$. Otherwise, replace N' by $\langle v \leq k \rangle$, where key k is the largest key less than q. (There must be such a key k. Node $\langle v < q \rangle$ is in T' but T' is irreducible, so there is a query, and hence a key k, smaller than q.) Since T' correctly classifies each query in $\mathcal{Q}$, so does T.

To finish, we note that the instance $\mathcal{I}'$ can be computed from $\mathcal{I}$ in $O(n \log n)$ time (by sorting the keys, under reasonable assumptions about $\mathcal{Q}$), and the second mapping (from T' to T) can be computed in $O(n \log n)$ time. Since $\mathcal{I}'$ has $\mathcal{K}' = \mathcal{Q}'$ and $\mathcal{C} = \{<\}$, it is known [10] to be equivalent to an instance of alphabetic encoding, which can be solved in $O(n \log n)$ time [4,7].

A Appendix

A.1 Python Code for Theorem 4 (GBST Algorithm of [9])

```
1  #!/usr/bin/env python3.4
2  import functools
3  memoize = functools.lru_cache(maxsize=None)
4
5  def huang1984(weights):
6      "Returns cost as computed by Huang and Wong's GBST algorithm (1984)."
7
8      n = len(weights)
9      beta = {i+1 : weights[key] for i, key in enumerate(sorted(weights.keys()))}
10
11     def is_legal(i, j, d): return 0 <= i <= j <= n and 0 <= d <= j -i
12
13     @memoize
14     def p_w_t(i, j, d):
15         "Returns triple: (cost p[i,j,d], weight w[i,j,d], deleted keys for t[i,j,d])."
16
17         interval = set(range(i+1, j+1))
18
19         if d == j-i: # base case
```

```
20              return (0, 0, interval)
21
22          def candidates(): # Lemma 4 recurrence from Huang et al.
23              for k in interval: # k = index of split key
24                  for m in range(d+2): # m = num. deletions from left subtree
25                      if is_legal(i, k-1, m) and is_legal(k-1, j, d-m+1):
26                          cost_l, weight_l, deleted_l = p_w_t(i, k-1, m)
27                          cost_r, weight_r, deleted_r = p_w_t(k-1, j, d-m+1)
28                          deleted = deleted_l .union( deleted_r )
29                          x = min(deleted, key = lambda h : beta[h])
30                          weight = beta[x] + weight_l + weight_r
31                          cost = weight + cost_l + cost_r
32                          yield cost, weight, deleted -set([x])
33
34          return min(candidates())
35
36      cost, weight, keys = p_w_t(0, n, 0)
37      return cost
38
39  weights = dict(b4=20,
40                 a3=20, v3=20,
41                 a2=20, p2=20, t2=20, x2=20,
42                 a1=20, d1=22, n1=20, q1=20, s1=20, u1=20, w1=20, y1=20,
43                 b0=10, c0= 5, d0=10, e0=10, n0=10, p0=10, q0=10, r0=10,
44                 s0=10, t0=10, u0=10, v0=10, w0=10, x0=10, y0=10, z0=10)
45
46  assert huang1984(weights) == 1763
    # Both assertions pass. The first is used in our Theorem\,{4}.
47
48  weights['d1'] += 0.99 # Increasing a weight cannot decrease the optimal cost, but
49  assert huang1984(weights) < 1763
    # in this case decreases the cost computed by the algorithm.
```

References

1. Anderson, R., Kannan, S., Karloff, H., Ladner, R.E.: Thresholds and optimal binary comparison search trees. J. Algorithms **44**, 338–358 (2002)
2. Chrobak, M., Golin, M., Munro, J.I., Young, N.E.: Optimal search trees with 2-way comparisons. CoRR, arXiv:1505.00357v4 [cs.DS] (2015). http://arxiv.org/abs/1505.00357v4
3. Comer, D.: A note on median split trees. ACM Trans. Program. Lang. Syst. **2**(1), 129–133 (1980)
4. Garsia, A.M., Wachs, M.L.: A new algorithm for minimum cost binary trees. SIAM J. Comput. **6**(4), 622–642 (1977)
5. Gilbert, E., Moore, E.: Variable-length binary encodings. Bell Syst. Tech. J. **38**(4), 933–967 (1959)
6. Hester, J.H., Hirschberg, D.S., Huang, S.H., Wong, C.K.: Faster construction of optimal binary split trees. J. Algorithms **7**(3), 412–424 (1986)
7. Hu, T.C., Tucker, A.C.: Optimal computer search trees and variable-length alphabetical codes. SIAM J. Appl. Math. **21**(4), 514–532 (1971)
8. Huang, S., Wong, C.: Optimal binary split trees. J. Algorithms **5**(1), 69–79 (1984)
9. Huang, S.-H., Wong, C.K.: Generalized binary split trees. Acta Inform. **21**(1), 113–123 (1984)
10. Knuth, D.E.: Optimum binary search trees. Acta Inform. **1**(1), 14–25 (1971)
11. Knuth, D.E.: The Art of Computer Programming. Sorting and Searching, vol. 3, 2nd edn. Addison-Wesley Publishing Company, Reading (1998)
12. Perl, Y.: Optimum split trees. J. Algorithms **5**, 367–374 (1984)
13. Sheil, B.A.: Median split trees: a fast lookup technique for frequently occuring keys. Commun. ACM **21**(11), 947–958 (1978)

14. Spuler, D.: Optimal search trees using two-way key comparisons. Acta Inform. **740**, 729–740 (1994)
15. Spuler, D.A.: Optimal search trees using two-way key comparisons. Ph.D. thesis, James Cook University (1994)
16. Yeung, R.: Alphabetic codes revisited. IEEE Trans. Inf. Theory **37**(3), 564–572 (1991)

Multidimensional Range Selection

Timothy M. Chan and Gelin Zhou(✉)

David R. Cheriton School of Computer Science,
University of Waterloo, Waterloo, Canada
{tmchan,g5zhou}@uwaterloo.ca

Abstract. We study the problem of supporting (orthogonal) *range selection* queries over a set of n points in constant-dimensional space. Under the standard word-RAM model with word size $w = \Omega(\lg n)$, we present data structures that occupy $O(n \cdot (\lg n / \lg\lg n)^{d-1})$ words of space and support d-dimensional range selection queries using $O((\lg n / \lg\lg n)^d)$ query time. This improves the best known data structure by a factor of $\lg\lg n$ in query time. To develop our data structures, we generalize the "parallel counting" technique of Brodal, Gfeller, Jørgensen, and Sanders (2011) for one-dimensional range selection to higher dimensions.

As a byproduct, we design data structures to support d-dimensional range counting queries within $O(n \cdot (\lg n / \lg w + 1)^{d-2})$ words of space and $O((\lg n / \lg w + 1)^{d-1})$ query time, for any word size $w = \Omega(\lg n)$. This improves the best known result of JaJa, Mortensen, and Shi (2004) when $\lg w \gg \lg\lg n$.

1 Introduction

Range searching is an important topic in data structures and computational geometry. Recently, there has been growing interest in so-called "range aggregate queries", where instead of reporting or counting points inside a query range, we want to compute some aggregate function over the weights of the points inside the query range. In this paper, we study the version of the problem for multidimensional orthogonal ranges (axis-aligned boxes), where the aggregate function is the median, or more generally, the k-th smallest element.

More precisely, we can formulate the *d-dimensional (orthogonal) range selection* problem as follows, by viewing the weights as an extra dimension. The coordinates of each input point p are represented as a $(d+1)$-tuple $(p_1, p_2, \ldots, p_d, p_{d+1})$. A query range is a d-dimensional rectangle $R = [a_1..b_1] \times [a_2..b_2] \times \cdots \times [a_d..b_d]$, and a range selection query asks for the point whose coordinate in the $(d+1)$-st dimension is the k-th smallest among all input points contained in $R \times (-\infty, \infty)$.

The underlying model of computation in this paper is the standard word-RAM model [4] with word size $w = \Omega(\lg n)$. Under this model, bitwise and arithmetic operations including multiplication can be performed over machine words in $O(1)$ time. Without loss of generality, coordinates of points are assumed

K. Elbassioni and K. Makino (Eds.): ISAAC 2015, LNCS 9472, pp. 83–92, 2015.
DOI: 10.1007/978-3-662-48971-0_8

to fit in *rank space* [5]. Coordinates can be replaced with their ranks in the point set, by increasing the query time by the cost of $O(1)$ predecessor searches.

The one-dimensional case of the range selection problem has been well studied [2,3,7,8]. Krizanc et al. [8] proposed the problem, and their structures required either super linear space or $O(n^\epsilon)$ query time for some constant $\epsilon > 0$. Brodal et al. [2] presented a linear space data structure with only $O(\lg n/\lg\lg n)$ query time, by a novel application of bit-level parallelism. As shown by Jørgensen and Larsen [7], Brodal et al.'s linear space structure achieved optimal worst-case query time for any data structure within $O(n \cdot \text{polylog}(n))$ bits of space. Jørgensen and Larsen [7] further designed an adaptive data structure for one-dimensional range selection queries, which occupied linear space and required only $O(\lg k/\lg\lg n + \lg\lg n)$ query time to select the k-th smallest element in the range. More recently, Chan and Wilkinson [3] reduced the query time to $O(\lg k/\lg\lg n + 1)$ using the same amount of space.[1] Shallow cutting [9] played a central role in designing these adaptive data structures.

For the case of higher dimensions, Brodal et al. [2] pointed out that a d-dimensional range selection query could be reduced to $O(\lg n)$ d-dimensional range counting queries. As shown by JaJa et al. [6], each d-dimensional range counting query requires $O((\lg n/\lg\lg n)^{d-1})$ query time. Thus the overall query time for a d-dimensional range selection query would be $O(\lg n \cdot (\lg n/\lg\lg n)^{d-1})$. To the best of our knowledge, this is the only known result for multidimensional range selection queries.

In this paper, we present data structures that support d-dimensional range selection queries using $O(n \cdot (\lg n/\lg\lg n)^{d-1})$ words of space and $O((\lg n/\lg\lg n)^d)$ query time, for any constant integer $d \geq 1$. This improves the straightforward solution by a factor of $\lg\lg n$ in query time. To develop our data structures, we generalize Brodal et al.'s "parallel counting" technique [2] into higher dimensions. In the search for the k-th smallest point, we keep solving subproblems of finding the first non-negative integer in an increasing array, where the length of an array is bounded above by $O(\lg^\epsilon n)$ for some constant $0 < \epsilon < 1$. Instead of performing binary search on each of these subproblems, we examine all integers in the array from the highest bits in a parallel fashion, to speed up the search. These integers are not stored explicitly and have to be retrieved at query time, where the retrievals are either multidimensional range counting queries or multidimensional "parallel counting" queries.

Along the way, we also improve JaJa et al.'s work for range counting queries [6] with some novel bit manipulation tricks, which may be of independent interest. Our data structures support d-dimensional range counting queries within $O((\lg n/\lg w + 1)^{d-1})$ query time and $O(n \cdot (\lg n/\lg w + 1)^{d-2})$ words of space, for any word size $w = \Omega(\lg n)$. When w is $\lg^{\omega(1)} n$, this improves JaJa et al.'s $O((\lg n/\lg\lg n)^{d-1})$ query and $O(n \cdot (\lg n/\lg\lg n)^{d-2})$ space bounds [6].

The rest of this paper is organized as follows. Section 2 contains preliminaries. Section 3 defines and solves a problem that abstracts the bottleneck of selection

[1] The conference version claimed $O(\lg k/\lg w + 1)$ query time but it would require non-standard word operations.

queries. In Sects. 4 and 5, we apply the "abstract" problem to range selection queries and present our data structures.

2 Preliminaries

Let $[a..b]$ denote the set of integers from a to b. For point $p = (p_1, p_2, \ldots, p_{d+1})$ and each $i \in [1..d+1]$, p_i is referred to as the i-th coordinate of p. For two points $p = (p_1, p_2, \ldots, p_{d+1})$ and $q = (q_1, q_2, \ldots, q_{d+1})$, p is said to be *dominated* by q if $p_i \leq q_i$ for each $i \in [1..d+1]$. Let σ be a fixed parameter, which will be set to be either $\lceil \lg^{\epsilon} n \rceil$ or $\lceil w^{\epsilon} \rceil$ for constant $0 < \epsilon < 1/d$. A point $p = (p_1, p_2, \ldots, p_{d+1})$ is said to be of *type* d' if $p_i \in [1..\sigma]$ for each $i \in [d'+1..d+1]$, i.e., the last $d - d' + 1$ coordinates fit in a narrow range $[1..\sigma]$. A set of m points is said to be of *type* d' if all these m points are of type d' and the i-th coordinates of points are in rank space for each $i \in [1..d']$, i.e., they are drawn from $[1..m]$ and pairwise different. The input point set is of type $d + 1$.

To exploit abilities of the word RAM, it is a standard technique to pack a short list of sufficiently small integers into a machine word. We divide a word into *subwords* of m bits, each storing the *two's complement representation* of a signed integer that ranges from -2^{m-1} to $2^{m-1} - 1$. With this representation, a set of operations can be performed in parallel to integers of the packed list in $O(1)$ time, provided that each of these integers in the input and the output fits in m bits: One can add a constant integer to, subtract a constant integer from, or bit shift all signed integers of a packed list. One can also add or subtract corresponding integers of two packed lists. One can even find the first non-negative integer or the last negative one in a packed list, given that multiplications are permitted [4].

3 The "Abstract" Problem

Let s, b, and t be parameters satisfying that $(s + b + 2)t < w$ and $b \ll s$. Intuitively, s denotes the "section size", and b denotes the number of "carry bits". The *j-th section* of an integer x is defined to be $\lfloor x/2^{sj} \rfloor \mod 2^s$.

Let $A[1..t]$ be an increasing sequence of w-bit signed integers, with $A[0] < 0$. The goal of our abstract problem is to find the smallest index $i^* > 0$ so that $A[i^* - 1] < 0 \leq A[i^*]$. However, the value of each $A[i]$ is not given explicitly; rather, each $A[i]$ is *decomposed* into a sequence of signed integers $A_0[i], A_1[i], \ldots$, satisfying the properties that $|A_j[i]| < 2^{s+b}$ and $A[i] = \sum_{j \geq 0} A_j[i] \cdot 2^{sj}$. (Note that for $b = 0$, the decomposition corresponds to precisely the sections of an integer, but the parameter b offers more flexibility, which will be needed in our applications later.) We can only access the sequence A using the following *oracles*:

- Given $1 \leq i \leq t$, return $A[i]$;
- Given $j \geq 0$, return the concatenation of the binary representations of $(A_j[1], \ldots, A_j[t])$, stored in a single word in which the i-th subword is equal to $A_j[i]$. (Because $(s + b + 2)t < w$, the result fits in a word.)

We resist to solve this problem with binary search directly, which would require $O(\lg t)$ time. Instead, we examine in a parallel manner the $A_j[i]$'s in decreasing order of j. The following lemma shows how to achieve a query time that is adaptive to the values of the $A[i]$'s.

Lemma 1. *The "abstract" problem described in this section can be solved within* $O(1 + \frac{1}{s} \cdot \lg \frac{A[t]-A[1]}{A[i^*]-A[i^*-1]})$ *word operations and oracle calls.*

Proof. Given an index p, we define $B_p[i] = \sum_{j \geq p} A_j[i] \cdot 2^{s(j-p)}$ for $1 \leq i \leq t$. Note that $B_p[1], \ldots, B_p[t]$ may not be in increasing order. However, as shown below, $B_p[i] \cdot 2^{sp}$ provides an approximation of $A[i]$:

$$\begin{aligned} \left|A[i] - B_p[i] \cdot 2^{sp}\right| &\leq \sum_{0 \leq j < p} 2^{s+b} \cdot 2^{sj} = 2^{s+b} \cdot \frac{2^{sp}-1}{2^s-1} \\ &< 2^{s+b} \cdot \frac{2^{sp}}{2^{s-1}} = 2^{sp+b+1}. \end{aligned} \quad (1)$$

We maintain a range $[\ell + 1..r]$ that contains i^*, as well as the concatenation of the binary representations of $(B_p[\ell], \ldots, B_p[r])$. We ensure the invariant that for $\ell \leq i \leq r$, $|B_p[i]| \leq 2^{b+1}$ before each iteration. Thus $(B_p[\ell], \ldots, B_p[r])$ can be packed into a single word in which the i-th subword equals to $B_p[i]$ for $\ell \leq i \leq r$, and the remaining bits are 0.

At the beginning of the algorithm, we compute $A[1]$ and $A[t]$ with two oracle calls, and set the initial value of p to be $p_0 = \lceil \frac{1}{s} \cdot \lg(A[t] - A[1]) \rceil$. We also set the initial value of $\ell = 1$ and $r = t$. Then for each i, $|A[i]| \leq A[t] - A[1] \leq 2^{sp_0}$. By Inequality 1,

$$|B_{p_0}[i] \cdot 2^{sp_0}| \leq |A[i] - B_{p_0}[i] \cdot 2^{sp_0}| + |A[i]| < 2^{sp_0+b+1} + 2^{sp_0}.$$

This implies that $|B_{p_0}[i]| \leq 2^{b+1}$. In addition, we observe that

$$B_{p_0}[i] = A_{p_0}[i] + A_{p_0+1}[i] \cdot 2^s + A_{p_0+2}[i] \cdot 2^{2s} + \cdots \equiv A_{p_0}[i] \pmod{2^s}.$$

We then have $B_{p_0}[i] + 2^{b+1} = (A_{p_0}[i] + 2^{b+1}) \mod 2^s$ since $|B_{p_0}[i]| \leq 2^{b+1}$ and $b \ll s$. This formula allows us to initialize $(B_{p_0}[1], ..., B_{p_0}[t])$ using one oracle call and $O(1)$ word operations.

In each iteration of the algorithm, we decrement the value of p and compute $(B_p[\ell], \ldots, B_p[r])$ using the following equation:

$$(B_p[\ell], \ldots, B_p[r]) = 2^s \cdot (B_{p+1}[\ell], \ldots, B_{p+1}[r]) + (A_p[\ell], \ldots, A_p[r]).$$

The computation requires $O(1)$ word operations and one oracle call. Note that for $\ell \leq i \leq r$, $B_p[i]$ fits in a subword, because $|B_{p+1}[i]| \leq 2^{b+1}$ and $|B_p[i]| \leq 2^{s+b+1} + 2^{s+b} < 2^{s+b+2}$. We then find the largest index ℓ' in $[\ell..r]$ with $B_p[\ell'] \leq -2^{b+1}$, and the smallest index r' in $[\ell..r]$ with $B_p[r'] \geq 2^{b+1}$. As described in Sect. 2, ℓ' and r' can be determined using $O(1)$ word operations [4].

By Inequality 1, we have $A[\ell'] < 0$ and $A[r'] > 0$. Thus subranges $[\ell..\ell']$ and $[r' + 1..r]$ can be discarded, and we know that i^* is contained in $[\ell' + 1..r']$.

We then evaluate $A[\ell'+1]$ and $A[r'-1]$ by two oracle calls. The algorithm terminates if one of the following conditions holds: $i^* = \ell'+1$ is returned if $A[\ell'+1] > 0$, or $i^* = r'$ is returned if $A[r'-1] < 0$. Otherwise, we reset $\ell = \ell'+1$ and $r = r'-1$. Note that $|B_p[\ell'+1]|, \ldots, |B_p[r'-1]| \le 2^{b+1}$, so the invariant is maintained, and we can continue on to the next iteration.

Now we analyze the running time of the algorithm. After each iteration before termination,

$$A[i^*] - A[i^*-1] \le A[r] - A[\ell] < (B[r] - B[\ell]) \cdot 2^{sp} + 2 \cdot 2^{sp+b+1}$$
$$\le 2 \cdot 2^{b+1} \cdot 2^{sp} + 2 \cdot 2^{sp+b+1} = 2^{sp+b+3}.$$

Thus, $p \ge \frac{1}{s} \cdot [\lg(A[i^*] - A[i^*-1]) - O(b)]$. The algorithm requires $O(1 + \frac{1}{s} \cdot \lg \frac{A[t]-A[1]}{A[i^*]-A[i^*-1]})$ oracle calls. □

4 Range Selection

In this section, we apply the above "abstract" problem to range selection queries. We use $t = \lceil \lg^{\epsilon} n \rceil$, section size $s = \lceil (1/2) \cdot \lg^{1-\epsilon} n \rceil$, and $b = \Theta(\lg\lg n)$ for constant $0 < \epsilon < 1/d$. We build a range tree over the $(d+1)$-st coordinates of points with branching factor t. Thus, the height of the range tree is $O(\lg n / \lg\lg n)$. Each node v in the range tree represents a range $[a_v..b_v]$ and the set $S(v)$ of points whose $(d+1)$-st coordinates are in $[a_v..b_v]$. The leaf nodes in the range tree each represent a single point.

To answer a given range selection query with query range R and rank k, we repeatedly solve subqueries of the following form: given an internal node v and its children $v_1, \ldots, v_t$ in the range tree, find the child v_{i^*} so that the desired answer is contained in $S(v_{i^*})$. To connect these subqueries with the "abstract" problem, we set $A[i] = N[i] - k$, where $N[i]$ is the number of points that fall into $R \times [a_{v_1}..b_{v_i}]$. We set $A_j[i] = N_j[i] - k_j$, where $N_0[i], N_1[i], \ldots$ is a sequence to be specified later that decomposes $N[i]$, and k_j is the j-th section of k.

To compute $N[i]$ and the $N_j[i]$'s, we define the following two kinds of queries over a point set S of type d with $\sigma = \lceil \lg^{\epsilon} n \rceil$, for which the support is summarized in Lemma 2. The proof of Lemma 2 is deferred to Sect. 5.

- *dominance counting* queries: given a query point $q = (q_1, q_2, \ldots, q_{d+1})$, return the number of points in S that are dominated by q;
- *parallel counting* queries: given a query $(q_1, q_2, \ldots, q_d, j)$ for some $j \ge 0$, return the concatenation of $(C_j[1], \ldots, C_j[t])$, where, for $1 \le i \le t$, $C_0[i], C_1[i], \ldots$ is a sequence that decomposes $C[i]$, the answer to the dominance counting query $(q_1, q_2, \ldots, q_d, i)$.

Lemma 2. *For any constant* $0 < \epsilon < 1/d$*, a point set* S *of size* $m \le n$ *and type* d' *with* $\sigma = \lceil \lg^{\epsilon} n \rceil$ *can be stored in* $O(m \lg\lg n \cdot (\lg n / \lg\lg n)^{d'-1})$ *bits of space, so that (a) dominance counting queries and (b) parallel counting queries can be answered in* $O((\lg n / \lg\lg n)^{d'-1})$ *query time.*

To facilitate the use of Lemma 2, we transform each $S(v)$ into a point set $D(v)$ of type d. For each point $p \in S(v)$, we replace the first d coordinates of p with their ranks in $S(v)$, and replace p_{d+1} with the index of v's child that represents a set containing p.

Given a d-dimensional query range R, we can express it as additions and subtractions of $2^d = O(1)$ d-dimensional dominance ranges. Let these ranges be $z_1, z_2, \ldots, z_{2^d}$. Computing $N[i]$ and $A[i]$, which is essentially a $(d+1)$-dimensional range counting query, can be reduced to dominance counting queries over $D(v)$ for ranges $z_1 \times [1..i], \ldots, z_{2^d} \times [1..i]$, and can be done by Lemma 2(a). Let $N[i, z_\ell]$ be the result for $z_\ell \times [1..i]$ for $1 \leq \ell \leq 2^d$. By Lemma 2(b) we can decompose $N[i, z_\ell]$ into a sequence $N_0[i, z_\ell], N_1[i, z_\ell], \ldots$ In addition, $(N_j[1, z_\ell], \ldots, N_j[t, z_\ell])$ can be computed for $j \geq 0$. We define $N_j[i]$ to be sum of $N_j[i, z_\ell]$ over all z_ℓ. Then the sequence $N_0[i], N_1[i], \ldots$ decomposes $N[i]$ and the sequence $A_0[i], A_1[i], \ldots$ decomposes $A[i]$, after increasing the parameter b by $\log(2^d) = O(1)$.

Now we can finally support range selection queries. Starting with the root node, we define and compute the oracles as described above. After determining i^*, the query algorithm recurses on v_{i^*} after setting $k = k - A[i^* - 1]$. We repeatedly apply Lemma 1 until we reach a leaf node v, and $a_v = b_v$ is the answer. The query algorithm requires solving $O(\lg n / \lg\lg n)$ "abstract" problems. We sum the cost of Lemma 1 over these $O(\lg n / \lg\lg n)$ subproblems. Observe that the sum of the logarithms of ratios in Lemma 1 is actually telescoping. The total number of oracle calls is thus $O(\lg n / \lg\lg n + \frac{1}{s} \cdot \lg n) = O(\lg n / \lg\lg n)$, each requiring $O((\lg n / \lg\lg n)^{d-1})$ time. We conclude:

Theorem 1. *Under the word RAM model with word size $w = \Omega(\lg n)$, d-dimensional range selection queries over a set of n points can be supported in $O((\lg n / \lg\lg n)^d)$ query time and $O(n \cdot (\lg n / \lg\lg n)^{d-1})$ words of space.*

5 Dominance Counting and Parallel Counting

Our method for dominance counting queries is similar to JaJa et al.'s work [6]. The major improvement is a novel algorithm to answer queries over a point set of size $\lceil w^{d\epsilon} \rceil$ and type 1 with $\sigma = \lceil w^\epsilon \rceil$ for any constant $0 < \epsilon < 1/d$ within $O(1)$ time and $O(\lg w)$ bits of space per point, which is presented in Lemma 4. This algorithm does not require a global lookup table, so it is able to handle larger word size $w = \omega(\lg n)$.

Lemma 3. *For any constant $0 < \epsilon < 1/d$, dominance counting queries over a point set S of size $m \leq n$ and type 1 with $\sigma = \lceil w^\epsilon \rceil$ can be supported using $O(m \lg w)$ bits of space and $O(1)$ query time.*

Proof. We sort all points of the point set in increasing order of the first coordinates, and divide the list into blocks of size $m_1 = w^2$. Then we divide each block into subblocks of size $m_2 = \lceil w^{d\epsilon} \rceil$. Each block/subblock is labeled with the largest first coordinate over the points inside the block/subblock. For each block β, we precompute a d-dimensional table F_β in which, for $1 \leq q_2, \ldots, q_{d+1} \leq$

$\lceil w^\epsilon \rceil$, the entry $F_\beta[q_2, \ldots, q_{d+1}]$ stores the number of points in S that are dominated by $(\mathtt{label}(\beta), q_2, \ldots, q_{d+1})$, where $\mathtt{label}(\beta)$ is the label of β. Similarly, for each subblock β' of β, we maintain a d-dimensional table $g_{\beta'}$ in which, for $1 \leq q_2, \ldots, q_{d+1} \leq \lceil w^\epsilon \rceil$, the entry $G_{\beta'}[q_2, \ldots, q_{d+1}]$ stores the number of points inside β that are dominated by $(\mathtt{label}(\beta'), q_2, \ldots, q_{d+1})$.

Given a dominance counting query $q = (q_1, q_2, \ldots, q_{d+1})$, we find the rightmost block β whose label is no greater than q_1. Then we find the rightmost subblock β' to the right of β whose label is no greater than q_1. Without loss of generality, we assume the existence of both β and β'. The other cases can be handled similarly. Thus the answer to the given dominance counting query can be expressed as $F_\beta[q_2, \ldots, q_{d+1}] + G_{\beta'}[q_2, \ldots, q_{d+1}] + h$, where h is the number of points in the subblock to the right of β' that are dominated by the given query.

Later in Lemma 4, we will show the computation of h requires $O(1)$ query time and $O(\lg w)$ bits of space per point. Thus, the overall query time for dominance counting queries over the point set of type 1 is $O(1)$. Finally we analyze the space cost. The tables for all blocks require $O((m/m_1) \times w^{d\epsilon} \times \lg m) = o(m)$ bits of space in total. The tables for all subblocks require $O((m/m_2) \times w^{d\epsilon} \times \lg m_1) = O(m \lg w)$ bits of space in total. Therefore the overall space cost is $O(m \lg w)$ bits. □

Lemma 4. *Dominance counting queries inside a subblock can be supported using $O(1)$ query time and $O(\lg w)$ bits of space per point.*

Proof. We divide a machine word into chunks of size $s_1 = d \cdot (\lceil \epsilon \lg w \rceil + 1)$ each. Each chuck is further divided into d subchunks of size $s_2 = \lceil \epsilon \lg w \rceil + 1$ each. We sort all points in increasing order of the first coordinates, and, for each point in the point set, we store its second coordinate to its $(d+1)$-st coordinate in a chunk γ. For $1 \leq \ell \leq d$, the $(\ell+1)$-st coordinate will be stored in the ℓ-th subchunk of γ. Note that these coordinates each fit in the lowest $\lceil \epsilon \lg w \rceil$ bits of a subchunk. The highest bit of the same subchunk, which is referred to as the flag bit, is set to be zero. Thus the space cost is $s_1 = O(\lg w)$ bits per point. Because each subblock consists of at most m_2 points and $m_2 \times s_1 = o(w)$, the chunks of all points in a subblock can fit in a single machine word.

Let $q = (q_1, q_2, \ldots, q_{d+1})$ be the query and β' be the rightmost subblock that intersects with q. We find the rank r of q_1 over the points of β', and copy the chunks of the first r points of β' into the first r chunks of a machine word $\mathcal{A}$. This requires only $O(1)$ time since these chunks are stored consecutively in memory. Then we store $q_2, q_3, \ldots, q_{d+1}$ duplicately in the first r chunks of another word $\mathcal{B}$. For each of these r chunks and each $1 \leq \ell \leq d$, $q_{\ell+1}$ is stored in the ℓ-th subchunk as the lowest $\lceil \epsilon \lg w \rceil$ bits, and the flag bit of the subchunk is set to be 1. The construction of $\mathcal{B}$ also requires $O(1)$ time.

We then compute $\mathcal{C} = \mathcal{B} - \mathcal{A}$, mask all bits of $\mathcal{C}$ to 0 except the flag bits of the subchunks in each of the first r chunks, and right-shift $\mathcal{C}$ by $s_2 - 1$ bits. It is not hard to see that a point is dominated by q iff the value the corresponding chunk represents is equal to $(2^{ds_2} - 1)/(2^{s_2} - 1)$.

To count the occurrences of that value, we create another word $\mathcal{D}$ so that each of the first r chunks represents $(2^{ds_2} - 1)/(2^{s_2} - 1) + 2^{ds_2 - 1}$. That is, the lowest bits of all subchunks and the flag bit of the d-th subchunk are set to be 1 in each of the first r chunks, and the other bit are set to be 0. We compute $\mathcal{E} = \mathcal{D} - \mathcal{C}$, and mask all bits of $\mathcal{E}$ to 0 except the highest bits of the first r chunks, i.e., the flag bits of the d-th subchunks. The highest bit of a chunk is 1 iff the corresponding point is dominated by q.

Finally we sum up the highest bits of the first r chunks. To achieve that, we right-shift $\mathcal{E}$ by $s_1 - 1$ bits, so that the highest bit of each chunk becomes the lowest one. Then we multiply the shifted word by $(2^{rs_1} - 1)/(2^{s_1} - 1)$ and the value stored in the r-th chunk will be the sum we need, which is also the answer to the query q. The whole algorithm requires $O(1)$ time and no table lookup. □

Lemma 5. *For any constant $0 < \epsilon < 1/d$, dominance counting queries over a point set S of size $m \leq n$ and type d' with $\sigma = \lceil w^{\epsilon} \rceil$ can be supported using $O(m \lg w \cdot (\lg n / \lg w + 1)^{d'-1})$ bits of space and $O((\lg n / \lg w + 1)^{d'-1})$ query time.*

Proof. The base case in which $d' = 1$ has been handled in Lemmas 3 and 4. We only show how to reduce the case of d' to that of $d' - 1$. We build a range tree over the d'-th coordinates of points with branching factor $\lceil w^{\epsilon} \rceil$. The height of the range tree is bounded above by $O(\lg n / \lg w + 1)$. Each node v in the range tree represents a range $[a_v..b_v]$ and the set $S(v)$ of points whose d'-th coordinates are in $[a_v..b_v]$. The leaf nodes in the range tree each represent a single point.

For each internal node v, we transform $S(v)$ into a point set $D(v)$ of type $d' - 1$. For any $\ell < d'$ and any point $p \in S(v)$, its ℓ-th coordinate p_ℓ is replaced with the rank of p_ℓ, i.e., the number of points in $S(v)$ whose ℓ-th coordinates are no greater than p_ℓ. In addition, the d'-th coordinate of p is replaced with an integer in $[1..t]$, which is the index of v's child that represents a set containing p. Queries over $D(v)$ can be supported recursively.

Inside each internal node v, for each dimension $1 \leq \ell \leq d'$ we write down a sequence $\mathcal{S}_{v,\ell}[1..|S(v)|]$. For each point $p \in S(v)$, $\mathcal{S}_{v,\ell}[p_\ell]$ is the integer that replaced the d'-th coordinate of p. We represent these sequences using the succinct data structures of Belazzougui and Navarro [1]. These data structures use $O(|\mathcal{S}_{v,\ell}| \lg w)$ bits of space, and support $\mathtt{rank}_i(\mathcal{S}_{v,\ell}, p_\ell)$ operations in $O(1)$ time, which count the occurrences of i's in $\mathcal{S}_{v,\ell}[1..p_\ell]$.

Let the given dominance counting query be $q = (q_1, q_2, \cdots, q_{d+1})$. Starting with the root node, we traverse the range tree from top to bottom. Let v be the root node and let $v_1, \ldots, v_t$ be the children of v from left to right. We find the largest i so that $b_{v_i} \leq q_{d'}$. Querying $D(v)$ recursively with $(q_1, \ldots, q_{d'-1}, i, q_{d'+1}, q_{d+1})$, we can find the number of points in the first i children of v that are dominated by q. Then we recursively query $S(v_{i+1})$ with $q' = (q'_1, q'_2, \ldots, q'_{d'}, q_{d'+1}, \ldots, q_{d+1})$, where $q'_\ell = \mathtt{rank}_{i+1}(\mathcal{S}_{v,\ell}, q_\ell)$ for $1 \leq \ell \leq d'$. We return the sum of the answers found.

This range tree is of height $O(\lg n / \lg w + 1)$, and a dominance counting query on a point set of type d' is reduced to $O(\lg n / \lg w + 1)$ queries on points sets of type $d' - 1$. Thus we achieve the desired bounds for query time and space cost. □

Remark. Lemma 5 is a stronger version of Lemma 2(a). By Lemma 5, one can support d-dimensional range counting queries within $O((\lg n/\lg w+1)^{d-1})$ query time and $O(n\cdot(\lg n/\lg w+1)^{d-2})$ words of space. This improves the data structures of JaJa et al. [6] when $w \geq \lg^{\omega(1)} n$.

Next we consider how to prove Lemma 2(b). Unlike the structures for dominance counting queries with $\sigma = \lceil w^{\epsilon} \rceil$, for parallel counting queries we can only set $\sigma = \lceil \lg^{\epsilon} n \rceil$.

Lemma 6. *For any constant $0 < \epsilon < 1/d$, parallel counting queries over a point set S of size $m \leq n$ and type 1 with $\sigma = \lceil \lg^{\epsilon} n \rceil$ can be supported using $O(m \lg\lg n)$ bits of space and $O(1)$ query time. In addition, we also need global lookup tables that occupy $o(n)$ bits of space in total.*

Proof. We sort all points of S in increasing order of the first coordinates. We divide S into blocks of size $n_1 = \lceil \lg^2 n \rceil$, and divide each block into subblocks of size $n_2 = \lceil \lg^{d\epsilon} n \rceil$. We still label each block/subblock with the largest first coordinate over the points inside the block/subblock.

For each block β we maintain a table D_β in which, for $j \in [0..\lceil (\lg n)/s \rceil]$ and $1 \leq q_2, q_3, \ldots, q_d \leq \lceil \lg^{\epsilon} n \rceil$, the entry $D_\beta[q_2, q_3, \ldots, q_d, j]$ stores the j-th sections of $f[1], f[2], \ldots, f[\lceil \lg^{\epsilon} n \rceil]$, where $f[i]$ is the number of points in S that are dominated by $(\texttt{label}(\beta), q_2, \ldots, q_d, i)$. As described in Sect. 3, we store the j-th section of each of these values in a subword of $s+b+2$ bits, and pack them into a single word. These table D_β's occupy $O(m/n_1) \times O(\lg^{(d-1)\epsilon} n) \times (\lceil \lg n/s \rceil + 1) \times O(\lg n) = O(m/\lg^{1-d\epsilon} n) = o(m)$ bits in total.

For each subblock β' of β we maintain a table $E_{\beta'}$ in which, for $1 \leq q_2, q_3, \ldots, q_d \leq \lceil \lg^{\epsilon} n \rceil$, the entry $E_{\beta'}[q_2, q_3, \ldots, q_d]$ stores the concatenation of $g[1], g[2], \ldots, g[\lceil \lg^{\epsilon} n \rceil]$, where $g[i]$ is the number of points inside β that are dominated by $(\texttt{label}(\beta'), q_2, \ldots, q_d, i)$. Each $g[i]$ can be represented in $\lceil \lg n_1 \rceil = O(\lg\lg n)$ bits. The overall space cost for all the tables $E_{\beta'}$ is $O(m/n_2) \times \lceil \lg^{d\epsilon} n \rceil \times O(\lg\lg n) = O(m \lg\lg n)$ bits. We further precompute a global lookup table X that, for each possible values of $g[1], g[2], \ldots, g[\lceil \lg^{\epsilon} n \rceil]$, stores a word in which the i-th subword is equal to $g[i]$. Clearly the lookup table X requires $o(n)$ bits of space.

We can encode each subblock in $O(m_2 \lg\lg n) = O((\lg^{d\epsilon} n)\cdot \lg\lg n)$ bits. Then we precompute another global lookup table Y that, for any possible encoding of a subblock β' and any $1 \leq q_1, q_2, \ldots, q_d \leq \lceil \lg^{\epsilon} n \rceil$, stores the concatenation of $h[1], h[2], \ldots, h[\lceil \lg^{\epsilon} n \rceil]$, where $h[i]$ is the number of points inside β' that are dominated by $(q_1, q_2, \ldots, q_d, i)$. The table Y also requires $o(n)$ bits of space since there are only $O(n^{1-\delta})$ possible encodings of subblocks for some $\delta > 0$.

Let $(q_1, q_2, \ldots, q_d, j)$ be a parallel counting query. We find the rightmost block β whose label is no greater than q_1, and the rightmost block β' to the right of β whose label is no greater than q_1. If $j > 0$, then we simply return $D_\beta[q_2, q_3, \ldots, q_d, j]$. If $j = 0$, then we further find the subblock β'' to the right of β' and the rank r of q_1 inside β''. The answer is the sum of $D_\beta[q_2, q_3, \ldots, q_d, 0]$, $E_{\beta'}[q_2, q_3, \ldots, q_d]$, and $Y[\texttt{enc}(\beta''), q_2, q_3, \ldots, q_d]$, where $\texttt{enc}(\beta'')$ is the encoding of β''. Note that we need X to transform the entry of $E_{\beta'}$. The overall query time is $O(1)$. □

Finally, following the same approach of the proof for Lemma 5 but using branching factor $\lceil \lg^{\epsilon} n \rceil$, we can prove Lemma 2(b). Since the answer is expressed as a sum of the j-th sections of $K = O((\lg n / \lg\lg n)^{d'})$ numbers, we need to set b larger than $\lg K = \Theta(\lg\lg n)$.

Acknowledgements. We thank the anonymous reviewers for their fruitful comments and suggestions.

References

1. Belazzougui, D., Navarro, G.: Optimal lower and upper bounds for representing sequences. ACM Trans. Algorithms (TALG) **11**(4), 31:1–31:21 (2015). Article 31
2. Brodal, G.S., Gfeller, B., Jørgensen, A.G., Sanders, P.: Towards optimal range medians. Theor. Comput. Sci. **412**(24), 2588–2601 (2011)
3. Chan, T.M., Wilkinson, B.T.: Adaptive and approximate orthogonal range counting. In: SODA, pp. 241–251 (2013)
4. Fredman, M.L., Willard, D.E.: Surpassing the information theoretic bound with fusion trees. J. Comput. Syst. Sci. **47**(3), 424–436 (1993)
5. Gabow, H.N., Bentley, J.L., Tarjan, R.E.: Scaling and related techniques for geometry problems. In: STOC, pp. 135–143 (1984)
6. JáJá, J., Mortensen, C.W., Shi, Q.: Space-efficient and fast algorithms for multidimensional dominance reporting and counting. In: Fleischer, R., Trippen, G. (eds.) ISAAC 2004. LNCS, vol. 3341, pp. 558–568. Springer, Heidelberg (2004)
7. Jørgensen, A.G., Larsen, K.G.: Range selection and median: tight cell probe lower bounds and adaptive data structures. In: SODA, pp. 805–813 (2011)
8. Krizanc, D., Morin, P., Smid, M.H.M.: Range mode and range median queries on lists and trees. Nord. J. Comput. **12**(1), 1–17 (2005)
9. Matousek, J.: Reporting points in halfspaces. Comput. Geom. **2**, 169–186 (1992)

Combinatorial Optimization and Approximation Algorithms I

On the Minimum Cost Range Assignment Problem

Paz Carmi and Lilach Chaitman-Yerushalmi(✉)

Department of Computer Science, Ben-Gurion University of the Negev,
Beersheba, Israel
{carmip,chaitman}@cs.bgu.ac.il

Abstract. We study the problem of assigning transmission ranges to radio stations placed in a d-dimensional (d-D) Euclidean space in order to achieve a strongly connected communication network with minimum total cost, where the cost of transmitting in range r is proportional to r^α. While this problem can be solved optimally in 1D, in higher dimensions it is known to be NP-hard for any $\alpha \geq 1$.

For the 1D version of the problem and $\alpha \geq 1$, we propose a new approach that achieves an exact $O(n^2)$-time algorithm. This improves the running time of the best known algorithm by a factor of n. Moreover, we show that this new technique can be utilized for achieving a polynomial-time algorithm for finding the minimum cost range assignment in 1D whose induced communication graph is a t-spanner, for any $t \geq 1$.

In higher dimensions, finding the optimal range assignment is NP-hard; however, it can be approximated within a constant factor. The best known approximation ratio is for the case $\alpha = 1$, where the approximation ratio is 1.5. We show a new approximation algorithm that breaks the 1.5 ratio.

1 Introduction

A wireless ad-hoc network is a self-organized decentralized network that consists of independent radio transceivers (transmitter/receiver) and does not rely on any existing infrastructure. The network nodes (stations) communicate over radio channels. Each node broadcasts a signal over a fixed range and any node within this transmission range receives the signal. Communication with nodes outside the transmission range is done using multi-hops, i.e., intermediate nodes pass the message forward and form a communication path from the source node to the desired target node. The twenty-first century witnesses widespread deployment of wireless networks for professional and private applications. The field of wireless communication continues to experience unprecedented market growth. For a comprehensive survey of this field see [11].

Let S be a set of points in the d-dimensional Euclidean space representing radio stations. A *range assignment* for S is a function $\rho : S \rightarrow \mathbb{R}^+$ that assigns

The research is partially supported by the Lynn and William Frankel Center for Computer Science and by grant 680/11 from the Israel Science Foundation (ISF).

K. Elbassioni and K. Makino (Eds.): ISAAC 2015, LNCS 9472, pp. 95–105, 2015.
DOI: 10.1007/978-3-662-48971-0_9

each point a transmission range (radius). The cost of a range assignment, is defined as $cost(\rho) = \sum_{v \in S}(\rho(v))^\alpha$ for some real constant $\alpha \geq 1$. In the case $\alpha \in (1, 6]$, the *cost* represents the power consumption of the network, where α varies depending on different environmental factors [11]. The linear case $\alpha = 1$, corresponds to minimizing the sum of ranges (radii).

A range assignment ρ induces a *directed communication graph* $G_\rho = (S, E_\rho)$, where $E_\rho = \{(u, v) : \rho(u) \geq |uv|\}$ and $|uv|$ denotes the Euclidean distance between u and v. A range assignment ρ is *valid* if the induced (communication) graph G_ρ is strongly connected. For ease of presentation, throughout the paper we refer to the terms 'assigning a range $|uv|$ to a point $u \in S$' and 'adding a directed edge (u, v)' as equivalent.

We consider the *d*-D Minimum Cost Range Assignment (MinRange) problem, that takes as an input a set S of n points in $\mathbb{R}^d$, and whose objective is finding a valid *range assignment* for S of minimum cost. This problem has been considered extensively, for different values of d and α, with additional requirements and modifications. Some of these works are mentioned in this section.

Kirousis et al. [9] considered the 1D MinRange problem (the radio stations are placed on a line) and showed an $O(n^4)$-time exact algorithm for the problem. Later, Das et al. [8] improved the running time to $O(n^3)$. Here, we propose an $O(n^2)$-time exact algorithm, this improves the running time of the best known algorithm by a factor of n without increasing the space complexity. The novelty of our method lies in separating the range assignment into two, *left* and *right*, assignments and restricting the algorithm search to optimal assignments that minimize a new evaluation function $cost'$, defined with respect to this separation. This counter intuitive approach reveals the existence of an optimal solution of a simple structure and allows us to achieve the aforementioned result. Moreover, it can be utilized to compute an optimal range assignment in 1D with the additional requirement that the induced graph is a t-spanner, for a given $t \geq 1$. Hopefully, our new technique will enable solving other range assignment variations as well.

A geometric directed graph $G = (S, E)$ is a t-spanner for a set S, if for every two points $u, v \in S$ there exists a path in G from u to v of length at most $t|uv|$, where the length of a path is defined as the sum of lengths of its edges. The importance of avoiding flooding the network when routing, was one of the reasons that led researchers to consider the combination of range assignment and t-spanners, e.g., [1,12–14], as well as the combination of range assignment and hop-spanners, e.g., [6,9]. While bounded-hop spanners bound the number of intermediate nodes forwarding a message, t-spanners bound the relative distance a message is forwarded. For the 1D bounded-hop range assignment problem, Clementi et al. [6] showed a 2-approximation algorithm whose running time is $O(hn^3)$. To the best of our knowledge, we are the first to show an algorithm that computes an optimal solution for the range assignment with the additional requirement that the induced graph (viewed as geometric graph) is a t-spanner.

While the 1D version of the MinRange problem can be solved optimally, for any $d \geq 2$ and $\alpha \geq 1$, it has been proven to be NP-hard (in [9] for $d \geq 3$ and $1 \leq \alpha < 2$ and later in [7] for $d \geq 2$ and $\alpha > 1$). However, some versions can be approximated within constant factor. For $\alpha = 2$ and any $d \geq 2$ Kirousis et al. [9] gave a 2-approximation algorithm based on the minimum spanning tree. The best known approximation ratio is for the case $\alpha = 1$, i.e., the cost function equals to the sum of radii, where the approximation ratio is 1.5 [3]. Minimization of the radii sum has been considered also in the context of other range assignment problems, such as set of circles connectivity [5] and circle coverage [2,10]. This linear model may be appropriate also for power consumption in future systems where the transmitting stations do not transmit in all directions simultaneously, but rather focus the transmission energy in a narrow angle beam whose direction changes according to the needs of the network, as predicted in [10].

We show a new approximation algorithm for the MinRange problem with $\alpha = 1$ that breaks the 1.5 ratio with a ratio of $1.5 - c$, for a suitable constant $c > 0$. for which the 1.5 ratio bound has not been breached, such as metric TSP, scheduling parallel jobs and minimum strongly connected sub-graph, or even have been proved to be the best one can hope for unless $P = NP$, such as the bin packing problem.

Due to space limitation, we omit an algorithm description (pseudo-code) and some figures and proofs; however, all of them are given in the full version of this manuscript [4].

2 Minimum Cost Range Assignment in 1D

In the 1D version of the MinRange problem, the input set $S = \{v_1, ..., v_n\}$ consists of points located on a line. For simplicity, we assume that the line is horizontal and for every $i < j$, v_i is to the left of v_j. Given two indices $1 \leq i < j \leq n$, we denote by $S_{i,j}$ the subset $\{v_i, ..., v_j\} \subseteq S$.

We present two polynomial-time algorithms for finding optimal range assignments, the first, in Sect. 2.1, for the basic 1D MinRange problem, and the second, in Sect. 2.2, subject to the additional requirement that the induced graph is a t-spanner (the 1D MinRangeSpanner problem). Our new approach for solving these problems requires introducing a variant of the *range assignment*. Instead of assigning each point in S a radius, we assign each point two directional ranges, *left range assignment*, $\rho^l : S \rightarrow \mathbb{R}^+$, and *right range assignment*, $\rho^r : S \rightarrow \mathbb{R}^+$. A pair of assignments (ρ^l, ρ^r) is called a *left-right assignment*. Assigning a point $v \in S$ a left range $\rho^l(v)$ and a right range $\rho^r(v)$ implies that in the induced graph, $G_{\rho^{lr}}$, v can reach every point to its left up to distance $\rho^l(v)$ and every point to its right up to distance $\rho^r(v)$. That is, $G_{\rho^{lr}}$, contains the directed edge (v_i, v_j) if and only if one of the following holds: (i) $i < j$ and $|v_i v_j| \leq \rho^r(v_i)$, or (ii) $j < i$ and $|v_i v_j| \leq \rho^l(v_i)$. The cost of an assignment (ρ^l, ρ^r), is defined as $cost(\rho^l, \rho^r) = \sum_{v \in S}(\max\{\rho^l(v), \rho^r(v)\})^\alpha$.

Our algorithms find a *left-right assignment* of minimum cost that can be converted into a *range assignment* ρ with the same cost by assigning each point $v \in S$ a range $\rho(v) = \max\{\rho^l(v), \rho^r(v)\}$. Note that any valid *range assignment* for S can be converted to a *left-right assignment* with the same cost, by assigning every point $v \in S$, $\rho^l(v) = \rho^r(v) = \rho(v)$. To be more precise, either $\rho^l(v)$ or $\rho^r(v)$

should be reduced to $|vu|$ where u is the farthest point in the directional range (for Lemma 1 to hold). Therefore, a minimum cost *left-right assignment* implies a minimum cost *range assignment.*

In addition to the *cost* function, we define $cost'(\rho^l, \rho^r) = \sum_{v \in S}((\rho^l(v))^\alpha + (\rho^r(v))^\alpha)$, and refine the term of *optimal solution* to include only solutions that minimize $cost'(\rho^l, \rho^r)$ among all solutions, (ρ^l, ρ^r), with minimum $cost(\rho^l, \rho^r)$.

2.1 An Exact Algorithm for the 1D MinRange Problem

Das et al. [8] state three basic lemmas regarding properties of an optimal range assignment. The following three lemmas are adjusted versions of these lemmas for a *left-right assignment.*

Lemma 1. *In an optimal solution* (ρ^l, ρ^r), *for every* $v_i \in S$, *either* $\rho^l(v_i) = 0$ *or* $\rho^l(v_i) = |v_i v_j|$; *similarly, either* $\rho^r(v_i) = 0$ *or* $\rho^r(v_i) = |v_i v_k|$, *for some* $j \leq i \leq k$.

Lemma 2. *Given indices* $1 \leq i < j < k \leq n$, *let* (ρ^l, ρ^r) *be an optimal solution for* $S_{i,k}$ *subject to the constraints* $\rho^l(v_j) \geq |v_i v_j|$ *and* $\rho^r(v_j) \geq |v_j v_k|$, *then,*
- *for all* $m = i, ..., j-1$, $\rho^r(v_m) = |v_m v_{m+1}|$ *and* $\rho^l(v_m) = 0$; *and*
- *for all* $m = j+1, ..., k$, $\rho^l(v_m) = |v_m v_{m-1}|$ *and* $\rho^r(v_m) = 0$.

Lemma 3. *In an optimal solution* (ρ^l, ρ^r), $\rho^l(v_1) = 0$ *and* $\rho^r(v_1) = |v_1 v_2|$.

Lemma 1 allows us to simplify the notation $\rho^x(v_i) = |v_i v_j|$ for $x \in \{l, r\}$ and $1 \leq i, j \leq n$, and write $\rho^x(i) = j$, for short. We use dynamic programming which exploits the special structure of an optimal solution according to our refined definition. An optimal solution consists of a 'division' of the interval $[v_1, v_n]$ into sub-intervals $[v_i, v_{k'}]$ having $i < k \leq k'$ with $\rho^l(k) = i$ and $\rho^r(k) = k'$, and thus Lemma 2 applies for each of them. The set of sub-intervals does not precisely admit a division since we allow each sub-interval $[v_i, v_{k'}]$ to share its endpoint and the point adjacent to it, i.e., $v_{k'}$ and $v_{k'-1}$, with the consecutive sub-interval.

Given $1 \leq i < n$, we denote by $OPT(i)$ the cost of an optimal solution for the sub-problem defined by the input $S_{i,n}$, subject to the constraint $\rho^r(i) = i+1$. Thus, the cost of an optimal solution for the whole problem is $OPT(1)$.

To guide the reader, we first present an algorithm with $O(n^3)$ running time Then, we reduce the running time to $O(n^2)$.

A Cubic-Time Algorithm. Our algorithm, to which we refer as 1DMinRA algorithm, computes the values $OPT(i)$ based on the recursive formula given in Lemma 4. This formula relies on the structure of an optimal solution described earlier.

Lemma 4. *Consider the value* $OPT(i)$, *then if* $i = n-1$, $OPT(i) = 2|v_{n-1}v_n|^\alpha$ *and if* $1 \leq i < n-1$,

$$OPT(i) = \min_{\substack{i<k<n \\ k<k'\leq n}} \left\{ \sum_{m=i}^{k'-2} |v_m v_{m+1}|^\alpha + OPT(k'-1) - |v_{k'-1}v_{k'}|^\alpha + \max\{|v_i v_k|^\alpha, |v_k v_{k'}|^\alpha\} \right\}.$$

Proof. Trivially, $OPT(n-1)$ equals $2|v_{n-1}v_n|^\alpha$. Let X_i denote the right side of the equation, we prove $OPT(i) = X_i$.

$OPT(i) \leq X_i$: We show that all costs that appear as min function arguments in X_i correspond to valid assignments and thus infer, by the optimality of $OPT(i)$, that the above inequality holds. Consider an argument with parameters k and k'. We associate it with an assignment (ρ^l, ρ^r) defined as follows. For $m \geq k'-1$ the assignment is inductively defined by $OPT(k'-1)$. For every $i \leq m < k$, $\rho^l(m) = m$ and $\rho^r(m) = m+1$, for every $k < m < k'$, $\rho^l(m) = m-1$ and $\rho^r(m) = m$ ($\rho^r(k'-1)$ is reassigned) and for k, $\rho^l(k) = i$, $\rho^r(k) = k'$. By the validity of $OPT(k'-1)$, every two points among $S_{k'-1,n}$ are (strongly) connected. By definition, $\rho^r(k'-1) = k'$ and our assignment for $S_{k,k'-1}$ (including reassigning $\rho^r(k'-1)$) ensures the existence of a path connecting $v_{k'-1}$ to $v_{k'}$ (passing through v_k). Moreover, our assignment for $S_{i,k'-1}$ guarantees the connectivity between every two points in $S_{i,k'-1}$, and altogether between every two points in $S_{i,n}$.

$OPT(i) \geq X_i$: Consider an optimal solution (ρ^l, ρ^r) for the points $S_{i,n}$ subject to the condition that $\rho^r(i) = i+1$. Let v_k be a point to the right of v_i with $\rho^l(k) = i$ and let $\rho^r(k) = k'$. Note that since v_i is the leftmost point and the induced graph is strongly connected, such a point necessarily exists.

Next we show that there is no edge directed either right or left connecting two points on different sides of $v_{k'}$ in $G_{\rho^{lr}}$, except for possibly an edge $(v_j, v_{k'-1})$ with $j > k'$. Assume towards contradiction that the former does not hold, i.e., there exists $i < t < k'$, with $\rho^r(t) \geq k'$; then, reassigning $\rho^r(k) = \max\{t, k\}$ maintains the connectivity, and reduces the value of *cost'* without increasing the value of *cost* in contradiction to the optimality of the solution. Now, let v_j be a point to the right of $v_{k'}$ with $\rho^l(j) = j' \in [i, k']$, we show that $j' \geq k'-1$. Consider a point v_t, $j' < t < k'$. As we have shown, $\rho^r(t) < k'$. By symmetric arguments we have $\rho^l(t) > j'$. Namely, there is no edge going out of the interval $(v_{j'}, v_{k'})$. Thus, connectivity can be achieved only if this interval is empty of vertices, i.e., either $j' = k'-1$ or $j' = k'$ (note that $k'-1 > i$).

The above observation allows us to divide the problem into two independent subproblems, one for the points $S_{i,k'-1}$ subject to the constraints $\rho^l(k) = i$ and $\rho^r(k) = k'$, and the other for the points $S_{k'-1,n}$ subject to the artificial constraint $\rho^r(k'-1) = k'$ that guarantees the existence of a path from $k'-1$ to k', due to the solution of the first subproblem, but should not be paid for. Note that the case where $j' = k'$ is covered by the choice of $v_{k'}$ as the point whose left range covers the leftmost point of the subproblem $S_{k'-1,n}$ (i.e., parameter k in the above formula). Regarding the first subproblem, by Lemma 2, in an optimal assignment, for every $i \leq m < k$, $\rho^l(m) = m$ and $\rho^r(m) = m+1$, and for every $k < m \leq k'-1$, $\rho^l(m) = m-1$ and $\rho^r(m) = m$. Thus, its cost is $\sum_{m=i}^{k'-2} |v_m v_{m+1}|^\alpha + \max\{|v_i v_k|^\alpha, |v_k v_{k'}|^\alpha\}$. The optimal cost of the second subproblem is $OPT(k'-1) - |v_{k'-1}v'_k|^\alpha$. Hence, the *cost* of an optimal solution to the whole problem is the sum of the above costs and the lemma follows.

The algorithm computes the values $OPT(i)$ according to the formula given in Lemma 4 for every $1 \leq i \leq n$ and stores them in a table T. During the i-th iteration $T[i'] = OPT(i')$ for every $i < i' < n$. Finally, it outputs T[1]. To reduce the running time of each computation it uses a 2-dimensional matrix storing for every $1 \leq i < j \leq n$ the sum $\sum_{m=i}^{j-1} |v_m v_{m+1}|^\alpha$.

The algorithm description (pseudo-code) is given in the full version [4].

While the table T maintains only the costs of the solutions, the optimal assignment can be easily retrieved by backtracking the entries leaded to the optimal cost and inferring the associated range assignment (as described in the proof of Lemma 4).

Complexity. The total running time is $O(n^3)$, since $O(n)$ iterations are performed during the algorithm and each iteration takes $O(n^2)$ time. Obviously, the algorithm requires $O(n^2)$ space (the same space complexity as in [8]).

Lemma 5. *Algorithm* 1DMinRA *runs in* $O(n^3)$ *time using* $O(n^2)$ *space.*

A Quadratic-Time Algorithm. We show how to reduce the running time of Algorithm 1DMinRA to $O(n^2)$. Consider the equality stated in Lemma 4. Observe that given fixed values i and k', the value k that minimizes the argument of the min function with respect to i and k' is simply the value k that minimizes $\max\{|v_i v_k|^\alpha, |v_k v_{k'}|^\alpha\}$. This value is simply the closest point to the midpoint of the segment $\overline{v_i v_{k'}}$, denoted by $c(i, k')$. Thus,

$$OPT(i) = \min_{i+1<k'\leq n} \left\{ \begin{array}{l} \sum_{m=i}^{k'-2} |v_m v_{m+1}|^\alpha + OPT(k'-1) - |v_{k'-1} v_{k'}|^\alpha \\ \qquad + \max\{|v_i v_{c(i,k')}|^\alpha, |v_{c(i,k')} v_{k'}|^\alpha\} \end{array} \right\}.$$

Consider Algorithm 1DMinRA after applying the above modification in the computation of $T[i]$. Since there are only $O(n)$ sub-problems to compute, each in $O(n)$ time, the running time reduces to $O(n^2)$ and Theorem 1 follows.

Theorem 1. *The* 1D MinRange *problem can be solved in* $O(n^2)$ *time using* $O(n^2)$ *space.*

2.2 An Exact Algorithm for the 1D MinRangeSpanner Problem

Given a set $S = \{v_1, .., v_n\}$ of points in $1D$ and a value $t \geq 1$, the 1D MinRangeSpanner problem aims to find a minimum cost range assignment for S, subject to the requirement that the induced graph is a t-spanner. (We view the induced graph as a geometric graph whose edges are line segments connecting pairs of input points.) We further utilize the technique presented in Sect. 2.1 to obtain a polynomial-time exact algorithm for the 1D MinRangeSpanner problem. This algorithm follows the same guidelines as Algorithm 1DMinRA and relies on arguments similar to those of Lemma 4.

We begin with providing the key notions required for understanding the correctness of the algorithm, followed by its description. The first and most crucial observation is that the problem can still be divided into two subproblems in the same way as in Algorithm 1DMinRA, by similar arguments to those of Lemma 4. In Lemma 4 we show that any assignment that does not satisfy the conditions required for the division can be adjusted to a new assignment with a lower value of *cost'* that preserves connectivity. The new assignment, however, preserves also the lengths of the shortest paths, which make the argument legitimate for this problem as well.

The two problems (MinRange and MinRangeSpanner) differ when it comes to solving each of the above subproblems. Consider the left subproblem, i.e., of the form described in Lemma 2. The optimal assignment for it is no longer necessarily the one stated in the lemma, since it does not ensure the existence of t-spanning paths. Therefore, our algorithm divides problems of this form into smaller subproblems handled recursively. Dealing with such subproblems requires defining new parameters: a rightmost input point v_j, and the length of the shortest paths connecting v_i to v_j, v_j to v_i and v_i to v_{i+1} not involving points in $S_{i,j}$ except for the endpoints, denoted by $\overrightarrow{\delta}$, $\overleftarrow{\delta}$, and δ^i, respectively. Regarding the computation of a subproblem, since points may be covered now by vertices outside the subproblem domain, we allow v_k to have either a right or a left range being 0 (in the terms of Algorithm 1DMinRA, either $k = i$ or $k = k'$).

Another key observation is that any directed graph G over S is a t-spanner for S if and only if for every $1 \leq i < n$ there exists a t-spanning path from v_i to v_{i+1} and from v_{i+1} to v_i. Moreover, given that G is strongly connected implies that the addition of an edge between consecutive points does not affect the length of the shortest path between any other pair of consecutive points. Therefore, for subproblems with $j = i + 1$ we assign $\rho^r(i) = i + 1$ (resp. $\rho^l(i + 1) = i$) if and only if $\overrightarrow{\delta}/|i, i+1| > t$ (resp. $\overleftarrow{\delta}/|i, i+1| > t$) and thus ensuring that the induced graph is a t-spanner.

Our algorithm may consider solutions in which an assignment to a node is charged more than once in the total cost; however, for every such solution, there exists an equivalent one in which the charging is done properly and is preferred by the algorithm due to its lower cost.

The description of our algorithm and its complexity given in [4] implies the following theorem.

Theorem 2. *The* $1D$ MinRangeSpanner *problem can be solved in* $O(n^7)$ *time using* $O(n^5)$ *space.*

3 The MinRange Problem in Higher Dimensions

In this section we focus on the MinRange problem for dimension $d \geq 2$ and $\alpha = 1$. As all the versions of the problem for $d \geq 2$ and $\alpha \geq 1$, it is known to be NP-hard. Currently, the algorithm achieving the best approximation ratio for

$\alpha = 1$ and $d \geq 2$ is the *Hub* algorithm with a ratio of 1.5. This algorithm was proposed by G. Calinescu, P.J. Wan, and F. Zaragoza for the general metric case, and analyzed by Ambühl et al. in [3] for the restricted Euclidean case. We show a new approximation algorithm and bound its approximation ratio from above by $1.5 - c$ for $c = 5/10^5$. Although in some cases our phrasing is restricted to the plane, all arguments hold for higher dimensions as well.

3.1 Our Approach

Presenting our approach requires acquaintance with two existing algorithms. The first is the *Hub* algorithm that finds the minimum enclosing disk C of S centered at point $hub \in S$. Then, it sets $\rho(hub) = r_{min}$ where r_{min} is C's radius. Finally, it directs the $MST(S)$ towards the *hub*. The cost of this assignment is $w(MST(S)) + r_{min} \leq w(MST(S)) + (w(MST(S)) + w(e_M))/2$, where e_M is the longest edge in $MST(S)$ and the weight function w is defined with respect to Euclidean lengths. The second algorithm is the algorithm for 1D MinRange problem by Kirousis et al. [9], to which we refer as the *1D RA* algorithm. We observe that this algorithm outputs an optimal solution for any ordered set $V = \{v_1, ..., v_n\}$ with distance function h that satisfies the following *line alike* condition: for every $1 \leq i \leq j < k \leq l \leq n$, it holds that $h(v_i, v_l) \geq h(v_j, v_k)$.

To guide the reader, we give an intuition and a rough sketch of our algorithm. The algorithm computes several solutions of four types and then chooses the minimum among them. Our analysis shows that at least one of the suggested solutions admits the required approximation. Those solutions roughly rely on two main methods. One uses a *hub* approach and achieves the required approximation for 'well spread' instances. The other uses more complicated techniques in order to achieve appropriate approximation for instances that roughly lie on a line. In order to distinguish between the two types of instances additional terminology is required. Given a graph G over S and two points $p, q \in S$, the *stretch factor* from p to q in G is $\delta_G(p,q)/|pq|$, where $\delta_G(p,q)$ denotes the Euclidean length of the shortest path between p and q in G. We use $\sim$*large* when referring to values greater than fixed thresholds, some with respect to $w(MST(S))$, defined later.

Consider $MST(S)$ and its longest path P_M. If one of the following conditions holds, then one of the first two solutions suggested by the algorithm, which use a *hub* approach, result in a better approximation than 1.5: (A1) there exists a $\sim$*large* edge in $MST(S)$; (A2) a $\sim$*large* fraction of P_M consists of disjoint sub-paths connecting pairs of points with $\sim$*large* stretch factor, not dominated by one sub-path of at least half the fraction; or (A3) the weight $w(MST(S) \setminus P_M)$ is $\sim$*large*.

Otherwise, there are three possible cases: (B1) S roughly lies on a line; (B2) there are two points in P_M with $\sim$*large* stretch factor, i.e., there is a $\sim$*large* 'hill' in P_M, and then either the optimal solution roughly consists of two independent subproblems, each roughly lies on a line (corresponds to a 'hill' side); or (B3) the optimal solution uses edges connecting the two sides of the 'hill', covering $\sim$*large* fraction of it.

The last three cases are approximated using the following technique. We consider every possible pair of *separation edges* of the optimal solution, i.e., edges that together cover a portion of the 'hill' and separate it into three sub-paths (the middle area covered by the edges and two uncovered areas), while the two uncovered areas are independent (namely, not connected by an edge). Note that such two edges exist (as shown later in our analysis). For each pair of separation edges we direct the middle covered sub-path to achieve a strongly connected subgraph via a cycle and solve the remaining two sub-paths separately, using two different techniques on each sub-path. In the first, we (i) 'flatten' the path, (ii) define a new distance function over the points of the sub-path, (iii) utilize the $1D$ RA algorithm to achieve an optimal assignment with respect to the defined distance function and then (iv) carefully transform it into a valid range assignment with respect to the Euclidean metric. Thus, we achieve a good approximation for input set that roughly lies on a line. In the second technique, we use the *Hub* algorithm. A $(1.5 - c)$-approximation is obtained for cases (B1) and (B2), using the first technique, and for case (B3), using the second technique.

3.2 The Approximation Algorithm

The algorithm uses the following three procedures that are defined precisely at the end of the algorithm's description.

- The *flatten* procedure f - a method performing shortcuts between pairs of points on a given path P resulting in a path without two points of stretch factor greater than c_s.
- The *distance function* h_S - a distance function defined for an ordered set $P \subseteq S$, satisfying the *line alike* condition.
- The *adjustment* transformation g - a function adjusting an optimal range assignment for an ordered set $P \subseteq S$ with distance function h, to a valid assignment for P.

Let R be the forest obtained by omitting from $MST(S)$ the edges of its longest path, P_M. Given a point $v \in P_M$, let $T(v)$ denote the tree of R rooted at v. For every $u \in T(v)$ let $r(u)$ denote the root of the tree in R containing u, namely, v. For a set of points $V \subset P_M$, let $T(V)$ denote the union $\bigcup_{v \in V} T(v)$. For ease of presentation, we assume the path P_M has a *left* and a *right* endpoints, thus, the *left* and *right* relations over P_M are naturally defined.

The Main Algorithm Scheme: Compute four solutions and return the one of minimum cost. In case of multiple assignments to a point in a solution, the maximum among the ranges counts.

Solution (i): apply the *Hub* algorithm.

Solution (ii): apply a variant of the *Hub* algorithm - find a point $p_h \in P_M$ that minimizes the value $r_h = \max\{|p_h\, p_1|, |p_h\, p_z|\}$, where p_1 and p_z are the endpoints of the path P_M. Assign p_h the range r_h, direct P_M towards p_h and bi-direct all edges in R.

(* *The rest of the algorithm handles cases (B1)-(B3) defined in Sect.* 3.1 *)

For every edge $e \in P_M$ do:

Let P_{e^l}, P_{e^r} be the paths of $P_M \backslash e$, to the left and to the right of e, respectively. Apply the *flatten procedure* f on P_{e^l} and P_{e^r} to obtain the paths $P_{l'} = (p_1, p_2, ..., p_m)$ and $P_{r'} = (p_{m+1}, p_{i+2}, ..., p_z)$, respectively.

(* *Note R has been changed during the flatten procedure* *)

For every 4 points $p_l, p_{l'}, p_{r'}, p_r$ with $l \leq l' \leq m < r' \leq r$ do:

In both solutions (iii) and (iv) direct the path $P_x = (p_l, ...p_m, p_{m+1}, ..., p_r)$ towards p_l and for each point p_i with $1 \leq i \leq z$ direct $T(p_i)$ towards p_i and assign p_i a range $w(T(p_i))$. Perform the least cost option among the following two, either add the edge (p_l, p_r), or add the two edges, one from u_l to $u_{r'}$ for $u_l \in T(p_l), u_{r'} \in T(p_{r'})$ of minimum length and the other from $u_{l'}$ to u_r for $u_{l'} \in T(p_{l'}), u_r \in T(p_r)$ of minimum length. As for the two paths $P_l = (p_1, p_2, ..., p_l)$ and $P_r = (p_r, p_{r+1}, ..., p_z)$, assign them ranges as follows:

Solution (iii): apply the *Hub* algorithm separately on each path.

Solution (iv): apply the *1D RA* algorithm separately on each path with respect to the *distance function* h_S.

The *Flatten* Procedure f. Let $c_s = 5/4$. Given a path $P = \{v_i, .., v_n\}$, set $Q_P = \{\}$. Let $j > i$ be the maximum index such that $\delta_P(v_i, v_j) > c_s|v_i v_j|$. If such index does not exist, let $j = i+1$. Else $(j > i+1)$, add the edge (v_i, v_j) to P, remove the edge (v_{j-1}, v_j) from P, move the sub-path $(v_i, .., v_{j-1})$ from P to the forest R, and update $Q_P = Q_P \cup \{(v_i, v_j)\}$. Finally, repeat with the sub-path $(v_j, .., v_n)$ without initializing Q_P.

The definitions for h_S and g are given with respect to the paths P_l and $P_{l'}$, the definitions for the path P_r and $P_{r'}$ are symmetric.

The Distance Function h_S. For every p_j, p_k with $1 \leq j \leq k \leq l$ we define,

$$h_S(p_j, p_k) = \min_{\substack{u \in T(p_{j'}), 1 \leq j' \leq j \\ v \in T(p_{k'}), k \leq k' \leq m}} |uv|.$$

The *Adjustment* Transformation g. Given an assignment $\rho' : P_l \rightarrow \mathbb{R}^+$, we transform it into an assignment $g(\rho') = \rho : P_{l'} \rightarrow \mathbb{R}^+$. First, we assign:

$$\rho(p_j) = \begin{cases} c_s \cdot \rho(p_j) + c_k \cdot T(p_j), 1 \leq j \leq l, \\ \qquad\qquad\quad c_k \cdot T(p_j), l < j \leq m, \end{cases}$$

where $c_k = 1 + 8(1 + c_s) = 19$. The multiplicity (by c_s) handles the gaps caused by points breaking the *line alike* condition with respect to the Euclidean metric. The role of the additive part, together with the second stage of the transformation, elaborated next, is to overcome the absence of points outside the path. In the second stage, for every p_j with $1 \leq j \leq m$, let $1 \leq j^- < j$ be the minimum index for which there exists $u \in T(p_{j^-})$ with $|p_j u| \leq c_k \cdot w(T(p_j))$, and let $j < j^+ \leq m$ be the maximum index for which there exists $u \in T(p_{j^+})$ with $|p_j u| \leq c_k \cdot w(T(p_j))$, direct the sub-path between p_{j^-} and p_{j^+} towards p_j.

The correctness of the algorithm, proved in [4], implies Theorem 3.

Theorem 3. *Given a set S of points in $\mathbb{R}^d$ for $d \geq 2$ and $\alpha = 1$, a minimum cost range assignment $(1.5 - c)$-approximation can be computed in polynomial time for S, where $c = \frac{5}{10^5}$.*

References

1. Abu-Affash, K., Aschner, R., Carmi, P., Katz, M.J.: Minimum power energy spanners in wireless ad-hoc networks. In: INFOCOM (2010)
2. Alt, H., Arkin, E.M., Brönnimann, H., Erickson, J., Fekete, S.P., Knauer, C., Lenchner, J., Mitchell, J.S.B., Whittlesey, K.: Minimum-cost coverage of point sets by disks. In: Proceedings of the Twenty-second Annual Symposium on Computational Geometry, SCG 2006, pp. 449–458 (2006)
3. Ambühl, C., Clementi, A.E.F., Penna, P., Rossi, G., Silvestri, R.: On the approximability of the range assignment problem on radio networks in presence of selfish agents. Theor. Comput. Sci. **343**(1–2), 27–41 (2005)
4. Carmi, P., Chaitman-Yerushalmi, L.: On the minimum cost range assignment problem. CoRR, abs/1502.04533 (2015)
5. Chambers, E.W., Fekete, S.P., Hoffmann, H.-F., Marinakis, D., Mitchell, J.S.B., Srinivasan, V., Stege, U., Whitesides, S.: Connecting a set of circles with minimum sum of radii. In: Dehne, F., Iacono, J., Sack, J.-R. (eds.) WADS 2011. LNCS, vol. 6844, pp. 183–194. Springer, Heidelberg (2011)
6. Clementi, A.E.F., Penna, P., Ferreira, A., Perennes, S., Silvestri, R.: The minimum range assignment problem on linear radio networks. Algorithmica **35**(2), 95–110 (2003)
7. Clementi, A.E.F., Penna, P., Silvestri, R.: On the power assignment problem in radio networks. Mob. Netw. Appl. **9**(2), 125–140 (2004)
8. Das, G.K., Ghosh, S.C., Nandy, S.C.: Improved algorithm for minimum cost range assignment problem for linear radio networks. Int. J. Found. Comput. Sci. **18**(3), 619–635 (2007)
9. Kirousis, L., Kranakis, E., Krizanc, D., Pelc, A.: Power consumption in packet radio networks. Theor. Comput. Sci. **243**(1–2), 289–305 (2000)
10. Lev-Tov, N., Peleg, D.: Polynomial time approximation schemes for base station coverage with minimum total radii. Comput. Netw. **47**(4), 489–501 (2005)
11. Pahlavan, K.: Wireless Information Networks. John Wiley, Hoboken (2005)
12. Shpungin, H., Segal, M.: Near optimal multicriteria spanner constructions in wireless ad-hoc networks. In: INFOCOM, pp. 163–171 (2009)
13. Wang, Y., Li, X.-Y.: Distributed spanner with bounded degree for wireless ad hoc networks. In: IPDPS 2002, pp. 120 (2002)
14. Wang, Y., Li, X.-Y.: Minimum power assignment in wireless ad hoc networks with spanner property. J. Comb. Optim. **11**(1), 99–112 (2006)

On the Approximability of the Minimum Rainbow Subgraph Problem and Other Related Problems

Sumedh Tirodkar(✉) and Sundar Vishwanathan

Department of Computer Science and Engineering, IIT Bombay, Mumbai, India
{sumedht,sundar}@cse.iitb.ac.in

Abstract. In this paper, we study the approximability of the Minimum Rainbow Subgraph (MRS) problem and other related problems. The input to the problem is an n-vertex undirected graph, with each edge colored with one of p colors. The goal is to find a subgraph on a minimum number of vertices which has one induced edge of each color. The problem is known to be NP-hard, and has an upper bound of $O(\sqrt{n})$ and a lower bound of $\Omega(\log n)$ on its approximation ratio.

We define a new problem called the Densest k Colored Subgraph problem, which has the same input as the MRS problem alongwith a parameter k. The goal is to output a subgraph on k vertices, which has the maximum number of edges of distinct colors. We give an $O(n^{1/3})$ approximation algorithm for it, and then, using that algorithm, give an $O(n^{1/3} \log n)$ approximation algorithm for the MRS problem. We observe that the MIN-REP problem is indeed a special case of the MRS problem. This also implies a combinatorial $O(n^{1/3} \log n)$ approximation algorithm for the MIN-REP problem. Previously, Charikar et al. [5] showed an ingenious LP-rounding based algorithm with an approximation ratio of $O(n^{1/3} \log^{2/3} n)$ for MIN-REP. It is quasi-**NP**-hard to approximate the MIN-REP problem to within a factor of $2^{\log^{1-\epsilon} n}$ [15]. The same hardness result now applies to the MRS problem. We also give approximation preserving reductions between various problems related to the MRS problem for which the best known approximation ratio is $O(n^c)$ where n is the size of the input and c is a fixed constant less than one.

1 Introduction

Given an input graph, to output an optimal subgraph satisfying some constraints is perhaps the most studied family of problems from an approximation perspective. Of late combinatorists have extensively studied such problems when the edges are colored, called Rainbow Subgraph problems. See [1,6,11,17,18] for a short representative list.

Our focus is arguably the simplest such computational problem, called the Minimum Rainbow Subgraph (MRS) problem. The input to the problem is an n-vertex undirected graph, with each edge colored with one of p colors. The goal is to find a subgraph on a minimum number of vertices which has one induced

K. Elbassioni and K. Makino (Eds.): ISAAC 2015, LNCS 9472, pp. 106–115, 2015.
DOI: 10.1007/978-3-662-48971-0_10

edge of each color. This was introduced in [4] and has been studied from the approximation viewpoint by [4,13,16]. The motivation for this problem comes from the Pure Parsimony Haplotyping (PPH) problem in computational biology. The reader is referred to [10] for a detailed description of the PPH problem. The MRS problem is known to be **NP**-hard.

There is a trivial $O(\sqrt{n})$ approximation algorithm for the MRS problem. Select one edge of each color and add its end points to the solution set. And this is the best known upper bound for this problem. The upper bound on the approximation ratio can be improved for bounded degree graphs. Camcho et al. [4] gave a $\frac{5}{6}\Delta$-approximation algorithm on graphs with maximum degree Δ, which was later improved by Katrenič et al. [13] to $\left(\frac{1}{2} + \left(\frac{1}{2} + \epsilon\right) \Delta\right)$. Katrenič et al. [13] also present an exact algorithm for the MRS problem that has a running time of $n^{O(1)} \cdot 2^p \cdot \Delta^{2p}$. Hüffner et al. [12] study the parameterized complexity of the MRS problem with different parameters.

We observe that the approximation ratio for the MRS problem achieved by the trivial algorithm may not be beaten using natural LP and SDP relaxations. We give an $\Omega(\sqrt{n})$ lower bound on the integrality gap for these natural relaxations (see Appendix B in the full version of this paper [19]). As the first idea towards an algorithm with an improved ratio, we define a new problem: the Densest k Colored Subgraph (DkCS) problem. The input to the DkCS problem consists of an undirected graph with each edge colored with one of p colors and a parameter k. The goal is to find a subgraph on k vertices which has the maximum number of edges with distinct colors. We show then that an f-approximation algorithm for the DkCS problem implies an $O(f \log n)$ approximation algorithm for the MRS problem (see Appendix A in the full version of this paper [19]).

Note that the well studied Densest k Subgraph (DkS) problem is a special case of the DkCS problem, in which every edge is colored with a different color. In addition to being **NP**-hard, the DkS problem has been shown not to admit a PTAS under various complexity theoretic assumptions [8,14]. The DkS problem is known to be notoriously hard to approximate. Breaking the $O(\sqrt{n})$ barrier, Feige et al. [9] gave an $O(n^{1/3-\epsilon})$ approximation algorithm, for some $\epsilon > 0$. In a remarkable paper, Bhaskara et al. [2] improve this to $O(n^{1/4+\epsilon})$, for any $\epsilon > 0$. There is a large gap between the known upper and lower bounds for the DkS problem. As evidence for the hardness of approximating the DkS problem within polynomial factors, a lower bound of $\Omega(n^{1/4}/\log^3 n)$ on the integrality gap for $\Omega(\log n/\log\log n)$ rounds of the Sherali-Adams relaxation for the DkS problem is shown in [3].

The introduction of colors (in the DkCS problem) intuitively seems to increase the difficulty. One difficulty, for instance is that exactly one edge of each color is of importance. In this paper, we give an $O(n^{1/3})$ approximation algorithm for the DkCS problem. Our algorithm builds on the one for the DkS problem in [9].

The MRS problem falls in a class of problems with the known upper bound on the approximation ratio $|I|^c$ where $|I|$ is the input size and c a constant less than one, and with the known lower bounds being smaller growing functions. Prior to a breakthrough result by Charikar et al. [5], several papers reduced the MIN-REP [15] problem (defined in Sect. 2.7) to other problems in order to obtain hardness results. It was conjectured that the MIN-REP problem has a lower bound $\Omega(\sqrt{n})$ on the approximation ratio, which was refuted by Charikar et al. [5]. They gave an LP-rounding algorithm with approximation ratio $O(n^{1/3}\log^{2/3} n)$. We observe that the MIN-REP problem is a special case of the MRS problem, and this gives a combinatorial $O(n^{1/3}\log n)$ approximation algorithm for MIN-REP. Note that an $o(n^{1/3})$ approximation algorithm for the MRS problem, implies an improved approximation ratio for the MIN-REP problem.

On the inapproximability side, a proof in [16] implies that it is quasi-**NP**-hard to approximate the MRS problem to within $\Omega(\log n)$. Kortsarz [15] showed that it is quasi-**NP**-hard to approximate the MIN-REP problem to within a factor of $2^{\log^{1-\epsilon} n}$, for any $\epsilon > 0$. The same hardness result applies to the MRS problem.

We present a randomized approximation preserving reduction from the DkS problem to the MRS problem (in Sect. 3). We also present approximation preserving reductions from the MRS problem to three problems, namely, the Red Blue Set Cover problem, the Power Dominating Set problem, and the Target Set Selection problem (See Appendix C in the full version of this paper [19] for problem definitions and reductions). We observe that there exists a PTAS for the MRS problem on planar graphs, and in general on minor free graphs (see Sect. 4 in the full version of this paper [19]).

2 An $O(n^{1/3})$ Approximation Algorithm for DkCS

Our algorithm follows the one in [9] to some extent. Some of the claims one can make in the uncolored case do not hold here and we need to overcome this. One difficulty is that exactly one edge of each color is of use in the optimal. The other difficulty is that we do not know which colors appear in the optimum and hence are "important". The basic idea in [9] is to pick the vertices in two phases. First pick a subset of vertices with a large number of edges incident on them and then to pick a subset with large number of edges incident on these and the first set.

Our algorithm A employs four different procedures, A_1, A_2, A_3, and A_4, each of which selects a dense colored subgraph. It returns the densest of the four colored subgraphs that are found.

2.1 Preliminaries

The *color degree* of a vertex is defined to be the number of distinct colors represented among edges incident on the vertex. The *average color degree* of a vertex in a set $S \subseteq V$ is the ratio of total number of distinct colors among the edges induced by the vertices in S to the size of S. For a set $S \subseteq V$ and a vertex v, the color degree of v *into* S is the color degree of v in the graph induced by $S \cup \{v\}$.

Let, for $1 \leq i \leq 4$, $A_i(G, k)$ denote the average color degree of the subgraph selected by the algorithm A_i.

One tool that we use repeatedly is the well known approximation algorithm for the unweighted maximum coverage problem. In this problem we are given a collection of subsets of a set and a positive integer k. The objective is to find k subsets which cover the maximum number of elements. This problem is known to be **NP**-hard. The greedy algorithm which chooses the subset which covers the most number of uncovered colors at each stage has an approximation ratio $1 - 1/e$, and no algorithm can do better [7].

For instance if we wish to determine k vertices which have the maximum number of edges with distinct colors incident on them, we can use the same greedy strategy and get an approximation ratio of $1 - 1/e$ on the maximum number of colors covered.

Proposition 1. *The greedy algorithm to find k vertices which have the maximum number of edges with distinct colors incident on them is* $1 - 1/e$ *approximate.*

Let the average color degree of the densest colored subgraph on k vertices in G be $d^*(G, k)$, or simply d^* when it is obvious from the context. Let G^* denote the optimum densest colored subgraph of k vertices.

2.2 Procedure A_1: A Trivial Procedure

Without loss of generality, we may assume that the graph G contains at least $k/2$ edges of distinct colors.

Procedure A_1. *Select* $k/2$ *edges of distinct colors from* G. *Return the set of vertices incident on these edges, adding arbitrary vertices to this set if its size is smaller than* k.

Clearly, $A_1(G, k) \geq 1$.

2.3 Procedure A_2: A Greedy Procedure

Our next procedure is a two step procedure. We first, greedily select a subset T of $k/2$ vertices to maximize the number of edges with distinct colors having at least on end-point in T. Later we again greedily pick $k/2$ vertices T' to maximize the number of edges with distinct colors covered by $T \cup T'$.

Procedure A_2. *Select the vertex of maximum color degree. Add it to* T. *Remove all edges of all colors incident on this vertex from* G. *Repeat this till* $|T| = k/2$. *After this is done, consider the original graph* G. *Select the vertex in* $G \setminus T$ *of maximum color degree into* T. *Add it to* T'. *Remove all edges of all colors incident on this vertex from the vertices in* $G \setminus T$. *Repeat this till* $|T'| = k/2$. *Return* $T \cup T'$.

Let $c(T)$ denote the number of distinct colors among edges incident on vertices in T. Let d_T denote the average color degree of a vertex in T. That is, $d_T = c(T)/(k/2)$.

Lemma 1. *Procedure A_2 returns a subgraph satisfying $A_2(G,k) \geq c_1 \frac{kd_T}{2n}$, for some constant $c_1 > 0$.*

Proof. Let m_1 denote the number of distinct colors among edges, both of whose endpoints lie in T. Then the number of distinct colors with one end point in T is $d_T|T| - 2m_1 = d_T k/2 - 2m_1 \geq 0$. The greedy strategy together with Proposition 1 ensures that at least $(1-\frac{1}{e})|T'|/|G \setminus T| > (1-\frac{1}{e})\frac{k}{2n}$ fraction of these distinct colored edges are contained in $T \cup T'$. Thus the total number of distinct colored edges in the subgraph induced by $T \cup T'$ is at least

$$\left(\frac{d_T k}{2} - 2m_1\right)\left(\left(1-\frac{1}{e}\right)\frac{k}{2n}\right) + m_1 \geq c_1 \frac{d_T k^2}{n}. \qquad \square$$

By Proposition 1, $d_T \geq (1-\frac{1}{e})\, d^*(G,k)$. Thus, this greedy procedure approximates $d^*(G,k)$ to within a ratio of at most $O(\frac{n}{k})$.

2.4 Colored Walks of Length 2

Our next procedure works when the color degree of vertices is small. Towards building intuition, consider G^*, the optimum densest k colored subgraph in G. In G^* every edge has a distinct color. Also, $|V(G^*)| = k$. Assume that d^* is at least $\sqrt{k}$. Then, if the graph expands (this term is used loosely), then there is a vertex v in $V(G^*)$ so that half the vertices in the optimum solution are at distance 2 from v in G^*. The idea is to look "greedily" in such neighborhoods. Suppose we knew this vertex v. (We try every vertex.) Then we first find vertices having maximum length two walks (with edges of distinct colors) to v and then vertices reachable by distinct colored edges having a large number of colored edges into the first set. Details follow.

For vertices u and v, let $W_2(u,v)$ denote the maximum number of colored length 2 walks from u to v, such that all edges in these paths are distinctly colored. Without colors, this number is easily determined. In the colored version, we need to invoke the algorithm for maximum matching. This number can be determined as follows. Any n-vertex graph can have at most n^2 distinct colors. Consider an auxillary graph G' on n^2 vertices, each vertex corresponding to one color. A length 2 walk between u and v is represented by an edge between two vertices in G' corresponding to the colors of the edges on that walk. Now, we find a maximum matching for the graph G'. Let $W_2(u,v)$ denote the set of colors corresponding to the maximum matching in G'.

Procedure A_3. *Construct a candidate graph $H(v)$ for every vertex v in G as follows. For every $w \in G$, compute $W_2(v,w)$. Select the vertex w for which $|W_2(v,w)|$ is maximum. Add it to $P(v)$. Remove all edges in G which are colored with the colors of edges incident on w that belong to $W_2(v,w)$. Repeat the procedure till $|P(v)| = k/2$. Now consider the original graph. Select a neighbor x of v which has maximum color degree into $P(v)$. Add x to $Q(v)$. Remove all edges of these colors from G. Repeat the procedure till $|Q(v)| = k/2$. Let $H(v)$ denote the subgraph induced on $P(v) \cup Q(v)$. (If $H(v)$ still contains less than k*

vertices, then it is completed to size k arbitrarily.) Among all vertices v, select the densest colored candidate graph H(v) as the output.

Let $cdeg^*(v)$ denote the color degree of v in G^*.

We now analyze the approximation ratio of this procedure. Let us first note that the number of colored length 2 walks within the optimum subgraph G^* is at least $k(d^*(G,k))^2$. This is because each $v \in G^*$ contributes $(cdeg^*(v))^2$ to this sum, and $\sum_{v \in G^*} (cdeg^*(v))^2 \geq k(d^*(G,k))^2$ by convexity.

It follows that there is a vertex v which is the endpoint of at least $(d^*(G,k))^2$ colored length-2 walks in G^*. By the greedy construction of $P(v)$ and Proposition 1, there are at least $(1 - \frac{1}{e})(d^*(G,k))^2/2$ walks of colored length 2 between this v and vertices of $P(v)$. There are at least $(1 - 1/e)(d^*(G,k))^2/2$ distinct colored edges between $Q(v)$ and $P(v)$ if $cdeg(v) \leq k/2$, and at least $(1 - \frac{1}{e})^2(d^*(G,k))^2k/4cdeg(v)$ distinct colored edges between $Q(v)$ and $P(v)$ otherwise. Since we do not require $P(v)$ and $Q(v)$ to be disjoint, each edge may have been counted twice. Hence, altogether, $H(v)$ contains at least $(1 - \frac{1}{e})^2 \min[(d^*(G,k))^2/4, (d^*(G,k))^2k/8\Delta_c(G))]$ edges, where $\Delta_c(G)$ denotes the maximum color degree in the graph.

This guarantees,

$$A_3(G,k) \geq \left(1 - \frac{1}{e}\right)^2 \frac{(d^*(G,k))^2}{2\max[k, 2\Delta_c(G))]}.$$

2.5 Procedure A_4: Another Greedy Procedure

This procedure is the key to handling colors. This complements Procedure A_3. In this procedure, we will pick a candidate subgraph with the following guarantee. Either the vertices in this subgraph will have high color degree, or the graph left after removing this subgraph has only vertices of low color degree.

So this procedure works in conjunction with Procedure A_3. In the uncolored case, a procedure like Procedure A_2 is enough to achieve this result. In this case it is tricky, and we need to get the algorithm just right. Details follow.

Procedure A_4. *Select a vertex u with maximum color degree. Add it to U. For the vertex u, arbitrarily keep only one edge of each color incident on it, and remove the rest. For every edge (u, v) colored c remove every edge colored c which is incident on v from the graph, except (u, v). Now, repeat this step till $|U| = k/2$.*

Now from the original graph (minus the edges removed in the above step), find a vertex of maximum color degree into U. Add this vertex to V. Call these colors as covered. Repeatedly choose the vertex which covers the maximum number of uncovered colors, and add it to V. Add a total of k/2 vertices to V this way. If $|U \cup V|$ is less then k, then add arbitrary vertices to it, to make it of size k.

Consider the vertices of U in the original graph. Find greedily the k/2 vertices that cover the most number of colors going into U as in Procedure A_2. Call this set V'.

Among the two subgraphs $U \cup V'$ and $U \cup V$, return the one which has more number of edges of distinct colors.

Let d_U denote the color degree of last vertex added to U.

Lemma 2. *If $k^2 \geq 4n$, then $A_4(G,k) \geq c_2 \frac{d_U}{k}$. Else, $A_4(G,k) \geq c_3 \frac{d_U k}{n}$.*

Proof. The total number of edges incident on the vertices U is at least $kd_U/2$. Let the number of colors covered after picking the ith vertex in V be f_i. Then the number of edges incident on vertices of U which are colored with one of these f_i colors is at most $kf_i/2$. This is because every vertex in U can have one edge of same color incident on it. Hence the total number of edges incident on vertices in U colored with uncovered colors is at least $kd_U/2 - kf_i/2$. Hence, the next vertex added to V will cover at least $(kd_U/2 - kf_i/2)/n$ colors. Hence, $f_{i+1} \geq f_i + (kd_U/2 - kf_i/2)/n$. By induction, the total number of colors covered in $U \cup V$ is $f_{\frac{k}{2}} \geq d_U \left(1 - (1 - \frac{k}{2n})^{\frac{k}{2}}\right)$. If $k^2 \geq 4n$, $f_{\frac{k}{2}} \geq d_U \left(1 - (\frac{1}{e})^{\frac{k^2}{4n}}\right) \geq d_U (1 - \frac{1}{e})$. Else, $f_{\frac{k}{2}} \geq d_U \left(1 - \left(1 - \binom{k/2}{1}\frac{k}{2n} + \binom{k/2}{2}\left(\frac{k}{2n}\right)^2\right)\right) \geq d_U \cdot \frac{k^2}{4n}$. □

2.6 Algorithm A

Algorithm A applies Procedures A_1, A_2, and A_4, on the graph G, and Procedure A_3 on the subgraph induced on $G_l = G \setminus U$, where U is the set of $k/2$ vertices chosen by Procedure A_4, and returns the densest colored subgraph of these.

Let an α fraction of the edges in the densest k colored subgraph be incident on the vertices of U. If $\alpha \leq \frac{1}{2}$, then clearly G_l has a densest k colored subgraph with average color degree $\Omega(d^*)$. Else, there are $d^*k/2$ edges of different colors from the optimum which are incident on the vertices of U. In A_4, we choose $k/2$ vertices having edges to U which cover the maximum number of colors (upto a constant factor). Thus, $A_4(G,k) = \Omega(d^*)$.

The performance guarantee of algorithm A is at least the geometric mean of the performance guarantee of any three of the Procedures A_1, A_2, A_3, and A_4. We look at three different cases.

1. If $k \geq d_U$, then $\Delta_c(G_l) \leq \frac{k}{2} + d_U \leq 2k$. Thus,

$$A(G,k) \geq \max\left[A_1(G,k), A_2(G,k), A_3(G_l,k)\right]$$
$$\geq \left(1 \cdot c_1 \frac{kd_T}{2n} \cdot c_4 \frac{(d^*(G,k))^2}{2\max[k, 2\Delta_c]}\right)^{1/3} \geq \frac{d^*(G,k)}{cn^{1/3}},$$

for some $c > 0$. Here the last inequality follows from the fact that $d_T \geq (1 - \frac{1}{e})d^*(G,k)$ (by Proposition 1).

For the remaining cases, we may assume $\Delta_c(G_l) \leq 2d_U$, because $d_U > k$.

2. If $k^2 \geq 4n$, then

$$A(G,k) \geq \max\left[A_2(G,k), A_4(G,k), A_3(G_l,k)\right]$$
$$\geq \left(c_1 \frac{kd_T}{2n} \cdot c_2 \frac{d_U}{k} \cdot c_4 \frac{(d^*(G,k))^2}{8d_U}\right)^{1/3} \geq \frac{d^*(G,k)}{cn^{1/3}},$$

for some $c > 0$. Here the last inequality follows from the fact that $d_T \geq (1 - \frac{1}{e})d^*(G,k)$ (by Proposition 1).

3. If $k^2 < 4n$, then

$$A(G,k) \geq \max\left[A_1(G,k), A_4(G,k), A_3(G_l,k)\right]$$
$$\geq \left(1 \cdot c_3 \frac{d_U k}{n} \cdot c_4 \frac{(d^*(G,k))^2}{8d_U}\right)^{1/3} \geq \frac{d^*(G,k)}{cn^{1/3}},$$

for some $c > 0$. Here the last inequality follows from the fact that $k \geq d^*(G,k)$.

This completes the proof for an $O(n^{1/3})$ approximation algorithm for the DkCS problem. The following theorem implies an $O(n^{1/3}\log n)$ approximation algorithm for the MRS problem.

Theorem 1. *If there is an f approximation algorithm for the DkCS problem then there is an $O(f\log n)$ approximation algorithm for the MRS problem.*

The proof of this theorem can be found in Appendix A in the full version of this paper [19].

2.7 An Algorithm for MIN-REP

The MIN-REP problem is a minimization version of the label cover problem [15]. The input consists of a bipartite graph $G = (A, B, E)$, where $|A| = |B| = n$, and equitable partitions of A and B into k sets of same size $q = n/k$. The bipartite graph and the partitions of A and B induce a "supergraph" H in the following way - The vertices of graph H are the equitable partitions of set A and B. Two vertices corresponding to sets A_i and B_j are adjacent by a "superedge" in H if and only if there exist $a_i \in A_i$ and $b_i \in B_i$ which are adjacent in G. The goal is to choose $A' \subset A$ and $B' \subset B$ such that the pairs (a,b), $a \in A'$ and $b \in B'$, cover all the superedges of H, while minimizing $|A'| + |B'|$.

Charikar et al. [5] gave an $O(n^{1/3}\log^{2/3} n)$ approximation algorithm for the MIN-REP problem using LP rounding. We observe that the MIN-REP problem is indeed a special case of the MRS problem. Consider an instance of the MIN-REP problem. Color all the edges between vertices of A_i and B_j with same color. Use different color for every pair A_i and B_j. Clearly, an f-approximation algorithm for the MRS problem implies an f-approximation algorithm for the MIN-REP problem. (Note that an LP based algorithm similar to [5] will not give a better approximation ratio for the MRS problem. This is evident from the $\Omega(\sqrt{n})$ integrality gap shown for $LP2$ in Appendix B in the full version of this paper [19].)

3 Reduction from the Densest k-Subgraph (DkS) Problem

The Densest k-Subgraph problem is a well studied problem [2,3,9]. Given a simple undirected graph, the goal is to output a subgraph on k vertices which has maximum number of edges. The best known hardness result for the DkS problem is due to Feige [8] and Khot [14], that the DkS problem does not have a PTAS. Feige proves the result assuming random 3-SAT formulas are hard to refute and Khot proves it assuming **NP** does not have randomized algorithms that run in subexponential time (i.e. that $\mathbf{NP} \not\subseteq \cap_{\epsilon>0} \mathbf{BPTIME}(2^{n^\epsilon})$). It is widely believed that the DkS problem has a lower bound on approximation ratio within a factor of n^c, for some $c > 0$.

Theorem 2. *If there is an f-approximation algorithm for the MRS problem, then there is a randomized $O(f^2 \cdot \log n)$-approximation algorithm for the DkS problem.*

Proof. We exhibit a randomized reduction from the DkS problem to the MRS problem. Consider an instance of the DkS problem. Assume that this graph G has a subgraph on k vertices with t edges such that t is maximum. Assume that t is known. This is a kosher assumption, since one can run the algorithm for each possible value of t. Color each edge of the graph independently and randomly with one of $t/(c \log t)$ colors. Let X be a random variable that denotes the number of edges needed to cover all the colors. The expected number of edges needed to cover all the colors (by the coupon collector argument) is

$$\left(\frac{t}{c \log t}\right) c' \log \left(\frac{t}{c \log t}\right) \leq \frac{t}{c''}$$

Then by Markov's inequality,

$$Pr(X > t) \leq \frac{\mathbb{E}[X]}{t} \leq \frac{1}{c''}$$

If c is chosen appropriately, then this probability will be small. This implies that there will be k vertices such that whp there is at least one edge of every color in the subgraph induced by these vertices. Suppose there exists an f approximation algorithm for the MRS problem. This algorithm will give a subgraph on at most $f \cdot k$ vertices and $t/(c \log t)$ edges. The density of this subgraph is $\left(\frac{t}{c \log t} \cdot \frac{1}{fk}\right)$. Select $k/2$ vertices of highest degree from this subgraph. Call this set U. Find $k/2$ vertices of highest degree into U from this subgraph. Call this set V. $U \cup V$ will have at least $\frac{t}{cf^2 \log t}$ edges. Thus, an f approximation algorithm for the MRS problem implies a randomized $O(f^2 \cdot \log n)$-approximation algorithm for the DkS problem. □

References

1. Alon, N., Jiang, T., Miller, Z., Pritikin, D.: Properly colored subgraphs and rainbow subgraphs in edge-colorings with local constraints. Random Struct. Algorithms **23**(4), 409–433 (2003)
2. Bhaskara, A., Charikar, M., Chlamtac, E., Feige, U., Vijayaraghavan, A.: Detecting High Log-Densities - an $O(n^{1/4})$ Approximation for Densest k-Subgraph. In: Proceedings of the Forty-second ACM Symposium on Theory of Computing, STOC 2010, pp. 201–210. ACM (2010)

3. Bhaskara, A., Charikar, M., Vijayaraghavan, A., Guruswami, V., Zhou, Y.: Polynomial integrality gaps for strong SDP relaxations of Densest k-subgraph. In: Proceedings of the Twenty-third Annual ACM-SIAM Symposium on Discrete Algorithms, SODA 2012, pp. 388–405. SIAM (2012)
4. Camacho, S.M., Schiermeyer, I., Tuza, Z.: Approximation algorithms for the minimum rainbow subgraph problem. Discrete Math. **310**(20), 2666–2670 (2010). Graph Theory - Dedicated to Carsten Thomassen on his 60th Birthday
5. Charikar, M., Hajiaghayi, M., Karloff, H.: Improved approximation algorithms for label cover problems. Algorithmica **61**(1), 190–206 (2011)
6. Erdős, P., Tuza, Z.: Rainbow subgraphs in edge-colorings of complete graphs. In: Quo Vadis, Graph Theory? A Source Book for Challenges and Directions, Annals of Discrete Mathematics, vol. 55, pp. 81–88. Elsevier (1993)
7. Feige, U.: A threshold of $\ln N$ for approximating set cover. J. ACM **45**(4), 634–652 (1998)
8. Feige, U.: Relations between average case complexity and approximation complexity. In: Proceedings of the Thiry-fourth Annual ACM Symposium on Theory of Computing, STOC 2002, pp. 534–543. ACM (2002)
9. Feige, U., Kortsarz, G., Peleg, D.: The dense k-subgraph problem. Algorithmica **29**(3), 410–421 (2001)
10. Gusfield, D.: Haplotype inference by pure parsimony. In: Baeza-Yates, R., Chávez, E., Crochemore, M. (eds.) CPM 2003. LNCS, vol. 2676, pp. 144–155. Springer, Heidelberg (2003)
11. Hahn, G., Thomassen, C.: Path and cycle sub-ramsey numbers and an edge-colouring conjecture. Discrete Math. **62**(1), 29–33 (1986)
12. Hüffner, F., Komusiewicz, C., Niedermeier, R., Rötzschke, M.: The parameterized complexity of the rainbow subgraph problem. Algorithms **8**(1), 60–81 (2015)
13. Katrenič, J., Schiermeyer, I.: Improved approximation bounds for the minimum rainbow subgraph problem. Inf. Process. Lett. **111**(3), 110–114 (2011)
14. Khot, S.: Ruling out PTAS for graph min-bisection, dense k-subgraph, and bipartite clique. SIAM J. Comput. **36**(4), 1025–1071 (2006)
15. Kortsarz, G.: On the hardness of approximating spanners. Algorithmica **30**, 2001 (1999)
16. Popa, A.: Approximating the rainbow – better lower and upper bounds. In: Gudmundsson, J., Mestre, J., Viglas, T. (eds.) COCOON 2012. LNCS, vol. 7434, pp. 193–203. Springer, Heidelberg (2012)
17. Rödl, V., Tuza, Z.: Rainbow subgraphs in properly edge-colored graphs. Random Struct. Algorithms, pp. 175–182 (1992)
18. Simonovits, M., Sós, V.T.: On restricted colourings of K_n. Combinatorica **4**(1), 101–110 (1984)
19. Tirodkar, S., Vishwanathan, S.: On the approximability of the minimum rainbow subgraph problem and other related problems (2015). https://www.cse.iitb.ac.in/internal/techreports/reports/TR-CSE-2015-75.pdf

General Caching Is Hard: Even with Small Pages

Lukáš Folwarczný(✉) and Jiří Sgall

Computer Science Institute of Charles University, Prague, Czech Republic
{folwar,sgall}@iuuk.mff.cuni.cz

Abstract. *Caching* (also known as *paging*) is a classical problem concerning page replacement policies in two-level memory systems. *General caching* is the variant with pages of different sizes and fault costs. The strong NP-hardness of its two important cases, the *fault model* (each page has unit fault cost) and the *bit model* (each page has the same fault cost as size) has been established. We prove that this already holds when page sizes are bounded by a small constant: The bit and fault models are strongly NP-complete even when page sizes are limited to $\{1, 2, 3\}$.

Considering only the decision versions of the problems, general caching is equivalent to the *unsplittable flow on a path problem* and therefore our results also improve the hardness results about this problem.

Keywords: General caching · Small pages · NP-hardness · Unsplittable flow on a path

1 Introduction

Caching (also known as *uniform caching* or *paging*) is a classical problem in the area of online algorithms and has been extensively studied since 1960s. It models a two-level memory system: There is the fast memory of size C (the *cache*) and a slow but large main memory where all data reside. The problem instance comprises a sequence of requests, each demanding a page from the main memory. No cost is incurred if the requested page is present in the cache (a *cache hit*). If the requested page is not present in the cache (a *cache fault*), the page must be loaded at the fault cost of one; some page must be evicted to make space for the new one when there are already C pages in the cache. The natural objective is to evict pages in such a way that the total fault cost is minimized. For a reference on classical results, see Borodin and El-Yaniv [7].

In 1990s, with the advent of World Wide Web, a generalized variant called *file caching* or simply *general caching* was studied [11,12]. In this setting, each page p has its $\text{SIZE}(p)$ and $\text{COST}(p)$. It costs $\text{COST}(p)$ to load this page into the cache and the page occupies $\text{SIZE}(p)$ units of memory there. Uniform caching is the special case satisfying $\text{SIZE}(p) = \text{COST}(p) = 1$ for every page p. Other important cases of this general model are

- the *cost model* (*weighted caching*): $\text{SIZE}(p) = 1$ for every page p;

A full version is available on arXiv. http://arxiv.org/abs/1506.07905.

K. Elbassioni and K. Makino (Eds.): ISAAC 2015, LNCS 9472, pp. 116–126, 2015.
DOI: 10.1007/978-3-662-48971-0_11

– the *bit model*: $\text{COST}(p) = \text{SIZE}(p)$ for every page p;
– the *fault model*: $\text{COST}(p) = 1$ for every page p.

Caching, as described so far, requires the service to load the requested page when a fault occurs, which is known as caching under the *forced policy.* Allowing the service to pay the fault cost without actually loading the requested page to the cache gives another useful and studied variant of caching, the *optional policy.*

Previous Work. In this article, we consider the problem of finding the optimal service in the *offline version* of caching when the whole request sequence is known in advance. Uniform caching is solvable in polynomial time with a natural algorithm known as Belady's rule [5]. Caching in the cost model is a special case of the *k-server problem* and is also solvable in polynomial time [8]. In late 1990s, the questions about the complexity status of general caching were raised. The situation was summed up by Albers et al. [1]: *"The hardness results for caching problems are very inconclusive. The* NP*-hardness result for the Bit model uses a reduction from* PARTITION, *which has pseudopolynomial algorithms. Thus a similar algorithm may well exist for the Bit model. We do not know whether computing the optimum in the Fault model is* NP*-hard."*

There was no improvement until a breakthrough in 2010 when Chrobak et al. [9] showed that general caching is strongly NP-hard, already in the case of the fault model as well as in the case of the bit model. General caching is usually studied under the assumption that the largest page size is very small in comparison with the total cache size, as is for example the case of the aforementioned article by Albers et al. [1]. Instances of caching with pages larger than half of the cache size (so called obstacles) are required in the proof given by Chrobak et al. Therefore, this hardness result is in fact still quite inconclusive.

Independently, a weak NP-hardness of the fault model was proven by Darmann et al. [10]. A substantially simpler proof of the strong NP-hardness of general caching was given by Bonsma et al. [6]; only pages of sizes in $\{1, 2, 3\}$ are needed in the proof, but they have many different costs: this means that the costs are far from the fault model and in fact they are also far from the bit model.

In the decision version, offline general caching is equivalent (together with interval scheduling/packing, resource allocation and other problems) to the *unsplittable flow on a path problem* (UFPP). An important parameter in the world of UFPP is *task density*, in the language of caching it is the fault cost divided by the page size. An approximation scheme with quasi-polynomial time complexity when the ratio of the maximum density to the minimum density is quasi-polynomial was given by Bansal et al. [2]. The first non-trivial instance of UFPP for which a PTAS was invented by Batra et al. [4] is the one when the task densities are in a constant range. The bit model of caching is equivalent to the case when all task densities are equal to one.

Our Contribution. We give a novel proof of strong NP-hardness for general caching which gives the first hardness result restricted to small pages in the fault and cost models:

Theorem 1.1. *General caching is strongly* NP*-hard even in the case when the page sizes are limited to* {*1, 2, 3*}*, for both the fault model and the bit model, and under each of the forced and optional policies.*

The proof of the result for general costs (and sizes $\{1, 2, 3\}$) is rather simple, in particular significantly simpler than the one given by Chrobak et al. [9] and at the same time it uses only two different costs, which is simpler compared to Bonsma et al. [6]. The reductions for the result in the fault and bit models are significantly more involved and require a non-trivial potential-function-like argument.

Open Problems. For the decision version, closing the remaining gap with small pages seems challenging and could provide new useful insights: *Is general caching also (strongly)* NP*-hard when page sizes are limited to* {*1, 2*}*? Can caching with page sizes* {*1, 2*} *be solved in polynomial time, at least in the bit or fault model?*

The currently best approximation algorithm is a 4-approximation by Bar-Noy et al. [3], while there is no hardness of approximation result. A better understanding of approximability of general caching remains a challenge.

Outline. Our main result – a polynomial-time reduction from independent set to caching in the fault model under the optional policy with page sizes restricted to {1, 2, 3} – is explained in Sect. 2 and its validity is proven in Sect. 3.

The remaining proofs are omitted; they are given in the full version available as arXiv:1506.07905. There, in section "Bit Model", we show how to modify the reduction so that it works for the bit model as well. In section "Forced Policy", we show how to obtain the hardness results also for the forced policy. Finally, in the appendix, we give a self-contained presentation of the simple proof of strong NP-hardness for general costs.

2 Reduction

The decision problem IndependentSet is well-known to be NP-complete. By 3Caching(forced) and 3Caching(optional) we denote the decision versions of caching under each policy with page sizes restricted to $\{1, 2, 3\}$.

Problem:	IndependentSet
Instance:	A graph G and a number K.
Question:	Is there an independent set of cardinality K in G?

Problem:	3Caching(*policy*)
Instance:	A universe of pages, a sequence of page requests, numbers C and L. For each page p it holds $\text{SIZE}(p) \in \{1, 2, 3\}$.
Question:	Is there a service under the policy *policy* of the request sequence using the cache of size C with a total fault cost of at most L?

We define 3Caching(Fault,*policy*) to be the problem 3Caching(*policy*) with the additional requirement that page costs adhere to the fault model. The problem 3Caching(bit,*policy*) is defined analogously.

In this section, we describe a polynomial-time reduction from INDEPENDENTSET to 3CACHING(FAULT,OPTIONAL). Informally, a set of pages of size two and three is associated with each edge and a page of size one is associated with each vertex. Each vertex-page is requested only twice while there are many requests on pages associated with edges. The request sequence is designed in such a way that the number of vertex-pages that are cached between the two requests in the optimal service is equal to the size of the maximum independent set.

We now show the request sequence of caching corresponding to the graph given in INDEPENDENTSET with a parameter H. In the next section, we prove that it is possible to set a proper value of H and a proper fault cost limit L such that the reduction becomes a valid polynomial-time reduction.

Reduction 2.1. Let $G = (V, E)$ be the instance of INDEPENDENTSET. The graph G has n vertices and m edges and there is an arbitrary fixed order of edges $e_1, \ldots, e_m$. Let H be a parameter bounded by a polynomial function of n.

A corresponding instance $\mathcal{I}_G$ of 3CACHING(FAULT,OPTIONAL) is an instance with the cache size $C = 2mH+1$ and the total of $6mH+n$ pages. The structure of the pages and the requests sequence is described below.

Pages. For each vertex v, we have a vertex-page p_v of size one. For each edge e, there are $6H$ edge-pages associated with it that are divided into H groups. The ith group consists of six pages $\bar{a}_i^e, \alpha_i^e, a_i^e, b_i^e, \beta_i^e, \bar{b}_i^e$ where pages α_i^e and β_i^e have size three and the remaining four pages have size two.

For a fixed edge e, let $\bar{a}^e$-pages be all pages $\bar{a}_i^e$ for $i = 1, \ldots, H$. Let also $\bar{a}$-pages be all $\bar{a}^e$-pages for $e = e_1, \ldots, e_m$. The remaining collections of pages (α^e-pages, α-pages, ...) are defined in a similar fashion.

Request Sequence. The request sequence of $\mathcal{I}_G$ is organized in phases and blocks. There is one phase for each vertex $v \in V$, we call such a phase the v-phase. There are exactly two requests on each vertex-page p_v, one just before the beginning of the v-phase and one just after the end of the v-phase; these requests do not belong to any phase. The order of phases is arbitrary. In each v-phase, there are $2H$ adjacent blocks associated with every edge e incident with v; the blocks for different incident edges are ordered arbitrarily. In addition, there is one initial block I before all phases and one final block F after all phases. Altogether, there are $d = 4mH + 2$ blocks.

Let $e = \{u, v\}$ be an edge, let us assume that the u-phase precedes the v-phase. The blocks associated with e in the u-phase are denoted by $B_{1,1}^e$, $B_{1,2}^e, \ldots, B_{i,1}^e, B_{i,2}^e, \ldots, B_{H,1}^e, B_{H,2}^e$, in this order, and the blocks in the v-phase are denoted by $B_{1,3}^e, B_{1,4}^e, \ldots, B_{i,3}^e, B_{i,4}^e, \ldots, B_{H,3}^e, B_{H,4}^e$, in this order. An example is given in Fig. 1.

Even though each block is associated with some fixed edge, it contains one or more requests to the associated pages for every edge e. In each block, we process the edges in the order $e_1, \ldots, e_m$ that was fixed above. Pages associated with the edge e are requested in two rounds. In each round, we process groups $1, \ldots, H$ in this order. When processing the ith group of the edge e, we request one or

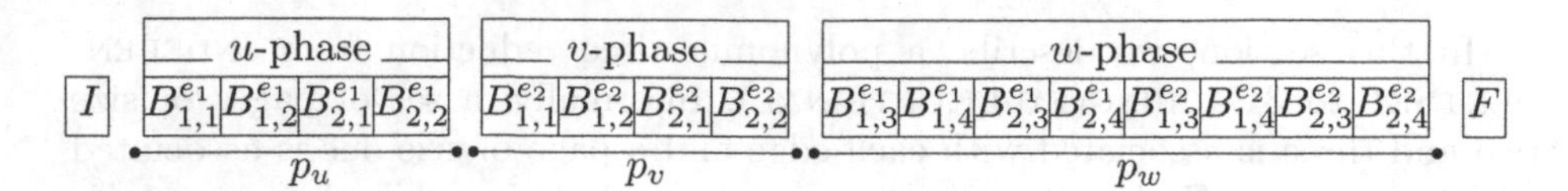

Fig. 1. An example of phases, blocks and requests on vertex-pages for a graph with three vertices u, v, w and two edges $e_1 = \{u, w\}$, $e_2 = \{v, w\}$ when $H = 2$

more pages of this group, depending on the block we are in. The following table determines which pages are requested.

Block	First round	•	Second round
before $B^e_{i,1}$	$\bar{a}^e_i$	•	
$B^e_{i,1}$	$\bar{a}^e_i$, α^e_i	•	b^e_i
$B^e_{i,2}$	α^e_i, a^e_i	•	b^e_i
between $B^e_{i,2}$ and $B^e_{i,3}$	a^e_i	•	b^e_i
$B^e_{i,3}$	a^e_i	•	b^e_i, β^e_i
$B^e_{i,4}$	a^e_i	•	β^e_i, $\bar{b}^e_i$
after $B^e_{i,4}$			$\bar{b}^e_i$

Reduction 2.1 is now complete. An example of requests on edge-pages associated with one edge e is depicted in a figure in the full version. Notice that the order of the pages associated with e is the same in all blocks; more precisely, in each block the requests on the pages associated with e form a subsequence of

$$\bar{a}^e_1\,\alpha^e_1\,a^e_1\,\ldots\,\bar{a}^e_i\,\alpha^e_i\,a^e_i\,\ldots\,\bar{a}^e_H\,\alpha^e_H\,a^e_H\;b^e_1\,\beta^e_1\,\bar{b}^e_1\,\ldots\,b^e_i\,\beta^e_i\,\bar{b}^e_i\,\ldots\,b^e_H\,\beta^e_H\,\bar{b}^e_H. \quad (1)$$

Preliminaries for the Proof. Instead of minimizing the service cost, we maximize the savings compared to the service which does not use the cache at all. This is clearly equivalent when considering the decision versions of the problems.

Without loss of generality, we assume that any page is brought into the cache only immediately before some request to that page and removed from the cache only after some (possibly different) request to that page; furthermore, the cache is empty at the beginning and at the end. That is, a page may be in the cache only between two consecutive requests to this page, and either it is in the cache for the whole interval or not at all.

Each page of size three is requested only twice in two consecutive blocks, and these blocks are distinct for all pages of size three. Thus, a service of edge-pages is valid if and only if at each time, at most mH edge-pages are in the cache. It is convenient to think of the cache as of mH *slots* for edge-pages.

As each vertex-page is requested twice, the savings on the n vertex-pages are at most n. Furthermore, a vertex-page can be cached if and only if during the phase it never happens that at the same time all slots for edge-pages are full and a page of size three is cached.

Let S_B denote the set of all edge-pages cached at the beginning of the block B and let S_B^e be the set of pages in S_B associated with the edge e. We use $s_B = |S_B|$ and $s_B^e = |S_B^e|$ for the sizes of the sets. Each edge-page is requested only in a contiguous segment of blocks, once in each block. It follows that the total savings on edge-pages are equal to $\sum_B s_B$ where the sum is over all blocks. In particular, the maximal possible savings on the edge-pages are $(d-1)mH$, using the fact that S_I is empty. We shall show that the maximum savings are $(d-1)mH + K$ where K is the size of the maximum independent set in G.

Almost-Fault Model. To understand the reduction, we consider what happens if we relax the requirements of the fault model and set the cost of each vertex-page to $1/(n+1)$ instead of 1 as required by the fault model.

In this scenario, the total savings on vertex-pages are $n/(n+1) < 1$ which is less than savings incurred by one edge-page. Therefore, edge-pages must be served optimally in the optimal service of the whole request sequence.

In this case, the reduction works already for $H = 1$. This leads to a quite short proof of the strong NP-hardness for general caching and we give this proof in an appendix of the full version. Now we show the main ideas that are important also for the design of our caching instance in the fault and bit models.

We first prove that for each edge e and each block $B \neq I$ we have $s_B^e = 1$ (see the appendix). Using this we show below that for each edge e, at least one of the pages α_1^e and β_1^e is cached between its two requests. This implies that the set of all vertices v such that p_v is cached between its two requests is independent.

For a contradiction, let us assume that for some edge e, neither of the pages α_1^e and β_1^e is cached between its two requests. Because pages α_1^e and β_1^e are forbidden, there is b_1^e in $S_{B_{1,2}^e}$ and a_1^e in $S_{B_{1,3}^e}$. Somewhere between these two blocks $B_{1,2}^e$ and $B_{1,3}^e$, we must switch from caching b_1^e to caching a_1^e. However, this is impossible, because the order of requests implies that we would have to cache both b_1^e and a_1^e at some moment (see Fig. 2). However, there is no place in the cache for such an operation, as $s_B^{e'} = 1$ for every e' and $B \neq I$.

In the fault model, the corresponding claim $s_B^e = H$ does not hold. Instead, we prove that the value of s_B^e cannot change much during the service and when we use H large enough, we still get a working reduction.

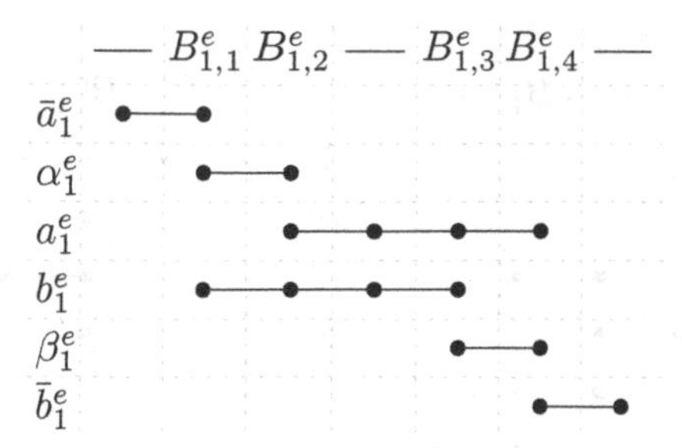

Fig. 2. Pages associated with one edge when $H = 1$

3 Proof of Correctness

In this section, we show that the reduction described in the previous section is indeed a reduction from IndependentSet set to 3Caching(fault,optional). We prove that there is an independent set of cardinality K in G if and only if there is a service of the caching instance $\mathcal{I}_G$ with the total savings of at least $(d-1)mH+K$. First the easy direction, which holds for any value of the parameter H.

Lemma 3.1. *Let G be a graph and $\mathcal{I}_G$ the corresponding caching instance from Reduction 2.1. Suppose that there is an independent set W of cardinality K in G. Then there exists a service of $\mathcal{I}_G$ with the total savings of at least $(d-1)mH+K$.*

Proof. For any edge e, denote $e=\{u,v\}$ so that the u-phase precedes the v-phase. If $u \in W$, we keep all $\bar{a}^e$-pages, b^e-pages, β^e-pages and $\bar{b}^e$-pages in the cache from the first to the last request on each page, but we do not cache a^e-pages and α^e-pages at any time. Otherwise, we cache all $\bar{a}^e$-pages, α^e-pages, a^e-pages and $\bar{b}^e$-pages, but do not cache b^e-pages and β^e-pages at any time. Figure 3 shows these two cases for the first group of pages. In both cases, at each time at most one page associated with each group of each edge is in the cache and the savings on those pages are $(d-1)mH$. We know that the pages fit in the cache because of the observations made in Sect. 2.

For any $v \in W$, we cache p_v between its two requests. To check that this is a valid service, observe that if $v \in W$, then during the corresponding phase no page of size three is cached. Thus, the page p_v always fits in the cache together with at most mH pages of size two. □

We prove the converse in a sequence of lemmata. In section "Bit Model" of the full version we will show how to reuse the proof for the bit model. To be able to do that, we list explicitly all the assumptions about the caching instance that are used in the following proofs.

Properties 3.2. Let $\mathcal{T}_G$ be an instance of general caching corresponding to a graph $G=(V,E)$ with n vertices, m edges $e_1,\ldots,e_m$, the same cache size and the same universe of pages as in Reduction 2.1. The request sequence is again

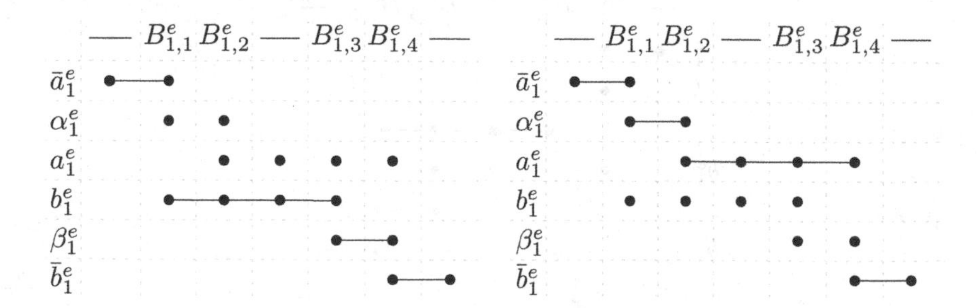

Fig. 3. The two ways of caching in Lemma 3.1

split into phases, one phase for each vertex. Each phase is again partitioned into blocks, there is one initial block I before all phases and one final block F after all phases. There is the total of d blocks.

The instance $\mathcal{T}_G$ is required to fulfill the following list of properties:

(a) Each vertex page p_v is requested exactly twice, right before the v-phase and right after the v-phase.
(b) The total savings incurred on edge-pages are equal to $\sum s_B$ (summing over all blocks).
(c) For each edge e, there are exactly H pages associated with e requested in I, all the $\bar{a}^e$-pages, and exactly H pages associated with e requested in F, all the $\bar{b}^e$-pages.
(d) In each block, pages associated with e_1 are requested first, then pages associated with e_2 are requested and so on up to e_m.
(e) For each block B and each edge e, all requests on a^e-pages and $\bar{b}^e$-pages in B precede all requests on $\bar{a}^e$-pages and b^e-pages in B.
(f) Let $e = \{u, v\}$ be an edge and p an α^e-page or β^e-page. Let B be the first block and $\overline{B}$ the last block where p is requested. Then B and $\overline{B}$ are either both in the u-phase or both in the v-phase. Furthermore, no other page of size three associated with e is requested in B, $\overline{B}$, or any block between them.

Lemma 3.3. *The instance from Reduction 2.1 satisfies Properties 3.2.*

Proof. All properties (a), (b), (c), (d), (f) follow directly from Reduction 2.1 and the subsequent observations. To prove (e), recall that the pages associated with an edge e requested in a particular block always follow the ordering (1). We need to verify that when the page a_i^e is requested, no page $\bar{a}_j^e$ for $j \leq i$ is requested and that when the page $\bar{b}_i^e$ is requested, no $\bar{a}^e$-page and no page b_j^e for $j \leq i$ is requested. This can be seen easily when we explicitly write down the request sequences for each kind of block. □

For the following claims, let $\mathcal{T}_G$ be an instance fulfilling Properties 3.2. We fix a service of $\mathcal{T}_G$ with the total savings of at least $(d-1)mH$.

Let $\mathcal{B}$ be the set of all blocks and $\overline{\mathcal{B}}$ the set of all blocks except for the initial and final one. For a block B, we denote the block immediately following it by B'.

We define two useful values characterizing the service for the block B: $\delta_B = mH - s_B$ (the number of free slots for edge-pages at the start of the service of the block) and $\gamma_B^e = |s_{B'}^e - s_B^e|$ (the change of the number of slots occupied by pages associated with e after requests from this block are served).

The first easy lemma says that only a small number of blocks can start with some free slots in the cache.

Lemma 3.4. *When summing over all blocks except for the initial one*

$$\sum_{B \in \mathcal{B} \setminus \{I\}} \delta_B \leq n.$$

Proof. Using the property (b) and $s_I = 0$, the savings on edge-pages are

$$\sum_{B \in \mathcal{B} \setminus \{I\}} s_B = (d-1)mH - \sum_{B \in \mathcal{B} \setminus \{I\}} \delta_B.$$

The total savings are assumed to be at least $(d-1)mH$. Due to the property (a), the savings on vertex-pages are at most n. Claim of the lemma follows. □

The second lemma states that the number of slots occupied by pages associated with a given edge does not change much during the whole service. The proof is postponed to the full version.

Lemma 3.5. *For each edge $e \in E$,*

$$\sum_{B \in \overline{\mathcal{B}}} \gamma_B^e \leq 6n.$$

For the rest of the proof, we set $H = 6mn + 3n + 1$. This enables us to show that the fixed service must cache some of the pages of size three.

Lemma 3.6. *For each edge $e \in E$, there is a block B such that some α^e-page or β^e-page is in S_B and $\delta_B = 0$.*

Proof. Fix an edge $e = e_k$. For each block B, we define

$$\varepsilon_B = \text{number of } \alpha^e\text{-pages and } \beta^e\text{-pages in } S_B.$$

Observe that due to the property (f), ε_B is always one or zero. We use a potential function

$$\Phi_B = \text{number of } a^e\text{-pages and } \bar{b}^e\text{-pages in } S_B.$$

Because there are only $\bar{a}$-pages in the initial block and only $\bar{b}$-pages in the final block (property (c)), we know

$$\Phi_{I'} = 0 \quad \text{and} \quad \Phi_F \geq H - \delta_F. \tag{2}$$

Now we bound the increase of the potential function as

$$\Phi_{B'} - \Phi_B \leq \delta_B + \sum_{\ell=1}^{k-1} \gamma_B^{e_\ell} + \varepsilon_B. \tag{3}$$

To justify this bound, we fix a block B and look at the cache state after requests on edges $e_1, \ldots, e_{k-1}$ are processed. How many free slots there can be in the cache? There are initial δ_B free slots in the beginning of the block B, and the number of free slots can be further increased when the number of pages in the cache associated with $e_1, \ldots, e_{k-1}$ decreases. This increase can be naturally bounded by $\sum_{\ell=1}^{k-1} \gamma_B^{e_\ell}$. Therefore, the number of free slots in the cache is at most $\delta_B + \sum_{\ell=1}^{k-1} \gamma_B^{e_\ell}$.

Because of the property (e), the number of cached a^e-pages and $\bar{b}^e$-pages can only increase by using the free cache space or caching new pages instead of α^e-pages and β^e-pages. We already bounded the number of free slots and ε_B is a natural bound for the increase gained on α^e-pages and β^e-pages. Thus, the bound (3) is correct.

Summing (3) over all $B \in \overline{\mathcal{B}}$, we have

$$\Phi_F - \Phi_{I'} = \sum_{B \in \overline{\mathcal{B}}} (\Phi_{B'} - \Phi_B) \leq \sum_{B \in \overline{\mathcal{B}}} \left(\delta_B + \sum_{\ell=1}^{k-1} \gamma_B^{e_\ell} + \varepsilon_B \right)$$

which we combine with (2) into

$$H - \delta_F \leq \sum_{B \in \overline{\mathcal{B}}} \left(\delta_B + \sum_{\ell=1}^{k-1} \gamma_B^{e_\ell} + \varepsilon_B \right),$$

and use Lemmata 3.4 and 3.5 to bound $\sum \varepsilon_B$ as

$$\begin{aligned} \sum_{B \in \overline{\mathcal{B}}} \varepsilon_B &\geq H - \delta_F - \sum_{B \in \overline{\mathcal{B}}} \left(\delta_B + \sum_{\ell=1}^{k-1} \gamma_B^{e_\ell} \right) \geq H - n - n - (k-1)6n \\ &\geq H - 6mn - 2n = n + 1. \end{aligned}$$

As there is at most one page of size three requested in each block (property (f)), the inequality $\sum \varepsilon_B \geq n+1$ implies that there are at least $n+1$ blocks where an α^e-page or a β^e-page is cached. At most n blocks have δ_B non-zero (Lemma 3.4); we are done. □

Lemma 3.7. *Suppose that there exists a service of $\mathcal{T}_G$ with the total savings of at least $(d-1)mH + K$. Then the graph G has an independent set W of cardinality K.*

Proof. Let W be a set of K vertices such that the corresponding page p_v is cached between its two requests. (There are at least K of them because the maximal savings on edge-pages are $(d-1)mH$.)

Consider an arbitrary edge $e = \{u, v\}$. Due to Lemma 3.6, there exists a block B such that $\delta_B = 0$ and some α^e-page or β^e-page is cached in the beginning of the block. This block B is either in the u-phase or in the v-phase, because of the statement of the property (f). This means that at least one of the two pages p_u and p_v is not cached between its two requests, because the cache is full. As a consequence, the set W is indeed independent. □

The value of H was set to $6mn + 3n + 1$, therefore Reduction 2.1 is indeed polynomial. Lemmata 3.1, 3.3 and 3.7 together imply that there is an independent set of cardinality K in G if and only if there is a service of the instance $\mathcal{I}_G$ with the total savings of at least $(d-1)mH + K$. We showed that the problem 3Caching(fault,optional) is indeed strongly NP-hard.

Acknowledgments. Partially supported by the Center of Excellence – ITI, project P202/12/G061 of GA ČR (J. Sgall) and by the project 14-10003S of GA ČR (L. Folwarczný). We are grateful to the anonymous reviewers, in particular for bringing the articles [6,10] to our attention.

References

1. Albers, S., Arora, S., Khanna, S.: Page replacement for general caching problems. In: Proceedings of the 10th Annual ACM-SIAM Symposium on Discrete Algorithms, pp. 31–40 (1999). http://dl.acm.org/citation.cfm?id=314500.314528
2. Bansal, N., Chakrabarti, A., Epstein, A., Schieber, B.: A quasi-PTAS for unsplittable flow on line graphs. In: Kleinberg, J.M. (ed.) Proceedings of the 38th Annual ACM Symposium on Theory of Computing, pp. 721–729. ACM (2006). http://doi.acm.org/10.1145/1132516.1132617
3. Bar-Noy, A., Bar-Yehuda, R., Freund, A., Naor, J., Schieber, B.: A unified approach to approximating resource allocation and scheduling. J. ACM **48**(5), 1069–1090 (2001). http://doi.acm.org/10.1145/502102.502107. A preliminary version appeared at STOC 2000
4. Batra, J., Garg, N., Kumar, A., Mömke, T., Wiese, A.: New approximation schemes for unsplittable flow on a path. In: Indyk, P. (ed.) Proceedings of the 26th Annual ACM-SIAM Symposium on Discrete Algorithms, pp. 47–58. SIAM (2015). http://dx.doi.org/10.1137/1.9781611973730.5
5. Belady, L.A.: A study of replacement algorithms for a virtual-storage computer. IBM Syst. J. **5**(2), 78–101 (1966). http://dx.doi.org/10.1147/sj.52.0078
6. Bonsma, P.S., Schulz, J., Wiese, A.: A constant-factor approximation algorithm for unsplittable flow on paths. SIAM J. Comput. **43**(2), 767–799 (2014). http://dx.doi.org/10.1137/120868360
7. Borodin, A., El-Yaniv, R.: Online Computation and Competitive Analysis. Cambridge University Press, New York (1998)
8. Chrobak, M., Karloff, H.J., Payne, T.H., Vishwanathan, S.: New results on server problems. SIAM J. Discrete Math. **4**(2), 172–181 (1991). http://dx.doi.org/10.1137/0404017. A preliminary version appeared at SODA 1990
9. Chrobak, M., Woeginger, G.J., Makino, K., Xu, H.: Caching is hard - even in the fault model. Algorithmica **63**(4), 781–794 (2012). http://dx.doi.org/10.1007/s00453-011-9502-9. A preliminary version appeared at ESA 2010
10. Darmann, A., Pferschy, U., Schauer, J.: Resource allocation with time intervals. Theoret. Comput. Sci. **411**(49), 4217–4234 (2010). http://dx.doi.org/10.1016/j.tcs.2010.08.028
11. Irani, S.: Page replacement with multi-size pages and applications to web caching. Algorithmica **33**(3), 384–409 (2002). http://dx.doi.org/10.1007/s00453-001-0125-4. A preliminary version appeared at STOC 1997
12. Young, N.E.: On-line file caching. Algorithmica **33**(3), 371–383 (2002). http://dx.doi.org/10.1007/s00453-001-0124-5. A preliminary version at SODA 1998

Randomized Algorithms I

The Secretary Problem with a Choice Function

Yasushi Kawase(✉)

Tokyo Institute of Technology, Tokyo, Japan
kawase.y.ab@m.titech.ac.jp

Abstract. In the classical secretary problem, a decision-maker is willing to hire the best secretary out of n applicants that arrive in a random order, and the goal is to maximize the probability of choosing the best applicant. In this paper, we introduce the secretary problem with a choice function. The choice function represents the preference of the decision-maker. In this problem, the decision-maker hires some applicants, and the goal is to maximize the probability of choosing the best set of applicants defined by the choice function. We see that the secretary problem with a path-independent choice function generalizes secretary version of the stable matching problem, the maximum weight bipartite matching problem, and the maximum weight base problem in a matroid. When the choice function is path-independent, we provide an algorithm that succeeds with probability at least $1/e^k$ where k is the maximum size of the choice, and prove that this is the best possible. Moreover, for the non-path-independent case, we prove that the success probability goes to arbitrary small for any algorithm even if the maximum size of the choice is 2.

1 Introduction

In the classical secretary problem, a decision-maker is willing to hire the best secretary out of n applicants that arrive in a random order, and the goal is to maximize the probability of choosing the best applicant. We call the probability *success probability*. As each applicant appears, it must be either selected or rejected, and the decision is irrevocable. It is assumed that the decision must be based only on the relative ranks of the applicants seen so far and the number of applicants n. It is well known that one can succeed with the optimal probability $1/e$ by the following algorithm: observe the first n/e applicants without selecting, and then select the next applicant who is the best among the observed applicants [8,15,21].

In this paper, we introduce the secretary problem with a choice function. In this model, the decision-maker may choose more than one applicant. As each applicant appears, it must be either selected or rejected, and the decision is irrevocable. It is assumed that the decision must be based only on the preference on the subsets of applicants seen so far and the number of applicants n. We assume that the preference of the decision-maker is represented by a *choice function*.

K. Elbassioni and K. Makino (Eds.): ISAAC 2015, LNCS 9472, pp. 129–139, 2015.
DOI: 10.1007/978-3-662-48971-0_12

A function $Ch : 2^S \to 2^S$ is called a choice function if $Ch(S') \subseteq S'$ holds for any $S' \subseteq S$. A choice function is called *path-independent* if it satisfies

$$Ch(Ch(X) \cup Y) = Ch(X \cup Y) \quad (X, Y \subseteq S).$$

A choice function Ch is said to be *consistent* (*irrelevance of rejected contracts*) if

$$Ch(Y) \subseteq X \subseteq Y \Longrightarrow Ch(X) = Ch(Y),$$

and *substitutable* (*comonotone*) if

$$X \subseteq Y \Longrightarrow Ch(Y) \cap X \subseteq Ch(X),$$

and *size-monotone* (*law of aggregate demand*) if

$$X \subseteq Y \Longrightarrow |Ch(X)| \leq |Ch(Y)|.$$

Aizerman and Malishevski [1] noted that a choice function is path-independent if and only if it satisfies the substitutability and the consistency. Moreover, if a choice function is substitutable and size-monotone, then it is consistent and path-independent. We define *width* of a choice function $Ch : 2^S \to 2^S$ as $\max_{S' \subseteq S} |Ch(S')|$.

Moreover, path-independent choice functions appear in the stable matching problem, the maximum weight bipartite matching problem, and the maximum weight base problem for a matroid, as shown in Sect. 2.

Related Work. Various variants of the secretary problem have been studied over several decades. An important generalization of the secretary problem is *the k-choice secretary problem.* In this setting, the decision-maker is allowed to accept k candidates. For an overview of the secretary problem, see the surveys [10,11,29].

Nikolaev [26] and Tamaki [30] considered the situation that the decision-maker is allowed to have two choices, and he must choose both the best and the second best out of n applicants. They proved that the asymptotic optimal probability is 0.2254 for the problem. Vanderbei [31] provided an asymptotic result for the case that k choices to select all of the k best candidates.

Buchbinder, Jain, and Singh [5] introduced *the J-choice, K-best secretary problem*, referred to as the (J, K)-secretary problem. Now, the decision-maker is allowed to accept J applicants, and the objective is to maximize the expected number of applicants selected among the best K. They formulated a linear programming (LP) that describe the optimal value for the problem. Chan, Chen, and Jiang [6] extended the LP to a continuous LP and analyze asymptotic behavior.

Recently, the secretary problem is generalized to online problems in the random order model. Now, we assume that each applicant has a value. The goal is to maximize the profit, i.e., the sum of the values of selected applicants. The quality of an online algorithm is usually measured by the *competitive ratio*, which is the ratio between the expected profit of the solution obtained by the online algorithm over the random arrival order and the optimal profit. Kleinberg [18] presented $\sqrt{k}/(\sqrt{k}-5)$ competitive algorithm for the sum of the value of k selected applicants. Babaioff, Immorlica, and Kleinberg [4] introduced the matroid secretary problem. In this model, the set of selected applicants must be an independent set in an underlying matroid. They conjectured that the matroid secretary problem is O(1)-competitive. Lachish [20] provided $O(\log\log r)$-competitive algorithm where r is the rank of given matroid, and Feldman, Svensson, and Zenklusen [9] presented a simpler algorithm with the same competitive ratio. Korula and Pál [19] presented 8-competitive algorithm for bipartite matching. Babaioff, Immorlica, Kempe, and Kleinberg [2] introduced *knapsack secretary problem* and proposed $10e$-competitive algorithm. For more details, see Dinitz [7] and Babaioff, Immorlica, Kempe, and Kleinberg [3].

Main Results Obtained in this Paper. In this paper, we discuss the optimal success probability of secretary problem with a choice function.

It is not difficult to see that selecting the best secretaries with size k is harder than simultaneously succeeds k classical secretary problems. Hence, the following theorem holds (the precise proof is given in Sect. 4).

Theorem 1. *For any $\varepsilon > 0$, there exists a path-independent and size-monotone instance (S, Ch) such that no algorithm succeeds with probability $1/e^{|Ch(S)|} + \varepsilon$.*

On the other hand, we can attain the probability if the given choice function satisfies the path-independency. Our algorithm is a simple threshold based one: observe the first n/e applicants without selecting, and then select if the next applicant is contained in the best set of applicants in the applicants seen so far.

Theorem 2. *There exists an algorithm that succeeds with probability $1/e^k$ for any path-independent instance (S, Ch), where k is the width of Ch, i.e., $k = \max_{S' \subseteq S} |Ch(S')|$.*

Moreover, we prove that if the given choice function does not satisfy the substitutability, there exists no constant success probability algorithm even if the width of the choice function is 2.

Theorem 3. *For any $\varepsilon > 0$, there exists consistent, size-monotone, and $|Ch(S)| \leq 2$ instance (S, Ch) such that no algorithm succeeds with probability ε.*

The Organization of the Paper. The rest of the paper is organized as follows. In Sect. 2, we formally define the secretary problem with a choice function. In Sect. 3, we provide a threshold algorithm and prove that it succeeds with a certain probability (Theorem 2). In Sect. 4, we show upper bounds for the problem (Theorems 1 and 3).

2 Model

Imagine that a decision-maker is willing to hire some secretaries out of n applicants S. The decision-maker knows the number n. The decision-maker has a choice function Ch on the set of applicants that represents her preference. Thus, the best set of secretaries is $Ch(S)$. The applicants are interviewed one-by-one in a random order. Immediately after the interview, an irrevocable decision must be made whether or not hire the applicant. Namely, if an applicant is hired, he will stay hired until the end of the process, and likewise if it is not. The decision must be based only on the number n and the value choice function Ch for the subsets of the applicants interviewed so far. The goal is to maximize the probability to hire $Ch(S)$.

Examples. The path-independency (substitutability) is not just a natural assumption as a preference but also it has some interesting applications. In fact, choice functions induced from M$^\natural$-concave value functions (and also quasi M$^\natural$-concave functions) are endowed with the substitutability and the size-monotonicity by Fujishige and Tamura [12] and Murota–Yokoi [25]. More specifically, the secretary problem with a path-independent (and size-monotone) choice function generalizes secretary version of the stable matching problem, the maximum weight bipartite matching problem, and the maximum weight base problem for a matroid, as we see below.

Stable Matching. Assume that the decision-maker is willing to hire secretaries at k positions P, out of n applicants A. Each position has a preference list that ranks applicants in strict order. Also, each applicant has a preference that ranks k positions in strict order. The lists are not necessarily complete to express unacceptable pairs. The applicants are interviewed one-by-one in a random order. The preferences related to an applicant is revealed when s/he arrives. Let $\mu : P \cup A \to A \cup P$ be a matching on positions and applicants. Then a pair $(p_i, a_j) \in P \times A$ is called a *blocking pair* for μ if p_i prefers a_j to $\mu(p_i)$ and a_j prefers p_i to $\mu(a_j)$. The matching μ is called *stable matching* if there exists no blocking pair for μ. It is well-known that the set of people who are matched is the same for all stable matching [14]. We define $Ch(X)$ for $X \subseteq A$ as the set of applicants who are matched in a stable matching for X to P. Then the width of Ch is at most k. Moreover, Ch is a size-monotone and path-independent choice function because the number of the "engaged pairs" is monotone increasing in the deferred-acceptance algorithm [13,23] and it produces the same stable matching no matter what the order of "proposal". See [16,22,28] for the details of the algorithm and the stable matching problem. If there is only one position ($k = 1$) and the preference lists are complete, the problem becomes the classical secretary problem.

Weighted Bipartite Matching. Assume that the decision-maker is willing to hire secretaries at k positions P, out of n applicants A. The worth of each assignment $a \in A$ to $p \in P$ is $w(a, p)$. The applicants are interviewed one-by-one in a random order. The worth of each assignment incident to an applicant is revealed when s/he arrives. The goal is to maximize the profit, which is the sum of the worth of assignments in a matching, i.e., the value $\sum_{(a,p)\in M} w(a, p)$ for the matching M. For simplicity, we assume that matchings have distinct profits. We define $Ch(X)$ for $X \subseteq A$ as the set of applicants who are matched in the maximum weight matching for X to P. Then the width of Ch is at most k.

Moreover, *Ch* is a path-independent and size-monotone choice function as we show below. Let $X \subseteq A$, $a^* \in A \setminus X$, and M and M' are the maximum matching for X to P and $X \cup \{a^*\}$ to P, respectively. Then the symmetric difference $M \triangle M'$ forms a path from a^* (or the empty set), since if not we can improve either M or M' [17]. Thus if $p \in P$ is matched in M then it is also matched in M', i.e., $\bigcup_{(a,p)\in M}\{p\} \subseteq \bigcup_{(a,p)\in M'}\{p\}$. Additionally, M' contains no unmatched applicant in M, i.e., if $a \in X$ satisfies $(a,p) \notin M$ for any $p \in P$ then $(a,p') \notin M'$ for any $p' \in P$. It implies that *Ch* is substitutable and size-monotone, and hence, it is path-independent and size-monotone.

We can also see the path-independency and the size-monotonicity by the facts that $\Gamma(X) = \max\{\sum_{(a,p)\in M} w(a,p) \mid M \text{ is a matching}, \bigcup_{(a,p)\in M}\{a\} = X\}$ is an M$^\natural$-concave function [24] and M$^\natural$-concave implies substitutable and size-monotone choice function [12,25].

If there is only one position ($k = 1$) and the preference lists are complete, the problem becomes the classical secretary problem.

Matroids. A *matroid* is a set system $(E, \mathcal{I})$, i.e. E is a finite set and $\mathcal{I}$ is a family of subsets of E, with the following properties:

(I1) $\emptyset \in \mathcal{I}$,
(I2) $J \subseteq I \in \mathcal{I} \Rightarrow J \in \mathcal{I}$,
(I3) $I, J \in \mathcal{I}$, $|J| < |I| \Rightarrow \exists v \in I \setminus J$ such that $J \cup \{v\} \in \mathcal{I}$.

Given a matroid $M = (E, \mathcal{I})$, a subset I of E is called *independent set* if I belongs to $\mathcal{I}$, and an inclusionwise maximal independent set is called a *base*. Any two bases of a matroid have the same number of elements, and the number is called *rank* of the matroid. For more details, see, e.g., [27]. Let $(E, \mathcal{I})$ be a matroid and each element $e \in E$ has a positive weight $w(e)$. For simplicity, we assume that independent sets have distinct weights.

Now we reinterpret the secretary problem in the auction context. Assume that the decision-maker is willing to sell some items to n bidders E. Each bidder e has a bid $w(e)$ to buy an item. There is a matroid structure $M = (E, \mathcal{I})$ on the bids, and a set of bids can be simultaneously accepted if and only if they are an independent set in the matroid. The bidders arrive one-by-one in a random order. Immediately after the arrival, an irrevocable decision must be made whether or not accept the bid. We define $Ch(X)$ for $X \subseteq E$ as the set $\operatorname{argmax}\{\sum_{e\in Y} w(e) \mid Y \subseteq X,\ Y \in \mathcal{I}\}$. Then the width of *Ch* is the rank of the matroid M. Moreover, *Ch* is a path-independent and size-monotone choice function [12,25]. If the matroid is a uniform matroid of rank 1, the problem becomes the secretary problem.

3 Algorithm

Assume that *Ch* is a path-independent choice function. We prove that the following simple algorithm succeeds with probability at least $1/e^k$ where k is the width of *Ch*, i.e., $k = \max_{S'\subseteq S} |Ch(S')|$. Let $S = \{1, \ldots, n\}$ and $\sigma(i)$ be the item given in the ith round. Denoted by B_i the best subset of applicants in $S_i^\sigma := \{\sigma(1), \ldots, \sigma(i)\}$, i.e., $B_i = Ch(S_i^\sigma)$. Note that $B_{i+1} = Ch(B_i \cup \{\sigma(i+1)\})$ holds since Ch is path-independent. Then our algorithm accept ith applicant when $\sigma(i) \in B_i$. The algorithm is summarized as Algorithm 1.

Algorithm 1. Threshold Choice Algorithm

1: $B_0 \leftarrow \emptyset$
2: **for all** applicant $\sigma(i)$ **do**
3: $\quad B_i \leftarrow Ch(B_{i-1} \cup \{\sigma(i)\})$
4: $\quad$ **if** $i > n/e$ and $\sigma(i) \in B_i$ **then** accept $\sigma(i)$
5: $\quad$ **else** reject $\sigma(i)$
6: **end for**

We prove Theorem 2 by showing that Algorithm 1 chooses the best set $Ch(S)$ with probability at least $1/e^k$ when Ch is a path-independent and width k choice function.

Proof of Theorem 2. Without loss of generality, we may assume that $n\,(=|S|)$ is a sufficiently large number, because we can add dummy applicants who are never selected by Ch without changing the width and the path-independency of Ch. We can also run the algorithm with dummy applicants in an online manner. Let $Ch(S) = \{\sigma(i_1), \sigma(i_2), \ldots, \sigma(i_t)\}$ ($i_1 < i_2 < \cdots < i_t$, $t \leq k$). Let A_j be the event that $Ch(S_{i_j-1}^\sigma) \cap (S_{i_j-1}^\sigma \setminus S_{i_{j-1}}^\sigma) = \emptyset$ where $i_0 = \lfloor n/e \rfloor$. The algorithm succeeds if $i_1 > n/e$ and all the events $A_1, \ldots, A_t$ occur, because it accepts all the applicants in $Ch(S)$ by the substitutability, and A_j means that the algorithm selects no applicants in $S_{i_j-1}^\sigma \setminus S_{i_{j-1}}^\sigma$ by the consistency. Thus, for a fixed sequence $i_1, \ldots, i_t$, the conditional probability of A_j given $A_{j+1}, \ldots, A_t$ is

$$\Pr\left[A_j \mid {}^{Ch(S)=\{\sigma(i_1),\ldots,\sigma(i_t)\}}_{A_{j+1},\ldots,A_t}\right] = \binom{i_{j-1}-(j-1)}{|Ch(S_{i_j-1}^\sigma)|+1-j} \Big/ \binom{i_j - j}{|Ch(S_{i_j-1}^\sigma)|+1-j}$$
$$\geq \binom{i_{j-1}-(j-1)}{k+1-j} \Big/ \binom{i_j - j}{k+1-j}$$

because $Ch(S_{i_j-1}^\sigma)$ is of the form

$$Ch(S_{i_j-1}^\sigma) = \{\sigma(i_1), \sigma(i_2), \ldots, \sigma(i_{j-1}), \sigma(r_j), \sigma(r_{j+1}), \ldots, \sigma(r_{t_j})\} \quad (t_j \leq k),$$

and A_j happens if $r_j, \ldots, r_{t_j} \in \{1, 2, \ldots, i_{j-1}\} \setminus \{i_1, i_2, \ldots, i_{j-1}\}$. Therefore, Algorithm 1 chooses the set $Ch(S)$ with probability at least

$$
\begin{aligned}
&\sum_{n/e<i_1<i_2<\cdots<i_t\le n} \frac{1}{\binom{n}{t}} \prod_{j=1}^{t} \Pr\left[A_j \mid {}^{Ch(S)=\{\sigma(i_1),\ldots,\sigma(i_t)\}}_{A_{j+1},\ldots,A_t}\right] \\
\ge &\sum_{n/e<i_1<i_2<\cdots<i_t\le n} \frac{1}{\binom{n}{t}} \prod_{j=1}^{t} \frac{\binom{i_{j-1}-(j-1)}{k+1-j}}{\binom{i_j-j}{k+1-j}} \\
\ge &\sum_{n/e<i_1<i_2<\cdots<i_t\le n} \frac{t!}{n^t} \left(\prod_{j=1}^{t-1} \frac{i_0-(j-1)}{i_j-k}\right)\left(\prod_{j=t}^{k} \frac{i_0-(j-1)}{i_t-j}\right) \\
\approx\ & t! \int_{1/e}^{1} dx_t \int_{1/e}^{x_t} dx_{t-1} \cdots \int_{1/e}^{x_3} dx_2 \int_{1/e}^{x_2} \left(\prod_{j=1}^{t-1} \frac{1/e}{x_j}\right)\left(\prod_{j=t}^{k} \frac{1/e}{x_t}\right) dx_1 \qquad (1) \\
=\ & t! \cdot \frac{1}{e^k} \int_{1/e}^{1} dx_t \int_{1/e}^{x_t} dx_{t-1} \cdots \int_{1/e}^{x_3} dx_2 \int_{1/e}^{x_2} \frac{dx_1}{x_1 \cdots x_{t-1} \cdot x_t^{k-t+1}} \\
=\ & t! \cdot \frac{1}{e^k} \int_{1/e}^{1} dx_t \int_{1/e}^{x_t} dx_{t-1} \cdots \int_{1/e}^{x_{l+1}} \frac{(\log(ex_l))^{l-1}}{(l-1)! \cdot x_l \cdots x_{t-1} \cdot x_t^{k-t+1}} dx_l \qquad (2) \\
=\ & t! \cdot \frac{1}{e^k} \cdot \int_{1/e}^{1} \frac{(\log(ex_t))^{t-1}}{(t-1)! \cdot x_t^{k-t+1}} dx_t \ge t! \cdot \frac{1}{e^k} \cdot \int_{1/e}^{1} \frac{(\log(ex_t))^{t-1}}{(t-1)! \cdot x_t} dx_t \qquad (3) \\
=\ & t! \cdot \frac{1}{e^k} \cdot \frac{(\log e)^t - (\log 1)^t}{t!} = \frac{1}{e^k}. \qquad (4)
\end{aligned}
$$

Equation (1) is a Riemann approximation where $i_j/n \to x_j$ as $n \to \infty$. The last four equalities (2)–(4) hold by

$$\frac{d}{dx}(\log(ex))^l = \frac{l(\log(ex))^{l-1}}{x}.$$

□

Remark 1. By (3) in the above proof, we obtain a success probability

$$\frac{1}{e^k} \cdot \int_{1/e}^{1} \frac{t(\log(ex_t))^{t-1}}{x^{k-t+1}} dx$$

when $|Ch(S)| = t$ and the width of Ch is k.

4 Upper Bounds

In this section, we prove upper bounds of the success probability.

We first show that for any $\varepsilon > 0$, there exists a path-independent and size-monotone instance (S, Ch) such that no algorithm succeeds with probability $1/e^{|Ch(S)|} + \varepsilon$ (Theorem 1). We use a choice function that implies k classical secretary problems at the same time.

Proof of Theorem 1. Let $S_i = \{(i-1)\cdot n+1, \ldots, (i-1)\cdot n+n\}$ for $i = 1, \ldots, k$ and $S = \bigcup_{i=1}^{k} S_i$. We consider the following choice function on S:

$$Ch(S') = \bigcup_{i=1}^{k} \operatorname{argmin}\{x \in S' \cap S_i\}.$$

Then, $Ch(S) = \{1, n+1, \ldots, (k-1)\cdot n+1\}$ and Ch is path-independent and size-monotone. Then, the success probability for any algorithm goes to at most $1/e^k$ as n goes to infinity because this situation is the classical secretary problem for each S_i. □

We note that the choice function in the above proof is implied by a stable matching with a partial list, a weighted bipartite matching, and a maximum weight base problem for a partition matroid.

Finally, we provide a proof of Theorem 3.

Proof of Theorem 3. Let Ch be a choice function on $S = \{1, 2, \ldots, 2m^2\}$ such that

$$Ch(S') = \begin{cases} \{m^2, m^2+1\} & (\{m^2, m^2+1\} \subseteq S'), \\ \operatorname{argmin}\{x \in S'\} & (\text{otherwise}). \end{cases}$$

Then, Ch is consistent, size-monotone, and $|\max_{S' \subseteq S} Ch(S')| \leq 2$, but not substitutable. This instance is the same as the classical secretary problem with $2m^2$ applicants, except for $\{m^2, m^2+1\}$. Thus, it is more difficult than choosing m^2 or m^2+1 in the classical setting, i.e., applicants are rankable, and the decision-maker chooses one applicant. Hence, it is sufficient to show that no algorithm can succeed with positive probability for the problem when m goes to infinity.

To provide an upper bound, we use a linear programming method introduced by Buchbinder, Jain, and Singh [5]. Consider the following linear programming:

$$\begin{aligned} \max\ & \textstyle\sum_{i=1}^{2m^2} \sum_{j=1}^{i} \frac{\binom{m^2-1}{j-1}\binom{m^2}{i-j} + \binom{m^2}{j-1}\binom{m^2-1}{i-j}}{\binom{2m^2}{i}} p_{ij} \\ \text{s.t. } & i \cdot p_{ij} \leq 1 - \textstyle\sum_{i'<i} \sum_{j'=1}^{i'} p_{i'j'} && (i \in [2m^2],\ j \in [i]), \\ & p_{ij} \geq 0 && (i \in [2m^2],\ j \in [i]) \end{aligned}$$

where $[n] = \{1, 2, \ldots, n\}$. We claim that the optimal value for the linear programming gives an upper bound of the success probability for the problem. Let p_{ij} be the probability of selecting the applicant relatively rank j in the ith round for $i \in [2m^2]$ and $j \in [i]$. Then the success probability is

$$\sum_{i=1}^{2m^2} \sum_{j=1}^{i} \frac{\binom{m^2-1}{j-1}\binom{m^2}{i-j} + \binom{m^2}{j-1}\binom{m^2-1}{i-j}}{\binom{2m^2}{i}} p_{ij}$$

because if ith applicant is relatively rank j, its true rank is k with probability

$$\binom{k-1}{j-1}\binom{2m^2-k}{i-j} \Big/ \binom{2m^2}{i}.$$

Also, p_{ij} must satisfy the following relation:

$$\begin{aligned}
p_{ij} &= \Pr[\sigma(i) \text{ is selected} \mid \sigma(i) \text{ is relatively rank } j] \cdot \Pr[\sigma(i) \text{ is relatively rank } j] \\
&\leq \Pr[\sigma(1), \ldots, \sigma(i-1) \text{ are not selected} \mid \sigma(i) \text{ is relatively rank} j] \cdot \frac{1}{i} \\
&= \Pr[\sigma(1), \ldots, \sigma(i-1) \text{ are not selected}] \cdot \frac{1}{i} = \frac{1}{i}\left(1 - \sum_{i'<i}\sum_{j'=1}^{i'} p_{i'j'}\right).
\end{aligned}$$

Thus, the linear programming present an upper bound of the success probability.

To evaluate the optimal value, now we consider the dual problem:

$$\begin{aligned}
&\min \textstyle\sum_{i=1}^{2m^2}\sum_{j=1}^{i} q_{ij} \\
&\text{s.t. } i \cdot q_{ij} + \textstyle\sum_{i'>i}\sum_{j'=1}^{i'} q_{i'j'} \geq \frac{\binom{m^2-1}{j-1}\binom{m^2}{i-j}+\binom{m^2}{j-1}\binom{m^2-1}{i-j}}{\binom{2m^2}{i}} \quad (i \in [2m^2],\ j \in [i]), \\
&\quad q_{ij} \geq 0 \quad (i \in [2m^2],\ j \in [i]).
\end{aligned}$$

Let q^* be

$$q^*_{ij} = \begin{cases} \frac{1}{i}\frac{\binom{m^2-1}{j-1}\binom{m^2}{i-j}+\binom{m^2}{j-1}\binom{m^2-1}{i-j}}{\binom{2m^2}{i}} & (2m^2 > i \geq 2m^2 - 2m,\ j \in [i]), \\ \frac{1}{m} + \frac{\binom{2m}{m}\binom{2m^2-2m}{m^2-m}}{\binom{2m^2}{m^2}} & (i = 2m^2,\ j \in \{m^2, m^2+1\}), \\ 0 & (\text{othwerwise}). \end{cases}$$

If q^* is a feasible solution for the dual LP, the success probability is at most

$$\begin{aligned}
&\frac{2}{m} + \frac{2\binom{2m}{m}\binom{2m^2-2m}{m^2-m}}{\binom{2m^2}{m^2}} + \sum_{i=2m^2-2m}^{2m^2-1}\sum_{j=1}^{i}\frac{1}{i}\frac{\binom{m^2-1}{j-1}\binom{m^2}{i-j}+\binom{m^2}{j-1}\binom{m^2-1}{i-j}}{\binom{2m^2}{i}} \\
&\leq \frac{2}{m} + \frac{2\binom{2m}{m}\binom{2m^2-2m}{m^2-m}}{\binom{2m^2}{m^2}} + \sum_{i=2m^2-2m}^{2m^2-1}\frac{2}{i} \leq \frac{2}{m} + \frac{2\binom{2m}{m}\binom{2m^2-2m}{m^2-m}}{\binom{2m^2}{m^2}} + \frac{4m}{2m^2-2m} \to 0
\end{aligned}$$

as m goes to infinity by Stirling's formula, which proves the theorem by weak duality of linear programming.

We finally claim that q^* is a feasible solution for the dual LP. For $2m^2 > i \geq 2m^2 - 2m$, it is clear that q^* satisfies the inequality. For $i = 2m^2$, q^* satisfies the inequality because the right-hand side value is one if $j = m^2, m^2+1$ and zero otherwise. For $i < 2m$, we can check as

$$\begin{aligned}
\frac{\binom{m^2-1}{j-1}\binom{m^2}{i-j}+\binom{m^2}{j-1}\binom{m^2-1}{i-j}}{\binom{2m^2}{i}} &\leq \frac{\frac{j}{m^2}\binom{m^2}{j}\binom{m^2}{i-j}+\frac{i-j+1}{m^2}\binom{m^2}{j-1}\binom{m^2}{i-j+1}}{\binom{2m^2}{i}} \\
&\leq \frac{j}{m^2} + \frac{i-j+1}{m^2} \leq \frac{2m}{m^2} \leq q^*_{2m^2,m^2} + q^*_{2m^2,m^2+1}.
\end{aligned}$$

For $2m \leq i < 2m^2 - 2m$, we have

$$\begin{aligned}
\frac{\binom{m^2-1}{j-1}\binom{m^2}{i-j} + \binom{m^2}{j-1}\binom{m^2-1}{i-j}}{\binom{2m^2}{i}} &\leq \max_{j'} \frac{2\binom{m^2}{j'}\binom{m^2}{i-j'}}{\binom{2m^2}{i}} = \max_{j'} \frac{2\binom{i}{j'}\binom{2m^2-i}{m^2-j'}}{\binom{2m^2}{m^2}} \\
&\leq \frac{2\binom{i}{\lfloor i/2 \rfloor}\binom{2m^2-i}{m^2-\lfloor i/2 \rfloor}}{\binom{2m^2}{m^2}} \leq \frac{2\binom{2m}{m}\binom{2m^2-2m}{m^2-m}}{\binom{2m^2}{m^2}} \\
&\leq q^*_{2m^2,m^2} + q^*_{2m^2,m^2+1}.
\end{aligned}$$

□

Intuitively, the dual solution q^* represents that the applicants with rank m^2 or m^2+1 appear after $(2m^2-2m)$th round with low probability and the success probability is low if an algorithm selects an applicant before $2m^2 - 2m$.

Acknowledgement. The author thanks Tomomi Matsui and Keisuke Bando for valuable comments and suggestions. This work was supported by JSPS KAKENHI Grant Number 26887014.

References

1. Aizerman, M., Malishevski, A.: General theory of best variants choice: some aspects. IEEE Trans. Autom. Control **26**, 1030–1040 (1981)
2. Babaioff, M., Immorlica, N., Kempe, D., Kleinberg, R.D.: A knapsack secretary problem with applications. In: Charikar, M., Jansen, K., Reingold, O., Rolim, J.D.P. (eds.) APPROX and RANDOM 2007. LNCS, vol. 4627, pp. 16–28. Springer, Heidelberg (2007)
3. Babaioff, M., Immorlica, N., Kempe, D., Kleinberg, R.: Online auctions and generalized secretary problems. ACM SIGecom Exch. **7**(2), 7:1–7:11 (2008)
4. Babaioff, M., Immorlica, N., Kleinberg, R.: Matroids, secretary problems, and online mechanisms. In: Proceedings of the Eighteenth Annual ACM-SIAM Symposium on Discrete Algorithms, pp. 434–443 (2007)
5. Buchbinder, N., Jain, K., Singh, M.: Secretary problems via linear programming. Math. Oper. Res. **39**(1), 190–206 (2014)
6. Chan, T.H.H., Chen, F., Jiang, S.H.C.: Revealing optimal thresholds for generalized secretary problem via continuous LP: impacts on online k-item auction and bipartite k-matching with random arrival order. In: Proceedings of the Eighteenth Annual ACM-SIAM Symposium on Discrete Algorithms, pp. 1169–1188 (2015)
7. Dinitz, M.: Recent advances on the matroid secretary problem. SIGACT News **44**(2), 126–142 (2013)
8. Dynkin, E.B.: The optimum choice of the instant for stopping a Markov process. Sov. Math. Dokl. **4**, 627–629 (1963)
9. Feldman, M., Svensson, O., Zenklusen, R.: A simple O(log log(rank))-completitive algorithm for the matroid secretary problem. In: Proceedings of the Twenty-Sixth Annual ACM-SIAM Symposium on Discrete Algorithms, pp. 1189–1201 (2015)
10. Ferguson, T.S.: Who solved the secretary problem? Statist. Sci. **4**(3), 282–289 (1989)
11. Freeman, P.R.: The secretary problem and its extensions: a review. Int. Statis. Rev. **51**(2), 189–206 (1983)

12. Fujishige, S., Tamura, A.: A general two-sided matching market with discrete concave utility functions. Discrete Appl. Math. **154**, 950–970 (2006)
13. Gale, D., Shapley, L.S.: College admissions and the stability of marriage. Am. Math. Mon. **69**, 9–14 (1962)
14. Gale, D., Sotomayor, M.: Some remarks on the stable matching problem. Discrete Appl. Math. **11**, 223–232 (1985)
15. Gilbert, J.P., Mosteller, F.: Recognizing the maximum of a sequence. J. Am. Statist. Assoc. **61**, 35–73 (1966)
16. Gusfield, D., Irving, R.W.: The Stable Marriage Problem: Structure and Algorithms. MIT Press, Boston (1989)
17. Khuller, S., Mitchell, S.G., Vazirani, V.V.: On-line algorithms for weighted bipartite matching and stable marriages. Theoret. Comput. Sci. **127**(2), 255–267 (1994)
18. Kleinberg, R.: A multiple-choice secretary algorithm with applications to online auctions. In: Proceedings of the Eighteenth Annual ACM-SIAM Symposium on Discrete Algorithms, pp. 630–631 (2005)
19. Korula, N., Pál, M.: Algorithms for secretary problems on graphs and hypergraphs. In: Albers, S., Marchetti-Spaccamela, A., Matias, Y., Nikoletseas, S., Thomas, W. (eds.) ICALP 2009, Part II. LNCS, vol. 5556, pp. 508–520. Springer, Heidelberg (2009)
20. Lachish, O.: O(log log rank) completitive-ratio for the matroid secretary problem. In: Proceedings of 55th Annual Symposium on Foundation of Compuster Science, pp. 326–335 (2014)
21. Lindley, D.V.: Dynamic programming and decision theory. Appl. Stat. **10**, 39–52 (1961)
22. Manlove, D.F.: Algorithmics of Matching Under Preferences. World Scientific, Singapore (2013)
23. McVitie, D.G., Wilson, L.B.: The stable marriage problem. Commun. ACM **14**(7), 486–490 (1971)
24. Murota, K.: Recent developments in discrete convex analysis. In: Cook, W.J., Lovász, L., Vygen, J. (eds.) Research Trends in Combinatorial Optimization, pp. 219–260. Springer, Heidelberg (2009)
25. Murota, K., Yokoi, Y.: On the lattice structure of stable allocations in a two-sided discrete-concave market. Math. Oper. Res. **40**, 460–473 (2015)
26. Nikolaev, M.L.: On a generalization of the best choice problem. Theor. Probab. Its Appl. **22**(1), 187–190 (1977)
27. Oxley, J.G.: Matroid Theory. Oxford University Press, New York (1992)
28. Roth, A.E., Sotomayor, M.: Two-Sided Matching: A Study in Game-Theoretic Modeling and Analysis. Cambridge University Press, Cambridge (1991)
29. Samuels, S.M.: Secretary problems. In: Ghosh, B.K., Sen, P.K. (eds.) Handbook of Sequential Analysis. Marcel Dekker, Boston (1991)
30. Tamaki, M.: Recognizing both the maximum and the second maximum of a sequence. J. Appl. Probab. **16**(4), 803–812 (1979)
31. Vanderbei, R.J.: The optimal choice of a subset of a population. Math. Oper. Res. **5**(4), 481–486 (1980)

The Benefit of Recombination in Noisy Evolutionary Search

Tobias Friedrich, Timo Kötzing(✉), Martin S. Krejca, and Andrew M. Sutton

Hasso Plattner Institute, Potsdam, Germany
timo.koetzing@hpi.de

Abstract. Practical optimization problems frequently include uncertainty about the quality measure, for example due to noisy evaluations. Thus, they do not allow for a straightforward application of traditional optimization techniques. In these settings meta-heuristics are a popular choice for deriving good optimization algorithms, most notably evolutionary algorithms which mimic evolution in nature. Empirical evidence suggests that genetic recombination is useful in uncertain environments because it can stabilize a noisy fitness signal. With this paper we want to support this claim with mathematical rigor.

The setting we consider is that of noisy optimization. We study a simple noisy fitness function that is derived by adding Gaussian noise to a monotone function. First, we show that a classical evolutionary algorithm that does not employ sexual recombination (the $(\mu+1)$-EA) cannot handle the noise efficiently, regardless of the population size. Then we show that an evolutionary algorithm which does employ sexual recombination (the Compact Genetic Algorithm, short: cGA) can handle the noise using a graceful scaling of the population.

1 Introduction

Heuristic optimization is widely used in practice for solving hard optimization problems for which no efficient problem-specific algorithm is known. Such problems are typically very large, noisy and constrained and cannot be solved by simple textbook algorithms. The inspiration for heuristic general-purpose problem solvers often comes from nature. A well-known example is simulated annealing, which is inspired from physical annealing in metallurgy. The largest and probably most successful class, however, are biologically-inspired algorithms, especially *evolutionary algorithms*.

Evolutionary and Genetic Algorithms. Evolutionary Algorithms (EAs) were introduced in the 1960s and have been successfully applied to a wide range of complex engineering and combinatorial problems [1,10,24]. Like Darwinian evolution in nature, evolutionary algorithms construct new solutions from old ones and select the fitter ones to continue to the next iteration. The construction of new solutions from old ones, so-called reproduction, can

K. Elbassioni and K. Makino (Eds.): ISAAC 2015, LNCS 9472, pp. 140–150, 2015.
DOI: 10.1007/978-3-662-48971-0_13

be *asexual* (mutation of a single individual) or *sexual* (crossover of several individuals). An EA that uses sexual reproduction is typically called *Genetic Algorithm* (GA). Since the beginning of EAs, it has been argued that GAs should be more powerful than pure EAs, which use only asexual reproduction [13]. This was debated for decades, but theoretical results and explanations on crossover are still scarce. There are some results for simple artificial test functions, where it was proven that a GA asymptotically outperforms an EA without crossover [16,17,20,25,31,35] and the other way around [30]. However, these artificial test functions are typically tailored to the specific algorithm and proof technique and the results give little insight into the advantage of sexual reproduction on realistic problems. There are also a few theoretical results for problem-specific algorithms and representations, namely coloring problems inspired by the Ising model [32] and the all-pairs shortest path problem [5]. For a nice overview of different aspects where populations and sexual recombination are beneficial for optimization of static fitness functions, see [29].

The underlying search space of many optimization problems is the set $\{0,1\}^n$ of all length-n bit strings. Many problems (including combinatorial ones such as the minimum spanning tree problem) have a straightforward formulation as an optimization problem on $\{0,1\}^n$. Many evolutionary algorithms are applicable to this search space without further modification adaption, and most formal analyses of evolutionary algorithms consider this search space. A popular simple fitness function on this search space is *OneMax*, which uses the number of 1s in a bit string as fitness value. A cornerstone of the analysis of any search heuristic is an analysis of its performance on the OneMax function [8,37], and studying the class of OneMax functions has also lead to several breakthroughs in the field of black-box complexity [4,9]. Finally, there are also works analyzing the use of crossover for the OneMax function [7,28,33].

Noisy Search. Heuristic optimization methods are typically not used for simple problems, but for rather difficult problems in *uncertain environments*. Evolutionary algorithms are very popular in settings including uncertainties; see [2] for a survey on examples in combinatorial optimization, but also [18] for an excellent survey also discussing different sources of uncertainty. Uncertainty can be modeled by a *probabilistic* fitness function, that is, a search point can have different fitness values each time it is evaluated. One way to deal with this is to replace fitness evaluations with an average of a (large) sample of fitness evaluations and then proceed as if there was no noise. In this work we show that generic GAs (with sexual reproduction) can overcome noise much more efficiently than using this naive approach. To do this in a rigorous manner, we assume *additive posterior noise*, that is, each time the fitness value of a search point is evaluated, we add a noise value drawn from some distribution. This model was studied in evolutionary algorithms without crossover in [6,11,12,14,34].

We will consider centered Gaussian noise with variance σ^2 and use OneMax as the underlying fitness function. Already such a seemingly simple setting poses difficulties to the analysis of evolutionary algorithms, as these algorithms are not developed with the analysis in mind. Particularly algorithms with sexual recombination have been resisting a mathematical analysis.

Our Results. We are interested in studying how well search heuristics can cope with noise, for which we use the concept of *graceful scaling* (Definition 2); intuitively, a search heuristic scales gracefully with noise if (polynomially) more noise can be compensated by (polynomially) more resources.

We first prove a sufficient condition for when a noise model is intractable for optimization by the classical $(\mu+1)$-EA (Theorem 4) and show that this implies that this simple asexual algorithm does not scale gracefully for Gaussian noise (Corollary 5). On the other hand, we study the compact GA (cGA), which models a genetic algorithm, and show how its gene-pool recombination operator is able to "smooth" the noise sufficiently to exhibit graceful scaling (Theorem 9).

We proceed in Sect. 2 by formalizing our setting and introducing the algorithms we consider. In Sect. 3 we give our results. Note that in this extended abstract, we omit many proof details and provide only proof sketches due to space constraints. We conclude the paper in Sect. 4.

2 Preliminaries

In the remainder of the paper, we will study a particular function class (OneMax) and a particular noise distribution (Gaussian, parametrized by the variance). Let $\sigma^2 \geq 0$. We define the noisy OneMax function $\text{OM}_{[\sigma^2]}\colon \{0,1\}^n \to \mathbb{R} := x \mapsto \|x\|_1 + Z$ where $\|x\|_1 := |\{i\colon x_i = 1\}|$ and Z is a normally distributed random variable $Z \sim \mathcal{N}(0, \sigma^2)$ with zero mean and variance σ^2.

The following proposition gives tail bounds for Z by using standard estimates of the complementary error function [36].

Proposition 1. *Let Z be a zero-mean Gaussian random variable with variance σ^2. For all $t > 0$ we have*

$$\Pr(Z < -t) = \frac{1}{2}\,\text{erfc}\left(\frac{t}{\sigma\sqrt{2}}\right) \leq \frac{1}{2}\mathrm{e}^{-t^2/(2\sigma^2)}$$

and asymptotically for large $t > 0$,

$$\Pr(Z < -t) = \frac{1}{1+o(1)}\frac{\sigma}{\sqrt{2\pi}t}\mathrm{e}^{-t^2/(2\sigma^2)}.$$

2.1 Algorithms

The $(\mu+1)$-EA, defined in Algorithm 1, is a simple mutation-only evolutionary algorithm that maintains a population of μ solutions and uses elitist survival selection. It derives its name from maintaining a population of μ individuals (randomly initialized) and generating one new individual each iteration by mutating a parent chosen uniformly at random from the current population. Then it evaluates the fitness of all individuals and chooses one with minimal value to be

Algorithm 1. The $(\mu + 1)$-EA

```
1 t ← 0;
2 P_t ← μ elements of {0,1}^n u.a.r.;
3 while termination criterion not met do
4     Select x ∈ P_t u.a.r.;
5     Create y by flipping each bit of x independently with probability 1/n;
6     Let z ∈ P_t ∪ {y} chosen s.t. ∀v ∈ P_t ∪ {y}: f(z) ≤ f(v);
7     P_{t+1} ← P_t ∪ {y} \ {z};
8     t ← t + 1;
```

removed from the population, so that again μ individuals proceed to the next generation.

The compact genetic algorithm (cGA) [15] is a genetic algorithm that maintains a population of size K *implicitly* in memory. Rather than storing each individual separately, the cGA only keeps track of population *allele frequencies* and updates these frequencies during evolution. Offspring are generated according to these allele frequencies, which is similar to what occurs in models of sexually-recombining natural populations. Indeed, the offspring generation procedure can be viewed as so-called *gene pool recombination* introduced by Mühlenbein and Paaß [23] in which all K members participate in uniform recombination. Since the cGA evolves a probability distribution, it is also a type of *estimation of distribution algorithm* (EDA). The correspondence between EDAs and models of sexually recombining populations has already been noted [22], and Harik et al. [15] demonstrate empirically that the behavior of the cGA is equivalent to a simple genetic algorithm at least on simple problems.

Algorithm 2. The compact GA

```
1  t ← 0;
2  p_{1,t} ← p_{2,t} ← ··· ← p_{n,t} ← 1/2;
3  while termination criterion not met do
4      for i ∈ {1,...,n} do
5          x_i ← 1 with probability p_{i,t}, x_i ← 0 with probability 1 − p_{i,t};
6      for i ∈ {1,...,n} do
7          y_i ← 1 with probability p_{i,t}, y_i ← 0 with probability 1 − p_{i,t};
8      if f(x) < f(y) then swap x and y;
9      for i ∈ {1,...,n} do
10         if x_i > y_i then p_{i,t+1} ← p_{i,t} + 1/K;
11         if x_i < y_i then p_{i,t+1} ← p_{i,t} − 1/K;
12         if x_i = y_i then p_{i,t+1} ← p_{i,t};
13     t ← t + 1;
```

The first rigorous analysis of the cGA is due to Droste [8] who gave a general runtime lower bound for all pseudo-Boolean functions, and a general upper bound for all linear pseudo-Boolean functions. Defined in Algorithm 2, the cGA maintains for all times $t \in \mathbb{N}_0$ a frequency vector $(p_{1,t}, p_{2,t}, \ldots, p_{n,t}) \in [0,1]^n$. In the t-th iteration, two strings x and y are sampled independently from this distribution where $\Pr(x = z) = \Pr(y = z) = \left(\prod_{i\colon z_i=1} p_{i,t}\right) \times \left(\prod_{i\colon z_i=0}(1 - p_{i,t})\right)$ for all $z \in \{0,1\}^n$. The cGA then compares the objective values of x and y, and updates the distribution by advancing $p_{i,t}$ toward the component of the winning string by an additive term. This small change in allele frequencies is equivalent to a population undergoing steady-state binary tournament selection [15].

Let F be a family of pseudo-Boolean functions $(F_n)_{n\in\mathbb{N}}$ where each F_n is a set of functions $f\colon \{0,1\}^n \to \mathbb{R}$. Let D be a family of distributions $(D_v)_{v\in\mathbb{R}}$ such that for all $D_v \in D$, $\mathrm{E}(D_v) = 0$. We define F with additive posterior D-noise as the set $F[D] := \{f_n + D_v \colon f_n \in F_n, D_v \in D\}$.

Definition 2. *An algorithm A* scales gracefully with noise *on $F[D]$ if there is a polynomial q such that, for all $g_{n,v} = f_n + D_v \in F[D]$, there exists a parameter setting p such that $A(p)$ finds the optimum of f_n using at most $q(n,v)$ calls to $g_{n,v}$.*

Algorithms that operate in the presence of noise often depend on *a priori* knowledge of the noise intensity (measured by the variance). In such cases, the following scheme can always be used to transform such algorithms into one that has no knowledge of the noise character. Suppose $A(\sigma^2)$ is an algorithm that solves a noisy function with variance at most σ^2 within $T_\delta(\sigma^2)$ steps with probability at least $1 - \delta$. A *noise-oblivious scheme* for A is in Algorithm 3.

Algorithm 3. Noise-oblivious scheme for A

```
1 i ← 0;
2 repeat until solution found
3 |   Run A(2^i) for T_δ(2^i) steps;
4 |   i ← i + 1;
```

If an algorithm A scales gracefully with noise, then the noise oblivious scheme for A scales gracefully with noise. The following proposition holds by a simple inductive argument.

Proposition 3. *Suppose $f_{n,v} \in F[D]$ is a noisy function with unknown variance v. Fixing n and assume that, for all $c > 0$ and all x, $cT_\delta(x) \leq T_\delta(cx)$. Then for any $s \in \mathbb{Z}^+$, the noise-oblivious scheme optimizes $f_{n,v}$ in at most $T_\delta(2^s v)$ steps with probability at least $1 - \delta^s$.*

3 Results

We derive rigorous bounds on the optimization time, defined as the first hitting time of the process to the true optimal solution (1^n) of $\mathrm{OM}_{[\sigma^2]}$, on a mutation-only based approach and the compact genetic algorithm.

3.1 Mutation-Based Approach

In this section we consider the $(\mu + 1)$-EA. We will first, in Theorem 4, give a sufficient condition for when a noise model is intractable for optimization by a $(\mu+1)$-EA. While uniform selection removes any individual from the population with probability $1/(\mu+1)$, the condition of Theorem 4 requires that the noise is strong enough so that the $(\mu + 1)$-EA will remove any individual with at least half that probability. Then we will show that, in the case of additive posterior noise sampled from a Gaussian distribution, this condition is fulfilled if the noise is large enough, showing that the $(\mu+1)$-EA cannot deal with arbitrary Gaussian noise (see Corollary 5).

Theorem 4. *Let $\mu \geq 1$ and D a distribution on $\mathbb{R}$. Let Y be the random variable describing the minimum over μ independent copies of D. Suppose*

$$\Pr(Y > D + n) \geq \frac{1}{2(\mu + 1)}.$$

Consider optimization of OneMax *with reevaluated additive posterior noise from D by $(\mu+1)$-EA. Then, for μ bounded from above by a polynomial, the optimum will* not *be evaluated after polynomially many iterations w.h.p.*

Proof Sketch. For all t and all $i \leq n$ let X_i^t be the random variable describing the *proportion of individuals* in the population of iteration t with exactly i 1s. The proof is by induction on t that

$$\forall t, \forall i \geq an \colon \mathrm{E}(X_i^t) \leq b^{an-i},$$

where a, b and c are specifically chosen constants. In other words, the expected number of individuals with i 1s is decaying exponentially with i after an. This will give the desired result with a simple union bound over polynomially many time steps. □

We apply Theorem 4 to show that large noise levels make it impossible for the $(\mu + 1)$-EA to efficiently optimize when the noise is significantly larger than the range of objective values. The proof is a simple exercise in bounding the tails of a Gaussian distribution using Proposition 1.

Corollary 5. *Consider optimization of* $\mathrm{OM}_{[\sigma^2]}$ *by $(\mu + 1)$-EA. Suppose $\sigma^2 \geq n^3$ and μ bounded from above by a polynomial in n. Then the optimum will* not *be evaluated after polynomially many iterations w.h.p.*

3.2 Compact GA

Let $T^\star$ be the optimization time of the cGA on $\mathrm{OM}_{[\sigma^2]}$, namely, the first time that it generates the underlying "true" optimal solution 1^n. We consider the stochastic process $X_t = n - \sum_{i=1}^{n} p_{i,t}$ and bound the optimization time by $T = \inf\{t \in \mathbb{N}_0 \colon X_t = 0\}$. Clearly $T^\star \leq T$ since the cGA produces 1^n in the T-th iteration almost surely. However, $T^\star$ and T can be infinite when there is a $t < T^\star$ where $p_{i,t} = 0$ since the process can never subsequently generate any string x with $x_i = 1$. To circumvent this, Droste [8] estimates $\mathrm{E}(T^\star)$ conditioned on the event that $T^\star$ is finite, and then bounds the probability of finite $T^\star$. In this paper, we will prove that as long as K is large enough, the optimization time is finite (indeed, polynomial) with high probability. To prove our result, we need the following drift theorem.

Theorem 6 (Tail Bounds for Multiplicative Drift [3,21]**).** *Let $\{X_t : t \in \mathbb{N}_0\}$ be a sequence of random variables over a set $S \subseteq \{0\} \cup [x_{\min}, x_{\max}]$ where $x_{\min} > 0$. Let T be the random variable that denotes the earliest point in time $t \geq 0$ such that $X_t = 0$. If there exists $0 < \delta < 1$ such that $\mathrm{E}(X_t - X_{t+1} \mid T > t, X_t) \geq \delta X_t$, then*

$$\Pr\left(T > \frac{\lambda + \ln(X_0/x_{\min})}{\delta} \,\middle|\, X_0\right) \leq e^{-\lambda} \text{ for all } \lambda > 0.$$

The following lemma bounds the drift on X_t, conditioned on the event that no allele frequency gets too small.

Lemma 7. *Consider the cGA optimizing $\mathrm{OM}_{[\sigma^2]}$ and let X_t be the stochastic process defined above. Assume that there exists a constant $a > 0$ such that $p_{i,t} \geq a$ for all $i \in \{1, \ldots, n\}$ and that $X_t > 0$, then $\mathrm{E}(X_t - X_{t+1} \mid X_t) \geq \delta X_t$ where $1/\delta = \mathcal{O}(\sigma^2 K \sqrt{n})$.*

Proof Sketch. Let x and y be the offspring generated in iteration t and $Z_t = \|x\|_1 - \|y\|_1$. Then $Z_t = Z_{1,t} + \cdots + Z_{n,t}$ where

$$Z_{i,t} = \begin{cases} -1 & \text{if } x_i = 0 \text{ and } y_i = 1, \\ 0 & \text{if } x_i = y_i, \\ 1 & \text{if } x_i = 1 \text{ and } y_i = 0. \end{cases}$$

Let $\mathcal{E}$ denote the event that in line 8, the evaluation of $\mathrm{OM}_{[\sigma^2]}$ correctly ranks x and y. Then

$$\mathrm{E}(X_t - X_{t+1} \mid X_t) = \frac{\mathrm{E}(|Z_t|)}{K} \left(1 - 2\Pr(\overline{\mathcal{E}})\right).$$

Using combinatorial arguments and properties of the Poisson-Binomial distribution, the expectation of $|Z_t|$ can be bounded from below by $aX_t\sqrt{2/n}$. The proof can then be completed by bounding the probability that x and y are incorrectly ranked, which is at most $\frac{1}{2}\left(1 - \Omega(\sigma^{-2})\right)$. This follows from straightforward deviation bounds on the normal distribution derived from Proposition 1. □

To use Lemma 7, we require that the allele frequencies stay large enough during the run of the algorithm. Increasing the effective population size K obviously translates to finer-grained allele frequency values, which means slower dynamics for $p_{i,t}$. Indeed, provided that K is set sufficiently large, the allele frequencies remain above an arbitrary constant for any polynomial number of iterations with very high probability. This is captured by the following lemma.

Lemma 8. *Consider the cGA optimizing* $\text{OM}_{[\sigma^2]}$ *with* $\sigma^2 > 0$. *Let* $0 < a < 1/2$ *be an arbitrary constant and* $T' = \min\{t \geq 0 \colon \exists i \in [n], p_{i,t} \leq a\}$. *If* $K = \omega(\sigma^2 \sqrt{n} \log n)$, *then for every polynomial* $\text{poly}(n)$, *n sufficiently large,* $\Pr(T' < \text{poly}(n))$ *is superpolynomially small.*

Proof Sketch. Let $i \in [n]$ be an arbitrary index. Let $\{Y_t \colon t \in \mathbb{N}_0\}$ be the tochastic process $Y_t = (1/2 - p_{i,t})\, K$. The proof begins by first showing that

$$\mathrm{E}(Y_t \mid Y_1, \ldots, Y_{t-1}) \leq Y_{t-1} - \Omega(\sigma^{-2}) \frac{1}{\sqrt{n}}.$$

The idea behind this claim is a follows. Obviously, $Y_t - Y_{t-1} \in \{-1, 0, 1\}$ and it suffices to bound the conditional expectation of this difference in one step. Again let x and y be the offspring generated in iteration t. The argument proceeds by considering the substrings of x and y induced by the remaining indexes (in $[n] \setminus \{i\}$), which are by definition statistically independent. Let $\mathcal{E}$ denote the event that these substrings are equal. If $x_i \neq y_i$, then whichever string contains a 1 in the i-th position has a strictly greater "true" fitness, and the change in Y_t with respect to Y_{t-1} depends only on the event that $\text{OM}_{[\sigma^2]}$ incorrectly ranks x and y. This probability can be bounded as in Lemma 7 and the $1/\sqrt{n}$ factor comes a bound on $\Pr(\mathcal{E})$ that arises from the fact that the number of positions $j \in [n] \setminus \{i\}$ where $x_j \neq y_j$ has a Poisson-Binomial distribution. It is then straightforward to show that the contributions to the expected difference conditioned on $\overline{\mathcal{E}}$ remains strictly negative. This is simply an exercise in checking the remaining possibilities and bounding their probability.

The proof is then finished by applying a refinement to the negative drift theorem of Oliveto and Witt [26,27] (cf. Theorem 3 of [19]). Implicitly ignoring self-loops in the Markov chain (which can only result in a slower process), we have $Y_1 = 0$ and $|Y_t - Y_{t+1}| \leq 1 < \sqrt{2}$, and thus for all $s \geq 0$,

$$\Pr(T' \leq s) \leq s \exp\left(-\frac{(1/2 - a)K|\epsilon|}{32}\right),$$

with $\epsilon = -\Omega(\sigma^{-2}/\sqrt{n})$. Since $K = \omega(\sigma^2 \sqrt{n} \log n)$, $\Pr(T' \leq s) = sn^{-\omega(1)}$.

So, for any polynomial $s = \text{poly}(n)$, with probability superpolynomially close to one, Y_s has not yet reached a state larger than $(1/2 - a)K$, and so $p_{i,t} > a$ for all $0 \leq t \leq s$. As this holds for arbitrary i, applying a union bound retains a superpolynomially small probability that any of the n frequencies have gone below a by $s = \text{poly}(n)$ steps. □

It is now straightforward to prove that the optimization time of the cGA is polynomial in the problem size and the noise variance. This is in contrast to the mutation-based $(\mu + 1)$-EA, which fails when the variance becomes large. This means the cGA scales gracefully with noise in the sense of Definition 2 applied to the $\text{OM}_{[\sigma^2]}$ noise model.

Theorem 9. *Consider the cGA optimizing* $\text{OM}_{[\sigma^2]}$ *with variance* $\sigma^2 > 0$. *If* $K = \omega(\sigma^2 \sqrt{n} \log n)$, *then with probability* $1 - o(1)$, *the cGA finds the optimum after* $\mathcal{O}(K\sigma^2 \sqrt{n} \log Kn)$ *steps.*

Proof. We will consider the drift of the stochastic process $\{X_t : t \in \mathbb{N}_0\}$ over the state space $S \subseteq \{0\} \cup [x_{\min}, x_{\max}]$ where $X_t = n - \sum_{i=1}^{n} p_{i,t}$. Hence, $x_{\min} = 1/K$.

Fix a constant $0 < a < 1/2$. We say the process has *failed* by time t if there exists some $s \leq t$ and some $i \in [n]$ such that $p_{i,s} \leq a$. Let $T = \min\{t \in \mathbb{N}_0 : X_t = 0\}$. Assuming the process never fails, by Lemma 7, the drift of $\{X_t : t \in \mathbb{N}_0\}$ in each step is bounded by $\mathrm{E}(X_t - X_{t+1} \mid X_t = s) \geq \delta X_t$ where $1/\delta = \mathcal{O}(\sigma^2 K \sqrt{n})$. By Theorem 6, $\Pr\left(T > (\ln(X_0/x_{\min}) + \lambda)/\delta\right) \leq \mathrm{e}^{-\lambda}$. Choosing $\lambda = d \ln n$ for any constant $d > 0$, the probability that $T = \Omega(K\sigma^2 \sqrt{n} \log Kn)$ is at most n^{-d}.

Letting $\mathcal{E}$ be the event that the process has not failed by $\mathcal{O}(K\sigma^2\sqrt{n}\log Kn)$ steps, by the law of total probability, the hitting time of $X_t = 0$ is bounded by $\mathcal{O}(K\sigma^2\sqrt{n}\log Kn)$ with probability $(1 - n^{-d})\Pr(\mathcal{E}) = 1 - o(1)$ where we can apply Lemma 8 to bound the probability of $\mathcal{E}$. □

4 Conclusions

In this paper we have examined the benefit of sexual recombination in evolutionary optimization on the fitness function $\text{OM}_{[\sigma^2]}$. The noise-free function ($\text{OM}_{[0]}$) is efficiently optimized by a simple hillclimber in $\Theta(n \log n)$ steps (this well-known statement follows from a coupon collector argument). Corollary 5 asserts that mutation-only (and by extension, simple hillclimbers) cannot optimize $\text{OM}_{[\sigma^2]}$ in polynomial time without using some kind of resampling strategy to reduce the variance. The intuitive reason for this is that the probability of generating and accepting a worse individual becomes larger than the probability of generating and accepting a better individual: mutation has a bias towards bit strings with about as many 0s as 1s, and for high noise the probability of accepting slightly worse individuals is about 1/2. Thus, mutation-only evolutionary algorithms do not scale gracefully in the sense that they cannot optimize noisy functions in polynomial time when the noise intensity is sufficiently high.

On the other hand, we proved that a genetic algorithm that uses gene pool recombination can always optimize noisy OneMax ($\text{OM}_{[\sigma^2]}$) in expected polynomial time, subject only to the condition that the noise variance σ^2 is bounded by some polynomial in n. Intuitively, the cGA can leverage the *sexual* operation of gene pool recombination to average out the noise and follow the underlying objective function signal.

Our results highlight the importance of understanding the influence of different search operators in uncertain environments, and suggest that algorithms such as the compact genetic algorithm that use some kind of recombination are able to scale gracefully with noise.

Acknowledgements. The research leading to these results has received funding from the European Union Seventh Framework Programme (FP7/2007–2013) under grant agreement no. 618091 (SAGE).

References

1. Bäck, T., Fogel, D.B., Michalewicz, Z. (eds.): Handbook of Evolutionary Computation, 1st edn. IOP Publishing Ltd., Bristol (1997)
2. Bianchi, L., Dorigo, M., Gambardella, L., Gutjahr, W.: A survey on metaheuristics for stochastic combinatorial optimization. Nat. Comput. **8**, 239–287 (2009)
3. Doerr, B., Goldberg, L.A.: Adaptive drift analysis. Algorithmica **65**, 224–250 (2013)
4. Doerr, B., Winzen, C.: Playing mastermind with constant-size memory. In: Proceedings of STACS 2012, pp. 441–452 (2012)
5. Doerr, B., Happ, E., Klein, C.: Crossover can provably be useful in evolutionary computation. Theor. Comput. Sci. **425**, 17–33 (2012a)
6. Doerr, B., Hota, A., Kötzing, T.: Ants easily solve stochastic shortest path problems. In: Proceedings of GECCO 2012, pp. 17–24 (2012b)
7. Doerr, B., Doerr, C., Ebel, F.: From black-box complexity to designing new genetic algorithms. Theor. Comput. Sci. **567**, 87–104 (2015)
8. Droste, S.: A rigorous analysis of the compact genetic algorithm for linear functions. Nat. Comput. **5**, 257–283 (2006)
9. Droste, S., Jansen, T., Wegener, I.: Upper and lower bounds for randomized search heuristics in black-box optimization. Theory Comput. Syst. **39**, 525–544 (2006)
10. Eiben, A.E., Smith, J.E.: Introduction to Evolutionary Computing. Springer, Heidelberg (2003)
11. Feldmann, M., Kötzing, T.: Optimizing expected path lengths with ant colony optimization using fitness proportional update. In: Proceedings of FOGA 2013, pp. 65–74 (2013)
12. Gießen, C., Kötzing, T.: Robustness of populations in stochastic environments. In: Proceedings of GECCO 2014, pp. 1383–1390 (2014)
13. Goldberg, D.E.: Genetic Algorithms in Search Optimization and Machine Learning. Addison-Wesley, Boston (1989)
14. Gutjahr, W., Pflug, G.: Simulated annealing for noisy cost functions. J. Global Optim. **8**, 1–13 (1996)
15. Harik, G.R., Lobo, F.G., Goldberg, D.E.: The compact genetic algorithm. IEEE Trans. Evol. Comp. **3**, 287–297 (1999)
16. Jansen, T., Wegener, I.: Real royal road functions–where crossover provably is essential. Discrete Appl. Math. **149**, 111–125 (2005)
17. Jansen, T., Wegener, I.: The analysis of evolutionary algorithms - a proof that crossover really can help. Algorithmica **34**, 47–66 (2002)
18. Jin, Y., Branke, J.: Evolutionary optimization in uncertain environments–a survey. IEEE Trans. Evol. Comp. **9**, 303–317 (2005)
19. Kötzing, T.: Concentration of first hitting times under additive drift. In: Proceedings of GECCO 2014, pp. 1391–1397 (2014)
20. Kötzing, T., Sudholt, D., Theile, M.: How crossover helps in pseudo-boolean optimization. In: Proceedings of GECCO 2011, pp. 989–996 (2011)
21. Lehre, P.K., Witt, C.: Concentrated hitting times of randomized search heuristics with variable drift. In: Ahn, H.-K., Shin, C.-S. (eds.) ISAAC 2014. LNCS, vol. 8889, pp. 686–697. Springer, Heidelberg (2014)
22. Mühlenbein, H., Paaß, G.: From recombination of genes to the estimation of distributions I. Binary parameters. In: Ebeling, W., Rechenberg, I., Voigt, H.-M., Schwefel, H.-P. (eds.) PPSN 1996. LNCS, vol. 1141, pp. 178–187. Springer, Heidelberg (1996)

23. Mühlenbein, H., Voigt, H.-M.: Gene pool recombination in genetic algorithms. In: Osman, I.H., Kelly, J.P. (eds.) Meta-Heuristics, pp. 53–62. Springer, New York (1996)
24. Neumann, F., Witt, C.: Bioinspired Computation in Combinatorial Optimization - Algorithms and Their Computational Complexity. Natural Computing Series. Springer, Heidelberg (2010)
25. Neumann, F., Oliveto, P.S., Rudolph, G., Sudholt, D.: On the effectiveness of crossover for migration in parallel evolutionary algorithms. In: Proceedings of GECCO 2011, pp. 1587–1594 (2011)
26. Oliveto, P.S., Witt, C.: Simplified drift analysis for proving lower bounds in evolutionary computation. Algorithmica **59**, 369–386 (2011)
27. Oliveto, P.S., Witt, C.: Erratum: simplified drift analysis for proving lower bounds in evolutionary computation (2012). arXiv:1211.7184 [cs.NE]
28. Oliveto, P.S., Witt, C.: Improved time complexity analysis of the simple genetic algorithm. Theoret. Comput. Sci. **605**, 21–41 (2015)
29. Prügel-Bennett, A.: Benefits of a population: five mechanisms that advantage population-based algorithms. IEEE Trans. Evol. Comp. **14**, 500–517 (2010)
30. Richter, J.N., Wright, A., Paxton, J.: Ignoble trails - where crossover is provably harmful. In: Rudolph, G., Jansen, T., Lucas, S., Poloni, C., Beume, N. (eds.) PPSN 2008. LNCS, vol. 5199, pp. 92–101. Springer, Heidelberg (2008)
31. Storch, T., Wegener, I.: Real royal road functions for constant population size. Theor. Comput. Sci. **320**, 123–134 (2004)
32. Sudholt, D.: Crossover is provably essential for the Ising model on trees. In: Proceedings of GECCO 2005, pp. 1161–1167 (2005)
33. Sudholt, D.: Crossover speeds up building-block assembly. In: Proceedings of GECCO 2012, pp. 689–702 (2012)
34. Sudholt, D., Thyssen, C.: A simple ant colony optimizer for stochastic shortest path problems. Algorithmica **64**, 643–672 (2012)
35. Watson, R.A., Jansen, T.: A building-block royal road where crossover is provably essential. In: Proceedings of GECCO 2007, pp. 1452–1459 (2007)
36. Weisstein, E.W.: Erfc, From MathWorld-A Wolfram Web Resource (2015). http://mathworld.wolfram.com/Erfc.html
37. Witt, C.: Optimizing linear functions with randomized search heuristics - the robustness of mutation. In: Proceedings of STACS 2012, pp. 420–431 (2012)

Algorithmic Learning for Steganography: Proper Learning of k-term DNF Formulas from Positive Samples

Matthias Ernst[1,2], Maciej Liśkiewicz[1], and Rüdiger Reischuk[1(✉)]

[1] Institut für Theoretische Informatik, Universität zu Lübeck, Lübeck, Germany
{ernst,liskiewi,reischuk}@tcs.uni-luebeck.de
[2] Graduate School for Computing in Medicine and Life Sciences, Universität zu Lübeck, Lübeck, Germany

Abstract. Proper learning from positive samples is a basic ingredient for designing secure steganographic systems for unknown covertext channels. In addition, security requirements imply that the hypothesis should not contain false positives. We present such a learner for k-term DNF formulas for the uniform distribution and a generalization to q-bounded distributions. We briefly also describe how these results can be used to design a secure stegosystem.

1 Introduction

Digital steganography is a fairly new field of modern computer science concerned with camouflaging the presence of secret data in legal communications. In the general setting, a sender, often called Alice or the *steganographer* wishes to send a hidden message to a recipient via a public channel, which is completely monitored by an adversary called Warden or *steganalyst*. Taking a "typical" document Alice tries to embed a secret message in it such that a steganalyst cannot determine whether the secret message is present or not. In particular, Warden should have little chances to distinguish original documents, called *coverdocuments*, from altered ones called *stegodocuments*. This implies in general that the distributions of coverdocuments and stegodocuments have to be fairly close.

A crucial component when modeling steganography and steganalysis is the *knowledge* of the parties involved about coverdocuments. Considering different levels of knowledge, various models have been defined and studied. For example, if both the steganographer and the steganalyst have perfect knowledge about the distribution of coverdocuments and these documents satisfy certain conditions, secure steganography can be modeled and investigated by means of information and coding theory, whereas steganalysis can be done by applying statistical detection theory. But, though well-understood, such models are quite artificial

M. Ernst—This work was supported by the Graduate School for Computing in Medicine and Life Sciences funded by Germany's Excellence Initiative [DFG GSC 235/2].

K. Elbassioni and K. Makino (Eds.): ISAAC 2015, LNCS 9472, pp. 151–162, 2015.
DOI: 10.1007/978-3-662-48971-0_14

and far away from reality (for more discussion, see [9]). The other extreme is to assume that the steganographer a priori has no knowledge whatsoever about typical documents and can only get information using a sampling oracle. Even if the steganalyst has full knowledge assuming the existence of secure cryptographic one-way functions, provably secure steganography is possible [7], but *any* secure steganographic system requires an exponential number of samples with respect to the message length [4]. Thus, steganography becomes highly inefficient.

To be closer to the real world, newer approaches to steganalysis and steganography assume some reasonable partial knowledge about the type of covertext channel. Then steganalysis can be formulated as a binary classification problem and examined using methods from machine learning. This line of research has currently received much attention (see e.g. [6,10,17]). However, learning approaches to steganography have not been studied systematically so far.

As in real applications of steganography we assume that Alice knows that the coverdocument distribution belongs to some class of distributions – she can choose the media where to embed into. Besides that, she can only use a sampling oracle to get information about the actual coverdocument distribution. Then the steganographic encoding can be stated as a two-stage problem (for a formal definition of steganography see Sect. 4):

(1) Algorithmic learning of the concrete distribution of coverdocuments and
(2) Generating a stegodocument that encodes a given piece of message.

Hence, the essential difficulties in constructing efficient algorithms arise because of two reasons. First, a standard PAC approach to model this situation typically fails because of a fundamental difference: only positive samples are available. Second, algorithms for random generation of combinatorial objects from a given (typically uniform) distribution, see e.g. [8], cannot be applied directly since the generated objects have to encode given messages.

Most recently Liśkiewicz et al. [12] have obtained several promising results in generating stegodocuments. They have considered three families of coverdocument channels described by monomials, by decision trees (DTs), and by DNF formulas, respectively, assuming uniform distribution of documents. The learning complexity of the corresponding concept classes in the general case ranges from low up to high (assuming $RP \neq NP$). For these families of channels efficient generic algorithms have been constructed that for a given description of the coverdocuments, suitably manipulate the documents to embed secret messages, even against a steganalyst with full knowledge. This solves Problem (2) above and allows secure steganography assuming the coverdocument distributions can be learned *properly*, i.e. such that the learning algorithm outputs a monomial, resp. a DT, or a DNF expression as its hypothesis, when learning from positive data only.

Notice the importance of the proper learning here. For example, it is well known that k-term DNF formulas can be learned efficiently from positive samples with respect to k-CNF formulas, i.e. such that the learning algorithm outputs a k-CNF formula for the concept represented by an unknown k-term DNF. However, such a k-CNF representation of coverdocuments is useless for stegodocuments generation, because one would have to find satisfying assignments for k-CNF formulas which cannot be done efficiently in general. Unlike monomials and k-CNF formulas, the problem whether DTs and DNF-formulas can be learned properly from positive samples in an efficient way, remains open even for simple probability distributions like the uniform one. This paper gives an affirmative answer to this question for k-term DNFs.

Learnability of k-term DNF: Known Results. For the notion of learnability, we loosely follow the PAC model. In the standard setting (i.e. with positive and negative samples) it is not feasible to learn k-term DNF formulas properly in a *distribution-free* sense for fixed $k \geq 2$ unless $RP = NP$. Learning k-term DNF concepts for $k \geq 4$ remains infeasible even if allowing as hypothesis $f(k)$-term DNF, for $f(k) \leq (2k - 3)$ [14]. For unrestricted DNF formulas, it is infeasible to learn with respect to DNF hypothesis, even if the number of terms in the hypotheses is arbitrary large [1]. Assuming that samples are drawn from specific distributions over the learning domain but still allowing positive and negative samples, the situation changes drastically. Flammini et al. [5] have shown that k-term DNF formulas are learnable (properly) in polynomial time using positive and negative samples drawn from q*-bounded* distributions (the ratio of the probabilities $D(x)/D(y)$ for elements in the support does not exceed q for some number $q \geq 1$). This class is a natural generalization of the uniform distribution.

If the number of terms of the DNFs may grow, from [19] we know that n-term DNF formulas over the uniform distribution can be learned using a polynomial number of samples in quasi-polynomial time. However, the hypothesis space has to be extended to $(n \cdot t)$-term DNF with t depending on the sample complexity.

Concerning steganographic applications one has to learn DNF formulas properly *and* from positive samples only. The next serious complication is to exclude false positives in order to achieve steganographic security. In the distribution free setting, this learning task can efficiently be mastered for 1-term DNF (monomials) [18]. But it becomes infeasible for k-term DNF, with $k \geq 2$, and log-term as well as for unrestricted DNF formulas [13]. There is a positive result for monotone DNF (MDNF) formulas over the uniform distribution. It is possible to learn log-term MDNF formulas from positive samples only [15]. The class of k-term MDNFs can even be learned over q-bounded distributions from positive samples [11,16]. Also, a method for positively learning 2-term DNF over q-bounded distributions is known [5]. Most recently De et al. [3] have shown that DNF formulas have efficient learning algorithms from uniformly distributed positive samples, but instead of a k-term DNF hypothesis the learner outputs a *sampler*. This model seems to be unsuitable for embedding secret messages efficiently, because it is unknown how coverdocuments can be modified to securely embed a given message without knowing an adequate k-term DNF hypothesis.

Our Contribution. The main result of this paper is an efficient learner without false positives for k-term DNF formulas from positive samples with hypothesis space identical to the concept class for arbitrary fixed k over q-bounded distributions. The major challenge already occurs for the uniform distribution: false positives cannot be tolerated at all. Our solution works in two phases. The learner switches from k-term DNF to k-CNF representation in phase 1 and then back in the second phase. In more details, in the first phase k-term DNF formulas are learned using k-CNF formulas with very high accuracy and without false positives using a first sequence of positive samples.

In phase 2, we construct a set of *maximal monomials* that should cover most of the k-CNF formula generated. The number of candidates for these monomials could be extremely large. Thus, we have to design a mechanism to select a suitable subset. This subset will still contain many more than k monomials. Finally, we apply tests with a second sequence of positive samples to select a subset of size at most k as final hypothesis.

As a negative result, we show that it is impossible to learn unrestricted DNF formulas without false positives. For q-bounded distributions learning n-term DNF formulas requires an exponential number of positive samples regardless of the hypothesis space. An overview of the current state of knowledge concerning DNF learning is given in Table 1.

Table 1. Positive and negative (unless $RP = NP$) results for learning DNF formulas from positive samples over several distributions in polynomial time.

Concept class	Distribution-free	Uniform/q-bounded
1-term DNF (monomials)	yes [18]	yes [18]
2-term DNF	no [13]	yes [5]
k-term DNF	no [13]	yes (Theorem 1)
log-term DNF	no [14]	open
unrestricted DNF	no [14]	no (Theorem 2)

2 Preliminaries

Let us start with some basic definitions. In the following, n will always denote the number of variables and $\mathcal{X} = \{0,1\}^n$ the set of binary strings of length n. For a distribution D over $\mathcal{X}$ let $\mathtt{sp}(D) := \{x \in \mathcal{X} \mid D(x) > 0\}$ denote the support of D. For $q \geq 1$ such a distribution is called *q-bounded* if $\max\{D(x) \mid x \in \mathtt{sp}(D)\} \leq q \cdot \min\{D(x) \mid x \in \mathtt{sp}(D)\}$.

For a Boolean formula φ let $\mathtt{sat}(\varphi) := \{x \in \mathcal{X} \mid \varphi(x)\}$ denote the set of assignments that satisfy φ; $\mathtt{sat}(\varphi)$ will also be called the *support* of φ. A k-CNF formula ψ is given by a conjunction of clauses each containing at most k literals. We may assume that ψ does not contain tautological clauses (having a variable and its negation simultaneously). A k-term DNF formula φ is a disjunction of at most k monomials. φ is called *non-redundant* if it does not contain monomials M such that removing M from φ does not change $\mathtt{sat}(\varphi)$, in particular there are no *identical* monomials (that means having the same set of literals) or *trivial* monomials with empty support (containing a variable and its negation). A monomial M will be called *shorter* than a monomial M' if it consists of less literals than M'; we call M *larger* than M' if $|\mathtt{sat}(M)| > |\mathtt{sat}(M')|$. In this paper we consider the family of concept classes $\{\mathtt{sat}(\varphi) \subseteq \mathcal{X} \mid \varphi \text{ is a } k\text{-term DNF formula}\}$ and proper learning of the classes from positive examples, i.e. we require that a learner seeing only satisfying assignments outputs a k-term DNF formula.

The reader is assumed to be familiar with the standard concepts of PAC theory (see e.g. [18]). Below we present only the definition of learnability of a concept C from positive examples. This can be modeled by the condition that the underlying distribution D on $\mathcal{X}$ fulfills $\mathtt{sp}(D) = C$. Allowing false positives makes the problem trivial because the hypothesis $H = \mathcal{X}$ would make errors $D(C \bigtriangleup H)$ with weight 0. We therefore define: *$\mathcal{A}$ learns $\mathcal{C}$ from positive samples without false positives* if for every pair (C, D) of a concept $C \in \mathcal{C}$ and distribution $D \in \mathcal{D}$ that fulfills $\mathtt{sp}(D) = C$ its hypothesis satisfies: $H \subseteq C$ and $\Pr[D(C \setminus H) \geq \varepsilon] \leq \delta$. A concept class $\mathcal{C}$ with a set $\mathcal{D}$ of q-bounded distributions can be learned efficiently if a learner exists with running time bounded by a polynomial in $(1/\varepsilon, 1/\delta, n, q)$.

3 Learning k-term DNF from Positive Samples

Flammini et al. [5] have presented a method for learning a k-term DNF formula φ for q-bounded distributions. In a first phase candidate monomials are generated from positive samples in such a way that all monomials of φ having enough assignments actually occur. But there are generally more, and some of these monomials may have assignments that do not belong to $\mathtt{sat}(\varphi)$. Therefore, in the second phase, combinations of at most k candidate monomials are tested against a set of positive and negative samples. If such a combination fulfills a specific error bound then it becomes the output. It has been shown that with high probability this yields an approximate hypothesis.

In the following we will develop a generalization of this method that is capable of positively learning k-term DNF formulas. The learner gets only positive samples and is not allowed to generate false positives.

Computing Maximal Monomials from CNF-Formulas. It is known how to learn a k-term DNF formula φ without false positives by using as hypothesis space k-CNF formulas. In this case $((2n)^{k+1} - \ln\delta)/\varepsilon$ positive samples are needed [2,14,18]. The learner starts with the conjunction of all possible non-tautological clauses of length at most k, of which there are at most $(2n)^{k+1}$. Then clauses not satisfied by positive samples are deleted.

Our first innovation will construct candidate monomials for φ by learning a k-CNF representation ψ for φ and extracting monomials from ψ afterwards. We choose monomials M with $\mathtt{sat}(M) \subseteq \mathtt{sat}(\psi)$ as large as possible. Generally, for $k \geq 3$ it is NP-hard to find a single satisfying assignment for a k-CNF formula. But here we already know a number of satisfying assignments, namely the positive samples used to create ψ. For this purpose, we define a criterion for potential candidate monomials generated from ψ and a sample $x \in \mathtt{sat}(\psi)$.

Definition 1. *Let ψ be a Boolean formula and $x \in \mathtt{sat}(\psi)$. A monomial M is* (ψ, x)-maximal *if $x \in \mathtt{sat}(M) \subseteq \mathtt{sat}(\psi)$ and there is no submonomial of M with this property (a submonomial is obtained by removing some literals from M).*

Algorithm 1 given below computes such maximal monomials. It starts with the monomial $M = 1$ and adds literals until $\mathtt{sat}(M) \subseteq \mathtt{sat}(\psi)$ is satisfied. We may assume that every clause of ψ does not contain any variable more than once.

Lemma 1. *For a k-CNF formula ψ and $x \in \mathtt{sat}(\psi)$ Algorithm 1 computes a (ψ, x)-maximal monomial. Its runtime is bounded by a polynomial $p_k(n)$. For every (ψ, x)-maximal monomial M there exists a sequence of literals selected in line 10 such that the algorithm outputs M.*

Input: k-CNF formula ψ without tautological clauses; assignment $x \in \mathtt{sat}(\psi)$
Output: some (ψ, x)-maximal monomial M

```
1  M ⟵ 1; remove every literal from ψ that is not satisfied by x;
2  while true do
3  |  foreach clause K in ψ do
4  |  |  if there is exactly one literal ℓ in K then
5  |  |  |  M ⟵ (M ∧ ℓ);
6  |  |  |  remove all clauses that contain ℓ from ψ;
7  |  |  end
8  |  end
9  |  if ψ is empty then return M;
10 |  select an arbitrary literal ℓ' from ψ;
11 |  remove ℓ' from every clause in ψ;
12 end
```

Algorithm 1. $\mathtt{MaxMonomial}(\psi, x)$

The learner to be defined below needs several (ψ, x)-maximal monomials, but at most $2^k - 1$ many. To get them one could perform a depth-first search over those literals that are selected and then deleted from ψ until enough maximal monomials have been found. However, different choices may lead to the same monomial eventually. In order to be efficient we need a suitable mechanism to prune the search tree. Our strategy and its analysis are quite involved; therefore, the details will be presented in a full version of this paper.

Learning Candidate Monomials. Considering every maximal monomial for each positive sample used to learn the k-CNF formula ψ, one might get a very large set of monomials. Thus, a new idea is needed to handle such a situation. To obtain a bounded number of candidates to continue with we try to prune the set of maximal monomials without losing too many satisfying assignments. To this aim every monomial of the unknown k-term DNF formula φ that has a large support should become a candidate monomial. On the other hand, monomials with a small support might be removed without losing much accuracy.

Let us start by considering the number of maximal monomials in case the k-CNF formula ψ is equivalent to the unknown k-term DNF formula φ. In general

$\mathtt{sat}(\psi)$ may cover only parts of the satisfying region of a monomial in a scattered way. Hence, there could exist many (ψ, x)-maximal monomials.

Definition 2. *Let $\varphi = M_1 \vee \cdots \vee M_k$ be a non-redundant k-term DNF formula, $x \in \mathtt{sat}(\varphi)$, and $I = \{i_1, \ldots, i_p\} \subseteq \{1, \ldots, k\}$ be a non-empty set of indices. A monomial $M_{I,x}$ is called (φ, I, x)-maximal if it is (φ, x)-maximal and $\mathtt{sat}(M_{I,x}) \subseteq \mathtt{sat}(M_{i_1} \vee \cdots \vee M_{i_p})$ and after removing any M_{i_j} from the right side this inclusion fails.*

Lemma 2. *For fixed φ, I, and x, a (φ, I, x)-maximal monomial $M_{I,x}$ is unique. If $y \in \mathtt{sat}(M_{i_1} \vee \cdots \vee M_{i_p})$ has a maximal monomial $M_{I,y}$ then $M_{I,y} = M_{I,x}$.*

This implies that the number of different (φ, I, x)-maximal monomials over all $x \in \mathtt{sat}(\varphi)$ and nonempty $I \subseteq \{1, \ldots, k\}$ is bounded by $2^k - 1$. Next we will derive a bound on the number of satisfying assignments for those maximal monomials that intersect potentially scattered regions of φ.

Lemma 3. *Let $\varphi = M_1 \vee \cdots \vee M_k$ be a non-redundant k-term DNF formula with monomials M_i ordered by increasing length. For $d \in \mathbb{N}$ let $\varphi_d = M_1 \vee \cdots \vee M_u$ be composed of all M_i with $|\mathtt{sat}(M_i)| \geq 2^d$. For a Boolean formula χ_d with $\mathtt{sat}(\chi_d) \subseteq \mathtt{sat}(M_{u+1} \vee M_{u+2} \vee \cdots \vee M_k)$ define $\psi_d := \varphi_d \vee \chi_d$, $\mathcal{M}^{[d]} := \{M \mid M is a (\psi_d, x) - max. monom. for some x \in (\mathtt{sat}(\chi_d) \setminus \mathtt{sat}(\varphi_d))\}$, and $\xi_d := \bigvee_{M \in \mathcal{M}^{[d]}} M$. Then it holds $|\mathtt{sat}(\xi_d)| \leq 2^{d+k-1}$.*

These notions provide the foundation for the learner specified in Algorithm 2 giving the following result.

Theorem 1. *For constant k, Algorithm 2 learns k-term DNF formulas without false positives over q-bounded distributions in polynomial time with respect to $(1/\varepsilon, 1/\delta, n, q)$ by drawing no more positive samples than*

$$\sigma(\varepsilon, \delta, n, k, q) \ := \ \varepsilon^{-1}\, q\, k\, 2^{3k+1}\left((2n)^{k+1} + \ln(2/\delta)\right) + 48\varepsilon^{-2}\, \ln\left(2^{k^2+2}/\delta\right).$$

Correctness Proof. We first show a bound on how much monomials may overlap (their $\mathtt{sat}$-regions have a nonempty intersection).

Lemma 4. *Let $\varphi = M_1 \vee \cdots \vee M_k$ be a non-redundant k-term DNF formula and φ_i equal φ without M_i. Then $|\mathtt{sat}(M_i) \setminus \mathtt{sat}(\varphi_i)| \geq |\mathtt{sat}(M_i)| \cdot 2^{-k+1}$.*

Next, let us estimate how well a k-CNF formula ψ can reconstruct the original monomials of the unknown k-term DNF φ.

Definition 3. *Let $g(\varphi, q, k) := q\, 2^k |\mathtt{sat}(\varphi)|$. For $\gamma > 0$ call a monomial M_i of φ γ-large if $|\mathtt{sat}(M_i)| \geq \gamma\, g(\varphi, q, k)$.*

Lemma 5. *Let $\varphi = M_1 \vee \cdots \vee M_k$ be a k-term DNF formula with monomials M_i and $\psi = K_1 \wedge \cdots \wedge K_p$ be a k-CNF formula with clauses K_j and $\mathtt{sat}(\psi) \subseteq \mathtt{sat}(\varphi)$. Let D be a q-bounded distribution with $\mathtt{sp}(D) = \mathtt{sat}(\varphi)$ and let $\gamma > 0$. If $D(\mathtt{sat}(\varphi) \setminus \mathtt{sat}(\psi)) < \gamma$ then for every γ-large M_i it holds $\mathtt{sat}(M_i) \subseteq \mathtt{sat}(\psi)$.*

Input: $\varepsilon, \delta, k, q$, sampling oracle EX
Output: hypothesis φ'
$\varepsilon_1 \longleftarrow \varepsilon\, q^{-1}\, k^{-1}\, 2^{-(3k+1)}$;
$N_1 \longleftarrow \varepsilon_1^{-1}\,((2n)^{k+1} + \ln(2/\delta))$;
draw N_1 samples $E = (e_1, \ldots, e_{N_1})$ using EX;
learn k-CNF formula ψ using samples in E;
$\mathcal{M} \longleftarrow \emptyset$;
for $j \longleftarrow 1$ **to** N_1 **do**
  let $\mathcal{M}_j$ denote all (ψ, e_j)-maximal monomials and $m_j := \min\{|\mathcal{M}_j|, 2^k - 1\}$;
  generate an arbitrary subset $\mathcal{M}'_j$ of $\mathcal{M}_j$ of size m_j;
  $\mathcal{M} \longleftarrow \mathcal{M} \cup \mathcal{M}'_j$;
end
reduce $\mathcal{M}$ to the $(2^k - 1)$-shortest monomials;
$N_2 \longleftarrow 48\,\varepsilon^{-2}\,\ln(2^{k^2+2}/\delta)$;
draw N_2 samples $S = (s_1, \ldots, s_{N_2})$ using EX;
foreach *subset* W *of* $\mathcal{M}$ *of size at most* k **do**
  $\varphi_W := \bigvee_{M \in W} M$;
  if φ_W *misclassifies less than* $3\varepsilon N_2/4$ *samples of* S **then return** $\varphi' := \varphi_W$;
end

Algorithm 2. Learn-k-Term-DNF$(\varepsilon, \delta, k, q, EX)$

Thus, if a CNF-formula ψ approximates a k-term DNF-formula φ quite well then every monomial of φ with large support is completely covered by ψ. Only monomials with small support may give rise to errors in the approximation.

Now we show that the set of candidate monomials $\mathcal{M}$ constructed by Algorithm 2 contains all large monomials.

Lemma 6. *Let $\varphi = M_1 \vee \cdots \vee M_k$ be a non-redundant k-term DNF formula. With probability at least $1 - \delta/2$, Algorithm 2 adds a monomial M'_i, with $\mathtt{sat}(M'_i) \supseteq \mathtt{sat}(M_i)$, to $\mathcal{M}$ for every $(\varepsilon_1 2^{2k})$-large M_i, where $\varepsilon_1 = \varepsilon\, q^{-1}\, k^{-1}\, 2^{-(3k+1)}$.*

Proof sketch. Let M_i be an $(\varepsilon_1 2^{2k})$-large monomial. Assume that the algorithm has learned a k-CNF formula ψ with $D(\mathtt{sat}(\varphi) \setminus \mathtt{sat}(\psi)) \leq \varepsilon_1$, which happens with probability at least $1 - \delta/2$. Then, using Lemmas 3, 4, and 5 one can show that the sample sequence E contains at least one element $e_j \in \mathtt{sat}(M_i)$, such that no (φ, e_j)-maximal monomial intersects with potential scattered regions of φ. Hence the number of (ψ, e_j)-maximal monomials can be bounded by Lemma 2 and some M'_i with $\mathtt{sat}(M'_i) \supseteq \mathtt{sat}(M_i)$ will be added to $\mathcal{M}$. All maximal monomials that intersect with scattered regions have less assignments than M_i by Lemmas 3 and 5. Thus M'_i is among the $2^k - 1$ shortest monomials in $\mathcal{M}$ by Lemma 2. □

From Lemma 6 one can conclude the correctness of Algorithm 2. The learning algorithm can be made applicable even if q is unknown (see [5]).

A Negative Result. Verbeurgt [19] has developed a method for learning poly(n)-term DNF over the uniform distribution from a polynomial number of positive and negative samples with a quasi-polynomial running time. In contrast, we can show (proof omitted):

Theorem 2. *For every q-bounded distribution D and every hypothesis space $\mathcal{H}$, learning n-term DNF formulas without false positives requires an exponential number of positive samples drawn according to D for $\varepsilon < 1/q$.*

4 Learning Documents for Steganography

We start this section with a short review of basic definitions similar to [7]. Let $\mathcal{X}$ denote the set of cover- or stegodocuments. A channel $\mathcal{C}$ is a mapping with domain $\mathcal{X}^*$ that for every sequence h of documents, called a *history*, defines a probability distribution $\mathcal{C}_h$ on $\mathcal{X}$.

A *sampling oracle* for $\mathcal{C}$ takes a history h as input and returns a random element according to $\mathcal{C}_h$. In order to generate a typical sequence of coverdocuments $c_1, c_2, \ldots$ of $\mathcal{C}$ one starts with the empty history and asks the sampling oracle for a first element c_1, then with history $h_1 = c_1$ a second element c_2 is requested, and so on. $\mathcal{C}$ is called *supuniform* if for every h, $\mathcal{C}_h$ is the uniform distribution on $\mathsf{sp}(\mathcal{C}_h)$.

A *stegosystem* for $\mathcal{X}$ is a pair of polynomial-time bounded probabilistic algorithms $\mathcal{S} = [SE, SD]$ such that, for a security parameter κ,

(1) the encoder SE having access to a sampling oracle for a channel $\mathcal{C}$ gets as input a history h (elements that have already been generated by $\mathcal{C}$), a secret key $K \in \{0,1\}^\kappa$, and a message $\mu \in \{0,1\}^m$ and returns a sequence of stegodocuments $s_1, s_2, \ldots$ that should look like typical elements of $\mathcal{C}$ starting with history h (the length of this sequence may depend on κ and m).
(2) The decoder SD takes as input a secret key K and a sequence of documents S and returns a string $\mu \in \{0,1\}^m$.

The *unreliability* of $\mathcal{S} = [SE, SD]$ with respect to a channel $\mathcal{C}$ is given by
$$\mathrm{UnRel}_{\mathcal{S},\mathcal{C}} := \max_{h,\mu\in\{0,1\}^m}\{\Pr_{K\in\{0,1\}^\kappa}[SD(K, SE(h,K,\mu)) \neq \mu]\}.$$
For security analysis we take as adversary a probabilistic machine W called a (t,ζ)-warden that can perform a chosen hiddentext attack:

- W can access a sampling oracle for the channel $\mathcal{C}$ that in the following will be called his *reference oracle*;
- W selects a history h and a message μ and queries a *challenge oracle* CH which is either $SE(h,K,\mu)$ or $\mathcal{C}(h,\mu)$, where $\mathcal{C}(h,\mu)$ returns a sequence of random elements of $\mathcal{C}$ with history h of the same length as $SE(h,\cdot,\mu)$;
- W runs in time t and can make up to ζ queries;
- with the help of the reference oracle $\mathcal{C}$ and the challenge oracle CH the warden $W^{\mathcal{C},CH}$ tries to distinguish stego- from coverdocuments.

His *advantage* over random guessing is defined as the difference

$$\mathrm{Adv}_{\mathcal{S},\mathcal{C}}(W) := \left|\Pr_{K\in\{0,1\}^\kappa}\left[W^{\mathcal{C},SE(\cdot,K,\cdot)} = 1\right] - \Pr\left[W^{\mathcal{C},\mathcal{C}(\cdot,\cdot)} = 1\right]\right|.$$

For a given family $\mathcal{F}$ of channels $\mathcal{C}$ the strongest notion of security for a stegosystem $\mathcal{S}$ is defined as $\mathrm{InSec}_{\mathcal{S},\mathcal{F}}(t,\zeta) := \sup_{\mathcal{C}\in\mathcal{F}} \sup_W \mathrm{Adv}_{\mathcal{S},\mathcal{C}}(W)$, where W runs over all (t,ζ)-wardens. Thus, if $\mathrm{InSec}_{\mathcal{S},\mathcal{F}}$ is small then for every channel $\mathcal{C}$ of $\mathcal{F}$ no W – even those having perfect knowledge about $\mathcal{C}$ – can detect the usage of $\mathcal{S}$ with significant advantage.

Now let us consider channels $\mathcal{C}$ over the document space $\mathcal{X} = \{0,1\}^n$ such that for every history h the support of $\mathcal{C}_h$ can be described by a k-term DNF formula. These will be called *k-term DNF channels*. In [12] a polynomial-time bounded embedding algorithm has been constructed that for a given string $\omega \in \{0,1\}^b$, an arbitrary key K, and a k-term DNF formula φ with sufficiently large support (depending on b) generates a document $s \in \mathtt{sat}(\varphi)$ that encodes ω. The distribution of these stegodocuments is uniform over $\mathtt{sat}(\varphi)$ where the probability is taken over random choices of K and the internal randomization of the algorithm. Assuming that the underlying k-term DNF channel $\mathcal{C}$ is known exactly – this means for every h a k-term DNF formula for $\mathtt{sp}(\mathcal{C}_h)$ – one can use this embedding procedure to construct an efficient stegosystem $\hat{\mathcal{S}}$ for the family $\mathcal{F}$ of all supuniform k-term DNF channels $\mathcal{C}$. It has both small unreliability and small insecurity.

Definition 4. *For $\eta \geq 1$ and an integer $k \geq 1$ let $\mathcal{F}_{k,\eta}$ be the set of all supuniform k-term DNF channels $\mathcal{C}$ such that for every history h it holds $|sp(\mathcal{C}_h)| \geq 2^\eta$.*

Let b denote the number of bits encoded per document and $m = \ell \cdot b$ the length of the secret message μ to be embedded. Combining the embedding technique of [12] with the results of the previous section we can show:

Theorem 3. *For the channel family $\mathcal{F}_{k,\eta}$ and given reliability parameters $\varepsilon, \delta > 0$ there exists a stegosystem $\mathcal{S}_k$ that for every $\mathcal{C} \in \mathcal{F}_{k,\eta}$ achieves the insecurity bound of $\hat{\mathcal{S}}$ and the unreliability bound* $\mathrm{UnRel}_{\mathcal{S}_k,\mathcal{C}} \leq 2\ell(\varepsilon+\delta) + 2em\left(k \cdot 2^{-\eta}/(1-\varepsilon)\right)^{(\log e)/b}$.

Trying to extend this result to q-bounded channels one faces the problem that the corresponding distributions are not efficiently learnable – their support can be learned, but not the individual probabilities which cannot even be specified in polynomial length in general. Thus, the stegoencoder cannot get complete knowledge about the channel and the same should hold for the steganalyst – otherwise he can easily detect any deviation from the channel distribution implying that secure and efficient steganography would be impossible. The analysis for this situation is given in a full version of this paper.

5 Conclusions

We have provided a polynomial-time algorithm for properly learning k-term DNF formulas from positive samples only. Further, we have shown that unrestricted DNF formulas cannot be learned from positive samples without false positives due to information theoretical reasons. Although the analogous learnability problem for log-term DNF formulas remains still open, the negative result for unrestricted DNF formulas shows that this new method for learning k-term DNF formulas is quite powerful.

Combining our learning algorithm with the embedding procedure of [12] we are able to construct an efficient and provably secure stegosystem for a family of channels that can be defined by k-term DNF formulas. This illustrates that methods of algorithmic learning are important for steganography. Here, however, both learning *and* embedding components are crucial. As an example, the embedding problem for supports represented by efficiently learnable k-CNF formulas seems to be infeasible.

References

1. Alekhnovich, M., Braverman, M., Feldman, V., Klivans, A.R., Pitassi, T.: The complexity of properly learning simple concept classes. J. Comput. Syst. Sci. **74**(1), 16–34 (2008)
2. Blumer, A., Ehrenfeucht, A., Haussler, D., Warmuth, M.K.: Learnability and the Vapnik-Chervonenkis dimension. J. ACM **36**(4), 929–965 (1989)
3. De, A., Diakonikolas, I., Servedio, R.A.: Learning from satisfying assignments. In: Indyk, P. (ed.) Proc. SODA, pp. 478–497. SIAM, Philadelphia (2015)
4. Dedić, N., Itkis, G., Reyzin, L., Russell, S.: Upper and lower bounds on black-box steganography. J. Cryptology **22**(3), 365–394 (2009)
5. Flammini, M., Marchetti-Spaccamela, A., Kučera, L.: Learning DNF formulae under classes of probability distributions. In: Proc. COLT, pp. 85–92. ACM, New York (1992)
6. Fridrich, J.: Steganography in digital media: principles, algorithms, and applications. Cambridge University Press, New York (2009)
7. Hopper, N., von Ahn, L., Langford, J.: Provably secure steganography. IEEE T. Comput. **58**(5), 662–676 (2009)
8. Jerrum, M.R., Valiant, L.G., Vazirani, V.V.: Random generation of combinatorial structures from a uniform distribution. Theor. Comput. Sc. **43**, 169–188 (1986)
9. Ker, A.D., Bas, P., Böhme, R., Cogranne, R., Craver, S., Filler, T., Fridrich, J., Pevný, T.: Moving steganography and steganalysis from the laboratory into the real world. In: Proc. IH&MMSec, pp. 45–58. ACM, New York (2013)
10. Kodovsky, J., Fridrich, J., Holub, V.: Ensemble classifiers for steganalysis of digital media. IEEE T. Inform. Forensics and Sec. **7**(2), 432–444 (2012)
11. Kucera, L., Marchetti-Spaccamela, A., Protasi, M.: On learning monotone DNF formulae under uniform distributions. Inform. Comput. **110**(1), 84–95 (1994)
12. Liśkiewicz, M., Reischuk, R., Wölfel, U.: Grey-box steganography. Theor. Comput. Sc. **505**, 27–41 (2013)
13. Natarajan, B.K.: Probably approximate learning of sets and functions. SIAM J. Comput. **20**(2), 328–351 (1991)
14. Pitt, L., Valiant, L.G.: Computational limitations on learning from examples. J. ACM **35**(4), 965–984 (1988)
15. Sakai, Y., Maruoka, A.: Learning monotone log-term DNF formulas under the uniform distribution. Theory of Comput. Syst. **33**(1), 17–33 (2000)
16. Sakai, Y., Maruoka, A.: Learning k-term monotone boolean formulae. In: Doshita, S., Furukawa, K., Jantke, K.P., Nishida, T. (eds.) ALT 1992. LNCS, vol. 743, pp. 195–207. Springer, Heidelberg (1993)

17. Schaathun, H.G.: Machine Learning in Image Steganalysis. Wiley-IEEE Press, Chichester (2012)
18. Valiant, L.G.: A theory of the learnable. CACM **27**(11), 1134–1142 (1984)
19. Verbeurgt, K.: Learning DNF under the uniform distribution in quasi-polynomial time. In: Proc. COLT, pp. 314–326. Morgan Kaufmann Publishers Inc., San Francisco (1990)

Combinatorial Optimization and Approximation Algorithms II

Obtaining a Triangular Matrix by Independent Row-Column Permutations

Guillaume Fertin[1], Irena Rusu[1], and Stéphane Vialette[2](✉)

[1] LINA UMR CNRS 6241, Université de Nantes, Nantes, France
{guillaume.fertin,irena.rusu}@univ-nantes.fr
[2] Université Paris-Est, LIGM (UMR 8049), CNRS, UPEM,
ESIEE Paris, ENPC, 77454 Marne-la-vallée, France
vialette@univ-mlv.fr

Abstract. Given a square (0, 1)-matrix A, we consider the problem of deciding whether there exists a permutation of the rows and a permutation of the columns of A such that, after these have been carried out, the resulting matrix is triangular. The complexity of the problem was posed as an open question by Wilf [6] in 1997. In 1998, DasGupta et al. [3] seemingly answered the question, proving it is **NP**-complete. However, we show here that their result is flawed, which leaves the question still open. Therefore, we give a definite answer to this question by proving that the problem is **NP**-complete. We finally present an exponential-time algorithm for solving the problem.

1 Introduction

In his contribution to the tribute to the late Professor Erdös [6], Wilf posed the following question: *"Let A be an $m \times n$ matrix of* 0's *and* 1's. *Consider the computational problem: do there exist permutations P of the rows of A, and Q, of the columns of A such that after carrying out these permutations, A is triangular? The question we ask concerns the complexity of the problem. Is this problem* **NP**-*complete? Or, does there exist a polynomial-time algorithm for doing it?"* As noted by Wilf, this problem is strongly related to job scheduling with precedence constraints, a well-known problem in theoretical computer science. The present paper is devoted to giving an answer to this question.

A square matrix is called *lower triangular* if all the entries above the main diagonal are zero. Similarly, a square matrix is called *upper triangular* if all the entries below the main diagonal are zero. A *triangular matrix* is one that is either lower triangular or upper triangular. Because matrix equations with triangular matrices are easier to solve, they are very important in linear algebra and numerical analysis. We refer the reader to [4] for an advanced discussion.

For an arbitrary square matrix A, it is well-known that there exists an invertible matrix S such that $S^{-1}AS$ is upper triangular. We focus here, however, on *permutation* matrices. Recall that a permutation matrix is a square matrix obtained from the same size identity matrix by a permutation of rows. A product of permutation matrices (resp. the inverse of a permutation matrix) is also a permutation matrix. In fact, for any permutation matrix P, $P^{-1} = P^T$.

K. Elbassioni and K. Makino (Eds.): ISAAC 2015, LNCS 9472, pp. 165–175, 2015.
DOI: 10.1007/978-3-662-48971-0_15

This paper is organized as follows. In Sect. 2, we provide the basic material needed for this paper. Section 3 is devoted to proving hardness of determining whether a square $(0, 1)$-matrix is permutation equivalent triangular, i.e. whether it can be transformed into a triangular matrix by independent row and column permutations. In Sect. 4, we give some properties of permutation equivalent triangular matrices, and present an exponential-time algorithm to determine whether a matrix is a permutation equivalent triangular matrix. The paper concludes with suggestions for further research directions.

2 Notations

For any positive integer n, denote $[n] = \{1, 2, \ldots, n\}$. Let $A = [a_{i,j}]$, $1 \leq i \leq m$ and $1 \leq j \leq n$, be a matrix of m rows and n columns. In the case that $m = n$, the matrix is *square* of *order* n. It is convenient to refer to either a row or a column of the matrix as a *line* of the matrix. We use the notation A^T for the *transpose* of matrix A. We always designate a zero matrix by $\mathbf{0}$, a matrix with every entry equal to 1 by J, and the identity matrix of order n by I. In order to emphasize the size of these matrices we sometimes include subscripts. Thus $J_{m,n}$ denotes the all 1's matrix of size m by n, and this is abbreviated to J_n if $m = n$. Notations $\mathbf{0}_{m,n}$, $\mathbf{0}_n$ and I_n are similarly defined. In displaying a matrix we often use $*$ to designate a submatrix of appropriate dimensions. Two matrices A and B are said to be *permutation equivalent* if there exist permutation matrices P and Q of suitable sizes such that $B = PAQ$.

We will be mostly concerned with matrices whose entries consist exclusively of the integers 0 and 1. Such matrices are referred to as $(0, 1)$-*matrices*. For a $(0, 1)$-matrix A, we let $\omega(A)$ stand for the number of 1's in A. A square matrix $A = [a_{i,j}]$ of order n is said to be *lower left triangular* (or *llt*, for short) if it has only 0's above the main diagonal (*i.e.* $a_{i,j} = 0$ for $1 \leq i < j \leq n$). We write $\triangle_n$ for the llt $(0, 1)$-matrix whose 0's are exclusively above the main diagonal. For two matrices $A = [a_{i,j}]$ and $B = [b_{i,j}]$ of size m by n, we write $A \leq B$ if $a_{i,j} \leq b_{i,j}$ for $1 \leq i \leq m$ and $1 \leq j \leq n$, so that a square matrix A of order n is llt if $A \leq \triangle_n$. In the context of permutation equivalent matrices, we will sometimes not be interested in any particular orientation of a triangular matrix and forget about any specific orientation such as "*lower left*". Furthermore, for readability, a matrix which is permutation equivalent to a triangular matrix is said to be a *pet matrix*. The *row sum vector* $\mathcal{R}(A) = [r_1 \, r_2 \ldots r_m]$ and the *column sum vector* $\mathcal{C}(A) = [c_1 \, c_2 \ldots c_n]$ of A are defined by $r_i = \sum_{1 \leq j \leq n} a_{i,j}$ for $1 \leq i \leq m$ and $c_j = \sum_{1 \leq i \leq m} a_{i,j}$ for $1 \leq j \leq n$. The row sum vector $\mathcal{R}(A)$ (resp. column sum vector $\mathcal{C}(A)$) is *stepwise bounded* if $|\{i : r_i \leq k\}| \geq k$ (resp. $|\{j : c_j \leq k\}| \geq k$) for $1 \leq k \leq n$. It is clear that if a $(0, 1)$-matrix A is a pet matrix then both $\mathcal{R}(A)$ and $\mathcal{C}(A)$ are stepwise bounded. The permanent of $A = [a_{i,j}]$ is defined as the number given by the formula $\text{per}(A) = \sum_{(j_1, j_2, \ldots, j_n) \in S_n} a_{1,j_1} \, a_{2,j_2} \ldots a_{n,j_n}$, where the summation is over all permutations $(j_1, j_2, \ldots, j_n)$ of $[n]$. Observe that, unlike the determinant, we do not put a minus sign in front of some of the terms in the summation. Of particular importance, the permanent does not change if the rows or columns of A are permuted.

For a set $K \subseteq [m]$ we will write $\overline{K}$ for the set $[m] \setminus K$. Let $K = \{i_1, i_2, \ldots, i_k\}$ be a set of k elements with $K \subseteq [m]$, and let $L = \{j_1, j_2, \ldots, j_l\}$ be a set of l elements with $L \subseteq [n]$. The sets K and L designate a collection of row indices and column indices, respectively, of the matrix A, and the k by l submatrix determined by them is denoted $A[K, L]$. Let $X = \{x_i : 1 \leq i \leq n\}$ be a non-empty set of n elements, that we call an *n-set.* Let $\mathcal{S} = (S_i : 1 \leq i \leq m)$ be m not necessarily distinct subsets of the n-set X. We refer to this collection of subsets of an n-set as a *configuration of subsets.* We set $a_{i,j} = 1$ if $x_j \in S_i$, and $a_{i,j} = 0$ if $x_i \notin S_i$. The resulting $(0,1)$-matrix $A = [a_{i,j}]$, $1 \leq i \leq m$ and $1 \leq j \leq n$ of size m by n is the *incidence matrix* for the configuration of subsets $\mathcal{S}$ of the n-set X. The 1s in row α_i of A display the elements in the subset S_i, and the 1's in column β_j display the occurrences of x_j among the subsets. Let $\mathcal{S} = (S_i : 1 \leq i \leq n)$ be a configuration of subsets of some ground n-set X. A bijective mapping $\varphi : \mathcal{S} \to [n]$ is said to be a *stepwise bounded labeling* (or sbl for short) of $\mathcal{S}$ if $\left|\bigcup_{\varphi(S_j) \leq i} S_j\right| \leq i$ for $1 \leq i \leq n$.

3 Answering Wilf's Question

We prove in this section that, given a square $(0,1)$-matrix A, deciding whether there exists a permutation matrix P and a permutation matrix Q of suitable size such that PAQ is triangular is **NP**-complete.

3.1 Disproving a Previous Related Result

Before giving our proof, it is worth mentioning that the following problem (called LBQIS(n, k) and rephrased to fit the context of this paper) is claimed to be **NP**-complete in [3]: Given a $(0,1)$-matrix of order n and a positive integer $k \leq n$, do there exist permutation matrices P and Q such that $PAQ = \begin{bmatrix} A_{1,1} & A_{1,2} \\ A_{2,1} & A_{2,2} \end{bmatrix}$ where $A_{1,2}$ is a square lower triangular matrix of size k by k? It is not very difficult to find a polynomial transformation from LBQIS to Wilf's question, which would prove the **NP**-completeness of the latter. Just add $n - k$ all-zero rows and $n - k$ all-zero columns to matrix A to obtain a new matrix A'. Now, notice that each submatrix $A_{1,2}$ in a solution for LBQIS may be completed with the $n - k$ all-zero rows put before row 1 of $A_{1,2}$ and with the $n - k$ all-zero columns put after column k of $A_{1,2}$ to yield a solution for the instance A' in Wilf's question, and viceversa.

Unfortunately, paper [3] contains a serious flaw in the proof. To fix things, note that in [3] LBQIS is stated in terms of bipartite graphs, for which matrix A is the reduced adjacency matrix. Then, LBQIS(n, k) is proved **NP**-complete by reduction from another problem on bipartite graphs called LBIS(n, k), using the so-called Rearrangement Lemma (Lemma 3.5 in [3]). However, the proof of this lemma is not correct, as shown by the two following counter-examples, which address two different assertions in the proof. Let G be the graph (input for LBIS) with vertices $U = \{i \mid 1 \leq i \leq 4\}$ and $V = \{i \mid 1 \leq i \leq 4\}$, whose edges are $(1,1), (2,1), (2,2), (3,2), (3,4), (4,3)$ and $(4,4)$. Thus, $n = 4$. Define $k = 1$. Let G' be the input graph for LBQIS built as in [3], and $k' = k^2 + k = 2$. In the proof of the Rearrangement Lemma, the first line claims that, given U' and V' with respective vertex orders σ_1 and σ_2 that realize an LBQIS of size k' for G', one may assume that the vertices in σ_1 and σ_2 (which are pairs of integers) are in non-decreasing order of their first integer. This is contradicted by the sets $U' = \{[2,4], [1,2]\}$ and $V' = \{[1,3], [2,2]\}$ which realize an LBQIS of size k' (*i.e.* 2) with the orders already indicated in U' and V', but which cannot be reordered as $U' = \{[1,2], [2,4]\}$ and $V' = \{[1,3], [2,2]\}$ since these new orders do not realize an LBQIS any longer. So the first assertion in the proof is false. Moreover, the vertex subset $U' \cup V'$ of G', with $U' = \{[1,1], [2,1]\}$ and $V' = \{[1,2], [1,3]\}$ is a solution of LBQIS of size k' for which the second assertion in the same lemma ("clearly $q_1 \leq p_1$") is also false. So, the proof of the Rearrangement Lemma is not correct, and consequently this also holds for the proof of the NP-completeness of LBQIS.

3.2 Our NP-completeness Proof for Wilf's Question

We present our results in terms of sbl for configurations of subsets. The rationale for considering sbl for configurations of subsets stems from the following lemma.

Lemma 1. *Let $\mathcal{S} = (S_i : 1 \leq i \leq n)$ be a configuration of subsets of some ground n-set, and let A be the corresponding incidence matrix. There exist permutation matrices P and Q of order n such that $PAQ \leq \triangle_n$ iff there exists an sbl of $\mathcal{S}$.*

We need to focus our attention on a special type of sbl. Call a bijective mapping $\varphi : \mathcal{S} \to [n]$ *normalized* if φ maps the identical subsets of elements of $\mathcal{S}$ to a set of consecutive integers. Most of the interest in normalized bijective labelings stems from the following intuitive lemma.

Lemma 2. *Let $\mathcal{S} = (S_i : 1 \leq i \leq n)$ be a configuration of subsets of some ground n-set. If there exists an sbl of $\mathcal{S}$ then there exists a normalized sbl of $\mathcal{S}$.*

We are now ready to prove that deciding whether there exists an sbl of some configuration of subsets is **NP**-complete thereby proving that deciding whether a square (0,1)-matrix is a pet matrix is **NP**-complete as well. The proof proceeds by a reduction from the **NP**-complete 3SAT problem [2]. Let an arbitrary instance of the 3SAT problem be given by a 3CNF formula $\phi = c_1 \vee c_2 \vee \ldots \vee c_m$ over variables $x_1, x_2, \ldots, x_n$. Our construction is divided into two steps: (1) construction of a (polynomial size) ground set $\mathbf{X}$ and (2) construction of a configuration of subsets C of the ground set $\mathbf{X}$. Throughout the proof, parts of the ground set $\mathbf{X}$ are written as capital bold letters $(\mathbf{V}, \mathbf{T}, \mathbf{F}, \ldots)$ and subsets of the configuration are written with capital calligraphic letters $(\mathcal{V}_i, \mathcal{T}_i, \mathcal{F}_i, \ldots)$.

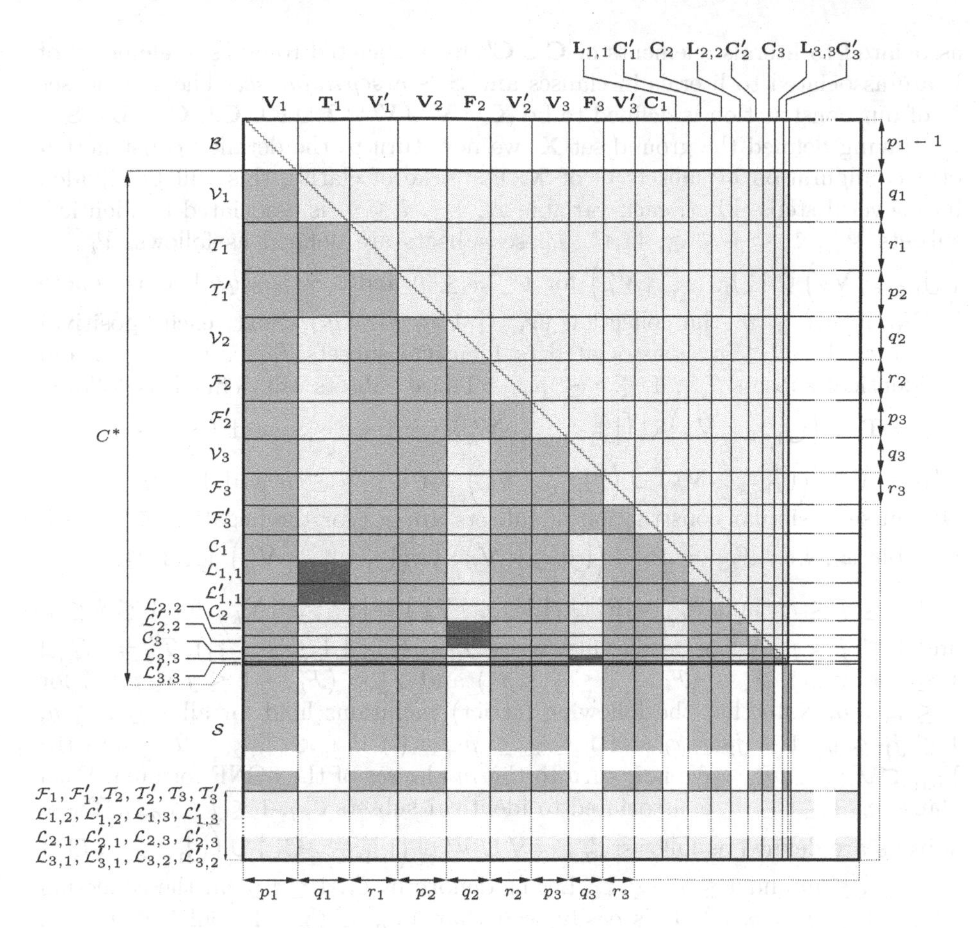

Fig. 1. Illustration of the construction for the 3CNF formula $\phi = (x_1 \vee x_2 \vee x_3) \wedge (\overline{x}_1 \vee \overline{x}_2 \vee x_3) \wedge (\overline{x}_1 \vee x_2 \vee \overline{x}_3)$. Identical subsets are not distinguishable in our representation. A satisfying truth assignment is given by $f(x_1) =$ TRUE, $f(x_2) =$ FALSE and $f(x_3) =$ FALSE. For sake of clarity, neither the ground set $\mathbf{X}$ nor the collection of subsets C is fully represented.

To begin with, define $p_i = 3(n+m+1-i)+2$, $q_i = 3(n+m+1-i)+1$ and $r_i = 3(n+m+1-i)$ for $1 \leq i \leq n+m$. Furthermore, define $p_{n+m+1} = 1$, $K = \sum_{i=1}^{n} q_i + 2\sum_{i=n+1}^{n+m} q_i$ and $L = \sum_{i=1}^{n+m}(p_{i+1}+r_i)$. Let us now define the ground set $\mathbf{X}$. Consider the pairwise disjoint sets defined as follows: $\mathbf{V}_i = \{v_{i,j} \mid 1 \leq j \leq p_i\}$, $\mathbf{V}'_i = \{v'_{i,j} \mid 1 \leq j \leq r_i\}$, $\mathbf{T}_i = \{t_{i,j} \mid 1 \leq j \leq q_i\}$, $\mathbf{F}_i = \{f_{i,j} \mid 1 \leq j \leq q_i\}$ for $1 \leq i \leq n$. Furthermore, define $\mathbf{C}_i = \{c_{i,j} \mid 1 \leq j \leq p_{n+i}\}$ $\mathbf{C}'_i = \{c'_{i,j} \mid 1 \leq j \leq r_{n+i}\}$ for $1 \leq i \leq m$, and $\mathbf{L}_{i,k} = \{\ell_{i,k,j} \mid 1 \leq j \leq q_{n+i}\}$ for $1 \leq i \leq m$ and $1 \leq k \leq 3$. Finally, define $\mathbf{S} = \{s\}$. For simplicity of notation, write $\mathbf{V} = \bigcup_{1 \leq i \leq n} \mathbf{V}_i$, $\mathbf{V}' = \bigcup_{1 \leq i \leq n} \mathbf{V}'_i$, $\mathbf{T} = \bigcup_{1 \leq i \leq n} \mathbf{T}_i$, $\mathbf{F} = \bigcup_{1 \leq i \leq n} \mathbf{F}_i$, $\mathbf{C} = \bigcup_{1 \leq i \leq m} \mathbf{C}_i$, $\mathbf{C}' = \bigcup_{1 \leq i \leq m} \mathbf{C}'_i$, and $\mathbf{L}_i = \bigcup_{1 \leq k \leq 3} \mathbf{L}_{i,k}$ for $1 \leq i \leq m$ and $\mathbf{L} = \bigcup_{1 \leq i \leq m} \mathbf{L}_i$. Informally, elements of $\mathbf{V} \cup \mathbf{V}'$ are associated to variables, elements of $\mathbf{T} \cup \mathbf{F}$ are associated to literals, elements of $\mathbf{C} \cup \mathbf{C}'$ are associated to clauses, elements of $\mathbf{L}$ are associated to literals in clauses and $\mathbf{S}$ is a *separator set*. The ground set $\mathbf{X}$ of our construction is defined to be $\mathbf{X} = \mathbf{V} \cup \mathbf{V}' \cup \mathbf{T} \cup \mathbf{F} \cup \mathbf{C} \cup \mathbf{C}' \cup \mathbf{L} \cup \mathbf{S}$.

Having defined the ground set $\mathbf{X}$, we now turn to the detailed construction of a configuration of subsets C of $\mathbf{X}$. For sake of clarity, this will be divided into several steps. First, each variable x_i, $1 \leq i \leq n$, is associated to identical subsets $\mathcal{V}_{i,j}$, $1 \leq j \leq q_i$, in C. These subsets are defined as follows: $\mathcal{V}_{i,j} = \left(\bigcup_{1 \leq k \leq i} \mathbf{V}_k\right) \cup \left(\bigcup_{1 \leq k \leq i-1} \mathbf{V}'_k\right)$ for $1 \leq i \leq n$ and $1 \leq j \leq q_i$. Let us denote by $\mathcal{V}_i$, $1 \leq i \leq n$, the collection $(\mathcal{V}_{i,j} \mid 1 \leq j \leq q_i)$. Next, each (positive) literal x_i, $1 \leq i \leq n$, is associated to identical subsets $\mathcal{T}_{i,j}$, $1 \leq j \leq r_i$, and to identical subsets $\mathcal{T}'_{i,j}$, $1 \leq j \leq p_{i+1}$. These subsets are defined as follows: $\mathcal{T}_{i,j} = \mathbf{T}_i \cup \left(\bigcup_{1 \leq k \leq i} \mathbf{V}_k\right) \cup \left(\bigcup_{1 \leq k \leq i-1} \mathbf{V}'_k\right)$ for $1 \leq i \leq n$ and $1 \leq j \leq r_i$, and $\mathcal{T}'_{i,j} = \mathbf{T}_i \cup \left(\bigcup_{1 \leq k \leq i} \mathbf{V}_k\right) \cup \left(\bigcup_{1 \leq k \leq i} \mathbf{V}'_k\right)$ for $1 \leq i \leq n$ and $1 \leq j \leq p_{i+1}$. Of course, a similar construction of subsets applies for the negation $\overline{x}_i$ of each variable x_i, *i.e.*, $\mathcal{F}_{i,j} = \mathbf{F}_i \cup \left(\bigcup_{1 \leq k \leq i} \mathbf{V}_k\right) \cup \left(\bigcup_{1 \leq k \leq i-1} \mathbf{V}'_k\right)$ for $1 \leq i \leq n$ and $1 \leq j \leq r_i$, and $\mathcal{F}'_{i,j} = \mathbf{F}_i \cup \left(\bigcup_{1 \leq k \leq i} \mathbf{V}_k\right) \cup \left(\bigcup_{1 \leq k \leq i} \mathbf{V}'_k\right)$ for $1 \leq i \leq n$ and $1 \leq j \leq p_{i+1}$. For readability, write $\mathcal{T}_i = (\mathcal{T}_{i,j} \mid 1 \leq j \leq r_i)$, $\mathcal{T}'_i = (\mathcal{T}'_{i,j} \mid 1 \leq j \leq p_{i+1})$, $\mathcal{F}_i = (\mathcal{F}_{i,j} \mid 1 \leq j \leq r_i)$ and $\mathcal{F}'_i = (\mathcal{F}'_{i,j} \mid 1 \leq j \leq p_{i+1})$ for $1 \leq i \leq n$. Note that the following (strict) inclusions hold for all $1 \leq i \leq n$, $1 \leq j_1 \leq q_i$, $1 \leq j_2 \leq r_i$ and $1 \leq j_3 \leq p_{i+1}$: (i) $\mathcal{V}_{i,j_1} \subset \mathcal{T}_{i,j_2} \subset \mathcal{T}'_{i,j_3}$ and (ii) $\mathcal{V}_{i,j_1} \subset \mathcal{F}_{i,j_2} \subset \mathcal{F}'_{i,j_3}$. We now turn to the m clauses of the 3CNF formula. Each clause c_i, $1 \leq i \leq m$, is associated to identical subsets $\mathcal{C}_{i,j}$, $1 \leq j \leq q_{n+i}$. These subsets are defined as follows: $\mathcal{C}_{i,j} = \mathbf{V} \cup \mathbf{V}' \cup \left(\bigcup_{1 \leq k \leq i} \mathbf{C}_k\right) \cup \left(\bigcup_{1 \leq k \leq i-1} \mathbf{C}'_k\right)$ for $1 \leq i \leq m$ and $1 \leq j \leq q_{n+i}$. Let us denote by $\mathcal{C}_i$, $1 \leq i \leq m$, the collection $(\mathcal{C}_{i,j} \mid 1 \leq j \leq q_{n+i})$. It is easily seen that $\mathcal{V}_{i,j_1} \subset \mathcal{C}_{k,j_2}$ for all $1 \leq i \leq n$, $1 \leq j_1 \leq q_i$, $1 \leq k \leq m$ and $1 \leq j_2 \leq q_{n+k}$.

Now, we consider the only part of the construction that depends on which literal occurs in which clauses. Denote by $\lambda_{i,k}$ the k-th literal of clause c_i, that is write $c_i = \lambda_{i,1} \vee \lambda_{i,2} \vee \lambda_{i,3}$ for $1 \leq i \leq m$, where each $\lambda_{i,k}$ is a variable or its negation. The k-th literal, $1 \leq k \leq 3$, of each clause c_i, $1 \leq i \leq m$, is associated to identical subsets $\mathcal{L}_{i,k,j}$, $1 \leq j \leq r_{n+i}$, and to identical subsets $\mathcal{L}'_{i,k,j}$, $1 \leq j \leq p_{n+i+1}$. These subsets are defined as follows: $\mathcal{L}_{i,k,j} = \mathbf{V} \cup \mathbf{V}' \cup \mathbf{A}_k \cup \mathbf{L}_{i,k} \cup \left(\bigcup_{1 \leq \ell \leq i} \mathbf{C}_\ell\right) \cup \left(\bigcup_{1 \leq \ell \leq i-1} \mathbf{C}'_\ell\right)$ for $1 \leq i \leq m$, $1 \leq j \leq r_{n+i}$ and $1 \leq k \leq 3$ and $\mathcal{L}'_{i,k,j} = \mathbf{V} \cup \mathbf{V}' \cup \mathbf{A}_k \cup \mathbf{L}_{i,k} \cup \left(\bigcup_{1 \leq \ell \leq i} \mathbf{C}_\ell\right) \cup \left(\bigcup_{1 \leq \ell \leq i} \mathbf{C}'_\ell\right)$ for $1 \leq i \leq m$, $1 \leq j \leq p_{n+i+1}$ and $1 \leq k \leq 3$, where $\mathbf{A}_k = \mathbf{T}_\ell$ if $\lambda_{i,k} = x_\ell$ and $\mathbf{A}_k = \mathbf{F}_\ell$ if $\lambda_{i,k} = \overline{x}_\ell$. For the sake of clarity, write $\mathcal{L}_{i,k} = (\mathcal{L}_{i,k,j} \mid 1 \leq j \leq r_{n+i})$ and $\mathcal{L}'_{i,k} = (\mathcal{L}'_{i,k,j} \mid 1 \leq j \leq p_{n+i+1})$ for $1 \leq i \leq m$ and $1 \leq k \leq 3$. Again, observe that $\mathcal{C}_{i,j_1} \subset \mathcal{L}_{i,k,j_2} \subset \mathcal{L}_{i,k,j_2}$ for all $1 \leq i \leq m$, $1 \leq j_1 \leq q_{n+i}$, $1 \leq j_2 \leq r_{n+i}$, $1 \leq j_3 \leq p_{n+i+1}$ and $1 \leq k \leq 3$.

Our construction ends with $p_1 + K - 1$ *utility subsets*. These subsets will be partitioned into two separate classes according to their intended function:

bootstrap subsets and *separator subsets.* First, C contains identical bootstrap subsets $\mathcal{B}_i$, $1 \leq i \leq p_1 - 1$, defined as follows: $\mathcal{B}_i = \emptyset$ for $1 \leq i \leq p_1 - 1$. The idea is to force any sbl to map the $p_1 - 1$ empty sets of $\mathcal{B}$ to the first $p_1 - 1 = 3(n+m)+1$ integers. Indeed, it is easily seen that all the above defined subsets of the configuration of subsets C but those of $\mathcal{B}$ contain at least p_1 elements and hence cannot be mapped to an integer $i \leq p_1 - 1$ in any sbl of C. Second, C contains identical separator subsets $\mathcal{S}_i, 1 \leq i \leq K$, defined by: $\mathcal{S}_i = \mathbf{V} \cup \mathbf{V}' \cup \mathbf{C} \cup \mathbf{C}' \cup \mathbf{S}$ for $1 \leq i \leq K$. The rationale of these subsets is that we need a separator between subsets in C corresponding to a satisfying truth assignment f for the 3CNF formula ϕ and garbage subsets of C, that is subsets not involved in the satisfying truth assignment f. For simplicity, let us denote by $\mathcal{B}$ the collection $(\mathcal{B}_i \mid 1 \leq i \leq p_1 - 1)$ and by $\mathcal{S}$ the collection $(\mathcal{S}_i \mid 1 \leq i \leq K)$. Clearly our construction can be carried out in polynomial time: indeed, we have $|\mathbf{X}| = O(m^2 + n^2)$ and $|C| = O(m^2 + n^2)$.

Lemma 3. *There exists a satisfying truth assignment f for ϕ iff there exists an sbl of the configuration of subsets C of the ground set $\mathbf{X}$.*

The key elements of the proof are as follows. First, it is crucial to focus on solutions that map identical subsets of elements of C to a set of consecutive elements (see Lemma 2). Second, the general shape of the solution is largely guided by the construction. Indeed, the empty subsets have to be placed first, followed by subsets corresponding to literals (either the positive or the negative literal of each variable is chosen) and next by subsets corresponding to clauses (one satisfying literal of each clause is chosen). Finally the separator subsets have to be placed, with the result that (thanks to the large polynomial number of such subsets) the remaining subsets can be placed in any order without violating the sought sbl property. The reader is invited to consider Fig. 1 for a schematic illustration of the reduction. We now briefly discuss, in an informal way, the two key arguments that are used in the proof. First, the whole procedure is, to some extent, similar to the accounting method used in amortized complexity analysis. Indeed, one might view the operation of placing a set (one after the other) as the process of charging some customer, the cost being the number of new elements that are introduced. With this metaphor in mind, notice that we do not charge when a subset does not introduce any new element, so that the leftover amount can be stored as "credit". When we place a new subset that does introduce some new elements, we can use the "credit" stored to pay for the cost of the operation. Second, when a subset uses the "credit" stored to pay the cost of introducing new elements, the following invariants can be shown to hold true: (i) it uses all the available credit and (ii) it does not allow to accumulate (it should be now clear that consecutive identical subsets do allow for accumulating credit) as much credit as it has consumed, thereby proving that subsets introduce less and less new elements as we progress adding subsets one after the other.

Theorem 1. *Let A be a $(0,1)$-matrix. Deciding whether A is a pet matrix is* **NP***-complete.*

4 Exponential-Time Algorithm

We present here an exponential-time algorithm for deciding whether a given a (0,1)-matrix A of order n is a pet matrix. We start by presenting some basic properties of square (0,1)-matrices that can be transformed into some triangular matrix by row and column independent permutations to help solving involved algorithmic issues. We of course focus of polynomial-time checkable properties.

We first focus on the permanent of a square (0,1)-matrix. A well-known result (see e.g. [1]) states that for a (0,1)-matrix A of order n, one has $\text{per}(A) = 1$ iff the lines of A may be permuted to yield a triangular matrix with 1's in the n main diagonal positions and 0's above the main diagonal. This theorem amounts to saying that $\text{per}(A) = 1$ iff there exist permutation matrices P and Q such that $I \leq PAQ \leq \triangle$. As shown in the following lemma, $\text{per}(A) = 1$ is certainly a threshold value in our context.

Lemma 4. *Let A be (0,1)-matrix. If A is a pet matrix then* $\text{per}(A) \leq 1$.

Notice that deciding $\text{per}(A) \leq 1$ for (0,1)-matrices of order n reduces to computing at most $n+1$ perfect matchings in bipartite graphs [1], and hence the above test is $O(n^3\sqrt{n})$ time as the Hopcroft–Karp algorithm for computing a maximum matching in a bipartite graph $B = (V, E)$ runs in $O(|E|\sqrt{|V|})$ [5].

Next, it is a simple matter to check that if a (0,1)-matrix A of order n is a pet matrix, then it contains at most $\frac{1}{2}n(n+1)$ 1's (*i.e.*, $\omega(A) \leq \frac{1}{2}n(n+1)$). The following lemma gives a lower bound.

Lemma 5. *Let A be (0,1)-matrix of order n, $n \geq 2$. If A contains at most $n+1$ 1's, then A is a pet matrix.*

Notice that, albeit not very impressive, Lemma 5 is tight as the square matrix $\begin{bmatrix} I_{n-2} & \mathbf{0}_{n-2,2} \\ \mathbf{0}_{2,n-2} & J_2 \end{bmatrix}$ of order n has $n-2+4 = n+2$ 1's and is not a pet matrix.

Finally, the following trivial lemma gives another condition that helps improving the running time of the algorithm in practice.

Lemma 6. *Let A be (0,1)-matrix of order n and $D(A)$ the directed graph associated to A (*i.e.*, the adjacency matrix of $D(A)$ is A). If $D(A)$ is acyclic (excluding self-loops), then A is a pet matrix.*

We now turn to presenting our exponential-time algorithm. The simplest exhaustive algorithm considers every possible pair of permutation matrices (P, Q) yielding an $O((n!)^2 \cdot \text{poly}(n))$ time algorithm. However, according to Lemma 1, it is enough to consider every permutation matrix P of order n and check whether the first i, $1 \leq i \leq n$, rows of PA have 1's in at most i columns. This observation yields an $O(n! \cdot \text{poly}(n))$ time algorithm. We propose here another exhaustive algorithm that improves on the $O(n! \cdot \text{poly}(n))$ time algorithm. The basic idea is to recursively split into smaller submatrices, instead of enumerating all permutations. For a (0,1)-matrix A of order n, we consider every possible set R of $\lceil n/2 \rceil$ rows of A and every possible set of $\lceil n/2 \rceil$ columns

$$PAQ = \begin{bmatrix} A_1 & \mathbf{0} \\ * & A_2 \end{bmatrix}$$
(a) Even

$$PAQ = \begin{bmatrix} A_1 & \mathbf{0} & \mathbf{0} \\ * & 1 & \mathbf{0} \\ & & A_2 \end{bmatrix}$$
(b) Odd and one 1

$$PAQ = \begin{bmatrix} A_1 & \mathbf{0} & \mathbf{0} \\ * & 0 & \mathbf{0} \\ & & A_2 \end{bmatrix}$$
(c) Odd and zero 1

Fig. 2. Obtaining a triangular (0,1)-matrix by recursively placing a **0** matrix in the upper right part.

C of A, and check whether these lines induce a zero matrix (or a matrix with at most one 1 in case the matrix has odd order; details follow).

If n is even, we let P and Q be two permutation matrices that put the rows in R at the first $\lceil n/2 \rceil$ positions and the columns in C at the last $\lceil n/2 \rceil$ positions. The key element for the improvement is that no specific order is required for the rows in R nor for the columns in C. The algorithm rejects the matrix A for the subsets R and C if $\omega(A[R,C]) > 1$, otherwise we can write PAQ as in Fig. 2(a), where A_1 and A_2 are matrices of order $\lceil n/2 \rceil = n/2$, and we proceed by recursively checking that both A_1 and A_2 are pet matrices. The case when n is odd is a bit more involved. First, the algorithm rejects matrix A for the subsets R and C if $\omega(A[R,C]) > 1$. Otherwise, we need to consider two (possibly positive) cases: $\omega(A[R,C]) = 1$ or $\omega(A[R,C]) = 0$. If $\omega(A[R,C]) = 1$, we let P and Q be two permutation matrices that put the rows in R at the first $\lceil n/2 \rceil$ positions and the columns of C at the last $\lceil n/2 \rceil$ positions (no specific order for the rows in R nor for the columns in C, except that the 1 of A is at row index $\lceil n/2 \rceil$ and at column index $\lceil n/2 \rceil$ in PAQ). We can write PAQ as in Fig. 2(b), where A_1 and A_2 are matrices of order $\lfloor n/2 \rfloor$, and we proceed by recursively checking that both A_1 and A_2 are pet matrices. Finally, if $\omega(A[R,C]) = 0$, for every row index $i \in R$ and every column index $j \in C$, we let P and Q be two permutation

Algorithm 1. Recognizing pet matrices.

1 **Algorithm:** permTriangular
  Data: A square matrix $A = [a_{i,j}]$ of order n
  Result: true if A is a pet matrix, false otherwise
2 **if** ($\omega(A) \leq n+1$) *or* (per(A) = 1) *or* *(A is stepwise bounded) or ($D(A)$ is acyclic)* **then return** true
3 **if** ($\omega(A) > \frac{n(n+1)}{2}$) *or* (per($A$) > 1) *or* *($\mathcal{R}(A)$ or $\mathcal{C}(A)$ is not stepwise bounded)* **then return** false
4 **for** *every subset* $R \subset [n]$ *of size* $\lceil \frac{n}{2} \rceil$ *and every subset* $C \subset [n]$ *of size* $\lceil \frac{n}{2} \rceil$ **do**
5   **if** *n is even* **then**
6   **if** permTriangularEven(A, R, C) **then return** true
7   **else**
8   **if** permTriangularOdd(A, R, C) **then return** true
9 **return** false

matrices that put the rows in R at the first $\lceil n/2 \rceil$ positions and the columns of C at the last $\lceil n/2 \rceil$ positions (no specific order for the rows in R nor for the columns in C except that row i in A is at row index $\lceil n/2 \rceil$ and column index j is at column index $\lceil n/2 \rceil$ in PAQ). We can write PAQ as in Fig. 2(c), where A_1 and A_2 are matrices of order $\lfloor n/2 \rfloor$, and we proceed by recursively checking that both A_1 and A_2 are pet matrices.

Algorithm 2. Subprocedure for recognizing pet matrices of even order.

1 **Algorithm:** permTriangularEven
  Data: A square matrix $A = [a_{i,j}]$ of even order n, and non-empty subsets $R \subset [n]$ and $C \subset [n]$, both of size $\frac{n}{2}$
  Result: true if A is a pet matrix with $A[R, C]$ as the upper right submatrix, false otherwise
2 **if** $\omega(A[R, C]) > 0$ **then return** false
3 Let $A_{\text{ul}} = A[R, \overline{C}]$ and $A_{\text{lr}} = A[\overline{R}, C]$
4 **return** permTriangular(A_{ul}) && permTriangular(A_{lr})

Algorithm 3. Subprocedure for recognizing pet matrices of odd order.

1 **Algorithm:** permTriangularOdd
  Data: A square matrix $A = [a_{i,j}]$ of odd order n, and non-empty subsets $R \subset [n]$ and $C \subset [n]$, both of size $\lceil \frac{n}{2} \rceil$
  Result: true if A is a pet matrix with $A[R, C]$ as the upper right submatrix, false otherwise
2 **if** $\omega(A[R, C]) > 1$ **then return** false
3 **if** $\omega(A) = 0$ **then**
4   **for** *every* $i \in R$ *and every* $j \in C$ **do**
5   Let $A_{\text{ul}} = A[R \setminus \{i\}, \overline{C}]$ and $A_{\text{lr}} = A[\overline{R}, C \setminus \{j\}]$
6   **if** permTriangular(A_{ul}) && permTriangular(A_{lr}) **then return** true
7   **return** false
8 **else**
9   Let i and j be the row and column indices of the unique 1 in $A[R, C]$
10   Let $A_{\text{ul}} = A[R \setminus \{i\}, \overline{C}]$ and $A_{\text{lr}} = A[\overline{R}, C \setminus \{j\}]$
11   **return** permTriangular(A_{ul}) && permTriangular(A_{lr})

A detailed description is given in Algorithms 1, 2 and 3. We now turn to evaluating the time complexity of this algorithm and we write $T(n)$ for the time complexity of calling permTriangular(A) for some $(0, 1)$-matrix A or order n. We have

$$T(n) \leq \begin{cases} (\lceil n/2 \rceil)^2 \binom{n}{\lceil n/2 \rceil}^2 (2T(\lfloor n/2 \rfloor) + 1) + O(n^3\sqrt{n}) & \text{if } n \text{ is odd} \\ 2\, (\lceil n/2 \rceil)^2 \binom{n}{\lceil n/2 \rceil}^2 T(\lfloor n/2 \rfloor) + O(n^3\sqrt{n}) & \text{if } n \text{ is even} \end{cases}$$

with $T(1) = O(1)$. The $O(n^3\sqrt{n})$ term is the time complexity for lines 2 and 3 in Algorithm 1. We also observe that the worst case occurs when

$n = 2^m - 1$ as $\lfloor n/2 \rfloor, \lfloor n/4 \rfloor, \ldots$ are odd integers. Looking for an asymptotic solution of the worst case, we thus write the following simplified recurrence: $T(2^m) = 2^{2m-2}\binom{2^m}{2^{m-1}}^2 (2T(2^{m-1}) + 1) + 2^{7m/6}$, with $T(1) = 1$. Now, write $\alpha(2^m) = 2^{2m-2}\binom{2^m}{2^{m-1}}^2$. Clearly, $\alpha(2^m) \geq 2^{7m/6}$, and hence we focus for now on on the recurrence $T(2^m) = 2\,\alpha(2^m)\,\left(T(2^{m-1}) + 1\right)$. A convenient non-recursive form of $T(2^m)$ is given in the following lemma.

Lemma 7. $T(2^m) = (2^m \prod_{i=1}^{m} \alpha(2^i)) + \left(\sum_{i=1}^{m} 2^{m-i+1} \prod_{j=i}^{m} \alpha(2^j)\right)$.

We now need the following lemma, in order to give an asymptotic solution for $T(n)$ in Proposition 1.

Lemma 8. $\sum_{i=1}^{m} 2^{m-i} \prod_{j=i}^{m} \alpha(2^j) = O\left(m\, 2^{2^{m+2}+m+1}\right)$.

Proposition 1. *Algorithm* permTriangular *runs in* $O\left(n\, 2^{4n}\, \pi^{-\log(n)}\right)$ *time.*

Proof. We have already observed that the worst case occurs for $n = 2^m - 1$. According to Lemma 8, we have $T(2^m) = O\left(2^{2^{m+2}+m-3}\,\pi^{-m}\right)$ and hence $T(n) = O\left(2^{2^{\log(n)+2}+\log(n)-3}\,\pi^{-\log(n)}\right) = O\left(n\, 2^{4n}\, \pi^{-\log(n)}\right)$. □

5 Conclusion

We suggest further research directions regarding the hardness of recognizing pet (0,1)-matrices. (i) What is the average running time of Algorithm permTriangular for pet matrices? (ii) A graph labeling strongly related to symmetric pet (0,1)-matrices can be defined as follows: Given a graph $G = (V, E)$ or order n, decide whether there exists a bijective mapping $f : V \to [n]$ such that $f(u) + f(v) > n$ for every edge $\{u, v\} \in E$ (*i.e.*, $PAP^T \leq \triangle_n$). Investigating the relationships between the two combinatorial problems is expected to yield fruitful results.

References

1. Brualdi, R.A., Ryser, H.J.: Combinatorial Matrix Theory. Cambridge University Press, New York (1991)
2. Cook, S.A.: The complexity of theorem-proving procedures. In: Proceeding 3rd Annual ACM Symposium on Theory of Computing, pp. 151–158. ACM, New York (1971)
3. DasGupta, B., Jiang, T., Kannan, S., Li, M., Sweedyk, E.: On the complexity and approximation of syntenic distance. Discrete Appl. Math. **88**(1–3), 59–82 (1998)
4. Golub, G.H., Van Loan, C.F.: Matrix Computations, 3rd edn. Johns Hopkins University Press, Baltimore and London (1996)
5. Hopcroft, J.E., Karp, R.M.: An $O(n^{2.5})$ algorithm for matching in bipartite graphs. SIAM J. Comput. **4**, 225–231 (1975)
6. Wilf, H.S.: On crossing numbers, and some unsolved problems. In: Bollobás, B., Thomason, A. (eds.) Combinatorics, Geometry and Probability: A Tribute to Paul Erdös, pp. 557–562. Cambridge University Press (1997)

Many-to-one Matchings with Lower Quotas: Algorithms and Complexity

Ashwin Arulselvan[1], Ágnes Cseh[2(✉)], Martin Groß[2], David F. Manlove[3], and Jannik Matuschke[2]

[1] Department of Management Science, University of Strathclyde, Glasgow, Scotland, UK
ashwin.arulselvan@strath.ac.uk

[2] Institute for Mathematics, TU Berlin, Berlin, Germany
{cseh,gross,matuschke}@math.tu-berlin.de

[3] School of Computing Science, University of Glasgow, Glasgow, Scotland, UK
David.Manlove@glasgow.ac.uk

Abstract. We study a natural generalization of the maximum weight many-to-one matching problem. We are given an undirected bipartite graph $G = (A \dot{\cup} P, E)$ with weights on the edges in E, and with lower and upper quotas on the vertices in P. We seek a maximum weight many-to-one matching satisfying two sets of constraints: vertices in A are incident to at most one matching edge, while vertices in P are either unmatched or they are incident to a number of matching edges between their lower and upper quota. This problem, which we call *maximum weight many-to-one matching with lower and upper quotas* (WMLQ), has applications to the assignment of students to projects within university courses, where there are constraints on the minimum and maximum numbers of students that must be assigned to each project.

In this paper, we provide a comprehensive analysis of the complexity of WMLQ from the viewpoints of classic polynomial time algorithms, fixed-parameter tractability, as well as approximability. We draw the line between NP-hard and polynomially tractable instances in terms of degree and quota constraints and provide efficient algorithms to solve the tractable ones. We further show that the problem can be solved in polynomial time for instances with bounded treewidth; however, the corresponding runtime is exponential in the treewidth with the maximum upper quota $u_{\max}$ as basis, and we prove that this dependence is necessary unless $\mathsf{FPT} = \mathsf{W}[1]$. Finally, we also present an approximation algorithm for the general case with performance guarantee $u_{\max} + 1$, which is asymptotically best possible unless $\mathsf{P} = \mathsf{NP}$.

Á. Cseh—Supported by the Deutsche Telekom Stiftung and by COST Action IC1205 on Computational Social Choice. Part of this work was carried out whilst visiting the University of Glasgow.

M. Groß—Supported by the DFG within project A07 of CRC TRR 154.

D.F. Manlove—Supported by EPSRC grant EP/K010042/1.

J. Matuschke—Supported by DAAD with funds of BMBF and the EU Marie Curie Actions.

K. Elbassioni and K. Makino (Eds.): ISAAC 2015, LNCS 9472, pp. 176–187, 2015.
DOI: 10.1007/978-3-662-48971-0_16

Keywords: Maximum matching · Many-to-one matching · Project allocation · Inapproximability · Bounded treewidth

1 Introduction

Many university courses involve some element of team-based project work. A set of projects is available for a course and each student submits a subset of projects as acceptable. For each acceptable student–project pair (s,p), there is a weight $w(s,p)$ denoting the *utility* of assigning s to p. The question of whether a given project can run is often contingent on the number of students assigned to it. Such quota constraints also arise in various other contexts involving the centralized formation of groups, including organizing activity groups at a leisure center, opening facilities to serve a community and coordinating rides within car-sharing systems. In these and similar applications, the goal is to maximize the utility of the assigned agents under the assumption that the number of participants for each open activity is within the activity's prescribed limits.

We model this problem using a weighted bipartite graph $G = (A \dot{\cup} P, E)$, where the vertices in A represent *applicants*, while the vertices in P are *posts* they are applying to. So in the above student–project allocation example, A and P represent the students and projects respectively, and E represents the set of acceptable student–project pairs. The edge weights capture the cardinal utilities of an assigned applicant–post pair. Each post has a lower and an upper quota on the number of applicants to be assigned to it, while each applicant can be assigned to at most one post. In a feasible assignment, a post is either *open* or *closed*: the number of applicants assigned to an open post must lie between its lower and upper quota, whilst a closed post has no assigned applicant. The objective is to find a maximum weight many-to-one matching satisfying all lower and upper quotas. We denote this problem by WMLQ.

In this paper, we study the computational complexity of WMLQ from various perspectives: Firstly, in Sect. 2, we show that the problem can be solved efficiently if the degree of every post is at most 2, whereas the problem becomes hard as soon as posts with degree 3 are permitted, even when lower and upper quotas are all equal to the degree and every applicant has a degree of 2. Furthermore, we show the tractability of the case of pair projects, i.e., when all upper quotas are at most 2. Then, in Sect. 3, we study the fixed parameter tractability of WMLQ. To this end, we generalize the known dynamic program for maximum independent set with bounded treewidth to WMLQ. The running time of our algorithm is exponential in the treewidth of the graph, with $u_{\max}$, the maximum upper quota of any vertex, as the basis. This yields a fixed-parameter algorithm when parameterizing by both the treewidth and $u_{\max}$. We show that this exponential dependence on the treewidth cannot be completely separated from the remaining input by establishing a $W[1]$-hardness result for WMLQ parameterized by treewidth. Finally, in Sect. 4, we discuss the approximability of the problem. We show that a simple greedy algorithm yields an approximation guarantee of $u_{\max}+1$ for WMLQ and $\sqrt{|A|}+1$ in the case of unit edge weights. We complement these results by showing that these approximation factors are asymptotically best possible, unless $\mathsf{P} = \mathsf{NP}$.

Related work

Among various applications of centralized group formation, perhaps the assignment of medical students to hospitals has received the most attention. In this context, as well as others, the underlying model is a bipartite matching problem involving lower and upper quotas. The *Hospitals/Residents problem with Lower Quotas* (HRLQ) [4,12] is a variant of WMLQ where applicants and posts have ordinal preferences over one another, and we seek a *stable matching* of residents to hospitals. Hamada et al. [12] considered a version of HRLQ where hospitals cannot be closed, whereas the model of Biró et al. [4] permitted hospital closures. Strategyproof mechanisms have also been studied in instances with ordinal preferences and no hospital closure [11].

The *Student/Project Allocation problem* [19, Sect. 5.6] models the assignment of students to projects offered by lecturers subject to upper and lower quota restrictions on projects and lecturers. Several previous papers have considered the case of ordinal preferences involving students and lecturers [1,14,20] but without allowing lower quotas. However two recent papers [15,21] do permit lower quotas together with project closures, both in the absence of lecturer preferences. Monte and Tumennasan [21] considered the case where each student finds every project acceptable, and showed how to modify the classical Serial Dictatorship mechanism to find a Pareto optimal matching. Kamiyama [15] generalized this mechanism to the case where students need not find all projects acceptable, and where there may be additional restrictions on the sets of students that can be matched to certain projects. This paper also permits lower quotas and project closures, but our focus is on cardinal utilities rather than ordinal preferences.

The unit-weight version of WMLQ is closely related to the *D-matching problem* [8,17,26], a variant of graph factor problems [24]. In an instance of the D-matching problem, we are given a graph G, and a domain of integers is assigned to each vertex. The goal is to find a subgraph G' of G such that every vertex has a degree in G' that is contained in its domain. Lovász [16] showed that the problem of deciding whether such a subgraph exists is NP-complete, even if each domain is either $\{1\}$ or $\{0,3\}$. On the other hand, some cases are tractable. For example, if for each domain D, the complement of D contains no consecutive integers, the problem is polynomially solvable [26]. As observed in [25], D-matchings are closely related to *extended global cardinality constraints* and the authors provide an analysis of the fixed-parameter tractability of a special case of the D-matching problem; see Sect. 3 for details.

The problem that we study in this paper corresponds to an optimization version of the D-matching problem. We consider the special case where G is bipartite and the domain of each applicant vertex is $\{0,1\}$, whilst the domain of each post vertex p is $\{0\} \cup \{\ell(p), \dots, u(p)\}$, where $\ell(p)$ and $u(p)$ denote the lower and upper quotas of p respectively. Since the empty matching is always feasible in our case, our aim is to find a domain-compatible subgraph G' such that the total weight of the edges in G' is maximum.

2 Degree- and Quota-restricted Cases

First, we provide a formal definition of the maximum weight many-to-one matching problem with lower quotas (WMLQ). Then, we characterize the complexity of the problem in terms of degree constraints on the two vertex sets: applicants and posts. At the end, we discuss the case of bounded upper quota constraints.

2.1 Problem Definition

In our problem, a set of applicants A and a set of posts P are given. A and P constitute the two vertex sets of an undirected bipartite graph $G = (V, E)$ with $V = A \dot{\cup} P$. For a vertex $v \in V$ we denote by $\delta(v) = \{\{v, w\} \in E : w \in V\}$ the set of edges incident to v and by $\Gamma(v) = \{w \in V : \{v, w\} \in E\}$ the *neighborhood* of v, i.e., the set of vertices that are adjacent to v. For a subset of vertices $V' \subset V$, we define $\delta(V') = \bigcup_{v \in V'} \delta(v)$. Each edge carries a *weight* $w : E \to \mathbb{R}_{\geq 0}$, representing the utility of the corresponding assignment. Each post is equipped with a *lower quota* $\ell : P \to \mathbb{Z}_{\geq 0}$ and an *upper quota* $u : P \to \mathbb{Z}_{\geq 0}$ so that $\ell(p) \leq u(p)$ for every $p \in P$. These functions bound the number of admissible applicants for the post (independent of the weight of the corresponding edges). Furthermore, every applicant can be assigned to at most one post. Thus, an *assignment* is a subset $M \subseteq E$ of the edges such that $|\delta(a) \cap M| \leq 1$ for every applicant $a \in A$ and $|\delta(p) \cap M| \in \{0, \ell(p), \ell(p) + 1, ..., u(p)\}$ for every $p \in P$. A post is said to be *open* if the number of applicants assigned to it is greater than 0, and *closed* otherwise. The *size* of an assignment M, denoted $|M|$, is the number of assigned applicants, while the *weight* of M, denoted $w(M)$, is the total weight of the edges in M, i.e., $w(M) = \sum_{e \in M} w(e)$. The goal is to find an assignment of maximum weight.

Remark 1. Note that when not allowing closed posts, the problem immediately becomes tractable. It is easy to see this in the unweighted case as any algorithm for maximum flow with lower capacities can be used to determine an optimal solution in polynomial time. This problem can be easily reduced to the classical maximum flow problem. The method can be naturally extended to the weighted case as the flow based linear program has integral extreme points due to its total unimodularity property.

Problem 1. WMLQ
Input: $\mathcal{I} = (G, w, \ell, u)$; *a bipartite graph* $G = (A \dot{\cup} P, E)$ *with edge weights* w.
Task: Find an assignment of maximum weight.
If $w = 1$ *for all* $e \in E$, *we refer to the problem as* MLQ.

Some trivial simplification of the instance can be executed right at start. If $u(p) > |\Gamma(p)|$ for a post p, then $u(p)$ can be replaced by $|\Gamma(p)|$. On the other hand, if $\ell(p) > |\Gamma(p)|$, then post p can immediately be deleted, since no feasible solution can satisfy the lower quota condition. Moreover, posts with $\ell(p) = 1$ behave identically to posts without a lower quota. From now on we assume that the instances have already been simplified this way.

2.2 Degree-Restricted Cases

In this subsection, we will consider WMLQ(i, j), a special case of WMLQ, in which we restrict us to instances in which every applicant submits at most i applications and every post receives at most j applications. In order to establish our first result, we reduce the maximum independent set problem (MIS) to MLQ. In MIS, a graph with n vertices and m edges is given and the task is to find an independent vertex set of maximum size. MIS is not approximable within a factor of $n^{1-\varepsilon}$ for any $\varepsilon > 0$, unless P = NP [29]. The problem remains APX-complete even for cubic (3-regular) graphs [2].

Theorem 1. *MLQ(2,3) is APX-complete.*

Proof. First of all, MLQ(2,3) is in APX because feasible solutions are of polynomial size and the problem has a 4-approximation (see Theorem 7).

To each instance $\mathcal{I}$ of MIS on cubic graphs we create an instance $\mathcal{I}'$ of MLQ such that there is an independent vertex set of size at least K in $\mathcal{I}$ if and only if $\mathcal{I}'$ admits an assignment of size at least $3K$, yielding an approximation-preserving reduction. The construction is as follows. To each of the n vertices of graph G in $\mathcal{I}$, a post with upper and lower quota of 3 is created. The m edges of G are represented as m applicants in $\mathcal{I}'$. For each applicant $a \in A$, $|\Gamma(a)| = 2$ and $\Gamma(a)$ comprises the two posts representing the two end vertices of the corresponding edge. Since we work on cubic graphs, $|\Gamma(p)| = 3$ for every post $p \in P$.

First we show that an independent vertex set of size K can be transformed into an assignment of at least $3K$ applicants. All we need to do is to open a post with its entire neighborhood assigned to it if and only if the vertex representing that post is in the independent set. Since no two posts stand for adjacent vertices in G, their neighborhoods do not intersect. Moreover, the assignment assigns exactly three applicants to each of the K open posts.

To establish the opposite direction, let us assume that an assignment of cardinality at least $3K$ is given. The posts' upper and lower quota are both set to 3, therefore, the assignment involves at least K open posts. No two of them can represent adjacent vertices in G, because then the applicant standing for the edge connecting them would be assigned to both posts at the same time.

The reduction given here is an L-reduction [23] with constants $\alpha = \beta = 3$. Since MLQ(2,3) belongs to APX and MIS is APX-complete in cubic graphs, it follows that MLQ(2,3) is APX-complete. □

So far we have established that if $|\Gamma(a)| \leq 2$ for every applicant $a \in A$ and $|\Gamma(p)| \leq 3$ for every post $p \in P$, then MLQ is NP-hard. In the following, we also show that these restrictions are the tightest possible. If $|\Gamma(p)| \leq 2$ for every post $p \in P$, then a maximum weight matching can be found efficiently, regardless of $|\Gamma(a)|$. Note that the case WMLQ$(1, \infty)$ is trivially solvable.

Theorem 2. WMLQ$(\infty, 2)$ *is solvable in* $O(n^2 \log n)$ *time, where* $n = |A| + |P|$.

Proof. After executing the simplification steps described after the problem definition, we apply two more changes to derive our helper graph H. Firstly, if $\ell(p) = 0$, $u(p) = 2$ and $|\Gamma(p)| = 2$, we separate p's two edges, splitting p into two posts with upper quota 1. After this step, all posts with $u(p) = 2$ also have $\ell(p) = 2$. All remaining vertices are of upper quota 1. Then, we substitute all edge pairs of posts with $\ell(p) = u(p) = 2$ with a single edge connecting the two applicants. This edge will carry the weight equal to the sum of the weights of the two deleted edges.

Clearly, any matching in H translates into an assignment of the same weight in G and vice versa. Finding a maximum weight matching in a general graph with n vertices and m edges can be done in $O(n(m + n \log n))$ time [10], which reduces to $O(n^2 \log n)$ in our case.

2.3 Quota-Restricted Cases

In this section, we address the problem of WMLQ with bounded upper quotas. Note that Theorem 1 already tells us that the case of $u(p) \leq 3$ for all posts $p \in P$ is NP-hard to solve. We will now settle the complexity of the only remaining case, where we have instances with every post $p \in P$ having an arbitrary degree and $u(p) \leq 2$. This setting models posts that need to be assigned to pairs of applicants.

The problem is connected to various known problems in graph theory, one of them being the *S-path packing problem*. In that problem, we are given a graph with a set of terminal vertices S. The task is to pack the highest number of vertex-disjoint paths so that each path starts and ends at a terminal vertex, while all its inner vertices are non-terminal. The problem can be solved in $O(n^{2.38})$ time [7,27] with the help of matroid matching [18]. An instance of MLQ with $\ell(p) = u(p) = 2$ for every post $p \in P$ corresponds to an S-path packing instance with $S = A$. The highest number of vertex-disjoint paths starting and ending in A equals half of the cardinality of a maximum assignment. Thus, MLQ with $\ell(p) = u(p) = 2$ can also be solved in $O(n^{2.38})$ time. On the other hand, there is no straightforward way to model posts with $u(p) = 1$ in S-path packing and introducing weights to the instances also seems to be a challenging task. Some progress has been made for weighted edge-disjoint paths, but to the best of our knowledge the question is unsettled for vertex-disjoint paths [13].

In the full version of the paper [3] we present a solution for the general case WMLQ with $u(p) \leq 2$. Our algorithm is based on f-factors of graphs [9].

Theorem 3. WMLQ *with* $u(p) \leq 2$ *for every* $p \in P$ *can be solved in* $O(nm + n^2 \log n)$ *time, where* $n = |V|$ *and* $m = |E|$.

3 Bounded treewidth graphs

In this section, we investigate WMLQ from the point of view of fixed-parameter tractability and analyze how efficiently the problem can be solved for instances with a bounded treewidth.

Fixed-parameter tractability. This field of complexity theory is motivated by the fact that in many applications of optimization problems certain input parameters stay small even for large instances. A problem, parameterized by a parameter k, is fixed-parameter tractable (FPT) if there is an algorithm solving it in time $f(k) \cdot \phi(n)$, where $f : \mathbb{R} \to \mathbb{R}$ is a function, ϕ is a polynomial function, and n is the input size of the instance. Note that this definition not only requires that the problem can be solved in polynomial time for instances where k is bounded by a constant, but also that the dependence of the running time on k is separable from the part depending on the input size. On the other hand, if a problem is shown to be W[1] – *hard*, then the latter property can only be fulfilled if FPT = W[1], which would imply $\mathsf{NP} \subseteq \mathsf{DTIME}(2^{o(n)})$. For more details on fixed-parameter algorithms see, e.g., [22].

Treewidth. In case of WMLQ we focus on the parameter *treewidth*, which, on an intuitive level, describes the likeness of a graph to a tree. A *tree decomposition* of graph G consists of a tree whose nodes—also called *bags*—are subsets of $V(G)$. These must satisfy the following three requirements.

1. Every vertex of G belongs to at least one bag of the tree.
2. For every edge $\{a, p\} \in E(G)$, there is a bag containing both a and p.
3. If a vertex in $V(G)$ occurs in two bags of the tree, then it also occurs in all bags on the unique path connecting them.

The *width* of a tree decomposition with a set of bags B is $\max_{b \in B} |b| - 1$. The *treewidth* of a graph G, $\text{tw}(G)$, is the smallest width among all tree decompositions of G. It is well known that a tree decomposition of smallest width can be found by a fixed-parameter algorithm when parameterized by $\text{tw}(G)$ [5].

In the following, we show that WMLQ is fixed-parameter tractable when parameterized simultaneously by the treewidth and $u_{\max}$, whereas it remains $W[1]$-hard when only parameterized by the treewidth. A similar study of the fixed-parameter tractability of the related *extended global cardinality constraint problem* (EGCC) was conducted in [25]. EGCC corresponds to the special case of the D-matching problem where the graph is bipartite and on one side of the bipartition all vertices have the domain $\{1\}$. Differently from WMLQ, EGCC is a feasibility problem (note that the feasibility version of WMLQ is trivial, as the empty assignment is always feasible). The authors of [25] provide a fixed-parameter algorithm for EGCC when parameterized simultaneously by the treewidth of the graph and the maximum domain size, and they show that the problem is W[1]-hard when only parameterized by the treewidth. These results mirror our results for WMLQ, and indeed both our FPT-algorithm for WMLQ and the one in [25] are extensions of the same classic dynamic program for the underlying maximum independent set problem. However, our hardness result uses a completely different reduction than the one in [25]. The latter makes heavy use of the fact that the domains can be arbitrary sets, whereas in WMLQ, we are confined to intervals.

Theorem 4. WMLQ *can be solved in time* $O(T + (u_{\max})^{3\,\mathrm{tw}(G)}|E|)$, *where* T *is the time needed for computing a tree decomposition of* G. *In particular,* WMLQ *can be solved in polynomial time when restricted to instances of bounded treewidth, and* WMLQ *parameterized by* $\max\{\mathrm{tw}(G), u_{\max}\}$ *is fixed-parameter tractable.*

The algorithmic proof of Theorem 4 can be found in the full version of the paper [3]. While our algorithm runs in polynomial time for bounded treewidth, the degree of the polynomial depends on the treewidth the algorithm only becomes a fixed-parameter algorithm when parameterizing by treewidth and $u_{\max}$ simultaneously. We will now show by a reduction from MINIMUM MAXIMUM OUTDEGREE that this dependence is necessary under the assumption that $\mathsf{FPT} \neq \mathsf{W}[1]$.

Problem 2. MINIMUM MAXIMUM OUTDEGREE
Input: A graph $G = (V, E)$, *edge weights* $w : E \to \mathbb{Z}_+$ *encoded in unary, a degree-bound* $r \in \mathbb{Z}_+$.
Task: Find an orientation D *of* G *such that* $\sum_{e \in \delta_D^+(v)} w(e) \leq r$ *for all* $v \in V$, *where* $\delta_D^+(v)$ *stands for the set of edges oriented so that their tail is* v.

Theorem 5 (Theorem 5 from [28]). MINIMUM MAXIMUM OUTDEGREE *is* $W[1]$*-hard when parameterized by treewidth.*

Theorem 6. MLQ *is* $\mathsf{W}[1]$*-hard when parameterized by treewidth, even when restricted to instances where* $\ell(p) \in \{0, u(p)\}$ *for every* $p \in P$.

Proof. Given an instance $(G = (V, E), w, r)$ of MINIMUM MAXIMUM OUTDEGREE, we construct an instance $(G' = (A \dot\cup P, E'), \ell, u)$ of MLQ as follows. For every vertex $v \in V$ we introduce a post $p_v \in P$ and let $\ell(p_v) = 0$ and $u(p_v) = r$. Furthermore, for every edge $e = \{v, v'\} \in E$, we introduce two posts $p_{e,v}$ and $p_{e,v'}$ with $\ell(p_{e,v}) = \ell(p_{e,v'}) = u(p_{e,v}) = u(p_{e,v'}) = w(e) + 1$, and $2w(e) + 1$ applicants $a_{e,v}^1, \ldots, a_{e,v}^{w(e)}, a_{e,v'}^1, \ldots, a_{e,v'}^{w(e)}, z_e$, for which we introduce the edges $\{p_v, a_{e,v}^i\}$, $\{a_{e,v}^i, p_{e,v}\}$, $\{p_{v'}, a_{e,v'}^i\}$, and $\{a_{e,v'}^i, p_{e,v'}\}$ for $i \in \{1, \ldots, w(e)\}$ as well as $\{p_{e,v}, z_e\}$ and $\{z_e, p_{e,v'}\}$.

We show that the constructed instance has a solution serving all applicants if and only if the MINIMUM MAXIMUM OUTDEGREE instance has an orientation respecting the bound on the outdegree.

First assume there is an orientation D of G with maximum outdegree at most r. Then consider the assignment that assigns for every oriented edge $(v, v') \in D$ the $w(e)$ applicants $a_{e,v}^i$ to p_v and the $w(e) + 1$ applicants $a_{e,v'}^i$ and z_e to $p_{e,v'}$. As the weighted outdegree of vertex v is at most r, every post p_v gets assigned at most $r = u(p_v)$ applicants.

Now assume M is a feasible assignment of applicants to posts serving every applicant. In particular, for every edge $e = \{v, v'\} \in E$, applicant z_e is assigned to either $p_{e,v}$ or $p_{e,v'}$ and exactly one of these two posts is open because the lower bound of $w(e) + 1$ can only be met if z_e is assigned to the respective post. If $p_{e,v}$ is open then all $w(e)$ applicants $a_{e,v'}^i$ are assigned to $p_{v'}$ and none of the applicants $a_{e,v}^i$ is assigned to p_v, and vice versa if $p_{e,v'}$ is open. Consider the orientation obtained by orienting every edge e from v to v' if and only if $p_{e,v}$ is open. By the above observations, the weighted outdegree of vertex v corresponds to the number of applicants assigned to post p_v, which is at most r.

Finally, note that G' can be constructed in time polynomial in the input size of the MINIMUM MAXIMUM OUTDEGREE instance as the weights are encoded in unary there. Furthermore, the treewidth of G' is at most $\max\{\text{tw}(G), 3\}$. To see this, start with a tree decomposition of G and identify each vertex $v \in V$ with the corresponding post p_v. For every edge $e = \{v, v'\} \in E$, there is a bag B with $p_v, p'_v \in B$. We add the new bag $B_e = \{p_v, p'_v, p_{e,v}, p_{e,v'}\}$ as a child to B. We further add the bags $B_{z_e} = \{p_{e,v}, p_{e,v'}, z_e\}$, $B_{a^i_{e,v}} = \{p_v, p_{e,v}, a^i_{e,v}\}$ and $B_{a^i_{e,v'}} = \{p_{v'}, p_{e,v'}, a^i_{e,v}\}$ for $i \in \{1, \dots, w(e)\}$ as children to B_e. Observe that the tree of bags generated by this construction is a tree decomposition. Furthermore, since we did not increase the size of any of the existing bags and added only bags of size at most 4 the treewidth of G' is at most $\max\{\text{tw}(G), 3\}$. □

4 Approximation

Having established the hardness of WMLQ even for very restricted instances in Theorem 1, we turn our attention towards approximability. In this section, we give an approximation algorithm and corresponding inapproximability bounds expressed in terms of $|A|, |P|$ and upper quotas in the graph.

The method, which is formally listed in Algorithm 1, is a simple greedy algorithm. We say a post p is *admissible* if it is not yet open and $|\Gamma(p)| \geq \ell(p)$. The algorithm iteratively opens an admissible post maximizing the assignable weight, i.e., it finds a post $p' \in P$ and a set A' of applicants in its neighborhood $\Gamma(p')$ with $\ell(p') \leq |A'| \leq u(p')$ such that $\sum_{a \in A'} w(a, p')$ is maximized among all such pairs. It then removes the assigned applicants from the graph (potentially rendering some posts inadmissible) and re-iterates until no admissible post is left.

Algorithm 1. Greedy algorithm for WMLQ

Initialize $P_0 = \{p \in P : |\Gamma(p)| \geq \ell(p)\}$.
Initialize $A_0 = A$.
while $P_0 \neq \emptyset$ **do**
  Find a pair $p' \in P_0$ and $A' \subseteq \Gamma(p')$ with $|A'| \leq u(p')$ such that $\sum_{a \in A'} w(a, p')$ is maximized among all such pairs.
  Open p' and assign all applicants in A' to it.
  Remove p' from P_0 and remove the elements of A' from A_0.
  for $p \in P_0$ with $\ell(p) > |\Gamma(p) \cap A_0|$ **do**
    Remove p from P_0.
  end for
end while

In the full version of the paper [3] we give a tight analysis of the algorithm, establishing approximation guarantees in terms of the number of posts $|P|$, number of applicants $|A|$, and the maximum upper quota $u_{\max} := \max_{p \in P} u(p)$ over all posts. We also provide two examples that show that our analysis of the greedy algorithm is tight for each of the described approximation factors. We further show there that the approximation ratios given above for WMLQ are almost tight from the point of view of complexity theory.

We point out a reduction from WMLQ to the set packing problem here. The elements in the universe of the set packing problem would be $A \cup P$. For each post p and for each subset $S \subset \Gamma(p)$, such that $l(p) \leq |S| \leq u(p)$, we create a set $S \cup \{p\}$ for the set packing instance. However, if the difference between upper and lower quota is not bounded, this would create an exponential sized input for the set packing problem and we could only employ an oracle based algorithm known for set packing problem to solve WMLQ. The greedy algorithm known for set packing problem [6] can be made to work in a fashion similar to the algorithm presented above.

Theorem 7. *Algorithm 1 is an α-approximation algorithm for* WMLQ *with* $\alpha = \min\{|P|, |A|, u_{\max} + 1\}$. *Furthermore, for* MLQ, *Algorithm 1 is a* $\sqrt{|A|} + 1$-*approximation algorithm. It can be implemented to run in time* $O(|E| \log |E|)$.

Theorem 8. MLQ *is not approximable within a factor of* $|P|^{1-\varepsilon}$ *or* $\sqrt{|A|}^{1-\varepsilon}$ *or* $u_{\max}^{1-\varepsilon}$ *for any* $\varepsilon > 0$, *unless* $\mathsf{P} = \mathsf{NP}$, *even when restricting to instances where* $\ell(p) = u(p)$ *for every* $p \in P$ *and* $|\Gamma(a)| \leq 2$ *for every* $a \in A$.

Acknowledgements. We would like to thank András Frank and Kristóf Bérczi for their observations that led us to Theorem 3 and the anonymous reviewers for their valuable comments, which have helped to improve the presentation of this paper.

References

1. Abraham, D.J., Irving, R.W., Manlove, D.F.: Two algorithms for the student-project allocation problem. J. Discrete Algorithms **5**(1), 79–91 (2007)
2. Alimonti, P., Kann, V.: Some APX-completeness results for cubic graphs. Theoret. Comput. Sci. **237**(1–2), 123–134 (2000)
3. Arulselvan, A., Cseh, Á., Groß, M., Manlove, D.F., Matuschke, J.: Many-to-one matchings with lower quotas: algorithms and complexity (2015). CoRR, abs/1412.0325

4. Biró, P., Fleiner, T., Irving, R.W., Manlove, D.F.: The College Admissions problem with lower and common quotas. Theoret. Comput. Sci. **411**, 3136–3153 (2010)
5. Bodlaender, H.L.: A linear-time algorithm for finding tree-decompositions of small treewidth. SIAM J. Comput. **25**(6), 1305–1317 (1996)
6. Chandra, B., Halldórsson, M.M.: Greedy local improvement and weighted set packing approximation. J. Algorithms **39**(2), 223–240 (2001)
7. Cheung, H.Y., Lau, L.C., Leung, K.M.: Algebraic algorithms for linear matroid parity problems. ACM Trans. Algorithms (TALG) **10**(3), 10:1–10:26 (2014)
8. Cornuéjols, G.: General factors of graphs. J. Comb. Theor. Ser. B **45**(2), 185–198 (1988)
9. Gabow, H.: An efficient reduction technique for degree-constrained subgraph and bidirected network flow problems. In: Proceedings of STOC 1983: the 15th Annual ACM Symposium on Theory of Computing, pages 448–456. ACM (1983)
10. Gabow, H.: Data structures for weighted matching and nearest common ancestors with linking. In: Proceedings of SODA 1990: the 1st ACM-SIAM Symposium on Discrete Algorithms, pp. 434–443. ACM-SIAM (1990)
11. Goto, M., Hashimoto, N., Iwasaki, A., Kawasaki, Y., Ueda, S., Yasuda, Y., Yokoo, M.: Strategy-proof matching with regional minimum quotas. In: Proceedings of the 2014 International Conference on Autonomous Agents and Multiagent Systems, AAMAS 2014, pp. 1225–1232, Richland (2014)
12. Hamada, K., Iwama, K., Miyazaki, S.: The Hospitals/Residents problem with lower quotas. Algorithmica, pp. 1–26 (2014)
13. Hirai, H., Pap, G.: Tree metrics and edge-disjoint S-paths. Math. Program. **147**(1–2), 81–123 (2014)
14. Iwama, K., Miyazaki, S., Yanagisawa, H.: Improved approximation bounds for the student-project allocation problem with preferences over projects. J. Discrete Algorithms **13**, 59–66 (2012)
15. Kamiyama, N.: A note on the serial dictatorship with project closures. Oper. Res. Lett. **41**, 559–561 (2013)
16. Lovász, L.: On the structure of factorizable graphs. Acta Mathematica Academiae Scientiarum Hungaricae **23**(1–2), 179–195 (1972)
17. Lovász, L.: Antifactors of graphs. Periodica Mathematica Hungarica **4**(2–3), 121–123 (1973)
18. Lovász, L.: Matroid matching and some applications. J. Comb. Theor. Ser. B **28**(2), 208–236 (1980)
19. Manlove, D.F.: Algorithmics of Matching Under Preferences. World Scientific, Singapore (2013)
20. Manlove, D.F., O'Malley, G.: Student project allocation with preferences over projects. J. Discrete Algorithms **6**, 553–560 (2008)
21. Monte, D., Tumennasan, N.: Matching with quorums. Econ. Lett. **120**, 14–17 (2013)
22. Niedermeier, R.: Invitation to Fixed-Parameter Algorithms. Oxford University Press, Oxford (2006)
23. Papadimitriou, C., Yannakakis, M.: Optimization, approximation and complexity classes. J. Comput. Syst. Sci. **43**(3), 425–440 (1991)
24. Plummer, M.: Graph factors and factorization: 1985–2003: a survey. Discrete Math. **307**(7–8), 791–821 (2007)
25. Samer, M., Szeider, S.: Tractable cases of the extended global cardinality constraint. Constraints **16**, 1–24 (2011)
26. Sebő, A.: General antifactors of graphs. J. Comb. Theor. Ser. B **58**(2), 174–184 (1993)

27. Sebő, A., Szegő, L.: The path-packing structure of graphs. In: Bienstock, D., Nemhauser, G.L. (eds.) IPCO 2004. LNCS, vol. 3064, pp. 256–270. Springer, Heidelberg (2004)
28. Szeider, S.: Not so easy problems for tree decomposable graphs (2011). CoRR, abs/1107.1177
29. Zuckerman, D.: Linear degree extractors and the inapproximability of max clique and chromatic number. Theor. Comput. **3**(6), 103–128 (2007)

Minimizing the Maximum Moving Cost of Interval Coverage

Haitao Wang[1](✉) and Xiao Zhang[2]

[1] Department of Computer Science, Utah State University, Logan, UT 84322, USA
haitao.wang@usu.edu

[2] Department of Computer Science, City University of Hong Kong, Kowloon Tong, Hong Kong
xiao.zhang@my.cityu.edu.hk

Abstract. In this paper, we study an interval coverage problem. We are given n intervals of the same length on a line L and a line segment B on L. Each interval has a nonnegative weight. The goal is to move the intervals along L such that every point of B is covered by at least one interval and the maximum moving cost of all intervals is minimized, where the moving cost of each interval is its moving distance times its weight. Algorithms for the "unweighted" version of this problem have been given before. In this paper, we present a first-known algorithm for this weighted version and our algorithm runs in $O(n^2 \log n \log \log n)$ time. The problem has applications in mobile sensor barrier coverage.

1 Introduction

In this paper, we consider an interval coverage problem, which has applications in mobile sensor barrier coverage in wireless sensor networks. For convenience, we introduce and discuss the problem from the barrier coverage point of view.

Let L be a line, say, the x-axis. Let B be a line segment on L, called a *barrier*. Denote by β the length of B. Without loss of generality, we assume B is the interval $[0, \beta]$ on L. Let $S = \{s_1, s_2, \ldots, s_n\}$ be a set of n sensors and each sensor s_i is a point on L with coordinate x_i. Each sensor s_i has a weight $w_i \geq 0$. All sensors have the same *sensing range* r. Namely, if a sensor is currently at a location x' on L, then all points of L in the interval $[x' - r, x' + r]$ is said to be *covered* by the sensor and the interval is called the *covering interval* of the sensor. The problem is to move each sensor s_i of S to a new location y_i on L such that every point of B is covered by at least one sensor of S and the value $\max_{1 \leq i \leq n} w_i \cdot |x_i - y_i|$ is minimized. For each sensor s_i, $w_i \cdot |x_i - y_i|$ is the *moving cost* of s_i. We call the problem the *weighted barrier coverage*, denoted by WBC. We assume $r \cdot n \geq \beta$ since otherwise a coverage of B would not be possible.

If all sensors have the same weight, then we refer to it as the "unweighted" version. An $O(n \log n)$ time algorithm has been given for the unweighted version

H. Wang was supported in part by NSF under Grant CCF-1317143. Part of the work by X. Zhang was carried out during his visit at Utah State University.

K. Elbassioni and K. Makino (Eds.): ISAAC 2015, LNCS 9472, pp. 188–198, 2015.
DOI: 10.1007/978-3-662-48971-0_17

by Chen et al. [5]. For the weighted version, to the best of our knowledge, we are not aware of any previous work. In this paper, we present an algorithm for the weighted version and our algorithm runs in $O(n^2 \log n \log\log n)$ time.

Related Work. Mobile Sensor Networks (MSNs) consist of a number of mobile wireless sensors, which have limited battery power and may have different energy dissipation ratio (characterized by the weights). The advantage of allowing the sensors to be mobile increases monitoring capability compared to those for which static wireless sensors are used. One of the most important applications in MSNs is to monitor a barrier to detect intruders in an attempt to cross a specific region.

If the sensors of S have different sensing ranges, we call the problem the "non-uniform case" (otherwise it is the "uniform case"). For the unweighted uniform case, Czyzowicz et al. [8] first gave an $O(n^2)$ time algorithm, and later, Chen et al. [5] solved the problem in $O(n \log n)$ time. For the unweighted non-uniform case, Chen et al. [5] presented an $O(n^2 \log n)$ time algorithm.

The *min-sum* unweighted version of the problem has also been studied, where the objective is to minimize the sum of the moving distances of all sensors. The non-uniform case of the problem is NP-hard [9]. For the uniform case, Czyzowicz et al. [9] gave an $O(n^2)$ time algorithm, and recently, Andrews and Wang [1] proposed an $O(n \log n)$ time solution. Another variation of the problem is the *min-num* version, where the goal is to move the minimum number of sensors to form a barrier coverage. Mehrandish et al. [12,13] proved the problem is NP-hard if sensors have different ranges and gave polynomial time algorithms otherwise.

Some problems on static sensors have also been considered. For example, Bar-Noy and Baumer [2] studied a problem of maximizing the lifetime of a network with static sensors, where the goal is to schedule the active time of sensors in a network so that the lifetime is maximized. A similar problem was considered in [3]. Fan et al. [10] studied a problem that aims to set an energy for each sensor to form a coverage such that the cost of all sensors is minimized.

Our Techniques. The unweighted uniform case of WBC is much easier due to an *order preserving property* [5,8]: There always exists an optimal solution in which the order of the sensors is the same as that in the input. However, the property no longer holds for the unweighted non-uniform case [5]. We can easily show that for the weighted version, the property does not hold even for the uniform case. This is one main difficulty for solving our problem WBC.

To solve the problem, we generalize the techniques in [5] for the unweighted non-uniform case. Specifically, let λ^* denote the maximum moving cost in an optimal solution of WBC. We first solve a *decision problem* in $O(n \log n)$ time to determine whether $\lambda \geq \lambda^*$ for any given value λ. If $\lambda \geq \lambda^*$, then our decision algorithm will find a "feasible solution" in which the order of the sensors will be determined. Further, with $O(n^2)$ time preprocessing, we can solve the decision problem in $O(n \log\log n)$ time for any λ (the $\log\log n$ factor is due to the van Emde Boas Tree [7]). For solving our original problem WBC (referred to as the *optimization problem* for differentiation from the decision problem), we use an approach similar in spirit to parametric search [6,11]. Namely, our optimization algorithm tries to "parameterize" the decision algorithm. Although we do not know the value λ^*, we will simulate the behavior of the decision algorithm on $\lambda = \lambda^*$, i.e., we determine the same order of the sensors as it would be obtained by the decision algorithm on $\lambda = \lambda^*$. To this end, our algorithm maintains an interval $(\lambda_1, \lambda_2]$ that contains λ^* and each step of the algorithm will shrink the interval by calling the decision algorithm on certain values λ. Unlike the traditional parametric search [6,11], our approach does not involve any parallel scheme and is actually quite practical.

The rest of the paper is organized as follows. In Sect. 2, we introduce some notation. In Sect. 3, we present our algorithm for the decision problem. Section 4 solves the optimization problem. Due to the space limit, some proofs are omitted but can be found in the full version of the paper.

2 Preliminaries

For ease of exposition, we assume the weight of each sensor of S is positive. We follow some terminologies in the previous work [5].

We use a *configuration* to refer to a specification on where each sensor $s_i \in S$ is located. For example, in the *input configuration*, each s_i is at x_i. We use C_I to denote the input configuration. Note that we can determine whether $\lambda^* = 0$ by checking whether the union of the covering intervals of all sensors in C_I contains B, which can be easily done in $O(n)$ time. If $\lambda^* = 0$, we do not need to move any sensor. Henceforth, we assume $\lambda^* > 0$.

For any sensor s_i, we use $I(s_i)$ to denote its covering interval. For any subset S' of sensors, with a little abuse of notation, we use $I(S')$ to denote the union of the covering intervals of all sensors in S'. For each sensor s_i, we call the left (resp., right) endpoint of $I(s_i)$ the *left (resp., right) extension* of s_i.

For convenience, for any point x on L, we also use x to denote its coordinate on L, and vice versa. For any point x on L, let $p^+(x)$ denote a point $x' \in L$ such that $x' > x$ and x' is infinitesimally close to x. Note that our algorithm never needs to find such a point $p^+(x)$ and we use $p^+(x)$ only for explaining our idea.

3 The Decision Problem

In this section, we consider the decision problem: given any value $\lambda > 0$, determine whether $\lambda \geq \lambda^*$.

We present an $O(n \log n)$ time algorithm that solves the decision problem. We call this algorithm the *decision algorithm*. Given any value λ, if $\lambda \geq \lambda^*$, then we say that λ is a *feasible value* and our decision algorithm will find a *feasible solution* in which B is covered and the moving cost of each sensor is at most λ.

Consider any value $\lambda > 0$. For any sensor $s_i \in S$, λ/w_i is the maximum distance s_i is allowed to move on L. Let $a_i = x_i - r - \lambda/w_i$ and $b_i = x_i + r + \lambda/w_i$. Note that a_i is the leftmost point and b_i is the rightmost point on L that can be covered by s_i with respect to λ. We call a_i (resp., b_i) the *leftmost (resp., rightmost)* λ*-coverable point* of s_i. Let $x_i^l = a_i + r$ and $x_i^r = b_i - r$. Namely, x_i^l (resp., x_i^r) is the leftmost (resp., rightmost) location that s_i is allowed to move, and we call it the *leftmost (resp., rightmost)* λ*-reachable location* of s_i.

In the following, in Sect. 3.1 we describe the algorithm while leaving the implementation details in Sect. 3.2. The correctness proof of the algorithm is omitted. The high-level scheme of our decision algorithm is similar to that for the unweighted non-uniform case in [5], but the low-level details are different.

3.1 The Algorithm Description

In the beginning, we move each sensor s_i to its rightmost λ-reachable location x_i^r. Let C_0 denote the resulting configuration. In C_0, for each sensor s_i, it is not allowed to move rightwards but can move leftwards by distance $2\lambda/w_i$.

If $\lambda \geq \lambda^*$, our algorithm will compute a subset S^c of sensors with their new locations such that B is covered by these sensors (i.e., $B \subseteq I(S^c)$) and the moving cost of each sensor of S^c is at most λ. For each sensor $s_i \in S \setminus S^c$, it can be anywhere in $[x_i^l, x_i^r]$, but in our solution it is at x_i^r (i.e., it does not move from its location in C_0). We call S^c a *solution subset.*

Consider a general step i with $i \geq 1$. Let C_{i-1} be the configuration right before the i-th step. Our algorithm maintains the following *invariants.* (1) We have a subset of sensors $S_{i-1} = \{s_{g(1)}, s_{g(2)}, \ldots, s_{g(i-1)}\}$, where for each $1 \leq j \leq i-1$, $g(j)$ is the index of the sensor $s_{g(j)}$ in S. Let $S_{i-1} = \emptyset$ for $i = 1$. (2) In C_{i-1}, for each sensor s_k of S, if s_k is in S_{i-1}, then s_k is at a new location $y_k \in [x_i^l, x_i^r]$; otherwise, it is still at x_k^r. (3) $B \cap I(S_{i-1})$, i.e., the intersection of B and the union of the covering intervals of all sensors of S_{i-1} in C_{i-1}, is an interval $[0, R_{i-1}]$ for some value $0 \leq R_{i-1} < \beta$. This means that the point $p^+(R_{i-1})$ is not covered by any sensor in S_{i-1}. Let $R_0 = 0$.

Initially when $i = 1$, we have $S_{i-1} = \emptyset$ and $R_0 = 0$, and thus all algorithm invariants hold for C_0. The i-th step of the algorithm will find a new sensor $s_{g(i)} \in S \setminus S_{i-1}$ and move it to a new location $y_{g(i)} \in [x_{g(i)}^l, x_{g(i)}^r]$ (and thus obtain a new configuration C_i). Let $R_i = y_{g(i)} + r$ and $S_i = S_{i-1} \cup \{s_{g(i)}\}$. We will show that $B \cap I(S_i) = [0, R_i]$. If $R_i \geq L$, then we have found a feasible

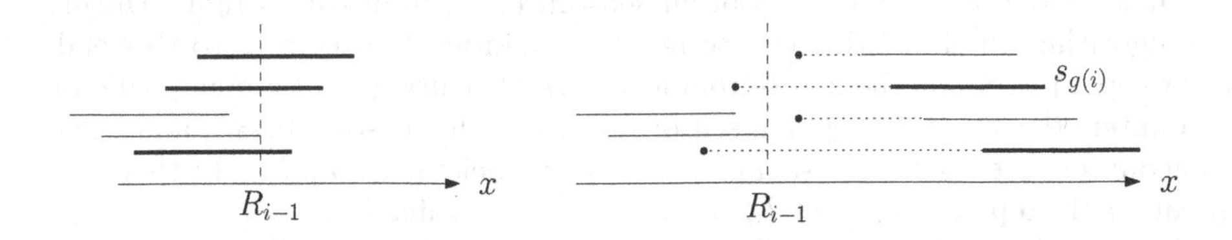

Fig. 1. Illustrating the two sets S_{i1} (left) and S_{i2} (right). The segments are the covering intervals of sensors. The thick segments correspond to the sensors in S_{i1} (left) and S_{i2} (right). The four black points in the right figure are the leftmost λ-coverable points of the four sensors to the right of R_{i-1}. The sensor $s_{g(i)}$ of S_{i2} is labeled (any sensor in S_{i1} can be $s_{g(i)}$).

solution with S_i as our solution subset. Otherwise, we proceed on the next step $i+1$ and all algorithm invariants are maintained. We give the details of the i-th step below. Note that the discussions are on the configuration C_{i-1}.

Since the interval $[0, R_{i-1}]$ is currently covered by the sensors of S_{i-1}, we need to find sensors in $S \setminus S_{i-1}$ to cover the rest of the barrier, i.e., $[R_{i-1}, \beta]$.

Define S_{i1} to be the set of sensors that cover the point $p^+(R_{i-1})$ in C_{i-1}, i.e., $S_{i1} = \{s_k \mid x_k^r - r \leq R_{i-1} < x_k^r + r\}$. According to the algorithm invariants, no sensor in S_{i-1} covers $p^+(R_{i-1})$. Thus, it holds that $S_{i1} \subseteq S \setminus S_{i-1}$.

If $S_{i1} \neq \emptyset$, then we choose an *arbitrary* sensor S_{i1} as $s_{g(i)}$[1] (e.g., see Fig. 1) and let $y_{g(i)} = x_{g(i)}^r$. We let $R_i = y_{g(i)} + r$ and $C_i = C_{i-1}$ (i.e., C_i is the same as C_{i-1} since $s_{g(i)}$ is not moved in this case).

If $S_{i1} = \emptyset$, then define S_{i2} as the set of sensors of S whose leftmost λ-coverable points are to the left of (or at) R_{i-1} and whose left extensions are strictly to the right of R_{i-1}, i.e., $S_{i2} = \{s_k \mid a_k \leq R_{i-1} < x_k^r - r\}$. Hence, for each $s_k \in S_{i2}$, s_k currently in C_{i-1} does not cover $p^+(R_{j-1})$ but we can move s_k leftwards for a distance at most $2\lambda/w_k$ to cover it.

If $S_{i2} \neq \emptyset$, then we choose the *leftmost* sensor of S_{i2} as $s_{g(i)}$ (e.g., see Fig. 1), and let $y_{g(i)} = R_{i-1} + r$. We move $s_{g(i)}$ to $y_{g(i)}$ to obtain the configuration C_i (where the right extension of $s_{g(i-1)}$ colocates with the left extension of $s_{g(i)}$).

If $S_{i2} = \emptyset$, then we conclude that $\lambda < \lambda^*$ and terminate the algorithm.

Hence, if $S_{i1} = S_{i2} = \emptyset$, the algorithm will stop and report $\lambda < \lambda^*$. Otherwise, a sensor $s_{g(i)}$ is found from either S_{i1} or S_{i2}, and it is moved to $y_{g(i)}$. In either case, according to our discussion, $R_i = y_{g(i)} + r$ and $[0, R_i] = I(S_i)$ in C_i (recall that $S_i = S_{i-1} \cup \{s_{g(i)}\}$). If $R_i \geq \beta$, then we terminate the algorithm and report $\lambda \geq \lambda^*$ and C_i as a feasible solution; otherwise, we proceed on the next step $i+1$ and all algorithm invariants have been maintained.

As there are n sensors in S, the algorithm will finish in at most n steps.

3.2 The Algorithm Implementation

We first give an implementation for our algorithm that runs in $O(n \log n)$ time. Later we will improve the implementation with certain preprocessing.

We first move each sensor s_i to x_i^r to obtain the initial configuration C_0. Then, we sort the $2n$ extensions of all sensors in C_0 from left to right. During the algorithm, in each i-th step, we need to maintain the set S_{i1}. To this end, we sweep a point p on the line L from left to right. During the sweeping, when p encounters the left extension of a sensor, we insert the sensor into S_{i1}, and when p encounters the right extension of a sensor, we delete it from S_{i1}. In this way, in each i-th step, when p is at R_{i-1}, the set S_{i1} is available.

If $S_{i1} \neq \emptyset$, then we arbitrarily pick a sensor in S_{i1} as $s_{g(i)}$. To store the set S_{i1}, since all sensor covering intervals have the same length, an easy observation is that the earlier a sensor is inserted into S_{i1}, the earlier it is deleted from S_{i1}. Therefore, we can simply use a first-in-first-out queue to store all sensors of S_{i1}

[1] It might be more natural to pick the rightmost sensor of S_{i1} as $s_{g(i)}$. In fact, an arbitrary one is sufficient.

such that each insertion and deletion can be done in constant time. To pick an arbitrary sensor in S_{i1}, we can always pick the sensor in the front of the queue.

If $S_{i1} = \emptyset$, then we need to determine whether $S_{i2} = \emptyset$. If yes, we terminate the algorithm and report $\lambda < \lambda^*$. Otherwise, we need to find the leftmost sensor of S_{i2} as $s_{g(i)}$. We assume that the sweeping point p is now at R_{i-1}. To maintain the set S_{i2} during the sweeping of p, we do the following.

In the beginning we sort the n leftmost λ-coverable points of all sensors of S along with the $2n$ extensions of all sensors in C_0. During the sweeping of p, if p encounters a leftmost λ-coverable point of some sensor s_k, then we insert s_k to S_{i2}. Further, if p encounters a left extension of some sensor s_k, then we delete s_k from S_{i2} (recall that this is also the moment we should insert s_k to S_{i1}).

In this way, when p is at R_{i-1}, S_{i2} is available. Since we need to find the leftmost sensor in S_{i2}, we use a balanced binary search tree T to store all sensors of S_{i2} where the "key" of each sensor s_k is the value x_k^r (which is its location in C_{i-1}). Clearly, T can support each of the following operations on S_{i2} in $O(\log n)$ time: inserting a sensor, deleting a sensor, finding the leftmost sensor.

After $s_{g(i)}$ is found, R_i can be computed immediately as discussed in the algorithm description. If $s_{g(i)}$ is from S_{i1}, then we do not need to actually move $s_{g(i)}$. We proceed to sweep p as usual. If $s_{g(i)}$ is from S_{i2}, we need to move $s_{g(i)}$ leftwards to $y_{g(i)} = R_{i-1} + r$. Since $s_{g(i)}$ is moved, we should also update the original sorted list of the $2n$ extensions of all sensors in C_0 to guide the future sweeping of p. We use the following approach to avoid the explicit update. We maintain a flag table for all sensor extensions in C_0. Initially, every table entry is *valid*. If $s_{g(i)}$ is moved, then we set the table entries of the two extensions of the sensor *invalid*, which can be done in constant time. Due to this extra table, during the sweeping of p, when p encounters a sensor extension, we first check the table to see whether the extension is still valid. If yes, then we proceed as usual; otherwise we ignore the event. This only cost an extra constant time at each event. In addition, after we proceed to sweep p from R_{i-1}, before processing the next event, we always check whether it is before p arrives at R_i (which is actually the right extension of $s_{g(i)}$). If yes, we proceed as usual; otherwise, we should process the event at R_i (i.e., determine the next sensor $s_{g(i+1)}$).

Hence, during the sweeping of p, each event can be handled in $O(\log n)$ time and each i-th step of the algorithm can be performed in $O(\log n)$ time. Since there are $O(n)$ events, the total running time of the algorithm is $O(n \log n)$.

Note that the space of the algorithm is $O(n)$.

Theorem 1. *Given any value λ, we can determine whether $\lambda \geq \lambda^*$ in $O(n \log n)$ time and $O(n)$ space.*

Our algorithm in Sect. 4 will call our decision algorithm many times, for which we have the following alternative result by improving the above implementation with certain preprocessing. The proof for Corrollary 1 is omitted.

Corollary 1. *With $O(n^2)$ time and $O(n^2)$ space preprocessing, we can determine whether $\lambda \geq \lambda^*$ in $O(n \log \log n)$ time for any given λ.*

4 The Optimization Problem

In this section, we give an $O(n^2 \log n \log\log n)$ time algorithm for solving the optimization problem. The goal is to compute λ^*, after which we can obtain an optimal solution by applying the decision algorithm on $\lambda = \lambda^*$.

The high-level scheme of our algorithm is different from that for the unweighted non-uniform case in [5] because the algorithmic scheme in [5] relies on another decision algorithm that can determine whether $\lambda^* = \lambda$ for any given λ while our algorithmic scheme does not need such a decision algorithm. The details of our algorithm are even more different from those in [5].

We assume that the preprocessing in Corollary 1 has been done. Unless otherwise stated, we always use Corollary 1 to implement our decision algorithm.

4.1 An Overview

As discussed before, a main difficulty is that we do not know the order of the sensors that cover B in an optimal solution. If we knew λ^*, then we could run our decision algorithm on $\lambda = \lambda^*$ to obtain an optimal solution in which the order of the sensors is also determined. We use an idea similar to the parametric search [6, 11]. We "parameterize" our decision algorithm with λ as a parameter. Although we do not know the value λ^*, we execute the decision algorithm in such a way that it will determine the same solution subset of sensors $s_{g(1)}, s_{g(2)}, \ldots$ in the same order as would be obtained if we ran the decision algorithm on $\lambda = \lambda^*$. To this end, we will use our decision algorithm to prune certain λ values.

Recall that for any value λ, step i of our decision algorithm determines the sensor $s_{g(i)}$ and obtains the set $S_i = \{s_{g(1)}, s_{g(2)}, \ldots, s_{g(i)}\}$ with $I(S_i) \cap B = [0, R_i]$ in the configuration C_i. In our following algorithm, we often consider λ as a variable rather than a fixed value. Thus, we will use $S_i(\lambda)$ (resp., $R_i(\lambda)$, $s_{g(i)}(\lambda)$, $C_i(\lambda)$, $x_i^r(\lambda)$) to refer to the corresponding S_i (resp., R_i, $s_{g(i)}$, C_i, x_i^r).

Our algorithm has at most $n+1$ steps. Consider a general i-th step for $i \geq 1$. Right before the step, we have an interval $(\lambda_{i-1}^1, \lambda_{i-1}^2]$ and a sensor set $S_{i-1}(\lambda)$, such that the following algorithm invariants hold.

1. $\lambda^* \in (\lambda_{i-1}^1, \lambda_{i-1}^2]$, i.e., λ^* is either equal to λ_{i-1}^2 or in $(\lambda_{i-1}^1, \lambda_{i-1}^2)$.
2. The set $S_{i-1}(\lambda)$ is the same for all values $\lambda \in (\lambda_{i-1}^1, \lambda_{i-1}^2)$. If $\lambda^* \neq \lambda_{i-1}^2$, then $S_{i-1}(\lambda)$ has the same sensors as $S_{i-1}(\lambda^*)$ with the same order.
3. $R_{i-1}(\lambda)$ on $\lambda \in (\lambda_{i-1}^1, \lambda_{i-1}^2)$ is a nondecreasing linear function and it has been explicitly computed.
4. $R_{i-1}(\lambda) < \beta$ for any $\lambda \in (\lambda_{i-1}^1, \lambda_{i-1}^2)$.

Initially when $i = 1$, we let $\lambda_0^1 = -\infty$ and $\lambda_0^2 = \infty$. Since $S_0(\lambda) = \emptyset$ and $R_0(\lambda) = 0$ for any λ, all invariants hold for $i = 1$.

The i-th step will either compute λ^*, or obtain a new interval $(\lambda_i^1, \lambda_i^2] \subseteq (\lambda_{i-1}^1, \lambda_{i-1}^2]$ and a sensor $s_{g(i)}(\lambda)$ with $S_i(\lambda) = S_{i-1}(\lambda) \cup \{s_{g(i)}(\lambda)\}$ such that all algorithm invariants hold on $S_i(\lambda)$ and $(\lambda_i^1, \lambda_i^2]$. We will show that the i-th step runs in $O(n \log n \log\log n)$ time. The details of the i-th step are given below.

4.2 A General i-th Step

We assume $\lambda^* \neq \lambda^2_{i-1}$ and thus λ^* is in $(\lambda^1_{i-1}, \lambda^2_{i-1})$. In fact, our following algorithm does not reply on this assumption, but only uses the set $S_{i-1}(\lambda)$, the interval $(\lambda^1_{i-1}, \lambda^2_{i-1})$, and the function $R_{i-1}(\lambda)$, which are all known. But we make the assumption only for explaining the rationale of our approach.

Since $\lambda^* \in (\lambda^1_{i-1}, \lambda^2_{i-1})$, by our algorithm invariants, for any $\lambda \in (\lambda^1_{i-1}, \lambda^2_{i-1})$, $S_{i-1}(\lambda)$ has the same sensors as $S_{i-1}(\lambda^*)$ with the same order. We simulate the decision algorithm on $\lambda = \lambda^*$. In order to determine the sensor $s_{g(i)}(\lambda^*)$, we first compute $S_{i1}(\lambda^*)$, i.e., the set of sensors covering the point $p^+(R_{i-1}(\lambda^*))$ in the configuration $C_{i-1}(\lambda^*)$, as follows.

Consider any sensor s_k in $S \backslash S_{i-1}(\lambda)$. Its position in the configuration $C_{i-1}(\lambda)$ is $x^r_k(\lambda) = x_k + \lambda/w_k$, which is an increasing function of λ. Thus, both the left and the right extensions of s_k in $C_{i-1}(\lambda)$ are increasing linear functions of λ. By our algorithm invariants, $R_{i-1}(\lambda)$ is a nondecreasing linear function. Suppose $f(\lambda)$ is the left or right extension of s_k in $C_{i-1}(\lambda)$. Unless the slope of $R_{i-1}(\lambda)$ is $1/w_k$, there is at most one value λ in $(\lambda^1_{i-1}, \lambda^2_{i-1})$ such that $R_{i-1}(\lambda) = f(\lambda)$. Let S' be the set of sensors s_k of $S \setminus S_{i-1}(\lambda)$ such that $1/w_k$ is not equal to the slope of $R_{i-1}(\lambda)$. We compute the set $S_{i1}(\lambda^*)$ as follows.

Suppose we increases λ from λ^1_{i-1} to λ^2_{i-1}. During the increasing of λ, we say that an "event" happens if $R_{i-1}(\lambda)$ is equal to the left or right extension value of a sensor $s_k \in S'$ at the current value of λ (called *event value*). It is not difficult to see that during the increasing of λ, the set $S_{i1}(\lambda)$ is fixed between any two adjacent events. In order to compute $S_{i1}(\lambda^*)$, we first compute all event values, which can be done in linear time by using the function $R_{i-1}(\lambda)$ and all left and right extension functions of the sensors in S'. Let Λ denote the set of all event values. In addition, we add λ^1_{i-1} and λ^2_{i-1} to Λ. Next we sort all values in Λ. Then, by using our decision algorithm, we do binary search on the sorted list of Λ to find two adjacent values λ_1 and λ_2 in the sorted list such that $\lambda^* \in (\lambda_1, \lambda_2]$. Note that $(\lambda_1, \lambda_2] \subseteq (\lambda^1_{i-1}, \lambda^2_{i-1}]$. Since $|\Lambda| = O(n)$, the binary search calls our decision algorithm $O(\log n)$ times, which takes $O(n \log n \log\log n)$ time in total.

We make another assumption that $\lambda^* \neq \lambda_2$. Again, this assumption is only for explaining the rationale of our approach, and the following algorithm does not rely on this assumption. Under the assumption, for any λ in (λ_1, λ_2), the set $S_{i1}(\lambda)$ is exactly $S_{i1}(\lambda^*)$. Hence, we can compute $S_{i1}(\lambda^*)$ by taking any $\lambda \in (\lambda_1, \lambda_2)$ and explicitly computing $S_{i1}(\lambda)$, which can be done in $O(n)$ time.

The above has computed $S_{i1}(\lambda^*)$ in $O(n \log n \log\log n)$ time (provided that our previous two assumptions are true). According to our decision algorithm, depending on whether $S_{i1}(\lambda^*) = \emptyset$, there are two cases.

If $S_{i1}(\lambda^*) \neq \emptyset$, we take any sensor of $S_{i1}(\lambda^*)$ as $s_{g(i)}(\lambda^*)$. Let $\lambda^1_i = \lambda_1$, $\lambda^2_i = \lambda_2$, and $S_i(\lambda) = S_{i-1}(\lambda) \cup \{s_{g(i)}(\lambda^*)\}$. We will show later that all algorithm invariants hold on $S_i(\lambda)$ and $(\lambda^1_i, \lambda^2_i]$ (further processing may be needed).

If $S_{i1}(\lambda^*) = \emptyset$, then according to our decision algorithm, we need to compute the set $S_{i2}(\lambda^*)$, i.e., the set of sensors of S whose leftmost λ-coverable points are to the left of (or at) $R_{i-1}(\lambda^*)$ and whose left extensions are strictly to the right of $R_{i-1}(\lambda^*)$. Under our previous two assumptions, we have $\lambda^* \in$

$(\lambda_1, \lambda_2) \subseteq (\lambda_{i-1}^1, \lambda_{i-1}^2)$. By our algorithm invariants, $R_{i-1}(\lambda)$ is a nondecreasing linear function on $\lambda \in (\lambda_1, \lambda_2)$. For each sensor $s_k \in S$, its leftmost λ-coverable point $a_k(\lambda) = x_k - \lambda/w_k - r$ is a decreasing linear function. Therefore, the interval (λ_1, λ_2) contains at most one value λ such that $R_{i-1}(\lambda) = a_k(\lambda)$.

Suppose we increases λ from λ_1 to λ_2. During the increasing of λ, we say that an "event" happens when $R_{i-1}(\lambda)$ is equal to $a_k(\lambda)$ for some sensor $s_k \in S$ at some *event value* λ. During the increasing of λ, the set $S_{i2}(\lambda)$ is fixed between any two adjacent events. This suggests the following way to compute $S_{i2}(\lambda^*)$.

First, we compute all event values by using $R_{i-1}(\lambda)$ and the functions $a_k(\lambda)$ of all sensors s_k of S. Let Λ contain all event values. We also add λ_1 and λ_2 to Λ. We sort all values of Λ. Then, by using our decision algorithm, we do binary search on the sorted list of Λ to find two adjacent values λ_1' and λ_2' in the sorted list such that $\lambda^* \in (\lambda_1', \lambda_2']$. Note that $(\lambda_1', \lambda_2'] \subseteq (\lambda_1, \lambda_2]$. Since $|\Lambda| = O(n)$, the above binary search calls the decision algorithm $O(\log n)$ time, which takes $O(n \log n \log\log n)$ time in total.

By our above analysis, the set $S_{i2}(\lambda)$ is fixed for all $\lambda \in (\lambda_1', \lambda_2')$. We take an arbitrary value $\lambda \in (\lambda_1', \lambda_2')$ and compute the set $S_{i2}(\lambda)$ explicitly, which can be done in $O(n)$ time. The proof of Lemma 1 is omitted.

Lemma 1. *If $S_{i2}(\lambda) = \emptyset$, then λ^* is in $\{\lambda_{i-1}^2, \lambda_2, \lambda_2'\}$.*

By Lemma 1, if $S_{i2}(\lambda) = \emptyset$, then λ^* is the smallest feasible value of $\{\lambda_{i-1}^2, \lambda_2, \lambda_2'\}$, which can be found by calling our decision algorithms on the three values respectively. Otherwise, we proceed as follows.

We make the third assumption that $\lambda^* \neq \lambda_2'$. Similarly, this assumption is only for explaining our approach, and the following algorithm does not rely on this assumption. Under the assumption, $\lambda^* \in (\lambda_1', \lambda_2')$. Hence, $S_{i2}(\lambda^*) = S_{i2}(\lambda)$. Next, we find the sensor $s_{g(i)}(\lambda^*)$, which is the leftmost sensor of $S_{i2}(\lambda^*)$. Although $S_{i2}(\lambda)$ is fixed for all $\lambda \in (\lambda_1', \lambda_2')$, the leftmost sensor of it may not be the same for all $\lambda \in (\lambda_1', \lambda_2')$. To find $s_{g(i)}(\lambda^*)$, we use the following approach.

For each sensor $s_k \in S_{i2}(\lambda)$, its location in the configuration $C_{i-1}(\lambda)$ is $x_k^r(\lambda) = x_k + \lambda/w$ for any $\lambda \in (\lambda_1', \lambda_2')$. Hence, $x_k^r(\lambda)$ is an increasing linear function of λ, which defines a line in the 2D coordinate system in which the x-coordinates correspond to the λ values and the y-coordinates correspond to x_k^r values. We consider the lower envelope $\mathcal{L}$ of the lines defined by all sensors of $S_{i2}(\lambda)$. For each point q of $\mathcal{L}$, suppose q lies on the line defined by the sensor s_k and q's x-coordinate is λ_q. Then, if $\lambda = \lambda_q$, the leftmost sensor of $S_{i2}(\lambda)$ is s_k. This means that each line segment of $\mathcal{L}$ corresponds to the same leftmost sensor of $S_{i2}(\lambda)$. Based on this observation, we proceed to compute $s_{g(i)}(\lambda^*)$ as follows.

We first compute the lower envelope $\mathcal{L}$, which can be done in $O(n \log n)$ time [4]. Then, let Λ be the set of the x-coordinates of the vertices of $\mathcal{L}$. Note that $|\Lambda| = O(n)$ since $\mathcal{L}$ is the lower envelope of at most n lines. We also add λ_1' and λ_2' to Λ. We sort all values of Λ. Then, by using our decision algorithm, we do binary search on the sorted list of Λ to find two adjacent values λ_1'' and λ_2'' such that $\lambda^* \in (\lambda_1'', \lambda_2'']$. Note that $(\lambda_1'', \lambda_2''] \subseteq (\lambda_1', \lambda_2']$. Since λ_1'' and λ_2'' are two adjacent values of the sorted Λ, by our above analysis, there is a sensor that is always the leftmost sensor of $S_{i2}(\lambda)$ for all $\lambda \in (\lambda_1'', \lambda_2'']$. To find the above leftmost sensor, we only need to take any value λ in $(\lambda_1'', \lambda_2'')$ and explicitly compute the locations of sensors in $S_{i2}(\lambda)$. The above algorithm finds $s_{g(i)}(\lambda^*)$ in $O(n \log n \log\log n)$ time, which is dominated by the binary search.

Further, we let $\lambda_i^1 = \lambda_1''$, $\lambda_i^2 = \lambda_2''$, and $S_i(\lambda) = S_{i-1}(\lambda) \cup \{s_{g(i)}(\lambda^*)\}$. We will show later that all algorithm invariants hold on $S_i(\lambda)$ and $(\lambda_i^1, \lambda_i^2]$ (further processing may be needed).

4.3 Maintaining the Algorithm Invariants

If λ^* has been computed above, then we terminate the algorithm. Otherwise, we have obtained an interval $(\lambda_i^1, \lambda_i^2] \subseteq (\lambda_{i-1}^1, \lambda_{i-1}^2]$ that contains λ^* and a sensor set $S_i(\lambda)$. Hence, the first algorithm invariant holds. In the sequel, we discuss the other three invariants.

According to our algorithm, $S_i(\lambda)$ is the same for all values λ in the open interval $(\lambda_i^1, \lambda_i^2)$. Further, assume $\lambda^* \neq \lambda_i^2$. Then, our above three assumptions (i.e., $\lambda^* \notin \{\lambda_{i-1}^2, \lambda_2, \lambda_2'\}$) are all true. By our algorithm invariants, $S_{i-1}(\lambda) = S_{i-1}(\lambda^*)$ holds for all $\lambda \in (\lambda_i^1, \lambda_i^2) \subseteq (\lambda_{i-1}^1, \lambda_{i-1}^2)$. Since our above algorithm for computing $s_{g(i)}(\lambda^*)$ is based on the above three assumptions and now that the assumptions are all true, the sensor $s_{g(i)}(\lambda^*)$ has been correctly computed above. Since $S_i(\lambda) = S_{i-1}(\lambda) \cup \{s_{g(i)}(\lambda^*)\}$ and $S_{i-1}(\lambda) = S_{i-1}(\lambda^*)$, $S_i(\lambda) = S_i(\lambda^*)$ holds for all $\lambda \in (\lambda_i^1, \lambda_i^2)$. This shows that the second invariant holds.

For the third invariant, recall that $R_i(\lambda)$ is equal to the right extension of $s_{g(i)}(\lambda)$ in $C_i(\lambda)$. Further, if $s_{g(i)}(\lambda) \in S_{i1}(\lambda)$, then $R_i(\lambda)$ is equal to $x_{g(i)} + r + \lambda/w_{g(i)}$, which is a nondecreasing linear function of λ. If $s_{g(i)}(\lambda) \in S_{i2}(\lambda)$, then $R_i(\lambda) = R_{i-1}(\lambda) + 2r$. Since $R_{i-1}(\lambda)$ is a nondecreasing linear function on $(\lambda_{i-1}^1, \lambda_{i-1}^2)$, $R_i(\lambda)$ is also a nondecreasing linear function on $(\lambda_i^1, \lambda_i^2) \subseteq (\lambda_{i-1}^1, \lambda_{i-1}^2)$. Therefore, in either case, the third algorithm invariant holds.

For the fourth invariant, if $R_i(\lambda) < \beta$ for all $\lambda \in (\lambda_i^1, \lambda_i^2)$, then the fourth invariant also holds. Otherwise, we have Lemma 2 whose proof is omitted.

Lemma 2. *If it is not true that $R_i(\lambda) < \beta$ for all $\lambda \in (\lambda_i^1, \lambda_i^2)$, then $R_i(\lambda)$ is a strictly increasing linear function on $(\lambda_i^1, \lambda_i^2)$ and there is a value $\lambda' \in (\lambda_i^1, \lambda_i^2)$ such that $R_i(\lambda') = \beta$.*

By Lemma 2, we compute the value $\lambda' \in (\lambda_i^1, \lambda_i^2)$ such that $R_i(\lambda') = \beta$. This means that B is covered by the sensors $S_i(\lambda')$ in the configuration $C_i(\lambda')$. Thus, λ' is a feasible value and $\lambda^* \in (\lambda_i^1, \lambda']$. Due to that $R_i(\lambda)$ is a strictly increasing function, $R_i(\lambda) < \beta$ for all $\lambda \in (\lambda_i^1, \lambda')$. We update λ_i^2 to λ'. Now the fourth invariant also holds. Note that all the first three invariants still hold with this updated (and smaller) interval $(\lambda_i^1, \lambda_i^2)$.

This completes the i-th step, which takes $O(n \log n \log\log n)$ time. If λ^* is not computed, all algorithm invariants hold and we proceed on the next step. We can prove that λ^* will be computed in at most $n+1$ steps.

Since each step of the algorithm takes $O(n \log n \log\log n)$ time, λ^* can be computed in $O(n^2 \log n \log\log n)$ time. The space complexity of the algorithm is $O(n^2)$, which is dominated by the preprocessing of Corollary 1. If we use Theorem 1 to implement the decision algorithm, then the total time for computing λ^* is $O(n^2 \log^2 n)$ and the space complexity is $O(n)$.

Theorem 2. *The problem WBC can be solved in $O(n^2 \log n \log\log n)$ time and $O(n^2)$ space; alternatively, it can be solved in $O(n^2 \log^2 n)$ time and $O(n)$ space.*

References

1. Andrews, A.M., Wang, H.: Minimizing the aggregate movements for interval coverage. In: Dehne, F., Sack, J.-R., Stege, U. (eds.) WADS 2015. LNCS, vol. 9214, pp. 28–39. Springer, Heidelberg (2015)
2. Bar-Noy, A., Baumer, B.: Average case network lifetime on an interval with adjustable sensing ranges. Algorithmica **72**, 148–166 (2015)
3. Bar-Noy, A., Rawitz, D., Terlecky, P.: Maximizing barrier coverage lifetime with mobile sensors. In: Bodlaender, H.L., Italiano, G.F. (eds.) ESA 2013. LNCS, vol. 8125, pp. 97–108. Springer, Heidelberg (2013)
4. de Berg, M., Cheong, O., van Kreveld, M., Overmars, M.: Computational Geometry – Algorithms and Applications, 3rd edn. Springer-Verlag, Berlin (2008)
5. Chen, D., Gu, Y., Li, J., Wang, H.: Algorithms on minimizing the maximum sensor movement for barrier coverage of a linear domain. Discrete Comput. Geom. **50**, 374–408 (2013)
6. Cole, R., Salowe, J., Steiger, W., Szemerédi, E.: An optimal-time algorithm for slope selection. SIAM J. Comput. **18**(4), 792–810 (1989)
7. Cormen, T., Leiserson, C., Rivest, R., Stein, C.: Introduction to Algorithms, 3rd edn. MIT Press, Cambridge (2009)
8. Czyzowicz, J., Kranakis, E., Krizanc, D., Lambadaris, I., Narayanan, L., Opatrny, J., Stacho, L., Urrutia, J., Yazdani, M.: On minimizing the maximum sensor movement for barrier coverage of a line segment. In: Ruiz, P.M., Garcia-Luna-Aceves, J.J. (eds.) ADHOC-NOW 2009. LNCS, vol. 5793, pp. 194–212. Springer, Heidelberg (2009)
9. Czyzowicz, J., Kranakis, E., Krizanc, D., Lambadaris, I., Narayanan, L., Opatrny, J., Stacho, L., Urrutia, J., Yazdani, M.: On minimizing the sum of sensor movements for barrier coverage of a line segment. In: Nikolaidis, I., Wu, K. (eds.) ADHOC-NOW 2010. LNCS, vol. 6288, pp. 29–42. Springer, Heidelberg (2010)
10. Fan, H., Li, M., Sun, X., Wan, P., Zhao, Y.: Barrier coverage by sensors with adjustable ranges. ACM Trans. Sens. Netw. **11**, 14 (2014)
11. Megiddo, N.: Applying parallel computation algorithms in the design of serial algorithms. J. ACM **30**(4), 852–865 (1983)
12. Mehrandish, M.: On routing, backbone formation and barrier coverage in wireless Ad Hoc and sensor networks. Ph.D. thesis, Concordia University, Montreal, Quebec, Canada (2011)
13. Mehrandish, M., Narayanan, L., Opatrny, J.: Minimizing the number of sensors moved on line barriers. In: Proceedings of the IEEE Wireless Communications and Networking Conference (WCNC), pp. 653–658 (2011)

Randomized Algorithms II

Heuristic Time Hierarchies via Hierarchies for Sampling Distributions

Dmitry Itsykson(✉), Alexander Knop(✉), and Dmitry Sokolov(✉)

Steklov Institute of Mathematics at St. Petersburg, Saint Petersburg, Russia
dmitrits@pdmi.ras.ru, {aaknop,sokolov.dmt}@gmail.com

Abstract. We introduce a new framework for proving the time hierarchy theorems for heuristic classes. The main ingredient of our proof is a hierarchy theorem for sampling distributions recently proved by Watson [11]. Class $\mathrm{Heur}_\epsilon\mathbf{FBPP}$ consists of functions with distributions on their inputs that can be computed in randomized polynomial time with bounded error on all except ϵ fraction of inputs. We prove that for every a, δ and integer k there exists a function $F : \{0,1\}^* \to \{0,1,\ldots,k-1\}$ such that $(F,U) \in \mathrm{Heur}_\epsilon\mathbf{FBPP}$ for all $\epsilon > 0$ and for every ensemble of distributions D_n samplable in n^a steps, $(F,D) \notin \mathrm{Heur}_{1-\frac{1}{k}-\delta}\mathbf{FBPTime}[n^a]$. This extends a previously known result for languages with uniform distributions proved by Pervyshev [9] by handling the case $k > 2$. We also prove that $\mathbf{P} \not\subseteq \mathrm{Heur}_{\frac{1}{2}-\epsilon}\mathbf{BPTime}[n^k]$ if one-way functions exist.

We also show that our technique may be extended for time hierarchies in some other heuristic classes.

1 Introduction

The time hierarchy theorem for a computational model states that given more time it is possible to solve more computational problems. For deterministic Turing machines this theorem was proved by Hartmanis and Stearns [4] by using diagonalization. To show that there exists a language that is decidable in $O(n^3)$ steps but not decidable in $O(n^2)$ one may consider a language that contains a string x if Turing machine M_x rejects x in n^2 steps. Time hierarchy theorems are known for all syntactic computational models (a model is syntactic if it is possible to enumerate all correct machines of this model). Standard diagonalization does not work if it is impossible to negate the answer of a machine in polynomial time (for example, we don't know whether $\mathbf{NP} = \mathbf{co\text{-}NP}$); but delayed diagonalization [13] works well for all syntactic models.

A computational model is semantic if it is impossible to enumerate correct machines. For example **BPTime**, **RTime**, **ZPTime** are semantic models; we cannot enumerate correct machines since they have to satisfy promises. There

The research is partially supported by the RFBR grant 14-01-00545, by the President's grant MK-2813.2014.1 and by the Government of the Russia (grant 14.Z50.31.0030).

K. Elbassioni and K. Makino (Eds.): ISAAC 2015, LNCS 9472, pp. 201–211, 2015.
DOI: 10.1007/978-3-662-48971-0_18

are no known tight time hierarchy theorems for any semantic model. The best current result for a time hierarchy for randomized computations with bounded error is superpolynomial: $\mathbf{BPTime}[n^{\log n}] \subsetneq \mathbf{BPTime}[2^{n^{\epsilon}}]$ [8]. However, we are not able to prove that $\mathbf{BPTime}[n] \subsetneq \mathbf{BPTime}[n^{100 \log n}]$.

The first advancement in that direction was a time hierarchy theorem for randomized classes with several bits of nonuniform advice [1,2], the latest results include a time hierarchy for classes with only one bit of advice: $\mathbf{BPTime}/1$ [2], $\mathbf{ZPTime}/1, \mathbf{MATime}/1$, etc. [10]. The idea of the proofs of time hierarchies with nonuniform advice from [2] (the similar idea was used in [1]) is based on the existence of an optimal algorithm for some **PSPACE**-complete language. The proof from [10] is based on a tricky delayed diagonalization.

Fortnow and Santhanam also proved the time hierarchy theorem for heuristic randomized algorithms with bounded error, such algorithms may give an incorrect answer (and also violate the promise) on a small fraction of inputs. Namely, there exists a language L that can be decided in $\mathrm{Heur}_{\frac{1}{n^c}}\mathbf{BPP}$, but cannot be decided in $\mathrm{Heur}_{\frac{1}{n^c}}\mathbf{BPTime}[n^a]$ with uniform distribution. This proof is also based on an optimal algorithm for a **PSPACE**-complete language. Pervyshev [9] simplifies and strengthens the time hierarchy theorem for heuristic **BPTime**: there exists a language L that can be decided in $\mathrm{Heur}_{\epsilon}\mathbf{BPP}$, but can not be decided in $\mathrm{Heur}_{\frac{1}{2}-\epsilon}\mathbf{BPTime}[n^a]$ with uniform distribution. Pervyshev used a delayed diagonalization against all randomized Turing machines. Delayed diagonalization require the ability to simulate a machine on the next input length. A randomized Turing machine may accept with any probability of error, thus it cannot be simulated with a bounded error. Let M be a randomized Turing machine that may violate a promise. Suppose we need to simulate it on an input x. Pervyshev suggested a method to simulate it heuristically: for every input x we put into the correspondence a set of strings $\{y_1, y_2, \dots, y_N\}$, where N is large enough. On every y_i we execute $M(x)$ many times and calculate the frequency of ones μ_i. We accept y_i if μ_i is greater than θ_{y_i}, where $\theta_{y_i} = \frac{2}{5} + \frac{i}{5N}$. Note that if $M(x)$ satisfies the promise of bounded error, then the answer of our simulation is the same for all y_i. And if $M(x)$ violates the promise, then our simulation may violate the promise only for a small fraction of y_i, namely for such y_i that θ_{y_i} is very close to $\Pr[M(x) = 1]$.

In this paper we consider k-valued functions and prove that for all a and δ and integer k there exists a function $F : \{0,1\}^* \to \{0, 1, \dots, k-1\}$ such that $(F, U) \in \mathrm{Heur}_{\epsilon}\mathbf{FBPP}$ for all $\epsilon > 0$ and for every ensemble of distributions D_n samplable in n^a steps $(F, D) \notin \mathrm{Heur}_{1-\frac{1}{k}-\delta}\mathbf{FBPTime}[n^a]$. So in case of k-valued functions we improve the fraction of hard instances from $\frac{1}{2} - \delta$ that was known for languages (i.e. $k = 2$) to $1 - \frac{1}{k} - \delta$. It is an interesting open question to prove something better than $\frac{1}{2} - \delta$ for languages. Note that for deterministic computations $\mathbf{P} \subsetneq \mathrm{Heur}_{1-\delta}\mathbf{DTime}[n^k]$ may be proved by the standard diagonalization.

Pervyshev's approach does not work for $k > 2$ by the following reason. Consider a function $F : \{0,1\}^n \to \{0, 1, \dots, k-1\}$ such that for some $a \in \{0, 1, \dots, k-1\}$ and $\delta > 0$, $\Pr_{x \leftarrow U_n}[F(x) = a] \geq 1 - \delta$. Assume that there is a function F' such that $\Pr_{x \leftarrow U_n}[F(x) \neq F'(x)] \leq 1 - \frac{1}{k} - 2\delta$, then $\Pr_{x \leftarrow U_n}[F'(x) = a] \geq \frac{1}{k} + \delta$. It is easy to see that if we have an access to F' then a can be computed for $k = 2$, while a cannot be determined for $k > 2$.

Our proof is based on the hierarchy for polynomial-time samplable distributions recently proved by Watson [11]. Watson proved that for any integer constant k, positive a and ϵ there exists a polynomial-time samplable ensemble of random variables γ_n that take values from the set $\{0, 1, \ldots, k-1\}$, such that for every samplable in n^a steps ensemble of random variables α_n with values in $\{0, 1, \ldots, k-1\}$ the statistical distance between α_n and γ_n is at least $1 - \frac{1}{k} - \epsilon$ for some n.

It is interesting to compare our proof for $k = 2$ with the proof of Pervyshev. In case of languages we need Watson's theorem only for $k = 2$; this particular case can be proved elementary by a delayed diagonalization. We define a language L_γ that is based on the ensemble γ_n; we prove that L_γ is solvable in $\text{Heur}_\epsilon\mathbf{BPP}$. If L_γ is solvable in randomized heuristic time n^a by an algorithm A, then A may be used to generate in n^a steps an ensemble of random variables α_n that is close to γ_n; the latter contradicts the theorem of Watson. In our proof for languages we separate the wheat from the chaff. For example, we don't care about machines that violate the promise, that's why we don't use the multithreshold trick described above. Instead we use similar but more simple observation: for polynomial-time samplable random variable $\gamma_n \in \{0, 1\}$ the language $\{r \mid \Pr[\gamma_{|r|} = 1] > 0.r\}$ is in $\text{Heur}_\epsilon\mathbf{BPP}$ for all ϵ.

This method can also be used to prove hierarchy theorems for other heuristic classes. Pervyshev proved the time hierarchy theorem for heuristic nondeterministic computations. This proof can also be formulated in our framework. To achieve that we extend the notion of samplability of random variables. We define a class of random variables (taking values in $\{0, 1\}$) that can be sampled in nondeterministic polynomial time: the sampling algorithm applies a function from **NP** to random bits. Watson's theorem for $k = 2$ also holds for nondeterministically samplable random variables. To prove a time hierarchy for heuristic **NP** we need a more accurate version of Watson's theorem, namely, we prove (only for $k = 2$) a hierarchy for a nondeterministic sampling for the case of a sampling algorithm that uses exactly n random bits, where n is the index of the random variable. In fact, this hierarchy was implicitly proved in [9]. It is an interesting open question to extend Watson's theorem for arbitrary k for nondeterministic sampling, the current proof does not work since it uses codes with good list decoding properties that are not monotone.

We also note that our method works for heuristic hierarchies for all classes $\mathfrak{C}\mathbf{Time}$ and $\mathbf{BP} \cdot \mathfrak{C}\mathbf{Time}$, where $\mathfrak{C}$ is a syntactic computational model that is closed under the application of majority. This observation for $\mathfrak{C}\mathbf{Time}$ was explicitly formulated in [9]. But the observation for $\mathbf{BP} \cdot \mathfrak{C}\mathbf{Time}$ has not appeared in [9]; the heuristic time hierarchy theorem for **AM** holds since $\mathbf{AM} = \mathbf{BP} \cdot \mathbf{NP}$, but Pervyshev proved it as a corollary of more complicated hierarchy for heuristic **MA**.

Conditional Results. There are several known conditions that imply the time hierarchy theorem for **BPTime**. The existence of a **BPP**-complete problem (under strong enough reductions) implies a time hierarchy theorem for **BPTime** (see for example [1]). The paper of Fortnow and Santhanam [2] implies that if it is possible to approximate the running time of the optimal algorithm for the **PSPACE**-complete language in polynomial time, then there exists a time hierarchy for **BPTime**. The time hierarchy theorem for **BPTime** also holds in case of the existence of a $\log(n) \to n$ pseudorandom generator (i.e. a generator mapping a seed of size $C \log n$ to n pseudorandom bits), since in that case $\mathbf{BPTime}[n^k] \subseteq \mathbf{DTime}[n^{k+\epsilon}]$ and the hierarchy follows from the deterministic time hierarchy. Such a pseudorandom generator exists if, for example, $\mathbf{E} \setminus \mathbf{Size}[2^{\epsilon n}] \neq \emptyset$ [6]. The existence of a $n \to \text{poly}(n)$ pseudorandom generator is not sufficient for a full derandomization, and thus **BPTime** hierarchy is not trivial in this case.

We prove that $\mathbf{P} \not\subseteq \text{Heur}_{\frac{1}{2}-\epsilon}\mathbf{BPTime}[n^k]$ under the existence of $n \to \text{poly}(n)$ a pseudorandom generator (that is equivalent to the existence of one-way functions). We also note that if $\mathbf{NP} \subseteq \mathbf{BPP}$ then $\mathbf{BPP} \not\subseteq \mathbf{BPTime}[n^k]$ for all k. In terms of Impagliazzo's worlds [5] the **BPTime** hierarchy theorem holds in Algorithmica and Criptomania worlds.

Organization of the Paper. In Sect. 3 we prove Pervyshev's result demonstrating our method; in Sect. 4 we prove **BPTime** hierarchy under the assumption of the existence of one-way functions; in Sect. 5 we prove heuristic hierarchy for k-valued functions; in Sect. 6 we prove our main result, extending the result from the previous section to arbitrary distributions. In Sect. 7 we show that it is possible to reprove the heuristic hierarchy for **NTime** in our framework.

2 Preliminaries

For two random variables χ_1, χ_2 with values from a set K the statistical distance between them is $\Delta(\chi_1, \chi_2) = \max_{S \subseteq K} |\Pr[\chi_1 \in S] - \Pr[\chi_2 \in S]|$.

An *ensemble of distributions* D is a family of distributions $\{D_n\}_{n=1}^{\infty}$, where D_n is a distribution on $\{0,1\}^n$. *A distributional problem* is a pair (L, D) of a language L and an ensemble of distributions D. Let U_n denote an ensemble of uniform distributions over $\{0,1\}^n$.

The class $\text{Heur}_{\delta(n)}\mathbf{BPTime}[f(n)]$ consists of distributional problems (L, D) such that there exists a probabilistic algorithm A that runs in at most $O(f(n))$ steps and for every n the following holds: $\Pr_{x \leftarrow D_n}[\Pr[A(x) = L(x)] \geq \frac{3}{4}] \geq 1 - \delta(n)$, where the inner probability is over random bits of the algorithm A. We also denote $\text{Heur}_{\delta(n)}\mathbf{BPP} = \bigcup_{k \geq 0} \text{Heur}_{\delta(n)}\mathbf{BPTime}[n^k]$. We also define classes for functions: $\text{Heur}_{\delta(n)}\mathbf{FBPTime}[f(n)]$ consists of pairs (F, D), where $F : \{0,1\}^* \to \{0,1\}^*$ is a function and D is an ensemble of distributions, such that there exists a probabilistic algorithm A that runs in at most $O(f(n))$ steps and for every n

the following holds: $\Pr_{x \leftarrow D_n}[\Pr[A(x) = F(x)] \geq \frac{3}{4}] \geq 1 - \delta(n)$, where the inner probability is over random bits of the algorithm A. We denote $\mathrm{Heur}_{\delta(n)}\mathbf{FBPP} = \bigcup_{k \geq 0} \mathrm{Heur}_{\delta(n)}\mathbf{FBPTime}[n^k]$.

In the following, we use some abuse of notation and omit uniform distributions. For example if we state that $L \in \mathrm{Heur}_{\delta(n)}\mathbf{BPP}$, we formally mean that $(L, U) \in \mathrm{Heur}_{\delta(n)}\mathbf{BPP}$.

We say that a distributional problem (L, D) is heuristically decidable in nondeterministic time $O(f(n))$ with an error $\delta(n)$ (we denote this as $(L, D) \in \mathrm{Heur}_{\delta(n)}\mathbf{NTime}[f(n)]$) iff there is a nondeterministic algorithm A that runs in at most $O(f(n))$ steps, such that for any n we have that $\Pr_{x \leftarrow D_n}[A(x) = L(x)] \geq 1 - \delta(n)$. We also define $\mathrm{Heur}_{\delta(n)}\mathbf{NP} = \bigcup_{k \geq 0} \mathrm{Heur}_{\delta(n)}\mathbf{NTime}[n^k]$.

3 Hierarchy for HeurBPP

Definition 1. *An ensemble of random variables γ_n is samplable in time $O(f(n))$ iff there exist a constant k and a deterministic algorithm A that on the input $(1^n, r)$ runs in $O(f(n))$ steps and $A(1^n, r)$ is distributed according to γ_n, where r is distributed uniformly over $\{0,1\}^{kf(n)}$. We denote the set of all ensembles samplable in time $O(f(n))$ as $\mathbf{DSamp}[f(n)]$.*

The following theorem is a particular case of a theorem from [11]. However the proof of this particular case is much simpler than the Watson's proof of his more general statement.

Theorem 1 [11]. *For every $a > 0$ and $\epsilon = \frac{1}{\mathrm{poly}(n)} > 0$, there exist $b > 0$ and an ensemble of random variables $\gamma_n \in \mathbf{DSamp}[n^b]$ that take values from $\{0,1\}$ such that for every ensemble $\alpha_n \in \mathbf{DSamp}[n^a]$ there exists an arbitrary large n_0 such that the statistical distance between α_{n_0} and γ_{n_0} is at least $\frac{1}{2} - \epsilon$.*

Let γ_n be an ensemble of random variables that take values from $\{0,1\}$. We denote $L_\gamma = \bigcup_n \{r \in \{0,1\}^n \mid \Pr[\gamma_n = 1] > 0.r\}$, where $0.r$ is a binary number.

Lemma 1. *For every polynomial-time samplable ensemble of random variables γ_n that take values from $\{0,1\}$ the language $L_\gamma \in \mathrm{Heur}_\epsilon\mathbf{BPP}$ for every $\epsilon = \frac{1}{\mathrm{poly}(n)} > 0$.*

Proof. Consider the following algorithm A: sample N independent instances of the random variable γ_n, let q be a fraction of 1s. If $q \geq 0.r$ then return 1 otherwise 0. By Chernoff bounds if $|0.r - \Pr[\gamma_n = 1]| > \epsilon/4$ then $\Pr[A(r) \neq L_\gamma(r)] < 2e^{-\frac{1}{8}\epsilon^2 N}$; it is less than $\frac{1}{4}$ for $N = O(\frac{1}{\epsilon^2})$. Note that $\Pr_r[|0.r - \Pr[\gamma_n = 1]| \leq \epsilon/4] \leq 2^{-n} + \epsilon/2$ that is less than ϵ for large enough n. □

Lemma 2. *Let L be a language such that for all n $|\Pr_{x \leftarrow U_n}[x \in L] - \Pr[\gamma_n = 1]| < \delta$ and $L \in \mathrm{Heur}_{\epsilon(n)}\mathbf{BPTime}[n^k]$ for some $\epsilon(n), \delta \geq 0$. Then there exists an* *ensemble of random variables* β_n *such that* $\beta_n \in \mathbf{DSamp}[n^{k+1}]$ *and* $\Delta(\beta_n, \gamma_n) \leq \epsilon(n) + \delta + \frac{1}{2^n}$.

Proof. Let E be a randomized algorithm that solves L in $\mathrm{Heur}_\epsilon\mathbf{BPTime}[n^k]$. Let $\hat{E}(x)$ execute $E(x)$ for $N = O(n)$ times and return the most frequent answer. Consider an ensemble of random variables α_n defined in the following way: sample a random element $x \in \{0,1\}^n$ and return $L(x)$. Then consider the following algorithm that samples β_n: sample a random element $x \in \{0,1\}^n$ and return $\hat{E}(x)$. Since $|\Pr_{x \leftarrow U_n}[x \in L] - \Pr[\gamma_n = 1]| < \delta$ we have that $\Delta(\alpha_n, \gamma_n) < \delta$. Let C be a set of all x such that $\Pr[E(x) = L(x)] \geq \frac{3}{4}$. Chernoff bounds imply that for $x \in C$ we have that $\Pr[\hat{E}(x) = L(x)] > 1 - \frac{1}{2^n}$.

$$\Delta(\beta_n, \gamma_n) = |\Pr_{x,r}[\hat{E}(x) = 1] - \Pr[\gamma_n = 1]|$$
$$\leq |\Pr_{x,r}[\hat{E}(x) = 1] - \Pr_x[x \in L]| + |\Pr_x[x \in L] - \Pr[\gamma_n = 1]| \leq \epsilon(n) + 2^{-n} + \delta.$$

□

Theorem 2 [9]. *For every* $b > 0$ *and* $\delta = \frac{1}{\mathrm{poly}(n)} > 0$ *there exists a language* L *such that* $L \notin \mathrm{Heur}_{\frac{1}{2}-\delta}\mathbf{BPTime}[n^b]$ *and for all* $\tau = \frac{1}{\mathrm{poly}(n)}$, $L \in \mathrm{Heur}_\tau\mathbf{BPP}$.

Proof. Let γ_n be an ensemble from Theorem 1 for $\epsilon = \delta/2$ and $a = b + 1$. By Lemma 1 $L_\gamma \in \mathrm{Heur}_\tau\mathbf{BPP}$. Assume that $L_\gamma \in \mathrm{Heur}_{\frac{1}{2}-\delta}\mathbf{BPTime}[n^b]$. Note that by construction of L_γ we have that $|\Pr_{x \leftarrow U_n}[x \in L_\gamma] - \Pr[\gamma_n = 1]| < \frac{1}{2^n}$. Hence by Lemma 2 there exists $\beta_n \in \mathbf{DSamp}[n^a]$ and $\Delta(\beta_n, \gamma_n) \leq \frac{1}{2} - \delta + \frac{1}{2^n} + \frac{1}{2^n} < \frac{1}{2} - \frac{\delta}{2}$ for n large enough. The latter contradicts Theorem 1. □

4 Conditional Hierarchy

Theorem 3. *Assume that one-way functions exist. Then for every* $\epsilon > 0$ *and* $a > 0$ *there exists a language* $L \in \mathbf{P}$ *such that* $L \notin \mathrm{Heur}_{\frac{1}{2}-\epsilon}\mathbf{BPTime}[n^a]$.

Proof. Consider the random variable γ_n from Theorem 1 and let S be a generator that generates γ_n. We assume that S gets random bits as the second input. Let S use $p(n)$ random bits. Let G be pseudorandom generator that maps n random bits to $p(n)$ pseudorandom ones. Consider the random variable $S(1^n, G(r))$, where $r \leftarrow U_n$. Since G is a pseudorandom generator we have that $\Delta(S(1^n, G(U_n)), \gamma_n) = \Delta(S(1^n, G(U_n)), S(1^n, U_{p(n)})) < \epsilon/4$ for all n large enough.

Consider the language $L = \bigcup_n \{r \in \{0,1\}^n \mid S(1^n, G(r)) = 1\}$. It is obvious that $L \in \mathbf{P}$. Lemma 2 and Theorem 1 implies $L \notin \mathrm{Heur}_{\frac{1}{2}-\epsilon}\mathbf{BPTime}[n^a]$. □

An anonymous reviewer noted that Theorem 3 also follows from the approach of Pervyshev.

We also show that the **BPTime** time hierarchy holds if all languages from **NP** are easy.

Theorem 4. *If* **NP** $\subseteq$ **BPP**, *then* **BPTime**$[n^k] \subsetneq$ **BPP** *for all* $k > 0$.

Proof. Assume, for the sake of contradiction, that **BPP** $\subseteq$ **BPTime**$[n^k]$. By the argument similar to Adleman's theorem we get **BPTime**$[n^k] \subseteq$ **Size**$[n^{2k+2}]$. Results of [12] implies that if **NP** $\subseteq$ **BPP**, then **PH** $\subseteq$ **BPP**. So if **BPP** $=$ **BPTime**$[n^k]$, then **PH** $\subseteq$ **Size**$[n^{2k+2}]$ that contradicts Kannan's theorem [7]. □

5 Hierarchy for k-valued Functions

In this section we extend heuristic time hierarchy theorem for functions that take constant number of values. Now we need the full power of Watson's theorem.

Theorem 5 [11]. *For every $a > 0$, $k > 0$ and $\epsilon = \frac{1}{\text{poly}(n)} > 0$, there exists $b > 0$ and an ensemble of random variables $\gamma_n \in$ **DSamp**$[n^b]$ that take values from $\{0, 1, \ldots, k-1\}$ such that for every ensemble $\alpha_n \in$ **DSamp**$[n^a]$ there exists n_0 such that the statistical distance between α_{n_0} and γ_{n_0} is at least $1 - \frac{1}{k} - \epsilon$.*

Let γ_n be an ensemble of random variables that take values from $\{0, 1, \ldots, k-1\}$. For every n we split the segment $[0, 1)$ on k disjoint parts: $I_0^{(n)} = \left[p_0^{(n)} = 0, p_1^{(n)}\right), I_1^{(n)} = \left[p_1^{(n)}, p_2^{(n)}\right), \ldots, I_{k-1}^{(n)} = \left[p_{k-1}^{(n)}, p_k^{(n)} = 1\right)$ (here we assume that $[a, a)$ is $\emptyset$). For all $i \in \{0, 1, \ldots, k-1\}$, $p_{i+1}^{(n)} - p_i^{(n)} = \Pr[\gamma_n = i]$. We define function $F_\gamma : \{0,1\}^* \to \{0, 1, \ldots, k-1\}$ such that for all $r \in \{0,1\}^n$, $F_\gamma(r) = i$ iff $0.r \in I_i^{(n)}$, where $0.r$ is a binary number. More formally $F_\gamma(r) = \min\{i \in \{0, 1, \ldots, k-1\} \mid \Pr[\gamma_n \in \{0, 1, \ldots, i\}] > 0.r\}$.

The following lemma is an extension of Lemma 1.

Lemma 3. *For every polynomial-time samplable ensemble of random variables γ_n that take values from $\{0, 1, \ldots, k-1\}$ the function $F_\gamma \in \text{Heur}_\epsilon$**FBPP** for every $\epsilon = \frac{1}{\text{poly}(n)} > 0$.*

Proof. Consider the following algorithm A: sample $N = O\left(\frac{k^4 \log(k)}{\epsilon^2}\right)$ independent instances of the random variable γ_n, let q_i be a fraction of is. Find minimum j such that $\sum_{i=1}^{j} q_i > 0.r$ and return j. By Chernoff bounds $\Pr[|q_i - \Pr[\gamma_n = i]| > \epsilon/2k^2] < 2e^{-\frac{1}{2k^4}\epsilon^2 N}$.

Assume that for all $i \in \{0, 1, \ldots, k-1\}$ we have that $|0.r - p_i^{(n)}| > \epsilon/2k$. In this case $A(r) \neq F_\gamma(r)$ only if for some $j \leq i$, $|q_j - \Pr[\gamma_n = j]| > \epsilon/2k^2$ that happens with probability at most $2e^{-\frac{1}{2k^4}k\epsilon^2 N} < \frac{1}{4}$ for $N = O\left(\frac{k^4 \log(k)}{\epsilon^2}\right)$. Hence $\Pr[A(r) \neq F_\gamma(r)] < \frac{1}{4}$.

Note that $\Pr_r[\forall i \in \{0, 1, \ldots, k-1\}\ |0.r - \Pr[\gamma_n = i]| \leq \epsilon/2k] \leq k2^{-n} + \epsilon/2$ that is less than ϵ for large enough n. □

Lemma 4. *Let F be a function $\{0,1\}^* \to \{0,1,\dots,k-1\}$ such that for all n statistical distance between $F(U_n)$ and γ_n is at most δ, where U_n is uniform distribution over $\{0,1\}^n$ and $F \in \mathrm{Heur}_{\epsilon(n)}\mathbf{FBPTime}[n^a]$ for some $\epsilon(n), \delta \geq 0$. Then there exists an ensemble of random variables β_n such that $\beta_n \in \mathbf{DSamp}[n^{a+1}]$ and $\Delta(\beta_n, \gamma_n) \leq \epsilon(n) + \delta + \frac{1}{2^n}$.*

Proof. The proof repeats the proof of Lemma 2. □

Theorem 6. *For every $b > 0$ and every $k > 0$ and every $\delta = \frac{1}{\mathrm{poly}(n)} > 0$ there exists a function $F\colon \{0,1\}^* \to \{0,1,\dots,k-1\}$ such that $F \notin \mathrm{Heur}_{1-\frac{1}{k}-\delta}\mathbf{FBPTime}[n^b]$ and for every $\tau = \frac{1}{\mathrm{poly}(n)} > 0$ $F \in \mathrm{Heur}_\tau\mathbf{FBPP}$.*

Proof. Let γ_n be an ensemble from Theorem 1 for $\epsilon = \frac{\delta}{2}$, and $a = b+1$. By Lemma 3 $F_\gamma \in \mathrm{Heur}_\tau\mathbf{FBPP}$. Assume that $F_\gamma \in \mathrm{Heur}_{1-\frac{1}{k}-\delta}\mathbf{FBPTime}[n^b]$. Note that by construction of F_γ for all $i \in \{0,1,\dots,k-1\}$ we have that $|\Pr_{x \leftarrow U_n}[F_\gamma(x) = i] - \Pr[\gamma_n = i]| < \frac{1}{2^n}$, hence $\Delta(F_\gamma(U_n), \gamma_n) \leq \frac{k}{2^{n+1}}$. Hence by Lemma 4 there exists $\beta_n \in \mathbf{DSamp}[n^a]$ and $\Delta(\beta_n, \gamma_n) \leq 1 - \frac{1}{k} - \delta + \frac{k}{2^{n+1}} + \frac{1}{2^n} < 1 - \frac{1}{k} - \frac{\delta}{2}$ for n large enough. The latter contradicts Theorem 5. □

6 Hierarchy for Arbitrary Distributions

In this section we strengthen the results from previous sections for arbitrary distributions. Now Watson's theorem is not sufficient for our goal, but actually Watson [11] proved slightly stronger statement:

Theorem 7. *For every $a > 0$, $k > 0$ and $\epsilon = \frac{1}{\mathrm{poly}(n)} > 0$, there exists $b > 0$ and an ensemble of random variables $\gamma_n \in \mathbf{DSamp}[n^b]$ that take values from $\{0,1,\dots,k-1\}$ such that for every ensemble $\alpha_n \in \mathbf{DSamp}[n^a]$ there exists $c \in \{0,1,\dots,k-1\}$ and arbitrary large n_0 such that $\Pr[\gamma_{n_0} = c] \geq 1 - \epsilon/2$ and $\Pr[\alpha_{n_0} = c] \leq \frac{1}{k} + \epsilon/2$.*

The proof of Theorem 7 for $k = 2$ repeats the proof of Theorem 1. For the proof of the general statement see [11]. Let γ_n be an ensemble of random variables that take values from $\{0,1,\dots,k-1\}$. As in previous section for every n we split the segment $[0,1)$ on k disjoint parts: $I_0^{(n)}, \dots, I_{k-1}^{(n)}$. We define function $H_\gamma : \{0,1\}^* \to \{0,1,\dots,k-1\}$ such that for all $r \in \{0,1\}^n$, $H_\gamma(r) = i$ iff $\Theta_r \in I_i^{(n)}$, where $\Theta_r = \frac{3}{8} + \frac{0.r}{4}$.

The proof of the next Lemma repeats the proof of Lemma 3 almost literally.

Lemma 5. *For every polynomial-time samplable ensemble of random variables γ_n that take values from $\{0,1,\dots,k-1\}$, $H_\gamma \in \mathrm{Heur}_\epsilon\mathbf{FBPP}$ for every $\epsilon = \frac{1}{\mathrm{poly}(n)} > 0$.*

Theorem 8. *For every $b > 0$ and every $k > 0$ and every $\delta = \frac{1}{\mathrm{poly}(n)} > 0$ there exists a function $H\colon \{0,1\}^* \to \{0,1,\dots,k-1\}$ such that for all $D \in \mathbf{DSamp}[n^b]$ $(H, D) \notin \mathrm{Heur}_{1-\frac{1}{k}-\delta}\mathbf{FBPTime}[n^b]$ and for every $\tau = \frac{1}{\mathrm{poly}(n)}$, $H \in \mathrm{Heur}_\tau\mathbf{FBPP}$.*

Proof. Let γ_n be an ensemble from Theorem 1 for $\epsilon = \frac{\delta}{2}$, and $a = b + 1$. By Lemma 5 $H_\gamma \in \mathrm{Heur}_\delta\mathbf{FBPP}$.

Assume that for some $D \in \mathbf{DSamp}[n^b]$, $(H_\gamma, D) \in \mathrm{Heur}_{1-\frac{1}{k}-\delta}\mathbf{FBPTime}[n^b]$. Let E be randomized algorithm that decides (H_γ, D) and G is a generator for D. Let $\hat{E}(x)$ execute $E(x)$ for $N = O(n)$ times and return the most frequent answer. Consider the following random variable α_n that is defined by the following algorithm: (1) Sample a random element $x \leftarrow G(1^n)$; (2) Return $\hat{E}(x)$. Note that α_n is samplable in $O(n^{b+1})$ steps.

By Theorem 7 applied to γ_n and α_n and $\epsilon = \frac{\delta}{2}$ there exists n_0 and $c \in \{0, 1, 2, \ldots, k-1\}$ such that $\Pr[\gamma_n = c] \geq 1 - \frac{\delta}{4}$ and $\Pr[\alpha_n = c] \leq \frac{1}{k} + \frac{\delta}{4}$. The size of $I_c^{(n_0)}$ is at least $\frac{3}{4}$ for $\delta < 1$, hence $\Theta_r \in I_c^{(n_0)}$ for all $r \in \{0,1\}^{n_0}$. Therefore $H_\gamma(x) = c$ for all $x \in \{0,1\}^{n_0}$ and by the definition of α_n and by Chernoff bounds $\Pr[\alpha_{n_0} = c] \geq \frac{1}{k} + \delta - 2^{-n_0} > \frac{1}{k} + \delta/2$ for large enough n_0 that contradicts Theorem 7. □

7 Hierarchy for HeurNP

In this section we show that our technique can also be used to prove a hierarchy theorem for heuristic **NP**.

Definition 2. *An ensemble of random variables γ_n is samplable in nondeterministic time $O(f(n))$ with n^k random bits iff there exists a nondeterministic algorithm A that on an input $(1^n, r)$ runs in $O(f(n))$ steps, and $A(1^n, r)$ is distributed according γ_n, where r is distributed uniformly over $\{0,1\}^{n^k}$. We denote the set of all ensembles samplable in time $O(f(n))$ with n^k random bits as $\mathbf{NSamp}_{n^k}[f(n)]$.*

In the proof of the time hierarchy theorem for heuristic **NTime** we use Boolean samplers from [3] instead of mixers used by Pervyshev. It is not very important whether to use samplers or mixers, but we prefer samplers since it makes the presentation slightly more elegant.

Definition 3. *A Boolean sampler is a randomized algorithm S, that takes on input an integer number n and rational numbers δ, ϵ. Algorithm S has an oracle access to a function $f\colon \{0,1\}^n \to \{0,1\}$; S makes several nonadaptive requests to the function f and outputs a number in the range $[0,1]$. Let us denote $\bar{f} = \frac{1}{2^n} \sum_{x \in \{0,1\}^n} f(x)$. For every function $f\colon \{0,1\}^n \to \{0,1\}$ the following inequality should be satisfied: $\Pr[|S^f(n, \epsilon, \delta) - \bar{f}| \geq \epsilon] < \delta$.*

A Boolean sampler is called averaging if it outputs the average value of requested values.

Theorem 9 [3]. *There is an averaging Boolean sampler S which uses n random bits, makes $q(n, \epsilon, \delta) = O(\frac{1}{\epsilon^2 \delta})$ requests to the function, and runs in time polynomial in n, $\frac{1}{\epsilon}$ and $\frac{1}{\delta}$.*

Corollary 1. *There exists an averaging Boolean sampler S that uses $n-1$ random bits, makes $O(\frac{1}{\epsilon^2\delta})$ requests to the function, and runs in time polynomial in n, $\frac{1}{\epsilon}$ and $\frac{1}{\delta}$.*

The following theorem is an analogue of Theorem 1 for distributions samplable by nondeterministic algorithms with fixed number of random bits. In the proof of Theorem 1 we may evaluate the frequency of random variable by sampling but now we use a Boolean sampler in order to save random bits.

Theorem 10. *For every $a > 0$ and $\epsilon > 0$ there exists $b > 0$ and an ensemble of random variables $\gamma_n \in \mathbf{NSamp}_n[n^b]$ that take values from $\{0,1\}$ such that for every ensemble $\alpha_n \in \mathbf{NSamp}_n[n^a]$ with values from $\{0,1\}$ there exists n such that the statistical distance between α_n and γ_n is at least $\frac{1}{2} - \epsilon$.*

Proof. We use delayed diagonalization. Let E_i be an enumeration of all nondeterministic algorithms (we interpret them as generators of random variables that use n random bits); we assume that E_i is supplied with an alarm clock that terminates its execution on an input $(1^n, r)$ after n^{a+1} steps. Let S be a Boolean sampler from Corollary 1. We define a sequence n_i as follows $n_1 = 1$, $n_{i+1} = n_i^* + 1$ and $n_i^* = 2^{n_i^{a+1}}$. We define γ_n by the following algorithm $\Gamma(1^n, r)$, where $r \leftarrow U_n$ is the string of random bits. For n such that $n_i \leq n \leq n_i^*$:

if $n = n_i^*$ then γ_n is concentrated on the element from $\{0,1\}$ that has the minimal probability according to $E_i(1^{n_i}, r)$, where r is uniformly distributed over $\{0,1\}^n$. This can be done by brute-force search in time poly(n_i^*);

if $n_i \leq n < n_i^* - 1$ then we execute $S^f(1^{n+1}, \frac{\epsilon}{2}, \frac{1}{4})$ using r as a random string, where $f : \{0,1\}^{n+1} \to \{0,1\}$ and $f(z) = E_i(1^{n+1}, z)$. Return 1 iff the result of the sampler exceeds $\frac{1}{2}$. Here we use that S is an averaging sampler and nondeterministic computations are closed under the application of majority.

Let α_n be generated by a nondeterministic algorithm $A(1^n, r)$ with n random bits in $O(n^a)$ steps, and A have number i in our enumeration. We prove by contradiction that there exists n $(n_i \leq n \leq n_i^*)$ such that $\Delta(\gamma_n, \alpha_n) > \frac{1}{2} - \epsilon$, where Δ denotes the statistical distance. Assume that $\Delta(\gamma_n, \alpha_n) \leq \frac{1}{2} - \epsilon$ for all n. Let b denote the element that has probability 1 according to $\gamma_{n_i^*}$ (by the construction $\Pr[E_i(1^{n_i}) = b] \leq \frac{1}{2}$). We prove by induction on k (for $0 \leq k \leq n_i^* - n_i$) that $\Pr[\gamma_{n_i^*-k} = b] > 1 - \frac{\epsilon}{2}$. The base $k = 0$ is trivial. Now we prove the induction step. By the induction hypothesis $\Pr[\alpha_{n_i^*-k} = b] \geq \Pr[\gamma_{n_i^*-k} = b] - \frac{1}{2} + \epsilon > 1 - \frac{\epsilon}{2} - \frac{1}{2} + \epsilon = \frac{1}{2} + \frac{\epsilon}{2}$. Hence by definition of a Boolean sampler $\Pr[\gamma_{n_i^*-k-1} = b] \geq 1 - \frac{\epsilon}{2}$. Finally we get a contradiction with $\Pr[\alpha_{n_i} = b] \leq \frac{1}{2}$. □

Theorem 11 [9]. *For every $b > 0$ and $\delta > 0$ there exists a language L such that $L \notin \mathrm{Heur}_{\frac{1}{2}-\delta}\mathbf{NTime}[n^b]$ and $L \in \mathbf{NP}$.*

Proof. Let γ_n be an ensemble from Theorem 10 for $\epsilon = \delta/2$ and $a = b + 1$, and S be a generator for this random variable. Consider the language $L = \{x | S(1^{|x|}, x) = 1\}$. It is easy to see that this language is in $\mathbf{NP}$. Let us prove that $L \notin \mathrm{Heur}_{\frac{1}{2}-\delta}\mathbf{NTime}[n^b]$. Assume the contrary and let nondeterministic algorithm A decide L in time n^b with error less than $\frac{1}{2} - \delta$. In this case for the random variable α_n that is distributed according to $A(x)$ for $x \leftarrow A_n$ we have that $\Delta(\alpha_n, \gamma_n) < \frac{1}{2} - \delta$ for all n. The latter contradicts Theorem 10. □

Further research. Is it possible to improve error from $\frac{1}{2} - \delta$ to $1 - \delta$ for Heur**NP** or Heur**BPP** hierarchy?

The second and third open questions are to prove **BPTime** hierarchy in Heuristica, where **NP** $\not\subseteq$ **BPP** but (**NP**, **PSamp**) $\subseteq$ Heur**BPP**, and in Pessiland, where (**NP**, **PSamp**) $\not\subseteq$ Heur**BPP** but there are no one-way functions.

Acknowledgments. The authors thank Edward A. Hirsch and anonimous reviewers for useful comments.

References

1. Barak, B.: A probabilistic-time hierarchy theorem for "slightly non-uniform" algorithms. In: Rolim, J.D.P., Vadhan, S.P. (eds.) RANDOM 2002. LNCS, vol. 2483, pp. 194–208. Springer, Heidelberg (2002)
2. Fortnow, L., Santhanam, R.: Hierarchy theorems for probabilistic polynomial time. In: FOCS, pp. 316–324 (2004)
3. Goldreich, O.: A sample of samplers: a computational perspective on sampling. In: Goldreich, O. (ed.) Studies in Complexity and Cryptography. LNCS, vol. 6650, pp. 302–332. Springer, Heidelberg (2011)
4. Hartmanis, J., Stearns, R.E.: On the computational complexity of algorithms. J. Symbolic Logic **32**(1), 120–121 (1967)
5. Impagliazzo, R.: A personal view of average-case complexity. In: SCT 1995: Proceedings of the 10th Annual Structure in Complexity Theory Conference (SCT 1995), p. 134. IEEE Computer Society, Washington, DC (1995)
6. Impagliazzo, R., Wigderson, A.: P = BPP if E requires exponential circuits: derandomizing the XOR lemma. In: Proceedings of the Twenty-Ninth Annual ACM Symposium on Theory of Computing, STOC 1997, pp. 220–229. ACM, New York (1997)
7. Kannan, R.: Circuit-size lower bounds and non-reducibility to sparse sets. Inf. Control **55**(1), 40–56 (1982)
8. Karpinski, M., Verbeek, R.: Randomness, provability, and the separation of Monte Carlo time and space. In: Börger, E. (ed.) Computation Theory and Logic. LNCS, vol. 270, pp. 189–207. Springer, Heidelberg (1987)
9. Pervyshev, K.: On heuristic time hierarchies. In: IEEE Conference on Computational Complexity, pp. 347–358 (2007)
10. van Melkebeek, D., Pervyshev, K.: A generic time hierarchy with one bit of advice. Comput. Complex. **16**(2), 139–179 (2007)
11. Watson, T.: Time hierarchies for sampling distributions. In: Innovations in Theoretical Computer Science, ITCS 2013, 9–12 January 2013, Berkeley, CA, USA, pp. 429–440 (2013)
12. Zachos, S.: Probabilistic quantifiers and games. J. Comput. Syst. Sci. **36**(3), 433–451 (1988)
13. Zak, S.: A turing machine time hierarchy. Theoret. Comput. Sci. **26**(3), 327–333 (1983)

Unbounded Discrepancy of Deterministic Random Walks on Grids

Tobias Friedrich[1(✉)], Maximilian Katzmann[2], and Anton Krohmer[1,2]

[1] Hasso Plattner Institute, Potsdam, Germany
{tobias.friedrich,anton.krohmer}@hpi.de
[2] Friedrich-Schiller-Universität Jena, Jena, Germany

Abstract. Random walks are frequently used in randomized algorithms. We study a derandomized variant of a random walk on graphs, called rotor-router model. In this model, instead of distributing tokens randomly, each vertex serves its neighbors in a fixed deterministic order. For most setups, both processes behave remarkably similar: Starting with the same initial configuration, the number of tokens in the rotor-router model deviates only slightly from the expected number of tokens on the corresponding vertex in the random walk model. The maximal difference over all vertices and all times is called single vertex discrepancy. Cooper and Spencer (2006) showed that on $\mathbb{Z}^d$ the single vertex discrepancy is only a constant c_d. Other authors also determined the precise value of c_d for $d = 1, 2$. All these results, however, assume that initially all tokens are only placed on one partition of the bipartite graph $\mathbb{Z}^d$. We show that this assumption is crucial by proving that otherwise the single vertex discrepancy can become arbitrarily large. For all dimensions $d \geq 1$ and arbitrary discrepancies $\ell \geq 0$, we construct configurations that reach a discrepancy of at least ℓ.

1 Introduction

Algorithms that are allowed to make random decision can solve many problems more efficiently than purely deterministic algorithms. One such example is the approximation of the volume of a convex body, where randomness gives a super-polynomial speed-up in computing power [11]. The first polynomial-time algorithm for this (and a number of other) problems is based on a certain random walk (e.g. [1]). Random walks appear to be powerful tools for designing efficient randomized algorithms.

Rotor-Router Model. The wide applicability of random walks raises the question what properties of the random walk are crucial and how much randomness is needed for this. To study this, we consider a derandomized variant of the random walk on the infinite grid $\mathbb{Z}^d$. In this *rotor-router model*, each vertex $\boldsymbol{x} \in \mathbb{Z}^d$ is equipped with a "rotor" together with a cyclic permutation (called a "rotor sequence") of the $2d$ cardinal directions of $\mathbb{Z}^d$. While the tokens performing a random walk leave a vertex in a random direction, in the rotor-router model the tokens deterministically go in the direction the rotor is pointing. After a token

K. Elbassioni and K. Makino (Eds.): ISAAC 2015, LNCS 9472, pp. 212–222, 2015.
DOI: 10.1007/978-3-662-48971-0_19

is sent, the rotor is rotated according to the fixed rotor sequence. This ensures that the tokens are distributed evenly among the neighbors.

Synonyms of the Rotor-Router Model. The rotor-router model was rediscovered independently several times in the literature. First under the name "Eulerian walker" [20], then as "edge ant walk" [22] and "whirling tour" [10]. It was later popularized by James Propp [16] and therefore also called "Propp machine" by Cooper and Spencer [6]. The same authors later also used the term "deterministic random walk" [4,8]. To emphasize the working principle, we only use the term "rotor-router model" in the rest of the paper.

Some Properties of the Rotor-Router Model. Many aspects of the model have been studied. The vertex and edge cover time of the rotor-router model can be asymptotically faster or slower as the classical random walk, depending on the topology [2,12,23]. Very precise bounds are also known if multiple tokens are deployed in parallel [7,15,17]. Our focus is on the *single-vertex discrepancy* with which we compare the rotor-router model and the expected behavior of the classical random walk. If particles are arbitrarily placed on the vertices and do a simultaneous walk in both models, we are interested in the maximal difference in the number of tokens between both models, at all times and on each vertex.

Known Results for the Single-Vertex Discrepancy. [6] proved that on $\mathbb{Z}^d$ the single vertex discrepancy is a constant c_d. For the case $d = 1$, that is, the graph being the infinite path, Cooper et al. [4] showed that $c_1 \approx 2.29$. For $d = 2$ the constant is $c_2 \approx 7.83$ for circular rotor sequences and $c_2 \approx 7.29$ otherwise [8]. It is further known that there is no such constant for infinite trees [5]. There are also (linear) upper and lower bounds for the discrepancy of finite graphs [14]. For some special finite graphs like hypercubes, stronger (i.e. polylogarithmic in the number of nodes) upper bounds are known [14].

Open Question. All three aforementioned results for the grid $\mathbb{Z}^d$ assume that the initial configuration is "even", that is, it only has tokens on one partition of the bipartite graph $\mathbb{Z}^d$. This assumption is, however, essential for achieving a constant discrepancy. Cooper et al. already pointed out for $d = 1$ that without this assumption their results "cannot be expected" [4, p. 2074]. We make this statement rigorous and present for each dimension d a configuration such that the single-vertex discrepancy on $\mathbb{Z}^d$ becomes arbitrarily large.

Results. To allow a direct comparison, let us first restate the result of Cooper and Spencer [6]. The mathematical notation is introduced in Sect. 2.

Theorem 1 ([6]). *For all $d \geq 1$ there is a constant $c_d \in \mathbb{R}_+$ such that for all* even *initial configurations, the single-vertex discrepancy on $\mathbb{Z}^d$ is bounded by c_d.*

Our main result is the following complement of the previous statement.

Theorem 2. *For all $d \geq 1$ and $\ell \in \mathbb{R}$ there is an initial configuration such that the single-vertex discrepancy on $\mathbb{Z}^d$ is at least ℓ.*

The reason for the unbounded discrepancy observed for non-even initial configurations is that the two partitions of $\mathbb{Z}^d$ subtly interfere with each other through the rotors. In every time step, all tokens switch back and forth between even and odd positions. In a random walk they are distributed independently, in the rotor-router model they follow the rotors, which exchange information between both partitions. This causes the unbounded discrepancy for appropriately set up initial configurations.

It should be noted that the discrepancy of ℓ in Theorem 2 already occurs for small configurations. In fact, Corollary 8 shows that a discrepancy of ℓ can be reached after $\Theta\left(\lceil \ell^2/d^2 \rceil\right)$ time steps with $\mathcal{O}(\lceil 1+\ell/d \rceil^{2d+1})$ tokens.

Techniques. For proving Theorem 2, we define a specific (infinitely large) initial configuration called (k,d)-wedge (cf. Definition 4), for which we study explicitly how it develops over time in the rotor-router and random walk model. We prove that this configuration is "stable" in the rotor-router model, that is, it stays unchanged after an even number of steps (cf. Lemma 6). The proof needs to consider 26 cases. We prove the cases using an automated theorem prover. Given this structural insight on the behavior of (k,d)-wedge, we calculate the resulting discrepancy (cf. Lemma 7). The proof makes use of the fact that the expected behavior of the d-dimensional random walk starting with a (k,d)-wedge can be decomposed into a collection of 1-dimensional random walks. To obtain a result for finite time and finite configurations, we observe that a subset of the (k,d)-wedge suffices to achieve a desired discrepancy (cf. Corollary 8).

2 Preliminaries

Random Walks. A random walk is a stochastic process that describes the movement of a number of tokens on a graph G. At each time step, each token at a vertex $\boldsymbol{x}$ chooses a neighbor independently and uniformly at random, and moves to that neighbor.

We consider simple random walks on an infinite d-dimensional grid $\mathbb{Z}^d$. A token at coordinate $\boldsymbol{x} = (x_1, \ldots, x_d)$ can move in the $2d$ cardinal directions, as given by the unit vectors: $\boldsymbol{e}_1 = (1,0,0\ldots), \boldsymbol{e}_2 = (0,1,0,\ldots), \ldots, -\boldsymbol{e}_1 = (-1,0,0,\ldots), -\boldsymbol{e}_2 = (0,-1,0,\ldots), \ldots, -\boldsymbol{e}_d = (0,\ldots,-1)$. We refer to this set of directions by E_{2d}. Following [18], we write Z_i for the direction that a token took at time step i. As all directions are equiprobable and independent, we have $\Pr[Z_i = \boldsymbol{e}_j] = \Pr[Z_i = -\boldsymbol{e}_j] = \frac{1}{2d}$ for all j. The position of a token after t steps can then be described as a sum of random variables $S_t = \boldsymbol{x} + Z_1 + Z_2 + \ldots + Z_t$.

We write $S_t^d(\boldsymbol{x})$ to express the probability that a d-dimensional random walk starting at the origin reaches vertex $\boldsymbol{x}$ after t steps. E.g., for dimension $d=1$ we obtain $S_t^1(x) = 2^{-t}\binom{t}{(t+x)/2}$.

We denote by $\bar{\boldsymbol{x}}$ the sum of the individual components of $\boldsymbol{x}$, i.e. $\bar{\boldsymbol{x}} := \boldsymbol{x}^T\mathbf{1} = \sum_{i=1}^d x_d$. Observe that the grid $\mathbb{Z}^d$ is a bipartite graph where all nodes with even $\bar{\boldsymbol{x}}$ form one partition, and nodes with odd $\bar{\boldsymbol{x}}$ form the other. With each time step, a token therefore switches the partition. To this end, we have $S_t^d(x) = 0$ if $(\bar{\boldsymbol{x}} - t \equiv 1) \mod 2$. We write $a \sim t$ to say that $(a \equiv t) \mod 2$, and we call a node $\boldsymbol{x}$ *even* if $\bar{\boldsymbol{x}} \sim 0$, and *odd* otherwise.

Rotor-Router Model. Let us now formally define the rotor-router model on the grid $\mathbb{Z}^d$. Each vertex $\boldsymbol{x}$ in this graph is equipped with a *rotor* $r_{\boldsymbol{x}} \in E_{2d}$. The *rotor sequence* for a vertex $\boldsymbol{x}$ is defined by a cyclic permutation $r_{\boldsymbol{x}}^{\rightarrow}: E_{2d} \rightarrow E_{2d}$.

At each time step t, all tokens at $\boldsymbol{x}$ do exactly one move as follows. A particular token moves in the direction of the rotor $r_{\boldsymbol{x}}$; and afterwards, the rotor is updated to point to $r_{\boldsymbol{x}}^{\rightarrow}(r_{\boldsymbol{x}})$. This is repeated until all tokens have been moved. Since tokens are not labeled, the order in which the tokens are passed to the rotor does not matter. All configurations of the rotor-router model are therefore fully defined by the initial placement of tokens, the initial rotor configurations $r_{\boldsymbol{x}}$ and the rotor sequences $r_{\boldsymbol{x}}^{\rightarrow}$ for all vertices $\boldsymbol{x} \in \mathbb{Z}^d$. If all tokens are initially on even vertices, we speak of an *even configuration.*

Single Vertex Discrepancy. When comparing the quality of the simulation of the rotor-router model, one often refers to the *single vertex discrepancy*, which is defined as follows. Let $f(\boldsymbol{x}, t): \mathbb{Z}^d \times \mathbb{N}_0 \rightarrow \mathbb{N}_0$ be the number of tokens at vertex $\boldsymbol{x}$ after t steps of the (deterministic) rotor-router model, and let $\mathbb{E}(\boldsymbol{x}, t): \mathbb{Z}^d \times \mathbb{N}_0 \rightarrow \mathbb{R}^+$ denote the expected number of tokens after t steps of a random walk with the same starting configuration $f(\boldsymbol{x}, 0)$. To compute $\mathbb{E}(\boldsymbol{x}, t)$ we determine for each $\boldsymbol{y} \in \mathbb{Z}^d$ the probability that a random walk starting at $\boldsymbol{y}$ reaches $\boldsymbol{x}$ after exactly t steps and multiply the result with the number of tokens that were at $\boldsymbol{y}$. Hence,

$$\mathbb{E}(\boldsymbol{x}, t) = \sum_{\boldsymbol{y} \in \mathbb{Z}^d} f(\boldsymbol{y}, 0) \cdot S_t^d(\boldsymbol{x} - \boldsymbol{y}). \tag{1}$$

Using this, we can define the *single vertex discrepancy.*

Definition 3. *Let $d \geq 1$ and an initial configuration $f(\boldsymbol{x}, 0)$ for all $x \in \mathbb{Z}^d$ be given. We call $\Delta(\boldsymbol{x}, t) = |f(\boldsymbol{x}, t) - \mathbb{E}(\boldsymbol{x}, t)|$ the* single vertex discrepancy at $\boldsymbol{x}$ after t steps. *Then, we define the* single vertex discrepancy Δ_d *as*

$$\Delta_d := \sup_{\boldsymbol{x} \in \mathbb{Z}^d, t \in \mathbb{N}} \Delta(\boldsymbol{x}, t). \tag{2}$$

3 Stable Configuration of the Rotor-Router Model

According to Theorem 1, the single vertex discrepancy is constant if we start with an even configuration. To prove that this condition is necessary, we construct the (k, d)-wedge, a starting configuration of tokens that ensures that there are effectively only two states of the rotor-router model.

The (k, d)-wedge intuitively forms a "peak" of tokens at the origin, and the rest of the graph is populated with tokens in a way that stabilizes the peak. In the random walk model, the expected number of nodes in the origin will decrease over time, while in the rotor-router model, the number of nodes always stays the same. The (k, d)-wedge is illustrated in Fig. 1 and formally defined as follows.

Definition 4. *Let $k, d \in \mathbb{N}$ be given, where k adjusts the vertex discrepancy. The rotor direction of vertex $\boldsymbol{x}$ at time t will be referred to by $r(\boldsymbol{x}, t): \mathbb{Z}^d \times \mathbb{N}_0 \rightarrow E_{2d}$.*

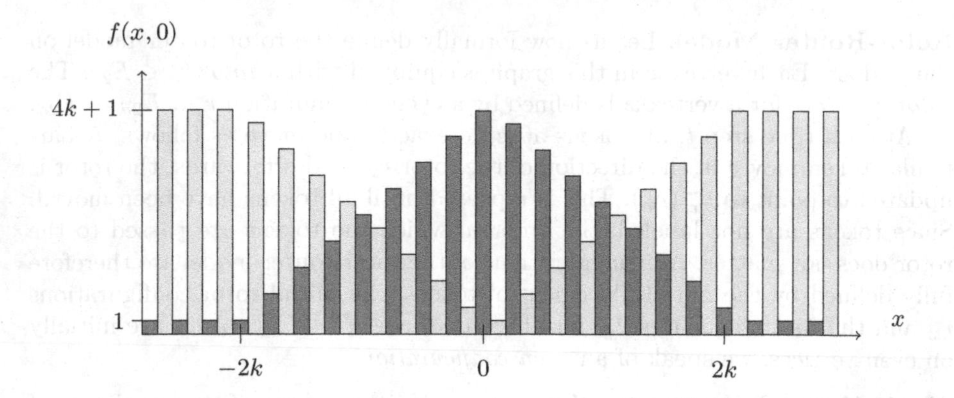

Fig. 1. Illustration of the $(k,1)$-wedge in dimension 1. The y-axis describes the number of tokens at position x. Dark colored bars show the even partition, light colored bars the odd one. This stable configuration is used to show our main result.

We define the (k,d)-wedge, *a starting configuration of the rotor-router model, as follows. For even vertices* $\boldsymbol{x}$ *with* $\bar{\boldsymbol{x}} \sim 0$, *we set*

$$f(\boldsymbol{x},0) := f_0(\bar{\boldsymbol{x}},0) := \begin{cases} d\cdot(4k+1+2\bar{\boldsymbol{x}}) & \text{if } \bar{\boldsymbol{x}} \in [-2k,0]\,, \\ d\cdot(4k+3-2\bar{\boldsymbol{x}}) & \text{if } \bar{\boldsymbol{x}} \in [1,2k]\,, \\ d & \text{otherwise.} \end{cases}$$

$$r(\boldsymbol{x},0) := r_0(\bar{\boldsymbol{x}},0) := \begin{cases} -\boldsymbol{e}_1 & \text{if } \bar{\boldsymbol{x}} \in [1,2k]\,, \\ \boldsymbol{e}_1 & \text{otherwise.} \end{cases}$$

For odd vertices $\boldsymbol{x}$ *with* $\bar{\boldsymbol{x}} \sim 1$, *we set*

$$f(\boldsymbol{x},0) := f_1(\bar{\boldsymbol{x}},0) := \begin{cases} d\cdot(1-2\bar{\boldsymbol{x}}) & \text{if } \bar{\boldsymbol{x}} \in [-2k,0]\,, \\ d\cdot(2\bar{\boldsymbol{x}}-1) & \text{if } \bar{\boldsymbol{x}} \in [1,2k]\,, \\ d\cdot(4k+1) & \text{otherwise.} \end{cases}$$

$$r(\boldsymbol{x},0) := r_1(\bar{\boldsymbol{x}},0) := \begin{cases} -\boldsymbol{e}_1 & \text{if } \bar{\boldsymbol{x}} \in [-2k,-1]\,, \\ \boldsymbol{e}_1 & \text{otherwise.} \end{cases}$$

The rotor sequences follow the order $\boldsymbol{e}_1,\ldots,\boldsymbol{e}_d,-\boldsymbol{e}_1,\ldots,-\boldsymbol{e}_d$.

Next, we show that the (k,d)-wedge is a stable configuration, meaning that the rotor-router model returns to the initial configuration every two steps. To this end, we introduce a function $g\colon \mathbb{Z}^d \times E_{2d} \times E_{2d} \times \mathbb{N} \to \mathbb{N}$, where $g(\boldsymbol{x},\pm\boldsymbol{e}_i,\pm\boldsymbol{e}_j,t)$ denotes the number of tokens that vertex $\boldsymbol{x}$ receives from vertex $\boldsymbol{x}\pm\boldsymbol{e}_i$ at time t when $r(\boldsymbol{x}\pm\boldsymbol{e}_i,t) = \pm\boldsymbol{e}_j$. Therefore,

$$g(\boldsymbol{x},\boldsymbol{e},\boldsymbol{f},t) = \begin{cases} \frac{f(\boldsymbol{x}+\boldsymbol{e},t)-d}{2d} & \text{if } \operatorname{sgn}(\boldsymbol{e}) = \operatorname{sgn}(\boldsymbol{f}), \\ \frac{f(\boldsymbol{x}+\boldsymbol{e},t)+d}{2d} & \text{otherwise,} \end{cases} \tag{3}$$

where $\operatorname{sgn}(-\boldsymbol{e}_i) = -1$ and $\operatorname{sgn}(+\boldsymbol{e}_i) = 1$ for all $i = 1, \ldots, d$. Then we can write

$$f(\boldsymbol{x}, t+1) = \sum_{i=1}^{d} g(\boldsymbol{x}, \boldsymbol{e}_i, r(\boldsymbol{x} + \boldsymbol{e}_i, t), t) + \sum_{i=1}^{d} g(\boldsymbol{x}, -\boldsymbol{e}_i, r(\boldsymbol{x} - \boldsymbol{e}_i, t), t), \tag{4}$$

which results from summing up the number of tokens that the neighbors of $\boldsymbol{x}$ pass to $\boldsymbol{x}$ at time step t. Recall that $f(\boldsymbol{x}, 0) = f(\bar{\boldsymbol{x}}, 0)$ and therefore $f(\boldsymbol{x} \pm \boldsymbol{e}_1, 0) = f(\bar{\boldsymbol{x}} \pm 1, 0)$. The same holds for $r(\boldsymbol{x}, 0)$. The definition of g in Eq. (3) can in this case be extended to $g(\bar{\boldsymbol{x}}, \pm 1, \pm \boldsymbol{e}_1, 0)$, and we can simplify Eq. (4) to

$$\begin{aligned} f(\bar{\boldsymbol{x}}, 1) &= \sum_{i=1}^{d} g(\bar{\boldsymbol{x}}, 1, r(\bar{\boldsymbol{x}} + 1, 0), 0) + \sum_{i=1}^{d} g(\bar{\boldsymbol{x}}, -1, r(\bar{\boldsymbol{x}} - 1, 0), 0) \\ &= d \cdot (g(\bar{\boldsymbol{x}}, 1, r(\bar{\boldsymbol{x}} + 1, 0), 0) + g(\bar{\boldsymbol{x}}, -1, r(\bar{\boldsymbol{x}} - 1, 0), 0)). \end{aligned} \tag{5}$$

To prove stability, it remains to show the following Lemmata.

Lemma 5. *Given a (k, d)-wedge, it holds*

$$r(\boldsymbol{x}, 1) = -r(\boldsymbol{x}, 0) \quad \textit{and} \quad f(\boldsymbol{x}, 1) = \begin{cases} f_1(\bar{\boldsymbol{x}}, 0) & \textit{if } \bar{\boldsymbol{x}} \sim 0, \\ f_0(\bar{\boldsymbol{x}}, 0) & \textit{if } \bar{\boldsymbol{x}} \sim 1. \end{cases}$$

Lemma 6. *Given a (k, d)-wedge, it holds $r(\boldsymbol{x}, 2) = r(\boldsymbol{x}, 0)$ and $f(\boldsymbol{x}, 2) = f(\boldsymbol{x}, 0)$.*

Lemma 5 states that the configuration of the rotor-router model after one step is again the (k, d)-wedge, except that it is shifted by one to the left. Furthermore, all rotors point in the opposite direction. By the same intuition, the next step undoes these changes and the configuration returns to the (k, d)-wedge after 2 steps, which is shown by Lemma 6.

These statements can be proven by a case distinction over Eq. (5). While none of the cases are mathematically challenging, there are 26 of them. Proving every case by hand is tedious and provides little to no further insight to the problem. Nevertheless, even small off-by-one errors break the stability of the (k, d)-wedge, which is why we wanted to convince ourselves that the (k, d)-wedge is indeed correct. To this end, we used the automated prover Isabelle/HOL [19] for the case distinction. Our code can be found in the long version of this paper.

Such provers excel at keeping track of all subgoals (i.e. cases) of a proof. Mostly, the proofs are not human readable, as they rely on internal proof routines. Automated proof systems like Isabelle/HOL, however, contain a certified kernel; so trusting the automated proof boils down to trusting the formalization of the problem and the correctness of the kernel. It is debated whether an automated proof can be considered correct or not—in our case, we believe that it is more reasonable to trust the correctness of Isabelle's kernel than to trust a lengthy and error-prone proof of 26 cases.

Discrepancy with Infinite Steps. If the rotor-router model is initialized with the (k,d)-wedge, the number of tokens stays the same at all vertices $\boldsymbol{x}$, independent of the number of steps the process is run (mod 2), as was shown above. In contrast, the expected number of tokens on the even partition decreases over time for the random walk. The reason for this is that at every time step and on every vertex the number of tokens is not a multiple of the number of neighboring vertices, ensuring that the rotor-router model cannot distribute the tokens equally to all neighbors as the random walk does. To show a lower bound on the discrepancy, we inspect the difference between the actual and the expected number of tokens at the origin after enough steps. We prove the following lemma.

Lemma 7. *If the rotor-router model is initialized with the (k,d)-wedge, we have*

$$\lim_{t\to\infty} \Delta(0,t) \geq 4dk.$$

Proof. Recall that $f(0,t)$ describes the number of tokens at $\boldsymbol{x} = 0$ when the rotor-router model is run, whereas $\mathbb{E}(0,t)$ describes the expected number of tokens at $\boldsymbol{x} = 0$ for the random walk after t steps. By Definition 3,

$$\Delta(0,t) = |f(0,t) - \mathbb{E}(0,t)|.$$

For the sake of brevity, we assume from now on that t is even; however, the statement holds for all t. Then, since the (k,d)-wedge was proven to be stable, we obtain $f(0,t) = d \cdot (4k+1)$.

The calculation of $\mathbb{E}(0,t)$ is more involved. According to Eq. (1),

$$\mathbb{E}(0,t) = \sum_{\boldsymbol{y}\in\mathbb{Z}^d} f(\boldsymbol{y},0)\cdot S_t^d(\boldsymbol{y}),$$

where $S_t^d(\boldsymbol{y})$ is the probability that a d-dimensional random walk that starts at $\boldsymbol{y} = (y_1,\ldots,y_d)$ ends at 0 after t steps. $S_t^d(\boldsymbol{y})$ admits simple formulas for $d \in \{1,2\}$, but there are no simple equations for $d \geq 3$ known to us.

To circumvent this problem, we show that the expected number of tokens $\mathbb{E}(\boldsymbol{x},t)$ is actually the same for all dimensions $d \geq 1$; if the starting configuration is the (k,d)-wedge.

Consider the expected number of tokens at a vertex $\boldsymbol{x}$ with respect to $\bar{\boldsymbol{x}} = x_1 + \ldots + x_d$. With one step, a token starting at $\boldsymbol{x}$ can only reach vertices $\boldsymbol{y}$ with $\bar{\boldsymbol{y}} \in \{\bar{\boldsymbol{x}} - 1, \bar{\boldsymbol{x}} + 1\}$. The probability that either happens is $1/2$, i.e.

$$\sum_{\substack{\boldsymbol{y}\in\mathbb{Z}^d\\ \bar{\boldsymbol{y}}=b}} S_1^d(\boldsymbol{x}-\boldsymbol{y}) = \begin{cases} \frac{1}{2}, & \text{if } b \in \{\bar{\boldsymbol{x}}-1, \bar{\boldsymbol{x}}+1\}\\ 0 & \text{otherwise.}\end{cases}$$

Consider now the following variation of a random walk on $\mathbb{Z}^d$, where each token can only move in one dimension, i.e.

$$\Pr[Z_i = \mathbf{e}_1] = \Pr[Z_i = -\mathbf{e}_1] = 1/2,$$
$$\Pr[Z_i = \mathbf{e}_j] = \Pr[Z_i = -\mathbf{e}_j] = 0 \quad \text{for all } j > 1.$$

In this setting, we obtain a collection of 1-dimensional random walks operating independently of each other. We write $\mathbb{E}'(\boldsymbol{x},t)$ to denote the expected number of tokens in this random walk; and we initialize $\mathbb{E}'(\boldsymbol{x},0)$ again with the (k,d)-wedge. Note that $\mathbb{E}'(\boldsymbol{x},t) = \mathbb{E}'(\bar{\boldsymbol{x}},t)$ again only depends on $\bar{\boldsymbol{x}}$ and t. By showing $\mathbb{E}'(\boldsymbol{x},t) = \mathbb{E}(\boldsymbol{x},t)$ we can analyze a 1-dimensional random walk and directly obtain results for d-dimensional random walks.

We prove $\mathbb{E}'(\boldsymbol{x},t) = \mathbb{E}(\boldsymbol{x},t)$ by induction over t. For the base case, we have $\mathbb{E}(\boldsymbol{x},0) = \mathbb{E}'(\boldsymbol{x},0)$ by definition. For the inductive step $t \to t+1$, we obtain

$$\begin{aligned}
\mathbb{E}(\boldsymbol{x},t) &= \sum_{\boldsymbol{y}\in\mathbb{Z}^d} \mathbb{E}(\boldsymbol{y},t-1)\cdot S_1^d(\boldsymbol{x}-\boldsymbol{y}) && (6)\\
&= \sum_{\substack{\boldsymbol{y}\in\mathbb{Z}^d\\ \bar{y}=\bar{x}+1}} \mathbb{E}'(\bar{y},t-1)\cdot S_1^d(\boldsymbol{x}-\boldsymbol{y}) + \sum_{\substack{\boldsymbol{y}\in\mathbb{Z}^d\\ \bar{y}=\bar{x}-1}} \mathbb{E}'(\bar{y},t-1)\cdot S_1^d(\boldsymbol{x}-\boldsymbol{y})\\
&= \mathbb{E}'(\bar{\boldsymbol{x}}+1,t-1)\cdot\frac{1}{2} + \mathbb{E}'(\bar{\boldsymbol{x}}-1,t-1)\cdot\frac{1}{2}\\
&= \mathbb{E}'(\bar{\boldsymbol{x}},t) = \mathbb{E}'(\boldsymbol{x},t), && (7)
\end{aligned}$$

where Eqs. (6) and (7) hold by the tower rule for expectation.

We now focus on the 1-dimensional random walk initialized with the (k,d)-wedge. Let $I_1 := [-2k,2k]$ and $I_2 := \mathbb{Z}\setminus I_1$. We know that $f(\boldsymbol{x},t) = d$ for all $x \in I_2$, $x \sim 0$. We denote the expected number of tokens that started in $S \subseteq \mathbb{Z}$ and arrive at the origin after $t \sim 0$ steps by $\mathbb{E}_S(0,t)$.

$$\mathbb{E}_{I_2}(0,t) = \sum_{\substack{x\in I_2\\ x\sim 0}} f(x,0)\cdot S_t^1(|x|) \le \sum_{\substack{x\in[-t,t]\\ x\sim 0}} d\cdot 2^{-t}\cdot\binom{t}{(t+|x|)/2}.$$

We now split the sum using that $S_t^1(x) = S_t^1(-x)$:

$$\mathbb{E}_{I_2}(0,t) \le \frac{d}{2^t}\cdot\left(\sum_{\substack{x=0\\ x\sim 0}}^{t}\binom{t}{(t+x)/2} + \sum_{\substack{x=2\\ x\sim 0}}^{t}\binom{t}{(t+x)/2}\right) = \frac{d}{2^t}\cdot\sum_{x=0}^{t}\binom{t}{x} = d.$$

This approximation shows that $\mathbb{E}_{I_2}(0,t) \le d$, which is obviously independent of the number of steps the process is run.

The number of expected tokens that started in I_1 and end at the origin after t steps will be approximated using the upper bound $\binom{t}{t/2} \le \sqrt{\frac{2}{\pi t}}\cdot 2^t\cdot e^{-\frac{18t-1}{72t^2+12t}}$ [21]. Then, $\mathbb{E}_{I_1}$ can be estimated the following way:

$$\begin{aligned}
\mathbb{E}_{I_1}(0,t) &= \sum_{i=1}^{k} S_t^1(2i)\cdot f(2i,0) + \sum_{i=0}^{k} S_t^1(2i)\cdot f(-2i,0)\\
&= d2^{-t}\left(\sum_{i=1}^{k}\binom{t}{\frac{t}{2}+i}\cdot(4k+3-4i) + \sum_{i=0}^{k}\binom{t}{\frac{t}{2}+i}\cdot(4k+1-4i)\right)
\end{aligned}$$

$$\leq \binom{t}{t/2} \cdot d2^{-t} \cdot \left(\sum_{i=1}^{k} (4k+3-4i) + \sum_{i=0}^{k} (4k+1-4i) \right)$$
$$= \binom{t}{t/2} \cdot d2^{-t} \cdot (2k+1)^2 \leq \sqrt{\frac{2}{\pi t}} \cdot e^{-\frac{18t-1}{72t^2+12t}} \cdot d \cdot (2k+1)^2.$$

Knowing $\mathbb{E}_{I_1}(0,t)$ and $\mathbb{E}_{I_2}(0,t)$, we compute $\mathbb{E}(0,t)$ by adding these terms and obtain $\mathbb{E}(0,t) \leq d + \sqrt{\frac{2}{\pi t}} \cdot e^{-\frac{18t-1}{72t^2+12t}} \cdot d \cdot (2k+1)^2$. This results in a discrepancy of

$$|f(0,t) - E(0,t)| \geq \max \left\{ 0, 4dk - \sqrt{\frac{2}{\pi t}} \cdot e^{-\frac{18t-1}{72t^2+12t}} \cdot d \cdot (2k+1)^2 \right\}. \quad (8)$$

For large enough t, this proves the claim. □

This means that by using the second partition of $\mathbb{Z}^d$ in the rotor-router model, it is possible to produce an arbitrarily large discrepancy of $\Omega(dk)$ which reveals that there is no constant bound for the single vertex discrepancy. Figure 2 illustrates the single vertex discrepancy in a $(k,1)$-wedge over time for $k \in \{16, 32, 64\}$.

Discrepancy Within Finite Steps. Lemma 7 shows that a discrepancy of $4dk$ can be reached if the processes are run for $t \to \infty$ steps. It is, however, possible to achieve high discrepancy using already few steps by investigating Eq. (8) more carefully. We show the following Corollary.

Corollary 8. *Given dimension $d \geq 1$ and a discrepancy $\ell \in \mathbb{R}_+$, there exists a (k,d)-wedge that reaches the discrepancy ℓ in $t \in \mathcal{O}\left(\lceil \ell^2/d^2 \rceil\right)$ steps using $\mathcal{O}(\lceil 1 + \ell/d \rceil^{2d+1})$ tokens.*

Proof. By Eq. (8), the number of steps that are needed to reach discrepancy ℓ with a (k,d)-wedge are

$$\ell \overset{!}{\leq} 4dk - \sqrt{\frac{2}{\pi t}} \cdot e^{-\frac{18t-1}{72t^2+12t}} \cdot d \cdot (2k+1)^2$$
$$\Leftarrow t \geq \frac{2}{\pi} \cdot \frac{d^2(2k+1)^4}{(4dk-\ell)^2}$$

Using standard analysis tools, we find that the minimum number of steps necessary to reach the given discrepancy ℓ is

$$t = \frac{2 \cdot d^2(\lceil \frac{d+\ell}{2d} \rceil + 1)^4}{\pi \cdot (2d+\ell)^2} \in \Theta\left(\left\lceil \frac{\ell^2}{d^2} \right\rceil\right)$$

when using a $(\lceil \frac{d+\ell}{2d} \rceil, d)$-wedge. As the process runs t steps, it visits $\Theta(t^d)$ positions of the grid $\mathbb{Z}^d$, each of which needs $\leq d \cdot (4k+1)$ tokens. Therefore, in total it needs at most $\mathcal{O}(\lceil 1 + \ell/d \rceil^{2d+1})$ tokens. □

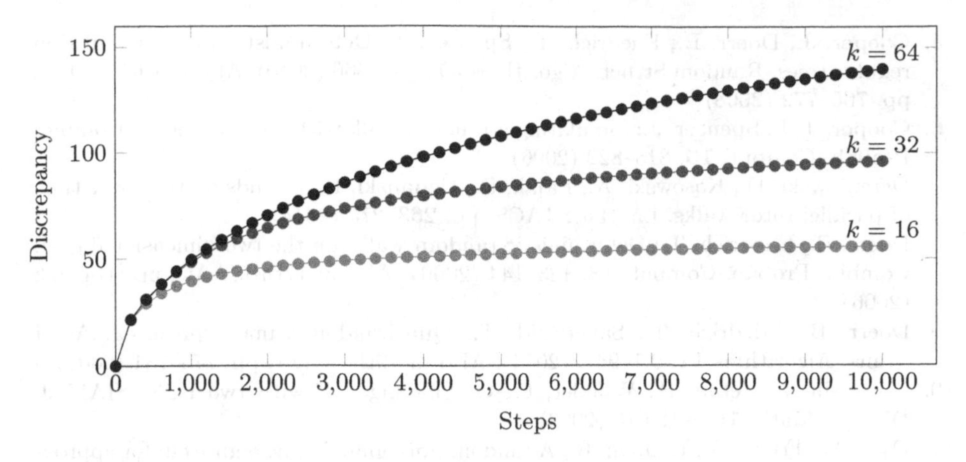

Fig. 2. The simulated single vertex discrepancies for different $(k, 1)$-wedges. The plots show that even for small t and k a high discrepancy can be achieved. This intuition is formalized in Corollary 8.

4 Conclusion

The rotor-router model is a derandomized variant of the classical random walk. It can be used algorithmically for example in broadcasting [9], external mergesort [3] and load balancing [13]. We study the similarity of the rotor-router model to the expected behavior of the random walk. It was observed and well studied that on grids the number of tokens only differs by some small constant at all times and on each vertex [4,6,8]. We closely look at the underlying assumptions of these results and prove that if tokens are allowed to start at an arbitrary position, both models can deviate arbitrarily far. Besides the revealed combinatorial structure, our result indicates that also in algorithmic applications the rotor-router model can deviate significantly from the expected behavior of the random walk, which should be studied further.

References

1. Aleliunas, R., Karp, R.M., Lipton, R.J., Lovasz, L., Rackoff, C.: Random walks, universal traversal sequences, and the complexity of maze problems. In: 20th FOCS, pp. 218–223 (1979)
2. Bampas, E., Gąsieniec, L., Hanusse, N., Ilcinkas, D., Klasing, R., Kosowski, A.: Euler tour lock-in problem in the rotor-router model. In: Keidar, I. (ed.) DISC 2009. LNCS, vol. 5805, pp. 423–435. Springer, Heidelberg (2009)
3. Barve, R.D., Grove, E.F., Vitter, J.S.: Simple randomized mergesort on parallel disks. Parallel Comput. **23**, 601–631 (1997). Also in 8th SPAA, pp. 109–118 (1996)
4. Cooper, J., Doerr, B., Spencer, J., Tardos, G.: Deterministic random walks on the integers. Eur. J. Combin. **28**, 2072–2090 (2007). Also in 3rd ANALCO, pp. 185–197 (2006)

5. Cooper, J., Doerr, B., Friedrich, T., Spencer, J.: Deterministic random walks on regular trees. Random Struct. Algorithms **37**, 353–366 (2010). Also in 19th SODA, pp. 766–772 (2008)
6. Cooper, J.N., Spencer, J.: Simulating a random walk with constant error. Combin. Probab. Comput. **15**, 815–822 (2006)
7. Dereniowski, D., Kosowski, A., Pajak, D., Uznanski, P.: Bounds on the cover time of parallel rotor walks. In: 31st STACS, pp. 263–275 (2014)
8. Doerr, B., Friedrich, T.: Deterministic random walks on the two-dimensional grid. Combin. Probab. Comput. **18**, 123–144 (2009). Also in 17th ISAAC, pp. 474–483 (2006)
9. Doerr, B., Friedrich, T., Sauerwald, T.: Quasirandom rumor spreading. ACM Trans. Algorithms **11**, 9:1–9:35 (2014). Also in 19th SODA, pp. 773–781 (2008)
10. Dumitriu, I., Tetali, P., Winkler, P.: On playing golf with two balls. SIAM J. Discrete Math. **16**, 604–615 (2003)
11. Dyer, M., Frieze, A., Kannan, R.: A random polynomial-time algorithm for approximating the volume of convex bodies. J. ACM **38**, 1–17 (1991). Also in 21st STOC, pp. 375–381 (1989)
12. Friedrich, T., Sauerwald, T.: The cover time of deterministic random walks. In: Thai, M.T., Sahni, S. (eds.) COCOON 2010. LNCS, vol. 6196, pp. 130–139. Springer, Heidelberg (2010)
13. Friedrich, T., Gairing, M., Sauerwald, T.: Quasirandom load balancing. SIAM J. Comput. **41**, 747–771 (2012). Also in 21st SODA, pp. 1620–1629 (2010)
14. Kijima, S., Koga, K., Makino, K.: Deterministic random walks on finite graphs. In: 9th ANALCO, pp. 16–25 (2012)
15. Klasing, R., Kosowski, A., Pajak, D., Sauerwald, T.: The multi-agent rotor-router on the ring: a deterministic alternative to parallel random walks. In: 32nd PODC, pp. 365–374 (2013)
16. Kleber, M.: Goldbug variations. The Math. Intell. **27**, 55–63 (2005)
17. Kosowski, A., Pająk, D.: Does adding more agents make a difference? a case study of cover time for the rotor-router. In: Esparza, J., Fraigniaud, P., Husfeldt, T., Koutsoupias, E. (eds.) ICALP 2014, Part II. LNCS, vol. 8573, pp. 544–555. Springer, Heidelberg (2014)
18. Lawler, G., Limic, V.: Random Walk: A Modern Introduction. Cambridge Studies in Advanced Mathematics. Cambridge University Press, Cambridge (2010)
19. Nipkow, T., Paulson, L.C., Wenzel, M. (eds.): Isabelle/HOL: A Proof Assistant for Higher-Order Logic. LNCS, vol. 2283. Springer, Heidelberg (2002)
20. Priezzhev, V.B., Dhar, D., Dhar, A., Krishnamurthy, S.: Eulerian walkers as a model of self-organized criticality. Phys. Rev. Lett. **77**, 5079–5082 (1996)
21. Robbins, H.: A remark on Stirling's formula. The Am. Math. Mon. **62**, 26–29 (1955)
22. Wagner, I.A., Lindenbaum, M., Bruckstein, A.M.: Distributed covering by ant-robots using evaporating traces. IEEE Trans. Rob. Autom. **15**, 918–933 (1999)
23. Yanovski, V., Wagner, I.A., Bruckstein, A.M.: A distributed ant algorithm for efficiently patrolling a network. Algorithmica **37**, 165–186 (2003)

Trading off Worst and Expected Cost in Decision Tree Problems

Aline Saettler[1], Eduardo Laber[1], and Ferdinando Cicalese[2](✉)

[1] PUC-Rio, Rio de Janeiro, Brazil
{asaettler,laber}@inf.puc-rio.br
[2] University of Verona, Verona, Italy
ferdinando.cicalese@univr.it

Abstract. We characterize the best possible trade-off achievable when optimizing the construction of a decision tree with respect to both the worst and the expected cost. It is known that a decision tree achieving the minimum possible worst case cost can behave very poorly in expectation (even exponentially worse than the optimal), and the vice versa is also true. Led by applications where deciding for the right optimization criterion might not be easy, recently, several authors have focussed on the bicriteria optimization of decision trees.

An unanswered fundamental question is about the best possible trade-off achievable. Here we are able to sharply define the limits for such a task. More precisely, we show that for every $\rho > 0$ there is a decision tree D with worst testing cost at most $(1+\rho)OPT_W + 1$ and expected testing cost at most $\frac{1}{1-e^{-\rho}}OPT_E$, where OPT_W and OPT_E denote the minimum worst testing cost and the minimum expected testing cost of a decision tree for the given instance. We also show that this is the best possible trade-off in the sense that there are infinitely many instances for which we cannot obtain a decision tree with both worst testing cost smaller than $(1+\rho)OPT_W$ and expected testing cost smaller than $\frac{1}{1-e^{-\rho}}OPT_E$.

1 Introduction

We consider a very general model of the decision tree construction problem: We have a set of objects $S = \{s_1, \ldots, s_n\}$ which is partitioned into m classes $C_1, \ldots, C_m$. Objects are characterized by the value they take with respect to a set of tests T. Each test $t \in T$ has a finite number of possible values, which we assume upper bounded by some fixed value ℓ. Each test t has also an associated rational positive cost $c(t)$ which has to be paid in order to use the test. The aim is to design a procedure which given an object identifies its class by adaptively using tests to acquire information on the object whose classification has to be discovered. Each new test restricts the set of possible classifications to those of the objects matching (or satisfying) the results of all the tests performed so far. The procedure stops when the objects agreeing with the results of the tests performed belong to the same class, which must also be the class of the object that had to be classified. We assume that the set of tests is complete, that is, for any pair of objects from distinct classes, there exists a test t which separates

K. Elbassioni and K. Makino (Eds.): ISAAC 2015, LNCS 9472, pp. 223–234, 2015.
DOI: 10.1007/978-3-662-48971-0_20

them, i.e., it has different values for the two objects. We also assume that a probability distribution $\mathbf{p}$ on the set of objects is provided, according to which the objects to be classified are believed to be chosen.

More formally, by an instance I we understand a quintuple $I = (S, \mathcal{C}, T, \mathbf{p}, \mathbf{c})$ where S is the set of objects, $\mathcal{C}$ is the family of classes defining the partition of S, T is the set of tests, $\mathbf{p}$ is the probability distribution on S and $\mathbf{c}$ is the cost function assigning a non-negative cost to each test in T.

As an example let us consider a diagnosis problem: the objects might be a set of possible diseases, e.g., $\{flu, dengue, cancer\}$ divided into infectious $\{flu\}$ and non-infectious $\{dengue, cancer\}$. The goal is to have a strategy for quickly/cheaply testing whether a patient allegedly ill might be infectious or not.

Any testing procedure can be represented by a *decision tree*, which is a tree where every internal node is associated with a test. The branches stemming out from a node are associated with the possible outcomes of the test associated with the node. Every leaf is associated with a set of objects that belong to the same class. More formally, a decision tree D over the set of objects S can be inductively defined as follows: (i) if every object of S belongs to the same class i, then D is a single leaf associated with class i; (ii) otherwise, the root r of D is associated with some test $t \in T$ and the children of r are decision trees for the non empty sets in $\{S_t^1, \ldots, S_t^\ell\}$, where S_t^i is the subset of S for which the test t takes value i.

Given a decision tree D, rooted at r, we can identify the class of an object s by following a path from r to a leaf as follows: first, we ask for the result of the test associated with r when performed on s; then, we follow the branch of r associated with the result of the test to reach a child r' of r; next, we apply the same steps recursively starting from r'. The procedure ends when a leaf is reached, which determines the class of s. We also say that this is the leaf associated to s.

We define $cost(D, s)$ as the sum of the tests' cost on the path from the root of D to the leaf associated with object s. Then, the *worst testing cost* and the *expected testing cost* of D are, respectively, defined as

$$cost_W(D) = \max_{s \in S}\{cost(D, s)\} \quad \text{and} \quad cost_E(D) = \sum_{s \in S} cost(D, s)p(s) \tag{1}$$

Most of the works on decision tree optimization focuses on building a decision tree that minimizes only one of the above measures [1,5,7,12–15,18,23]. From an application point of view, the choice of the optimization criterion reflects different assumptions on the data model: a more optimistic perspective on the knowledge of the underlying distribution might elicit the minimization of the expected testing cost; a more pessimistic perspective might prefer the more conservative minimization of the worst case testing cost.

However, the two different optimization criteria can lead to very different trees: a decision tree minimizing the expected cost for a very skewed distribution can have a skewed shape with a very high worst case cost even exponentially bigger than the worst cost of a decision tree optimized with respect to the worst testing cost. Conversely optimizing with respect to the worst testing cost can lead to a tree with poor performance in expectation. The choice of the "wrong" optimization criterion might have serious consequences in practical applications (see, e.g., [17] for such a study in the economics literature).

These arguments have motivated recent work on decision tree constructions optimizing both worst and the expected testing cost [3,9,16,19]. In [9], which was our starting point in this line of research, we provided an algorithm which builds a decision tree guaranteeing simultaneously $O(\log n)$-approximation for both worst and expected testing cost—which is the best possible approximation achievable for either criterion under standard complexity assumptions.

Here, we address, and, surprisingly, exactly answer, a more fundamental question regarding the existence of arbitrarily good trade-offs between expected and worst cost: Does there exist in general (asymptotically for any instance) a decision tree with worst testing cost and expected testing cost arbitrarily close, respectively, to the optimal worst testing cost and the optimal expected testing cost? Or, otherwise, what is the threshold for the best trade-off we can hope for?

Our Results. In Sect. 2, we show that for every $\rho > 0$ and every instance I there exists a decision tree D with worst testing cost at most $(1+\rho)OPT_W(I)+1$ and expected testing cost at most $\frac{1}{1-e^{-\rho}}OPT_E(I)$, where $OPT_W(I)$ (resp. $OPT_E(I)$) denote the cost of the decision tree with minimum worst testing cost (resp. minimum expected testing cost) for the instance I.

We then show, in Sect. 3, that this is a sharp characterization of the best possible trade-off attainable, in the sense that there are infinitely many instances for which we cannot obtain a decision tree with both worst testing cost smaller than $(1+\rho)OPT_W(I)$ and expected testing cost smaller than $\frac{1}{1-e^{-\rho}}OPT_E(I)$.

To obtain the upper bound, we present a procedure that given a parameter $\rho > 0$, a decision tree D_W with worst testing cost W and a decision tree D_E with expected testing cost E, produces a decision tree D with worst testing cost at most $(1+\rho)W+1$ and expected testing cost at most $\frac{1}{1-e^{-\rho}}E$. For the analysis of our procedure we employ techniques from non-linear programming (NLP) [4]. Although we had reached our upper bound in an independent way, we shall mention that our techniques are similar to those employed in [2,25] to obtain tight trade-offs between the minimization of the expected completion time and the makespan for scheduling problems.

For the lower bound, we then make use of the probability distribution used in the analysis of the upper bound—obtained by the optimal solution of the NLP—as a starting point for constructing non-trivial instances that guarantee that the upper bound is tight. In this case, the results of [2,25] give no clue how to obtain a tight lower bound for our problem.

Related Work. There are some studies related to the simultaneous minimization of the expected testing cost and the worst testing cost for the prefix code problem [6,11,20–22]. The problem of constructing a prefix code is a particular case of decision tree optimization in which each object belongs to a distinct class, the testing costs are uniform and the set of tests is in one to one correspondence with the set of all binary strings of length n so that the test corresponding to a binary string b outputs 0 (1) for object s_i if and only if the ith bit of b is 0 (1).

In contrast to our present findings, the results of Milidiu and Laber [22] imply that in the case of the prefix code problem, asymptotically, there exists a decision tree that is arbitrarily close to the optimum with respect to both expected and worst cost. More precisely, for every instance I with n objects and any $\rho > 0$, there is a decision tree D such that $cost_W(D)/OPT_W(I) \leq (1+\rho)$ and $cost_E(D)/OPT_E(I) \leq 1/\psi^{\rho \log n - 1}$, where ψ is the golden ratio $(1+\sqrt{5})/2$.

Some proofs are deferred to the full version of the paper.

2 Trade-off: Upper Bound

In this section, we show our upper bound on the achievable trade-off between worst and expected testing cost for the decision tree optimization problem. Our proof will be constructive, that is, we will show a procedure for constructing a decision tree guaranteeing the desired trade off.

Given a positive number j, and two decision trees D_E and D_W for instance I, the procedure `CombineTrees`(D_E, D_W, j) (See Algorithm 1) constructs a new decision tree D^j for I whose worst testing cost is increased by at most j w.r.t the worst testing cost of D_W, i.e., $cost_W(D^j) \leq j + cost_W(D_W)$. Our algorithm uses the definition of a *j-replaceable node*, by which we mean a node v in D such that the total cost of the tests on the path from the root of D to v (including v) is at least j and the cost of the path from the root of D to the parent of v is smaller than j. The procedure `Trade-Off` repeatedly uses `CombineTrees` to create several decision trees with increasingly worst testing cost and chooses the one with the best expected testing cost. We will show that this way it can guarantee the best possible trade off.

Proposition 1. *The decision tree D^j returned by* `CombineTrees` *has worst testing cost at most $j + cost_W(D_W)$.*

Now we analyze the decision tree $D = D^{j^*}$ output by `Trade-Off`(D_E, D_W, C), where C is an integer parameter. Notice that D is the decision tree with minimum

---

Algorithm 1. Computes trade off tree between D_W and D_E

Procedure `CombineTrees`(D_E, D_W, j)

1: $D^j \leftarrow D_E$
2: Traverse D^j and construct $R = \{v \mid v \text{ is a } j\text{-replaceable node of } D^j\}$
3: **for** each $v \in R$ **do**
4:     Replace in D^j the subtree rooted at v with D_W
5: **return** D^j

Procedure `Trade-Off`(D_E, D_W, C)

1: **for** $j = 0, \ldots, C$ **do**
2:     $D^j \leftarrow$ `CombineTrees`(D_E, D_W, j)
3: $j^* \leftarrow \arg\min_{0 \leq j \leq C} cost_E(D^j)$
4: **return** D^{j^*}

---

expected testing cost among the decision trees $D^0, D^1, D^2, \ldots, D^C$, where D^j is the decision tree returned by `CombineTrees`(D_E, D_W, j). It follows from the previous proposition that $cost_W(D) \leq C + cost_W(D_W)$.

The analysis of the expected testing cost of D is more involved. In order to simplify the notation we will let $W = cost_W(D_W)$. We also assume for simplicity in the following that test costs are integers. Given a decision tree D' and an object/leaf $s \in S$ with $cost(D', s) = \kappa$ we will say that s has cost κ in D'.

Let p_i, with $i = 1, \ldots, C$, be the sum of the probabilities of objects with cost i in D_E and p_{C+1} be the sum of the probabilities of the objects with cost larger than C in D_E. Clearly $cost_E(D_E) \geq \sum_{i=1}^{C+1} p_i \cdot i$. Furthermore, for $j = 0, \ldots, C$, we have that $cost_E(D^j) \leq \sum_{i=1}^{j} p_i \cdot i + \left((j+W) \sum_{i=j+1}^{C+1} p_i\right)$ because the objects whose cost in D_E is larger than j have cost at most $j + W$ in D^j. Thus,

$$\frac{cost_E(D)}{cost_E(D_E)} = \min_{j=0,\ldots,C} \frac{cost_E(D^j)}{cost_E(D_E)} \leq \min_{j=0,\ldots,C} \left\{ \frac{\sum_{i=1}^{j} p_i \cdot i + (j+W) \sum_{i=j+1}^{C+1} p_i}{\sum_{i=1}^{C+1} p_i \cdot i} \right\}$$

$$\leq \max_{\mathbf{p} \in \mathcal{P}} \min_{j=0,\ldots,C} \left\{ \frac{\sum_{i=1}^{j} p_i \cdot i + (j+W) \sum_{i=j+1}^{C+1} p_i}{\sum_{i=1}^{C+1} p_i \cdot i} \right\},$$

where $\mathcal{P} = \{(p_1, p_2, \ldots, p_{C+1}) | \sum_{i=1}^{C+1} p_i = 1 \text{ and } p_1, p_2, \ldots, p_{C+1} \geq 0\}$.

Thus, we can conclude that $cost_E(D)/cost_E(D_E) \leq z^*$, where z^* is the maximum achieved by the following non linear program (NLP):

$$z^* = \max z \quad \text{s. t.} \tag{2}$$

$$z\left(\sum_{i=1}^{C+1} i \cdot p_i\right) - \sum_{i=1}^{j} i \cdot p_i - (j+W)\left(\sum_{i=j+1}^{C+1} p_i\right) \leq 0, \; j = 0, \ldots, C \tag{3}$$

$$\sum_{i=1}^{C+1} p_i = 1 \tag{4}$$

$$p_i \geq 0, \; i = 1, \ldots, C+1 \tag{5}$$

Perharps surprisingly we can prove that the optimal solution of the NLP is given by $\mathbf{p}^* = (p_1^*, p_2^*, \ldots, p_{C+1}^*, z^*)$, where for $i = 1, \ldots, C$,

$$p_i^* = \frac{(W-1)^{i-1}}{W^i}, \text{ and } p_{C+1}^* = \frac{(W-1)^C}{W^C} \text{ and } z^* = \frac{1}{\left(1 - \left(\frac{W-1}{W}\right)^{C+1}\right)}. \tag{6}$$

The proof consists of showing that the functions that define the $C+1$ non linear constraints are convex in the polyhedra $\mathcal{P}$ and the point p^* satisfies the Karush-Kuhn-Tucker conditions [4]. Thus, by setting C equal to an integer such that $(C-1)/W < \rho \leq C/W$, we get the following theorem.

Theorem 1. *Fix an instance I of the decision tree optimization problem and let D_E be a decision tree such that $cost_E(D_E) = OPT_E(I)$. For every $\rho > 0$ there exists a decision tree D such that*

$$cost_W(D) \leq (1+\rho)OPT_W(I) + 1 \quad and \quad cost_E(D) \leq \left(\frac{1}{1-e^{-\rho}}\right) OPT_E(I).$$

3 Trade-off: Lower Bound

In this section we show that the previous theorem is essentially tight by proving the following result.

Theorem 2. *Fix positive integers W and C. There exists an instance I such that the following hold: 1. $OPT_W(I) \leq W$. 2. $OPT_E(I) \leq W\left(1 - \left(\frac{W-1}{W}\right)^C\right) + \lfloor \log W \rfloor \left(\frac{W-1}{W}\right)^C$. 3. If a decision tree D for I is such that $cost_W(D) \leq W + C$ then it holds that $cost_E(D) \geq W$.*

Corollary 1. *For any fixed $\rho > 0$ and $\epsilon > 0$, there are infinitely many instances I of the decision tree problem such that no decision tree can simultaneously guarantee worst testing cost smaller than $OPT_W(I)(1+\rho)$ and expected testing cost smaller $OPT_E(I)\left(\frac{1}{1-e^{-\rho}}\right) - \epsilon$.*

Proof. Pick an integer W and $C = \lfloor \rho W \rfloor$. Then, consider the instance I given by the previous theorem. It follows that every decision tree D, with $cost_W(D) \leq W + C \leq (1+\rho)W$, satisfies $cost_E(D) \geq W$. Thus,

$$\frac{cost_E(D)}{OPT_E(I)} \geq \frac{1}{1 - \left(\frac{W-1}{W}\right)^C + \frac{\lfloor \log W \rfloor}{W}\left(\frac{W-1}{W}\right)^C} \geq \frac{1}{1 - \left(\frac{W-1}{W}\right)^{\rho W} + \frac{\lfloor \log W \rfloor}{W}\left(\frac{W-1}{W}\right)^{\rho W}}$$

It is easy to see that the right hand side expression goes to $\frac{1}{1-e^{-\rho}}$ as W goes to ∞. Thus, for every W larger than a certain integer, say W_ϵ we get that right hand side is larger than $\frac{1}{1-e^{-\rho}} - \epsilon$ so that there are infinitely many instances with the required property.

3.1 The Structure of the Instance I in Theorem 2

For positive integers W and C we define the following instance $I = (S, T, \mathcal{C}, \mathbf{p}, \mathbf{c})$.

The Set of Objects S. For the sake of simplifying notation, let $L_W = \lfloor \log W \rfloor$. The set of objects is divided into the objects of type i (for each $i = 1, \ldots, C + L_W$) and *light* objects with almost zero probability mass. For each $i = 1, \ldots, C + L_W$ there are 2^i objects of type i, which we denote by $S^{(i)} = \{o_1^{(i)}, \ldots, o_{2^i}^{(i)}\}$.

For each $i = 1, \ldots, C$ and $j = 1, \ldots, 2^i$, the probability of $o_j^{(i)}$ is $\frac{(W-1)^{i-1}}{2^i W^i}$. Hence, the total probability of objects of type i is $p(S^{(i)}) = \frac{(W-1)^{i-1}}{W^i}$. Note that this is exactly the probability distribution of the optimal solution of the NLP presented in the previous section.

For each $i = C+1, \ldots, C+L_W$ and $j = 1, \ldots, 2^i$, the probability of $o_j^{(i)}$ is $\left(\frac{W-1}{W}\right)^C \frac{1}{2^C(2^{L_W+1}-2)}$. Hence, for the total cumulative probability of objects of type larger than C we have $p(S^{(C+1)} \cup \cdots \cup S^{(C+L_W)}) = \left(\frac{W-1}{W}\right)^C$.

Finally, there are 2^{C+L_W+1} light objects, each of which has the same probability which is *very close to zero*. We denote by $S^L = \{o_j^L \mid j = 1, \ldots, 2^{C+L_W+1}\}$ the set of the light objects, and we set $p(S^L) = \epsilon' \to 0$.[1]

The Partition into Classes $\mathcal{C}$. Each object belongs to a different class.

A Canonical Representation of the Objects. For later purposes it is convenient to visualize the set of objects as a complete binary tree $\mathcal{T}$ of depth $C + L_W + 1$. By the ith level of $\mathcal{T}$ we understand the set of nodes at distance i from the root.

For $i = 1, \ldots, C+L_W$ the objects of type i are identified with the nodes at level i of $\mathcal{T}$. Therefore, for $i = 1, \ldots, C+L_W$ and $j = 1, \ldots, 2^i$, the jth node (counting from left to right) in level i is identified with object $o_j^{(i)}$ of $S^{(i)}$. We use $O_j^{(i)}$ to denote the set of objects of the subtree of $\mathcal{T}$ rooted at $o_j^{(i)}$.

The light objects are identified with the nodes at level $C + L_W + 1$ of $\mathcal{T}$. Therefore, for $j = 1, \ldots, 2^{C+L_W+1}$, the jth node (counting from left to right) in level $C + L_W + 1$ of $\mathcal{T}$ is identified with object o_j^L of S^L. We shall note that the root of $\mathcal{T}$ is not associated with an object.

The Set of Tests T. The set T of available tests is easily explained with reference to the canonical representation of the objects presented above. The values taken by a test can be interpreted as a partition of the set of objects, each value corresponding to the subset of objects for which the test has that value. Therefore, we describe a test by the way it partitions or splits the set of objects.

There is one test of type 1, which we denote with $t_1^{(1)}$ and splits the objects as follows: (i) the single object $\{o_1^{(1)}\}$; (ii) the single object $\{o_2^{(1)}\}$; (iii) the set $O_1^{(1)} - \{o_1^{(1)}\}$; (iv) the set $O_2^{(1)} - \{o_2^{(1)}\}$.

For each $i = 2, \ldots, C+L_W$ and $j = 1, \ldots, 2^{i-2}$ the set T includes a test $t_j^{(i)}$ which splits the set of objects into 4 parts as follows: (i) the single object $\{o_{2j-1}^{(i)}\}$; (ii) the single object $\{o_{2j}^{(i)}\}$; (iii) $O_1^{(1)} - \{o_1^{(1)}, o_{2j-1}^{(i)}\} - O_{2j}^{(i)}$; (iv) $O_2^{(1)} \cup \{o_1^{(1)}\} \cup \left(O_{2j}^{(i)} \setminus \{o_{2j}^{(i)}\}\right)$.

For each $i = 2, \ldots, C+L_W$ and $j = 2^{i-2}+1, \ldots, 2^{i-1}$ the set T includes a test $t_j^{(i)}$ which splits the set of objects into 4 parts as follows: (i) the single object $\{o_{2j-1}^{(i)}\}$; (ii) the single object $\{o_{2j}^{(i)}\}$; (iii) $O_1^{(1)} \cup \{o_2^{(1)}\} \cup \left(O_{2j-1}^{(i)} \setminus \{o_{2j-1}^{(i)}\}\right)$; (iv) $O_2^{(1)} - \{o_2^{(1)}, o_{2j}^{(i)}\} - O_{2j-1}^{(i)}$.

---

[1] The probability of all the other objects should be multiplied by $1-\epsilon$. For simplifying the notation we shall simply assume that the light objects have zero probability.

We can try to visualize the split produced by test $t_j^{(i)}$, with $i > 1$, as follows: we first separate the objects in the subtrees of $\mathcal{T}$ rooted at the children of $o_j^{(i-1)}$; then we separate the roots of these two subtrees from the remaining objects. This way we create 4 groups, two of them being singletons. Then, one of the non-singleton groups gets the remaining objects from the set $O_1^{(1)} - \{o_1^{(1)}\}$ while the other gets those from $O_2^{(1)} - \{o_2^{(1)}\}$. Finally, we add $\{o_1^{(1)}, o_2^{(1)}\}$ to one of the non-singleton groups according to whether the object $o_j^{(i-1)}$ is in $O_1^{(1)}$ or in $O_2^{(1)}$. For each $i = 1, \ldots, C + L_W$, we will refer to tests $\{t_j^{(i)} \mid j = 1, \ldots, 2^{i-1}\}$ as the tests of type i.
Finally, T includes a test denoted by t^* which separates each single object.

The Cost of the Tests. The tests of type $i = 1, \ldots, C + L_W$ have cost 1 while the test t^* has cost W. We will refer to test t^* as *the costly test.*

3.2 Proof of Theorem 2

Proof of Item 1. The first item of Theorem 2 follows because a tree with the costly test t^* at the root has worst testing cost W.

Proof of Item 2. To prove the second item we construct a decision tree $D^C(I)$ for instance I, which we call the Canonical Decision Tree, and we evaluate its expected testing cost.

If we ignore the leaves—which can be added in the *natural* way—the structure of the nodes associated with tests in the canonical decision tree $D^C(I)$ can be obtained as follows: start with the canonical tree of objects $\mathcal{T}$; replace the root of $\mathcal{T}$ with the test $t_1^{(1)}$ and each node $o_j^{(i)}$ with the test $t_j^{(i+1)}$, for $i = 1, \ldots C + L_W - 1$. Finally, each node on level $C + L_W$ is replaced by the costly test t^*.

It is not to hard to verify that

$$cost_E(D^C(I)) \leq \sum_{j=1}^{C} j\frac{(W-1)^{j-1}}{W^j} + (C + L_W)\left(\frac{W-1}{W}\right)^C \tag{7}$$

$$= W\left(1 - \left(\frac{W-1}{W}\right)^C\right) + L_W\left(\frac{W-1}{W}\right)^C. \tag{8}$$

Inequality (7) follows by observing that in the canonical decision tree every object of type larger than C has cost at most $C + L_W$. Moreover, we use $\sum_{j=1}^{C} j\frac{(W-1)^{j-1}}{W^j} = W - (C + W)\left(\frac{W-1}{W}\right)^C$ to obtain (8). Thus, we have established the item 2 of Theorem 2.

Proof of Item 3. To establish the item 3, we need some additional notation. By the test associated with a non-light object o we mean the non-costly test that separates the two children of o, that is, for $o = o_j^{(i)}$ the test associated to o is $t_j^{(i+1)}$. For a decision tree D, we use $Obj(\nu)$ to denote the set of objects associated with the leaves in the subtree of D rooted at ν. We say that a test t occurs at cost level κ in a decision tree D if the total cost of tests on the path from the root of D to t (excluding t) is κ.

The proof of the following propositions are deferred to the full version.

Proposition 2. *Let D be a decision tree for instance I. Let ν be an internal node of D, such that $Obj(\nu)$ includes non-light objects. Then there are two sibling nodes/objects of the canonical tree of objects $\mathcal{T}$, name them x_1 and x_2, such that each object in $Obj(\nu)$ is a descendant of either x_1 or x_2 in $\mathcal{T}$.*

Proposition 3. *The following inequalitiy holds: $Pr[O_k^{(i)} - o_k^{(i)}] \leq (W-1)Pr[o_k^{(i)}]$, for any $1 \leq i \leq C + L_W$ and $1 \leq k \leq 2^i$.*

Lemma 1. *Let D be a decision tree for the instance I such that $cost_W(D) \leq W + C$. Then $cost_E(D) \geq W$.*

Proof. Let D be a decision tree with minimum expected testing cost among all decision trees for I with worst testing cost not larger than $W + C$.

First, we argue that every non-costly test in D with at least one non-light object as a descendant occurs at cost level at most $C - 1$. For the sake of contradiction, let us assume that some non-costly test that has at least one non-light object as a descendant occurs at cost level larger than or equal to C. Let ν be the node of D corresponding to such a test and let o be a non-light object in $Obj(\nu)$. Assume that $o \in O_1^{(1)}$ (the proof for the other case is analogous so that we omit it). We have two cases: o is of type $i > 1$ and o is of type 1. If o is of type larger than 1 then the two light objects identified with the two nodes in the leftmost path of the subtree rooted at o in $\mathcal{T}$ are also in $Obj(\nu)$ because the only tests that separate them from o are the costly test and the test corresponding to the parent of o. However, none of these tests can be an ancestor of ν in D, for otherwise we would have $o \notin Obj(\nu)$. Thus, in order to separate these light objects we need a costly test in the subtree of D rooted at ν. This implies that $cost_W(D) > C + W$, which is a contradiction. If o has type 1 then the argument is the same except for the fact that we consider the two light objects identified with the two nodes in the *righmost* path of the subtree rooted at o in $\mathcal{T}$.

Now, we argue that there exists a tree $\tilde{D}$ with worst testing cost at most $C + W$ and expected testing cost not larger than that of D such that all costly tests in $\tilde{D}$, which are ancestors of at least one non-light object, occur at cost level C. For that, let ν be an internal node of D associated with a costly test that occurs at cost level smaller than C and such that $Obj(\nu)$ contains non-light objects. By Proposition 2 the set of objects $Obj(\nu)$ can be partitioned into three parts $\{o_{2j-1}^{(s)}\}$, $\{o_{2j}^{(s)}\}$ and $Obj(\nu) \setminus \{o_{2j-1}^{(s)}, o_{2j}^{(s)}\} \subseteq \left(O_{2j-1}^{(s)} \cup O_{2j}^{(s)}\right) - \{o_{2j-1}^{(s)}, o_{2j}^{(s)}\}$, for some $1 \leq s \leq C + L_W$. and some $1 \leq j \leq 2^s$. From Proposition 3 we have

$$Pr[Obj(\nu) \setminus \{o_{2j-1}^{(s)}, o_{2j}^{(s)}\}] \leq Pr[\{O_{2j-1}^{(s)}\} \setminus \{o_{2j-1}^{(s)}\}] + Pr[\{O_{2j}^{(s)}\} \setminus \{o_{2j}^{(s)}\}] \quad (9)$$

$$\leq Pr[\{o_{2j-1}^{(s)}, o_{2j}^{(s)}\}](W-1). \quad (10)$$

Let D' be the decision tree obtained by replacing the costly test at ν with the test $t_j^{(s)}$ and then using costly tests on the two branches not leading to leaves. This reduces the cost of the leaves associated with $o_{2j-1}^{(s)}$ and $o_{2j}^{(s)}$ by $W-1$ and increases by 1 the cost of the leaves associated to the objects in $Obj(\nu) \setminus \{o_{2j-1}^{(s)}, o_{2j}^{(s)}\}$. In formulas, using (9)-(10), we have

$$cost_E(D') = cost_E(D) - (W-1)Pr[\{o_{2j-1}^{(s)}, o_{2j}^{(s)}\}] + Pr[Obj(\nu) \setminus \{o_{2j-1}^{(s)}, o_{2j}^{(s)}\}] \leq cost_E(D).$$

Repeated application of the above transformation gives a decision tree $\tilde{D}$ such that: $cost_W(\tilde{D}) = C + W$; $cost_E(\tilde{D}) \leq cost_E(D)$ and for each node ν of $\tilde{D}$ associated with a costly test, either $Obj(\nu)$ contains only light objects or ν occurs at cost level C.

For each $\ell = 1, 2, \ldots, C$ let $\tilde{p}_\ell$ be the probability of the *non-light* objects whose cost in $\tilde{D}$ is ℓ. Note that every non-costly test splits the set of non-light objects into exactly four parts, with exactly two of them associated to leaves. Thus, it follows that in $\tilde{D}$, there are at most 2^ℓ leaves at level ℓ associated with non-light objects (such objects having cost ℓ). Therefore, for each $k = 1, \ldots, C$.

$$\sum_{\ell=1}^{k} \tilde{p}_\ell \leq \sum_{\ell=1}^{k} \frac{(W-1)^{\ell-1}}{W^\ell}. \tag{11}$$

In fact for each k there are at most $2^{k+1}-2$ leaves associated with non-light objects in the first k levels. In addition the set of $2^{k+1}-2$ objects of largest probability in I is given by the set of objects of type $1, \ldots, k$, whose cumulative probability coincides with the right-hand-side expression.

Then, ignoring the contribution of the light objects, we can write

$$cost_E(\tilde{D}) = \sum_{j=1}^{C} \left(1 - \sum_{\ell=1}^{j-1} \tilde{p}_\ell\right) + W\left(1 - \sum_{\ell=1}^{C} \tilde{p}_\ell\right) \tag{12}$$

$$\geq \sum_{j=1}^{C} \left(1 - \sum_{\ell=1}^{j-1} \frac{(W-1)^{\ell-1}}{W^\ell}\right) + W\left(1 - \sum_{\ell=1}^{C} \frac{(W-1)^{\ell-1}}{W^\ell}\right) \tag{13}$$

$$= \sum_{j=1}^{C} \frac{(W-1)^{j-1}}{W^{j-1}} + W\left(\frac{(W-1)^C}{W^C}\right) = W \tag{14}$$

where (12) is a rewriting of $cost_E(\tilde{D})$ in terms of the contribution of the internal nodes/tests by cost level; and (13) follows from (12) because of (11). By the construction of $\tilde{D}$ we finally have the desired result $cost_E(D) \geq cost_E(\tilde{D}) \geq W$.

4 Open Problems

An interesting open question regards the case of uniform testing costs. We ask whether for every $\epsilon > 0$, there is some integer n_0 such that every instance I with uniform testing costs and with more than n_0 objects, admits a decision tree D such that $cost_E(D) \leq (1+\epsilon)OPT_E(I)$ and $cost_W(D) \leq (1+\epsilon)OPT_W(I)$. Notice that this result holds for the special case of decision tree construction that we have in prefix code problem [22].

References

1. Adler, M., Heeringa, B.: Approximating optimal binary decision trees. In: Goel, A., Jansen, K., Rolim, J.D.P., Rubinfeld, R. (eds.) APPROX and RANDOM 2008. LNCS, vol. 5171, pp. 1–9. Springer, Heidelberg (2008)
2. Aslam, J.A., Rasala, A., Stein, C., Young, N.E.: Improved bicriteria existence theorems for scheduling. In: SODA, pp. 846–847 (1999)
3. Alkhalid, A., Chikalov, I., Moshkov, M.: A tool for study of optimal decision trees. In: Yu, J., Greco, S., Lingras, P., Wang, G., Skowron, A. (eds.) RSKT 2010. LNCS, vol. 6401, pp. 353–360. Springer, Heidelberg (2010)
4. Bazaraa, M.S., Sherali, H.D., Shetty, C.M.: Nonlinear Programming: Theory and Algorithms, 2nd edn. John Wiley, New York (1993)
5. Bellala, G., Bhavnani, S.K., Scott, C.: Group-based active query selection for rapid diagnosis in time-critical situations. IEEE-IT **58**(1), 459–478 (2012)
6. Buro, M.: On the maximum length of huffman codes. IPL **45**, 219–223 (1993)
7. Chakaravarthy, V.T., Pandit, V., Roy, S., Awasthi, P., Mohania, M.: Decision trees for entity identification: approximation algorithms and hardness results. In: Proceedings of PODS 2007, pp. 53–62 (2007)
8. Cicalese, F., Jacobs, T., Laber, E., Molinaro, M.: On greedy algorithms for decision trees. In: Cheong, O., Chwa, K.-Y., Park, K. (eds.) ISAAC 2010, Part II. LNCS, vol. 6507, pp. 206–217. Springer, Heidelberg (2010)
9. Cicalese, F., Laber, E., Saettler, A.: Diagnosis determination: decision trees optimizing simultaneously worst and expected testing cost. In: ICML 2014, pp. 414–422 (2014)
10. Garey, M.R.: Optimal binary identification procedures. SIAM J. Appl. Math. **23**(2), 173–186 (1972)
11. Garey, M.R.: Optimal binary search trees with restricted maximal depth. SIAM J. Comput. **3**(2), 101–110 (1974)
12. Golovin, D., Krause, A., Ray, D.: Near-optimal bayesian active learning with noisy observations. In: Advances in Neural Information Processing Systems, vol. 23, pp. 766–774 (2010)
13. Guillory, A., Bilmes, J.: Average-case active learning with costs. In: Gavaldà, R., Lugosi, G., Zeugmann, T., Zilles, S. (eds.) ALT 2009. LNCS, vol. 5809, pp. 141–155. Springer, Heidelberg (2009)
14. Guillory, A., Bilmes, J.: Interactive submodular set cover. In: Proceedings of ICML 2010, pp. 415–422 (2010)
15. Gupta, A., Nagarajan, V., Ravi, R.: Approximation algorithms for optimal decision trees and adaptive TSP problems. In: Abramsky, S., Gavoille, C., Kirchner, C., Meyer auf der Heide, F., Spirakis, P.G. (eds.) ICALP 2010. LNCS, vol. 6198, pp. 690–701. Springer, Heidelberg (2010)
16. Hussain, S.: Relationships among various parameters for decision tree optimization. In: Faucher, C., Jain, L.C. (eds.) Innovations in Intelligent Machines-4. SCI, vol. 514, pp. 393–410. Springer, Heidelberg (2014)

17. Kelle, P., Schneider, H., Yi, H.: Decision alternatives between expected cost minimization and worst case scenario in emergency supply second revision. Int. J. Prod. Econ. **157**, 250–260 (2014)
18. Kosaraju, S.R., Przytycka, T.M., Borgstrom, R.: On an optimal split tree problem. In: Dehne, F., Gupta, A., Sack, J.-R., Tamassia, R. (eds.) WADS 1999. LNCS, vol. 1663, pp. 157–168. Springer, Heidelberg (1999)
19. Krause, A.: Optimizing Sensing: Theory and Applications. Ph.D. thesis, Carnegie Mellon University, December 2008
20. Larmore, L.L.: Height restricted optimal binary trees. SICOMP **16**(6), 1115–1123 (1987)
21. Larmore, L.L., Hirschberg, D.S.: A fast algorithm for optimal length-limited huffman codes. J. ACM **37**(3), 464–473 (1990)
22. Milidi, R.L., Laber, E.S.: Bounding the inefficiency of length-restricted prefix codes. Algorithmica **31**(4), 513–529 (2001)
23. Moshkov, M.J.: Greedy algorithm with weights for decision tree construction. Fundam. Inform. **104**(3), 285–292 (2010)
24. Papadimitriou, C.H., Yannakakis, M.: On the approximability of trade-offs and optimal access of web sources. In: FOCS 2000, pp. 86–92 (2000)
25. Rasala, A., Stein, C., Torng, E., Uthaisombut, P.: Existence theorems, lower bounds and algorithms for scheduling to meet two objectives. In: SODA 2002, pp. 723–731 (2002)

Graph Algorithms and FPT I

Sliding Token on Bipartite Permutation Graphs

Eli Fox-Epstein[1](✉), Duc A. Hoang[2], Yota Otachi[2], and Ryuhei Uehara[2]

[1] Brown University, Providence, USA
ef@cs.brown.edu
[2] JAIST, Nomi, Japan
{hoanganhduc,otachi,uehara}@jaist.ac.jp

Abstract. Sliding Token is a natural reconfiguration problem in which vertices of independent sets are iteratively replaced by neighbors. We develop techniques that may be useful in answering the conjecture that Sliding Token is polynomial-time decidable on bipartite graphs. Along the way, we give efficient algorithms for Sliding Token on bipartite permutation and bipartite distance-hereditary graphs.

1 Introduction

Reconfiguration problems have been subject to much recent attention and study. We focus on just one reconfiguration problem, Sliding Token, which is a natural reconfiguration problem over independent sets on graphs. Recall that an *independent set* of a graph is a subset of its vertices such that no two are adjacent. A vertex in an independent set is called a *token*. Intuitively, one "slides" tokens across edges to form new independent sets.

For independent sets I and J, we write $I \overset{G}{\leftrightarrow} J$ if $|I| = |J|$ and there exists an edge $uv \in E(G)$ where $I \triangle J = \{u, v\}$, where $\triangle$ denotes symmetric difference. A *reconfiguration sequence* is a sequence of independent sets $\langle I_1, I_2, \ldots, I_k \rangle$ such that $I_i \overset{G}{\leftrightarrow} I_{i+1}$ for all $1 \leq i < k$. For independent sets I and J on graph G, the binary relation $I \overset{G}{\leftrightsquigarrow} J$ denotes that a reconfiguration sequence containing both I and J exists. "$\overset{G}{\leftrightsquigarrow}$" partitions independent sets into equivalence classes: let $[I]_G = \{J \mid I \overset{G}{\leftrightsquigarrow} J\}$ be the equivalence class of I (with the subscript omitted when implied from context). A yes-instance of Sliding Token is a graph G and independent sets I and J where $I \overset{G}{\leftrightsquigarrow} J$.

Hearn and Demaine [3] show Sliding Token is PSPACE-complete. Kamiński et al. [5] give a linear-time algorithm for Sliding Token on cographs. There are also polynomial-time algorithms on trees and claw-free graphs for Sliding Token [1,2]. On graphs of bounded bandwidth (and thus treewidth), Sliding Token remains PSPACE-complete [9]. Sliding Token is $W[1]$-hard parameterized only by the length of the reconfiguration sequence [4,6].

1.1 Preliminaries

Let G be a graph with vertex set $V(G)$ (with $n = |V(G)|$) and edge set $E(G)$, and S a subset of its vertices. $G[S]$ is the subgraph induced by S: the graph

K. Elbassioni and K. Makino (Eds.): ISAAC 2015, LNCS 9472, pp. 237–247, 2015.
DOI: 10.1007/978-3-662-48971-0_21

with vertex set S and edge set $E(G) \cap (S \times S)$. Define $G \setminus S$ as $G[V(G) \setminus S]$. $N_G(v)$ is the set of all vertices adjacent to v in G and $N_G[v] = N_G(v) \cup \{v\}$. $N_G[S] = \cup_{v \in S} N_G[v]$ for vertex-subset S. When the graph is unambiguous, it is omitted from the notation.

Let $R(G,I) = \{v \mid v \in \cap_{I' \in [I]_G} I'\}$ be the subset of I containing all of the tokens v such that $v \in I'$ for all $I' \in [I]_G$. Vertices in $R(G,I)$ are called *rigid* with respect to G and I. An independent set I is *unlocked* if $R(G,I) = \emptyset$.

Because we frequently form sets that are just slight modifications of others, we write $A + x$ to be $A \cup \{x\}$ and $A - x$ to be $A \setminus \{x\}$.

A graph is a *permutation* graph if and only if there is a bijection between the vertices and a set of line segments between two parallel vertical lines such that two vertices are adjacent if and only if their corresponding segments intersect. A *bipartite permutation* graph is a permutation graph that has no odd-length cycles.

Given an ordering $\langle v_1, \ldots, v_n \rangle$ of the vertices of a graph, let $N_G^+(v_i) = N_G(v_i) \cap \{v_{i+1}, \ldots, v_n\}$. Similarly, define $N_G^-(v_i) = N_G(v_i) \cap \{v_1, \ldots, v_{i-1}\}$.

The following is easily derived from e.g. [7,8]:

Proposition 1. *Each connected bipartite permutation graph G has an ordering $\langle v_1, v_2, \ldots, v_n \rangle$ to $V(G)$ such that*

1. *for all $j > 1$, $N(v_j) \not\subset N(v_1)$,*
2. *for all $i \leq j \leq k$, every path from v_i to v_k contains some vertex in $N_G[v_j]$*
3. *$v_2 \in N(v_1)$ if $n > 1$,*
4. *v_2 is a pendant only if $n = 2$,*
5. *for all i and j where $1 \leq i < j \leq n$, v_i's distance to v_1 is at most v_j's distance to v_1, and*
6. *for all i and j where $1 \leq i < j \leq n$ and v_i and v_j have equal distance to v_1, $N_G^-(v_j) \subseteq N_G^-(v_i)$ and $N_G^+(v_i) \subseteq N_G^+(v_j)$, and*
7. *$N_G^-(v_i) \neq \emptyset$ for all $1 < i \leq n$.*

Such an ordering can be found in linear time. □

Bipartite permutation graphs may seem somewhat arbitrary; however, their many definitions make them a compelling class to study. For example, they are also characterized as bipartite AT-free graphs, bipartite bounded tolerance graphs, bipartite tolerance graphs, bipartite trapezoid graphs, and unit interval bigraphs. They are well studied (see e.g. [7]) and SLIDING TOKEN is PSPACE-complete on some slight non-bipartite generalizations (e.g. AT-free, perfect [5]).

We present an algorithm to efficiently decide SLIDING TOKEN on bipartite permutation graphs. Our main theorem is:

Theorem 1. SLIDING TOKEN *can be decided in polynomial time on bipartite permutation graphs of n vertices.*

This result bounds the diameter of the "reconfiguration graph" for SLIDING TOKEN on a bipartite permutation graph; the algorithm produces a sequence of length quadratic in the number of tokens if any sequence exists. Because of this, determining if there exists a reconfiguration sequence of length at most k is in NP.

To prove the main result, we first give some results about general and biparite graphs in Sects. 2 and 3. We prove our main result in Sect. 4 and then briefly show how techniques developed within can be applied to other classes of bipartite graphs.

2 Coping with Rigid Tokens

In general, tokens may be confined to specific areas of the graph. For example, in the PSPACE-hardness reduction for SLIDING TOKEN given by Demaine and Hearn [3], no token can ever slide out of its specific gadget (see e.g. Theorem 23 in [3]). Rigidity is a much stricter form of confinement; easing proof of strong statements about it, and for the purposes of SLIDING TOKEN on bipartite permutation graphs, it is not too restrictive. Once identified, rigid vertices and their neighborhoods can be deleted. This allows algorithms to only consider instances without rigid vertices, which, in this case, significantly simplifies them.

Proposition 2. *If G' is an induced subgraph of G and $I \overset{G'}{\leftrightsquigarrow} J$, then $I \overset{G}{\leftrightsquigarrow} J$ via the same reconfiguration sequence.* □

Proposition 3. *$I \overset{G}{\leftrightsquigarrow} J$ if and only if $I - v \overset{G \setminus N[v]}{\leftrightsquigarrow} J - v$ for any $v \in R(G, I) \cap R(G, J)$.*

Proof. First, assume $I \overset{G}{\leftrightsquigarrow} J$. Fix a reconfiguration sequence $\langle I = I_0, I_1, \ldots, I_k = J\rangle$. $v \in I_j$ and $N(v) \cap I_j = \emptyset$ for $0 \leq j \leq k$. Therefore, simply remove v from all I_j, $0 \leq j \leq k$, and remove $N_G[v]$ from G: the sets remain independent and do not use deleted vertices.

Next, suppose $I - v \overset{G \setminus N[v]}{\leftrightsquigarrow} J - v$. Proposition 2 gives $I - v \overset{G}{\leftrightsquigarrow} J - v$. Modify the reconfiguration sequence by inserting v into each independent set. This maintains independence: no vertex in $N_G(v)$ is in the reconfiguration sequence as those vertices do not exist in the induced subgraph. □

Proposition 4. *$I \overset{G}{\leftrightsquigarrow} J$ if and only if $R(G, I) = R(G, J)$ and $I \setminus R(G, I) \overset{G \setminus N[R(G,I)]}{\leftrightsquigarrow} J \setminus R(G, I)$.*

Proof. By definition of rigidity, if $R(G, I) \neq R(G, J)$ then $J \notin [I]_G$. Repeated application of Proposition 3 implies the other direction. □

Proposition 5. *Let I be an independent set and $S \subseteq I$. If, for all $w \in N(S)$, $|N(w) \cap S| > 1$, then $S \subseteq R(G, I)$.* □

3 An Algorithm on Bipartite Graphs

In this section, we show that it is relatively straightforward to manipulate the tokens of an independent set in a bipartite graph in a number of ways to e.g. find rigid tokens. In general graphs, identifying $R(G, I)$ is PSPACE-complete; a proof is briefly sketched here. Given an instance $\langle G, I, J\rangle$ of SLIDING TOKEN, we modify G by adding two vertices to produce G': vertex v adjacent to all vertices in $V(G) \setminus J$ and vertex w adjacent to only v. A token is added to I on w. Clearly, $w \notin R(G', I \cup \{w\})$ if and only if $I \overset{G}{\leftrightsquigarrow} J$.

Algorithm 1. SWITCHSIDES(A, B, E, I_0)

Input: Bipartite graph $G = (A \cup B, E)$, independent set I_0
Output: Reconfiguration sequence $\langle I_0, \ldots, I_k\rangle$ where $I_k \cap A = R(G, I_0) \cap A$ and $k = |I_0| - |R(G, I_0) \cap A|$

```
1  M ← ∅                                          // Will hold available slides
2  C ← table from vertices to subsets of vertices
   // Initialize M
3  foreach vertex u ∈ B do
4      C_u ← N(u) ∩ I_0
5      if |C_u| = 1 then
6          M ← M ∪ {u}
7  k ← 0
8  while |M| > 0 do
9      k ← k + 1
10     u ← remove an arbitrary element u from M         // u ∈ B will be in I_k
11     v ← remove the unique vertex v from C_u                 // v ∈ I_{k-1}
12     I_k ← I_{k-1} − v + u
13     foreach vertex w ∈ N(v) do
14         C_w ← C_w − v
15         if |C_w| = 1 then
16             M ← M ∪ {w}
17 return ⟨I_0, I_1, ..., I_k⟩
```

Algorithm 2. WIGGLE(A, B, E, I_0)

Input: Bipartite graph $G = (A \cup B, E)$, independent set I_0
Output: Reconfiguration sequence $\langle I_0, \ldots, I_k\rangle$ with $k \leq 4|I_0|$ such that for all $v \in I_0 \setminus R(G, I_0)$, there is some j where $I_j \setminus I_{j-1} = \{v\}$

```
1 ⟨I_0, ..., I_{k1}⟩ ← SwitchSides(A, B, E(G), I_0)
2 ⟨I_0 = I'_0, ..., I'_{k2}⟩ ← SwitchSides(B, A, E(G), I_0)
3 return ⟨I_0, ..., I_{k1}, I_{k1−1}, ..., I_0, I'_1, ..., I'_{k2}, I'_{k2−1}, ..., I_0⟩
```

Proposition 6. *Given a bipartite graph $G = (A \cup B, E)$ and an independent set I_0, in linear time a reconfiguration sequence $\langle I_0, \ldots, I_k \rangle$ can be computed where $I_k \cap A = R(G, I_0) \cap A$ and $k = |I_0| - |R(G, I_0 \cap A)|$.*

Proof. We analyze Algorithm 1.

Runtime. The first loop, when processing u, charges its work to all the incident edges to u. Charge each iteration of the inner loop (lines 14–16) to the edge vw and charge the work on lines 9–12 to the vertex v. No edge or vertex is charged more than twice, and each charge takes $O(1)$ time.

Correctness. Let C_u^t (M^t) be the state of C_u (resp., M) at the top of the tth execution of the while loop (i.e. at line 9 when k is incremented to be t).

The while loop of Algorithm 1 maintains these properties going into the tth iteration: (P1) $C_u^t = N(u) \cap I_{t-1}$ for all vertices $u \in B$ and (P2) $M^t = \{u \in I_0 : |C_u^t| = 1\}$.

The output is a valid reconfiguration sequence because (1) I_k and I_{k-1} differ by adjacent vertices (line 12) and (2) P1 guarantees that each set is independent.

Next, we prove $I_0 \cap I_k \cap A = R(G, I_0) \cap A$. As only vertices in $I_0 \cap A$ are removed from an independent set during the reconfiguration sequence, both I_0 and I_k contain $I_0 \cap B$. Since it is a valid reconfiguration sequence, we know $R(G, I_0) \subseteq I_0 \cap I_k$. Thus, $R(G, I_0) \subseteq I_0 \cap I_k$ and it remains to be shown that no non-rigid vertices of $I_0 \setminus (R(G, I_0) \cap A)$ are in I_k. Since M is empty at the end of the algorithm, $|C_u| \neq 1$ for all $u \in I_k$. Consider $S = I_0 \cap I_k \cap A$. Any $w \in N(S)$ must have $|C_w| > 1$ by property (P1), so Proposition 5 with G and S shows $S \subseteq R(G, I_0)$. Thus, $I_0 \cap I_k \cap A = R(G, I_0) \cap A$.

Finally, we show that the length of the reconfiguration sequence, k, is as promised. For all $0 < j \leq k$, we have that $|I_j \cap I_0| = |I_{j-1} \cap I_0| - 1$, so $|I_0| - |R(G, I_0) \cap A|$ is an upper bound on k. To lower-bound k, it takes k slides to reconfigure k vertices out of I_0. □

Algorithm 2 applies Algorithm 1 twice to produce a sequence that starts and ends with the same sequence but ensures that each token not in $R(G, I)$ slides exactly twice.

Lemma 1. *Given bipartite graph $G = (A \cup B, E)$, and independent set I_0 in linear time Algorithm 2 finds a reconfiguration sequence of length at most $4|I_0|$ in which each token of $I_0 \setminus R(G, I_0)$ slides exactly twice.* □

Lemma 2. *Let $G = (A \cup B, E)$ be a bipartite graph and I be an independent set of G. In linear time, $R(G, I)$ can be computed.*

Proof. Invoke Algorithm 2. By the post-condition promises, the tokens that never slid in the output sequence are exactly $R(G, I)$. □

Lemma 3. *Let $G = (A \cup B, E)$ be a connected bipartite graph and I an unlocked independent set. Then for any $v \in V(G)$, in linear time, one can find a reconfiguration sequence $\langle I = I_0, I_1, \ldots, I_k = J \rangle$ where $v \in J$, $v \notin I_{k-1}$, and k is at most $|I|$ plus the distance between v and the closest token of I.*

Proof. We distinguish 3 cases:

(1) If $v \in I$, the entire sequence is just $\langle I \rangle$.

(2) If there is a unique closest token w in I to v, the reconfiguration sequence repeatedly replaces that token with a vertex that is one closer to v. Let u be any vertex in $N(v)$ where some shortest path from w to v passes through u. Since w is uniquely closest to v among all tokens in I, it must be the case that $N(u) \cap I = \{w\}$. So update construct $I' = I - w + u$; u is now uniquely closest in I' to v, so this process can be repeated.

(3) Otherwise, let S be the set of all closest vertices to v at distance d. Without loss of generality assume $S \subseteq A$. By the correctness of Algorithm 1, there is a $J \in [I]$ where $J \subseteq B$. Consider a reconfiguration sequence $\langle I = I_0, I_1, \ldots, I_k = J \rangle$ from I to J. There must be an index j, with $j \leq k \leq |I|$, where I_j has a unique closest token to v as either some token will first move to be distance $d-1$ away from v, or all but one token will slide to be at least distance $d+1$ away. Then, from I_j, the reconfiguration sequence is as described in case (2). □

We write I_v^G (with the graph usually omitted) to indicate an independent set resulting in invoking Lemma 3 on G and I to place a vertex on v. This produces some reconfiguration sequence of linear length from I to I_v, in which I_v is the only independent set containing v.

We are able to simplify instances with the following lemma:

Lemma 4. *Let I be an unlocked independent set in bipartite graph G.*

(1) If $N[v] \cap I = \emptyset$, then $R(G \setminus \{v\}, I) = \emptyset$.

(2) If $N[N[v]] \cap I \subseteq \{v\}$, then $R(G \setminus N[v], I - v) = \emptyset$.

Proof. Invoke Algorithm 2 on G. Since all tokens move no farther than to their neighbors, both cases immediately follow. □

Proposition 7. *Suppose $N_G(u) = N_G(v)$. For any unlocked independent sets I and J, $I \overset{G}{\leftrightsquigarrow} J$ if and only if $I_u^G \overset{G \setminus \{v\}}{\leftrightsquigarrow} J_u^G$.*

Proof. The "if" direction is trivial. For the "only if" direction, assume $I \overset{G}{\leftrightsquigarrow} J$. This implies $I_u^G \overset{G}{\leftrightsquigarrow} J_u^G$. First, observe that no unlocked set contains both u and v. Each set S in any reconfiguration sequence from I_u^G to J_u^G on G that contains v can be replaced by $S - v + u$. Now this sequence only uses vertices only in $G \setminus \{v\}$, so $I_u^G \overset{G \setminus \{v\}}{\leftrightsquigarrow} J_u^G$. □

4 Sliding Token on Bipartite Permutation Graphs

Throughout the section, let G be a bipartite permutation graph with vertices $\langle v_1, v_2, \ldots, v_n \rangle$ ordered as described previously.

Proposition 8. *Assume $R(G, I) = R(G, J)$. If $v_i \in R(G, I)$, then each component of $G \setminus N[v_i]$ is a bipartite permutation graph and $I \overset{G}{\leftrightsquigarrow} J$ if and only if, for each component C of $G \setminus N[v_i]$, we have $I \cap C \overset{G[C]}{\leftrightsquigarrow} J \cap C$.*

Proof. First, note that an induced subgraph of a bipartite permutation graph is still a bipartite permutation graph. Now, we appeal to Proposition 4. □

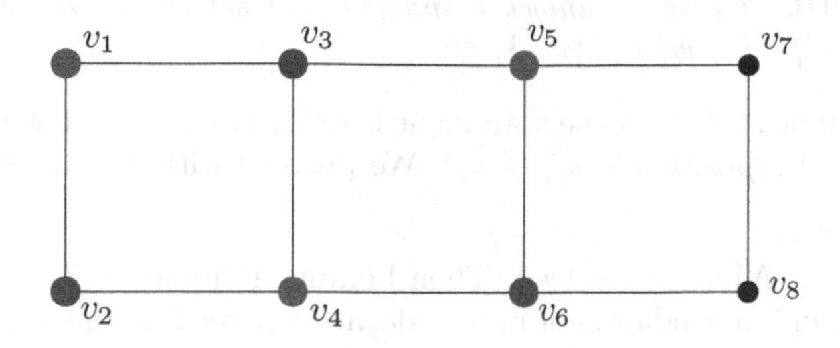

Fig. 1. Two unlocked independent sets in different equivalence classes: $\{v_1, v_4, v_5\}$ and $\{v_2, v_3, v_6\}$.

Lemma 2 locates rigid vertices in linear time and Proposition 8 permits treating each component independently after deleting rigid vertices and their neighborhoods. We assume $R(G, I) = \emptyset$ for the remainder of the section. Using Proposition 7 allows us to assume that each vertex has a distinct neighborhood.

In each equivalence class over $\overset{G}{\leftrightsquigarrow}$, we will pick a representative independent set as the lexicographically least element (i.e. the independent set I minimizing $\sum_{v_i \in I} 2^i$). We write I_+ to indicate the representative of the equivalence class to which some independent set I belongs. Then, deciding if $I \overset{G}{\leftrightsquigarrow} J$ is equivalent to determining if $I_+ = J_+$.

To give some intuition on why finding I_+ is nontrivial, Fig. 1 illustrates two unlocked independent sets in different equivalence classes.

Fix some I and let w_j^+ be the jth least token of I_+. The algorithm relies on two vital observations: first, that there are only two possibilities for where the token of least index in I will reside in I_+ and second, that I_+ can be assembled one vertex at a time.

Proposition 9. $|\{v_1, v_2\} \cap I_+| = 1$. *If* $|I_+| \geq 2$ *and* $v_2 \in I_+$ *then* $|N(v_1) \cap I_+| \geq 2$.

Proof. First, we prove $|\{v_1, v_2\} \cap I_+| = 1$. Suppose not: that $w_1^+ = v_i$ for some $i > 1$. There are two cases to consider:

(1) Assume $v_i \in N(v_1)$. Use Lemma 3 to place a token on v_1 and obtain a reconfiguration sequence $\langle I_+ = I_0, I_1, \ldots, I_k \rangle$. Recall that $v_i \in I_j$ for all $j < k$. Consider the sequence $\langle I_0, \ldots, I_k, I_{k-1} - v_i + v_2, I_{k-1} - v_i + v_2, I_{k-2} - v_i + v_2, \ldots, I_0 - v_i + v_2 \rangle$. This sequence is valid, so $I_+ - v_i + v_2 \in [I_+]$. But $I_+ - v_i + v_2$ is lexicographically less than I_+, a contradiction.

(2) Now assume $v_i \notin N(v_1)$. Again use Lemma 3 to place a token on v_1. Similarly, the sequence can be unrolled in reverse, except this time leaving a token on v_1.

Now we prove if $v_2 \in I_+$ then $|N(v_1) \cap I_+| \geq 2$. Suppose not: that $v_2 \in I_+$ but $N(v_1) \cap I_+ = \{v_2\}$. Then $I_+ \overset{G}{\leftrightarrow} I_+ - v_2 + v_1$ is legal and lexicographically less, a contradiction. □

Proposition 10. *If I is an unlocked independent set containing w_1^+ and w_2^+ then $R(G \setminus N[w_1^+], I - w_1^+) \subseteq \{w_2^+\}$.*

Proof. We assume $|I| > 2$ as the statement is otherwise trivial. Let $v_j = w_2^+$ for some $j > 2$ (by Proposition 9, $w_2^+ \neq v_2$). We proceed with case analysis:

1. Assume $v_1 \in I$.
   (a) Assume $N[N[v_1]] \cap I = \{v_1\}$. Then Lemma 4 applies to I.
   (b) Assume v_1 is a pendant. For v_1 to slide, at some set I' in the reconfiguration sequence given by Algorithm 2, $N(v_2) \cap I' = \{v_1\}$. Lemma 4 applies to I'.
   (c) Assume no neighbor of v_1 has v_1 as its only neighboring token. $N(v_1) \subseteq N(v_j)$ (otherwise, we fall into one of the previous cases) so the token on v_1 cannot slide until v_j slides. Once v_j slides, $N[N[v_1]] = \{v_1\}$ and Lemma 4 completes the proof.
   (d) Otherwise, observe that $N(v_2) \cap I = \{v_1\}$. Let L_i be the set of vertices distance i away from v_1.
   If any two vertices v_a, v_b in $I \cap L_2$ have $N^-(u) = N^-(v)$, then Algorithm 1 slides all vertices of L_2 with index at least b (assuming $a < b$) into L_3. Notice that it suffices to show that $a = j$.
   In I_+, there must be a k where $k > j$ and $N^-(v_k) = N^-(v_j)$; (otherwise $I_+ - v_1 + v_2 - v_j + v_i$ is lexicographically less than I_+). However, if $v_k \notin I$, more argument is required. Consider any reconfiguration sequence from I to I_+. Let I' be the last independent set in the sequence containing v_k. In I', the token of second-least index cannot be in $N^-(v_k)$ but must be in L_2. We show this gives a contradiction to I_+'s lexicographical minimality: since the token on v_k does not slide for the remainder of the reconfiguration sequence, the two first tokens are able to reconfigure from v_1 and v_j in I_+ to a lesser configuration.
2. Assume $v_2 \in I$. By Proposition 9, $v_j \in N(v_1)$. Thus, $N(v_2) \subseteq N(v_1)$. Consider a reconfiguration sequence in which v_2 eventually slides, e.g. the one generated by Lemma 3 to produce $I_{v_1}^G$. In this, v_j must slide before v_1. Let I' be the independent set immediately after v_j slides. $N[N[v_i]] \cap I' = \{v_i\}$, so Lemma 4 applies.

□

Proposition 11. *$I_+ - w_1^+$ is lexicographically minimal on $G \setminus N[w_1^+]$.* □

We find a reconfiguration sequence between I and I_+ using dynamic programming over vertex index with a table $T[\cdot]$. For notational convenience, we define $J^{i,k} = \{v_j \in J \mid i \leq j \leq k\}$ for any independent set J. Let G_i be the unique component of $G \setminus N[v_i]$ containing vertices of higher index. $T[i]$ will be assigned some $J = \arg\max_{J \in [I]: J \ni v_i} |J^{0,i}|$. As a base case, set $T[0] = I$.

Define

$$W(i,j) = \begin{cases} T[j] & \text{if } v_i \in T[j] \\ T[j]^{0,k} \cup (T[j]^{j+1,n})_{v_i}^{G_j} & \text{if } R(G_j, T[j]^{j+1,n}) = \emptyset \\ \text{“}invalid\text{”} & \text{otherwise.} \end{cases}$$

(Recall, the notation in the middle case invokes Lemma 3.) Say $W(i,j)$ is *valid* if $0 \le j < i$ and $W(i,j)$ is an independent set and not "invalid". Among the valid $W(i,j)$ that maximize $|W(i,j)^{0,i}|$, set $T[i]$ to be the $W(i,j)$ where j is least.

Lemma 5. *If $v_i \in I_+$ then $T[i]^{0,i} = I_+^{0,i}$.*

Proof. Using Propositions 10 and 11, this follows from a simple induction on the size of I_+. □

Theorem 2. *Given a connected bipartite permutation graph G and an unlocked independent set I, there is a cubic-time algorithm to find I_+.*

Proof. Given the dynamic programming table $T[\cdot]$, find the least index i where $|T[i]^{0,i}| = |I|$ and report $I_+ = T[i]$; by Lemma 5, this is correct.

In total, $O(n^2)$ sets $W(i,j)$ are computed, each of which takes linear time, giving cubic runtime. □

Given this, proving the main theorem is straightforward:

Proof (of Theorem 1). As input, we are given a bipartite permutation graph G and two independent sets I and J. If $R(G,I) \neq R(G,J)$, then output "no". Otherwise, form $G' = G \backslash N[R(G,I)]$. For each C component of G', find $I' = I \cap C$ and $J' = J \cap C$; then find I'^{+} and J'^{+} using Theorem 2. If in any component, I'^{+} and J'^{+} differ, then output "no". Otherwise, it must be that $I \overset{G}{\leftrightsquigarrow} J$. □

5 Sliding Token on Bipartite Distance-Hereditary Graphs

In this section, we give an additional application of the techniques built in Sect. 3. A graph is *distance-hereditary* if the distance between two vertices in any connected induced subgraph is exactly the distance in the original graph. One characterization of bipartite distance-hereditary graphs is graphs obtainable from a single vertex by repeatedly picking a vertex v in the graph and then adding a new vertex w with either $N(w) = \{v\}$ (pendant) or $N(w) = N(v)$ (twin).

Theorem 3. *There is a polynomial-time algorithm to decide* Sliding Token *on bipartite distance-hereditary graphs.*

Proof. Let I_0 and J_0 be independent sets of the same cardinality on bipartite distance-hereditary graph G. We analyze the following algorithm.

We can assume, using Lemma 2 and Proposition 4 that $R(G,I) = R(G,J) = \emptyset$. Repeatedly:

1. If $N(v) = N(w)$ for any v, w, use Lemma 3 to place a token on v in I and in J, and then delete w.
2. Else, if there is a pendant v whose neighbor w has degree 2, use Lemma 3 to place a token from I and from J on v, then delete $N(v)$.
3. Otherwise, compute a sequence of operations used to construct the graph and look at the last twin operation used. At least one of the two involved vertices must have a pendant. Use Lemma 3 to place a token from I and from J on the pendant and delete it and its neighborhood.

Bipartite distance-hereditary graphs are closed under vertex deletion, so after each iteration the graph remains bipartite distance-hereditary. Suppose that before an iteration, $R(G, I) = \emptyset$. Let G', I', J' be the graph and independent sets after the iteration. We show that $R(G', I') = R(G', J') = \emptyset$.

In case (1), since $N_G(v) = N_G(w)$ and $v \in I' \cap J'$, we have $N_G(w) \cap I' = N_G(w) \cap J' = \emptyset$. Lemma 4 implies $R(G', I') = R(G', J') = \emptyset$. In cases (2) and (3), if there is a token on any neighbor u of w besides v in I', then after invoking Algorithm 2, there must be an intermediate independent set I'' where $v \in I''$ but $u \notin I''$. From I'', Lemma 4 completes the proof. □

6 Discussion

We show that SLIDING TOKEN can be efficiently decided on bipartite permutation graphs and bipartite distance-hereditary graphs. The results of [5] show that SLIDING TOKEN is PSPACE-hard on AT-free graphs, which are a natural generalization of bipartite permutation graphs to non-bipartite graphs. This suggests that bipartitedness is closely related to the complexity of SLIDING TOKEN.

The complexity of SLIDING TOKEN on bipartite graphs remains a compelling topic for future research; the tools developed here tackle rigidity but need strengthening to be able to decide SLIDING TOKEN when dynamic programming does not fit as naturally.

Acknowledgements. E.F. partially supported by an NSF EAPSI fellowship and NSF grants CCF-09-64037 and CCF-14-09520.

References

1. Bonsma, P., Kamiński, M., Wrochna, M.: Reconfiguring independent sets in claw-free graphs. In: Ravi, R., Gørtz, I.L. (eds.) SWAT 2014. LNCS, vol. 8503, pp. 86–97. Springer, Heidelberg (2014)
2. Demaine, E.D., Demaine, M.L., Fox-Epstein, E., Hoang, D.A., Ito, T., Ono, H., Otachi, Y., Uehara, R., Yamada, T.: Polynomial-time algorithm for sliding tokens on trees. In: Ahn, H.-K., Shin, C.-S. (eds.) ISAAC 2014. LNCS, vol. 8889, pp. 389–400. Springer, Heidelberg (2014)
3. Hearn, R.A., Demaine, E.D.: PSPACE-completeness of sliding-block puzzles and other problems through the nondeterministic constraint logic model of computation. Theor. Comput. Sci. **343**(1–2), 72–96 (2005)

4. Ito, T., Kamiński, M., Ono, H., Suzuki, A., Uehara, R., Yamanaka, K.: On the parameterized complexity for token jumping on graphs. In: Gopal, T.V., Agrawal, M., Li, A., Cooper, S.B. (eds.) TAMC 2014. LNCS, vol. 8402, pp. 341–351. Springer, Heidelberg (2014)
5. Kamiński, M., Medvedev, P., Milani, M.: Complexity of independent set reconfigurability problems. Theor. Comput. Sci. **439**, 9–15 (2012)
6. Mouawad, A.E., Nishimura, N., Raman, V., Wrochna, M.: Reconfiguration over tree decompositions. In: Cygan, M., Heggernes, P. (eds.) IPEC 2014. LNCS, vol. 8894, pp. 246–257. Springer, Heidelberg (2014)
7. Spinrad, J., Brandstädt, A., Stewart, L.: Bipartite permutation graphs. Discrete Appl. Math. **18**(3), 279–292 (1987)
8. Sprague, A.P.: Recognition of bipartite permutation graphs. Congressus Numerantium **62**, 151–161 (1995)
9. Wrochna, M.: Reconfiguration in bounded bandwidth and treedepth. CoRR, abs/1405.0847 (2014)

Output-Polynomial Enumeration on Graphs of Bounded (Local) Linear MIM-Width

Petr A. Golovach[1(✉)], Pinar Heggernes[1], Mamadou Moustapha Kanté[2], Dieter Kratsch[3], Sigve H. Sæther[1], and Yngve Villanger[1]

[1] Department of Informatics, University of Bergen, Bergen, Norway
{petr.golovach,pinar.heggernes,sigve.sether,yngve.villanger}@ii.uib.no
[2] Clermont-Université, Université Blaise Pascal, LIMOS, CNRS, Aubiére, France
mamadou.kante@isima.fr
[3] Université de Lorraine, LITA, Metz, France
dieter.kratsch@univ-lorraine.fr

Abstract. The linear maximum induced matching width (LMIM-width) of a graph is a width parameter based on the maximum induced matching in some of its subgraphs. In this paper we study output-polynomial enumeration algorithms on graphs of bounded LMIM-width and graphs of bounded local LMIM-width. In particular, we show that all 1-minimal (σ, ρ)-dominating sets, and hence all minimal dominating sets, of graphs of bounded LMIM-width can be enumerated with polynomial (linear) delay using polynomial space. Furthermore, we show that all minimal dominating sets of a unit square graph can be enumerated in incremental polynomial time.

1 Introduction

Enumeration is at the heart of computer science and combinatorics. Enumeration algorithms for graphs and hypergraphs typically deal with listing all vertex subsets or edge subsets satisfying a given property. As the size of the output is often exponential in the size of the input, it is customary to measure the running time of enumeration algorithms in the size of the input plus the size of the output. If the running time of an algorithm is bounded by a polynomial in the size of the input plus the size of the output, then the algorithm is called output-polynomial. A large number of such algorithms have been given over the last 30 years; many of them solving problems on graphs and hypergraphs [7–9,13,19–21,23]. It is also possible to show that certain enumeration problems have no output-polynomial time algorithm unless P = NP [19–21].

Recently Kanté et al. showed that the famous longstanding open question whether there is an output-polynomial algorithm to enumerate all minimal transversals of a hypergraph is equivalent to the question whether there is an output-polynomial algorithm to enumerate all minimal dominating sets of a

The research leading to these results has received funding from the European Research Council under the European Union's Seventh Framework Programme (FP/2007-2013)/ERC Grant Agreement n. 267959.

K. Elbassioni and K. Makino (Eds.): ISAAC 2015, LNCS 9472, pp. 248–258, 2015.
DOI: 10.1007/978-3-662-48971-0_22

graph [14]. Although the main question remains open, a large number of results have been obtained on graph classes. Output-polynomial algorithms to enumerate all minimal dominating sets exist for graphs of bounded treewidth and of bounded clique-width [6], interval graphs [7], strongly chordal graphs [7], planar graphs [9], degenerate graphs [9], split graphs [14], path graphs [15], permutation graphs [16], line graphs [10,15,18], chordal bipartite graphs [12], chordal graphs [17] and graphs of girth at least 7 [10].

In this paper, we extend the above results to graphs of bounded *linear maximum induced matching width* (LMIM-width), which is a linearized version of the notion of *maximum induced matching width* introduced by Vatshelle [25]. Belmonte and Vatshelle showed that several important graph classes, among them interval, circular-arc and permutation graphs, have bounded LMIM-width [1]. Polynomial-time algorithms solving optimization problems on such graph classes have been studied in [4,25].

In this paper, we study two ways of using bounded LMIM-width in enumeration algorithms. In Sect. 3 we study the enumeration problem corresponding to an extended and colored version of the well-known (σ, ρ)-domination problem, asking to enumerate all 1-minimal **Red** (σ, ρ)-dominating sets. This includes the enumeration of all minimal (total) dominating sets on graphs of bounded LMIM-width. We establish as our main result an enumeration algorithm with polynomial (linear) delay and polynomial space for this problem. Our algorithm uses the enumeration (and counting) of paths in directed acyclic graphs. In Sect. 4 we study the enumeration of all minimal dominating sets in unit square graphs. We first show that any r-neighborhood in such graphs have LMIM-width bounded by $O(r^2)$. Then we show how to adapt the so-called flipping method developed by Golovach et al. [10] to enumerate all minimal dominating sets of a unit square graph in incremental polynomial time. Due to space constraints, various proofs are omitted in this extended abstract. The full version of the paper is available in [11].

2 Definitions and Preliminaries

Graphs. The power set of a set V is denoted by 2^V. For two sets A and B we let $A \setminus B$ be the set $\{x \in A \mid x \notin B\}$, and if X is a subset of a ground set V, we let $\bar{X}$ be the set $V \setminus X$. We often write x to denote the singleton set $\{x\}$. We denote by $\mathbb{N}$ the set of positive or null integers, and let $\mathbb{N}^*$ be $\mathbb{N} \setminus \{0\}$.

A graph G is a pair $(V(G), E(G))$ with $V(G)$ its set of vertices and $E(G)$ its set of edges. An edge between two vertices x and y is denoted by xy (respectively yx). The subgraph of G induced by a subset X of its vertex set is denoted by $G[X]$. The set of vertices that is adjacent to x is denoted by $N_G(x)$, and we let $N_G[x]$ be the set $N_G(x) \cup \{x\}$. For $U \subseteq V(G)$, $N_G[U] = \bigcup_{v \in U} N_G[v]$ and $N_G(U) = N_G[U] \setminus U$. For a vertex x and a positive integer r, $N_G^r[x]$ denotes the set of vertices at distance at most r from x. For two disjoint subsets A and B of $V(G)$, let $G[A, B]$ denote the graph with vertex set $A \cup B$ and edge set $\{uv \in E(G) \mid u \in A, v \in B\}$. Clearly, $G[A, B]$ is a bipartite graph and $\{A, B\}$

is its bipartition. Recall that a set of edges M is an *induced matching* if the end-vertices of distinct edges of M are different and not adjacent. We denote by $\mathbf{mim}_G(A, B)$ the size of a maximum induced matching in $G[A, B]$.

Let G be a graph, and let $\mathbf{Red}, \mathbf{Blue} \subseteq V(G)$ such that $\mathbf{Red} \cup \mathbf{Blue} = V(G)$. We refer to the vertices of **Red** as the *red* vertices, the vertices of **Blue** as the *blue* vertices, and we say that G together with given sets **Red** and **Blue** is a *colored graph.* For simplicity, whenever we say that G is a colored graph, it is assumed that the sets **Red** and **Blue** are given. Notice that **Red** and **Blue** are not necessarily disjoint. In particular, it can happen that $\mathbf{Red} = \mathbf{Blue} = V(G)$; a non-colored graph G can be seen as a colored graph with $\mathbf{Red} = \mathbf{Blue} = V(G)$.

A graph G is an *(axis-parallel) unit square* graph if it is an intersection graph of squares in the plane with their sides parallel to the coordinate axis. These graphs also are known as the graphs of cubicity 2. We use the following equivalent definition, see e.g. [5], in which each vertex v of G is represented by a point in $\mathbb{R}^2$. A graph G is a unit square graph if there is a function $f\colon V(G) \to \mathbb{R}^2$ such that two vertices $u, v \in V(G)$ are adjacent in G if and only if $\|f(u) - f(v)\|_\infty < 1$, where the norm $\|\|_\infty$ is the L_∞ norm. For a vertex $v \in V(G)$, we let $x_f(v)$ and $y_f(v)$ denote the x and y-coordinate of $f(v)$ respectively. We say that the point $(x_f(v), y_f(v))$ *represents* v. The function f is called a *realization* of the unit square graph. It is straightforward to see that for any unit square graph G, there is a realization $f\colon V(G) \to \mathbb{Q}^2$. We always assume that a unit square graph is given with its realization. Indeed, it is NP-hard to recognize unit square graphs [3]. We refer to the survey of Brandstädt, Le and Spinrad [2] for the definitions of all other graph classes mentioned in our paper.

Enumeration. Let $\mathcal{D}$ be a family of subsets of the vertex set of a given graph G on n vertices and m edges. An *enumeration algorithm* for $\mathcal{D}$ lists the elements of $\mathcal{D}$ without repetitions. The running time of an enumeration algorithm $\mathcal{A}$ is said to be *output polynomial* if there is a polynomial $p(x, y)$ such that all the elements of $\mathcal{D}$ are listed in time bounded by $p((n+m), |\mathcal{D}|)$. Assume now that $D_1, \ldots, D_\ell$ are the elements of $\mathcal{D}$ enumerated in the order in which they are generated by $\mathcal{A}$. Let us denote by $T(\mathcal{A}, i)$ the time $\mathcal{A}$ requires until it outputs D_i, also $T(\mathcal{A}, \ell+1)$ is the time required by $\mathcal{A}$ until it stops. Let $delay(\mathcal{A}, 1) = T(\mathcal{A}, 1)$ and $delay(\mathcal{A}, i) = T(\mathcal{A}, i) - T(\mathcal{A}, i-1)$. The *delay* of $\mathcal{A}$ is $\max\{delay(\mathcal{A}, i)\}$. Algorithm $\mathcal{A}$ runs in *incremental polynomial* time if there is a polynomial $p(x, i)$ such that $delay(\mathcal{A}, i) \le p(n+m, i)$. Furthermore $\mathcal{A}$ is a *polynomial delay* algorithm if there is a polynomial $p(x)$ such that the delay of $\mathcal{A}$ is at most $p(n+m)$. Finally $\mathcal{A}$ is a *linear delay* algorithm if $delay(\mathcal{A}, 1)$ is bounded by a polynomial in $n+m$ and $delay(\mathcal{A}, i)$ is bounded by a linear function in $n+m$.

Linear induced matching width. The notion of the *maximum induced matching width* was introduced by Vatshelle [25] (see also [1]). We will give the definition in terms of colored graphs and restrict ourselves to the case of linear maximum induced matching width. Let G be a colored n-vertex graph with $n \ge 2$ and let $x_1, \ldots, x_n$ be a linear ordering of its vertex set. For each $1 \le i \le n$, we let $A_i = \{x_1, x_2, \ldots x_i\}$ and $\bar{A}_i = \{x_{i+1}, x_{i+2}, \ldots x_n\}$. The *maximum induced matching width* (*MIM-width*) of $x_1, \ldots, x_n$ is

$\max\{\max\{\mathbf{mim}_G(A_i \cap \mathbf{Red}, \bar{A}_i \cap \mathbf{Blue}), \mathbf{mim}_G(A_i \cap \mathbf{Blue}, \bar{A}_i \cap \mathbf{Red})\} \mid 1 \leq i \leq n\}$.

The *linear maximum induced matching width* (*LMIM-width*) of G, denoted by $\mathbf{lmimw}(G)$, is the minimum value of the MIM-width taken over all linear orderings of G.

Belmonte and Vatshelle [1] proved that several important graph classes have bounded linear maximum induced matching width. For example, the LMIM-width of an interval graph is 1 and the LMIM-width of a permutation graph is at most 2.

(σ, ρ)-domination. The *(σ, ρ)-dominating set* notion was introduced by Telle and Proskurowski [24] as a generalization of dominating sets. Indeed, many NP-hard domination type problems such as the problems d-Dominating Set, Independent Dominating Set and Total Dominating Set are special cases of the (σ, ρ)-Dominating Set Problem. See [4, Table 1] for more examples. For technical reasons, we introduce **Red** (σ, ρ)-domination. Let σ and ρ be finite or co-finite subsets of $\mathbb{N}$. We say that a set $D \subseteq V(G)$ *(σ, ρ)-dominates* $U \subseteq V(G)$ if it (σ, ρ)-dominates every $u \in U$, *i.e.*, for each $u \in U$, $|N_G(u) \cap D| \in \sigma$ if $u \in D$, otherwise $|N_G(u) \cap D| \in \rho$.

Let G be a colored graph. A set of vertices $D \subseteq \mathbf{Red}$ is a **Red** *(σ, ρ)-dominating set* if D (σ, ρ)-dominates **Blue**. If **Red** $=$ **Blue** $= V(G)$, then a **Red** (σ, ρ)-dominating set is a *(σ, ρ)-dominating set.*

Notice that if $\sigma = \mathbb{N}$ and $\rho = \mathbb{N}^*$, then a set $D \subseteq V(G)$ (σ, ρ)-dominates a vertex u if $u \in D$ or u is adjacent to a vertex of D, i.e., the notion of (σ, ρ)-domination coincides with the classical domination in this case. Whenever we consider this case, we simply write that a set D dominates a vertex or set and D is a (**Red**) dominating set omitting (σ, ρ). We are interested in **Red** dominating sets because that is what we actually need in Sect. 4.

A **Red** (σ, ρ)-dominating set D of a graph G is said *minimal* if for any proper subset $D' \subset D$, D' is not a **Red** (σ, ρ)-dominating set, and we say that D is 1-*minimal* if for each vertex x in D, $D \setminus x$ is not a **Red** (σ, ρ)-dominating set. Clearly, every minimal **Red** (σ, ρ)-dominating set is 1-minimal, but the converse is not true for arbitrary σ and ρ.

Because our aim is to enumerate 1-minimal **Red** (σ, ρ)-dominating sets, we need some certificate that a considered set is 1-minimal. Let D be a **Red** (σ, ρ)-dominating set of a colored graph G. For a vertex $u \in D$, the vertex $v \in \mathbf{Blue}$ is its *certifying vertex* (or a *certificate*) if v is not (σ, ρ)-dominated by $D \setminus \{u\}$.

Notice that because D is a **Red** (σ, ρ)-dominating set, if v is a certificate for u, then $v \in N_G[u]$. Observe also that, for some pairs (σ, ρ), a vertex may be a certificate for many vertices and it can be a certificate for itself. Notice that in the case of the classical domination, certificates are usually called *privates* because they are certificates for exactly one vertex, including itself. It is straightforward to show the following.

Lemma 1. *A set $D \subseteq$ **Red** is a 1-minimal **Red** (σ, ρ)-dominating set of a colored graph G if and only if each vertex $u \in D$ has a certificate.*

Lemma 2. *Let D be a* **Red** *(σ,ρ)-dominating set of G. If v is a certificate for $u \in D$, then $v = u$ or v is a certificate for all vertices of $N_G(v) \cap D$.*

3 Enumerations for Graphs of Bounded LMIM-width

In this section we prove the following, which generalizes the results in [16].

Theorem 1. *Let (σ,ρ) be a pair of finite or co-finite subsets of $\mathbb{N}$ and let c be a positive integer. For a colored graph G given with a linear ordering of $V(G)$ of MIM-width at most c, one can count in time bounded by $O(n^c)$, and enumerate with linear delay, all 1-minimal* **Red** *(σ,ρ)-dominating sets of G.*

Corollary 1. *Let (σ,ρ) be a pair of finite or co-finite subsets of $\mathbb{N}$. Then, for every colored graph G in one of the following graph classes, we can count in polynomial time, and enumerate with linear delay all 1-minimal* **Red** *(σ,ρ)-dominating sets of G: interval graphs, permutation graphs, circular-arc graphs, circular permutation graphs, trapezoid graphs, convex graphs, and for fixed k, k-polygon graphs, Dilworth-k graphs and complements of k-degenerate graphs.*

The following corollary improves some known results in the enumeration of minimal transversals of interval and circular-arc hypergraphs where only an incremental polynomial time algorithm was known (see e.g. [22]).

Corollary 2. *For every hypergraph $\mathcal{H}$ being an interval hypergraph or a circular-arc hypergraph one can count in polynomial time, and enumerate with linear delay, all minimal transversals of $\mathcal{H}$.*

The remaining part of the section is devoted to the main ideas of the proof of Theorem 1. Throughout this section let (σ,ρ) be a fixed pair of finite or co-finite subsets of $\mathbb{N}$ and let G be a fixed n-vertex colored graph with $n \geq 2$. Let $x_1,\dots,x_n$ be a fixed linear ordering of the vertex set of G such that the maximum induced matching width of $x_1,\dots,x_n$ is bounded by a constant c. Furthermore, for all $i \in \{1,2,\dots,n\}$, we let $A_i = \{x_1,x_2,\dots x_i\}$ and $\bar{A}_i = \{x_{i+1},x_{i+2},\dots x_n\}$.

Let $d(\mathbb{N}) = 0$. For every finite set $\mu \subseteq \mathbb{N}$, let $d(\mu) = 1 + \max\{a \mid a \in \mu\}$, and for every co-finite set $\mu \subseteq \mathbb{N}$, let $d(\mu) = 1 + \max\{a \mid a \in \mathbb{N} \setminus \mu\}$. For finite or co-finite subsets σ and ρ of $\mathbb{N}$, we let $d(\sigma,\rho) = \max(d(\sigma),d(\rho))$. As pointed out in [4] given a subset D of **Red**, we can check if D is a **Red** (σ,ρ)-dominating set by computing $|D \cap N_G(x)|$ up to $d(\sigma,\rho)$ for each vertex x in **Blue**. We define $\sigma^* = \sigma \setminus \rho$ and $\rho^* = \rho \setminus \sigma$. Let also $\sigma^- = \{i \in \sigma \mid i-1 \notin \sigma\}$, $\rho^- = \{i \in \rho \mid i-1 \notin \rho\}$.

Lemma 3. *The sets σ^*,ρ^*,σ^- and ρ^- are finite or co-finite. Also, $d(\sigma^*,\rho^*) \leq d(\sigma,\rho)$ and $d(\sigma^-,\rho^-) \leq d(\sigma,\rho)+1$.*

Lemma 4. *Let D be a* **Red** *(σ,ρ)-dominating set of G and let $u \in D$. The vertex u is a certificate for itself if and only if $u \in$* **Blue** *and $|N_G(u) \cap D| \in \sigma^*$. A vertex $v \in N_G(u) \cap$* **Blue** *is a certificate for u if and only if D (σ^-,ρ^-)-dominates v.*

Let $d \in \mathbb{N}$ and let $A \subseteq V(G)$. Two red subsets X and Y of A are *d-neighbor equivalent w.r.t.* A, denoted by $X \equiv_A^d Y$, if $\min(d, |X \cap N_G(x)|) = \min(d, |Y \cap N_G(x)|)$ for all $x \in \bar{A} \cap \mathbf{Blue}$. It is clear that $\equiv_A^d$ is an equivalence relation and we let $nec(\equiv_A^d)$ be its number of equivalence classes.

Lemma 5 ([1]). *Let $d \in \mathbb{N}$ and let $A \subseteq V(G)$. Then $nec(\equiv_A^d) \leq n^{d \cdot c}$.*

We will follow the same idea as in [4] where a minimum (or a maximum) (σ, ρ)-dominating set is computed. For every $i \in \{1, \ldots, n\}$ and every subset X of $A_i \cap \mathbf{Red}$, we denote by $rep_{A_i}^d(X)$ the lexicographically smallest set $R \subseteq A_i \cap \mathbf{Red}$ such that $|R|$ is minimised and $R \equiv_{A_i}^d X$. Notice that it can happen that $R = \varnothing$.

Lemma 6 ([4]). *For every $i \in \{1, \ldots, n\}$, one can compute a list LR_i containing all representatives* w.r.t. $\equiv_{A_i}^d$ *in time $O(nec(\equiv_{A_i}^d) \cdot \log(nec(\equiv_{A_i}^d)) \cdot n^2)$. One can also compute a data structure that given a set $X \subseteq A_i \cap \mathbf{Red}$ in time $O(\log(nec(\equiv_{A_i}^d)) \cdot |X| \cdot n)$ allows us to find a pointer to $rep_{A_i}^d(X)$ in LR_i. Similar statements hold for the list $LR_{\bar{i}}$ containing all representatives* w.r.t. $\equiv_{\bar{A}_i}^d$.

Our goal now is to define a DAG, denoted by $DAG(G)$, the maximal paths of which correspond exactly to the 1-minimal **Red** (σ, ρ)-dominating sets.

For $1 \leq j \leq n$ and $C \subseteq A_j \cap \mathbf{Blue}$ (or $C \subseteq \bar{A}_j \cap \mathbf{Blue}$) we denote by $\mathcal{SG}_j(C)$ (or by $\mathcal{GG}_j(C)$) the set X obtained from C if we we initially set $X = C$ and recursively apply the following rule: let x be the greatest (or smallest) vertex in X such that $N(X \setminus \{x\}) \cap (\bar{A}_j \cap \mathbf{Red}) = N(X) \cap (\bar{A}_j \cap \mathbf{Red})$ (or $N(X \setminus \{x\}) \cap (A_j \cap \mathbf{Red}) = N(X) \cap (A_j \cap \mathbf{Red})$) and set $X = X \setminus \{x\}$. Notice that $\mathcal{SG}_j(C)$ and $\mathcal{GG}_j(C)$ are both uniquely determined, and both have sizes bounded by c from [1, Lemma 1]. Observe also that if $C \subseteq A_j \cap \mathbf{Blue}$ (or $C \subseteq \bar{A}_j \cap \mathbf{Blue}$), then $\mathcal{SG}_\ell(C \cup \{x_\ell\}) = \mathcal{SG}_\ell(\mathcal{SG}_j(C) \cup \{x_\ell\})$ for all $\ell > j$ (or $\mathcal{GG}_\ell(C \cup \{x_\ell\}) = \mathcal{GG}_\ell(\mathcal{GG}_j(C) \cup \{x_\ell\})$ for all $\ell \leq j$).

Let $1 \leq j < n$ and let $(R_j, R'_j, C_j, C'_j) \in LR_j \times LR_{\bar{j}} \times 2^{A_j \cap \mathbf{Blue}} \times 2^{\bar{A}_j \cap \mathbf{Blue}}$ and $(R_{j+1}, R'_{j+1}, C_{j+1}, C'_{j+1}) \in LR_{j+1} \times LR_{\overline{j+1}} \times 2^{A_{j+1} \cap \mathbf{Blue}} \times 2^{\bar{A}_{j+1} \cap \mathbf{Blue}}$. There is an *ε-arc-1* from (R_j, R'_j, C_j, C'_j) to $(R_{j+1}, R'_{j+1}, C_{j+1}, C'_{j+1})$ if

(1.1) $R_j \equiv_{A_{j+1}}^d R_{j+1}$ and $R'_j \equiv_{\bar{A}_j}^d R'_{j+1}$, and

(1.2) if ($x_{j+1} \notin \mathbf{Blue}$ or ($x_{j+1} \in \mathbf{Blue}$ and $|N(x_{j+1}) \cap (R_j \cup R'_{j+1})| \in \rho$ and $|N(x_{j+1}) \cap (R_j \cup R'_{j+1})| \notin \rho^-$)) then ($C_{j+1} = \mathcal{SG}_{j+1}(C_j)$ and $C'_j = \mathcal{GG}_j(C'_{j+1})$), otherwise we should have ($|N(x_{j+1}) \cap (R_j \cup R'_{j+1})| \in \rho^-$) and

  (1.2.a) if $N(x_{j+1}) \cap (\bar{A}_{j+1} \cap \mathbf{Red}) \neq \varnothing$, then $C_{j+1} = \mathcal{SG}_{j+1}(C_j \cup \{x_{j+1}\})$, else $C_{j+1} = \mathcal{SG}_{j+1}(C_j)$, and

  (1.2.b) if $N(x_{j+1}) \cap (A_j \cap \mathbf{Red}) \neq \varnothing$, then $C'_j = \mathcal{GG}_j(C'_{j+1} \cup \{x_{j+1}\})$, else $C'_j = \mathcal{GG}_j(C'_{j+1})$.

There is an *ε-arc-2* from (R_j, R'_j, C_j, C'_j) to $(R_{j+1}, R'_{j+1}, C_{j+1}, C'_{j+1})$ if

(2.1) $R_{j+1} \equiv_{A_{j+1}}^d (R_j \cup \{x_{j+1}\})$, $R'_j \equiv_{\bar{A}_j}^d (R'_{j+1} \cup \{x_{j+1}\})$, $x_{j+1} \in \mathbf{Red}$, ($|N(x_{j+1}) \cap (R_j \cup R'_{j+1})| \in \sigma$ if $x_{j+1} \in \mathbf{Blue}$), and

(2.2) if ($x_{j+1} \notin$ **Blue** or ($x_{j+1} \in$ **Blue** and $|N(x_{j+1}) \cap (R_j \cup R'_{j+1})| \notin \sigma^-$)), then ($C_{j+1} = \mathcal{SG}_{j+1}(C_j)$ and $C'_j = \mathcal{GG}_j(C'_{j+1})$), otherwise we should have ($|N(x_{j+1}) \cap (R_j \cup R'_{j+1})| \in \sigma^-$) and
  (2.2.a) if $N(x_{j+1}) \cap (\bar{A}_{j+1} \cap \mathbf{Red}) \neq \emptyset$, then $C_{j+1} = \mathcal{SG}_{j+1}(C_j \cup \{x_{j+1}\})$, else $C_{j+1} = \mathcal{SG}_{j+1}(C_j)$, and
  (2.2.b) if $N(x_{j+1}) \cap (A_j \cap \mathbf{Red}) \neq \emptyset$, then $C'_j = \mathcal{GG}_j(C'_{j+1} \cup \{x_{j+1}\})$, else $C'_j = \mathcal{GG}_j(C'_{j+1})$, and
(2.3) either ($N(x_{j+1}) \cap (C_j \cup C'_{j+1}) \neq \emptyset$) or (($x_{j+1} \in$ **Blue** and $|N(x_{j+1}) \cap (R_j \cup R'_{j+1})| \in \sigma^*$).

The nodes of $DAG(G)$**.** $(R, R', C, C', i) \in LR_i \times LR_{\bar{i}} \times 2^{A_i \cap \mathbf{Blue}} \times 2^{\bar{A}_i \cap \mathbf{Blue}} \times [n]$ is a node of $DAG(G)$ whenever $x_i \in$ **Red**, $C = \mathcal{SG}_i(C)$ and $C' = \mathcal{GG}_i(C')$. We call i the *index* of (R, R', C, C', i). Finally $s = (\emptyset, \emptyset, \emptyset, \emptyset, 0)$ is the *source node* and $t = (\emptyset, \emptyset, \emptyset, \emptyset, n+1)$ is the *terminal node* of $DAG(G)$.

The arcs of $DAG(G)$. There is an arc from the node $(R_0, R'_0, C_0, C'_0, j)$ to the node $(R_p, R'_p, C_p, C'_p, j+p)$ with $1 \leq j < j+p \leq n$ if there exist tuples $(R_1, R'_1, C_1, C'_1), \ldots, (R_{p-1}, R'_{p-1}, C_{p-1}, C'_{p-1})$ such that (1) for each $1 \leq i \leq p-1$, $(R_i, R'_i, C_i, C'_i) \in LR_{j+i} \times LR_{\overline{j+i}} \times 2^{A_{j+i} \cap \mathbf{Blue}} \times 2^{\bar{A}_{j+i} \cap \mathbf{Blue}}$ and there is an ε-arc-1 from $(R_{i-1}, R'_{i-1}, C_{i-1}, C'_{i-1})$ to (R_i, R'_i, C_i, C'_i), and (2) there is an ε-arc-2 from $(R_{p-1}, R'_{p-1}, C_{p-1}, C'_{p-1})$ to (R_p, R'_p, C_p, C'_p).

There is an arc from the source node to a node (R, R', C, C', j) if ($S = \{x \in (A_j \cap \mathbf{Blue}) \setminus \{x_j\} \mid N(x) \cap (\bar{A}_j \cap \mathbf{Red}) \neq \emptyset$ and $|N(x) \cap (\{x_j\} \cup R')| \in \rho^-\}$)

(S1) $\{x_j\} \equiv^d_{A_j} R$ and $(\{x_j\} \cup R')$ (σ, ρ)-dominates $A_j \cap$ **Blue**,
(S2) if ($x_j \in$ **Blue** and $|N(x_j) \cap R'| \in \sigma^-$) then $C = \mathcal{SG}_j(S \cup \{x_j\})$, otherwise $C = \mathcal{SG}_j(S)$, and
(S3) either ($N(x_j) \cap (C' \cup C) \neq \emptyset$) or ($x_j \in$ **Blue** and $|N(x_j) \cap R'| \in \sigma^*$).

There is an arc from a node (R, R', C, C', j) to the terminal node if

(T1) $|N(x) \cap R| \in \rho$ for each $x \in \bar{A}_{j+1} \cap$ **Blue**, and
(T2) $C' = \mathcal{GG}_j(\{x \in \bar{A}_j \cap \mathbf{Blue} \mid N(x) \cap (A_j \cap \mathbf{Red}) \neq \emptyset$ and $|N(x) \cap R| \in \rho^-\})$.

If $P = (s, v_1, v_2, \ldots, v_p, t)$ is a path in $DAG(G)$, then the *trace of* P, denoted by **trace**(P), is defined as $\{x_{j_1}, x_{j_2}, \ldots, x_{j_p}\}$ where for all $i \in \{1, 2, \ldots, p\}$, j_i is the index of the node v_i. We have the following lemmas.

Lemma 7. *Let* $\mathcal{P}$ *be the set of paths in* $DAG(G)$ *from the source node to the terminal node. The mapping which associates with every* $P \in \mathcal{P}$ **trace**(P) *is a one-to-one correspondence with the set of* 1*-minimal* **Red** (σ, ρ)*-dominating sets.*

Lemma 8. $DAG(G)$ *is a DAG and can be constructed in time* $O(n^{c \cdot d})$.

We can now prove Theorem 1. By Lemma 8 $DAG(G)$ is a DAG and can be constructed in time $O(n^{c \cdot d})$. By Lemma 7 it is sufficient to count and enumerate the maximal paths in $DAG(G)$, and since we can count the maximal paths and enumerate them with linear delay (see for instance [16]), this concludes the proof.

4 Enumeration of Minimal Dominating Sets for Unit Square Graphs

Let G be a unit square graph and suppose that $f\colon V(G) \to \mathbb{Q}^2$ is a realization of G. (See Sect. 2 for more details on the point model of unit square graphs used in our paper.) For a vertex $v \in V(G)$, $\mathbf{frac}(v) = x_f(v) - \lfloor x_f(v) \rfloor$ is the fractional part of the x-coordinate of the point representing v. Let $v_1, \ldots, v_n$ be a linear ordering of the vertex set of G such that $\mathbf{frac}(v_i) \leq \mathbf{frac}(v_j)$ for all $j > i$. We prove that the MIM-width of $v_1, \ldots, v_n$ is bounded by $O(diam^2)$ where *diam* is the diameter of G, which we state in the following.

Theorem 2. *For a unit square graph G, $u \in V(G)$ and a positive integer r,* $\mathbf{lmimw}(G[N_G^r[u]]) = O(r^2)$. *Moreover, if a realization $f\colon V(G) \to \mathbb{Q}^2$ of G is given, then a linear ordering of vertices of MIM-width $O(r^2)$ can be constructed in polynomial time.*

We will now explain how to use this property and Theorem 1, to obtain an incremental polynomial time enumeration algorithm for the minimal dominating sets of G. To do it, we use a variant of the *flipping* method proposed in [10].

Given a minimal dominating set D^*, the flipping operation replaces an isolated vertex of $G[D^*]$ with its neighbor outside of D^*, and, if necessary, adds or deletes some vertices to obtain new minimal dominating sets D, such that $G[D]$ has more edges compared to $G[D^*]$. The enumeration algorithm starts with enumerating all maximal independent sets of the input graph G using the algorithm of Johnson, Papadimitriou, and Yannakakis [13], which gives the initial minimal dominating sets. Then the flipping operation is applied to every appropriate minimal dominating set found, to find new minimal dominating sets inducing subgraphs with more edges.

Let G be a graph. Let also $D \subseteq V(G)$. For $u \in D$, $C_D[u] = \{v \in V(G) \mid v \in N_G[u] \setminus N_G[D \setminus \{u\}]\}$ and $C_D(u) = \{v \in V(G) \mid v \in N_G(u) \setminus N_G[D \setminus \{v\}]\} = C_D[u] \setminus \{u\}$. Observe that if D is a minimal dominating set, then $C_D[u]$ is the set of certificates for a vertex $u \in D$.

Let us describe the variant of the flipping operation from [10], that we use. Let G be the input graph; we fix an (arbitrary) order of its vertices: $v_1, \ldots, v_n$. Suppose that D' is a dominating set of G. We say that the minimal dominating set D is obtained from D' by *greedy removal of vertices (with respect to order $v_1, \ldots, v_n$)* if we initially let $D = D'$, and then recursively apply the following rule: *If D is not minimal, then find a vertex v_i with the smallest index i such that $D \setminus \{v_i\}$ is a dominating set in G, and set $D = D \setminus \{v_i\}$.* Clearly, when we apply this rule, we never remove vertices of D' that have certificates. Whenever greedy removal of vertices of a dominating set is performed, it is done with respect to this ordering.

Let D be a minimal dominating set of G such that $G[D]$ has at least one edge uw. Then the vertex $u \in D$ is dominated by the vertex $w \in D$. Therefore, $C_D[u] = C_D(u) \neq \varnothing$. Let X be a non-empty inclusion-maximal independent set such that $X \subseteq C_D(u)$. Consider the set $D' = (D \setminus \{u\}) \cup X$. Notice that

D' is a dominating set in G, since all vertices of $C_D(u)$ are dominated by X by the maximality of X and u is dominated by w, but D' is not necessarily minimal, because it can happen that X dominates all the certificates of some vertex of $D \setminus \{u\}$. We apply greedy removal of vertices to D' to obtain a minimal dominating set. Let Z be the set of vertices that are removed by this to ensure minimality. Observe that $X \cap Z = \varnothing$ and $u \notin Z$ by the definition of these sets; in fact there is no edge between a vertex of X and a vertex of Z. Finally, let $D^* = ((D \setminus \{u\}) \cup X) \setminus Z$.

It is important to notice that $|E(G[D^*])| < |E(G[D])|$. Indeed, to construct D^*, we remove the endpoint u of the edge $uw \in E(G[D])$ and, therefore, reduce the number of edges. Then we add X but these vertices form an independent set in G and, because they are certificates for u with respect to D, they are not adjacent to any vertex of $D \setminus \{u\}$. Therefore, $|E(G[D^*])| \leq |E(G[D'])| < |E(G[D])|$.

The *flipping* operation is exactly the *reverse* of how we generated D^* from D; i.e., it replaces a non-empty independent set X in $G[D^*]$ such that $X \subseteq G[D^*] \cap N_G(u)$ for a vertex $u \notin D^*$ with their neighbor u in G to obtain D. In particular, we are interested in all minimal dominating sets D that can be generated from D^* in this way. Given D and D^* as defined above, we say that D^* is a *parent of D with respect to flipping u and X*. We say that D^* is a *parent* of D if there is a vertex $u \in V(G)$ and an independent set $X \subseteq N_G(u)$ such that D^* is a parent with respect to flipping u and X. It is important to note that each minimal dominating set D such that $E(G[D]) \neq \varnothing$ has a unique parent with respect to flipping of any $u \in D \cap N_G[D \setminus \{u\}]$ and a maximal independent set $X \subseteq C_D(u)$, as Z is lexicographically selected by a greedy algorithm. Similarly, we say that D is a *child* of D^* (with respect to flipping u and X) if D^* is the parent of D (with respect to flipping u and X). The proof of the following lemma is implicit in [10].

Lemma 9 [10]. *Suppose that for a graph G, all independent sets $X \subseteq N_G(u)$ for a vertex u can be enumerated in polynomial time. Suppose also that there is an enumeration algorithm $\mathcal{A}$ that, given a minimal dominating set D^* of a graph G such that $G[D^*]$ has an isolated vertex, a vertex $u \in V(G) \setminus D^*$ and a non-empty independent set X of $G[D^*]$ such that $X \subseteq D^* \cap N_G(u)$, generates with polynomial delay a family of minimal dominating sets $\mathcal{D}$ with the property that $\mathcal{D}$ contains all minimal dominating sets D that are children of D^* with respect to flipping u and X. Then all minimal dominating sets of G can be enumerated in incremental polynomial time.*

To obtain our main result, we will show that there is indeed an algorithm as algorithm $\mathcal{A}$ described in the statement of Lemma 9 when the input graph G is a unit square graph. We show that we can construct $\mathcal{A}$ by reduction to the enumeration of minimal **Red** dominating sets in an auxiliary colored induced subgraph of $G[N_G^3[u]]$. Let D^* be a minimal dominating set of a graph G such that $G[D^*]$ has an isolated vertex. Let also $u \in V(G) \setminus D^*$ and X is a non-empty independent set of $G[D^*]$ such that $X \subseteq D^* \cap N_G(u)$. Consider the set $D' = (D \setminus X) \cup \{u\}$. Denote by **Blue** the set of vertices that are not dominated by D'. Notice that **Blue** $\subseteq N_G(X) \setminus N_G[u]$. Therefore, **Blue** $\subseteq N_G^2[u]$. Let **Red** $= N_G($**Blue**$) \setminus N_G[X]$. Clearly, **Red** $\subseteq N_G^3[u]$. We construct the colored graph $H = G[$**Red** $\cup$ **Blue**$]$. Let $\mathcal{A}'$ be an algorithm that enumerates minimal **Red** dominating sets in H. Assume that if **Blue** $= \emptyset$, then $\mathcal{A}'$ returns $\emptyset$ as the unique **Red** dominating set. We construct $\mathcal{A}$ as follows.

Step 1. If $\mathcal{A}'$ returns an empty list of sets, then $\mathcal{A}$ returns an empty list as well.
Step 2. For each **Red** dominating set R of H, consider $D'' = D' \cup R$ and construct a minimal dominating set D from D'' by greedy removal.

Lemma 10. *If $\mathcal{A}'$ lists all minimal* **Red** *dominating sets with polynomial delay, then $\mathcal{A}$ generates with polynomial delay a family of minimal dominating sets $\mathcal{D}$ with the property that $\mathcal{D}$ contains all minimal dominating sets D that are children of D^* with respect to flipping u and X.*

Now we are ready to prove the main result of the section.

Theorem 3. *For a unit square graph G given with its realization f, all minimal dominating sets of G can be enumerated in incremental polynomial time.*

Proof. It is straightforward to observe that for a vertex u of a unit square graph G, any independent set $X \subseteq N_G(u)$ has at most 4 vertices. Hence, all independent sets $X \subseteq N_G(u)$ for a vertex u can be enumerated in polynomial time. By combining Theorems 1 and 2, and Lemmas 9 and 10, we obtain the claim. □

References

1. Belmonte, R., Vatshelle, M.: Graph classes with structured neighborhoods and algorithmic applications. Theor. Comput. Sci. **511**, 54–65 (2013)
2. Brandstädt, A., Le, V.B., Spinrad, J.P.: Graph classes: a survey. SIAM Monographs on Discrete Mathematics and Applications, SIAM, Philadelphia (1999)
3. Breu, H.: Algorithmic aspects of constrained unit disk graphs. PhD thesis, The University of British Columbia (1996)
4. Bui-Xuan, B.M., Telle, J.A., Vatshelle, M.: Fast dynamic programming for locally checkable vertex subset and vertex partitioning problems. Theor. Comput. Sci. **511**, 66–76 (2013)
5. Chandran, L.S., Francis, M.C., Sivadasan, N.: On the cubicity of interval graphs. Graphs and Combinatorics **25**(2), 169–179 (2009)
6. Courcelle, B.: Linear delay enumeration and monadic second-order logic. Discrete Appl. Math. **157**, 2675–2700 (2009)
7. Eiter, T., Gottlob, G.: Identifying the minimal transversals of a hypergraph and related problems. SIAM J. Comput. **24**, 1278–1304 (1995)
8. Eiter, T., Gottlob, G.: Hypergraph transversal computation and related problems in logic and AI. In: Flesca, S., Greco, S., Leone, N., Ianni, G. (eds.) JELIA 2002. LNCS (LNAI), vol. 2424, pp. 549–564. Springer, Heidelberg (2002)
9. Eiter, T., Gottlob, G., Makino, K.: New results on monotone dualization and generating hypergraph transversals. SIAM J. Comput. **32**, 514–537 (2003)

10. Golovach, P.A., Heggernes, P., Kratsch, D., Villanger, Y.: An incremental polynomial time algorithm to enumerate all minimal edge dominating sets. Algorithmica **72**, 836–859 (2015)
11. Golovach, P.A., Heggernes, P., Kanté, M.M., Kratsch, D., Sæther, S.H., Villanger, Y.: Output-Polynomial Enumeration on Graphs of Bounded (Local) Linear MIM-Width, CoRR, abs/1509.03753 (2015)
12. Golovach, P.A., Heggernes, P., Kante, M., Kratsch, D., Villanger, Y.: Enumerating minimal dominating sets in chordal bipartite graphs. Discrete Applied Mathematics, to appear. doi:10.1016/j.dam.2014.12.010
13. Johnson, D.S., Papadimitriou, C.H., Yannakakis, M.: On generating all maximal independent sets. Inf. Process. Lett. **27**(3), 119–123 (1988)
14. Kanté, M.M., Limouzy, V., Mary, A., Nourine, L.: On the enumeration of minimal dominating sets and related notions. SIAM J. Discrete Math. **28**, 1916–1929 (2014)
15. Kanté, M.M., Limouzy, V., Mary, A., Nourine, L.: On the neighbourhood helly of some graph classes and applications to the enumeration of minimal dominating sets. In: Chao, K.-M., Hsu, T., Lee, D.-T. (eds.) ISAAC 2012. LNCS, vol. 7676, pp. 289–298. Springer, Heidelberg (2012)
16. Kanté, M.M., Limouzy, V., Mary, A., Nourine, L., Uno, T.: On the enumeration and counting of minimal dominating sets in interval and permutation graphs. In: Cai, L., Cheng, S.-W., Lam, T.-W. (eds.) Algorithms and Computation. LNCS, vol. 8283, pp. 339–349. Springer, Heidelberg (2013)
17. Kanté, M.M., Limouzy, V., Mary, A., Nourine, L., Uno, T.: A Polynomial Delay Algorithm for Enumerating Minimal Dominating Sets in Chordal Graphs. In: Proceedings of WG 2015 (2014). arxiv:1407.2036 (to appear)
18. Kanté, M.M., Limouzy, V., Mary, A., Nourine, L., Uno, T.: Polynomial delay algorithm for listing minimal edge dominating sets in graphs. In: Dehne, F., Sack, J.-R., Stege, U. (eds.) WADS 2015. LNCS, vol. 9214, pp. 446–457. Springer, Heidelberg (2015)
19. Khachiyan, L., Boros, E., Borys, K., Elbassioni, K.M., Gurvich, V.: Generating all vertices of a polyhedron is hard. Discrete Comput. Geom. **39**, 174–190 (2008)
20. Khachiyan, L., Boros, E., Elbassioni, K.M., Gurvich, V.: On enumerating minimal dicuts and strongly connected subgraphs. Algorithmica **50**, 159–172 (2008)
21. Lawler, E.L., Lenstra, J.K., Rinnooy Kan, A.H.G.: NP-hardness and polynomial-time algorithms: generating all maximal independent sets. SIAM J. Comput. **9**, 558–565 (1980)
22. Rauf, I.: Polynomially Solvable Cases of Hypergraph Transversal and Related Problems. PhD thesis, Saarland University (2011)
23. Tarjan, R.E.: Enumeration of the elementary circuits of a directed graph. SIAM J. Comput. **2**, 211–216 (1973)
24. Telle, J.A., Proskurowski, A.: Algorithms for vertex partitioning problems on partial k-trees. SIAM J. Discrete Math. **10**(4), 529–550 (1997)
25. Vatshelle, M.: New width parameters of graphs. PhD thesis, University of Bergen (2012)

Minimum Degree Up to Local Complementation: Bounds, Parameterized Complexity, and Exact Algorithms

David Cattanéo[1] and Simon Perdrix[2](✉)

[1] LIG, University of Grenoble, Grenoble, France
[2] CNRS, Inria Project Team Carte, LORIA, Nancy, France
simon.perdrix@loria.fr

Abstract. The local minimum degree of a graph is the minimum degree that can be reached by means of local complementation. For any n, there exist graphs of order n which have a local minimum degree at least $0.189n$, or at least $0.110n$ when restricted to bipartite graphs. Regarding the upper bound, we show that the local minimum degree is at most $\frac{3}{8}n+o(n)$ for general graphs and $\frac{n}{4}+o(n)$ for bipartite graphs, improving the known $\frac{n}{2}$ upper bound. We also prove that the local minimum degree is smaller than half of the vertex cover number (up to a logarithmic term). The local minimum degree problem is NP-Complete and hard to approximate. We show that this problem, even when restricted to bipartite graphs, is in W[2] and FPT-equivalent to the EvenSet problem, whose W[1]-hardness is a long standing open question. Finally, we show that the local minimum degree is computed by a $\mathcal{O}^*(1.938^n)$-algorithm, and a $\mathcal{O}^*(1.466^n)$-algorithm for the bipartite graphs.

1 Introduction

Notations. Given a graph $G = (V, E)$, $\sim_G$ denotes the neighbourhood relation of G i.e., $\forall u, v \in V$, $u \sim_G v \Leftrightarrow \{u, v\} \in E$. We consider simple ($\forall u \in V, u \not\sim u$), undirected ($u \sim v \Leftrightarrow v \sim u$) graphs. The set $N_G(u) = \{v \mid u \sim_G v\}$ is the neighbourhood of u and its size $\delta_G(u) = |N_G(u)|$ is the degree of u. $\delta(G) = \min_{u\in V} \delta_G(u)$ is the minimum degree of G and $\tau(G)$ is the vertex cover number i.e., the size of the smallest set S such that if $u \sim v$, then $u \in S$ or $v \in S$. For any $D \subseteq V$, $Odd_G(D) = \Delta_{u\in D} N_G(u) = \{v \in V \mid |N_G(v) \cap D| = 1 \bmod 2\}$ is the odd-neighbourhood of D, where Δ denotes the symmetric difference.

Local Complementation. Local complementation of a graph with respect to one of its vertices consists in complementing the neighbourhood of this vertex:

Definition 1. *The local complementation of a graph G with respect to one of its vertices u is the graph $G \star u$ such that $v \sim_{G\star u} w$ iff $(v \sim_G w)$ xor $(u \sim_G v \wedge u \sim_G w)$.*

The local complementation is an involution ($G \star u \star u = G$). Two graphs are LC-equivalent if there exists a sequence of local complementation transforming one into the other: $G \equiv_{LC} H \Leftrightarrow \exists u_0, \dots u_k, G \star u_0 \dots \star u_k = H$.

K. Elbassioni and K. Makino (Eds.): ISAAC 2015, LNCS 9472, pp. 259–270, 2015.
DOI: 10.1007/978-3-662-48971-0_23

Local complementation has been introduced by Kotzig [20]. The study of this quantity is motivated by several applications: Bouchet [4,5] and de Fraysseix [9] used local complementation to give a characterization of circle graphs, and Oum [22] links the notion of vertex minor of a graph to LC-equivalence. A noticeable property of local complementation proved by Bouchet [2] is that LC-equivalence of graphs can be decided in time polynomial in the order of the graphs.

Cut Rank. Local complementation is related to the cut-rank function[1] [2,22]: given a graph G and a bipartition $(A, V\backslash A)$ of its vertices, $\text{cutrk}_G(A)$ is the rank of the linear map $L_A : 2^A \to 2^{V\backslash A} = X \mapsto Odd_G(X) \cap (V\backslash A)$. L_A is linear with respect to the symmetric difference: $L_A(X \Delta Y) = L_A(X) \Delta L_A(Y)$. The cut-rank can equivalently be defined as the rank of the cut-matrix, a sub-matrix of the adjacency matrix. Notice that for any A, $\text{cutrk}_G(A) = \text{cutrk}_G(V\backslash A)$.

LC-equivalent graphs have the same cutrank ($\text{cutrk}_G(\cdot) = \text{cutrk}_{G\star u}(\cdot)$) [3], however the converse which was conjectured in [2], has been disproved by Fon deer Flaass [12]: the counterexample involves two isomorphic Petersen graphs which have the same cut-rank but which are not LC-equivalent.

LU-equivalence. More recently, local complementation has emerged as a key operation in the field of quantum information theory. The graph state formalism consists in representing a quantum state using a graph (see [15] for details). This powerful formalism provides a graphical representation of quantum entanglement: each vertex represent a quantum bit (qubit) and the edges represent intuitively the entanglement between the qubits. Since entanglement is a non local property, the strength of the entanglement can only decrease when *local* operations are applied on the quantum state, and as a consequence the entanglement is invariant by *local reversible* operations. In the field of quantum information theory this intuition is captured by the LU-equivalence of quantum states: two quantum states have the same entanglement if and only if they are LU-equivalent i.e., there is a local unitary operation transforming one state into the other. LU-equivalence of quantum states can be naturally lifted to graphs as follows: two graphs are LU-equivalent if and only if the corresponding quantum states are LU-equivalent. Van den Nest [27] proved that LC-equivalent graphs are LU-equivalent. Moreover Hein et al. [15] proved that LU-equivalent graphs have the same cutrank. Thus LU-equivalence is weaker than LC-equivalence but stronger than the cut-rank equivalence. Using Fon der Flaass's counterexample based on the Petersen graph, one can show that there exist pairs of graphs which are not LU-equivalent but which have the same cutrank [15]. LC- and LU-equivalences were conjectured to coincide [25]. Indeed, LC- and LU-equivalence actually coincide for several families of graphs [26,28], however a counterexample of order 27 has been discovered using computer assisted methods [19].

[1] It was used by Bouchet [2] and others under the name *connectivity function*, and coined the cut-rank by Oum [22].

Local Minimum Degree. In this paper we will focus on the minimum degree up to local complementation called local minimum degree:

Definition 2. *Given a graph G, the* local minimum degree *of G is*

$$\delta_{loc}(G) = \min_{H \equiv_{LC} G} \delta(H)$$

The local minimum degree has been used to bound the rate of some quantum codes obtained by graph concatenation [1]. This quantity has also been used to characterise the complexity of preparation of graph states [16] which are used as a resource in measurement-based quantum computation [24] (a model of quantum computation which is very promising in terms of physical implementation), as well as blind quantum computation [6] for instance. The local minimum degree is also used to bound the optimal threshold that can be achieved by graph-based quantum secret sharing [13,21].

The local minimum degree is related to the cut-rank function and the smallest set of the form $D \cup Odd_G(D)$:

Property 1 [16]. Given a graph $G = (V, E)$,

$$\delta_{loc}(G) + 1 = \min_{\emptyset \subset D \subseteq V} |D \cup Odd_G(D)| = \min\{|A| : A \subseteq V \wedge \text{cutrk}_G(A) < |A|\}$$

The second equation provides a cut-rank characterisation of the local minimum degree which implies that two graphs which have the same cut-rank have the same local minimum degree. As a consequence, since LU-equivalent graphs have the same cut-rank function, they have the same local minimum degree, too. Thus the local minimum degree is invariant for the three closely related, albeit distinct, classes of equivalence based respectively on local complementation, local unitary operations, and cut-rank functions.

Bounds on the Local Minimum Degree. The local minimum degree has been studied for several families of graphs: the local minimum degree of the hypercube is at least logarithmic in the order of the hypercube [16]; the local minimum degree of a Paley graph $\mathcal{P}_n$ of order n is at least $\sqrt{n}$. There is no known specific upper bound on the local minimum degree of Paley graphs except that not all Paley graphs can have a linear local minimum degree (i.e., $\delta_{loc}(\mathcal{P}_n) = \Theta(n)$), and the existence of an infinite number of Paley graphs with a linear local minimum degree would imply the Bazzi-Mitter conjecture on elliptic curves [17,18].

There is no known explicit construction which leads to a local minimum degree greater than the square root of the order of the graph, however using probabilistic methods, it has been proven that there exist graphs of order n which have a local minimum degree larger than $0.189n$ [18]. There are even bipartite graphs with a linear local minimum degree: for any n there exists a bipartite graph of order n and local minimum degree at least $0.110n$ [18].

Regarding the upper-bounds, Property 1 implies that the local minimum degree is at most half of the order of the graph, since no set larger than half of the vertices can have a full cut-rank. In Sect. 2, we improve this upper bound, proving that for any graph of order n, its local minimum degree is at most $\frac{3}{8}n + o(n)$, and $\frac{n}{4} + o(n)$ for bipartite graphs. We also prove that the local minimum degree is smaller than half of the vertex cover number (up to a logarithmic term).

Complexity of the Local Minimum Degree. One motivation for studying the complexity of computing the local minimum degree comes from the problem of producing graphs with a 'large' local minimum degree. Indeed, there is no known explicit construction of graphs with a local minimum degree linear in the order of the graph, but a random graph has such a 'large' local minimum degree with high probability. So to produce a graph with a large local minimum degree, one can pick a graph at random and then double check that the local minimum degree is actually 'large'. However, computing the local minimum degree is hard, even for bipartite graphs: the associated decision problem is NP-Complete [18] and hard to approximate [18].

In Sect. 3, we investigate the parameterized complexity of the local minimum degree problem and its restriction to bipartite graphs. We show that both problems are FPT-equivalent to the so-called EvenSet problem, implying their W[2]-membership. However, it does not imply any hardness result since the W[1]-hardness of EvenSet is long standing open question [11].

In Sect. 4, we introduce exponential algorithms for computing the local minimum degree, mainly based on the improved upper bounds. We show that the local minimum degree of any graph of order n can be computed in time $\mathcal{O}^*(1.938^n)$ and more interestingly that the local minimum degree of bipartite graphs can be computed in time $\mathcal{O}^*(1.466^n)$.

2 Upperbounds on the Local Minimum Degree

For improving the known bounds on the local minimum degree, we use as a routine the fact that in any bipartite graph $G = (V_1, V_2, E)$, there exists a non empty subset of V_1 which oddly dominates at most $\frac{|V_2|}{2(1-2^{-|V_1|})}$ vertices, so roughly speaking as long as V_1 is not too small with respect to V_2 there is a non empty subset of V_1 which oddly dominates at most half of the vertices of V_2. This fact is a direct consequence of the so called Plotkin bound [23] on linear codes:

Lemma 1. *For any bipartite graph $G = (V_1, V_2, E)$, there exists a non empty set $D \subseteq V_1$ s.t.*

$$|Odd_G(D)| \leq \frac{|V_2|}{2(1 - 2^{-|V_1|})}$$

Proof. $C := \{Odd_G(D) : D \subseteq V_1\}$ is a linear binary code of length $n = |V_2|$ and rank $k = |V_1|$, where $Odd_G(D)$ is identified with its indicator vector in V_2. According to the Plotkin bound [23], the minimum distance d of C is at most $n/(2(1 - 2^{-k}))$, thus there exists a non empty set $D \subseteq V_1$ such that $|Odd_G(D)| \leq |V_2|/(2(1 - 2^{-|V_1|}))$. □

The local minimum degree can be bounded by the vertex cover number as follows:

Lemma 2. *Given a graph G of order n and vertex cover number $\tau(G) > 0$,*

$$2\delta_{loc}(G) \leq \tau(G) + \log_2(\tau(G)) + 1$$

Proof. Let $G = (V, E)$ be a graph of order n, and let S be an independent set of size $\alpha = n - \tau(G)$, and $R \subseteq S$ a subset of size k to be fixed later. Let $G' = (R, (V \backslash S) \cup R, E')$ be a bipartite graph s.t. for any $u \in R$, $N_{G'}(u) = \{u\} \cup N_G(u)$. Notice that there are two copies of R in G', one on each side of the bipartite graph: there is a matching between these two copies of R, the other edges of G' are those of G between R and $V \setminus S$. According to Lemma 1 there exists $D \subseteq R'$ s.t

$$|Odd_{G'}(D)| \leq \frac{|V| - |S| + |R|}{2(1 - 2^{-|R|})} = \frac{\tau(G) + k}{2(1 - 2^{-k})}$$

The odd-neighbourhood of D in G' is related to the odd-neighbourhood of D in G as follows: $Odd_{G'}(D) = \Delta_{u \in D} N_{G'}(u) = \Delta_{u \in D}(\{u\} \cup N_G(u)) = D \Delta Odd_G(D)$. Thus $|Odd_{G'}(D)| = |D \cup Odd_G(D)|$. As a consequence, $\delta_{loc}(G) + 1 \leq \frac{\tau(G)+k}{2(1-2^{-k})}$.

- If $\lceil \log_2(\tau(G) + 1) \rceil \leq n - \tau(G)$, then we fix $k = \lceil \log_2(\tau(G) + 1) \rceil$:

$$\delta_{loc}(G) + 1 \leq \frac{\tau(G) + \lceil \log_2(\tau(G) + 1) \rceil}{2(1 - 2^{-\lceil \log_2(\tau(G)+1) \rceil})} < \frac{1}{2}(\tau(G) + \log_2(\tau(G))) + 1 \quad (1)$$

  To prove the second inequality of Eq. (1), let $\tau(G) = 2^r + y$ with $y < 2^r$. Notice that $\lceil \log_2(\tau(G) + 1) \rceil = r + 1$, thus

$$\delta_{loc}(G) + 1 \leq \frac{2^r + y + r + 1}{2(1 - 2^{-r-1})}$$

  Moreover, standard calculation shows that $\frac{2^r+y+r+1}{1-2^{-r-1}} < 2^r + y + \log_2(2^r + y) + 2$ when $r > 0$. Thus $2\delta_{loc}(G) + 2 < \tau(G) + \log_2(\tau(G)) + 2$. When $r = 0$, $\tau(G) = 1$, thus G is a star (and possibly some isolated vertices), so $2\delta_{loc}(G) \leq 2 = \tau(G) + \log_2(\tau(G)) + 1$.
- If $\lceil \log_2(\tau(G) + 1) \rceil > n - \tau(G)$, then it is enough to prove that $2\delta_{loc}(G) \leq n$ since $\tau(G) + \log_2(\tau(G)) + 1 \geq \tau(G) + \lceil \log_2(\tau(G) + 1) \rceil > n$. For any set S of size $\lfloor \frac{n}{2} \rfloor + 1$, $\text{cutrk}_G(S) < |S|$ since $|V \setminus S| < |S|$, thus according to property 1, $\delta_{loc}(G) < \lfloor \frac{n}{2} \rfloor + 1 \leq n/2$. □

Remark 1. In Lemma 2, the condition $\tau(G) > 0$ only excludes the empty graph and is used to guarantee that the logarithm is well defined. The bound is tight for star graphs: $\delta_{loc}(S_n) = 1$ and $\tau(S_n) = 1$. This is the only tight case and when $\tau(G) > 1$, the proof can be modified to prove the following statement where the constant factor is removed: if $\tau(G) > 1$, $2\delta_{loc}(G) \leq \tau(G) + \log_2(\tau(G))$.

The vertex cover number-based bound on the local minimal degree leads to an improved general upper bound for bipartite graphs:

Theorem 1. *For any bipartite graph G of order $n > 0$,*

$$\delta_{loc}(G) < \frac{n}{4} + \log_2 n$$

Proof. If $n \leq 2$, the property is satisfied. Otherwise, since G is bipartite $\tau(G) \leq \lfloor \frac{n}{2} \rfloor$, so according to Lemma 2, $\delta_{loc}(G) \leq \frac{1}{2}(\tau(G) + \log_2(\tau(G)) + 1) \leq \frac{n}{4} + \frac{1}{2}\log_2(n/2) + \frac{1}{2} \leq \frac{n}{4} + \frac{1}{2}\log_2 n < \frac{n}{4} + \log_2 n$. □

Contrary to the bipartite case, the bound involving the vertex cover number does not lead to an improved upper bound for non-bipartite graphs. However, we prove that the local minimum degree of a graph of order n is at most $\frac{3}{8}n + o(n)$ exploiting the structure of the kernels of the linear maps associated with the cuts of the graph:

Theorem 2. *For any graph G of order $n > 0$,*

$$\delta_{loc}(G) < \frac{3}{8}n + \log_2 n$$

Proof. For any integer $0 < k < n/2$, let S be a subset of $\lfloor n/2 \rfloor + k$ vertices. Let $L : S \to V \setminus S$ be the map $D \mapsto Odd_G(D) \setminus S$ which is linear for the symmetric difference, i.e. $L(D_1 \Delta D_2) = L(D_1)\Delta L(D_2)$. Notice that for any $D \in Ker(L)$, $D \cup Odd(D) \subseteq S$. According to the rank nullity theorem, $dim(Ker(L)) \geq 2k - 1$. Let $R \subseteq S$ be a basis for $Ker(L)$. Let $G' = (R, S \times \{1,2,3\}, E')$ be a bipartite graph s.t. for any $D \in R, N_{G'}(D) = D \times \{1\} \cup Odd_G(D) \times \{2\} \cup (Odd_G(D)\Delta D) \times \{3\}$: the neighbourhood of D in G' is the disjoint union of D, $Odd_G(D)$ and $D\Delta Odd_G(D)$. Notice that $|R| \geq 2k - 1$ and $|S \times \{1,2,3\}| = 3(\lfloor n/2 \rfloor + k)$, so according to Lemma 1, there exists a non empty $R_0 \subseteq R$ such that $|Odd_{G'}(R_0)| \leq \left\lfloor \frac{3}{2} . \frac{\lfloor n/2 \rfloor + k}{1 - 2^{-2k+1}} \right\rfloor$.

Let $F := \Delta_{D \in R_0} D$. Since R is a basis and $R_0 \neq \emptyset$, $F \neq \emptyset$. Moreover $Odd_{G'}(R_0) = \Delta_{D \in R_0} N_{G'}(D) = \Delta_{D \in R_0}(D \times \{1\} \cup Odd_G(D) \times \{2\} \cup (Odd_G(D)\Delta D) \times \{3\}) = F \times \{1\} \cup Odd_G(F) \times \{2\} \cup (F\Delta Odd_G(F)) \times \{3\}$. Thus $|Odd_{G'}(R_0)| = |F| + |Odd_G(F)| + |F\Delta Odd_(F)| = 2|F \cup Odd_G(F)|$. As a consequence,

$$|F \cup Odd_G(F)| \leq \left\lfloor \frac{1}{2} \left\lfloor \frac{3}{2} . \frac{\lfloor n/2 \rfloor + k}{1 - 2^{1-2k}} \right\rfloor \right\rfloor \tag{2}$$

We choose $k = \lfloor 4 \log_2(n)/3 \rfloor$ to guarantee $|F \cup Odd_G(F)| \leq \frac{3}{8}n + \log_2(n) + O(1)$. More precisely, notice that $|F \cup Odd_G(F)| \leq \frac{3}{8} . \frac{n + 2\lfloor 4\log_2(n)/3 \rfloor}{1 - 2 \times 2^{-2\lfloor 4 \log_2(n)/3 \rfloor}} \leq \frac{3}{8} . \frac{n + 8 \log_2(n)/3}{1 - 8.n^{-8/3}}$ which is strictly smaller than $\frac{3}{8}n + \log_2 n + 1$ when $n > 60$. For $2 < n \leq 61$, one can double check by direct calculation that the bound in Eq. 2 is actually strictly smaller than $\frac{3}{8}n + \log_2(n) + 1$. Thus for any $n > 2$, $\min_{D \neq \emptyset} |D \cup Odd_G(D)| < \frac{3}{8}n + \log_2 n + 1$, so $\delta_{loc}(G) < \frac{3}{8}n + \log_2 n$. Finally, it is easy to check that $\delta_{loc}(G) < \frac{3}{8}n + \log_2 n$ also holds for $n \leq 2$. □

Remark 2. Choosing $k = \lfloor \log_2(n)/2 \rfloor$ in the proof of Theorem 2 gives an asymptotically slightly better bound: $\delta_{loc}(G) \leq 3/8n + 3/4 \log_2(n) + O(1)$.

3 Parameterized Complexity

The decision problem associated with the local minimum degree is known to be NP-complete and hard to approximate: there exists no k-approximation algorithm for this problem for any constant k unless P=NP [18]. In this section we consider the parameterized complexity of this problem, and its bipartite version. Please refer to [10] for an introduction to parameterized complexity.

Local Minimum Degree :
input: A graph G
parameter: An integer k
question: Is $\delta_{loc}(G) \leq k$?

Bipartite Local Minimum Degree :
input: A bipartite graph G
parameter: An integer k
question: Is $\delta_{loc}(G) \leq k$?

We show that both problems are FPT-equivalent to the EvenSet problem [11]:

EvenSet:
input: A bipartite graph $G = (R, B, E)$
parameter: An integer k
question: Is there a non empty $D \subseteq R$, such that $|D| \leq k$ and $Odd_G(D) = \emptyset$ i.e., every vertex in B has an even number of neighbours in D?

To prove the FPT-equivalence of these three problems, first we prove that EvenSet is harder than Local Minimum Degree, and then that Bipartite Local Minimum Degree is harder than EvenSet.

Theorem 3. EvenSet *is* FPT*-reducible to* Local Minimum Degree.

Proof. Given an instance (G, k) of Local Minimum Degree, let (G', k') be an instance of EvenSet where:
$G' = (A_1 \cup A_2, \cup A_3, A_4 \cup A_5, E_1 \cup E_2 \cup E_3)$, $k' = 2k+2$

$\forall i \in [1, 5], A_i = \{a_{i,u}, \forall u \in V(G)\}$
$E_1 = \{(a_{1,u}, a_{4,u}), \forall u \in V(G)\}$,
$E_2 = \{(a_{i,u}, a_{5,u}), \forall i \in \{2, 3\}, \forall u \in V(G)\}$
$E_3 = \{(a_{2,u}, a_{i,v}), \forall i \in \{4, 5\}, \forall \{u, v\} \in E(G)\}$

In other words, G' consists of 5 copies A_is of $V(G)$, there is a matching between A_1 and A_4, and between A_3 and A_5. Moreover, the subgraph induced by $A_2 \cup A_4$ is the bipartite double of G, whereas subgraph induced by $A_2 \cup A_5$ the bipartite double of G augmented with a matching.

– If (G, k) is a positive instance of Local Minimum Degree with a non empty $D \subseteq V(G)$ such that $|D \cup Odd_G(D)| \leq k+1$. Let $D' = \{a_{1,u} \mid u \in Odd_G(D)\} \cup \{a_{2,u} \mid u \in D\} \cup \{a_{3,u} \mid u \in Odd_G(D)\Delta D\}$, thus D' is composed of the copy of D in A_2, the copy of $Odd_G(D)$ in A_1 and the copy of $D\Delta Odd_G(D)$ in A_3. Notice that $Odd_{G'}(D') = \emptyset$, and $D' \neq \emptyset$ since $D \neq \emptyset$. Moreover $|D'| = |Odd_G(D)| + |D| + |D\Delta Odd_G(D)| = 2|D \cup Odd_G(D)| \leq 2k + 2 = k'$. Thus D' makes (G', k') a positive instance of EvenSet.

– If (G', k') is a positive instance of EvenSet with a non empty $D \subseteq A_1 \cup A_2 \cup A_3$ of size at most k' such that $Odd_{G'}(D) = \emptyset$. For $i \in [1,3]$, let $D_i = \{u \in V(G) \mid a_{i,u} \in D\}$. Notice that $D_1 = Odd_G(D_2)$ and $D_3 = Odd_G(D_2) \Delta D_2$. $D \neq \emptyset$ implies $D_2 \neq \emptyset$, moreover $|D_2 \cup Odd_G(D_2)| = \frac{1}{2}(|D_2| + |Odd_G(D_2)| + |Odd_G(D_2) \Delta D_2|) = \frac{1}{2}|D| \leq \frac{1}{2}k' = k+1$, so D_2 makes (G, k) a positive instance of Local Minimum Degree. □

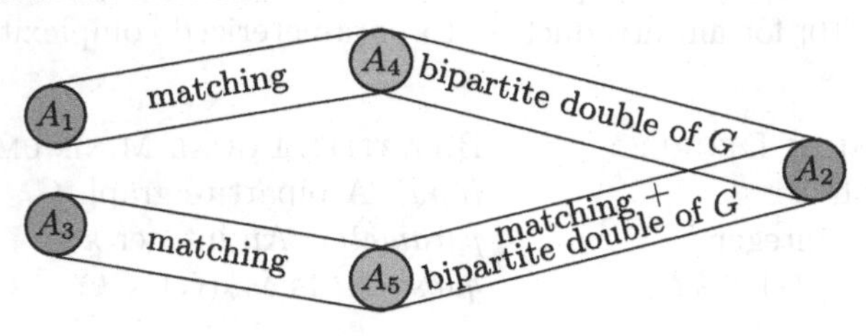

Corollary 1. Local Minimum Degree *is in* W[2].

W[2]-membership of Local Minimum Degree is not surprising in the sense that not only EvenSet but all similar problems of graph domination with parity conditions are known to be in W[2] [8]. We refine this W[2]-membership by proving that both Local Minimum Degree and Bipartite Local Minimum Degree are FPT-equivalent to EvenSet. They form a peculiar subclass of W[2] for which no hardness results are known: the W[1]-hardness of EvenSet is a long standing open question in parameterized complexity [11]. This contrasts with the subclass of problems FPT-equivalent to the W[1]-hard OddSet problem which contains problems like Weak Odd Domination and Quantum Threshold [7,14].

Theorem 4. Bipartite Local Minimum Degree *is* FPT*-reducible to* EvenSet.

Proof. If $(G{=}(R, B, E), k)$ is a positive instance of EvenSet, then it is also a positive instance of Bipartite Local Minimum Degree. But if (G, k) is a positive instance of Bipartite Local Minimum Degree, it may fail to be a positive instance of EvenSet mainly for two reasons:

(i) A set D such that $|D \cup Odd_G(D)| \leq k+1$ may not be a subset of R
(ii) For solving EvenSet, one wants to guarantee that $Odd_G(D) = \emptyset$.

Regarding the first point, a gadget with a local minimum degree larger than $k+1$ is attached to each vertex in B to guarantee that no vertex of B can occur in a set D such that $|D \cup Odd(D)| \leq k+1$. Concretely we can use a Paley graph P_q which vertices are $\{0, \ldots, q-1\}$ for $q = 1 \bmod 4$ a power of prime, and (i, j) is an edge iff $\exists x, i - j = x^2 \bmod q$. The local minimal degree of a Paley graph is at least square root of its order. However to keep the bipartiteness of the graph we use the bipartite double of a Paley graph rather than a Paley graph. Indeed, it is known that the local minimum degree of a bipartite double graph is as large as the local minimum degree of the original graph ($\delta_{loc}(G^{\oplus 2}) \geq \delta_{loc}(G)$ [17]).

Regarding the second point, each vertex of B is duplicated k times in such a way that for any $D \subseteq R$ if a vertex $v \in B$ is in the odd neighbourhood of D than its k copies are also in the odd-neighbourhood which contradicts the fact that $|D \cup Odd(D)|$ is at most $k+1$.

Concretely, let q be a prime number such that $q \geq k^2+1$ and $q = 1 \bmod 4$, let (G', k) be an instance of BIPARTITE LOCAL MINIMUM DEGREE such that $G' = (R \cup P', P, E_G \cup E_{\text{Paley}})$, where $P = \cup_{b \in B, i \in [0,k]} P_{b,i}$, $P' = \cup_{b \in B, i \in [0,k]} P'_{b,i}$ $P_{b,i} = \{p_{b,i,r}, \forall r \in [0, q-1]\}$, $P'_{b,i} = \{p'_{b,i,r}, \forall r \in [0, q-1]\}$ $E_{\text{Paley}} = \cup_{b \in B, i \in [0,k]} E^{(b,i)}_{\text{Paley}}$ and $E^{(b,i)}_{\text{Paley}} = \{(p_{b,i,r}, p'_{b,i,r'}), \forall r, r' \in [0, q-1]$ *s.t.* $\exists \ell \in [0, q-1], \ell^2 = r - r' \bmod q\}$.

– If (G, k) is a positive instance of EVENSET with $D \subseteq E$ s.t. $Odd_G(D) = \emptyset$ then $Odd_{G'}(D) = \emptyset$ so (G', k) is a positive instance of BIPARTITE LOCAL MINIMUM DEGREE.

– If (G', k) is a positive instance of BIPARTITE LOCAL MINIMUM DEGREE with D s.t. $|D \cup Odd_{G'}(D)| \leq k+1$. For any $b \in B, i \in [0,k]$, let $D'_{b,i} = D \cap (P_{b,i} \cup P'_{b,i})$, in the subgraph induced by $P_{b,i} \cup P'_{b,i}$ $|D' \cup Odd_{G'[P_{b,i} \cup P'_{b,i}]}(D)| \leq k+1$, thus $D'_{n,i} = \emptyset$ since $\delta_{loc}(\text{Paley}_{k^2+1}) > k$. So $D \subseteq R$. Moreover if there exists $p_{b,i,0} \in Odd_{G'}(D)$ then $\forall j \in [0,k], p_{b,j,0} \in Odd_{G'}(D)$, so $|D \cup Odd_{G'}(D)| > k+1$, so by contradiction $Odd_{G'}(D) = \emptyset$. Thus (G, k) is a positive of EVENSET. □

Corollary 2. BIPARTITE LOCAL MINIMUM DEGREE *and* LOCAL MINIMUM DEGREE *are* FPT*-equivalent to* EVENSET.

W[1]-hardness of EVENSET is a long standing open problem, the FPT-equivalence with (BIPARTITE) LOCAL MINIMUM DEGREE might give some more insights and open new perspectives on the parameterized complexity of EVENSET.

4 Exponential Algorithms

In this section we introduce exact exponential algorithms for computing the local minimum degree of a graph.

Property 2. The local minimum degree of a graph of order n can be computed in time $\mathcal{O}^*(1.938^n)$.

Proof. Thanks to Property 1 and Theorem 2, $\delta_{loc}(G) + 1 = \min\{|A| : |A| \leq \frac{3}{8}n + \log_2(n) \wedge \text{cutrk}_G(A) < |A|\}$. The algorithm consists in enumerating all subsets of at most $\frac{3}{8}n + \log_2(n)$ vertices and computing its cut-rank. The cut-rank can be computed in polynomial time, so the complexity of this algorithm is $\mathcal{O}^*(2^{H(\frac{3}{8})n})$ where $H(x) = -x\log_2 x - (1-x)\log_2(1-x)$ is the binary entropy function. □

Regarding the bipartite case, enumerating all the subsets of size at most $\frac{n}{4} + \log_2(n)$ leads to a $\mathcal{O}^*(1.755^n)$ algorithm. This naive algorithm can be improved:

Theorem 5. *The local minimum degree of a bipartite graph of order n can be computed in time $\mathcal{O}^*(1.466^n)$.*

Proof. We use the following property of bipartite graphs: given a bipartite graph $G = (V_1, V_2, E)$, $\delta_{loc}(G) + 1 = \min_{\emptyset \subset D \subseteq V_1 \text{ or } \emptyset \subset D \subseteq V_2} |D \cup Odd_G(D)|$. Indeed, for any $D \subseteq V_1 \cup V_2$, both $(D \cap V_1) \cup Odd_G(D \cap V_1)$ and $(D \cap V_2) \cup Odd_G(D \cap V_2)$ are subsets of $D \cup Odd_G(D)$. Let $|V_1| = \alpha n$ and $|V_2| = (1 - \alpha)n$. We assume w.l.o.g. that $\alpha \leq 1/2$. Since V_1 is a vertex cover set, according to Lemma 2, $\delta_{loc}(G) \leq \frac{\alpha}{2}n + \frac{\log_2(\alpha n)}{2}$. Thus to compute the local miminum degree, it is enough to enumerate all sets D of size at most $\frac{\alpha}{2}n + \frac{\log_2(\alpha n)}{2}$ in both V_1 and V_2 and to compute their odd neighbourhood – which can be done in time polynomial in n. There are $\binom{\alpha n}{\frac{\alpha}{2}n + \frac{\log_2(\alpha n)}{2}} + \binom{(1-\alpha)n}{\frac{\alpha}{2}n + \frac{\log_2(\alpha n)}{2}} = \mathcal{O}^*(2^{(1-\alpha)nH(\frac{\alpha}{2(1-\alpha)})})$ sets to enumerate. Notice that $\alpha \mapsto (1 - \alpha)H(\frac{\alpha}{2(1-\alpha)})$ is maximal for $\alpha_0 = 0.3885$, and $2^{(1-\alpha_0)H(\frac{\alpha_0}{2(1-\alpha_0)})} = 1.46557$. □

5 Conclusion

After having shown that the local minimum degree is smaller than half of the vertex cover number (up to a logarithmic term), we have improved the best known upper bound on the local minimum degree, proving that it is at most $\frac{3}{8}n + o(n)$ and $\frac{n}{4} + o(n)$ for bipartite graphs. Moreover, we have investigated the parametrized complexity of the problem, showing its W[2]-membership and its FPT-equivalence with the EvenSet problem, even when restricted to bipartite graphs. Finally, we have introduced a $\mathcal{O}^*(1.938^n)$-algorithm – $\mathcal{O}^*(1.466^n)$-algorithm for the bipartite graphs – for computing the local minimum degree.

This is noticeable that the bipartite case evolves quite similarly to the general case: same parameterized complexity, and upper bound and algorithm slightly better in the bipartite case. It would be interesting to investigate other families of graphs, in particular those defined by excluded vertex minors, in order to identify a family of graphs which local minimum is large but easy to compute or to approximate.

Acknowledgments. We would like to thank Emmanuel Jeandel and Mehdi Mhalla for several helpful discussions. This work has been partially funded by the ANR-10-JCJC-0208 CausaQ grant and by région Rhône-Alpes (ADR Cible R637).

References

1. Beigi, S., Chuang, I., Grassl, M., Shor, P., Zeng, B.: Graph concatenation for quantum codes. J. Math. Phys. **52**(2), 022201 (2011)
2. Bouchet, A.: Graphic presentations of isotropic systems. J. Comb. Theory Ser. A **45**, 58–76 (1987)
3. Bouchet, A.: Connectivity of isotropic systems. In: Proceedings of the third international conference on Combinatorial mathematics, pp. 81–93. New York Academy of Sciences (1989)
4. Bouchet, A.: κ-transformations, local complementations and switching. In: NATO Advance Research Workshop, vol. C, pp. 41–50 (1990)

5. Bouchet, A.: Circle graph obstructions. J. Comb. Theor. Ser. B **60**(1), 107–144 (1994)
6. Broadbent, A., Fitzsimons, J., Kashefi, E.: Universal blind quantum computation. In: 50th Annual IEEE Symposium on Foundations of Computer Science, FOCS 2009 (2009)
7. Cattanéo, D., Perdrix, S.: Parameterized complexity of weak odd domination problems. In: Gąsieniec, L., Wolter, F. (eds.) FCT 2013. LNCS, vol. 8070, pp. 107–120. Springer, Heidelberg (2013)
8. Cattanéo, D., Perdrix, S.: The parameterized complexity of domination-type problems and application to linear codes. In: Gopal, T.V., Agrawal, M., Li, A., Cooper, S.B. (eds.) TAMC 2014. LNCS, vol. 8402, pp. 86–103. Springer, Heidelberg (2014)
9. de Fraysseix, H.: Local complementation and interlacement graphs. Discrete Math. **33**(1), 29–35 (1981)
10. Downey, R.G., Fellows, M.R.: Parameterized Complexity. Springer, New York (1999)
11. Downey, R.G., Fellows, M.R., Vardy, A., Whittle, G.: The parameterized complexity of some fundabmental problems in coding theory. CDMTCS Research Report Series (1997)
12. Fon-Der-Flaasss, D.G.: Local complementations of simple and directed graphs. Discrete Analysis and Operations Research, pp. 15–34. Springer, Netherlands (1996)
13. Gravier, S., Javelle, J., Mhalla, M., Perdrix, S.: Quantum secret sharing with graph states. In: Kučera, A., Henzinger, T.A., Nešetřil, J., Vojnar, T., Antoš, D. (eds.) MEMICS 2012. LNCS, vol. 7721, pp. 15–31. Springer, Heidelberg (2013)
14. Gravier, S., Javelle, J., Mhalla, M., Perdrix, S.: On weak odd domination and graph-based quantum secret sharing. Theor. Comput. Sci. **598**, 129–137 (2015)
15. Hein, M., Eisert, J., Briegel, H.J.: Multi-party entanglement in graph states. Phys. Rev. A **69**, 062311 (2004)
16. Høyer, P., Mhalla, M., Perdrix, S.: Resources required for preparing graph states. In: Asano, T. (ed.) ISAAC 2006. LNCS, vol. 4288, pp. 638–649. Springer, Heidelberg (2006)
17. Javelle, J.: Cryptographie Quantique, Protocoles et Graphes. PhD thesis, Grenoble University (2014)
18. Javelle, J., Mhalla, M., Perdrix, S.: On the minimum degree up to local complementation: bounds and complexity. In: Golumbic, M.C., Stern, M., Levy, A., Morgenstern, G. (eds.) WG 2012. LNCS, vol. 7551, pp. 138–147. Springer, Heidelberg (2012)
19. Ji, Z., Chen, J., Wei, Z., Ying, M.: The lu-lc conjecture is false (2007)
20. Kotzig, A.: Eulerian lines in finite 4-valent graphs and their transformations. In: Colloqium on Graph Theory Tihany 1966, pp. 219–230. Academic Press (1968)
21. Markham, D., Sanders, B.C.: Graph states for quantum secret sharing. Phys. Rev. A **78**, 042309 (2008)
22. Oum, S.: Approximating rank-width and clique-width quickly. ACM Trans. Algorithms **5**(1), 10 (2008)
23. Plotkin, M.: Binary codes with specified minimum distance. IRE Trans. Inf. Theor. **6**(4), 445–450 (1960)
24. Raussendorf, R., Briegel, H.J.: A one-way quantum computer. Phys. Rev. Lett. **86**, 5188–5191 (2001)
25. Schlingemann, D.: Local equivalence of graph states. In: Krueger, O., Werner, R.F. (eds.) Some Open Problems in Quantum Information Theory (2005). arXiv:quant-ph/0504166

26. Van den Nest, M.: Local equivalence of stabilizer states and codes. PhD thesis, Faculty of Engineering, K. U. Leuven, Belgium, May 2005
27. Van den Nest, M., Dehaene, J., De Moor, B.: Graphical description of the action of local clifford transformations on graph states. Phys. Rev. A **69**, 022316 (2004)
28. Zeng, B., Chung, H., Cross, A.W., Chuang, I.L.: Local unitary versus local clifford equivalence of stabilizer and graph states. Phys. Rev. A **75**(3), 032325 (2007)

Exact and FPT Algorithms for Max-Conflict Free Coloring in Hypergraphs

Pradeesha Ashok, Aditi Dudeja, and Sudeshna Kolay(✉)

Institute of Mathematical Sciences, Chennai, India
{pradeesha,aditid,skolay}@imsc.res.in

Abstract. Conflict-free coloring of hypergraphs is a very well studied question of theoretical and practical interest. For a hypergraph $H = (U, \mathcal{F})$, a conflict-free coloring of H refers to a vertex coloring where every hyperedge has a vertex with a unique color, distinct from all other vertices in the hyperedge. In this paper, we initiate a study of natural maximization version of this problem, namely, Max-CFC: For a given hypergraph H and a fixed $r \geq 2$, color the vertices of U using r colors so that the number of hyperedges that are conflict-free colored is maximized. By previously known hardness results for conflict-free coloring, this maximization version is NP-hard.

We study this problem in the context of both exact and parameterized algorithms. In the parameterized setting, we study this problem with respect to the natural parameter, the solution size. In particular, we study the following question: p-CFC: For a given hypergraph, can we conflict-free color at least k hyperedges with at most r colors, the parameter being k. We show that this problem is FPT by designing an algorithm with running time $2^{\mathcal{O}(k \log\log k + k \log r)}(n+m)^{\mathcal{O}(1)}$ using a novel connection to the Unique Coverage problem and applying the method of color coding in a non-trivial manner. For the special case for hypergraphs induced by graph neighbourhoods we give a polynomial kernel. Finally, we give an exact algorithm for Max-CFC running in $\mathcal{O}(2^{n+m})$ time. All our algorithms, with minor modifications, work for a stronger version of conflict-free coloring, Unique Maximum Coloring.

1 Introduction

A hypergraph H is a pair $(U, \mathcal{F})$ where U is a set of n vertices and $\mathcal{F}$ contains m subsets of U. We call these subsets *hyperedges*. Thus a general graph is a hypergraph where every hyperedge contains exactly two vertices. A k-vertex-coloring of H, for $k \in \mathbb{N}$ is a function $c : U \to \{1, 2, \ldots, k\}$. A coloring is called a *proper* coloring if none of the hyperedges are monochromatic, i.e. all the vertices of the hyperedge are not of the same color. We look at a stricter version of coloring called conflict-free coloring.

The research leading to these results has received partial funding from the European Research Council under the European Union's Seventh Framework Programme (FP7/2007-2013)/ERC grant agreement no. 306992.

K. Elbassioni and K. Makino (Eds.): ISAAC 2015, LNCS 9472, pp. 271–282, 2015.
DOI: 10.1007/978-3-662-48971-0_24

Definition 1. *A vertex coloring $c : U \rightarrow \{1, 2, \ldots, k\}$ of a hypergraph $H(U, \mathcal{F})$ is said to be* conflict-free, *if for every $F \in \mathcal{F}, \exists v \in F$ such that $\forall u \in F$, $u \neq v$ implies $c(u) \neq c(v)$. In other words, every hyperedge has a uniquely colored vertex.*

The minimum number of colors required to conflict-free color the vertices of a hypergraph H is called the *conflict-free chromatic number* of H and is represented as $\chi_{cf}(H)$. For a given hypergraph H, the *minimum conflict-free coloring problem* refers to computing the value of $\chi_{cf}(H)$.

The concept of conflict-free coloring was introduced for hypergraphs induced by geometric regions, motivated by the frequency allocation problem in cellular networks [4]. This problem also found applications in areas like Radio Frequency Identification and Robotics. Conflict-free coloring question has been extensively studied for hypergraphs induced by various geometric regions [1,8,13].

Pach and Tardos [12] initiated the study of conflict-free coloring for general hypergraphs and gave an upper bound of $O(\sqrt{m})$ on the conflict-free chromatic number. On the algorithmic side, the minimum conflict-free coloring problem for a general hypergraph is NP-hard by results shown in [4,7]. [12] also studied the conflict-free coloring of hypergraphs induced by graph neighborhoods. Here the vertex set of the hypergraph corresponds to vertex set of a general graph $G = (V, E)$ and the hyperedges are defined by the neighborhoods (open or closed) of the vertices in G. [12] showed an upperbound of $O(\log^2 n)$ and a lower bound of $\Omega(\log n)$ for this problem. Gargano and Rescigno [7] studied the minimum conflict-free coloring of these hypergraphs and showed NP- completeness. [7] also showed that the minimum conflict-free coloring problem for these graphs becomes tractable when parameterized by the vertex cover or the neighborhood diversity number of the graph. Specifically, they gave an algorithm that decides whether a hypergraph induced by neighborhoods of a graph G can be conflict-free colored using k colors. This algorithm runs in time $2^{\mathcal{O}(kt \log k)}$ where t represents the neighborhood diversity number of G. Note that this also implies an algorithm to solve the minimum conflict-free coloring problem in hypergraphs induced by graph neighbourhoods, which runs in $\mathcal{O}(n^n)$ time.

In this paper, we initiate a study a maximization version of the Minimum Conflict-Free Coloring problem.

> Maximum Conflict-free Coloring(Max-CFC)
> **Input:** A hypergraph $(U, \mathcal{F})$ on n vertices and m hyperedges, and an integer $r \geq 2$.
> **Output:** A maximum-sized subfamily of hyperedges that can be conflict-free colored with r colors.

The NP-hardness of this problem follows from the NP-hardness reductions shown in [7]. We give an exact algorithm for this problem that runs in $\mathcal{O}(2^{m+n}) \cdot n^{\mathcal{O}(1)}$ time. As a corollary, we obtain an exact algorithm, of running time $\mathcal{O}(4^n) \cdot n^{\mathcal{O}(1)}$, for hypergraphs induced by neighbourhoods in graphs. We also define a stronger variant of conflict-free coloring namely, *unique-maximum coloring* [3].

Definition 2. *A vertex coloring* $c : U \rightarrow \{1, 2, \ldots, k\}$ *is said to be* unique-maximum, *if for every* $F \in \mathcal{F}, \exists v \in F$ *such that* $\forall u \in F$, $u \neq v$ *implies* $c(u) < c(v)$. *In other words, the maximum color occuring in a hyperedge occurs uniquely. The minimum number of colors required to unique-maximum color H is called the unique-maximum chromatic number of H.*

For a given hypergraph H, the *minimum unique-maximum coloring problem* refers to computing the minimum number of colors required to unique-maximum color H.

Similar to the definition of Max-CFC, we can define Maximum Unique-Maximum Coloring (Max-UMC) to take as input a hypergraph H and a positive integer $r \geq 2$, and output the largest subfamily of hyperedges that has a unique-maximum coloring with r colors. Our algorithms for Max-CFC, with some modification, also works for Max-UMC.

In the parameterized setting, we study Max-CFC parameterized by solution size.

p-CFC	**Parameter:** k
Input: A hypergraph $(U, \mathcal{F})$ on n vertices and m hyperedges, and positive integers $r \geq 2$ and k.	
Question: Is there a subfamily of at least k hyperedges that can be conflict-free colored using r colors?	

We also study this problem when we restrict the input hypergraph to that induced by the closed/open neighbourhood of a graph G.

Our Results and Methods. In the realm of parameterized algorithm we obtain the following result.

1. We show that the problem is FPT by designing a kernel with at most 4^k vertices and $\mathcal{O}(k \log k)$ hyperedges. The kernel is obtained by finding a novel connection to Unique Coverage problem [10]. We use this one way connection to either say that the given instance for p-CFC is a YES instance or conclude that the number of hyperedges is upper bounded by $\mathcal{O}(k \log k)$. Finally, using extremal results on set-family we bound the number of vertices (elements) to 4^k. Moreover, when we restrict the input hypergraph to that induced by the closed/open neighbourhood of a graph G, then the above imply polynomial kernels for these variants.
2. A direct consequence of our kernel is an $r^{4^k}(n+m)^{\mathcal{O}(1)}$ algorithm for p-CFC. We exploit the fact that the number of hyperedges is at most $\mathcal{O}(k \log k)$ in the reduced instance to design an FPT algorithm with running time $2^{\mathcal{O}(k \log \log k + k \log r)}(n+m)^{\mathcal{O}(1)}$. We arrive at the required algorithm by combining the fact that we have small number of hyperedges and using the technique of color coding introduced in [2] in a non-trivial manner.

Finally, we design an exact algorithm that solves the Max-CFC problem for general hypergraphs. This algorithm exploits structural properties of a YES instance for Max-CFC. Our algorithm runs in $\mathcal{O}(2^{m+n})$ time. The algorithm also works for the Minimum Conflict-Free coloring problem. In particular, for hypergraphs induced by graph neighbourhoods, our algorithm runs in time $\mathcal{O}(4^n)$ which is a non-trivial improvement over the best known exact algorithm that runs in $\mathcal{O}(n^n)$ time [7]. The algorithm is based on dynamic programming combined with an application of subset-convolution. We refer to [6] for a more detailed introduction to exact algorithms. Some minor modifications to our algorithm give an exact algorithm for Unique Maximum Coloring.

2 Preliminaries

A set of consecutive integers $\{1, 2, \ldots n\}$ will be written as $[n]$ in short. We denote the hypergraph as $H = (U, \mathcal{F})$. We refer to the objects in the universe U by either vertices or elements. Furthermore, for a vertex $v \in U$, $\mathsf{deg}(v)$ denotes the number of hyperedges v is part of. The neighbourhood of a vertex $v \in U$, denoted by $N(v)$, is the subfamily of hyperedges in $\mathcal{F}$ that contain v.

Parameterized Algorithms. The instance of a parameterized problem is a pair containing the actual problem instance of size n and a positive integer called a parameter, usually represented as k. The problem is said to be in FPT if there exists an algorithm that solves the problem in $f(k)n^{\mathcal{O}(1)}$ time, where f is a computable function. The problem is said to admit a $g(k)$-sized kernel, if there exists an polynomial time algorithm that converts the actual instance to a reduced instance of size $g(k)$, while preserving the answer. When g is a polynomial function, then the problem is said to admit a polynomial kernel. A reduction rule is a polynomial time procedure that changes a given instance I_1 of a problem Π to another instance I_2 of the same problem Π. We say that the reduction rule is *safe* when I_1 is a YES instance of Π if and only if I_2 is a YES instance. Readers are requested to refer [5] for more details.

Exact Algorithms. Although all NP-Complete problems can be solved by some brute-force algorithm, the running time of these algorithms can be extremely large even for some small input. However, for some these problems, we can design super-polynomial algorithms which are considerably faster than brute-force. Such algorithms which solve NP-Complete problems optimally are called exact algorithms. At times, these may even be practical for moderate or small instance sizes.

3 FPT Algorithm for p-CFC

We are given a hypergraph $\mathcal{H} = (U, \mathcal{F})$ as input and two positive integers, k and r. In this section, we give an FPT algorithm for p-CFC on hypergraphs, parameterized by k. In other words, we wish to find out if k hyperedges can be conflict-free colored using r colors. *For simplicity, throughout this section, we assume that we are given a simple hypergraph, that is no hyperedges are repeated.* We first give a kernel and then use this kernel to get the desired FPT algorithm.

Kernel for P-CFC. We begin with a simple observation that if $r > k$, then we can conflict-free color any subfamily of k edges with r colors. Thus, for the remaining section, we assume that $r \leq k$.

We can also preprocess the input instance to detect simple YES instances of the problem, by applying the following reductions to the instance.

Reduction 1. *If there is a vertex $v \in U$ such that* $\mathsf{deg}(v)$ *is at least k say* YES.

Lemma 1 (†).[1] *Reduction Rule 1 is safe.*

Next, we draw a connection between P-CFC and the UNIQUE HITTING SET (UHS) problem. In UHS, we take a hypergraph H and a positive integer k as input. The question is to decide whether there is a set S of vertices and a subfamily $\mathcal{F}'$ of size at least k such that each hyperedge in $\mathcal{F}'$ contains exactly 1 vertex from S i.e., each hyperedge of $\mathcal{F}'$ needs to be uniquely hit by S.

Observation 1 (†). *Given a hypergraph H and a positive integer k, if (H, k) is a* YES *instance for* UHS, *then $(H, k, r = 2)$ is a* YES *instance for* P-CFC.

The UHS problem, in turn, is related to the UNIQUE COVERAGE (UC) problem. In UC, we take a hypergraph H and a positive integer k as input. The question is to decide whether there is a subfamily $\mathcal{F}'$ of hyperedges and a set S of at least k vertices such that each vertex in S belongs to exactly 1 hyperedge of $\mathcal{F}'$. In other words, each vertex of S needs to be uniquely covered by $\mathcal{F}'$.

Lemma 2 (†). *An instance $(H = (U, \mathcal{F}), k)$ of* UHS *has an equivalent instance $(H' = (\hat{U}, \hat{\mathcal{F}}), k)$ of* UC, *where the parameter remains the same, and $|U| = |\hat{\mathcal{F}}|, |\hat{U}| = |\mathcal{F}|$.*

The UC problem has been studied in the field of parameterized complexity. When k, the number of vertices to be uniquely covered, is the parameter, the problem was shown to be in FPT in [10]. The following Proposition was proved in [10], and we will shortly show how this is useful to us.

Proposition 1 [10, Lemma 17]. *Let $(H = (U, \mathcal{F}), k)$ be an instance of* UC *such that every hyperedge has size at most $k - 1$. Then there exists a constant α_{uc} such that if $|U| \geq \alpha_{uc} k \log k$ then $(H = (U, \mathcal{F}), k)$ is a* YES *instance and furthermore in polynomial time, it is possible to find a subfamily covering at least k elements uniquely.*

We use Proposition 1 to bound the universe size for P-CFC.

Lemma 3 (†). *Let $(H = (U, \mathcal{F}), k, r)$ be an instance of* P-CFC. *Then in polynomial time, either we can conclude that (H, k, r) is a* YES *instance of* P-CFC *or $|\mathcal{F}| \leq \alpha_{uc} k \log k$.*

Thus, from now onwards, we assume our instance to have at most $\mathcal{O}(k \log k)$ hyperedges. Using an extremal result on set systems [9, Theorem 8.12], we obtain the following.

---

[1] Proofs labelled with † can be found in the full version.

Theorem 1 **(†).** P-CFC *has a kernel with at most* 4^k *vertices and* $\mathcal{O}(k \log k)$ *sets.*

Corollary 1. P-CFC *for hypergraphs induced by graph neighborhoods admits polynomial kernels.*

Corollary 1 follows from Lemma 3 and the fact that the number of hyperedges are same as the number of vertices in hypergraphs induced by graph neighborhoods.

Theorem 1 immediately implies that P-CFC is FPT. Given an instance $(H = (U, \mathcal{F}), k, r)$ of P-CFC, by using Theorem 1, we either conclude that $(H = (U, \mathcal{F}), k, r)$ is a YES instance of P-CFC or we have that $|U| \leq 4^k$. Now we look at every r-partition of U and check whether there are k hyperedges that are conflict-free colored. If we succeed for any partition then we return YES, else we conclude that the given instance is a NO instance. The running time of this algorithm is upper bounded by $r^{4^k}(|U| + |\mathcal{F}|)^{\mathcal{O}(1)}$.

Faster FPT algorithm for P-CFC**.** Let $N = |U| + |\mathcal{F}|$. In this section, we give the full description of an FPT algorithm for P-CFC that runs in $2^{\mathcal{O}(k \log\log k + k \log r)} \cdot N^{\mathcal{O}(1)}$ time. We will assume that our input instance contains at most $\mathcal{O}(k \log k)$ hyperedges and 4^k vertices.

Towards this we first define some concepts. Given a set $S \subseteq U$, a subfamily $\mathcal{F}'$, and a coloring $\Gamma : U \rightarrow [r]$, we say that S is a *cfc-solution* if each hyperedge h in $\mathcal{F}'$ is conflict-free colored and a uniquely colored vertex of h belongs to S. Furthermore, given such a set S and a hyperedge h, let $\mathsf{unicolelt}_S(h)$ denote the uniquely colored vertex of h that belongs to S. In what follows we define an auxiliary problem and give an FPT algorithm for this problem. Finally, we reduce our problem to this one with some guesses and by using the color coding technique, introduced by Alon et al. in [2], to obtain the desired algorithm for P-CFC.

PARTITIONED P-CFC **Parameter:** $r + p + \vert\mathcal{F}\vert$
Input: A hypergraph $(U = U_1 \uplus U_2 \cdots U_p, \mathcal{F})$, a function $\Psi_{\mathsf{family}} : \mathcal{F} \rightarrow [r]$, $\Psi_{\mathsf{parts}} : [p] \rightarrow [r]$, a subset $U' \subseteq U$ and a coloring function $\Gamma' : U' \rightarrow [r]$, for every $v \in U - U'$, a list $L_v \subseteq [r]$ **Question:** Does there exist a coloring function $\Gamma : U \rightarrow [r]$ such that: Each hyperedge is conflict-free colored, $\Gamma(U') = \Gamma'(U')$. For each $v \in U - U', \Gamma(v) \in L_v$. Also, there exists a cfc-solution set S of size exactly p, for all $i \in [p]$, $\vert S \cap U_i\vert = 1$ and for every $h \in \mathcal{F}$, $\mathsf{unicolelt}_S(h) \in \bigcup_{j \in \Psi^{-1}_{\mathsf{parts}}(\Psi_{\mathsf{family}}(h))} U_j$?

In simple words, the problem definition can be explained as follows. We are given a partitioning of the universe U into p-parts and a partial coloring function Γ' on a subset U'. We are looking for a coloring $\Gamma : U \rightarrow [r]$ which extends Γ'. Each vertex v in $U - U'$ has a list of admissible colors, and Γ must choose a color from L_v. Also, due to Γ, each of the hyperedge is conflict free colored and there exists a cfc-solution set S such that it contains exactly one vertex from each part. Suppose the hypothetical set S be $\{x_1, x_2, \ldots, x_p\}$ (think of x_i as some kind of variables) where $x_i \in U_i$. The function Ψ_{parts} is used to guess the color of x_i in Γ. The function Ψ_{family} divides the family $\mathcal{F}$ into r *chunks* (not to be confused with parts and coloring). The idea is that the uniquely colored vertex of $h \in \mathcal{F}$, say x_j, has been assigned the same color by Γ as h has been assigned to the chunk number by Ψ_{family}, i.e., $\Gamma(x_j) = \Psi_{\mathsf{family}}(h)$. Next we show how we can solve the PARTITIONED P-CFC problem.

Given an instance $((U = U_1 \uplus U_2 \cdots U_p, \mathcal{F}'), \Psi_{\mathsf{family}}, \Psi_{\mathsf{parts}}, U', \Gamma', \{L_v \subseteq [r] | v \in U - U'\})$ of PARTITIONED P-CFC, we first do a polynomial time preprocessing of the instance. For all $v \in U'$, we must set $\Gamma(v) = \Gamma'(v)$. In the following Reduction Rules, we show that the input functions Ψ_{family} and Ψ_{parts} allow us to prune the list of some of the vertices. The first reduction rule deals with hyperedges h where $|\Gamma'^{-1}(\Psi_{\mathsf{family}}(h)) \cap h| = 1$

Reduction 2. *Suppose there is a hyperedge h containing $w \in U'$ such that $\Psi_{\mathsf{family}}(h) = \Gamma'(w)$. Then, for every $v \in h - \{w\}$ we delete $\Psi_{\mathsf{family}}(h)$ from L_v. We delete h from $\mathcal{F}$.*

Lemma 4 (†). *Reduction Rule 2 is safe.*

Reduction 3. *If there is a vertex $v \in U_i, i \in [p]$, and $h \in \mathcal{F}$, such that $v \in h$, $\Psi_{\mathsf{family}}(h) \neq \Psi_{\mathsf{parts}}(i)$, then we remove the color $\Psi_{\mathsf{family}}(h)$ from the list of v.*

Lemma 5 (†). *Reduction Rule 3 is safe.*

The next rule deals with hyperedges h where $|\Gamma'^{-1}(\Psi_{\mathsf{family}}(h)) \cap h| \geq 2$.

Reduction 4. *If there are two vertices $v, w \in U'$ and a hyperedge $h \in \mathcal{F}$, such that $\Psi_{\mathsf{family}}(h) = \Gamma'(v) = \Gamma'(w)$, then we say* NO.

Lemma 6 (†). *Reduction Rule 4 is safe.*

Reduction 5. *Suppose there is a vertex $w \in U - U'$ with $L_w = \{c\}$, then we put w in U' and set $\Gamma'(w) = c$. If there is a vertex v where $L_v = \emptyset$, then we say* NO.

Lemma 7 (†). *Reduction Rule 5 is safe.*

Given an instance $((U = U_1 \uplus U_2 \cdots U_p, \mathcal{F}'), \Psi_{\mathsf{family}}, \Psi_{\mathsf{parts}}, U', \Gamma', \{L_v \subseteq [r] | v \in U - U'\})$ of PARTITIONED P-CFC, we apply Reduction Rules 2, 3, 4, 5 exhaustively. If in the process we infer that the given instance is a NO instance then we return the same. It could also happen that we get $\mathcal{F} = \emptyset$. In this case for every vertex $v \in U - U'$, Γ assigns to v an element of $L(v)$ arbitrarily. Thus, from now onwards we assume that neither we obtain that the given instance is a NO instance nor that $\mathcal{F} = \emptyset$. We call an instance of PARTITIONED P-CFC *reduced* if Reduction Rules 2, 3, 4, 5 are not applicable. For simplicity, let $((U = U_1 \uplus U_2 \cdots U_p, \mathcal{F}'), \Psi_{\mathsf{family}}, \Psi_{\mathsf{parts}}, U', \Gamma', \{L_v \subseteq [r] | v \in U - U'\})$ denote the reduced instance of PARTITIONED P-CFC. Observe that the reduced instance have the following properties:

1. For every vertex v, $|L_v| \geq 2$.
2. For every hyperedge h, $|\Gamma'^{-1}(\Psi_{\mathsf{family}}(h)) \cap h| = 0$.

We define the set $V_i \subseteq U - U'$ as the set of vertices that have i in their list of admissible colors. Then there are two kinds of vertices in V_i: It could be that the vertex v has $i \in L_v$ and $\exists h \in \mathcal{F}$, $v \in U_j \cap h$ such that $\Psi_{\mathsf{family}}(h) = i, \Psi_{\mathsf{parts}}(j) = i$. Or, the vertex v has $i \in L_v$. Also, for any h with $\Psi_{\mathsf{family}}(h) = i, v \notin h$.

To solve the reduced instance of PARTITIONED P-CFC, we will solve some r instances of an even more specialized problem that we define now. Let PARTITIONED UHS be the problem of determining, for a given partition $U_1 \uplus \ldots \uplus U_q$ of the universe and a family $\mathcal{F}$, whether there is a set S of vertices that uniquely hits all hyperedges of the input hypergraph (that is, for all $h \in \mathcal{F}$, $|h \cap S| = 1$) and where $\forall i \in [q]$, $|U_i \cap S| = 1$. Now we define some sets based on $V_i \subseteq U$:

1. For every $j \in [r]$, and $x \in \Psi^{-1}_{\mathsf{parts}}(j)$ let $Z^x_j = U_x \cap V_j$ and $Z_j = \bigcup_{x \in \Psi^{-1}_{\mathsf{parts}}(j)} Z^x_j$.
2. For every $j \in [r]$, and $h \in \Psi^{-1}_{\mathsf{family}}(j)$ let $h_j = h \cap V_j$ and $\mathcal{F}_j = \{h_j \mid h \in \Psi^{-1}_{\mathsf{family}}(j)\}$.

Next we relate the instance of PARTITIONED P-CFC to PARTITIONED UHS.

Lemma 8. *Let* $((U = U_1 \uplus U_2 \cdots U_p, \mathcal{F}'), \Psi_{\mathsf{family}}, \Psi_{\mathsf{parts}}, U', \Gamma', \{L_v \subseteq [r] | v \in U - U'\})$ *denote the reduced instance of* PARTITIONED P-CFC. *Then it is a* YES *instance of* PARTITIONED P-CFC *if and only if for all* $j \in [r]$, $(\biguplus_{x \in \Psi^{-1}_{\mathsf{parts}}(j)} Z^x_j, \mathcal{F}_j)$ *is a* YES *instance of* PARTITIONED UHS.

Proof. First, suppose that $((U = U_1 \uplus U_2 \cdots U_p, \mathcal{F}'), \Psi_{\mathsf{family}}, \Psi_{\mathsf{parts}}, U', \Gamma', \{L_v \subseteq [r] | v \in U - U'\})$ is a YES instance of PARTITIONED P-CFC. Then there is a satisfying assignment Γ such that each hyperedge is conflict-free colored, $\Gamma'(U') = \Gamma(U')$. For each $v \in U - U', \Gamma(v) \in L_v$. Also, there exists a cfc-solution set $S = \{v_1, \ldots, v_p\}$ such that for all $i \in [p]$, $|S \cap U_i| = 1$. In the reduced instance, for all h, $|\Gamma'^{-1}(\Psi_{\mathsf{family}}(h)) \cap h| = 0$. Thus, $S \cap U' = \emptyset$. For each $i \in [r]$, we look at $S \cap V_i$. By definition of Z_i, every vertex in $S \cap V_i$ must belong to a part in Z_i. In particular, every vertex of $S \cap \Gamma^{-1}(i)$ must belong to a part in Z_i. Also, since every vertex of S belongs to a unique part of $U_1 \uplus U_2 \cdots U_p$, there is exactly one vertex in $S \cap Z^x_i$, for each $Z^x_i \in Z_i$. Also, we know that for every $h \in \mathcal{F}$, if $\mathsf{unicolelt}_S(h) = v_j$, then $\Gamma(v_j) = \Psi_{\mathsf{family}}(h)$. Thus, for each hyperedge $h \in \mathcal{F}_i$, $\mathsf{unicolelt}_S(h) \in S \cap \Gamma^{-1}(i)$. F or every other vertex $u \in h - \mathsf{unicolelt}_S(h)$, $\Gamma(u) \neq i$ and therefore $u \notin S \cap \Gamma^{-1}(i)$. Thus, for every $i \in [r]$, $S_i = S \cap \Gamma^{-1}(i)$ is a unique hitting set of $\mathcal{F}_j$ with the property that $\forall x \in \Psi^{-1}_{\mathsf{parts}}(i)$, $|Z^x_i \cap S_i| = 1$. Thus, $(\biguplus_{x \in \Psi^{-1}_{\mathsf{parts}}(j)} Z^x_j, \mathcal{F}_j)$ is a YES instance of PARTITIONED UHS.

In the reverse direction, suppose $(\biguplus_{x \in \Psi^{-1}_{\mathsf{parts}}(j)} Z^x_j, \mathcal{F}_j)$ is a YES instance of PARTITIONED UHS. Then a solution set S_i is a unique hitting set of $\mathcal{F}_j$ with the property that $\forall x \in \Psi^{-1}_{\mathsf{parts}}(i)$, $|Z^x_i \cap S_i| = 1$. By definition, $S_i \subseteq Z_i \subseteq V_i$. First, for each vertex $v \in S_i$, we assign $\Gamma(v) = i$. For each $w \in U'$, must we set $\Gamma(w) = \Gamma'(w)$. Now, we look at a vertex $w \in V_i - S_i$. Look at the colors in $L_w - \{i\}$. In the reduced instance, it must be the case that, for any h with $\Psi_{\mathsf{family}}(h) \neq i$, $w \notin h$. For a vertex $w \in (U - U') - \bigcup_{j \in [r]} S_i$, we arbitrarily pick a color $c \in L_w - \{i\}$ and set $\Gamma(w) = c$. Every hyperedge h has exactly one vertex in the color class $\Psi_{\mathsf{family}}(h)$, namely the vertex in $S_{\Psi_{\mathsf{family}}(h)} \cap h$ that uniquely hit h. Thus, Γ is a satisfying assignment and $((U = U_1 \uplus U_2 \cdots U_p, \mathcal{F}'), \Psi_{\mathsf{family}}, \Psi_{\mathsf{parts}}, U', \Gamma', \{L_v \subseteq [r] | v \in U - U'\})$ is a YES instance of PARTITIONED P-CFC. $\square$

Lemma 8 allows us to reduce an instance of the PARTITIONED P-CFC problem to r instances of PARTITIONED UHS. Next, we design an algorithm for PARTITIONED UHS.

Lemma 9 (†). PARTITIONED UHS, *where the number of hyperedges is* m, *the universe size is* n *and a* $q \leq m$ *partitioning of the universe is given, is* FPT *parameterized by* m. *The running time of the algorithm is* $4^m \cdot (n+m)^{\mathcal{O}(1)}$.

Lemmata 8, 9 and safeness of the Reduction Rules 2, 3, 4, 5 together result in the following algorithm for PARTITIONED P-CFC.

Lemma 10. PARTITIONED P-CFC *can be solved in time* $2^{p+|\mathcal{F}|} \cdot N^{\mathcal{O}(1)}$.

We give an algorithm for P-CFC using Lemma 10 and the method of color coding technique of [2]. For this we need the notion of a Perfect Hash Family. A Perfect Hash Family is a family of functions, whose domain is a universe U of n elements and range is a set of k elements, and with the following property: for every k-sized subset $S \subseteq U$, there is a function ζ in the family that maps S to the range injectively. That is, every element of S maps to a different number in $[k]$. The following Proposition shows that such families are constructive [11].

Proposition 2. *For any* n *and* $k \leq n$, *a* (n,k)*-Perfect Hash Family of size* $e^k k^{\mathcal{O}(\log k)} \log n$ *can be deterministically computed in time* $e^k k^{\mathcal{O}(\log k)} n \log n$.

Our main theorem is the following.

Theorem 2. P-CFC *can be solved in time* $2^{\mathcal{O}(k \log\log k + k \log r)} \cdot N^{\mathcal{O}(1)}$.

Proof. Let $((U,\mathcal{F}),k,r)$ be an instance of P-CFC. Recall that $|U| = n$, $|\mathcal{F}| = m$ and $N = n + m$. Given an instance we first apply Theorem 1 and obtain an equivalent instance with at most 4^k vertices and $\mathcal{O}(k \log k)$ hyperedges. We run through all $p \leq k$. Since the number of hyperedges in the input instance is $\alpha_{uc} k \log k$, the number of subfamilies of size k is $\binom{\alpha_{uc} k \log k}{k} \leq \frac{(\alpha_{uc} k \log k e)}{k}^k \leq (\alpha_{uc} \log k)^k$. We guess a subfamily $\mathcal{F}'$ of hyperedges that will be conflict free colored. That is, we are trying to find a coloring $\Gamma : U \to [r]$ such that each hyperedge h in $\mathcal{F}'$ is conflict-free colored. Let S be a hypothetical cfc-solution corresponding to it. In other words, for each hyperedge h in $\mathcal{F}'$, a uniquely colored vertex of h (with respect to Γ) belongs to S. We guess the size of $|S|$, say $p \leq k$. For a fixed p, let $\mathfrak{F}$ be the family of (n,p)-Perfect Hash Family of size $e^p p^{\mathcal{O}(\log p)} \log n$. By the property of $\mathfrak{F}$, we know that there exists a function $\zeta \in \mathfrak{F}$ that maps S to $[p]$ injectively. Let $U_1, \ldots, U_p$ denote the partition of U given by ζ. Observe that after this we will be seeking for a cfc-solution S such that $|S \cap U_i| = 1$ for all $i \in [p]$.

Next for each hyperedge h in $\mathcal{F}'$, we guess the color of a vertex in h that is uniquely colored by Γ. There are r^k such guesses. Thus, after this guess, we define a function $\Psi_{\mathsf{family}} : \mathcal{F}' \to [r]$ such that h is assigned the color of the vertex in h that will be uniquely colored by Γ. Finally, for the potential solution set S we guess the color of each vertex given by Γ. Since we are looking for a cfc-solution set S, such that $\forall i \in [p], |U_i \cap S| = 1\}$ it is equivalent to say that we guess an r partitioning of the p parts in $U = (U_1, \ldots, U_p)$. That is, the vertex of $S \in U_i$ will be assigned to each color by Γ. To express this guess, we define another function $\Psi_{\mathsf{parts}} : [p] \to [r]$ such that $\Psi_{\mathsf{parts}}(j) = i$ if the vertex x in $S \cap U_j$ will have $\Gamma(x) = i$. Thus, there are r^p guesses for the coloring of the potential solution set S by Γ. At the end of this sequence of guesses, we have fixed a choice of hyperedges that are to be r conflict-free colored, a coloring of the potential solution set S (without actually knowing the vertices of S, this essentially means a partitioning of the parts of U) and a partitioning of the hyperedges according to which color of Γ will determine that the hyperedge is conflict-free colored. This results in the following instance of PARTITIONED P-CFC: $((U = U_1 \uplus U_2 \cdots U_p, \mathcal{F}'), \Psi_{\mathsf{family}}, \Psi_{\mathsf{parts}}, U' = \emptyset, (\forall v \in U : L_v = [r]))$. By Lemma 10 we know that we can solve this in time $2^{p+k} \cdot N^{\mathcal{O}(1)} \leq 4^k \cdot N^{\mathcal{O}(1)}$. Thus the overall running time for P-CFC is upper bounded by the number of guesses and the running time of an algorithm for PARTITIONED P-CFC. Thus, the running time of the algorithm is upper bounded by:

$$\binom{\alpha_{uc} k \log k}{k} \times k \times |\mathfrak{F}| \times r^k \times r^k \times 4^k \cdot N^{\mathcal{O}(1)} = 2^{\mathcal{O}(k \log\log k + k \log r)} \cdot N^{\mathcal{O}(1)}.$$

□

4 Exact Algorithm for Max-Conflict Free Coloring

In this section, we give an exact algorithm for solving MAX-CFC for hypergraphs. We give a recurrence on subproblems, using which we can give a dynamic programming algorithm to solve the problem. However, a much faster algorithm can be designed using the technique of subset convolutions on functions.

Theorem 3 (†). MAX-CFC *for hypergraphs can be solved by an exact algorithm that runs in* $\mathcal{O}(2^{(m+n)})$ *time.*

Proof Outline. Let $H = (U, \mathcal{F})$ be the input hypergraph. Suppose, for a given hypergraph, there is a procedure to decide whether there exists an r-coloring that is conflict-free. Then, we can generate all subsets $\mathcal{F}'$ of $\mathcal{F}$, such that there exists an r-coloring of vertices of $(U(\mathcal{F})', \mathcal{F}')$ that is conflict-free, by running this procedure for all subsets. Then solving the MAX-CFC problem reduces to picking the maximum sized subsets among those.

We now give a procedure to find the minimum number of colors required to conflict-free color a given hypergraph, $(U', \mathcal{F}')$. Let χ' be a r-coloring on U' and let $\mathcal{F}'$ be conflict-free colored by χ'. Then χ' partitions U' into r partitions, $U_1, U_2, \ldots, U_r$, such that the following property is true.

$$\forall F \in \mathcal{F}', \exists i \in [r] \text{ such that } |F \cap U_i| = 1.$$

Let $\mathcal{F}_1$ be the set of hyperedges such that $\forall F \in \mathcal{F}_1, |F \cap U_1| = 1$ i.e., all the hyperedges in $\mathcal{F}_1$ have a unique vertex colored by color 1. Then, if we correctly guessed U_1 then solving whether $\mathcal{F}'$ has an r conflict-free coloring in U is equivalent to solving whether $\mathcal{F}' \setminus \mathcal{F}_1$ has an $r - 1$ conflict-free coloring in $U \setminus U_1$.

Let $\mathcal{C}(X, \mathcal{E})$ be the minimum number of colors needed to conflict-free color the hypergraph $(X, \mathcal{E})$. We give the following recurrence relation to find $\mathcal{C}(X, \mathcal{E})$.

$$\mathcal{C}(X, \mathcal{E}) = \begin{cases} \min\limits_{X' \subseteq X : \exists h \in \mathcal{E}, |h \cap X'| = 1} \{1 + \mathcal{C}(X \setminus X', \mathcal{E} \setminus \mathcal{E}')\}, & \text{if } X \neq \phi \\ 0, & \text{if } X = \phi \end{cases} \tag{1}$$

where $\mathcal{E}' = \{h \in \mathcal{E} | |h \cap X'| = 1\}$. We use the above recurrence to get first an algorithm with running time $\mathcal{O}(3^n 2^m)$ and then using the method of subset-convolution speed up the running time to $\mathcal{O}(2^{n+m})$. □

It is to be noted that by setting $r = n$, $\mathcal{C}(V, \mathcal{F})$ returns the minimum number of colors required to conflict-free color the given hypergraph.

Corollary 2. *Given a hypergraph H, $\chi_{cf}(H)$ can be found in $\mathcal{O}(2^n 2^m)$ time.*

Corollary 3. *The* Max-CFC *problem on hypergraphs induced by neighbourhoods of graphs can be solved in $\mathcal{O}(4^n)$ time.*

References

1. Ajwani, D., Elbassioni, K., Govindarajan, S., Ray, S.: Conflict-free coloring for rectangle ranges using $O(n^{.}382)$ colors. Discrete Comput. Geom. **48**(1), 39–52 (2012)
2. Alon, N., Yuster, R., Zwick, U.: Color-coding. J. ACM **42**(4), 844–856 (1995)
3. Cheilaris, P., Tóth, G.: Graph unique-maximum and conflict-free colorings. J. Discrete Algorithms **9**(3), 241–251 (2011)
4. Even, G., Lotker, Z., Ron, D., Smorodinsky, S.: Conflict-free colorings of simple geometric regions with applications to frequency assignment in cellular networks. In: 2002 Proceedings of the 43rd Annual IEEE Symposium on Foundations of Computer Science, pp. 691–700 (2002)
5. Flum, J., Grohe, M.: Parameterized Complexity Theory. Texts in Theoretical Computer Science. An EATCS Series. Springer-Verlag, Berlin (2006)
6. Fomin, F.V., Kratsch, D.: Exact Exponential Algorithms. Springer, New York (2010)
7. Gargano, L., Rescigno, A.A.: Complexity of conflict-free colourings of graphs. Theoret. Comput. Sci. **566**, 39–49 (2015)
8. Har-Peled, S., Smorodinsky, S.: Conflict-free coloring of points and simple regions in the plane. Discrete Comput. Geom. **34**(1), 47–70 (2005)
9. Jukna, S.: Extremal Combinatorics: With Applications in Computer Science. Springer Science & Business Media, New York (2011)
10. Misra, N., Moser, H., Raman, V., Saurabh, S., Sikdar, S.: The parameterized complexity of unique coverage and its variants. Algorithmica **65**(3), 517–544 (2013)

11. Naor, M., Schulman, L.J., Srinivasan, A.: Splitters and near-optimal derandomization. In: 36th Annual Symposium on Foundations of Computer Science, 23–25 October 1995, Milwaukee, Wisconsin, pp. 182–191 (1995)
12. Pach, J., Tardos, G.: Conflict-free colourings of graphs and hypergraphs. Comb. Probab. Comput. **18**(05), 819–834 (2009)
13. Smorodinsky, S.: On the chromatic number of geometric hypergraphs. SIAM J. Discrete Math. **21**(3), 676–687 (2007)

Computational Geometry II

Geometric Matching Algorithms for Two Realistic Terrains

Sang Duk Yoon(✉), Min-Gyu Kim, Wanbin Son, and Hee-Kap Ahn

Department of Computer Science and Engineering, POSTECH, Pohang, Korea
{egooana,alsrbbk,mnbiny,heekap}@postech.ac.kr

Abstract. We consider a geometric matching of two realistic terrains, each of which is modeled as a piecewise-linear bivariate function. For two realistic terrains f and g where the domain of g is relatively larger than that of f, we seek to find a translated copy f' of f such that the domain of f' is a sub-domain of g and the L_∞ or the L_1 distance of f' and g restricted to the domain of f' is minimized. In this paper, we show a tight bound on the number of different combinatorial structures that f and g can have under translation in their projections on the xy-plane. We give a deterministic algorithm and a randomized algorithm that compute an optimal translation of f with respect to g under L_∞ metric. We also give a deterministic algorithm that computes an optimal translation of f with respect to g under L_1 metric.

1 Introduction

In the terrain matching problem, we are given two terrains and the goal is to measure the similarity between two terrains. Terrain matching has been extensively used for various applications to locate the exact position of objects such as aircrafts [5,10,15], cruise missiles [3,6], underwater vehicles [13,16,17], rockets and robots for space missions [7,14].

In these applications, terrain matching is used to specify the location of an object by constructing local terrain data around the object and finding the most similar sub-terrain in the existing global terrain data. A typical method to find the most similar sub-terrain is feature matching. Well known examples of features are linear edges, $2D$ curves, contour lines and Gaussian curvatures [5,6,10]. These features describe some characteristics of a terrain, but may not fully reflect the geometric properties of the terrain.

Moroz and Aronov [12] and Agarwal et al. [1] dealt with terrain matching as a geometric matching problem. They defined a terrain f as a piecewise-linear bivariate function $f : \mathbb{D}_f \rightarrow \mathbb{R}$, where $\mathbb{D}_f$ is a triangulated domain of f in the xy-plane. For each vertex of the triangulation, the function value $f(v)$ is given, and the other values are given by the linear interpolation within each triangle. They

This work was supported by the National Research Foundation of Korea(NRF) grant funded by the Korea government(MSIP) (No. 2011-0030044).

K. Elbassioni and K. Makino (Eds.): ISAAC 2015, LNCS 9472, pp. 285–295, 2015.
DOI: 10.1007/978-3-662-48971-0_25

gave algorithms that compute exact L_∞, L_1 (under vertical scaling and translation) and L_2 distances between two terrains, respectively. These algorithms only handle two input terrains defined on the same domain, and to the best of our knowledge, there is no research about matching two triangulated terrains defined on different domains.

We deal with a geometric matching problem concerning two terrains defined by the piecewise-linear bivariate functions on triangulated domains, but in this paper we do not require that two terrains have the same domain. Let $\mathbb{D}_f + t = \{p+t \mid p \in \mathbb{D}_f\}$ be a translated image of $\mathbb{D}_f$ by a translation vector $t \in \mathbb{R}^2$. We define the distance between two terrains as follows.

Definition 1. *Let $f : \mathbb{D}_f \to \mathbb{R}$ and $f' : \mathbb{D}_{f'} \to \mathbb{R}$ be two terrains such that $\mathbb{D}_{f'} = \mathbb{D}_f + t$ for a translation vector $t \in \mathbb{R}^2$. The distance $d_\infty(f, f')$ between f and f' under L_∞ metric is*

$$\min_{h\in\mathbb{R}} \max_{p\in\mathbb{D}_f} |(f(p)+h) - f'(p+t)|$$

and the distance $d_1(f, f')$ between f and f' under L_1 metric is

$$\min_{h\in\mathbb{R}} \iint_{p\in\mathbb{D}_f} |(f(p)+h) - f'(p+t)|dp.$$

We use two different distances to measure the similarity between two terrains. The distance d_∞ measures the min-max vertical distance under vertical translation h and the distance d_1 measures the minimum volume under vertical translation h.

Our problem can be stated as follows:

Problem (Terrain Matching). Given two terrains $f : \mathbb{D}_f \to \mathbb{R}$ and $g : \mathbb{D}_g \to \mathbb{R}$, find an optimal translation vector t^* such that $\mathbb{D}^* = \mathbb{D}_f + t^* \subset \mathbb{D}_g$ and the distance between f and $g \restriction_{\mathbb{D}^*}$ is minimized, where $g \restriction_{\mathbb{D}^*}$ denotes the restriction of g to $\mathbb{D}^*$.

In many computational geometric problems, there is a certain gap between the worst-case computational complexity of an algorithm and the actual running time of the algorithm on inputs from real world applications [11]. The same phenomenon happens for the terrain matching problem. So, it is an important issue to develop an algorithm that is efficient for a realistic input. We assume that our input terrains satisfy some realistic constraints. A *realistic terrain* is a terrain with three additional constraints. In the following definition, k and r are assumed to be positive constants.

Definition 2 [11]. *A terrain $f : \mathbb{D}_f \to \mathbb{R}$ is a realistic terrain if it satisfies the followings:*

(a) *The triangulation of $\mathbb{D}_f$ is a k-low-density triangulation.*
(b) *For the smallest rectangle that contains $\mathbb{D}_f$, the ratio of the length of a short side to the length of a long side is $1 : r$.*
(c) *The longest edge in the triangulation of $\mathbb{D}_f$ is at most constant times as long as the shortest one.*

A planar triangulation $\mathcal{T}$ is called a *k-low-density triangulation* if for any axis-aligned square R with side length s, the number of edges of $\mathcal{T}$ with length greater than or equal to s that intersect R is at most k. Also, by Definition 2(*b*), we assume that the domain of a realistic terrain is an axis-aligned rectangle with constant side-length ratio.

1.1 Our Results

We present first algorithms for matching two triangulated terrains with different domains and also show geometric properties between two realistic terrains. To solve the terrain matching problem under L_∞ metric, we first gather two-dimensional translation vectors t such that $\mathbb{D}_f + t \subset \mathbb{D}_g$. Then, we subdivide the set of translation vectors into a partition such that the translation vectors t of a cell of the partition correspond to "one combinatorial structure" between two triangulations of $\mathbb{D}_f + t$ and $\mathbb{D}_g$. For each cell of the partition, we can find a translation vector that minimizes the distance among the translation vectors in the cell by reducing it to a linear programming problem.

To solve the terrain matching problem under L_1 metric, we need to treat the amount of vertical translation h of f explicitly. After we concatenate h as a third coordinate of the two-dimensional translation vectors $t = (t_x, t_y)$, we subdivide the set of three-dimensional translation vectors (t_x, t_y, h) into a partition such that the volume function between f and g for the translation vectors (t_x, t_y, h) of a cell can be expressed by a single formula. For each cell of the partition, we find a three-dimensional translation vector that minimizes the distance among the translation vectors of the cell by using numerical methods.

Our results are twofold: Let m (resp. n) be the number of triangles in the triangulation of $\mathbb{D}_f$ (resp. $\mathbb{D}_g$), assuming that $m \leq n$. The side lengths of $\mathbb{D}_f$ (resp. $\mathbb{D}_g$) are a and $\mathsf{a}r$ (resp. a' and $\mathsf{a}'r'$) for a positive constant r (resp. r'). Let $A = m + (\frac{\mathsf{a}}{\mathsf{a}'})^2 n$.

1. Under L_∞ metric,
   - We show that the number of different combinatorial structures between the triangulation of $\mathbb{D}_f$ and the triangulation of $\mathbb{D}_g$ is $O(nmA)$, and this bound is tight if $\mathsf{a}' > 2\mathsf{a}$.
   - We present a deterministic algorithm and a randomized algorithm for the terrain matching problem. The deterministic algorithm runs in $O(nmA^{4/3+\delta})$ time using $O(n + A^2)$ space for a fixed $\delta > 0$, and the randomized algorithm runs in $O(nmA \log n \log^2 A \log^2 \log A)$ expected time using $O(n+A^2)$ space. The randomized algorithm outperforms the deterministic algorithm when $m = \Omega(\log^3 A)$.

  - The time complexity of the randomized algorithm is near linear to the number of different combinatorial structures. It seems hard to avoid searching the whole combinatorial structures, so our algorithms run reasonably fast.
2. Under L_1 metric,
  - We show that the number of different formulae of the volume function defined between f and g is $O(nmA^4)$.
  - We present a deterministic algorithm for the terrain matching problem that runs in $O(nmA^4)$ time using $O(n + A^3)$ space.

The factor $A = (m + (\frac{\mathsf{a}}{\mathsf{a}'})^2 n)$ is $O(m)$ when $m : n \approx \mathsf{a}^2 : \mathsf{a}'^2$. This condition holds when edge lengths of triangles in both realistic terrains are asymptotically same. Many real world applications use terrains with this condition to match. With $A = O(m)$, the running times and the space complexities of our algorithms are linear to n (except the randomized algorithm); it means that our algorithms can be used as a query algorithm for finding the most similar part of large terrain data g for query terrain data f which runs in time linear to the size of the database.

Due to lack of space, all proofs are omitted but can be found in the full version.

2 Translation Space Under L_∞ Metric

Let $\mathcal{S}$ be a set of translation vectors t that satisfies $\mathbb{D}_f + t \subset \mathbb{D}_g$, i.e., $\mathcal{S} = \{t \in \mathbb{R}^2 \mid \mathbb{D}_f + t \subset \mathbb{D}_g\}$. We call $\mathcal{S}$ the *translation space* of f and g. As mentioned before, our goal is to find an optimal translation vector in $\mathcal{S}$. To find it among infinitely many translation vectors in $\mathcal{S}$, we need to investigate geometric properties of input terrains.

Let the triangulations of $\mathbb{D}_f$ and $\mathbb{D}_g$ be $\mathcal{T}_f$ and $\mathcal{T}_g$, respectively. In this section, we show that $\mathcal{S}$ can be subdivided into the finite number of cells such that the interior of each cell induces the same combinatorial structure of the overlay of the triangulations. Then we show a tight upper bound on the number of the combinatorially different sets of translation vectors.

2.1 Candidate Pairs Defining the Distance Between Two Terrains

For a translation vector t, let $\mathcal{O}(f, g, t)$ be the overlay of $\mathcal{T}_f + t$ and $\mathcal{T}_g$ where $\mathcal{T}_f + t = \{p + t \mid p \in \mathcal{T}_f\}$ be a translated image of $\mathcal{T}_f$ (Fig. 1). The following lemma shows that there is a vertex $v \in \mathcal{O}(f, g, t)$ that realizes the distance between f and $g \restriction_{\mathbb{D}_f + t}$, i.e., $d_\infty(f, g \restriction_{\mathbb{D}_f + t}) = |(f(v - t) + h') - g(v)|$, where $h' = \operatorname*{argmin}_{h \in \mathbb{R}} \max_{p \in \mathbb{D}_f} |(f(p) + h) - g(p + t)|$.

Lemma 1. *There is a vertex of $\mathcal{O}(f, g, t)$ that realizes $d_\infty(f, g \restriction_{\mathbb{D}_f + t})$ for any translation vector $t \in \mathcal{S}$.*

Each vertex of $\mathcal{O}(f, g, t)$ corresponds to a vertex-triangle pair or an edge-edge pair of two triangulations $\mathcal{T}_f + t$ and $\mathcal{T}_g$. We define the *combinatorial structure* $\mathcal{C}(t)$ between $\mathcal{T}_f$ and $\mathcal{T}_g$ at $t \in \mathcal{S}$ as the set of these pairs between $\mathcal{T}_f + t$ and $\mathcal{T}_g$.

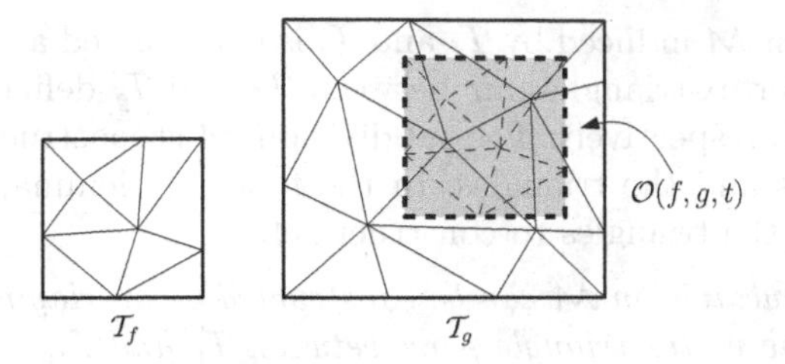

Fig. 1. An example of the overlay $\mathcal{O}(f,g,t)$.

2.2 Subdividing Translation Space

Now we describe how to subdivide $\mathcal{S}$ into cells such that $\mathcal{C}(t) = \mathcal{C}(t')$ for any two translation vectors t and t' in the interior of a cell. We denote by $\mathcal{M}$ the subdivision. The combinatorial structure corresponding to an edge or a vertex of $\mathcal{M}$ is the union of the combinatorial structures of the adjacent cells of the edge or the vertex in $\mathcal{M}$.

Let us consider the different combinatorial structures induced by two triangles $\triangle$ and $\triangle'$ from $\mathcal{T}_f$ and $\mathcal{T}_g$, respectively (Fig. 2(a)). Let $\mathcal{S}_\triangle = \{t \in \mathbb{R}^2 \mid (\triangle + t) \cap \triangle' \neq \emptyset\}$ (the gray region in Fig. 2(b)). For an edge-edge pair (e, e') where e and e' are edges of $\triangle$ and $\triangle'$, respectively, the set of translation vectors t such that $e + t$ and e' intersect forms a parallelogram in $\mathcal{S}_\triangle$ (Fig. 2(b)). For a vertex-triangle pair $(\triangle, v)$ where v is a vertex of $\triangle'$, the set of translation vectors t such that $\triangle + t$ and v intersect forms a triangle. Similarly, for a vertex-triangle pair $(v, \triangle')$ where v is a vertex of $\triangle$, the set of translation vectors such that $v + t$ and $\triangle'$ intersect forms a triangle. We say that the parallelogram in $\mathcal{S}_\triangle$ is defined by an edge-edge pair and the triangle in $\mathcal{S}_\triangle$ is defined by a vertex-triangle pair. The overlay of the parallelograms and triangles defined by all edge-edge and vertex-triangle pairs between $\triangle$ and $\triangle'$, respectively, subdivides $\mathcal{S}_\triangle$ into cells such that the interior of each cell corresponds to exactly one combinatorial structure between $\triangle$ and $\triangle'$ (Fig. 2(c)).

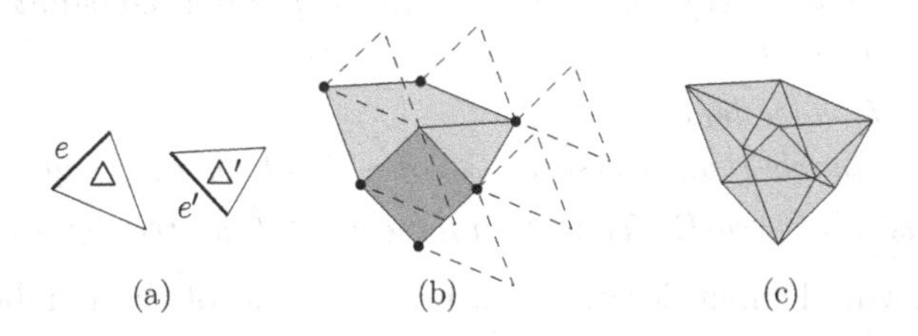

Fig. 2. (a) Two triangles $\triangle$ and $\triangle'$, and two edges e and e' of them. (b) The set of translation vectors t such that $e + t \cap e' \neq \emptyset$ in $\mathcal{S}_\triangle$ (darker parallelogram). (c) The resulting subdivision of $\mathcal{S}_\triangle$.

The subdivision $\mathcal{M}$ induced by $\mathcal{T}_f$ and $\mathcal{T}_g$ is constructed as follows. An edge-edge pair and a vertex-triangle pair between $\mathcal{T}_f$ and $\mathcal{T}_g$ define a parallelogram and a triangle in $\mathcal{S}$, respectively. The subdivision $\mathcal{M}$ is constructed by overlaying the parallelograms and the triangles. In the following lemma, we show that it suffices to overlay the triangles to construct $\mathcal{M}$.

Lemma 2. *The subdivision $\mathcal{M}$ can be constructed by overlaying the triangles in $\mathcal{S}$ defined by all the vertex-triangle pairs between $\mathcal{T}_f$ and $\mathcal{T}_g$.*

We propose the simplified way to construct $\mathcal{M}$ as follows. The triangle in $\mathcal{S}$ defined by a vertex of $\mathcal{T}_f$ and a triangle $\triangle$ of $\mathcal{T}_g$ is a translated copy of $\triangle$. For a vertex v of $\mathcal{T}_f$, the set of triangles in $\mathcal{S}$ that are defined by v and the triangles in $\mathcal{T}_g$ forms a translated copy of $\mathcal{T}_g$. Let $\mathcal{T}_g(v)$ be the translated copy of $\mathcal{T}_g$ formed by a vertex v of $\mathcal{T}_f$. The triangle in $\mathcal{S}$ defined by a vertex of $\mathcal{T}_g$ and a triangle $\triangle$ of $\mathcal{T}_f$ is a translated copy of $-\triangle$, where $-\triangle$ is the reflection through the origin of $\triangle$. For a vertex u of $\mathcal{T}_g$, the set of triangles in $\mathcal{S}$ that are defined by u and the triangles in $\mathcal{T}_f$ forms a translated copy of $-\mathcal{T}_f$, where $-\mathcal{T}_f$ is a set of $-\triangle$ for all $\triangle \in \mathcal{T}_f$. Let $\mathcal{T}_f(u)$ be the translated copy of $-\mathcal{T}_f$ formed by a vertex u of $\mathcal{T}_g$. By Lemma 2, $\mathcal{M}$ is the overlay of $\mathcal{T}_f(u)$ and $\mathcal{T}_g(v)$ for all vertices u of $\mathcal{T}_g$ and v of $\mathcal{T}_f$, restricted to $\mathcal{S}$.

2.3 Complexity of the Subdivision $\mathcal{M}$

We analyze the number of cells in $\mathcal{M}$ for two terrains. Since $\mathcal{M}$ is a planar subdivision, we can bound the number of cells by bounding the number of vertices of $\mathcal{M}$.

For two terrains, not necessarily realistic, the subdivision of their translation space has $O(n^2m^2)$ cells; $\mathcal{M}$ is the overlay of $\mathcal{T}_f(u)$ and $\mathcal{T}_g(v)$ for all vertices u of $\mathcal{T}_g$ and v of $\mathcal{T}_f$, restricted to $\mathcal{S}$. So, $\mathcal{M}$ is the overlay of $O(nm)$ edges and has $O(n^2m^2)$ cells. It can be easily shown that this bound is tight. However, the worst case scenario hardly happens in real world applications. We are going to present a better bound for realistic terrains. First of all, we introduce some properties of a realistic terrain.

Lemma 3 [11]. *Let $f : \mathbb{D}_f \to \mathbb{R}$ be a realistic terrain such that $\mathcal{T}_f$ has n triangles and the side lengths of $\mathbb{D}_f$ are a and $\mathsf{a}r$ for a positive constant r. Then the following conditions hold.*

- *All edges in $\mathcal{T}_f$ have length $\Theta(\frac{\mathsf{a}}{\sqrt{n}})$.*
- *Let R be a rectangle that intersects $\mathcal{T}_f$, of which both side lengths are $\Omega(\frac{\mathsf{a}}{\sqrt{n}})$, and that has total area $\mathcal{R}$. Then R intersects $O(\frac{\mathcal{R}}{\mathsf{a}^2}n)$ triangles of $\mathcal{T}_f$.*

Next, the following lemma describes an upper bound of the number of cells in $\mathcal{M}$ when f and g are realistic terrains.

Lemma 4. *The number of cells in the subdivision $\mathcal{M}$ is $O(nm(m+(\frac{\mathsf{a}}{\mathsf{a}'})^2n))$, and the number of pairs in a combinatorial structure is $O(m + (\frac{\mathsf{a}}{\mathsf{a}'})^2n)$. When edge lengths of $\mathcal{T}_f$ and $\mathcal{T}_g$ are asymptotically same, i.e., $\frac{\mathsf{a}}{\sqrt{m}} = \Theta(\frac{\mathsf{a}'}{\sqrt{n}})$, the number of cells becomes $O(nm^2)$.*

If f and g are realistic terrains and $\mathbb{D}_g$ is "larger enough" than $\mathbb{D}_f$, we can show that the bound in Lemma 4 is tight by a simple example.

Theorem 1. *Let $f : \mathbb{D}_f \to \mathbb{R}$ and $g : \mathbb{D}_g \to \mathbb{R}$ be two realistic terrains such that $\mathcal{T}_f$ (resp. $\mathcal{T}_g$) has m (resp. n) triangles and the side lengths of $\mathbb{D}_f$ (resp. $\mathbb{D}_g$) are a and $\mathsf{a}r$ (resp. a' and $\mathsf{a}'r'$) for a positive constant r (resp. r'). The number of different combinatorial structures between $\mathcal{T}_f$ and $\mathcal{T}_g$ is $O(nmA)$, where $A = m + (\frac{\mathsf{a}}{\mathsf{a}'})^2 n$. This bound is tight if $\mathsf{a}' > 2\mathsf{a}$.*

3 Geometric Matching Algorithms Under L_∞ Metric

In this section, we first show how to compute $\mathcal{M}$ and the combinatorial structures corresponding to the interiors of the cells. We find an optimal translation vector $t^* \in c$ for each cell c in $\mathcal{M}$ by considering the combinatorial structure corresponding to the interior of c. For two translation vectors t in the interior of c and t' on the boundary of c, every vertex of $\mathcal{O}(f, g, t')$ is an intersection induced by an edge-edge pair or a vertex-triangle pair in $\mathcal{C}(t)$, so it is enough to consider the combinatorial structure corresponding to the interior of c. From now on, the term "combinatorial structure of a cell" means $\mathcal{C}(t)$ for a translation vector t in the interior of the cell.

We observe that the combinatorial structures of the adjacent cells of $\mathcal{M}$ have only $O(1)$ different edge-edge or vertex-triangle pairs, so they can be computed efficiently. We present a deterministic algorithm and a randomized algorithm to compute an optimal translation vector.

3.1 Construction of $\mathcal{M}$

The subdivision $\mathcal{M}$ is constructed by overlaying $\mathcal{T}_f(u)$ and $\mathcal{T}_g(v)$ for all vertices $v \in \mathcal{T}_f$ and $u \in \mathcal{T}_g$ restricted to $\mathcal{S}$ as explained in Sect. 2.2. The total number of the edges to overlay is $O(nm)$. The overlay of N edges can be constructed in $O(N \log N + K)$ time using $O(N + K)$ space [2], where K is the number of intersections. In our case, $N = O(nm)$ and $K = O(nm(m + (\frac{\mathsf{a}}{\mathsf{a}'})^2 n))$ by Theorem 1, so $\mathcal{M}$ can be constructed in $O(nm(m + (\frac{\mathsf{a}}{\mathsf{a}'})^2 n + \log n))$ time using $O(nm(m + (\frac{\mathsf{a}}{\mathsf{a}'})^2 n))$ space.

However, it is not necessary to maintain the whole subdivision $\mathcal{M}$; after finding an optimal translation vector among the translation vectors in a cell of $\mathcal{M}$, the cell is not necessary anymore. So we only maintain a small part of $\mathcal{M}$ to reduce the space. We divide $\mathcal{S}$ into a regular grid of length ℓ and then compute cells of the subdivision $\mathcal{M}$ that intersect each of the grid cells, one by one.

Lemma 5. *For two realistic terrains, $\mathcal{M}$ can be reported cell by cell of a grid of length $\ell = \Theta(\frac{\mathsf{a}}{\sqrt{m}} + \frac{\mathsf{a}'}{\sqrt{n}})$ in $O(nmA)$ time using $O(n + A^2)$ space, where $A = m + (\frac{\mathsf{a}}{\mathsf{a}'})^2 n$.*

After constructing the cells of $\mathcal{M}$ in a grid cell, we compute the corresponding combinatorial structure of each cell of $\mathcal{M}$. The straightforward way is computing the combinatorial structure of each cell separately, but this is inefficient because the combinatorial structures of the adjacent cells of $\mathcal{M}$ are similar as in the following.

Let c_i and c_j be two adjacent cells of $\mathcal{M}$ with a common edge e, and $\mathcal{C}_i$ and $\mathcal{C}_j$ be the corresponding combinatorial structures, respectively. Without loss of generality, we can say that e is a part of an edge of a triangle in $\mathcal{S}$ defined by a vertex v of $\mathcal{T}_f$ and a triangle $\triangle$ of $\mathcal{T}_g$. Let t be a translation vector in the interior of e. In the overlay of $\mathcal{T}_f + t$ and $\mathcal{T}_g$, $v + t$ lies on the interior of an edge of $\triangle$ and v is the unique vertex that lies on an edge of the overlay of $\mathcal{T}_f + t$ and $\mathcal{T}_g$ by the construction. It means that $\mathcal{C}_i$ and $\mathcal{C}_j$ have at most $O(1)$ different pairs because of the k-low-density assumption (Definition 2(a)) which implies that the degree of each vertex of $\mathcal{T}_f$ and $\mathcal{T}_g$ is at most k.

Observation 1. *The combinatorial structures of two adjacent cells of $\mathcal{M}$ have $k = O(1)$ different pairs.*

Note that the sequence of cells can be obtained by a standard DFS(depth first search) scheme.

3.2 A Deterministic Geometric Matching Algorithm

We first consider a deterministic algorithm to compute an optimal translation vector of a cell in $\mathcal{M}$. Computing an optimal translation vector of a cell can be reduced to a linear programming in $\mathbb{R}^4$ as follows. We use w to represent the fourth coordinate of $\mathbb{R}^4$.

Let $f[\mathbb{D}] = \{(x, y, f(p)) \mid p = (x, y) \in \mathbb{D} \subseteq \mathbb{D}_f\}$, and $g[\mathbb{D}] = \{(x, y, g(p)) \mid p = (x, y) \in \mathbb{D} \subseteq \mathbb{D}_g\}$. For an edge-edge pair (e, e') of a combinatorial structure of a cell c in $\mathcal{M}$, let $(0, 0, v(t, h))$ be a vertical translation vector with respect to $t = (t_x, t_y) \in \mathbb{R}^2$ and $h \in \mathbb{R}$ such that $f[e] + (t_x, t_y, h) + (0, 0, v(t, h))$ and $g[e']$ are contained in a common plane in $\mathbb{R}^3$. For a translation vector $t \in c$, let p be the intersection point of $e + t$ and e'. Then, $|v(t, h)| = |(f(p) + h) - g(p + t)|$. The set of points $(t_x, t_y, h, |v(t, h)|)$ in $\mathbb{R}^4$ is the upper envelope of two hyperplanes which consist of two sets of points $(t_x, t_y, h, v(t, h))$ and $(t_x, t_y, h, -v(t, h))$ for $t = (t_x, t_y) \in \mathbb{R}^2$ and $h \in \mathbb{R}$, respectively. For a vertex-triangle pair of a combinatorial structure, we can construct the upper envelope of two hyperplanes in $\mathbb{R}^4$ analogously.

For a cell c in $\mathcal{M}$ and a translation vector $t \in c$, we construct a set of hyperplanes in $\mathbb{R}^4$ corresponding to the pairs in a combinatorial structure $\mathcal{C}(t)$ as described. The problem of finding an optimal translation vector t for c reduces to the linear programming problem of finding a point q of the smallest w-coordinate in the upper envelope of the hyperplanes in $\mathbb{R}^4$ for $(q_x, q_y) \in c$, where q_x and q_y are the x- and y-coordinates of q, respectively. Note that the restriction $(q_x, q_y) \in c$ can be described by a set of linear inequality constraints. The translation vector (q_x, q_y) is an optimal translation vector for c.

Matoušek and Schwarzkopf [9] gave a dynamic data structure supporting a linear programming in $\mathbb{R}^4$. With N hyperplanes, the data structure can be constructed in $O(N^{4/3+\delta})$ deterministic time and space for any fixed $\delta > 0$. Also, both update time (insertions and deletions of hyperplanes) and query time are $O(N^{1/3+\delta})$ amortized time.

The overall strategy is as follows. We subdivide $\mathcal{S}$ by a regular grid of length $\ell = \Theta(\frac{\mathsf{a}}{\sqrt{m}} + \frac{\mathsf{a}'}{\sqrt{n}})$ and treat the cells one by one. For a grid cell, we compute the corresponding part of $\mathcal{M}$ and a sequence of adjacent cells in $\mathcal{M}$ as described in Sect. 3.1. Next, we build the data structure [9] with hyperplanes in $\mathbb{R}^4$ for the combinatorial structure of the first cell of the sequence. We find the lowest point in the upper envelope of the hyperplanes. For the rest cells in the sequence, we update the data structure for the corresponding combinatorial structure, and then find the lowest point. Note that the number of updates for a cell is $O(1)$ by Observation 1 if we follow the sequence of adjacent cells. We repeat this until the end of the sequence. By Lemma 4, the number of edge-edge and vertex-triangle pairs in a combinatorial structure is $O(m + (\frac{\mathsf{a}}{\mathsf{a}'})^2 n)$ and the number of different combinatorial structure is $O(nm(m + (\frac{\mathsf{a}}{\mathsf{a}'})^2 n))$. The following theorem summarizes the overall time and space complexity.

Theorem 2. *Let $f : \mathbb{D}_f \to \mathbb{R}$ and $g : \mathbb{D}_g \to \mathbb{R}$ be the two realistic terrains such that $\mathcal{T}_f$ (resp. $\mathcal{T}_g$) has m (resp. n) triangles and the side lengths of $\mathbb{D}_f$ (resp. $\mathbb{D}_g$) are a and $\mathsf{a}r$ (resp. a' and $\mathsf{a}'r'$) for a positive constant r (resp. r'). We can compute an optimal translation vector t^* such that $\mathbb{D}_f + t^* \subset \mathbb{D}_g$ and the distance under L_∞ metric between f and $g \restriction_{\mathbb{D}^*}$ is minimized in $O(nmA^{4/3+\delta})$ time with $O(n + A^2)$ space for any fixed $\delta > 0$ and $A = m + (\frac{\mathsf{a}}{\mathsf{a}'})^2 n$.*

3.3 A Randomized Geometric Matching Algorithm

In this section, we propose a randomized algorithm for the reduced problem described in Sect. 3.2. We first present a decision algorithm to decide the existence of a point of the upper envelope of hyperplanes in $\mathbb{R}^4$ such that the w-coordinates of the point is smaller than an input value. Next, we propose a randomized approach to compute an optimal translation vector.

As described in Sect. 3.2, we construct the set of hyperplanes in $\mathbb{R}^4$ corresponding to the combinatorial structure of a cell c of $\mathcal{M}$ and the set of hyperplanes corresponding to the boundary edges of c. A decision version of finding the lowest point in the upper envelope of hyperplanes can be stated as follows: given $\delta > 0$, is there a point in the upper envelope of hyperplanes whose w-coordinate is smaller than δ?

We check whether there is such a point in the upper envelope by introducing a new hyperplane $H_\delta : w = \delta$. If the upper envelope has such a point then the intersections of H_δ and each 'upper half-space' of the hyperplanes have a non-empty common intersection in H_δ. This problem can be reduced to a linear programming in $\mathbb{R}^3$, and we use a semi-online data structure for a linear programming in $\mathbb{R}^3$ [4].

Lemma 6. *For given $\delta > 0$, we can solve the decision problem for each cell of $\mathcal{M}$ in $O(\log^2 A \log^2 \log A)$ amortized time where $A = m + (\frac{\mathsf{a}}{\mathsf{a}'})^2 n$.*

Therefore, we can find translation vectors t *such that* $d_\infty(f, g\restriction_{\mathbb{D}_f+t}) \le \delta$ *in* $O(nmA\log^2 A\log^2\log A)$ *time.*

Next, we present how to use the solutions of the decision problems to find an optimal translation vector. First we randomly choose one cell c of $\mathcal{M}$ and compute the minimum distance δ between f and g for the combinatorial structure of c by solving a linear programming [8]. With this δ, we solve the decision problem for each cells of $\mathcal{M}$ and find cells which realize the distance smaller than δ. We can expect that only constant fraction of the number of the cells realize the distance smaller than δ. We repeat this procedure recursively with a new distance δ' computed from one of the cells decided as 'yes' from the previous recursive step until we find a cell which realizes the minimum distance. The expected number of recursion is $O(\log(\#\text{cells})) = O(\log(nm(m + (\frac{\mathsf{a}}{\mathsf{a}'})^2 n))) = O(\log n)$.

Theorem 3. *Let* $f : \mathbb{D}_f \to \mathbb{R}$ *and* $g : \mathbb{D}_g \to \mathbb{R}$ *be the two realistic terrains such that* $\mathcal{T}_f$ *(resp.* $\mathcal{T}_g$*) has* m *(resp.* n*) triangles and the side lengths of* $\mathbb{D}_f$ *(resp.* $\mathbb{D}_g$*) are* a *and* $\mathsf{a}r$ *(resp.* a' *and* $\mathsf{a}'r'$*) for a positive constant* r *(resp.* r'*). We can compute an optimal translation vector* t^* *such that* $\mathbb{D}_f + t^* \subset \mathbb{D}_g$ *and the distance under* L_∞ *metric between* f *and* $g\restriction_{\mathbb{D}^*}$ *is minimized in* $O(nmA\log n\log^2 A\log^2\log A)$ *expected time with* $O(n + A^2)$ *space where* $A = m + (\frac{\mathsf{a}}{\mathsf{a}'})^2 n$.

4 Geometric Matching Algorithm Under L_1 Metric

In this section, we solve the terrain matching problem under L_1 metric. To compute the distance between two terrains under L_1 metric, we need to compute the volume function between two terrains. The volume function of two terrains is a sum of volume functions between pairs of triangles, so we need to compute the volume functions between pairs of triangles.

To get a single formula for the volume function, we need to consider the amount of vertical translation h of f along with the translation vector $t = (t_x, t_y) \in \mathcal{S}$. If we concatenate h to the translation vector $t = (t_x, t_y)$ as a third coordinate, then a vertical prism $\{(t_x, t_y, h) | h \in \mathbb{R}\}$ in $\mathbb{R}^3$ over each cell of $\mathcal{M}$ can be seen as all possible vertical translations.

The basic idea is as follows. We first compute $\mathcal{M}$ as described in Sect. 3.1. For each cell c of $\mathcal{M}$, we subdivide the translation space of $\mathbb{R}^3$ with domain c such that the volume function between each pair of triangles is described as a single formula within a cell of the subdivision. We compute an optimal translation vector for each cell by using numerical methods.

Theorem 4. *Let* $f : \mathbb{D}_f \to \mathbb{R}$ *and* $g : \mathbb{D}_g \to \mathbb{R}$ *be the two realistic terrains such that* $\mathcal{T}_f$ *(resp.* $\mathcal{T}_g$*) has* m *(resp.* n*) triangles and the side lengths of* $\mathbb{D}_f$ *(resp.* $\mathbb{D}_g$*) are* a *and* $\mathsf{a}c$ *(resp.* a' *and* $\mathsf{a}'c'$*) for a positive constant* c *(resp.* c'*). We can compute an optimal translation vector* t^* *such that* $\mathbb{D}_f + t^* \subset \mathbb{D}_g$ *and the distance under* L_1 *metric between* f *and* $g\restriction_{\mathbb{D}^*}$ *is minimized in* $O(nmA^4)$ *time using* $O(n + A^3)$ *space, where* $A = m + (\frac{\mathsf{a}}{\mathsf{a}'})^2 n$.

References

1. Agarwal, P.K., Aronov, B., van Kreveld, M., Löffler, M., Silveira, R.I.: Computing correlation between piecewise-linear functions. SIAM J. Comput. **42**(5), 1867–1887 (2013)
2. Balaban, I.J.: An optimal algorithm for finding segments intersections. In: Proceedings of the 11th Symposium on Computational Geometry, pp. 211–219 (1995)
3. Carr, J.R., Sobek, J.S.: Digital scene matching area correlator (DSMAC). In: Proceedings on the Society of Photo-Optical Instrumentation Engineers, vol. 0238, pp. 36–41 (1980)
4. Eppstein, D.: Dynamic three-dimensional linear programming. ORSA J. Comput. **4**(4), 360–368 (1992)
5. Ernst, M.D., Flinchbaugh, B.E.: Image/map correspondence using curve matching. In: Proceedings on AAAI Symposium on Robot Navigation, pp. 15–18 (1989)
6. Golden, J.P.: Terrain contour matching (TERCOM): a cruise missile guidance aid. In: Proceedings on the Society of Photo-Optical Instrumentation Engineers, vol. 0238, pp. 10–18 (1980)
7. Johnson, A.E., Ansar, A., Matthies, L.H., Trawny, N., Mourikis, A.I., Roumeliotis, S.I.: A general approach to terrain relative navigation for planetary landing (2007)
8. Matoušek, J., Sharir, M., Welzl, E.: A subexponential bound for linear programming. Algorithmica **16**(4–5), 498–516 (1996)
9. Matoušek, J., Schwarzkopf, O.: Linear optimization queries. In: Proceedings of the Eighth Annual Symposium on Computational Geometry, SCG 1992, pp. 16–25. ACM (1992)
10. Medioni, G., Nevatia, R.: Matching images using linear features. IEEE Trans. Pattern Anal. Mach. Intell. (PAMI) **6**(6), 675–685 (1984)
11. Moet, E., van Kreveld, M., van der Stappen, A.F.: On realistic terrains. Comput. Geom. **41**(12), 48–67 (2008)
12. Moroz, G., Aronov, B.: Computing the distance between piecewise-linear bivariate functions. In: Proceedings of the Twenty-third Annual ACM-SIAM Symposium on Discrete Algorithms, SODA 2012, pp. 288–293. SIAM (2012)
13. Newman, P., Durrant-Whyte, H.: Using sonar in terrain-aided underwater navigation. In: Proceedings on IEEE International Conference on Robotics and Automation, vol. 1, pp. 440–445 (1998)
14. Olson, C., Matthies, L.: Maximum likelihood rover localization by matching range maps. In: Proceedings on IEEE International Conference on Robotics and Automation, vol. 1, pp. 272–277 (1998)
15. Rodriquez, J.J., Aggarwal, J.K.: Matching aerial images to 3-D terrain maps. IEEE Trans. Pattern Anal. Mach. Intell. **12**(12), 1138–1149 (1990)
16. Sistiaga, M., Opderbecke, J., Aldon, M., Rigaud, V.: Map based underwater navigation using a multibeam echosounder. In: Proceedings on OCEANS, vol. 2, pp. 747–751 (1998)
17. Williams, S., Dissanayake, G., Durrant-Whyte, H.: Towards terrain-aided navigation for underwater robotics. Adv. Robot. **15**(5), 533–549 (2001)

Size-Dependent Tile Self-Assembly: Constant-Height Rectangles and Stability

Sándor P. Fekete[1], Robert T. Schweller[2], and Andrew Winslow[3](✉)

[1] TU Braunschweig, Braunschweig, Germany
s.fekete@tu-bs.de
[2] University of Texas–Pan American, Edinburg, TX, USA
rtschweller@utpa.edu
[3] Université Libre de Bruxelles, Brussels, Belgium
awinslow@ulb.ac.be

Abstract. We introduce a new model of algorithmic tile self-assembly called *size-dependent assembly*. In previous models, supertiles are stable when the total strength of the bonds between any two halves exceeds some constant temperature. In this model, this constant temperature requirement is replaced by an nondecreasing *temperature function* $\tau : \mathbb{N} \rightarrow \mathbb{N}$ that depends on the size of the smaller of the two halves. This generalization allows supertiles to become unstable and break apart, and captures the increased forces that large structures may place on the bonds holding them together.

We demonstrate the power of this model in two ways. First, we give fixed tile sets that assemble constant-height rectangles and squares of arbitrary input size given an appropriate temperature function. Second, we prove that deciding whether a supertile is stable is coNP-complete. Both results contrast with known results for fixed temperature.

1 Introduction

In this paper, we introduce the *size-dependent tile self-assembly model*, a natural extension of the well-studied *two-handed tile assembly model* or *2HAM* [4]. As in the 2HAM, a size-dependent system consists of a collection of square Wang tiles [17,21] with an associated *bond strength* assigned to each tile edge color. In the 2HAM, self-assembly proceeds by repeatedly combining any two previously assembled *supertiles* into a new *stable* supertile provided the total bond strength between the supertiles meets or exceeds some positive integer called the *temperature*.

Although the 2HAM is both simple and natural, the model does not capture the intuition that two large assemblies should require more bond strength to be stable than two very small assemblies. As an analogy, a single staple is sufficient to attach two pieces of paper or to attach a sheet of paper to the hull of a battleship. However, a staple is too weak to amalgamate together two battleships.

The size-dependent self-assembly model generalizes the 2HAM by replacing the fixed, integer temperature parameter τ of the 2HAM with a nondecreasing

K. Elbassioni and K. Makino (Eds.): ISAAC 2015, LNCS 9472, pp. 296–306, 2015.
DOI: 10.1007/978-3-662-48971-0_26

temperature function $\tau(n)$ that specifies a required threshold of bond strength when given the size of the smaller of two supertiles under consideration. A set of tile types and temperature function together define a size-dependent self-assembly system.

Our results. We first consider efficiently assembling fixed-height rectangles and squares in the size-dependent self-assembly model. We prove that there exists a fixed tile set assembling a $k \times 3$ rectangle for every $k \geq 7$ given an appropriate temperature function. This tile set is extended to obtain a matching result for $k \times k$ squares. These results demonstrate that size-dependent temperature functions can, in theory, direct assembly in the spirit of temperature programming [11,20], concentration programming [3,7,12], and staging [5]. Unlike these other methods, size-dependence is present in all physical systems, but has not be demonstrated to be programmable. Thus these constructions demonstrate that this ubiquitous aspect of physical systems can (and likely already does) direct assembly in dramatic ways, regardless of whether they can be implemented physically.

In addition to the design of systems that assemble rectangles and squares, we consider the complexity of determining if a supertile is *stable*, i.e. cannot break apart due to insufficient bond strength. Determining the stability of an supertile is a fundamental problem for design, simulation, and analysis of tile self-assembly systems. This problem enjoys a straightforward, polynomial-time solution in the 2HAM. In contrast, we prove that the problem is coNP-complete in the size-dependent model, even for temperature functions with just two distinct temperatures.

Reversibility. A key feature of size-dependence is *reversibility*: the possibility of breaking bonds. Our rectangle and square constructions make critical use of reversibility to beat tile type lower bounds in similar models (see [18]), and our hardness result proves that this mechanism is capable of complex behaviors.

Reversibility has been more directly incorporated into a number of other self-assembly models via glues that repel [8,16] or deactivate [10,13,14], tiles that dissolve [1], and temperatures that change over time [2,20]. Reversibility in these models has yielded a number of new functionalities, including replication [1,13], fuel-efficient computation [14,19], shape identification [15], and efficient small-scale assembly of general shapes [6]. We believe that further study of the ubiquitous but indirect form of reversibility found in size-dependent self-assembly may yield similar functionality.

2 Definitions

The first three subsections define the 2HAM, giving definitions equivalent to those in prior work, e.g. [4]. The final section describes the differences between the two-handed and size-dependent models.

2.1 Tiles, Assemblies, and Supertiles

A *tile type* is a quadruple (g_N, g_E, g_S, g_W) of *glues* from a fixed alphabet Σ. Each glue $g_i \in \Sigma$ has an associated non-negative integer *strength*, denoted by $\text{str}(g_i)$.[1] An instance of a tile type, called a *tile*, is an axis-aligned unit square with center in $\mathbb{Z}^2$. The edges of a tile are labeled with the glues of the tile's type (e.g. g_N, g_E, g_S, g_W) in clockwise order, starting with the edge with normal vector $\langle 0, 1 \rangle$. Two tiles are *adjacent* if their centers have distance 1.

An *assembly* α is a partial mapping $\alpha : \mathbb{Z}^2 \to T$ from tile locations to a set of tile types T, also called a *tile set.* The domain of this partial function is denoted by $\text{dom}(\alpha)$. Each assembly has a dual *bond graph*: a grid graph with vertex set $\text{dom}(\alpha)$ and an edge between every pair of adjacent tiles that form a bond. An edge cut of the bond graph of an assembly is also called a *cut* of the assembly, and the total strength of the bonds of the edges in the cut is the *strength* of the cut. An assembly is τ*-stable* if every cut of the assembly has strength at least τ.

For an assembly $\alpha : \mathbb{Z}^2 \to T$ and vector $\boldsymbol{u} = \langle x, y \rangle$ with $x, y \in \mathbb{Z}^2$, the assembly $\alpha + \boldsymbol{u}$ denotes the assembly consisting of the tiles in α, each translated by $\boldsymbol{u}$. For two assemblies α and β, β is a *translation* of α, written $\beta \simeq \alpha$, provided that there exists a vector $\boldsymbol{u}$ such that $\beta = \alpha + \boldsymbol{u}$. The *supertile* of α is the set $\tilde{\alpha} = \{\beta : \alpha \simeq \beta\}$. A supertile $\tilde{\alpha}$ is τ*-stable* provided that the assemblies it contains are τ-stable. The *size* of a supertile is denoted by $|\tilde{\alpha}|$ and is equal to the size of an assembly in $\tilde{\alpha}$ (and not the cardinality of $\tilde{\alpha}$, which is always $\aleph_0$).

2.2 The Assembly Process

Two assemblies α and β are *disjoint* if $\text{dom}(\alpha) \cap \text{dom}(\beta) = \varnothing$. The *union* of two disjoint assemblies α and β, denoted by $\alpha \cup \beta$, is the partial function $\alpha \cup \beta : \mathbb{Z}^2 \to T$ defined as $(\alpha \cup \beta)(x, y) = \alpha(x, y)$ if $(x, y) \in \text{dom}(\alpha)$ and $(\alpha \cup \beta)(x, y) = \beta(x, y)$ if $(x, y) \in \text{dom}(\beta)$. Two supertiles $\tilde{\alpha}$ and $\tilde{\beta}$ can *combine* into a supertile $\tilde{\gamma}$ provided:

- There exist disjoint assemblies $\alpha \in \tilde{\alpha}$ and $\beta \in \tilde{\beta}$.
- $\alpha \cup \beta = \gamma \in \tilde{\gamma}$ and the cut partioning $\text{dom}(\gamma)$ into $\text{dom}(\alpha)$ and $\text{dom}(\beta)$ has strength at least τ (equivalently, γ is τ-stable).

The set of all combinations of $\tilde{\alpha}$ and $\tilde{\beta}$ at temperature τ is denoted by $C^{\tau}_{\tilde{\alpha},\tilde{\beta}}$.

2.3 Two-Handed Tile Assembly Systems

A *two-handed tile assembly system* or *two-handed system* is a pair $\mathcal{T} = (T, \tau)$, where T is a *tile set* and $\tau \in \mathbb{N}$ is a *temperature.* Given a system $\mathcal{T} = (T, \tau)$, a supertile $\tilde{\alpha}$ is *producible*, written $\tilde{\alpha} \in \mathcal{A}[\mathcal{T}]$, provided that either $|\tilde{\alpha}| = 1$ or $\tilde{\alpha}$ is a combination of two other producible supertiles of $\mathcal{T}$. A supertile $\tilde{\alpha}$ is *terminal* provided that for all producible supertiles $\tilde{\beta}$, $C^{\tau}_{\tilde{\alpha},\tilde{\beta}} = \varnothing$. A system is *directed* or *deterministic* provided that it has only one terminal supertile.

[1] In later sections, glues with strength 0 are treated as non-existent.

Given a *shape* $P \subseteq \mathbb{Z}^2$, we say a system $\mathcal{T}$ *self-assembles* P, provided that every terminal supertile $\tilde{\alpha}$ of $\mathcal{T}$ has an assembly $\alpha \in \tilde{\alpha}$ such that $\mathrm{dom}(\alpha) = P$. That is, every terminal supertile has shape P, up to translation. A shape P is a $w \times h$ *rectangle* provided that $P = \{x+1, x+2, \ldots, x+w\} \times \{y+1, y+2, \ldots, y+h\}$ for some $x, y, w, h \in \mathbb{Z}$. If $w = h$, then the rectangle is a *square*.

2.4 Size-Dependent Systems

A *size-dependent two-handed tile assembly system* or *size-dependent system* $\mathcal{S} = (T, \tau)$ is a generalization of a two-handed tile assembly system. Two-handed and size-dependent systems are identical, except for the definition of τ. Recall that in two-handed systems, $\tau \in \mathbb{N}$ determines the bond strength needed for two supertiles to combine and for a supertile to be τ-stable.

In size-dependent systems, τ is not an integer temperature, but rather a non-decreasing *temperature function* $\tau : \mathbb{N} \to \mathbb{N}$. An assembly γ is τ*-stable* provided any cut partioning $\mathrm{dom}(\gamma)$ into two assemblies $\mathrm{dom}(\alpha)$, $\mathrm{dom}(\beta)$ has strength at least $\tau(\min(|\alpha|, |\beta|))$. A supertile $\tilde{\gamma}$ is τ*-stable* provided the assemblies in $\tilde{\gamma}$ are τ-stable. Also, two supertiles $\tilde{\alpha}$ and $\tilde{\beta}$ can *combine* into a supertile $\tilde{\gamma}$ provided that:

- There exist disjoint assemblies $\alpha \in \tilde{\alpha}$ and $\beta \in \tilde{\beta}$.
- $\alpha \cup \beta = \gamma \in \tilde{\gamma}$ and the cut partioning $\mathrm{dom}(\gamma)$ into $\mathrm{dom}(\alpha)$ and $\mathrm{dom}(\beta)$ has strength at least $\tau(\min(|\alpha|, |\beta|))$.

For a given temperature function $\tau : \mathbb{N} \to \mathbb{N}$, the set of all combinations of $\tilde{\alpha}$ and $\tilde{\beta}$ is denoted by $C^{\tau}_{\tilde{\alpha},\tilde{\beta}}$. Note that the second condition is not equivalent to γ being τ-stable. Figure 1 illustrates an example: a cut in a supertile has sufficient strength, but combining with another supertile causes increased size that causes the cut to become insufficiently strong. So $\tilde{\alpha}$, $\tilde{\beta}$ may be τ-stable while their combination $\tilde{\gamma}$ is τ*-unstable.*

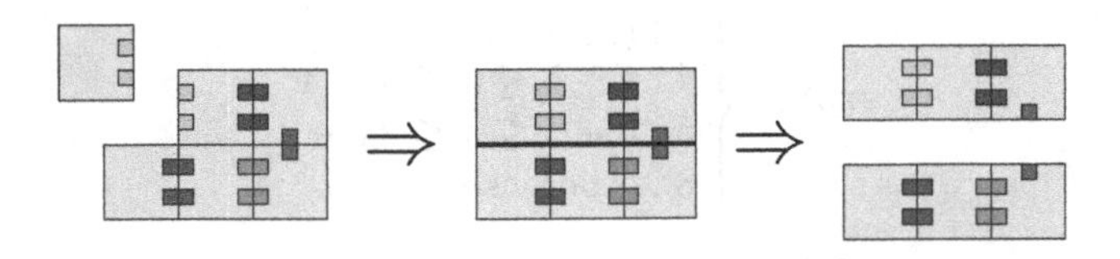

Fig. 1. Three steps of size-dependent self-assembly with glue function $\tau(n) = n - 1$. The addition of a new tile (left) causes the supertile to have a strength-1 cut partioning it into two supertiles of 3 tiles each (center). Because $\tau(3) = 2 > 1$, the supertile can then break (right).

Supertiles that are τ-unstable can also "break" into smaller supertiles. A supertile $\tilde{\gamma}$ can *break* into $\tilde{\alpha}$ and $\tilde{\beta}$ provided that:

- There exist disjoint assemblies $\alpha \in \tilde{\alpha}$ and $\beta \in \tilde{\beta}$ with connected bond graphs.
- $\alpha \cup \beta = \gamma \in \tilde{\gamma}$ and the strength of the cut partioning γ into α and β is less than $\tau(\min(|\alpha|, |\beta|))$.

A cut between two supertiles resulting from a break is called a *break cut.* For a given temperature function $\tau : \mathbb{N} \to \mathbb{N}$, the set of all supertiles resulting from breaks of $\tilde{\gamma}$ is denoted by $B^{\mathcal{T}}_{\tilde{\gamma}}$. Given a size-dependent system $\mathcal{T} = (T, \tau)$, a supertile $\tilde{\alpha}$ is *producible* provided either:

- $|\tilde{\alpha}| = 1$.
- $\tilde{\alpha}$ is the combination of two other producible supertiles.
- $\tilde{\alpha}$ is the result of a break of a producible supertile.

A producible supertile $\tilde{\alpha}$ is *terminal* provided $C^{\mathcal{T}}_{\tilde{\alpha},\tilde{\beta}} = \varnothing$ and $B^{\mathcal{T}}_{\tilde{\alpha}} = \varnothing$.

Note that the conditions on supertiles combining and breaking do *not* imply that combining supertiles or supertiles resulting from a break are τ-stable. This allows for systems with an infinite number of producible supertiles and a unique terminal supertile, including those described in this work.

3 Constant-Height Rectangles

Here we prove that there exists a single set of tiles that can be used to self-assemble constant-height rectangles of arbitrary width using an appropriate choice of temperature function. Such a result contrasts with the polynomial number of tiles required to assemble a constant-height rectangle in an assembly system with constant temperature [2].

Theorem 1. *There exists a tile set T such that for every $k \geq 7$, there exists a size-dependent system with tile set T that self-assembles a $k \times 3$ rectangle.*

Proof. The temperature function used is:

$$\tau(n) = \begin{cases} 3 : n \leq k - 6 \\ 4 : k - 5 \leq n \leq k + 3 \\ 5 : k + 4 \leq n \leq 2k - 2 \\ 8 : \text{otherwise} \end{cases}$$

The tile set consists of three tile types and two *blocks*: supertiles with unique internal glues and strength 8, the maximum temperature of the system. The tiles and blocks are listed and named in Fig. 2.

The system works by assembling a unique terminal $k \times 3$ supertile in three phases. First, top filler tiles and top bases combine into arbitrarily wide height-2 supertiles. These undergo at least two breaks to form *top half* supertiles of size $2k - 3$. Second and separately, bottom filler tiles and bottom bases combine to form *bottom half* supertiles of size approximately $k + 3$. Finally, these two halves combine into a terminal $k \times 3$ supertiles shown in Fig. 3. It can easily be verified that this supertile is a terminal supertile of the system; it remains to be shown

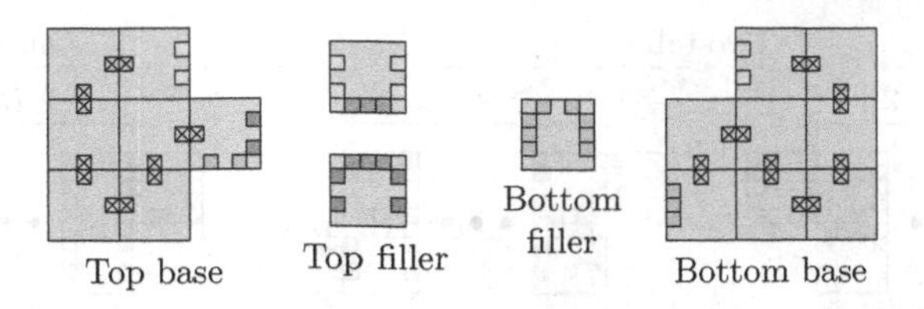

Fig. 2. The tile types and blocks for the constant-height rectangle construction. The gray glues are unique and strength at least 8 (Color figure online).

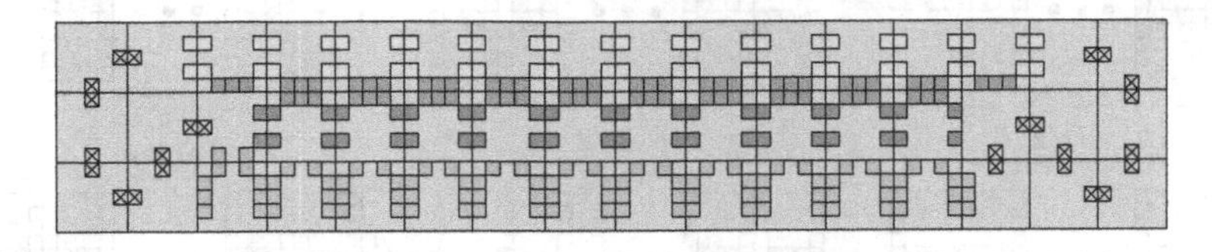

Fig. 3. The unique terminal supertile of the constant-height rectangle construction.

that no other terminal supertiles of the system exist (necessary for the system to *self-assemble* a $k \times 3$ rectangle).

Top Filler Supertiles. To start, consider the producible supertiles consisting of only top filler tiles, called *top filler* supertiles. Because $\tau(n) > 2$ for all n, upper and lower top filler tiles must first combine into size-2 supertiles before combining with other top filler supertiles into height-2 rectangular supertiles (lower right supertile in Fig. 4). These rectangular supertiles break along 2-edge and 3-edge cuts into the remaining supertiles seen in Fig. 4.

Because $k \geq 7$, any partition of the lower right supertile in Fig. 4 either has a part that is a single tile or uses a strength-4 cut of at least 2 edges and thus both parts have size at least $k + 3 \geq 10$. Therefore, the remaining 8 types of supertiles in Fig. 4 have at least 4 columns of 2 tiles each.

The width bounds seen in the figure are computed by considering how the supertiles are created. If the supertile is the result of a break, it must satisfy the size bound for the strength of the cut used in the break. If it is the result of a combination, it must be larger than the total sizes of the combined supertiles.[2]

We designate three types of top filler supertiles as seen in Fig. 4. As already proven, breaks only result in single tiles or supertiles of size 10 and larger. Any two-tab (one-tab) supertile can break into a one-tab (tabless) supertile and a single tile, and these are the only breaks that use cuts of strength at most 3. Then any other break uses a cut of strength 4 or more, and so results in supertiles of size at least $k + 4$. Thus any combination of two-tab and one-tab supertiles has size at least $2(k + 4)$. A two-tab supertile can also be the result of a break using a cut of strength 7 and thus have size at least $2k - 3$ and, because two-tab supertiles have even size, $2k - 2$. Because $\min(2(k + 4), 2k - 2) - 1 = 2k - 3$ and $k \geq 7$, $2k - 3 \geq k + 4$ and a break of a two-tab supertile into a single tile

[2] An upper bound is also implied by τ, but this is ignored here.

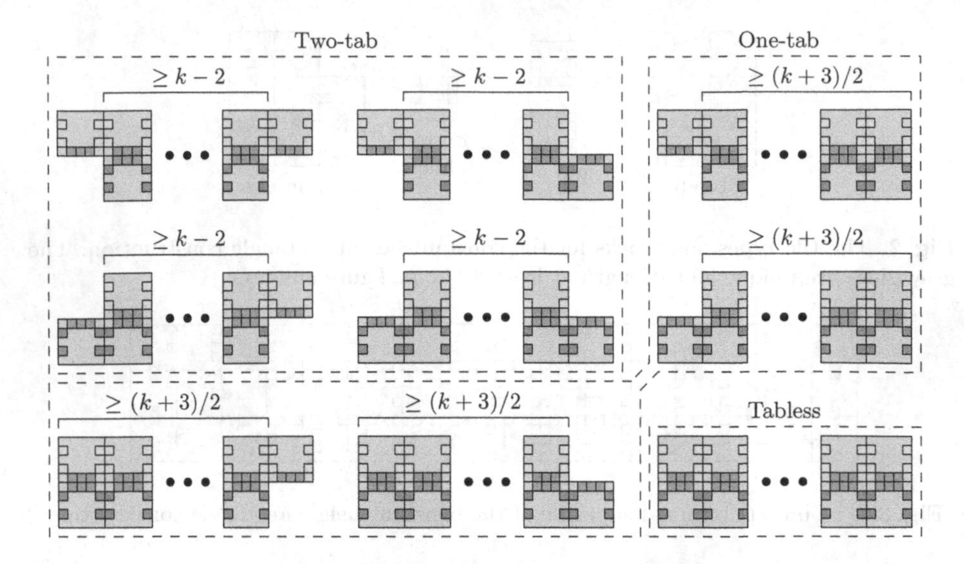

Fig. 4. The producible top filler supertiles.

and one-tab supertile cannot yield a one-tab supertile smaller than $k + 4$. In conclusion, one-tab and two-tab supertiles have size at least $k + 4$ and $2k - 2$, respectively, implying the bounds seen in Fig. 4.

Top and Bottom Halves. Top filler supertiles cannot combine with other supertiles, except for a complete top base to form a *top half* supertile (upper supertile in Fig. 5). Top half supertiles may combine with top filler supertiles and break into top half and top filler supertiles. A top half supertile with a single upper filler tile in the rightmost column is *ready.* Because ready top half supertiles are two-tab top filler supertiles that have combined with a top base, they have size at least $2k - 3$ and thus width at least $k - 2$.

Independently of top halves, bottom filler tiles combine into arbitrarily wide height-1 supertiles called a *bottom filler* supertile. These supertiles also combine with bottom bases at various stages of assembly. A *bottom half* supertile contains bottom filler tiles and a completed bottom base. If the number of bottom filler tiles in a bottom half is at least $2k-18$ (and there exists a 1-edge strength-3 cut partitioning the supertile into two of size at least $k - 5$), the bottom half can break into a bottom half and bottom filler supertile.

Combining Halves. The only shared glues between top and bottom tiles are the strength-2 glues on the south of the top base and west of the bottom base (turquoise and yellow in Fig. 2). Thus a supertile consisting of bottom tiles cannot combine with a supertile consisting of top tiles, unless the supertiles are bottom and top halves.

A bottom half and top half can combine, provided they have the same width and the top half is ready (and thus has width at least $k - 2$. Moreover, because

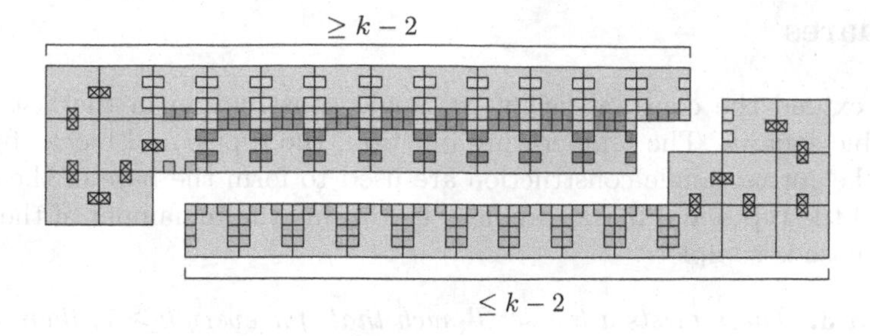

Fig. 5. The top half and bottom half supertiles. The bottom half λ can be arbitrarily large, but the upper bound follows from the requirement that to combine, $\tau(|\lambda|) \leq 4$ and thus $|\lambda| \leq k$.

the maximum strength of the bonds between the bottom and top halves is 4, they can only combine only if the smaller supertile, necessarily the bottom half, has size at most $k+3$ and thus width at most $k-2$. Thus, the bottom and top halves combine provided they both have width exactly $k-2$, forming a terminal supertile of width exactly k.

No Waste. Although it is not required by the definition of self-assembly, this system also has the property that every supertile may undergo a sequence of breaks and combinations to become terminal. In other words, the system has no "waste" supertiles. This can be seen by noting that supertiles not found within the (unique) terminal supertile, i.e. top filler supertiles wider than $k-4$, top halves wider than $k-2$, bottom filler supertiles of width more than $k-5$, and bottom halves of width more than $k-2$ can repeatedly break into smaller supertiles that *are* found in the terminal supertile. □

The temperature functions used in the previous construction all have a maximum bounded above by the constant 8. Next, we prove that any set of temperature functions used to assemble arbitrarily large constant-height rectangles are similarly bounded above by a constant.

Theorem 2. *Let T be a tile set and $\tau_1, \tau_2, \ldots$ be an infinite sequence of temperature functions such that the size-dependent system (T, τ_i) assembles a $k_i \times O(1)$ rectangle and all k_i are distinct. Let $f(n) = \min_{i \in \mathbb{N}}(\tau_i(n))$. Then $f(n) = O(1)$.*

Proof. Let $c \in \mathbb{N}$ be the maximum height of a rectangle assembled by a system (T, τ_i). Let $g_{\max}$ be the maximum strength of a glue in T. Let $\tilde{\gamma}$ be a terminal assembly of (T, τ_i) and thus a rectangle with width k_i. For any $n \leq k_i/2$, there exists a cut of $\tilde{\gamma}$ into supertiles $\tilde{\alpha}$, $\tilde{\beta}$ such that $n = |\tilde{\alpha}| \leq |\tilde{\beta}|$ and the cut contains at most $c+1$ edges. Then since $\tilde{\gamma}$ is stable, $f(n) \leq \tau_i(n) \leq (c+1)g_{\max}$ for all $n \leq k_i/2$. Because there exist infinitely many k_i, every n has $n \leq k_i/2$ for large enough k_i and we conclude that $f(n) \leq (c+1)g_{\max}$ for all $n \in \mathbb{N}$. □

4 Squares

Here we extend the constant-height rectangle construction in the last section to assemble squares. The temperature function, tile types, and blocks from the constant-height rectangle construction are used to form the base of the square; additional tile types and blocks are used to "fill in" the remainder of the square once the base is complete.

Theorem 3. *There exists a tile set T such that, for every $k \geq 7$, there exists a size-dependent system with tile set T that self-assembles a $k \times k$ square.*

The constant-height rectangle construction used as the basis for the construction of Theorem 3 result in temperature functions that are bounded above by a constant. We conjecture that there exists a square construction that uses temperature functions that all scale as $\Omega(\sqrt{n})$, and prove that no better lower bound is possible:

Theorem 4. *Let T be a tile set and $\tau_1, \tau_2, \ldots$ be an infinite sequence of temperature functions such that the size-dependent system (T, τ_i) assembles a $k_i \times k_i$ square and k_i are all distinct. Let $f(n) = \min_{i \in \mathbb{N}}(\tau_i(n))$, the minimum of all temperature functions for size n. Then $f(n)$ is not $\omega(\sqrt{n})$.*

5 τ-stabilility is coNP-complete

In two-handed tile assembly systems that are not size-dependent, determining whether a supertile is τ-stable amounts to determining if there exists a cut of the bond graph of weight less than τ, a problem decidable in polynomial time. In contrast, we prove that the same problem is coNP-complete for size-dependent systems, even when restricted to constant-time-computable temperature functions with just two distinct temperatures.

The reduction is from maximum independent set in Hamiltonian cubic (3-regular) planar graphs, proved NP-hard in [9]. The constructed assembly contains vertex gadgets arranged horizontally along a line bisecting the assembly. Gadgets are connected by zero-strength cuts mirroring the edges of the input graph, and have two possible cuts through them: *include* or *exclude*. The include path has slightly lower strength, but intersects the zero-strength cuts connecting the vertex gadget to the gadgets of adjacent vertices in the input graph. The temperature function requires that any cut passes through all vertex gadgets, does not use the include cuts of the gadgets of two adjacent vertices in the graph, and does not use too many exclude paths. Thus an independent set of at least some size exists if and only if there exists a sufficiently larger independent set of vertices.

Theorem 5. *Given a temperature function $\tau : \mathbb{N} \rightarrow \mathbb{N}$ and supertile, determining whether the supertile is τ-stable is coNP-complete.*

6 Open Problems

The rectangle and square constructions in this work use artificial temperature functions engineered in tandem with the tile sets. A central open question is whether physically implementable families of temperature functions (e.g. $\tau(n) = cn^\delta$ for varying $c, \delta > 0$) are similarly capable of such control. We conjecture that the design of such systems is possible but difficult; consider the lengthy analysis of the construction in Sect. 3 with just 5 components. Alternatively, temperature functions may be given as input along with shapes, with the goal of designing systems that assemble shapes *despite* the temperature functions.

The difficulty of system design is supported by the coNP-hardness of determining stability. Proving the PSPACE-hardness of predicting a system outcomes, such as whether a unique terminal supertile exists, would give even further evidence of this difficulty.

As previously discussed, *reversibility* is a key feature of size-dependent systems. Reversibility has been more directly incorporated into algorithmic design in other tile assembly models, leading to functionality not found in irreversible models. For instance, *replication* of shapes and patterns [1,13], *fuel-efficient* systems [14,19], and assembly of arbitrary shapes using a small, bounded scale factor [6]. Can any of these be achieved with size-dependent systems?

Acknowledgements. This work began at the Bellairs Workshop on Self-Assembly and Computational Geometry, March 21–28th, 2014. We thank the other participants for a productive and positive atmosphere, in particular, Alexandra Keenan and the co-organizers, Erik Demaine and Godfried Toussaint.

References

1. Abel, Z., Benbernou, N., Damian, M., Demaine, E.D., Demaine, M.L., Flatland, R., Kominers, S.D., Schweller, R.: Shape replication through self-assembly and RNAse enzymes. In: Proceedings of the 21st ACM-SIAM Symposium on Discrete Algorithms (SODA), pp. 1045–1064 (2010)
2. Aggarwal, G., Cheng, Q., Goldwasser, M., Kao, M., de Espanes, P., Schweller, R.: Complexities for generalized models of self-assembly. SIAM J. Comput. **34**(6), 1493–1515 (2005)
3. Becker, F., Rapaport, I., Rémila, É.: Self-assemblying classes of shapes with a minimum number of tiles, and in optimal time. In: Arun-Kumar, S., Garg, N. (eds.) FSTTCS 2006. LNCS, vol. 4337, pp. 45–56. Springer, Heidelberg (2006)
4. Cannon, S., Demaine, E.D., Demaine, M.L., Eisenstat, S., Patitz, M.J., Schweller, R.T., Summers, S.M., Winslow, A.: Two hands are better than one (up to constant factors): self-assembly in the 2HAM vs. aTAM. In: STACS 2013. LIPIcs, vol. 20, pp. 172–184. Schloss Dagstuhl (2013)
5. Demaine, E.D., Demaine, M.L., Fekete, S.P., Ishaque, M., Rafalin, E., Schweller, R.T., Souvaine, D.L.: Staged self-assembly: nanomanufacture of arbitrary shapes with $O(1)$ glues. Nat. Comput. **7**(3), 347–370 (2008)

6. Demaine, E.D., Patitz, M.J., Schweller, R.T., Summers, S.M.: Self-assembly of arbitrary shapes using RNAse enzymes: meeting the Kolmogorov bound with small scale factor (extended abstract). In: Proceedings of the 28th International Symposium on Theoretical Aspects of Computer Science (STACS 2011) (2011)
7. Doty, D.: Randomized self-assembly for exact shapes. SIAM J. Comput. **39**(8), 3521–3552 (2010)
8. Doty, D., Kari, L., Masson, B.: Negative interactions in irreversible self-assembly. In: Sakakibara, Y., Mi, Y. (eds.) DNA 16 2010. LNCS, vol. 6518, pp. 37–48. Springer, Heidelberg (2011)
9. Fleischner, H., Sabidussi, G., Sarvanov, V.I.: Maximum independent sets in 3- and 4-regular Hamiltonian graphs. Discrete Math. **310**, 2742–2749 (2010)
10. Hendricks, J., Padilla, J.E., Patitz, M.J., Rogers, T.A.: Signal transmission across tile assemblies: 3D static tiles simulate active self-assembly by 2D signal-passing tiles. In: Soloveichik, D., Yurke, B. (eds.) DNA 2013. LNCS, vol. 8141, pp. 90–104. Springer, Heidelberg (2013)
11. Kao, M.Y., Schweller, R.: Reducing tile complexity for self-assembly through temperature programming. In: Proceedings of the 17th Annual ACM-SIAM Symposium on Discrete Algorithms (SODA), pp. 571–580 (2006)
12. Kao, M.-Y., Schweller, R.T.: Randomized self-assembly for approximate shapes. In: Aceto, L., Damgård, I., Goldberg, L.A., Halldórsson, M.M., Ingólfsdóttir, A., Walukiewicz, I. (eds.) ICALP 2008, Part I. LNCS, vol. 5125, pp. 370–384. Springer, Heidelberg (2008)
13. Keenan, A., Schweller, R., Zhong, X.: Exponential replication of patterns in the signal tile assembly model. In: Soloveichik, D., Yurke, B. (eds.) DNA 2013. LNCS, vol. 8141, pp. 118–132. Springer, Heidelberg (2013)
14. Padilla, J.E., Patitz, M.J., Pena, R., Schweller, R.T., Seeman, N.C., Sheline, R., Summers, S.M., Zhong, X.: Asynchronous signal passing for tile self-assembly: fuel efficient computation and efficient assembly of shapes. In: Mauri, G., Dennunzio, A., Manzoni, L., Porreca, A.E. (eds.) UCNC 2013. LNCS, vol. 7956, pp. 174–185. Springer, Heidelberg (2013)
15. Patitz, M.J., Summers, S.M.: Identifying shapes using self-assembly. Algorithmica **64**(3), 481–510 (2012)
16. Reif, J.H., Sahu, S., Yin, P.: Complexity of graph self-assembly in accretive systems and self-destructible systems. In: Carbone, A., Pierce, N.A. (eds.) DNA 2005. LNCS, vol. 3892, pp. 257–274. Springer, Heidelberg (2006)
17. Robinson, R.M.: Undecidability and nonperiodicity for tilings of the plane. Inventiones Mathematicae **12**, 177–209 (1971)
18. Rothemund, P.W.K., Winfree, E.: The program-size complexity of self-assembled squares (extended abstract). In: Proceedings of ACM Symposium on Theory of Computing (STOC), pp. 459–468 (2000)
19. Schweller, R.T., Sherman, M.: Fuel efficient computation in passive self-assembly. In: Proceedings of the 24th Annual ACM-SIAM Symposium on Discrete Algorithms, pp. 1513–1525 (2013)
20. Summers, S.M.: Reducing tile complexity for the self-assembly of scaled shapes through temperature programming. Algorithmica **63**(1), 117–136 (2012)
21. Wang, H.: Proving theorems by pattern recognition–II. Bell Syst. Tech. J. **40**(1), 1–41 (1961)

The 2-Center Problem in a Simple Polygon

Eunjin Oh[1(✉)], Jean-Lou De Carufel[2], and Hee-Kap Ahn[1]

[1] Pohang University of Science and Technology, Pohang, Korea
{jin9082,heekap}@postech.ac.kr
[2] University of Ottawa, Ottawa, Canada
jdecaruf@uottawa.ca

Abstract. The geodesic k-center problem in a simple polygon with n vertices consists in the following. Find k points, called *centers*, in the polygon to minimize the maximum geodesic distance from any point of the polygon to its closest center. In this paper, we focus on the case where $k = 2$ and present an exact algorithm that returns an optimal geodesic 2-center in $O(n^2 \log^2 n)$ time.

1 Introduction

The geodesic k-center problem in a simple polygon P consists in the following. Find a set S of k points in P that minimizes $\max_{p \in P} \min_{s \in S} d(s, p)$, where $d(x, y)$ is the length of the shortest path between x and y lying in P (also called *geodesic path*). Geometrically, this is equivalent to find k smallest geodesic disks with the same radius whose union contains P.

The k-center problem in the 2-dimensional Euclidean space is the same as the one for a simple polygon, except that, the distance between two points, x and y, is their Euclidean distance. For a finite point set, there have been a lot of results on the k-center problem in the 2-dimensional Euclidean space. For a set of n points in the plane, the 1-center problem can be solved in linear time [6]. Chan showed that the 2-center problem can be solved in $O(n \log^2 n \log^2 \log n)$ deterministic time and in $O(n \log n)$ expected time [4]. The k-center problem can be solved in $O(n^{O(\sqrt{k})})$ time [9]. Kim and Shin considered the problem of covering a convex polygon with two congruent disks and presented a $O(n \log^3 n \log \log n)$-time algorithm for the problem [10].

The k-center problem has also been studied under the geodesic metric inside a simple polygon for the special case $k = 1$. Asano and Toussaint studied the geodesic 1-center problem and presented the first algorithm for the problem which returns the geodesic center of a polygon with n vertices in $O(n^4 \log n)$ time [2]. In 1989, the running time was improved to $O(n \log n)$ time by Pollack et al. [13]. They first triangulate the polygon and find the triangle that contains the center in $O(n \log n)$ time. Then they subdivide the triangle further and find the region containing the center such that any point inside the region has the

This work was supported by the NRF grant 2011-0030044 (SRC-GAIA) funded by the government of Korea.

K. Elbassioni and K. Makino (Eds.): ISAAC 2015, LNCS 9472, pp. 307–317, 2015.
DOI: 10.1007/978-3-662-48971-0_27

combinatorially equivalent shortest path tree from the point. Then the problem is reduced to find the lowest point of the upper envelope of a family of functions, which was dealt in [12] due to Megiddo. Very recently, the running time was improved by Ahn et al. to $O(n)$ [1], which is optimal. They construct a set of $O(n)$ chords instead of triangulating the polygon and find a triangle that is bounded by at most three chords in the set and contains the center in linear time. Afterwards they find the lowest point of the upper envelope of a family of functions inside the triangle, using an algorithm similar to the one in [12].

Surprisingly, there has been no result for the geodesic k-center problem for $k > 1$, except the one by Vigan [14]. They gave an exact algorithm for the geodesic 2-center problem in a simple polygon with n vertices, which runs in $O(n^8 \log n)$ time. The algorithm follows the framework of Kim and Shin [10], which returns a pair of congruent disks of smallest radius whose union contains a convex polygon in the Euclidean plane. However, the algorithm does not seem to work as it is because of the following reasons. First, they claim that the decision version of the geodesic 2-center problem in a simple polygon can be solved using a technique similar to the one by Kim and Shin without providing any detailed argument. The decision algorithm by Kim and Shin, however, does not seem to work for an arbitrary simple polygon unless it is modified to handle a simple polygon. Second, they apply parametric search using their decision algorithm. Again, they do not describe how their parallel algorithm works. The parallel algorithm by Kim and Shin [10] does not seem to extend for this problem.

Our Results. We present an $O(n^2 \log^2 n)$-time exact algorithm for the geodesic 2-center problem in a simple polygon with n vertices, which is the first correct algorithm for this problem. To be more specific, we first observe that a simple polygon P can always be partitioned into two regions by a geodesic path connecting two boundary points x and y of P such that the radius of an optimal geodesic 2-center is the larger of the radii of the optimal geodesic 1-centers of the regions. Then we consider $O(n)$ candidate pairs of polygon edges one of whose element (e, e') satisfies $x \in e$ and $y \in e'$ and present a procedure that finds $O(n)$ such pairs in $O(n^2)$ time. Finally, we present an algorithm that computes an optimal 2-center restricted to such a pair of edges in $O(n \log^2 n)$ time using the parametric search by Megiddo with a decision and a parallel algorithms.

2 Preliminary

A polygon is *simple* if every vertices are distinct and edges intersect only at common endpoints. A polygon P is *weakly simple* if, for any $\varepsilon > 0$, there is a simple polygon Q such that the Fréchet distance between P and Q is at most ε [5]. Most algorithms designed for simple polygons also work for weakly simple polygons including the algorithms we use in this paper.

The vertices of a simple polygon P with n vertices are labeled $v_1, \ldots, v_n$ in clockwise order along the boundary of P. We set $v_{n+k} = v_k$ for all $1 \leq k < n$. An edge whose endpoints are v_i and v_{i+1} is denoted by e_i.

For any two points x and y lying inside a (weakly) simple polygon P, the *geodesic path* between x and y, denoted by $\pi(x, y)$, is the shortest path between x and y inside P. The length of $\pi(x, y)$ is called the *geodesic distance* between x and y, denoted by $d(x, y)$. In this paper, "distance" refers to the geodesic distance unless specified otherwise.

For a set X (for instance a polygon or a disk), ∂X denotes the boundary of X. For any points u and w on ∂P, let $C[u, w]$ be the part of ∂P in clockwise order from u to w. The subpolygon of P bounded by $\pi(u, w)$ and $C[u, w]$ is denoted by $P[u, w]$. Note that $P[u, w]$ may not be simple, but it is always weakly simple. Indeed, consider the Euclidean disks centered at points in $\pi(u, w)$ with radius $\varepsilon > 0$. There exists a simple curve connecting u and w that lies in the union of these disks and which does not cross $C[w, u]$. A region bounded by that simple curve and $C[w, u]$ is a simple polygon whose Fréchet distance from P is at most ε. The *radius* of P is defined as $\max_{p \in P} d(c, p)$ and is denoted by $r(P)$, where c is the optimal geodesic 1-center of P. We set $r(\alpha, \beta) = r(P[\alpha, \beta])$ for any points $\alpha, \beta \in \partial P$ for brevity.

The geodesic disk centered at v with radius r, denoted by $D_r(v)$, is the set of points whose geodesic distance from v is at most r. The boundary of a geodesic disk inside P consists of disjoint polygonal chains of ∂P and $O(n)$ circular arcs [3].

We call a pair of points (c_1, c_2) in P a *2-center*. The radius of a 2-center (c_1, c_2) in P is defined as $\max_{p \in P} \min\{d(c_1, p), d(c_2, p)\}$. If the radius of a 2-center is the minimum over all 2-centers in P, the 2-center is said to be *optimal*. Note that $D_r(c_1) \cup D_r(c_2)$ contains P for a 2-center (c_1, c_2) and a radius r at least as large as the radius of (c_1, c_2).

For any two points x and $y \in P$, the *bisector* of x and y is defined as the set of points in P equidistant from x and y. A bisector of two points may contain a two-dimensional region if there is a vertex of P equidistant from x and y. To avoid this, we define the *bisecting curve* of x and y, denoted by $b(x, y)$, to be the maximal curve which contains the midpoint of $\pi(x, y)$ and does not contain any point on ∂P in its interior.

All missing proofs can be found in the full version of this paper.

Lemma 1. *The bisecting curve is well-defined for any two points in P.*

3 The Partition by an Optimal 2-center

Let (c_1^*, c_2^*) be an optimal geodesic 2-center and r^* be the radius of (c_1^*, c_2^*). For any two points α and β on ∂P, let $r_{\max}(\alpha, \beta) = \max\{r(\alpha, \beta), r(\beta, \alpha)\}$.

Lemma 2. *If P is covered by two geodesic disks centered at points in P with radius r, then there are two points $x, y \in \partial P$ with $r_{\max}(x, y) \leq r$.*

For any 2-center (c_1, c_2) in P and radius r, we call a pair of points $\alpha, \beta \in \partial P$ a *point-partition* of P with respect to the tuple (c_1, c_2, r) if and only if $d(c_1, x) \leq r$ and $d(c_2, y) \leq r$ for all points $x \in P[\alpha, \beta]$ and $y \in P[\beta, \alpha]$. Note

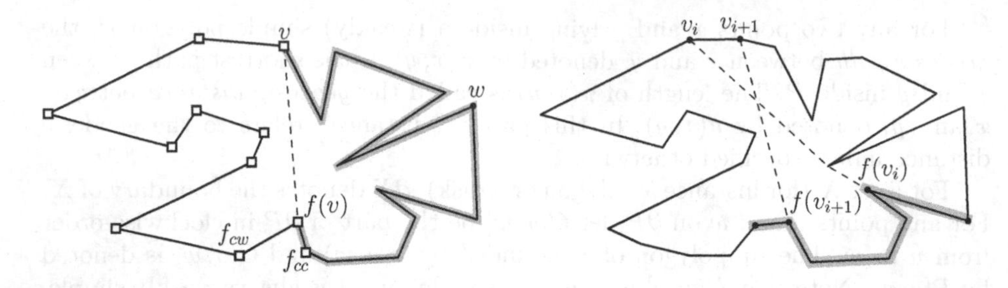

Fig. 1. (Left) The gray chain indicates $C[v, f(v)]$ and the points marked with squares are possible positions of $f(w)$. (Right) The edges on the gray chain are the candidate edges of $e_i = v_i v_{i+1}$.

that a point-partition with respect to (c_1, c_2, r) does not exist if r is less than the radius of (c_1, c_2). A pair of edges (e, e') is called an *edge-partition* with respect to (c_1, c_2, r) if and only if $x \in e$ and $y \in e'$ for some point-partition (x, y) with respect to (c_1, c_2, r). In particular, a point-partition and an edge-partition with respect to (c_1^*, c_2^*, r^*) are called *an optimal point-partition* and *an optimal edge-partition*, respectively. By Lemma 2, there always exist an optimal point-partition and an optimal edge-partition in a simple polygon. Note that a point-partition and an edge-partition with respect to (c_1, c_2, r) are not necessarily unique if $\min\{d(c_1, \alpha), d(c_1, \beta)\} < r$, where α and β are the two endpoints of $b(c_1, c_2)$. If an optimal point-partition (α, β) of P is given, we can compute an optimal 2-center in linear time using the algorithm in [1].

We first compute a set of edge pairs, which we call *candidate edge pairs*, containing at least one optimal edge-partition. For each candidate edge pair (e_i, e_j), we compute an *optimal 2-center* (c_1, c_2) *restricted to* (e_i, e_j), that is, c_1 and c_2 are 1-centers of $P[\alpha, \beta]$ and $P[\beta, \alpha]$, respectively, where (α, β) is the pair realizing $\inf_{(x,y) \in e_i \times e_j} r_{\max}(x, y)$.

3.1 Computing a Set of Candidate Edge Pairs

In this section, we define *candidate edge pairs* and describe how to find a set of all candidate edge pairs in $O(n^2)$ time. We define the function $f(v)$ which maps each vertex v of P to *the first clockwise vertex* v' of P from v that minimizes $r_{\max}(v, v')$.

Lemma 3. *Let v be a vertex of P and w be a vertex lying on $C[v, f(v)]$. Then $f(w) \in C[f_{cc}, v]$, where f_{cc} is the counterclockwise neighbor vertex of $f(v)$.*

An edge $e_j = v_j v_{j+1}$ is called a *candidate edge* of $e_i = v_i v_{i+1}$ if and only if (1) $f(v_{i+1}) = v_j$ and $f(v_i) = v_{j+1}$, or (2) v_j or v_{j+1} lies on $C[f(v_i), f(v_{i+1})]$ when $f(v_i)$ appears earlier than $f(v_{i+1})$ when we traverse the boundary of P in clockwise order from v_i (See Fig. 1 (Right)). A pair of edges (e_i, e_j) is called *a candidate edge pair* if e_j is a candidate edge of e_i.

Lemma 4. *There is an optimal edge-partition in the set of all candidate edge pairs.*

Lemma 5. *The number of candidate edge pairs is* $O(n)$.

Now we give a procedure that finds the set of all candidate edge pairs. First, we compute the vertex $f(v_1)$ by traversing all vertices of P in clockwise order. Afterwards, we find $f(v_i)$ for all $i \in [2, n]$. Suppose that we have already computed $f(v_{i-1})$ and we want to find $f(v_i)$. By Lemma 3, we do not need to consider the vertices lying in the interior of $C[v_{i-1}, f_{cc}]$, where f_{cc} is the counterclockwise neighbor vertex of $f(v_{i-1})$. Thus we traverse the vertices from f_{cc} in clockwise order and check whether the current vertex is $f(v_i)$. To check this, we consider three vertices: the current vertex v_c and the two neighbor vertices v_{n_1}, v_{n_2} of v_c. If $r_{\max}(v_c, v_i) \leq \min\{r_{\max}(v_{n_1}, v_i), r_{\max}(v_{n_2}, v_i)\}$, $f(v_i)$ is the current vertex v_c by the monotonicity of the functions $r(v, \cdot)$ and $r(\cdot, v)$, where v is a fixed vertex of P. Otherwise, v_c is not $f(v_i)$, so we move to the vertex next to v_c. We can find all $f(v_i)$ for $i \in [1, n]$ by traversing ∂P twice. For each vertex we visit during the traversal, we compute $r_{\max}(\alpha, \beta)$ for three different pairs (α, β) of vertices of P, each of which takes $O(n)$ time by the algorithm in [1].

Afterwards, we compute the set of all candidate edge pairs based on the information we just computed. For each edge e_i, we traverse the edges lying between $f(v_i)$ and $f(v_{i+1})$. It takes time proportional to the number of candidate edge pairs, which is $O(n)$ by Lemma 5.

Lemma 6. *The set of all candidate edge pairs can be computed in* $O(n^2)$ *time.*

4 A Decision Algorithm for a Candidate Edge Pair

We say that a point-partition (α, β) is *restricted* to (e_i, e_j) if $\alpha \in e_i$ and $\beta \in e_j$. We say that a tuple (c_1, c_2, r) consisting of a 2-center (c_1, c_2) and a radius r is *restricted* to (e_i, e_j) if some point-partitions with respect to (c_1, c_2, r) are restricted to (e_i, e_j). We can view $r_{\max}(\alpha, \beta)$ as the function whose variables are $\alpha \in e_i$ and $\beta \in e_j$. Since the function is continuous and the domain is bounded, there exist two points $\alpha^* \in e_i, \beta^* \in e_j$ that minimize the function. We call (c_1^*, c_2^*) *an optimal 2-center restricted* to (e_i, e_j) if c_1^* and c_2^* are the optimal 1-centers of $P[\alpha^*, \beta^*]$ and $P[\beta^*, \alpha^*]$, respectively.

In this section, we present a decision algorithm for a candidate edge pair (e_i, e_j). Let r_{ij}^* be the radius of an optimal 2-center restricted to (e_i, e_j). Let r be an input of the algorithm. The decision algorithm decides whether $r_{ij}^* > r$ or not.

We assume that $r(v_{i+1}, v_j) \leq r \leq r(v_i, v_{j+1})$ and $r(v_{j+1}, v_i) \leq r \leq r(v_j, v_{i+1})$, since the other cases can be handled easily.

The decision algorithm first assumes that $r \geq r_{ij}^*$ and constructs a 2-center restricted to (e_i, e_j) with radius r. The 2-center produced by the algorithm is valid if and only if $r \geq r_{ij}^*$. Therefore, the algorithm can then decide whether $r \geq r_{ij}^*$ by checking whether the 2-center is valid. Thus, from now on, we assume that $r > r^*_{ij}$. Let (c_1, c_2, r) be a tuple consisting of a 2-center (c_1, c_2) and radius r which is restricted to (e_i, e_j), and (α, β) be a point-partition with respect to (c_1, c_2, r) which is restricted to (e_i, e_j). Without loss of generality, we assume that $D_r(c_1)$ covers $P[\beta, \alpha]$ and $D_r(c_2)$ covers $P[\alpha, \beta]$.

4.1 Computing the Intersection of Geodesic Disks

The first step of the decision algorithm is to compute $I_1 = \cap_{k=j+1}^{i} D_r(v_k)$ and $I_2 = \cap_{k=i+1}^{j} D_r(v_k)$ in $O(n \log n)$ time. Clearly, $c_1 \in I_1$ and $c_2 \in I_2$. The following lemmas show a few properties of ∂I_1 and ∂I_2 which will be used by our algorithm.

Lemma 7. *The number of circular arcs in ∂I_1 and ∂I_2 is $O(n)$.*

Lemma 8. *Let $\mathcal{D} = \{D_1, \ldots, D_k\}$ be a set of geodesic disks with the same radius and let I be the intersection of all disks in $\mathcal{D}$. Let $S = < s_1, \ldots, s_t >$ be the cyclic sequence of circular arcs of ∂I along its boundary in clockwise order. For any $i \in [1, k]$, the circular arcs in $\partial I \cap \partial D_i$ are consecutive in S.*

Let S_1 and S_2 be the closures of $\partial I_1 \setminus \partial P$ and $\partial I_2 \setminus \partial P$, respectively. Note that ∂I_1 and ∂I_2 consist of $O(n)$ circular arcs and (possibly incomplete) edges of ∂P. By the following lemma, it is sufficient to choose two points, one from S_1 and one from S_2, in order to find a 2-center restricted to (e_i, e_j) with radius r.

Lemma 9. *If $r \geq r^*_{ij}$, there is a tuple (c_1, c_2, r) restricted to (e_i, e_j) such that $c_1 \in S_1$ and $c_2 \in S_2$.*

4.2 Subdividing the Edges and the Chains

We choose any two points $w_1 \in \partial I_1$ and $w_2 \in \partial I_2$ which are endpoints of some circular arcs of ∂I_1 and ∂I_2, respectively. We use them as reference points for ∂I_1 and ∂I_2. To find a 2-center, we traverse ∂I_1 and ∂I_2 from w_1 and w_2, respectively. We write $p \prec q$ for any two points $p \in \partial I_t$ and $q \in \partial I_t$, if p comes earlier than q when we traverse ∂I_t in clockwise order from the reference point w_t for $t = 1, 2$. We consider ∂I_1 and ∂I_2 as chains starting from w_1 and w_2, respectively.

We compute the shortest path maps SPM_{v_i} and $\text{SPM}_{v_{i+1}}$, where SPM_x is the graph which is obtained by extending the edges of the shortest path tree rooted at a vertex x towards their descendants [8]. By overlaying them with ∂I_1, we obtain the set of $O(n)$ *finer* arcs of ∂I_1 in linear time.

We also subdivide the polygon edge e_i into $O(n)$ *subedges* by overlaying the extensions of the edges in the shortest path trees rooted at v_i and v_{i+1} towards the parents of their endpoints with e_i. Let $\mathcal{L}_i$ be the set of intersections of the extensions of the edges in the shortest path trees with e_i. We define $\mathcal{L}_j$ similarly.

For any point x on a finer circular arc and any point p on a subedge, the combinatorial structures of $\pi(x, p)$ are the same if the geodesic path $\pi(x, p)$ consists of more than two segments for all x, p. Otherwise, there are at most three distinct combinatorial structures of $\pi(x, p)$ depending on x and p.

4.3 Four Coverage Functions and Their Extrema

We define four functions $\phi_t(x)$ and $\psi_t(x)$ for $t = 1, 2$ as follows. We represent each point $p \in e_i$ (and $q \in e_j$) as a real number $\|v_i - p\|/\|v_i - v_{i+1}\|$ (and $\|v_j - q\|/\|v_j - v_{j+1}\|$) in $[0, 1]$, where $\|x - y\|$ is the Euclidean distance between the points x and y. We use a real number in $[0, 1]$ and its corresponding point interchangeably. We define the function $\phi_1 : S_1 \to [0, 1]$ (and $\phi_2 : S_2 \to [0, 1]$) that maps $x \in S_1$ (and $x \in S_2$) to the supremum (and the infimum) of the numbers which represent the points in $D_r(x) \cap e_i$. Similarly, we define a function $\psi_1 : S_1 \to [0, 1]$ (and $\psi_2 : S_2 \to [0, 1]$) that maps $x \in S_1$ (and $x \in S_2$) to the infimum (and the supremum) of the numbers which represent the points in $D_r(x) \cap e_j$.

We need to split S_t (for $t = 1, 2$) in subsections such that ϕ_t and ψ_t are monotone when restricted to each subsection. However, S_t is not necessarily a connected subset of ∂I_t. Thus, to simplify the description of the split, we define four continuous functions $\phi'_t, \psi'_t : \partial I_t \to [0, 1]$ by interpolating ϕ_t and ψ_t on ∂I_t:

$$\phi'_t(x) = \begin{cases} \phi_t(x) & \text{if } x \in S_t \\ \frac{d_c(x,x_1)}{d_c(x_1,x_2)}\phi_t(x_1) + \frac{d_c(x,x_2)}{d_c(x_1,x_2)}\phi_t(x_2) & \text{otherwise,} \end{cases}$$

where x_1 and x_2 are the first and the last point of S_t along ∂I_t from x in clockwise order, respectively, and $d_c(x', y')$ denotes the length of a chain $C[y', x']$. The function ψ'_t is defined similarly. Recall that we let w_t be the reference point for ∂I_t. As we traverse ∂I_t, we eventually come back to w_t. Therefore, w_t can be thought of as the first and the last point of ∂I_t. The functions ϕ'_t and ψ'_t are *well-defined* in the sense that each of $\phi'_t(w_t)$ and $\psi'_t(w_t)$ takes the same value, no matter if w_t is the first or the last point of ∂I_t.

Lemma 10. *The functions ϕ'_t and ψ'_t (for $t = 1, 2$) are well-defined. Excluding w_t, each of ϕ'_t and ψ'_t has at most one local maximum and at most one local minimum. Moreover, these local extrema lie on S_t.*

We can compute the local extrema of ϕ'_t and ψ'_t in $O(\log^3 n)$ time. These local extrema subdivide ∂I_1 into at most five subchains $c_{1,k}$ for $k \in \{1, 2, \ldots, 5\}$ as follows. Let x_1, x_2, x_3 and x_4 be the local maxima and the local minima of ϕ'_1 and ψ'_1 with $x_1 \prec x_2 \prec x_3 \prec x_4$. The subchain $c_{1,k}$ is the set of points $x \in \partial I_1$ with $x_{k-1} \prec x \prec x_k$ for $k \in \{1, 2, \ldots, 5\}$, where we set $x_0 = x_5 = w_1$. Similarly, the local extrema of ϕ'_2 and ψ'_2 divide the chain ∂I_2 into five subchains $c_{2,\ell}$ ($\ell \in \{1, 2, \ldots, 5\}$). After subdividing ∂I_1, ϕ_1 and ψ_1 are monotone when the domain is restricted to $c_{1,k} \cap S_1$ ($k \in \{1, 2, \ldots, 5\}$). Similarly, ϕ_2 and ψ_2 restricted to $c_{2,\ell} \cap S_2$ ($\ell \in \{1, 2, \ldots, 5\}$) are monotone.

4.4 Computing a 2-center for a Pair of Subchains

We consider a pair of subchains $(c_{1,k}, c_{2,\ell})$ ($k \in \{1, 2, \ldots, 5\}$ and $\ell \in \{1, 2, \ldots, 5\}$). Let $s_{1,k} = S_1 \cap c_{1,k}$ and $s_{2,\ell} = S_2 \cap c_{2,\ell}$. We find a 2-center with radius r that is restricted to (e_i, e_j), if it exists, where one center is on

$s_{1,k}$ and the other is on $s_{2,\ell}$. Assume that ϕ_1 and ψ_1 (respectively ϕ_2 and ψ_2) are decreasing when their domain is restricted to $s_{1,k}$ (respectively to $s_{2,\ell}$). The other cases can be handled in a similar way.

We define two new functions $\mu_1 : s_{1,k} \to s_{2,\ell}$ and $\mu_2 : s_{1,k} \to s_{2,\ell}$. For a point $x \in s_{1,k}$, $\mu_1(x)$ (and $\mu_2(x)$) is the first clockwise point (and the last clockwise point) in $s_{2,\ell}$ which is covered by $D_r(\phi_1(x))$ (and $D_r(\psi_1(x))$). If every point in $s_{2,k}$ is covered by $D_r(\phi_1(x))$, then $\mu_1(x)$ is defined as the last clockwise point of $s_{2,l}$. Notice that $\mu_1(x)$ and $\mu_2(x)$ are increasing on $s_{1,k}$. Then, if there is a point $x \in s_{1,k}$ such that $\mu_1(x) \prec \mu_2(x)$, $(x, \mu_1(x), r)$ is restricted to (e_i, e_j).

To decide whether this is the case, we traverse $c_{1,k}$ and $c_{2,\ell}$ and $(x, \mu_1(x), r)$ is restricted to (e_i, e_j) for the current points $x \in c_{1,k}$ and $\mu_1(x) \in c_{2,\ell}$. As we scan the subchains, we pick $O(n)$ points, which are called *event points.* Then we compute $\mu_1(x)$ and $\mu_2(x)$ for all event points in $s_{1,k}$ in linear time. Then we traverse the two subchains and find a 2-center using the information we just computed.

We explain how we compute the event points on $c_{1,k}$. The set of *event points* of $c_{1,k}$ is the subset of $c_{1,k}$ consisting of points belonging to one of the three types defined below.

- (T1) The subchain $c_{1,k} \subseteq \partial I_1$ consists of circular arcs and line segments. It was subdivided by two SPM's in Sect. 4.2. The endpoints of all these circular arcs, line segments and subdivisions are the event points of type T1.
- (T2) the points $x \in s_{1,k}$ such that $d(x,p) = r$ for some $p \in \mathcal{L}_i$.
- (T3) the points $x \in s_{1,k}$ such that $d(x,p) = r$ for some $p \in \mathcal{L}_j$.

Let $\mathcal{E}_1$, $\mathcal{E}_2$ and $\mathcal{E}_3$ be the sets of points of types T1, T2 and T3, respectively. Let $\mathcal{E} = \mathcal{E}_1 \cup \mathcal{E}_2 \cup \mathcal{E}_3$. We say $\eta \in \mathcal{E}$ is *caused* by p if $d(\eta, p) = r$ for $p \in \mathcal{L}_i \cup \mathcal{L}_j$. We do not need to compute $\mathcal{E}_1$, since we already maintain the arcs of $c_{1,k}$ in clockwise order. Recall that $\mathcal{L}_i$ is the set of intersection points of the extensions of the edges in the two shortest path trees of v_i, v_{i+1} with e_i. Let $\mathcal{L}_i = \{p_1, \ldots, p_m\}$ be sorted along e_i in clockwise order from v_i.

We explain how to compute $\mathcal{E}_2$ by traversing $c_{1,k}$ from its starting point. Assume that we have reached an event point $\eta \in \mathcal{E}_1 \cup \mathcal{E}_2$ and have already computed all T2 points on the subchain lying before η. Let η' be the T1 point next to η. It is sufficient to show that we can find all T2 points on the subchain lying between η and η' by walking the subchain from η to η' once. If $\eta \in c_{1,k} \backslash s_{1,k}$, let $h(\eta)$ be the last T2 point in $s_{1,k}$ in clockwise order with $h(\eta) \prec \eta$. Otherwise, let $h(\eta) = \eta$. While computing all T2 points, we also compute $\phi_1(h(\eta))$ and $\pi(\eta, \phi_1(h(\eta)))$ for every event point $\eta \in \mathcal{E}_1 \cup \mathcal{E}_2$.

If the subchain of $c_{1,k}$ connecting η' and η lies in ∂P, there are two possible cases: $\eta' \notin s_{1,k}$ or $\eta' \in s_{1,k}$. In both cases, there is no T2 point between η' and η. Thus η' is the event point next to η and it is sufficient to compute $\phi_1(h(\eta))$ and $\pi(\eta, \phi_1(h(\eta)))$. If $\eta' \notin s_{1,k}$ then $h(\eta') = h(\eta)$. Thus, we compute $\pi(\eta', \phi_1(h(\eta)))$, which takes constant time. If $\eta' \in s_{1,k}$, then $h(\eta') = \eta'$. To compute $\pi(\eta', \phi_1(\eta'))$, we first compute $d(\eta', p_{i'})$, where $p_{i'}p_{i'+1}$ is the subedge of e_i which contains $\phi_1(h(\eta))$. Since ϕ_1 is decreasing, $\phi_1(\eta')$ lies on $C[v_i, \phi_1(h(\eta))]$. Thus, if $d(\eta', p_{i'}) > r$, then $\phi_1(h(\eta))$ does not lie on $C[p_{i'}, \phi_1(h(\eta))]$, so we can skip $p_{i'}$. We check each subedge of e_i from $p_{i'}$ in counterclockwise order until it contains $\phi_1(\eta')$, and compute the geodesic path $\pi(\eta', \phi_1(\eta'))$ for $\phi_1(\eta')$ on the subedge. It takes time proportional to the number of subedges we traverse on e_i.

Otherwise, the subchain of $c_{1,k}$ connecting η' and η lies in $s_{1,k}$. In this case, we first compute $d(\eta', p_{i'})$, where $p_{i'}p_{i'+1}$ is the subedge of e_i which contains $\phi_1(h(\eta))$. If $d(\eta', p_{i'})$ is smaller than r, then $\phi_1(\eta')$ lies between $\phi_1(\eta)$ and $p_{i'}$, and it can be computed in constant time. In this case, there is no T2 point between η and η'. If $d(\eta', p_{i'})$ is greater than or equal to r, then there is an event point caused by $p_{i'}$ lying between η and η'. It can be computed in constant time. Moreover, it is the first T2 point from η.

Until now, we have computed the set $\mathcal{E}_1 \cup \mathcal{E}_2$ and the values $\phi_1(\eta)$ for all $\eta \in \mathcal{E}_1 \cup \mathcal{E}_2$ lying in $s_{1,k}$. Similarly, we also compute $\mathcal{E}_3$ and ϕ_1 for all event points in $\mathcal{E}_3$. Moreover, we can define the event points in $c_{2,\ell}$ and compute all event points similarly. Using this information, we can compute $\mu_1(x)$ and $\mu_2(x)$ for all $x \in \mathcal{E}$ in linear time.

By the following lemma, we can find a 2-center restricted to (e_i, e_j) with radius r by traversing $c_{1,k}$ once.

Lemma 11. *Let η and η' in $\mathcal{E}$ be two event points adjacent along $s_{1,k}$. We can determine whether there is a point $\eta \prec x \prec \eta'$ such that $\mu_1(x) \prec \mu_2(x)$ in time proportional to the number of event points between $\mu_2(\eta')$ and $\mu_1(\eta)$.*

If $r^*_{ij} \geq r$, there exists a 2-center (c_1, c_2) with radius r such that $c_1 \in S_1$ and $c_2 \in S_2$ (see Lemma 9). We have $\mu_1(c_1) \prec c_2 \prec \mu_2(c_1)$. Thus, the algorithm always considers the subarc pair, one containing c_1 and the other containing c_2.

4.5 The Analysis of the Decision Algorithm

The first step, computing the intersection of the geodesic disks, takes linear time once the farthest-point geodesic Voronoi diagrams of the vertices of $C[v_{j+1}, v_i]$ and of the vertices of $C[v_{i+1}, v_j]$ have been constructed. The second step, subdividing the edges and the chains with the SPM's, takes $O(n)$ time. The third step, finding the local extrema of ϕ'_t and ψ'_t (for $t = 1, 2$), takes $O(\log^3 n)$ time.

In the last step, we consider $O(1)$ subchain pairs. For a given subchain pair $(c_{1,k}, c_{2,\ell})$, we compute all event points on the subchains. The set $\mathcal{E}_1$ of T1 points has already been given. The sets $\mathcal{E}_2$ and $\mathcal{E}_3$ of T2 and T3 points, respectively, can be computed in $O(n)$ time by the following lemma.

Lemma 12. *Let η be a T1 or T2 point that the algorithm has computed. If we have already computed $\pi(\eta, \phi_1(\eta))$ and the subedge $p_{i'}p_{i'+1}$ of e_i containing $\phi_1(\eta)$, we can compute $d(\eta', p_{i'})$ in constant time, where η' is the T1 event point next to η.*

Lemma 13. *For a candidate edge pair (e_i, e_j) and a radius r, we can decide whether $r \geq r^*_{ij}$ in $O(n)$ time, once the farthest-point geodesic Voronoi diagrams of the vertices of $C[v_{j+1}, v_i]$ and of the vertices of $C[v_{i+1}, v_j]$ are constructed.*

5 An Optimization Algorithm for a Candidate Edge Pair

Using the parametric searching technique [11], our decision algorithm can be extended into an optimization algorithm as follows.

Instead of computing $\partial I_1(r_{ij}^*)$ and $\partial I_2(r_{ij}^*)$ explicitly, we compute the combinatorial structures of $\partial I_1(r_{ij}^*)$ and $\partial I_2(r_{ij}^*)$. There is an interval $[r_1, r_2]$ which contains r_{ij}^* and such that the combinatorial structures of $I_1(r)$ and $I_2(r)$ are the same for any $r \in [r_1, r_2]$. We first find such an interval by binary search on a set of candidate radii using our decision algorithm. Let r_L and r_U be the radii which we obtain by binary search. We have $r_{ij}^* \in [r_L, r_U]$. We construct the combinatorial structures of $I_1(r_1)$ and $I_2(r_1)$.

Then we subdivide $\partial I_1(r_{ij}^*)$ (and $\partial I_2(r_{ij}^*)$) into $O(n)$ *finer arcs* by overlaying the shortest path maps of v_i, v_{i+1}, v_j, and v_{j+1} with $\partial I_1(r_{ij}^*)$ (and $\partial I_2(r_{ij}^*)$). To be specific, we compute the combinatorial structure of the subdivisions of $\partial I_1(r_{ij}^*)$ and $\partial I_2(r_{ij}^*)$. Let $g_1(r), \ldots, g_m(r)$ (and $g'_1(r), \ldots, g'_{m'}(r)$) be the algebraic functions of constant degree which represent the endpoints of the finer arcs of $\partial I_1(r)$ (and $\partial I_2(r)$) for $r \in [r_L, r_U]$.

Lemma 14. *The set of endpoints of the finer arcs of $\partial I_1(r)$ and $\partial I_2(r)$ can be computed in $O(n \log^2 n)$ time for $r \in [r_L, r_U]$.*

5.1 Computing the Coverage Function Values

Lemma 15. *For an index $k \in [1, m]$, the subedge of e_i containing $\phi_1(g_k(r_{ij}^*))$ can be computed in $O(n \log n)$ time.*

We can compute the subedge of e_i containing $g_k(r_{ij}^*)$ for all indices $k \in [1, m]$ in $O(n^2 \log n)$ time. To compute them efficiently, we parallelize this procedure using $O(n)$ processors. The details can be found in the full version of the paper. Afterwards, we compute the algebraic functions $\phi_1(g_k(r))$ and $\phi_2(g'_{k'}(r))$ for all $k \in [1, m]$ and all $k' \in [1, m']$. Then we sort the points in $\mathcal{L}_i$ and the points $\phi_1(g_k(r_{ij}^*)), \phi_2(g'_{k'}(r_{ij}^*))$ for all $k \in [1, m]$ and all $k' \in [1, m']$ in $O(n \log^2 n)$ time using Cole's parallelized sorting algorithm [7] and the decision algorithm in Sect. 4.

Lemma 16. *The points $\phi_1(g_k(r_{ij}^*)), \phi_2(g'_k(r_{ij}^*))$ for all indices $k \in [1, m]$ and the points in $\mathcal{L}_i$ can be sorted in $O(n \log^2 n)$ time.*

5.2 Constructing 4-Tuples Consisting of Two Cells and Two Subedges

Consider a 4-tuple (x_1, x_2, y_1, y_2), where x_t is a finer arc of $\partial I_t(r_{ij}^*)$ for $t = 1, 2$ and y_1, y_2 are subedges in e_i, e_j, respectively. We say the 4-tuple (x_1, x_2, y_1, y_2) is *optimal* if there is an optimal 2-center (c_1^*, c_2^*) such that $c_1^* \in x_1, c_2^* \in x_2$ and $\alpha \in y_1, \beta \in y_2$ for some point-partition (α, β) with respect to (c_1^*, c_2^*, r_{ij}^*). If an optimal 4-tuple is given, then we can compute c_1^* and c_2^* in constant time.

Lemma 17. *Given an optimal 4-tuple* (x_1, x_2, y_1, y_2)*, an optimal 2-center* (c_1, c_2) *restricted to the candidate edge pair* (e_i, e_j) *can be computed in constant time.*

Instead of considering all 4-tuples, we construct a set of 4-tuples with size $O(n)$ containing at least one optimal 4-tuple by parallelizing the procedure in Sect. 4.4 after modifying it.

Lemma 18. *An optimal 2-center restricted to a given candidate edge pair can be computed in* $O(n \log^2 n)$ *time.*

Theorem 1. *An optimal 2-center can be computed in* $O(n^2 \log^2 n)$ *time.*

References

1. Ahn, H.-K., Barba, L., Bose, P., De Carufel, J.-L., Korman, M., Oh, E.: A linear-time algorithm for the geodesic center of a simple polygon. In: Arge, L., Pach, J. (eds.) 31st International Symposium on Computational Geometry (SoCG 2015). Leibniz International Proceedings in Informatics (LIPIcs), vol. 34, pp. 209–223. Schloss Dagstuhl-Leibniz-Zentrum fuer Informatik, Dagstuhl (2015)
2. Asano, T., Toussaint, G.T.: Computing geodesic center of a simple polygon. Technical Report SOCS-85.32, McGill University (1985)
3. Borgelt, M.G., Van Kreveld, M., Luo, J.: Geodesic disks and clustering in a simple polygon. Int. J. Comput. Geom. Appl. **21**(06), 595–608 (2011)
4. Chan, T.M.: More planar two-center algorithms. Comput. Geom. **13**(3), 189–198 (1999)
5. Chang, H.-C., Erickson, J., Xu, C.: Detecting weakly simple polygons. In: Proceedings of the Twenty-Sixth Annual ACM-SIAM Symposium on Discrete Algorithms (SODA 2015), pp. 1655–1670. SIAM (2015)
6. Chazelle, B., Matoušek, J.: On linear-time deterministic algorithms for optimization problems in fixed dimension. J. Algorithms **21**(3), 579–597 (1996)
7. Cole, R.: Parallel merge sort. SIAM J. Comput. **17**(4), 770–785 (1988)
8. Guibas, L., Hershberger, J., Leven, D., Sharir, M., Tarjan, R.: Linear-time algorithms for visibility and shortest path problems inside triangulated simple polygons. Algorithmica **2**(1–4), 209–233 (1987)
9. Hwang, R., Lee, R., Chang, R.: The slab dividing approach to solve the Euclidean p-center problem. Algorithmica **9**(1), 1–22 (1993)
10. Kim, S.K., Shin, C.-S.: Efficient algorithms for two-center problems for a convex polygon. In: Du, D.-Z., Eades, P., Sharma, A.K., Lin, X., Estivill-Castro, V. (eds.) COCOON 2000. LNCS, vol. 1858, pp. 299–309. Springer, Heidelberg (2000)
11. Megiddo, N.: Applying parallel computation algorithms in the design of serial algorithms. J. ACM **30**(4), 852–865 (1983)
12. Megiddo, N.: On the ball spanned by balls. Discrete Comput. Geom. **4**(1), 605–610 (1989)
13. Pollack, R., Sharir, M., Rote, G.: Computing the geodesic center of a simple polygon. Discrete Comput. Geom. **4**(1), 611–626 (1989)
14. Vigan, I.: Packing and covering a polygon with geodesic disks. Technical report, CoRR abs/1311.6033 (2013)

Choice Is Hard

Esther M. Arkin[1], Aritra Banik[2], Paz Carmi[2], Gui Citovsky[1], Matthew J. Katz[2](✉), Joseph S.B. Mitchell[1], and Marina Simakov[2]

[1] Department of Applied Mathematics and Statistics, Stony Brook University, Stony Brook, USA
{estie,gcitovsk,jsbm}@ams.stonybrook.edu

[2] Department of Computer Science, Ben-Gurion University, Beersheba, Israel
aritrabanik@gmail.com, carmip@gmail.com, {matya,simakov}@cs.bgu.ac.il

Abstract. Let $P = \{C_1, C_2, \ldots, C_n\}$ be a set of color classes, where each color class C_i consists of a pair of objects. We focus on two problems in which the objects are points on the line. In the first problem (*rainbow minmax gap*), given P, one needs to select exactly one point from each color class, such that the maximum distance between a pair of consecutive selected points is minimized. This problem was studied by Consuegra and Narasimhan, who left the question of its complexity unresolved. We prove that it is NP-hard. For our proof we obtain the following auxiliary result. A 3-SAT formula is an LSAT formula if each clause (viewed as a set of literals) intersects at most one other clause, and, moreover, if two clauses intersect, then they have exactly one literal in common. We prove that the problem of deciding whether an LSAT formula is satisfiable or not is NP-complete. We present two additional applications of the LSAT result, namely, to *rainbow piercing* and *rainbow covering*.

In the second problem (*covering color classes with intervals*), given P, one needs to find a minimum-cardinality set $\mathcal{I}$ of intervals, such that exactly one point from each color class is covered by an interval in $\mathcal{I}$. Motivated by a problem in storage systems, this problem has received significant attention. Here, we settle the complexity question by proving that it is NP-hard.

1 Introduction

A multiple choice problem consists of a set of color classes $P = \{C_1, C_2, \ldots, C_n\}$, where each color class C_i consists of a pair of objects. When the underlying objects are points (resp., intervals) on the x-axis, we say that P is a set of point (resp., interval) color classes. Consider a set P of point color classes. We call an interval on the x-axis that contains at most one point from each color class a *conflict-free* interval (or CF-interval for short). Given a set of color classes P and a set $Q \subseteq \cup_{i=1}^{n} C_i$, we say that Q is a *rainbow* if it contains at most one object

Research supported by US-Israel Binational Science Foundation (project 2010074). E. Arkin and J. Mitchell are partially supported by the National Science Foundation (CCF-1526406).

K. Elbassioni and K. Makino (Eds.): ISAAC 2015, LNCS 9472, pp. 318–328, 2015.
DOI: 10.1007/978-3-662-48971-0_28

from each color class. The first problem that we study (rainbow minmax gap) is mentioned in a recent paper by Consuegra and Narasimhan [6].

Rainbow Minmax Gap (Decision Version): Given a set P of n point color classes and a value $d > 0$, determine whether there exists a rainbow Q of size n with *max gap* at most d, where the max gap of Q is the maximum distance between a pair of consecutive points in Q.
This problem is the 1-dimensional version of a more general 2-dimensional problem. Consider a set of agents (represented by points in the plane) where each agent provides a certain service, and for each of these services, there are several agents in the set providing this service. The goal is to compute a minimum bottleneck spanning tree consisting of exactly one agent for each of the available services. In [6], the authors present a 2-approximation algorithm for rainbow minmax gap, but leave the question whether the problem is NP-hard or not open. In Sect. 3 we prove that the problem is NP-hard.

In order to obtain this result we define a new and especially simple satisfiability problem, which we call *linear SAT* (or LSAT for short), and prove that it is still NP-complete. A 3-SAT formula is an LSAT formula if each clause (viewed as a set of literals) intersects at most one other clause, and, moreover, if two clauses intersect, then they have exactly one literal in common. An LSAT formula can be depicted as a set of disjoint semi-closed intervals on a line, see Fig. 1. We prove that the problem of deciding whether an LSAT formula is satisfiable or not is NP-complete. This is quite surprising, since the satisfiability problem for the class of formulas that can be depicted as disjoint closed intervals on a line is already polynomially solvable. We believe that the NP-completeness of LSAT may be useful in deriving other hardness results. In particular, we use LSAT to prove NP-hardness of the following two multiple choice problems, see Sect. 3.

Rainbow Piercing: Given a set P of point color classes and a set of intervals $\mathcal{I}$ on the x-axis, determine whether there exists a rainbow Q that is a *piercing set* for $\mathcal{I}$ (i.e., each interval in $\mathcal{I}$ is pierced by at least one point in Q).

Rainbow Covering: Given a set P of interval color classes, i.e., where each color class C_i is a pair of intervals on the x-axis, and a set of points S on the x-axis, determine whether there exists a rainbow Q that *covers* S (i.e., each point in S is covered by at least one interval in Q).

A fascinating related problem is: cover exactly one point from each color class using a minimum number of (arbitrary) intervals. This problem is motivated by the following problem. Consider a storage system where each item is stored in multiple places, and the objective is to retrieve all items with a minimum number of contiguous read operations. The problem is formally defined as follows.

Covering Color Classes with Intervals of Arbitrary Length: Given a set P of point color classes, find a minimum-cardinality set $\mathcal{I}$ of intervals of arbitrary length, such that exactly one point from each color class is covered by an interval in $\mathcal{I}$.

In Sect. 4 we show that this problem is NP-hard, by first showing that the following simpler problem is NP-hard.

Covering Color Classes with Unit Length Intervals: Given a set P of point color classes, decide whether or not there exists a set of unit length intervals, $\mathcal{I}$, such that exactly one point from each color class is covered. Assuming a feasible solution exists, minimize the cardinality of $\mathcal{I}$.

Related Work. As far as we know, the first to consider a "multiple-choice" problem of this kind were Gabow et al. [9], who studied the following problem. Given a directed acyclic graph with two distinguished vertices s and t and a set of k pairs of vertices, determine whether there exists a path from s to t that uses at most one vertex from each of the given pairs. They showed that the problem is NP-complete. A sample of additional graph problems of this kind can be found in [2,10,13]. The first to consider a problem of this kind in a geometric setting were Arkin and Hassin [3], who studied the following problem. Given a set V and a collection of subsets of V, find a cover of minimum diameter, where a cover is a subset of V containing at least one representative from each subset. They also considered the multiple-choice dispersion problem, which asks one to maximize the minimum distance between any pair of elements in the cover. They proved that both problems are NP-hard. Recently, Arkin et al. [1] considered the following problem. Given a set S of n pairs of points in the plane, color the points in each pair by red and blue, so as to optimize the radii of the minimum enclosing disk of the red points and the minimum enclosing disk of the blue points. In particular, they consider the problems of minimizing the maximum and minimizing the sum of the two radii. In another recent paper, Consuegra and Narasimhan [6] consider several problems of this kind, including the rainbow minmax gap problem, for which they present a 2-approximation algorithm (and we prove NP-hardness).

2 A New Satisfiability Result

In the boolean satisfiability problem (SAT), one is given a formula in conjunctive normal form and the goal is to determine whether it is satisfiable or not. SAT is one of the first problems that was shown to be NP-complete (by Cook [7]). Subsequently, many variants of SAT were shown to be NP-complete, including the variant known as 3-SAT, in which each clause consists of at most three literals [5,11,14]. Some restricted variants of SAT can be solved in polynomial time [4,8,12]. In this section we define an especially simple variant of 3-SAT, which we call *linear SAT* (LSAT for short), and prove that it is NP-complete. A 3-SAT formula is an LSAT formula if each clause (viewed as a set of literals) intersects at most one other clause. Moreover, if two clauses intersect, then they have exactly one literal in common. Let F be an LSAT formula and let T be its corresponding set of literals, then F can be depicted as in Fig. 1. That is, one can sort the literals in T, such that (i) each clause of F corresponds to at most three consecutive literals in the sorted list, and (ii) each clause shares at most one of its literals with another clause, in which case this literal is extreme in both clauses.

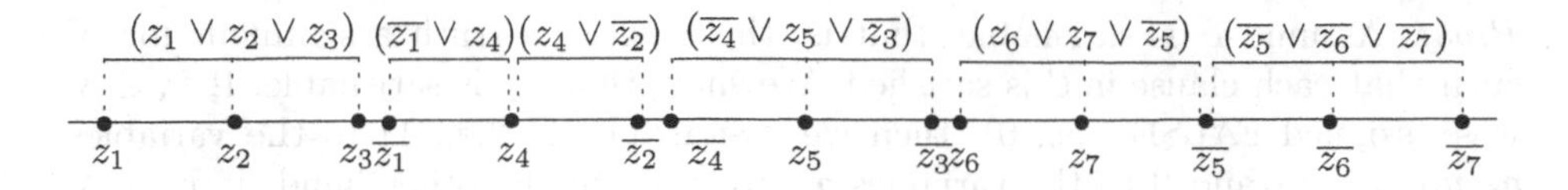

Fig. 1. An example of an LSAT formula.

Observe that if the clauses of a 3-SAT formula F are pairwise disjoint, then one can determine in polynomial time whether F is satisfiable or not, by determining whether the corresponding bipartite graph in which there is an edge between clause C and variable x if and only if either x or $\overline{x}$ appear in C contains a perfect matching. It is therefore somewhat surprising that LSAT is NP-complete, since the clauses of an LSAT formula are almost pairwise disjoint. We now prove that LSAT is NP-complete by a reduction from 3,4-SAT. A 3-SAT formula is a 3,4-SAT formula if each variable appears in at most 4 clauses, either negated or unnegated. 3,4-SAT was shown to be NP-complete by Tovey [14].

Let F be a 3,4-SAT formula, let X be the underlying set of variables, and let $\mathcal{C}$ be the set of clauses of F. Without loss of generality, we assume that each variable $x_i \in X$ appears unnegated (i.e., as x_i) in at most three clauses and negated (i.e., as $\overline{x_i}$) in at most two clauses. We construct an LSAT formula $F_L = (X_L, \mathcal{C}_L)$ from F, and show that there is a truth assignment for X such that each clause in $\mathcal{C}$ is satisfied if and only if there is a truth assignment for X_L such that each clause in $\mathcal{C}_L$ is satisfied. We construct X_L from X as follows. For each variable $x_i \in X$ we add to X_L the variables $x_i, a_i, y_{i1}, y_{i2}, y_{i3}, z_{i1}, z_{i2}$. Also, for each variable x_i we add the following clauses to $\mathcal{C}_L$:

1. $(y_{i1} \vee x_i)$ 2. $(x_i \vee a_i)$ 3. $(y_{i2} \vee \overline{a_i})$ 4. $(\overline{a_i} \vee y_{i3})$ 5. $(z_{i1} \vee \overline{x_i})$ 6. $(\overline{x_i} \vee z_{i2})$.

Observe that clause 1 and clause 2 share x_i, clause 3 and 4 share $\overline{a_i}$, and clause 5 and 6 share $\overline{x_i}$. Now, for each clause $C_i \in \mathcal{C}$, we add a clause to $\mathcal{C}_L$ as follows. For each variable x_i that appears in C_i, if x_i appears unnegated, then we replace it by $\overline{y_{i1}}$, $\overline{y_{i2}}$, or $\overline{y_{i3}}$, depending on whether this is the first, second, or third occurrence of x_i, and if x_i appears negated, we replace it by $\overline{z_{i1}}$ or $\overline{z_{i2}}$, depending on whether this is the first or second occurrence of $\overline{x_i}$. For example, given the formula

$$(x_1 \vee x_2 \vee \overline{x_3}) \wedge (x_1 \vee x_4 \vee \overline{x_2}) \wedge (x_1 \vee \overline{x_4} \vee \overline{x_2}),$$

we create the following three clauses (in addition to the six clauses that are created for each of the variables x_1, x_2, x_3, x_4).

$$(\overline{y_{11}} \vee \overline{y_{21}} \vee \overline{z_{31}}), (\overline{y_{12}} \vee \overline{y_{41}} \vee \overline{z_{21}}), (\overline{y_{13}} \vee \overline{z_{41}} \vee \overline{z_{22}}).$$

It is easy to see that the obtained formula, F_L, is indeed an LSAT formula, since each clause in $\mathcal{C}_L$ that was obtained from a clause in $\mathcal{C}$ by replacement does not share any of its literals with another clause in $\mathcal{C}_L$.

Theorem 1. *F is satisfiable if and only if F_L is satisfiable.*

Proof. Assume F is satisfiable, that is, there exists a truth assignment for X such that each clause in $\mathcal{C}$ is satisfied. We show that F_L is satisfiable. If $x_i \in X$ was assigned FALSE (i.e., 0), then we assign TRUE (i.e., 1) to the variables $a_i, y_{i1}, y_{i2}, y_{i3}$ and 0 to the variables x_i, z_{i1}, z_{i2}. On the other hand, if $x_i \in X$ was assigned 1, then we assign 0 to $a_i, y_{i1}, y_{i2}, y_{i3}$ and 1 to x_i, z_{i1}, z_{i2}. We claim that this truth assignment to the variables of X_L satisfies F_L. Observe first that all the $6|X|$ clauses created for the variables in X are satisfied, since each of them consists of two literals which assume opposite values. It remains to show that the assignment satisfies the clauses consisting of three literals. But this is obvious, since for any clause $C \in \mathcal{C}$ and any literal t of C, the value of t and the value of t' are equal, where t' is the literal replacing t in F_L. (That is, if $t = x_i$, then $t' = \overline{y_{ik}}$, for some $k \in \{1, 2, 3\}$, and $x_i = 1$ if and only if $\overline{y_{ik}} = 1$, and if $t = \overline{x_i}$, then $t' = \overline{z_{ik}}$, for some $k \in \{1, 2\}$, and $\overline{x_i} = 1$ if and only if $\overline{z_{ik}} = 1$.)

We now prove that if F_L is satisfiable, then so is F. Consider any truth assignment for X_L that satisfies F_L. This truth assignment (restricted to X) also satisfies F. This is true, since, as can easily be verified, $\overline{y_{ik}} = 1 \implies x_i = 1$ and $\overline{z_{ik}} = 1 \implies \overline{x_i} = 1$. (For example, if $\overline{y_{i2}} = 1$, then since $(y_{i2} \vee \overline{a_i})$ is satisfied, we deduce that $\overline{a_i} = 1$ and therefore, $a_i = 0$. But now, since $(x_i \vee a_i)$ is satisfied, we deduce that $x_i = 1$). □

We conclude that

Theorem 2. *LSAT is NP-complete.*

3 Applications of LSAT to Rainbow Problems

In this section we prove that the rainbow problems mentioned in the introduction are NP complete, by devising reductions from LSAT. Specifically, we first prove that (the decision version) of minmax gap is NP-complete, and then we show that rainbow piercing and rainbow covering are NP-complete.

3.1 Rainbow Minmax Gap (Decision Version) is NP-complete

Let P be a set of n color classes, where each color class C_i is a pair of points $\{p_i, \overline{p_i}\}$ on the x-axis, and let $d > 0$. We prove that the decision version of rainbow minmax gap is NP-complete, that is, it is NP-complete to determine whether there exists a rainbow $Q \subset \cup_{i=1}^{n} C_i$ of size n, such that the maximum gap between a pair of consecutive points in Q is at most d.

We present a reduction from LSAT. Let F be an LSAT formula, and let X be the underlying set of variables and $\mathcal{B}$ be the set of clauses of F. Let k be the number of clauses in $\mathcal{B}$ that do not intersect any other clause in $\mathcal{B}$. Place the points $q_1, q_2, \ldots, q_{k+1}$ on the x-axis, from left to right, such that the distance between any two consecutive points is $d + \frac{d}{4}$. Now, for each clause B_i of these k clauses, place three additional points between q_i and q_{i+1}, one for each of its literals. For example, if $B_i = (x_a \vee x_b \vee \overline{x_c})$, then we place the points p_a, p_b, and $\overline{p_c}$ such that p_b is at the middle of the interval $\overline{q_i q_{i+1}}$ and p_a and $\overline{p_c}$ are to its left and right, respectively, at distance $\frac{d}{8}$ from p_b (see Fig. 2(a)).

Next, consider the pairs of clauses that have a single literal in common, and let l be their number. Place the points $q_{k+2}, \ldots, q_{k+l+1}$, from left to right, such that the distance between q_{k+i} and q_{k+i+1} is $2d$, for $i = 1, \ldots, l$. Now, for the i'th pair B, B' of these l pairs of clauses, place five additional points between q_{k+i} and q_{k+i+1}. For example, if $B = (x_a \vee \overline{x_b} \vee x_c)$ and $B' = (x_c \vee x_d \vee \overline{x_e})$, then we place the points $p_a, \overline{p_b}, p_c, p_d, \overline{p_e}$ such that the distance between q_{k+i} and the first point (p_a), as well as the distance between q_{k+i+1} and the last point ($\overline{p_e}$), is $\frac{d}{2}$ and the distance between any two consecutive points is $\frac{d}{4}$ (see Fig. 2(b)). Finally, place the points $\overline{q_1}, \ldots, \overline{q_{k+l+1}}$ such that distance between any two consecutive points, including the distance between q_{k+l+1} and $\overline{q_1}$, is $d + \epsilon$, for some $\epsilon > 0$. See Fig. 3 for a complete example.

Notice that in our reduction, we have assumed that each clause in F consists of three literals. However, we can adapt the reduction to fit formulas containing two literal clauses.

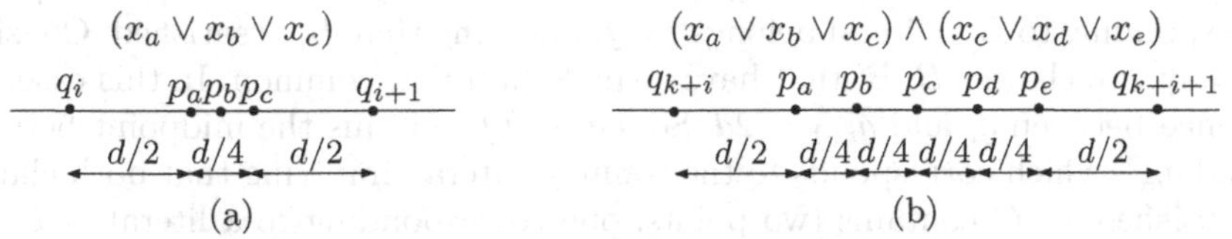

$(x_a \vee x_b \vee x_c)$ — q_i, $p_a p_b p_c$, q_{i+1} — $d/2$, $d/4$, $d/2$ — (a)

$(x_a \vee x_b \vee x_c) \wedge (x_c \vee x_d \vee x_e)$ — q_{k+i}, p_a, p_b, p_c, p_d, p_e, q_{k+i+1} — $d/2$, $d/4$, $d/4$, $d/4$, $d/4$, $d/2$ — (b)

Fig. 2. The reduction from LSAT to the decision version of minmax gap.

Lemma 1. *Let P be the resulting set of color classes (i.e., $P = \{\{q_1, \overline{q_1}\}, \ldots, \{q_{k+l+1}, \overline{q_{k+l+1}}\}, \{p_a, \overline{p_a}\}, \{p_b, \overline{p_b}\}, \ldots\}$). F is satisfiable if and only if there exists a rainbow Q consisting of one point from each color class, such that the maximum gap between a pair of consecutive points in Q is at most d.*

Proof. Assume F is satisfiable, and consider the rainbow Q that is obtained as follows. First, add the points $q_1, \ldots, q_{k+l+1}$ to Q. Next, for each variable x_i appearing in F, if x_i was assigned TRUE, then add the point p_i to Q; otherwise, add the point $\overline{p_i}$ to Q. Obviously, Q consists of exactly one point from each color class. We claim that the distance between any two consecutive points in Q is at most d. To see this, it is enough to examine the situation between q_i and q_{i+1}, for $i = 1, \ldots, k + l$. If $i \leq k$, then since the clause, B_i, corresponding to this interval is satisfied, one of its literals is true and the point corresponding to it was added to Q. Clearly, the distance between this point and q_i (alternatively, q_{i+1}) is at most $\frac{3d}{4}$ (see Fig. 2(a)). If $k + 1 \leq i \leq k + l$, then consider the pair of clauses B, B' corresponding to this interval. Since both are satisfied, then either the literal that is common to both is true, or each of them has a unique literal that is true. In the former case, Q contains the midpoint between q_i and q_{i+1}, whose distance from q_i (alternatively, q_{i+1}) is exactly d, and in the latter case, Q contains two points p_1, p_2, such that the distance between them is at most d and the distance between p_1 and q_i, as well as the distance between p_2 and q_{i+1} is at most $\frac{3d}{4}$ (see Fig. 2(b)). We have shown that Q is as required.

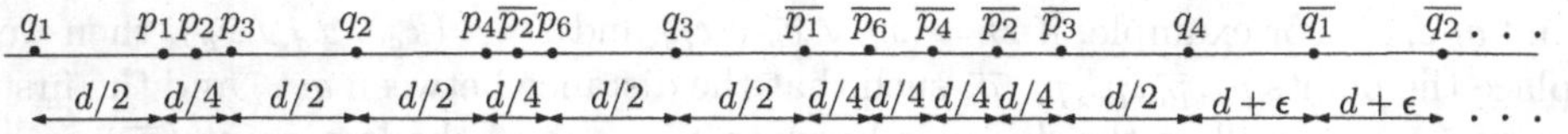

Fig. 3. A complete example: $F = (x_1 \vee x_2 \vee x_3) \wedge (x_4 \vee \overline{x_2} \vee x_6) \wedge (\overline{x_1} \vee \overline{x_6} \vee \overline{x_4}) \wedge (\overline{x_4} \vee \overline{x_2} \vee \overline{x_3})$.

Assume now that there exists a rainbow Q consisting of one point from each color class, such that the maximum gap between a pair of consecutive points in Q is at most d. Observe that Q cannot contain any of the points $\overline{q_1}, \ldots, \overline{q_{k+l+1}}$, so it must contain all the points $q_1, \ldots, q_{k+l+1}$. We assign values to the variables appearing in F is follows. If $p_i \in Q$, then set $x_i = 1$, and if $\overline{p_i} \in Q$, then set $x_i = 0$. We claim that this assignment satisfies each of the clauses of F. Consider any clause B_i that does not intersect any other clause in $\mathcal{B}$. Since the distance between q_i and q_{i+1} is greater than d, at least one of the three points corresponding to B_i's literal belongs to Q, implying that B_i is satisfied. Consider now any two clauses B, B' that have a single literal in common. In this case, the distance between q_i and q_{i+1} is $2d$. So, either Q contains the midpoint between q_i and q_{i+1} which corresponds to the common literal, implying that both clauses are satisfied, or Q contains two points, one corresponding to a literal of B and one to a literal of B', again implying that both clauses are satisfied. Thus, we have shown that F is satisfiable. □

Hence, we have proved the following theorem.

Theorem 3. *The decision version of rainbow minmax gap is NP-complete.*

Corollary 1. *Rainbow minmax gap is NP-hard.*

3.2 Rainbow Piercing and Rainbow Covering are NP-complete

In the full version of this paper, we prove the following two theorems:

Theorem 4. *Rainbow piercing is NP-complete.*

Theorem 5. *Rainbow covering is NP-complete.*

4 Exact Coverage of Color Classes

Let $P = \{C_1, C_2, \ldots, C_n\}$ be a set of n color classes, where each color class C_i is a pair of points $\{p_i, \overline{p_i}\}$ on the x-axis. We consider coverage problems where the goal is to use intervals on the x-axis to cover *exactly one* point from each color class. We now prove that the following three problems are NP-hard; the decision versions are easily seen to be in NP. Note that in each of these problems, it is implied that the intervals are conflict-free (no interval can contain two points from the same color class). In this section, we represent point pairs in Figs. 4, 5 and 6 as the tips of a ⊓ shape or the tips of a ⊔ shape. Certain pairs in these figures are drawn in color in order to help explain the constructions.

Problem 1 (Covering color classes with unit length intervals). *Decide whether or not there exists a set of unit length intervals, $\mathcal{I}$, such that exactly one point from each color class is covered by an interval in $\mathcal{I}$.*

Problem 2 (Covering color classes with the fewest unit length intervals). *Find a minimum-cardinality set $\mathcal{I}$ of unit length intervals (assuming a feasible solution exists), such that exactly one point from each color class is covered by an interval in $\mathcal{I}$.*

Problem 3 (Covering color classes with intervals of arbitrary length). *Find a minimum-cardinality set $\mathcal{I}$ of intervals of arbitrary length, such that exactly one point from each color class is covered by an interval in $\mathcal{I}$.*

4.1 Unit Intervals

Theorem 6. *Problem 1 is NP-Complete.*

Proof. Problem 1 is clearly in NP because we can check whether or not exactly one point from each color class is covered in polynomial time. The reduction is from 3-SAT. Given n variables $\{x_1, x_2, x_3, \ldots, x_n\}$, and m clauses $\{c_1, c_2, c_3, \ldots, c_m\}$, we design the following gadgets.

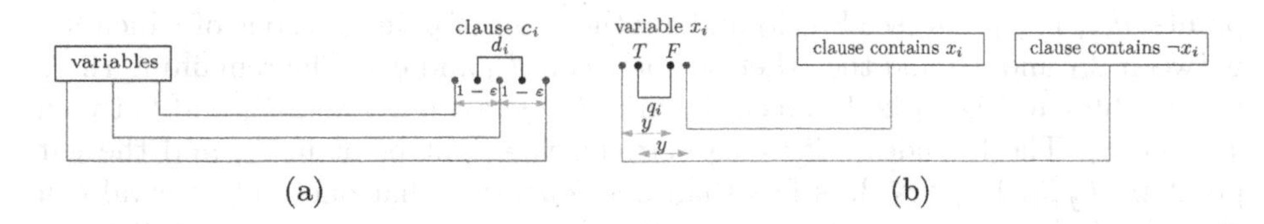

Fig. 4. Clause and variable gadgets for Problem 1 (Color figure online).

Each clause gadget (Fig. 4(a)) consists of five points. It contains a pair of points d_i (represented by a ⊓ shape in Fig. 4(a)), interleaved with three blue points; each of the three paired to a point in a variable gadget (these blue pairs are represented by a ⊔ shape in Fig. 4(a)). In a clause gadget, the Euclidean distance between any two consecutive (blue) points that are paired to variables is $1 - \varepsilon$ for $\varepsilon > 0$ (ε should be bounded above; $\varepsilon < 1/3$ suffices). Each variable gadget (Fig. 4(b)) consists of a consecutive pair of points, q_i, surrounded by blue points on each side. If variable x_i (resp. $\overline{x_i}$) appears in clause c_i, then one blue point will be placed to the right (resp. left) of q_i and this point will be paired to a blue point in c_i. The blue points that surround q_i are placed a distance of y from their respective farthest points in q_i. We set $y < 1$, ensuring that any unit interval that covers a point in q_i must also cover either the surrounding blue points to the left or right of q_i. Setting x_i to FALSE is equivalent to covering the right point of q_i. Setting x_i to TRUE is equivalent to covering the left point of q_i. We line up all of the variables, followed by all of the clauses, so that each consecutive gadget is spaced farther than unit distance apart.

If a clause evaluates to FALSE, each of the three blue points in a clause cannot be covered. Pair d_i will now be left uncovered because we cannot cover a point in d_i with a unit interval without covering one of the blue points in the clause. If a clause evaluates to TRUE, then a point from d_i can always be covered. Therefore, there exists a satisfying truth assignment in 3-SAT if and only if there exists a covering with unit length intervals such that exactly one point from each color class is covered. □

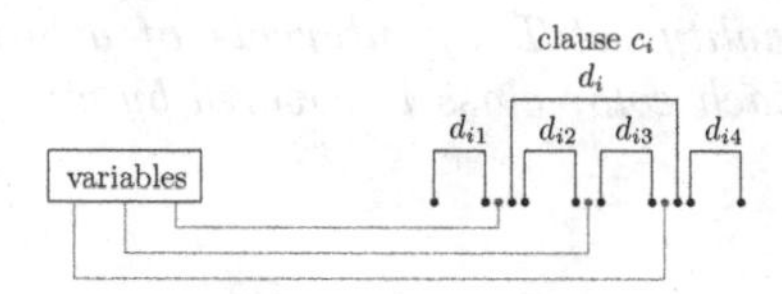

Fig. 5. Clause gadget for Problem 2 (Color figure online).

We now suppose that there indeed exists a set of unit length intervals $\mathcal{I}$ such that exactly one point from each color class is covered by an interval in $\mathcal{I}$. We show that finding such a set of minimum-cardinality is NP-hard. The reduction is from 3-SAT. We use the same variable gadgets and modify the clause gadgets. A clause gadget, c_i, contains 13 points. It contains four consecutive pairs of points, d_{ij}, $1 \leq j \leq 4$ (see Fig. 5) and another pair of points, d_i, one of which lies between d_{i1} and d_{i2} and the other lies between d_{i3} and d_{i4}. The remaining three points (blue in Fig. 5) lie between d_{i1} and d_i, between d_{i2} and d_{i3} and between d_{i3} and d_i. The Euclidean distance between the right point in d_{ij} and the left point in d_{ij+1}, $1 \leq j \leq 3$, is less than one, ensuring that one unit interval can cover both d_{ij} and d_{ij+1}. The two points that define d_{ij} are spaced unit distance apart. In the full version of this paper, we prove the following theorem using the gadget described above.

Theorem 7. *Problem 2 is NP-Hard.*

4.2 Arbitrary Length Intervals

With intervals of arbitrary length, there always exists a solution that gives complete coverage. We show that finding such a solution of minimum is NP-hard.

Theorem 8. *Problem 3 is NP-Hard.*

Proof. The reduction is again from 3-SAT. In this case, spacing of points is irrelevant. Variable gadgets are set up very similarly to the unit interval version. This time, in order to ensure that in a minimum-cardinality cover, the blue points (either to the left or to the right of q_i) in a variable gadget are covered with the same interval that covers q_i, we enclose pair q_i with a 'safety' pair s_i (see Fig. 6). We will see that covering a point in q_i and not using the same interval to cover a point in s_i would be too costly. Clause gadgets are set up in the same way as the unit interval, optimization problem (Fig. 5).

We break the set of points in the construction into two halves, H_1 which contains the variable and clause gadgets, and H_2 which contains another gadget which will be described soon (see Fig. 6). Surrounding each variable and each clause we place a cluster of $M >> n + 3m$ points. Note that the points in a cluster are laid out side-by-side (rather than on the same x-coordinate). In H_2, we create M groups of points, where each group is made up of $n + m + 1$ points, one paired to each cluster in H_1. Surrounding these groups are pairs of consecutive points, g_1 and g_2. Pair g_1 lies to the left of the first group and pair g_2 lies to the right of the last group. The gadget in H_2 will help us isolate all of the variable and clause gadgets in H_1.

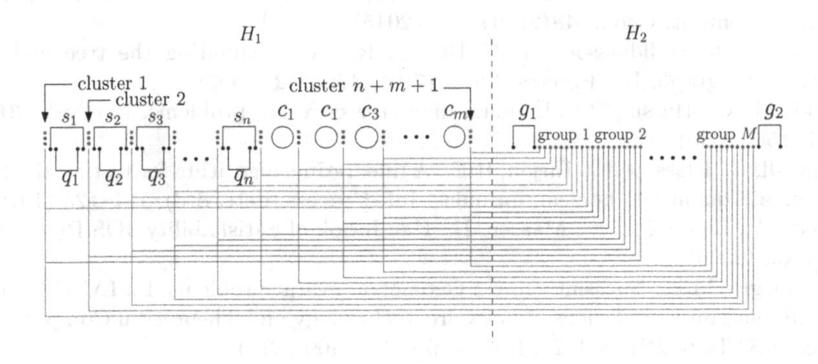

Fig. 6. Arbitrary length intervals – the big picture (Color figure online).

First, we show that any feasible solution uses at least $n + 3m + 1$ intervals.

Case 1: No cluster in H_1 is completely covered. The variable and clause gadgets are now isolated. We need at least n intervals to cover the variable gadgets and at least $3m$ intervals to cover the clause gadgets. At least one more interval is needed to cover the remaining points in H_2.

Case 2: At least one cluster in H_1 is completely covered. If any cluster is completely uncovered then at least M intervals will be needed in H_2. If all clusters are "touched" by an interval then at least $n + 3m + 1$ intervals will be used in H_1 (at least $n+m+1$ intervals touch a cluster and at least $2m$ intervals are needed to finish covering the clauses). At least one more interval is needed to cover points in H_2. At least $n + 3m + 2$ intervals are used in total.

Now we claim that there exists a satisfying truth assignment in 3-SAT if and only if a minimum cover uses $n + 3m + 1$ intervals. Suppose there exists a satisfying truth assignment. Any feasible solution must use at least $n + 3m + 1$ intervals. This lower bound can be achieved by covering pairs in H_1 the same way as in the unit interval optimization problem construction and using one more interval in H_2 to cover g_1, all groups, and g_2.

Now suppose that a minimum cover uses $n+3m+1$ intervals. By Case 2, we know that no cluster in H_1 can be completely covered. Therefore, all variable and clause gadgets are isolated the same way they were in the unit interval version. Recall that in the variable gadgets, a 'safety' pair s_i encloses the set of blue points that extend to clause gadgets. If the interval used to cover q_i does not also cover pair s_i, then an extra interval will be needed in the covering; this would be one interval too many. Therefore, we now see that variable gadgets work the same way as in the unit interval version. This means that if any clause would have evaluated to FALSE then at least $n + 3m + 2$ intervals would have been needed. □

References

1. Arkin, E.M., Díaz-Báñez, J.M., Hurtado, F., Kumar, P., Mitchell, J.S.B., Palop, B., Pérez-Lantero, P., Saumell, M., Silveira, R.I.: Bichromatic 2-center of pairs of points. Comput. Geom. **48**(2), 94–107 (2015)
2. Arkin, E.M., Halldórsson, M.M., Hassin, R.: Approximating the tree and tour covers of a graph. Inf. Process. Lett. **47**(6), 275–282 (1993)
3. Arkin, E.M., Hassin, R.: Minimum-diameter covering problems. Networks **36**(3), 147–155 (2000)
4. Aspvall, B., Plass, M.F., Tarjan, R.E.: A linear-time algorithm for testing the truth of certain quantified boolean formulas. Inf. Process. Lett. **8**(3), 121–123 (1979)
5. Biere, A., Heule, M., van Maaren, H.: Handbook of Satisfiability. IOS Press, Amsterdam (2009)
6. Consuegra, M.E., Narasimhan, G.: Geometric avatar problems. In: IARCS Annual Conference on Foundations of Software Technology and Theoretical Computer Science, FSTTCS 2013, vol. 24. LIPIcs, pp. 389–400 (2013)
7. Cook, S.A.: The complexity of theorem-proving procedures. In: Proceedings of the 3rd Annual ACM Symposium on Theory of Computing, pp. 151–158 (1971)
8. Dowling, W.F., Gallier, J.H.: Linear-time algorithms for testing the satisfiability of propositional horn formulae. J. Logic Program. **1**(3), 267–284 (1984)
9. Gabow, H.N., Maheshwari, S.N., Osterweil, L.J.: On two problems in the generation of program test paths. IEEE Trans. Softw. Eng. **2**(3), 227–231 (1976)
10. Hudec, O.: On alternative p-center problems. Zeitschrift fur Operations Research **36**(5), 439–445 (1992)
11. Knuth, D.: Sat11 and sat11k. In: Proceedings of SAT Competition 2013; Solver and Benchmark Descriptions, p. 32 (2013)
12. Knuth, D.E.: Nested satisfiability. Acta Informatica **28**(1), 1–6 (1990)
13. Tanimoto, S.L., Itai, A., Rodeh, M.: Some matching problems for bipartite graphs. J. ACM (JACM) **25**(4), 517–525 (1978)
14. Tovey, C.A.: A simplified NP-complete satisfiability problem. Discrete Appl. Math. **8**(1), 85–89 (1984)

Graph Algorithms and FPT II

Fully Dynamic Betweenness Centrality

Matteo Pontecorvi and Vijaya Ramachandran(✉)

Computer Science Department, University of Texas, Austin, TX 78712, USA
{cavia,vlr}@cs.utexas.edu

Abstract. We present fully dynamic algorithms for maintaining betweenness centrality (BC) of vertices in a directed graph $G = (V, E)$ with positive edge weights. BC is a widely used parameter in the analysis of large complex networks. We achieve an amortized $O(\nu^{*2} \cdot \log^3 n)$ time per update with our basic algorithm, and $O(\nu^{*2} \cdot \log^2 n)$ time with a more complex algorithm, where $n = |V|$, and ν^* bounds the number of distinct edges that lie on shortest paths through any single vertex. For graphs with $\nu^* = O(n)$, our algorithms match the fully dynamic all pairs shortest paths (APSP) bounds of Demetrescu and Italiano [8] and Thorup [28] for unique shortest paths, where $\nu^* = n - 1$. Our first algorithm also contains within it, a method and analysis for obtaining fully dynamic APSP from a decremental algorithm, that differs from the one in [8].

1 Introduction

Betweenness centrality (BC) is a widely-used measure in the analysis of large complex networks, and is defined as follows. Given a directed graph $G = (V, E)$ with $|V| = n$, $|E| = m$ and positive edge weights, let σ_{xy} denote the number of shortest paths (SPs) from x to y in G, and $\sigma_{xy}(v)$ the number of SPs from x to y in G that pass through v, for each pair $x, y \in V$. Then, $BC(v) = \sum_{s \neq v, t \neq v} \frac{\sigma_{st}(v)}{\sigma_{st}}$.

The measure $BC(v)$ is often used as an index that determines the relative importance of v in G, and is computed for all $v \in V$. Some applications of BC include analyzing social interaction networks [13], identifying lethality in biological networks [20], and identifying key actors in terrorist networks [6,15]. In the static case, the widely used algorithm by Brandes [5] runs in $O(mn+n^2 \log n)$ on weighted graphs. Several approximation algorithms are available: [1,24] for static computation and, recently, [3,4] for dynamic computation. Heuristics for dynamic betweenness centrality with good experimental performance are given in [10,16,26], but none provably improve on Brandes. The only earlier exact dynamic BC algorithms that provably improve on Brandes on some classes of graphs are the recent separate incremental and decremental[1] algorithms in [18, 19]. Table 1 contains a summary of these results.

This work was supported in part by NSF grants CCF-0830737 and CCF-1320675.

[1] Incremental/decremental refer to the insertion/deletion of a vertex or edge; the corresponding weight changes that apply are weight decreases/increases, respectively.

K. Elbassioni and K. Makino (Eds.): ISAAC 2015, LNCS 9472, pp. 331–342, 2015.
DOI: 10.1007/978-3-662-48971-0_29

In this paper, we present two results for fully dynamic exact betweenness centrality: a basic algorithm that provably improves over Brandes for dense graphs (where m is close to n^2) with succinct single-source SP dags, and a faster algorithm that is considerably more complicated.

Our techniques recompute the BC scores using certain data structures related to shortest paths extensions (see Sect. 2), which are generalizations of similar ones introduced by Demetrescu and Italiano in [8] for fully dynamic all pairs shortest paths (APSP) (the `DI` method), where only one SP is maintained for each pair of vertices. To compute BC, however, we need *all* the SPs for each pair of vertices (*all pairs all shortest paths – APASP*). Our fully dynamic algorithms build on our recent work (with Nasre) [19] on decremental APASP (the `NPRdec` method), which generalizes the `DI` data structures to represent all of the multiple SPs for every pair of vertices using a *tuple-system* (see Sect. 3.1 (A System of Tuples)).

Table 1. Related results (DR stands for Directed and UN for Undirected)

Paper	Year	Time	Weights	Update type	DR/UN	Result
Brandes [5]	2001	$O(mn)$	NO	Static Alg.	Both	Exact
Brandes [5]	2001	$O(mn + n^2 \log n)$	YES	Static Alg.	Both	Exact
Geisberger et al. [9]	2007	Heuristic	YES	Static Alg.	Both	Approx.
Riondato et al. [24]	2014	depends on ϵ	YES	Static Alg.	Both	ϵ-Approx.
			Semi Dynamic			
Green et al. [10]	2012	$O(mn)$	NO	Edge Inc.	Both	Exact
Kas et al. [12]	2013	Heuristic	YES	Edge Inc.	Both	Exact
NPR [18]	2014	$O(\nu^* \cdot n)$	YES	Vertex Inc.	Both	Exact
NPRdec [19]	2014	$O(\nu^{*2} \cdot \log n)$	YES	Vertex Dec.	Both	Exact
Bergamini et al. [4]	2015	depends on ϵ	YES	Batch (edges) Inc.	Both	ϵ-Approx.
			Fully Dynamic			
Lee et al. [16]	2012	Heuristic	NO	Edge Update	UN	Exact
Singh et al. [26]	2013	Heuristic	NO	Vertex Update	UN	Exact
Kourtellis+ [14]	2014	$O(mn)$	NO	Edge Update	Both	Exact
Bergamini et al. [3]	2015	depends on ϵ	YES	Batch (edges)	UN	ϵ-Approx.
This paper (Basic)	2015	$O(\nu^{*2} \cdot \log^3 n)$	YES	Vertex Update	Both	Exact
This paper (FFD)	2015	$O(\nu^{*2} \cdot \log^2 n)$	YES	Vertex Update	Both	Exact

Our Results. Let ν^* be the maximum number of distinct edges that lie on shortest paths through any given vertex in G; we assume $\nu^* = \Omega(n)$. Both of our BC algorithms are obtained through fully dynamic all pairs all shortest paths. The first APASP algorithm FULLY-DYNAMIC matches the `DI` APSP bound (which computes unique SPs) for graphs with $\nu^* = O(n)$; FULLY-DYNAMIC generalizes `DI`, though it is somewhat different from `DI` even for unique SPs, and its analysis is quite different from `DI`. The second APASP algorithm FFD is a generalization of Thorup [28] (the `Thorup` method) for APASP and matches its bound for APSP when $\nu^* = O(n)$; the main challenge here is to generalize the 'level graphs' of

Thorup to the case when SPs for a given vertex pair can be distributed across multiple levels. Both APASP algorithms lead to fully dynamic BC algorithms as follows:

Theorem 1. *Let Σ be a sequence of $\Omega(n)$ fully dynamic vertex updates on a directed n-node graph $G = (V, E)$ with positive edge weights. Let ν^* bound the number of distinct edges that lie on shortest paths through any single vertex in any of the updated graphs or their vertex induced subgraphs. Then, all BC scores (and APASP) can be maintained in amortized time:*

(1) $O(\nu^{*2} \cdot \log^3 n)$ per update with algorithm FULLY-DYNAMIC*,*
(2) $O(\nu^{*2} \cdot \log^2 n)$ per update with algorithm FFD*.*

Discussion of the parameters m^* and ν^*. Let m^* be the number of distinct edges in G that lie on shortest paths; ν^*, defined above, is the maximum number of distinct edges on shortest paths through a single vertex. Clearly, $\nu^* \leq m^* \leq m$.

- m^* vs m: In many cases, $m^* \ll m$: as noted in [11], in a complete graph ($m = \Theta(n^2)$) where edge weights are chosen from a large class of probability distributions, $m^* = O(n \log n)$ with high probability.
- ν^* vs m^*: Clearly, $\nu^* = O(n)$ in any graph with only a constant number of SPs between every pair of vertices. These graphs are called k-geodetic [23] (when at most k SPs exists between two nodes), and are well studied in graph theory [2,17,27]. In fact $\nu^* = O(n)$ even in some graphs that have an exponential number of SPs between some pairs of vertices. In contrast, m^* can be $\Theta(n^2)$ even in some graphs with unique SPs, for example the complete unweighted graph K_n.

Another type of graph with $\nu^* \ll m^*$ is one with large clusters of nodes (e.g., as described by the *planted ℓ-partition model* [7,25]). Consider a graph H with k clusters of size n/k (for some constant $k \geq 1$) with $\delta < w(e) \leq 2\delta$, for some constant $\delta > 0$, for each edge e in a cluster; between the clusters is a sparse interconnect. Then $m^* = \Omega(n^2)$ but $\nu^* = O(n)$.

For the above classes of graphs, both of our BC algorithms will run in amortized $\tilde{O}(n^2)$ time per update ($\tilde{O}$ hides polylog factors). More generally we have:

Theorem 2. *Let Σ be a sequence of $\Omega(n)$ updates on graphs with $O(n)$ distinct edges on shortest paths through any single vertex in any vertex-induced subgraph. Then, all BC scores (and APASP) can be maintained in amortized time $O(n^2 \cdot \log^2 n)$ per update.*

In this extended abstract, we present the key features of our results; details and full proofs are on arXiv [21,22]. Our algorithms use $\widetilde{O}(m \cdot \nu^*)$ space, extending the $\widetilde{O}(mn)$ result in DI for APSP. Brandes uses only linear space, but all known dynamic algorithms require at least $\Omega(n^2)$ space.

Overview of the Paper. In Sect. 2 we describe our fully dynamic BC algorithm that uses the data structures maintained by our APASP algorithms. In Sect. 3 we review the NPRdec and DI algorithms. In Sect. 4 we describe our fully dynamic approach, and in Sect. 5 we present our first algorithm FULLY-DYNAMIC and establish its amortized time bound of $O(\nu^{*2} \cdot \log^3 n)$. In Sect. 6 we briefly describe our faster algorithm FFD, and we conclude with Sect. 7.

2 The Fully Dynamic Betweenness Centrality Algorithm

The static Brandes algorithm [5] computes BC scores in a two phase process. The first phase (implicitly) computes the SP out-dag for every source through n applications of Dijkstra's algorithm. The second phase uses an 'accumulation' technique that computes all BC scores using these SP dags in $O(n \cdot \nu^*)$ time.

In our fully dynamic algorithm, we will leave the second phase unchanged. For the first phase, we will use the approach in the incremental BC algorithm in [18], which maintains the SP dags using a very simple and efficient incremental algorithm. For decremental and fully dynamic updates, the corresponding dynamic APASP algorithms to maintain the SP dags are more involved. Neither the decremental nor our new fully dynamic APASP algorithms maintain the SP dags explicitly, instead they maintain data structures to update a collection of *tuples* (see Sect. 3.1 (A System of Tuples)). We now describe a very simple method to construct the SP dags from these data structures (this step is not addressed in the decremental APASP algorithm in [19]).

For every vertex pair x, y, the following sets $R^*(x, y)$, $L^*(x, y)$ are maintained in NPRdec, and in both of our fully dynamic algorithms (a restricted version of these sets was introduced for APSP in DI) :

- $R^*(x, y)$ contains all nodes y' such that every shortest path $x \rightsquigarrow y$ in G can be extended with the edge (y, y') to generate another shortest path $x \rightsquigarrow y \rightarrow y'$.
- $L^*(x, y)$ contains all nodes x' such that every shortest path $x \rightsquigarrow y$ in G can be extended with the edge (x', x) to generate another shortest path $x' \rightarrow x \rightsquigarrow y$.

These sets allow us to construct the SP dag for each source s using the following algorithm BUILD-DAG. In our fully dynamic algorithms R^* and L^* will be supersets of the exact collections of nodes defined above, but the check in Step 3 will ensure that only the correct SP dag edges are included. The combined sizes of these R^* and L^* sets is $O(n \cdot \nu^* \cdot \log n)$ in our fully dynamic algorithms, hence the amortized time bound for the overall fully dynamic BC algorithm is dominated by the time bound for fully dynamic APASP.

Algorithm 1. BUILD-DAG$(G, s, \mathbf{w}, D)$ (w is the weight function; D is the distance matrix)

```
1: for each t ∈ V do
2:     for each u ∈ R*(s,t) do
3:         if D(s,t) + w(t,u) = D(s,u) then add the edge (t,u) to dag(s)
```

3 Background

3.1 The NPR Decremental APASP Algorithm [19]

The decremental algorithm `NPRdec` for APASP builds on the key concept of a *locally shortest path (LSP)* in a graph, introduced in the `DI` method [8]. A path p in G is an LSP if the path p' obtained by removing the first edge from p and the path p'' obtained by removing the last edge from p are both SPs in G. For APASP, we need to maintain all shortest paths, and G can have an exponential (in n) number of SPs. Thus the `DI` method is not feasible for APASP since it maintains each SP (and LSP) separately. In order to succinctly maintain all SPs and LSPs in a manner suitable for efficient decremental updates, `NPRdec` developed the tuple-system described below.

A System of Tuples. Since a graph could have an exponential number of shortest paths, `NPRdec` introduced the compact tuple-system described below. Let $\mathbf{w}$ be the edge weight function in G, and let $d(x, y)$ denote the shortest path length from x to y. A *tuple*, $\tau = (xa, by)$, represents a set of paths in G, all with the same weight, and all of which use the same first edge (x, a) and the same last edge (b, y). If the paths in τ are LSPs, then τ is an LST (locally shortest tuple), and the weight of every path in τ is $\mathbf{w}(x, a) + d(a, b) + \mathbf{w}(b, y)$. If $d(x, y) = \mathbf{w}(x, a) + d(a, b) + \mathbf{w}(b, y)$, then τ is a *shortest path tuple (ST)*.

A *triple* $\gamma = (\tau, wt, count)$ represents the tuple $\tau = (xa, by)$ that contains *count* paths from x to y, each with weight wt. We use triples to succinctly store all LSPs and SPs for each vertex pair in G. For $x, y \in V$, we define:

$$P(x,y) = \{((xa,by), wt, count)\colon (xa,by) \text{ is an LST from } x \text{ to } y \text{ in } G\}$$
$$P^*(x,y) = \{((xa,by), wt, count)\colon (xa,by) \text{ is an ST from } x \text{ to } y \text{ in } G\}.$$

A *left tuple* (or ℓ-tuple), $\tau_\ell = (xa, y)$, represents the set of LSPs from x to y, all of which use the same first edge (x, a). A *right tuple* (r-tuple) $\tau_r = (x, by)$ is defined analogously. For a shortest path r-tuple $\tau_r = (x, by)$, $L(\tau_r)$ is the set of vertices which can be used as pre-extensions to create LSTs in G, and for a shortest path ℓ-tuple $\tau_\ell = (xa, y)$, $R(\tau_\ell)$ is the set of vertices which can be used as post-extensions to create LSTs in G. Hence:

$$L(x,by) = \{x' : (x',x) \in E(G) \text{ and } (x'x,by) \text{ is an LST in } G\}$$
$$R(xa,y) = \{y' : (y,y') \in E(G) \text{ and } (xa,yy') \text{ is an LST in } G\}.$$

For $x, y \in V$, $L^*(x, y)$ denotes the set of vertices which can be used as pre-extensions to create shortest path tuples in G; $R^*(x, y)$ is defined symmetrically:

$$L^*(x,y) = \{x' : (x',x) \in E(G) \text{ and } (x'x,y) \text{ is a } \ell\text{-tuple representing SPs in } G\}$$
$$R^*(x,y) = \{y' : (y,y') \in E(G) \text{ and } (x,yy') \text{ is an } r\text{-tuple representing SPs in } G\}.$$

Data Structures. The `NPRdec` algorithm uses priority queues for P and P^*, and balanced search trees for L^*, L, R^* and R, as well as for a set Marked-Tuples that is specific only to one update. It also uses priority queues H_c and H_f for the cleanup and fixup procedures, respectively.

Lemma 1. *[19] Let $G = (V, E)$ be a directed graph with positive edge weights. The number of LSTs (or triples) that contain a vertex v in G is $O(\nu^{*2})$, and the total number of LSTs (or triples) in G is bounded by $O(m^* \cdot \nu^*)$.*

The `NPRdec` algorithm maintains all STs and LSTs in the current graph, and for each tuple, it maintains the L, R, L^* and R^* sets. To execute a new update to a vertex v, `NPRdec` (similar to `DI`) first calls an algorithm cleanup on v which removes all STs and LSTs that contain v. This is followed by a call to algorithm fixup on v which computes all STs and LSTs in the updated graph that are not already present in the system. The overall algorithm update consists of cleanup followed by fixup. If the updates are all decremental then `NPRdec` maintains exactly all the SPs and LSPs in the graph in $O(\nu^{*2} \cdot \log n)$ amortized time per update. Several challenges to adapting the techniques in the `DI` decremental method to the tuple-system are addressed in [19]. The analysis of the amortized time bound is also more involved since with multiple shortest paths it is possible for the dynamic APASP algorithm to examine a tuple and merely change its count; in such a case, the `DI` proof method of charging the cost of the examination to the new path added to or removed from the system does not apply.

3.2 The DI Fully Dynamic APSP Algorithm [8]

The `DI` method first gives a decremental APSP algorithm, and shows that this is also a correct, though inefficient, fully dynamic APSP algorithm. The inefficiency arises because under incremental updates the method may maintain some old SPs and their combinations that are not currently SPs or LSPs; such paths are called historical shortest paths (HPs) and locally historical paths (LHPs). To obtain an efficient fully dynamic algorithm, the `DI` method introduces 'dummy updates' into the update sequence. A dummy update performs cleanup and fixup on a vertex that was updated in the past. Using a strategically chosen sequence of dummy updates, it is established in [8] that the resulting APSP algorithm runs in amortized time $O(n^2 \cdot \log^3 n)$ per real update. The `DI` method continues to use the notation P^*, L^*, etc., even though these are supersets of the defined sets in a fully dynamic setting. We will do the same in our fully dynamic algorithms.

4 Overview of Our Fully Dynamic APASP Approach

A natural approach to obtain a fully dynamic APASP algorithm would be to convert the `NPRdec` decremental APASP algorithm to an efficient fully dynamic APASP algorithm by using dummy updates, similar to `DI`. There are two steps in this process, and each has challenges (the second step is more challenging).

Step 1: *Converting `NPRdec` to a correct (but inefficient) fully dynamic APASP algorithm.* Recall that the decremental APSP algorithm in `DI` stores old or 'historical' SPs (i.e., HPs) in the P^* sets if it is used as a fully dynamic algorithm.

Historical paths arise due to the following reason: When incremental updates are interleaved with decremental ones, a path placed previously in a P^* may cease to be an SP and become a *historical SP (or HP)* if a shorter path for the same vertex pair is created by an incremental update. However, `DI` show that their decremental algorithm remains correct if HPs remain in P^*.

For the APASP case, the decremental `NPRdec` algorithm is not correct when used with fully dynamic updates. To see this, let us extend the notion of historical paths to historical tuples (HT and LHT) in the natural way. Using `NPRdec`, we could have an HT τ in P^* which is no longer an ST, but is an LST. Now, if additional paths are added to τ in the next update, then `NPRdec` will treat this tuple as an LST and update its count in P but not in P^*. If later, τ is restored as an ST (through a decremental update), it will have an incorrect lower count in P^* which will not be detected (this can never happen in `DI` since it assumes unique SPs). Additional issues occur in `NPRdec` that need to be addressed (see [21]).

Our first step in developing a fully dynamic APASP algorithm is to update the `NPRdec` algorithm so that the resulting algorithm FULLY-UPDATE remains correct under fully dynamic updates. This algorithm and its analysis are available in [21]; the details are technical and are omitted here. Algorithm FULLY-UPDATE matches the amortized bound in `NPRdec` for decremental updates while being correct for fully dynamic updates. However, as with `DI`, it is inefficient as a fully dynamic algorithm.

Lemma 2. *Consider a sequence of r calls to* FULLY-UPDATE *on a graph with n vertices. Let C be the maximum number of tuples in the tuple-system that can contain a path through a given vertex, and let D be the maximum number of tuples that can be in the tuple-system at any time. Then* FULLY-UPDATE *executes the r updates in $O((r \cdot (n^2 + C) + D) \cdot \log n)$ time.*

Lemma 3. *Suppose every HT in the tuple-system is an ST in one of z different n-node graphs, and every LHT is formed from these HTs. Then,*

1. *The number of LHTs in G's tuple-system is at most $O(z \cdot m \cdot \nu^*)$.*
2. *If all HTs that contain a given vertex u lie within $z' \leq z$ of the z graphs, then the number of LHTs that contain u is $O((z + z'^2) \cdot \nu^{*2})$.*

The proof of Lemma 2 adapts the `NPRdec` analysis to FULLY-UPDATE; Lemma 3 follows from basic properties of the tuple-system (see [21] for both proofs).

Step 2. *Obtaining a good dummy sequence for efficient fully dynamic APASP.* The `DI` method uses 'dummy updates', where a vertex updated at time t is also given a 'dummy' update at steps $t + 2^i$, for each $i > 0$ (this update is performed along with the real update at step $t + 2^i$). The effect of a dummy update on a vertex v is to remove any HP or LHP that contains v, thereby streamlining the collection of paths maintained. Further, with unique SPs, each HP in $P^*(x, y)$ for a given pair x, y will have a different weight. An $O(\log n)$ bound on the number of HPs in a $P^*(x, y)$ is established in `DI` as follows. Let the current time step be t, and consider an HP τ last updated at $t' < t$. Let us denote the smallest i such that $t' + 2^i > t$ as the dummy-index for τ. By observing that different HPs for x, y must have different dummy-indices, it follows that their number is $O(\log t)$, which is $O(\log n)$ since the data structure is reconstructed after $O(n)$ updates.

If we try to apply the DI dummy sequence to APASP, we are faced with the issue that a new ST for x, y (with the same weight) could be created at each update in a long sequence of successive updates. Then, an incremental update could transform all of these STs into HTs. If this happens, then several HTs for x, y, all with the same weight, could have the same dummy-index (in DI only one HP can be present for this entire collection due to unique SPs). Thus, the DI approach of obtaining an $O(\log n)$ bound for the number of HPs for each vertex pair does not work for HTs in our tuple-system.

Our method for Step 2 is to use a different dummy sequence, and a completely different analysis that obtains an $O(\log n)$ bound for the number of different 'PDGs' (a PDG is a type of derived graph defined in Sect. 5) that can contain the HTs. Our new dummy sequence is inspired by the 'level graph' method introduced in Thorup [28] to improve the amortized bound for fully dynamic APSP to $O(n^2 \cdot \log^2 n)$, saving a log factor over DI. The Thorup method is complex because it maintains $O(\log n)$ levels of data structures for suitable 'level graphs'. Our first algorithm FULLY-DYNAMIC does not maintain these level graphs (though our second algorithm FFD does). Instead, FULLY-DYNAMIC performs exactly like the fully dynamic algorithm in DI, except that it uses this alternate dummy update sequence, and it calls FULLY-UPDATE for APASP instead of the DI update algorithm for APSP. Our change in the update sequence requires a completely new proof of the amortized bound which we sketch in the next section (Sect. 5). We consider this to be a contribution of independent interest: If we replace FULLY-UPDATE by the DI update algorithm in FULLY-DYNAMIC, we get a new fully dynamic APSP algorithm which is as simple as DI, with a new analysis. The full details of algorithm FULLY-DYNAMIC are in [21].

In Sect. 6 we briefly describe the second algorithm FFD, which achieves an $O(\log n)$ improvement over the amortized bound for FULLY-DYNAMIC. This algorithm overcomes some technical challenges in order to generalize the Thorup method to APASP, and is considerably more complicated than FULLY-DYNAMIC.

5 Algorithm FULLY-DYNAMIC

Algorithm FULLY-DYNAMIC applies FULLY-UPDATE (see Sect. 4, Step 1) to vertex v with the new weight function $\mathbf{w}'$ for the t-th update. Then it executes dummy updates on a sequence $\mathcal{N}$ of the most recently updated vertices as specified in Steps 2-5. The length of this sequence of vertices is determined by the position k of the lsb set to 1 in the bit representation $B = b_{r-1} \cdots b_0$ of t.

Let G_t be the graph after the t-th update, with G_0 the initial graph. Thus, $G = G_{t-1}$ in Algorithm 2, and the updated graph is G_t. For each i such that $b_i = 1$, we let $time_t(i)$ be the earlier update step t' whose bit representation matches B in positions $b_{r-1} \cdots b_i$ and has zeros elsewhere. We define $Prior\text{-}times(t) = \{time_t(i) \mid b_i = 1\}$. Note that $|Prior\text{-}times(t)| = O(\log t)$. The following lemma

Algorithm 2. FULLY-DYNAMIC$(G, v, \mathbf{w}', t)$

1: FULLY-UPDATE$(v, \mathbf{w}')$
2: $k \leftarrow$ position of the least significant bit set in the representation $b_{r-1} \cdots b_0$ of t
3: $\mathcal{N} \leftarrow$ set of vertices updated at steps $t-1, \cdots, t-(2^k-1)$
4: **for** each $u \in \mathcal{N}$ in decreasing order of update time **do**
5: FULLY-UPDATE$(u, \mathbf{w}')$ (dummy updates)

follows from the fact that a vertex updated at $t' \notin$ *Prior-times*(t) would have been updated by a more recent dummy update (see [21] for the proof).

Lemma 4. *For every vertex v in G_t, the step t_v of the most recent update to v is in Prior-times(t).*

The Prior Deletion Graph (PDG). For $t' < t$, let W be the set of vertices that are updated in the interval of steps $[t'+1, t]$. We define the *prior deletion graph (PDG) $\Gamma_{t',t}$* as the induced subgraph of $G_{t'}$ on the vertex set $V(G_{t'}) - W$. If t is the current update step, then we simply use $\Gamma_{t'}$ instead of $\Gamma_{t',t}$.

We say that a path p *is present in both $G_{t'}$ and G_t* if no call to FULLY-UPDATE is made on any vertex in p in the interval $[t'+1, t]$. The following lemma follows from a PDG $\Gamma_{t',t}$ being the result of applying a sequence of decremental updates to $G_{t'}$. Thus, an ST in $G_{t'}$ is an ST in $\Gamma_{t',t}$ if it is present in it.

Lemma 5. *1. If τ is an ST in $G_{t'}$ then τ continues to be an ST in every PDG $\Gamma_{t',t}$ with $t \geq t'$ in which τ is present.*
2. For any $\hat{t} \geq t'$, if τ is an ST in $G_{\hat{t}}$ then τ is an ST in every PDG $\Gamma_{t',t''}$, $t'' \geq \hat{t}$, in which τ is present.

PDGs for Update t: We will associate with the current update step t, the set of PDGs $\Gamma_{t'}$, for $t' \in$ *Prior-times*(t). These PDGs are similar to the *level graphs* maintained in Thorup, but we choose to give them a different name since we do not maintain these graphs; we only use them here to analyze the performance of our algorithm. We rebuild the tuple-system after $2n$ updates, so $t \leq 2n$.

Lemma 6. *Each HT in the tuple-system for G_t is an ST in at least one of the $\Gamma_{t',t}$ for $t' \in$ Prior-times(t). Further $z = O(\log n)$ in Lemma 3 for G_t.*

Proof. Consider an HT $\tau = (xa, by)$ in G_t. Let the most recently updated vertex in τ be v, and let its update step be $t_v \leq t$. By definition of HT, τ is an ST in some t' in $[t_v, t]$, hence by Lemma 5, part 2, using $\hat{t} = t' = t_v$ and $t'' = t$, we have τ an ST in Γ_{t_v}. Further, by Lemma 4, $t_v \in$ *Prior-times*(t). Finally, since $|$*Prior-times*$(t)| = O(\log t) = O(\log n)$ for any t, $z = O(\log n)$ in Lemma 3. ∎

We will now use the above lemma to establish the amortized time bound.

Lemma 7. *Algorithm 2 executes a sequence Σ of n real updates on an n-node graph in $O(\nu^{*2} \cdot \log^3 n)$ amortized time per update.*

Proof. (Sketch) We apply Lemma 2. By Lemma 6 we have $z = O(\log n)$ in Lemma 3, hence $D = O(m \cdot \nu^* \cdot \log n)$ in Lemma 2. Let C_1 and C_2 be the cost of a cleanup for a real and dummy update, respectively. Then, we use $z' = z$ in Lemma 3 for the real updates, so $C_1 = O(\nu^{*2} \cdot \log^2 n)$.

It is readily seen that there are $O(n \log n)$ dummy updates performed during the n real updates. At the real update step t, when a dummy update is performed on vertex u (last updated at time t_u), only PDGs Γ_t and Γ_{t_u} contain u, hence $z' = 2$ in Lemma 3. Thus $C_2 = O(\nu^{*2} \cdot \log n)$. Hence, by Lemma 2, the total time for the n real updates and $n \log n$ dummy updates is $O(\ (n \cdot (n^2 + C_1) + n \log n \cdot (n^2 + C_2) + D) \cdot \log n) = O(n \cdot \nu^{*2} \cdot \log^3 n + m \cdot \nu^* \cdot \log^2 n)$. Since we assume $\nu^* = \Omega(n)$, we have $m = O(n \cdot \nu^*)$, and we obtain the desired amortized cost for each of the n real updates. ■

6 Algorithm FFD

We give a very brief overview of Algorithm FFD, deferring the details to [22].

Background. For unique SPs, `Thorup` uses a *level system* of decremental-only graphs, with updates being insertion or deletion of a node with incident edges. The PDGs in Sect. 5 are an abstract representation of the graphs maintained in Thorup's level system. Every path maintained by `Thorup` is an SP or LSP in some level graph (i.e., PDG), and when a node is removed from the current graph, it is also removed from every PDG that contains it. This saves a log factor in the amortized time bound over the DI bound.

Algorithm FFD**.** In our algorithm FFD for fully dynamic APASP, we explicitly maintain the PDGs of Sect. 5 using 'local' data structures (see [22] for a detailed description of the structures). A level i PDG is *active* at time t if i is the lsb set in some $t' \in$ *Prior-times*(t); we say *level*$(t') = i$. Each path p in a tuple is centered in level $k =$ *level*(t'), where t' is the most recent step in which p entered the tuple system in a fixup step. Thus, the paths represented by a tuple are spread across the active levels at which these paths are centered; this avoids copying over all data structures each time a new level is activated, which would be very expensive. To keep track of this distribution of paths, we associate an $O(\log n)$-size array C_γ with each tuple γ that stores the number of paths in γ centered at each level.

We face several challenges when we try to extend the `Thorup` method to APASP. Here we briefly describe a major challenge, which we call the *partial extension problem (PEP)* (see [22] for a detailed example). This arises when a collection of HTs for x, y are restored as STs due to a decremental update. A tuple τ in this collection may have its correct extensions in the local structures L_i^* and R_i^* in level i, but its extensions in a more recent level j may not be in L_j^* and R_j^* if τ is not an ST in that level, and is instead an HT. Thus, when the algorithm processes τ as an ST after the current decremental update, it needs to generate the correct extensions in L_j^* and R_j^* since they are not currently present in these sets, but to maintain efficiency, it should not try to generate extensions in L_i^* and R_i^*, since they are already present there. Neither Thorup nor FULLY-DYNAMIC need to distinguish between these two cases. In FULLY-DYNAMIC, algorithm FULLY-UPDATE (called in Steps 1 and 5) creates LHTs by combining every pair of compatible HTs, hence these LHTs will always be available in the corresponding tuple-system. This problem is not an issue in Thorup either, due to the assumption of unique SPs: Thorup can afford to look at all HPs, since there are only $O(n^2 \cdot \log n)$ of them. Algorithm FFD maintains HTs (since it maintains APASP), and their number can be much larger.

In order to maintain both correctness and efficiency in the PEP scenario for APASP, we introduce two new data structures: (1) the historical distance matrices DL that allow us to efficiently determine the most recent level graph in which an HT was an ST, and (2) data structures LN and RN that allow us to efficiently identify exactly those new extensions that need to be performed.

7 Conclusion

We conclude with a possible avenue for improving the amortized bound. Instead of the tuples we maintain in our tuple systems, we could have maintained left and right tuples (see Sect. 3.1 (A System of Tuples)). This would reduce the space usage from $\tilde{O}(m \cdot \nu^*)$ to $\tilde{O}(mn)$. This improved space bound is achievable with $\tilde{O}(\nu^{*2})$ amortized time (details omitted). The number of left or right tuples that contain a given vertex is only $\tilde{O}(n \cdot \nu^*)$, but the time bound does not improve with our current method. Is there an improved method that achieves $\tilde{O}(n \cdot \nu^*)$ amortized time?

References

1. Bader, D.A., Kintali, S., Madduri, K., Mihail, M.: Approximating betweenness centrality. In: Bonato, A., Chung, F.R.K. (eds.) WAW 2007. LNCS, vol. 4863, pp. 124–137. Springer, Heidelberg (2007)
2. Bandelt, H.-J., Mulder, H.M.: Interval-regular graphs of diameter two. Discrete Math. **50**, 117–134 (1984)
3. Bergamini, E., Meyerhenke, H.: Fully-dynamic approximation of betweenness centrality (2015). arXiv:1504.07091 [cs.DS]
4. Bergamini, E., Meyerhenke, H., Staudt, C.L.: Approximating betweenness centrality in large evolving networks. In: Proceedings of ALENEX 2015, ch. 11, pp. 133–146. SIAM (2015)
5. Brandes, U.: A faster algorithm for betweenness centrality. J. Math. Sociol. **25**(2), 163–177 (2001)
6. Coffman, T., Greenblatt, S., Marcus, S.: Graph-based technologies for intelligence analysis. Commun. ACM **47**(3), 45–47 (2004)
7. Condon, A., Karp, R.M.: Algorithms for graph partitioning on the planted partition model. Random Struct. Algorithms **18**(2), 116–140 (2001)
8. Demetrescu, C., Italiano, G.F.: A new approach to dynamic all pairs shortest paths. J. ACM **51**(6), 968–992 (2004)
9. Geisberger, R., Sanders, P., Schultes, D.: Better approximation of betweenness centrality. In: Proceedings of ALENEX 2008, ch. 8, pp. 90–100. SIAM (2008)

10. Green, O., McColl, R., Bader, D.A.: A fast algorithm for streaming betweenness centrality. In: Proceedings of 4th PASSAT, pp. 11–20 (2012)
11. Karger, D.R., Koller, D., Phillips, S.J.: Finding the hidden path: time bounds for all-pairs shortest paths. SIAM J. Comput. **22**(6), 1199–1217 (1993)
12. Kas, M., Wachs, M., Carley, K.M., Carley, L.R.: Incremental algorithm for updating betweenness centrality in dynamically growing networks. In: Proceedings of ASONAM (2013)
13. Kourtellis, N., Alahakoon, T., Simha, R., Iamnitchi, A., Tripathi, R.: Identifying high betweenness centrality nodes in large social networks. SNAM **3**, 899–914 (2013)
14. Kourtellis, N., Morales, G.D.F., Bonchi, F.: Scalable online betweenness centrality in evolving graphs. IEEE Trans. Knowl. Data Eng. **27**(9), 2494–2506 (2015)
15. Krebs, V.: Mapping networks of terrorist cells. Connections **24**(3), 43–52 (2002)
16. Lee, M.-J., Lee, J., Park, J.Y., Choi, R.H., Chung, C.-W.: Qube: a quick algorithm for updating betweenness centrality. In: Proceedings of the 21st WWW Conference, pp. 351–360 (2012)
17. Mulder, H.M.: Interval-regular graphs. Discrete Math. **41**(3), 253–269 (1982)
18. Nasre, M., Pontecorvi, M., Ramachandran, V.: Betweenness centrality – incremental and faster. In: Csuhaj-Varjú, E., Dietzfelbinger, M., Ésik, Z. (eds.) MFCS 2014, Part II. LNCS, vol. 8635, pp. 577–588. Springer, Heidelberg (2014)
19. Nasre, M., Pontecorvi, M., Ramachandran, V.: Decremental all-pairs ALL shortest paths and betweenness centrality. In: Ahn, H.-K., Shin, C.-S. (eds.) ISAAC 2014. LNCS, vol. 8889, pp. 766–778. Springer, Heidelberg (2014)
20. Pinney, J.W., McConkey, G.A., Westhead, D.R.: Decomposition of biological networks using betweenness centrality. In: Proceedings of 9th RECOMB (2005)
21. Pontecorvi, M., Ramachandran, V.: Fully dynamic all pairs all shortest paths (2014). http://arxiv.org/abs/1412.3852v2
22. Pontecorvi, M., Ramachandran, V.: A faster algorithm for fully dynamic betweenness centrality (2015). http://arxiv.org/abs/1506.05783
23. Ramos, J.S.R.M., Ramos, M.T.: A generalization of geodetic graphs: K-geodetic graphs. Inverstigacin Operativa **1**, 85–101 (1998)
24. Riondato, M., Kornaropoulos, E.M.: Fast approximation of betweenness centrality through sampling. In: Proceedings of the 7th ACM WSDM, pp. 413–422. ACM (2014)
25. Schaeffer, S.E.: Survey: graph clustering. Comput. Sci. Rev. **1**(1), 27–64 (2007)
26. Goel, K., Singh, R.R., Iyengar, S., Sukrit, : A faster algorithm to update betweenness centrality after node alteration. In: Bonato, A., Mitzenmacher, M., Prałat, P. (eds.) WAW 2013. LNCS, vol. 8305, pp. 170–184. Springer, Heidelberg (2013)
27. Srinivasan, N., Opatrny, J., Alagar, V.: Bigeodetic graphs. Graphs and Combinatorics **4**(1), 379–392 (1988)
28. Thorup, M.: Fully-dynamic all-pairs shortest paths: faster and allowing negative cycles. In: Hagerup, T., Katajainen, J. (eds.) SWAT 2004. LNCS, vol. 3111, pp. 384–396. Springer, Heidelberg (2004)

When Patrolmen Become Corrupted: Monitoring a Graph Using Faulty Mobile Robots

Jurek Czyzowicz[1], Leszek Gasieniec[2], Adrian Kosowski[3], Evangelos Kranakis[4](✉), Danny Krizanc[5], and Najmeh Taleb[4]

[1] Dépt. d'informatique, Univ. du Québec en Outaouais, Gatineau, Québec, Canada
[2] Department of Computer Science, University of Liverpool, Liverpool, UK
[3] LIAFA, Inria and Université Paris Diderot, Paris, France
[4] School of Computer Science, Carleton University, Ottawa, Canada
kranakis@scs.carleton.ca
[5] Department of Mathematics and Computer Science, Wesleyan University, Middletown, CT, USA

Abstract. A team of k mobile robots is deployed on a weighted graph whose edge weights represent distances. The robots perpetually move along the domain, represented by all points belonging to the graph edges, not exceeding their maximal speed. The robots need to patrol the graph by regularly visiting all points of the domain. In this paper, we consider a team of robots (patrolmen), at most f of which may be unreliable, i.e. they fail to comply with their patrolling duties.

What algorithm should be followed so as to minimize the maximum time between successive visits of every edge point by a reliable patrolmen? The corresponding measure of efficiency of patrolling called *idleness* has been widely accepted in the robotics literature. We extend it to the case of untrusted patrolmen; we denote by $\Im_k^f(G)$ the maximum time that a point of the domain may remain unvisited by reliable patrolmen. The objective is to find patrolling strategies minimizing $\Im_k^f(G)$.

We investigate this problem for various classes of graphs. We design optimal algorithms for line segments, which turn out to be surprisingly different from strategies for related patrolling problems proposed in the literature. We then use these results to study the case of general graphs. For Eulerian graphs G, we give an optimal patrolling strategy with idleness $\Im_k^f(G) = (f+1)|E|/k$, where $|E|$ is the sum of the lengths of the edges of G. Further, we show the hardness of the problem of computing the idle time for three robots, at most one of which is faulty, by reduction from 3-edge-coloring of cubic graphs — a known NP-hard problem. A byproduct of our proof is the investigation of classes of graphs minimizing idle time (with respect to the total length of edges); an example of such a class is known in the literature under the name of Kotzig graphs.

Keywords: Fault tolerant · Idleness · Kotzig graphs · Patrolling

K. Elbassioni and K. Makino (Eds.): ISAAC 2015, LNCS 9472, pp. 343–354, 2015.
DOI: 10.1007/978-3-662-48971-0_30

1 Introduction

Patrolling occurs in many activities of everyday life whenever it is required to monitor a specific region, for example, the perimeter of a piece of land or a building, so as to investigate a feature of interest for purposes of surveillance. Typically, in such a setting patrolmen are assigned to monitor specified regions by moving perpetually at regular intervals through areas assigned to them.

In this paper, we are interested in patrolling when some of the patrolmen may be unreliable (faulty) in that they fail to report their monitoring activities. More specifically, we model and study the following problem: We are given a team of robot patrolmen and a domain to be monitored. Assume that some of the patrolmen may be unreliable. We want to design a strategy constructing perpetual patrolmen trajectories, so that, independently of which subset of them (of a given size) will turn out to be faulty, no point of the environment will ever be left unvisited by some reliable robot longer than the allowed *idle time*.

Preliminaries and Notation. We are given a connected topological graph $G = (V, E)$ with V being its set of vertices and E its set of edges. In the sequel we define several useful concepts.

The Jordan arc representing each edge $e \in E$ of the graph $G = (V, E)$ is modeled as a smooth continuous and rectifiable curve of arbitrary positive length represented by its edge weight $w(e)$. We may suppose that the graph is embedded in $3D$ space, with no edge crossings. By $|E|$ we denote the sum of the lengths of the edges of G.

At any time a robot may occupy any point belonging to edge e (so the sum of its distances from both endpoints of e sums up to $w(e)$). We denote by $\mathcal{D}_G$ the domain (the union of edges) along which the robots walk. We assume a continuous traversal model, whereby the movement of the i-th robot within $\mathcal{D}_G$ follows a continuous function of time $\pi_i : [0, \infty) \to \mathcal{D}_G$, for each $i = 1, 2, \ldots, k$. Hence, $\pi_i(t)$ denotes the position in $\mathcal{D}_G$ of the i-th robot at time t. Each robot may move in any direction along an edge not exceeding the maximum (unit) speed so within time interval $[t_1, t_2]$ each robot may travel a distance of at most $t_2 - t_1$. We also suppose that when walking at maximum speed, a robot travels the unit distance in unit time, so that time and distance travelled are commensurable. By *patrolling strategy* we understand the set $\mathcal{P} = \{\pi_1, \pi_2, \ldots, \pi_k\}$ of infinite trajectories of k robots in $\mathcal{D}_G$, where $\pi_i(t)$ is the point of $\mathcal{D}_G$ occupied by the i-th robot at time t.

The performance of the patrolling strategy is evaluated by using a measure of *idleness*, widely used in robotics literature. Suppose that we design the patrolling strategy $\mathcal{P} = \{\pi_1, \pi_2, \ldots, \pi_k\}$ for k robots moving in the domain of a geometric graph G when each robot is reliable. Then the idleness of strategy $\mathcal{P}$ for graph G (or its *idle time*), denoted by $\Im_k^f(G, \mathcal{P})$ is the supremum of the lengths of time intervals between two consecutive visits to the same point of $\mathcal{D}_G$ (supremum taken over time and all points of $\mathcal{D}_G$). When up to f robots may be faulty, we assume that the adversary, knowing our strategy, may choose a set F of f

faulty robots, a point p of the domain and a time moment $t \geq 0$. The idleness of the strategy is the supremum (taken over all such adversarial choices) of time intervals T such that point p is not visited during the time interval $[t, t+T]$ by any reliable robot. Finally, the idleness of a graph G for k robots, at most f of which may be faulty, is denoted by $\Im_k^f(G) := \inf_{\mathcal{P}} \Im_k^f(G, \mathcal{P})$. Hence $\Im_k^f(G)$ is the lower bound of idleness over all possible patrolling strategies. When there are no faulty robots (i.e., $f = 0$) we use the notation $\Im_k(G, \mathcal{P}) := \Im_k^0(G, \mathcal{P})$.

Consider a walk of a robot within the segment, which starts at one of its endpoints, walks to the other endpoint and returns to the initial one. Such a cyclic path around the segment has length equal to twice its size. By an Eulerian tour of the segment by r robots we mean a perpetual movement of these robots, which are equally spaced around such a cyclic path, and walking in the same cyclic direction with the same speed. By $CPT(G)$ we denote the length of a Chinese Postman Tour on the graph G.

Related Work. Patrolling has been defined as the act of surveillance consisting in walking perpetually around an area in order to protect or supervise it. It is useful in monitoring and locating objects or humans that need to be rescued from a disaster, in ecological monitoring or detecting intrusion. Network administrators may use mobile agent patrols to detect network failures or to discover web pages which need to be indexed by search engines, cf. [22]. Patrolling has been recently intensively studied in robotics (cf. [6,13,14,17,22,30]) where it is often viewed as a version of terrain *coverage*, a central task in robotics.

Boundary and area patrolling have been studied in [1,13,14,26] with approaches placing more emphasis on experimental results. The accepted measure of the algorithmic efficiency of patrolling is called *idleness* and it is related to the frequency with which the points of the environment are visited (cf. [6,13,14,22]); this criterion was first introduced in [22]. Depending on the requirements, idleness may sometimes be viewed as the average [13], worst-case [30], probabilistic [1] or experimentally verified [22] time elapsed since the last visit of a node (cf. also [6]). In some papers the terms of *blanket time* [30] or *refresh time* [26] have been used instead.

A survey of diverse approaches to patrolling based on the idleness criteria can be found in [27]. In [3–5] patrolling is studied as a game between patrollers and the intruder. Some papers consider the patrolling problem based on swarm or ant-based algorithms [15,24,30]. In these approaches robots are memoryless (or having small memory), decentralized [24] with no explicit communication permitted either with other robots or the central station, with local sensing capabilities (e.g., [15]). Ant-like algorithms usually mark the visited nodes of the graph. [30] presents an evolutionary process and shows that a team of memoryless robots, by leaving marks at the nodes while walking through them, after relatively short time stabilizes to the patrolling scheme in which the frequency of the traversed edges is uniform to a factor of two (i.e., the number of traversals of the most often visited edge is at most twice the number of traversals of the least visited one).

A theoretical analysis of approaches to patrolling in graph-based models can be found in [6]. The two basic methods are referred to as *cyclic strategies*, where a single cycle spanning the entire graph is constructed with the robots assigned to consecutively traverse this cycle in the same direction, and as *partition-based strategies*, where the region is split into a number of either disjoint or overlapping portions to be patrolled by subsets of robots assigned to these regions. The environment and the time considered in the models studied are usually discrete in an underlying graph environment. In [26], polynomial-time patrolling solutions for lines and trees are proposed. For the case of cyclic graphs, [26] proves the NP-hardness of the problem and a constant-factor approximation is proposed.

Patrolling with robots that do not necessarily have identical speeds has been initiated in [9]. As shown in [12,20] it offers several surprises both in terms of the difficulty of the problem as well as in terms of the algorithmic results obtained. In particular, no optimal patrolling strategy involving more than three robots has yet been proposed.

Fault tolerance related to mobile robots has been considered for several problems in distributed computing with failures occurring either to the environment (nodes or links) or to the robots themselves. The cases of faulty robots were often studied for the robot gathering problem under various assumptions of faults (crash and Byzantine), e.g., [2,10,11]. Other studies concerned the problem of convergence, e.g. [7], flocking, e.g., [29] and many other ones. Several papers, e.g., [8,19,28] concerned unreliable or inaccurate robot sensing devices, rather than the robots themselves. Experimental papers related to unreliable robots performing patrolling were considered in the robotics literature [13,14,17,23]. To the best of our knowledge, the theoretical study, considered in our paper, concerning optimally patrolling a connected graph in the presence of faulty robots has not been investigated in the past.

Outline and Results of the Paper. In Sect. 2, we provide optimal patrolling strategies for line segments. These non-intuitive strategies rely on a decomposition of the set of robots into three groups with different patrolling tours, in a way dependent on k and f. Next we employ these results in Sect. 3 as building blocks to provide strategies for general graphs. In particular, for any Eulerian graph G we show that the idleness satisfies $\Im_k^f(G) = (f+1)|E|/k$. In Sect. 3.2, we analyze the hardness of the problem of computing the idle time on a specific class of graphs (derived from the class of Kotzig graphs) by showing that if the idle time could be computed optimally then we could solve 3-edge-coloring of cubic graphs, a well known NP-hard problem (see [16]). Finally, in Sect. 4 we conclude with a summary of our results and mention additional work and various related open problems. All missing proofs can be found in the full paper.

2 Idleness of Line Segments

In this section we study exclusively the idleness of the line segment and provide upper and lower bounds for idleness. Without loss of generality we assume that we have to patrol the unit-length segment, represented by the interval $I = [0, 1]$.

However, the results can be easily reformulated for segments of any given length. Throughout the main part of this section we assume that most of the robots are reliable, more precisely that $f < \frac{k-2}{2}$. We first give a patrolling strategy and analyze its performance. Then we analyze the lower bound for segment idleness showing that our strategy is optimal for odd f and almost optimal for even f.

2.1 The Upper Bound

The idea of the strategy is the following. We partition the segment I into three subsegments I_L (left), I_R (right), and I_M (middle), where I_M does not contain any endpoint of I. Two subsets of robots will follow Eulerian tours of I_L and I_R and the remaining robots are assigned to do the Eulerian tour of the entire I. We show that by choosing sizes of the segments of the partition as well as the number of robots assigned to each Eulerian tour we obtain an efficient strategy. We have the following theorem:

Theorem 1. *Consider k robots patrolling segment $I = [0, 1]$, with at most f of them faulty where $k > 2$ and $f < \frac{k}{2} - 1$. There exists a patrolling strategy $\mathcal{P}$ of I whose idleness satisfies $\Im_k^f(I, \mathcal{P}) \leq \frac{2\lfloor f/2 \rfloor + 2}{k - 2\lceil f/2 \rceil}$*

Proof. First we give explicitly the patrolling strategy.

1. Decompose the unit interval I into three segments I_L, I_M and I_R with pairwise disjoint interiors:

$$I_L := \left[0, \frac{\lceil f/2 \rceil}{k-2\lceil f/2 \rceil}\right], \; I_M := \left[\frac{\lceil f/2 \rceil}{k-2\lceil f/2 \rceil}, 1 - \frac{\lceil f/2 \rceil}{k-2\lceil f/2 \rceil}\right], \; I_R := \left[1 - \frac{\lceil f/2 \rceil}{k-2\lceil f/2 \rceil}, 1\right]$$

2. For each of the segments I_L, I_R assign $\lceil f/2 \rceil$ equally spaced robots to perform an Eulerian tour of this segment.
3. The remaining $k - 2\lceil f/2 \rceil$ robots perform an Eulerian tour of the entire segment I. These robots are also equally spaced around I.

Observe first that the subsegments I_L, I_R, I_M are well defined. Indeed, as f is an integer $f < \frac{k}{2} - 1$ implies $f \leq \frac{k-3}{2}$. Hence $2\lceil f/2 \rceil \leq f + 1 \leq \frac{k-3}{2} + 1 = \frac{k-1}{2}$. However, for any integer $k > 2$ we have $\frac{k-1}{2} \leq k - 2$ which implies $2\lceil f/2 \rceil \leq k - 2$, so the denominator of the fractions in the definitions of the segments I_L, I_M, I_R is not zero. Moreover, the point $1/2$ belongs to I_M, hence all these segments are well defined.

Denote by S_L (respectively S_R, S_I) the set of robots executing an Eulerian tour of I_L (respectively I_R, I). Observe that the distance d between two consecutive robots of S_I, computed around this Eulerian tour, equals $d = \frac{2}{k-2\lceil f/2 \rceil}$. We now prove the correctness of the upper bound on the idleness of any point $p \in I$. We consider two cases: when p belongs to an extremal subsegment I_L or I_R and when p is in the middle segment I_M.

Case 1: Point p is in extremal subsegment (by symmetry we may assume without loss of generality that $p \in I_L$). Suppose first that at least one robot

$r_i \in S_L$ is not faulty. Then r_i revisits every point of I_L at time intervals of at most $2|I_L|$. Hence the idleness of $p \in I_L$ (maximized at endpoints of I_L) is bound by $\Im_k^f(I,\mathcal{P}) \leq 2|I_L| = \frac{2\lceil f/2\rceil}{k-2\lceil f/2\rceil} < \frac{2\lfloor f/2\rfloor+2}{k-2\lceil f/2\rceil}$ When all $\lceil f/2\rceil$ robots of S_L are faulty, the idle time is maximized for $p = 0$, while the adversary chooses the remaining $\lfloor f/2\rfloor$ faulty robots to form a stream of consecutive robots of S_M. Then the time between visits of point $p = 0$ by two reliable robots (i.e. one preceding and one following such a stream) equals $d(\lfloor f/2\rfloor + 1)$ and we have

$$\Im_k^f(I,\mathcal{P}) \leq d(\lfloor f/2\rfloor + 1) = \frac{2(\lfloor f/2\rfloor + 1)}{k - 2\lceil f/2\rceil} \tag{1}$$

again verifying the claim of the theorem. The argument is entirely symmetric when $p \in I_R$ and is therefore omitted.

Case 2: $p \in I_M$. The visits to this point are made exclusively by the robots from S_I. Point p is being visited by two streams of robots executing the Eulerian tour of I, one walking over p from left to right and the other one from right to left (clearly, in the Eulerian cycle the robots are moving in the same direction, but from the "point of view of the point p" the traversal is in opposite directions). Each of these streams may have several faulty robots, and the idle time at p depends on the distance between the two reliable robots starting and ending such faulty streams.

Consider first the case when the two faulty-robots streams, visiting p at the same time, are disjoint, i.e. separated by at least one reliable robot (cf. Fig. 1 (a), where ∘ denotes a reliable robot and • a faulty robot).

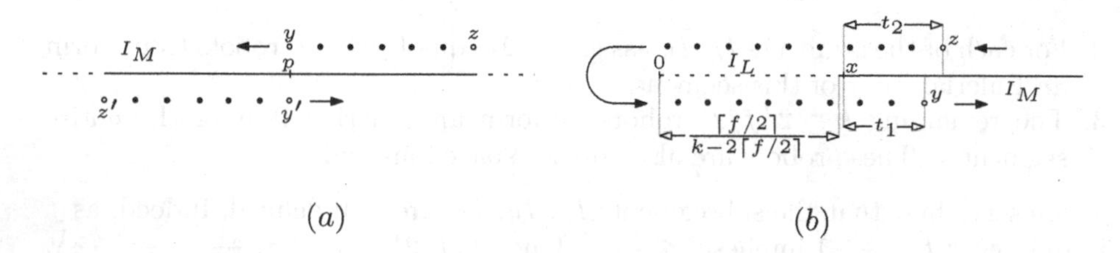

Fig. 1. Illustration of the patrolling strategy for a point p on the central part of the line segment.

To maximize the time while point p remains unvisited by reliable robots, the adversary has to make faulty two sequences of consecutive robots (i.e. those belonging to both left-to-right and right-to-left streams) arriving at p at the same time. The idle time is then determined by the length of the shorter of the two sequences of consecutive faulty robots, which in the worst-case contains $\lfloor f/2\rfloor$ robots. Then the claim of the theorem is again satisfied by Eq. (1).

Consider now the case when there is a single faulty-robot stream visiting p in both directions (see Fig. 1 (b) which depicts a time moment t when this happens). In the worst case this stream may contain f robots. Let t_1 be the time since the last visit of p by a reliable robot y and t_2 - the time when the next reliable robot z visits p. As $x > |I_L| = \frac{\lfloor f/2 \rfloor}{k-2\lceil f/2 \rceil}$ and all robots present within I_L at time t are faulty, as well as the distance between y and z around the Eulerian cycle is at most $d(f+1)$, we have

$$\Im_k^f(I,\mathcal{P}) = t_1 + t_2 \leq d(f+1) - 2|I_L| = \frac{2(f+1)}{k-2\lceil f/2 \rceil} - \frac{2\lceil f/2 \rceil}{k-2\lceil f/2 \rceil} = \frac{2\lfloor f/2 \rfloor + 2}{k-2\lceil f/2 \rceil}.$$

This completes the proof of Theorem 1. ∎

2.2 The Lower Bound

We first show the following lemma, which applies for general graphs.

Lemma 1. *Consider a patrolling strategy $\mathcal{P}$ of graph G. Let E' be a subset of segments of edges of G, such that starting from some time moment of the strategy, in the union of the interiors of all elements of E' there are always at most r robots. Then $\Im_k^f(\mathcal{I},\mathcal{P}) \geq \frac{(f+1)|E'|}{r}$, where $|E'|$ denotes the sum of lengths of segments of E'.*

The next theorem proves that the patrolling strategy from the previous section is optimal for odd f and almost optimal for even f.

Theorem 2. *For any k and f such that $f < k/2 - 1$ we have $\Im_k^f(\mathcal{I}) \geq \frac{f+1}{k-f-1}$.*

Proof. Partition the unit interval into the following three segments

$$I_L := \left[0, \frac{f+1}{2(k-f-1)}\right], \quad I_M := \left(\frac{f+1}{2(k-f-1)}, 1 - \frac{f+1}{2(k-f-1)}\right), \quad I_R := \left[1 - \frac{f+1}{2(k-f-1)}, 1\right].$$

By the condition $f < k/2 - 1$ in the hypothesis of the theorem the three sub-segments should not have a non-trivial overlap. Before proving the theorem we derive a crucial claim.

Claim. If $\Im_k^f(\mathcal{I}) < \frac{f+1}{k-f-1}$ then at each time moment during the patrolling there must be at least $f+1$ robots in each of the segments I_L and I_R.

By symmetry it is sufficient to prove this claim for the segment I_L. Suppose that $\Im_k^f(\mathcal{I}) < \frac{f+1}{k-f-1}$ and assume on the contrary that at some time, say t_0, we have at most f robots in the segment I_L. If an adversary makes all of these robots faulty then it would follow that no reliable robot could visit the endpoint 0 during the entire time interval $[t_0 - |I_L|, t_0 + |I_L|]$, where $|I_L|$ denotes the length of the interval I_L. Therefore the idle time at the endpoint 0 would be larger than $2|I_L| = \frac{f+1}{k-f-1}$, which contradicts the hypothesis of the claim.

From the Claim above we see that at all times each of the two intervals I_L and I_R contains at least $f+1$ robots. Since at each time moment at least $2f+2$ robots must be visiting I_L, I_R, the open interval I_M must always contain at most $k-2f-2$ robots. Applying Lemma 1 to the set E' consisting of the segment I_M, since $|I_M| = (1 - \frac{f+1}{k-f-1})$ we have $\Im_k^f(\mathcal{I}) \geq \frac{(f+1)|I_M|}{(k-2f-2)} = \frac{(f+1)(1-\frac{f+1}{k-f-1})}{(k-2f-2)} = \frac{f+1}{k-f-1}$. ∎

3 Idleness of Arbitrary Graphs

In this section we study upper and lower bounds for patrolling times on general graphs. First we prove a theorem associating the patrolling time to the length of a Chinese Postman Tour on the graph. Next we use the results in Sect. 2 concerning line segments so as to determine asymptotic bounds on the patrolling time for arbitrary graphs. The efficiency of the proposed strategy is arbitrarily close to the optimal one when k is sufficiently large.

3.1 A General Result and Algorithm

First we prove the following theorem and approximation patrolling strategy on arbitrary graphs.

Theorem 3. *For any connected graph G, and $k \geq 2$ robots, at most f of which are faulty, ($f \leq k-1$), we have that* $\frac{(f+1)|E|}{k} \leq \Im_k^f(G) \leq \frac{(f+1)CPT(G)}{k}$.

Proof. The upper bound is implied from the following patrolling algorithm:

1. Select any Chinese Postman Tour of G.
2. Have the robots patrol the graph by placing them equidistant along the Chinese Postman Tour.

It is clear that the respective distances between consecutive robots will be $\frac{CPT(G)}{k}$. The worst case idle time occurs when we have f consecutive faulty robots. In this case the resulting idle time will never exceed $\frac{(f+1)CPT(G)}{k}$, which proves the upper bound.

The lower bound follows directly from Lemma 1 applied to E' being the set of all edges of G. ∎

As a corollary of Theorem 3 we obtain the following tight (and simple) expression for the value of the idleness for Eulerian graphs.

Corollary 1 (Idleness for Connected Eulerian Graphs). *For any connected Eulerian graph G, and $k \geq 2$ robots, at most f of which are faulty ($f \leq k-1$), we have that* $\Im_k^f(G) = \frac{(f+1)|E|}{k}$.

The claim is immediate since in this case $|E| = CPT(G)$.

3.2 Hardness of Computing the Idleness

To show the hardness of our problem in general graphs, we restrict ourselves to the special case of $k = 3$ robots with exactly $f = 1$ fault. We will now prove that the problem of computing the idleness $\Im_3^1(G)$ is NP-hard for general graphs with unit-length edges. The proof proceeds by reduction from 3-Edge-Coloring in Cubic Graphs ($3ECC$), a well-known NP-complete problem (see [18]).

First we show the following auxiliary result which partially characterizes graphs having minimum possible idleness $\Im_3^1$ with respect to the total length of their edges. For a graph $H = (V, E)$ and $E' \subseteq E$, denote by $H[E'] \subseteq H$ the connected subgraph of H with edge set E'.

Lemma 2. *Let $H = (V, E)$ be a graph with unit-length edges.*

(i) If $\Im_3^1(H) = \frac{2}{3}E(H)$, then there exists a partition of the edge set $E = E_1 \cup E_2 \cup E_3$ such that each of the graphs $H[E \setminus E_i]$, $1 \le i \le 3$, is semi-Eulerian (i.e., connected and with at most two vertices of odd degree).

(ii) Conversely, if there exist a decomposition of the edge set: $E = E_1 \cup E_2 \cup E_3$, such that $|E_1| = |E_2| = |E_3| = \frac{1}{3}E(H)$ and each of the graphs $H[E \setminus E_i]$, $1 \le i \le 3$, is Eulerian, then $\Im_3^1(H) = \frac{2}{3}E(H)$.

We use Lemma 2 to show the following theorem.

Theorem 4 (Hardness of Computing the Idleness). *It is NP-hard to decide whether for a given graph H with unit-length edges we have $\Im_3^1(H) = \frac{2}{3}E(H)$.*

The above theorem shows that the problem of computing the optimal idle time for patrolling with unreliable robots is NP-hard in general. For an unbounded number of robots k (i.e., when k is treated as part of the input) and graphs with edges of integer length, the decision problem belongs to PSPACE, but we do not know whether it belongs to NP. We leave this as an open problem.

3.3 Characterizing Graphs with Minimum Idle Time

We close this section by considering some properties which hold for graphs with small idle time in fault-tolerant patrolling, and giving examples of classes of such graphs.

For the case of 3 robots, some classes of graphs minimizing idle time $\Im_3^1$ (with respect to the total length of edges) are given by Lemma 2. An example of such a class is known in the literature under the name of Kotzig graphs [21]. A graph is *Kotzig* if it is 3-regular and admits a decomposition into three matchings M_1, M_2, M_3 such that $E = M_1 \cup M_2 \cup M_3$ and for each pair $i \neq j$, the union $M_i \cup M_j$ forms a Hamiltonian cycle of the graph. By Lemma 2(*ii*), we immediately obtain that Kotzig graphs have the minimum possible idleness $\Im_3^1$ in the class of cubic graphs, i.e., $\Im_3^1 = \frac{2}{3}E = \frac{2}{3} \cdot \frac{3}{2}n = n$. Interestingly, this idleness $\Im_3^1$ is also best possible in the sense that we cannot obtain better idle time if we know beforehand which of the three robots is faulty, and attempt to solve the problem only for two non-faulty robots: for Kotzig graphs, we have $\Im_2^0 = \frac{CPT}{2} = \frac{2n}{2} = n$.

Corollary 2 (Idleness of Kotzig Graphs). $\Im_3^1(G) = \Im_2^0(G) = n$, *for n-vertex Kotzig graphs G.*

We remark that we do not know of a complete structural characterization of all graphs having minimum idleness $\Im_3^1 = \frac{2}{3}E$. The characterization from Lemma 2 is only partial, and the distinction between semi-Eulerian and Eulerian graphs in claims (*i*) and (*ii*) of the Lemma 2 is important. For example, when the patrolled graph is a cycle with 3 unit-length edges, we have $\Im_3^1 = \frac{2}{3}E = 2$, whereas this graph does not admit a decomposition $E = E_1 \cup E_2 \cup E_3$ into non-empty sets E_1, E_2, E_3 such that each of the graphs induced by $E \setminus E_i$ is Eulerian.

On the other hand, when the patrolled graph is a star with 3 unit-length edges, this graph admits a decomposition $E = E_1 \cup E_2 \cup E_3$ into single edges such that each of the graphs induced by $E \setminus E_i$ is semi-Eulerian, but this graph does not minimize idle time: by Theorem 3, $\Im_3^1 \geq \frac{CPT}{2} = 3 > \frac{2}{3}E$.

Variants of Lemma 2 can also be obtained for a larger number of robots. For k even, classes of graphs having minimum possible idle time $\Im_k^1 = \frac{2}{k}E$ include *Hamiltonian Decomposable Graphs*, i.e., k-regular graphs whose edge set can be partitioned into $k/2$ edge-disjoint Hamiltonian cycles [25].

4 Conclusion and Open Problems

We gave optimal fault-tolerant patrolling strategy for segments (for odd f) and Eulerian graphs. In all proposed strategies the collection of patrolmen is divided into sub-collections, each of the sub-collections, forming a "cycle" of equally spaced robots walking around a portion of the graph (with some portions being covered by more than one sub-collection). Somewhat surprisingly, for a graph as simple as a segment, the optimal strategy consists of two sub-collections patrolling small sub-segments and the third sub-collection patrolling the entire segment (hence the points close to the endpoints being visited by the robots belonging to two sub-collections). We also proved that for some graphs finding an optimal patrolling strategy is NP-hard.

While optimal strategies for Eulerian graphs work for any ratio of faulty patrolmen, the strategies for segments assume the maximal faulty robots ratio to be slightly smaller than half the total of all robots. One open question is to give optimal patrolling strategies for segments when the faulty robot ratio is high. There are plenty of open questions concerning different models of patrolling: robot failures may be dynamic, failures may happen with given probability, robots may have non-zero visibility radii, or may be allowed to communicate. Some questions, like robots with distinct patrolling speeds and two-dimensional domains may be hard.

Acknowledgements. This work was partially supported by NSERC grants. Research on this problem was initiated at the MITACS "International Problem Solving Workshop" held on July 16–20, 2012, in Vancouver, BC, Canada. The authors would like to express their deepest appreciation for the generous support of MITACS.

References

1. Agmon, N., Kraus, S., Kaminka, G.A.: Multi-robot perimeter patrol in adversarial settings. In: ICRA, pp. 2339–2345 (2008)
2. Agmon, N., Peleg, D.: Fault-tolerant gathering algorithms for autonomous mobile robots. SIAM J. Comput. **36**(1), 56–82 (2006)
3. Alpern, S., Morton, A., Papadaki, K.: Optimizing Randomized Patrols. Operational Research Group, London School of Economics and Political Science (2009)
4. Alpern, S., Morton, A., Papadaki, K.: Patrolling games. Oper. Res. **59**(5), 1246–1257 (2011)

5. Amigoni, F., Basilico, N., Gatti, N., Saporiti, A., Troiani, S.: Moving game theoretical patrolling strategies from theory to practice: an USARSim simulation. In: ICRA, pp. 426–431 (2010)
6. Chevaleyre, Y.: Theoretical analysis of the multi-agent patrolling problem. In: IAT, pages 302–308 (2004)
7. Cohen, R., Peleg, D.: Convergence properties of the gravitational algorithm in asynchronous robot systems. SIAM J. Comput. **41**(1), 1516–1528 (2005)
8. Cohen, R., Peleg, D.: Convergence of autonomous mobile robots with inaccurate sensors and movements. SIAM J. Comput. **38**(1), 276–302 (2008)
9. Czyzowicz, J., Gąsieniec, L., Kosowski, A., Kranakis, E.: Boundary patrolling by mobile agents with distinct maximal speeds. In: Demetrescu, C., Halldórsson, M.M. (eds.) ESA 2011. LNCS, vol. 6942, pp. 701–712. Springer, Heidelberg (2011)
10. Défago, X., Gradinariu, M., Messika, S., Raipin-Parvédy, P.: Fault-tolerant and self-stabilizing mobile robots gathering. In: Dolev, S. (ed.) DISC 2006. LNCS, vol. 4167, pp. 46–60. Springer, Heidelberg (2006)
11. Dieudonné, Y., Pelc, A., Peleg, D.: Gathering despite mischief. In: Proceedings of the Twenty-Third Annual ACM-SIAM Symposium on Discrete Algorithms, pp. 527–540. SIAM (2012)
12. Dumitrescu, A., Ghosh, A., Tóth, C.D.: On fence patrolling by mobile agents. Electr. J. Comb. **21**(3), P3.4 (2014)
13. Elmaliach, Y., Agmon, N., Kaminka, G.A.: Multi-robot area patrol under frequency constraints. Ann. Math. Artif. Intell. **57**(3–4), 293–320 (2009)
14. Elmaliach, Y., Shiloni, A., Kaminka, G.A.: A realistic model of frequency-based multi-robot polyline patrolling. AAMAS **1**, 63–70 (2008)
15. Elor, Y., Bruckstein, A.M.: Autonomous multi-agent cycle based patrolling. In: Dorigo, M., Birattari, M., Di Caro, G.A., Doursat, R., Engelbrecht, A.P., Floreano, D., Gambardella, L.M., Groß, R., Şahin, E., Sayama, H., Stützle, T. (eds.) ANTS 2010. LNCS, vol. 6234, pp. 119–130. Springer, Heidelberg (2010)
16. Garey, M., Johnson, D.: Computers and Intractability, vol. 174. Freeman, San Francisco (1979)
17. Hazon, N., Kaminka, G.A.: On redundancy, efficiency, and robustness in coverage for multiple robots. Robot. Auton. Syst. **56**, 1102–1114 (2008)
18. Holyer, I.: The NP-completeness of edge-coloring. SIAM J. Comput. **10**(4), 718–720 (1981)
19. Izumi, T., Souissi, S., Katayama, Y., Inuzuka, N., Défago, X., Wada, K., Yamashita, M.: The gathering problem for two oblivious robots with unreliable compasses. SIAM J. Comput. **41**(1), 26–46 (2012)
20. Kawamura, A., Kobayashi, Y.: Fence patrolling by mobile agents with distinct speeds. In: Chao, K.-M., Hsu, T., Lee, D.-T. (eds.) ISAAC 2012. LNCS, vol. 7676, pp. 598–608. Springer, Heidelberg (2012)
21. Kotzig, A.: Hamilton graphs and hamilton circuits. In: Theory of Graphs and its Applications, Proceedings of the Symposium of Smolenice, pp. 63–82. Publ. House Czechoslovak Acad. Sci. (1964)
22. Machado, A., Ramalho, G.L., Zucker, J.-D., Drogoul, A.: Multi-agent patrolling: an empirical analysis of alternative architectures. In: Sichman, J.S., Bousquet, F., Davidsson, P. (eds.) MABS 2002. LNCS (LNAI), vol. 2581, pp. 155–170. Springer, Heidelberg (2003)
23. Marino, A., Parker, L., Antonelli, G., Caccavale, F., Chiaverini, S.: A fault-tolerant modular control approach to multi-robot perimeter patrol. In: Robotics and Biomimetics (ROBIO), pp. 735–740 (2009)

24. Marino, A., Parker, L.E., Antonelli, G., Caccavale, F.: Behavioral control for multi-robot perimeter patrol: a finite state automata approach. In: ICRA, pp. 831–836 (2009)
25. Park, J.-H., Kim, H.-C.: Dihamiltonian decomposition of regular graphs with degree three. In: Widmayer, P., Neyer, G., Eidenbenz, S. (eds.) WG 1999. LNCS, vol. 1665, pp. 240–249. Springer, Heidelberg (1999)
26. Pasqualetti, F., Franchi, A., Bullo, F.: On optimal cooperative patrolling. In: CDC, pp. 7153–7158 (2010)
27. Portugal, D., Rocha, R.: A survey on multi-robot patrolling algorithms. In: Camarinha-Matos, L.M. (ed.) Technological Innovation for Sustainability. IFIP AICT, vol. 349, pp. 139–146. Springer, Heidelberg (2011)
28. Souissi, S., Défago, X., Yamashita, M.: Gathering asynchronous mobile robots with inaccurate compasses. In: Shvartsman, M.M.A.A. (ed.) OPODIS 2006. LNCS, vol. 4305, pp. 333–349. Springer, Heidelberg (2006)
29. Yang, Y., Souissi, S., Défago, X., Takizawa, M.: Fault-tolerant flocking for a group of autonomous mobile robots. J. Syst. Softw. **84**(1), 29–36 (2011)
30. Yanovski, V., Wagner, I.A., Bruckstein, A.M.: A distributed ant algorithm for efficiently patrolling a network. Algorithmica **37**(3), 165–186 (2003)

Cops and Robbers on String Graphs

Tomáš Gavenčiak[1], Przemysław Gordinowicz[2], Vít Jelínek[3], Pavel Klavík[3](✉), and Jan Kratochvíl[1]

[1] Department of Applied Mathematics, Faculty of Mathematics and Physics, Charles University in Prague, Prague, Czech Republic
{gavento,honza}@kam.mff.cuni.cz

[2] Institute of Mathematics, Technical University of Lodz, Łódź, Poland
pgordin@p.lodz.pl

[3] Computer Science Institute, Charles University in Prague, Prague, Czech Republic
{jelinek,klavik}@iuuk.mff.cuni.cz

Abstract. The game of cops and robber, introduced by Nowakowski and Winkler in 1983, is played by two players on a graph. One controls k cops and the other a robber. The players alternate and move their pieces to the distance at most one. The cops win if they capture the robber, the robber wins by escaping indefinitely. The cop number of G is the smallest k such that k cops win the game.

We extend the results of Gavenčiak et al. [ISAAC 2013], investigating the maximum cop number of geometric intersection graphs. Our main result shows that the maximum cop number of string graphs is at most 15, improving the previous bound 30. We generalize this approach to string graphs on a surface of genus g to show that the maximum cop number is at most $10g + 15$, which strengthens the result of Quilliot [J. Combin. Theory Ser. B 38, 89–92 (1985)]. For outer string graphs, we show that the maximum cop number is between 3 and 4. Our results also imply polynomial-time algorithms determining the cop number for all these graph classes.

1 Introduction

The Cops and Robber game on graphs has been introduced by Winkler and Nowakowski [12] and independently by Quilliot [14]. In this paper, we investigate this game on the classes of geometric intersection graphs.

Rules of the Game. In this game two players alternate their moves. The first player (called "the cops") places k cops on the vertices of a graph G. Then the second player (called "the robber") chooses a vertex for the robber. Then the players alternate. In the cops' move, every cop either stays in its vertex or moves

P. Gordinowicz and V. Jelínek—Supported by CE-ITI (P202/12/G061 of GAČR). For the full version, see [5].

T. Gavenciak, P. Klavík and J. Kratochvíl—Supported by Charles University as GAUK 196213.

K. Elbassioni and K. Makino (Eds.): ISAAC 2015, LNCS 9472, pp. 355–366, 2015.
DOI: 10.1007/978-3-662-48971-0_31

to one of its neighbors. More cops may occupy the same vertex. In the robber's move, the robber either stays in its vertex, or goes to a neighboring vertex.

The game ends when the robber is *captured* which happens when a cop occupies the same vertex as the robber. The cops wins if he is able to capture the robber. The robber wins if he is able to escape indefinitely.

Definition 1.1. *For a graph G, its* cop number $\mathrm{cn}(G)$ *is the least number k such that the cops have a winning strategy on G with k cops. For a class of graphs $\mathcal{C}$, the* maximum cop number *max-cn*$(\mathcal{C})$ *is the maximum cop number* $\mathrm{cn}(G)$ *of a connected graph $G \in \mathcal{C}$, possibly $+\infty$.*

The restriction to connected graphs is standard. The reason is that if G has connected components $C_1, \dots, C_k$, then $\mathrm{cn}(G) = \sum_{i=1}^{k} \mathrm{cn}(C_i)$.

Known Results. Graphs of the cop number one were characterized already by Quilliot [14]. These are the graphs whose vertices can be linearly ordered $v_1, v_2, \dots, v_n$ so that each v_i for $i \geq 2$ is a corner of $G[v_1, \dots, v_i]$, i.e., v_i has a neighbor v_j for some $j < i$ such that v_j is adjacent to all other neighbors of v_i.

For k part of the input, deciding whether the cop number of a graph is at most k has been shown to be NP-hard (2010) [4], PSPACE-hard (2013) [10] and very recently (2015) even EXPTIME-complete [8], confirming a conjecture of Goldstein and Reingold (1995) [7]. In order to test whether k cops suffice to capture the robber on an n-vertex graph, we can search the game graph which has $\mathcal{O}(n^{k+1})$ vertices to find a winning strategy for cops. In particular, if k is a fixed constant, this algorithm runs in polynomial time.

For general graphs on n vertices, it is known that at least $\sqrt{n}$ cops may be needed (e.g., for the incidence graph of a finite projective plane). Meyniel's conjecture states that the cop number of a connected n-vertex graph is $\mathcal{O}(\sqrt{n})$. For more details and results, see the recent book [2].

Geometrically Represented Graphs. Figure 1 shows that the geometry of a graph class heavily influences the maximum cop number. For planar graphs, the classical result of Aigner and Fromme [1] shows that the maximum cop number is 3. This result was generalized to graphs of bounded genus by Quilliot [14] and improved by Schroeder [15]. However, while for planar graphs (genus 0) the maximum cop number is equal 3, already for toroidal graphs (genus 1) the exact value is not known.

We study *intersection representations* in which a graph G is represented by a map $\varphi : V \to 2^X$ for some ground set X such that the edges of G are described by the intersections: $uv \in E \iff \varphi(u) \cap \varphi(v) \neq \emptyset$. The ground set X and the images of φ are usually somehow restricted to get particular classes of intersection graphs. For example, the well-known interval graphs have $X = \mathbb{R}$ and every $\varphi(v)$ a closed interval.

All of these classes admit large cliques, so their genus is unbounded and the above bound of the maximum cop number does not apply. On the other hand existence of large cliques does not imply big maximum cop number since only one cop can guard a maximal clique. It was shown by Gavenčiak et al. [6] that for most of these intersection graph classes, the maximum cop-number is bounded.

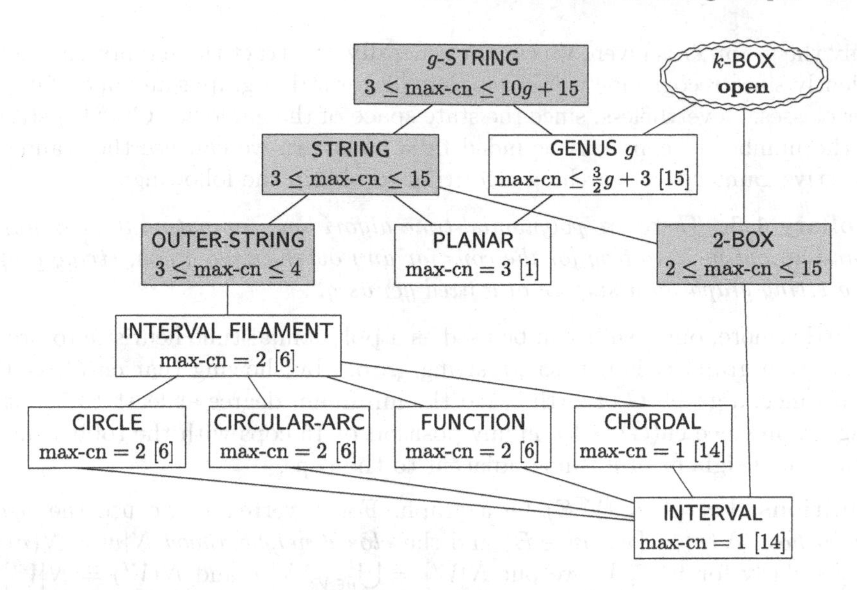

Fig. 1. The Hasse diagram of inclusions between important classes of geometrically represented graphs with known bounds on the maximum cop number. The bounds presented in this paper are in gray.

In particular, it is shown in [6] that the maximum cop number of string graphs is at most 30. The class of *string graphs* (STRING) is the class of intersection graphs of *strings*: $X = \mathbb{R}^2$ and every $\varphi(v)$ is required to be a finite curve that is a continuous image of the interval $[0, 1]$ in $\mathbb{R}^2$. It is known that every intersection graph of *arc-connected sets* in the plane (i.e., connected regions bounded by closed simple Jordan curves) is a string graph, so the above bound applies to most classes of intersection graphs in the plane. For instance, *boxicity k graphs* (k-BOX) are intersection graphs of k-dimensional intervals in $\mathbb{R}^k$ and they are string graphs for $k \leq 2$.

Let $\mathbf{S}$ be an arbitrary surface of genus g. We consider a generalization of string graphs for $X = \mathbf{S}$, and we denote this class by g-STRING. It is known that every graph embeddable to a genus g can be represented by a contact representation of disks on a suitable Riemann surface of genus g; so GENUSg$\subsetneq$g-STRING. The class of *outer-string graphs* (OUTER-STRING) consists of all string graphs having string representations with each string in the upper half-plane, intersecting the x-axis in exactly one point, which is an endpoint of this string.

Theorem 1.2. *We show the following bounds for the maximum cop number:*

(i) $3 \leq$ max-cn(OUTER-STRING) ≤ 4.
(ii) $3 \leq$ max-cn(STRING) ≤ 15.
(iii) $3 \leq$ max-cn(g-STRING) $\leq 10g + 15$.

We note that the strategies of cops in all upper bounds are geometric and their description is constructive, using an intersection representation of G.

If only the graph G is given, we cannot generally construct these representations efficiently since recognition is NP-complete [9] for string graphs and open for the other classes. Nevertheless, since the state space of the game has $\mathcal{O}(n^{k+1})$ states and the number of cops k is bounded by a constant, we can use the standard exhaustive game space searching algorithm to obtain the following:

Corollary 1.3. *There are polynomial-time algorithms computing the cop number and an optimal strategy for the cops for any outer-string graph, string graph and a string graph on a surface of a fixed genus g.*

Furthermore, our results can be used as a polynomial-time heuristic to prove that a given graph G is not, say, a string graph, by showing that $\text{cn}(G) > 15$. For instance, a graph G of girth 5 and the minimum degree at least 16 is not a string graph since $\text{cn}(G) > 15$: in any position of 15 cops with the robber on v, at least one neighbor of v is non-adjacent to the cops.

Definitions. Let $G = (V, E)$ be a graph. For a vertex v, we use the *open neighborhood* $N(v) = \{u : uv \in E\}$ and the *closed neighborhood* $N[v] = N(v) \cup \{v\}$. Similarly for $V' \subseteq V$, we put $N[V'] = \bigcup_{v \in V'} N[v]$ and $N(V') = N[V'] \setminus V'$. For $V' \subseteq V$, we denote by $G|_{V'}$ the subgraph of G induced by V'. For assumptions for string representations, see the full version.

2 Outer-String Graphs

In this section, we prove that the maximum cop number of outer string graphs is between 3 and 4, thus establishing Theorem 1.2(i).

Proof (Theorem 1.2 (i), sketch). Figure 2 shows a connected outer string graph requiring three cops. It remains to show the four cops are always sufficient.

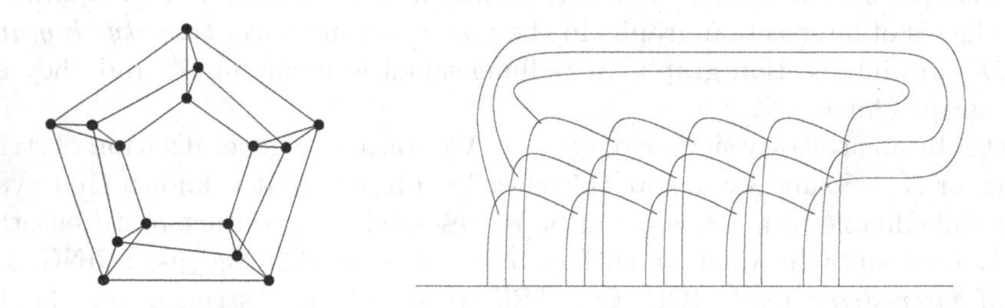

Fig. 2. The 3-by-5 toroidal grid G and one of its outer-string representations. Clearly $\text{cn}(G) = 3$.

We quickly sketch the strategy using Fig. 3, for the details, see the full version. Two cops are called *guards* and in each phase they guard two strings s_i and s_j such that the robber is confined under them. Two other cops called *hunters* travel along strings $p_0, \ldots, p_k$ covering the top of the confined area, always one is on p_i and the other on p_{i+1}. When the robber is confined under p_i and p_{i+1}, the guards move to p_i and p_{i+1} and the next phase begins. □

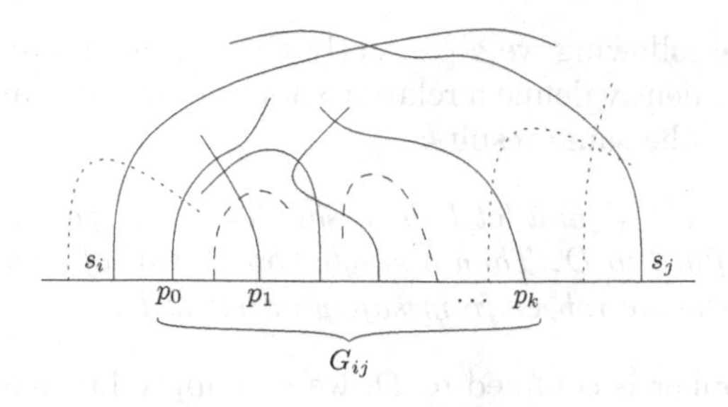

Fig. 3. An overview of the strategy.

3 Guarding Shortest Paths and Curves in String Graphs

Shortest Paths. We recall a lemma by Aigner and Fromme [1] giving us a strategy to prevent the robber to enter any given shortest path using only one cop in general graphs.

Lemma 3.1 ([1], **Lemma 4**). *Let $P = \{u = p_0, p_1, \ldots, p_k = v\}$ be any shortest $u - v$ path. Then a single cop C can, after a finite number of moves (used to move cop C to an appropriate position on P), prevent the robber from safely entering P. That is, if the robber ever moved on P, he would be captured in the next move.*

This result is particularly useful for planar graphs where one can cut the graph by protecting several shortest paths. For intersection graphs, forbidding the robber to visit vertices of P is not sufficient to prevent him from moving from one side of the part to the other. We need a stronger tool to geometrically restrict the robber. We get this by showing that in general graphs we can protect the closed neighborhood of a given shortest path using five cops, preventing the robber from safely stepping on any string even crossing the protected path.

We first need one additional generalization of Lemma 3.1 – we protect paths which are not necessarily shortest in G, but are shortest from the point of the robber within a region he is already confined to. Below we combine this generalization with guarding path neighborhood. We believe that these tools may be of some further interest. We say that an *$u - v$ path P is shortest relative to $D \subseteq V$* if there is no shorter $u - v$ path in G using at least one vertex of D. Note that P itself may or may not go through D.

Confinement, Safe Moves and Guarding. When our strategy makes sure that any time in the future of the game, whenever the robber leaves $D \subseteq V$ he is captured immediately, we say that the robber is *confined to* D. Note that this includes the case when the robber cannot even get outside D without being captured. If the robber can be immediately captured by moving to a vertex v, we say that the robber *cannot safely move to* v. When the strategy makes sure that the robber may not safely move to any $v \in P$, we say that *P is guarded.*

Note that in the following we get exactly the original statement for $D = V$. Also, we could equivalently define a relative shortest path by having no shortcuts contained in D with the same results.

Lemma 3.2. *Let $u, v \in V$ and let P be a shortest $u - v$ path relative to $D \subseteq V$ with the robber confined to D. Then a single cop C can, after a finite number of initial moves, prevent the robber from safely entering P.*

Proof. Since the robber is confined to D, we can apply Lemma 3.1 to $G|_{P \cup D}$. □

Lemma 3.3 ([6], **Lemma 6**). *Let $u, v \in V$ and let P be a shortest $u - v$ path relative to $D \subseteq V$ with the robber is confined to D. Then five cops $C_{-2}, C_{-1}, C_0, C_1, C_2$ can, after a finite number of initial moves, guard $N[P]$.*

In the following, when we say "start guarding a path", we do not explicitly mention the initial time required to position the five cops onto the path and assume that the strategy waits for enough turns.

Shortest Curves. Since our strategy for string graphs is partially geometric, we introduce the concept of shortest curves as particular curves through the string representation of a shortest path. Note that below we consider any curves sharing only their endpoints to be disjoint.

Let G be a string graph together with a fixed string representation φ, robber confined to $D \subseteq V$ and P a shortest $u - v$ relative to D. Suppose that we choose and fix two points $\pi_u \in \varphi(u)$ and $\pi_v \in \varphi(v)$. Let $\pi_{uv} \subseteq \varphi(P)$ be a curve from π_u to π_v such that $\pi_{uv} \subseteq \bigcup_{p \in P} \varphi(p)$ and for every $p \in P$ π_{uv} has a connected intersection with $\phi(p)$ and these correspond to the points of P in the same order. We call π_{uv} a *shortest curve of P (relative to D)* with endpoints π_u and π_v. A curve π is called a *shortest curve (relative to D)* if it is a shortest curve of some shortest path. We leave out D if $D = V$ or it is clear from the context.

The shortest path in the graph corresponding to a shortest curve π is uniquely defined by the sequence of strings that intersect π on a substring of non-zero length. To *guard a shortest curve* π means to guard its corresponding shortest path. The number of its strings is the *length* of π. Note that the Euclidean length of π plays no role in this paper.

Corollary 3.4. *Let G be a string graph together with a string representation φ and let π be a shortest curve relative to D such that the robber is confined to D. Then five cops can (after a finite number of initial moves) prevent the robber from entering any string intersecting π.*

Proof. Let P be the shortest path such that π is a shortest curve of P. By guarding $N[P]$, the cops prevent from entering strings intersecting π. □

Lemma 3.5. *Any sub-curve (continuous part) of a shortest curve (relative to D) is also a shortest curve (relative to D).* □

4 Capturing Robber in String Graphs

In this section we show that the number of cops sufficient to capture a robber on a connected string graphs is bounded by 15. The idea of the proof of Theorem 1.2(ii) is inspired by the proof of Aigner and Fromme for planar graphs [1]. Before we prove Theorem 1.2(ii), we introduce several notions used below.

Segments, Faces and Regions. For a given string graph G and its string representation φ let the *faces* (of φ) be the open arc-connected regions of $\mathbb{R}^2 \setminus \varphi(G)$, and let a *closed face* be the closure of a face. As we assume that the number of intersections of φ is finite, the number of faces is also finite. Note that every face is an open set.

A *segment* of string π is a part of the string not containing any intersection with another string between either two intersections, an intersection and an endpoint, or two endpoints. Note that the number of segments is also finite. A *region* is a closed subset of $\mathbb{R}^2$ obtained as a closure of a union of some of the faces.

Let $\mathrm{clos}(X)$ denote the topological closure of a set X and $\mathrm{int}(X)$ the topological interior of X. A vertex v is *internal to* B (also *contained in* B) if $\varphi(v) \subseteq \mathrm{int}(B)$. Denote all vertices internal to a region B by $V_B = \{v \in V \mid \varphi(v) \subseteq \mathrm{int}(B)\}$. In the next section we will use the following topological result, following from the previous section and Corollary 3.4.

Proposition 4.1. *If there is $D \subseteq V$ such that the cops guard disjoint shortest curves π_1 and π_2 (relative to D) between points π_u to π_v such F is the closed face of $\mathbb{R}^2 \setminus (\pi_1 \cup \pi_2)$ containing the robber's string and D is the component of V_F containing the robber, then the robber may not safely leave D.*

Additionally, we use the following topological lemma, for the proof see the full version.

Lemma 4.2. *Given two disjoint simple $\pi_u - \pi_v$ curves π_1 and π_2 in $\mathbb{R}^2$, with π_u, π_v different points, let F be one of the closed faces of $\mathbb{R}^2 \setminus (\pi_1 \cup \pi_2)$. For any simple $\pi_u - \pi_v$ curve π_3 contained in F and going through at least one of its inner points we have that every face of $F \setminus (\pi_1 \cup \pi_2 \cup \pi_3)$ is bounded by simple and disjoint curves π_i' and π_3' with $\pi_i' \subseteq \pi_i$, $\pi_3' \subseteq \pi_3$ and $i \in \{1, 2\}$.*

Restricted Graphs and Strategies. Given a closed region $B \subseteq \mathbb{R}^2$, let G *restricted to* B, denoted $G|_B$, be the intersection graph of the curves of $\varphi \cap B$. This operation may remove vertices (for entire strings outside B), remove edges (crossings outside B) and it also splits each vertex v whose string $\varphi(v)$ leaves and then reenters B at least once. In the last case, every arc-connected part of $\varphi(v) \cap B$ spans a new vertex v_i. The new vertices are also called the *splits* of v. The new graph is again a string graph with representation denoted $\varphi|_B$ directly derived from φ. Note that this operation preserves the faces and strings in $\mathrm{int}(B)$ and all representation properties assumed above, namely the vertex set of $G|_B$ is finite. Also, the number of segments does not increase.

Lemma 4.3. *Let B be a region. If π is a shortest curve (optionally relative to D with the robber confined to D) and $\pi' \subseteq \pi$ is a sub-curve with $\pi' \subseteq B$, then π' is a shortest curve (relative to D) in $G|_B$ and $\varphi|_B$.*

Proof. This follows from Proposition 3.5 and the fact that underlying path of π' is preserved (and if any $p \in P'$ got split into $\{p_i\}$ we use the p_i intersecting π') and no path (e.g. through D) can get shortened by a restriction. □

We now show that a strategy for a restricted graph may be used in the original graph.

Lemma 4.4. *Let B be a region. If there is a cop's strategy $\mathcal{S}'$ eventually capturing a robber in $G|_B$ confining him to V_B then there is a strategy $\mathcal{S}$ for the same number of cops capturing the robber on G confining him to V_B.*

Proof. The strategy $\mathcal{S}$ plays out as $\mathcal{S}'$ except when $\mathcal{S}'$ would move a cop to a split $v_i \in V_{G|_B}$ of $v \in V_G$, $\mathcal{S}$ moves the cop to v. Note that all such moves are possible. Robber's choices while internal to B are not extended in any way.

Assume the robber moves from internal u to non-internal v, which is split to $v_1, \ldots v_k$ in $G|_B$. Note that at least one of $vv_1, \ldots v_k$, say v_i, is adjacent to u in $G|_B$, as $\varphi(u)$ has to intersect $\varphi(v)$ in $\mathrm{int}(B)$. Let $\mathcal{S}$ play as $\mathcal{S}'$ would if the robber moved to v_i, capturing him with this move as assumed in the statement. □

The Strategy for 15 Cops. Our strategy proceeds in phases, monotonously shrinking the safe area of the robber. Slightly informally, in every phase the robber is confined to $D \subseteq V$ by either (A) a single cop guarding a cut-vertex separating D from the rest of the graph or (B) two squads of cops guarding two shortest curves forming a simple closed circle. Then we show that we can decrease either the number of the segments of φ or size of D while not increasing the other and get one of the cases again.

Note that it is important that in case (B) the curves form a simple (not self-intersecting) cycle (as the general case has many technical issues). Also, the reader can assume that the robber is always *inside* the cycle, e.g. using circular inversion of φ if not, but it is not necessary.

Proof (Theorem 1.2(ii)). In any situation let $\widetilde{P}$ be the union of currently guarded paths and vertices. Let D be the component of $V \setminus N[\widetilde{P}]$ containing the robber and let $Q = N[\widetilde{P}] \cap N[D]$. Recall that whenever the robber would be in $N[\widetilde{P}]$ he would be immediately captured, so D is well defined. Let s be the number of segments of φ.

We build a strategy that confines the robber to D for the rest of the game. Therefore we may assume that $V = D \cup Q \cup \widetilde{P}$ as the vertices outside $N[D]$ are irrelevant for the robber and unused by our strategy.

Claim. Let $V = D \cup Q \cup \widetilde{P}$, the robber be on $r \in D$ and one of the following:

(A) One cop guards a vertex c, $|\widetilde{P}| = 1$.
(B) The cops guard two shortest curves π_1 and π_2 (relative to D) between points π_u to π_v such that $\pi_1 \cup \pi_2$ forms a simple cycle, $|\widetilde{P}| \geq 2$ and additionally $G = G|_F$ where F is the closed face of $\mathbb{R}^2 \setminus (\pi_1 \cup \pi_2)$ containing $\varphi(r)$.

Then 15 cops have a strategy to capture the robber confining him to D.

Proof (Claim). We prove this by induction on s and $|D|$, the claim obviously holds for $s \leq 1$ and $|D| = 0$. We distinguish two cases:

Case A. If $Q = \{q\}$ then start guarding q, stop guarding c and let $G' = G - c$ while also leaving out any irrelevant vertices to have $V' = D' \cup Q' \cup \{q\}$ as above. We then use claim case (a) for G' with both smaller s' and $D' \subsetneq D$.

If $Q = \{q_1, \ldots q_k\}$, $k \geq 2$, let $G' = G - c$ and let π_{q_i} be any point of $\varphi(c) \cap \varphi(q_i)$. Now let π_1 be a shortest curve between some π_{q_i} and π_{q_j}. We let $\pi_2 \subseteq \varphi(c)$ be the part of $\varphi(c)$ between π_{q_i} and π_{q_j}.

However, $\pi_1 \cup \pi_2$ may not be a simple cycle. Let $\pi_u = \pi_{q_i}$ and let π_v to be the first point of $\pi_1 \cup \pi_2$ along π_2 going from π_u. Note that if there is no other intersection then $\pi_v = \pi_{q_j}$. Now let π_1' and π_2' be the parts of π_1 and π_2 between π_u and π_v, forming a simple cycle.

Let $G'' = G|_F$ where F is the closed face of $\mathbb{R}^2 \setminus (\pi_1 \cup \pi_2)$ containing $\varphi(r)$. Remove any irrelevant vertices from G'' to have $V'' = D'' \cup Q'' \cup \widetilde{P}''$ as above and use claim case (b) for smaller $D'' \subsetneq D$ (as P_1 has a neighbor in D) and not increased $|s''|$. Note that $\widetilde{P}''$ uses at least one vertex other than c.

Case B. If there is no $\pi_u - \pi_v$ path through a vertex of D then, according to Menger's theorem, there must be a cut-vertex $c \in \widetilde{P} \cup Q$ separating D from $\widetilde{P}$. Let one cop guard c and then stop guarding $\widetilde{P}$. Let $G' = G \setminus (\widetilde{P} - c)$ while also leaving out irrelevant vertices to have $V' = D' \cup Q' \cup \{c\}$ as above. We then use claim case (a) for G' with smaller s' and $D' \subseteq D$.

If there is a $\pi_u - \pi_v$ path through a vertex of D, let π_3 be shortest such curve. Note that it is a shortest curve relative to D. Let five cops start guarding π_3 and then let F be the closed face of $\mathbb{R}^2 \setminus (\pi_1 \cup \pi_2 \cup \pi_3)$ containing the robber string. According to Lemma 4.2 we have that F is delimited by disjoint π_i' and π_j' where $i = 3$ or $j = 3$ and $\pi_i' \cup \pi_j'$ form a simple cycle. We let the cops stop guarding π_k where $k \notin \{i, j\}$ and restrict the guarding of π_i and π_j to π_i' and π_j' as in Proposition 3.5. $|\widetilde{P}| \geq 2$ as one vertex string can not form a closed loop.

Let $G' = G|_F$ while also removing any irrelevant vertices from G' to have $V' = D' \cup Q' \cup \widetilde{P}'$ as above. We then use claim case (b) for G' with non-increased s and $D' \subsetneq D$. ◇

Having proven the claim, the theorem then follows by guarding an arbitrary vertex c with one cop so $\widetilde{P} = \{c\}$, defining D and Q as before the claim, and discarding irrelevant vertices to get $V' = D \cup Q \cup \widetilde{P}$. We then use claim case (a) with $G' = G|_{V'}$. □

5 String Graphs on Bounded Genus Surfaces

In this section, we generalise the results of the previous section and prove that $10g + 15$ cops are sufficient to catch the robber on graphs having a string representation on a surface of genus g.

We assume familiarity with basic topological concepts related to curves on surfaces, such as genus, non-contractible closed curves and the fundamental group of surfaces; otherwise see [13]. Specifically, we use the following topological lemma, which directly follows from the properties of the fundamental group.

Lemma 5.1. *Let π_1, π_2 and π_3 be three curves on a surface* $\mathbf{S}$, *all sharing the same endpoints x and y and oriented from x to y. If the closed curve $\pi_1 - \pi_2$ is non-contractible, then at least one of $\pi_1 - \pi_3$ and $\pi_2 - \pi_3$ is non-contractible as well.* □

Let G be a graph with a string representation φ on a surface $\mathbf{S}$. We represent the combinatorial structure of φ by an auxiliary multigraph $A(G)$ embedded on $\mathbf{S}$ and defined as follows: the vertices of $A(G)$ are the endpoints of the strings of φ and the intersection points of pairs of strings of φ, and the edges of $A(G)$ correspond to segments of strings of φ connecting pairs of vertices appearing consecutively on a string of φ. By representing φ by $A(G)$, we will be able to apply the well-developed theory of graph embeddings on surfaces.

We say that a (closed) walk $W = w_0, w_1, \dots w_k$ in G *imitates* a (closed) curve $\pi \subseteq \varphi[G]$ on the surface $\mathbf{S}$ if π can be partitioned into a sequence of consecutive segments $\pi_0, \pi_1, \dots, \pi_k$ of positive length, such that $\pi = \sum_{i=0}^{k} \pi_i$ and $\pi_i \subseteq \varphi(w_i)$ for each $i = 0, \dots, k$. A closed walk W *imitates a non-contractible curve* if there is a non-contractible curve $\pi \subseteq \varphi[G]$ imitated by W.

Lemma 5.2. *Let φ be a string representation of a connected graph G on a surface* $\mathbf{S}$ *of genus $g > 0$ and let W be a closed walk in G imitating a non-contractible curve. Then every connected component of the graph $G' = G - N[W]$ has a string representation on a surface of genus at most $g - 1$.*

Lemma 5.3. *If a graph G has no string representation in the plane, then for every string representation φ of G on a surface* $\mathbf{S}$ *there is a closed walk W in G imitating a non-contractible curve.*

Proof. Let $A(G)$ be the auxiliary multigraph corresponding to the string representation φ. Since $A(G)$ is not planar, the embedding of $A(G)$ contains a non-contractible cycle (see [11, Chapter 4.2]), which corresponds to a noncontractible curve on $\mathbf{S}$. This curve is imitated by a closed walk W of G. □

Lemma 5.4. *On a graph G with a string representation φ on a surface* $\mathbf{S}$ *and a shortest closed walk W imitating a non-contractible curve, 10 cops have a strategy to guard $N[W]$ after a finite number of initial moves – that is capture a robber immediately after he enters a vertex of $N[W]$.*

Proof (Theorem 1.2(iii)). Let G be a connected graph with a string representation φ on a surface $\mathbf{S}$ of the smallest possible genus g. We want to show that $10g+15$ cops have a strategy to capture the robber on G. We proceed by induction on the genus g. If $g=0$, we use Theorem 1.2(ii).

Let $g>0$, and fix a string representation of G on a surface of genus g. Let W be a shortest closed walk in G imitating a non-contractible curve.

By Lemma 5.4, 10 cops may, after a finite amount of moves, prevent the robber from entering $N[W]$. The first part of the cops' strategy is to designate a group of 10 cops that will spend the entire game guarding $N[W]$. Thus, after a finite number of moves the robber will remain confined to a single connected component K of the graph $G' = G - N[W]$.

By Lemma 5.2, the graph K has a string representation on a surface of genus at most $g-1$, and by induction, $10(g-1)+15$ cops have a strategy to capture the robber on K. Thus, $10g+15$ cops will capture the robber on G. □

6 Conclusions

In this paper, we improve the bound on the maximum cop number of string graphs and also generalize this bound for string graphs on arbitrary surfaces. It remains open whether other intersection classes of special and higher dimensional sets have bounded maximum cop number. In particular:

Problem 6.1. Is the maximum cop number of k-BOX bounded?

We note that bounded genus graphs have bounded boxicity [3]. If the answer is positive, it implies another strengthening of [14,15].

References

1. Aigner, M., Fromme, M.: Game of cops and robbers. Discrete Appl. Math. **8**(1), 1–12 (1984)
2. Bonato, A., Nowakowski, R.J.: The Game of Cops and Robbers on Graphs. American Mathematical Society, Providence (2011)
3. Esperet, L., Joret, G.: Boxicity of graphs on surfaces. Graphs and Combinatorics **29**(3), 417–427 (2013)
4. Fomin, F.V., Golovach, P.A., Kratochvíl, J., Nisse, N., Suchan, K.: Pursuing a fast robber on a graph. Theor. Comput. Sci. **411**(7–9), 1167–1181 (2010)
5. Gavenčiak, T., Gordinowicz, P., Jelínek, V., Klavík, P., Kratochvíl, J.: Cops and robbers of intersection graphs (in preparation, 2015)
6. Gavenčiak, Tomás, Jelínek, Vít, Klavík, Pavel, Kratochvíl, Jan: Cops and robbers on intersection graphs. In: Cai, Leizhen, Cheng, Siu-Wing, Lam, Tak-Wah (eds.) Algorithms and Computation. LNCS, vol. 8283, pp. 174–184. Springer, Heidelberg (2013)
7. Goldstein, A.S., Reingold, E.M.: The complexity of pursuit on a graph. Theor. Comput. Sci. **143**(1), 93–112 (1995)
8. Kinnersley, W.B.: Cops and robbers is exptime-complete. J. Comb. Theor. Ser. B **111**, 201–220 (2015)

9. Kratochvíl, J.: String graphs. II. recognizing string graphs is NP-hard. J. Comb. Theor. Ser. B **52**(1), 67–78 (1991)
10. Mamino, M.: On the computational complexity of a game of cops and robbers. Theor. Comput. Sci. **477**, 48–56 (2013)
11. Mohar, B., Thomassen, C.: Graphs on Surfaces. The John Hopkins University Press, Baltimore (2001)
12. Nowakowski, R., Winkler, P.: Vertex-to-vertex pursuit in a graph. Discrete Math. **43**, 235–239 (1983)
13. Prasolov, V.: Elements of Combinatorial and Differential Topology. Graduate Studies in Mathematics. American Mathematical Soc., Providence (2006)
14. Quilliot, A.: A short note about pursuit games played on a graph with a given genus. J. Combin. Theory Ser. B **38**, 89–92 (1985)
15. Schroeder, B.S.W.: The copnumber of a graph is bounded by 3/2 genus(g) + 3. Trends Math., pp. 243–263. Birkhäuser, Boston (2001)

Min-Power Covering Problems

Eric Angel[1], Evripidis Bampis[2], Vincent Chau[3,4(✉)], and Alexander Kononov[5]

[1] IBISC, Université d'Évry Val d'Essonne, Évry, France
[2] Sorbonne Universités, UPMC Univ. Paris 06, UMR 7606, LIP6, Paris, France
[3] Department of Computer Science, City University of Hong Kong, Kowloon Tong, Hong Kong
[4] Department of Computer Science, Hong Kong Baptist University, Kowloon Tong, Hong Kong
vincchau@comp.hkbu.edu.hk
[5] Sobolev Institute of Mathematics, Novosibirsk, Russia

Abstract. In the classical vertex cover problem, we are given a graph $G = (V, E)$ and we aim to find a minimum cardinality cover of the edges, i.e. a subset of the vertices $C \subseteq V$ such that for every edge $e \in E$, at least one of its extremities belongs to C. In the MIN-POWER-COVER version of the vertex cover problem, we consider an edge-weighted graph and we aim to find a cover of the edges and a valuation (power) of the vertices of the cover minimizing the total power of the vertices. We say that an edge e is covered if at least one of its extremities has a valuation (power) greater than or equal than the weight of e. In this paper, we consider MIN-POWER-COVER variants of various classical problems, including vertex cover, min cut, spanning tree and path problems.

1 Introduction

In the classical vertex cover problem, we are given a graph $G = (V, E)$ and we aim to find a minimum cardinality cover of the edges, i.e. a subset of the vertices $C \subseteq V$ such that for every edge $e \in E$, at least one of its extremities belongs to C. In the MIN-POWER-COVER version of the vertex cover problem, we consider an edge-weighted graph and we aim to find a cover of the edges and a valuation (power) of the vertices of the cover minimizing the total power of the vertices. We say that an edge e is covered if at least one of its extremities has a valuation (power) greater than or equal than the weight of e. A motivating example is related to the installation of security cameras in the crossroads of a town. Here the graph represents the road network of the town, where the vertices represent the crossroads and the edges the roads. Each edge is associated with a weight which represents the cost of the camera that is needed in order to guarantee the adequate visibility. The bigger the length of the road-segment the bigger the cost of the camera that is needed for offering good quality images. The objective is to determine a choice and a placement of cameras minimizing the overall cost, i.e. a cover of minimum power (cost). We call this problem the MIN-POWER-COVER vertex cover problem. More generally, we are interested in

K. Elbassioni and K. Makino (Eds.): ISAAC 2015, LNCS 9472, pp. 367–377, 2015.
DOI: 10.1007/978-3-662-48971-0_32

the MIN-POWER-COVER variants of classical graph problems, including asymmetric vertex cover, minimum cut, spanning tree and (s,t)-path. In the classical setting, the solution of all these problems is just a subgraph of the initial graph respecting some particular structural property. In their MIN-POWER-COVER variants, we aim in determining subgraphs whose edges can be covered with the minimum total power. A natural question is then to know what is the difficulty of the MIN-POWER-COVER variants compared to the corresponding classical graph problems. In this paper, we investigate this question. Interestingly, some of these problems have the same difficulty as their classical variant, while others become much more difficult in the MIN-POWER-COVER setting. As a warm-up take for instance the maximum matching problem, one can easily see that the classical version coincides with its MIN-POWER-COVER variant: a matching whose edges can be covered with the minimum total power is just a minimum weighted maximum matching. More generally, it is not difficult to see that if the subgraph that we are interested to determine has a maximum degree of Δ then a ρ-approximation algorithm for the classical problem gives directly a $(\Delta \cdot \rho)$-approximation for its MIN-POWER-COVER variant. As a corollary, there is a 3-approximation algorithm for the MIN-POWER-COVER variant of the metric Traveling Salesman Problem (TSP) and a 2-approximation algorithm for the MIN-POWER-COVER variant of the cycle cover problem.

Notice that our model differs from the MIN-POWER model studied in [1–4] where for covering a given edge it is required that the power associated to both of its extremities is at least equal to the edge cost.

Our Contribution. In Sect. 2, we study the MIN-POWER-COVER variant of the vertex cover problem. We show that there is a 2-approximation algorithm for both the symmetric and the asymmetric versions of the problem for general graphs. We also prove that the problem can be solved in polynomial time for bipartite graphs. In Sect. 3, we show how to solve the MIN-POWER-COVER variant of the (s,t)–cut problem both for the directed and the undirected case. In Sect. 4, we show that the MIN-POWER-COVER spanning tree problem is as hard to approximate as the classical dominating set problem. Finally in Sect. 5, we propose a simple variation of Dijkstra's algorithm for solving the MIN-POWER-COVER variant of the shortest path problem.

2 Min-Power-Cover Vertex Cover

We consider two versions of the problem, the asymmetric and the symmetric version. In the asymmetric version, we are given a graph $G = (V, E)$ with two weights $w_{u,v}$ and $w_{v,u}$ on each edge $(u, v) \in E$. We aim to find a non negative power p_v for each vertex $v \in V$ such that for all $(u, v) \in E$, either $w_{v,u} \leq p_v$ or $w_{u,v} \leq p_u$, and such that $\sum_{v \in V} p_v$ is minimized. Intuitively, $w_{u,v}$ (resp. $w_{v,u}$) denotes the minimum power that have to be associated to vertex u (resp. v) in order to cover the edge (u, v). An instance of the minimum power vertex cover is called *symmetric* if one has $w_{u,v} = w_{v,u}$ for all edges $(u, v) \in E$.

Notice that the MIN-POWER-COVER vertex problem is NP-hard since for the special case in which $w_{u,v} = w_{v,u} = 1$ for all edges $(u, v) \in E$ we get the classical VERTEX COVER problem.

2.1 A 2-Approximation Algorithm for the Asymmetric Case

If (u, v) is an edge of G, we use $u \sim v$ to denote that u and v are adjacent. We give an algorithm inspired by the local ratio method in approximation algorithms (see for example [5] for a survey):

Algorithm 1. Algorithm for the asymmetric case of MIN-POWER-COVER Vertex Cover

---

$p_v \leftarrow 0$ for all $v \in V$
while $E \neq \emptyset$ **do**
  Select $(u, v) \in E$ such that $\varepsilon \leftarrow \min\{w_{u,v}, w_{v,u}\}$ is minimum.
  Update p_u and p_v:
  $p_u \leftarrow p_u + \varepsilon$
  $p_v \leftarrow p_v + \varepsilon$
  Update all adjacent edges with u and v:
  $\forall u' \sim u,\ w_{u,u'} \leftarrow w_{u,u'} - \varepsilon$
  $\forall v' \sim v,\ w_{v,v'} \leftarrow w_{v,v'} - \varepsilon$
  Delete all edges $(a, b) \in E$ if $w_{a,b} = 0$ including edge (u, v).
end while

Theorem 1. *This algorithm is a polynomial time 2-approximation algorithm for the asymmetric* MIN-POWER-COVER *vertex cover problem.*

Proof. It is easy to see that the algorithm returns a feasible solution. Now we want to analyze its approximation ratio. The algorithm works in several iterations. At each iteration we select an edge, we increase the power at its extremities, and we simplify the instance by decreasing some weights and removing some edges (at least one). Let denote by I_k the instance at the end of the k-th iteration (or the beginning of the $k+1$-th iteration), and let denote by $OPT(k)$ the value of an optimal solution for the instance I_k. The algorithm ends when E becomes empty. This means that if the while loop is executed K times, the instance I_K has no edge and therefore $OPT(K) = 0$. Notice that $OPT(0) = OPT$ is the cost of the optimal solution for the initial instance.

At each iteration of the algorithm, we can consider that we make a payment. Let $PAY(k) = 2\varepsilon$ with ε corresponding to the selected edge (u, v) at the k-th iteration. We are going to prove that we pay at most twice the decrease in cost of the optimal solution, i.e. $OPT(k) - OPT(k+1) \geq \varepsilon = PAY(k)/2$ for $1 \leq k \leq K-1$. For this, we construct a feasible solution for the instance I_{k+1} from the solution $OPT(k)$. There are three cases to consider according to whether edge (u, v) is covered only by u, or only by v, or by both u and v, in solution $OPT(k)$. For the first case, $p_u^*(k)$ (the assigned power for u in $OPT(k)$) is at least ε. We construct a solution as follows: $p_u(k+1) \leftarrow p_u^*(k) - \varepsilon$ and $p_v(k+1) \leftarrow p_v^*(k)\ \forall v \neq u$. This is a feasible solution for I_{k+1} of cost $OPT(k) - \varepsilon$, and thus $OPT(k+1) \leq OPT(k) - \varepsilon$. Similarly for the second case, we have also $OPT(k+1) \leq OPT(k) - \varepsilon$. For the third case, we have $p_u^*(k) \geq \varepsilon$ and $p_v^*(k) \geq \varepsilon$, and the solution such that $p_u(k+1) \leftarrow p_u^*(k) - \varepsilon$, $p_v(k+1) \leftarrow p_v^*(k) - \varepsilon$ and $p_w(k+1) \leftarrow p_w^*(k)\ \forall w \neq u, v$, shows that $OPT(k+1) \leq OPT(k) - 2\varepsilon$. So in any case we always have $OPT(k) - OPT(k+1) \geq \varepsilon$.

Notice that the cost of the solution returned by the algorithm is $cost(ALG) = \sum_{k=1}^{K} PAY(k)$. Since we have $OPT(k) - OPT(k+1) \geq \varepsilon\ \forall k$, we obtain that $OPT(0) - OPT(K) \geq \sum_k PAY(k)/2 = cost(ALG)/2$. Finally $2\,OPT \geq cost(ALG)$. Since at each iteration at least one edge is removed from E, the time complexity is $\mathcal{O}(|E|)$. □

2.2 A Faster 2-Approximation Algorithm for Symmetric Case

We give an algorithm which is faster for the symmetric case:

Algorithm 2. Algorithm for the symmetric case of MIN-POWER-COVER Vertex Cover

```
p_v ← 0 for all v ∈ V
while E ≠ ∅ do
  Select (u, v) ∈ E such that w_{u,v} is maximum.
  Update p_u and p_v:
  p_u ← w_{u,v}
  p_v ← w_{u,v}
  Delete (u, v) and all adjacent edges with u and v:
  ∀u' ∼ u, w_{u,u'} ← 0
  ∀v' ∼ v, w_{v,v'} ← 0
  Delete all edges (a, b) ∈ E if w_{a,b} = 0 including edge (u, v).
end while
```

Theorem 2. *This algorithm is a polynomial time 2-approximation algorithm for the* SYMMETRIC MIN POWER VERTEX COVER *problem.*

Proof. Let M be the set of edges selected by the algorithm. It is easy to see that M is a matching of the graph. Indeed, when selecting an edge, the algorithm removes all its adjacent edges. The optimal solution must cover every edge in M. So, it must include at least one of the endpoints of each edge $\in M$, where no two edges in M share an endpoint. Hence, the cost of an optimal solution satisfy $OPT \geq \sum_{e \in M} w_e$. But the algorithm returns a power vertex cover of cost $2 \times \sum_{e \in M} w_e$, so we have $cost(ALG) = 2 \times \sum_{e \in M} w_e \leq 2 \times OPT$. □

2.3 A Polynomial Case: Bipartite Graphs

If the graph is bipartite we show that the asymmetric MIN-POWER-COVER vertex cover problem can be solved in polynomial time.

We state the problem as a integer linear program and then we show that the matrix of constraints is totally unimodular. We assume that the edges incident to a vertex $v \in V$ are sorted in non decreasing order of their weights (with respect to the vertex v), so we have $w_v^1 \le w_v^2 \le \ldots \le w_v^{d(v)}$, with $w_v^1 := \min_{z \mid (v,z) \in E} w_{vz}$, and $d(v)$ the degree of vertex v. We define $\delta_{v,1} := w_v^1$ and $\delta_{v,i} := w_v^i - w_v^{i-1}$ for $2 \le i \le d(i)$. We denote by $f(e,v)$ the index of edge e according to vertex v in this ordering. So one has $w_v^{f(e,v)} = w_{v,u}$ if $e = (v,u)$. Notice that $f(e,u)$ can be different of $f(e,v)$, with $e = (u,v)$. The integer linear program can be written as follows:

$$\begin{array}{llr}
\min \sum_v \sum_{i=1}^{d(v)} \delta_{v,i}\, x_{v,i} & & \\
\text{s.t.} \quad x_{v,i-1} - x_{v,i} \ge 0 & \forall v \in V,\ \forall i \in \{2,\ldots,d(v)\} & (1) \\
x_{u,f(e,u)} + x_{v,f(e,v)} \ge 1 & \forall e = (u,v) \in E & (2) \\
x_{v,i} \in \{0,1\} & \forall v \in V,\ \forall i \in \{1,\ldots,d(v)\} & (3)
\end{array}$$

In this integer linear program, $x_{v,i} = 1$ means that we assign to vertex v a power which is at least the i-th weight w_v^i. Constraints (1) ensure the coherence between the different values $x_{v,i}$ for $i = 1,\ldots,d(v)$. Constraints (2) ensure that each edge $e \in E$ is covered.

Proposition 1. *The matrix of constraints of the above integer linear program is totally unimodular.*

Proof. We use the Hoffman's conditions which are sufficient for a matrix A to be totally unimodular. The rows of A can be partitioned into two disjoint sets B and C such that:

- Every column of A contains at most two non-zero entries;
- Every entry in A is 0, +1, or −1;
- If two non-zero entries in a column of A have opposite signs, then the rows of both are in B, or both in C;
- If two non-zero entries in a column of A have the same sign, then the row of one is in B, and the other in C.

The two first conditions can be easily verified. Moreover every variable $x_{v,i}$ $\forall i$ are in the same set thanks to the first property. Finally, since we have a bipartite graph, we have two sets B and C such that $B \cup C = V$ and $B \cap C = \emptyset$, $\forall u \in V$ either $u \in B$, or $u \in C$. Then, it is easy to see that constraints (2) verify condition 4. Indeed, it corresponds to an edge (u,v) and because it is a bipartite graph, $x_{u,f(e,u)}$ and $x_{v,f(e,v)}$ are not in the same set. □

3 Min-Power-Cover Cut

We study two versions of the MIN-POWER-COVER cut problem. In the first version we study the directed case while in the second one, we consider the undirected case.

3.1 Directed Graphs

Given a directed graph $G = (V, E)$, with a weight $w_{u,v}$ on each arc $(u, v) \in E$, we want to find a minimum cost $s - t$ cut $C = (S, T)$ of G with $S, T \subset V$, $S \cap T = \emptyset$, such that $s \in S$ and $t \in T$, where s and t are two vertices of G. The cost is given by the sum of powers that we have to assign to each vertex in order to cover all the arcs belonging to the cut, i.e. the arcs (u, v) such that $u \in S$ and $v \in T$. The problem can be viewed as finding the minimum total power to disconnect the graph between s and t. This value could serve as an indication of the robustness of the graph against attacks. To disconnect an arc (u, v), we have to set a power p_u on the source u of the arc such that $w_{u,v} \le p_u$. The objective function is $\min_{v\in V} \sum p_v$.

We show that we can solve in polynomial time this problem on graph G by solving the classical minimum (s, t)-cut problem on a related graph G'. We transform the graph G into a graph G' in the following way:

Let $O(v)$ be the set of outcoming arcs of vertex v. We sort $O(v)$ in non increasing order of their weights. Let $O(v, i)$ be the i-th outcoming arc of vertex v according to the order of weights defined previously. Thus we have $w_{O(v,1)} \geq w_{O(v,2)} \geq \ldots \geq w_{O(v,|O(v)|)}$. We denote by $g_{v,i}$ a vertex in G' which corresponds to arc $O(v, i)$ in G. For each vertex v in G, we create $|O(v)| + 1$ new vertices forming a path in the following order: $v, g_{v,1}, \ldots, g_{v,|O(v)|}$. Then we set the cost of arcs as follows: the cost of arc $(v, g_{v,1})$ is $w_{O(v,1)}$, and the cost of arc $(g_{v,i}, g_{v,i+1})$ is $w_{O(v,i+1)}$ for $i = 1, \ldots, |O(v)| - 1$. Finally for each arc $(v, u) = O(v, i)$, we create an arc $(g_{v,i}, u)$ with infinite cost. This construction is depicted in Fig. 1.

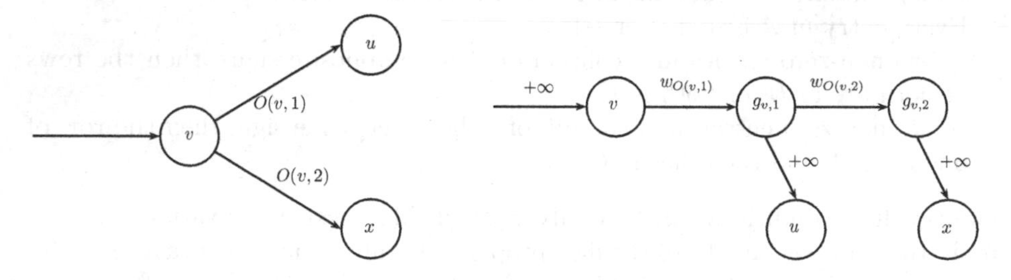

Fig. 1. Illustration of the transformation of a vertex v from graph G to G'

Definition 1. *We call a path in G' a* valid *path if all its arcs have a finite cost, except the last one.*

Lemma 1. *(u, v) is an arc in G if and only if there is a valid path from u to v in G'.*

Proposition 2. *The* MIN-POWER-COVER *cut problem on graph G can be solved by finding a classical minimum cut in graph G'.*

Proof. Let OPT be the cost of an optimal solution for the MIN-POWER-COVER cut problem on graph G, and let OPT' be the cost of an optimal solution for the MIN-CUT problem in graph G'.

We first show that $OPT \geq OPT'$. We show that an optimal solution for the MIN-POWER-COVER cut problem can be transformed to a feasible solution for the MIN-CUT problem in graph G' with the same cost. Let p_v^* the assigned power at vertex v in OPT. We may assume that the value of an assigned power for a vertex v corresponds to a weight of an outcoming edge of v. Indeed, if $w_{O(v,i+1)} < p_v^* < w_{O(v,i)}$, then p_v^* can be lowered to $w_{O(v,i+1)}$, which leads to a better solution, in contradiction with the fact that OPT was optimal. Let $\mathcal{P}$ be the set of vertices v in G such that $p_v^* > 0$ in solution OPT. For all vertices $v \in \mathcal{P}$, we delete all outcoming arcs e of v such that $w_e \leq p_v^*$. Moreover, for all vertices $v \in \mathcal{P}$, we delete an arc in G'. Since the affected power p_v^* on a vertex v is equal to some $w_{O(v,i)}$, then we can delete the arc $(g_{v,i-1}, g_{v,i})$ in G'. If there is some arc with the same value, then cut the arc with the smallest index.

Let S and T with $\mathcal{P} \subseteq S$ the two sets of vertices according to OPT.

Let us consider any two vertices $u \in S$ and $v \in T$. By OPT, we know that there is no arc (u, v) since we already deleted these arcs by assigning power to vertices. Then there is no path from u to v in G' thanks to Lemma 1.

Thus we have a feasible solution for the MIN-CUT problem in graph G' with the same cost.

We show now that $OPT \leq OPT'$. We show that an optimal solution for the MIN-CUT problem in graph G' can be transformed into a feasible solution for the MIN-POWER-COVER cut problem in G with the same cost.

Let $\mathcal{A}$ be the set of arcs that disconnect G'. Each arc $a \in \mathcal{A}$ corresponds to some arc $(g_{v,i-1}, g_{v,i})$ (otherwise it would be an infinite arc), then for each arc $a \in \mathcal{A}$, we set the power of vertex v in G to $w_{v,i}$. Then we can delete all outcoming arcs e of v such that $w_e \leq p_v^*$ in G. Since there is no valid path from u to v in OPT' such that $u \in S \cap V$ and $v \in T \cap V$ without crossing a cut arc, then there is no arc (u, v) in G. Thus, it is a feasible solution for the MIN-POWER CUT problem with the same cost. □

3.2 Undirected Graph

Given an undirected graph $G = (V, E)$, we want to find an $s - t$ cut $C = (S, T)$ of G such that $s \in S$ and $t \in T$, where s and t are two given vertices.

The cost is given by the sum of power that we assign to each vertex in order to disconnect the graph. To disconnect an edge, we have to set a power on one of its two extremities forming the edge, i.e. if (u, v) is the edge, then we must have either $w_{u,v} \leq p_u$, or $w_{u,v} \leq p_v$. The objective function is $\min \sum p_v$.

As for the directed case, we transform the graph G into a graph G' in the following way:

Let $E(v)$ the set of adjacent edges of vertex v. We sort, in non increasing order of weights, the set $E(v)$. Let $E(v,i)$ be the i-th incident edge to vertex v according to the order of weights previously defined. Thus we have $w_{E(v,1)} \geq w_{E(v,2)} \geq \cdots \geq w_{E(v,|E(v)|)}$.

We denote by $g_{v,i}$ and $g'_{v,i}$ a vertex in G' which corresponds to the edge $e_{v,i}$ in G. For each vertex v in G, we create $2|E(v)|+1$ new vertices forming a path in the following order: $g'_{v,|E(v)|}, \ldots, g'_{v,1}, v, g_{v,1}, \ldots, g_{v,|E(v)|}$.

Then, we set the cost of arcs as follows: the cost of $(g'_{v,i+1}, g'_{v,i})$ is $w_{E(v,i+1)}$ for $i = 1, \ldots, |E(v)|-1$, the cost of $(g'_{v,1}, v)$ is $w_{E(v,1)}$, the cost of $(v, g_{v,1})$ is $w_{E(v,1)}$, and the cost of $(g_{v,i}, g_{v,i+1})$ is $w_{E(v,i+1)}$ for $i = 1, \ldots, |E(v)|-1$. This construction is depicted in Fig. 2.

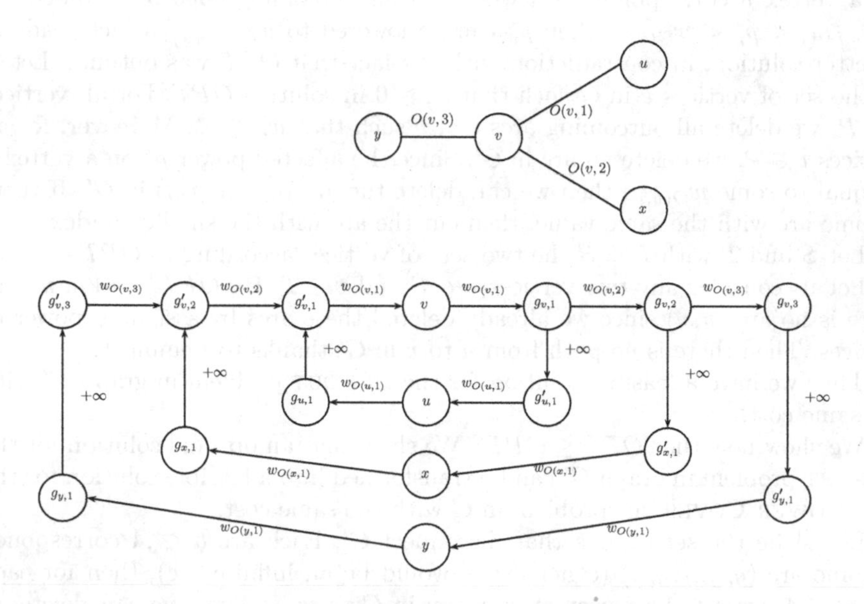

Fig. 2. Illustration of the transformation of a vertex v from graph G to G'

Then for each arc $(v,u) = E(v,i) = E(u,j)$, we create an arc $(g'_{v,i}, g_{u,j})$ and an arc $(g'_{u,j}, g_{v,i})$ with infinite cost.

Proposition 3. *The* MIN-POWER-COVER *cut problem on an undirected graph can be solved using the classical* MIN-CUT *problem in graph* G'.

Proof. Let OPT be the optimal solution for the MIN-POWER-COVER cut problem and OPT' be the optimal solution for the MIN-CUT problem in graph G'.

Let us first show that $OPT \geq OPT'$.

We show that an optimal solution for the MIN-POWER-COVER cut problem can be transformed to a feasible solution for the MIN-CUT problem in graph G'

with the same cost. Let p_v^* the affected power at vertex v in OPT. We may assume that the value of the affected power for a vertex v corresponds to a weight of an adjacent edge of v. Indeed, if $w_{E(v,i+1)} < p_v^* < w_{E(v,i)}$, then p_v^* can be reduced to $w_{O(v,i+1)}$ and we have a better solution which contradicts the fact that OPT was optimal.

Let $\mathcal{P}$ be the set of vertices v in OPT in G such that $p_v^* > 0$.

For all vertices $v \in \mathcal{P}$, we delete all adjacent edges e of v such that $w_e \leq p_v^*$. Moreover, for all vertices $v \in \mathcal{P}$, we delete an arc in G'. Since the affected power p_v^* on a vertex v is equal to some $w_{O(v,i)}$, then we can delete the arc $(g_{v,i-1}, g_{v,i})$ or $(g'_{v,i}, g'_{v,i-1})$ in G' according to whether $v \in S$ or $v \in T$. If there is some arc with the same value, then we cut the arc with the smallest index.

Let S and T be the two sets of vertices according to OPT.

Consider two vertices $u \in S$ and $v \in T$. By OPT, we know that there is no edge (u, v) since we already deleted these edges by affecting power to the vertices. Then (u, v) cannot be a simple path in G' thanks to Lemma 1.

Thus we have a feasible solution for the MIN-CUT problem in graph G' with the same cost.

Now, let us show that $OPT \leq OPT'$.

We show that an optimal solution for the MIN-CUT problem in graph G' can be transformed to a feasible solution for the MIN-POWER-COVER cut problem in G with the same cost.

Let $\mathcal{A}$ be the set of arcs that disconnect G'. Each arc $a \in \mathcal{A}$ corresponds to some $(g_{v,i-1}, g_{v,i})$ or $(g'_{v,i}, g'_{v,i-1})$ (otherwise it is an infinite arc). Then for each arc $a \in \mathcal{A}$, we set the power of vertex v in G to $w_{O(v,i)}$. Then we can delete all adjacent edges e of v such that $w_e \leq p_v^*$ in G. Since there is no simple path (u, v) (resp. (v, u)) in OPT' such that $u \in S \cap V$ and $v \in T \cap V$ without crossing a cut arc $(g_{v,i-1}, g_{v,i})$ (resp. $(g'_{v,i}, g'_{v,i-1})$), then there is no arc (u, v) (resp. (v, u)) in G. Thus, we have a feasible solution for the MIN-POWER-COVER cut problem with the same cost. □

4 Min-Power-Cover Spanning Tree

In this section, we will show that the MIN-POWER-COVER spanning tree problem is as hard to approximate as the DOMINATING SET problem.

Given an edge-weighted graph $G = (V, E)$, the objective is to find a spanning tree and to assign power to vertices such that all the edges of the spanning tree are covered. An edge is covered if the assigned power in one of its extremities is at least the weight of that edge. We want to minimize the sum of powers over all vertices. W.l.o.g. we assume that G is connected.

Recall that a *dominating set* is a subset of vertices $S \subseteq V$ such that any vertex outside S is adjacent with at least one vertex inside S. The MIN DOMINATING SET problem consists in finding a dominating set S such that $|S|$ is minimum.

In the following we assume that each edge of the graph has a unit weight. Let $MPST^*$ (resp. DOM^*) be the cost of an optimal solution for the MIN-POWER-COVER spanning tree (resp. MIN DOMINATING SET) problem. We denote by T^* the corresponding spanning tree with respect to $MPST^*$.

Proposition 4. *One has,* $DOM^* \leq MPST^* \leq 2\,DOM^* - 1$.

Proof. Since all edges of the graph have a unit weight, it is easy to see that in an optimal solution for MIN-POWER-COVER spanning tree, each vertex is either assigned a power of 1 or 0. Let $L(T^*)$ be the set of vertices which are assigned a power of 1. Therefore, $MPST^* = |L(T^*)|$. Let us consider a vertex $v \in V$, since T^* is a spanning tree, v belongs to some edge $e = (u, v) \in T^*$, and since e is covered, either $u \in L(T^*)$ or $v \in L(T^*)$. It means that $L(T^*)$ is a dominating set, and therefore $DOM^* \leq MPST^*$.

We prove now that $MPST^* \leq 2\,DOM^* - 1$. Let $V(DOM^*)$ be the dominating set such that $|V(DOM^*)| = DOM^*$. We denote by $\partial V(DOM^*) := \{e = (u, v) \in E \mid u \text{ or } v \in DOM^*\}$ the set of edges which are adjacent with a vertex in $V(DOM^*)$. If the graph $G(V, \partial V(DOM^*))$ is connected then we can select a spanning tree $T(V, E')$ of $G(V, E)$ such that $E' \subseteq \partial V(DOM^*)$ and such that $L(T) = DOM^*$. In this case, $MPST^* \leq DOM^*$. Let us assume now that there are $k \geq 2$ connected components in $G(V, \partial V(DOM^*))$, and let denote by K_i the i-th connected component for $1 \leq i \leq k$. Notice that $k \leq DOM^*$. We show how to transform $V(DOM^*)$ into a feasible solution for the MIN-POWER-COVER spanning tree problem: we add a vertex v in $V(DOM^*)$ such that $v \in K_i$ and $u \in K_j$ with $i \neq j$ and such that $(u, v) \in E$. Thus we merge two connected components into one. We iterate, and thus we have to add at most $k-1$ vertices to $V(DOM^*)$ in order to obtain a feasible solution for the MIN-POWER-COVER spanning tree problem. Therefore, $MPST^* \leq DOM^* + k - 1 \leq 2\,DOM^* - 1$. □

5 Min-Power-Cover Path

Given a graph $G = (V, E)$, the objective is to find a path from a vertex s to a vertex t with the minimum power cost. We consider the asymmetric case. If an edge (u, v) is covered, then either $w_{u,v} \leq p_u$, or $w_{u,v} \leq p_v$. We say that a path is valid if each edge of the path is covered by the power affectation on the vertices. Then the objective function is $\min \sum p_v$.

For each vertex v we consider a cost function $C(v, p)$ which is the cost of a path from s to v if the power in vertex v is equal to p. Let δ be the degree of vertex v. In each vertex we consider at most δ different values of power. Let Q be a set of pairs (v, p). We consider the following modification of Dijkstra's Algorithm.

Theorem 3. *The algorithm solves the symmetric* MIN-POWER-COVER *path problem optimally.*

Sketch of Proof. In order to prove the correctness of the algorithm we note that the following statements hold each time that (2) is executed in the algorithm: (a) For all $(v, p) \in R$ and all $(u, k) \in Q : C(v, p) \leq C(u, k)$. (b) For all $(v, p) \in R$: If the power in vertex v is equal to p, $C(v, p)$ is the cost of a cheapest $s - v$-path in G. In particular, (b) holds when the algorithm terminates, so we obtain the optimal solution. The rest of the proof can be found in Sect. 7 of [6]. Since the number of pairs is at most $n + 2m$ then the running time of the algorithm is $O(m^2)$, where $m = |E(G)|$.

Algorithm 3. Algorithm for the Min-Power-Cover Path

1: Set $C(s,p) = p$. Set $C(v,p) = \infty$ for all v and p. Set $R = \emptyset$.
2: Find a pair $(v,p) \in Q \setminus R$ such that $C(v,p) = \min_{(u,k)\in Q\setminus R} C(u,k)$.
3: Set $R = R \cup (v,p)$.
4: **for** $(u,k) \in Q \setminus R$ such that $(v,u) \in E$ and $w_{v,u} \leq p$ or $w_{u;v} \leq k$ **do**
5:    **if** $C(u,k) > C(v,p) + k$ **then**
6:       set $C(u,k) = C(v,p) + k$
7:       $path(u,k) = (v,p)$
8:    **end if**
9: **end for**
10: If $R \neq Q$ then go to 2.
11: For each vertex v take a minimum cost $C(v,p)$ over all p.

6 Concluding Remarks

We have introduced a new variant of classical graph problems, namely the MIN-POWER-COVER variant. We have shown that while the difficulty of the problem for the power-cover variants of the min-cut, path and vertex cover problems remains the same, the power-cover variant of the spanning tree problem is hard to approximate. An interesting open question is to study the complexity and the approximability of the power-cover variants of other problems. In particular, it would be interesting to know whether there is an approximation algorithm for the MIN-POWER-COVER metric Traveling Salesman Problem with an approximation ratio better than 3.

References

1. Althaus, E., Călinescu, G., Mandoiu, I.I., Prasad, S.K., Tchervenski, N., Zelikovsky, A.: Power efficient range assignment for symmetric connectivity in static ad hoc wireless networks. Wireless Netw. **12**(3), 287–299 (2006)
2. Byrka, J., Grandoni, F., Rothvoß, T., Sanità, L.: An improved lp-based approximation for steiner tree. In: Schulman, L.J. (ed.) STOC, pp. 583–592. ACM (2010)
3. Grandoni, F.: On min-power steiner tree. In: Epstein, L., Ferragina, P. (eds.) ESA 2012. LNCS, vol. 7501, pp. 527–538. Springer, Heidelberg (2012)
4. Kirousis, L.M., Kranakis, E., Krizanc, D., Pelc, A.: Power consumption in packet radio networks. Theor. Comput. Sci. **243**(1–2), 289–305 (2000)
5. Bar-Yehuda, R., Bendel, K., Freund, A., Rawitz, D.: Local ratio: a unified framework for approximation algorithms in memoriam: shimon even 1935–2004. ACM Comput. Surv. **36**(4), 422–463 (2004)
6. Korte, B., Vygen, J.: Combinatorial Optimization. Springer, Heidelberg (2002)

Computational Geometry III

Minimizing the Diameter of a Spanning Tree for Imprecise Points

Chih-Hung Liu[1(✉)] and Sandro Montanari[2(✉)]

[1] Department of Computer Science, University of Bonn, Bonn, Germany
chliu@uni-bonn.de

[2] Department of Computer Science, ETH Zurich, Zürich, Switzerland
sandro.montanari@inf.ethz.ch

Abstract. We study the diameter of a spanning tree, i.e., the length of its longest simple path, under the *imprecise points* model, in which each point is assigned an own occurrence disk instead of an exact location. We prove that the minimum diameter of a spanning tree for n points each of which is selected from its occurrence disk can be computed in $O(n^9)$ time for arbitrary disks and in $O(n^6)$ time for unit ones. If the disks are disjoint, we improve the run-time respectively to $O(n^8 \log^2 n)$ and $O(n^5)$. These results contrast with the fact that minimizing the sum of the edge lengths of a spanning tree for imprecise points is NP-hard.

1 Introduction

Given a set of points, a *spanning tree* is a tree connecting all points, its *diameter* is the length of its longest simple path, and a *minimum diameter spanning tree* is a spanning tree with the smallest diameter. We consider the *imprecise points* model in which each point is assigned an own occurrence disk instead of an exact location, and attempt to place one point in each disk such that the diameter of a minimum diameter spanning tree of the resulting point set is minimized.

The motivation for the study of imprecise points comes from data imprecision in real-world applications. In practical applications, the exact locations of points are often unknown because the measurements of an instrument have some error interval or the corresponding objects may move or fluctuate. An important task for imprecise points is to compute the minimum and maximum values that certain geometric measures on the point set can attain. Geometry measures such as the smallest bounding box [10], the smallest enclosing circle [8,10], the farthest pair [10], the width [10], the closest pair [5,10], or the area and perimeter of the convex hull [9] have been studied for various shapes of occurrence regions.

For points whose locations are all exactly known, Ho *et al.* [7] proved that there always exists a minimum diameter spanning tree that is either *monopolar*, i.e., it contains a point, called a *pole*, linked to all the remaining points, or *bipolar*, i.e., it contains two poles such that all remaining points are linked to one of the two, and developed an algorithm for computing such a tree in $O(n^3)$ time.

This work has been partially supported by DFG, DACH grant Kl 655/19-1.

K. Elbassioni and K. Makino (Eds.): ISAAC 2015, LNCS 9472, pp. 381–392, 2015.
DOI: 10.1007/978-3-662-48971-0_33

Chan [2] further improved the time bound to $o(n^{17/6+\alpha})$ for any fixed $\alpha > 0$. Since the best known exact algorithm still takes near-cubic time, Gudmundsson *et al.* [6] and Spriggs *et al.* [12] proposed fast approximation algorithms.

Although to the best of our knowledge there is no study on minimum diameter spanning trees for imprecise points, there is a considerable amount of work concerning minimum cost spanning trees, where the cost is the sum of edge lengths. Löffler and van Kreveld [9] proved that it is NP-hard to minimize the cost of a spanning tree for imprecise points in intersecting disks or squares. The problem is NP-hard even for *disjoint* disks [4] and for axis-aligned segments [3].

1.1 Our Results

Theorem 1. *The minimum diameter of a spanning tree of n points each of which is selected from a respective disk region can be computed in* $O(n^9)$ *time for arbitrary disks and in* $O(n^6)$ *time for unit disks. If the disks are disjoint, the required computation time decreases respectively to* $O(n^8 \log^2 n)$ *and* $O(n^5)$.

Theorem 1 reveals a stark contrast: while for imprecise points a minimum diameter spanning tree can be computed much faster than a minimum cost spanning tree (polynomial run-time versus NP-hardness), the reverse is true for exact points (almost cubic versus $O(n \log n)$ run-time). The difference in the imprecise points model comes from the fact that there always exists a minimum diameter spanning tree whose longest simple path consists of at most three edges (monopolar or bipolar). Therefore, the diameter can be attained by at most four disks, while the cost depends instead on all the n disks.

There are two main challenges for achieving this polynomial run-time. First, although there are only linearly many monopolar tree topologies, the number of bipolar tree topologies is exponential in the number of points. It is not clear how to reduce the number of topologies to be considered without the exact locations of points. Second, even for a fixed tree topology, it is unknown how to place points in their respective disks efficiently in order to minimize the diameter.

For the first challenge, we use the arrangements formed by bisectors among disks to prove that it is sufficient to consider $O(n^7)$ bipolar tree topologies. For unit disks, we further show that it is sufficient to consider only $O(n^4)$ bipolar tree topologies admitting a straight line separating the centers of the disks linked to one pole from the centers of the disks linked to the other pole.

For the second challenge, we employ farthest-disk Voronoi diagrams for the bipolar case and refined second-order farthest-disk Voronoi diagrams for the monopolar case to analyze the locations of poles, and to compute optimum locations for them. For disjoint disks, we develop a sequential search achieving a faster run-time than in the intersecting case. For unit disks, we further improve the run-time by computing farthest-disk Voronoi diagrams in a batch.

The paper is organized as follows. In Sect. 2 we introduce notation, definitions, and farthest-disk Voronoi diagrams. In Sects. 3 and 4 we respectively compute optimal placements for monopolar and bipolar topologies. In Sect. 5 we develop the sequential search for disjoint disks and in Sect. 6 we discuss unit disks.

2 Preliminary

Let $\mathfrak{D} = \{D_1, \ldots, D_n\}$ be a set of distinct disks in the plane with centers $c_i = (h_i, k_i)$ and radius r_i for $1 \leq i \leq n$; we use $\mathfrak{C}$ to denote the set of all centers. A point set $P = \{p_1, \ldots, p_n\}$ is called *placement* if $p_i \in D_i$ for $1 \leq i \leq n$. For a tree T, let δ_T denote its (Euclidean) diameter, and for a placement P, let δ_P denote the diameter of its minimum diameter spanning tree. A placement P is called *minimum* if δ_P is smallest among all placements. We say that $\mathfrak{D}$ *admits monopolarity* if there exists a minimum placement which ensures the existence of a monopolar minimum diameter spanning tree. Throughout the paper, we assume the intersection of all disks in $\mathfrak{D}$ to be empty. Otherwise, the problem can be solved trivially by selecting the same point in the intersection of all disks.

For a point x and a disk D, we define the distance $d(x, D)$ between x and D as $\min_{y \in D} |\overline{xy}|$. Given a compact, connected subset $A \subset \mathbb{R}^2$, we use cl$A$ and ∂A to indicate respectively the closure and the boundary of A. The following two facts indicate the run-time of two basic operations used in our computation and implemented by the Newton-Raphson method.

Fact 1. *For two disks D_p, D_q and an edge e that is a line segment or a circular or hyperbolic arc, a point $t \in cl\ e$ minimizing $d(t, D_p) + d(t, D_q)$ can be computed in $O(1)$ time.*

Fact 2. *For two disks D_p, D_q and two circular arcs e_1, e_2, the points $s \in cl\ e_1, t \in cl\ e_2$ minimizing $d(s, D_p) + |\overline{st}| + d(t, D_q)$ can be computed in $O(1)$ time.*

Farthest-disk Voronoi diagrams. For a set $\mathfrak{D}$ of disjoint disks, the farthest-disk Voronoi diagram FV($\mathfrak{D}$) of $\mathfrak{D}$ is a planar subdivision such that all points in a region share the same farthest disk among $\mathfrak{D}$, and FVR($D, \mathfrak{D}$) represents the Voronoi region associated with a disk $D \in \mathfrak{D}$. The common boundary between two Voronoi regions is called a Voronoi edge, and the common vertex among more than two Voronoi regions is called a Voronoi vertex. A Voronoi region FVR($D, \mathfrak{D}$) consists of $O(n)$ disjoint faces.

FV($\mathfrak{D}$) can also be defined by bisectors among disks in $\mathfrak{D}$. For two disks D and D', their bisector $B(D, D') = \{x \in \mathbb{R}^2 \mid d(x, D) = d(x, D')\}$ partitions the plane into two connected regions $H(D, D') = \{x \in \mathbb{R}^2 \mid d(x, D) < d(x, D')\}$ and $H(D', D) = \{x \in \mathbb{R}^2 \mid d(x, D') < d(x, D)\}$. We then have

$$\mathrm{FVR}(D, \mathfrak{D}) = \bigcap_{D' \in \mathfrak{D} \setminus \{D\}} H(D', D) \text{ and } \mathrm{FV}(\mathfrak{D}) = \bigcup_{D \in \mathfrak{D}} \partial \mathrm{FVR}(D, \mathfrak{D}).$$

A Voronoi edge between FVR($D, \mathfrak{D}$) and FVR($D', \mathfrak{D}$) is a part of $B(D, D')$.

Note that, if D and D' intersect, $B(D, D')$ contains a two-dimensional face corresponding to $D \cap D'$, which makes the definition of FV($\mathfrak{D}$) ambiguous. To remove the ambiguity, we amend the bisector system as follows. If $D' \subseteq D$, we let $H(D, D')$ be $\mathbb{R}^2$, and both $B(D, D')$ and $H(D', D)$ be empty. If $D \subseteq D'$, the treatment is symmetric. Otherwise, let y, z be the two intersection points between ∂D and $\partial D'$; we redefine $B(D, D')$ as $\{x \in \mathbb{R}^2 \setminus D \cap D' \mid d(x, D) = d(x, D')\} \cup \overline{yz}$. That is, we replace the part of $B(D, D')$ within $D \cap D'$ with $\overline{yz}$. The sets $H(D, D')$ and $H(D', D)$ become the connected regions induced by the resulting curve and containing their original parts.

The *refined* second order farthest-disk Voronoi diagram $\mathcal{FV}_2(\mathfrak{D})$ is generated from $\mathrm{FV}(\mathfrak{D})$ by further partitioning each non-empty $\mathrm{FVR}(D, \mathfrak{D})$ with $\mathrm{FV}(\mathfrak{D} \setminus \{D\})$. We use $\mathcal{FVR}_2((D, D'), \mathfrak{D})$ to denote the intersection between $\mathrm{FVR}(D, \mathfrak{D})$ and $\mathrm{FVR}(D', \mathfrak{D} \setminus \{D\})$. Then $\mathcal{FV}_2(\mathfrak{D}) = \bigcup_{D, D' \in \mathfrak{D}} \partial \mathcal{FVR}_2((D, D'), \mathfrak{D})$, and all points in $\mathcal{FVR}_2((D, D'), \mathfrak{D})$ share the same farthest disk D and the same second farthest disk D' in $\mathfrak{D}$. The following theorems provide structural and computational properties of $\mathrm{FV}(\mathfrak{D})$ and of $\mathcal{FV}_2(\mathfrak{D})$.

Theorem 2. *Both $FV(\mathfrak{D})$ and $\mathcal{FV}_2(\mathfrak{D})$ have $O(n)$ faces, and $FV(\mathfrak{D})$ is a tree.*

Theorem 3. *$FV(\mathfrak{D})$ and $\mathcal{FV}_2(\mathfrak{D})$ can be computed respectively in $O(n \log^2 n)$ time and $O(n^2 \log^2 n)$ time. If $\mathfrak{D}$ consists of unit disks, both $FV(\mathfrak{D})$ and $\mathcal{FV}_2(\mathfrak{D})$ can be computed in $O(n \log n)$ time.*

3 Monopolar Case

We consider the case where $\mathfrak{D}$ admits monopolarity, and compute a minimum placement P with a monopolar minimum diameter spanning tree T. The general idea is to compute an optimum pole p_i for each disk $D_i \in \mathfrak{D}$, and select the one resulting in the smallest δ_T. To minimize δ_T, for each disk $D_j \in \mathfrak{D} \setminus \{D_i\}$ it is sufficient to place p_j at the first intersection between $\overrightarrow{p_i c_j}$ and D_j.

The following fact indicates that δ_T is determined by the first and second farthest disks of p_i, and relates p_i to the refined second-order farthest Voronoi diagram $\mathcal{FV}_2(\mathfrak{D})$. Using this fact, we prove that it is sufficient to consider the location of p_i either on ∂D_i, or on an edge of $\mathcal{FV}_2(\mathfrak{D})$.

Fact 3. *For a pole $p_i \in D_i$, it holds that*

$$\delta_T = d(p_i, D_j) + d(p_i, D_k),$$

where D_j and D_k are the first and second farthest disks of p_i in $\mathfrak{D}$.

Lemma 1. *If $\mathfrak{D}$ admits monopolority, there exists a minimum placement with a monopolar minimum diameter spanning tree T whose pole p_i belongs to ∂D_i or $\mathcal{FV}_2(\mathfrak{D}) \cap D_i$.*

Proof. Following the notation of Fact 3, it holds that $p_i \in \mathcal{FVR}_2((D_j, D_k), \mathfrak{D})$, and we further let p'_i be the first intersection between $\partial(\mathcal{FVR}_2((D_j, D_k), \mathfrak{D}) \cap D_i)$ and $\overrightarrow{p_i c_j}$. Since D_j is the farthest disk for all points in $\overline{p_i p'_i}$, the segment $\overline{p_i p'_i}$ does not intersect D_j; otherwise all disks contain the intersection point, contradicting the assumption that the intersection of all disks is empty. Since moving p_i toward c_j will not increase $d(p_i, D_j) + d(p_i, D_k)$ unless p_i enters D_j (no matter whether p_i belongs to D_k or not), replacing p_i with p'_i results in a placement satisfying the statement. Note that if $\mathcal{FVR}_2((D_j, D_k), \mathfrak{D}) \cap D_i = \emptyset$, then $\partial(\mathcal{FVR}_2((D_j, D_k), \mathfrak{D}) \cap D_i) = \partial D_i$. □

By Lemma 1, for computing an optimal pole in D_i, it is sufficient to consider a planar graph formed by combining ∂D_i with the edges of $\mathcal{FV}_2(\mathfrak{D}) \cap D_i$. For each edge e of the planar graph, we find the optimum location of p_i on cl e, and select the one resulting in the smallest δ_T. We then conclude the following theorem.

Theorem 4. *If $\mathfrak{D}$ admits monopolarity, a minimum placement for $\mathfrak{D}$ can be computed in $O(n^2 \log^2 n)$ time, and in $O(n^2)$ time when $\mathfrak{D}$ are all unit disks.*

Proof. Consider a disk D_i and an edge e of the planar graph induced by ∂D_i and $(\mathcal{FV}_2(\mathfrak{D}) \cap D_i)$. If $e \subset \text{cl}\mathcal{FVR}_2((D_j, D_k), \mathfrak{D})$, by Fact 3, the optimum location $p_i \in \text{cl } e$ minimizes $d(p_i, D_j) + d(p_i, D_k)$. Furthermore, if e belongs to ∂D_i, then e is a circular arc; otherwise, e belongs to a bisector between two disks, and consists of at most three parts, each of which is a line segment or a hyperbolic arc. Therefore, by Fact 1 p_i can be computed in $O(1)$ time. By Theorem 2 the planar graph has $O(n)$ edges, and thus an optimal location $p_i \in D_i$ can be computed in $O(n)$ time, resulting in overall $O(n^2)$ time for all n disks. Furthermore, by Theorem 3 $\mathcal{FV}_2(\mathfrak{D})$ can be computed in $O(n^2 \log^2 n)$ time for general disks and $O(n \log n)$ time for unit disks, leading to the statement. □

4 Bipolar Case

We now describe how to compute a minimum placement P in the case where $\mathfrak{D}$ does not admit monopolarity. It is known [7] that in this case there exists a bipolar minimum diameter spanning tree T of P. Throughout this section, we use $p_i \in D_i$ and $p_j \in D_j$ to denote the two poles of T, and $\mathfrak{D}_i$ and $\mathfrak{D}_j$ to denote the disks of $\mathfrak{D} \setminus \{D_i, D_j\}$ whose points in P are linked in T respectively to p_i and p_j. We also call D_i and D_j a pair of *polar disks*, and a configuration consisting of D_i, D_j, $\mathfrak{D}_i$, and $\mathfrak{D}_j$ a *bipolar tree topology.* If p_i and p_j are fixed, to minimize δ_T, for each disk $D_k \in \mathfrak{D}_i$ and for each $D_l \in \mathfrak{D}_j$, it is sufficient to place p_k and p_l such that $d(p_i, p_k)$ and $d(p_j, p_l)$ are minimized; that is, p_k and p_l are the first intersections respectively between $\overrightarrow{p_i c_k}$ and D_k and between $\overrightarrow{p_j c_l}$ and D_l.

The idea of our algorithm is to compute a minimum placement P for every pair of polar disks D_i and D_j and every *essential* 2-partition $\mathfrak{D}_i, \mathfrak{D}_j$ of $\mathfrak{D} \setminus \{D_i, D_j\}$, and select the one resulting in the minimum δ_T. We use the arrangement generated by the bisectors among disks to show that, for a fixed pair of polar disks, the number of essential 2-partitions to consider is $O(n^5)$ (i.e., $O(n^7)$ topologies in overall). We then use FV($\mathfrak{D}_i$) and FV($\mathfrak{D}_j$) to prove that there exists a minimum placement where p_i and p_j lie on the boundaries of D_i and D_j, respectively, and to compute the optimum locations in $O(n^2)$ time, resulting in an overall run-time of $O(n^9)$. These results rely on the following fact coming from the "stability condition" of Ho *et al.* [7], i.e., a longest simple path of a bipolar minimum diameter spanning tree contains the edge between two poles.

Fact 4. *Given a bipolar minimum diameter spanning tree with poles $p_i \in D_i$ and $p_j \in D_j$, it holds that*

$$\delta_T = d(p_i, D_k) + |\overline{p_i p_j}| + d(p_j, D_l),$$

where D_k and D_l are respectively the farthest disks of p_i in $\mathfrak{D}_i$ and of p_j in $\mathfrak{D}_j$. Thus, p_i belongs to $FVR(D_k, \mathfrak{D}_i)$ and p_j belongs to $FVR(D_l, \mathfrak{D}_j)$.

4.1 Bipolar Tree Topologies

In order to bound the number of bipolar tree topologies we look at the situation where the poles p_i, p_j and the 2-partition $\mathfrak{D}_i, \mathfrak{D}_j$ are fixed. If D_k is the farthest disk of p_i in $\mathfrak{D}_i$, by Fact 4 moving any disk $D \in \mathfrak{D}_j$ satisfying $d(p_i, D) \leq d(p_i, D_k)$ into $\mathfrak{D}_i$ will not increase δ_T. Therefore, it is sufficient to consider a bipolar tree topology where $\mathfrak{D}_i$ contains every disk in $\mathfrak{D} \setminus \{D_i, D_j\}$ whose distance to p_i is at most $d(p_i, D_k)$ and $\mathfrak{D}_j$ contains all the remaining ones.

More formally, for a pole $p_i \in D_i$ we define a sequence $(B_1, B_2, \ldots, B_{n-2})$ of $\mathfrak{D} \setminus \{D_i, D_j\}$ to be the *nearest ordered sequence* of p_i if B_m is the mth nearest disk of p_i in $\mathfrak{D} \setminus \{D_i, D_j\}$ for $1 \leq m \leq n-2$ (with ties broken arbitrarily). If the farthest disk D_k of p_i in $\mathfrak{D}_i$ is B_m, replacing $\mathfrak{D}_i$ with $\{B_1, \ldots, B_m\}$ and $\mathfrak{D}_j$ with $\{B_{m+1}, \ldots, B_{n-2}\}$ (and moving the points in the placement accordingly) does not increase δ_T. By this argument, once the pole p_i is fixed (regardless of p_j), the number of different candidates for $\mathfrak{D}_i$ and $\mathfrak{D}_j$ that need to be considered is $n-2$, i.e., $\mathfrak{D}_i = \{B_1, \ldots, B_m\}$ and $\mathfrak{D}_j = \{B_{m+1}, \ldots, B_{n-2}\}$ for $1 \leq m \leq n-2$.

As a result, for a pair of polar disks D_i and D_j, if we could bound the total number of different nearest ordered sequences of $\mathfrak{D} \setminus \{D_i, D_j\}$ for all points $p_i \in D_i$ by a polynomial in n, the number of bipolar tree topologies we would need to consider is therefore equal to the polynomial times $(n-2)$. This polynomial bound is proven in the following lemma using the arrangement formed by bisectors among the disks in $\mathfrak{D} \setminus \{D_i, D_j\}$.

Lemma 2. *For a pair of polar disks D_i and D_j it is sufficient to consider $O(n^5)$ 2-partitions, and those partitions can be enumerated in $O(n^5)$ time.*

Proof. (Sketch) We consider the arrangement formed by the $\binom{n-2}{2}$ bisectors between all pairs of disks in $\mathfrak{D} \setminus \{D_i, D_j\}$. Observe that all points in a face of the arrangement share the same nearest ordered sequence of $\mathfrak{D} \setminus \{D_i, D_j\}$, because there is no bisector passing through it. Since any two bisectors intersect at most four times (even for modified bisectors), the arrangement has $O(n^4)$ faces. Therefore, the total number of different nearest ordered sequences of $\mathfrak{D} \setminus \{D_i, D_j\}$ for all points in the plane is $O(n^4)$, and the number of 2-partitions that need to be considered is $O(n^4) \cdot (n-2) = O(n^5)$. The arrangement can be computed in $O(n^4)$ time by incrementally inserting disks since the kth insertion forms $O((k-1)^3)$ intersections. For each face of the arrangement intersecting D_i, we generate the corresponding nearest nearest order sequence and the $(n-2)$ candidates for $\mathfrak{D}_i$, resulting in a total run-time of $O(n^4) \cdot (n-2) = O(n^5)$. □

4.2 Locations of an Optimum Placement

The following lemma employs FV($\mathfrak{D}_i$) and FV($\mathfrak{D}_j$) to analyze the optimum locations of $p_i \in D_i$ and $p_j \in D_j$ for fixed $\mathfrak{D}_i$ and $\mathfrak{D}_j$.

Lemma 3. *If $\mathfrak{D}$ does not admit monopolarity, there exists a minimum placement with a bipolar minimum diameter spanning tree with poles $p_i \in D_i$ and $p_j \in D_j$ such that $p_i \in \partial D_i$ and $p_j \in \partial D_j$.*

Proof. We assume that in a minimum placement p_i and p_j do not belong to $D_i \cap D_j$. Otherwise, by triangle inequality placing p_i on p_j results in a minimum placement admitting a monopolar minimum diameter spanning tree. If $p_i \notin \partial D_i$, we move p_i along the direction $\overrightarrow{p_i p_j}$ until we reach ∂D_i. While moving p_i, even if its farthest disk D_k in $\mathfrak{D}_i$ changes, $d(p_i, D_k) + |\overline{p_i p_j}|$ will not increase, and thus the diameter will not increase. If also $p_j \notin \partial D_j$, we move p_j to ∂D_j in a symmetric way without increasing the diameter, leading to this lemma. □

For a pair of polar disks D_i and D_j and a 2-partition of $\mathfrak{D} \setminus \{D_i, D_j\}$ into $\mathfrak{D}_i$ and $\mathfrak{D}_j$, we compute a minimum placement as follows. We first compute $\mathrm{FV}(\mathfrak{D}_i)$ and $\mathrm{FV}(\mathfrak{D}_j)$ and use them to partition respectively ∂D_i and ∂D_j into arcs. Then, for each arc $a_k \in \partial D_i \setminus \mathrm{FV}(\mathfrak{D}_i)$ and each arc $a_l \in \partial D_j \setminus \mathrm{FV}(\mathfrak{D}_j)$, we find a point $p'_i \in \mathrm{cl}\ a_k$ and a point $p'_j \in \mathrm{cl}\ a_l$ minimizing the corresponding diameter. Finally, we select p_i and p_j as the pair p'_i and p'_j resulting in the smallest diameter.

Theorem 5. *If $\mathfrak{D}$ does not admit monopolarity, a minimum placement for $\mathfrak{D}$ can be computed in $O(n^9)$ time.*

Proof. For each pair of polar disks D_i, D_j we consider by Lemma 2 $O(n^5)$ 2-partitions of $\mathfrak{D} \setminus \{D_i, D_j\}$. For each 2-partition $\mathfrak{D}_i, \mathfrak{D}_j$, we compute by Theorem 3 $\mathrm{FV}(\mathfrak{D}_i)$ and $\mathrm{FV}(\mathfrak{D}_j)$ in $O(n \log^2 n)$ time. Since by Theorem 2 both diagrams have $O(n)$ edges, both $\partial D_i \setminus \mathrm{FV}(\mathfrak{D}_i)$ and $\partial D_j \setminus \mathrm{FV}(\mathfrak{D}_j)$ have $O(n)$ arcs, and $O(n^2)$ pairs of arcs will be dealt with. We process each pair of arcs by Fact 2 in $O(1)$ time and thus compute a minimum placement for fixed D_i, D_j and $\mathfrak{D}_i, \mathfrak{D}_j$ in $O(n^2)$ time. The total run-time is then $O(n^9)$. □

5 Disjoint Disks

Since the run-time is dominated by the bipolar case, the improvements presented in Sects. 5 and 6 only consider the bipolar case. If the disks in $\mathfrak{D}$ are pair-wise disjoint, we propose a linear-time sequential search to compute a minimum placement for fixed D_i, D_j and $\mathfrak{D}_i, \mathfrak{D}_j$, leading to $O(n^8 \log^2 n)$ run-time (dominated by the construction of $\mathrm{FV}(\mathfrak{D}_i)$ and $\mathrm{FV}(\mathfrak{D}_j)$). The key observation in order to obtain this improvement is that, for $p_i \in \partial D_i$ and $p_j \in \partial D_j$, a local minimum placement is also a global minimum one.

We reformulate δ_T as a function of $p_i \in \partial D_i$ and $p_j \in \partial D_j$ by parametrizing p_i and p_j for two angles θ_i and θ_j such that $p_i = (r_i \cos\theta_i + h_i, r_i \sin\theta_i + k_i)$ and $p_j = (r_j \cos\theta_j + h_j, r_j \sin\theta_j + k_j)$. By Fact 4, δ_T is attained by a path from p_k to p_l passing through p_i and p_j, for some $D_k \in \mathfrak{D}_i$ and $D_l \in \mathfrak{D}_j$ depending on the locations of p_i and p_j. Thus, we define a periodic function

$$f_{k,l}(\theta_i, \theta_j) = d(p_i, D_k) + |\overline{p_i p_j}| + d(p_j, D_l) = |\overline{c_k p_i}| + |\overline{p_i p_j}| + |\overline{p_j c_l}| - r_k - r_l,$$

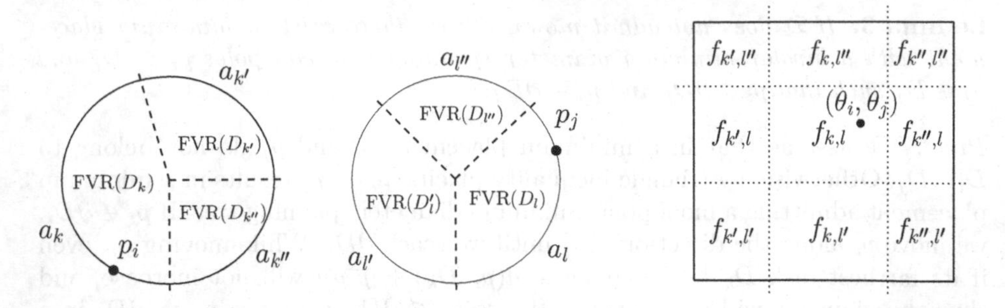

Fig. 1. Geometric interpretation of the functions $f_{k,l}$ for a pair of disks D_i and D_j.

representing the length of such a simple path. Then, regardless of the locations of p_i and p_j, we can express δ_T as a function

$$f(\theta_i, \theta_j) = \max_{D_k \in \mathfrak{D}_i, D_l \in \mathfrak{D}_j} f_{k,l}(\theta_i, \theta_j).$$

The following lemmas indicate that a local minimal point of $f_{k,l}$ is a global minimum one, and that the same property also holds for f.

Lemma 4. *If $f_{k,l}$ has a local minimum at θ_i and θ_j, the point is also a global minimum.*

Lemma 5. *If f has a local minimum at θ_i and θ_j, the point is also a global minimum.*

Proof. Assume that f has a global minimum at θ_i' and θ_j'. Let D_k and $D_{k'}$ be the farthest disks respectively of p_i and p_i' in $\mathfrak{D}_i$, and D_l and $D_{l'}$ be the farthest disks respectively of p_j and p_j' in $\mathfrak{D}_j$. By Lemma 4, the definition of f, and the fact that f has a global minimum at θ_i' and θ_j', it holds that

$$f_{k,l}(\theta_i, \theta_j) \leq f_{k,l}(\theta_i', \theta_j') \leq f_{k',l'}(\theta_i', \theta_j') \leq f_{k,l}(\theta_i, \theta_j).$$

Thus, $f_{k,l}(\theta_i, \theta_j) = f_{k',l'}(\theta_i', \theta_j')$ and f has a global minimum at θ_i, θ_j. □

Corollary 1. *If f has a local minimum at θ_i for fixed θ_j (resp. at θ_j for fixed θ_i), the point is also a global minimum for fixed θ_j (resp. for fixed θ_i).*

We re-interpret f geometrically to present the sequential search as follows. We view $(\theta, \phi, f_{k,l}(\theta, \phi))$ as a point in three dimension, so that $f_{k,l}$ is a 3D surface, and f is the upper envelope of all $f_{k,l}$. If we assign different colors to different $f_{k,l}$, the vertical projection of f onto the xy-plane is a planar subdivision consisting of axis-parallel rectangles, where each rectangle is associated with a function $f_{k,l}$ such that $f_{k,l}(\theta_i, \theta_j) = f(\theta_i, \theta_j)$ if (θ_i, θ_j) belongs to the rectangle. Therefore, a point (θ_i, θ_j) in a rectangle associated with a function $f_{k,l}$ corresponds to a point p_i on the arc $a_k = \mathrm{FVR}(D_k, \mathfrak{D}_i) \cap \partial D_i$ and a point p_j on the arc $a_l = \mathrm{FVR}(D_l, \mathfrak{D}_j) \cap \partial D_j$. Searching the minimal point of $f_{k,l}$ inside the corresponding rectangle is equivalent to searching the minimum δ_T for a pair of poles, p_i and p_j on the corresponding arcs. Figure 1 illustrates the geometric interpretation.

We can sequentially search the rectangles until we find a point (θ_i, θ_j) with minimum $f(\theta_i, \theta_j)$. We begin with an arbitrary rectangle and compute the minimal value inside it, i.e., the minimum δ_T achieved by a pair of points on the corresponding arcs. If the minimal point occurs in the interior of the rectangle, by Lemma 5 we have found a global minimum. Otherwise, we compute the minimum values for the at most three other rectangles adjacent to the minimal point and select the smallest one. If the smallest value of the selected adjacent rectangle is not smaller than that of the previous one, then the previous point was a global minimum. Otherwise, we continue searching in the selected rectangle. The following lemma indicates the run-time of this sequential search. Note that the geometric interpretation is presented only for easier understanding; we do not construct the corresponding planar subdivision due to its quadratic size.

Lemma 6. *For fixed D_i, D_j and $\mathfrak{D}_i, \mathfrak{D}_j$, a pair of poles $p_i \in \partial D_i$ and $p_j \in \partial D_j$ with minimum δ_T can be computed in $O(|\partial D_i \setminus FV(\mathfrak{D}_i)| + |\partial D_j \setminus FV(\mathfrak{D}_j)|)$ time.*

Proof. Moving from one rectangle to the next one, at least one of the corresponding arcs is replaced with its adjacent arc. By Corollary 1, if we leave one arc it will not be considered again. Therefore, the number of pairs of tested arcs (i.e., tested rectangles) is $O(|\partial D_i \setminus \mathrm{FV}(\mathfrak{D}_i)| + |\partial D_j \setminus \mathrm{FV}(\mathfrak{D}_j)|)$. By Fact 2 a solution for a pair of arcs can be computed in $O(1)$ time, concluding this lemma. □

Since $O(|\partial D_i \setminus \mathrm{FV}(\mathfrak{D}_i)| + |\partial D_j \setminus \mathrm{FV}(\mathfrak{D}_j)|) = O(n)$, the time for computing a minimum placement for fixed D_i, D_j and $\mathfrak{D}_i, \mathfrak{D}_j$ is bounded by the $O(n \log^2 n)$ construction time of $\mathrm{FV}(\mathfrak{D}_i)$ and $\mathrm{FV}(\mathfrak{D}_j)$. There are $O(n^2)$ pairs of disks and we consider $O(n^5)$ topologies per pair, resulting in the following overall run-time.

Theorem 6. *If all disks are pairwise disjoint, a minimum placement for $\mathfrak{D}$ can be computed in $O(n^8 \log^2 n)$ time.*

6 Unit Disks

If all disks in $\mathfrak{D}$ are unit, we improve the run-time to $O(n^6)$ for the intersecting case and to $O(n^5)$ for the disjoint case. We obtain these improvements by reducing the number of bipolar topologies that need to be considered, and by decreasing the overall time required to compute all the Voronoi diagrams.

6.1 Separable Bipolar Tree Topologies

The key property for reducing the number of bipolar tree topologies to be considered is proven by the following lemma.

Lemma 7. *If all disks in $\mathfrak{D}$ are unit, there exists a minimum placement admitting a bipolar minimum diameter spanning tree with poles $p_i \in D_i$ and $p_j \in D_j$ such that the centers of $\mathfrak{D}_i$ and the centers of $\mathfrak{D}_j$ can be separated by a straight line, or admitting a monopolar minimum diameter spanning tree.*

Proof. Consider a minimum placement P for $\mathfrak{D}$ with a bipolar minimum diameter spanning tree T whose poles are $p_i \in D_i$ and $p_j \in D_j$. We modify $\mathfrak{D}_i$ and $\mathfrak{D}_j$ to satisfy the statement without increasing δ_T. Let B_i and B_j denote the smallest balls centered respectively at p_i and p_j and enclosing respectively the centers of $\mathfrak{D}_i$ and $\mathfrak{D}_j$. Let R_i and R_j be the radii of B_i and B_j, respectively.

If $R_i < 1$, all disks in $\mathfrak{D}_i$ contain p_i, and all points for $\mathfrak{D}_i$ are placed on p_i, so the distance between p_i and its farthest disk in $\mathfrak{D}_i$ is 0; if $R_i \geq 1$, such distance is $R_i - 1$. Hence we can reformulate this distance as $\max\{R_i - 1, 0\}$. By symmetry for R_j and Fact 4, it holds that $\delta_T = \max\{R_i - 1, 0\} + \max\{R_j - 1, 0\} + |\overline{p_i p_j}|$.

If B_i and B_j do not intersect or are tangent, the first statement trivially holds. Otherwise, we distinguish the relation between B_i and B_j into two cases.

If $B_j \subset B_i$, for all disks $D_l \in \mathfrak{D}_j$, we replace $p_l \in D_l$ with the first intersection between $\overrightarrow{p_i c_l}$ and D_l and link p_l to p_i instead of p_j. These operations result in a monopolar spanning tree whose diameter is not greater than δ_T, leading to the second statement. The symmetric holds if B_i is completely contained inside B_j.

If $B_i \cap B_j \neq \emptyset$ but neither ball is completely contained inside the other, we construct another minimum placement for $\mathfrak{D}$ admitting a bipolar minimum diameter spanning tree satisfying the first statement as follows. Let L be the line passing through the two intersections between ∂B_i and ∂B_j, and H_i and H_j be the two connected regions of $\mathbb{R}^2$ separated by L, where $B_i \setminus B_j \subset H_i$ and $B_j \setminus B_i \subset H_j$. We move each $D_k \in \mathfrak{D}_i$ whose center c_k belongs to H_j from $\mathfrak{D}_i$ to $\mathfrak{D}_j$, and move each disk $D_l \in \mathfrak{D}_j$ whose center c_l belongs to H_i from $\mathfrak{D}_j$ to $\mathfrak{D}_i$. Since all centers of the moved disks belong to $B_i \cap B_j$, the movement will not increase neither R_i nor R_j, leading the first statement. □

An essential 2-partition of $\mathfrak{D} \setminus \{D_i, D_j\}$ thus corresponds to a 2-partition of the centers $\mathfrak{C} \setminus \{c_i, c_j\}$ that can be separated by a straight line. The following well-known lemma implies that, for a pair of polar disks D_i and D_j, the number of 2-partitions to be considered is $O(n^2)$, and they can be computed in $O(n^2)$ time.

Lemma 8. *For a set S of n points, there are $O(n^2)$ 2-partitions of S into S_1 and S_2 such that there exists a line separating S_1 and S_2, and those 2-partitions can be generated in $O(n^2)$ time.*

6.2 Computing Farthest Voronoi Diagrams in a Batch

To further improve the run-time, for a fixed pair of polar disks D_i, D_j we compute all $O(n^2)$ diagrams $FV(\mathfrak{D}_i)$ and $FV(\mathfrak{D}_j)$ in a batch. Since for unit disks it holds that $\text{FV}(\mathfrak{D}) = \text{FV}(\mathfrak{C})$, we can use two well-known facts: the farthest-site Voronoi diagram of n point sites can be computed in $O(n)$ time if their convex hull is precomputed [1], and we can dynamically update convex hulls in $O(\log^2 n)$ time [11]. To apply these facts, we first generate a tour of all the 2-partitions such that any two consecutive 2-partitions differ by one disk. This tour is computed in $O(n^2)$ time by applying the central point-line duality on $\mathfrak{C} \setminus \{c_i, c_j\}$, and its length is $O(n^2)$. We then construct the convex hulls of the two sets of centers in the first 2-partition and start following the tour. When moving from a 2-partition to the next one, we update the convex hulls by inserting one center point into one hull and deleting it from the other one in $O(\log^2 n)$ time. While visiting a 2-partition, we compute the two farthest-site Voronoi diagrams from the respective hulls in $O(n)$ time. Since there are $O(n^2)$ 2-partitions, we conclude the following lemma.

Lemma 9. *If $\mathfrak{D}$ are all unit disks, for a pair of polar disks D_i and D_j all the $O(n^2)$ 2-partitions of $\mathfrak{D} \setminus \{D_i, D_j\}$ satisfying Lemma 7 and the corresponding farthest-disk Voronoi diagrams can be computed in $O(n^3)$ time.*

For a pair of polar unit disks D_i and D_j, we find a minimum placement by computing all the $O(n^2)$ 2-partitions of $\mathfrak{D} \setminus \{D_i, D_j\}$ together with the corresponding farthest Voronoi diagrams in $O(n^3)$ time. For each 2-partition, a minimum placement can be found in $O(n^2)$ and $O(n)$ time respectively for intersecting and for disjoint disks, leading to $O(n^3)$ and $O(n^4)$ time respectively for a pair of polar unit disks. Thus, we conclude the following overall run-time.

Theorem 7. *If all disks in $\mathfrak{D}$ are unit, a minimum placement can be computed in $O(n^6)$ time and $O(n^5)$ time respectively for intersecting and for disjoint disks.*

References

1. Aggarwal, A., Guibas, L.J., Saxe, J.B., Shor, P.W.: A linear-time algorithm for computing the voronoi diagram of a convex polygon. Discrete Comput. Geom. **4**, 591–604 (1989)
2. Chan, T.M.: Semi-online maintenance of geometric optima and measures. SIAM J. Comput. **32**(3), 700–716 (2003)
3. Disser, Y., Mihalák, M., Montanari, S., Widmayer, P.: Rectilinear shortest path and rectilinear minimum spanning tree with neighborhoods. In: Fouilhoux, P., Gouveia, L.E.N., Mahjoub, A.R., Paschos, V.T. (eds.) ISCO 2014. LNCS, vol. 8596, pp. 208–220. Springer, Heidelberg (2014)
4. Dorrigiv, R., Fraser, R., He, M., Kamali, S., Kawamura, A., López-Ortiz, A., Seco, D.: On minimum-and maximum-weight minimum spanning trees with neighborhoods. In: Erlebach, T., Persiano, G. (eds.) WAOA 2012. LNCS, vol. 7846, pp. 93–106. Springer, Heidelberg (2013)
5. Fiala, J., Kratochvíl, J., Proskurowski, A.: Systems of distant representatives. Discrete Appl. Math. **145**(2), 306–316 (2005)
6. Gudmundsson, J., Haverkort, H.J., Park, S., Shin, C., Wolff, A.: Facility location and the geometric minimum-diameter spanning tree. Comput. Geom. **27**(1), 87–106 (2004)
7. Ho, J., Lee, D.T., Chang, C., Wong, C.K.: Minimum diameter spanning trees and related problems. SIAM J. Comput. **20**(5), 987–997 (1991)
8. Jadhav, S., Mukhopadhyay, A., Bhattacharya, B.K.: An optimal algorithm for the intersection radius of a set of convex polygons. J. Algorithms **20**(2), 244–267 (1996)
9. Löffler, M., van Kreveld, M.J.: Largest and smallest convex hulls for imprecise points. Algorithmica **56**(2), 235–269 (2010)

10. Löffler, M., van Kreveld, M.J.: Largest bounding box, smallest diameter, and related problems on imprecise points. Comput. Geom. **43**(4), 419–433 (2010)
11. Overmars, M.H., van Leeuwen, J.: Maintenance of configurations in the plane. J. Comput. Syst. Sci. **23**(2), 166–204 (1981)
12. Spriggs, M.J., Keil, J.M., Bespamyatnikh, S., Segal, M., Snoeyink, J.: Computing a (1+ϵ)-approximate geometric minimum-diameter spanning tree. Algorithmica **38**(4), 577–589 (2004)

Model-Based Classification of Trajectories

Maike Buchin and Stef Sijben(✉)

Ruhr-Universität Bochum, Bochum, Germany
{Maike.Buchin,Stef.Sijben}@rub.de

Abstract. We present algorithms for classifying trajectories based on a movement model parameterized by a single parameter, like the Brownian bridge movement model. Classification is the problem of assigning trajectories to classes of similar movement characteristics. For instance, the set of trajectories might be the subtrajectories resulting from segmenting a trajectory, thus identifying movement phases. We give an efficient algorithm to compute the optimal classification for a discrete set of parameter values. We also show that classification is NP-hard if the parameter values are allowed to vary continuously and present an algorithm that solves the problem in polynomial time under mild assumptions on the input.

1 Introduction

Recent advances in tracking technology lead to increasing amounts of movement data. For instance, animals, vehicles, and people are tracked to analyze their movement. Movement data is typically recorded as a sequence of time-stamped positions, called a *trajectory*. To analyze these data requires efficient algorithms, a task addressed by the emerging field of computational movement analysis [7].

Here we study the fundamental analysis task of classifying trajectory data. Classification asks to group trajectories (or trajectory pieces) into classes of similar trajectories. That is, a *classification* of a set of trajectories $\mathcal{T}$ is a partition of $\mathcal{T}$ into disjoint *classes*. If the trajectories correspond to periods of homogeneous behaviour (e.g. they are the segments produced by a segmentation algorithm), the classification can detect when behavioural states recur.

Our Approach. The motivation for classification is typically to make inferences about the underlying movement process. Hence it is only natural to take a statistical perspective on this analysis tasks: As we describe in more detail below, we see trajectory classification as fitting a parameterized movement model to the data. Taking such an approach is essential when designing algorithms for applications –as in ecology– that use movement data in a statistical analysis.

In ecology, movement models are used to infer a continuous motion from discrete samples of the movement path. Mostly random movement models, like the Brownian bridge movement model (BBMM) [3,5,9] and variants of it, e.g. [11], Lévy walks [10] and behavioural change point analysis [8] are used.

In these movement models, a link l, i.e. the part of the trajectory between two consecutive observations, has an associated log-likelihood function $L_l(x)$ as

K. Elbassioni and K. Makino (Eds.): ISAAC 2015, LNCS 9472, pp. 393–403, 2015.
DOI: 10.1007/978-3-662-48971-0_34

a function of the model parameter x, indicating how well the model fits the data for each possible value of x. The log-likelihood of a parameter value x for a trajectory τ is given by $L_\tau(x) = \sum_{l \in \tau} L_l(x)$. A classification $\mathcal{C}$ has an associated log-likelihood, which is defined as $L_\mathcal{C} = \sum_{C \in \mathcal{C}} \sum_{\tau \in C} L_\tau(x(C))$. That is, the log-likelihood of each class C is the sum over the log-likelihoods of its elements (i.e., trajectories) τ, evaluated at the parameter value $x(C)$ assigned to C. The log-likelihood of $\mathcal{C}$ is the sum of the log-likelihoods of the classes. Our algorithms assume that the log-likelihood functions are bitonic. We think this assumption is justified, since only bitonic log-likelihood functions are encountered with the movement models used in practice.

We could now define an optimal classification as one that maximizes the log-likelihood, but then it would be optimal to put each trajectory into its own class, resulting in the largest possible number of degrees of freedom for the model. One solution is to fix the number of classes, but typically the number of classes is not known beforehand. To determine a good number of classes an information criterion like the Bayesian information criterion (BIC) can be used [8,11].

To facilitate multi-scale analysis, we use a more general notion of an *information criterion* (IC) to define the optimal classification. An IC assigns a value to each classification based on its likelihood and the complexity of the model (that is, the number of classes). In particular we consider ICs of the form $\text{IC}(\mathcal{C}) = -2L_\mathcal{C} + |\mathcal{C}| \cdot p$, where $L_\mathcal{C}$ is the log-likelihood of the model instance and $|\mathcal{C}|$ is the number of classes. The number p is a penalty factor for adding complexity to the model that counteracts overfitting. We now define an optimal classification to be one that minimizes the value of the IC. p may be chosen simply as $\log k$ to obtain the BIC, where k is the number of trajectories, or one may use a stability diagram [1] to select a good value. Our algorithms can be adapted to produce stability diagrams.

Problem Statement. Given a set of trajectories $\mathcal{T}$, an *optimal classification* $\mathcal{C}_{opt}$ is the classification $\{C_1, \ldots, C_\ell\}$ and selection of model parameters for the classes $x(C_i)$, $1 \leq i \leq \ell$, that achieves the minimum value for the information criterion among all classifications and parameter values for $\mathcal{T}$. Our goal is to compute optimal classifications. Note that an optimal classification asks both for the choice of classes and their model parameters.

Related Work. In movement ecology, classification algorithms have been used to identify behavioural states from acceleration data [14,15]. Also, criteria-based segmentation can be used for classification, by using multiple criteria, one for each class. This setting has been successfully applied to data of migrating geese [4]. This work extends our previous work [2] where we use a model-based approach for trajectory segmentation and classification.

Similar problems are studied for time series data such as audio. However, the nature and models of these data differ from ours. It would be interesting to explore whether our approach can be extended to these by appropriate modelling.

Results and Overview. First we consider a discrete setting where we assume that the parameter values are drawn from a finite set of candidate values and that the log-likelihood functions are given by listing the values they take on these. We give a dynamic programming algorithm for this case in Sect. 2. Next we consider the case where the parameter values are drawn from a continuous domain $\mathcal{D}$, which is usually an interval on the real line. Here, we first show that the decision problem becomes NP-hard in Sect. 3. Then, we give a polynomial time algorithm under mild assumptions on the input in Sect. 4.

Figure 1 shows an example of our method. Two movement tracks, collected from a fisher (animal) [12,13], were segmented based on the BBMM [2]. Each trajectory produced 5 segments, and these 10 segments were given as input to the discrete classification algorithm, which was implemented in R, again using likelihoods based on the BBMM and using the BIC penalty factor. This resulted in two classes, indicating that the animal was alternating between just two behavioural states.

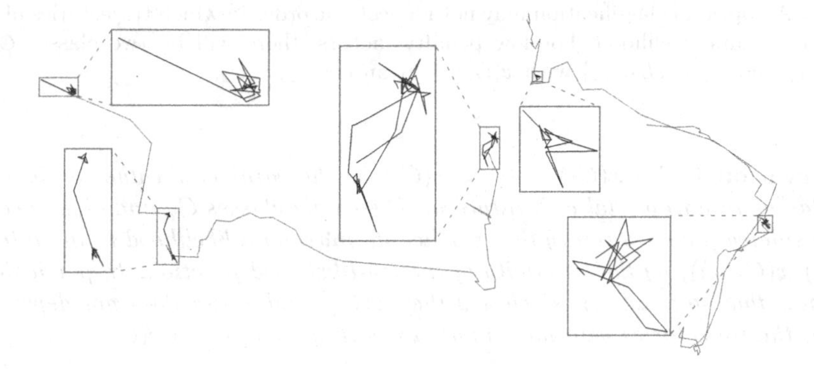

Fig. 1. Examples of classification of fisher tracks. Classes indicated by colour (Color figure online).

Preliminaries. We assume a *trajectory* τ is given by a sequence of n time-stamped positions. When classifying a set of trajectories $\mathcal{T}$, we use ℓ to denote the number of classes and $C_1, \ldots, C_\ell \subseteq \mathcal{T}$ to denote the classes. Each class C_i is assigned a value of the model parameter $x(C_i)$. Each trajectory $\tau_i \in \mathcal{T}$ has an associated log-likelihood function $L_i(x)$, which is the sum of the log-likelihood functions of the links in τ_i. We assume that the log-likelihood functions L_i are bitonic, i.e. that they have one maximum and are increasing before that and decreasing after. We also assume w.l.o.g. that the functions are given in increasing order by the parameter value at which they reach their maximum. That is, if we define $M_i := \arg\max_x(L_i(x))$, then $i < j \Rightarrow M_i \leq M_j$. We represent a classification $\mathcal{C}$ for $L_1, \ldots, L_k$ by an array of length k where $\mathcal{C}[i]$ is the parameter value that the classification assigns to the class of L_i, or $\mathcal{C}[i] = \text{NIL}$ if $\mathcal{C}$ is a partial classification that does not classify L_i. For the discrete classification, we assume we are given a set of model parameters $\{x_1, \ldots, x_m\}$ in sorted order.

2 Dynamic Programming for Discrete Parameter

First we present a dynamic programming algorithm for the case that a discrete set of parameter values is given. A natural approach would be to process the trajectories in the order they are given. However it is not necessarily the case that if $M_i < M_j$ that this order is also reflected in the classes they are associated with. Figure 2 shows an example. However, we can use the following property to efficiently compute the optimal classification.

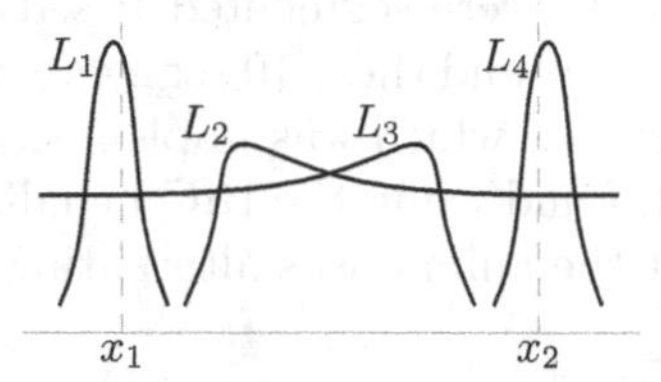

Fig. 2. An optimal classification may not respect the order in which trajectories obtain their maximum likelihood. For low penalty factors, there will be two classes $C_1 = \{L_1, L_3\}$ and $C_2 = \{L_2, L_4\}$ with $x(C_1) = x_1$ and $x(C_2) = x_2$.

Observation 1. *Let $x(C_1) < \cdots < x(C_\ell)$ be the parameter values assigned to the classes in an optimal classification. Then only classes C_j and C_{j+1} need to be considered for a trajectory that reaches its maximum likelihood in the interval $[x(C_j), x(C_{j+1}))$, by the bitonicity of the log-likelihood function. In particular if we know that some $x(C_i)$ is selected then $x(C_j)$ with $j < i$ does not depend on any of the trajectories with maximum larger than or equal to $x(C_i)$.*

Using this observation, the optimal classification can be efficiently computed. For a problem instance $(L_1, \ldots, L_k)$, $(x_1, \ldots, x_m)$, add dummy values $x_0 := -\infty$ and $x_{m+1} := \infty$. Let $\mathcal{L}_{i,j} := \{L_l \in \{L_1, \ldots, L_k\} \mid x_i \leq M_l < x_j\}$, with $0 \leq i < j \leq m+1$, denote the set of functions that reach their maximum value between x_i and x_j. For $i \in \{1, \ldots, m\}$, let $Opt_i = \{C_1, \ldots, C_\ell\}$ be the optimal classification of $\mathcal{L}_{0,i}$, conditioned on $x(C_\ell) = x_i$, even if C_ℓ is empty. Let $\mathcal{C}_i$ denote the complete classification that is obtained from Opt_i by assigning all trajectories in $\mathcal{L}_{i,m+1}$ to C_ℓ. The following lemma suggests a means to efficiently compute Opt_i.

Lemma 2. *Opt_i consists of an optimal classification of $\mathcal{L}_{0,j}$ for some $j \in \{0, \ldots, i-1\}$ extended with a class with parameter value x_i, by assigning each trajectory in $\mathcal{L}_{j,i}$ the parameter value x_j or x_i that has the highest log-likelihood.*

Proof. Let $x(C_1) < \cdots < x(C_{\ell-1}) < x(C_\ell) = x_i$ be the parameter values for the classes in Opt_i. If $\ell = 1$, all trajectories are assigned to one class, which we view as an extension of the empty classification Opt_0. Otherwise $x(C_{\ell-1}) = x_j$ for some $j < i$, since the parameter values come from $\{x_1, \ldots, x_m\}$. By Observation 1, Opt_i consists of an optimal classification of $\mathcal{L}_{0,j}$ (that uses x_j), i.e. Opt_j, and the trajectories in $\mathcal{L}_{j,i}$ are independently assigned to either $C_{\ell-1}$ or C_ℓ. □

```
DiscreteClassification((L_1, ..., L_k), (x_1, ..., x_m))
 1  Opt_0 ← An array of length n with all elements set to NIL
 2  C ← An arbitrary (complete) classification of L_1, ..., L_k
 3  for i ← 1 to m
 4        do for j ← 0 to i − 1
 5              do O_j ← Opt_j
 6                 for L_l ∈ 𝓛_{j,i}
 7                      do O_j[l] ← argmax_{x∈{x_j,x_i}} (L_l(x))
 8           Opt_i ← argmin_{{O_j | 0≤j<i}} IC_i(O_j)
 9           C_i ← Opt_i with all NIL s replaced by x_i
10           f IC(C_i) < IC(C)
11              then C ← C_i
12  return C
```

Algorithm 1. Discrete classification.

See Algorithm 1 for the pseudocode of the algorithm. IC is the function that computes the value of the information criterion for a given (partial) classification. IC_i computes the IC under the assumption that x_i is used, even if no trajectory has been assigned to x_i. We do this to take into account that we assign the remaining trajectories to x_i in the next step. Comparing values of the IC for partial classifications is useful only if they assign values to exactly the same trajectories, as is the case for all O_j in a single iteration of the algorithm.

Theorem 3. *Algorithm 1 computes the optimal classification of k trajectories with respect to an information criterion in $O(km^2)$ time and $O(mk)$ space, where m is the number of candidate parameter values.*

Proof. For the correctness observe that the optimal classification $\mathcal{C}$ is one of the $\{\mathcal{C}_i \mid 1 \leq i \leq m\}$ computed by the algorithm: Let x_i be the largest value that is selected in $\mathcal{C}$. Then all trajectories in $\mathcal{L}_{i,m+1}$ are assigned to x_i, and the other trajectories are assigned according to an optimal classification of $\mathcal{L}_{0,i}$ conditioned on using x_i, i.e. Opt_i. This is exactly the $\mathcal{C}_i$ computed in the algorithm.

The runtime and space use are as follows. The outer loop has m iterations. The middle and inner loops have $O(m)$ and $O(k)$ iterations respectively. These middle and inner loops with all the operations in them take $O(mk)$ time. Computing the IC of $O(m)$ classifications, each of size k, takes $O(mk)$ time too, and all the other operations need less time than that, so the running time of the algorithm is $O(km^2)$. At any time, the algorithm stores $O(m)$ (partial) classifications, each of size k, leading to a space use of $O(mk)$. □

By reusing values computed in previous iterations of the outer loop, the running time can be improved to $O(m^2 + mk(\log m + \log k))$.

3 NP-hardness for Continuous Parameter

We briefly sketch a reduction from SET COVER, showing that the decision version of continuous classification is NP-hard. Let $\mathcal{S} = \{S_1, \dots, S_n\}$ be a family of subsets of a universe $\mathcal{U} = \{1, \dots, m\}$ such that $\bigcup_{i=1}^{n} S_i = \mathcal{U}$, and assume that an integer $k < n$ is given. Given this instance of SET COVER, we construct a set of log-likelihood functions, a penalty factor p and a constant c such that a classification $\mathcal{C}$ with $\text{IC}(\mathcal{C}) \leq c$ exists if and only if $\mathcal{S}$ has a cover of size $\leq k$.

For each $i \in \mathcal{U}$, we construct a pair of log-likelihood functions, L_i^- and L_i^+, over the domain $[0, 2^n)$, which encode for each subset of $\mathcal{S}$ whether it covers i. The sum of all these functions is the number of elements covered by the subsets under consideration. Furthermore there are two likelihood functions which encode the number of subsets selected. Only if a cover of size $\leq k$ exists, there is a value x in the domain such that a single class with parameter value x has a sufficiently large log-likelihood. A large penalty factor prevents using two or more classes.

4 Polynomial-Time Algorithm for Realistic Inputs

We now present an algorithm that runs in polynomial time for realistic inputs. For reasonably homogeneous trajectories (and a standard movement model) the likelihood functions are sufficiently smooth. In this case the number of classifications we need to consider reduces; essentially, this is because the reordering illustrated in Fig. 2 does not happen too often. At the end of this section, we give the precise assumptions we need to make on the likelihood functions.

For a set of input functions $\mathcal{L} := \{L_1, \dots, L_k\}$, let Opt_i denote the optimal classification with exactly i classes. Assume that we can optimize the value of the model parameter and compute the log-likelihood for a fixed class of size n in $F(n)$ time. Let C_j^i denote the jth class in Opt_i (sorted by parameter value).

The algorithm iteratively computes Opt_{i+1} from Opt_i, for $i \in \{1, \dots, k-1\}$. There is only one classification with one class (ignoring the parameter value), so $Opt_1 = \{C_1^1\} = \{\{L_1, \dots, L_k\}\}$. $x(C_1^1)$ can be computed in $O(F(k))$ time.

x_j^i be a shorthand notation for $x(C_j^i)$. Recall that a classification is stored in an array specifying for each function the parameter value of the class to which it is assigned. Recall that M_i is the parameter value at which function L_i reaches its maximum likelihood. In constructing Opt_{i+1}, we use the following properties, where C_0^i and C_{i+1}^i are empty dummy classes:

Lemma 4. *Let $Opt_i = \{C_1^i, \dots, C_i^i\}$ and $Opt_{i+1} = \{C_1^{i+1}, \dots, C_{i+1}^{i+1}\}$ be the optimal classifications of a set of functions $\mathcal{L} = \{L_1, \dots, L_k\}$ with i and $i+1$ classes respectively. Let $x(C_0^i) := M_1$ and $x(C_{i+1}^i) := M_k$. Then, the jth class of Opt_{i+1} has a smaller parameter value than the jth class in Opt_i, but no smaller than the $j-1$th class in Opt_i. That is,*

$$\forall j \in \{1, \dots, i+1\} : x_{j-1}^i \leq x_j^{i+1} \leq x_j^i \ .$$

Lemma 5. *Functions shift to a class with a larger index as the number of classes increases, but no more than one class at a time. That is,*

$$\forall j \in \{1, \ldots, k\} : L_j \in C_m^i \Rightarrow L_j \in C_m^{i+1} \cup C_{m+1}^{i+1} \ .$$

Proof. Consider a function $L_l \in C_m^i$. Then, $M_l \in (x_{m-1}^i, x_{m+1}^i)$ and $L_l(x_{m-1}^i) \leq L_l(x_m^i) \geq L_l(x_{m+1}^i)$, and the first inequality is strict if $M_l < x_m^i$. If $M_l \geq x_m^{i+1}$, $L_l \notin C_p^{i+1}$ with $p \leq m-1$ by Observation 1. If $M_l < x_m^{i+1}$, $L_l(x_m^{i+1}) \geq L_l(x_m^i) > L_l(x_{m-1}^i) \geq L_l(x_p^{i+1})$ for any $p \leq m-1$ by Lemma 4 and thus $L_l \notin C_p^{i+1}$. Similarly, $L_l \notin C_p^{i+1}$ for any $p > m+1$ and thus $L_l \in C_m^{i+1}$ or $L_l \in C_{m+1}^{i+1}$. □

When Opt_i is computed, we can use it to compute Opt_{i+1}. By Lemmas 4 and 5, a function can be in one of only two classes of Opt_{i+1} and the parameter value x_j^{i+1} for a class C_j^{i+1} must be in a particular interval $[x_{j-1}^i, x_j^i]$.

For a function $L_l \in C_j^i$, whether $L_l \in C_j^{i+1}$ or $L_l \in C_{j+1}^{i+1}$ is determined by the parameter values x_j^{i+1} and x_{j+1}^{i+1} for these classes. The parameter values for two consecutive classes can be represented as a point inside a rectangle $\mathcal{R}_j^i \subseteq \mathcal{D}^2$:

$$(x_j^{i+1}, x_{j+1}^{i+1}) \in [x_{j-1}^i, x_j^i] \times [x_j^i, x_{j+1}^i] =: \mathcal{R}_j^i \ .$$

Let $P_l := \{(p,q) \in \mathcal{D}^2 \mid L_l(p) = L_l(q) \wedge p < q\}$ be the set of points that represent pairs of parameter values at which L_l has the same value. If L_l is continuous and strictly bitonic (i.e. strictly increasing and decreasing), P_l is a bimonotone continuous curve that separates the region $\{(x,y) \in \mathcal{D}^2 \mid x < y\}$ into two parts. For any point (p,q) in the region below P_l, we have $L_l(p) < L_l(q)$. Thus, if the two candidate classes for L_l have parameter values x_j^{i+1} and x_{j+1}^{i+1}, and $(x_j^{i+1}, x_{j+1}^{i+1})$ lies below P_l, then L_l is in C_{j+1}^{i+1} rather than C_j^{i+1}. Similarly, if $(x_j^{i+1}, x_{j+1}^{i+1})$ lies above P_l, then L_l is in C_j^{i+1}. These curves are illustrated in Fig. 3. If L_l is not strictly bitonic or discontinuous at a finite number of points, P_l is not a well-defined Jordan curve. In this case, we relax the definition of P_l to be the continuous, bimonotone curve separating the region with $L_l(p) < L_l(q)$ from that with $L_l(q) \leq L_l(p)$ (note that the latter inequality is no longer strict). This curve always exists and suffices to find the optimal classification.

If the intersection of P_l with $\mathcal{R}_j^i$ for some $L_l \in C_j^i$ is empty, L_l is assigned to a fixed class in Opt_{i+1} regardless of the choice of parameter values. Otherwise, P_l divides $\mathcal{R}_j^i$ into two contiguous regions: one in which $L_l \in C_j^{i+1}$ and one in which $L_l \in C_{j+1}^{i+1}$. We first show how to compute Opt_2 from Opt_1 using these curves, then how to generalize this to arbitrary i.

Computing Opt_2. The set $\{P_1, P_2, \ldots, P_k\}$ defines an arrangement of curves $\mathcal{A}$. This is illustrated in Fig. 3. Let $\mathcal{A}_1^1$ be the part of $\mathcal{A}$ inside $\mathcal{R}_1^1$. Each face of $\mathcal{A}_1^1$ defines a partition of $C_1^1 = \{L_1, L_2, \ldots, L_k\}$ into two classes, fixing the structure of the classification.

Lemma 6. *One of the faces f in $\mathcal{A}_1^1$ represents Opt_2, in the sense that $C_1^2 = \{L_l \mid P_l$ lies below $f\}$ and $C_2^2 = \{L_l \mid P_l$ lies above $f\}$.*

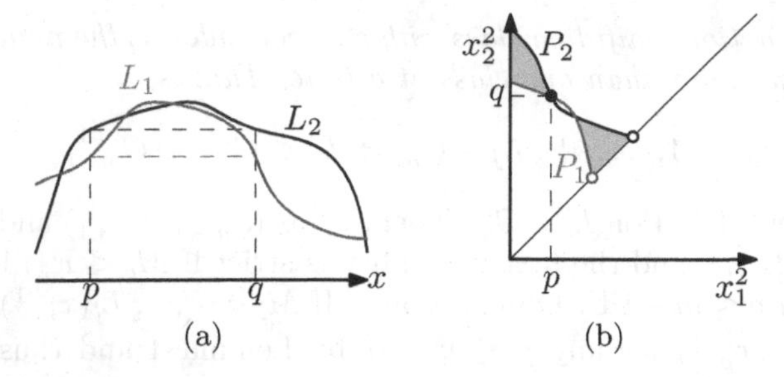

Fig. 3. A pair of log-likelihood functions L_1, L_2 (a) and the arrangement $\mathcal{A}$ induced by the curves P_1, P_2 (b). Both functions have $L(p) = L(q)$, and thus the arrangement has a vertex at (p, q). $\mathcal{A}$ has 5 faces, with the shaded faces corresponding to the same classification $C_1^2 = \{L_1\}$, $C_2^2 = \{L_2\}$.

Proof. By Lemma 4 and the definition of $\mathcal{R}_1^1$, we know that $p := (x_1^2, x_2^2) \in \mathcal{R}_1^1$. Let f be the face of $\mathcal{A}_1^1$ that contains p. If Opt_2 is not the classification represented by f there is a function L_l such that $L_l \in C_1^2$, but p lies below P_l or vice versa. However, if p is below P_l, $L_l(x_1^2) < L_l(x_2^2)$ and the classification is improved by assigning L_l to C_2^2, contradicting the optimality of Opt_2. Thus, the classification represented by f is precisely Opt_2. □

For each face, the parameter values of the two classes are optimized separately and the resulting classification with the highest likelihood is selected as Opt_2.

Computing Opt_i ($i > 2$). In computing Opt_3 from Opt_2 (or generally Opt_{i+1} from Opt_i for $i \geq 2$), we have to deal with the fact that both $\mathcal{A}_1^2$ and $\mathcal{A}_2^2$ produce candidates for subsets of C_2^3. We could consider each pair of faces $(f_1, f_2) \in \mathcal{A}_1^2 \times \mathcal{A}_2^2$ and optimize parameter value for the three classes represented by f_1 and f_2, but that does not generalize to arbitrary i, since we would have to consider every i-tuple of faces in $\mathcal{A}_1^i \times \mathcal{A}_2^i \times \cdots \times \mathcal{A}_i^i$, which is exponential in i.

Instead, the optimal classification can be computed by computing a longest path on a polynomial-size directed acyclic graph G_i. The vertices of G_i are organized in $i+2$ levels, with a source vertex r at level 0 and a target vertex t at level $i+1$. For each face in $\mathcal{A}_j^i$, G_i has a vertex at level j representing it. A vertex v at level j, or face in $\mathcal{A}_j^i$, corresponds to a partition of C_j^i into two parts, say λ_v and ρ_v, which are candidates to become part of C_j^{i+1} and C_{j+1}^{i+1} respectively. For each $j \in \{0, \ldots, i\}$, G_i contains an edge from each vertex at level j to each vertex at level $j+1$. See Fig. 4 for an illustration of this construction.

An edge from a vertex u at level $j-1$ to a vertex v at level j fixes a set of functions $\rho_u \cup \lambda_v$ that is a candidate for C_j^{i+1}, with $\rho_r = \lambda_t = \emptyset$, i.e. r and t do not contribute to C_1^{i+1} or C_{i+1}^{i+1}. For the candidate class represented by an edge, the parameter value is optimized and its log-likelihood is set as the edge weight.

An r–t path in G_i visits exactly one vertex at each level, and this path represents a particular classification, with each edge corresponding to a class.

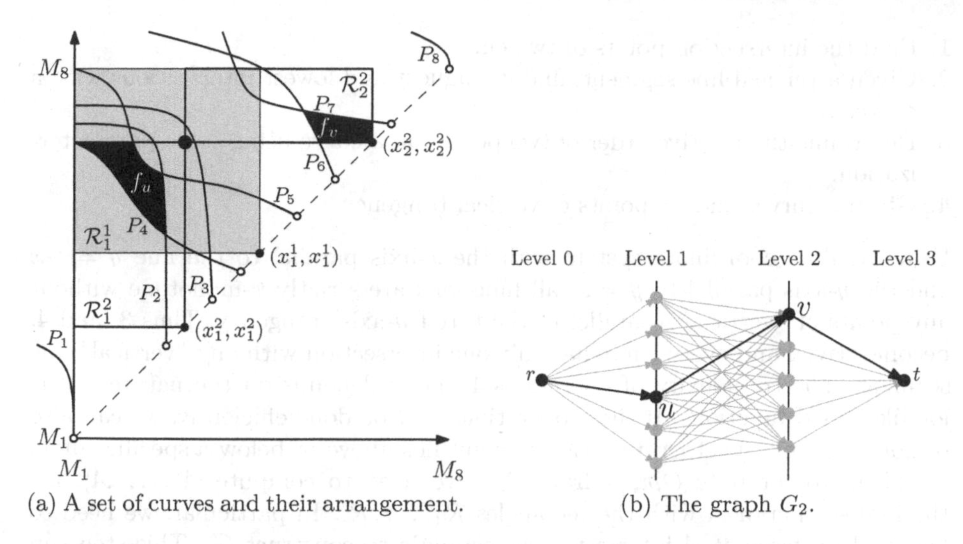

(a) A set of curves and their arrangement. (b) The graph G_2.

Fig. 4. Illustration of the algorithm for a set of curves. The coloured path through u and v represents the coloured faces $f_u \subseteq \mathcal{R}_1^2$ and $f_v \subseteq \mathcal{R}_2^2$, yielding a candidate for Opt_3 with $C_1 = \{L_1, L_4\}$, $C_2 = \{L_2, L_3, L_5, L_6\}$ and $C_3 = \{L_7, L_8\}$ (Color figure online).

The log-likelihood of the classification is the sum of the weights of the edges in the path, i.e. the path's length.

Lemma 7. *Opt_{i+1} corresponds to the longest r–t path in G_i.*

We omit the proof for space reasons, but note that an optimal classification must correspond to an r–t path using similar reasoning to the proof of Lemma 6.

The use of $\mathcal{A}$ and G_i is illustrated in Fig. 4. For the value of x_1^1, the shaded rectangle is $\mathcal{R}_1^1$. Suppose that the parameter values for Opt_2 are (x_1^2, x_2^2) at the big red dot. Then $C_1^2 = \{L_1, L_2, L_4, L_5\}$ and $C_2^2 = \{L_3, L_6, L_7, L_8\}$, i.e. the curves lying below and above this point. The red rectangles are $\mathcal{R}_1^2$ and $\mathcal{R}_2^2$ and the graph G_2 has six vertices at level 1 corresponding to the six faces of $\mathcal{A}_1^2$ and five vertices at level 2. A path through G_2 corresponds to selecting one face in each of the rectangles, thus fixing the classification.

Analysis. $\mathcal{A}$ has $O(sk^2)$ faces, where s is the maximal number of intersections between a pair of curves in $\mathcal{A}$. Thus, a trivial bound on the complexity of G_i is obtained by observing that each level of G_i has at most $O(sk^2)$ vertices and thus the number of vertices in G_i is $O(sk^3)$. A more careful analysis shows that G_i has $O(sk^2)$ vertices and $O(s^2k^4)$ edges.

The curves P_i and the arrangement $\mathcal{A}$ can be computed in $O(k\lambda_{s+2}(k))$ time [6], where $\lambda_s(n)$ is the maximum length of an (n, s) Davenport-Schinzel sequence and the constant in the big-O notation depends on the input functions. The algorithm described in [6] requires that the following operations can be performed efficiently on the curves:

1. Find the intersection points of two curves.
2. Given a vertical line segment, find its highest and lowest intersections with a curve.
3. Determine the relative order of two points on a curve using some parameterization.
4. Given a curve, find its points of vertical tangency.

Using a tilted coordinate system with the x-axis parallel to the line $y = -x$ and the y-axis parallel to $y = x$, all functions are strictly x-monotone without any points of "vertical" (parallel to the tilted y-axis) tangency. Thus 3. and 4. become trivial and every curve has only one intersection with any "vertical" line segment. The complexity of operations 1 and 2 depends on the nature of the log-likelihood functions. If these operations can be done efficiently, we can also decide efficiently whether a particular point lies above or below a specific curve.

Then, to compute Opt_{i+1} from Opt_i we need to compute $\mathcal{A}_1^i, \ldots, \mathcal{A}_i^i$, i.e. the intersection of $\mathcal{A}$ with the rectangles $\mathcal{R}_1^i, \ldots, \mathcal{R}_i^i$. In particular, we need to know which faces of $\mathcal{A}$ intersect each rectangle to construct G_i. This step can be performed in $O(sk^2)$ time using a depth-first search on the dual graph of $\mathcal{A}$.

Computing the length of an edge in G_i is assumed to take $O(F(n))$ time, where n is the size of the class represented by the edge. Then, computing all edge lengths takes $O(s^2k^4F(k))$ time. Since G_i is a DAG, the longest path can be computed in $O(|V| + |E|) = O(s^2k^4)$ time.

Thus, given Opt_i and $\mathcal{A}$, computing Opt_{i+1} takes $O(s^2k^4F(k))$ time. To find the optimal classification, we have to compute all classifications up to Opt_k and select the one among those with minimum IC for the given penalty factor.

Theorem 8. *The optimal classification of k trajectories can be computed in time $O(s^2k^5F(k) + k\lambda_{s+2}(k))$, where $F(n)$ is the time required to optimize the likelihood for a set of n trajectories, s is the maximum number of pairwise intersections between the curves P_i and the constant in the big O-notation depends on the input functions.*

For the algorithm to run in polynomial time, we need to be able to efficiently compute the arrangement $\mathcal{A}$, and to optimize the likelihood for a set of trajectories (expressed as $F(n)$). Furthermore, the complexity of the algorithm depends on the complexity of $\mathcal{A}$ and the graphs G_i, which are polynomial if s is polynomially bounded. For the Brownian bridge movement model, the curves intersect at most once in practice, and many pairs do not intersect. For example, in an arrangement for 20 fisher trajectories similar to those in Fig. 1 there were only 4 intersections. The arrangement often has linear complexity, reducing $|G_i|$ by a factor k^2. The parameter value for a set of trajectories can be optimized to a fixed precision in time linear in the total length of the trajectories. Exact solutions can be obtained using root-finding techniques for polynomials, where the degree is linear in the total length of the trajectories. So, the algorithm computes an optimal classification of k trajectories of total length n in $O(k^3n)$ for the BBMM in practice.

Acknowledgements. We would like to thank Sander Alewijnse, Kevin Buchin, and Michel Westenberg for discussing and sharing ideas on this topic.

References

1. Alewijnse, S.P.A., Buchin, K., Buchin, M., Kölzsch, A., Kruckenberg, H., Westenberg, M.A.: A framework for trajectory segmentation by stable criteria. In: Proceedings of the 22nd International Conference on Advances in Geographic Information Systems (ACM GIS), pp. 351–360. ACM (2014)
2. Alewijnse, S.P.A., Buchin, K., Buchin, M., Sijben, S., Westenberg, M.A.: Model-based segmentation and classification of trajectories. In: Proceedings of the 30th European Workshop on Computational Geometry (2014)
3. Buchin, K., Sijben, S., Arseneau, T.J., Willems, E.P.: Detecting movement patterns using Brownian bridges. In: Proceedings of the 20th International Conference on Advances in Geographic Information Systems (ACM GIS), pp. 119–128. ACM (2012)
4. Buchin, M., Kruckenberg, H., Kölzsch, A.: Segmenting trajectories by movement states. In: Timpf, S., Laube, P. (eds.) Advances in Spatial Data Handling, pp. 15–25. Springer, Heidelberg (2013)
5. Bullard, F.: Estimating the Home Range of an Animal: A Brownian Bridge Approach. Master's thesis, The University of North Carolina (1999)
6. Edelsbrunner, H., Guibas, L., Pach, J., Pollack, R., Seidel, R., Sharir, M.: Arrangements of curves in the plane-topology, combinatorics, and algorithms. Theoret. Comput. Sci. **92**(2), 319–336 (1992)
7. Gudmundsson, J., Laube, P., Wolle, T.: Computational movement analysis. In: Kresse, W., Danko, D.M. (eds.) Springer Handbook of Geographic Information, pp. 423–438. Springer, Heidelberg (2012)
8. Gurarie, E., Andrews, R.D., Laidre, K.L.: A novel method for identifying behavioural changes in animal movement data. Ecol. Lett. **12**(5), 395–408 (2009)
9. Horne, J., Garton, E., Krone, S., Lewis, J.: Analyzing animal movements using Brownian bridges. Ecology **88**(9), 2354–2363 (2007)
10. de Jager, M., Weissing, F.J., Herman, P.M.J., Nolet, B.A., van de Koppel, J.: Lévy walks evolve through interaction between movement and environmental complexity. Science **332**(6037), 1551–1553 (2011)
11. Kranstauber, B., Kays, R., LaPoint, S.D., Wikelski, M., Safi, K.: A dynamic brownian bridge movement model to estimate utilization distributions for heterogeneous animal movement. J. Anim. Ecol. **81**(4), 738–746 (2012)
12. LaPoint, S., Gallery, P., Wikelski, M., Kays, R.: Animal behavior, cost-based corridor models, and real corridors. Landscape Ecol. **28**(8), 1615–1630 (2013)
13. LaPoint, S., Gallery, P., Wikelski, M., Kays, R.: Data from: animal behavior, cost-based corridor models, and real corridors. Movebank Data Repository (2013)
14. Nathan, R., Spiegel, O., Fortmann-Roe, S., Harel, R., Wikelski, M., Getz, W.M.: Using tri-axial acceleration data to identify behavioral modes of free-ranging animals: general concepts and tools illustrated for griffon vultures. J. Exp. Biol. **215**(6), 986–996 (2012)
15. Shamoun-Baranes, J., Bom, R., van Loon, E.E., Ens, B.J., Oosterbeek, K., Bouten, W.: From sensor data to animal behaviour: an oystercatcher example. PLoS ONE **7**(5), e37997 (2012)

Linear-Time Algorithms for the Farthest-Segment Voronoi Diagram and Related Tree Structures

Elena Khramtcova and Evanthia Papadopoulou(✉)

Faculty of Informatics, Università della Svizzera italiana (USI), Lugano, Switzerland
{elena.khramtcova,evanthia.papadopoulou}@usi.ch

Abstract. We present linear-time algorithms to construct tree-like Voronoi diagrams with disconnected regions after the sequence of their faces along an enclosing boundary (or at infinity) is known. We focus on the farthest-segment Voronoi diagram, however, our techniques are also applicable to constructing the order-$(k+1)$ subdivision within an order-k Voronoi region of segments and updating a nearest-neighbor Voronoi diagram of segments after deletion of one site. Although tree-structured, these diagrams illustrate properties surprisingly different from their counterparts for points. The sequence of their faces along the relevant boundary forms a Davenport-Schinzel sequence of order ≥ 2. Once this sequence is known, we show how to compute the corresponding Voronoi diagram in linear time, expected or deterministic, augmenting the existing linear-time frameworks for points in convex position with the ability to handle non-point sites and multiple Voronoi faces.

1 Introduction

It is well known that the Voronoi diagram of points in convex position can be computed in linear time, given the order of their convex hull [1]. Linear-time constructions also exist for a class of related diagrams such as the farthest-point Voronoi diagram, computing the medial axis of a convex polygon, and deleting a point from the nearest-neighbor Voronoi diagram. In an abstract setting, a *Hamiltonian abstract Voronoi diagram* can be computed in linear time [9], given the order of Voronoi regions along an unbounded simple curve, which visits each region exactly once and can intersect each bisector only once. This construction has been extended recently to include forest structures [5] under similar conditions where no region can have multiple faces within the domain enclosed by the curve. The medial axis of a simple polygon can also be computed in linear time [8]. It is therefore natural to ask what other types of Voronoi diagrams can be constructed in linear time.

Classical variants of Voronoi diagrams such as higher-order Voronoi diagrams for sites other than points, had surprisingly been ignored in the literature of

Research supported in part by the Swiss National Science Foundation, project 20GG21-134355, under the ESF EUROCORES program EuroGIGA/VORONOI.

K. Elbassioni and K. Makino (Eds.): ISAAC 2015, LNCS 9472, pp. 404–414, 2015.
DOI: 10.1007/978-3-662-48971-0_35

computational geometry until recently [4,13]. Given a set S of n simple geometric objects in the plane, called sites, the *order-k Voronoi diagram* of S is a partitioning of the plane into regions such that every point within a region has the same k nearest sites. For $k = 1$, this is the *nearest-neighbor Voronoi diagram* and for $k = n - 1$ it is the *farthest-site Voronoi diagram* of S. Despite similarities, these diagrams for non-point sites, e.g., line segments, illustrate fundamental structural differences from their counterparts for points, such as the presence of disconnected regions (see also [2,6,10]). This had been a gap in the computational geometry literature, until recently, as segment Voronoi diagrams are fundamental to problems involving proximity among polygonal objects. This paper contributes further in closing this gap. For more information on Voronoi diagrams see the book of Aurenhammer et al. [3]. For application examples of higher order segment Voronoi diagrams see, e.g., [11] and references therein.

In this paper we give linear-time algorithms (expected and deterministic) for constructing tree-like Voronoi diagrams with disconnected regions, after the sequence of their faces within an enclosing boundary (or at infinity) is known. We focus on the farthest-segment Voronoi diagram, however, the same techniques are applicable to constructing the order-$(k+1)$ subdivision within a given order-k segment Voronoi region, and updating in linear time the nearest-neighbor segment Voronoi diagram after the deletion of one site. Interestingly, the latter two problems require computing initially two different tree-like diagrams. A major difference from the respective problems for points is that the sequence of faces along the relevant enclosing boundary forms a Davenport-Schinzel sequence of order at least two,[1] in contrast to the case of points, where no repetition can exist. Repetition introduces several complications, including the fact that the sequence of Voronoi faces along the relevant boundary for a subset of the original segments, $S' \subset S$, is not a subsequence of the respective sequence for S. In addition, such a subsequence may not even correspond to a Voronoi diagram. Thus, the intermediate diagrams computed by our algorithms are interesting on their own right. They have the structural properties of the relevant segment Voronoi diagram, however, they do not correspond to such a diagram nor are they instances of abstract Voronoi diagrams.

The purpose of this paper is to extend the paradigm of the existing linear constructions for tree-structured diagrams beyond the case of points in convex position [1]. Our goal is to generalize fundamental techniques known for points to more general objects so that the computation of their basic diagrams can be unified, despite their structural differences. As a byproduct we also improve the time complexity of the basic iterative approach to construct the order-k segment Voronoi diagram to $O(k^2 n + n \log n)$ from the standard $O(k^2 n \log n)$ [13], and also updating a nearest neighbor diagram after deletion of one site in time proportional to the number of updates performed in the diagram.

[1] Order-3 for the farthest-segment Voronoi diagram [2,12], order-4 for the order-k segment Voronoi diagram (easy to derive from [13]), order-2 for disjoint segments or the corresponding abstract Voronoi diagrams [10,13].

2 Preliminaries and Definitions

Let S be a set of arbitrary line segments in $\mathbb{R}^2$; segments in S may intersect or touch at a single point. The distance between a point q and a line segment s_i is $d(q, s_i) = \min\{d(q, y) \mid y \in s_i\}$, where $d(q, y)$ denotes the ordinary distance between two points q, y in the L_2 (or the L_p) metric. The bisector of two segments $s_i, s_j \in S$ is $b(s_i, s_j) = \{x \in \mathbb{R}^2 \mid d(x, s_i) = d(x, s_j)\}$. For disjoint segments, $b(s_i, s_j)$ is an unbounded curve that consists of a constant number of pieces, where each piece is a portion of an elementary bisector between the endpoints and open portions of s_i, s_j. If two segments intersect at point p, their bisector consists of two such curves intersecting at p.

The farthest Voronoi region of a segment s_i is $\mathit{freg}(s_i) = \{x \in \mathbb{R}^2 \mid d(x, s_i) > d(x, s_j), 1 \leq j \leq n, j \neq i\}$. For disjoint line segments or line segments that intersect but do not touch at endpoints, the order-k Voronoi region of a set H, where $H \subset S$, $|H| = k$, and $1 \leq k \leq n-1$, is $k\text{-}\mathit{reg}(H) = \{x \mid \forall s \in H, \forall t \in S \setminus H\ d(x, s) < d(x, t)\}$. For an extension of this definition to line segments forming a *planar straight-line graph*, see [13]. Note, for $k = n-1$, $\mathit{freg}(s_i) = k\text{-}\mathit{reg}(S \setminus \{s_i\})$. The (non-empty) farthest (resp., order-k) Voronoi regions of the segments in S, together with their bounding edges and vertices, define a partition of the plane, called the *farthest-segment Voronoi diagram*, denoted FVD(S), see Fig. 1(a) (resp., *order-k Voronoi diagram*). Any maximally connected subset of a Voronoi region is called a *face*.

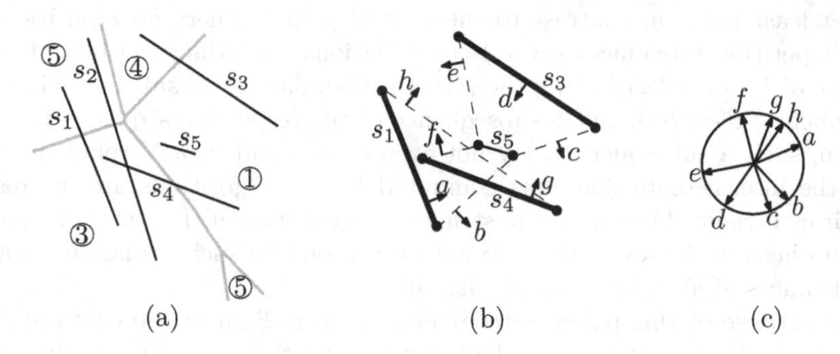

Fig. 1. [12] (a) FVD(S), $S = \{s_1, \ldots, s_5\}$; (b) its farthest hull; (c) Gmap(S)

A farthest Voronoi region $\mathit{freg}(s_i)$ is non-empty and unbounded in direction ϕ if and only if there exists an open halfplane, normal to ϕ, which intersects all segments in S but s_i [2]. The line ℓ, normal to ϕ, bounding such a halfplane, is called a *supporting line*. The direction ϕ (normal to ℓ) is referred to as the *hull direction* of ℓ and it is denoted by $\nu(\ell)$. An unbounded Voronoi edge separating $\mathit{freg}(s_i)$ and $\mathit{freg}(s_j)$ is a portion of $b(p, q)$, where p, q are endpoints of s_i and s_j, such that the line through $\overline{pq}$ induces an open halfplane that intersects all segments in S, except s_i, s_j (and possibly except additional segments incident to p, q). Segment $\overline{pq}$ is called a *supporting segment*; the direction normal to it pointing to the inside of this halfplane is denoted by $\nu(\overline{pq})$ and is called the *hull direction* of $\overline{pq}$. A segment $s_i \in S$ such that the line ℓ through s_i is supporting, is called a *hull segment*; its *hull direction* is $\nu(s_i) = \nu(\ell)$, normal to ℓ. The closed polygonal line obtained by following the supporting and hull segments in the angular order of their hull directions is called the *farthest hull.* Figures 1(a) and (b) illustrate a farthest-segment Voronoi diagram and its hull respectively. In Fig. 1(b), supporting segments are shown in dashed lines, and hull segments are shown in bold. Arrows indicate the hull directions of all supporting and hull segments. For more information see [12].

The Gaussian map of FVD(S), denoted Gmap(S), (see Fig. 1(c)) provides a correspondence between the faces of FVD(S) and a *circle of directions* K [12]. K can be assumed to be a unit circle, where each point x on K corresponds to a direction as indicated by the radius of K at x. Each Voronoi face is mapped to an arc on K, which represents the set of directions along which the face is unbounded. An arc is delimited by two consecutive hull directions of supporting segments. The Gmap(S) can be viewed as a cyclic sequence of consecutive arcs on K, where each arc corresponds to one face of FVD(S). Two neighboring arcs α, γ are separated by the hull direction $\nu(\alpha, \gamma)$ of a supporting segment $\overline{pq}$ ($\nu(\alpha, \gamma) = \nu(\overline{pq})$); $\nu(\alpha, \gamma)$ is the direction towards infinity of the relevant portion of bisector $b(p, q)$. The arc of a hull segment is called a *segment arc* and consists of two sub-arcs separated by the hull direction ν of the segment, where each sub-arc corresponds to an endpoint of the hull segment. An arc that corresponds to a single endpoint of a segment is called a *single-vertex* arc. The Gmap(S) can be computed in $O(n \log n)$ time (or output-sensitive $O(n \log h)$ time, where $h = |\text{Gmap}(S)|$) [12].

The standard point-line duality transformation T offers a correspondence between the faces of FVD(S) and envelopes of *wedges* [2]. A segment $s_i = uv$ corresponds to a *lower wedge*, defined by the lower envelope of $T(u)$ and $T(v)$ (see, e.g., Fig. 5), and to an *upper wedge* defined as the area above the upper envelope of $T(u)$, $T(v)$. Let E (resp., E') be the boundary of the union of the lower (resp., upper) wedges. The faces of FVD(S) correspond exactly to the edges of E and E' [2]. Let the upper and lower Gmap be the portion of Gmap(S) above and below the horizontal diameter of K respectively.

There is a clear correspondence between E (resp., E') and the upper (resp., lower) Gmap: the vertices of E are exactly the hull directions of supporting segments on the upper Gmap and the apexes of wedges in E are exactly the hull directions of hull segments [12]. In fact, any x-monotone path π in the arrangement of upper (resp., lower) wedges can be transformed into a sequence of arcs in the portion of K above (resp., below) its horizontal diameter. Each edge of π, portion of $T(u)$, corresponds to an arc on K for u, and each vertex of π, which is an intersection point $T(u) \cap T(v)$, corresponds to the hull direction $\nu(\overline{uv})$ of the supporting segment $\overline{uv}$.

Throughout this paper, given an arc α, let s_α denote the segment in S that induces α.

3 The Farthest Voronoi Diagram of a Sequence

Let G be a sequence of arcs on the circle of directions K, corresponding to a pair of x-monotone paths in the dual space, one in the arrangement of upper (resp. lower) wedges. No arcs in the sequence can overlap and no gaps are allowed. We call G an *arc sequence*. Consecutive arcs of the same segment in G are assumed unified into a single maximal arc.

In the following we define the farthest Voronoi diagram of such an arc sequence G, FVD(G). For G = Gmap(S), FVD(G) = FVD(S). The diagrams of such sequences appear as intermediate diagrams in the process of computing FVD(S), however, they do not correspond to any type of segment Voronoi diagram. We first define such a diagram and then present an arc deletion and arc insertion operation, which constitute the basis for our algorithms.

Given an arc $\alpha \in G$ and a point $x \in \mathbb{R}^2$, $x \notin s_\alpha$, let $r(x, s_\alpha)$ denote the ray emanating from x in the direction $\overrightarrow{px}$, where p is the point in s_α closest to x (see Fig. 2). We say that x is *attainable* from α if the direction of $r(x, s_\alpha)$ is contained in α. A point x in the interior of s_α is attainable from α if $\nu(s_\alpha)$ is in α (i.e., if α is a segment arc). An endpoint of s_α is attainable from all its corresponding arcs (see Sect. 2).

Let $d(x, \alpha) = d(x, s_\alpha)$, if x is attainable from α, and let $d(x, \alpha) = -\infty$, otherwise. The locus of points attainable from arc α is called the *attainable region* of α, $R(\alpha)$. Figure 2 illustrates the attainable regions of arcs α_1, α_2, and β, shaded. Intuitively, an arc α *exists* only for points within its attainable region (i.e., α is relevant exclusively within $R(\alpha)$ and it should not be considered outside).

Remark 1. For arcs $\alpha_1, \alpha_2 \in G$ of the same segment s_α, $R(\alpha_1) \cap R(\alpha_2) \setminus \{s_\alpha\} = \emptyset$.

Given two arcs α, β ($s_\alpha \neq s_\beta$) we define their *arc bisector* by $b(\alpha, \beta) = b(s_\alpha, s_\beta) \cap R(\alpha) \cap R(\beta)$. If $s_\alpha = s_\beta$ and α, β are consecutive, then $b(\alpha, \beta) = R(\alpha) \cap R(\beta)$ is called the *artificial bisector* of α, β. The farthest Voronoi region of an arc α is now defined in the ordinary way

$$freg(\alpha) = \{x \in \mathbb{R}^2 \mid d(x, \alpha) > d(x, \gamma), \forall \text{ arc } \gamma \in G, \gamma \neq \alpha\}.$$

The subdivision of the plane derived by the farthest regions of all arcs in G and their boundaries, is called the *farthest Voronoi diagram* of G, denoted FVD(G). The closure of $freg(\alpha)$ is denoted by $\overline{freg}(\alpha)$.

Definition 1. *Let $\mathcal{T}(G) = \mathbb{R}^2 \setminus \cup_{\alpha \in G} \overline{freg}(\alpha)$. If all edges of $\mathcal{T}(G)$ are portions of arc bisectors, then G, $\mathcal{T}(G)$, and* FVD(G) *are all called* proper.

For a proper sequence, $\mathcal{T}(G)$ is simply the graph structure of FVD(G). The diagrams and sequences produced by our algorithms are always proper. Note, however, that for an arbitrary arc sequence, $\mathcal{T}(G)$ may contain boundaries of attainable regions and even two-dimensional regions. Figure 3(a) illustrates FVD(G) for a proper arc sequence G, which consists of three maximal arcs of segments s_1, s_4, and s_5 and is derived from Gmap(S) of Fig. 1. Ray r indicates an *artificial bisector* between two consecutive arcs of s_5 (which have been unified into a single maximal arc for s_5). Figure 3(b) illustrates FVD(G''), where G'' contains an additional arc β of segment s_3 ($G'' = G \oplus \beta$).

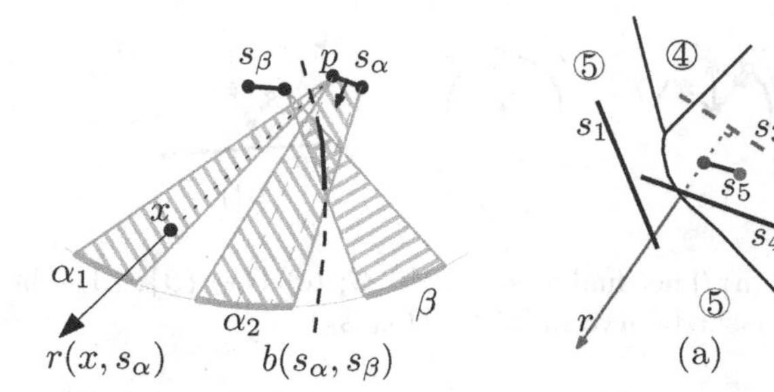

Fig. 2. Attainable regions $R(\alpha_1)$, $R(\alpha_2)$, $R(\beta)$

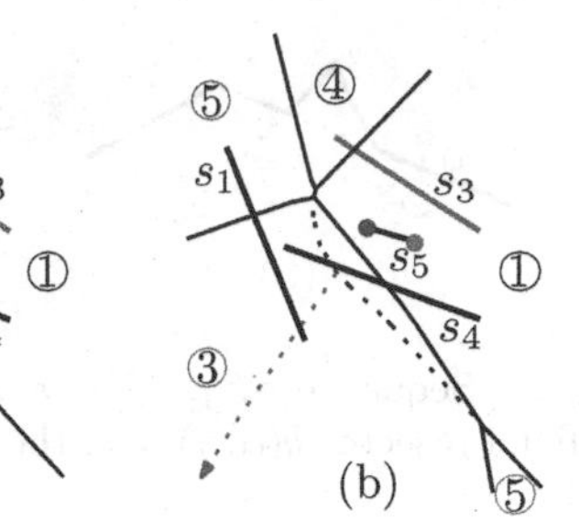

Fig. 3. FVD(G) for an arc sequence of Fig. 1; (a) FVD(G); (b) FVD(G''), $G'' = G \oplus \beta$.

Lemma 1. *For a proper arc sequence G, $\mathcal{T}(G)$ is a tree.*

Proof (Sketch). Since G is proper, all the edges of $\mathcal{T}(G)$ are portions of arc bisectors. Let x be a point on $\mathcal{T}(G)$ along arc bisector $b(\alpha, \beta)$. We first prove that the entire ray $r(x, s_\alpha)$ must be enclosed in $\overline{freg}(\alpha)$, i.e., regions are unbounded. This is because no arc bisector involving α can bound $r(x, s_\alpha)$ as we walk on it starting at x, unless an arc δ suddenly becomes attainable because $r(x, s_\alpha)$ intersects $R(\delta)$ at point z and $d(z, \delta) > d(z, \alpha)$; but then $z \in \mathcal{T}(G)$ without being on an arc bisector, a contradiction. It remains to show that $\mathcal{T}(G)$ is connected. If $\mathcal{T}(G)$ contained two different components, there would be a face of a segment s_α inducing two non consecutive arcs in G, α_1 and α_2. But then *freg*(α_1) and *freg*(α_2) would be neighboring, contradicting Remark 1. □

An arc sequence G is called a *subsequence* of Gmap(S) if every arc of G entirely contains a corresponding arc of Gmap(S) induced by the same segment. The arcs in G are simply expanded versions of the arcs in Gmap(S). The arcs in Gmap(S) as well as their expanded versions in G are called *original arcs.* A sequence G' is called an *augmented subsequence* of Gmap(S) if G' contains at least one arc of Gmap(S) for every segment with an arc in G'. An augmented subsequence consists of *original* arcs, which are expanded versions of the arcs in Gmap(S), and *new* arcs, which do not correspond to arcs of Gmap(S). An augmented subsequence G', which has the same original arcs as G, is said to be *corresponding to G.* Note that in the dual space, G and G' no longer correspond to envelopes of wedges, but to x-monotone paths that contain portions of these envelopes. The intermediate sequences of diagrams produced by our algorithms are always augmented subsequences of Gmap(S).

3.1 Deletion and Insertion of Arcs

Throughout our algorithms we use a deletion and a re-insertion operation for original arcs in sequences derived from Gmap(S). The deletion operation produces subsequences of Gmap(S) that are not necessarily proper. As a result, the insertion operation introduces new arcs, creating augmented subsequences, which are always proper. Let $G \ominus \beta$ (resp., $G \oplus \beta$) denote the arc sequence derived from G after deleting from it (resp., inserting to it) arc β.

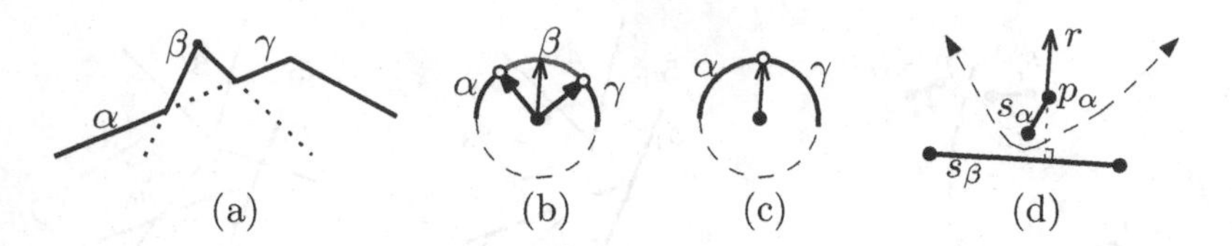

Fig. 4. Sequence $\alpha\beta\gamma$, $s_\alpha = s_\gamma$. (a) The dual wedges; (b) G; (c) $G \ominus \{\beta\}$; (d) The artificial bisector $b(\alpha, \gamma) = r$; the dashed curve indicates $b(s_\alpha, s_\beta)$.

Arc Deletion. A subsequence G is derived from Gmap(S) by deleting arcs. When an arc β is deleted from G, the neighboring arcs α and γ *expand* over β (see Fig. 4(a)–(c)). Either both α and γ expand (see Figs. 4 and 5(a) illustrating segments in the dual space) or one expands while the other shrinks (see Fig. 5(b)). During the expansion, α and γ may change from being a single-vertex arc to a segment arc. Since α and γ are original, they both remain present in $G\ominus\{\beta\}$, and their common endpoint becomes $\nu(\alpha, \gamma)$. Assuming $s_\alpha \neq s_\gamma$, $\nu(\alpha, \gamma)$ corresponds to bisector $b(\alpha, \gamma)$ as obtained from $b(s_\alpha, s_\gamma)$. If $s_\alpha = s_\gamma$, we let α and γ expand until they reach $\nu(s_\beta)$, i.e., $\nu(\alpha, \gamma) = \nu(s_\beta)$. If $s_\alpha = s_\beta$ then α expands to cover the entire β and $\nu(\alpha, \gamma) = \nu(\beta, \gamma)$.

Remark 2. The artificial bisector $b(\alpha, \gamma)$ (for $s_\alpha = s_\gamma$) is (or contains) a ray perpendicular to s_β, emanating from the relevant endpoint of s_α and extending away from s_β (see Fig. 4(d)).

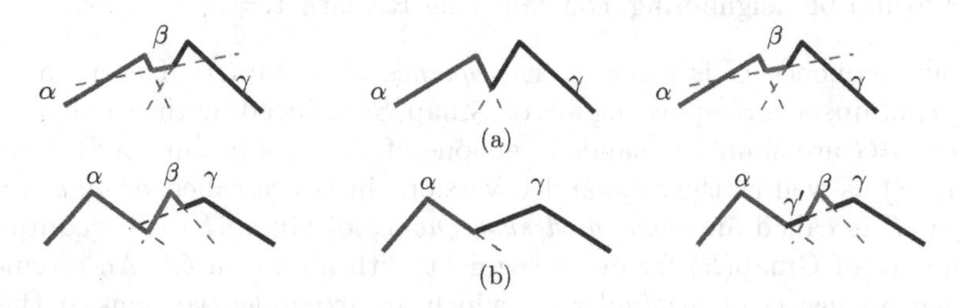

Fig. 5. Deleting and re-inserting β in sequence $\alpha\beta\gamma$. (a) α and γ enlarge; (b) γ enlarges, α shrinks. From left to right: the initial sequence; after deleting β; after re-inserting β.

Arc Insertion. Let G' be a proper augmented subsequence of Gmap(S) and let β be an original arc, $\beta \notin G'$. Let α, γ be two consecutive original arcs in G', such that β is between α, γ in Gmap(S). A number of new arcs may lie between α, γ in G'. To insert β in G' there are several cases to consider. The insertion of arc β in G' corresponds to inserting *freg*(β) in FVD(G'). Figure 5 illustrates in dual space the deletion and re-insertion of an arc β in a sequence $\alpha\beta\gamma$.

Basic cases are as follows (assuming for simplicity that α, γ are consecutive in G'): (1) s_α, s_β, and s_γ are all distinct, and $\nu(\alpha, \gamma)$ is in β. This is the standard case, resulting in $\alpha\beta\gamma$, see Fig. 5(a). (2) $s_\alpha = s_\gamma$. Then β is inserted over $\nu(s_\beta) = \nu(\alpha, \gamma)$, resulting in $\alpha\beta\gamma$, and *freg*(β) is inserted over the artificial bisector $b(\alpha, \gamma)$ in FVD(G'), see Fig. 3. (3) Arc γ (equiv. α), as it appears expanded in G', entirely contains β, see Fig. 5(b) (note that α had shrunk during the deletion of β). Then the insertion of β splits γ in two arcs resulting in $\alpha\gamma'\beta\gamma$, where γ' is a new arc. In FVD(G'), *freg*(β) splits *freg*(γ) into *freg*(γ) and *freg*(γ'). (4) $s_\alpha = s_\beta$ (equiv. $s_\beta = s_\gamma$). Then α is split by $\nu(\alpha, \beta)$ and one part becomes β. Note that $\nu(\alpha, \beta)$ is determined when α and β became consecutive in a deletion operation, and that α, β cannot be neighbors in Gmap(S).

If α and γ are not consecutive in G', a number of new arcs may be traced to find the actual entry point for β between α and γ. The insertion of β may delete a series of such consecutive new arcs. Assuming that G' is proper, it is not hard to show that $G'' = G' \oplus \beta$ is also a proper augmented subsequence of Gmap(S).

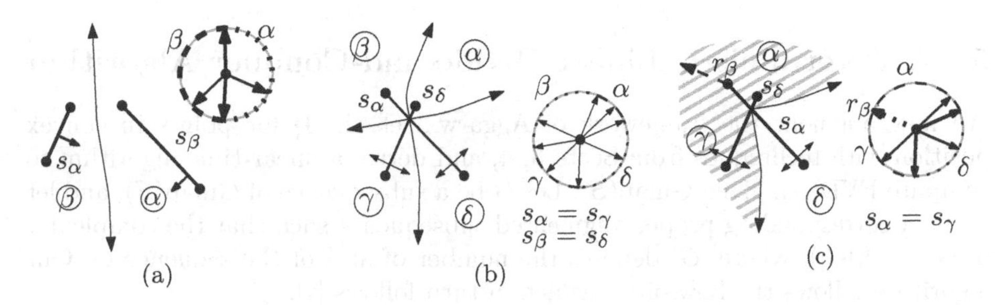

Fig. 6. FVD(G) and G. (a) G = Gmap(S) for $S = \{s_\alpha, s_\beta\}$; (b) G = Gmap(S) for $S = \{s_\alpha, s_\delta\}$; (c) $G = \alpha\gamma\delta$ = Gmap(S) $\ominus \beta$ where $S = \{s_\alpha, s_\delta\}$

Note that arc sequences defined by two segments are always proper. Figure 6 illustrates such sequences and their Voronoi diagrams. Figures 6(a) and (b) show FVD(S) and Gmap(S) for two disjoint and intersecting segments respectively. Figure 6(c) illustrates FVD(G) for G = Gmap(S) $\ominus \beta$, where S is the same as in Fig. 6(b). In the latter figure, arcs α and γ ($s_\alpha = s_\gamma$) become neighbors inducing one maximal arc $\alpha\gamma$; region *freg*($\alpha\gamma$) is shown shaded; it is split into *freg*(α) and *freg*(γ) by the artificial bisector $b(\alpha, \gamma) = r_\beta \cup s_\alpha$.

4 A Randomized Linear Construction

We sketch an expected linear-time algorithm to compute FVD(S), given Gmap(S). It is inspired by the simple two-phase randomized approach of [7] for points in convex position and uses the concepts of Sect. 3. Let $\alpha_1, \alpha_2, \ldots, \alpha_h$ be a random permutation of arcs in Gmap(S), and let $A_i = \{\alpha_1, \alpha_2, \ldots, \alpha_i\}$, $1 \leq i \leq h$, be the set of the first i arcs in this order. Let t be the largest index such that $\alpha_1, \ldots, \alpha_t$ consists of arcs of only two segments that form exactly two maximal arcs.

The algorithm proceeds in two phases. Phase 1 computes the subsequence G_i, $t \leq i < h$, where G_h = Gmap(S), and G_i is obtained from G_{i+1} by deleting arc α_{i+1} as described in Sect. 3. The two neighbors of α_{i+1} in G_{i+1} are recorded as a tentative re-entry point for α_{i+1} during phase 2. Note that both neighbors may correspond to the same segment or the segment of one neighbor may coincide with $s_{\alpha_{i+1}}$. In phase 2, the algorithm computes incrementally G'_i and FVD(G'_i), for $t < i \leq h$, starting with FVD(G'_t), $G'_t = G_t$. G'_t is proper as it consists of exactly two maximal arcs. G'_{i+1} is obtained from G'_i by inserting back α_{i+1} ($G'_{i+1} = G'_i \oplus \alpha_{i+1}$). During the re-entry of α_{i+1} a *new* arc may be created, thus, $G'_{i+1} \neq G_{i+1}$. The entry point for α_{i+1} is either an unbounded bisector (regular or artificial) or an arc σ that entirely contains α_{i+1}. In the latter case, a new arc is created. At the end of phase 2 we obtain FVD(G'_h) = FVD(S) ($G'_h = G_h$).

In the full paper we prove: (1) the complexity of G'_i is $O(i)$ despite the new arcs; (2) the expected number of new arcs traced in a step of phase 2 is constant. Then using backwards analysis we can derive the following theorem.

Theorem 1. *Given* Gmap(S), *the* FVD(S) *can be computed in expected* $O(h)$ *time, where* h *is the complexity of* FVD(S).

5 A Deterministic Linear Divide-and-Conquer Algorithm

We now augment the framework of Aggarwal et al. [1] for points in convex position with techniques from Sects. 3, 4, and derive a linear-time algorithm to compute FVD(S), given Gmap(S). Let G be a subsequence of Gmap(S), and let G' be a corresponding proper augmented subsequence such that the complexity of G' is $O(|G|)$, where $|G|$ denotes the number of arcs of the sequence G. Our algorithm follows the flow of [9], which in turn follows [1].

1. Unite consecutive arcs of the same segment in G into single maximal arcs.
2. Color each arc of G *red* or *blue* by applying the following two rules:
   (a) For each 5-tuple F of consecutive arcs $\alpha\beta\gamma\delta\epsilon$ in G, compute FVD(F') as follows: start with the sequence $\gamma\delta$, and consecutively insert the arcs β, ϵ, α (in this order) resulting in FVD(F'). (F' is a possibly augmented version of F.) In FVD(F'), if *freg*(γ) does not neighbor any region of segments s_α and s_ϵ, color γ red; else color γ blue.
   (b) For each series of consecutive blue arcs, color red every other arc, except the last one.
3. Let B (blue) be the sequence obtained from G by deleting all the red arcs. Recursively compute FVD(B'). (B' is a possibly augmented version of B.)
4. Partition the red arcs into *crimson* and *garnet*: Re-color as *crimson* at least a constant fraction of the red arcs, such that for any two crimson arcs, if they were inserted in FVD(B'), their Voronoi regions would not touch.
5. Insert the crimson arcs one by one in FVD(B') resulting in FVD(V').
6. Let Gr (garnet) be the sequence obtained from G by deleting all blue and crimson arcs. Recursively compute FVD(Gr').
7. Merge FVD(V') and FVD(Gr') into FVD(G') so that $|G'|$ is $O(|G|)$.
8. For any arcs united in Step 1, subdivide their regions in FVD(G') into finer parts by inserting the corresponding artificial bisectors.

The recursion ends when the number of maximal arcs in G is at most five. Then FVD(G') can be directly computed in $O(1)$ time and also enhanced as indicated in Step 8. If all arcs in G are of the same segment, no diagram is generated but instead G is returned as a list of arcs. In this case, in Step 7, we obtain FVD(G') by inserting this list of arcs in FVD(V') one by one.

Step 2. Rules 2a and 2b guarantee that no two consecutive arcs in G are red and no three consecutive arcs in G are blue. The insertion order in Rule 2a guarantees that γ neighbors at most one new arc.

Step 4. To choose the crimson arcs we apply the *combinatorial lemma* of [1] on (a modified) $\mathcal{T}(B')$. The lemma states that for a binary tree T with n leaves embedded in $\mathbb{R}^2$, if each leaf of T is associated with a subtree of T and if for any two successive leaves these subtrees are disjoint, then in $O(n)$ time we can choose a set of leaves, whose number is at least a constant fraction of n and whose subtrees are pairwise disjoint. We associate each red arc β in G with a unique leaf of $\mathcal{T}(B')$, which would be the entry point for β in FVD(B'). If the insertion of β splits an arc of B' in two, then we also add an artificial bisector to $\mathcal{T}(B')$ to serve as an entry point for β. The leaf in $\mathcal{T}(B')$ associated with β is in turn associated with the incident subtree of $\mathcal{T}(B')$, which would be intersected by *freg*(β), if β were inserted in FVD(B'). The modified $\mathcal{T}(B')$ satisfies the requirements of the combinatorial lemma, and has complexity proportional to $|B'|$ plus the number of red arcs $|R|$.

Step 7. We obtain G' and FVD(G') by merging FVD(V') and FVD(Gr'). To keep the complexity of G' within $O(|G|)$, we merge the two diagrams while discarding parts that are guaranteed to contain no original arcs. Merging is done in two steps: (1) identify starting points for the *merge curves* between the two diagrams, and (2) trace the *merge curves*. Here, we identify starting points only for the merge curves that are related to original arcs. Skipping a merge curve has the effect of discarding the portion of one diagram that is bounded by it. This can be safely done because any portions of the diagram that are associated with only new arcs can not appear in FVD(S). G' contains all the original arcs of V' and Gr'; furthermore, $|G'|$ is $O(|V'| + |Gr'|)$. Since G' contains all the original arcs of G, it is an augmented subsequence of Gmap(S) corresponding to G. It is not hard to prove that G' is proper.

G' is an augmented subsequence of Gmap(S) corresponding to G, and the recursive algorithm starts with $G =$ Gmap(S). Thus, at the end of the algorithm, the resulting arc sequence must be $G' =$ Gmap(S) (easy to see in dual space).

Lemma 2. *$|G'|$ is $O(|G|)$.*

Proof. Let $m = |G|$ and $S(m) = |G'|$. Since Step 4 is performed by applying the combinatorial lemma of [1], $|Gr| \leq q|R|$, where $0 < q < 1$ and $|R|$ is the number of red arcs ($|R| = |G| - |B|$). Thus, (following [1,9]) there exist positive constants q_1 and q_2, $q_1 + q_2 < 1$, such that $|B| \leq q_1|G|$ and $|Gr| \leq q_2|G|$. At Step 4, at most one new arc is generated for every crimson arc inserted in B', thus, $|V'| = S(q_1m) + O(m)$. At Step 7, $|G'| \leq |V'| + |Gr'| + O(m)$. Thus, $|G'| \leq S(q_1m) + S(q_2m) + O(m)$. Hense, $S(m) = O(m)$. □

Since the size of the augmented subsequences is always kept bounded, the time complexity can be analyzed similarly to [1]. We conclude:

Theorem 2. *Given* Gmap(S), *the* FVD(S) *can be computed in $O(h)$ time, where h is the combinatorial complexity of* FVD(S).

Concluding Remarks

Theorems 1 and 2 apply also to computing the order-(k+1) subdivision within an order-k Voronoi region in time proportional to the complexity of the region's boundary. It also applies to updating a nearest-neighbor segment Voronoi diagram after the deletion of one segment in time proportional to the number of updates in the diagram. In this paper we considered line segments, however, the presented techniques are not specific to them. For example, the constructions can be easily adapted for the respective farthest abstract Voronoi diagram (to be described in the full paper). Note that the farthest abstract Voronoi diagram can be constructed in expected $O(n \log n)$ time by a randomized incremental construction [10], which is not related to the randomized linear-time approach in this paper.

References

1. Aggarwal, A., Guibas, L., Saxe, J., Shor, P.: A linear-time algorithm for computing the Voronoi diagram of a convex polygon. Discrete Comput. Geom. **4**, 591–604 (1989)
2. Aurenhammer, F., Drysdale, R., Krasser, H.: Farthest line segment Voronoi diagrams. Inform. Process. Lett. **100**, 220–225 (2006)
3. Aurenhammer, F., Klein, R., Lee, D.T.: Voronoi Diagrams and Delaunay Triangulations. World Scientific, Singapore (2013)
4. Bohler, C., Cheilaris, P., Klein, R., Liu, C.-H., Papadopoulou, E., Zavershynskyi, M.: On the complexity of higher order abstract Voronoi diagrams. In: Kwiatkowska, M., Peleg, D., Fomin, F.V., Freivalds, R.U. (eds.) ICALP 2013, Part I. LNCS, vol. 7965, pp. 208–219. Springer, Heidelberg (2013)
5. Bohler, C., Klein, R., Liu, C.: Forest-like abstract Voronoi diagrams in linear time. In: Proceedings of the 26th CCCG (2014)
6. Cheong, O., Everett, H., Glisse, M., Gudmundsson, J., Hornus, S., Lazard, S., Lee, M., Na, H.: Farthest-polygon Voronoi diagrams. Comput. Geom. **44**(4), 234–247 (2011)
7. Chew, L.P.: Building Voronoi diagrams for convex polygons in linear expected time. Technical report, Dartmouth College, Hanover, USA (1990)
8. Chin, F., Snoeyink, J., Wang, C.A.: Finding the medial axis of a simple polygon in linear time. Discrete Comput. Geom. **21**(3), 405–420 (1999)
9. Klein, R., Lingas, A.: Hamiltonian abstract Voronoi diagrams in linear time. In: Du, D.-Z., Zhang, X.-S. (eds.) ISAAC 1994. LNCS, vol. 834, pp. 11–19. Springer, Heidelberg (1994)
10. Mehlhorn, K., Meiser, S., Rasch, R.: Furthest site abstract Voronoi diagrams. Int. J. Comput. Geom. Ap. **11**(6), 583–616 (2001)
11. Papadopoulou, E.: Net-aware critical area extraction for opens in VLSI circuits via higher-order Voronoi diagrams. IEEE T. Comput. Aid. D. **30**(5), 704–716 (2011)
12. Papadopoulou, E., Dey, S.K.: On the farthest line-segment Voronoi diagram. Int. J. Comput. Geom. Ap. **23**(6), 443–459 (2013)
13. Papadopoulou, E., Zavershynskyi, M.: The higher-order Voronoi diagram of line segments. Algorithmica (2014). doi:10.1007/s00453-014-9950-0

Unfolding Orthogonal Polyhedra with Linear Refinement

Yi-Jun Chang[1] and Hsu-Chun Yen[2(✉)]

[1] Department of EECS, University of Michigan, Ann Arbor, MI 48109, USA
[2] Department of Electrical Engineering, National Taiwan University, Taipei 10617, Taiwan
yen@cc.ee.ntu.edu.tw

Abstract. An unfolding of a polyhedron is a single connected planar piece without overlap resulting from cutting and flattening the surface of the polyhedron. Even for orthogonal polyhedra, it is known that *edge-unfolding*, i.e., cuts are performed only along the edges of a polyhedron, is not sufficient to guarantee a successful unfolding in general. However, if additional cuts parallel to polyhedron edges are allowed, it has been shown that every orthogonal polyhedron of genus zero admits a *grid-unfolding* with *quadratic refinement*. Using a new unfolding technique developed in this paper, we improve upon the previous result by showing that *linear refinement* suffices. Our approach not only requires fewer cuts but is also much simpler.

1 Introduction

The study of folding and unfolding in computational geometry can be traced back to several hundred years ago [4], reflecting the fact that it is natural for human to construct a polyhedron by folding from its unfolding. The interested reader is referred to [3] for a nice introduction to this field. Specifically, an *unfolding* of a polyhedron is a single connected planar piece resulting from cutting and flattening its surface. Ideally, we hope to only cut the edges of a polyhedron when producing an unfolding. In reality, however, even for orthogonal polyhedra it is known that cutting along edges is not sufficient to guarantee an unfolding in general. Such a negative result gives rise to the so-called *grid-unfolding*, in which new edges (which are allowed to be cut) are added by intersecting the polyhedron with all coordinate planes passing through a vertex. Even with such a relaxation in unfolding, it remains a major open problem to decide whether every orthogonal polyhedron admits a grid-unfolding [5]. Up to this point, only a few restricted classes of orthogonal polyhedra, such as orthotubes, are known to admit grid-unfoldings [5].

Y.-J. Chang—Supported by NSF grant CCF-1514383.
H.-C. Yen—Supported in part by Ministry of Science and Technology, Taiwan, under grant MOST 103-2221-E-002-154-MY3.

K. Elbassioni and K. Makino (Eds.): ISAAC 2015, LNCS 9472, pp. 415–425, 2015.
DOI: 10.1007/978-3-662-48971-0_36

The notion of a *refinement* under grid-unfolding was proposed in [2], in which each rectangular face of a polyhedron under grid-unfolding is further refined to an $a \times b$ grid, where all grid lines are allowed to be cut. Such an unfolding style is called an $(a \times b)$-*grid-unfolding.* Using the *epsilon-unfolding* algorithm [2], all orthogonal polyhedra of genus zero can be grid-unfolded with exponential refinement, i.e., $(O(2^n) \times O(2^n))$-grid-unfolded. Just recently, an $(O(n^2) \times O(n^2))$-grid-unfolding algorithm (called *delta-unfolding*) was proposed [1], adapting the heavy-path decomposition technique for balancing trees in data structure design in conjunction with the idea behind the strategy of [2].

In this paper, we give a new grid-unfolding algorithm that only requires linear refinement (i.e., $(O(n) \times O(n))$-grid-unfolding), yielding an improvement over the delta-unfolding method which needs quadratic refinement. Our algorithm is based on some new ideas which differ from [1,2]:

- Instead of insisting on keeping a partial unfolding in a single piece throughout the procedure, we allow the presence of many (up to $O(n)$) pieces during the process, and they are linked together in the last step. This prevents the complicated back and forth spiraling for each component.
- The strategy behind [1,2] relies on an unfolding tree based on the adjacency relationship among components of a polyhedron. We introduce a new type of an unfolding tree, which attempts to establish better "bridges" between (not necessarily adjacent) components of a polyhedron.
- In our method, we first identify and unfold the so-called *backbone* of a polyhedron into a y-convex polygon on the plane; and the remaining faces are stitched to the unfolded backbone afterward.

2 Preliminaries

An *orthogonal polyhedron* is a polyhedron with all edges parallel to axes of the Cartesian coordinate system in 3D. Throughout the paper, we write O to denote an orthogonal polyhedron. We always assume that O is of genus zero. We let $Y_0, Y_1, \ldots$ be the planes orthogonal to the y-axis containing some vertex in O. These planes are ordered in a way that $y_a > y_b$ for any $a > b$, if we let $y = y_i$ be the plane for Y_i. We also call Y_i the *layer* i. The portion of O within the two layers Y_{i-1} and Y_i consists of some disjoint connected parts, each of which is called a *component* of layer i. Viewing in the $y+$ direction (i.e., increasing y coordinates), each component C (of layer i) has two *rims.* The *back rim* (resp., *front rim*) is the portion of the surface of O that is located in both C and Y_i (resp., C and Y_{i-1}).

The part of the surface of C not located in Y_{i-1} or Y_i consists of possibly several connected pieces, which are called *bands.* The outmost one that surrounds the entire C is called *protrusion*, denoted as *prot*(C). The other ones that surround holes of C are called *dents.* However, as observed in [2], when O is of genus zero, all the dents can be "popped out" to become protrusions of other components. As a result, we assume that there is no dent.

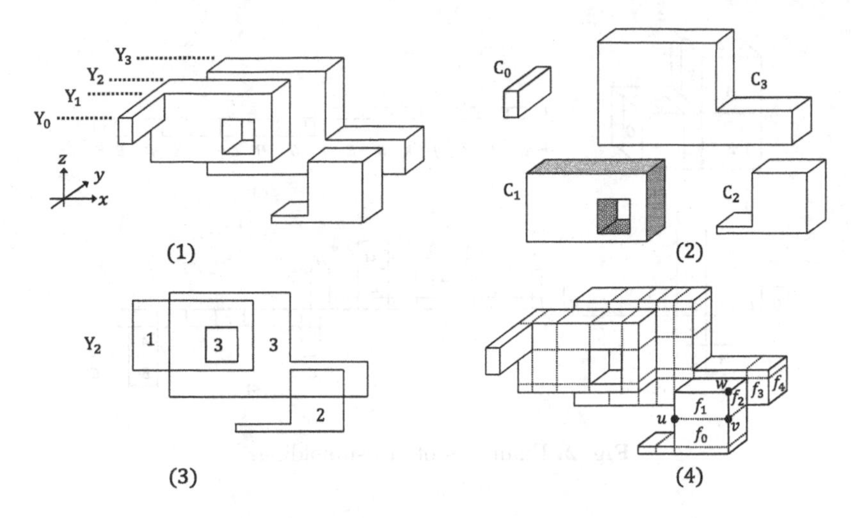

Fig. 1. Illustrations of some basic concepts.

Recall that an *edge-unfolding* [5] cuts a polyhedron along some (not necessarily all) of its edges. In the setting of a *grid-unfolding* [5], it is allowed to cut along additional edges formed by intersecting the polyhedron with coordinate planes passing through vertices of the polyhedron. Unless stated otherwise, a polyhedron is referred to one with such additional edges in our subsequent discussion. With this assumption in mind, it is not hard to see that all the faces are axis-parallel rectangles in a polyhedron (see Fig. 1(4)).

Let G be the graph formed by the vertices and the edges of a polyhedron. The set of vertices, edges, and faces in the polyhedron are denoted as $V(G), E(G)$, and $F(G)$, respectively. Let G^* be the dual graph of G in which each node corresponds to a face in the polyhedron such that $\{f_1, f_2\} \in E(G^*)$ if f_1 and f_2 are neighboring faces. As there is a natural isomorphism between $E(G)$ and $E(G^*)$, with a slight abuse of notation we simply write $\{v_1, v_2\} = \{f_1, f_2\}$, where $v_1, v_2 \in V(G)$ and $f_1, f_2 \in V(G^*)$, if $\{v_1, v_2\}$ is the edge shared between two rectangular faces f_1 and f_2. A set of faces $P = \{f_1, f_2, ..., f_k\}$ in G^* is called a *straight* path if faces in P can be unfolded to a rectangle that is an $1 \times k$ grid. We sometimes write $P = (f_1, f_2, ..., f_k)$ to denote the sequence of faces along a straight path.

An $(a \times b)$-refinement of a grid-unfolding (also called $(a \times b)$-grid-unfolding), where $a, b \in Z^+$, is to refine each face (which is a rectangle) into an $a \times b$ grid, and in the unfolding process, all grid lines are allowed to be cut [5].

Figure 1(1) is a polyhedron having layers Y_0, Y_1, Y_2 and Y_3. Figure 1(2) shows its four components. The dotted and the dashed portions of the surface of C_1 are its dent and protrusion, respectively. Figure 1(3) shows the portion of the surface of O in Y_2. The portion 1, 2, and 3 are the back rim of C_1, the back rim of C_2, and the front rim of C_3, respectively. Figure 1(4) shows the vertices,

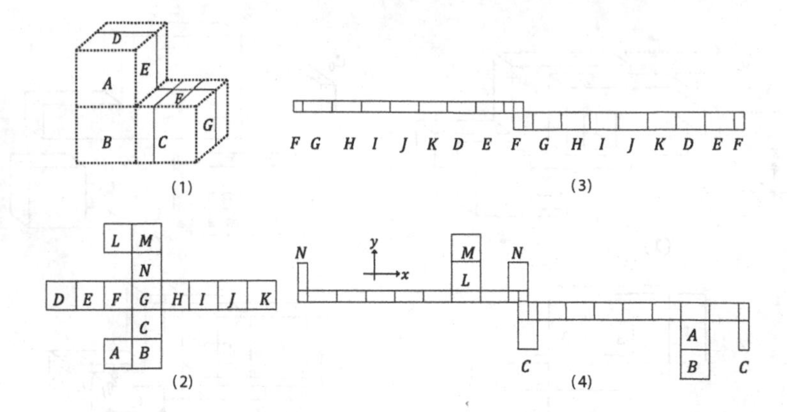

Fig. 2. Examples of grid-unfolding.

edges, and faces of the polyhedron. Note that (f_1, f_2, f_3, f_4) is a straight path, while (f_0, f_1, f_2, f_3) is not. We have $\{u, v\}, \{v, w\} \in E(G)$ and $\{v, w\} = \{f_1, f_2\}$.

The components of a polyhedron naturally form a tree structure [2]. We define the *component tree* T_0 to be the tree of components such that $\{C, C'\} \in E(T_0)$ iff there are some faces $f \in C$ and $f' \in C'$ such that $\{f, f'\} \in E(G^*)$. For the polyhedron in Fig. 1(1), its component tree is the path (C_0, C_1, C_3, C_2).

An *orthogonal polygon* is a polygon with edges parallel to either the x-axis or the y-axis of the Cartesian coordinate system in 2D. A polygon Q is called *y-convex* if the intersection of any straight line parallel to the y-axis and the interior of Q is a single connected line segment.

Figure 2(2) shows an (1×1)-grid unfolding for the polyhedron depicted in Fig. 2(1). The y-convex polygon in Fig. 2(4) is a (2×3)-grid unfolding resulting from cutting along the solid lines on the polyhedron in Fig. 2(1).

3 The Unfolding Tree

In this section we introduce a tree structure which is a key element behind our linear refinement unfolding algorithm. Before proceeding further, we require the definition of a *bridge*. A bridge between components C and C' is a straight path $P = (f_0, f_1, \dots, f_k)$ such that

- f_0 (resp., f_k) belongs to the protrusion of C (resp., C'), and f_0 and f_k are normal to the $z+$ or $z-$ directions.
- $f_1, \dots, f_{k-1}$ belong to the same layer, and hence, they are all normal to either $y+$ or $y-$.

A bridge is said to be in layer Y_i if $f_1, \dots, f_{k-1}$ above are in layer Y_i.

An *unfolding tree* T_U is a tree of components of the orthogonal polyhedron O returned by the following procedure. The procedure starts with $V(T_U)$ = the set of all components, and $E(T_U) = \emptyset$ initially. It adds an edge $e = \{C, C'\}$ to

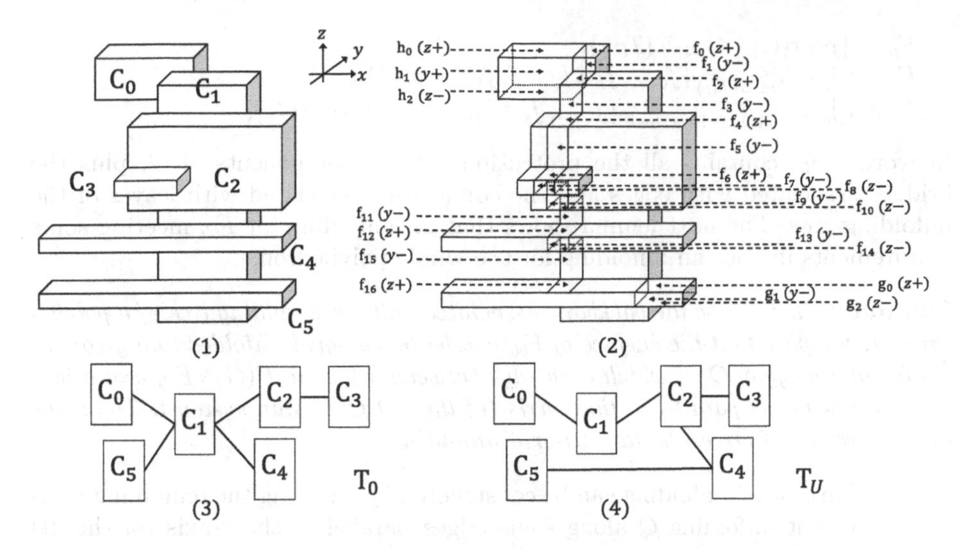

Fig. 3. Illustrations of an unfolding tree, bridges, and connectors. Here $f(d+)$ (resp., $f(d-)$) indicates that face f is normal to the $d+$ (resp., $d-$) direction, where $d \in \{x, y, z\}$.

$E(T_U)$ iteratively provided that there is a bridge between C and C', and adding e does not create a cycle. The procedure ends when no more edges can be added.

The bridge under consideration when adding edge $\{C, C'\}$ is denoted as $br(C, C')$ (also called the bridge associated with $\{C, C'\}$).

Using the bridges $br(C_0, C_1) = (f_0, f_1, f_2)$, $br(C_1, C_2) = (f_2, f_3, f_4)$, $br(C_2, C_3) = (f_4, f_5, f_6)$, $br(C_2, C_4) = (f_{10}, f_{11}, f_{12})$, $br(C_4, C_5) = (f_{14}, f_{15}, f_{16})$, the unfolding tree of Fig. 3(1) is shown in Fig. 3(4). Note that the unfolding tree, depending on the choices of the bridges, is not unique. We also note that the so-called unfolding tree in [2] is actually the component tree in our terminology (see T_0 in Fig. 3(3), for instance).

With respect to an unfolding tree, we have the following lemmas:

Lemma 1. *The graph T_U computed by the aforementioned procedure is a tree.*

Lemma 2. *For each leaf C of T_U, there must be one rim R_C of C such that all faces in the rim are not adjacent to any face in another component C' ($C \neq C'$).*

We associate each leaf C of an unfolding tree T_U with a *connector* (denoted as $con(C)$), which is a straight path $P = (f_0, f_1, \ldots, f_k)$ such that f_0 and f_k are normal to z+ or z− directions, and all of $f_1, \ldots, f_{k-1}$ belong to the rim R_C guaranteed by Lemma 2. In Fig. 3, we may set (h_0, h_1, h_2), (f_6, f_7, f_8), and (g_0, g_1, g_2) as the connector of C_0, C_3, and C_5, respectively. Note that h_1 (normal to the $y+$ direction) belongs to the back rim of component C_0.

Suppose T_U is an unfolding tree of an orthogonal polyhedron O, we consider the following set F_{bb} of faces (called the *backbone*) which plays a critical role in our subsequent unfolding process. $F_{bb} = F_p \cup F_b \cup F_c$, where

- $F_p = \{prot(C) \mid C \in V(T_U)\}$,
- $F_b = \bigcup_{(C,C') \in E(T_U)} \{f_0, .., f_k \mid (f_0, .., f_k) = br(C, C')\}$,
- $F_c = \bigcup_{C \in V(T_U), C \text{ is a leaf}} \{f_0, .., f_k \mid (f_0, .., f_k) = con(C)\}$,

In words, F_{bb} contains all the protrusions of the components of O, plus the bridges associated with edges and the connectors associated with leaves of the unfolding tree. The next lemma shows that an unfolding for F_{bb} meeting some requirements implies an unfolding for the whole polyhedron.

Lemma 3. *Let F_{bb} be the backbone associated with an unfolding tree of a polyhedron O. Suppose that the surface of F_{bb} can be $(a \times b)$-grid-unfolded to a y-convex orthogonal polygon Q in which each edge between a face in $F(G) \setminus F_{bb}$ and a face in a protrusion is parallel to the x-axis (of the 2D Cartesian system). Then, the entire surface of O can be $(a \times b)$-grid-unfolded.*

Proof. The desired unfolding can be constructed by stitching the remaining faces to the current unfolding Q along some edges parallel to the x-axis (of the 2D Cartesian system), just as how the unfolding in Fig. 2(4) is extended from the partial unfolding in Fig. 2(3). □

4 The Linear Refinement Unfolding Algorithm

In this section we present a grid-unfolding algorithm that only requires linear refinement for orthogonal polyhedra. Specifically, the required refinement is $(2|\text{leaves}(T_U)| \times 4|\text{leaves}(T_U)|)$. Note that $|\text{leaves}(T_U)|$ is $O(|V(G)|)$. In the subsequent discussion, Lemmas 1 and 2 are applied implicitly.

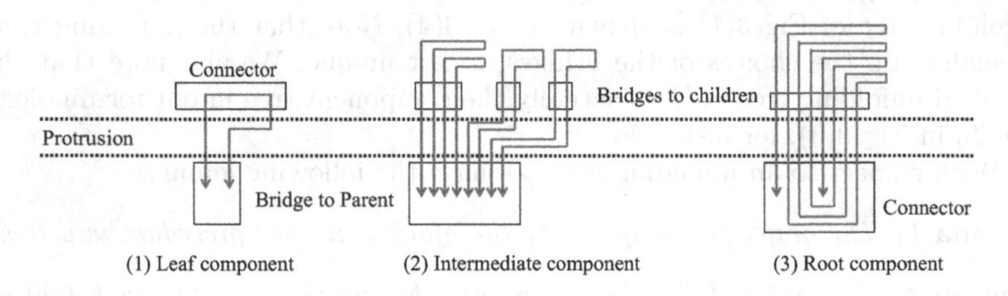

Fig. 4. An overview of the unfolding algorithm.

We designate a leaf of T_U as the root to make the tree directed. For each component $C \in V(T_U)$, we write T_C to denote the subtree rooted at C. Our approach is based on a bottom-up procedure operating on the unfolding tree T_U. A high-level overview of our algorithm is illustrated in Fig. 4, which involves three types of operations:

- For each leaf component, a strip is created by unfolding its connector and its protrusion (Fig. 4(1)).

- For each intermediate component, we gather all strips from its children subtrees, extend them to cover the entire protrusion, and pass them to its parent (Fig. 4(2)).
- Finally, at the root, $|\text{leaves}(T_U)|$ strips are concatenated to form a desired partial unfolding meeting the condition of Lemma 3 (Fig. 4(3)).

Note that bridges and connectors play important roles in our algorithm. Bridges allow us to extend strips from one component to another, and connectors allow us to create and concatenate the strips.

For each C which is not a root, let $F_{bb}(C)$ be the part of the backbone containing only the protrusions and connectors of components in T_C, and the bridges associated with each links in $E(T_C)$. Suppose $(f_0, f', ...)$ is the bridge linking C to its parent, where f_0 is in C. We define an invariant $\mathbb{I}$ as follows:

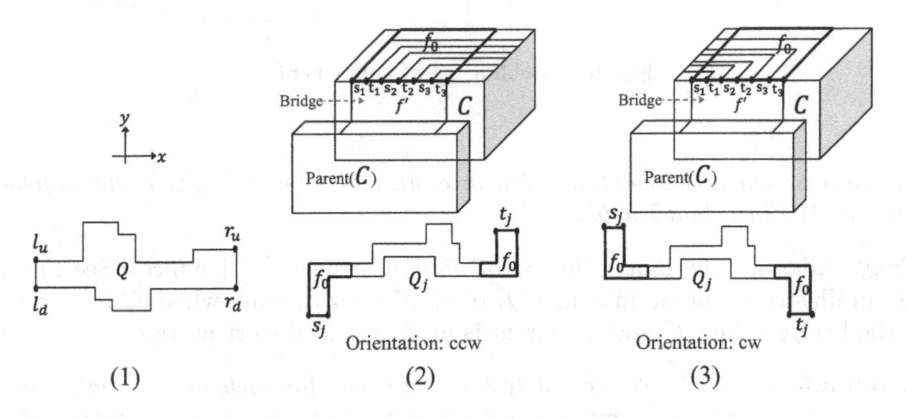

Fig. 5. Requirement for each Q_j.

Invariant $\mathbb{I}$. Given a component C and an orientation of either *clockwise* (*cw*, for short) or *counter-clockwise* (*ccw*, for short), let $r = |\text{leaves}(T_C)|$ be the number of leaf nodes in T_C. The surface of $F_{bb}(C)$ can be $(2r \times 4r)$-grid-unfolded into $Q_1, Q_2, \ldots, Q_r$ y-convex orthogonal polygons such that

1. the edge between f_0 and f' is divided into $2r$ segments: $\{f_0, f'\} = (s_1, t_1, s_2, t_2, \ldots, s_r, t_r)$. Both s_j and t_j, $1 \le j \le r$, are parallel to the x-axis in the polygon Q_j. See Fig. 5(2-3).
2. for each Q_j, suppose (l_u, l_d) and (r_u, r_d) are the left-most side and the right most side of Q_j, respectively, as depicted in Fig. 5(1).
   - If the given orientation is *ccw*, Q_j is of the shape depicted in the lower figure of Fig. 5(2), in which the left (resp., right) endpoint of s_j (resp., t_j) is l_d (resp., r_u). Figure 5(2) also displays the 3D view of the unfolding.
   - If the given orientation is *cw*, Q_j is of the shape depicted in the lower figure of Fig. 5(3), in which the left (resp., right) endpoint of s_j (resp., t_j) is l_u (resp., r_d). Figure 5(3) also displays the 3D view of the unfolding.

We call each Q_j a *strip* in the subsequent discussion. In words, Invariant $\mathbb{I}$ says that for a component C and a chosen orientation (ccw or cw), $F_{bb}(C)$ can be unfolded into r strips (i.e., y-convex orthogonal polygons) all of which respect the same orientation associated with C. Furthermore, the r strips meet at the beginning of the bridge linking C to its parent.

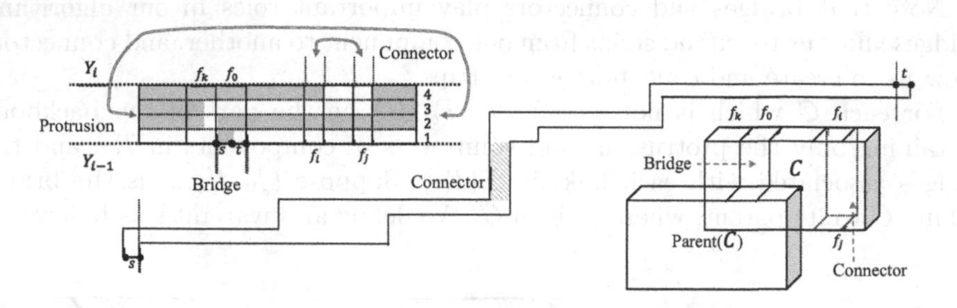

Fig. 6. Unfolding a leaf component.

Lemma 4. *For each orientation (cw or ccw), and for each leaf C in the unfolding tree T_U, Invariant $\mathbb{I}$ holds.*

Proof. Adapting the *Single Box Spiral Path* described in [2] suffices. See Fig. 6 for an illustration of an unfolding Q_1 of $F_{bb}(C)$, for the case where C is in layer i, the bridge linking C and its parent is in Y_{i-1}, and the orientation is ccw. □

Lemma 5. *For each orientation (cw or ccw), and for each intermediate component C (i.e., neither a root nor a leaf) in the unfolding tree T_U, Invariant $\mathbb{I}$ holds.*

Proof. The proof is done by induction on $|V(T_C)|$. The desired unfolding of $F_{bb}(C)$ is constructed by extending the strips of $F_{bb}(C')$ for each child C' of C. We assume that C is in layer i, and we only deal with the case that the bridge linking C to its parent is in Y_{i-1} and the chosen orientation is ccw, as the other cases are similar. We denote B_0 as the bridge linking C to its parent, and $B_1, B_2, ..., B_d$ the bridges linking the d children of C to C.

By Lemma 4 and by induction hypothesis, for each child C' of C, we assume that $F_{bb}(C')$ is already unfolded into $|\text{leaves}(T_{C'})|$ strips meeting Invariant $\mathbb{I}$. If $br(C', C)$ is in Y_{i-1} (resp., Y_i), we assume that $F_{bb}(C')$ is unfolded in the cw (resp., ccw) orientation, respectively.

We let $(f_0, f_1, \ldots, f_k)$ be the faces in the protrusion of C in a counterclockwise ordering with respect to the y+ direction, where f_0 belongs to B_0. $B_j, 1 \leq j \leq d$, can be divided into two groups, B^1 and B^2, such that $B^1 = \{B_j \mid B_j \text{ is in } Y_{i-1}\}$ and $B^2 = \{B_j \mid B_j \text{ is in } Y_i\}$.

Let $\{g_1, \ldots, g_a\}$ ($a = |B^1|$) be the set of faces of the protrusion of C associated with B^1. We further assume that if $g_i = f_{m_i}$ and $g_j = f_{m_j}$, $i < j$ implies $m_i < m_j$ ($g_1, \ldots, g_a$ are listed in increasing order with respect to the index of f_{m_j}).

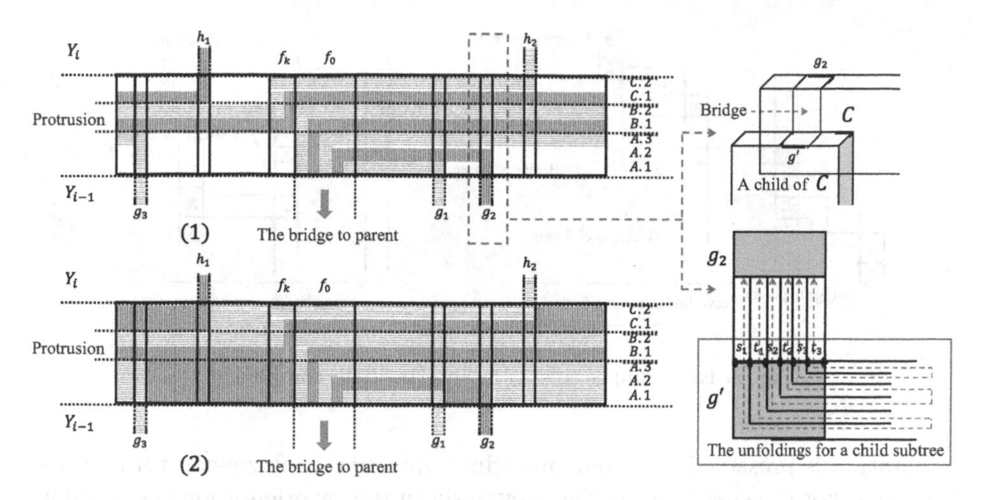

Fig. 7. Unfolding an intermediate component.

Similarly, let $\{h_1, \ldots, h_b\}$ ($b = |B^2|$) be the set of faces of the protrusion of C associated with B^2 listing in decreasing order with respect to $(m_j - k) \bmod k$, for $f_{m_j} = h_j$.

The protrusion of C is partitioned into three regions, A, B and C. Region A is divided into $|B^1|$ levels; and both of B and C are divided into $|B^2|$ levels. See Fig. 7.

Next, we show how to extend the strips (associated with $F_{bb}(C')$) along $br(C', C)$ and $prot(C)$ to the face f_0, the beginning of B_0 (the bridge linking C and its parent). We have the following two cases:

- (Case 1: $br(C', C) = (\ldots, g_j) \in B^1$) We extend the strips coming from the bridge $(\ldots, g_j) \in B^1$, $1 \leq j \leq a$, by going up along the bridge to level $A.j$, turning left until reaching f_0.
- (Case 2: $br(C', C) = (\ldots, h_j) \in B^2$) We extend the strips coming from the bridge $(\ldots, h_j) \in B^2$, $1 \leq j \leq b$, by going down along the bridge to level $C.j$, turning right until reaching f_k, going down to level $B.j$, turning right, and then going straight until reaching f_0.

What remains is to extend some strips in either the up or the down direction, if needed, to unfold the entire protrusion of C.

Figure 7(1) shows the result after carrying out the procedure described in Cases 1 and 2 for each child; and Fig. 7(2) shows the final result after extending some strips to "fill" the entire protrusion.

To see that each Q_j is unfolded in the ccw orientation, see Fig. 8 (note that the circled region indicates the portion of Q_j before the procedure). For the case that the strip (before the procedure) is coming here via a bridge in Y_i (which ends at a face h_m, for some m), then according to our choice of the orientation, Q_j is in the ccw orientation (before the procedure). After the procedure, the ccw

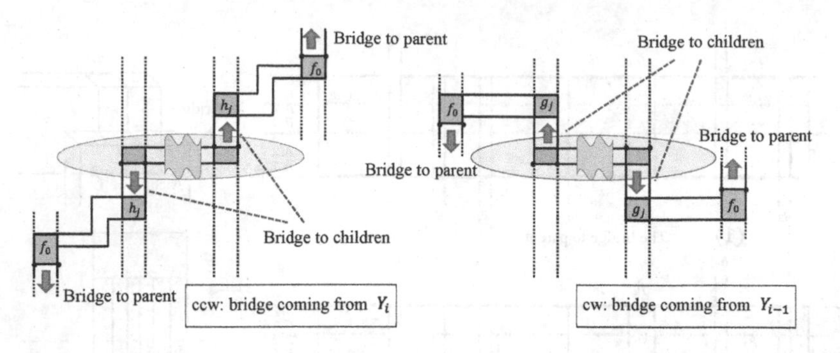

Fig. 8. Extending strips at an intermediate component.

orientation is preserved since our procedure only add some zig-zag turns in its two ends. For the case that Q_j was previously in the cw orientation (i.e. coming from Y_{i-1}), a U-turn is made to change it to ccw after the procedure.

Note that the amount of refinement is at most $(2r \times 4r)$, which can be reached at f_0 when $a = 0$ (i.e. when there is no g_j). □

We are now in a position to prove the main theorem:

Theorem 1. *Every orthogonal polyhedron O can be $(2r \times 4r)$-grid-unfolded, where $r = |leaves(T_U)|$.*

Proof. Let C be the root of T_U, and let C' be the child of C, and let $br(C, C')$ ends at the face f_0. We assume that C is in layer i and that $br(C, C')$ is in Y_{i-1}, as the other cases are similar.

In view of Lemmas 4 and 5, we assume that $F_{bb}(C')$ is already unfolded into $|\text{leaves}(T_{C'})| = r$ strips in the ccw orientation.

Let $(f_0, f_1, \ldots, f_k)$ be the faces in the protrusion in a counter-clockwise ordering with respect to the y+ direction, where f_i and $f_j, i < j$, are the two faces belonging to the connector of C. We divide the protrusion into two regions A and B.

Our task is to link these strips together to form a single piece of partial unfolding meeting the condition of Lemma 3 using $br(C, C')$, $prot(C)$, and $con(C)$:

1. For each strip Q_j, we extend its two ends by going down along the bridge to layer B, turning right until reaching f_1, going down to layer A, turning right and going straight until reaching f_i.
2. Now, the edge $\{f_j, f_{j+1}\}$ in portion B is divided into $2r$ segments: $L(Q_1)$, $R(Q_1)$, $L(Q_2)$, $R(Q_2)$, ..., $L(Q_r)$, $R(Q_r)$, where $L(Q_j)$ (resp., $R(Q_j)$) stands for the left-most (resp., right-most) edge of the strip Q_j. We connect all the strips via the connector and the path $(f_i, \ldots, f_j)$ in the protrusion as follows: $R(Q_1)$ and $R(Q_r)$ are linked, $L(Q_2)$ and $L(Q_r)$ are linked, ... etc. See Fig. 9(2) for the case when $r = 3$.
3. Finally, as in the intermediate case, we finish the rest of the unfolding for the protrusion by extending some strips in either the up or the down direction.

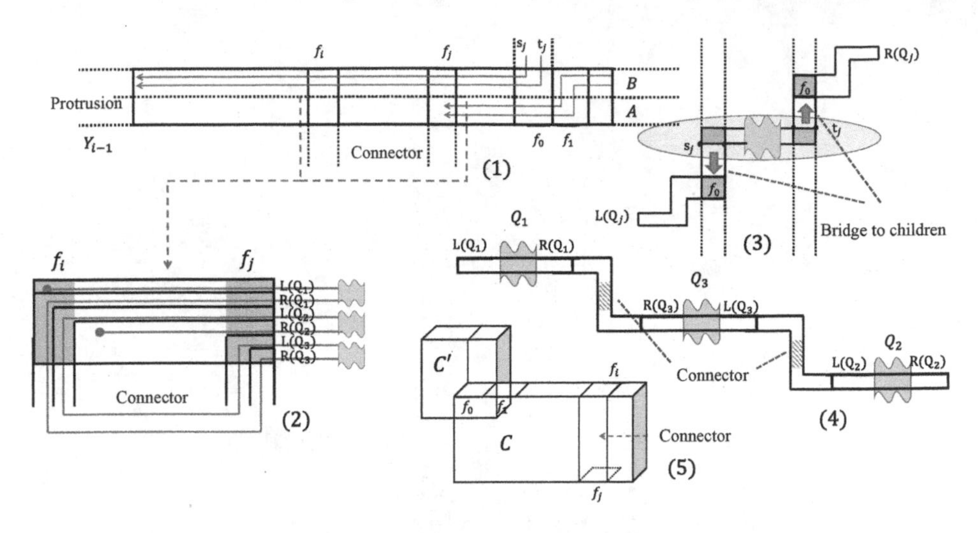

Fig. 9. Unfolding the root component.

Figure 9(1) and (3) show the first step; Fig. 9(2) and (4) show the second step; Fig. 9(5) illustrates the component C in the polyhedron. The required refinement is $(2r \times 4r)$. By Lemma 3, the theorem is concluded. □

References

1. Damian, M., Demaine, E.D., Flatland, R.: Unfolding orthogonal polyhedra with quadratic refinement: the delta-unfolding algorithm. Graphs and Combinatorics **30**(1), 125–140 (2014)
2. Damian, M., Flatland, R., O'Rourke, J.: Epsilon-unfolding orthogonal polyhedra. Graphs and Combinatorics **23**(1), 179–194 (2007)
3. Demaine, E.D., O'Rourke, J.: Geometric Folding Algorithms: Linkages, Origami, Polyhedra. Cambridge University Press, Cambridge (2007)
4. Dürer, A.: Unterweysung der Messung mit dem Zirkel und Richtscheyt, in Linien Ebnen und gantzen Corporen, 1525. Reprinted 2002, Verlag Alfons Uhl, Nördlingen; translated as the Painter's Manual, Abaris Books, New York (1977)
5. O'Rourke, J.: Unfolding orthogonal polyhedra. In: Goodman, J.E., Pach, J., Pollack, R. (eds.) Surveys on Discrete and Computational Geometry: Twenty Years Later, pp. 231–255. American Mathematical Society, New York (2008)

Combinatorial Optimization and Approximation Algorithms III

Colored Non-crossing Euclidean Steiner Forest

Sergey Bereg[1], Krzysztof Fleszar[2(✉)], Philipp Kindermann[2], Sergey Pupyrev[3,4], Joachim Spoerhase[2], and Alexander Wolff[2]

[1] University of Texas, Dallas, USA
[2] Lehrstuhl für Informatik I, Universität Würzburg, Würzburg, Germany
krzysztof.fleszar@uni-wuerzburg.de
[3] Department of Computer Science, University of Arizona, Tucson, AZ, USA
[4] Institute of Mathematics and Computer Science, Ural Federal University, Yekaterinburg, Russia

Abstract. Given a set of k-colored points in the plane, we consider the problem of finding k trees such that each tree connects all points of one color class, no two trees cross, and the total edge length of the trees is minimized. For $k = 1$, this is the well-known Euclidean Steiner tree problem. For general k, a $k\rho$-approximation algorithm is known, where $\rho \leq 1.21$ is the Steiner ratio.

We present a PTAS for $k = 2$, a $(5/3 + \varepsilon)$-approximation for $k = 3$, and two approximation algorithms for general k, with ratios $O(\sqrt{n} \log k)$ and $k + \varepsilon$.

1 Introduction

Steiner tree is a fundamental problem in combinatorial optimization. Given an edge-weighted graph and a set of vertices called *terminals*, the task is to find a minimum-weight subgraph that connects the terminals. For *Steiner forest*, the terminals are colored, and the desired subgraph must connect, for each color, the terminals of that color.

In this paper, we consider a geometric variant of Steiner forest where we add the constraint of planarity and require that terminals with distinct colors lie in distinct connected components. More precisely, we consider the problem of computing, for a k-colored set of points in the plane (which we also call *terminals*), k pairwise non-crossing planar Euclidean Steiner trees, one for each color. Such trees exist for every given set of points. We call the problem of minimizing the total length of these trees *k-Colored Non-Crossing Euclidean Steiner Forest* (k-CESF). Figure 1 shows some instances.

The problem is motivated by a method that Efrat et al. [7] suggested recently for visualizing embedded and clustered graphs. They visualize clusters by regions in the plane that enclose related graph vertices. Their method attempts to reduce visual clutter and optimize "convexity" of the resulting regions by reducing the amount of "ink" necessary to connect all elements of a cluster. Efrat et al. [7] proposed the problem k-CESF and provided a simple $k\rho$-approximation algorithm, where ρ is the *Steiner ratio*, that is, the supremum, over all finite point

K. Elbassioni and K. Makino (Eds.): ISAAC 2015, LNCS 9472, pp. 429–441, 2015.
DOI: 10.1007/978-3-662-48971-0_37

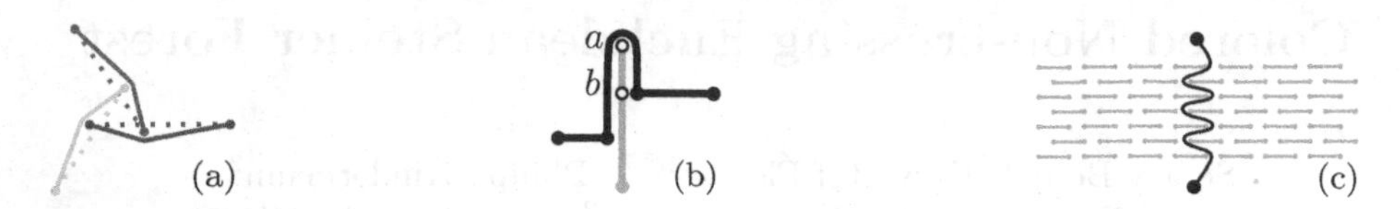

Fig. 1. Difficult examples for k-CESF. (a) The optimum contains no straight-line edge. (b) Segment ab is used twice by the black curve. (c) The black curve can be made arbitrarily longer than the corresponding straight-line segment (Gray segments represent different colors.) (Color figure online).

sets in the plane, of the ratio of the total edge length of a minimum spanning tree over the total edge length of a Euclidean Steiner tree (EST). Chung and Graham [6] showed that $\rho \leq 1.21$.

Our Contribution. The middle column of Table 1 shows our results. For k-CESF, we present a deterministic $(k + \varepsilon)$- and a randomized $O(\sqrt{n} \log k)$-approximation algorithm; see Sect. 2. The main result of our paper is that 2-CESF admits a polynomial-time approximation scheme (PTAS); see Sect. 3. By a non-trivial modification of the PTAS, we prove that 3-CESF admits a $(5/3+\varepsilon)$-approximation algorithm; see Sect. 4.

Table 1. Known and new results for k-CESF (hardness and approximation ratios)

k	k-CESF	planar graph
1	EST: NP-hard [10], $1 + \varepsilon$ [1,15]	ST: NP-hard [10], $1 + \varepsilon$ [4]
2	$1 + \varepsilon$ (Theorem 4)	
3	$5/3 + \varepsilon$ (Theorem 5)	
general k	$k + \varepsilon$ (Theorem 1), $O(\sqrt{n} \log k)$ (Theorem 3)	k const.-size nets on 2 faces, exact [12]
$n/2$	NP-hard [2], $O(\sqrt{n} \log n)$ [5]	k size-2 nets on h faces, exact [9]

Our PTAS for 2-CESF uses some ideas of Arora's algorithm [1] for EST, which is equivalent to 1-CESF. Since, in a solution to 2-CESF, the two trees are not allowed to cross, our approach differs from Arora's algorithm in several respects. We use a different notion of *r-lightness*, and by a *portal-crossing reduction* we achieve that each portal is crossed at most three times. More care is also needed in the perturbation step and in the base case of the dynamic program.

Related Work. Apart from the result of Efrat et al. [7], so far the only two variants of k-CESF that have been studied are those with extreme values of k. As mentioned above, 1-CESF is the same as EST, which is NP-hard [10]. Arora [1] and Mitchell [15] showed independently that EST admits a PTAS. The other extreme value of k, for which k-CESF has been considered, is $k = n/2$. This is the problem of joining specified pairs of points via non-crossing curves of minimum total length. Liebling et al. [13] gave some heuristics for this problem. Bastert and Fekete [2] claimed that $(n/2)$-CESF is NP-hard, but their proof has not been formally published. Recently, Chan et al. [5] considered $(n/2)$-CESF in the context of embedding planar graphs at fixed vertex locations. They gave an $O(\sqrt{n}\log n)$-approximation algorithm based on an idea of Liebling et al. [13] for computing a short non-crossing tour.

There is substantial work on the case where there are obstacles in the plane. Note that, in contrast to k-CESF, a valid solution may not exist in that setting. For a single color (that is, 1-CESF with obstacles), Müller-Hannemann and Tazari [17] give a PTAS. Papadopoulou [18] gave an algorithm for finding minimum-length non-crossing paths joining pairs of points (that is, $n/2$-CESF) on the boundary of a single polygon. A practical aspect of the problem—computing non-crossing paths of specified thickness—was studied by Polishchuk and Mitchell [19]. Their algorithm computes a representation of the thick paths inside a simple polygon; they also show how to find shortest thick disjoint paths joining endpoints on the boundaries of polygonal obstacles (with exponential dependence on the number of obstacles). The main difficulty with multiple obstacles is deciding which homotopy class of the paths gives the minimum length. If the homotopy class of the paths is specified, then the problem is significantly easier [8,20]. Hurtado et al. [11] studied a set visualization problem where points can be blue or red or both, and the points of either color must be connected. Their aim was to minimize the total length of the network. The blue and red subgraphs may intersect.

The graph version of the problem has been studied in the context of VLSI design. Given an edge-weighted plane graph G and a family of k vertex sets (called nets), the goal is to find a set of k non-crossing Steiner trees interconnecting the nets such that the total weight is minimized. The problem is clearly NP-hard, as the special case $k = 1$ is the graph Steiner tree problem (ST), which is known to be NP-hard [10]. ST admits a PTAS [4]. On planar graphs, k-CESF can be solved in $O(2^{O(h^2)} n \log k)$ time [9] for k terminal pairs (that is, size-2 nets) if all terminals lie on h faces of the given n-vertex graph and in $O(n \log n)$ time for $h = 2$ and k constant-size nets [12]. We list these results in Table 1; many entries are still open.

In the *group Steiner tree* problem, one is given a k-colored point set and the task is to find a minimum-length tree that connects at least one point of each color. The problem is discussed in a survey by Mitchell [16]. Another related problem is that of constructing a minimum-length non-crossing path through a given sequence of points in the plane. Its complexity status remains open [14].

2 Algorithms for k-CESF

Despite its simple formulation, the k-CESF problem seems to be rather difficult. There are instances where the optimum contains no straight-line edges or contains paths with repeated line segments; see Figs. 1(a) and (b). This shows that obvious greedy algorithms fail to find an optimal solution, as Liebling et al. [13] observed. They also provided an instance of the problem in a unit square for $k = n/2$ in which the length of an optimal solution is in $\Omega(n\sqrt{n})$, whereas the trivial lower bound (the sum of lengths of straight-line segments connecting the pairs of terminals) is only $O(n)$. The example is based on the existence of expander graphs with a quadratic number of edge crossings. In Fig. 1(c), we provide an example in which one of the curves in the optimal solution can be arbitrarily longer than the trivial lower bound for the corresponding color.

Efrat et al. [7] suggested an approximation algorithm for k-CESF. The key ingredient of their algorithm is the following observation, which shows how to make a pair of given trees non-crossing: reroute one of the trees using a "shell" around the other tree. For any geometric graph G, we denote its total edge length by $|G|$.

Lemma 1 (Efrat et al. [7]). *Let R and B be two trees in the plane spanning red and blue terminals, respectively. Then, there exists a tree R' spanning the red terminals such that (i) R' and B are non-crossing and (ii) $|R'| \leq |R| + 2|B|$.*

Efrat et al. [7] start with k (possibly intersecting) minimum spanning trees, one for each color. Then, they iteratively go through these trees in order of increasing length. In every step, they reroute the next tree by laying a shell around the current solution as in Lemma 1. Their algorithm has approximation factor $k\rho$. In the full version of the paper [3], we show that the algorithm even yields approximation factor $k + \varepsilon$ if we use a PTAS for EST for the initial solution to each color.

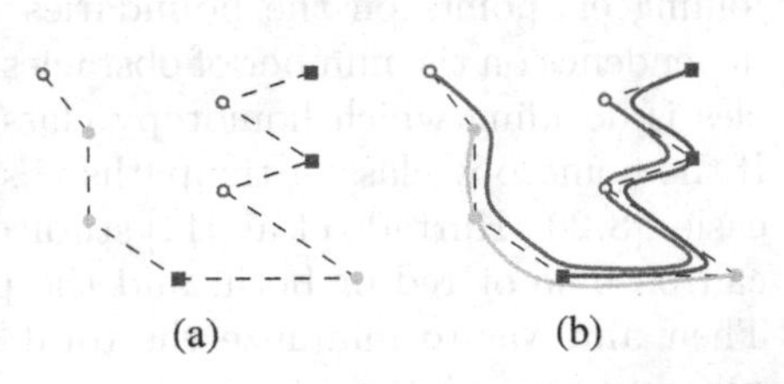

Fig. 2. (a) A low-stretch curve C through the terminals; (b) a 3-CESF solution to the instance reated by wrapping paths around C (Color figure online).

Theorem 1. *For every $\varepsilon > 0$, there is a $(k + \varepsilon)$-approximation algorithm for k-CESF.*

For even k, we can slightly improve on this by using our PTAS for 2-CESF (Theorem 4).

Theorem 2. *For every $\varepsilon > 0$, there is a $(k - 1 + \varepsilon)$-approximation algorithm for k-CESF if k is even.*

Next, we present an approximation algorithm for k-CESF whose ratio depends only logarithmically on k, but also depends on $\sqrt{n}$. The algorithm employs a space-filling curve through a set of given points. The curve was utilized in a heuristic for $(n/2)$-CESF by Liebling et al. [13]. Recently, Chan et al. [5] showed that the approach yields an $O(\sqrt{n} \log n)$-approximation for $(n/2)$-CESF. We show that similar arguments yield approximation ratio $O(\sqrt{n} \log k)$ for general k.

Theorem 3. *k-CESF admits a (randomized) $O(\sqrt{n}\log k)$-approximation algorithm.*

Proof. Chan et al. [5] gave a randomized algorithm to construct a curve C through the given set P of n points. Their curve has small *stretch*, that is, the ratio between the Euclidean distance $d(p,q)$ of two points $p,q \in P$ and their distance $d_C(p,q)$ along the curve is small. Assuming that the points are scaled to lie in a unit square, Chan et al. showed, for a fixed pair of points $p,q \in P$, $\mathbb{E}[d_C(p,q)] \leq O(\sqrt{n}\log(\frac{1}{d(p,q)})) \cdot d(p,q)$. Using C, we construct a solution to k-CESF so that, for every color, the terminals are visited in the order given by the curve; and thus, the solution to every color is a path. All paths can be wrapped around the curve without intersecting each other; see Fig. 2.

If the order of the points along the curve for a specific color i is $p_1^i, \ldots, p_{n_i}^i$, then the length of the corresponding path is $\sum_{j=1}^{n_i-1} d_C(p_j^i, p_{j+1}^i) = d_C(p_1^i, p_{n_i}^i)$. Let $\bar{d} = \sum_{i=1}^{k} d(p_1^i, p_{n_i}^i)/k$. The total (expected) length of the solution is

$$\mathrm{ALG} = \sum_{i=1}^{k} \mathbb{E}[d_C(p_1^i, p_{n_i}^i)] \leq \sum_{i=1}^{k} O(\sqrt{n}\log(1/d(p_1^i, p_{n_i}^i))) \cdot d(p_1^i, p_{n_i}^i).$$

Given that log is concave, this expression is bounded by $\sum_{i=1}^{k} O(\sqrt{n}\log(1/\bar{d})) \cdot \bar{d}$; see Chan et al. [5]. Since the optimal solution to P connects all pairs of terminals of the same color (possibly using non-straight-line curves), $\mathrm{OPT} \geq \sum_{i=1}^{k} d(p_1^i, p_{n_i}^i) = k\bar{d}$. Hence,

$$\mathrm{ALG} \leq \sum_{i=1}^{k} O(\sqrt{n}\log(k/\mathrm{OPT})) \cdot \mathrm{OPT}/k \leq O(\sqrt{n}\log k)\,\mathrm{OPT}.$$

□

3 PTAS for 2-CESF

In this section, we show that 2-CESF admits a PTAS. We follow Arora's approach for computing EST [1], which consists of the following steps. First, Arora performs a recursive geometric partitioning of the plane using a quadtree and snaps the input points to the corners of the tree. Next, he defines an *r-light* solution, which is allowed to cross an edge of a square in the quadtree at most r times and only at so-called *portals.* Then he builds an optimal *portal-respecting* solution using dynamic programming (DP), and finally trims the edges of the solution to get the result. To get an algorithm for 2-CESF, we modify these steps as follows:

(i) The perturbation step, which snaps the terminal to a grid, is modified to avoid crossings between trees. Similarly, the reverse step transforming a perturbed instance solution into one to the original instance is different; see Lemmas 2 and 3.

(ii) We use a different notion of an r-light solution in which every portal is crossed at most r times. We devise a *portal-crossing reduction* that reduces the number of crossings to $r = 3$; see Lemma 5.
(iii) The base case of the DP needs a special modification; it computes a set of crossing-free Steiner trees of minimum total length (see Lemma 6).

We assume that the bounding rectangles of the two sets of input terminals overlap; otherwise, we can use a PTAS for the Steiner tree of each input set individually. We first snap the instance to an $(L \times L)$-grid with $L = O(n)$. We proceed as follows. Let L_0 be the diameter of the smallest bounding box of the given 2-CESF instance. We place an $(L \times L)$-grid of granularity (grid cell size) $g = L_0/L$ inside the bounding box. By scaling the instance appropriately, we can assume that the granularity is $g = 1$. We move each terminal of one color to the nearest grid point in an even row and column, and each terminal of the other color to the nearest grid point in an odd row and column. Thus, the grid point for each terminal is uniquely defined, and no terminals of different color end up at the same location. If there are more terminals of the same color on a grid point, we remove all but one of them. We call the resulting instance a *perturbed* instance.

Lemma 2. *Let* OPT_I *be the length of an optimal solution to a 2-CESF instance* I *of* n *terminals and let* $\varepsilon > 0$*. There is an* $(L\times L)$*-grid with* $L = O(n/\varepsilon)$ *such that* $\mathrm{OPT}_{I^*} \leq (1+\varepsilon)\,\mathrm{OPT}_I$*, where* OPT_{I^*} *is the length of an optimal solution to the perturbed instance* I^**.*

Proof. Choose L to be a power of 2 within the interval $[3\sqrt{2}n/\varepsilon, 6\sqrt{2}n/\varepsilon]$ and perturb the instance as described above. Consider an optimal solution to I. Iteratively, we connect every terminal in I^* to the optimum solution as follows: Connect the terminal to the closest point of the tree in the optimum solution that has the same color. If this line segment crosses the tree of the other color, then reroute this tree around the line segment by using two copies of the line segment. Two copies suffice even if the other tree is crossed more than once since all crossing edges can be connected to the two new line segments. The distance between the terminal and the tree is at most the distance between the terminal and the corresponding terminal in I, which is bounded by $\sqrt{2}$ as we are assuming the unit grid. Hence, we pay at most $3\sqrt{2}$ for connecting the terminal. Since the bounding rectangles of the input terminals overlap, $\mathrm{OPT}_I \geq L$. Thus, the additional length of an optimal solution to I^* is

$$\mathrm{OPT}_{I^*} - \mathrm{OPT}_I \leq 3\sqrt{2}n \leq \varepsilon \cdot \mathrm{OPT}_I\,. \qquad \square$$

The next lemma, proven analogously to Lemma 2, shows that we can transform a solution to the perturbed instance into one to the original instance.

Lemma 3. *Given a solution* $\mathcal{T}$ *to the perturbed instance as defined in Lemma 2, we can transform* $\mathcal{T}$ *into a solution to the original instance, increasing its length by at most* $\varepsilon \cdot \mathrm{OPT}_I$*.*

In the following, we assume that the instance is perturbed. We place a quadtree in dependence of two integers $a, b \in [0, \ldots, L-1]$ that we choose independently uniformly at random. We place the origin of the coordinate system on the bottom left corner of the bounding box of our instance. Then we take a box B whose width and height is twice the width and height of the bounding box. We place it such that its bottom left corner has coordinates $(-a, -b)$. Note that the bounding box is inside B. We extend the $(L \times L)$-grid to cover B. Thus, we have an $(L' \times L')$-grid with $L' = 2L$.

Then we partition B with a quadtree along the $(L' \times L')$-grid. The partition is stopped when the current quadtree box coincides with a grid cell. We define the *level of a quadtree square* to be its depth in the quadtree. Thus, B has level 0, whereas the level of a leaf is bounded by $\log L' = \log(2L) = O(\log n)$. Then, for each grid line ℓ, we define its *level* as the highest (that is, of minimum value) level of all the quadtree squares that touch ℓ (but which are not crossed by it).

Let $m = \lceil 4 \log L'/\varepsilon \rceil$. On each grid line ℓ of level i, we place $2^i \cdot m$ equally spaced points. We call these points *portals*. Thus, each square contains at most m portals on each of its edges. A solution that crosses the grid lines only at portals is called *portal-respecting*. We show that there is a close-to-optimal portal-respecting solution. Note that, in contrast to Arora, we first make the solution portal-respecting before reducing the number of crossings on each grid line. The proof of the following lemma is similar to the Arora's prove and is provided in the full version of the paper [3].

Lemma 4. *Let* OPT_I *be the length of an optimal solution to a 2-CESF instance* I*, and let* $\varepsilon > 0$ *be as in the definition of* m*. Then, there exists a position of the quadtree and a portal-respecting solution to* I *of length at most* $(1+\varepsilon)\,\mathrm{OPT}_I$.

The last ingredient of our DP is to reduce the number of crossings in every portal. We call a solution *r-light* if each portal is crossed at most r times.

In the following, we explain an operation which we call a *portal-crossing reduction.* We are given a portal-respecting solution consisting of two Steiner trees R and B (red and blue) and we want to reduce (that is, modify without increasing its length) it such that R and B pass through each portal at most three times in total.

Lemma 5. *Every portal-respecting solution of 2-CESF can be transformed into a 3-light portal-respecting solution without increasing its length.*

Proof. Consider a sequence of passes through a portal. We assume that there are no terminals in the portals. If two adjacent passes belong to the same tree, then we can eliminate one of them by snapping it to the other one. Note that this may create cycles, but they can be broken by removing the longest part of each cycle. Therefore, we can assume that the passes form an alternating sequence. It suffices to show that any alternating sequence of four passes can be reduced to two passes by shortening the trees. Let a, b, c, and d be such a sequence as shown

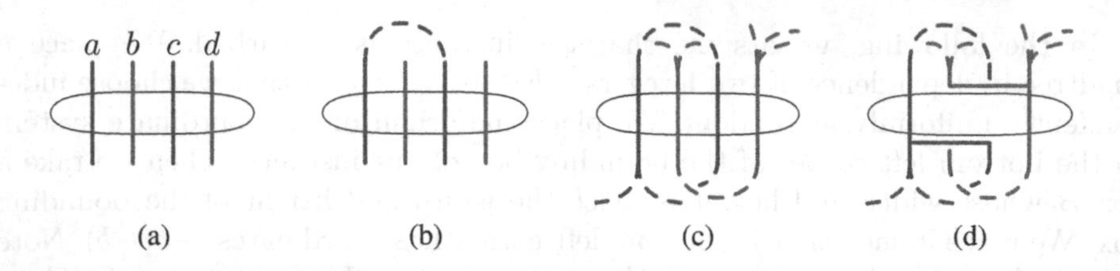

Fig. 3. A portal modification for four passes (Color figure online).

in Fig. 3a, where a and c belong to B and b and d to R. We cut the passes b and c. This results in two connected components in each tree. W.l.o.g., a and the upper part of c belong to the same connected component; see Fig. 3b. Otherwise, we can change the colors because (i) a and the lower part of c are connected, and (ii) the upper part of b and d are connected.

Since R and B are disjoint, d and the lower part of b are in the same connected component; see Fig. 3c. Then, we connect the component as shown in Fig. 3d and shorten the trees (e.g., the lower part of b can be reduced to a terminal of R). Note that the passes a and d remain in the solution, while the passes b and c are eliminated. We repeat the procedure for the remaining passes, until there are at most three passes left. The length of the solution does not increase because the portal has zero width. □

With the next Lemma 6, we show how to find a close-to-optimal 3-light portal-respecting solution to the perturbed instance. We assume that an appropriate quadtree (as defined in Lemma 4) is given.

Lemma 6. *Let $\varepsilon > 0$ be as above. Given a perturbed instance I^* of an n-terminal 2-CESF instance, we can compute, in time $O(n^{O(1/\varepsilon)})$ and $O(n^{O(1/\varepsilon)})$ space, a solution of length at most $(1+\varepsilon)\,\mathrm{OPT}_{I^*}$, where OPT_{I^*} is the length of an optimal 3-light portal-respecting solution to I^*.*

Proof. We use DP with a subproblem consisting of (a) a square of the quadtree, (b) a sequence of up to three red and blue points on each portal on the border of the square, and (c) a *non-crossing* partition of these points into sets of the same color. A partition of these points is non-crossing if for no four points a, b, c, d, occurring in that order on the boundary of the square, it holds that a and c belong to one set of the partition, and b and d to another one. The goal is to find an optimal collection of crossing-free red and blue Steiner trees, such that each set of the partition and each terminal inside the square is contained in a tree of the same color.

The base case of our DP is a unit square, which is either empty or contains terminals only at corners of the square. If the square is empty, we consider each set of the partition as an instance of 1-CESF and solve it by the PTAS for EST [1]. For each point set, we force its Steiner tree to lie inside its convex hull, by projecting any part outside the convex hull to its border. Since the partition is non-crossing, the convex hulls of its point sets are pairwise disjoint. Therefore, the Steiner trees and their union is also a close-to-optimal solution to the base case. If the square contains (up to four) terminals at the corners, these terminals are treated in a similar way as portals.

For composite squares in the quadtree, we proceed as follows. For the four squares that subdivide the composite square, we consider all combinations of all possible (b) and (c) that match together and match the subproblem. In the DP, we already have computed a close-to-optimal solution to every choice of (b) and (c) of each of the four squares; taking the best combination gives a close-to-optimal solution.

The size of the DP table is proportional to the number of subproblems, that is, (a) $\times$ (b) $\times$ (c). There are $O(n^2)$ squares in the quadtree in total. Each square contains at most $m = O(\log n/\varepsilon)$ portals. For each portal, there is a constant number of possible sequences of up to three colored points. Thus, there are $2^{O(\log n/\varepsilon)} = n^{O(1/\varepsilon)}$ possibilities for (b). Since the number of non-crossing partitions of a set of k elements is the k'th Catalan number C_k, we have $C_{O(\log n/\varepsilon)} < 2^{O(\log n/\varepsilon)} = n^{O(1/\varepsilon)}$ possibilities for (c). In total, we consider $n^{O(1/\varepsilon)}$ subproblems in the DP.

The running time to solve an instance of the base case is polynomial in $m = O(\log n/\epsilon)$. The running time to handle a composite square is polynomial in $(n^{O(1/\epsilon)})^4$, which is $n^{O(1/\epsilon)}$. Thus, the total running time is bounded by $n^{O(1/\epsilon)}$. □

Now we prove the main result of this section.

Theorem 4. *2-CESF admits a PTAS.*

Proof. Consider a 2-CESF instance I. Let OPT be the length of an optimum solution. For any $\varepsilon > 0$, by Lemmas 2, 4 and 5, the length, OPT′, of an optimal 3-light portal-respecting solution to the perturbed version of I is a most $(1 + \varepsilon)\,\mathrm{OPT}$. Using Lemma 6, we find a 3-light portal-respecting solution to the perturbed instance of length at most $(1 + \varepsilon)\,\mathrm{OPT}' \leq (1 + \varepsilon)(1 + \varepsilon)\,\mathrm{OPT}$. By Lemma 3, we transform the solution into a solution to I by increasing its length by at most $\varepsilon \cdot \mathrm{OPT}$. Therefore, for every $\varepsilon' > 0$, we can construct a solution to I of length $(1 + \varepsilon)(1 + \varepsilon)\,\mathrm{OPT} + \varepsilon \cdot \mathrm{OPT} \leq (1 + \varepsilon')\,\mathrm{OPT}$ by choosing $\varepsilon > 0$ appropriately. □

4 Algorithm for 3-CESF

The above approach for 2-CESF cannot be directly applied to 3-CESF since optimal trees may need to pass portals many times. For example, the three paths crossing the portal in Fig. 4 are difficult because we cannot locally reroute to make them $O(1)$-light as in Lemma 5.

Instead, we now improve the approximation ratio of $3 + \varepsilon$ (from Theorem 1) to $5/3 + \varepsilon$. We re-use some ideas of the approach for 2-CESF.

To this end, take an optimal solution T for 3-CESF. The terminals are red, green, and blue; we call the corresponding trees R, G, and B. We assume that B

is the cheapest among the three trees. In the beginning, we construct a quadtree partitioning the plane and choose the portals, for a given ε, as described in Sect. 3. We then make the solution portal-respecting, which results in a solution T^* consisting of trees R^*, G^*, and B^*. In expectation, this increases the length of each of the trees (and hence, of T) by a factor of at most $1+\varepsilon$.

First, we show that we have few portal passes if the blue and the green tree do not *meet* at any portal, that is, no blue and green passes are adjacent.

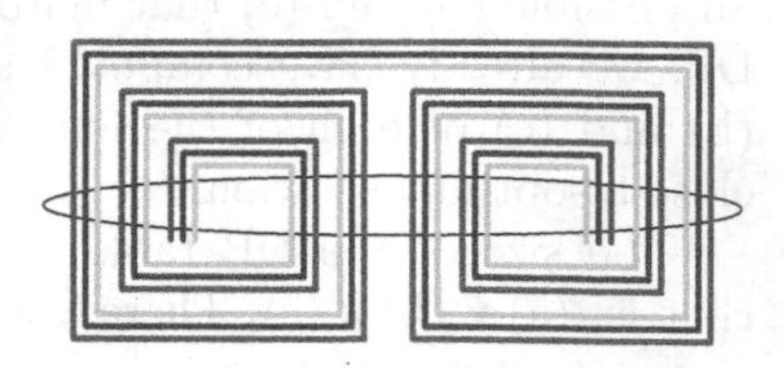

Fig. 4. A difficult portal crossing of a 3-CESF instance (Color figure online).

Lemma 7. *Consider a portal-respecting solution T^* to 3-CESF consisting of trees R^*, G^*, B^*. If B^* and G^* do not meet at any portal, then T^* can be transformed into a 7-light portal-respecting solution.*

Proof. Apply the portal-crossing reduction from Lemma 5 and consider a portal. Recall that, after this operation, there are no *rbrb* and *rgrg* subsequences in the passes of the portal. Here, r, b, and g correspond to the passes of the trees R^*, B^*, and G^*, respectively. If the portal has only one blue or one green pass, then the solution is already 7-light at the portal (with the longest possible sequences *rgrbrgr* and *rbrgrbr*, respectively). Otherwise, it contains at least two blue and at least two green passes. Notice that the sequence of passes must be r-alternate, that is, of the form $\ldots r \circ r \circ r \ldots$ since blue and green do not meet. Thus, a sequence of more than 7 passes must contain a subsequence *grbrgrb* (or a symmetric one, *brgrbrg*). These subsequences are reducible. See Fig. 5 for one of the possible cases, the other cases are analogous. □

Now, we show that T^* can be transformed into a 10-light portal-respecting solution T' of length at most $|R^*|+|G^*|+3|B^*|$.

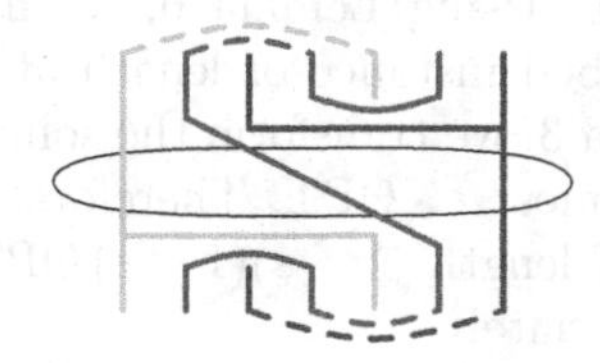

Fig. 5. Constructing a 7-light solution to an instance without adjacent blue-green passes (one of several possible cases) (Color figure online).

Lemma 8. *A portal-respecting solution T^* to 3-CESF, consisting of trees B^*, R^*, and G^*, can be transformed into a portal-respecting solution T' such that*

(i) T' passes at most 10 times through each portal, and
(ii) $|T'| \leq |R^*| + |G^*| + 3|B^*|$.

Proof sketch. We define a BG-*solution*; informally, this is a solution in which we are allowed to connect green branches to the blue tree (if they never meet, we can apply Lemma 7). We prove the lemma in two steps. First, we show that T can be transformed to a portal-respecting BG-solution T^{BG} with at most 6 passes per portal having the same (or smaller) length. In this solution, B^* remains connected and passes each portal at most twice. Then, we further modify T^{BG} to get a portal-respecting solution T^* with at most 10 passes per portal and the desired length by laying a shell around B^* to reroute G^*. The full proof is given in the full version of the paper [3]. □

Before we describe our approximation algorithm, we first need to discuss the perturbation step. The perturbation itself is the same as in Sect. 3: we move each terminal to a uniquely defined closest grid point (we assign the grid points of even row and odd column to the third color) and merge terminals of the same color to one terminal. However, we need a different technique to transform a solution to the original instance into a solution to the perturbed instance and vice versa.

Lemma 9. *Let I be a 3-CESF instance with n terminals, let OPT_I be the length of an optimal solution to I, and let $\varepsilon > 0$. Then, we can place an $(L\times L)$-grid with $L = O(n/\varepsilon)$ such that, for the perturbed instance I^* of I, $\mathrm{OPT}_{I^*} \leq (1+\varepsilon)\,\mathrm{OPT}_I$.*

Proof sketch. We proceed similar to the proof of Lemma 2 by connecting each terminal of I^* to the nearest point of its corresponding tree. Since this connection can cross segments of two colors, we have to be more careful with the rerouting. The full proof is given in the full version of the paper [3]. □

Analogously to the proof of Lemma 9, we transform a solution to a perturbed instance back into one to the original instance not increasing the length by much. Then, we combine the lemmas to prove the main result of this section.

Lemma 10. *We can transform a solution $\mathcal{T}$ to the perturbed instance I^* into a solution to the original instance I, increasing the length by at most $\varepsilon\,\mathrm{OPT}_I$.*

Proof. Iteratively connect each terminal of the original instance to the solution $\mathcal{T}$ analogously to the proof of Lemma 9. □

Theorem 5. *For every $\varepsilon > 0$, 3-CESF admits a $(5/3+\varepsilon)$-approximation algorithm.*

Proof. Let $\varepsilon' = \sqrt[3]{1+3\varepsilon/5} - 1$. Let T be an optimal solution to a 3-CESF instance I with trees R, G and B. W.l.o.g., assume that $|B| \leq |R|, |G|$. Denote by $\mathrm{OPT}_I = |R|+|G|+|B|$ the length of T. We first construct a portal-respecting solution T^* of length $|T^*| = |R^*|+|G^*|+|B^*| \leq (1+\varepsilon')(|R|+|G|+|B|)$. Then, Lemma 8 yields an optimal 10-light portal-respecting solution T' of length

$$\begin{aligned}|T'| \leq |R^*|+|G^*|+3|B^*| \leq 5/3\cdot|T^*| &\leq 5/3\cdot(1+\varepsilon')\cdot(|R|+|G|+|B|)\\ &= 5/3\cdot(1+\varepsilon')\cdot\mathrm{OPT}_I\,.\end{aligned}$$

Using a DP similar to the one described in Sect. 3 and using Lemma 9, we find a 10-light portal-respecting solution of length $(1+\varepsilon')|T'|$ to the perturbed instance I^* of I. By Lemma 10, we can transform our solution to I^* into a solution to I whose total length is bounded by

$$(1+\varepsilon')^2|T'| \leq 5/3(1+\varepsilon')^3\,\mathrm{OPT}_I < (5/3+\varepsilon)\,\mathrm{OPT}_I\,.$$

□

5 Conclusion

We have presented approximation algorithms for k-CESF. We leave the following questions open. Is k-CESF APX-hard for some $k \geq 3$? Can we improve the running time of the PTAS for 2-CESF from $O(n^{O(1/\varepsilon)})$ to $O(n(\log n)^{O(1/\varepsilon)})$ as Arora [1] did for EST?

Currently, we are studying an "anchored" version of k-CESF where the only allowed Steiner points are input points of a different color. Any α-approximation for k-CESF yields an $\alpha(1+\sqrt{3})/2$- approximation for the anchored version.

Acknowledgments. We are grateful to Alon Efrat, Jackson Toeniskoetter, and Thomas van Dijk for the initial discussion of the problem.

References

1. Arora, S.: Polynomial time approximation schemes for Euclidean traveling salesman and other geometric problems. J. ACM **45**(5), 753–782 (1998)
2. Bastert, O., Fekete, S.P.: Geometric wire routing. Technical report. 96.247, Universität zu Köln (1998). http://e-archive.informatik.uni-koeln.de/247
3. Bereg, S., Fleszar, K., Kindermann, P., Pupyrev, S., Spoerhase, J., Wolff, A.: Colored non-crossing Euclidean steiner forest. CoRR abs/1509.05681 (2015). http://arxiv.org/abs/1509.05681
4. Borradaile, G., Klein, P., Mathieu, C.: An $O(n \log n)$ approximation scheme for Steiner tree in planar graphs. ACM Trans. Algorithms **5**(3), 31 (2009)
5. Chan, T.M., Hoffmann, H.-F., Kiazyk, S., Lubiw, A.: Minimum length embedding of planar graphs at fixed vertex locations. In: Wismath, S., Wolff, A. (eds.) GD 2013. LNCS, vol. 8242, pp. 376–387. Springer, Heidelberg (2013)
6. Chung, F.R.K., Graham, R.L.: A new bound for Euclidean Steiner minimal trees. Ann. New York Acad. Sci. **440**(1), 328–346 (1985)
7. Efrat, A., Hu, Y., Kobourov, S.G., Pupyrev, S.: MapSets: visualizing embedded and clustered graphs. In: Duncan, C., Symvonis, A. (eds.) GD 2014. LNCS, vol. 8871, pp. 452–463. Springer, Heidelberg (2014)
8. Efrat, A., Kobourov, S.G., Lubiw, A.: Computing homotopic shortest paths efficiently. Comput. Geom. Theory Appl. **35**(3), 162–172 (2006)
9. Erickson, J., Nayyeri, A.: Shortest non-crossing walks in the plane. In: Proceedings of ACM-SIAM Symposium on Discrete Algorithms (SODA 2011), pp. 297–308 (2011)
10. Garey, M.R., Johnson, D.S.: Computers and Intractability: A Guide to the Theory of NP-Completeness. W. H. Freeman & Co., New York (1979)
11. Hurtado, F., Korman, M., van Kreveld, M., Löffler, M., Sacristán, V., Silveira, R.I., Speckmann, B.: Colored spanning graphs for set visualization. In: Wismath, S., Wolff, A. (eds.) GD 2013. LNCS, vol. 8242, pp. 280–291. Springer, Heidelberg (2013)
12. Kusakari, Y., Masubuchi, D., Nishizeki, T.: Finding a noncrossing Steiner forest in plane graphs under a 2-face condition. J. Combin. Optim. **5**, 249–266 (2001)
13. Liebling, T.M., Margot, F., Müller, D., Prodon, A., Stauffer, L.: Disjoint paths in the plane. ORSA J. Comput. **7**(1), 84–88 (1995)

14. Löffler, M.: Existence and computation of tours through imprecise points. Int. J. Comput. Geom. Appl. **21**(1), 1–24 (2011)
15. Mitchell, J.S.: Guillotine subdivisions approximate polygonal subdivisions: a simple polynomial-time approximation scheme for geometric TSP, k-MST, and related problems. SIAM J. Comput. **28**(4), 1298–1309 (1999)
16. Mitchell, J.S.: Geometric shortest paths and network optimization. In: Urrutia, J., Sack, J.R. (eds.) Handbook of Computational Geometry, chap. 15, pp. 633–701. North-Holland (2000)
17. Müller-Hannemann, M., Tazari, S.: A near linear time approximation scheme for Steiner tree among obstacles in the plane. Comput. Geom. Theory Appl. **43**(4), 395–409 (2010)
18. Papadopoulou, E.: k-pairs non-crossing shortest paths in a simple polygon. Int. J. Comput. Geom. Appl. **9**(6), 533–552 (1999)
19. Polishchuk, V., Mitchell, J.S.B.: Thick non-crossing paths and minimum-cost flows in polygonal domains. In: Proceedings of the ACM Symposium on Computational Geometry (SoCG 2007), pp. 56–65 (2007)
20. Verbeek, K.: Homotopic $\mathcal{C}$-oriented routing. In: Didimo, W., Patrignani, M. (eds.) GD 2012. LNCS, vol. 7704, pp. 272–278. Springer, Heidelberg (2013)

On a Generalization of Nemhauser and Trotter's Local Optimization Theorem

Mingyu Xiao(✉)

School of Computer Science and Engineering,
University of Electronic Science and Technology of China, Chengdu, China
myxiao@gmail.com

Abstract. The Nemhauser and Trotter's theorem applies to the famous VERTEX COVER problem and can obtain a 2-approximation solution and a problem kernel of $2k$ vertices. This theorem is a famous theorem in combinatorial optimization and has been extensively studied. One way to generalize this theorem is to extend the result to the BOUNDED-DEGREE VERTEX DELETION problem. For a fixed integer $d \geq 0$, BOUNDED-DEGREE VERTEX DELETION asks to delete at most k vertices of the input graph to make the maximum degree of the remaining graph at most d. VERTEX COVER is a special case that $d = 0$. Fellows, Guo, Moser and Niedermeier proved a generalized theorem that implies an $O(k)$-vertex kernel for BOUNDED-DEGREE VERTEX DELETION for $d = 0$ and 1, and for any $\varepsilon > 0$, an $O(k^{1+\varepsilon})$-vertex kernel for each $d \geq 2$. In fact, it is still left as an open problem whether BOUNDED-DEGREE VERTEX DELETION parameterized by k admits a linear-vertex kernel for each $d \geq 3$. In this paper, we refine the generalized Nemhauser and Trotter's theorem. Our result implies a linear-vertex kernel for BOUNDED-DEGREE VERTEX DELETION parameterized by k for each $d \geq 0$.

1 Introduction

VERTEX COVER, to find a minimum set of vertices in a graph such that each edge in the graph is incident on at least one vertex in this set, is one of the most fundamental problems in graph algorithms, graph theory, parameterized algorithms, theories of NP-completeness and many others. Nemhauser and Trotter [22] proved a famous theorem (NT-Theorem) for VERTEX COVER.

Theorem 1 [NT – Theorem]. *For an undirected graph $G = (V, E)$ of $n = |V|$ vertices and $m = |E|$ edges, there is an $O(\sqrt{n}m)$-time algorithm to compute two disjoint vertex subsets C and I of G such that for any minimum vertex cover K' of the induced subgraph $G[V \setminus (C \cup I)]$, $K' \cup C$ is a minimum vertex cover of G and*

$$|K'| \geq \frac{|V \setminus (C \cup I)|}{2}.$$

M. Xiao—Supported by NFSC of China under the Grant 61370071.

K. Elbassioni and K. Makino (Eds.): ISAAC 2015, LNCS 9472, pp. 442–452, 2015.
DOI: 10.1007/978-3-662-48971-0_38

This theorem provides a polynomial-time algorithm to reduce the size of the input graph by possibly finding partial solution. It turns out that NT-Theorem has great applications in approximation algorithms [5,17,19] and parameterized algorithms [2,7]. We can see that $V \setminus I$ is a 2-approximation solution and $G[V \setminus (C \cup I)]$ is a $2k$-vertex kernel of the problem taking the size of the solution as the parameter k. Lokshtanov et al. [21] also apply NT-Theorem to branching algorithms for Vertex Cover and some other related problems. Due to NT-Theorem's practical usefulness and theoretical depth in graph theory, it has attracted numerous further studies and follow-up work [2,4,9,14]. Bar-Yehuda, Rawitz and Hermelin [4] extended NT-Theorem for a generalized vertex cover problem, where edges are allowed not to be covered at a certain predetermined penalty. Fellows, Guo, Moser and Niedermeier [14] extended NT-Theorem for Bounded-Degree Vertex Deletion.

In this paper, we are interested in Bounded-Degree Vertex Deletion. A *d-degree deletion set* of a graph G is a subset of vertices, whose deletion leaves a graph of maximum degree at most d. For each fixed d, Bounded-Degree Vertex Deletion is to find a d-degree deletion set of minimum size in an input graph. Bounded-Degree Vertex Deletion and its "dual problem" to find maximum s-plexes have applications in computational biology [8,14] and social network analysis [3,24]. There is a substantial amount of theoretical work on this problem [20,23,24], specially in parameterized complexity [6,8,14].

Since Vertex Cover is a special case of Bounded-Degree Vertex Deletion, we are interested in finding a local optimization theorem similar to NT-Theorem for Bounded-Degree Vertex Deletion. Fellows, Guo, Moser and Niedermeier [14] made a great progress toward to this interesting problem by giving the following theorem.

Theorem 2 [14]. *For an undirected graph $G = (V, E)$ of $n = |V|$ vertices and $m = |E|$ edges, any constant $\varepsilon > 0$ and any integer $d \geq 0$, there is an $O(n^4 m)$-time algorithm to compute two disjoint vertex subsets C and I of G such that for any minimum d-degree deletion set K' of the induced subgraph $G[V \setminus (C \cup I)]$, $K' \cup C$ is a minimum d-degree deletion set of G, and*

$$|K'| \geq \frac{|V \setminus (C \cup I)|}{d^3 + 4d^2 + 6d + 4} \quad \text{for } d \leq 1, \qquad \text{and}$$

$$|K'|^{1+\varepsilon} \geq \frac{|V \setminus (C \cup I)|}{c} \quad \text{for } d \geq 2,$$

where c is a function of d and ε.

In this theorem, for $d \geq 2$, the number of remaining vertices in $V \setminus (C \cup I)$ is not bounded by a constant times of the solution size $|K'|$ of $G[V \setminus (C \cup I)]$. This is a significant difference between this theorem and the NT-Theorem for Vertex Cover. In terms of parameterized algorithms, Theorem 2 cannot get a linear-vertex kernel for Parameterized Bounded-Degree Vertex Deletion (with parameter k being the solution size) for each $d \geq 2$. In fact, in an initial version [15] of Fellows, Guo, Moser and Niedermeier's paper, a better result was claimed, which can get a linear-vertex kernel for PARAMETERIZED BOUNDED-DEGREE VERTEX DELETION for each $d \geq 0$. Unfortunately, the proof in [15] is incomplete. We also note that Chen et al. [8] proved a $37k$-vertex kernel for BOUNDED-DEGREE VERTEX DELETION for $d = 2$. However, whether BOUNDED-DEGREE VERTEX DELETION for each $d \geq 3$ allows a linear-vertex kernel is not known. In this paper, based on Fellows, Guo, Moser and Niedermeier's work [15], we close the above gap by proving the following theorem for BOUNDED-DEGREE VERTEX DELETION.

Theorem 3 [Our result]. *For an undirected graph $G = (V, E)$ of $n = |V|$ vertices and $m = |E|$ edges and any integer $d \geq 0$, there is an $O(n^{5/2}m)$-time algorithm to compute two disjoint vertex subsets C and I of G such that for any minimum d-degree deletion set K' of the induced subgraph $G[V \setminus (C \cup I)]$, $K' \cup C$ is a minimum d-degree deletion set of G and*

$$|K'| \geq \frac{|V \setminus (C \cup I)|}{d^3 + 4d^2 + 5d + 3}.$$

From this version of the generalized Nemhauser and Trotter's theorem, we can get a $(d^3 + 4d^2 + 5d + 3)k$-vertex kernel for BOUNDED-DEGREE VERTEX DELETION parameterized by the size k of the solution, which is linear in k for any constant $d \geq 0$. There is no difference between the cases that $d \leq 1$ and $d \geq 2$ anymore. For the special case that $d = 0$, our theorem specializes a $3k$-vertex kernel for VERTEX COVER, while Theorem 2 provides a $4k$-vertex kernel and NT-Theorem provides a $2k$-vertex kernel. For the special case that $d = 1$, our theorem provides a $13k$-vertex kernel and Theorem 2 provides a $15k$-vertex kernel. For the special case that $d = 2$, our theorem obtains a $37k$-vertex kernel, the same result obtained by Chen et al. [8].

Recently, Dell and van Melkebeek [12] showed that unless the polynomial-time hierarchy collapses, PARAMETERIZED BOUNDED-DEGREE VERTEX DELETION does not have kernels consisting of $O(k^{2-\epsilon})$ edges for any constant $\epsilon > 0$, which implies that linear size would be the best possible bound on the number of vertices in any kernel for this problem. It has also been proved by Fellows, Guo, Moser and Niedermeier [14] that when d is not bounded, PARAMETERIZED BOUNDED-DEGREE VERTEX DELETION is W[2]-hard. Then unless FPT = W[2], it is impossible to remove d from the size function of any kernel of this problem. These two hardness results also imply that our result is 'tight' in some sense.

The framework of our algorithm follows that of Fellows, Guo, Moser and Niedermeier's algorithm [14]. But we still need some new and nontrivial ideas to get our result. For the purpose of presentation, we will define a decomposition, called 'd-bounded decomposition' to prove Theorem 3 and construct our algorithms. This decomposition can be regarded as an extension of the crown decomposition for VERTEX COVER [1,10], but more sophisticated. To compute C and I in Theorem 3, we will change to compute a proper d-bounded decomposition. Some similar ideas in construction of crown decompositions as in Fellows, Guo, Moser and Niedermeier's algorithm for Theorem 2 [14] are used to construct our decomposition. The detailed differences between our and previous algorithms will be addressed in Sect. 4. Before introducing the decompositions, we first give the notation system in this paper. Proofs of some lemmas are omitted due to space limitation, which can be found in the full version of this paper.

2 Notation System

Let $G = (V, E)$ stand for a simple undirected graph with a set V of $n = |V|$ vertices and a set E of $m = |E|$ edges. For simplicity, we may denote a singleton set $\{v\}$ by v. For a vertex subset V', a vertex in V' is denoted by *V'-vertex*. The graph induced by V' is denoted by $G[V']$. We also use $N(V')$ to denote the set of vertices in $V \setminus V'$ adjacent to some vertices in V' and let $N[V'] = N(V') \cup V'$. The vertex set and edge set of a graph G' are denoted by $V(G')$ and $E(G')$, respectively. A bipartite graph with two parts of vertices A and B and edge set E_H is denoted by $H = (A, B, E_H)$.

For an integer $d' \geq 1$, a star with $d'+1$ vertices is called a *d'-star*. For $d' > 1$, the unique vertex of degree > 1 in a d'-star is called the *center* of the star and all other degree-1 vertices are called the *leaves* of the star. For a 1-star, any vertex can be regarded as a *center* and the other vertex as a *leaf*. A star with a center v is also called a star *centered at* v. For two disjoint vertex sets V_1 and V_2, a set of stars is *from* V_1 *to* V_2 if the centers of the stars are in V_1 and leaves are in V_2. A *$\leq d'$-star* is a star with at most d' leaves. A *d'-star packing* (resp., *$\leq d'$-star packing*) is a set of vertex-disjoint d'-stars (resp., $\leq d'$-stars). We will use $\alpha(G)$ to denote the size of a minimum d-degree deletion set of a graph G.

3 The Decomposition Techniques

Crown decomposition is a powerful tool to obtain kernels for Vertex Cover. This technique was firstly introduced in [1,10] and found to be very useful in designing kernelization algorithms for Vertex Cover and related problems [2,9,26].

Definition 1 [**Crown Decomposition**]. *A crown decomposition of a graph G is a partition of the vertex set of G into three sets I, C and J such that*

(1) I is an independent set,
(2) there are no edges between I and J, and
(3) there is a matching M on the edges between I and C such that all vertices in C are matched.

See Fig. 1(a) for an illustration for crown decompositions. In some references, $I \neq \emptyset$ is also required in the definition of crown decompositions. Here we allow $I = \emptyset$ for the purpose of presentation. It is known that

Lemma 1 [1]. *Let (I, C, J) be a crown decomposition of G. Then (I, C) satisfies the local optimality condition in Theorem 1, i.e., $K' \cup C$ is a minimum vertex cover of G for any minimum vertex cover K' of the induced subgraph $G[V \setminus (I \cup C)]$.*

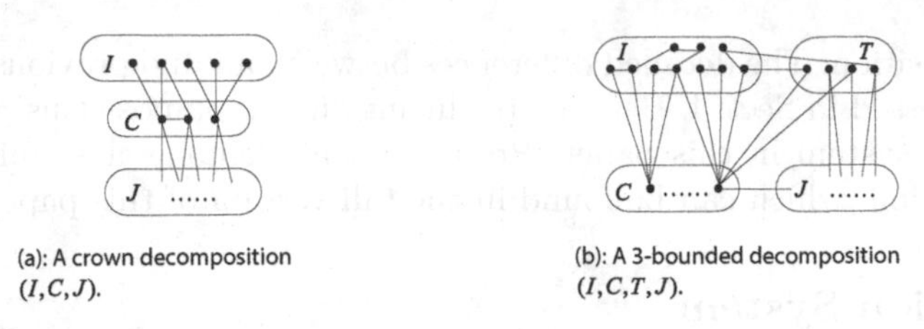

(a): A crown decomposition (I,C,J).

(b): A 3-bounded decomposition (I,C,T,J).

Fig. 1. Decompositions

By this lemma, we can reduce the instance of VERTEX COVER by removing $I \cup C$ of a crown decomposition. There are some methods that find certain crown decompositions of a graph and result in a linear-vertex kernel for VERTEX COVER [2].

In this paper, we will use *d-bounded decomposition*, which extends the definition of crown decompositions and Lemma 1. Let A and B be two disjoint vertex subsets of a graph G. A *full d'-star packing from A to B* is a set of $|A|$ vertex-disjoint d'-stars with centers in A and leaves in B. The third item in Definition 1 means that there is a full 1-star packing from C to I. We define the following decomposition.

Definition 2 [d**–BoundedDecomposition**]. *A d-bounded decomposition of a graph $G = (V, E)$ is a partition of the vertex set of G into four sets I, C, T and J such that*

(1) any vertex in $I \cup T$ is of degree $\leq d$ in the induced subgraph $G[V \setminus C]$,
(2) there are no edges between I and J, and
(3) there is a full $(d+1)$-star packing from C to I.

An illustration for d-bounded decompositions is given in Fig. 1(b). We have the following Lemma 2 for d-bounded decompositions. This lemma can be derived from the lemmas in [14], although d-bounded decomposition is not formally defined in [14].

Lemma 2. *Let (I, C, T, J) be a d-bounded decomposition of G. Then (I, C) satisfies the local optimality condition in Theorem 3, i.e., $K' \cup C$ is a minimum d-degree deletion set of G for any minimum d-degree deletion set K' of the induced subgraph $G[V \setminus (I \cup C)]$.*

By Lemma 2, we can reduce an instance by removing $I \cup C$ if the graph has a d-bounded decomposition (I, C, T, J). This is the main idea how we get Theorem 3 and kernels for our problem. Here arises a problem how to find a d-bounded decomposition (I, C, T, J) of a graph such that $I \neq \emptyset$ if it exists. First, we give a simple observation.

Observation 1. *Let R be a set of vertices v such that any vertex in $N[v]$ is of degree $\leq d$. Then $(I = R, C = \emptyset, T = N(R), J = V \setminus (I \cup T))$ is a d-bounded decomposition of G.*

By Lemma 2 and Observation 1, we can reduce an instance by removing from the graph the set B of vertices v such that any vertex in $N[v]$ is of degree $\leq d$. We will introduce an algorithm that can find more d-bounded decompositions.

4 Algorithms

We first introduce an algorithm to find d-bounded decompositions of graphs, based on which we can easily get an algorithm for the generalization of NT-theorem in Theorem 3.

4.1 The Algorithm for Decompositions

First of all, we give the main idea of our algorithm to find a d-bounded decomposition (I, C, T, J) of a graph $G = (V, E)$. It contains three major phases.
Phase 1: find a partition (X, Y) of the vertex set V such that the maximum degree in $G[Y]$ is at most d.
Phase 2: find two subsets $C' \subseteq X$ and $I' \subseteq Y$ satisfying *Basic Condition*: there is a full $(d+1)$-star packing from C' to I' and there is no edge between I' and $X \setminus C'$.
Phase 3: iteratively move some vertices out of I' and some vertices out of C' to make $(I', C', T' = N(I') \setminus C', J' = V \setminus (I' \cup C' \cup T'))$ a d-bounded decomposition.

In fact, the first two phases of our algorithm are almost the same as that of Fellows, Guo, Moser and Niedermeier's algorithm [14]. However, in Phase 3, our algorithm uses a different method to compute I' and C'. This is critical for us to get an improvement.

Phase 1. For Phase 1, we can find a maximal $(d+1)$-star packing S and let $X = V(S)$. By the maximality of S, we know that X is a d-degree deletion set and $G[Y]$ has no vertex of degree $> d$. Then the partition (X, Y) satisfies the condition in Phase 1. In order to obtain a good performance, our algorithm may not use an arbitrary maximal $(d+1)$-star packing S. When we obtain a new $(d+1)$-star packing S' such that $|S'| > |S|$ in our algorithm, we will update X by letting $X = V(S')$.

Phase 2. After obtaining (X, Y) in Phase 1, our algorithm finds two special sets $C' \subseteq X$ and $I' \subseteq Y$ in Phase 2. To find C' and I' satisfying Basic Condition, we need to find a special $\leq(d+1)$-star packing from X to Y, which can be computed by the algorithms for finding maximum matchings in bipartite graphs. Note that the idea of computing $\leq(d+1)$-stars from X and Y has been used to solve some other problems in references [11,16,25].

We consider the bipartite graph $H = (X, Y, E_H)$ with edge set E_H being the set of edges between X and Y in G, and are going to find a $\leq(d+1)$-star packing from X to Y in H. Note that a Y-vertex no adjacent to any vertex in X will become a degree-0 vertex in H. We construct an auxiliary bipartite graph $H' = (X_1 \cup X_2 \cup \ldots X_{d+1}, Y, E'_H)$, where each X_i $(i = 1, 2, \ldots, d+1)$ is a copy of X and a vertex $v_i \in X_i$ is adjacent to a vertex $u \in Y$ if and only if the corresponding vertex $v \in X$ is adjacent to u in H. For a vertex $v \in X$, we may use v_i to denote its corresponding vertex in X_i.

We find a maximum matching M' in H' by using a $O(n^{1/2}m)$-time algorithm [13,18]. Let M be the set of edges in H corresponding to the matching M', i.e., an edge uv ($u \in Y$ and $v \in X$) of H is in M if and only if uv_i is in M' for some v_i corresponding to v. Edges in M are called *marked* and others are called *unmarked.* Observe that since M' is a matching in H', we have that $|M| = |M'|$. The set of marked edges in H forms a $_{\leq}(d+1)$-star packing $S_{\leq d+1}$. This is the $_{\leq}(d+1)$-star packing we are seeking for. It is also easy to observe that

Lemma 3. *Graph H has a $_{\leq}(d+1)$-star packing containing t edges if and only if H' has a matching of size t.*

Next, we analyze some properties of $S_{\leq d+1}$ and find C' and I' satisfying Basic Condition based on these properties.

Let S_{d+1} denote the set of $(d+1)$-stars in $S_{\leq d+1}$. An X-vertex in a star in S_{d+1} is *fully tagged.* Then $X \cap V(S_{d+1})$ is the set of fully tagged vertices. A Y-vertex is *untagged* if it is adjacent to at least one vertex in X in H but not contained in any star in $S_{\leq d+1}$. A path P in H that alternates between edges not in M and edges in M is called an *M-alternating path.*

Lemma 4. *If there is an M-alternating path P from an untagged vertex $u \in Y$ to a vertex $v \in X$ in H, then v is fully tagged.*

Next, we are going to set C' and I'. If there is no untagged vertex, let $C' = \emptyset$. Otherwise let C' be the set of X-vertices connected with at least one untagged vertex by an M-alternating path in H. Let $X' = X \setminus C'$. Let Y' be the set of Y-vertices that is a leaf of a $_{\leq}(d+1)$-star in $S_{\leq d+1}$ that is centered at a vertex in X', and $I' = Y \setminus Y'$.

Lemma 5. *The two sets C' and I' obtained above satisfy Basic Condition.*

We describe the above progress to compute C' and I' as an algorithm $\texttt{basic}(G, X, Y)$ in Fig. 2, which will be used as a subalgorithm in our main algorithm.

Lemma 6. *Algorithm* $\texttt{basic}(G, X, Y)$ *runs in $O(n^{1/2}m)$ time.*

Note that all untagged vertices will be in I'. So if the size of Y is large, for example $|Y| > (d+1)|X|$, we can guarantee that there is always some untagged vertices and the set I' returned by $\texttt{basic}(G, X, Y)$ is not an empty set.

Phase 3. After obtaining (C', I') from Phase 2, we look at the partition $\mathcal{P} = (I', C', T' = N(I') \setminus C', J' = V \setminus (I' \cup C' \cup T'))$. Since there is no edge between I' and $X' = X \setminus C'$, we know that $T' \subseteq Y$ and $X' \subseteq J'$. Then there is no edge between I' and J'. The partition $\mathcal{P}$ satisfies Conditions (2) and (3) in Definition 2 for d-bounded decompositions. Next, we consider Condition (1). Let $G^* = G[V \setminus C']$. Any vertex in I' is of degree $\leq d$ in G^*, because $G[Y] = G[V \setminus X]$ has maximum degree $\leq d$ and I'-vertices are not adjacent to any vertex in $X \setminus C'$.

Input: A graph $G = (V, E)$ and a partition (X, Y) of the vertex set V.
Output: Two sets $C' \subseteq X$ and $I' \subseteq Y$ satisfying the Basic Condition.

1. Compute the bipartite graph H and the auxiliary bipartite graph H'.
2. Compute a maximum matching M' in H' and the corresponding edge set M and the $\leq(d+1)$-star packing $S_{\leq d+1}$ in H.
3. Let C' be $\emptyset$ if there is no untagged vertex, and the set of X-vertices connected with at least one untagged vertex by an M-alternating path in H otherwise. Let $X' \leftarrow X \setminus C'$. Let Y' be the set of Y-vertices each of which is a leaf of a $\leq(d+1)$-star centered at a vertex in X' and let $I' \leftarrow Y \setminus Y'$.
4. Return (C', I').

Fig. 2. Algorithm $\texttt{basic}(G, X, Y)$

Although $T' = N(I') \setminus C' \subseteq Y$, vertices in T' is possible to be of degree $> d$ in G^*. In fact, we only know that each vertex in T' is of degree $\leq d$ in $G[Y]$. But in G^*, every T'-vertex is adjacent to some vertices in $X' = X \setminus C'$ and thus can be of degree $> d$. So Condition (1) may not hold. We will move some vertices out of C' and I' to make the decomposition satisfying Condition (1).

Let B be the set of T'-vertices that are of degree $> d$ in G^*. Note that any vertex in B is adjacent to some vertices in X. We call vertices in $N_{I'}(B) = N(B) \cap I'$ *bad* vertices. Note that B is not an empty set if and only if $N_{I'}(B)$ is not an empty set. If $B = \emptyset$, then Condition (1) holds directly. For the case that $B \neq \emptyset$, i.e., $N_{I'}(B) \neq \emptyset$, our idea is to update I' by removing $N_{I'}(B)$ out of I'. However, after moving some vertices out of I', there may not be a full $(d+1)$-star packing from C' to I' anymore. So after moving $N_{I'}(B)$ out of I' we invoke the algorithm $\texttt{basic}(G[C' \cup I'], C', I')$ for Phase 2 on the subgraph $G[C' \cup I']$ to find new C' and I', and then check whether there are new bad vertices or not. We do these iteratively until we find a d-bounded decomposition, where no bad vertex exists. In the returned d-bounded decomposition, I' and C' may become empty. However, we can guarantee $I' \neq \emptyset$ when the size of the graph satisfies some conditions. We analyze this after describing the whole algorithm.

The Whole Algorithm for Decomposition. Our algorithm $\texttt{decomposition}$ (G) presented in Fig. 3 is to compute two subsets of vertices C and I of the input graph G such that $(I, C, T = N(I) \setminus C, J = V \setminus (I \cup C \cup T))$ is a d-bounded decomposition of G.

Steps 3, 4 and 6 in $\texttt{decomposition}(G)$ are the same steps in $\texttt{basic}(G, X, Y)$. Here we add Step 5 into these steps, which is used to update the $(d+1)$-star packing S. In $\texttt{decomposition}(G)$, Steps 1, 2 and 5 are corresponding to Phase 1, Steps 3, 4 and 6 are corresponding to Phase 2, and Steps 7 and 8 are corresponding to Phase 3. Note that Step 8 will also invoke $\texttt{basic}(G, X, Y)$.

Input: A graph $G = (V, E)$.
Output: Two subsets of vertices C and I such that $(I, C, T = N(I) \setminus C, J = V \setminus (I \cup C \cup T))$ is a d-bounded decomposition.

1. Find a maximal $(d+1)$-star packing S in G.
2. $X \leftarrow V(S)$ and $Y \leftarrow V \setminus X$.
3. Compute the bipartite graph H and the auxiliary bipartite graph H'.
4. Compute a maximum matching M' in H' and the corresponding edge set M and the $\leq(d+1)$-star packing $S_{\leq d+1}$ in H.
5. Let S_{d+1} be the set of $(d+1)$-stars in $S_{\leq d+1}$.
   If $\{|S_{d+1}| > |S|\}$,
   then $S \leftarrow S_{d+1}$ and **goto** Step 2.
6. Let C' be $\emptyset$ if there is no untagged vertex, and be the set of X-vertices connected with at least one untagged vertex by an M-alternating path in H otherwise. Let $X' \leftarrow X \setminus C'$. Let Y' be the set of leaves of $\leq(d+1)$-stars in $S_{\leq d+1}$ centered at vertices in X' and let $I' \leftarrow Y \setminus Y'$.
7. Compute the set $N_{I'}(B)$ of bad vertices based on C' and I'.
8. **If** $\{N_{I'}(B) \neq \emptyset\}$,
   then $I' \leftarrow I' \setminus N_{I'}(B)$, $(C', I') \leftarrow$ `basic`$(G[C' \cup I'], C', I')$, and **goto** Step 7.
9. **Return** $(C = C', I = I')$.

Fig. 3. Algorithm `decomposition`(G)

Lemma 7. *The two vertex sets C and I returned by* `decomposition`(G) *make* $(I, C, T = N(I) \setminus C, J = V \setminus (I \cup C \cup T))$ *a d-bounded decomposition.*

We can prove the following two important lemmas.

Lemma 8. *Algorithm* `decomposition`(G) *runs in $O(n^{3/2}m)$ time and returns (C, I) such that (I, C, T, J) is a d-bounded decomposition of G, where $T = N(I) \setminus C$ and $J = V(G) \setminus (I \cup C \cup T)$.*

Lemma 9. *Algorithm* `decomposition`(G) *returns (C, I) such that*

$$|V \setminus (C \cup I)| \leq (d^3 + 4d^2 + 5d + 3)\alpha(G).$$

4.2 The Algorithm for Theorem 3

Lemma 9 can get the size condition in Theorem 3 directly. We use the following algorithm in Fig. 4 for Theorem 3.

From the second iteration of Step 2 in `BDD`(G), each execution of $I \leftarrow I \cup I'$ will include at least one new vertex to I. So `decomposition`$(G[V \setminus (C \cup I)])$ will be called for at most $n + 1$ times. Algorithm `BDD`(G) runs in $O(n^{5/2}m)$ time. Furthermore, if `decomposition`$(G' = G[V \setminus (C \cup I)])$ returns two empty sets, then by Lemma 9 we have $|V(G')| = |V(G') \setminus (C \cup I)| \leq (d^3 + 4d^2 + 5d + 3)\alpha(G')$. These together with Lemmas 8 and 9 imply Theorem 3.

Input: A graph $G = (V, E)$.
Output: Two subsets of vertices C and I satisfying the conditions in Theorem 3.

1. $C, I \leftarrow \emptyset$.
2. **Do** { $(C', I') \leftarrow$ `decomposition`$(G[V \setminus (C \cup I)])$, $C \leftarrow C \cup C'$ and $I \leftarrow I \cup I'$ } **while** $I' \neq \emptyset$.
3. **Return** (C, I).

Fig. 4. Algorithm BDD(G)

5 Concluding Remarks

In this paper, we provide a refined version of the generalized Nemhauser-Trotter-Theorem, which applies to Bounded-Degree Vertex Deletion and for any $d \geq 0$ can get a linear-vertex problem kernel for the problem parameterized by the solution size. This is the first linear-vertex kernel for the case that $d \geq 3$. Our algorithms and proofs are based on extremal combinatorial arguments, while the original NT-Theorem uses linear programming relaxations [22]. It seems no way to generalize the linear programming relaxations used for the original NT-Theorem to Bounded-Degree Vertex Deletion [14]. A crucial technique in this paper is the d-bounded decomposition. To find such kinds of decompositions, we follow the ideas to find crown decompositions [2] and the algorithmic strategy in [14]. However, we use more ticks and can finally obtain the linear size condition.

As pointed out by Fellows et al. [14], the results for Bounded-Degree Vertex Deletion in this paper can be modified for the problem of packing stars. We believe that the new decomposition technique can be used to get local optimization properties and kernels for more deletion and packing problems.

References

1. Abu-Khzam, F.N., Collins, R.L., Fellows, M.R., Langston, M.A., Suters, W.H., Symons, C.T.: Kernelization algorithms for the vertex cover problem: theory and experiments. In: ALENEX 2004, pp. 62–69. ACM/SIAM (2004)
2. Abu-Khzam, F.N., Fellows, M.R., Langston, M.A., Suters, W.H.: Crown structures for vertex cover kernelization. Theor. Comput. Syst. **41**(3), 411–430 (2007)
3. Balasundaram, B., Butenko, S., Hicks, I.V.: Clique relaxations in social network analysis: the maximum k-plex problem. Oper. Res. **59**(1), 133–142 (2011)
4. Bar-Yehuda, R., Rawitz, D., Hermelin, D.: An extension of the Nemhauser & Trotter theorem to generalized vertex cover with applications. SIAM J. Discrete Math. **24**(1), 287–300 (2010)
5. Bar-Yehuda, R., Even, S.: A local-ratio theorem for approximating the weighted vertex cover problem. Ann. Discrete Math. **25**, 27–45 (1985)
6. Betzler, N., Bredereck, R., Niedermeier, R., Uhlmann, J.: On bounded-degree vertex deletion parameterized by treewidth. Discrete Appl. Math. **160**(1–2), 53–60 (2012)

7. Chen, J., Kanj, I.A., Jia, W.: Vertex cover: further observations and further improvements. J. Algorithms **41**(2), 280–301 (2001)
8. Chen, Z.-Z., Fellows, M., Fu, B., Jiang, H., Liu, Y., Wang, L., Zhu, B.: A linear kernel for co-path/cycle packing. In: Chen, B. (ed.) AAIM 2010. LNCS, vol. 6124, pp. 90–102. Springer, Heidelberg (2010)
9. Chlebík, M., Chlebíková, J.: Crown reductions for the minimum weighted vertex cover problem. Discrete Appl. Math. **156**, 292–312 (2008)
10. Chor, B., Fellows, M., Juedes, D.W.: Linear kernels in linear time, or how to save k colors in $O(n^2)$ steps. In: Hromkovič, J., Nagl, M., Westfechtel, B. (eds.) WG 2004. LNCS, vol. 3353, pp. 257–269. Springer, Heidelberg (2004)
11. Cygan, M., Pilipczuk, M., Pilipczuk, M., Wojtaszczyk, J.O.: An improved FPT algorithm and a quadratic kernel for pathwidth one vertex deletion. Algorithmica **64**(1), 170–188 (2012)
12. Dell, H., van Melkebeek, D.: Satisfiability allows no nontrivial sparsification unless the polynomial-time hierarchy collapses. J. ACM **61**(4), 23:1–23:27 (2014)
13. Even, S., Tarjan, R.E.: An $O(n^{2.5})$ algorithm for maximum matching in general graphs. In: FOCS 1975, pp. 100–112 (1975)
14. Fellows, M.R., Guo, J., Moser, H., Niedermeier, R.: A generalization of Nemhauser and Trotter's local optimization theorem. J. Comput. Syst. Sci. **77**, 1141–1158 (2011)
15. Fellows, M.R., Guo, J., Moser, H., Niedermeier, R.: A generalization of Nemhauser and Trotter's local optimization theorem. In: STACS 2009, pp. 409–420. IBFI Dagstuhl (2009)
16. Fomin, F.V., Lokshtanov, D., Misra, N., Philip, G., Saurabh, S.: Hitting forbidden minors: approximation and kernelization. In: STACS 2011, pp. 189–200 (2011)
17. Hochbaum, D.S.: Approximation algorithms for the set covering and vertex cover problems. SIAM J. Comput. **11**(3), 555–556 (1982)
18. Hopcroft, J., Karp, R.M.: An $O(n^{2.5})$ algorithm for maximum matching in bipartite graphs. SIAM J. Comput. **2**(4), 225–231 (1973)
19. Khuller, S.: The vertex cover problem. SIGACT News **33**(2), 31–33 (2002)
20. Komusiewicz, C., Huffner, F., Moser, H., Niedermeier, R.: Isolation concepts for efficiently enumerating dense subgraphs. Theoret. Comput. Sci. **410**(38–40), 3640–3654 (2009)
21. Lokshtanov, D., Narayanaswamy, N.S., Raman, V., Ramanujan, M.S., Saurabh, S.: Faster parameterized algorithms using linear programming. ACM T. Algorithms **11**(2), 15 (2014)
22. Nemhauser, G.L., Trotter, L.E.: Vertex packings: structural properties and algorithms. Math. Program. **8**, 232–248 (1975)
23. Nishimura, N., Ragde, P., Thilikos, D.M.: Fast fixed-parameter tractable algorithms for nontrivial generalizations of vertex cover. Discrete Appl. Math. **152**(1–3), 229–245 (2005)
24. Seidman, S.B., Foster, B.L.: A graph-theoretic generalization of the clique concept. J. Math. Sociol. **6**, 139–154 (1978)
25. Thomassé, S.: A $4k^2$ kernel for feedback vertex set. ACM T. Algorithms **6**(2), 32:1–32:28 (2010)
26. Xiao, M.: A note on vertex cover in graphs with maximum degree 3. In: Thai, M.T., Sahni, S. (eds.) COCOON 2010. LNCS, vol. 6196, pp. 150–159. Springer, Heidelberg (2010)

Approximation Algorithms in the Successive Hitting Set Model

Sabine Storandt(✉)

Department of Computer Science, University of Freiburg,
Freiburg im Breisgau, Germany
storandt@informatik.uni-freiburg.de

Abstract. We introduce the successive Hitting Set model, where the set system is not given in advance but a set generator produces the sets that contain a specific element from the universe on demand. Despite incomplete knowledge about the set system, we show that several approximation algorithms for the conventional Hitting Set problem can be adopted to perform well in this model. We describe, and experimentally investigate, several scenarios where the new model is beneficial compared to the conventional one.

1 Introduction

The Hitting Set problem is a classical NP-complete problems, with applications in various areas as computational geometry [1], sensor networks [2], or route planning [3]. The unweighted Hitting Set problem (HS) is defined as follows.

Definition 1 (Hitting Set). *Given a set system* $(U, \mathfrak{S})$ *with* U *being a universe of elements and* $\mathfrak{S}$ *a collection of subsets of* U*, the Hitting Set problem demands to find a smallest subset of the universe* $H \subseteq U$ *such that all sets in* $\mathfrak{S}$ *are hit by* H*, i.e.* $\forall S \in \mathfrak{S} : S \cap H \neq \emptyset$.

In the weighted version, additionally a weight function $w : U \rightarrow \mathbb{R}^+$ is given. The goal is then to find the cheapest $H \subseteq U$ which hits all sets in $\mathfrak{S}$.

Both problem versions are not only NP-hard but exhibit also an inapproximability bound of $\ln(m)(1 - o(1))$ with $m = |\mathfrak{S}|$ as the dual of the Hitting Set problem, the Set Cover problem, was proven to be $\ln(n)(1-o(1))$ inapproximable with n being the number of elements in the universe [4]. Hitting Set problems are often tackled in practice with the greedy algorithm [5], as it provides an asymptotically optimal approximation guarantee of $\ln(m) + \Theta(1)$. For more refined Hitting Set problem versions, or with a priori knowledge about the set system, better approximations and custom-tailored heuristics are possible.

The main obstacle for solving Hitting Set type problems in practice is that with $\mathfrak{S} \subseteq \mathcal{P}(U)$ the number of sets might be significantly larger than the number of elements in the universe U. Therefore storing $\mathfrak{S}$ explicitly can demand enormous space, and operations on the complete set system are extremely expensive.

K. Elbassioni and K. Makino (Eds.): ISAAC 2015, LNCS 9472, pp. 453–464, 2015.
DOI: 10.1007/978-3-662-48971-0_39

This limits the applicability of greedy and other approximation algorithms to rather small instances.

A natural question is, whether we could solve the Hitting Set problem without having to store and investigate whole $\mathfrak{S}$ at once. Obviously, this demands that we can access certain subsets of $\mathfrak{S}$ efficiently. We formalize this idea into a new model, which we call the successive Hitting Set (SHS) model.

Definition 2 (Successive Hitting Set Model). *Given a universe of elements U, and a deterministic set generator $G : U \to \mathcal{P}(U)$. The set generator called for $u \in U$ reveals the collection of sets that contain u. The universe together with $\cup_{u \in U} G(u)$ forms the set system.*

The Hitting Set problem in the successive model remains basically unchanged. And, of course, we could just call the generator for all elements first, and then run the conventional Hitting Set algorithms to compute H. But the scope of the paper is to design (approximation) algorithms that issue calls to the generator in a way that the number of known sets (i.e. sets that have to be explicitly stored) at any point in time is significantly lower than $|\mathfrak{S}|$.

1.1 Related Work

Hitting Set problems or Set Cover problems with the set system being not fully provided a priori were tackled before in the context of an on-line model [6]. In this model, sets or elements are revealed to the algorithm in some unpredictable order and have to be handled immediately. On-line algorithms are analyzed by bounding the competitive ratio, that is the solution cost in the on-line model divided by the solution cost in the conventional off-line model. This differs significantly from our successive setting, as here the sets are not revealed by some 'adversary' but it is part of our envisioned approximation algorithms to call the set generator for elements in the universe wisely.

Furthermore, there is a wide range of heuristics for the conventional Hitting Set problem which aim at compressing the set system or avoid its explicit construction in order to be able to tackle large instances or to accelerate the computation [7–10]. Most often these heuristics are custom-tailored for certain kinds of set systems, and the focus is rather on providing good solutions in practice than on investigating theoretical approximation guarantees. We will provide successive algorithms in the following which exhibit good approximation guarantees and perform well in practice at the same time.

1.2 Contribution

- We adopt the standard greedy algorithm to work in the SHS model with an approximation guarantee of $\ln(m) + 2$.
- We show that the k-approximation for the k-Hitting Set problem via the pricing method carries over to our new model.

- We prove that for set systems with VC-dimension d, a $2dc\log(dc)$ approximation is possible in the SHS model. In the conventional model, the guarantee is $dc\log(dc)$.
- We investigate several applications where the successive model leads to a considerably reduced space consumption and/or faster computation times compared to the conventional model. Furthermore, we show that in practice our devised approximation algorithms achieve close-to-optimal solutions.

2 Preliminaries

In this paper we restrict ourselves to algorithms where the generator is only called once per element in the universe. Otherwise, every time an operation needs to be conducted in a conventional Hitting Set algorithm, the respective part of the set system is generated (if it fits in memory) and simply forgotten afterwards. But this potentially leads to a very high number of calls to the generator. Hence the time spend on set generation might dominate the total runtime, which is not what we aim for. So the paradigm in this paper is that a set once generated can only be forgotten after it was hit. This also provides us with an easy correctness prove for all our algorithms: If the generator was called for every element in the universe and the set system is empty, a feasible Hitting Set is at hand.

We use the following notation. With c we denote the size of the optimal solution H^*. We refer to the underlying set of elements for a collection of sets $\mathfrak{S}$ as $\underline{\mathfrak{S}} = \cup_{S \in \mathfrak{S}} S$. We assume the sets in $\mathfrak{S}$ to be closed under intersection, that is we cannot divide $\mathfrak{S}$ into two partitions $\mathfrak{S}_A$ and $\mathfrak{S}_B$ with $\underline{\mathfrak{S}_A} \cap \underline{\mathfrak{S}_B} = \emptyset$. This of course is only a technical restriction. If $\mathfrak{S}$ is not closed under intersection, we could define independent subproblems and solve them individually. In our algorithms, whenever a temporary set system runs empty because no sets intersect with previously chosen ones, we just call the generator for some arbitrary element (for which the generator was not already called), and proceed from there.

3 Greedy Algorithm for General Set Systems

The classical greedy algorithm for the Hitting Set problem works as follows. In every round of the algorithm, the element $u \in U$ is selected which hits most so far unhit sets. Or, in the weighted case, the element u which minimizes $w(u)/|\{S \in \mathfrak{S} : S \ni u\}|$. Then u is added to the Hitting Set H, and all newly hit sets are removed from the system. The algorithm proceeds until $\mathfrak{S}$ runs empty.

The computation of the best hitter in every round and the removal of the newly hit sets induce a complete sweep over all elements in so far unhit sets. This makes the execution of greedy quite expensive, especially in early rounds.

The greedy algorithm guarantees a $\ln(b) + \Theta(1)$ approximation with b being the size of the largest subset of $\mathfrak{S}$ that can be hit with a single element from U. As this subset potentially contains (almost) all sets from $\mathfrak{S}$, we have a $\ln(m) + \Theta(1)$ approximation with $m = |\mathfrak{S}|$.

3.1 Successive Greedy

In the SHS setting, we proceed as follows. We start with an arbitrary subset $\mathfrak{S}' \subseteq \mathfrak{S}$. (If no such set is specified, we call the generator G for an arbitrary element in U and refer to the resulting set as $\mathfrak{S}'$.) We select the best hitter for $\mathfrak{S}'$, add it to the solution H and remove all hit sets from $\mathfrak{S}'$ just like in the conventional greedy algorithm. But now, for every set that was hit, we call the set generator for all contained elements. (Of course, we never call the set generator twice for an element during the course of the algorithm, and we discard sets immediately that are already hit.) The generated sets are added to $\mathfrak{S}'$. Then the whole process is repeated. The algorithm stops after the set generator was called for every element in U, and $\mathfrak{S}'$ ran empty.

3.2 Approximation Quality

The successive greedy algorithm has very limited knowledge about the set system in every round. This is the very opposite of the way conventional greedy works, as it always selects the best hitter *globally*. Nevertheless, we will prove that the approximation guarantee of the successive greedy algorithm is quite close to the guarantee in the conventional model, as specified in the following Theorem.

Theorem 1. *Successive greedy computes a Hitting Set H with the property $|H| \leq c \cdot (\ln m + 2)$ where c denotes the optimal solution size and m being the number of sets in the complete set system.*

Proof. Let h be a hitter in the optimal solution, and $\mathfrak{S}(h)$ the collection of sets hit by h. With $s = |\mathfrak{S}(h)|$ we denote the number of sets hit by h, i.e. every set in $\mathfrak{S}(h)$ is bought at cost $w(h)/s$ (in the unweighted case $w(h) = 1$). We will argue that the total costs for $\mathfrak{S}(h)$ in the successive greedy algorithm are lower or equal to $w(h)(H_{s-1}+1)$ with H_{s-1} indicating the $(s-1)^{th}$ harmonic number.

Let h_0 be the first hitter in the course of the successive greedy algorithm which hits a set S in $\mathfrak{S}(h)$. Obviously, S is bought at cost $\leq w(h)$, as h would have been a possible choice as well. After h_0 is added to H, all sets are generated which intersect with S. Therefore, in the next round of the successive algorithm all sets $\mathfrak{S}(h) \setminus S$ are available in the temporary set system. (Of course, h_0 might hit more than one set in $\mathfrak{S}(h)$, but this would only reduce the total costs for $\mathfrak{S}(h)$.) The successive greedy algorithm could now choose h as the next hitter for $\mathfrak{S}(h)$, with a cost ratio of $w(h)/(s-1)$. So the only reason why the algorithm decides for another element h_1 is, that its ratio is even better or equal to $w(h)/(s-1)$. This ratio determines the cost for the next hit set in $\mathfrak{S}(h)$. Then we can apply the same argument recursively, providing us with a cost ratio of $\leq w(h)/(s-i)$ for hitter h_i, until $i = s-1$ and all sets in $\mathfrak{S}(h)$ are hit. Hence the total cost for all sets in $\mathfrak{S}(h)$ can be expressed as:

$$w(h) + \sum_{i=1}^{s-1} \frac{w(h)}{s-i} = w(h) + w(h) \sum_{i=1}^{s-1} \frac{1}{i} = w(h)(H_{s-1}+1)$$

Using $H_n < \ln n + 1$, we can upper bound the costs for $\mathfrak{S}(h)$ by $w(h)(\ln(s)+2)$. So compared to the costs of $w(h)$ for $\mathfrak{S}(h)$ in the optimal solution, we pay more by at most a factor of $\ln(s) + 2$. As this is true for every hitter in the optimal solution, and $s \leq m$, successive greedy has an approximation guarantee of $\ln(m) + 2$. ■

Arguing more precisely, the standard greedy algorithm exhibits an approximation guarantee of H_b with b being the size of the largest subset of $\mathfrak{S}$ that can be hit with a single element from U. This term converges to $\ln(b) + \gamma$ for growing b, with γ denoting the Euler-Mascheroni constant ($\gamma \approx 0.57721$). In the successive model, the approximation guarantee is $H_{b-1}+1 = H_b+1-1/(b-1)$.

4 Pricing Method for the K-Hitting Set Problem

We now consider the k-Hitting Set problem. Here, all sets in the collection $\mathfrak{S}$ contain at most k elements. For this special kind of Hitting Set problem, the general inapproximability bound does not apply. In fact, there exists a k-approximation algorithm which is an instance of the primal-dual method. The algorithm is called the pricing method as it assigns prices p_S to sets in the system. Initially all prices are zero, so $\forall S \in \mathfrak{S} : p_S = 0$. For every element $u \in U$, the following constraint yields $\sum_{S \in \mathfrak{S}(u)} p_S \leq w(u)$ with $\mathfrak{S}(u)$ being the collection of sets that contain u. If equality holds, the element u is called tight. The pricing method operates in rounds. In every round, a set S from $\mathfrak{S}$ is selected which contains only elements that are not tight. Then the price of the set p_S is increased as much as possible without violating any constraint. This leads to at least one of the elements in S becoming tight. The algorithm exits as soon as every set in $\mathfrak{S}$ contains some tight element. All tight elements form then the Hitting Set H.

4.1 Successive Algorithm

Again, we start with some arbitrary set $\mathfrak{S}' \subseteq \mathfrak{S}$. We select a set S from $\mathfrak{S}'$ in every round to make one of the contained elements tight. But we have to be very careful about not violating any constraints when increasing the price of S. Therefore, we maintain potential weights w' for every $u \in U$. In the beginning, we have $w'(u) = w(u)$ for all elements. Then after selecting S, we compute $\Delta = \min_{u \in S} w'(u)$. We increase the price p_S by Δ and at the same time decrease all potential weights from elements in S by Δ. At least one of those elements will have a potential weight of 0 afterwards. All elements with w' being 0 are added to H, and hit sets are removed from $\mathfrak{S}'$. Note, that it does not matter how we issue calls to the generator. We can just select some arbitrary element u in every round (for which the generator was not already called); and add the respective sets $G(u)$ to $\mathfrak{S}'$. Again, the algorithm exits as soon as the generator was called for all elements and $\mathfrak{S}'$ is empty.

4.2 Correctness and Analysis

Correctness of the successive pricing method is obvious, as the algorithm only terminates when every set in the system contains some tight element; and the tight elements coincide with the Hitting Set.

The quality analysis works similar to the conventional pricing method analysis. The only point we have to assure is not to violate any constraint. Every time we increase the price of a set, we decrease the potential weights of all contained elements. Therefore the summed prices of sets that contain a specific element are always equal to the original potential weight minus the final potential weight of the element. As potential weights never drop below 0, the summed prices of sets that contain element u are bounded by $w(u)$. Therefore all constraints are satisfied at any point in time. Accordingly, the quality analysis for the conventional pricing method can be applied, proving that the successive pricing method also has an approximation guarantee of k.

Theorem 2. *The successive pricing method returns a solution H for an instance of the k-Hitting-Set problem with $|H| \leq c \cdot k$ where c denotes the optimal solution size.*

The advantage of the successive approach is again that it requires only a very small subset of $\mathfrak{S}$ to be explicitly stored. More precisely, a *single* set being available in each round suffices for correctness and to achieve the desired approximation quality. Hence, linear space in the size of U would be enough for the pricing method to work, if our set generator can be instrumented to produce the sets per element one by one. For comparison, whole $\mathfrak{S}$ might require space in the order of $k \cdot |U|^k$.

5 Concatenated Hitting Sets

Lets assume we have some algorithm $\mathcal{A}$ for HS which provides a better approximation guarantee than the generally tight $\ln(m)$ bound by making use of characteristics of the underlying set system. A famous incarnation of such $\mathcal{A}$ is the algorithm by Brönimann and Goodrich [11] which provides for set systems with VC-dimension d a solution within $dc \log(dc)$. In this context, the VC-dimension can be regarded as a complexity measure for the set system. Low VC-dimensions are exhibited e.g. by many set systems on geometric objects [12]. We now describe a successive scheme which can exploit such algorithms $\mathcal{A}$ to find good approximate solutions while only operating on subsets of $\mathfrak{S}$.

5.1 Successive Algorithm

Like before, we start with some arbitrary subset $\mathfrak{S}_1 \subseteq \mathfrak{S}$. We apply algorithm $\mathcal{A}$ to $\mathfrak{S}_1$ conventionally. This provides us with an initial Hitting Set H_1. Then we delete the sets in $\mathfrak{S}_1$ from the system but add all sets that have a non-empty intersection with a set in $\mathfrak{S}_1$ (by calling the generator for all elements in $\underline{\mathfrak{S}_1}$).

Of course, we never add already hit sets to the system at any point in time. For the newly generated set system $\mathfrak{S}_2$, we again apply $\mathcal{A}$ which leads to a second Hitting Set H_2. We repeat the process until the generator was called for every element in U and the set system is empty. The final Hitting Set returned is the union of $H_1, H_2, \cdots, H_k$.

5.2 Correctness and Approximation Quality

As every set in $\mathfrak{S}$ is generated at some point and only deleted after it was hit, correctness of the successive algorithm is obvious.

It remains to analyze the solution quality. We make the following two simple but crucial observations:

Observation 3. *The optimal solution size c_i for $\mathfrak{S}_i$ is smaller or equal to the optimal solution size $c = |H|$ for $\mathfrak{S}$ reduced to $\mathfrak{S}_i$, i.e. $c_i \leq |H \cap \underline{\mathfrak{S}_i}|$.*

Observation 4. $\underline{\mathfrak{S}_i} \cap \underline{\mathfrak{S}_j} = \emptyset$ *if $|i - j| > 1$, because all sets intersecting with $\mathfrak{S}_i$ are either already contained in $\mathfrak{S}_{i-1}$ or are created in $\mathfrak{S}_{i+1}$ and therefore hit and deleted before the construction of $\mathfrak{S}_{i+2}$.*

The second observation tells us that $\mathfrak{S}_{odd} = \mathfrak{S}_1, \mathfrak{S}_3, \cdots$ is a collection of pairwise intersection free instances, and the same is true for $\mathfrak{S}_{even} = \mathfrak{S}_2, \mathfrak{S}_4, \cdots$. According to the first observation, the optimal solution for whole $\mathfrak{S}$ requires at least as many hitters for $\mathfrak{S}_i$ as the individual optimal solution. As the union of intersection free instances can not lead to any redundant hitters, we conclude $c \geq \sum_{i=1}^{k/2} c_{2i}$ and $c \geq \sum_{i=1}^{k/2} c_{2i-1}$. In total we get $2c \geq \sum_{i=1}^{k} c_i$.

Now, we consider $\mathcal{A}$ with an approximation guarantee of $d \log(dc)$ with d denoting the VC-dimension of the set system. For any subset of $\mathfrak{S}$, the VC-dimension can not be higher than d. So we have $|H_i| \leq dc_i \log(dc_i)$. The optimal solution for any $\mathfrak{S}_i$ is smaller or equal to the global solution, i.e. $c_i \leq c$. Hence we can upper bound $|H_i|$ by $dc_i \log(dc)$. Then we can upper bound the size of the solution resulting from combining all individual H_i by:

$$\sum_{i=1}^{k} |H_i| \leq \sum_{i=1}^{k} dc_i \log(dc) = d \log(dc) \sum_{i=1}^{k} c_i \leq 2dc \log(dc)$$

The last inequality uses our lower bound for the optimal solution c as constructed above.

There are other set systems which even exhibit constant approximations [13] or PTAS [1]. The respective approximation algorithms can easily be plugged into our successive scheme and the analysis is quite similar.

Theorem 5. *For an approximation algorithm $\mathcal{A}$ which computes a Hitting Set H with the guarantee $|H| \leq f(c)$, the successive variant of $\mathcal{A}$ exhibits an approximation guarantee of $2f(c)$.*

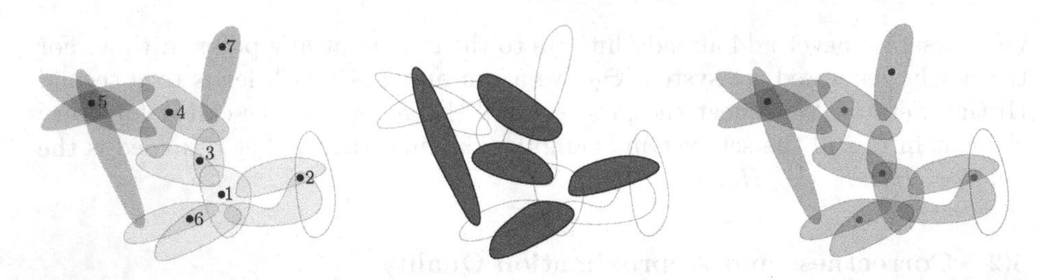

Fig. 1. Left: Illustration of the successive greedy algorithm. Hitters are indicated by black dots, the enumeration reflects the order in which they were chosen. Middle: Independent collection of sets picked by the successive pricing method. All elements in the lilac sets form the Hitting Set. Right: Illustration of the concatenation algorithm. The first instance is given by the blue sets, the second by the red sets and the third by the green sets (together with a Hitting Set per instance). The green and the blue instances are intersection-free. The colorless set on the right is not in the green instance as it is already hit by the red Hitting Set (Colour figure online).

So our successive scheme produces a solution with an approximation guarantee which is only worse by a factor of 2 compared of the original approximation guarantee. At the same time, our algorithm only requires the storage of the actual $\mathfrak{S}_i$, and operations to compute H are only performed on sub-instances.

For improving the solution quality in practice, we can apply a backwards pruning strategy. At the moment we constructed the Hitting Set H_i for $\mathfrak{S}_i$, we can check if elements in H_{i-1} become superfluous due to H_i. For that purpose, we sweep over the sets in $\mathfrak{S}_{i-1}$ that are *not* hit by H_i and only maintain their hitters in H_{i-1}.

6 Applications

Our theoretical investigations showed that operating in the successive model leads to the same approximation guarantees in Big-O-notation than in the conventional model. But the question remaining is, if there are really applications where the intermediate sizes of the set system known to the successive approximation algorithms are considerably smaller than $|\mathfrak{S}|$.

Figure 1 illustrates all three introduced algorithms (successive greedy, successive pricing method and concatenation). For the k-Hitting Set problem we observed that a single known set suffices for the pricing method to work correctly. But for greedy and the concatenated algorithm, the existence of elements $u \in U$ with $\overline{G(u)}$ containing a significant fraction of the elements in U possibly leads to set systems with their size comparable to $\mathfrak{S}$. So in that case our successive algorithms are not advantageous.

In this section, we will describe applications where the successive model is intuitively beneficial compared to the conventional one.

Set Systems with Efficient Generators. The efficiency of our successive algorithms relies on how quick they can operate on the temporary set system as well as on how quick they can generate the next required sets. We will describe an exemplary application in the following, where efficient generators are easily available. So at the latest when the complete set system would no longer fit in memory, the successive algorithm will outperform the conventional one.

Example 1 (Hitting k-Paths or Shortest Paths). Given a graph $G(V, E)$, the objective is to hit all simple paths in G which contain at least k nodes or have a length exceeding some bound B (when additionally given a cost function $f : E \to \mathbb{R}$). The efficient construction of all paths that contain a certain vertex v can be accomplished using a breadth-first-search or Dijkstra based approach [8]. Note that the successive framework only makes sense when k or B are chosen as a small fraction of the diameter of the graph.

Incomplete Knowledge. In some applications, it might not even be possible to call the generator G a priori for every element, as necessary information might be missing. This is typically the case in AI applications, where e.g. mobile robots have to explore unknown terrain. Let, for example, the task of the robot be to physically mark every square of side length a which contains a certain amount of items in some finite area. Of course, the robot might explore the whole area first, then compute the set of all relevant squares, identify the respective marker positions, and then drive back to place them. But in the spirit of our successive scheme, it always could explore areas next that intersect with the ones just hit by driving in an a-tube around them. Then the set of squares it has to remember and that are used for computation of the next marker position(s) is smaller, and potentially the robot has to drive less of a detour to place the marker.

Solving Conventional Instances in the SHS Model. Even if the set system is explicitly available and fits in memory, it might be beneficial in terms of runtime to use the successive version of the greedy algorithm. Think of a set system where the best hitter hits only a very small fraction of all sets. Therefore the number of sets will decrease slowly in the greedy algorithm. But every round requires a complete scan over all remaining sets, so the computation gets quite expensive. In the successive algorithm, we could define $\mathfrak{S}'$ as a collection of sufficiently small sets in $\mathfrak{S}$. Then, computing the initial hitter can be made as cheap as desired. If every element can hit only a small fraction of sets in $\mathfrak{S}$, also the increase in the set system size by calling the generator is moderate.

An efficient generator for explicitly available set systems is easy to design. For example, one could store $\mathfrak{S}$ as an array of sets and keep for every element $u \in U$ a list of corresponding set indices.

7 Experiments

We implemented the standard greedy algorithm and the described successive greedy variant in C++ and evaluated them in terms of quality, space consumption and runtime. The timings were measured on a single core of an Intel i5-4300U CPU with 1.90 GHz and 12 GB RAM.

Table 1. Comparison of greedy and successive greedy on several benchmarks. 'k' equals 10^3. 'lb' stands for lower bound, T(ext) is time for set extraction, T(hit) for hitting set computation, T(total) or T for complete execution time. Timings are given in seconds (s), minutes (m) or hours (h).

		Greedy						Successive Greedy		
#nodes	lb	$\lvert\mathfrak{S}\rvert$	space	T(ext)	T(hit)	T(total)	$\lvert H\rvert$	max $\lvert\mathfrak{S}'\rvert$	T	$\lvert H\rvert$
100 k	1,857	731 k	0.17 GB	12 s	47 s	59 s	3,302	15 k	21 s	4,411
500 k	8,959	4,431 k	1.08 GB	99 s	19 m	21 m	16,423	72 k	457 s	21,313
996 k	18,862	8,033 k	1.98 GB	172 s	76 m	79 m	33,909	126 k	18 m	41,073
6,611 k	96,468	–	–	–	–	–	–	749 k	10 h	246,370
21,945 k	274,981	–	–	–	–	–	–	2251 k	37 h	691,513

As example application we chose the construction of Hitting Sets on shortest paths in a graph (Example 1 in Sect. 6) as it is of theoretical and practical interest (see [3,8,14]).

We extracted real-world road networks from OSM[1] to model the graphs. We chose networks with the number of nodes increasing from about 100,000 to 20 million. The number of edges in our test graphs is about twice the number of nodes. We demanded to find a Hitting Set for each of the graphs which hits every shortest path with a length exceeding 1000 m. The results for greedy and successive greedy are provided in Table 1. We observe that the greedy algorithm can only provide solutions for the instances with up to one million nodes in the graph. For larger benchmarks, the space consumption of the set system exceeds our hardware capabilities. The successive greedy algorithm on the other hand leads to results on all benchmarks. The solution quality is naturally worse compared to the classical greedy solution (about 25 % on average in our experiments). To make statements about the solution quality compared to the optimum, we computed simple lower bounds along by selecting a collection of pair-wise independent sets in the system. Comparing the solutions found by successive greedy to those lower bounds (provided in Table 1, second column), we see that they are never more than a factor of 3 apart. So the approximation ratio of successive greedy is quite good in our setting.

If we compare the size of the complete set system $|\mathfrak{S}|$ to the maximum number of sets in the temporary set system (max $|\mathfrak{S}'|$) maintained by successive greedy, we observe a drastic reduction. For example, for the 996k instance, $\mathfrak{S}'$ has at most 1.5 % of the size of $\mathfrak{S}$. The space consumption of $\mathfrak{S}'$ was comparable to the space consumption of the input graph for all instances. This is also reflected in the computation times. The extraction times of the set system are negligible compared to the times for the Hitting Set computation. For successive greedy the total time is always smaller than for conventional greedy, e.g. by a factor of 4.28 for the 996 k instance. For larger instances this effect expectedly would be even more pronounced.

[1] openstreetmap.org.

So in concurrency with out theoretical investigations, successive greedy turns out to be a useful tool to construct Hitting Sets in practice – especially on large instances.

8 Conclusions and Future Work

We introduced the successive Hitting Set model and designed algorithms that work in this model with good theoretical approximation guarantees – close to the guarantees in the conventional model. The experimental study confirmed that there are indeed applications where algorithms in the successive model lead to less space consumption and better computation times than the conventional algorithms. In future work, memory consumption could be turned into a hard constraint. We observed that the successive pricing method works if only a single set is available in every round. For successive greedy and the concatenation algorithm, the 'wavefront' of sets might become huge, though. Therefore it would be interesting to study algorithms in the successive model with the number of sets in the temporary system being restricted a priori – either instance-dependent (e.g. considering the maximal number of sets that can be hit by a single element) or completely ad hoc.

References

1. Mustafa, N.H., Ray, S.: Ptas for geometric hitting set problems via local search. In: Proceedings of the Twenty-Fifth Annual Symposium on Computational Geometry, pp. 17–22. ACM (2009)
2. Hefeeda, M., Bagheri, M.: Randomized k-coverage algorithms for dense sensor networks. In: IEEE 26th IEEE International Conference on Computer Communications, INFOCOM 2007, pp. 2376–2380. IEEE (2007)
3. Eisner, J., Funke, S.: Transit nodes-lower bounds and refined construction. In: ALENEX, SIAM, pp. 141–149 (2012)
4. Feige, U.: A threshold of ln n for approximating set cover. J. ACM (JACM) **45**(4), 634–652 (1998)
5. Slavík, P.: A tight analysis of the greedy algorithm for set cover. In: Proceedings of the Twenty-Eighth Annual ACM Symposium on Theory of Computing, pp. 435–441. ACM (1996)
6. Alon, N., Awerbuch, B., Azar, Y.: The online set cover problem. In: Proceedings of the Thirty-Fifth Annual ACM Symposium on Theory of Computing, pp. 100–105. ACM (2003)
7. Vinterbo, S., Øhrn, A.: Minimal approximate hitting sets and rule templates. Int. J. Approximate Reasoning **25**(2), 123–143 (2000)
8. Funke, S., Nusser, A., Storandt, S.: On k-path covers and their applications. Proc. VLDB Endow. **7**(10), 893–902 (2014)
9. Funke, S., Nusser, A., Storandt, S.: Placement of loading stations for electric vehicles: no detours necessary! In: Twenty-Eighth AAAI Conference on Artificial Intelligence (2014)
10. Delling, D., Goldberg, A.V., Pajor, T., Werneck, R.F.: Robust distance queries on massive networks. In: Schulz, A.S., Wagner, D. (eds.) ESA 2014. LNCS, vol. 8737, pp. 321–333. Springer, Heidelberg (2014)

11. Brönnimann, H., Goodrich, M.T.: Almost optimal set covers in finite VC-dimension. Discrete Comput. Geom. **14**(1), 463–479 (1995)
12. Matoušek, J., Seidel, R., Welzl, E.: How to net a lot with little: small epsilon-nets for disks and halfspaces. In: Proceedings of the Sixth Annual Symposium on Computational Geometry, pp. 16–22. ACM (1990)
13. Even, G., Rawitz, D., Shahar, S.M.: Hitting sets when the VC-dimension is small. Inf. Process. Lett. **95**(2), 358–362 (2005)
14. Abraham, I., Delling, D., Fiat, A., Goldberg, A.V., Werneck, R.F.: VC-dimension and shortest path algorithms. In: Aceto, L., Henzinger, M., Sgall, J. (eds.) ICALP 2011. LNCS, vol. 6755, pp. 690–699. Springer, Heidelberg (2011)

Randomized Algorithms III

Generating Random Hyperbolic Graphs in Subquadratic Time

Moritz von Looz(✉), Henning Meyerhenke, and Roman Prutkin

Karlsruhe Institute of Technology (KIT), Karlsruhe, Germany
{moritz.looz-corswarem,meyerhenke,roman.prutkin}@kit.edu

Abstract. Complex networks have become increasingly popular for modeling various real-world phenomena. Realistic generative network models are important in this context as they simplify complex network research regarding data sharing, reproducibility, and scalability studies. *Random hyperbolic graphs* are a very promising family of geometric graphs with unit-disk neighborhood in the hyperbolic plane. Previous work provided empirical and theoretical evidence that this generative graph model creates networks with many realistic features.
In this work we provide the first generation algorithm for random hyperbolic graphs with subquadratic running time. We prove a time complexity of $O((n^{3/2} + m) \log n)$ with high probability for the generation process. This running time is confirmed by experimental data with our implementation. The acceleration stems primarily from the reduction of pairwise distance computations through a polar quadtree, which we adapt to hyperbolic space for this purpose and which can be of independent interest. In practice we improve the running time of a previous implementation (which allows more general neighborhoods than the unit disk) by at least two orders of magnitude this way. Networks with billions of edges can now be generated in a few minutes.

Keywords: Complex networks · Hyperbolic geometry · Efficient range query · Polar quadtree · Generative graph model

1 Introduction

The algorithmic analysis of *complex networks* is a highly active research area since complex networks are increasingly used to represent phenomena as varied as the WWW, social relations, protein interactions, and brain topology [18]. Complex networks have several non-trivial topological features: They are usually *scale-free*, which refers to the presence of a few high-degree vertices (hubs) among many low-degree vertices. A heavy-tail degree distribution that occurs frequently in practice follows a power law [18, Chap. 8.4], i. e. the number of vertices with

This work is partially supported by SPP 1736 *Algorithms for Big Data* of the German Research Foundation (DFG) and by the Ministry of Science, Research and the Arts Baden-Württemberg (MWK) via project *Parallel Analysis of Dynamic Networks*.

K. Elbassioni and K. Makino (Eds.): ISAAC 2015, LNCS 9472, pp. 467–478, 2015.
DOI: 10.1007/978-3-662-48971-0_40

degree k is proportional to $k^{-\gamma}$, for a fixed exponent $\gamma > 0$. Moreover, complex networks often have the *small-world property*, i. e. the distance between any two vertices is surprisingly small, regardless of network size.

Generative network models play a central role in many complex network studies for several reasons: Real data often contains confidential information; it is then desirable to work on similar synthetic networks instead. Quick testing of algorithms requires small test cases, while benchmarks and scalability studies need bigger graphs. Graph generators can provide data at different user-defined scales for this purpose. Also, transmitting and storing a generative model and its parameters is much easier than doing the same with a gigabyte-sized network. A central goal for generative models is to produce networks which replicate relevant structural features of real-world networks [9]. Finally, generative models are an important theoretical part of network science, as they can improve our understanding of network formation. The most widely used graph-based system benchmark in high-performance computing, Graph500 [5], is based on R-MAT [10]. This model is efficiently computable, but has important drawbacks concerning realism and preservation of properties over different graph sizes [14].

Random hyperbolic graphs (RHGs), introduced by Krioukov et al. [15], are a very promising graph family in this context: They yield a provably high clustering coefficient (a measure for the frequency of triangles) [12], small diameter [7] and a power-law degree distribution with adjustable exponent. They are based on *hyperbolic geometry*, which has negative curvature and is the basis for one of the three isotropic spaces. (The other two are Euclidean (flat) and spherical geometry (positive curvature).) In the generative model, vertices are distributed randomly on a hyperbolic disk of radius R and edges are inserted for every vertex pair whose distance is below R.[1] This family of graphs has been analyzed well theoretically [7,8,12,13] and Krioukov et al. [15] showed that complex networks have a natural embedding in hyperbolic geometry. Calculating all pairwise distances in the generation process has quadratic time complexity. This impedes the creation of massive networks and is likely the reason previously published networks based on hyperbolic geometry have been in the range of at most 10^5 vertices. A faster generator is necessary to use this promising model for networks of interesting scales.

Proofs, details, further experiments, pseudocode and visualizations omitted due to space constraints can be found in the full version [24].

Outline and Contribution. We develop, analyze, and implement a fast, subquadratic generation algorithm for random hyperbolic graphs.

To lay the foundation, Sect. 2 discusses other generative network models and introduces fundamentals of hyperbolic geometry. The main technical part starts with Sect. 3, in which we use the Poincaré disk model to relate hyperbolic to

[1] We consider the name "hyperbolic unit-disk graphs" as more precise, but we use "random hyperbolic graphs" to be consistent with the literature. More general neighborhoods are possible [15] but not considered here since most theoretical works [7,8,12] are for unit-disk neighborhoods.

Euclidean geometry. This allows the use of a new spatial data structure, namely a polar quadtree adapted to hyperbolic space, to reduce both asymptotic complexity and running time of the generation. We further prove the time complexity of our generation process to be $O((n^{3/2} + m) \log n)$ with high probability (whp, i.e. $\geq 1 - 1/n$) for a graph with n vertices, m edges, and sufficiently large n. Our experimental results in Sect. 4 confirm our theoretical bound for the running time. A graph with 10^7 vertices and 10^9 edges can be generated with our shared-memory parallel implementation in about 8 min. The generator code is available in the network analysis toolkit NetworKit [22].

2 Related Work and Preliminaries

To make the following discussions clearer, we first introduce some network terminology. The *clustering coefficient* measures how likely two vertices with a common neighbor are to be connected. Different definitions exist, we use the global clustering coefficient, the fraction of closed triplets to connected triplets. Many real networks have multiple *connected components*, yet one large component is usually dominant. The *diameter* is the longest shortest path in the graph, which is often surprisingly small in complex networks. Complex networks also often exhibit a *community structure*, i.e. dense subgraphs with sparse connections between them.

2.1 Existing Graph Generators

The Barabasi-Albert model [2] is a preferential attachment model, designed to replicate the growth of real complex networks. The probability that a new vertex will be attached to an existing vertex v is proportional to v's degree, which results in a power-law degree distribution. While the distribution's exponent is constant for the basic model, generalizations for arbitrary exponents exist (see e.g. [18, Chap. 14]). Preferential attachment processes can be implemented with a running time in $O(n + m)$ [6].

The Dorogovtsev-Mendes model [11] is designed to model network growth with a fixed average degree. It is very fast in theory ($\Theta(n)$) and practice, but accepts only the vertex count as parameter and is thus inflexible.

The Recursive Matrix (R-MAT) model [10] was proposed to recreate properties of complex networks including a power-law degree distribution, the small-world property and self-similarity. Design goals also include few parameters and high generation speed. The R-MAT generator recursively subdivides the initially empty adjacency matrix into quadrants and drops edges into it according to given probabilities. It has $\Theta(m \log n)$ asymptotic complexity and is fast in practice. However, at least the R-MAT parameters used by the Graph500 system benchmark [5] lead to an insignificant community structure and clustering coefficients, as no incentive to close triangles exists.

Given a degree sequence *seq*, the Chung-Lu (CL) model [1] adds edges (u, v) with a probability of $p(u, v) = \frac{seq(u)seq(v)}{\sum_k seq(k)}$, recreating *seq* in expectation.

The model can be conceived as a weighted version of the well-known Erdős-Rényi (ER) model and has similar capabilities as the R-MAT model [21]. Implementations exist with $\Theta(n+m)$ time complexity [17]. LFR [16], in turn, was designed as a benchmark generator for community detection algorithms. Usually the user specifies vertex degrees and community sizes. However, the implementation is also able to handle other parameters.

BTER [14] is a two-stage structure-driven model. It uses the standard ER model to form relatively dense subgraphs and thus distinct communities. Afterwards, the CL model is used to add edges, matching the desired degree distribution in expectation [20]. This is done in $\Theta(n + m \log d_{\max})$, where $d_{\max}$ is the maximum vertex degree.

To summarize, all these models have their characteristics and also deficiencies. While some seem more preferable than others, no model is widely accepted as suitable for all (or at least most) possible scenarios or applications. Thus, the promising previous results described next motivate a deeper investigation of RHGs.

2.2 Graphs in Hyperbolic Geometry

Kriokouv et al. [15] introduced the family of random hyperbolic graphs and showed how they naturally develop a power-law degree distribution and other properties of complex networks. In the generative model, vertices are generated as points in polar coordinates (ϕ, r) on a disk of radius R in the hyperbolic plane with curvature $-\zeta^2$. We denote this disk with $\mathbb{D}_R$. The angular coordinate ϕ is drawn from a uniform distribution over $[0, 2\pi]$. The probability density for the radial coordinate r is given by [15, Eq. (17)] and controlled by a growth parameter α:

$$f(r) = \alpha \frac{\sinh(\alpha r)}{\cosh(\alpha R) - 1} \tag{1}$$

For $\alpha = 1$, this yields a uniform distribution on hyperbolic space within $\mathbb{D}_R$. For lower values of α, vertices are more likely to be in the center, for higher values more likely at the border of $\mathbb{D}_R$.

We denote the hyperbolic distance between two points p_1 and p_2 with $\text{dist}_{\mathcal{H}}(p_1, p_2)$. In the model, any two vertices u and v are connected by an edge if their hyperbolic distance $\text{dist}_{\mathcal{H}}(u, v)$ is below R. The neighborhood of a point (= vertex) thus consists of the points lying in a hyperbolic circle around it. (Krioukov et al. also present a more general model in which edges are inserted with a probability depending on hyperbolic distance. For this purpose, they define a family of monotonically falling functions parametrized by a temperature T. As noted earlier, we are in line with the theoretical works [7,8,12] and discuss unit-disk neighborhoods only. The latter can be considered as the special case $T = 0$.) Several works have analyzed the properties of the resulting graphs theoretically. Krioukov et al. show that for $\alpha/\zeta > \frac{1}{2}$, the degree distribution follows a power law with exponent $2 \cdot \alpha/\zeta + 1$ [15, Eq. (29)]. Gugelmann et al. [12] prove non-vanishing clustering and a low variation of the clustering coefficient.

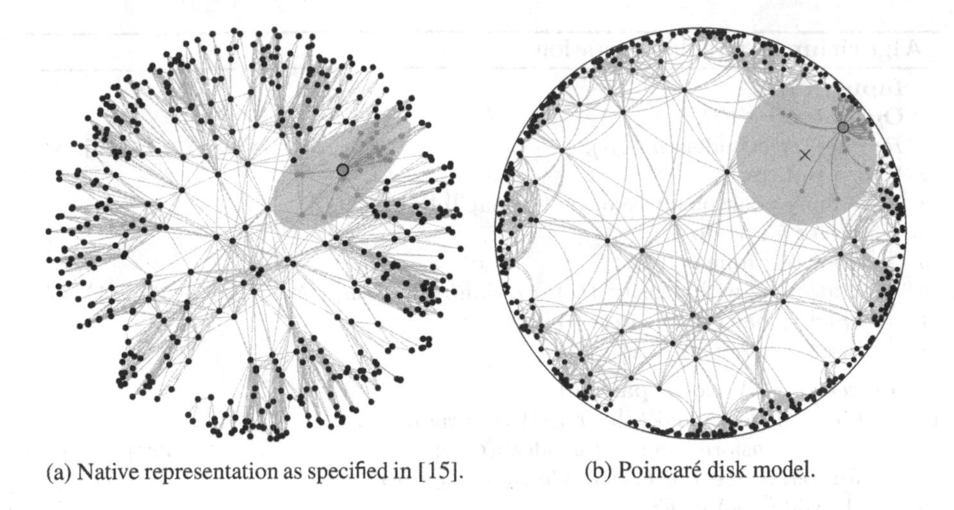

(a) Native representation as specified in [15]. (b) Poincaré disk model.

Fig. 1. Comparison of geometries. Neighbors of the bold blue vertex are in the hyperbolic respective Euclidean circle. The center of the Euclidean circle is marked with $\times$ (Colour figure online).

Bode et al. [7] discuss the size of the giant component and the probability that the graph is connected [8]. They also show [8] that the curvature parameter ζ can be fixed while retaining all degrees of freedom, we thus assume $\zeta = 1$ from now on. Kiwi and Mitsche [13] bound the diameter asymptotically almost surely for $\frac{1}{2} < \alpha < 1$. The average degree $\overline{k}$ of a random hyperbolic graph is controlled with the radius R, using an approximation given by [15, Eq. (22)]. An example graph with 500 vertices, $R \approx 5.08$ and $\alpha = 0.8$ is shown in Fig. 1a. For the purpose of illustration in the figure, we choose a vertex u (the bold blue vertex) and add edges (u, v) for all vertices v where $\text{dist}_{\mathcal{H}}(u, v) \leq 0.2 \cdot R$. The neighborhood of u then consists of vertices within a hyperbolic circle (marked in blue).

A previous generator implementing the extended model and with quadratic complexity is available [3]. We show in Sect. 4.1 that for the random hyperbolic graphs described above, our implementation is at least two orders of magnitude faster in practice.

2.3 Poincaré Disk Model

The Poincaré disk model is one of several representations of hyperbolic space within Euclidean geometry and maps the hyperbolic plane onto the Euclidean unit disk $D_1(0)$. The hyperbolic distance between two points $p_E, q_E \in D_1(0)$ is then given by the Poincaré metric [4]:

$$\text{dist}_{\mathcal{H}}(p_E, q_E) = \text{acosh}\left(1 + 2\frac{||p_E - q_E||^2}{(1 - ||p_E||^2)(1 - ||q_E||^2)}\right). \tag{2}$$

Algorithm 1. Graph Generation

```
   Input: n, k̄, α
   Output: G = (V, E)
 1 R = getTargetRadius(n, k̄, α);                                  /* Eq.(22) [15] */
 2 V = n vertices;
 3 T = empty polar quadtree of radius mapToPoincare(R);
 4 for vertex v ∈ V do
 5     draw φ[v] from U[0, 2π);
 6     draw r_H[v] with density f(r) = α sinh(αr)/(cosh(αR) − 1);   /* Eq.(1) */
 7     r_E[v] = mapToPoincare(r_H[v]);
 8     insert v into T at (φ[v], r_E[v]);
 9 for vertex v ∈ V do in parallel
10     C_H = circle around (φ[v], r_H[v]) with radius R;
11     C_E = transformCircleToEuclidean(C_H);                         /* Prop. 1 */
12     for vertex w ∈ T.getVerticesInCircle(C_E) do
13         add (v, w) to E;
14 return G;
```

Figure 1b shows the same graph as in Fig. 1a, but translated into the Poincaré model. This model is conformal, i. e. it preserves angles. More importantly for us, it maps hyperbolic circles onto Euclidean circles.

3 Fast Generation of Graphs in Hyperbolic Geometry

We proceed by showing how to relate hyperbolic to Euclidean circles. Using this transformation, we are able to partition the Poincaré disk with a polar quadtree that supports efficient range queries. We adapt the network generation algorithm to use this quadtree and prove subsequently that it achieves subquadratic generation time.

3.1 Generation Algorithm

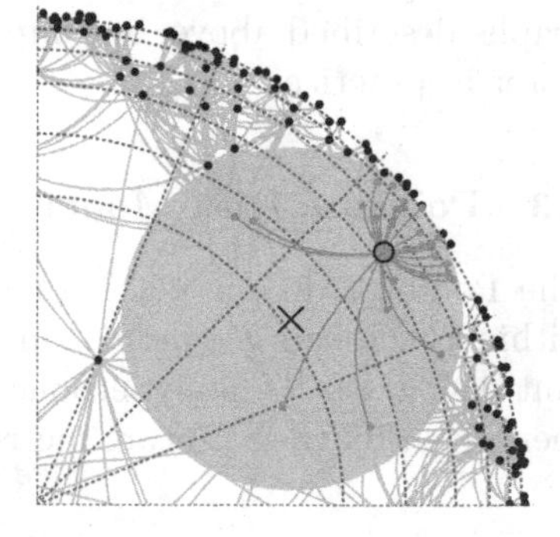

Fig. 2. Polar quadtree

Transformation from Hyperbolic Geometry. Neighbors of a query point $u = (\phi_h, r_h)$ lie in a hyperbolic circle around u with radius R. This circle, which we denote as H, corresponds to a Euclidean circle E in the Poincaré disk. The center E_c and radius rad_E of E are in general different from u and R. All points on the boundary of E in the Poincaré disk are also on the boundary of H and thus have hyperbolic distance R from u. Among these points, the two on the ray from the origin through u are straightforward to construct by keeping the angular coordinate fixed and choosing the radial coordinates to match the hyperbolic distance: (ϕ_h, r_{e_1}) and (ϕ_h, r_{e_2}), with $r_{e_1}, r_{e_2} \in [0,1)$, $r_{e_1} \neq r_{e_2}$ and $\text{dist}_{\mathcal{H}}(E_c, (\phi_h, r_e)) = R$ for $r_e \in \{r_{e_1}, r_{e_2}\}$. It follows:

Proposition 1. *E_c is at $(\phi_h, \frac{2r_h}{ab+2})$ and* rad_E *is* $\sqrt{\left(\frac{2r_h}{ab+2}\right)^2 - \frac{2r_h^2 - ab}{ab+2}}$, *with* $a = \cosh(R) - 1$ *and* $b = (1 - r_h^2)$.

Algorithm. The generation of $G = (V, E)$ with n vertices and average degree k is shown in Algorithm 1. As in previous efforts [3], vertex positions are generated randomly (lines 5 and 6). We then map these positions into the Poincaré disk (line 7) and, as a new feature, store them in a polar quadtree (line 8). For each vertex u the hyperbolic circle defining the neighborhood is mapped into the Poincaré disk according to Proposition 1 (lines 10–11) – also see Fig. 1b, where the neighborhood of u consists of exactly the vertices in the light blue Euclidean circle. Edges are then created by executing a Euclidean range query with the resulting circle in the polar quadtree (lines 12–13). We use the same probability distribution for the node positions and add an edge (u, v) exactly if the hyperbolic distance between u and v is less than R. This leads to the following proposition:

Proposition 2. *Algorithm 1 generates random hyperbolic graphs as defined in Sect. 2.2.*

Data Structure. As mentioned above, our central data structure is a polar quadtree on the Poincaré disk. While Euclidean quadtrees are common [19], we are not aware of previous adaptations to hyperbolic space. A node in the quadtree is defined as a tuple $(\min_\phi, \max_\phi, \min_r, \max_r)$ with $\min_\phi \leq \max_\phi$ and $\min_r \leq \max_r$. It is responsible for a point $p = (\phi_p, r_p) \in D_1(0)$ iff $(\min_\phi \leq \phi_p < \max_\phi)$ and $(\min_r \leq r_p < \max_r)$. Figure 2 shows a section of a polar quadtree, where quadtree nodes are marked by dotted red lines. We call the geometric region corresponding to a quadtree node its *quadtree cell*. When a point is to be inserted into an already full leaf node, the node is split into four children. Splitting in the angular direction is straightforward as the angle range is halved: $\text{mid}_\phi := \frac{\max_\phi + \min_\phi}{2}$. For the radial direction, we choose the splitting radius to result in an equal division of probability mass:

$$\text{mid}_{r\mathcal{H}} := \text{acosh}\left(\frac{\cosh(\alpha \max_{r\mathcal{H}}) + \cosh(\alpha \min_{r\mathcal{H}})}{2}\right) / \alpha \tag{3}$$

(Note that Eq. (3) uses radial coordinates in the native representation, which are converted back to coordinates in the Poincaré disk.) This leads to two lemmas useful for establishing the time complexity of the main quadtree operations:

Lemma 1. *Let $\mathbb{D}_R$ be a hyperbolic disk of radius R, p a point in $\mathbb{D}_R$ which is chosen according to the distribution discussed in Sect. 2.2, and T be a polar quadtree on $\mathbb{D}_R$. Let C be a quadtree cell at depth i. Then, the probability that p is in C is 4^{-i}.*

Lemma 2. *Let R and $\mathbb{D}_R$ be as in Lemma 1. Let T be a polar quadtree on $\mathbb{D}_R$ containing n points distributed according to Sect. 2.2. Then, for n sufficiently large,* $\text{height}(T) \in O(\log n)$ *whp.*

3.2 Time Complexity

The time complexity of the generator is determined by the quadtree operations.

Quadtree Insertion. For the amortized analysis, we consider each element's initial and final position during the insertion of n elements. Let $h(T)$ be the final height of quadtree T, let $h(i)$ be the final depth of element i and let $t(i)$ be the depth of i when it was inserted. During insertion of element i, $t(i)$ quadtree nodes are visited until the correct leaf for insertion is found, the cost for this is linear in $t(i)$. When a leaf cell is full, it splits into four children and the depth of each element in the leaf increases by one. Over the course of inserting all n elements, element i thus moves $h(i)-t(i)$ times due to leaf splits. To reach its final position at depth $h(i)$, element i accrues cost of $O(t(i)+h(i)-t(i)) = O(h(i)) \subseteq O(h(T))$, which is $O(\log n)$ whp due to Lemma 2. The amortized time complexity for a node insertion is then: $T(\text{Insertion}) \in O(\log n)$ whp.

Quadtree Range Query. Neighbors of a vertex u are the vertices within a Euclidean circle constructed according to Proposition 1. Let $\mathcal{N}(u)$ be this neighborhood set in the final graph, thus $\deg(u) := |\mathcal{N}(u)|$. We denote leaf cells that do not have non-leaf siblings as *bottom leaf cells.*

Lemma 3. *Let T and n be as in Lemma 2. A range query on T returning a point set A will examine at most $O(\sqrt{n} + |A|)$ bottom leaf cells with probability at least $1 - \frac{1}{n^2}$.*

Due to Lemma 3, the number of examined bottom leaf cells for a range query around u is in $O(\sqrt{n} + \deg(u))$ with probability at least $1 - \frac{1}{n^2}$. The query algorithm traverses T from the root downward. For each bottom leaf cell b, $O(h(T))$ inner nodes and non-bottom leaf cells are examined on the path from the root to b. Due to Lemma 2, $h(T)$ is in $O(\log n)$ whp. The time complexity to gather the neighborhood of a vertex u with degree $\deg(u)$ is thus: $T(\text{RQ}(u)) \in O\left((\sqrt{n} + \deg(u)) \cdot \log n\right)$ whp.

Graph Generation. To generate a graph G from n points, the n positions need to be generated and inserted into the quadtree. The time complexity of this is $O(n) + n \cdot O(\log n) = O(n \log n)$ whp. In the next step, neighbors for all points are extracted. This has a complexity of

$$T(\text{Edges}) = \sum_{v} O\left((\sqrt{n} + \deg(v)) \cdot \log n\right) = O\left(\left(n^{3/2} + m\right) \log n\right) \text{ whp.} \quad (4)$$

The complexity bounds for each of the n range queries hold with probability at least $1 - \frac{1}{n^2}$, with a union bound we get a probability of at least $1 - 1/n$ for the above complexity. This dominates the quadtree operations and thus total running time. We conclude:

Theorem 1. *Generating random hyperbolic graphs can be done in* $O((n^{3/2} + m) \log n)$ *time whp for sufficiently large* n, *i. e. with probability* $\geq 1 - 1/n$.

A more thorough discussion of the probabilities can be found in the full version.

4 Experimental Evaluation

We compare the output and running times of our implementation and the implementation of Aldecoa et al. [3] for the same parameters. Please note that the implementation at [3] supports a more general model including six different regimes, among them the generalized model of Krioukov et al. described in Sect. 2.2. In practice, they support a variable clustering coefficient and non-powerlaw degree distributions. The random hyperbolic graphs we consider, where two nodes are connected exactly if their hyperbolic distance is at most R, correspond to a finite γ and a temperature of zero.

Implementation. Our implementation uses the NetworKit toolkit [22] and is written in C++ 11. The code is compiled with GCC 4.8 and parallelized with OpenMP. The parallelization over the range queries is straightforward as they are independent.

Several optimizations improve performance. Sorting the points by angular coordinates before generating the graph improves cache locality, since points close to each other also have similar neighbors and thus similar access patterns to the quadtree data structure. The number of memory reallocations while constructing the neighborhood of a vertex v can be reduced by pre-allocating memory according to the expected degree of v, which is approximated by [15, Eq. (12)]. This is especially useful in a parallel setting with a global lock for memory allocations. The effect of these optimizations depend on the density of the generated graph.

Experimental Setup. In the comparison with the implementation of [3], the generated graphs could not be compared directly as both implementations sample random graphs. In its output files, the implementation of [3] does provide the hyperbolic coordinates of the generated points. Yet, since the distance threshold R is computed non-deterministically with a Monte Carlo process and not written to the log file, we do not have all necessary information to recreate the graphs exactly. Instead, we generate series of graphs with 10000 nodes, average degree $\overline{k}$ between 4 and 256 and degree distribution exponent γ between 2.2 and 7. We then compare the average properties of the generated graphs.

Running time measurements were made on a server with 256 GB RAM and 2x8 Intel Xeon E5-2680 cores at 2.7 GHz. With hyperthreading, we use 32 virtual threads for our parallel implementation. The implementation of [3] is sequential.

4.1 Results

Qualitative Comparison. Figures showing properties of the generated graphs in comparison with the implementation of [3] can be found in the full paper [24].

Some random fluctuations are visible, but for almost all properties the averages of our implementation are very similar to the implementation of [3]. The measured values of γ for thin graphs and various target γs differ from the previous implementation, but the fluctuation within the measurements of each implementation are sufficiently strong that it leads us to assume some measurement noise. The differences between the implementations are smaller than the variations within one implementation. Both generators create graphs with at times high diameters, about 600 for graphs with 10000 nodes, $\overline{k} = 16$ and $\gamma = 7$. At first glance, this seems to contradict the theoretical bound in [13]. However, their result is only for $\frac{1}{2} < \alpha < 1$, while $\gamma = 7$ corresponds to a value of 3 for α.

Running Time. Figure 3 shows the running times for networks with 10^4-10^7 vertices and up to $1.2 \cdot 10^9$ edges. We achieve a throughput of up to 13 million edges/s. Even at only 10^4 vertices, our implementation is two orders of magnitude faster than the quadratic-time implementation of [3] for the same parameters – where only one order of magnitude stems from parallelization (the typical parallel speedup values at this scale range between 8 and 12). For graphs with 10^5 vertices, we already see an improvement of three orders of magnitude and, due to our algorithm's smaller asymptotic complexity of $O((n^{3/2}+m)\log n)$, this gap grows with increasing graph sizes. Note that the proof of this complexity bound (stated in Sect. 3) is supported by the measurements, as illustrated by the lines for the theoretical fit in Fig. 3.

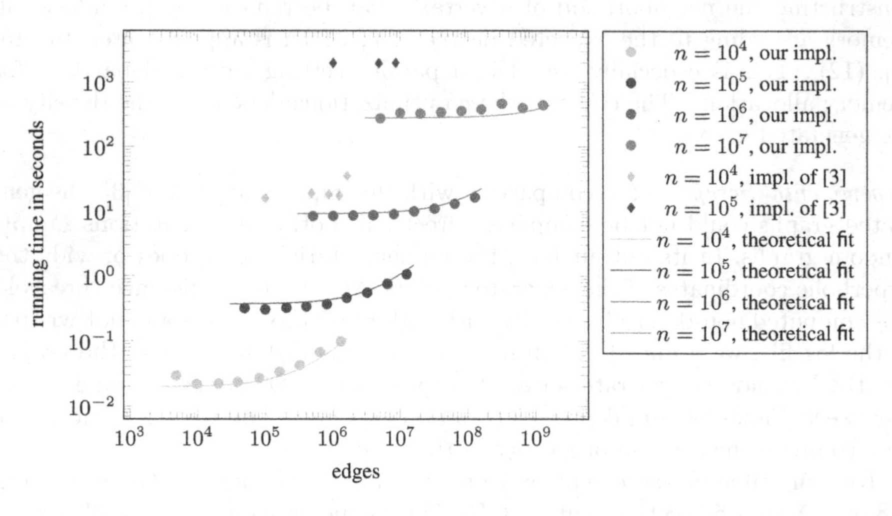

Fig. 3. Comparison of running times to generate networks with 10^4-10^7 vertices, $\alpha = 1$ and varying $\overline{k}$. Circles represent running times of our implementation, diamonds the running times of the implementation of [3]. Our running times are fitted with the equation $T(n,m) = \left(\left(3.8 \cdot 10^{-7} n + 1.14 \cdot 10^{-9} n^{3/2} + 1.38 \cdot 10^{-8} m\right) \log\, n\right)$ seconds.

5 Conclusions

In this work we have provided efficient range queries in the hyperbolic plane – based on a polar quadtree, which we have adapted to hyperbolic space and which can thus be of independent interest. We have further shown that the fast range queries facilitate the generation of random hyperbolic graphs (RHGs) in running time $O((n^{3/2}+m)\log n)$ time whp. In practice our parallel generator constructs RHGs with billions of edges in a few minutes, about two orders of magnitude faster than the quadratic-time algorithm.

Previous work (both theoretical and empirical) has already shown that the generated graphs are complex networks with many properties also found in real-world instances [15, Sect. 4]. Thus, RHGs constitute a promising model for complex network research that deserve even further attention. In the context of our paper, future work will investigate the incremental quadtree construction in order to admit a dynamic model with vertex movement; this deserves a more thorough treatment than possible here given the space constraints. An extension to generate random hyperbolic graphs with non-zero temperatures can be found in our manuscript about *probabilistic neighborhood queries* [23].

References

1. Aiello, W., Chung, F., Lu, L.: A random graph model for massive graphs. In: Proceedings of the 32nd ACM Symposium on Theory of Computing, pp. 171–180. ACM (2000)
2. Albert, R., Barabási, A.L.: Statistical mechanics of complex networks. Rev. Mod. Phys. **74**(1), 47 (2002)
3. Aldecoa, R., Orsini, C., Krioukov, D.: Hyperbolic graph generator. Comput. Phys. Commun. **196**, 492–496 (2015). doi:10.1016/j.cpc.2015.05.028. http://www.sciencedirect.com/science/article/pii/S0010465515002088
4. Anderson, J.W.: Hyperbolic Geometry. Springer Undergraduate Mathematics Series, 2nd edn. Springer, Berlin (2005)
5. Bader, D.A., Berry, J., Kahan, S., Murphy, R., Riedy, E.J., Willcock, J.: Graph 500 benchmark 1 ("search"), version 1.1. Technical report, Graph 500 (2010)
6. Batagelj, V., Brandes, U.: Efficient generation of large random networks. Phys. Rev. E **71**(3), 036113 (2005)
7. Bode, M., Fountoulakis, N., Müller, T.: On the giant component of random hyperbolic graphs. In: The Seventh European Conference on Combinatorics, Graph Theory and Applications. CRM Series, vol. 16, pp. 425–429. Scuola Normale Superiore (2013)
8. Bode, M., Fountoulakis, N., Müller, T.: The probability that the hyperbolic random graph is connected (2014). http://web.mat.bham.ac.uk/N.Fountoulakis/BFM.pdf. Preprint
9. Chakrabarti, D., Faloutsos, C.: Graph mining: laws, generators, and algorithms. ACM Comput. Surv. (CSUR) **38**(1), 2 (2006)
10. Chakrabarti, D., Zhan, Y., Faloutsos, C.: R-MAT: a recursive model for graph mining. In Proceedings of the 4th SIAM International Conference on Data Mining (SDM), Orlando, FL. SIAM, April 2004

11. Dorogovtsev, S.N., Mendes, J.F.F.: Evolution of Networks: from Biological Nets to the Internet and WWW. Oxford University Press, Oxford (2003)
12. Gugelmann, L., Panagiotou, K., Peter, U.: Random hyperbolic graphs: degree sequence and clustering. In: Czumaj, A., Mehlhorn, K., Pitts, A., Wattenhofer, R. (eds.) ICALP 2012, Part II. LNCS, vol. 7392, pp. 573–585. Springer, Heidelberg (2012)
13. Kiwi, M., Mitsche, D.: A bound for the diameter of random hyperbolic graphs. In: 2015 Proceedings of the Twelfth Workshop on Analytic Algorithmics and Combinatorics (ANALCO), pp. 26–39. SIAM, January 2015
14. Kolda, T.G., Pinar, A., Todd, P., Seshadhri, C.: A scalable generative graph model with community structure. SIAM J. Sci. Comput. **36**(5), C424–C452 (2014)
15. Krioukov, D., Papadopoulos, F., Kitsak, M., Vahdat, A., Boguñá, M.: Hyperbolic geometry of complex networks. Phys. Rev. E **82**(3), 036106 (2010)
16. Lancichinetti, A., Fortunato, S., Radicchi, F.: Benchmark graphs for testing community detection algorithms. Phys. Rev. E **78**(4), 046110 (2008)
17. Miller, J.C., Hagberg, A.: Efficient generation of networks with given expected degrees. In: Frieze, A., Horn, P., Prałat, P. (eds.) WAW 2011. LNCS, vol. 6732, pp. 115–126. Springer, Heidelberg (2011)
18. Newman, M.: Networks: An Introduction. Oxford University Press, Oxford (2010)
19. Samet, H.: Foundations of Multidimensional and Metric Data Structures. Morgan Kaufmann Publishers Inc., San Francisco (2005)
20. Seshadhri, C., Kolda, T.G., Pinar, A.: Community structure and scale-free collections of Erdős-Rényi graphs. Phys. Rev. E **85**(5), 056109 (2012)
21. Seshadhri, C., Pinar, A., Kolda, T.G.: The similarity between stochastic Kronecker and Chung-Lu graph models. In: Proceedings of the 2012 SIAM International Conference on Data Mining (SDM), pp. 1071–1082 (2012)
22. Staudt, C.L., Sazonovs, A., Meyerhenke, H.: NetworKit: an interactive tool suite for high-performance network analysis (2014). arXiv preprint arXiv:1403.3005
23. von Looz, M., Meyerhenke, H.: Querying probabilistic neighborhoods in spatial data sets efficiently, September 2015. ArXiv preprint arXiv:1509.01990
24. von Looz, M., Meyerhenke, H., Prutkin, R.: Generating random hyperbolic graphs in subquadratic time, September 2015. ArXiv preprint arXiv:1501.03545

Provable Efficiency of Contraction Hierarchies with Randomized Preprocessing

Stefan Funke[1](✉) and Sabine Storandt[2]

[1] FMI, University of Stuttgart, Stuttgart, Germany
funke@fmi.uni-stuttgart.de
[2] Department of Computer Science, University of Freiburg, Freiburg, Germany
storandt@informatik.uni-freiburg.de

Abstract. We present a new way of analyzing Contraction Hierarchies (CH), a widely used speed-up technique for shortest path computations in road networks. In previous work, preprocessing and query times of deterministically constructed CH on road networks with n nodes were shown to be polynomial in n as well as the highway dimension h of the network and its diameter D. While h is conjectured to be polylogarithmic for road networks, a tight bound remains an open problem. We rely on the empirically justifiable assumption of the road network exhibiting small growth. We introduce a method to construct randomized Contraction Hierarchies on road networks as well as a probabilistic query routine. Our analysis reveals that randomized CH lead to sublinear search space sizes in the order of $\sqrt{n}\log\sqrt{n}$, auxiliary data in the order of $n\log^2\sqrt{n}$, and correct query results with high probability after a polynomial time preprocessing phase.

1 Introduction

Contraction Hierarchies (CH) [1] are a preprocessing based technique to accelerate shortest path computations in road networks. The basic idea behind CH is to augment the network with shortcut edges which allow to settle less nodes in a Dijkstra run without compromising correctness of the result. CH are widely used on real-world instances, as they provide an excellent trade-off between the amount of auxiliary data (only doubling the network size) and speed-up (about three orders of magnitude compared to a plain Dijkstra). But these values are solely empirical, based on experiments on real-world networks [2]. Theoretical explanations for this good empirical behaviour are still not fully satisfying.

1.1 Related Work

In [3], the notion of the *highway dimension* h of a network was introduced to explain the practical performance of CH and other speed-up techniques (on undirected networks). A small highway dimension indicates that shortest paths in the road network longer than a parameter r can be hit by a set S of nodes with S being locally sparse. Here, locally sparse means that the intersection

K. Elbassioni and K. Makino (Eds.): ISAAC 2015, LNCS 9472, pp. 479–490, 2015.
DOI: 10.1007/978-3-662-48971-0_41

of a ball of radius r with S contains at most h elements. Assuming optimal preprocessing, it was shown that $\mathcal{O}(nh \log D)$ shortcut edges are added to the original network, with n being the number of nodes in the network and $D \leq n$ the network diameter. The number of nodes settled in a CH-Dijkstra run was shown to be in $\mathcal{O}(h \log D)$. As optimal CH preprocessing is NP-hard (using Hitting Set computations as a subroutine), they also study a polynomial time approximation version. This adds another factor of $\log n$ to the auxiliary data size and the query time. While h is conjectured to be polylogarithmic for road networks, the problem of proving h to be small is still open (for grids it is known that $h \in \Theta(\sqrt{n})$). Moreover, h-values for real-world networks are unknown (as its computation is NP-hard as well) and the preprocessing methods introduced to study CH theoretically are too slow to be practical for large road networks [4]. Hence validating whether the theoretical results reflect real-world behavior is difficult.

In [5], CH were studied based on the topology of the network. It was shown that for planar graphs, CH preprocessing based on nested dissection leads to auxiliary data in the order of $\mathcal{O}(n \log n)$. For minor-closed graphs with balanced $\mathcal{O}(\sqrt{n})$ separators, search spaces are shown to be in $\mathcal{O}(\sqrt{n})$, for graphs with treewidth k in $\mathcal{O}(k \log n)$. For graphs with highway dimension h, results matching those in [3] were reported assuming edge costs that maximize h. An implementation of CH based on nested dissection [6] showed that it leads to good performance in practice. Nevertheless a real comparison to the theoretical results again is hardly possible due to h being unknown. Moreover all results so far heavily use Big-O-Notation, making it difficult to tell whether the observed behaviour in practice is due to asymptotics or due to hidden constants.

1.2 Contribution

We exhibit a so far unexplored connection of CH to Skip Lists [7], a data structure for fast search within ordered sets of elements. Based on the model of randomized Skip List construction, we describe a CH variant with randomized preprocessing for some probability parameter $p \in]0,1[$. We prove the expected number of shortcuts to be at most $n(1-p)(0.5 \cdot \log^2_{1/p} \sqrt{n} + (1-p^2)^{-1})$, and the expected search space size to be $(6 \ln {}^1\!/\!_p \log \sqrt{n} + 2)\sqrt{n}$ (no O-notation here!). Preprocessing is in polynomial time. For our results to be valid, we rely on a simple and intuitive bound on the growth rate of the underlying metric. Moreover, we prove our theoretical bounds to be meaningful by comparing them to experimental results on real-world road networks. Surprisingly, the randomized construction shares certain characteristics with a common heuristic CH construction scheme. For this simple heuristic construction, no theoretical guarantees of any kind are known. While heuristically constructed CH naturally outperform CH with randomized preprocessing, our studies are the first to give some insight in the theoretical auxiliary data size and search space parameters for this heuristic construction.

2 Preliminaries

In the following, we describe preprocessing and query answering for conventional CH carefully. We then briefly review randomized Skip Lists to show their potential to serve as model for randomized CH construction. Finally, we provide some details on the graph model that is used in our analysis.

2.1 Contraction Hierarchies

Given a road network $G(V, E)$ and edge costs $c : E \to \mathbb{R}^+$, the preprocessing phase of CH works as follows: Every node $v \in V$ gets assigned a level $l : V \to \mathbb{N}$ inducing a (not necessarily total) order on the nodes. Then a CH-graph $G'(V, E \cup E^+)$ is constructed upon this order where E^+ denotes the set of shortcut edges. For a pair of nodes v, w a shortcut edge $e = (v, w)$ is added to E^+ if all nodes on the shortest path between v and w exhibit a level smaller than $\min\{l(v), l(w)\}$. The cost of e is set to the shortest path distance between v and w. Determining node levels that minimizes $|E^+|$ is APX-hard [8,9].

In practice, though, there exist heuristics which construct CH-graphs with small sets of shortcut edges very efficiently. The most common heuristic is based on the *node contraction* operation [1]. Here, a node v and all its adjacent edges are removed from the current graph, and shortcut edges are inserted between any pair of neighbors u, w of v if u, v, w was the shortest path from u to w. Nodes are contracted one-by-one and their rank in the contraction order is used as node level. If the goal is keeping $|E^+|$ small, a good candidate for the next node to contract is the one with minimal *edge difference* (ED), which denotes the number of added shortcuts if v is contracted minus the number of edges that are currently adjacent to v (some heuristics also consider a linear combination of the ED and other values). So after contraction of v, the current graph has one node less and $ED(v)$ more edges (note that $ED(v)$ can be negative). The ED-values need to be continuously updated, as contracting a node influences the ED-values of its neighbors. It was noted, though, that independent sets of nodes can be contracted at once without violating correctness. So nodes in the current graph are first sorted increasingly by their ED-value. Then an independent set of nodes is chosen greedily considering the nodes in the ED-order (with all nodes in the set receiving the same level). This approach allows to construct the CH-graph in very few contraction rounds and leads to few added shortcuts in practice.

The final CH-graph (original graph plus all shortcuts) has the following nice property (no matter how l was chosen): Between any pair of nodes s, t there exists a shortest path on which the node levels at first monotonously increase and then monotonously decrease (so the path is unimodal wrt. l). Therefore queries can be answered via a bi-directional Dijkstra computation that only relaxes upward edges (v, w) with $l(v) \leq (w)$ in the forward run and accordingly downward edges in the backwards run. By construction, the forward and the backward run both settle the node(s) with the highest level on the shortest path from s to t. Hence a node that minimizes the forward plus the backwards distance yields the optimal shortest path distance.

2.2 Randomized Skip Lists

Skip Lists are a data structure for efficient search and maintenance of an ordered set of elements. Skip Lists consist of layers of linked lists. The bottom layer is a linked list that contains all n elements in sorted order. Each element gets assigned a height h. To determine the height values randomly, first a probability $p \in]0,1[$ is chosen. Then for every element a coin with probability p for *HEAD* is flipped until *TAIL* shows up. The number of times the coin showed *HEAD* marks the height. The maximum height among all elements determines the number of additional layers in the Skip List data structure. Each list i contains only links between elements with a height $\geq i$ and 'skips' over the others. In expectation, the maximum height is $\log_{1/p} n$ and therefore the total space consumption is in $\mathcal{O}(n \log_{1/p} n)$. Searching for an element using a suitable query algorithm that works its way from the topmost layer down demands $1/p \log_{1/p} n$. Choosing different values of p allows to trade search costs against storage costs. We will use randomized Skip List construction as model for randomized CH construction by interpreting the road network as the bottom layer.

2.3 Our Model: Graph Metrics with Bounded Growth

Like [3] we need to make some assumptions about the structure of our road networks for an analysis to succeed. In typical representations of a road network (as for example derived from data of the OpenStreetMap project) as graphs $G(V, E, c)$ with edge costs $c : E \to \mathbb{N}$, edges represent road segments of rather uniform length, so we can replace edge costs by respective sequences of unit-cost edges without blowing up the size of the graph by more than a constant factor. So from now on we will focus on graphs with unit edge costs. Our crucial assumption on the structure of G can be stated as follows: For any node $v \in V$, the number of nodes w at distance k is bounded by $g \cdot k$ for some constant $g \geq 1$, that is

$$|\{w \in V : d(v, w) = k\}| \leq g \cdot k$$

We have verified this condition to hold for small values of g for several real-world networks. It also implies $|\{w \in V : d(v, w) \leq k\}| \leq gk(k+1)/2$ – which mimics the area growth in R^2 when increasing the radius of a circle. To keep the presentation simpler, we assume in the following $g = 1$, but it is easy to see that the parameter g could be carried along all following calculations.

Our condition has some connection to already existing characterizations of graph metrics. For example, demanding that the number of nodes at distance k is *exactly* $g \cdot k$ implies an *expansion rate* of 4 according to the definition of [10] as well as constant doubling dimension [11]. On the other hand there are metrics with an unbounded expansion rate yet satisfying our condition.

3 Randomized Contraction Hierarchies

3.1 Preprocessing

We start the preprocessing phase by assigning levels $l : V \to \mathbb{N}$ by coin tosses in the same way as for Skip Lists (with a probability p for *HEAD*). So $l(v)$ is an integer greater than zero with $P(l(v) \geq L) = p^{L-1}$.

To complete the preprocessing, we need to compute the set of shortcuts resulting from our randomized choice of node levels. To that end, we run a Dijkstra computation from each node v until on every active path in the search tree there is a node with a level $\geq l(v)$. For every first node w on a shortest path from v with $l(w) \geq l(v)$ we insert the shortcut $e = (v, w)$ with $c(e) = d_v(w)$ in the CH-graph (avoiding multi edges).

The preprocessing obviously demands only polynomial time. We expect a maximum node level of $\mathcal{O}(\log n)$, so assigning levels to n nodes can be done in expected $\mathcal{O}(n \log n)$ time. The n Dijkstra runs in the second phase require $\mathcal{O}(n^2 \log n + nm)$ time and dominate the overall runtime.

3.2 Analysis

Let us now analyze our CH construction. The two key performance indicators are the *total number of shortcuts* and the *number of settled nodes* in a query. Ideally, both of these values should be small in order to guarantee a space-efficient CH-graph and a good speed-up compared to plain Dijkstra's algorithm.

Throughout the analysis, log always refers to $\log_{1/p}$.

Total Number of Shortcuts. To bound the total number of shortcuts we first bound the number of upward edges emanating from some node v. We provide two such bounds, one being stronger for nodes v with small levels, the other being stronger for nodes v with large levels.

Lemma 1. *The expected number of upwards edges (original or shortcut) emanating from a node v with level L is bounded by p^{1-L}.*

Proof. A shortcut (v, w) from v to a node w with a shortest path $v \rightsquigarrow w$ of length k exists if and only if the level of w is at least L while the level of all $k-1$ nodes inbetween on the shortest path from v to w is less than L. So the probability for the shortcut (v, w) to exist can be expressed as $P(l(w) \geq L) \cdot P(l < L)^{k-1}$. Due to our condition we have at most k nodes at distance k, hence the total number of upward edges can be bounded as:

$$E(X) \leq \sum_{k=1}^{D} k \cdot P(l \geq L) \cdot P(l < L)^{k-1} = \sum_{k=1}^{D} kp^{L-1}(1 - p^{L-1})^{k-1}$$

Here D is the diameter of the graph, $D \leq n$. We then substitute $1 - p^{L-1}$ with q and end up with:

$$E(X) \leq \frac{1-q}{q}\sum_{k=1}^{D} kq^k < \frac{1-q}{q}\sum_{k=0}^{\infty} kq^k = \frac{1-q}{q}\cdot\frac{q}{(1-q)^2} = \frac{1}{p^{L-1}} \qquad \square$$

Lemma 2. *The expected number $E(X)$ of upwards edges (original or shortcut) emerging from a node v with level L is bounded by np^{L-1}.*

Proof. Upwards shortcuts demand the target node to have a level $\geq L$. Therefore the total number of shortcuts emerging from a node with level L is bounded by the expected number of nodes with a level $\geq L$ in the network. As the expected number of nodes with level L equals $n(1-p)p^{L-1}$, the expected number of nodes with a level at least L is np^{L-1}. $\square$

We observe that the bound by Lemma 1 is tighter for $L \leq \log\sqrt{n}$ and the bound by Lemma 2 for $L > \log\sqrt{n}$.

Theorem 1. *The expected number of upwards edges in the CH-graph is bounded by $n(1-p)(0.5\cdot\log^2\sqrt{n} + (1-p^2)^{-1})$.*

Proof. Using Lemma 1 and the fact that we expect $np^{L-1}(1-p)$ nodes at level L we bound the number of outgoing edges from nodes with level $\leq \log\sqrt{n}$ by

$$\sum_{L=1}^{\log\sqrt{n}} n(1-p)p^{L-1}p^{1-L} \leq n(1-p)\cdot 0.5\cdot\log^2\sqrt{n}$$

and the number of edges from nodes with higher level using Lemma 2 by

$$\sum_{L=\log\sqrt{n}+1}^{\infty} n(1-p)p^{L-1}np^{L-1} = \frac{n^2(1-p)}{p^2}\cdot\sum_{L=\log\sqrt{n}+1}^{\infty} p^{2L}$$

$$= \frac{n^2(1-p)}{p^2}\cdot\frac{p^2(p^{\log n}-p^{2n})}{1-p^2} = \frac{n^2(1-p)(1/n-p^{2n})}{1-p^2} \leq \frac{n(1-p)}{1-p^2} \qquad \square$$

The analysis for the number of downwards edges can be done analogously. So the final number of expected edges in the CH-graph is $n(1-p)(0.5\cdot\log^2\sqrt{n} + (1-p^2)^{-1})$ for undirected networks (summing up the two bounds in the Theorem) and twice this number, i.e., $n(1-p)(\log^2\sqrt{n} + 2(1-p^2)^{-1})$ for directed networks.

Search Space Analysis. We define the search space $SS(v)$ for a node $v \in V$ as the number of nodes that are pushed into the priority queue (PQ) during a CH-Dijkstra run from v (relaxing only upwards edges). We will first analyze the *direct search space (DSS)* of v. A node w is in $DSS(v)$ if on the shortest path from v to w all nodes have levels $\leq l(w)$. Therefore, w will be settled with the correct distance $d(v,w)$ in the CH-Dijkstra run. Unfortunately, $SS(v)$ is typically a superset of $DSS(v)$ as also nodes on monotonously increasing but non-shortest paths are considered. We will modify the query algorithm to bound the number of such nodes.

Lemma 3. *The expected size of $DSS(v)$ is bounded by $(1+p)\sqrt{n}$.*

Proof. We can assume that all nodes with a level $l > \log\sqrt{n}$ are always in $DSS(v)$. In expectation, there are $\sum_{L=\log\sqrt{n}+1}^{\infty} np^L(1-p) = \sqrt{n}$ such nodes in the network. A node w is in $DSS(v)$ if on the shortest path from v to w all nodes have a level of at most $l(w)$. The expected number of such nodes with $l(w) \leq \log\sqrt{n}$ can be bounded by

$$\sum_{L=1}^{\log\sqrt{n}} P(l=L) \sum_{k=1}^{D} kP(l \leq L)^{k-1} = \sum_{L=1}^{\log\sqrt{n}} p^{L-1}(1-p) \sum_{k=1}^{D} k(1-p^{L-1})^{k-1}.$$

As the last sum can be bounded by p^{-2L+2}, we get:

$$\sum_{L=1}^{\log\sqrt{n}} p^{-L+1}(1-p) = p(\sqrt{n}-1)$$

Together with the at most $\sqrt{n}$ nodes in $DSS(v)$ with a level greater than $\log\sqrt{n}$, the size of $DSS(v)$ is bounded by $(1+p)\sqrt{n}$. □

To characterize and reduce the number of nodes in $SS(v) \setminus DSS(v)$, we need the following properties about nodes in $DSS(v)$.

Lemma 4. *The probability for a node w at distance k from v to be in $DSS(v)$ but exhibiting a level $l(w) < \log k - \log(c\ln(1/p)\log k)$ is bounded by k^{-c}.*

Proof. If $w \in DSS(v)$, all k nodes on the shortest path from v to w have a level of at most $l(w)$. As $l(w) < \log k - \log(c\ln(1/p)\log k)$ the same needs to hold for all nodes on this shortest path. The probability for that can be expressed as:

$$\left(1 - p^{\log k - \log(c\ln(1/p)\log k)}\right)^k = \left(1 - \frac{c\ln(1/p)\log k}{k}\right)^k$$

Using $(1+x) \leq e^x$ with $x = -c\ln(1/p)\log k \cdot k^{-1}$, we can upper bound the above formula by

$$\left(e^{-c\ln(1/p)\log k \cdot k^{-1}}\right)^k = e^{-c\ln(1/p)\log k} = e^{-c\ln(1/p)\ln(k)/ln(1/p)} = k^{-c}. \quad \square$$

Applying the above Lemma we show that with high probability a node in $DSS(v)$ whose shortest path from v is at least $n^{1/4}$ long does not have too small a level.

Lemma 5. *A node w at shortest path distance $k > n^{1/4}$ from v is in $DSS(v)$ and exhibits a level $l(w) \geq \log k - \log(c\ln(1/p)\log k)$ with probability $\geq 1 - n^{-c/4}$.*

Proof. The probability $P = P(l(w) < \log k - \log(c\ln(1/p)\log k))$ is bounded by k^{-c} (according to Lemma 4). So the larger k the smaller the probability. For $k > n^{1/4}$ we get $P < n^{-c/4}$. Hence we can lower bound the probability of the counter-event by $1 - n^{-c/4}$. □

Armed with this insight, we modify our query algorithm such that nodes with too small a level relative to their distance are discarded during the exploration. That is, during the run of CH-Dijkstra, we discard a node w from further consideration (not pushing it into the PQ) if $d(w) > n^{1/4}$ and $l(w) < \min(\log\sqrt{n}, \log d(w) - \log(c\ln(1/p)\log d(w)))$, where $d(w)$ denotes the current distance label of w in the CH-Dijkstra run. The following theorem shows that for appropriate choice of c, this leads to small search spaces and with high probability to the correct result.

Theorem 2. *Our modified query algorithm has an expected search space size of at most* $\sqrt{n}(2 + c\ln(1/p)\sqrt{2}\log\sqrt{n})$ *and computes the correct result with probability* $\geq 1 - 2n^{\frac{-c+4}{4}}$.

Proof. We know that always $d(v, w) \leq d(w)$ has to be true, where $d(v, w)$ is the true distance from v to w. Therefore, the number of nodes with $d(v, w) \leq d(w) \leq n^{1/4}$ can be bounded by $\sum_{i=1}^{n^{1/4}} k \leq \sqrt{n}$. The number of nodes with $l(w) \geq \log d(w) - \log(c\ln(1/p)\log d(w))$ can be bounded by

$$\sum_{k=1}^{D} x_k P(l \geq \log k - \log(c\ln(1/p)\log k)$$

with $x_k \leq k$, $\forall k = 1, \ldots, D$ and $\sum x_k = n$. As $P(l \geq \log k - \log(c\ln(1/p)\log k)$ decreases with growing k, this sum can be upper bounded by:

$$\sum_{k=1}^{\sqrt{2n}} kP(l \geq \log k - \log(c\ln(1/p)\log k) = \sum_{k=1}^{\sqrt{2n}} k\frac{c\ln(1/p)\log(k)}{k}$$

$$= c\ln(1/p)\sum_{k=1}^{\sqrt{2n}} \log k \leq c\ln(1/p)\sqrt{2n}\log\sqrt{n}$$

Together with the at most $\sqrt{n}$ nodes above level $\log\sqrt{n}$ in expectation, our search space size does not exceed $\sqrt{n}(2 + c(\ln(1/p)\sqrt{2}\log\sqrt{n})$ nodes.

It remains to show that queries are answered correctly with high probability. Queries are answered correctly for sure if $SS(v) \supseteq DSS(v)$. According to Lemma 5, a node $w \in DSS(v)$ at distance $k > n^{1/4}$ is not contained in our pruned search space with probability at most $n^{-c/4}$. We are interested in an upper bound for the probability that at least one of the nodes in $DSS(v)$ is not in our search space. We simply apply the union bound upper bounding the probability that one or more nodes of $DSS(v)$ do not have large enough level by $n \cdot n^{-c/4} = n^{(-c+4)/4}$. So with probability $\geq 1 - n^{\frac{-c+4}{4}}$, all nodes of $DSS(v)$ are actually in the search space of v and with the same argument holding for the (reverse) search space of the target, we arrive at the bound for the correctness of the query result. □

The above Theorem implies for $c = 4 + \alpha$, $\alpha > 0$ our query routine produces the correct result with probability $\geq 1 - 2n^{-\alpha/4}$. Choosing for example $c = 6$ we have a success probability of $1 - 2/\sqrt{n}$ and expected search space sizes for source and target of less than $\sqrt{n}(2 + 6\log\sqrt{n})$ for $p = 1/2$.

4 Experimental Results

We implemented randomized CH construction and the proposed query answering algorithm in C++. We also implemented the heuristic CH construction based on iterative contraction of independent sets as described in Sect. 2.1. Experiments were conducted on a single core of an Intel i5-4300U CPU with 1.90 GHz and 12 GB RAM. We used the OSM road network data of a cut-out of Germany with 2, 275, 793 nodes and 4, 637, 537 directed edges for evaluation.

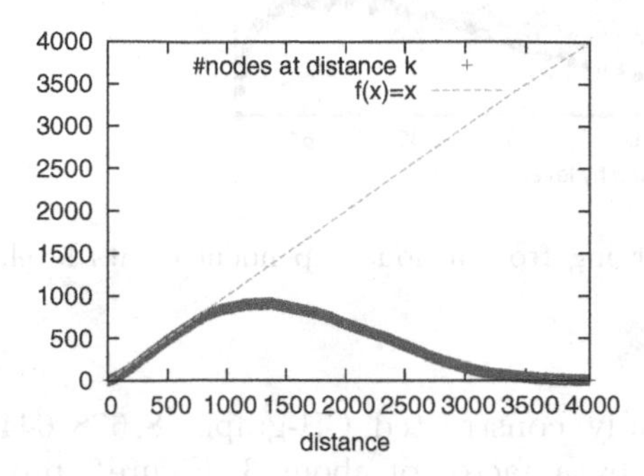

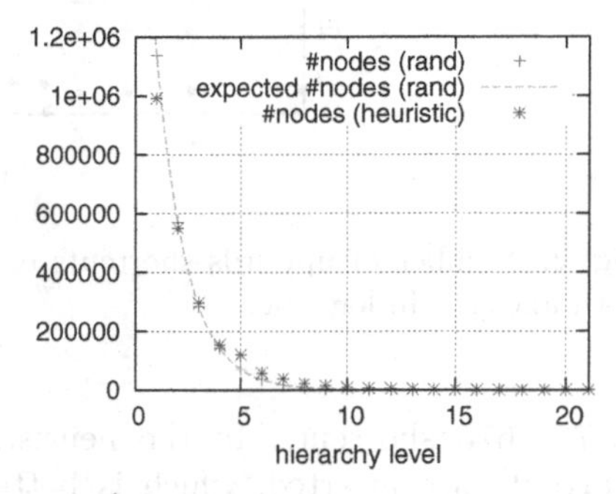

Fig. 1. Left: Our model predicts the number of nodes with distance k to be beneath the green line. The red boxes indicate the real distance dependent node distribution based on Dijkstra search trees from 1,000 randomly chosen source nodes. Right: Node level distribution resulting from heuristic and randomized CH construction (Color figure online).

We first validate our chosen model. In Fig. 1, left, we compare the average number of nodes at distance k from a source in our real network (with euclidean distances as cost metric) to the prediction according to the model. We observe that up to distance about 1000 there are indeed almost exactly k nodes with distance k. Then the number declines.

Unless mentioned otherwise, the following experiments are conducted using $p = 1/2$ for randomized CH. We first want to evaluate the CH preprocessing. Figure 1, right, shows that the expected number of nodes per level reflects the real node levels quite perfectly. While this is not surprising for the randomized construction, it is indeed for the heuristic construction. So basically in every contraction round, about half of the remaining nodes form an independent set and get contracted at once, leading to the same node level distribution as for our Skip List based randomized levels. The maximum level in the heuristic construction was 99, though, and therefore about a factor of 5 higher than in the randomized CH, but this is due to the fact that towards the end of the contraction process the remaining nodes form clique-like structures which only allow for contracting a single element as independent set.

For randomized CH, we expect the number of shortcuts to be less than 41 million for $n = 2, 275, 793$. Averaged over three runs, our CH-graph contained

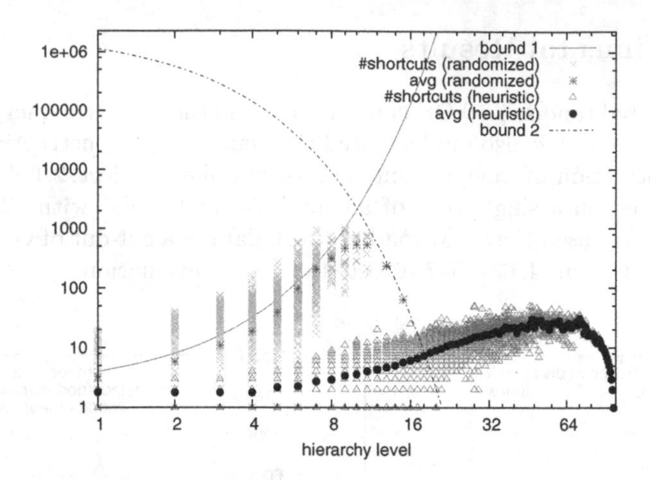

Fig. 2. Number of upwards shortcuts emerging from a node dependent on its level. Both axes are in logscale.

23, 758, 675 shortcuts. In the heuristically constructed CH-graph, 8, 678, 644 shortcuts are inserted, which is better by a factor of about 3. Figure 2 provides a detailed overview of the number of upwards shortcuts emerging from a node in dependency of its level. We observe that bound 1 and 2 resulting from Lemmas 1 and 2 are almost perfect predictions for the average value per level. For the heuristic construction the curve is stretched and exhibits a lower peak. This is a result of a wider range of node levels and a lower number of total shortcuts (due to the ED-related node contraction order).

Finally, we evaluated search spaces and query times for randomized and heuristic CH. In a query answered in the heuristically constructed CH-graph, 497 nodes were settled on average. Query times were in the order of a half microsecond which results in a speed-up of factor 500 compared to plain Dijkstra. With randomized CH, the predicted search space size for $c > 4$ is over 66,000. The number of actually settled nodes in our experiments was only 5,241, showing that some of our upper bound assumptions in the analysis are too pessimistic. All queries were answered correctly in our experiments. Query times were in the order of 20 ms, yielding only a speed-up of 10 compared to plain Dijkstra. Still, the fact that there is speed-up at all using a randomized construction shows that our results have some degree of practical justification. Moreover, it follows straight from the Skip List like construction that the expected number of nodes on the optimal CH-path from s to t with $d(s,t) = k$ is $2 \log k$. Evaluating 1,000 example queries, we observed that this result is matched accurately – indeed for both, randomized and heuristically constructed CH.

To study the influence of p, we ran the same experiments with $p = 1/4$ and $p = 3/4$. For $p = 1/4$, the expected maximum number of shortcuts is 32 million, the real number was 16, 941, 163. For $p = 3/4$ our upper bound implies no more than 49 million expected edges in the CH-graph, and we ended up with 28, 526, 499 in the experiments. The search space sizes and query times were slightly worse for both $p = 1/4$ and $p = 3/4$ compared to $p = 1/2$. If levels are more spread due to higher p they are also more connected due to more shortcuts. If levels are less spread there are large connected components of nodes with the same level which are explored until our level-distance-bound kicks in. Hence $p = 1/2$ seems to be a suitable choice, not only because it leads to a close relation with the heuristic CH construction but also since the auxiliary data size vs search space size trade-off is good.

5 Conclusions

It is rather surprising that it is possible to construct CH via a level assignment that does not take into account the structure of the graph. This is in stark contrast to common heuristic construction schemes as well as the approaches in [3,5] where the construction process is heavily guided by the graph structure. Furthermore, our randomized construction scheme shares some natural characteristics with the common heuristic construction scheme – both in theory as well as in empirical evaluations. Our results should be seen as a step towards a better understanding of the good performance of contraction hierarchies in practice.

References

1. Geisberger, R., Sanders, P., Schultes, D., Vetter, C.: Exact routing in large road networks using contraction hierarchies. Transp. Sci. **46**(3), 388–404 (2012)
2. Bast, H., Delling, D., Goldberg, A., Müller-Hannemann, M., Pajor, T., Sanders, P., Wagner, D., Werneck, R.F.: Route planning in transportation networks. arXiv preprint (2015). arXiv:1504.0514
3. Abraham, I., Fiat, A., Goldberg, A.V., Werneck, R.F.: Highway dimension, shortest paths, and provably efficient algorithms. In: Proceedings of 21st Annual ACM-SIAM symposium on Discrete Algorithms, pp. 782–793 (2010)
4. Abraham, I., Delling, D., Fiat, A., Goldberg, A.V., Werneck, R.F.: Highway dimension and provably efficient shortest path algorithms. Microsoft Research, USA, Tech. report 9 (2013)
5. Bauer, R., Columbus, T., Rutter, I., Wagner, D.: Search-space size in contraction hierarchies. In: Fomin, F.V., Freivalds, R., Kwiatkowska, M., Peleg, D. (eds.) ICALP 2013, Part I. LNCS, vol. 7965, pp. 93–104. Springer, Heidelberg (2013)
6. Dibbelt, J., Strasser, B., Wagner, D.: Customizable contraction hierarchies. In: Gudmundsson, J., Katajainen, J. (eds.) SEA 2014. LNCS, vol. 8504, pp. 271–282. Springer, Heidelberg (2014)
7. Pugh, W.: Skip lists: a probabilistic alternative to balanced trees. Commun. ACM **33**(6), 668–676 (1990)
8. Bauer, R., Columbus, T., Katz, B., Krug, M., Wagner, D.: Preprocessing speed-up techniques is hard. In: Calamoneri, T., Diaz, J. (eds.) CIAC 2010. LNCS, vol. 6078, pp. 359–370. Springer, Heidelberg (2010)
9. Milosavljević, N.: On optimal preprocessing for contraction hierarchies. In: Proceedings of the 5th ACM SIGSPATIAL International Workshop on Computational Transportation Science, pp. 33–38. ACM (2012)

10. Karger, D.R., Ruhl, M.: Finding nearest neighbors in growth-restricted metrics. In: Proceedings of 34th Annual ACM Symposium on Theory of Computing, STOC 2002, pp. 741–750. ACM, New York (2002)
11. Gupta, A., Krauthgamer, R., Lee, J.R.: Bounded geometries, fractals, and low-distortion embeddings. In: Proceedings of 44th Symposium on Foundations of Computer Science (FOCS 2003), pp. 534–543. IEEE Computer Society (2003)

Randomized Minmax Regret for Combinatorial Optimization Under Uncertainty

Andrew Mastin[1(✉)], Patrick Jaillet[2], and Sang Chin[3]

[1] Lawrence Livermore National Laboratory, Livermore, CA 94550, USA
mastin1@llnl.gov
[2] Massachusetts Institute of Technology, Cambridge, MA 02139, USA
jaillet@mit.edu
[3] Draper Laboratory, Cambridge, MA 02139, USA
schin@draper.com

Abstract. The minmax regret problem for combinatorial optimization under uncertainty can be viewed as a zero-sum game played between an optimizing player and an adversary, where the optimizing player selects a solution and the adversary selects costs with the intention of maximizing the regret of the player. The conventional minmax regret model considers only deterministic solutions/strategies, and minmax regret versions of most polynomial solvable problems are NP-hard. In this paper, we consider a randomized model where the optimizing player selects a probability distribution (corresponding to a mixed strategy) over solutions and the adversary selects costs with knowledge of the player's distribution, but not its realization. We show that under this randomized model, the minmax regret version of any polynomial solvable combinatorial problem becomes polynomial solvable. This holds true for both interval and discrete scenario representations of uncertainty. Using the randomized model, we show new proofs of existing approximation algorithms for the deterministic model based on primal-dual approaches. We also determine integrality gaps of minmax regret formulations, giving tight bounds on the limits of performance gains from randomization. Finally, we prove that minmax regret problems are NP-hard under general convex uncertainty.

Keywords: Robust optimization · Approximation algorithms · Game theory

1 Introduction

Many optimization applications involve cost coefficients that are not fully known. When probability distributions are available for cost coefficients (e.g. from

Research supported in part by NASA ESTOs Advanced Information System Technology (AIST) program under grant number NNX12H81G. Also supported by NSF grant 1029603, ONR grant N00014-12-1-0033, and AFOSR grant FA9550-12-1-0136. Lawrence Livermore National Laboratory is operated by Lawrence Livermore National Security, LLC, for the U.S. Department of Energy, National Nuclear Security Administration under Contract DE-AC52-07NA27344.

K. Elbassioni and K. Makino (Eds.): ISAAC 2015, LNCS 9472, pp. 491–501, 2015.
DOI: 10.1007/978-3-662-48971-0_42

historical data or other estimates), stochastic programming is often an appropriate modeling choice [13,26]. In other cases, costs may only be known to be contained in intervals (i.e. each cost has a known lower and upper bound) or to be a member of a finite set of scenarios, and one is more interested in worst-case performance. Robust optimization formulations are desirable here as they employ a minmax-type objective [12,17,23].

In a general robust optimization problem with cost uncertainty, one must select a set of items from some feasible *solution set* such that item costs are contained in some *uncertainty set*. The basic problem of selecting an optimal solution from the solution set when costs are known is referred to as the *nominal problem*. When only the uncertainty set is known, the goal under the *minmax* criterion (also referred to as absolute robustness) is to select a solution that gives the best upper bound on objective cost over all possible costs from the uncertainty set [27] (assuming that the nominal problem is a minimization problem). That is, one must select the solution that, when item costs are chosen to maximize the cost of the selected solution, is minimum. Under the *minmax regret* criterion (sometimes called the robust deviation model), the goal is instead to select the solution that minimizes the maximum possible regret, defined as the difference between the cost of the selected solution and the optimal solution [25].

A problem under the minmax regret criterion can be viewed as a two-stage game played between an optimizing player and an adversary. In the first stage, the optimizing player selects a deterministic solution. In the second stage, an adversary observes the selected solution and chooses values/costs from the uncertainty set with the intention of maximizing the player's regret. The goal of the optimizing player is thus to select a solution that least allows the adversary to generate regret. For both interval and discrete scenario representations of cost uncertainty, the minmax regret versions of most polynomial solvable problems are NP-hard [1,2,5,8,23,28]. A variation on this model, first suggested by Bertsimas et al. [10] for minmax robust optimization, is to allow the optimizing player to select a probability distribution over solutions and require the adversary to select costs based on knowledge of the player's distribution, but not its realization. In this paper, we study this randomized model under the minmax regret criterion instead of the minmax criterion.

We show that under this randomized model, the minmax regret version of any polynomial solvable 0–1 integer linear programming problem is polynomial solvable. This holds true for both interval and discrete scenario representations of uncertainty. Our observation is that the randomized model corresponds to the linear programming relaxation of the mixed integer program for the deterministic model. While the relaxation may have an exponential number of constraints, an efficient separation oracle is given by the nominal problem.

The linear programming formulation leads to further insights. We show that currently known approximation algorithms for deterministic minmax regret problems [3,19], which have been proved using combinatorial arguments, can be proved using simpler primal-dual methods. This analysis also yields integrality gaps for deterministic minmax regret problems. The integrality gaps are shown to be equal to k for discrete scenario uncertainty, where k is the number of scenarios, and equal to 2 for interval uncertainty. Both gaps match the ratios of the approximation algorithms, showing that these algorithms are optimal. The integrality gaps also establish lower bounds on performance when moving from the deterministic model to the randomized model. Letting Z_{D} and Z_{R} denote the deterministic and randomized minmax regret values for a common nominal problem, we effectively show that $Z_{\mathrm{D}}/k \leq Z_{\mathrm{R}} \leq Z_{\mathrm{D}}$ for discrete scenario uncertainty and $Z_{\mathrm{D}}/2 \leq Z_{\mathrm{R}} \leq Z_{\mathrm{D}}$ for interval uncertainty.

Given that the randomized model makes many minmax regret problems polynomial solvable for interval uncertainty and discrete scenario uncertainty, it is natural to ask if polynomial solvability remains in the presence of slightly more elaborate uncertainty sets. We show that for general convex uncertainty sets, however, deterministic and randomized minmax regret problems are NP-hard, even for polynomial solvable nominal problems.

The paper is structured as follows. In the remainder of this section we review related work. Section 2 introduces notation and definitions. Section 3 highlights our most important results for discrete scenario uncertainty, interval uncertainty, and convex uncertainty. A full development of these topics, with proofs, is contained in the full version of the paper [24]. A conclusion is given in Sect. 4.

Related Work. One of the first studies of minmax regret from both an algorithmic and complexity perspective was that of Averbakh [7]. He looked at the minmax regret version of the simple problem of selecting k items out of n total items, where the cost of each item is uncertain and the goal is to select the set of items with minimum total cost. For interval uncertainty, he derived a polynomial-time algorithm based on interchange arguments. He demonstrated that for the discrete scenario representation of uncertainty, however, the minmax regret problem becomes NP-hard, even for the case of only two scenarios. It is interesting to contrast these results with the case of general minmax regret linear programming which, as shown by Averbakh and Lebedev [9], is NP-hard for interval uncertainty but polynomial solvable for discrete scenario uncertainty.

Apart from the item selection problem, most polynomial solvable minmax regret combinatorial problems are NP-hard, both for interval and discrete scenario uncertainty [1,2,8,23,28]. The survey paper of Aissi et al. [5] provides a comprehensive summary of results related to both minmax and minmax regret combinatorial problems. For problems that are already NP-complete, most of their minmax regret versions are Σ_2^p-complete [16]. To solve minmax regret problems in practice, the book by Kasperski reviews standard mixed integer program (MIP) formulations for both interval and discrete scenario uncertainty [17].

General approximation algorithms for deterministic minmax regret problems are known for both types of uncertainty. Kasperski and Zieliński [19] proved a 2-approximation algorithm based on midpoint costs under interval uncertainty, and Aissi et al. [3] gave a k-approximation algorithm using mean costs under discrete scenario uncertainty, where k is the number of scenarios. Under interval uncertainty, fully polynomial-time approximation schemes are known for many problems [17,20]. For discrete scenario uncertainty, Kasperski et al. [18] looked at the minmax regret item selection problem, which models special cases of many combinatorial problems. They showed that for a non-constant number of scenarios, the problem is not approximable within any constant factor unless P = NP. If the number of scenarios is constant, fully polynomial-time approximation schemes are known for some problems [4,6].

The application of a game-theoretic model with mixed strategies to robust optimization problems was introduced by Bertsimas et al. [10]. They focused on the minmax robust model, and their analysis was motivated by adversarial models used for online optimization algorithms. They showed that if it is possible to optimize over both the solution set and the uncertainty set in polynomial time, then an optimal mixed strategy solution can be calculated in polynomial time, and that the expected cost under the randomized model is no greater than the cost for the deterministic model. They also gave bounds on the improvement gained from randomization for various uncertainty sets. Our work is similar to theirs, but we focus on the minmax regret criterion instead of the minmax criterion.

Other related areas of research are Stackelberg security games [21,22], network interdiction games [11], and dueling algorithms [15]. A common feature of many of these works, as well as ours, is that they involve games that at first glance have exponential size but can be solved efficiently using the appropriate reductions.

2 Definitions

We consider a general combinatorial optimization problem where we are given a set of n items $E = \{e_1, e_2, \ldots, e_n\}$ and a set $\mathcal{F}$ of feasible subsets of E. Each item $e \in E$ has a cost $c_e \in \mathbb{R}$. Given the vector $c = (c_1, \ldots, c_n)$, the goal of the optimization problem is to select a feasible subset of items that minimizes the total cost; we refer to this as the *nominal problem*:

$$F^*(c) := \min_{T \in \mathcal{F}} \sum_{e \in T} c_e. \tag{1}$$

Let $x = (x_1, \ldots, x_n)$ be a characteristic vector for some set T, so that $x_e = 1$ if $e \in T$ and $x_e = 0$ otherwise. Also let $\mathcal{X} \subseteq \{0,1\}^n$ denote the set of all characteristic vectors corresponding to feasible sets $T \in \mathcal{F}$. We assume that $\mathcal{X}$ is described in size m (e.g. with m linear inequalities). We can equivalently write the nominal problem with a linear objective function

$$F^*(c) = \min_{x \in \mathcal{X}} \sum_{e \in E} c_e x_e. \tag{2}$$

Throughout the paper, we use both set notation and characteristic vectors for ease of presentation.

We review the conventional definitions for the deterministic minmax regret framework and then present the analogous definitions for our randomized model.

For some cost vector $c \in \mathcal{C}$, the deterministic cost of a solution $T \in \mathcal{F}$ is

$$F(T, c) := \sum_{e \in T} c_e. \tag{3}$$

The regret of a solution T under some cost vector c is the difference between the cost of the solution and the optimal cost:

$$R(T, c) := F(T, c) - F^*(c). \tag{4}$$

The *maximum regret problem* for a solution T is

$$R_{\max}(T) := \max_{c \in \mathcal{C}} R(T, c) = \max_{c \in \mathcal{C}} (F(T, c) - F^*(c)). \tag{5}$$

The *deterministic minmax regret problem* is then

$$Z_{\mathrm{D}} := \min_{T \in \mathcal{F}} R_{\max}(T) = \min_{T \in \mathcal{F}} \max_{c \in \mathcal{C}} (F(T, c) - F^*(c)). \tag{6}$$

We abuse the notation $F(\cdot, c)$, $R(\cdot, c)$ and $R_{\max}(\cdot)$ by replacing set arguments with vectors (e.g. $F(x, c)$ in place of $F(T, c)$), but we generally use capital letters for sets and lowercase letters for vectors.

We now move to the randomized framework, where the optimizing player selects a distribution over solutions and the adversary selects a distribution over costs. Starting with the optimizing player, for some set $T \in \mathcal{F}$, let y_T denote the probability that the optimizing player selects set T. Let $y = (y_T)_{T \in \mathcal{F}}$ be the vector of length $|\mathcal{F}|$ specifying the set selection distribution; we refer to y simply as a *solution*. Define the feasible region for y as

$$\mathcal{Y} := \{y \mid y \geq \mathbf{0}, \mathbf{1}^\top y = 1\}, \tag{7}$$

where the notation $\mathbf{0}$ and $\mathbf{1}$ indicates a full vector of zeros and ones, respectively. We similarly define a distribution over costs for the adversary. The set $\mathcal{C}$ may in general be infinite, but we only consider strategies with finite support; for now we assume that such strategies are sufficient. Thus consider a finite set $\mathcal{C}_f \subseteq \mathcal{C}$, and for some $c \in \mathcal{C}_f$, let w_c denote the probability that the adversary selects costs c. Then let $w = (w_c)_{c \in \mathcal{C}_f}$ and define the feasible region

$$\mathcal{W} := \{w \mid w \geq \mathbf{0}, \mathbf{1}^\top w = 1\}. \tag{8}$$

We are interested in succinct descriptions of strategies for both players. We define for the optimizing player a *mixed strategy encoding* $\mathcal{M} = (\Theta, Y)$ as a set of deterministic solutions $\Theta = \{T_i \in \mathcal{F} \mid i = 1, \ldots, \mu\}$ that should be selected with nonzero probability and the corresponding probabilities $Y = \{y_{T_i} \in [0, 1] \mid i = 1, \ldots, \mu\}$ that satisfy $\sum_{i=1}^{\mu} y_{T_i} = 1$. Here μ is the support size of the mixed strategy (i.e. the number of deterministic solutions with nonzero probability). Likewise, define an adversarial mixed strategy encoding $\mathcal{L} = (C, W)$ as a set of costs $C = \{c^j \in \mathcal{C}_f \mid j = 1, \ldots, \eta\}$ to be selected with corresponding probabilities $W = \{w_{c^j} \in [0, 1] \mid j = 1, \ldots, \eta\}$ satisfying $\sum_{j=1}^{\eta} w_{c^j} = 1$.

The expected regret under y and w is simply

$$\overline{R}(y,w) := \sum_{T\in\mathcal{F}}\sum_{c\in\mathcal{C}_f} y_T w_c R(T,c) = \sum_{T\in\mathcal{F}}\sum_{c\in\mathcal{C}_f} y_T w_c (F(T,c) - F^*(c)). \quad (9)$$

For a given y, the *maximum expected regret problem* is

$$\overline{R}_{\max}(y) := \max_{w\in\mathcal{W}} \sum_{c\in\mathcal{C}_f} w_c \sum_{T\in\mathcal{F}} y_T R(T,c) = \max_{c\in\mathcal{C}_f} \sum_{T\in\mathcal{F}} y_T R(T,c). \quad (10)$$

The above equality follows from a standard observation in game theory: the optimization of $w \in \mathcal{W}$ is maximization of the function $G(y,c) = \sum_{T\in\mathcal{F}} y_T R(T,c)$ over the convex hull of $\mathcal{C}_f$, which is equivalent to optimizing over $\mathcal{C}_f$ itself. The minmax expected regret problem, which we refer to as the *randomized minmax regret problem*, is

$$Z_{\mathrm{R}} := \min_{y\in\mathcal{Y}} \overline{R}_{\max}(y) = \min_{y\in\mathcal{Y}} \max_{c\in\mathcal{C}} \left(\sum_{T\in\mathcal{F}} y_T (F(T,c) - F^*(c)) \right), \quad (11)$$

where we have replaced $\mathcal{C}_f$ with $\mathcal{C}$ under the assumption that $\mathcal{C}_f$ contains the maximizing cost vector.

3 Results

We only consider the perspective of the optimizing player here; analogous results for the adversary are given in the full version of the paper [24].

3.1 Discrete Scenario Uncertainty

Under discrete scenario uncertainty, we are given a finite set $\mathcal{S}$ of $|\mathcal{S}| = k$ scenarios. For each $S \in \mathcal{S}$, there exists a cost vector $c^S = (c_e^S)_{e\in E}$. Our solvability result for the optimizing player is the following.

Theorem 1. *For discrete scenario uncertainty, if the nominal problem $F^*(c)$ can be solved in time polynomial in n and m, then the corresponding randomized minmax regret problem $\min_{y\in\mathcal{Y}} \max_{S\in\mathcal{S}}(\overline{F}(y,c^S) - F^*(c^S))$ can be solved in time polynomial in n, m, and k.*

Recall that the feasible region $\mathcal{X}$ is described in size m. The algorithm for determining the optimizing player's mixed strategy, shown in Algorithm 1, solves two linear programs. The first is a linear programming relaxation of the deterministic minmax regret problem,

$$\begin{aligned} \min_{p,z} \quad & z \\ \text{s.t.} \quad & \sum_{e\in E} c_e^S p_e - F^*(c^S) \le z, \quad \forall S \in \mathcal{S}, \qquad \text{(LPD)} \\ & p \in \mathrm{CH}(\mathcal{X}),\ z \text{ free}, \end{aligned}$$

where $\mathrm{CH}(\mathcal{X})$ denotes the convex hull of $\mathcal{X}$ and $p \in [0,1]^n$. We refer to the vector $p = (p_1, \ldots, p_n)$ as the *marginal probability vector*; it indicates the total probability that each item should be selected in the mixed strategy. Given the optimal vector p^* from solving (LPD), the second linear program maps the marginal probabilities to an optimal mixed strategy:

$$\begin{aligned} \max_{u,w} \quad & w - \sum_{e \in E} p_e u_e \\ \text{s.t.} \quad & w - \sum_{e \in T} u_e \leq 0, \qquad \forall T \in \mathcal{F}, \\ & u, w \text{ free}, \end{aligned} \tag{LPM}$$

where $u = (u_1, \ldots, u_n)$. While both (LPD) and (LPM) potentially have an exponential number of constraints, a separation oracle is given by solvability of the nominal problem. Solvability of the latter program (LPM) is a known result from [14].

Algorithm 1. Rand-Minmax-Regret (discrete scenario uncertainty)

Input: Nominal combinatorial problem, cost vectors $(c^S)_{S \in \mathcal{S}}$
Output: Optimizing player's optimal mixed strategy $\mathcal{M}^* = (\Theta^*, Y^*)$ where $\Theta^* = (T_1, \ldots, T_\mu)$ and $Y^* = (y_{T_1}, \ldots, y_{T_\mu})$

1: Solve linear program (LPD) to get probability vector $p^* = (p_1^*, \ldots, p_n^*)$.
2: Solve linear program (LPM) with $p = p^*$ to generate constraints indexed $i = 1, \ldots, \mu$. Each constraint i corresponds to a set $T_i \in \mathcal{F}$ and dual variable y_{T_i}, indicating that T_i is an element in the optimal mixed strategy and has probability y_{T_i}.

A k-approximation algorithm for the deterministic minmax regret problem was introduced by Aissi et al. [3] and is shown in Algorithm 2. Using a new primal-dual interpretation with the formulation (LPD), as well as some arguments from [3], we show a simple proof of Theorem 2.

Algorithm 2. Mean-Cost-Approximation (Aissi et al. [3])

Input: Nominal combinatorial problem, cost vectors $(c^S)_{S \in \mathcal{S}}$
Output: Feasible solution $M \in \mathcal{F}$ satisfying $R_{\max}(M) \leq k Z_{\mathrm{D}}$.

1: Determine mean costs for each item: $d_e \leftarrow \frac{1}{k} \sum_{S \in \mathcal{S}} c_e^S, \quad \forall e \in E$.
2: Solve nominal problem with mean costs: $M \leftarrow \operatorname{argmin}_{T \in \mathcal{F}} \sum_{e \in T} d_e$.

Theorem 2. *For discrete scenario uncertainty, the solution to the nominal problem with mean costs is a k-approximation for the deterministic minmax regret problem.*

Since the randomized minmax regret problem corresponds to a linear programming relaxation of the deterministic minmax regret problem (specifically, the deterministic formulation is given by replacing the constraint $p \in \mathrm{CH}(\mathcal{X})$ with $x \in \mathcal{X}$ in (LPD)), it follows that $Z_{\mathrm{R}} \leq Z_{\mathrm{D}}$. Additionally, the primal-dual interpretation allows us to prove a new lower bound on Z_{R}, stated in Theorem 3 below. We show that this bound is tight.

Theorem 3. *For discrete scenario uncertainty and all nominal problems,*

$$Z_{\mathrm{R}} \geq \frac{Z_{\mathrm{D}}}{k}, \tag{12}$$

where $k = |S|$ is the number of scenarios. Equivalently, the integrality gap of the mixed integer program corresponding to (LPD) *is equal to k.*

3.2 Interval Uncertainty

For interval uncertainty, each item cost is independently contained within known lower and upper bounds:

$$c_e \in [c_e^-, c_e^+], \quad \forall e \in E. \tag{13}$$

Define the region

$$\mathcal{I} := \{c \mid c_e \in [c_e^-, c_e^+], e \in E\}. \tag{14}$$

Our solvability result for interval uncertainty is the following.

Theorem 4. *For interval uncertainty, if the nominal problem $F^*(c)$ can be solved in time polynomial in n and m, then the corresponding randomized minmax regret problem $\min_{y \in \mathcal{Y}} \max_{c \in \mathcal{I}}(\overline{F}(y, c) - F^*(c))$ can be solved in time polynomial in n and m.*

The algorithm for determining the optimizing player's mixed strategy is shown in Algorithm 3. This is the same algorithm that is used for the discrete scenario uncertainty case, except the linear program (LPI) is used instead of (LPD),

$$\begin{aligned} \min_{p,z} \quad & z \\ \text{s.t.} \quad & \sum_{e \in E \setminus T} c_e^+ p_e - \sum_{e \in T} c_e^- (1 - p_e) \leq z, \qquad \forall T \in \mathcal{F}, \\ & p \in \mathrm{CH}(\mathcal{X}), \ z \text{ free.} \end{aligned} \tag{LPI}$$

For the deterministic minmax regret problem under interval uncertainty, the known 2-approximation algorithm of Kasperski and Zieliński [19] uses midpoint costs and is shown in Algorithm 4. Using primal-dual methods, we show a new proof for this algorithm as stated by Theorem 5. We also prove Theorem 6, establishing the integrality gap for interval uncertainty, and we show that the corresponding bound is tight.

Algorithm 3. RAND-MINMAX-REGRET (interval uncertainty)

Input: Nominal combinatorial problem, item cost bounds (c_e^-, c_e^+), $e \in E$.
Output: Optimizing player's optimal mixed strategy $\mathcal{M}^* = (\Theta^*, Y^*)$ where $\Theta^* = (T_1, \ldots, T_\mu)$ and $Y^* = (y_{T_1}, \ldots, y_{T_\mu})$

1: Solve linear program (LPI) to get probability vector $p^* = (p_1^*, \ldots, p_n^*)$.
2: Solve linear program (LPM) with $p = p^*$ to generate constraints indexed $i = 1, \ldots, \mu$. Each constraint i corresponds to a set $T_i \in \mathcal{F}$ and dual variable y_{T_i}, indicating that T_i is an element in the optimal mixed strategy and has probability y_{T_i}.

Algorithm 4. MIDPOINT-COST-APPROXIMATION (Kasperski and Zieliński [19])

Input: Nominal combinatorial problem, item cost bounds (c_e^-, c_e^+), $e \in E$.
Output: Feasible solution $M \in \mathcal{F}$ satisfying $R_{\max}(M) \leq 2Z_{\mathrm{D}}$.

1: Determine midpoint costs for each item: $d_e \leftarrow \left(\dfrac{c_e^- + c_e^+}{2}\right)$, $\quad \forall e \in E$.
2: Solve nominal problem with midpoint costs: $M \leftarrow \operatorname*{argmin}_{T \in \mathcal{F}} \sum_{e \in T} d_e$.

Theorem 5. *For interval uncertainty, the solution to the nominal problem with midpoint costs is a 2-approximation for the deterministic minmax regret problem.*

Theorem 6. *For interval uncertainty and all nominal problems,*

$$Z_{\mathrm{R}} \geq \frac{Z_{\mathrm{D}}}{2}. \tag{15}$$

Equivalently, the integrality gap of the mixed integer program corresponding to (LPI) is equal to 2.

3.3 Convex Uncertainty

If the uncertainty set $\mathcal{C}$ is allowed to be a general nonnegative convex set and the nominal problem is polynomial solvable, we show that the maximum expected regret problem becomes NP-hard. This result implies that both randomized and deterministic minmax regret problems are NP-hard under convex uncertainty, since both are at least as hard as the maximum expected regret problem.

Theorem 7. *For polynomial solvable nominal problems* $F^*(c) = \min_{x \in \mathcal{X}} \sum_{e \in E} c_e x_e$ *and nonnegative convex uncertainty sets* $\mathcal{C}$*, the maximum expected regret problem* $\max_{c \in \mathcal{C}} \left(\sum_{e \in E} c_e p_e - F^*(c)\right)$ *where* $p \in \mathrm{CH}(\mathcal{X})$ *is NP-hard.*

4 Conclusion

Our results on lower bounds for randomized minmax regret in relation to deterministic minmax regret, specifically Theorems 3 and 6, have important implications for approximating deterministic minmax regret problems. Theorem 3 indicates that the integrality gap for the minmax regret problem under discrete scenario uncertainty is equal to k, and it is easy to create instances of nearly all nominal problems that achieve this gap. This also holds true for the integrality gap of 2 under interval uncertainty. In Kasperski [17], it is posed as an open problem whether or not there exist approximation algorithms under interval uncertainty that, for some specific nominal problems, achieve an approximation ratio better than 2. We have answered this question in the negative for approximation schemes based on our linear programming relaxations.

An important future step with randomized minmax regret research is to develop approximation algorithms for dealing with nominal problems that are already NP-hard. This problem is non-trivial: an algorithm with an approximation factor α for a nominal problem does not immediately yield an algorithm to approximate the randomized minmax regret problem with a factor α.

References

1. Aissi, H., Bazgan, C., Vanderpooten, D.: Complexity of the min-max and min-max regret assignment problems. Oper. Res. Lett. **33**(6), 634–640 (2005)
2. Aissi, H., Bazgan, C., Vanderpooten, D.: Complexity of the min-max (regret) versions of cut problems. In: Deng, X., Du, D.-Z. (eds.) ISAAC 2005. LNCS, vol. 3827, pp. 789–798. Springer, Heidelberg (2005)
3. Aissi, H., Bazgan, C., Vanderpooten, D.: Approximating min-max (regret) versions of some polynomial problems. In: Chen, D.Z., Lee, D.T. (eds.) COCOON 2006. LNCS, vol. 4112, pp. 428–438. Springer, Heidelberg (2006)
4. Aissi, H., Bazgan, C., Vanderpooten, D.: Approximation of min-max and min-max regret versions of some combinatorial optimization problems. Eur. J. Oper. Res. **179**(2), 281–290 (2007)
5. Aissi, H., Bazgan, C., Vanderpooten, D.: Min-max and min-max regret versions of combinatorial optimization problems: a survey. Eur. J. Oper. Res. **197**(2), 427–438 (2009)
6. Aissi, H., Bazgan, C., Vanderpooten, D.: General approximation schemes for min-max (regret) versions of some (pseudo-) polynomial problems. Discrete Optim. **7**(3), 136–148 (2010)
7. Averbakh, I.: On the complexity of a class of combinatorial optimization problems with uncertainty. Math. Program. **90**(2), 263–272 (2001)
8. Averbakh, I., Lebedev, V.: Interval data minmax regret network optimization problems. Discrete Appl. Math. **138**(3), 289–301 (2004)
9. Averbakh, I., Lebedev, V.: On the complexity of minmax regret linear programming. Eur. J. Oper. Res. **160**(1), 227–231 (2005)
10. Bertsimas, D., Nasrabadi, E., Orlin, J.B.: On the power of nature in robust discrete optimization (2013, working papers)
11. Bertsimas, D., Nasrabadi, E., Orlin, J.B.: On the power of randomization in network interdiction. arXiv preprint (2013). http://arxiv.org/abs/1312.3478
12. Bertsimas, D., Sim, M.: The price of robustness. Oper. Res. **52**(1), 35–53 (2004)
13. Birge, J.R., Louveaux, F.V.: Introduction to Stochastic Programming. Springer, New York (1997)
14. Grötschel, M., Lovász, L., Schrijver, A.: The ellipsoid method and its consequences in combinatorial optimization. Combinatorica **1**(2), 169–197 (1981)

15. Immorlica, N., Kalai, A.T., Lucier, B., Moitra, A., Postlewaite, A., Tennenholtz, M.: Dueling algorithms. In: Proceedings of the Forty-Third annual ACM Symposium on Theory of Computing, pp. 215–224. ACM (2011)
16. Johannes, B., Orlin, J.B.: Minimax regret problems are harder than minimax problems (2012, working papers)
17. Kasperski, A.: Discrete Optimization with Interval Data. STUDFUZZ, vol. 228. Springer, Heidelberg (2008)
18. Kasperski, A., Kurpisz, A., Zieliński, P.: Approximating the min-max (regret) selecting items problem. Inf. Process. Lett. **113**(1), 23–29 (2013)
19. Kasperski, A., Zieliński, P.: An approximation algorithm for interval data minmax regret combinatorial optimization problems. Inf. Process. Lett. **97**(5), 177–180 (2006)
20. Kasperski, A., Zieliński, P.: On the existence of an fptas for minmax regret combinatorial optimization problems with interval data. Oper. Res. Lett. **35**(4), 525–532 (2007)
21. Kiekintveld, C., Jain, M., Tsai, J., Pita, J., Ordóñez, F., Tambe, M.: Computing optimal randomized resource allocations for massive security games. In: Proceedings of The 8th International Conference on Autonomous Agents and Multiagent Systems, vol. 1, pp. 689–696. International Foundation for Autonomous Agents and Multiagent Systems (2009)
22. Korzhyk, D., Conitzer, V., Parr, R.: Complexity of computing optimal stackelberg strategies in security resource allocation games. In: Twenty-Fourth AAAI Conference on Artificial Intelligence (2010)
23. Kouvelis, P., Yu, G.: Robust Discrete Optimization and Its Applications. Springer, New York (1997)
24. Mastin, A., Jaillet, P., Chin, S.: Randomized minmax regret for combinatorial optimization under uncertainty. arXiv preprint (2014). http://arxiv.org/abs/1401.7043
25. Savage, L.J.: The theory of statistical decision. J. Am. Stat. Assoc. **46**(253), 55–67 (1951)
26. Shapiro, A., Dentcheva, D., Ruszczyński, A.P.: Lectures on Stochastic Programming: Modeling and Theory. SIAM, Philadelphia (2009)
27. Wald, A.: Contributions to the theory of statistical estimation and testing hypotheses. Ann. Math. Stat. **10**(4), 299–326 (1939)
28. Yu, G., Yang, J.: On the robust shortest path problem. Comput. Oper. Res. **25**(6), 457–468 (1998)

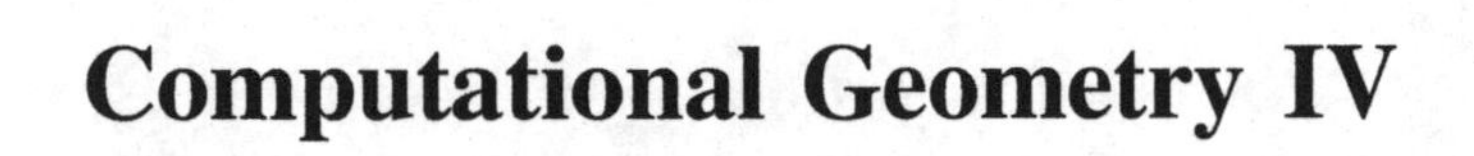

Computational Geometry IV

An Optimal Algorithm for Reconstructing Point Set Order Types from Radial Orderings

Oswin Aichholzer[1], Vincent Kusters[2(✉)], Wolfgang Mulzer[3], Alexander Pilz[1], and Manuel Wettstein[2]

[1] Institute for Software Technology, Graz University of Technology, Graz, Austria
{oaich,apilz}@ist.tugraz.at
[2] Department of Computer Science, ETH Zürich, Zurich, Switzerland
{vincent.kusters,manuelwe}@inf.ethz.ch
[3] Institut für Informatik, Freie Universität Berlin, Berlin, Germany
mulzer@inf.fu-berlin.de

Abstract. Given a set P of n labeled points in the plane, the *radial system* of P describes, for each $p \in P$, the radial ordering of the other points around p. This notion is related to the *order type* of P, which describes the orientation (clockwise or counterclockwise) of every ordered triple of P. Given only the order type of P, it is easy to reconstruct the radial system of P, but the converse is not true. Aichholzer et al. (*Reconstructing Point Set Order Types from Radial Orderings*, in Proc. ISAAC 2014) defined $T(R)$ to be the set of order types with radial system R and showed that sometimes $|T(R)| = n - 1$. They give polynomial-time algorithms to compute $T(R)$ when only given R.

We describe an optimal $O(n^2)$ time algorithm for computing $T(R)$. The algorithm constructs the convex hulls of all possible point sets with the given radial system, after which sidedness queries on point triples can be answered in constant time. This set of convex hulls can be found in $O(n)$ time. Our results generalize to *abstract order types*.

1 Introduction

Let P be a set of n labeled points in the plane. The *chirotope* of P is a function that indicates the orientation of each triple of P (clockwise, counterclockwise, or collinear). Throughout this paper, we consider only point sets in *general position*, that is, without collinear triples. Two labeled point sets have the same *order type* if they have the same chirotope or if one chirotope is the negation of the other. Many problems on planar point sets do not depend on the exact coordinates of the points but only on their order type. Examples include computing the convex hull and determining whether two segments with endpoints in the point set intersect. A *generalized configuration of points* is a labeled point set and an arrangement of pseudo-lines such that each pair of points is on a pseudo-line and each pseudo-line contains exactly two points [5]. By the containment in semispaces defined by these supporting pseudo-lines, orientations of point triples are defined analogously to point sets: if a point c is to the left of

K. Elbassioni and K. Makino (Eds.): ISAAC 2015, LNCS 9472, pp. 505–516, 2015.
DOI: 10.1007/978-3-662-48971-0_43

the pseudo-line through a and b when going from a to b, then the triple (a, b, c) is oriented *counterclockwise*. *Abstract order types* are the generalization of point set order types to generalized configurations of points. For most combinatorial purposes, generalized configurations of points behave like point sets; their convex hull is the intersection of those halfspaces bounded by the pseudolines that contain all the points and determines a cycle of directed arcs. Their chirotope determines whether two arcs defined by pairs of points cross. We refer to the work of Goodman and Pollack (see, e.g., [6]) and to a book by Knuth [7] (who calls abstract order types "CC systems") for more details. In this paper, we will be solely concerned with abstract order types. As opposed to many other publications on the subject, we stress that we consider *labeled* abstract order types here (and not abstract order type isomorphism classes). That is, we say that two abstract order types are *equivalent* when the bijection between them is fixed and they have the same chirotope, or one chirotope is the negation of the other.

Radial Systems. The *counterclockwise radial system* R_χ of an abstract order type χ on a set P defines, for each $p \in P$, the counterclockwise order $R_\chi(p)$ of the elements in $P \setminus \{p\}$ around p. We call each $R_\chi(p)$ a *counterclockwise radial ordering*. When χ is realizable as a point set, then $R_\chi(p)$ can be found by sweeping a ray around p in counterclockwise direction. Given a function U, we write $U \sim R_\chi$ when, for all $p \in P$, it holds that $U(p)$ is equal to $R_\chi(p)$ or the reverse of $R_\chi(p)$. Thus, in a sense, the relation $\sim$ "forgets" the clockwise/counterclockwise direction of each individual $R_\chi(p)$. We call U an *undirected radial system* and each $U(p)$ an *undirected radial ordering*. When we say *radial system*, we always mean counterclockwise radial system. It is possible to recover R_χ from U (all omitted proofs can be found in the full version of the paper):

Theorem 1.1. *Let χ be an abstract order type on V with $|V| = n$ and let $U \sim R_\chi$. Then U uniquely determines R_χ (up to complete reversal) and we can recover R_χ from U by reporting the direction of every $U(v)$ in $O(n)$ time.*

Aichholzer et al. [1] investigated under which circumstances the undirected radial system U of a generalized configuration of points P uniquely determines the abstract order type χ. They show that if P has a convex hull with at least four points, then U uniquely determines χ. More precisely: let $T(U)$ be the set of abstract order types with undirected radial system U (i.e., the sequences in U are known to originate from an abstract order type). We have

Theorem 1.2. ([1, Theorems 1 and 2]). *Consider an abstract order type χ on a set V with $n = |V| \geq 5$ and let $U \sim R_\chi$. Let $H \subseteq V$ be the points of the convex hull of χ. Then we can compute $|H|$ from U in polynomial time. Further, (i) if $|H| \neq 3$, then $T(U) = \{\chi\}$ and we can compute χ from U in polynomial time; and (ii) if $|H| = 3$, then $|T(U)| \leq n - 1$; all elements of $T(U)$ have convex hull size 3; and we can compute $T(U)$ from U in polynomial time.*

In the full version of [1] it is shown that (i) can be implemented in $O(n^3)$ time. There exist counterclockwise radial systems R with $|T(R)| = n - 1$. Hence, it is not possible to improve the bound on $|T(U)|$ in (ii), even if we consider counterclockwise radial systems instead of undirected radial systems [1].

Although U does not always uniquely determine χ, the pair (U, H), where H is the set of points on the convex hull, always suffices [1]. Thus, the abstract order types in $T(U)$ all have different convex hulls. Given an undirected radial system U on a set V, we say that a subset $H \subseteq V$ is *important* if H is the convex hull of some abstract order type in $T(U)$. An *important triangle* is an important set of size 3. Important sets are interrelated as follows.

Theorem 1.3 ([1, Propositions 1–4]). *Consider a radial system R on a set V with $n = |V| \geq 5$. If V has more than two important triangles, then all important triangles must have an element $v^* \in V$ in common. Thus, in general, exactly one of the following cases applies:*

(1) there is exactly one important set, and it has size at least four; or
(2) all important sets are triangles, there are at most $n-1$ of them, and they all share an element $v^* \in V$; or
(3) there are exactly two important sets, and they are disjoint triangles.

For cases (2) and (3), there exists actually a complete characterization of the important triangles. For an abstract order type $\chi \in T(U)$, an *inner* important triangle of χ is an important triangle of U that is not equal to the convex hull of χ. The following lemma reformulates the fact that an inner important triangle is not contained in a convex quadrilateral [1,9].

Lemma 1.4 ([1,9]). *Let χ be an abstract order type on a set P. A triangle $\langle a, b, c\rangle$ of χ is an inner important triangle iff the following conditions hold.*

(1) It is empty of points of P.
(2) It partitions $P \setminus \{a, b, c\}$ into three subsets P_a, P_b, and P_c, such that P_a is to the left of the directed line ba and to the right of ca, and P_b and P_c are defined analogously.
(3) For any two points $v, w \in P_a$, the pseudo-line vw intersects the edge bc; and similarly for points in P_b and P_c.

In this context, we mention that, if R is the radial system of some point set order type, then every abstract order type with radial system R can be realized as a point set [9, Theorem 27]. We do not consider realizability of abstract order types as point sets in this work. In the following, with a *realization of a radial system R*, we mean an abstract order type whose radial system is R.

Interestingly, realizability of radial systems cannot be decided by checking realizability of all induced radial systems up to any fixed constant size. Figure 1 shows a construction which is not realizable as an abstract order type, while every radial system induced by any strict subset of the vertices can be realized, even as a point set order type.

Theorem 1.5. *For any $k \geq 3$, there exists a radial system R_k over $n = 2k + 1$ vertices that is not realizable as an abstract order type, but that becomes realizable as a point set order type when removing any point.*

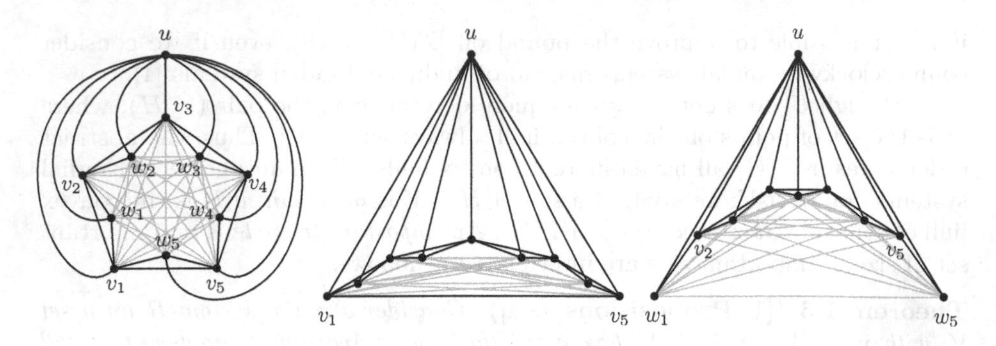

Fig. 1. The construction of R_5 on the left, and point set order type realizations of two induced radial systems after removing either w_5 or v_1 on the right.

Good Drawings. A *good drawing* (sometimes also called *simple topological graph*) of a graph is a drawing in the plane or on the sphere where each vertex is represented by a distinct point, and each edge is represented by a Jordan arc between its two vertices; any two such arcs intersect in at most one point, which is either a common endpoint or a proper crossing. The *rotation* of a vertex v in a good drawing is the cyclic order of the edges incident to v. The *rotation system* of a good drawing is the set of the rotations of its vertices. The radial system of a point set P is equivalent to the rotation system of the complete geometric graph on P. A generalized configuration of points Q defines a good drawing of K_n where the vertices are embedded on the points of Q and every edge is a segment of a pseudo-line in Q. The radial system of Q is equivalent to the rotation system of this good drawing. In a good drawing of K_n, the rotation system determines which edges cross. Therefore, it fixes the drawing up to the ordering of the crossings; in particular, we can find out whether two edges cross by locally inspecting the rotations for the four vertices involved [8]. We will use good drawings as a tool to maintain important sets in our algorithm.

Related Work. Variations on the notion of radial systems have been studied in many contexts. A prime example are *local sequences*, which are obtained by sweeping a line (instead of a ray) around each point. Goodman and Pollack [6] show that they determine the order type of P. Pilz and Welzl [9] describe a hierarchy on order types based on crossing edges in which two order types are considered equivalent iff they have the same radial system. We refer to Aichholzer et al. [1] for a more complete list of related work.

Our Results. For a given undirected radial system U on n vertices (which has size $\Theta(n^2)$), we provide an algorithm to direct the n radial orderings in $O(n)$ time (Theorem 1.1). Our main algorithm identifies the set of convex hulls of all abstract order types consistent with the given radial system in $O(n)$ time (provided that the input is the radial system of an abstract order type). This set allows for constant-time queries to the chirotope for any of these abstract order types. Hence, this is a means of reporting an explicit representation of $T(U)$ in $O(n)$ time, significantly improving Theorem 1.2. We remark that this can be shown to be optimal, as an adversary can use any unconsidered point in a suitable example to alter $|T(n)|$ (e.g., by using it to "destroy" a top triangle as defined in Sect. 2.1). If we do not know that the set of permutations provided as input is indeed the radial system of an abstract order type, we show how to verify this in $O(n^2)$ time. A straight-forward adversary argument shows that $\Omega(n^2)$ time (i.e., reading practically the whole input) is necessary to verify whether $|T(n)| = 0$. In this sense, our algorithm is optimal.

For radial systems as a data structure, we require that we can obtain the relative order of three elements in a radial ordering in constant time. This can be done by storing not only the radial ordering, but also the rank of each element within some linear order defined by the radial ordering around each vertex, when considering the n elements to be identified by their index in $\{1, \ldots, n\}$.

2 Obtaining Chirotopes from Radial Systems

Let R be the radial system for which we want to obtain the set $T(R)$ of abstract order types that realize it. (This set may be empty.) Our algorithm for computing $T(R)$ (conceptually) constructs a good drawing of a plane graph on the sphere by adding the vertices one-by-one and maintaining the faces that are candidates for the convex hull. We will see later that this actually boils down to maintaining at most two sequences of vertices plus one special vertex. Throughout the description, we assume that the radial orderings indeed correspond to the radial system of an abstract order type. If any of the assumptions is not fulfilled, we know that there is no abstract order type for the given set of radial orderings. If R can be realized as an abstract order type, then the plane graph is the subdrawing of a drawing weakly isomorphic (cf. [8]) to the complete graph on any generalized configuration of points that realizes that abstract order type.

For a plane cycle $C = \langle c_0, \ldots, c_{m-1} \rangle$ of m vertices (which we think of as counterclockwise with its *interior* to its left) in a good drawing of the complete graph, we say that an edge $c_i v$ *emanates to the outside* of the cycle at c_i if we encounter v in a counterclockwise sweep in $R(c_i)$ from c_{i-1} to c_{i+1}.[1] Otherwise, $c_i v$ *emanates to the inside*. If cv emanates to the outside for all $c \in C$, then v *covers* the cycle. If cv emanates to the inside for all $c \in C$, then we say that v is *inside* the cycle, and *outside* otherwise. If v neither is inside C nor covers C, then the good drawing restricted to C plus all edges from vertices of C to v is not plane. We call a cycle $\langle c_0, \ldots, c_{m-1} \rangle, m \geq 4$, *compact* if it is plane and, for each c_i, the edges $c_i c_{i+2}, c_i c_{i+3}, \ldots, c_i c_{i-2}$ all emanate to the inside (i.e., its rotation system corresponds to the radial system of m points in convex position).

Observation 2.1. *In any realization of a radial system, a compact cycle corresponds to a set of points in convex position.*

[1] We consider all indices modulo the length of the corresponding sequence.

Lemma 2.2. *Consider a radial system R. If Γ is a good drawing of the complete graph whose rotation system corresponds to R, then no element of an important set is inside of a compact cycle in Γ. In particular, no edge crosses an edge of the cell in Γ that defines the convex hull of a realization.*

Lemma 2.2 is closely related to Lemma 1.4 (see also [2, Theorem 3.2]). Consider a radial system R and a directed edge ab. Assume that ab is an edge of the convex hull of an abstract order type χ with $R_\chi \sim R$ (i.e., a *realization*) so that all other points of χ are to the left of ab. It is easy to see that the edge ab and R together uniquely determine the convex hull of our abstract order type. Hence, there is only one abstract order type realizing R with such an edge. We re-state the following well-known fact.

Lemma 2.3. *Given the radial system and a directed convex hull edge of an abstract order type, the orientation of a triple can be reported in constant time.*

2.1 Obtaining Hull Edges

Let P be a set of n points (or a generalized configuration of points), and let R be the radial system of the abstract order type χ of P. We assume that there is at least one abstract order type realizing R. The goal is to find a set of $O(n)$ *candidate edges* that may appear on the convex hull of a realization (i.e., the edges of the convex hull of P if there is no other realization of R or the union of the edges of all important triangles). Our algorithm incrementally builds a "hull structure" (defined below) for P. Before step k, we have a current set $P_{k-1} \subseteq P$ of $k-1$ points and a hull structure Z_{k-1} that represents the candidate edges for P_{k-1}. The algorithm selects a point $p_k \in P \setminus P_{k-1}$, adds it to P_{k-1}, and updates Z_{k-1}. A careful choice of p_k allows for updates in constant amortized time.

We begin with the description of the hull structure. Let $P_k \subseteq P$ be a set of k points ($k \geq 4$). The kth *hull structure* Z_k is an abstract representation of a graph with vertex set $V_k \subseteq P_k$ that is embedded on the sphere. That is, Z_k stores the incidences between the vertices, edges, and faces, but it does not assign coordinates to the points. Hull structures come in three types (see Fig. 2), which correspond in one-to-one-fashion to the three possible configurations of important sets in Theorem 1.3:

Type 1: Z_k is a compact cycle (recall that therefore, R restricted to V_k represents a convex $|V_k|$-gon with $|V_k| \geq 4$).

Type 2: Z_k consists of a compact cycle C and a *top vertex* t that covers C. The 3-cycles incident to t are called *top triangles.* A top triangle τ is marked either *unexamined, dirty,* or *empty.* Initially, τ is unexamined. Later, τ is marked either dirty or empty. "Dirty" indicates that τ cannot contain a convex hull vertex in its interior. "Empty" means that τ is a candidate for an important triangle. We orient each top triangle so that all other vertices of Z_k are to the exterior.

Type 3: Z_k is the union of two vertex-disjoint 3-cycles T_1 and T_2, called *independent triangles.* T_1 and T_2 are directed so that each has all of P_k to the interior. Moreover, the edges between the vertices of T_1 and T_2 appear as in Fig. 2.

Let R_k be the restriction of R to P_k. We maintain the following invariant: (a) if R_k has exactly one important set of size at least four, Z_k is of Type 1 and represents the counterclockwise convex hull boundary; (b) if R_k has two disjoint important triangles, Z_k is of Type 3, and the important triangles are exactly the independent triangles; (c) if R_k has several important triangles with a common vertex, Z_k is of Type 2 and all important triangles appear as top triangles; (d) if R_k has exactly one important triangle, Z_k is of Type 2 or 3, with the important triangle as a top triangle (Type 2) or as an independent triangle (Type 3). Furthermore, if Z_k is of Type 2, no convex hull vertex for P lies inside a dirty triangle, and each point of P_k lies either in C or in a dirty triangle.

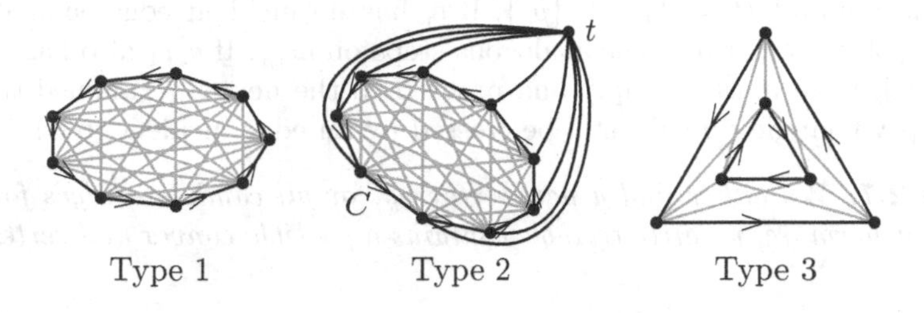

Fig. 2. The three different types of hull structures.

Initially, we pick 5 arbitrary points from P. Among those, there must be a compact 4-cycle Z_4 (e.g., [1, Figure 4]), which can be found in constant time. Our initial hull structure Z_4 is of Type 1, with vertex set $V_4 = P_4$. We next describe the insertion step for each possible type. For the running time analysis, we subdivide the algorithm into *phases*. Each phase is of Type 1, 2, or 3, and a new phase begins each time the type of the hull structure changes.

Type 1. We take an arbitrary vertex c of Z_{k-1} and check in constant time whether c has an incident edge in R emanating to the outside of Z_{k-1}. If not, the edges incident to c in Z_{k-1} are on the convex hull of P, and we are done; see below. Otherwise, let $p_k \in P \setminus P_{k-1}$ be the endpoint of such an edge. We set $P_k = P_{k-1} \cup \{p_k\}$, and we walk along Z_{k-1} (starting at c) to find the interval I of vertices for which the edge to p_k emanates to the outside. There are two cases: (i) if $I = Z_{k-1}$ (i.e., p_k covers Z_{k-1}), then Z_k is the hull structure of Type 2 with compact cycle Z_{k-1}, top vertex p_k, and all top triangles marked unexamined; (ii) if $I = \langle c_i, \dots, c_j \rangle$ is a proper subinterval of Z_{k-1}, the next hull structure Z_k is of Type 1 with vertex sequence $\langle p_k, c_j, \dots, c_i \rangle$ (R is realizable, so $c_j \neq c_i$).

Lemma 2.4. *We either obtain an edge from which the convex hull can be determined uniquely, or Z_k is a valid hull structure for P_k.*

Lemma 2.5. *A Type 1 phase that begins with a hull structure of size m and lasts for ℓ insertions takes $O(m + \ell)$ time. Furthermore, the next phase (if any) is of Type 2, beginning with a hull structure of size at most $m + \ell$.*

Type 2. We begin with a simple observation.

Observation 2.6. *Let Z_{k-1} be a Type 2 hull structure with compact cycle C and top vertex t. The vertices of C appear in their circular order in the clockwise radial ordering around t.*

We need to identify a suitable vertex p_k to insert. For this, we select an unexamined top triangle $\tau = \langle t, c_{i+1}, c_i \rangle$ and test whether c_i has an incident edge that emanates to the inside of τ. If yes, let $v \in P \setminus P_{k-1}$ be an endpoint of such an edge and check whether $c_i v$ crosses the edge tc_{i+1}. If so, then by Lemma 2.2 the vertices of τ lie inside a convex quadrilateral and there is no convex hull vertex inside τ. We mark τ dirty and proceed to the next unexamined triangle. If not, we set $p_k = v$ and $P_k = P_{k-1} \cup \{p_k\}$. If c_i has no incident edge emanating to the inside of τ, we perform the analogous steps on c_{i+1}. If c_{i+1} also has no such incident edge, we mark τ empty and proceed to the next unexamined triangle. (The empty triangle τ might still be crossed by an edge incident to t.)

Lemma 2.7. *We either find a new vertex p_k, or all candidate edges for P lie in Z_k. Furthermore, no dirty triangle contains a possible convex hull vertex of P.*

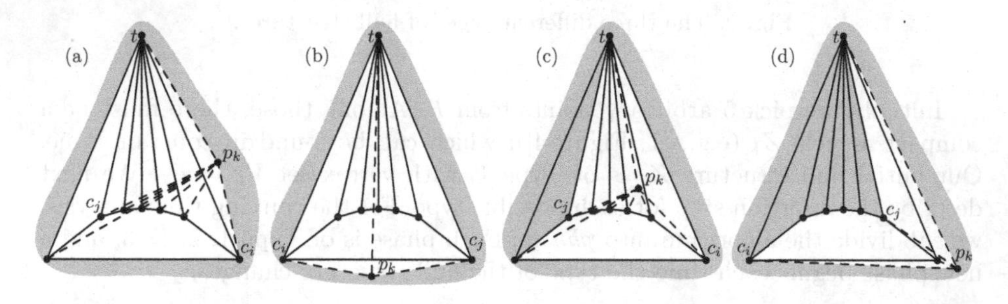

Fig. 3. Z_{k-1} is of Type 2 and p_k is not covering: if p_k forms a non-crossed 4-cycle, Z_k is of Type 1 (a, b); if not, Z_k is of Type 2 with p_k on the compact cycle (c, d). The algorithm will later discover that the triangle $\langle t, c_{j+1}, c_j \rangle$ in (c) is not important since it is inside a convex quadrilateral (Color figure online).

With p_k at hand, we inspect the boundary of C to find the interval I of vertices for which the edge to p_k emanates to the outside of C. First, if p_k does not cover C, i.e., $I = \langle c_i, \ldots, c_j \rangle$ is a proper subinterval of C, then p_k must lie between c_{i-1} and c_{j+1} in the clockwise order around t, as in any realization one of the cases in Fig. 3 applies. If p_k is between c_{i-1} and c_i or between c_j and c_{j+1}, then either $\langle p_k, t, c_{i-1}, c_i \rangle$ or $\langle t, p_k, c_j, c_{j+1} \rangle$ is a compact 4-cycle containing P_k, and we make it the next hull structure Z_k of Type 1; see Fig. 3(a). The green areas in the figures are the only regions where we might still find candidate edges. Otherwise, if $i + 1 = j$ and the edge tp_k crosses $c_i c_{i+1}$, the compact 4-cycle $\langle t, c_j, p_k, c_i \rangle$ contains P_k and becomes the next Type 1 hull structure Z_k; see Fig. 3(b). In any other case (i.e., p_k lies between c_i and c_j in clockwise order around t and if $i+1=j$ then tp_k does not cross c_ic_{i+1}), Z_k is of Type 2 and obtained from Z_{k-1} by removing the top triangles between c_i and c_j and adding the top triangles $\langle t, p_k, c_i\rangle$ and $\langle t, c_j, p_k\rangle$; see Fig. 3(c) and (d). If c_ip_k intersects an edge of Z_{k-1}, then $\langle t, p_k, c_i\rangle$ lies in a compact 4-cycle and is marked dirty. Otherwise, it is marked unexamined. We handle $\langle t, c_j, p_k\rangle$ similarly.

Second, suppose p_k covers C and let i, j be so that p_k is between c_i and c_{i+1} in clockwise order around t and t lies between c_j and c_{j+1} in clockwise order around p_k. Observation 2.6 ensures that these edges are well-defined; see Fig. 4. Now there are three cases. First, if $i=j$, then one of $\langle c_i, c_{i+1}, t, p_k\rangle$ or $\langle c_i, c_{i+1}, p_k, t\rangle$ defines a compact 4-cycle containing P_k, so Z_k is of Type 1 and consists of this cycle; see Fig. 5(a). Second, if $\{i, i+1\} \cap \{j, j+1\} = \emptyset$, then Z_k is of Type 3, with independent triangles $\langle p_k, c_i, c_{i+1}\rangle$ and $\langle t, c_j, c_{j+1}\rangle$; see Fig. 5(b). Third, suppose that $j=i+1$ or $i=j+1$, say, $j=i+1$. Then Z_k is of Type 2, with top vertex c_j and compact cycle $\langle t, p_k, c_i, c_{j+1}\rangle$. The top triangle $\langle c_j, c_{j+1}, c_i\rangle$ is dirty, the other top triangles are unexamined; see Fig. 5(c–d).

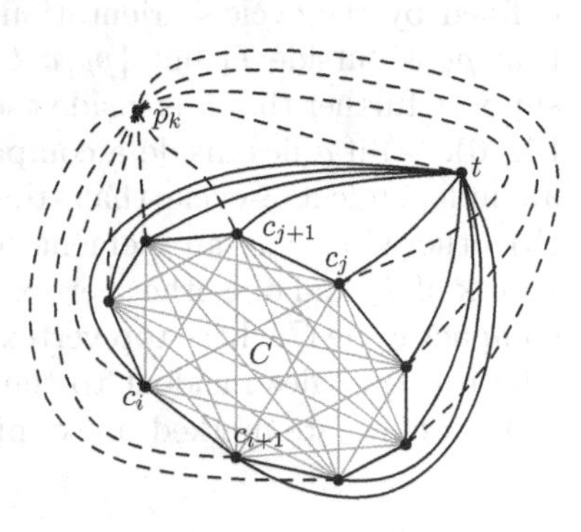

Fig. 4. Z_{k-1} is of Type 2 and p_k is covering.

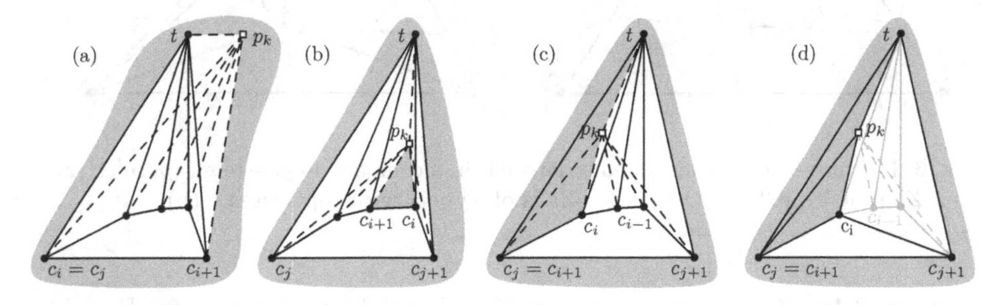

Fig. 5. Z_{k-1} is of Type 2 and p_k (box) is covering: if t and p_k are between the same vertices in each other's rotation, Z_k is of Type 1 (a); if these vertices are disjoint, Z_k is of Type 3 (b); if t and p_k have a common neighbor c_j in the other's rotation (c), the new top vertex c_j of Z_k structure requires the construction of a new compact cycle (d).

Lemma 2.8. *The resulting hull structure is valid for* P_k.

Lemma 2.9. *A Type 2 phase that begins with a hull structure of size* m *and lasts for* ℓ *insertions takes* $O(\ell + m)$ *time. Furthermore, if the next phase (if any) is of Type 1, it begins with a hull structure of size at most* 4.

Type 3. Let $T_1 = \langle a, b, c \rangle$ and $T_2 = \langle a', c', b' \rangle$ be the two independent triangles of Z_{k-1}, and let p_k be an arbitrary vertex of $P \setminus P_{k-1}$. We set $P_k = P_{k-1} \cup \{p_k\}$, and we distinguish three cases. First, if p_k is inside both T_1 and T_2, then $Z_k = Z_{k-1}$. Second, suppose that p_k is outside, say, T_1, and that $\{p_k, a, b, c\}$ forms a compact 4-cycle C. (Hence, p_k is inside T_2; recall that "inside" and "outside" is defined by the cycle's orientation.) Then $Z_k = C$ is of Type 1. Third, suppose that p_k is outside T_1 but $\{p_k, a, b, c\}$ does not form a compact 4-cycle. W.l.o.g., suppose further that a is inside the triangle $\langle p_k, b, c \rangle$. There are two subcases (see Fig. 6): (a) if a lies inside a compact 4-cycle, we replace a by p_k in T_1 to obtain an independent 3-cycle that, together with T_2, defines Z_k, again of Type 3; (b) otherwise, a is an element of a compact 4-cycle C that involves p_k, one vertex of T_2 and one other vertex of T_1. Then, Z_k is a Type 2 hull structure with compact cycle C whose top vertex is the vertex of T_1 that is not an element of C. The top triangles incident to the vertex of T_2 are marked dirty, the remaining top triangles are marked unexamined.

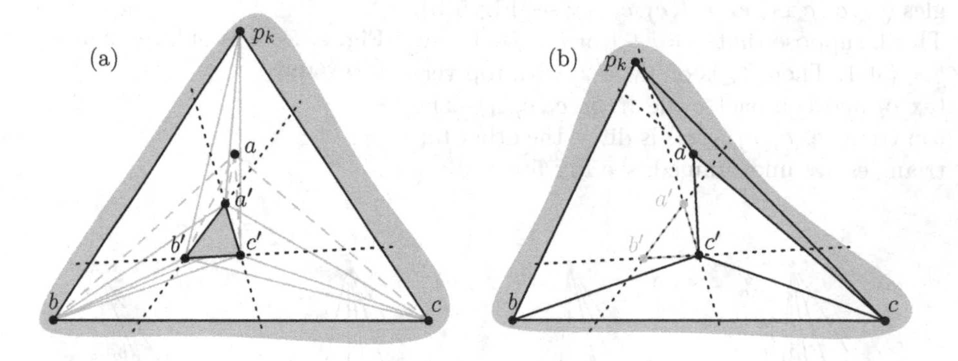

Fig. 6. If a vertex of an independent triangle is in a compact 4-cycle (e.g., $\langle p_k, a', c', c \rangle$), then Z_k if of Type 3 (a). Otherwise, Z_k is of Type 2 with top vertex c (b).

Lemma 2.10. *The resulting structure Z_k is a valid hull structure for P_k.*

Observation 2.11. *A Type 3 phase with ℓ insertions takes $O(\ell)$ time. If the next phase (if any) is of Type 2, it begins with a hull structure with at most 5 vertices, if it is of Type 1, it begins with a hull structure of size 4.*

To wrap up, we get the following lemma:

Lemma 2.12. *The final hull structure Z_n contains all candidate edges for R, and it can be obtained in $O(n)$ time.*

2.2 Obtaining the Actual Hulls from a Hull Structure

After having obtained Z_n, it remains to identify the faces that are important sets. If Z_n is of Type 1, then it is the only important set of R. If this is not the case, we want to obtain all the important triangles of R, i.e., all convex hulls of abstract order types realizing the radial system.

Lemma 2.13. *Given a Type 2 hull structure, we can decide in linear time which top triangles are important triangles of* R.

Lemma 2.14. *For a Type 3 hull structure, we can decide in linear time which of the two independent triangles are important triangles of* R.

For each important set we obtained for the radial system R, its chirotope is now given by Lemma 2.3.

Theorem 2.15. *Given a radial system* R *of an abstract order type, we can answer queries to the chirotopes of* $T(R)$ *in constant time, after* $O(n)$ *preprocessing time.*

Recall that we assumed that there is at least one realization of R. We can now check this assumption in the following way. We build the dual pseudo-line arrangement using an arbitrary chirotope we obtained for R using Lemma 2.3. This whole process takes $O(n^2)$ time [3,4]. If it fails then R has no realization. Otherwise, the dual pseudo-line arrangement explicitly gives the rotation system of the corresponding abstract order type, which we now compare to R.

Corollary 2.16. *Testing whether a set of radial orderings is the radial system of an abstract order type can be done in* $O(n^2)$ *time.*

We can apply our insights to obtain all important sets of a given chirotope.

Theorem 2.17. *Given an abstract order type, a hull structure of its radial system can be found in* $O(n \log n)$ *time. Further, the faces in the hull structure that can become convex hulls can be reported in the same time.*

Acknowledgments. This work was initiated during the *ComPoSe Workshop on Order Types and Rotation Systems* held in February 2015 in Strobl, Austria. We thank the participants for valuable discussions.

O.A. and A.P. are partially supported by the ESF EUROCORES programme EuroGIGA - ComPoSe, Austrian Science Fund (FWF): I 648-N18. W.M. is supported in part by DFG grants MU-3501/1 and MU-3501/2.

References

1. Aichholzer, O., Cardinal, J., Kusters, V., Langerman, S., Valtr, P.: Reconstructing point set order types from radial orderings. In: Ahn, H.-K., Shin, C.-S. (eds.) ISAAC 2014. LNCS, vol. 8889, pp. 15–26. Springer, Heidelberg (2014)
2. Balko, M., Fulek, R., Kynčl, J.: Crossing numbers and combinatorial characterization of monotone drawings of K_n. Discrete Comput. Geom. **53**(1), 107–143 (2015)
3. Chazelle, B., Guibas, L.J., Lee, D.T.: The power of geometric duality. BIT **25**(1), 76–90 (1985)

4. Edelsbrunner, H., O'Rourke, J., Seidel, R.: Constructing arrangements of lines and hyperplanes with applications. SIAM J. Comput. **15**(2), 341–363 (1986)
5. Goodman, J.E.: Proof of a conjecture of Burr, Grünbaum, and Sloane. Discrete Math. **32**(1), 27–35 (1980)
6. Goodman, J.E., Pollack, R.: Semispaces of configurations, cell complexes of arrangements. J. Combin. Theor. Ser. A **37**(3), 257–293 (1984)
7. Knuth, D.E.: Axioms and Hulls. LNCS, vol. 606. Springer, Heidelberg (1992)
8. Kynčl, J.: Simple realizability of complete abstract topological graphs in P. Discrete Comput. Geom. **45**(3), 383–399 (2011)
9. Pilz, A., Welzl, E.: Order on order types. In: Proceedings 31st International Symposium on Computational Geometry (SOCG 2015), pp. 285–299. LIPICS (2015)

Improved Approximation for Fréchet Distance on c-packed Curves Matching Conditional Lower Bounds

Karl Bringmann[1] and Marvin Künnemann[2,3](✉)

[1] Institute of Theoretical Computer Science, ETH Zurich, Zurich, Switzerland
karlb@inf.ethz.ch
[2] Max Planck Institute for Informatics, Saarbrücken, Germany
marvin@mpi-inf.mpg.de
[3] Saarbrücken Graduate School of Computer Science,Saarbrücken, Germany

Abstract. The Fréchet distance is a well-studied and popular measure of similarity of two curves. The best known algorithms have quadratic time complexity, which has recently been shown to be optimal assuming the Strong Exponential Time Hypothesis (SETH) [Bringmann FOCS'14]. To overcome the worst-case quadratic time barrier, restricted classes of curves have been studied that attempt to capture realistic input curves. The most popular such class are c-packed curves, for which the Fréchet distance has a $(1 + \varepsilon)$-approximation in time $\mathcal{O}(cn/\varepsilon + cn \log n)$ [Driemel et al. DCG'12]. In dimension $d \geqslant 5$ this cannot be improved to $\mathcal{O}((cn/\sqrt{\varepsilon})^{1-\delta})$ for any $\delta > 0$ unless SETH fails [Bringmann FOCS'14]. In this paper, exploiting properties that prevent stronger lower bounds, we present an improved algorithm with time complexity $\mathcal{O}(cn \log^2(1/\varepsilon)/\sqrt{\varepsilon} + cn \log n)$. This improves upon the algorithm by Driemel et al. for any $\varepsilon \ll 1/\log n$, and matches the conditional lower bound (up to lower order factors of the form $n^{o(1)}$).

1 Introduction

The Fréchet distance is a popular measure of similarity of curves and has two classic variants. Roughly speaking, the *continuous Fréchet distance* of two curves π, σ is the minimal length of a leash required to connect a dog to its owner, as they walk without backtracking along π and σ, respectively. In the *discrete Fréchet distance* we replace the dog and its owner by two frogs – in each time step each frog can jump to the next vertex along its curve or stay where it is.

In a seminal paper in 1991, Alt and Godau introduced the continuous Fréchet distance to computational geometry [2,13]. For polygonal curves π and σ with n and m vertices, respectively, (we always assume $m \leqslant n$) they presented an $\mathcal{O}(nm \log n)$ algorithm. The discrete Fréchet distance was defined by Eiter and Mannila [12], who presented an $\mathcal{O}(nm)$ algorithm. Since then, Fréchet distance

An extended version of this article can be accessed at http://arxiv.org/abs/1408.1340.

K. Elbassioni and K. Makino (Eds.): ISAAC 2015, LNCS 9472, pp. 517–528, 2015.
DOI: 10.1007/978-3-662-48971-0_44

has become a rich field of research, with many variants and related problems being studied (see, e.g., the references in [6]). As a natural measure for curve similarity, the Fréchet distance has found applications in various areas such as signature verification (see, e.g., [16]), map-matching tracking data (see, e.g., [5]), and moving objects analysis (see, e.g., [7]).

Apart from log-factor improvements [1,8] the quadratic complexity of the classic algorithms for the continuous and discrete Fréchet distance are still the state of the art. In fact, the first author recently showed a conditional lower bound: Assuming the Strong Exponential Time Hypothesis (SETH) there is no algorithm for the (continuous or discrete) Fréchet distance in time $\mathcal{O}((nm)^{1-\delta})$ for any $\delta > 0$, so apart from lower order factors of the form $n^{o(1)}$ the classic algorithms are optimal [6].

In attempts to obtain faster algorithms for realistic inputs, various restricted classes of curves have been considered, such as backbone curves [4], κ-bounded and κ-straight curves [3], and ϕ-low density curves [11]. The most popular model of realistic inputs are *c-packed curves.* A curve π is c-packed if for any point $z \in \mathbb{R}^d$ and any radius $r > 0$ the total length of π inside the ball $B(z,r)$ is at most cr, where $B(z,r)$ is the ball of radius r around z. This model has been used for several generalizations of the Fréchet distance [9,10,14,15]. Driemel et al. [11] introduced c-packed curves and presented a $(1+\varepsilon)$-approximation for the continuous Fréchet distance in time $\mathcal{O}(cn/\varepsilon + cn\log n)$, which works in any $\mathbb{R}^d$, $d \geqslant 2$. Assuming SETH, the following lower bounds have been shown for c-packed curves: (1) For sufficiently small constant $\varepsilon > 0$ there is no $(1+\varepsilon)$-approximation in time $\mathcal{O}((cn)^{1-\delta})$ for any $\delta > 0$ [6]. (2) In any dimension $d \geqslant 5$ and for varying $\varepsilon > 0$ there is no $(1+\varepsilon)$-approximation in time $\mathcal{O}((cn/\sqrt{\varepsilon})^{1-\delta})$ for any $\delta > 0$ [6].

In this paper we are interested in better-than-constant approximation algorithms. Specifically, for any constant $0 < \beta < 1$ we set $\varepsilon = n^{-\beta}$ and ask what running time is necessary to compute a $(1+\varepsilon)$-approximation of the Fréchet distance on c-packed curves. Note that in this regime the known upper and lower bounds differ by a factor $\sqrt{\varepsilon}^{-1-o(1)} = n^{\beta/2+o(1)}$. We improve upon the algorithm by Driemel et al. [11] by presenting an algorithm that matches the conditional lower bound of [6].

Theorem 1.1. *For any $\varepsilon \in (0,1]$ we present a $(1+\varepsilon)$-approximation on c-packed curves for the continuous and discrete Fréchet distance in time $\tilde{\mathcal{O}}(cn/\sqrt{\varepsilon})$.*

We use the $\tilde{\mathcal{O}}$-notation to hide polylogarithmic factors in n and $1/\varepsilon$. Specifically, our running time is $\mathcal{O}(\frac{cn}{\sqrt{\varepsilon}}\log(1/\varepsilon) + cn\log n)$ for the discrete variant and $\mathcal{O}(\frac{cn}{\sqrt{\varepsilon}}\log^2(1/\varepsilon) + cn\log n)$ for the continuous variant. This improves upon the algorithm by Driemel et al. for any $\varepsilon \ll 1/\log n$. While our algorithm might be too complex to speed up the algorithm by Driemel et al. in practical situations, it clarifies the optimal asymptotic dependence on c,n, and ε – apart from lower order factors $n^{o(1)}$, in dimension $d \geqslant 5$, and unless SETH fails [6]. Specifically, for any constants $\alpha, \beta \geqslant 0$ even after restricting $c = n^{\alpha+o(1)}$ and $\varepsilon = n^{-\beta+o(1)}$, any algorithm takes time $\Omega(\min\{cn/\sqrt{\varepsilon}, n^2\}^{1-o(1)})$ in dimension $d \geqslant 5$ [6], i.e., any algorithm is at most a factor of $n^{o(1)}$ faster than the better of our proposed algorithm ($\tilde{\mathcal{O}}(cn/\sqrt{\varepsilon})$) and the exact algorithm ($\tilde{\mathcal{O}}(n^2)$).

Organization. After setting up notation in Sect. 2, we present a new approximate decision procedure in Sect. 3. This procedure reduces the problem to one-dimensional curves, which we study in Sect. 4. Our exposition is limited to the continuous Fréchet distance; it is straightforward to obtain a similar algorithm for the discrete variant. For reasons of space, proofs, illustrations and basic technical lemmas for compositions of curves are deferred to the full version.

2 Preliminaries

Throughout the paper we fix the dimension $d \geqslant 2$. We let $B(z, r)$ be the ball of radius r around $z \in \mathbb{R}^d$. For integers $i \leqslant j$, we let $[i..j] := \{i, i+1, \ldots, j\}$, which is not to be confused with the real interval $[i, j]$. A (polygonal) curve π is defined by its vertices $(\pi_1, \ldots, \pi_n)$ with $\pi_p \in \mathbb{R}^d$, $p \in [1..n]$. We let $|\pi| = n$ be the number of vertices of π and $\|\pi\|$ be its total length $\sum_{i=1}^{n-1} \|\pi_i - \pi_{i+1}\|$, where $\|.\|$ denotes the Euclidean norm. We write $\pi_{p..b}$ for the subcurve $(\pi_p, \pi_{p+1}, \ldots, \pi_b)$. We can also view π as a continuous function $\pi\colon [1, n] \to \mathbb{R}^d$ with $\pi_{p+\lambda} = (1 - \lambda)\pi_p + \lambda\pi_{p+1}$ for $p \in [1..n-1]$ and $\lambda \in [0, 1]$. For the second curve $\sigma = (\sigma_1, \ldots, \sigma_m)$ we will use indices of the form $\sigma_{q..d}$ for the reader's convenience.

Variants of the Fréchet Distance. Let Φ_n be the set of all continuous and non-decreasing functions ϕ from $[0, 1]$ onto $[1, n]$. The *continuous Fréchet distance* between two curves π, σ with n and m vertices, respectively, is defined as

$$d_{\mathrm{F}}(\pi, \sigma) := \inf_{\phi_1 \in \Phi_n, \phi_2 \in \Phi_m} \max\nolimits_{t \in [0,1]} \|\pi_{\phi_1(t)} - \sigma_{\phi_2(t)}\|.$$

We call $\phi := (\phi_1, \phi_2)$ a (continuous) *traversal* of (π, σ), and say that it has *width* $\max_{t \in [0,1]} \|\pi_{\phi_1(t)} - \sigma_{\phi_2(t)}\|$.

In the discrete case, we let Δ_n be the set of all non-decreasing functions ϕ from $[0, 1]$ onto $[1..n]$. We obtain the *discrete Fréchet distance* $d_{\mathrm{dF}}(\pi, \sigma)$ by replacing Φ_n and Φ_m by Δ_n and Δ_m. We obtain an analogous notion of a (discrete) *traversal* and its *width*. Note that any $\phi \in \Delta_n$ is a staircase function attaining all values in $[1..n]$. Hence, $(\phi_1(t), \phi_2(t))$ changes only at finitely many points in time t. At any such *time step*, we jump to the next vertex in π or σ or both.

Free-Space Diagram. The discrete *free-space* of curves π, σ is defined as $\mathcal{D}^d_{\leqslant\delta}(\pi, \sigma) := \{(p, q) \in [1..n] \times [1..m] \mid \|\pi_p - \sigma_q\| \leqslant \delta\}$. Note that any discrete traversal of π, σ of width at most δ corresponds to a monotone sequence of points in the free-space where at each point in time we increase p or q or both. Because of this property, the free-space is a standard concept used in many algorithms for the Fréchet distance.

The continuous free-space is defined as $\mathcal{D}_{\leqslant\delta}(\pi, \sigma) := \{(p, q) \in [1, n] \times [1, m] \mid \|\pi_p - \sigma_q\| \leqslant \delta\}$. Again, a monotone path from $(1, 1)$ to (n, m) in $\mathcal{D}_{\leqslant\delta}(\pi, \sigma)$ corresponds to a traversal of width at most δ. It is well-known [2,13] that each *free-space cell* $C_{i,j} := \{(p,q) \in [i,i+1] \times [j,j+1] \mid \|\pi_p - \sigma_q\| \leqslant \delta\}$ (for $i \in [1..n-1], j \in [1..m-1]$) is convex, specifically it is the intersection of an ellipse with $[i,i+1] \times [j,j+1]$. In particular, the intersection of the free-space with any interval $[i,i+1] \times \{j\}$ (or $\{i\} \times [j,j+1]$) is an interval $I^h_{i,j}$ (or $I^v_{i,j}$), and for any such interval the subset that is reachable by a monotone path from $(1,1)$ is an interval $R^h_{i,j}$ (or $R^v_{i,j}$). Moreover, in constant time one can solve the following *free-space cell problem*: Given intervals $R^h_{i,j} \subseteq [i,i+1] \times \{j\}, R^v_{i,j} \subseteq \{i\} \times [j,j+1]$, determine the intervals $R^h_{i,j+1} \subseteq [i,i+1] \times \{j+1\}, R^v_{i+1,j} \subseteq \{i+1\} \times [j,j+1]$ consisting of all points that are reachable from a point in $R^h_{i,j} \cup R^v_{i,j}$ by a monotone path within the free-space cell $C_{i,j}$. Solving this problem for all cells from lower left to upper right we determine whether (n,m) is reachable from $(1,1)$ by a monotone path and thus decide whether the Fréchet distance is at most δ.

From Approximate Deciders to Approximation Algorithms. An *approximate decider* is an algorithm that, given curves π, σ and $\delta > 0, 0 < \varepsilon \leqslant 1$, returns one of the outputs (1) $d_F(\pi,\sigma) > \delta$ or (2) $d_F(\pi,\sigma) \leqslant (1+\varepsilon)\delta$. In particular, if $\delta < d_F(\pi,\sigma) \leqslant (1+\varepsilon)\delta$ the algorithm may return either of the two outputs.

Let $D(\pi,\sigma,\delta,\varepsilon)$ be the running time of an approximate decider and set $D(\pi,\sigma,\varepsilon) := \max_{\delta>0} D(\pi,\sigma,\delta,\varepsilon)$. We assume polynomial dependence on ε, i.e., that there are constants $0 < c_1 < c_2 < 1$ such that for any $0 < \varepsilon \leqslant 1$ we have $c_1 D(\pi,\sigma,\varepsilon/2) \leqslant D(\pi,\sigma,\varepsilon) \leqslant c_2 D(\pi,\sigma,\varepsilon/2)$. Driemel et al. [11] gave the following construction of a $(1+\varepsilon)$-approximation for the Fréchet distance.

Lemma 2.1. *For any approximate decider with running time $D(\pi,\sigma,\varepsilon)$ we can construct a $(1+\varepsilon)$-approximation running in time $\mathcal{O}\big(D(\pi,\sigma,\varepsilon)+D(\pi,\sigma,1)\log n\big)$.*

3 The Approximate Decider

By Lemma 2.1 it suffices to give an improved approximate decider with running time $\mathcal{O}(\frac{cn}{\sqrt{\varepsilon}} \log^2(1/\varepsilon))$ for the Fréchet distance to prove Theorem 1.1.

Long Segments and Pieces. We first partition our curves into subcurves, each of which is either a *long segment*, i.e., a single segment of length at least $\Lambda = \Theta(\sqrt{\varepsilon}\,\delta)$, or a *piece*, i.e., a subcurve staying in the ball of radius Λ around its initial vertex. More formally, we modify the given curve π by introducing new vertices as follows. Start at the initial vertex π_1. If the segment following the current vertex has length at least $\Lambda = \Lambda_{\varepsilon,\delta} := \min\{\frac{1}{2}\sqrt{\varepsilon}, \frac{1}{4}\} \cdot \delta$ then mark this segment as *long* and proceed to the next vertex. Otherwise follow π from the current vertex π_x to the first point π_y such that $\|\pi_x - \pi_y\| = \Lambda$ (or until we reach the end of π). If π_y is not a vertex, but lies on some segment of π, then introduce a new vertex at π_y. Mark $\pi_{x..y}$ as a *piece* of π and set π_y as current vertex. Repeat until π is completely traversed. Since this procedure introduces at most $|\pi|$ new vertices and does not change the shape of π, with slight abuse of notation we call the resulting curve again π and set $n := |\pi|$. This partitions π into subcurves $\pi^1, \ldots, \pi^k$, with $\pi^s = \pi_{p_s..b_s}$, where every part π^s is either

- a *long segment*: $b_s = p_s + 1$ and $\|\pi_{p_s} - \pi_{b_s}\| \geqslant \Lambda$, or
- a *piece*: $\|\pi_{p_s} - \pi_{b_s}\| = \Lambda$ and $\|\pi_{p_s} - \pi_x\| < \Lambda$ for all $x \in [p_s, b_s)$.

Note that the last piece actually might have distance $\|\pi_{p_s} - \pi_{b_s}\|$ less than Λ, however, for simplicity we assume equality for all pieces. Similarly, we introduce new vertices on σ and partition it into subcurves $\sigma^1, \ldots, \sigma^\ell$, with $\sigma^t = \sigma_{q_t..d_t}$, each of which is a long segment or a piece. Let $m := |\sigma|$.

Free-Space Regions. We follow the usual approach of exploring the reachable free-space (see Sect. 2). However, we treat regions spanned by a piece π' of π and a piece σ' of σ in a special way. Typically, if π', σ' consist of n', m' segments then their free-space would be resolved in time $\mathcal{O}(n'm')$, by resolving each of the $n'm'$ induced free-space cells in constant time. We show that by resorting to approximation we can reduce this running time to $\tilde{\mathcal{O}}(n' + m')$. This is made formal by the following subproblem and lemma.

Problem 3.1 (Free-space region problem). Given $\delta > 0$, $0 < \varepsilon \leqslant 1$, curves π, σ with n and m vertices, and *entry intervals* $\tilde{R}^h_{i,1} \subseteq [i, i+1] \times \{1\}$ for $i \in [1..n)$ and $\tilde{R}^v_{1,j} \subseteq \{1\} \times [j, j+1]$ for $j \in [1..m)$, compute *exit intervals* $\tilde{R}^h_{i,m} \subseteq [i, i+1] \times \{m\}$ for $i \in [1..n)$ and $\tilde{R}^v_{n,j} \subseteq \{n\} \times [j, j+1]$ for $j \in [1..m)$ such that (1) the exit intervals contain all points reachable from the entry intervals by a monotone path in $\mathcal{D}_{\leqslant\delta}(\pi, \sigma)$ and (2) all points in the exit intervals are reachable from the entry intervals by a monotone path in $\mathcal{D}_{\leqslant(1+\varepsilon)\delta}(\pi, \sigma)$. Here and in the remainder of the paper, we denote reachable intervals by $\tilde{R}$ instead of R to stress that we work with approximations.

Lemma 3.1. *If π and σ are pieces then the free-space region problem can be solved in time $\mathcal{O}((n + m)\log^2 1/\varepsilon)$.*

We will prove this lemma in Sects. 3.1 and 4. Using the algorithm of the above lemma, we obtain an approximate decider for the Fréchet distance as follows.

Algorithm 1. We consider all regions $r_{s,t} = [p_s, b_s] \times [q_t, d_t]$ spanned by parts π^s and σ^t. With each region $r_{s,t}$ we store the entry intervals $\tilde{R}^h_{i,q_t} \subseteq [i, i+1] \times \{q_t\}$ for $i \in [p_s..b_s)$ and $\tilde{R}^v_{p_s,j} \subseteq \{p_s\} \times [j, j+1]$ for $j \in [q_t..d_t)$. We correctly initialize the outer reachability intervals $\tilde{R}^h_{i,1}$ and $\tilde{R}^v_{1,j}$. Then we enumerate all regions sorted by increasing layer $s + t$, and among all regions with equal $s + t$ sorted by s. For each region $r_{s,t}$ we resolve its free-space region: (1) If both π^s, σ^t are long segments, we can resolve the free-space cell $r_{s,t}$ in constant time, (2) if π^s is a piece and σ^t is a long segment, we sequentially resolve the free-space cells $[i, i+1] \times [q_t, d_t]$ for $i = p_s, \ldots, b_s - 1$ (and symmetrically if π^s is a long segment and σ^t a piece), and (3) if both π^s, σ^t are pieces we solve the corresponding free-space region problem using Lemma 3.1. Finally, we return $d_F(\pi, \sigma) \leqslant (1 + \varepsilon)\delta$ if $(n, m) \in \tilde{R}^h_{n-1,m}$ and $d_F(\pi, \sigma) > \delta$ otherwise.

Observe that instead of enumerating *all* regions, we may enumerate only *reachable* regions, i.e., regions where some stored entry interval is non-empty. Indeed, if the reachable regions in layer L are $r_{s_1,L-s_1}, \ldots, r_{s_a,L-s_a}$, sorted by

$s_1 \leqslant \ldots \leqslant s_a$, then the reachable regions in layer $L+1$ are among $\{r_{s_i+1,L-s_i}, r_{s_i,L-s_i+1} \mid 1 \leqslant i \leqslant a\}$, and we can check for each such region in constant time whether it is reachable, so we can efficiently enumerate all reachable regions in layer $L+1$, again sorted by s. This trick was also used in [11, Lemma 3.1].

To bound the running time of this algorithm, we charge the time spent for region $r_{s,t}$ to appropriate segments of π^s or σ^t. Then we argue that the number of charges of a segment $\pi_{i..i+1}$ gives a lower bound on the length of σ in a small ball around $\pi_{i..i+1}$, which in turn is bounded since σ is c-packed. This analysis is similar to [11] and yields Lemma 3.2. Combining Lemma 3.2 with Lemma 2.1 yields Theorem 1.1.

Lemma 3.2. *Algorithm 1 is a correct approximate decider with running time* $\mathcal{O}(\frac{cn}{\sqrt{\varepsilon}} \log^2 1/\varepsilon)$ *on c-packed curves.*

3.1 Solving the Free-Space Region Problem on Pieces

It remains to prove Lemma 3.1. Let $(\pi, \sigma, \delta, \varepsilon)$ be an instance of the free-space region problem, i.e., π and σ are curves that stay within distance $\Lambda = \Lambda_{\varepsilon,\delta} = \Theta(\sqrt{\varepsilon}\,\delta)$ of their initial vertices. We reduce this instance to the free-space region problem on *one-dimensional separated* curves, i.e., curves $\hat{\pi}, \hat{\sigma}$ in $\mathbb{R}$ such that all vertices of $\hat{\pi}$ lie above 0 and all vertices of $\hat{\sigma}$ lie below 0.

Consider the line L containing the starting points of π and σ. Denote by $\Pi\colon \mathbb{R}^d \to L$ the projection onto L. By projecting π and σ we obtain *one-dimensional* curves $\hat{\pi} := \Pi(\pi) = (\Pi(\pi_1), \ldots, \Pi(\pi_n))$ and $\hat{\sigma} := \Pi(\sigma) = (\Pi(\sigma_1), \ldots, \Pi(\sigma_m))$. Moreover, if the initial vertices are within distance $\|\pi_1 - \sigma_1\| \leqslant \delta - 2\Lambda$ then all pairs of points in π, σ are within distance δ, since π, σ stay within distance Λ of their initial vertices. In this case, the free-space region problem is trivial. Otherwise, π and σ are sufficiently far apart so that their projections $\hat{\pi}, \hat{\sigma}$ are *separated*, i.e., there is a point $z \in L$ such that $\hat{\pi}$ and $\hat{\sigma}$ lie on different sides of z on L. Thus, $\hat{\pi}, \hat{\sigma}$ are one-dimensional separated curves, and after rotation and translation we can assume that they lie on $\mathbb{R}$ and are separated by 0.

We show that it suffices to solve the free-space region problem on $\hat{\pi}$, $\hat{\sigma}$, with parameters $\hat{\delta} := \delta$ and $\hat{\varepsilon} := \frac{1}{2}\varepsilon$ (with the same entry intervals as for π, σ).

Lemma 3.3. *Any solution to the free-space region problem on* $(\hat{\pi}, \hat{\sigma}, \hat{\delta}, \hat{\varepsilon})$ *solves the free-space region problem on* $(\pi, \sigma, \delta, \varepsilon)$.

In the proof, we crucially use that since π, σ stay within distance $\Lambda = \Theta(\sqrt{\varepsilon}\,\delta)$ of their initial vertices, the projection does not change distances between π and σ significantly – it follows from the Pythagorean theorem that any distance of approximately δ is changed by less than $\varepsilon\delta$. Thus, we can replace π, σ by $\hat{\pi}, \hat{\sigma}$ without introducing too much error.

In the following section, we show how to solve the free-space region problem on one-dimensional separated curves (Lemma 4.1). Together with the above Lemma 3.3 this concludes the proof of Lemma 3.1.

Remark. We distilled the property that we use in Lemma 3.3 from the conditional lower bound [6]. In [6], two curves π, σ are constructed for which it is hard to decide whether $d_F(\pi, \sigma)$ is at most δ or at least $(1+\varepsilon)\delta$. In this construction, for some consecutive points π_i, π_{i+1} and σ_j, σ_{j+1} we want to force any algorithm reaching (π_i, σ_j) to make a simultaneous step to $(\pi_{i+1}, \sigma_{j+1})$, i.e., we want that (i) $\|\pi_i - \sigma_j\|, \|\pi_{i+1} - \sigma_{j+1}\| \leqslant \delta$ and (ii) $\|\pi_i - \sigma_{j+1}\|, \|\pi_{i+1} - \sigma_j\| > (1+\varepsilon)\delta$. By elementary geometric arguments, (i) and (ii) imply $\|\pi_i - \pi_{i+1}\|, \|\sigma_j - \sigma_{j+1}\| \geqslant \sqrt{\varepsilon}\delta$. Thus, we cannot "compress" the curves π and σ too well (in terms of c-packedness), resulting in the factor $1/\sqrt{\varepsilon}$ in the lower bound of [6]. This bottleneck is closely related to Lemma 3.3.

4 On One-Dimensional Separated Curves

This section is dedicated to proving the following lemma.

Lemma 4.1. *The free-space region problem on one-dimensional separated curves can be solved in time $\mathcal{O}((n+m)\log^2 1/\varepsilon)$.*

To this end, we reduce our problem to the following simpler problem by subdividing the curves and performing parallel binary searches (the details are deferred to the extended version). In the remainder of this section we prove Lemma 4.2.

Problem 4.1 (Reduced free-space problem). Given $\delta > 0$ and $0 < \varepsilon \leqslant 1$, given one-dimensional separated curves π, σ with n, m vertices and all vertex coordinates being multiples of $\frac{1}{4}\varepsilon\delta$, and given an entry set $E \subseteq [1..n]$, compute the exit set $F^\pi \subseteq [1..n]$ consisting of all points f such that $d_{dF}(\pi_{e..f}, \sigma) \leqslant \delta$ for some $e \in E$ and the exit set $F^\sigma \subseteq [1..m]$ consisting of all points f such that $d_{dF}(\pi_{e..n}, \sigma_{1..f}) \leqslant \delta$ for some $e \in E$.

Lemma 4.2. *The reduced free-space problem has a $\mathcal{O}((n+m)\log 1/\varepsilon)$ algorithm.*

4.1 Greedy Decider for One-Dimensional Separated Curves

Before solving the reduced free-space problem, let us consider the simpler problem of deciding $d_{dF}(\pi, \sigma) \leqslant \delta$ for one-dimensional separated curves π, σ. In this section, we present a near-linear time algorithm for this problem, by walking along π and σ with *greedy steps* to either find a feasible traversal or bottleneck subcurves. We are not the first to have this observation[1], which in any case is not the focus of this work. Instead, we are interested in the (quite complex) extension of this result to the reduced free-space problem, which we use as a subroutine for our main result on c-packed curves, see Sect. 4.2.

In the remainder of the paper all indices of curves will be integral. Let $\pi = (\pi_1, \dots, \pi_n)$ and $\sigma = (\sigma_1, \dots, \sigma_m)$ be two separated curves in $\mathbb{R}$, i.e., $\pi_i \geqslant 0 \geqslant \sigma_j$. For indices $1 \leqslant i \leqslant n$ and $1 \leqslant j \leqslant m$, define $\mathrm{vis}_\sigma(i,j) := \{k \mid k \geqslant j \text{ and } \sigma_k \geqslant$

[1] We thank Wolfgang Mulzer for pointing us to this (unpublished) result by Matias Korman and Sergio Cabello (personal communication).

$\pi_i - \delta\}$ as the index set of vertices on σ that are later in sequence than σ_j and are in distance δ to π_i (i.e., *seen* by π_i). Symmetrically, $\text{vis}_\pi(i,j) := \{k \mid k \geqslant i \text{ and } \pi_k \leqslant \sigma_j + \delta\}$. Note that $\pi_i \leqslant \pi_{i'}$ implies that $\text{vis}_\sigma(i,j) \supseteq \text{vis}_\sigma(i',j)$, however the converse does not necessarily hold. Moreover, we set $\text{reach}_\sigma(i,j) := [j+1..j+k]$ where k is maximal with $[j+1..j+k] \subseteq \text{vis}_\sigma(i,j)$. Note that for any $j' \in \text{reach}_\sigma(i,j)$ we can reach $(\pi_i, \sigma_{j'})$ starting from (π_i, σ_j) and staying in π_i, since $|\pi_i - \sigma_{j'}| \leqslant \delta$ for any $j' \in \text{reach}_\sigma(i,j)$. We define $\text{reach}_\pi(i,j)$ symmetrically.

These visibility sets enable us to define the greedy algorithm for the Fréchet distance of π and σ given in Algorithm 2. Let $1 \leqslant p \leqslant n$ and $1 \leqslant q \leqslant m$ be arbitrary indices on σ and π. We say that p' is a *greedy step on* π *from* (p,q), written $p' \leftarrow \text{GREEDYSTEP}_\pi(\pi_{p..n}, \sigma_{q..m})$, if $p' \in \text{reach}_\pi(p,q)$ and $\text{vis}_\sigma(i,q) \subseteq \text{vis}_\sigma(p',q)$ holds for all $p \leqslant i \leqslant p'$. Symmetrically, $q' \in \text{reach}_\sigma(p,q)$ is a *greedy step on* σ *from* (p,q), if $\text{vis}_\pi(p,j) \subseteq \text{vis}_\pi(p,q')$ for all $q \leqslant j \leqslant q'$. In pseudocode, $\text{GREEDYSTEP}_\pi(\pi_{p..n}, \sigma_{q..m})$ denotes a function that returns an arbitrary greedy step p' on π from (p,q) if such an index exists and returns an error otherwise. We remark that greedy steps do not imply monotonicity of the coordinates π_p, because $\pi_{i'} > \pi_i$ might have equal visibility sets $\text{vis}_\sigma(i',j) = \text{vis}_\sigma(i,j)$. Greedy steps have, however, a certain monotonicity in visibility sets, see Lemma 4.3.

Algorithm 2. Greedy decider for separated curves $\pi_{1..n}$ and $\sigma_{1..m}$ in $\mathbb{R}$

1: $p \leftarrow 1, q \leftarrow 1$
2: **repeat**
3: **if** $p' \leftarrow \text{GREEDYSTEP}_\pi(\pi_{p..n}, \sigma_{q..m})$ **then** $p \leftarrow p'$
4: **if** $q' \leftarrow \text{GREEDYSTEP}_\sigma(\pi_{p..n}, \sigma_{q..m})$ **then** $q \leftarrow q'$
5: **until** no greedy step was found in the last iteration
6: **if** $p = n$ and $q = m$ **then return** $d_{\text{dF}}(\pi,\sigma) \leqslant \delta$
7: **else return** $d_{\text{dF}}(\pi,\sigma) > \delta$

Theorem 4.1. *Let* π *and* σ *be one-dimensional separated curves and* $\delta > 0$. *Algorithm 2 decides whether* $d_{\text{dF}}(\pi,\sigma) \leqslant \delta$ *in time* $\mathcal{O}((n+m)\log(nm))$.

Correctness. We call the indices (p,q) of point pairs considered in some iteration of Algorithm 2 (for any choice of greedy steps, if more than one exists) *greedy point pairs*. The following useful monotonicity property holds: If some greedy point on π sees a point on σ that is yet to be traversed, all following greedy points on π will see it *until it is traversed.*

Lemma 4.3. *For any greedy point pair* (p,q) *we have (1)* $\text{vis}_\sigma(\ell,q) \subseteq \text{vis}_\sigma(p,q)$ *for all* $1 \leqslant \ell \leqslant p$, *and symmetrically, (2)* $\text{vis}_\pi(p,\ell) \subseteq \text{vis}_\pi(p,q)$ *for all* $1 \leqslant \ell \leqslant q$.

We exploit this monotonicity to prove that if Algorithm 2 gets stuck then no feasible traversal of π and σ exists. We derive an even stronger statement using the following notion: For a greedy point pair (p,q), define $\text{STOP}_\pi(\pi_{p..n}, \sigma_{q..m}) := \max(\text{reach}_\pi(p,q) \cup \{p\}) + 1$ as the index of the first point after π_p on π which is not seen by σ_q, or $n+1$ if no such index exists. Let STOP_σ be defined symmetrically.

Lemma 4.4 (Correctness of Algorithm 2). *Let (p,q) be a greedy point pair, $p_{\text{stop}} := \text{STOP}_\pi(\pi_{p..n}, \sigma_{q..m})$ and $q_{\text{stop}} := \text{STOP}_\sigma(\pi_{p..n}, \sigma_{q..m})$. If on both curves no greedy step from (p,q) exists, then $d_{\text{dF}}(\pi,\sigma) > \delta$. Moreover, if $q_{\text{stop}} \leqslant m$ then $d_{\text{dF}}(\pi_{1..p'}, \sigma_{1..q_{\text{stop}}}) > \delta$ for all $p' \in [1..n]$, and if $p_{\text{stop}} \leqslant n$ then $d_{\text{dF}}(\pi_{1..p_{\text{stop}}}, \sigma_{1..q'}) > \delta$ for all $q' \in [1..m]$.*

Our Building Blocks and Their Implementation. A greedy step on a curve is not uniquely defined. We choose to implement the function GREEDYSTEP_π such that among all greedy steps that maximize the visibility region, we return the longest possible step on the curve. This property will be exploited by our algorithms in Sect. 4.2. More formally, let $p' \in \text{reach}_{\pi'}(p,q)$ be such that (i) p' is the largest index in $\text{reach}_\pi(p,q)$ with $|\text{vis}_\sigma(p',q)| = \max\{|\text{vis}_\sigma(z,q)| \mid z \in \text{reach}_\pi(p,q)\}$ and (ii) $\text{vis}_\sigma(p',q) \supseteq \text{vis}_\sigma(p,q)$. If p' exists, MAXGREEDYSTEP_π returns this value, otherwise it reports that no such index exists. We show the following lemma by a reduction to range searching on the point sets $P = \{(i,\pi_i) \mid i \in [1..n]\}$ and $Q = \{(i,\sigma_i) \mid i \in [1..m]\}$.

Lemma 4.5. *For one-dimensional separated curves $\pi = \pi_{1..n}$ and $\sigma = \sigma_{1..m}$, MAXGREEDYSTEP_π and STOP_π can be implemented to run in time $\mathcal{O}(\log nm)$ after $O((n+m)\log nm)$ preprocessing. If π, σ are input curves of the reduced free-space problem then these procedures can be implemented in time $\mathcal{O}(\log 1/\varepsilon)$ after $O((n+m)\log 1/\varepsilon)$ preprocessing.*

4.2 Solving the Reduced Free-Space Problem

In this section, we solve the reduced free-space problem, given entries E on π. We first solve the problem of determining the exit set F^σ assuming $E = \{1\}$ (single entry). Then we show for general $E \subseteq [1..n]$ how to compute F^π and F^σ.

Algorithm 3. Special Case: Single entry

```
1: function FIND-σ-EXITS(π_{p..b}, σ_{q..d})
2:     if q = d then
3:         if STOP_π(π_{p..b}, σ_q) = b + 1 then return {q}   ▷ End of π is reachable
4:         else return ∅
5:     if p' ← MAXGREEDYSTEP_π(π_{p..b}, σ_{q..d}) then
6:         return FIND-σ-EXITS(π_{p'..b}, σ_{q..d})
7:     else if q' ← GREEDYSTEP_σ(π_{p..b}, σ_{q..d}) then
8:         return FIND-σ-EXITS(π_{p..b}, σ_{q..q'-1}) ∪ FIND-σ-EXITS(π_{p..b}, σ_{q'..d})
9:     else return FIND-σ-EXITS(π_{p..b}, σ_{q..d-1})          ▷ No greedy step possible
```

Single Entry. The above recursive algorithm computes F^σ for curves π, σ if we have only one entry $E = \{1\}$, i.e., any traversal has to start in $(1,1)$. By

Lemma 4.4, whenever a greedy step in π is possible we can perform this greedy step and still reach all exits in F^σ. When a greedy step in σ is possible (from q to q'), then one can show that it is a greedy step with respect to $\pi_{p..b}, \sigma_{q..d'}$ for any $q' \leqslant d' \leqslant d$. Thus by Lemma 4.4, after the greedy step we can still reach all exits $F^\sigma \cap [q', \infty)$. This explains the split into $\sigma_{q..q'-1}$ and $\sigma_{q'..d}$. Finally, if no greedy step is possible, then Lemma 4.4 implies that $d_{\mathrm{dF}}(\pi_{p..b}, \sigma_{q..d}) > \delta$, so that d is no exit, and we can decrement d.

It is crucial that we use MaxGreedyStep$_\pi$: Since there can be no two consecutive maximal greedy steps in π, we can charge greedy steps in π to the operations in σ, whose number can be easily seen to be bounded by $\mathcal{O}(m)$.

Lemma 4.6. Find-σ-Exits$(\pi_{1..n}, \sigma_{1..m})$ *correctly identifies* F^σ *given the single entry* $E = \{1\}$ *and runs in time* $\mathcal{O}(m \log 1/\varepsilon)$.

Note that symmetrically, we can implement Find-π-Exits$(\pi_{1..n}, \sigma_{1..m})$ returning F^π given the single entry $E = \{1\}$ on π in time $\mathcal{O}(n \log 1/\varepsilon)$.

Entries on π, Exits on π. We now want to compute F^π given a set of entries E on π. To obtain near-linear time, it is essential to avoid computing the exits by iterating over every entry. We show how to divide π into disjoint subcurves that can be solved by a single call to Find-π-Exits each.

Assume we want to traverse $\pi_{p..n}$ and $\sigma = \sigma_{1..m}$ starting in π_p and σ_1. Let $u(p) := \max\{p' \in [p..n] \mid \exists 1 \leqslant q' \leqslant m : d_{\mathrm{dF}}(\pi_{p..p'}, \sigma_{1..q'}) \leqslant \delta\}$ be the last point on π that is reachable while traversing an arbitrary subcurve of σ that starts in σ_1. We prove that all exits reachable from p are contained in the interval $[p..u(p)]$ and that all entries in $(p..u(p)]$ can be ignored, as their exits can also be reached from p. Thus, we can determine the exits reachable from p by calling Find-π-Exits$(\pi_{p..u(p)}, \sigma)$.

Lemma 4.7. *The exits set* F^σ *can be computed in time* $\mathcal{O}((n + m) \log 1/\varepsilon)$.

Entries on π, Exits on σ. Similarly to F^π, we show how to compute the exits F^σ given entries E on π, by reducing the problem to calls of Find-σ-Exits. This time, however, the task is more intricate, as it is much more complex to define which entries can be ignored. Our solution works as follows. For any index p on π, let $Q(p) := \min\{q \mid d_{\mathrm{dF}}(\pi_{p..n}, \sigma_{1..q}) \leqslant \delta\}$ be the endpoint of the shortest initial fragment of σ such that the remaining part of π can be traversed together with this fragment. Let $P(p) := \min\{p' \mid d_{\mathrm{dF}}(\pi_{p..p'}, \sigma_{1..Q(p)}) \leqslant \delta\}$ be the endpoint of the shortest initial fragment of π, such that $\sigma_{Q(p)}$ can be reached by a feasible traversal. We show that we can ignore all entries in $(p..P(p)]$ – which makes the algorithm quite complex. We remark that $Q(p)$ and $P(p)$ are defined as a minimum, making it necessary to also implement *minimal* greedy steps.

Lemma 4.8. *The exit set* F^σ *can be computed in time* $\mathcal{O}((n + m) \log 1/\varepsilon)$.

Lemmas 4.8 and 4.7 yield an algorithm for the reduced free-space problem, proving Lemma 4.2. This concludes the proof of Theorem 1.1.

5 Conclusion

We presented an improved $(1+\varepsilon)$-approximation algorithm for the Fréchet distance on c-packed curves running in time $\tilde{O}(cn/\sqrt{\varepsilon})$. While our running time improves the state of the art for $\varepsilon \ll 1/\log n$, we suspect that our algorithm is too complex to speed up Fréchet distance computation in practical situations, unless ε is very small. Our running time matches a conditional lower bound, so that it is asymptotically optimal, up to lower order factors of the form $n^{o(1)}$, in dimension $d \geqslant 5$, and unless the Strong Exponential Time Hypothesis fails. We leave it as open problems to (1) find simpler and more practical algorithms with the same asymptotic guarantees as ours, (2) improve our $\log n$ and $\log 1/\varepsilon$-factors, and (3) determine the correct asymptotic behaviour in dimension $d = 2, 3, 4$.

References

1. Agarwal, P., Avraham, R.B., Kaplan, H., Sharir, M.: Computing the discrete Fréchet distance in subquadratic time. In: Proceedings of the 24th Annual ACM-SIAM Symposium on Discrete Algorithms (SODA 2013), pp. 156–167 (2013)
2. Alt, H., Godau, M.: Computing the Fréchet distance between two polygonal curves. Internat. J. Comput. Geom. Appl. **5**(1–2), 78–99 (1995)
3. Alt, H., Knauer, C., Wenk, C.: Comparison of distance measures for planar curves. Algorithmica **38**(1), 45–58 (2004)
4. Aronov, B., Har-Peled, S., Knauer, C., Wang, Y., Wenk, C.: Fréchet distance for curves, revisited. In: Azar, Y., Erlebach, T. (eds.) ESA 2006. LNCS, vol. 4168, pp. 52–63. Springer, Heidelberg (2006)
5. Brakatsoulas, S., Pfoser, D., Salas, R., Wenk, C.: On map-matching vehicle tracking data. In: Proceedings of the 31st International Conference on Very Large Data Bases (VLDB 2005), pp. 853–864 (2005)
6. Bringmann, K.: Why walking the dog takes time: Fréchet distance has no strongly subquadratic algorithms unless SETH fails. In: Proceedings of the 55th Annual IEEE Symposium on Foundations of Computer Science (FOCS 2014), pp. 661–670 (2014)
7. Buchin, K., Buchin, M., Gudmundsson, J., Löffler, M., Luo, J.: Detecting commuting patterns by clustering subtrajectories. Internat. J. Comput. Geom. Appl. **21**(3), 253–282 (2011)
8. Buchin, K., Buchin, M., Meulemans, W., Mulzer, W.: Four soviets walk the dog - with an application to Alt's conjecture. In: Proceedings of the 25th Annual ACM-SIAM Symposium on Discrete Algorithms (SODA 2014), pp. 1399–1413 (2014)
9. Chen, D., Driemel, A., Guibas, L.J., Nguyen, A., Wenk, C.: Approximate map matching with respect to the Fréchet distance. In: Proceedings of 13th Workshop on Algorithm Engineering and Experiments (ALENEX 2011), pp. 75–83 (2011)
10. Driemel, A., Har-Peled, S.: Jaywalking your dog: computing the Fréchet distance with shortcuts. SIAM J. Comput. **42**(5), 1830–1866 (2013)
11. Driemel, A., Har-Peled, S., Wenk, C.: Approximating the Fréchet distance for realistic curves in near linear time. Discrete Comput. Geom. **48**(1), 94–127 (2012)
12. Eiter, T., Mannila, H.: Computing discrete Fréchet distance. Technical report. CD-TR 94/64, Christian Doppler Laboratory for Expert Systems, TU Vienna, Austria (1994)

13. Godau, M.: A natural metric for curves - computing the distance for polygonal chains and approximation algorithms. In: Jantzen, M., Choffrut, C. (eds.) STACS 1991. LNCS, vol. 480, pp. 127–136. Springer, Heidelberg (1991)
14. Gudmundsson, J., Smid, M.: Fréchet queries in geometric trees. In: Bodlaender, H.L., Italiano, G.F. (eds.) ESA 2013. LNCS, vol. 8125, pp. 565–576. Springer, Heidelberg (2013)
15. Har-Peled, S., Raichel, B.: The Fréchet distance revisited and extended. In: Proceedings of 27th Annual Symposium on Computational Geometry (SoCG 2011), pp. 448–457 (2011)
16. Munich, M.E., Perona, P.: Continuous dynamic time warping for translation-invariant curve alignment with applications to signature verification. In: Proceedings of the 7th International Conference on Computer Vision (ICCV 1999), pp. 108–115 (1999)

Computing the Gromov-Hausdorff Distance for Metric Trees

Pankaj K. Agarwal[1], Kyle Fox[1(✉)], Abhinandan Nath[1], Anastasios Sidiropoulos[2], and Yusu Wang[2]

[1] Duke University, Durham, USA
{pankaj,kylefox,abhinath}@cs.duke.edu
[2] Ohio State University, Columbus, USA
sidiropoulos.1@osu.edu, yusu@cse.ohio-state.edu

Abstract. The Gromov-Hausdorff distance is a natural way to measure distance between two metric spaces. We give the first proof of hardness and first non-trivial approximation algorithm for computing the Gromov-Hausdorff distance for geodesic metrics in trees. Specifically, we prove it is NP-hard to approximate the Gromov-Hausdorff distance better than a factor of 3. We complement this result by providing a polynomial time $O(\min\{n, \sqrt{rn}\})$-approximation algorithm where r is the ratio of the longest edge length in both trees to the shortest edge length. For metric trees with unit length edges, this yields an $O(\sqrt{n})$-approximation algorithm.

1 Introduction

The Gromov-Hausdorff distance (or GH distance for brevity) [9] is one of the most natural distance measures between metric spaces, and has been used, for example, for matching deformable shapes [4,14] and for analyzing hierarchical clustering trees [6]. Informally, the Gromov-Hausdorff distance measures the *additive* distortion suffered when mapping one metric space into another using a correspondence between their points. Multiple approaches have been proposed to estimate the Gromov-Hausdorff distance or provide alternatives to its computation [4,13,14].

Despite much effort, the problem of computing, either exactly or approximately, GH distance has remained elusive. On one hand, the problem is not known to be NP-hard, and on the other hand no polynomial-time approximation algorithm exists for graphic metrics[1] unless the graph isomorphism problem

Work on this paper by P. K. Agarwal, K. Fox and A. Nath was supported by NSF under grants CCF-09-40671, CCF-10-12254, CCF-11-61359, and IIS-14-08846, and by Grant 2012/229 from the U.S.-Israel Binational Science Foundation. A. Sidiropoulos was supported by NSF under grants CAREER-1453472 and CCF-1423230. Y. Wang was supported by NSF under grant CCF–1319406.

[1] A graphic metric measures the shortest path distance between vertices of a graph with unit length edges.

K. Elbassioni and K. Makino (Eds.): ISAAC 2015, LNCS 9472, pp. 529–540, 2015.
DOI: 10.1007/978-3-662-48971-0_45

is in P. (The metrics for two graphs have GH distance 0 if and only if the two graphs are isomorphic.) Motivated by this trivial hardness result, it is natural to ask whether GH distance becomes easier in more restrictive settings such as geodesic metrics over trees.

Our Results. In this paper, we give the first non-trivial results on approximating the GH distance between metric trees. First, we prove (in Sect. 3) that the problem remains NP-hard even for metric trees via a reduction from 3-PARTITION. In fact, we show that there exists no algorithm with approximation ratio less than 3 unless P = NP. As noted above, we are not aware of any result that shows the GH distance problem being NP-hard even for general graphic metrics.

To complement our hardness result, we give an $O(\sqrt{n})$-approximation algorithm for the GH distance between metric trees with n nodes and *unit length* edges. Our algorithm works with arbitrary edge lengths as well; however, the approximation ratio becomes $O(\min\{n, \sqrt{rn}\})$ where r is the ratio of the longest edge length in both trees to the shortest edge length. Even achieving the $O(n)$-approximation ratio present here for arbitrary r is a non-trivial task.

Our algorithm uses a reduction, described in Sect. 4, to the similar problem of computing the *interleaving distance* [15] between two *merge trees*. Given a function $f : \mathbb{X} \to \mathbb{R}$ over a topological space $\mathbb{X}$, the merge tree T_f describes the connectivity between components of the sublevel sets of f. Morozov et al. [15] proposed the interleaving distance as a way to compare merge trees and their associated functions[2]. To take advantage of our reduction from GH distance, we describe, in Sect. 5, an $O(\min\{n, \sqrt{rn}\})$-approximation algorithm for interleaving distance between merge trees. Due to lack of space, most of the proofs have been provided in the full version [1].

Related Work. Most work on associating points between two metric spaces involves *embedding* a given high dimensional metric space into an infinite host space of lower dimensional metric spaces. However, there is some work on finding a bijection between points in two given finite metric spaces that minimizes typically multiplicative distortion of distances between points and their images, with some limited results on additive distortion. See [10,12,16] for recent surveys.

The interleaving distance between merge trees [15] was proposed as a measure to compare functions over topological domains that is stable to small perturbations in a function. Distances for the more general Reeb graphs are given in [3,8]. These concepts are related to the GH distance (Sect. 4), which we will leverage to design an approximation algorithm for the GH distance for metric trees.

2 Preliminaries

Metric Spaces and the Gromov-Hausdorff Distance. A *metric space* $\mathcal{X} = (X, \rho)$ consists of a (potentially infinite) set X and a function $\rho : X \times X \to$

[2] In fact, our hardness result can be easily extended to the GH distance between discrete tree metrics and the interleaving distance between merge trees.

$\mathbb{R}_{\geq 0}$ such that the following hold: $\rho(x,y) = 0$ iff $x = y$; $\rho(x,y) = \rho(y,x)$; and $\rho(x,z) \leq \rho(x,y) + \rho(y,z)$.

Given sets A and B, a *correspondence* between A and B is a set $\mathcal{C} \subseteq A \times B$ such that: (i) $\forall a \in A, \exists b \in B$ such that $(a,b) \in \mathcal{C}$; and (ii)$\forall b \in B, \exists a \in A$ such that $(a,b) \in \mathcal{C}$. We use $\Pi(A,B)$ to denote the set of all correspondences between A and B.

Let $\mathcal{X}_1 = (X_1, \rho_1)$ and $\mathcal{X}_2 = (X_2, \rho_2)$ be two metric spaces. The *distortion* of a correspondence $\mathcal{C} \in \Pi(X_1, X_2)$ is defined as:

$$\mathrm{Dist}(\mathcal{C}) = \sup_{(x,y),(x',y') \in \mathcal{C}} |\rho_1(x,x') - \rho_2(y,y')| \ .$$

The *Gromov-Hausdorff distance* [13], d_{GH}, between $\mathcal{X}_1$ and $\mathcal{X}_2$ is defined as:

$$d_{GH}(\mathcal{X}_1, \mathcal{X}_2) = \frac{1}{2} \inf_{\mathcal{C} \in \Pi(X_1,X_2)} \mathrm{Dist}(\mathcal{C}) \ .$$

Intuitively, d_{GH} measures how close can we get to an *isometric* (distance-preserving) embeddding between two metric spaces. We note that there are different equivalent definitions of the Gromov-Hausdorff distance; see e.g., Theorem 7.3.25 of [5] and Remark 1 of [13].

Given a tree $T = (V,E)$ and a length function $l : E \to \mathbb{R}_{\geq 0}$, we associate a metric space $\mathcal{T} = (|T|, d)$ with T as follows. $|T|$ is a geometric realization of T. The metric space is extended to points in an edge such that each edge of length l is isometric to the interval $[0,l]$. For $x, y \in |T|$, define $d(x,y)$ to be the length of the path $\pi(x,y) \in |T|$ which is simply the sum of the lengths of the restrictions of this path to edges in T. It is clear that d is a metric. The metric space thus obtained is a *metric tree.* We denote $\mathcal{T} = (T,d)$, treating T as the same as $|T|$.

Merge Trees and the Interleaving Distance. Let $f : \mathbb{X} \to \mathbb{R}$ be a continuous function from a connected topological space $\mathbb{X}$ to the set of real numbers. The *sublevel set* at a value $a \in \mathbb{R}$ is defined as $F_{\leq a} = \{x \in \mathbb{X} \mid f(x) \leq a\}$. A *merge tree* T_f captures the evolution of the topology of the sublevel sets as the function value is increased continuously from $-\infty$ to $+\infty$. Formally, it is obtained as follows. Let $\mathrm{epi}f = \{(x,y) \in \mathbb{X} \times \mathbb{R} \mid y \geq f(x)\}$. Let $\bar{f} : \mathrm{epi}f \to \mathbb{R}$ be such that $\bar{f}((x,y)) = y$. We may say $\bar{f}((x,y))$ is the *height* of point $(x,y) \in \mathbb{X}\times\mathbb{R}$. For two points (x,y) and (x',y') in $\mathbb{X} \times \mathbb{R}$ with $y = y'$, let $(x,y) \sim (x',y')$ denote them lying in the same component of $\bar{f}^{-1}(y)(= \bar{f}^{-1}(y'))$. Then $\sim$ is an equivalence relation, and the merge tree T_f is defined as the quotient space $(\mathbb{X}\times\mathbb{R})/\sim$.

Since two components at a certain height can only merge at a higher height and a component can never split as height increases, we get a rooted tree where the internal nodes represent the points where two components merge and the leaves represent the birth of a new component at a local minimum. Figure 1 shows an example of a merge tree for a 1-dimensional function. Note that the merge tree extends to a height of ∞, and our assumption that $\mathbb{X}$ is connected implies we have only one component in $F_{\leq \infty}$. We define the *root* of merge tree T_f to be the node with the highest function value.

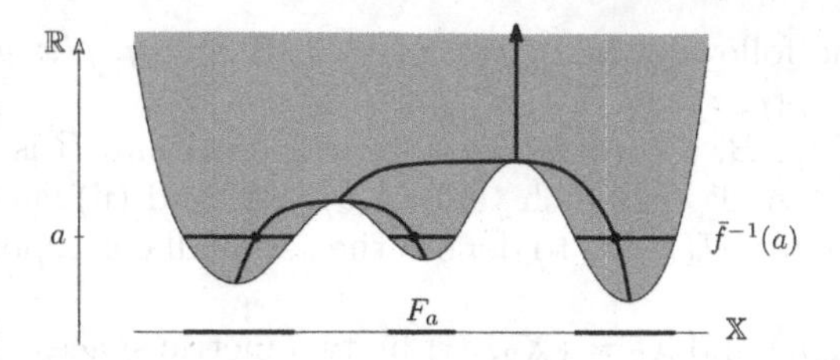

Fig. 1. Merge tree for a function from $\mathbb{R} \to \mathbb{R}$ (image by Morozov et al. [15]).

Since each point $x \in T_f$ represents a component of a sublevel set at a certain height, we can associate a height value $\hat{f}(x)$ with x. Given a merge tree T_f and $\epsilon \geq 0$, an ϵ-shift map $i^\epsilon : T_f \to T_f$ maps a point in the tree to its ancestor at height ϵ higher. We thus have $\hat{f}(i^\epsilon(x)) = \hat{f}(x) + \epsilon$. Given $\epsilon \geq 0$ and merge trees T_f and T_g with the associated shift maps i^ϵ and j^ϵ respectively, two continuous maps $\alpha^\epsilon : T_f \to T_g$ and $\beta^\epsilon : T_g \to T_f$ are said to be ϵ-compatible if they satisfy the following conditions

$$\hat{g}(\alpha^\epsilon(x)) = \hat{f}(x) + \epsilon, \forall x \in T_f \; ; \qquad \hat{f}(\beta^\epsilon(y)) = \hat{g}(y) + \epsilon, \forall y \in T_g \; ;$$
$$\beta^\epsilon \circ \alpha^\epsilon = i^{2\epsilon} \; ; \qquad \alpha^\epsilon \circ \beta^\epsilon = j^{2\epsilon} \; .$$

The *interleaving distance* [15] is then defined as

$$d_I(T_f, T_g) = \inf\{\epsilon \mid \text{there exist } \epsilon\text{-compatible maps } \alpha^\epsilon \text{ and } \beta^\epsilon\}.$$

Remark. We can relax the requirements on α^ϵ and β^ϵ as follows. Instead of requiring *exact* value changes, we require $\hat{f}(x) \leq \hat{g}(\alpha^\epsilon(x)) \leq \hat{f}(x) + \epsilon$ and $\hat{g}(y) \leq \hat{f}(\beta^\epsilon(y)) \leq \hat{g}(y) + \epsilon$. In addition, as x moves toward the root of T_f, $\alpha^\epsilon(x)$ must move toward the root of T_g (although $\alpha^\epsilon(x)$ may remain constant for a range of x values) and we do not need them to be continuous. A similar rule applies for β^ϵ. Finally, $\beta^\epsilon(\alpha^\epsilon(x))$ must go to an ancestor of x and $\alpha^\epsilon(\beta^\epsilon(y))$ must go to an ancestor of y. Both definitions of interleaving distance are equivalent, and we may use either based on which is more convenient.

As shown in [15], the interleaving distance is a metric and has the desirable properties of being both stable to small function perturbations and more discriminative than the popular bottleneck distance between persistence diagrams [7].

In the remainder of the paper, we will frequently drop the superscript ϵ when it is clear from the context. Also, we may stop alluding to the underlying functions f and g of the merge trees T_f and T_g and simply refer to them as T_1 and T_2. We may also use f and g to sometimes denote the height of the points in the trees themselves.

3 Hardness of Approximation

We show a reduction from the following decision problem called UNRESTRICTED-PARTITION: given a multiset of positive integers $X = \{a_1, \ldots, a_n\}$ with

$n = 3m$, is it possible to partition them into m multisets $\{X_1, \ldots, X_m\}$ such that all the elements in each multiset sum to the same quantity $S = \left(\sum_{i=1}^{n} a_i\right)/m$. This problem can be proved to be strongly NP-complete by a reduction from 3-Partition (see [1] for proof), so we can assume that the size of the integers is polynomial in the input.

We construct two trees T_1 and T_2 as follows. Let A and B be two sufficiently large numbers. Let $T_{s,t}$ denote a star graph having t edges of length s. T_1 consists of a node r_1 incident to an edge (r_1, r_1') of length B and to n edges $\{(r_1, p_1), \ldots, (r_1, p_n)\}$ of length 1, where p_i is the center of a copy of T_{A,a_i}. T_2 consists of a node r_2 incident to an edge (r_2, r_2') of length B and to m edges $\{(r_2, q_1), \ldots, (r_2, q_m)\}$ of length 2, where each q_i is the center of a distinct copy of $T_{A+1,S}$. See Fig. 2 for an illustration. Let $\mathcal{T}_1$ and $\mathcal{T}_2$ denote the metric trees associated with T_1 and T_2 respectively. Clearly, this construction can be done in polynomial time.

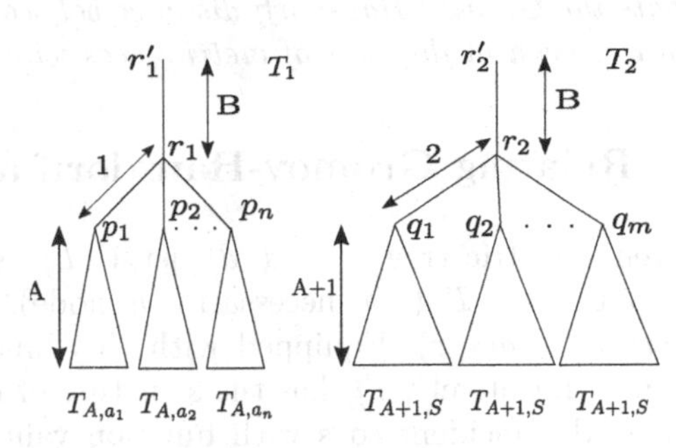

Fig. 2. The trees T_1 and T_2.

Lemma 1. *If the given instance of* UNRESTRICTED-PARTITION *is a* yes *instance, then* $d_{GH}(\mathcal{T}_1, \mathcal{T}_2) \leq 1$. *Otherwise,* $d_{GH}(\mathcal{T}_1, \mathcal{T}_2) \geq 3$.

Proof. (*Yes* instance) We construct a correspondence $\mathcal{C}$ between $\mathcal{T}_1$ and $\mathcal{T}_2$ with distortion at most 2, hence distance at most 1. A linearly interpolated bijection between the points of edges (r_1, r_1') and (r_2, r_2'), with r_1 mapping to r_2 and r_1' mapping to r_2', is added to $\mathcal{C}$. If a_i is assigned to X_j, the linearly interpolated bijection between edges (r_1, p_i) and (r_2, q_j) is added to $\mathcal{C}$. Also, the leaves of T_{A,a_i} are each mapped to a distinct leaf of $T_{A+1,S}$ attached to q_j such that there is a bijection between the leaves of T_1 and T_2 – this can be done since we have a *yes* instance. The interior points of the edges are mapped using linear interpolation. It can be easily verified that the distortion induced by $\mathcal{C}$ is at most 2.

(*No* instance) We show that any correspondence induces a distortion of at least 6, hence distance at least 3. Assume A and B are large enough so that for any correspondence with distortion ≤ 6, we can construct a bijection between the leaf edges of T_1 and T_2 such that two leaf edges are related if the correspondence sends the leaf of one edge to points on the other edge, with (r_1, r_1') mapping to (r_2, r_2'). Since we have a *no* instance, either no such bijection exists or there exists an i such that two leaves of T_{A,a_i} map to points inside leaf edges in $T_{A+1,S}$ attached to q_{j_1} and q_{j_2}, for some $j_1 \neq j_2$. Then the corresponding leaves attached to q_{j_1} and q_{j_2} (say l_1 and l_2 resp.) must map to points l_1' and l_2' inside T_{A,a_i} in T_1. We then have $d_1(l_1', l_2') \leq 2A$ while $d_2(l_1, l_2) = 2A + 6$. The distortion is at least 6.

We may also apply the reduction to metric trees with unit edge lengths by subdividing longer edges with an appropriate number of vertices. We thus have the following theorem.

Theorem 1. *Unless* P = NP, *there is no polynomial-time algorithm to approximate the Gromov-Hausdorff distance between two metric trees to a factor better than 3, even in the case of metric trees with unit edge lengths.*

4 Relating Gromov-Hausdorff and Interleaving Distances

Given a metric tree $\mathcal{T} = (T, d)$, let $V(T)$ denote the nodes of the tree. Given a point $s \in T$ (not necessarily a node), let $f_s : T \to \mathbb{R}$ be defined as $f_s(x) = -\ d(s, x)$. Equipped with this function, we obtain a merge tree T_{f_s} from $\mathcal{T}$. Intuitively, T_{f_s} has the structure of rooting T at s, and then adding an extra edge incident to s with function value extending to $+\infty$. The following theorem, proved in [1], connects the GH distance and the interleaving distance.

Theorem 2. *Let* $\gamma = \min_{u\in V(T_1), v\in V(T_2)} d_I(T_{1f_u}, T_{2f_v})$. *Then*

$$\tfrac{1}{2}d_{GH}(\mathcal{T}_1, \mathcal{T}_2) \le \gamma \le 10 d_{GH}(\mathcal{T}_1, \mathcal{T}_2)\ .$$

Corollary 1. *If there is a c-approximation algorithm for the interleaving distance between two merge trees, then there is a* 20*c-approximation algorithm for the Gromov-Hausdorff distance between two metric trees.*

5 Computing the Interleaving Distance

We propose algorithms for the decision version of the interleaving distance problem, which is stated as follows: *Given two merge trees* T_1 *and* T_2 *and a value* $\epsilon \ge 0$, *compute an* ϵ*-compatible map between them if such a map exists; otherwise report that no such map exists.*

Given two merge trees T_1 and T_2, a *c-approximate decision procedure* for any $c \ge 1$ does the following: if $d_I(T_1, T_2) \le \epsilon$, it returns a pair of $c\epsilon$-compatible maps between T_1 and T_2; if $d_I(T_1, T_2) > \epsilon$ it will either return a pair of $c\epsilon$-compatible maps between T_1 and T_2 or report that no such maps exist. Using binary search, this gives us a c-approximation to $d_I(T_1, T_2)$.

If we know $\alpha^\epsilon(x)$ for a point x at height h, then we can compute $\alpha^\epsilon(y)$ for any ancestor y of x at height $h' \ge h$ by simply putting $\alpha^\epsilon(y) = j^{h'-h} \circ \alpha^\epsilon(x)$. A similar claim holds for β^ϵ. Thus specifying the maps for the leaves of the trees suffices, because any point in the tree is the ancestor of at least one of the leaves. Hence, these maps have a representation that requires linear space in the size of the trees.

We define the *length* of any edge in a merge tree other than the edge to infinity to be the height difference between its two end points. Given a parameter $\epsilon > 0$, an edge is called ϵ-*long*, or *long* for brevity, if its length is greater than 2ϵ.

We first describe an exact decision procedure if all edges in both trees are long, and then describe an approximate decision procedure when there are short edges. Finally, we combine the two procedures to handle arbitrary merge trees.

Algorithm for Trees with Long Edges. A *subtree* rooted at a point x in a merge tree T, denoted T^x, includes all the points in the merge tree that are descendants of x and an edge from x that extends upwards to height ∞. For every $x \in T$, the nearest descendant of x (including x) that is in $V(T)$, say $\tau(x)$, is the only node such that $T^x = T^{\tau(x)}$. For $u \in V(T)$, let $C(u)$ denote the children of u.

Assume $d_I(T_1, T_2) \leq \epsilon$, and let $\alpha : T_1 \to T_2$ and $\beta : T_2 \to T_1$ be a pair of ϵ-compatible maps. We define an indicator function $\Phi : T_1 \times T_2 \to \{0,1\}$ such that $\Phi(u,v) = 1$ if $d_I(T_1^u, T_2^v) \leq \epsilon$ and 0 otherwise. We propose an algorithm to compute $\Phi(u,v)$ for all $u \in V(T_1), v \in V(T_2)$. If $\Phi(u,v) = 1$, the algorithm also computes a pair of ϵ-compatible maps between T_1^u and T_2^v. We are interested in $\Phi(r_1, r_2)$, where r_1 (resp. r_2) is the root of T_1 (resp. T_2).

Lemma 2. *If all the edges are long, the maps α and β induce a bijection between the subtrees rooted at the nodes of T_1 and the nodes of T_2.*

Proof. We define $\Psi_1 : V(T_1) \to V(T_2)$ as follows. Let $u \in V(T_1)$, and let u_p be its parent (for $u = r_1$ we set u_p to be an artificial node at height ∞ above r_1). Let u' be the ancestor of u at height $f(u_p) - 2\epsilon - \epsilon_0$ where ϵ_0 is such that all the children of u_p have height less than $f(u')$ and $\alpha(u') \notin V(T_2)$. We may use the same ϵ_0 for all $u \in V(T_1)$. Set $\Psi_1(u) = \tau(\alpha(u'))$. We prove that $|f(u) - g(\Psi_1(u))| \leq \epsilon$. This is true because all the points in T_1^u map to points in $T_2^{\Psi_1(u)}$ and vice versa, hence $d_I\left(T_1^u, T_2^{\Psi_1(u)}\right) \leq \epsilon$. If $|f(u) - g(\Psi_1(u))| > \epsilon$, the roots of T_1^u and $T_2^{\Psi_1(u)}$ are more than ϵ apart and at least one edge e incident to one of the roots will not be in the image of the corresponding ϵ-compatible map. However, the composition map applied to the lower node incident to e must map it to a point inside e (since the edges are longer than 2ϵ), a contradiction. Define Ψ_2 similarly.

We now prove $\Psi_2(\Psi_1(u)) = u$ for all $u \in V(T_1)$. We know $\beta(\alpha(u'))$ lies on the edge (u_p, u), because $f(u') < f(u_p) - 2\epsilon$. Therefore, $\beta(\Psi_1(u))$ is a descendant of u_p. Because $g(\Psi_1(u)) \geq f(u) - \epsilon$, we further conclude $\beta(\Psi_1(u))$ is an ancestor of u and $\Psi_2(\Psi_1(u))$ is an ancestor of u as well. Since $|f(u) - g(\Psi_1(u))| \leq \epsilon$ and $|g(v) - f(\Psi_2(v))| \leq \epsilon$ for all $u \in V(T_1)$ and $v \in V(T_2)$, we have $|f(u) - f(\Psi_2(\Psi_1(u))| \leq 2\epsilon$. All the edges are longer than 2ϵ, so $\Psi_2(\Psi_1(u)) = u$. We conclude Ψ_1 is a surjection, with Ψ_2 as its inverse. By symmetry, Ψ_2 must be surjective as well, making Ψ_1 a bijection.

Lemma 3. *Suppose all the edges in T_1 and T_2 are long. For any pair of nodes $u \in V(T_1), v \in V(T_2)$, $\Phi(u,v) = 1$ iff all of the following hold: (i) $|f(u) - g(v)| \leq \epsilon$; (ii) $|C(u)| = |C(v)|$; (iii) Let $C(u) = \{u_1, \ldots u_k\}$ and $C(v) = \{v_1, \ldots, v_k\}$, then there exists a permutation π of $[1:k]$ such that $\Phi(u_i, v_{\pi(i)}) = 1$ for all $i \in [1:k]$.*

See [1] for a proof. Using Lemma 3, we compute $\Phi(u,v)$ in a bottom-up manner. Suppose we have computed $\Phi(u_i, v_j)$ for all $u_i \in C(u)$ and $v_j \in C(v)$.

We compute $\Phi(u, v)$ as follows. If (i) or (ii) of Lemma 3 does not hold for u and v, then we return $\Phi(u, v) = 0$. Otherwise we construct the bipartite graph $G_{uv} = \{C(u) \cup C(v), E = \{(u_i, v_j) \mid \Phi(u_i, v_j) = 1\}\}$ and determine in $O(k^{5/2})$ time whether G_{uv} has a perfect matching, using the algorithm by Hopcroft and Karp [11]. If G_{uv} has a perfect matching $M = \{(u_1, v_{\pi(1)}), \ldots, (u_k, v_{\pi(k)})\}$, we set $\Phi(u, v) = 1$, else we set $\Phi(u, v) = 0$. If $\Phi(u, v) = 1$, we use the ϵ-compatible maps for $T_{u_i}, T_{v_{\pi(i)}}, i \in [1 : k]$, to compute a pair of ϵ-compatible maps between T_1^u and T_2^v, as discussed in the proof of Lemma 3. The theorem below follows (see [1] for the runtime analysis).

Theorem 3. *Given two merge trees T_1 and T_2 and a parameter $\epsilon > 0$ such that all edges of T_1 and T_2 are ϵ-long, then whether $d_I(T_1, T_2) \leq \epsilon$ can be determined in $O(n^{5/2})$ time. If the answer is yes, a pair of ϵ-compatible maps between T_1 and T_2 can be computed within the same time.*

Algorithm for Short Edges. Given two merge trees, a naive map is to map the lowest among all the leaves in both the trees to a point at height equal to the height of the higher root (see Fig. 3). Thus, all the points in one tree will be mapped to the infinitely long edge on the other tree. This map produces a distortion equal to the height of the trees, which can be arbitrarily larger than the optimum. Nevertheless, this simple idea leads to an approximation algorithm.

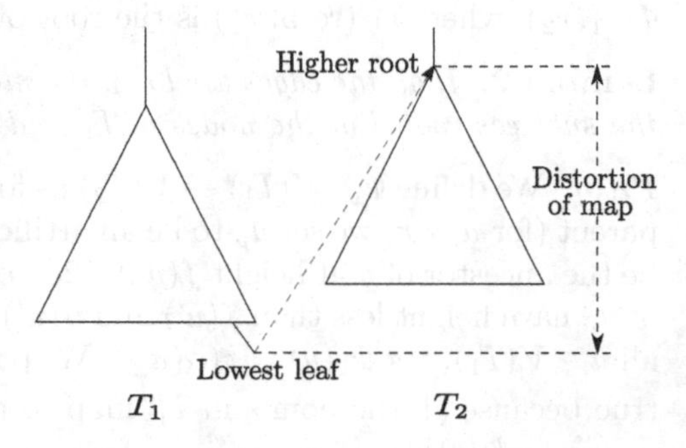

Fig. 3. A naive map (Color figure online).

Here is an outline of the algorithm. After carefully *trimming* off short subtrees from the input trees, the algorithm decomposes them into two kinds of regions – those with nodes and those without nodes. If the interleaving distance between the input trees is small, then there exists an isomorphism between trees induced by the regions without nodes. Using this isomorphism, the points in the nodeless regions are mapped without incurring additional distortion. Using a counting argument and the naive map described above, it is shown that the distortion incurred while mapping the regions with nodes and the trimmed regions is bounded.

More precisely, given T_1, T_2 and $\epsilon > 0$, define the *extent* $e(x)$ of a *point* x (which is not necessarily a tree node) in T_1 or T_2 as the maximum height difference between x and any of its descendants. Suppose each edge is at most $s\epsilon$ long. Let T_1' and T_2' be subsets of T_1 and T_2 consisting only of points with extent at least $2\sqrt{ns}\epsilon$, adding nodes to the new leaves of

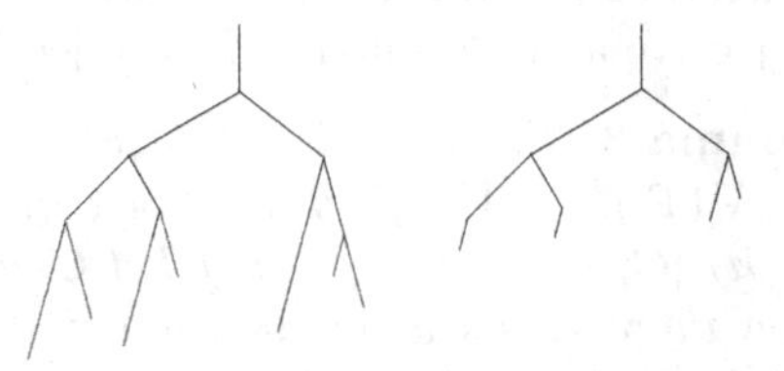

Fig. 4. Tree after trimming red points (Color figure online).

T_1' and T_2' as necessary. Note that T_1' and T_2' themselves are trees. For example, in Fig. 4 the red points in the left tree are those with extent less than a fixed value, and the right tree is obtained after trimming the red points.

Lemma 4. *If $d_I(T_1, T_2) \leq \epsilon$, then $d_I(T_1', T_2') \leq \epsilon$.*

See [1] for a proof of the above lemma. We now define *matching points* in T_1' and T_2'. A point x in T_1' is a matching point if there exists a branching node x' in T_1' or y' in T_2' with function value $f(x)$ and there exist no branching nodes nor leaves in T_1' or T_2' with function value in the range $(f(x), f(x) + 2\epsilon]$. Matching points on T_2' are defined similarly. By this definition, no two matching points share a function value within 2ϵ of each other unless they share the exact same function value. Furthermore, if x is a matching point, then all points with the same function value as x on both T_1' and T_2' are matching points. There are at most $O(n^2)$ matching points.

Suppose $d_I(T_1', T_2') \leq \epsilon$, and let $\alpha' : T_1' \to T_2'$ and $\beta' : T_2' \to T_1'$ be a pair of ϵ-compatible functions for T_1' and T_2'. Call a matching point x in T_1' and a matching point y in T_2' with $f(x) = g(y)$ *matched* if $\alpha'(x)$ is an ancestor of y.

Lemma 5. *Let x be any matching point in T_1'. The matched relation is a bijective function between matching points in T_1' with function value $f(x)$ and matching points in T_2' with function value $f(x)$.*

See [1] for a proof. Let T_1^m be a rooted tree consisting of one node per matching point on T_1'. Let $p(v)$ be the matching point for node v. Tree T_1^m has node v as an ancestor of node u if $p(v)$ is an ancestor of $p(u)$ (see Fig. 5). Define T_2^m similarly. The size of T_1^m and T_2^m is $O(n^2)$.

Intuitively, T_1^m and T_2^m represent the trees induced by matching points. By the definition of interleaving distance and Lemma 5, T_1^m and T_2^m are isomorphic if T_1' and T_2' have interleaving distance at most ϵ.

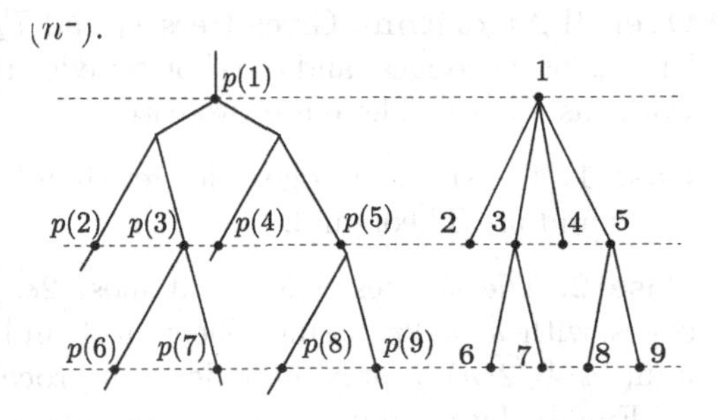

Fig. 5. The left tree shows matching points on tree T_1' and the right tree shows T_1^m.

Our algorithm finds an isomorphism between T_1^m and T_2^m in linear time [2]. If one does not exist, then the interleaving distance between T_1' and T_2' must be greater than ϵ; by Lemma 4, it thus reports that T_1 and T_2 have interleaving distance greater than ϵ.

If an isomorphism between T_1^m and T_2^m does exist, then the following functions $\alpha : T_1 \to T_2$ and $\beta : T_2 \to T_1$ are returned. For each matched pair of matching points x and y, the algorithm sets $\alpha(x) = y$ and $\beta(y) = x$. Now, let (f_1, f_2) be any maximal range of function values without any branching points in T_1' or T_2' where $f_2 - f_1 > 2\epsilon$. Let x' be any point in T_1' with $f(x') \in (f_1, f_2)$. Point x' has a unique matching point descendant x. The algorithm sets $\alpha(x')$

to the point y' in T_2' where y' is the ancestor of $\alpha(x)$ with $g(y') = f(x')$, and it sets $\beta(y') = x'$. For every remaining point x'' in T_1', the algorithm sets $\alpha(x'')$ to $\alpha(x)$ where x is the lowest matching point *ancestor* of x''. Assignment $\beta(y'')$ is defined similarly for remaining points y'' in T_2'. We call such points x'' and y'' *lazily assigned.* Finally, each point x''' in $T_1 - T_1'$ has $\alpha(x''')$ set to $\alpha(x)$ where x is the lowest ancestor of x''' on T_1'. Similar assignments are done for points in $T_2 - T_2'$.

One can verify that α and β meet all their desired properties except for how much a point's function value can change going from one tree to the other. A counting argument gives the following lemma, proved in [1].

Lemma 6. *For each lazily assigned point x'' in T_1', we have $g(\alpha(x'')) \leq f(x'') + 2\sqrt{ns\epsilon}$.*

Theorem 4. *Let T_1 and T_2 be two merge trees and $\epsilon > 0$ a parameter. There is an $O(n^2)$ time algorithm that returns a pair of $4\sqrt{ns\epsilon}$-compatible maps between T_1 and T_2, if $d_I(T_1, T_2) \leq \epsilon$ and the maximum length of a tree edge is $s\epsilon$. If $d_I(T_1, T_2) > \epsilon$, then the algorithm may return no or return a pair of $4\sqrt{ns\epsilon}$-compatible maps.*

Proof. By Lemma 6 and the symmetric lemma for T_2', each point in T_1' and T_2' has its function value changed by at most $2\sqrt{ns\epsilon}$. Points outside T_1' and T_2' have their function value changed by at most $2 \cdot 2\sqrt{ns\epsilon}$.

Remark. If $s = \Omega(n)$, we modify the above algorithm slightly – we skip the trimming step, but keep the rest same. It can be shown, as in Lemma 6, that the height of a point and its image differ by at most $2n\epsilon$.

Overall Algorithm. Given trees T_1 and T_2, let r denote the ratio between the lengths of the longest and the shortest edge in both trees. Our decision procedure works as follows. There are two cases –

Case 1. The shortest edge is longer than 2ϵ. We invoke the procedure for long edges and use Theorem 3.

Case 2. The shortest edge is at most 2ϵ. We invoke the procedure for short edges with $s = 2r$. Using Theorem 4 and the remark following it, we get a $\min(2n, 4\sqrt{2rn})$-approximate decision procedure.

Finally, by plugging this decision into a binary search over all possible candidate values for ϵ, we obtain an approximation algorithm for the interleaving distance. The following lemma, proved in [1], states that the number of candidate values for ϵ is only $O(n^2)$. Thus binary search takes $O(\log n)$ time, and Theorem 5 follows.

Lemma 7. *Let T_1 and T_2 be two merge trees with internal nodes I_1 and I_2 resp. and leaves L_1 and L_2 resp. Then the value of $d_I(T_1, T_2)$ is either*

(i) $|f(u) - g(v)|$ for some pair $(u, v) \in I_1 \times I_2 \cup L_1 \times L_2$, or
(ii) $\frac{1}{2}|f(u) - f(u')|$ for some $u \in L_1$, where u' is an ancestor node of u, or

(iii) $\frac{1}{2}|f(v) - f(v')|$ *for some* $v \in L_2$*, where* v' *is an ancestor node of* v*.*

Theorem 5. *Given two merge trees* T_1 *and* T_2 *with a total of* n *vertices, there exists an* $O(n^{5/2} \log n)$ *time* $O(\min\{n, \sqrt{rn}\})$*-approximation algorithm for computing the interleaving distance between them, where* r *is the ratio between the lengths of the longest and the shortest edge in both trees.*

Combining Theorem 5 with Corollary 1, we have:

Corollary 2. *Given two metric trees* T_1 *and* T_2 *with a total of* n *vertices, there exists an* $O(n^{7/2} \log n)$ *time* $O(\min\{n, \sqrt{rn}\})$*-approximation algorithm for computing the Gromov-Hausdorff distance between them, where* r *is the ratio between the lengths of the longest and the shortest edge in both trees.*

6 Conclusion

We have presented the first hardness results for computing the Gromov-Hausdorff distance between metric trees. We have also given a polynomial time approximation algorithm for the problem. But the current gap between the lower and upper bounds on the approximation factor is polynomially large. It would be very interesting to close this gap. In general, we hope that our current investigation will stimulate more research on the theoretical and algorithmic aspects of embedding or matching under additive metric distortion.

References

1. Agarwal, P.K., Fox, K., Nath, A., Sidiropoulos, A., Wang, Y.: Computing the Gromov-Hausdorff distance for metric trees (2015). CoRR, abs/1509.05751
2. Aho, A.V., Hopcroft, J.E., Ullman, J.D.: The Design and Analysis of Computer Algorithms. Addison-Wesley, Reading (1974)
3. Bauer, U., Ge, X., Wang, Y.: Measuring distance between Reeb graphs. In: 30th Annual Symposium on Computational Geometry, p. 464 (2014)
4. Bronstein, A.M., Bronstein, M.M., Kimmel, R.: Efficient computation of isometry-invariant distances between surfaces. SIAM J. Sci. Comput. **28**(5), 1812–1836 (2006)
5. Burago, D., Burago, Y., Ivanov, S.: A Course in Metric Geometry. American Mathematical Society, Providence (2001)
6. Carlsson, G., Mémoli, F.: Characterization, stability and convergence of hierarchical clustering methods. J. Mach. Learn. Res. **11**, 1425–1470 (2010)
7. Cohen-Steiner, D., Edelsbrunner, H., Harer, J.: Stability of persistence diagrams. Disc. Comput. Geom. **37**(1), 103–120 (2007)
8. de Silva, V., Munch, E., Patel, A.: Categorification of Reeb graphs, Preprint (2014)
9. Gromov, M.: Metric Structures for Riemannian and Non-Riemannian Spaces. Birkhäuser Basel, Basel (2007)
10. Hall, A., Papadimitriou, C.: Approximating the distortion. In: Chekuri, C., Jansen, K., Rolim, J.D.P., Trevisan, L. (eds.) APPROX 2005 and RANDOM 2005. LNCS, vol. 3624, pp. 111–122. Springer, Heidelberg (2005)

11. Hopcroft, J.E., Karp, R.M.: An $n^{5/2}$ algorithm for maximum matchings in bipartite graphs. SIAM J. Comp. **2**(4), 225–231 (1973)
12. Kenyon, C., Rabani, Y., Sinclair, A.: Low distortion maps between point sets. SIAM J. Comp. **39**(4), 1617–1636 (2009)
13. Memoli, F.: On the use of Gromov-Hausdorff distances for shape comparison. In: Eurographics Symposium on Point-Based Graphics (2007)
14. Mémoli, F., Sapiro, G.: A theoretical and computational framework for isometry invariant recognition of point cloud data. Found. Comput. Math. **5**(3), 313–347 (2005)
15. Morozov, D., Beketayev, K., Weber, G.H.: Interleaving distance between merge trees. In: Workshop on Topological Methods in Data Analysis and Visualization: Theory, Algorithms and Applications (2013)
16. Papadimitriou, C., Safra, S.: The complexity of low-distortion embeddings between point sets. In: 16th Annual ACM-SIAM Symposium on Discrete Algorithms, pp. 112–118 (2005)

The VC-Dimension of Visibility on the Boundary of a Simple Polygon

Matt Gibson[1], Erik Krohn[2], and Qing Wang[1](✉)

[1] Department of Computer Science,
University of Texas at San Antonio, San Antonio, TX, USA
uvg160@my.utsa.edu
[2] Department of Computer Science,
University of Wisconsin - Oshkosh, Oshkosh, WI, USA

Abstract. In this paper, we prove that the VC-Dimension of visibility on the boundary of a simple polygon is exactly 6. Our result is the first tight bound for any variant of the VC-Dimension problem regarding simple polygons. Our upper bound proof is based off several structural lemmas which may be of independent interest to researchers studying geometric visibility.

1 Introduction

Geometric covering problems have been a focus of research for decades. Here we are given some set of points P and a set S where each $s \in S$ can cover some subset of P. The subset of P is generally induced by some geometric object. For example, P might be a set of points in the plane, and s consists of the points contained within some disk in the plane. The goal is to choose the smallest number of elements in S to cover all of the points in P. Most variants of the problem are NP-hard, and therefore most research on geometric set cover focuses on designing polynomial-time approximation algorithms whose approximation ratio is as good as possible. For most variants, the problem can easily be reduced to an instance of the combinatorial set cover problem which has a polynomial-time $O(\log n)$-approximation algorithm, which is the best possible approximation under standard complexity assumptions [3,5,9,11,12]. The main question therefore is to determine for which variants of geometric set cover can we obtain polynomial-time approximation algorithms with approximation ratio $o(\log n)$, as any such algorithm must exploit the geometry of the problem to achieve the result. This area has been studied extensively, see for example [1,2,14], and much progress has been made utilizing algorithms that are based on solving the standard linear programming relaxation.

Unfortunately this technique has severe limitations for some variants of geometric set cover, and new ideas are needed to make progress on these variants. In particular, the techniques are lacking when the points P we wish to cover is the interior or boundary of a simple polygon, and we wish to place the smallest number of points in P that collectively "see" the polygon. This problem is classically

K. Elbassioni and K. Makino (Eds.): ISAAC 2015, LNCS 9472, pp. 541–551, 2015.
DOI: 10.1007/978-3-662-48971-0_46

referred to as the *art gallery problem* as an art gallery can be modeled as a polygon and the points placed by an algorithm represent cameras that can "guard" the art gallery. This has been one of the most well-known problems in computational geometry for many years, yet still to this date the best polynomial-time approximation algorithm for this problem is the $O(\log n)$ approximation algorithm that uses no geometric information. The key issue is a fundamental lack of understanding of the combinatorial structure of visibility inside simple polygons. It seems that in order to develop powerful approximation algorithms for this problem, the community first needs to better understand the underlying structure of such visibility.

VC-Dimension. An interesting measure of the complexity of a set system is the notion of *VC-dimension.* To define this in the context of a simple polygon P, we say that a finite set of points G in P is *shattered* if for every subset of $G' \subseteq G$ there exists some point $v \in P$ such that v sees every point in G' and does not see any point in $G \setminus G'$. In this context, we call v a *viewpoint.* See Fig. 1, where the red points are G and the green points are a set of viewpoints that shatter G. The VC-dimension is the largest d such that there exists some simple polygon P and point set G of size d that can be shattered.

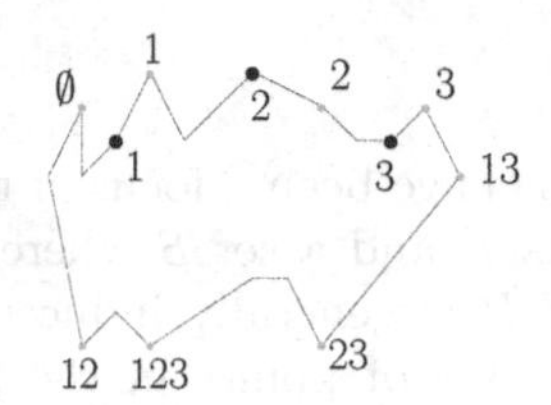

Fig. 1. Three points are shattered. The label on green points denotes subset of red points that it sees (Color figure online).

Brönnimann and Goodrich give a polynomial-time $O(\log OPT)$-approximation algorithm for any set system with constant VC-dimension [4], establishing a connection between the VC-Dimension problem and set cover. The VC-dimension problem has received a lot of recent attention due to this connection with set cover, but the problem is quite interesting on its own. In 1998, Valtr showed that the VC-dimension of the visibility in a simple polygon is between 6 and 23 [13]. The lower bound of 6 is still the best known lower bound, and the upper bound was not improved until very recently by Gilbers and Klein who give an upper bound of 14 [8]. The gap here is still extremely large, as 2^6 viewpoints are needed for the lower bound and if the upper bound were able to be realized then 2^{14} viewpoints would be needed. Gilbers and Klein suggest that the actual VC-dimension is likely to be closer to the lower bound of 6 rather than the upper bound of 14. Regardless, a lack of knowledge of the structure of visibility in simple polygons has prevented the community from tightening this gap.

Due to the complexity of the VC-Dimension problem in simple polygons, other special cases have been considered. King [10] showed that the VC-dimension of visibility on x-monotone terrains is 4, where an x-monotone terrain is a polyginal curve such that any vertical line intersects the chain in at most one point. In the context of polygons, research has considered special cases, both by restricting the class of polygon considered or by restricting the location of G and/or the viewpoints. A polygon is *x-monotone* (or simply *monotone*) if any vertical line intersects the boundary in at most two points. Gibson, Krohn, and Wang [6] recently showed that the VC-dimension on the boundary of a monotone polygon (that is, when G and all viewpoints are required lie on the boundary) is exactly 6. The lower bound is particularly surprising given that it seems that the monotonicity property would be quite restrictive relative to simple polygons, yet the lower bound matches the best known lower bound for simple polygons (when G and the viewpoints can lie anywhere inside of the polygon). Note that this lower bound of 6 also applies to simple polygons. Recently Gilbers [7] proved an upper bound of 7 on the VC-dimension of visibility on the boundary of a simple polygon, but it was not clear if the actual bound should be 6 or 7.

Our Contribution. Given the seemingly restrictive nature of the monotonicity constraint for the lower bound of Gibson, Krohn, and Wang, it would be reasonable to think that a lower bound of 7 might be possible. Our main result is to improve the upper bound on the VC-Dimension of visibility on the boundary of a simple polygon to 6, and thus we obtain the following theorem.

Theorem 1. *The VC-Dimension of visibility on the boundary of a simple polygon is 6.*

To achieve the tight upper bound, we first give a set of structural lemmas regarding visibility of points on the boundary of a simple polygon. We believe these lemmas may be of general interest to researchers studying visibility problems in simple polygons. Our result is the first tight bound for any variant of the VC-Dimension problem regarding simple polygons, and we hope this result will serve as a springboard for further research on the general VC-Dimension problem and other visibility-based problems for simple polygons.

2 Preliminaries

Let a and b denote two points on the boundary of a simple polygon P. We let ∂P denote the boundary of the polygon, and we let $\partial(a, b)$ denote subset of ∂P obtained by walking "clockwise" along ∂P from a to b but not including a or b. If we wish to include a or b in the subset then we denote this $\partial[a, b)$ or $\partial(a, b]$ respectively. See Fig. 2(a) and (b). For any set of $k \geq 3$ points on ∂P, their *clockwise ordering* is the order in which we visit them when walking clockwise around ∂P. For example, if a, b, and c are three points on ∂P, then if $b \in \partial(a, c)$ then their clockwise ordering is (a, b, c) (or equivalently (b, c, a) and (c, a, b)).

For any two points a and b on ∂P, we say *a sees b* if the line segment $\overline{ab}$ does not go outside of P. We call the line segment connecting two points on ∂P

that see each other a *good line segment*, and we call the line segment connecting two points on ∂P that do not see each other a *bad line segment.* In the figures throughout the paper, good lines are represented as solid, green lines and bad lines are represented as dashed, red lines. If a and b do not see each other, then there must be a point p on ∂P that "cuts through" $\overline{ab}$. See Fig. 2(c). We say p is a *blocker* for a and b, or equivalently we say *p blocks a from seeing b.* Note that p can block a from seeing b and not see a or b. See Fig. 2(d). Note that two points a and b do not see each other if and only if there is some point in $\partial(a, b)$ that blocks a from seeing b or there is a point in $\partial(b, a)$ that blocks a from seeing b.

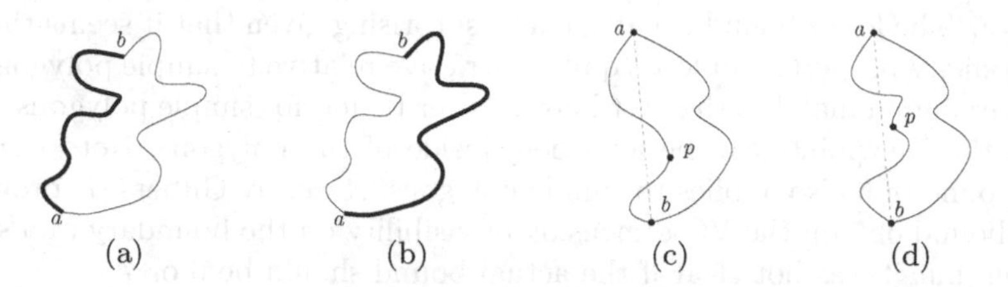

Fig. 2. (a) Bold is $\partial(a, b)$. (b) Bold is $\partial(b, a)$. (c) p blocks a from seeing b. (d) p is a blocker but does not see a or b (Color figure online).

Structural Lemmas. Our proof technique is to consider sequences of G and viewpoints in clockwise order and to consider whether or not this ordering is realizable. Intuitively, one can view the problem as placing the points on a circle according to the clockwise sequence and then determining if it is possible to "bend" the boundary of the circle into a simple polygon so that the required visibility constraints are met. See Fig. 3 for an illustration. Part (a) shows some ordering of G and viewpoints, and part (b) shows that it is possible to block all of the bad lines without blocking any good lines. Part (c) however is not realizable, because we cannot block the bad line without also blocking one of the good lines.

In this section, we give a set of lemmas which help us show that certain sequences are not realizable. All lemmas are stated for a sequence of points in clockwise order; however, it is easy to see that symmetric versions of all lemmas apply (when the sequence of points is in "counterclockwise order"). The first two lemma play a major role in our upper bound proof by restricting the location of possible blockers for a bad line.

Lemma 1. *Let $a, b,$ and c denote three points on the boundary of a simple polygon in clockwise order and b sees c. If a sees some point p in $\partial(b, c)$ then no point in $\partial(a, c)$ can block a from seeing c.*

Proof. First note that no point in $\partial(b, c)$ can block a from seeing c or else it would also block b from seeing c. Moreover, if a sees p in $\partial(b, c)$ then no point in $\partial(a, p)$ can block a from seeing c or else a could not see p, and therefore we have that no point in $\partial(a, p) \cup \partial(b, c)$ can block a from seeing c. The Lemma follows as $\partial(a, c) = \partial(a, p) \cup \partial(b, c)$. See Fig. 4(a). □

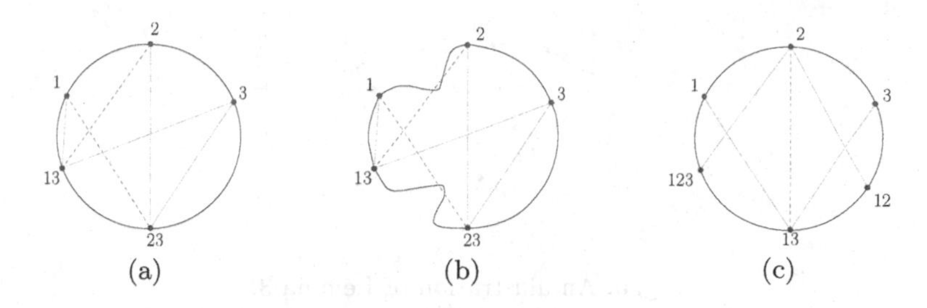

Fig. 3. An illustration of the proof technique.

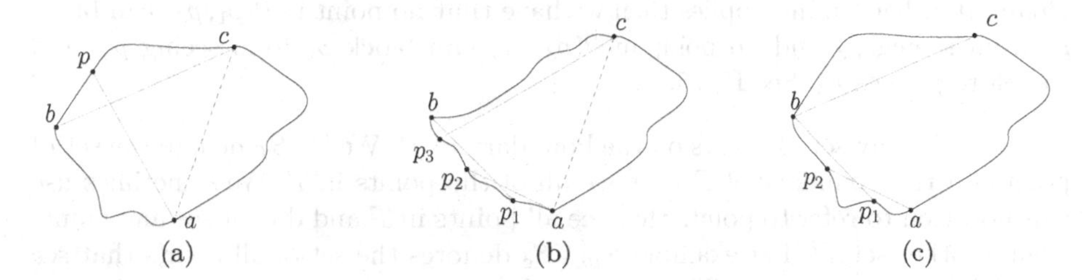

Fig. 4. An illustration of Lemmas 1 and 2.

Lemma 2. *Let a, b, and c denote three points on the boundary of a simple polygon in clockwise order. If there are two points p_1 and p_2 such that $p_1 \in \partial(a, b)$, $p_2 \in \partial(p_1, b)$, p_1 sees b, and p_2 sees a, then no point from $\partial(a, c)$ can block a from seeing c if: (1) there is a point in $\partial(a, b)$ that sees c, or (2) b sees c and does not block a from seeing c.*

Proof. First note that due to the visibility assumptions and positioning of p_1 and p_2, we have that no point in $\partial(a, b)$ can block a from seeing b by Lemma 1. This necessarily implies that no point in $\partial(a, b)$ can block a from seeing c. Therefore if any point in $\partial(a, c)$ blocks a from seeing c then it must be in $\partial[b, c)$. Suppose condition (1) occurs, and let p_3 be a point in $\partial(a, b)$ that sees c. Then no point in $\partial(p_3, c)$ can block a from seeing c or it would also block p_3 from seeing c, and therefore no point in $\partial(a, b) \cup \partial(p_3, c) = \partial(a, c)$ can block a from seeing c. See Fig. 4(b). Now suppose condition (2) occurs. Since b sees c, we have that no point in $\partial(b, c)$ can block a from seeing c or it would also block b from seeing c, and b does not block a from seeing c by assumption. Therefore no point in $\partial(a, b) \cup \partial[b, c) = \partial(a, c)$ can block a from seeing c. See Fig. 4(c). □

The following lemma plays a large role in our upper bound proof.

Lemma 3. *Let p_1, p_2, p_3, and p_4 denote four points on the boundary of a simple polygon in clockwise order, p_2 sees p_3, and p_4 sees p_3. Then if Lemma 1 or 2 applies with $a := p_1$, $b := p_2$, and $c := p_3$ and Lemma 1 or 2 applies with $a := p_1$, $b := p_4$, and $c := p_3$ then we have that p_1 sees p_3.*

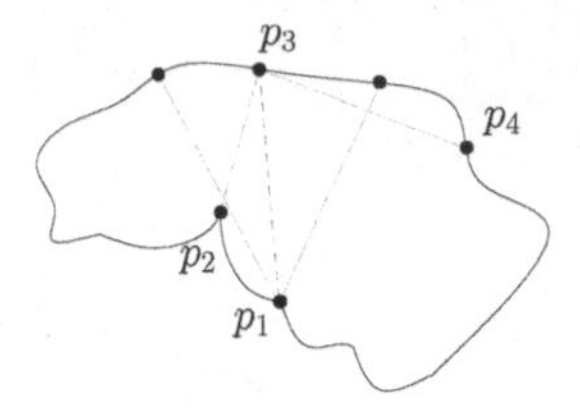

Fig. 5. An illustration of Lemma 3.

Proof. If either lemma applies then we have that no point in $\partial(p_1, p_3)$ can block p_1 from seeing p_3 and no point in $\partial(p_3, p_1)$ can block p_1 from seeing p_3, and therefore p_1 sees p_3. See Fig. 5. □

Let T be any set of points on the boundary of P. We let S_T denote the set of points on the boundary of P that see all of the points in T. We sometimes use this notation to refer to points that see all points in T and do not see any points from another set T'. For example $S_{ab} \setminus S_c$ denotes the set of all points that see a and b but do not see c. The next two lemmas are helpful for restricting the location of viewpoints in certain scenarios. The proofs of Lemmas 4, 5 and 6 are ommitted due to lack of space.

Lemma 4. *Let a, b, c, d denote four points on the boundary of a simple polygon in clockwise order such that $b \in S_d$ and $a \notin S_d$, and suppose a is blocked from seeing d by a point in $\partial(a, d)$. If there is a point $v \in S_{ac}$ that is in $\partial(a, b)$, then every point $z \in \partial(v, d)$ satisfies the following: no point of $S_{az} \setminus \{v\}$ is in $\partial(a, d)$.*

Lemma 5. *Let a, b, c, d, e, and f denote six points on the boundary of a simple polygon in clockwise order. Suppose $a \in S_{ce} \setminus S_d$, d is blocked from seeing a by a point in $\partial(a, d)$, and d is blocked from seeing a by a point in $\partial(d, a)$ as well. Then $S_{bdf} = \emptyset$.*

The next lemma serves as a "sub-lemma" for the final three lemmas.

Lemma 6. *Let a, b, c, and d denote four points on the boundary of a simple polygon in clockwise order such that $b \in S_a \setminus S_d$, $c \in S_d \setminus S_a$, and suppose a sees a point $x \in \partial(c, d]$. Then if b blocks a from seeing c then it must be that c does not block b from seeing d.*

The upper bound proof relies upon the final three lemmas heavily.

Lemma 7. *Let a, b, c, d, p_1 and p_2 denote six points on the boundary of a simple polygon in clockwise order. Let p_1 and p_2 be points such that $p_1 \in S_{ac} \setminus S_b$ and $p_2 \in S_{bd} \setminus S_c$. Then at most one viewpoint in S_{bc} can go in $\partial(b, c)$ and the rest must go in $\partial(p_1, p_2)$.*

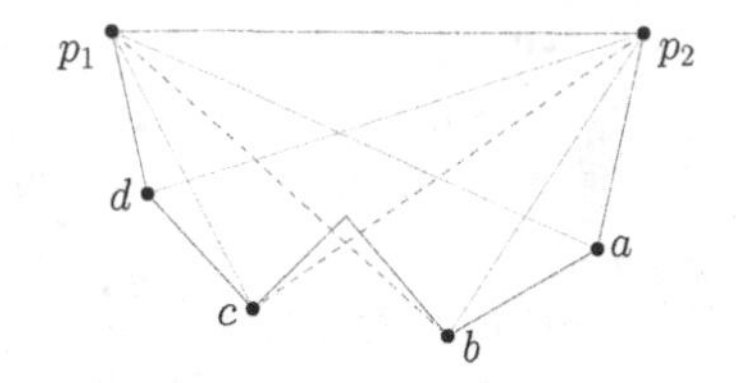

Fig. 6. An illustration of Lemma 7.

Proof. Let v be some point in S_{bc}. First note that if $v \in \partial(p_2, b)$ we cannot block c from seeing p_2 by Lemma 3, and similarly if $v \in \partial(c, p_1)$ we cannot block b from seeing p_1 by Lemma 3. By Lemma 6, we cannot have that c blocks b from seeing p_1 and simultaneously b blocks c from seeing p_2. Without loss of generality assume that b does not block c from seeing p_2, and suppose for the sake of contradiction we have two viewpoints $v_1, v_2 \in S_{bc}$ in $\partial(b, c)$. Then we have that no point in $\partial(p_2, c)$ can block c from seeing p_2 by the second condition of Lemma 2. Therefore we have that c sees p_2 by Lemma 3. Finally if $b \in S_{bc}$ or if $c \in S_{bc}$, then we have that b and c see each other, and it follows that c sees p_2 by Lemma 3. See Fig. 6. □

The next lemma is similar to Lemma 7, but the positions of a and p_2 flip (Fig. 7).

Lemma 8. *Let a, p_2, b, c, d, p_1, and x denote seven points on the boundary of a simple polygon in clockwise order such that $p_1 \in S_{ac} \setminus S_b$ and $p_2 \in S_{bdx} \setminus S_c$. If p_2 does not block b from seeing p_1, then at most one viewpoint in S_{bc} can go in $\partial(b, c)$ and the rest must go in $\partial(p_1, p_2)$.*

Proof. Let v be some point in S_{bc}. First note that if $v \in \partial(p_2, b)$ we cannot block c from seeing p_2 by Lemma 3. Since p_2 does not block b from seeing p_1, then if $v \in \partial(c, p_1)$ we cannot block b from seeing p_1 by Lemma 3. Now assume for the sake of contradiction that we have two viewpoints $v_1, v_2 \in S_{bc}$ in $\partial(b, c)$. First suppose that b does not block c from seeing p_2. We have nothing can block c from seeing p_2 in $\partial(c, p_2)$ by Lemma 1, and nothing can block c from seeing p_2 in $\partial(p_2, c)$ by Lemma 2. Therefore we have c sees p_2 by Lemma 3. So now suppose that b does block c from seeing p_2. By Lemma 6, we have that c cannot block b from seeing p_1. By Lemma 2 we have that no point in $\partial(b, p_1)$ can block b from seeing p_1, and therefore must be blocked by a point in $\partial(p_1, b)$. By Lemma 1, we cannot block p_1 from seeing p_2 with a point in $\partial(p_1, p_2)$, and therefore we cannot block b from p_1 with a point in $\partial(p_1, p_2)$. Also we cannot block b from seeing p_1 with a point in $\partial(p_2, b)$ or else b would not see p_2. Therefore we must have p_2 blocks b from seeing p_1. But if p_2 blocks b from seeing p_1, and b blocks c from seeing p_2, then Lemma 6 implies that c cannot see p_1, a contradiction. □

Our final lemma has points in the same setup as Lemma 8, except here we assume that p_2 does block b from seeing p_1.

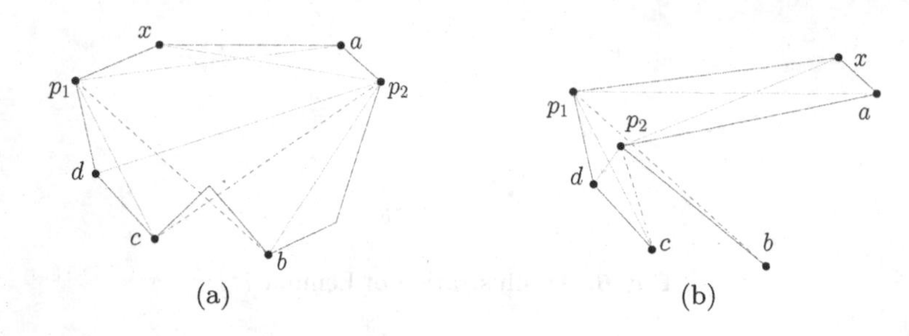

Fig. 7. An illustration of Lemmas 8 and 9.

Lemma 9. *Let a, p_2, b, c, d, p_1, and x denote seven points on the boundary of a simple polygon in clockwise order such that $p_1 \in S_{ac} \setminus S_b$ and $p_2 \in S_{bdx} \setminus S_c$. If p_2 blocks b from seeing p_1, then at most one viewpoint in S_{bc} can go in $\partial(b, c)$ and the rest must go in $\partial(c, p_1)$.*

Proof. Let v be some point in S_{bc}. First note that if $v \in \partial(p_2, b)$ we cannot block c from seeing p_2 by Lemma 3. Now suppose $v \in \partial(p_1, p_2)$. If $v \in \partial(p_1, a)$ then we have that no point in $\partial(p_1, b)$ can block b from seeing p_1 by Lemma 1 which contracts that p_2 is blocking b from seeing p_1. If $v \in \partial(x, p_2)$, then by Lemma 1 we have that no point in $\partial(x, b)$ can block b from seeing x, which contracts that p_2 is blocking b from seeing p_1.

We now show that at most one point in $S_{b,c}$ can go in $\partial(b, c)$. Since p_2 is blocking b from seeing p_1, Lemma 4 implies that b cannot block c from seeing p_2. Therefore if there are distinct points $v_1, v_2 \in S_{bc}$ in $\partial(b, c)$, Lemma 2 implies that no point in $p(p_2, c)$ blocks c from seeing p_2 which implies that c sees p_2 by Lemma 3, a contradiction. Finally if $b \in S_{bc}$ or if $c \in S_{bc}$, then we have that b and c see each other, and it follows that c sees p_2 by Lemma 3. □

3 VC-Dimension Upper Bound Proof

In this section, we prove the following theorem. Combined with the lower bound result of [6], this completes the proof of Theorem 1.

Theorem 2. *The VC-Dimension of visibility on the boundary of a simple polygon is at most 6.*

We prove Theorem 2 by showing that there is no simple polygon that contains a set G of 7 points on its boundary that can be shattered. Let G be any set of 7 points on the boundary of a simple polygon P denoted $g_1, \ldots, g_7$ in clockwise order. For any subset $T \subseteq G$, consider the viewpoint that sees every point in T and no point in $G \setminus T$. We denote this viewpoint based off the indices of the points in T. For example, v_{167} is the viewpoint that sees g_1, g_6, and g_7 and does not see g_2, g_3, g_4, and g_5. Similarly, we let S_{24} denote the set of all viewpoints that see g_2 and g_4.

We remind the reader that our proof technique is based on considering different clockwise sequences of points and showing that there is no simple polygon that can satisfy that particular sequence. The viewpoints v_{1357} (that sees exactly the odd indexed points) and v_{246} (that sees exactly the even indexed points) play a particularly important role in the proof. Intuitively, these points cause issues for lower bound constructions due to the fact that they (roughly) alternate which points they see in G. We consider four main cases based on the location of v_{1357}: (1) $v_{1357} \in \partial(g_7, g_1)$, (2) $v_{1357} \in \partial(g_1, g_2)$, (3) $v_{1357} \in \partial(g_2, g_3)$, and (4) $v_{1357} \in \partial(g_3, g_4)$. Note that other locations of v_{1357} is symmetric to one of the four considered cases (e.g., $v_{1357} \in \partial(g_4, g_5)$ is symmetric to Case 4). We prove Case 1 in the paper, and the proofs for the final three cases are ommitted due to lack of space. These proofs are similar in nature to that of Case 1.

Case 1: $v_{1357} \in \partial(g_7, g_1)$
In this case, we can assume without loss of generality that g_4 is blocked from seeing v_{1357} by a point in $\partial(v_{1357}, g_4)$. Moreover, there must be a blocking point in $\partial[g_3, g_4)$. If not, then there would be a blocker $p \in \partial(v_{1357}, g_3)$ and g_3 would not be a blocker, which implies that p would block g_3 from seeing v_{1357}. We will show that it follows that there cannot be a point in $\partial(g_4, v_{1357})$ that also blocks g_4 from seeing v_{1357}. If there is such a blocking point, then there must be one in $\partial(g_4, g_5]$ by a symmetric argument to why there must be a blocking point in $\partial[g_3, g_4)$, and then Lemma 5 implies that we cannot place any points in S_{147}. See Fig. 8(a). Now consider v_{246}. Note that it cannot be in $\partial(v_{1357}, g_3)$ as Lemma 1 with $a = g_4, b = g_3$, and $c = v_{1357}$ would contradict the assumption that there is a point in $\partial(v_{1357}, g_4)$ that blocks g_4 from seeing v_{1357}, and so v_{246} must be in $\partial(g_3, v_{1357})$.

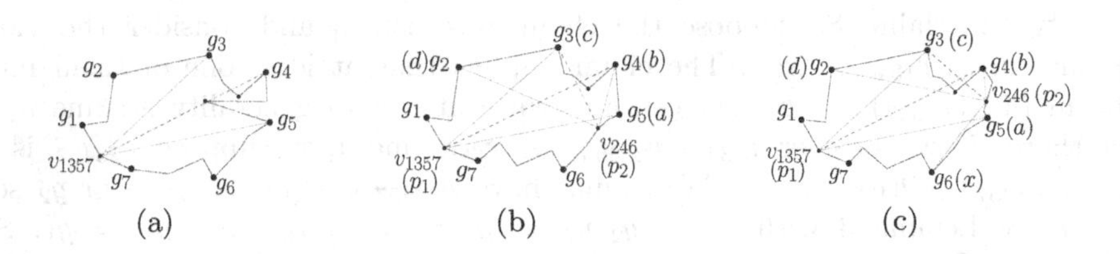

Fig. 8. An illustration of Case 1.

We claim that at most one viewpoint in S_{34} can go in $\partial(g_3, g_4)$ and the rest must be in $\partial(g_4, v_{1357}) \cap \partial(v_{246}, v_{1357})$. In other words when walking clockwise around ∂P starting at g_3, we should reach at most one viewpoint of S_{34} before having reached both of g_4 and v_{246}, and then we will reach the rest of the points in S_{34} prior to reaching v_{1357}. We prove the claim by considering the possible locations of v_{246}.

- If $v_{246} \in \partial(g_5, v_{1357})$ then the claim is true by Lemma 7 with $a = g_5, b = g_4, c = g_3, d = g_2, p_1 = v_{1357}$, and $p_2 = v_{246}$. See Fig. 8(b).

- If $v_{246} \in \partial(g_4, g_5)$ then the claim is true by Lemma 8 with $a = g_5, b = g_4, c = g_3, d = g_2, x = g_6, p_1 = v_{1357}$, and $p_2 = v_{246}$. Note we apply Lemma 8 here and not Lemma 9 since no point in $\partial(g_4, v_{1357})$ blocks g_4 from seeing v_{1357}. See Fig. 8(c).
- If $v_{246} \in \partial(g_3, g_4)$, then the claim follows by applying Lemma 4 with $a = g_4, b = g_3$, and $c = g_2$, and $d = v_{1357}$.

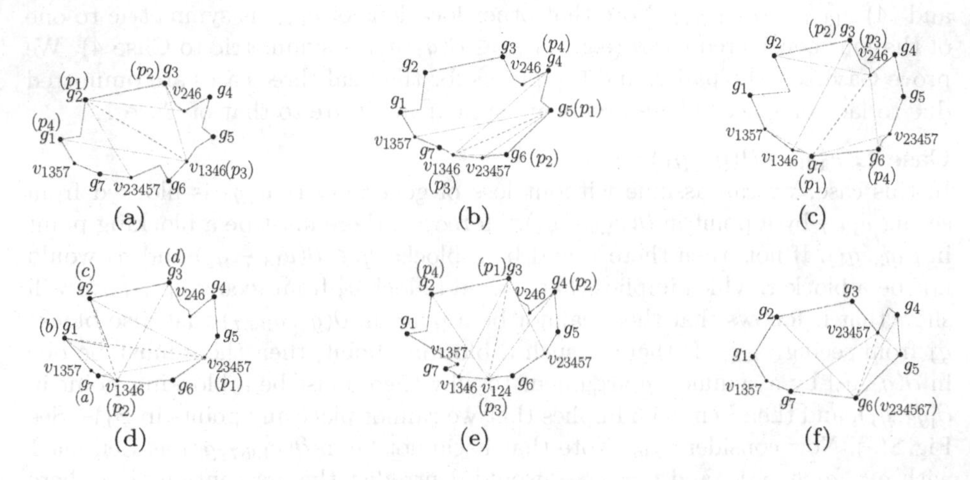

Fig. 9. An illustration of $v_{1357} \in \partial(g_7, g_1)$.

If there is a simple polygon that shatters G, the ordering of S_{34} must satisfy the claim. So suppose the claim is satisfied, and consider the viewpoints $v_{1346}, v_{13467} \in S_{34}$. The claim implies that at least one of them must be in $\partial(g_4, v_{1357}) \cap \partial(v_{246}, v_{1357})$, and without loss of generality assume v_{1346} is there. Now consider $v_{23457}, v_{234567} \in S_{34}$, and first suppose v_{23457} is in $\partial(g_4, v_{1357}) \cap \partial(v_{246}, v_{1357})$. We cannot have $v_{23457} \in \partial(v_{1346}, v_{1357})$ or g_2 sees v_{1346} by Lemma 3 with $p_1 = g_2, p_2 = g_3, p_3 = v_{1346}$, and $p_4 = g_1$. See Fig. 9(a). So assume we have $v_{23457} \in \partial(g_4, v_{1346})$. If we additionally have $v_{23457} \in \partial(g_6, v_{1357})$, then g_5 sees v_{1346} by Lemma 3 with $p_1 = g_5, p_2 = g_6, p_3 = v_{1346}$, and $p_4 = g_4$. See Fig. 9(b). So assume that $v_{23457} \in \partial(g_4, g_6)$. We will now show that v_{1346} cannot be in $\partial(g_7, v_{1357})$, and therefore it must be in $\partial(v_{23457}, g_7)$. If $v_{1346} \in \partial(g_7, v_{1357})$, then we have that g_7 sees v_{246} by Lemma 3 with $p_1 = g_7, p_2 = g_3, p_3 = v_{246}$, and $p_4 = g_6$. See Fig. 9(c). So now assume that $v_{23457} \in \partial(g_4, g_6)$ and $v_{1346} \in \partial(v_{23457}, g_7)$, and consider the viewpoints in S_{12}. By Lemma 7 with $a = g_7, b = g_1, c = g_2, d = g_3, p_1 = v_{23457}$, and $p_2 = v_{1346}$, at most one of them can be in $\partial(g_1, g_2)$ and the rest must be in $\partial(v_{23457}, v_{1346})$. See Fig. 9(d). So consider the viewpoints v_{124} and v_{1246}. At least one must be in $\partial(v_{23457}, v_{1346})$ and without loss of generality assume v_{124} is there. Then g_3 sees v_{124} by Lemma 3 with $p_1 = g_3, p_2 = g_4, p_3 = v_{124}$, and $p_4 = g_2$. See Fig. 9(e).

So we now have that v_{23457} cannot be in $\partial(g_4, v_{1357}) \cap \partial(v_{246}, v_{1357})$ and therefore must be in $\partial(g_3, g_4)$. The claim then implies that $v_{234567} \in \partial(g_4, v_{1357}) \cap \partial(v_{246}, v_{1357})$. The analysis in the previous paragraph can also show that v_{234567} is not in $\partial(g_4, g_6)$ or $\partial(g_6, v_{1357})$, but it could be that $g_6 = v_{234567}$. But if $v_{23457} \in \partial(g_3, g_4)$ and $g_6 = v_{234567}$ then g_6 sees v_{23457} by Lemma 3 with $p_1 = g_6, p_2 = g_7, p_3 = v_{23457}$, and $p_4 = g_5$. See Fig. 9(f).

References

1. Aloupis, G., Cardinal, J., Collette, S., Langerman, S., Orden, D., Ramos, P.: Decomposition of multiple coverings into more parts. In: Proceedings of the Twentieth Annual ACM-SIAM Symposium on Discrete Algorithms, SODA 2009, pp. 302–310. Society for Industrial and Applied Mathematics, Philadelphia, PA (2009)
2. Aronov, B., Ezra, E., Sharir, M.: Small-size epsilon-nets for axis-parallel rectangles and boxes. SIAM J. Comput. **39**(7), 3248–3282 (2010)
3. Bellare, M., Goldwasser, S., Lund, C., Russell, A.: Efficient probabilistically checkable proofs and applications to approximations. In: Proceedings of the Twenty-fifth Annual ACM Symposium on Theory of Computing, STOC 1993, pp. 294–304. ACM, New York (1993)
4. Brönnimann, H., Goodrich, M.: Almost optimal set covers in finite VC-dimension. Discrete Comput. Geom **14**, 463 (1995)
5. Feige, U., Halldórsson, M.M., Kortsarz, G., Srinivasan, A.: Approximating the domatic number. SIAM J. Comput. **32**(1), 172–195 (2003)
6. Gibson, M., Krohn, E., Wang, Q.: On the VC-dimension of visibility in monotone polygons. In: 26th Canadian Conference on Computational Geometry (CCCG) (2014)
7. Gilbers, A.: VC-dimension of perimeter visibility domains. Inf. Process. Lett. **114**(12), 696–699 (2014)
8. Gilbers, A., Klein, R.: A new upper bound for the VC-dimension of visibility regions. Comput. Geom. **47**(1), 61–74 (2014)
9. Johnson, D.S.: Approximation algorithms for combinatorial problems. In: Proceedings of the Fifth Annual ACM Symposium on Theory of Computing, STOC 1973, pp. 38–49. ACM, New York (1973)
10. King, J.: VC-dimension of visibility on terrains. In: CCCG (2008)
11. Lund, C., Yannakakis, M.: On the hardness of approximating minimization problems. J. ACM **41**(5), 960–981 (1994)
12. Raz, R., Safra, S.: A sub-constant error-probability low-degree test, and a sub-constant error-probability PCP characterization of NP. In: Proceedings of the Twenty-Ninth Annual ACM Symposium on Theory of Computing, STOC 1997, pp. 475–484. ACM, New York (1997)
13. Valtr, P.: Guarding galleries where no point sees a small area. Israel J. Math. **104**(1), 1–16 (1998)
14. Varadarajan, K.R.: Epsilon nets and union complexity. In: Symposium on Computational Geometry, pp. 11–16 (2009)

Computational Complexity I

Quantum Bit Commitment with Application in Quantum Zero-Knowledge Proof (Extended Abstract)

Jun Yan[1(✉)], Jian Weng[1], Dongdai Lin[2], and Yujuan Quan[1]

[1] Jinan University, Guangzhou 510632, China
tjunyan@jnu.edu.cn
[2] State Key Laboratory of Information Security, Institute of Information Engineering, Chinese Academy of Sciences, Beijing 100093, China

Abstract. In this work, we study formalization and construction of non-interactive statistically binding quantum bit commitment scheme (QBC), as well as its application in quantum zero-knowledge (QZK) proof. We explore the fully quantum model, where both computation and communication could be quantum. While most of the proofs here are straightforward based on previous works, we have two technical contributions. First, we show how to use reversibility of quantum computation to construct non-interactive QBC. Second, we identify new issue caused by quantum binding in security analysis and give our idea to circumvent it, which may be found useful elsewhere.

Keywords: Bit commitment · Zero-knowledge proof · Quantum cryptography · Quantum complexity theory

1 Introduction

Bit commitment scheme (BC) is a two-stage protocol between sender and receiver; it can be viewed as the electronic implementation of a "locked box". Intuitively, in the first stage of BC, sender conveys a locked box which contains a bit to receiver, who cannot tell whether the bit is 0 or 1 (it is hidden inside the box); this is known as *hiding* property of BC. In the second stage, sender sends a key to receiver, who can then open the box to see the bit. In this stage, we require that the revealed bit is the same as the one locked in the first stage; that is, sender commits himself to a single bit value. This is known as the *binding* property of BC. Unfortunately, unconditional BC does not exist. As a compromise, we can base BC on some plausible complexity assumptions such as one-way function [14]. In modern cryptography, BC serves as a cryptographic primitive and is widely used in various contructions. Interested readers can refer to standard textbook such as [7] for a formal introduction of BC.

Of particular interest to this work is *quantum bit commitment scheme* (QBC) — bit commitment scheme implemented with quantum mechanism that

K. Elbassioni and K. Makino (Eds.): ISAAC 2015, LNCS 9472, pp. 555–565, 2015.
DOI: 10.1007/978-3-662-48971-0_47

is also (naturally required) *secure against quantum attack*. We highlight that like (classical) BC, QBC still aims to secure classical information, i.e., implement commitment to a bit rather than a qubit. One should also be careful about the difference between the (general) QBC we are studying here and the *bit commitment scheme against quantum attack* [1]. For the former, both the computation and communication, i.e. the construction of the scheme, could be quantum, whereas for the latter, the construction is restricted to be classical. Nevertheless, they both provides security against quantum adversary. Hence, the latter can be viewed as a special case of the former. We further remark that BC against quantum attack is often called QBC in *post-quantum cryptography*.

Like in classical setting, there are also two flavors of QBC: statistically-binding QBC and statistically-hiding QBC[1]. In this work, we restrict our study to the former, and call it QBC for short henceforth. Statistically-hiding QBC raises more issues when generalized from its classical counterpart [2] and is beyond the scope of this work.

In the past two decades, the study of QBC has attracted a great deal of attentions. However, we notice that most of them only focus on QBC per se, failing to consider it as a building block (primitive) to construct larger protocols. In this work, we study a natural application of QBC: quantum zero-knowledge (QZK) proof, an important notion in both complexity theory and cryptography.

Informally speaking, a zero-knowledge (ZK) proof is an interactive proof between two parties, prover and verifier, such that at the end of the proof, verifier is convinced of, say, the membership of the input instance of an **NP** language, but learns nothing else; in particular, the witness is *not* leaked to verifier. Quantum zero-knowledge proof is zero-knowledge proof realized using quantum mechanism that is secure against quantum attack. Again, one should differentiate (general) QZK and classical ZK secure against quantum attack (which is also called QZK in post-quantum cryptography).

Due to space reasons, in this extended abstract we just state our results, ideas, and sketch proofs. Details are referred to the full paper [12].

1.1 Our Contribution

We have three main results. *First*, we propose a *formalization* of non-interactive statistically-binding QBC (Sect. 3). Here "non-interactive" means in both commit and reveal stages, there is only one direction of message from sender to receiver. We define *weakest* quantum binding property, i.e., honest commitment to 0 cannot be opened as 1, and vice versa; it turns out that this binding suffices for the purpose of constructing QZK proof later. We argue that our formalization is robust, conceptually simple, and enjoys some nice properties; in particular, it *composes in parallel*. We also give a construction of *non-interactive* statistically-binding QBC based on pesudorandom generator (PRG) secure against quantum distinguisher, improving its classical counterpart [14] regarding round complexity.

[1] Hiding an binding properties cannot be simultaneously information-theoretic secure either [13].

Second, it may come as no surprise but we show that just like in the classical setting, statistically-binding QBC can be used to construct QZK proof for **NP** languages (Sect. 4). We remark that the actual proof is not as straightforward as it first appears.

Theorem 1. *All **NP** languages have quantum computational zero-knowledge proof given access to non-interactive computationally hiding, statistically binding quantum bit commitment scheme.*

Combining it with QBC generalized from Naor [14], we have:

Theorem 2. *If pseudorandom generator secure against quantum distinguisher exists, then all languages in **NP** have quantum computational zero-knowledge proof.*

We remark that in [20], Watrous constructed classical ZK proof against quantum attack using classical BC secure against quantum attack, in contrast to the quantum construction here. One advantage of our QZK proof over Watrous is gained from the QBC we used: our QBC relies on a complexity assumption (PRG against quantum attack) believed to be weaker than quantum one-way permutation assumed in [20].

Finally, we generalize the classical unconditional (not relying on complexity assumptions) study of zero-knowledge [16,18] to the quantum setting. First, similar to classical setting, we introduce *instance-dependent* QBC associated with a promise problem A: the scheme is constructed from instances of A such that when it is yes instance, the scheme is hiding; when it is no instance, the scheme is binding. Then, we can prove the following equivalence between instance-dependent QBC and QZK:

Theorem 3. *For every language $A \in$ **NP**, A has quantum statistical (resp. computational) zero-knowledge proof if and only if A has an instance-dependent non-interactive quantum bit commitment scheme that is statistically (resp. computationally) hiding on* YES *instances and statistically binding on* NO *instances.*

Such equivalence not only implies statistically-binding QBC is the minimum assumption for QZK proof, but also useful in proving many properties of QZK proof *unconditionally*.

These results can also be viewed as the follow-up works of [10,19,21] on unconditional study of QZK proof.

1.2 Techniques and Proof Overview

Formalization of non-interactive QBC. Motivated by Watrous' construction of complete problem for quantum statistical zero-knowledge proof [19], we formalize non-interactive QBC in terms of a pair of quantum circuits (Q_0, Q_1) (See Definition 2). The QBC proceeds as illustrated in Fig. 1, where quantum register B contains the value to open that is to be announced in the reveal stage;

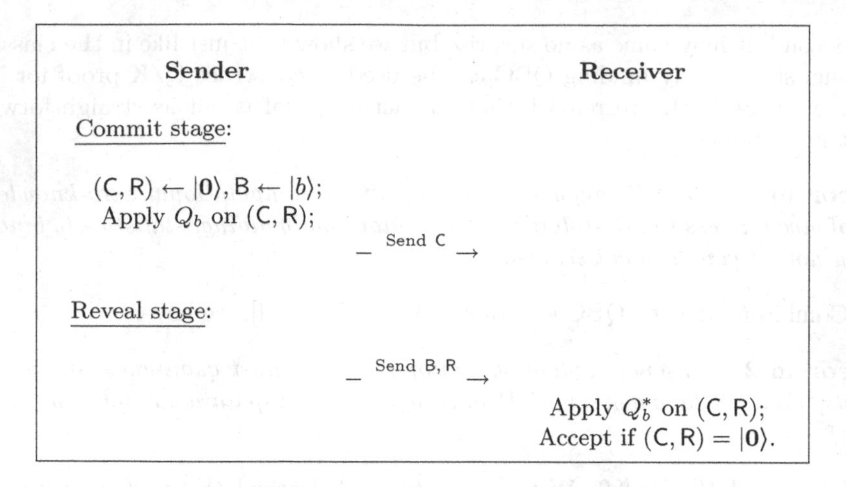

Fig. 1. Formalization of non-interactive QBC.

quantum register C contains bit commitment, and register R which is typically entangled with C, is used to open the commitment. We remark that here we make an essential use of the *reversibility* of quantum computation (quantum circuit Q^*); similar technique is also used in [4]. Inspired by our formalization, we *quantize* Naor's construction of BC from pseudorandom generator [14], obtaining a *non-interactive* (as opposed to the original two-message) QBC.

From QBC to QZK. We construct QZK proof for all **NP** languages given access to QBC. To this end, we construct QZK proof for **NP**-complete languages Hamiltonian Cycle. Our constructions are of GMW-type, almost the same as their classical counterparts [8] and [3], respectively, except that now we plug in QBC instead. For the security proof, completeness is trivial; QZK property follows from quantum hiding property of QBC by applying Watrous' rewinding technique [20] directly.

The nontrivial part lies in the soundness. When we try to generalize classical soundness analysis to the quantum setting, we encounter new difficulty raised by quantum binding. We shall discuss the difficulty in more detail shortly.

We highlight that a technical lemma (Lemma 1) we use in soundness analysis gives two characterizations of a bunch of subspaces with small "angles", which could be found useful elsewhere. We note that this lemma is similar to [17, Lemma 6]; the advantage of ours is its ability to handle multiple projectors/subspaces rather than two. Thus, it can be used to establish soundness of protocols with more than one challenge bit, e.g. protocol for Graph 3-Coloring [8].

Unconditional Study of QZK Proof. Like in classical setting [16,18], we show that from any QZK proof for problem $\mathsf{A} \in \mathbf{NP}$, we can construct an instance-dependent statistically binding QBC associate with it. Our basic construction is borrowed from Watrous [19], which is incomparable to its classical counterpart. Conversely, from QBC to QZK, we construct QZK proof for A by plugging instance-dependent QBC into the GMW-type zero-knowledge protocol.

This equivalence between QBC and QZK implies that every promise problem in **NP** with QZK proof also has a GMW-type QZK proof. For space reasons, we shall give no further detail about this result in this extended abstract.

1.3 New Issues Raised by Quantum Binding in Security Analysis

It is well-known that quantum binding is weaker than its classical counterpart [6]. In more detail, recall that classical binding property of BC roughly says sender can only open a bit commitment as one value, 0 or 1, but exclusively. This property is pretty easy to use in showing the security of larger protocol when BC is used as a building block. For example, soundness of classical ZK protocol can be reduced to the binding property of underlying BC (see, e.g. [7]). However, regarding quantum binding, malicious sender is no longer bound to a single bit value. Explanation follows.

Using our formalization of non-interactive QBC describe in Fig. 1, honest sender prepares quantum state $Q_b|\mathbf{0}\rangle_{CR}$ to commit bit $b \in \{0,1\}$. Then malicious sender is certainly legal to prepare a quantum state like

$$\sqrt{p_0} \cdot |0\rangle_B \otimes Q_0|\mathbf{0}\rangle_{CR} + \sqrt{p_1} \cdot |1\rangle_B \otimes Q_1|\mathbf{0}\rangle_{CR}, \quad (1)$$

where real numbers p_0, p_1 satisfy $p_0 + p_1 = 1$. In this case, it is easy to see that receiver can open the bit commitment successfully, but open as 0 with probability p_0 and 1 with p_1. Actually, malicious sender can tweak the probabilities p_0 and p_1 arbitrarily as he wants; he can even deviate from the ways as described in expression (1) as well. Therefore, by quantum binding we cannot guarantee malicious sender be bound to a single bit value.

More issues arise in the security analysis of protocols within which QBC is composed in parallel to commit a string. In more detail, suppose sender first commits to a string bit by bit using QBC. Later, sender will announce some *classical information* to tell receiver *which bit commitments and what values* are to open. But these classical information could be in superposition and entangled with commitments in a similar way as register B in expression (1). If receiver measures these classical information, then they will collapse, distributed according to some probability distributions, meanwhile the original quantum commitments will be disturbed due to entanglement. What making things complicated is, typically, the kinds of classical information vary according to the random coins of the outer protocol (refer to Sect. 4 to see an concrete QZK protocol); their probability distributions will be correlated through the original quantum commitments which, unfortunately, has been disturbed in different ways (according to the random coins). This makes it hard to analyze receiver's accepting probability (averaging over all random coins of the outer protocol).

We finally remark that in a *special case* where QBC is actually *classical*, that is, classical BC secure against quantum attack, things go back to the easy (classical) case. To see this, now we can let (honest) receiver measure everything upon its arrival, including quantum register C (which is not allowed with general QBC) to obtain the commitment. Then it follows from statistically binding property that (quantum) commitment (malicious sender could still be quantum) will collapse to classical commitment to either 0 or 1. Once the superposition collapses, subsequent security analysis follows almost the same lines of analysis in the classical setting. We point out that the triviality of soundness in [20] relies on classical BC against quantum attack being used underly.

Our Solution. The idea of our solution is easy: we pretend receiver does not measure classical information, but rather takes a "big" *binary measurement* over the whole workspace to determine whether to accept or not. In this way, quantum commitments will be fixed in the subsequent analysis. Each binary measurement (corresponding to the random coins of the outer protocol) will induce an accepting subspace. Intuitively, if we can show that these accepting subspaces are pairwisely (almost) orthogonal, then we can conclude that receiver cannot accept with high probability. This is because there cannot exist a quantum state which has large projections (almost 1) on each orthogonal subspace. To have a glimpse at how this geometric picture is related to quantum (statistically) binding, observe that honest quantum bit commitments to 0 and 1 (two quantum states) are (almost) orthogonal.

1.4 Related Work

Non-interactive perfectly binding BC secure against quantum attack can be based on quantum one-way permutation [1]. Our construction of non-interactive statistically binding QBC generalized from Naor [14] has the advantage of relying on a seemingly weaker complexity assumption, but at the cost of quantum construction. Moreover, while it is folklore that Naor's two-message construction can be trivially generalized to the quantum setting, as far as we know, our way of using reversible computation to realize non-interactive QBC is surprisingly new.

Compared with Kobayashi's work on QZK proof [11], here we restrict to **NP** languages, which is of the most interest from cryptographic view: we expect (honest) prover can be implemented in *polynomial time* (given access to a witness). However, some of transformations in [11] may not preserve prover's complexity. We also prove many properties of QZK proof that are similar to Kobayashi, but in a completely different approach that just meets the call for unconditional study of QZK proof in [11].

To the best of our knowledge, the only previous work using (general) QBC as a building block is [5], where the security analysis based on quantum (computationally) binding is much different from us.

The remainder of this extended abstract is organized as follows. In Sect. 2, we introduce some terminologies and notations. Next in Sect. 3, we give a formalization of QBC, followed a construction. Finally in Sect. 4, we show that QBC suffices for QZK.

2 Preliminaries

Most terminologies and notations of quantum information we are using here are standard. To save notations, given a projector Π, we also use Π to denote the *subspace* on which it projects.

We assume readers are familiar with classical GMW-type zero-knowledge protocol and its security analysis [7].

We adopt definitions of *quantum (in)distinguishability* and *quantum zero-knowledge* from Watrous [20], which are straightforward generalizations of their classical counterparts [7].

Quantum circuit is composed of quantum gates chosen from some fixed universal, finite, and *unitary* quantum gate set [15,19]. To save notations, we overload notations to use Q and Q^* to denote the *unitary transformation* induced by quantum circuit Q and its inverse, respectively. Quantum algorithm can be formalized in terms of *uniformly generated* quantum circuit family; that is, the description of each quantum circuit in this family can be output by *a single classical algorithm* given as input the index of this circuit in the family.

3 Formalization and Construction of QBC

In this section, we shall give a formalization of non-interactive statistically-binding QBC as motivated by [19]. Then we generalize Naor's construction [14] to the quantum setting inspired by our formalization.

We first recall a definition in [19].

Definition 1 (Quantum state defined by quantum circuit). *Quantum circuit Q operates on a pair of quantum registers (C, R). We can view Q encodes a quantum state in the following way: first initialize (C, R) in state $|\mathbf{0}\rangle$, and then apply Q; we call the resulting state of C, i.e. $Tr_{\mathsf{R}}(Q|\mathbf{0}\rangle_{\mathsf{CR}}\langle\mathbf{0}|Q^*)$, the quantum state defined by quantum circuit Q.*

Our formalization of QBC is based on the definition above.

Definition 2. *A non-interactive quantum bit commitment scheme (QBC) is a two-party, two-stage protocol. It can be represented by an ensemble of quantum circuit pair $\{(Q_0(n), Q_1(n))\}_n$ that are uniformly generated in polynomial time. Specifically,*

- *The protocol consists of two parties, sender and receiver, proceeding in two stages: first a commit stage and later a reveal stage.*
- *In commit stage, to commit bit $b \in \{0, 1\}$, sender applies quantum circuit Q_b on quantum registers (C, R) that are initialized in all $|0\rangle$'s state. Then sender sends quantum register C, whose state we denote by ρ_b, to receiver.*
- *In reveal stage, sender announces b, and sends quantum register R to receiver. Receiver then applies Q_b^* on (C, R), accepts if (C, R) return to all $|0\rangle$'s state.*

The execution of QBC is depicted in Fig. 1. We are next to define hiding (or concealing) and binding properties of QBC.

- ***Hiding**. We say the scheme is computationally hiding if quantum state ensembles $\{\rho_0(n)\}_n$ and $\{\rho_1(n)\}_n$ are computationally indistinguishable.*
- *Statistically $\epsilon(n)$-**binding**. We say the scheme is statistically $\epsilon(n)$-binding if fidelity $F(\rho_0(n), \rho_1(n)) < \epsilon(n)$. For cryptographic applications, we usually require $\epsilon(n)$ be negligible, or even exponentially small.*

A discussion about the definition above is referred to the full paper, where we argue that our formalization is robust, conceptually simple, and without loss of generality. Hereafter, when we talk about QBC, we are referring to the formalization given in Definition 2 (rather than any specific constructions).

We highlight that we are defining the *weakest* quantum binding here (see the full paper for detail), which enjoys a very appealing property: it *composes in parallel.* This makes it easy to extend QBC to quantum string commitment by *committing a binary string bit by bit.* Formal definition of quantum string commitment and proof of the following theorem is referred to the full paper.

Theorem 4. *Suppose* $\{(Q_0(n), Q_1(n))\}_n$ *is a non-interactive statistically* $\epsilon(n)$*-binding QBC. Then given a binary string* $s \in \{0,1\}^m$*, committing each bit* s_i *(*$i = 1, 2, \ldots, m$*) of* s *using QBC results in a non-interactive statistically* $\epsilon(n)$*-binding quantum string commitment scheme, which can be represented by quantum circuit* $Q_s(n) = \bigotimes_{i=1}^{m} Q_{s_i}(n)$*.*

We next consider how to construct non-interactive statistically-binding QBC from plausible complexity assumption. Our idea is inspired by our formalization of QBC: it suffices for us to generate an ensemble of quantum states $\{(\rho_0(n), \rho_1(n))\}_n$ such that $\{\rho_0(n)\}_n$ and $\{\rho_1(n)\}_n$ are *computationally indistinguishable* but statistically distinguishable. To this end, we adapt Naor's idea [14]: let $\rho_0(n)$ be the density operator corresponding to a pseudorandom distribution (against quantum distinguisher), and $\rho_1(n)$ corresponding to a truly random (i.e. uniform) distribution. The commit and open procedures follow the generic one as given in Definition 2. Formally, we prove the following theorem, whose proof is referred to the full paper.

Theorem 5. *If classical pseudorandom generator against quantum distinguisher exists, then we have (computationally hiding) statistically binding quantum non-interactive bit commitment scheme.*

We remark that if [9] can be generalized to the quantum setting, which is widely believed, then the assumption of the theorem above and in turn Theorem 2 can be replaced with quantum one-way function.

4 From QBC to QZK

We finally consider how to use QBC to construct QZK proof for all **NP** languages. An immediate idea is to plug QBC into GMW-type zero-knowledge protocols for some **NP**-complete languages (e.g. [3,8]). This idea is natural, but we have to reconsider its security, i.e. soundness and quantum zero-knowledge; we know that showing security against quantum adversary is not always immediate even given classical security.

Indeed, quantum zero-knowledge property is not trivial, since commonly used classical rewinding technique cannot be applied in quantum setting generally.

Fortunately, Watrous [20] develops a quantum rewinding technique that is suitable in some special case. After a careful examination, it turns out that this technique works equally well in our case with QBC. Hence, quantum zero-knowledge is checked.

Soundness is not clear with QBC either. Quantum binding raises several new issues that never happen with classical BC (even against quantum attack), as generally discussed in "Introduction". In the remainder of this section, we explore a specific protocol, GMW-type ZK protocol for Hamiltonian Cycle problem [3], showing that it is sound given quantum binding.

We first briefly recap this protocol and set up some notations for our exposition. In the GMW-type ZK protocol for Hamiltonian Cycle, prover, who plays the role of sender in QBC, first sends a message or quantum register A that is expected to contain the quantum string commitment (using QBC to commit bit by bit) of $\pi(G)$ (encoded as an n^2-bit string), where G is the input graph of n vertices and π a random permutation over vertex set $\{1, 2, \ldots, n\}$. After receiving the (quantum) commitment, verifier, who plays the role of receiver in QBC, sends back a challenge bit, 0 or 1. Then depending on this challenge bit, prover sends his second message or quantum registers (B, C), trying to either open the whole quantum string commitment as $\pi(G)$, or n (out of n^2) positions of the string commitment as n 1's corresponding to a Hamiltonian cycle. Here, quantum register C is to store classical information such as permutation π or position of Hamiltonian cycle c; they will determine which positions and what values of bit commitments are to open. Register B is just used for opening commitments.

In soundness analysis, G is assumed not Hamiltonian and prover could be malicious. To estimate verifier's accepting probability, as mentioned in "Introduction", our idea is to pretend receiver does not measure π or c, thus avoiding perturbing quantum commitment or register A. Instead, verifier will apply some "big" binary measurements $\{P_0, \mathbb{1} - P_0\}$ or $\{P_1, \mathbb{1} - P_1\}$ on the combined quantum system (A, B, C) received from prover, where

$$P_0 = \sum_{\pi} |\pi\rangle_C\langle\pi| \otimes Q_{\pi(G)}|\mathbf{0}\rangle_{AB}\langle\mathbf{0}|Q^*_{\pi(G)}, \tag{2}$$

$$P_1 = \sum_{c} |c\rangle_C\langle c| \otimes (Q_c \otimes \mathbb{1}_{\bar{c}})|\mathbf{0}\rangle_{AB}\langle\mathbf{0}|(Q^*_c \otimes \mathbb{1}_{\bar{c}}). \tag{3}$$

Here, quantum circuit $Q_{\pi(G)}$ denotes the quantum circuit to commit graph $\pi(G)$, and Q_c denotes the quantum circuit to commit n 1's at position c corresponding to a Hamiltonian cycle. It is easy to write out expressions of $Q_{\pi(G)}$ and Q_c in terms of (Q_0, Q_1) according to Theorem 4.

Note that projectors P_0 and P_1 induce accepting subspaces corresponding to verifier's challenge bit 0 and 1, i.e. succeeding in opening graph $\pi(G)$ and Hamiltonian cycle c, respectively. To prove soundness, we are to show that whatever quantum state (malicious) prover may prepare as commitment, its projections on subspaces P_0 and P_1 cannot be close to 1 simultaneously. This in turn suffices to show that subspaces P_0 and P_1 are (almost) orthogonal. To see this, note that for each pair of π, c, vectors $Q_{\pi(G)}|\mathbf{0}\rangle$ and $Q_c|\mathbf{0}\rangle$ are purifications of honest quantum commitments (within A component) to string $\pi(G)$ and n 1's, respectively. Since G does not have Hamiltonian cycle, the substring of $\pi(G)$ at position c cannot be all 1's — this is the very argument used in classical soundness analysis. Then quantum (statistically) binding implies that P_0 and P_1, which sum over all π's and c's respectively, are (almost) orthogonal. We further remark that the formal proof is not as straightforward as the intuition would suggest; some technical work is required to deal with errors (in case of statistical binding as opposed to perfect binding), which if we are not careful, may blow up exponentially due to exponential terms ($\approx n!$) within expressions of P_0 and P_1.

Formally, to relate the (almost) orthogonality of subspaces P_0 and P_1 with quantum binding, we need a technical lemma as below; it is stated in a *contrapositive* way for our purpose of proof by contradiction. In geometric picture, this lemma gives two characterizations of a bunch of subspaces with small "angles" (orthogonality corresponds to angle $\pi/2$) pairwisely.

Lemma 1. *Let $\mathcal{X}, \mathcal{Y}$ be two complex Euclidean spaces, and $P_1, \ldots, P_m$ be projectors on $\mathcal{X} \otimes \mathcal{Y}$. If there exists a vector $|\psi\rangle \in \mathcal{X} \otimes \mathcal{Y}$ and unitary transformations $U_1, \ldots, U_m \in U(\mathcal{Y})$ such that $\sum_i \|P_i U_i |\psi\rangle\|^2/m \geq 1 - \delta$ for some $0 \leq \delta \leq 1$, then there exists unitary transformations $U'_1, \ldots, U'_m \in U(\mathcal{Y})$ satisfying $\|P_1 U'_1 \cdots P_m U'_m |\psi\rangle\| \geq 1 - m\sqrt{\delta}$.*

Further expositions and details of the proof are referred to the full paper.

Acknowledgement. We thank Yi Deng for helpful discussion during the progress of this work. Thanks also go to Dominique Unruh and anonymous referees of several conferences for their invaluable insights and comments.

Jun Yan is supported in part by the Fundamental Research Funds for the Central Universities (21615317), by the Open Project Program of the State Key Laboratory of Information Security (2015-MS-08), by the PhD Start-up Fund of Natural Science Foundation of Guangdong Province, China (2014A030310333), and by the National Natural Science Foundation of China (61501207). Jian Weng is supported in part by National Science Foundation of China (61272413, 61133014, 61272415 and 61472165), by Fok Ying Tong Education Foundation (131066), by Program for New Century Excellent Talents in University (NCET-12-0680), by Research Fund for the Doctoral Program of Higher Education of China (20134401110011), by Foundation for Distinguished Young Talents in Higher Education of Guangdong (2012LYM 0027), and by China Scholarship Council. Dongdai Lin is supported in part by National Science Foundation of China (61379139) and by the Strategic Priority Research Program of the Chinese Academy of Science (XDA06010701). Yujuan Quan is supported in part by Special Project on the Integration of Industry, Education and Research of Guangdong Province (2013B090500030) and by Key Technology R&D Program of Guangzhou,China (2014Y2-00133).

References

1. Adcock, M., Cleve, R.: A quantum goldreich-levin theorem with cryptographic applications. In: Alt, H., Ferreira, A. (eds.) STACS 2002. LNCS, vol. 2285, pp. 323–334. Springer, Heidelberg (2002)

2. Ambainis, A., Rosmanis, A., Unruh, D.: Quantum attacks on classical proof systems: the hardness of quantum rewinding. In: FOCS, pp. 474–483 (2014)
3. Blum, M.: How to prove a theorem so no one else can claim it. In: Proceedings of the International Congress of Mathematicians, vol. 1, p. 2 (1986)
4. Chailloux, A., Kerenidis, I., Rosgen, B.: Quantum commitments from complexity assumptions. In: Aceto, L., Henzinger, M., Sgall, J. (eds.) ICALP 2011, Part I. LNCS, vol. 6755, pp. 73–85. Springer, Heidelberg (2011)
5. Crépeau, C., Dumais, P., Mayers, D., Salvail, L.: Computational collapse of quantum state with application to oblivious transfer. In: Naor, M. (ed.) TCC 2004. LNCS, vol. 2951, pp. 374–393. Springer, Heidelberg (2004)
6. Dumais, P., Mayers, D., Salvail, L.: Perfectly concealing quantum bit commitment from any quantum one-way permutation. In: Preneel, B. (ed.) EUROCRYPT 2000. LNCS, vol. 1807, pp. 300–315. Springer, Heidelberg (2000)
7. Goldreich, O.: Foundations of Cryptography, Basic Tools, vol. I. Cambridge University Press, Cambridge (2001)
8. Goldreich, O., Micali, S., Wigderson, A.: Proofs that yield nothing but their validity for all languages in NP have zero-knowledge proof systems. J. ACM **38**(3), 691–729 (1991)
9. Håstad, J., Impagliazzo, R., Levin, L.A., Luby, M.: A pseudorandom generator from any one-way function. SIAM J. Comput. **28**(4), 1364–1396 (1999)
10. Kobayashi, H.: Non-interactive quantum perfect and statistical zero-knowledge. In: Ibaraki, T., Katoh, N., Ono, H. (eds.) ISAAC 2003. LNCS, vol. 2906, pp. 178–188. Springer, Heidelberg (2003)
11. Kobayashi, H.: General properties of quantum zero-knowledge proofs. In: Canetti, R. (ed.) TCC 2008. LNCS, vol. 4948, pp. 107–124. Springer, Heidelberg (2008)
12. Lin, D., Quan, Y., Weng, J., Yan, J.: Quantum bit commitment with application in quantum zero-knowledge proof. Cryptology ePrint Archive, Report 2014/791, this is a preliminary version; the final full version is in preparation
13. Mayers, D.: Unconditionally secure quantum bit commitment is impossible. Phys. Rev. Lett. **78**(17), 3414–3417 (1997)
14. Naor, M.: Bit commitment using pseudorandomness. J. Cryptology **4**(2), 151–158 (1991)
15. Nielsen, M.A., Chuang, I.L.: Quantum computation and Quantum Informatioin. Cambridge University Press, Cambridge (2000)
16. Ong, S.J., Vadhan, S.P.: An equivalence between zero knowledge and commitments. In: Canetti, R. (ed.) TCC 2008. LNCS, vol. 4948, pp. 482–500. Springer, Heidelberg (2008)
17. Unruh, D.: Quantum proofs of knowledge. In: Pointcheval, D., Johansson, T. (eds.) EUROCRYPT 2012. LNCS, vol. 7237, pp. 135–152. Springer, Heidelberg (2012)
18. Vadhan, S.P.: An unconditional study of computational zero knowledge. SIAM J. Comput. **36**(4), 1160–1214 (2006)
19. Watrous, J.: Limits on the power of quantum statistical zero-knowledge. In: FOCS, pp. 459–468 (2002)
20. Watrous, J.: Zero-knowledge against quantum attacks. SIAM J. Comput. **39**(1), 25–58 (2009). preliminary version appears in STOC 2006
21. Yan, J.: Complete problem for perfect zero-knowledge quantum proof. In: Bieliková, M., Friedrich, G., Gottlob, G., Katzenbeisser, S., Turán, G. (eds.) SOFSEM 2012. LNCS, vol. 7147, pp. 419–430. Springer, Heidelberg (2012)

Effectiveness of Structural Restrictions for Hybrid CSPs

Vladimir Kolmogorov[1], Michal Rolínek[1], and Rustem Takhanov[2](✉)

[1] IST Austria, Klosterneuburg, Austria
{vnk,michal.rolinek}@ist.ac.at
[2] Nazarbayev University, Astana, Kazakhstan
takhanov@mail.ru

Abstract. Constraint Satisfaction Problem (CSP) is a fundamental algorithmic problem that appears in many areas of Computer Science. It can be equivalently stated as computing a homomorphism $\mathbf{R} \to \mathbf{\Gamma}$ between two relational structures, e.g. between two directed graphs. Analyzing its complexity has been a prominent research direction, especially for the *fixed template CSPs* where the right side $\mathbf{\Gamma}$ is fixed and the left side $\mathbf{R}$ is unconstrained.

Far fewer results are known for the *hybrid* setting that restricts both sides simultaneously. It assumes that $\mathbf{R}$ belongs to a certain class of relational structures (called a *structural restriction* in this paper). We study which structural restrictions are *effective*, i.e. there exists a fixed template $\mathbf{\Gamma}$ (from a certain class of languages) for which the problem is tractable when $\mathbf{R}$ is restricted, and NP-hard otherwise. We provide a characterization for structural restrictions that are *closed under inverse homomorphisms*. The criterion is based on the *chromatic number* of a relational structure defined in this paper; it generalizes the standard chromatic number of a graph.

As our main tool, we use the algebraic machinery developed for fixed template CSPs. To apply it to our case, we introduce a new construction called a "lifted language". We also give a characterization for structural restrictions corresponding to minor-closed families of graphs, extend results to certain Valued CSPs (namely conservative valued languages), and state implications for (valued) CSPs with ordered variables and for the maximum weight independent set problem on some restricted families of graphs.

1 Introduction

The *Constraint satisfaction problems (CSPs)* and the valued constraint satisfaction problems (VCSP) provide a powerful framework for analysis of a large set of computational problems arising in propositional logic, combinatorial optimization, artificial intelligence, graph theory, scheduling, biology, computer vision etc. Traditionally CSP is formalized either as a problem of (a) finding an assignment of values to a given set of variables, subject to constraints on the values that can be assigned simultaneously to specified subsets of variables, or as problem

K. Elbassioni and K. Makino (Eds.): ISAAC 2015, LNCS 9472, pp. 566–577, 2015.
DOI: 10.1007/978-3-662-48971-0_48

of (b) finding a homomorphism between two finite relational structures A and B (e.g., two oriented graphs). These two formulations are polynomially equivalent under the condition that the input constraints in the first case or input relations in the second case are given by lists of their elements. Soft version of CSP, that is VCSP, generalizes the CSP by replacing crisp constraints with cost functions applied to tuples of variables. In the VCSP we require to find the maximum (or minimum) of a sum of cost functions applied to corresponding variables.

The CSPs have been the cutting edge research field of theoretical computer science since the 70s, and recently this interest has been expanded to VCSP. One of the themes that revealed rich logical and algebraic structure of the CSPs was the question of classification of the problem's computational complexity when constraint relations are restricted to a given set of relations or, alternatively, when the second relational structure is some fixed $\mathbf{\Gamma}$. Thus, this problem is parameterized by $\mathbf{\Gamma}$, denoted as CSP($\mathbf{\Gamma}$) and called a fixed template CSP with a template $\mathbf{\Gamma}$ (another name is a non-uniform CSP). E.g., if the domain set is boolean and $\mathbf{\Gamma}$ is a relational structure with four ternary predicates $x \vee y \vee z, \overline{x} \vee y \vee z, \overline{x} \vee \overline{y} \vee z, \overline{x} \vee \overline{y} \vee \overline{z}$, CSP($\mathbf{\Gamma}$) models 3-SAT which is historically one the first NP-complete problems [8]. At the same time, if we restrict $\mathbf{\Gamma}$ to binary predicates, then we obtain tractable 2-SAT. Generally, Schaeffer proved [23] that for any template $\mathbf{\Gamma}$ over the boolean set, CSP($\mathbf{\Gamma}$) is either in P or NP-complete, and any tractable constraint language belongs to one of 6 classes (0 or 1-preserving, binary, horn, anti-horn and linear subspaces). When $\mathbf{\Gamma}$ contains only one graph (irreflexive symmetric predicate) Hell and Nešetřil [14] proved an analogous statement, by showing that only for bipartite graphs the problem is tractable. Feder and Vardi [10] found that all fixed template CSPs can be expressed as problems in a fragment of SNP, called Monotone Monadic SNP (MM SNP). They introduced this class as a natural restriction of SNP for which Ladner's argument about the existence of problems with intermediate complexity between P and NP-hard could not be applied. Moreover, they showed that all problems in MM SNP can be reduced with respect to Turing reduction to fixed template CSPs and, thus, non-uniform CSPs complexity classification would lead to a classification of MM SNP problems. This result placed fixed-template CSPs into a broad logical context that naturally lead to a conjecture that such CSPs are either tractable or NP-hard, the so called dichotomy conjecture.

In [16] Jeavons observed that any predicate given by primitive positive formula using predicates of the template $\mathbf{\Gamma}$, when added to $\mathbf{\Gamma}$, does not change the complexity of CSP($\mathbf{\Gamma}$). This result clarified that the computational complexity of CSP($\mathbf{\Gamma}$) is fully defined by the minimal predicate clone that contains predicates of $\mathbf{\Gamma}$. In universal algebra, it has long been known that the predicate clones are dual to the so called functional clones [11,19,22]. Specifically, it implies that the complexity of CSP($\mathbf{\Gamma}$) is defined by the set of polymorphisms of $\mathbf{\Gamma}$. The last was the main motive for subsequent research. Intensive studies in this direction lead to a conjectured algebraic description of all tractable templates made by Bulatov et al. [6], with subsequent reformulations of this conjecture by Maroti and McKenzie [20]. In the long run it was shown by Siggers [24] that if Bulatov-Jeavons-Krokhin characterization of tractable templates is correct, then the tractable core structures can be characterized as those that admit a single 6-ary polymorphism that satisfies a certain equality. The last fact will serve as a key ingredient for one of our results.

Besides fixed template CSPs, another parameterization of CSP concerns restrictions on the left relational structure of the input. If we restrict the left structure of the input to some specified set $\mathcal{H}$ and impose no restriction on the right relational structure, then the problem is called CSP with structural restrictions $\mathcal{H}$. For example, if $\mathcal{H}$ is a set of graphs with treewidth less or equal to $k \in \mathbb{N}$, then the problem can be solved in polynomial time. It was found by Grohe [12] that any structural restriction $\mathcal{H}$ that defines tractable CSP should be of bounded treewidth modulo homomorphic equivalence.

Related Work. Since many (V)CSP instances do not fall into any of the tractable classes offered by one of the previous approaches, there has been growing interest in the so-called hybrid restrictions. That is when the input is restricted to a subset of all input pairs $(\mathbf{R}, \mathbf{\Gamma})$. One approach to this problem is to construct a new structure for any input $(\mathbf{R}, \mathbf{\Gamma})$, $G_{\mathbf{R},\mathbf{\Gamma}}$, and shift the analysis to $G_{\mathbf{R},\mathbf{\Gamma}}$. In case of binary CSPs (i.e. when all predicates of an input are binary) it is natural to define $G_{\mathbf{R},\mathbf{\Gamma}}$ as a microstructure graph [17] of a template $(\mathbf{R}, \mathbf{\Gamma})$. Thereby, a set of inputs for which certain local substructures in $G_{\mathbf{R},\mathbf{\Gamma}}$ are forbidden form a parametrized problem. Cooper and Živný [9] investigated this formulation and found examples of specific forbidden substructures that result in tractable hybrid CSPs. Microstructure graphs also naturally appear in the context of fixed template CSPs. Specifically, all templates $\mathbf{\Gamma}$ with binary predicates that define fixed template CSPs for which local consistency preprocessing of the input results in a perfect microstructure graph were completely classified in [26].

Our Results. The main topic of our paper is a hybrid framework for (V)CSP, when left structures are restricted to some set $\mathcal{H}$ and combined with a fixed right structure $\mathbf{\Gamma}$ (corresponding CSP is denoted as $\text{CSP}_{\mathcal{H}}(\mathbf{\Gamma})$). The difficulty of applying known algebraic machinery to this framework is due to the fact that the closure operator, analogous to the minimal containing clone, cannot depend on $\mathbf{\Gamma}$ only. Therefore, in an algebraic theory of hybrid CSPs an analogue of primitive positive formula should depend on both input structures. In our approach we define for any $\mathbf{R} \in \mathcal{H}$ and $\mathbf{\Gamma}$ a set of predicates $\mathbf{\Gamma}_{\mathbf{R}}$ that we call a "lifted" language. Our key idea is that the closures $\langle \mathbf{\Gamma}_{\mathbf{R}} \rangle$ for $\mathbf{R} \in \mathcal{H}$, under certain conditions, could maintain the information on the tractability of $\text{CSP}_{\mathcal{H}}(\mathbf{\Gamma})$. In this paper, by that "certain conditions" we understand the property that $\mathcal{H}$ is closed under inverse homomorphisms. We are especially interested in a classification of structural restrictions $\mathcal{H}$ closed under inverse homomorphisms for which we could find a template $\mathbf{\Gamma}$ (in a certain class of templates $\mathcal{C}$) that defines tractable $\text{CSP}_{\mathcal{H}}(\mathbf{\Gamma})$, whereas a $\text{CSP}(\mathbf{\Gamma})$ is NP-hard. We call such restrictions effective for a class $\mathcal{C}$. Our key results are formulated for 2 cases: the class of BJK languages, that is, the class of templates that are either tractable or have core a without a Siggers polymorphism, and a class of conservative valued templates.

Specifically, we prove that if $\mathcal{H}$ is a set of binary structures closed under inverse homomorphisms, it is effective for BJK languages if and only if $\{\chi(\mathbf{R}) \mid \mathbf{R} \in \mathcal{H}\}$ is bounded, where $\chi(\mathbf{R})$ is a chromatic number of $\mathbf{R}$ (considered as a graph). The last result is extended to the nonbinary case, with natural generalization of the chromatic number to arbitrary relational structures. A notable corollary of this result is that the set of acyclic digraphs is an ineffective structural restriction for BJK languages. This explains why NP-hardness arguments for certain fixed templates of digraph homomorphism problem can be extended to a case when the input digraph is acyclic [25]. Less straightforward corollary: let $\mathcal{H}$ be a set of binary structures such that their "graph copies" forbid specific minors, then $\mathcal{H}$ is effective for BJK languages if and only if $\{\chi(\mathbf{R}) \mid \mathbf{R} \in \mathcal{H}\}$ is bounded. The last statement does not require that $\mathcal{H}$ is closed under inverse homomorphisms.

For $\mathrm{VCSP}_{\mathcal{H}}(\mathbf{\Gamma})$ we prove an analogue of our previous result for a class $\mathcal{C}$ of all conservative valued templates. We obtain as a corollary that the maximum weight independent set problem is still NP-hard in some graph classes.

2 Preliminaries

Throughout the paper we assume $P \neq NP$. A problem is called *tractable* if it can be solved in polynomial time.

The symbol $[n]$ will denote the set $\{1, \dots, n\}$, and $\overline{\mathbb{Q}} = \mathbb{Q} \cup \{\infty\}$ the set of rational numbers with (positive) infinity. Also D will stand for a finite set.

We will denote the tuples in lowercase boldface such as $\mathbf{a} = (a_1, \dots, a_k)$. Also for mappings $h \colon A \to B$ and tuples $\mathbf{a} = (a_1, \dots, a_k)$, where $a_j \in A$ for $j = 1, \dots, k$, we will write $\mathbf{b} = (h(a_1), \dots, h(a_k))$ simply as $\mathbf{b} = h(\mathbf{a})$. Relational structures will be denoted in uppercase boldface as $\mathbf{R} = (R, r_1, \dots, r_k)$.

Finally let $\mathrm{ar}(\varrho)$, $\mathrm{ar}(\mathbf{a})$, and $\mathrm{ar}(f)$ stand for arity of a relation ϱ, size of a tuple $\mathbf{a}$, and arity (number of parameters) of a function f, respectively.

2.1 Fixed Template CSP

We will first formulate the general CSP in an algebraic way as a decision problems whether there exists a homomorphism between certain relational structures.

Definition 1. *Let $\mathbf{R} = (R, r_1, \dots, r_k)$ and $\mathbf{R}' = (R', r_1', \dots, r_k')$ be relational structures with a common signature (that is $\mathrm{ar}(r_i) = \mathrm{ar}(r_i')$ for every $i = 1, \dots, k$). A mapping $h \colon R \to R'$ is called a* homomorphism *from $\mathbf{R}$ to $\mathbf{R}'$ if for each $i = 1, \dots, k$, whenever $(x_1, \dots, x_{\mathrm{ar}(r_i)}) \in r_i$, then $((h(x_1), \dots, h(x_{\mathrm{ar}(r_i')})) \in r_i'$. In that case, we write $\mathbf{R} \xrightarrow{h} \mathbf{R}'$ or sometimes just $\mathbf{R} \to \mathbf{R}'$.*

Definition 2 (General CSP). *The general CSP is the following decision problem. Given a pair of relational structures with common signature $\mathbf{R} = (V, r_1, \dots, r_k)$ and $\mathbf{\Gamma} = (D, \varrho_1, \dots, \varrho_k)$, decide whether $\mathbf{R} \to \mathbf{\Gamma}$. Equivalently, decide whether there is a mapping $h : V \to D$ that satisfies*

$$\bigwedge_{(\varrho, \mathbf{v}) \in T} [h(\mathbf{v}) \in \varrho] \tag{1}$$

where $T = \{(\varrho_i, \mathbf{v}) \mid i \in [k], \mathbf{v} \in r_i\}$ specifies the set of constraints.

The set V represents the set of *variables* and we will only consider V finite, similarly D is the *domain set* or the set of *labels* for variables. The relations r_1, ..., r_k specify the tuples of V constrained by relations $\varrho_1, \ldots, \varrho_k$, respectively.

As we mentioned in the introduction, one natural way to restrict the general CSP is to fix the constraint types. A finitary relational structure $\mathbf{\Gamma} = (D, \varrho_1, \ldots, \varrho_k)$ over a fixed finite domain D will be called a constraint language. For such $\mathbf{\Gamma}$ we will denote by Γ (without boldface) the set of relations $\{\varrho_1, \ldots, \varrho_k\}$; with some abuse of terminology set Γ will also be called a constraint language. (Note that both views are used in the literature).

Definition 3 (Fixed template CSP). *Let D be a finite set and $\mathbf{\Gamma}$ a constraint language over D. Then the decision problem* CSP($\mathbf{\Gamma}$) *is defined as follows: given a relational structure $\mathbf{R} = (V, r_1, \ldots, r_k)$ of the same signature as $\mathbf{\Gamma}$, decide whether $\mathbf{R} \to \mathbf{\Gamma}$.*

We will usually write CSP(Γ) instead of CSP($\mathbf{\Gamma}$). Although there are multiple relational structures $\mathbf{\Gamma}$ that correspond to the same set Γ, it can be seen that all choices give equivalent problems; this justifies the notation CSP(Γ).

2.2 Fixed Template VCSP

A more general framework operates with *cost functions* $f : D^n \to \overline{\mathbb{Q}}$ instead of relations $\varrho \subseteq D^n$. This idea leads to the notion of valued CSP.

Definition 4. *We denote the set of all functions $f : D^n \to \overline{\mathbb{Q}}$ by $\mathbf{\Phi}_D^{(n)}$ and let $\mathbf{\Phi}_D = \bigcup_{n \geq 1} \mathbf{\Phi}_D^{(n)}$. We will often call the functions in $\mathbf{\Phi}_D$* cost functions *over D. For every cost function $f \in \mathbf{\Phi}_D^{(n)}$, let* $\operatorname{dom} f = \{x \mid f(x) < \infty\}$. *Note that* $\operatorname{dom} f$ *can be considered both as an n-ary relation and as an n-ary function such that* $\operatorname{dom} f(x) = 0$ *if and only if $f(x)$ is finite.*

We will say that the cost functions in $\mathbf{\Phi}_D$ take *values*. Note that in some papers on VCSP cost functions are called weighted relations.

Definition 5. *An instance of the* valued constraint satisfaction problem *(VCSP) is specified by finite sets D, V and a function from D^V to $\overline{\mathbb{Q}}$ given by*

$$f_{\mathcal{I}}(h) = \sum_{(f,\mathbf{v}) \in T} w(f, \mathbf{v}) f(h(\mathbf{v})), \tag{2}$$

where V is a finite set of variables, $w(f, \mathbf{v})$ are positive numbers,[1] *and T is a finite set of constraints of the form $(f, \mathbf{v})$ where $f \in \mathbf{\Phi}_D$ is a cost function and $\mathbf{v} \in V^{\operatorname{ar}(f)}$ is a tuple of variables of size* $\operatorname{ar}(f)$. *The goal is to find an* assignment *(or* labeling*) $h \in D^V$ that minimizes $f_{\mathcal{I}}$.*

---

[1] We will allow two possibilities: (i) weights are positive integers, and the length of the description of $\mathcal{I}$ grows linearly with $w(f, \mathbf{v})$; (ii) weights are positive rationals. All our statements for VCSPs will hold under both models. Note that in the literature weights $w(f, \mathbf{v})$ are usually omitted, and T is allowed to be a multiset rather than a set; this is equivalent to model (i). Including weights will be convenient for hybrid VCSPs.

Note that $f_{\mathcal{I}}$ can also be looked at as a cost function over the variable set V.

Definition 6. *A* valued constraint language *over D is either a tuple $\mathbf{\Gamma} = (D, f_1, \ldots, f_k)$ with $f_1, \ldots, f_k \in \mathbf{\Phi}_D$ or the corresponding finite set $\Gamma = \{f_1, \cdots, f_k\} \subseteq \mathbf{\Phi}_D$. We will denote by* VCSP(Γ) *the class of all VCSP instances in which the cost functions are all contained in Γ.*

This framework subsumes many other frameworks studied earlier and captures many specific well-known problems, including k-SAT, GRAPH k-COLOURING, MAX CUT, MIN VERTEX COVER, and others (see [15]).

A function $f \in \mathbf{\Phi}_D^{(n)}$ that takes values in $\{0, \infty\}$ is called *crisp*. We will often view it as a relation in D^n, and vice versa (this should be clear from the context). If language Γ is crisp (i.e. it contains only crisp functions), then VCSP(Γ) is a pure feasibility problem corresponding to CSP(Γ). Note, however, that according to our definitions there is a slight difference between the two: CSP(Γ) is a decision problem while VCSP(Γ) asks to compute a solution explicitly if it exists.

The dominant research line in this area is to classify the complexity of problems VCSP(Γ). Sometimes, problems CSP(Γ) and VCSP(Γ) are defined also for infinite languages Γ and then VCSP(Γ) is called tractable if for each finite $\Gamma' \subseteq \Gamma$, VCSP(Γ') is tractable. Also, VCSP(Γ) is called NP-hard if for some finite $\Gamma' \subseteq \Gamma$, VCSP(Γ') is NP-hard. In turn, we will focus purely on finite languages Γ.

2.3 Polymorphisms

Let $\mathcal{O}_D^{(m)}$ denote the set of all operations $g : D^m \to D$ and let $\mathcal{O}_D = \bigcup_{m \geq 1} \mathcal{O}_D^{(m)}$. When D is clear from the context, we will sometimes write simply $\mathcal{O}^{(m)}$ and $\mathcal{O}$.

Any language Γ defined on D can be associated with a set of operations on D, known as the polymorphisms of Γ, defined as follows.

Definition 7. *An operation $g \in \mathcal{O}_D^{(m)}$ is a* polymorphism *of a cost function $f \in \mathbf{\Phi}_D$ if for any $\mathbf{x}^1, \ldots, \mathbf{x}^m \in \operatorname{dom} f$, we have that $g(\mathbf{x}^1, \ldots, \mathbf{x}^m) \in \operatorname{dom} f$ where g is applied component-wise.*

For any valued constraint language Γ over a set D, we denote by Pol(Γ) *the set of all operations on D that are polymorphisms of every $f \in \Gamma$.*

Clearly, if g is a polymorphism of a cost function f, then g is also a polymorphism of dom f. For $\{0, \infty\}$-valued functions, which naturally correspond to relations, the notion of a polymorphism defined above coincides with the standard notion of a polymorphism for relations. Note that the projections, i.e. operations of the form $e_n^i(x_1, \ldots, x_n) = x_i$, are polymorphisms of all valued constraint languages. Polymorphisms play the key role in the algebraic approach to the CSP. For VCSPs more general constructs called *fractional polymorphisms* are necessary. We refer to the full version of the paper [18] for further background on this topic.

2.4 Algebraic Dichotomy Conjecture

The condition for tractability of CSPs was first conjectured by Bulatov et al. [6], and a number of equivalent formulations was later given in [2,20,24]. We will use the formulation by Siggers [24]; it will be important for our purposes that Siggers polymorphisms have a fixed arity six and so for example on a fixed finite domain D there is only a finite number of them.

Definition 8. *An operation* $s\colon D^6 \to D$ *is called a Siggers operation on* D *if it is idempotent (i.e.* $s(x,x,x,x,x,x) = x$ *for all* $x \in D$*), and also* $s(x,x,x,x,y,y) = s(x,y,x,y,x,x)$ *and* $s(y,y,x,x,x,x) = s(x,x,y,x,y,x)$ *for all* $x, y \in D$*.*

The conjecture is usually stated for *core* languages. To reduce the number of definitions, we will give an alternative formulation that avoids cores. For a language Γ on D and a domain $D' \subseteq D$ let $\Gamma[D']$ be the language obtained from Γ by restricting each function to the domain D'.

Definition 9. *Tuple* (g,s) *will be called a* Siggers pair *on a domain* D *if* g *is a unary operation on* D *satisfying* $g \circ g = g$ *and* s *is a Siggers operation on* $g(D) \subseteq D$*. We say that a crisp language* Γ *on domain* D *admits* (g,s) *if* g *is a unary polymorphism of* Γ *and* s *is a 6-ary polymorphism of* $\Gamma[g(D)]$*.*

Theorem 1 ([24]). *A crisp constraint language* Γ *that does not admit a Siggers pair is NP-Hard.*

Conjecture 1 (A version of the Algebraic Dichotomy Conjecture). If a crisp language Γ admits a Siggers pair, then CSP(Γ) is tractable.

There has been remarkable progress on this conjecture. It has been verified for domains of size 2 [23] and 3 [4], or for languages containing all unary relations on D [5]. It has also been shown that it is equivalent to its restriction for directed graphs (that is when Γ contains a single binary relation ϱ) [7]. Further, the conjecture holds if ϱ corresponds to a directed graph with no sources and sinks [3]. Nevertheless, in the general case the conjecture remains open.

Definition 10. *A crisp language* Γ *is called a* BJK *language if it satisfies one of the following:*

- *CSP(Γ) is tractable*
- *Γ does not admit a Siggers pair.*

Conjecture 2 (Another version of the Algebraic Dichotomy Conjecture). Every crisp language Γ is a BJK language.

2.5 Hybrid (V)CSP Setting

Definition 11. *Let us call a family* $\mathcal{H}$ *of relational structures with a common signature a structural restriction. If all the relations in* $\mathcal{H}$ *are unary, we call* $\mathcal{H}$ all-unary.

Definition 12 (Hybrid CSP). *Let D be a finite domain, $\mathbf{\Gamma}$ a constraint language over D, and $\mathcal{H}$ a structural restriction of the same signature as $\mathbf{\Gamma}$. We define* $\mathrm{CSP}_{\mathcal{H}}(\mathbf{\Gamma})$ *as the following decision problem: given a relational structure $\mathbf{R} \in \mathcal{H}$ as input, decide whether $\mathbf{R} \to \mathbf{\Gamma}$.*

Definition 13 (Hybrid VCSP). *Let D be a finite domain, $\mathbf{\Gamma} = (D, f_1, \dots, f_k)$ a valued constraint language over D, and $\mathcal{H}$ a structural restriction of the same signature as $\mathbf{\Gamma}$. We define* $\mathrm{VCSP}_{\mathcal{H}}(\mathbf{\Gamma})$ *as the class of instances of the following form.*

An instance is a function from D^V to $\overline{\mathbb{Q}}$ given by

$$f_{\mathcal{I}}(h) = \sum_{(f,\mathbf{v}) \in T} w(f,\mathbf{v}) f(h(\mathbf{v})), \tag{3}$$

where V is a finite set of variables, $w(f,\mathbf{v})$ are positive numbers and T is a finite set of constraints determined by some relational structure $\mathbf{R} = (V, r_1, \dots, r_k) \in \mathcal{H}$ as follows: $T = \{(f_i, \mathbf{v}) \mid i \in [k], \mathbf{v} \in r_i\}$. The goal is to find an assignment *(or* labeling*) $h \in D^V$ that minimizes $f_{\mathcal{I}}$.*

Definition 14. *A structural restriction $\mathcal{H}$ is called* effective *for a class of (valued) languages $\mathcal{C}$ if there is a language $\mathbf{\Gamma}$ with $\Gamma \in \mathcal{C}$, of the same signature as $\mathcal{H}$, such that* (V)CSP(Γ) *is NP-Hard, whereas* (V)$\mathrm{CSP}_{\mathcal{H}}(\mathbf{\Gamma})$ *is tractable.*

$\mathcal{H}$ is called ineffective *for $\mathcal{C}$ if for every $\mathbf{\Gamma}$ with $\Gamma \in \mathcal{C}$, of the same signature as $\mathcal{H}$,* (V)CSP(Γ) *and* (V)$\mathrm{CSP}_{\mathcal{H}}(\mathbf{\Gamma})$ *are either both tractable or both NP-hard.*

Note, some structural restrictions could potentially be neither effective nor ineffective for a given $\mathcal{C}$ (since there exist intermediate complexity classes between NP-hard and tractable problems).

Example 1. Let us give some examples of effective restrictions for the class $\mathcal{C}$ of all crisp languages.

Let $\mathcal{H}$ be the set of k-colorable graphs for $k > 2$. Note that k-colorable graphs are exactly those that map homomorphically to the complete graph K_k. Therefore for the language $\Gamma = \{\neq_D\}$ on domain D with $|D| > 2$, we get that $\mathrm{CSP}_{\mathcal{H}}(\mathbf{\Gamma})$ is tractable (with a constant time algorithm that outputs YES), whereas CSP(Γ) is NP-Hard.

Similarly, also restricting to the class of planar graphs or perfect graphs is effective, since planar graphs are 4-colorable [1], and for perfect graphs the GRAPH k-COLOURING problem is known to be solvable in polynomial time [13].

3 Our Results

Most of our results will apply to structural restrictions $\mathcal{H}$ that are *up-closed.*

Definition 15. *A family of relational structures $\mathcal{H}$ is called* closed under inverse homomorphisms *(or* up-closed *for short) if whenever $\mathbf{R}' \to \mathbf{R}$ and $\mathbf{R} \in \mathcal{H}$, then also $\mathbf{R}' \in \mathcal{H}$.*

As examples of up-closed relational structures, let us mention directed acyclic graphs or k-colorable graphs. The proofs are straightforward. On the other hand, many natural graph classes do not possess this property, e.g. planar graphs and perfect graphs.

We introduce a notion of a chromatic number of relational structures that generalizes the usual chromatic number of graphs.

Definition 16. *Let* $\mathbf{R} = (V, r_1, \ldots, r_k)$ *be a relational structure. A coloring of* $\mathbf{R}$, *that is a mapping* $c\colon V \to [m]$, *is* improper *if there is a color* $j \in [m]$ *such that for each* $i \in [k]$, *the relation* r_i *contains a monochromatic tuple of the color* j. *A coloring that is not improper is called* proper.

We define the chromatic number $\chi(\mathbf{R})$ *of* $\mathbf{R}$ *to be the smallest number of colors that can yield a proper coloring of* $\mathbf{R}$. *(If no proper coloring exists, we set* $\chi(\mathbf{R}) = \infty$; *this will happen if e.g.* $\mathbf{R}$ *contains only one unary relation). Also, we define the* chromatic number $\chi(\mathcal{H})$ *of a structural restriction as*

$$\chi(\mathcal{H}) = \sup\{\chi(\mathbf{R}) : \mathbf{R} \in \mathcal{H}\}.$$

Theorem 2. *A structural restriction* $\mathcal{H}$ *with* $\chi(\mathcal{H}) < \infty$ *that is not all-unary is effective for the class of BJK languages.*

Theorem 3. *An up-closed structural restriction* $\mathcal{H}$ *with* $\chi(\mathcal{H}) = \infty$ *is ineffective for the class of BJK languages.*

In particular, Theorem 3 means that the Algebraic Dichotomy Conjecture would imply that up-closed structural restrictions $\mathcal{H}$ with $\chi(\mathcal{H}) = \infty$ are ineffective for the class of all CSP languages. Next, we state our results for valued languages.

Definition 17. *A valued language is called* conservative *if it contains all unary* $\{0, 1\}$*-valued cost functions.*

Definition 18. *We say that a relational structure* $\mathcal{H}$ does not restrict unaries *if for each* $\mathbf{R} \in \mathcal{H}$ *of the form* $\mathbf{R} = (V, r_1, \ldots, r_{i-1}, r_i, r_{i+1}, \ldots, r_k)$ *with* $\mathrm{ar}(r_i) = 1$ *and for each unary relation* $r'_i \subseteq V$, *we have* $\mathbf{R}' \in \mathcal{H}$, *where* $\mathbf{R}' = (V, r_1, \ldots, r_{i-1}, r'_i, r_{i+1}, \ldots, r_k)$.

Theorem 4. *An up-closed structural restriction* $\mathcal{H}$ *with* $\chi(\mathcal{H}) = \infty$ *that does not restrict unaries is ineffective for the class of conservative valued languages.*

Below we list three implications of our theorems. All missing proofs can be found in the full version of the paper [18].

3.1 Implications of Theorems 2, 3 and 4

Ordered CSP. One natural structural restriction to fixed template CSP is to introduce ordering of variables and request the constraints to respect the ordering.

Definition 19. *We call a relational structure* $(V, r_1, \ldots, r_k)$ ordered *if, after some identification of* V *with* $[n]$ *for* $n = |V|$*, whenever* $(v_1, \ldots, v_{\mathrm{ar}(r_j)}) \in r_j$ *for some* $j = 1, \ldots, k$*, then* $v_1 < \cdots < v_{\mathrm{ar}(r_j)}$.

Theorem 5. *Let* $\mathcal{H}$ *be the set of all ordered relational structures of some fixed signature. Such structural restriction* $\mathcal{H}$ *is ineffective for BJK languages and for conservative valued languages.*

This has an interesting consequence for graph homomorphism problems.

Corollary 1. *For the class of directed acyclic graphs* $\mathcal{H}$*, algebraic dichotomy conjecture implies that for every language* $\mathbf{\Gamma} = (D, \varrho)$ *with a binary relation* ϱ*,* $\mathrm{CSP}(\Gamma)$ *is tractable if and only if* $\mathrm{CSP}_{\mathcal{H}}(\mathbf{\Gamma})$ *is tractable.*

Minor-Closed Families of Graphs. It is known that a minor-closed family of undirected graphs has either bounded chromatic number or contains all graphs (see [21, Lemma 2]). Using this result and Ramsey's Theorem, we can show the following.

Theorem 6. *Let the structural restriction* $\mathcal{H}$ *be a family of directed graphs such that the underlying family of undirected graphs is minor-closed. Then* $\mathcal{H}$ *is effective for BJK languages if and only if* $\chi(\mathcal{H}) < \infty$.

Maximum Independent Set. Although Theorem 4 is formulated for conservative languages, it also gives implications for some optimization problems corresponding to non-conservative languages. Namely, the following can be shown.

Theorem 7. *Let* $\mathcal{G}$ *be a family of undirected graphs with* $\chi(\mathcal{G}) = \infty$ *that is closed under inverse homomorphisms (i.e. if* G, G' *are undirected graphs such that* $G \in \mathcal{G}$ *and* G' *maps homomorphically to* G *then* $G' \in \mathcal{G}$*). Then the* MAX WEIGHT INDEPENDENT SET *problem (with positive node weights) is NP-hard even when restricted to graphs in* $\mathcal{G}$.

Acknowledgements. We thank Andrei Krokhin for helpful comments on the manuscript. This work was supported by the European Research Council under the European Unions Seventh Framework Programme (FP7/2007–2013)/ERC grant agreement no 616160.

References

1. Appel, K., Haken, W.: Every planar map is four colorable. Part i: discharging. Illinois J. Math. **21**(3), 429–490 (1977)
2. Barto, L., Kozik. M.: New conditions for Taylor varieties and CSP. In: Proceedings of the 25th Annual IEEE Symposium on Logic in Computer Science, LICS 2010, 11–14 July 2010, Edinburgh, UK, pp. 100–109 (2010)
3. Barto, L., Kozik, M., Niven, T.: The CSP dichotomy holds for digraphs with no sources and no sinks (a positive answer to a conjecture of Bang-Jensen and Hell). SIAM J. Comput. **38**(5), 1782–1802 (2009)

4. Bulatov, A.: A dichotomy theorem for constraint satisfaction problems on a 3-element set. J. ACM **53**(1), 66–120 (2006)
5. Bulatov, A.: Complexity of conservative constraint satisfaction problems. ACM Trans. Comput. Logic, 12(4) (2011). Article 24
6. Bulatov, A., Krokhin, A., Jeavons, A.: Classifying the complexity of constraints using finite algebras. SIAM J. Comput. **34**(3), 720–742 (2005)
7. Bulín, J., Delić, D., Jackson, M., Niven, T.: On the reduction of the CSP dichotomy conjecture to digraphs. In: Schulte, C. (ed.) CP 2013. LNCS, vol. 8124, pp. 184–199. Springer, Heidelberg (2013)
8. Cook, S.A.: The complexity of theorem-proving procedures. In: Proceedings of the Third Annual ACM Symposium on Theory of Computing, STOC 1971, pp. 151–158. ACM, New York (1971)
9. Cooper, M.C., Živný, S.: Hybrid tractability of valued constraint problems. Artif. Intell. **175**(9–10), 1555–1569 (2011)
10. Feder, T., Vardi, M.Y.: The computational structure of monotone monadic SNP and constraint satisfaction: a study through datalog and group theory. SIAM J. Comput. **28**(1), 57–104 (1998)
11. Geiger, D.: Closed systems of functions and predicates. Pacific J. Math. **27**(1), 95–100 (1968)
12. Grohe, M.: The complexity of homomorphism and constraint satisfaction problems seen from the other side. J. ACM **54**(1), 1:1–1:24 (2007)
13. Grötschel, M., Lovász, L., Schrijver, A.: Geometric Algorithms and Combinatorial Optimization. Springer-Verlag, New York (1988)
14. Hell, P., Nešetřil, J.: On the complexity of h-coloring. J. Comb. Theory, Series B **48**(1), 92–110 (1990)
15. Jeavons, P., Krokhin, A., Živný, S.: The complexity of valued constraint satisfaction. Bull. EATCS **113**, 21–55 (2014)
16. Jeavons, P.: On the algebraic structure of combinatorial problems. Theor. Comput. Sci. **200**(1–2), 185–204 (1998)
17. Jégou, P.: Decomposition of domains based on the micro-structure of finite constraint-satisfaction problems. In: AAAI, pp. 731–736 (1993)
18. Kolmogorov, V., Rolínek, M., Takhanov, R.: Effectiveness of structural restrictions for hybrid CSPs (2015). arXiv1504.07067
19. Kuznetsov, A.V.: Algebra of logic and their generalizations. In: Mathematics in USSR for 40 years, vol. 1, pp. 105–115. Fizmatgiz Moscow (1959)
20. Maróti, M., McKenzie, R.: Existence theorems for weakly symmetric operations. Algebra Universalis **59**(3–4), 463–489 (2008)
21. Nešetřil, J., Ossona de Mendez, P.: Colorings and homomorphisms of minor closed classes. In: Aronov, B., Basu, S., Pach, J., Sharir, M. (eds.) Discrete and Computational Geometry. Algorithms and Combinatorics, vol. 25, pp. 651–664. Springer, Heidelberg (2003)
22. Post, E.L.: On The Two-Valued Iterative Systems of Mathematical Logic. Princeton University Press, Princeton (1941)
23. Schaefer, T.J.: The complexity of satisfiability problems. In: Proceedings of the 10th Annual ACM Symposium on Theory of Computing (STOC), pp. 216–226 (1978)
24. Siggers, M.H.: A strong Mal'cev condition for locally finite varieties omitting the unary type. Algebra Universalis **64**(1–2), 15–20 (2010)

25. Swarts, J.: The complexity of digraph homomorphisms: Local tournaments, injective homomorphisms and polymorphisms. Ph. D. thesis, University of Victoria, Canada (2008)
26. Takhanov, R.S.: A dichotomy theorem for the general minimum cost homomorphism problem. In: Proceedings of the 27th International Symposium on Theoretical Aspects of Computer Science (STACS), pp. 657–668 (2010)

Polynomial-Time Isomorphism Test of Groups that are Tame Extensions (Extended Abstract)

Joshua A. Grochow[1] and Youming Qiao[2](✉)

[1] Santa Fe Institute, Santa Fe, NM, USA
jgrochow@santafe.edu
[2] Centre for Quantum Computation and Intelligent Systems, University of Technology, Sydney, Australia
jimmyqiao86@gmail.com

Abstract. We give new polynomial-time algorithms for testing isomorphism of a class of groups given by multiplication tables (GPI). Two results (Cannon & Holt, J. Symb. Comput. 2003; Babai, Codenotti & Qiao, ICALP 2012) imply that GPI reduces to the following: given groups G, H with characteristic subgroups of the same type and isomorphic to $\mathbb{Z}_p^d$, and given the coset of isomorphisms $\mathrm{Iso}(G/\mathbb{Z}_p^d, H/\mathbb{Z}_p^d)$, compute $\mathrm{Iso}(G, H)$ in time $\mathrm{poly}(|G|)$. Babai & Qiao (STACS 2012) solved this problem when a Sylow p-subgroup of $G/\mathbb{Z}_p^d$ is trivial. In this paper, we solve the preceding problem in the so-called "tame" case, i.e., when a Sylow p-subgroup of $G/\mathbb{Z}_p^d$ is cyclic, dihedral, semi-dihedral, or generalized quaternion. These cases correspond exactly to the group algebra $\overline{\mathbb{F}}_p[G/\mathbb{Z}_p^d]$ being of tame type, as in the celebrated tame-wild dichotomy in representation theory. We then solve new cases of GPI in polynomial time.

Our result relies crucially on the divide-and-conquer strategy proposed earlier by the authors (CCC 2014), which splits GPI into two problems, one on group actions (representations), and one on group cohomology. Based on this strategy, we combine permutation group and representation algorithms with new mathematical results, including bounds on the number of indecomposable representations of groups in the tame case, and on the size of their cohomology groups.

Finally, we note that when a group extension is *not* tame, the preceding bounds do not hold. This suggests a precise sense in which the tame-wild dichotomy from representation theory may also be a key barrier to cross to put GPI into P.

1 Introduction

The group isomorphism problem (GPI) is to decide whether two finite groups, given by their multiplication tables, are isomorphic. It is one of the few natural problems not known to be in P, and unlikely to be NP-complete, as it reduces to Graph Isomorphism (GRAPHI; see, e.g., [27]). In addition to being intrinsically interesting, resolving the exact complexity of GPI is thus a tantalizing question. Further, there is a surprising connection between GPI and the Geometric

K. Elbassioni and K. Makino (Eds.): ISAAC 2015, LNCS 9472, pp. 578–589, 2015.
DOI: 10.1007/978-3-662-48971-0_49

Complexity Theory program (see, e. g., [31] and references therein): Techniques from GPI were used to solve cases of LIE ALGEBRA ISOMORPHISM that have applications in Geometric Complexity Theory [21]. In a survey article [2] in 1995, after enumerating several isomorphism-type problems including GRAPHI and GPI, Babai expressed the belief that GPI might be the only one expected to be in P.[1] Despite its connection with GRAPHI, P seems an achievable goal for GPI, as there are many reasons GPI seems easier than GRAPHI (see, e. g., the introduction to [22] for an overview of these reasons).

As a group of order n can be generated by $\lceil \log n \rceil$ elements, GPI is solvable in time $n^{\log n + O(1)}$ [17,30].[2] The only improvement for the general case was Rosenbaum's recent $n^{0.5 \log n + O(1)}$-time algorithm [34]. However, there have been more significant improvements for special group classes, representing a more structural approach to the problem. Isomorphism of Abelian groups was recognized as easy quite early [35,38], leading to an $O(n)$-time algorithm [26]. Since 2009, there have been several non-trivial polynomial-time algorithms for much more complicated group classes: groups with no Abelian normal subgroups [3,4], groups with Abelian Sylow towers [5,28,32], and p-groups of genus 2 [10,29].

Partly motivated to distill a common pattern from the three then-recent major polynomial-time algorithms [4,5,29], the authors proposed [22] a divide-and-conquer strategy for GPI based on the extension theory of groups. This strategy is crucial for Theorem 1. Before getting to the details of this strategy, let us first examine an approach for GPI that motivates the problem that we study.

In 2003, Cannon and Holt [12] suggested the following outline for GPI. First, they introduce a natural sequence of characteristic subgroups: $G = G_0 \rhd G_1 \rhd \cdots \rhd G_\ell = \mathrm{id}$, where $G_1 = \mathrm{Rad}(G)$ is the solvable radical of G—the largest solvable normal subgroup—and G_i/G_{i+1} is elementary Abelian for all $1 \le i \le \ell - 1$. This filtration is easily computed, and for each factor we know how to test isomorphism: $G/\mathrm{Rad}(G)$ has no Abelian normal subgroups, so is handled by [4].

Given two groups G and H, after computing these filtrations of G and H, the strategy is to first test isomorphisms of the corresponding factors, which is necessary for G and H to be isomorphic. Then, starting from $G_0/G_1 (= G/\mathrm{Rad}(G))$, proceed inductively along this filtration. Note that for G_0/G_1, not only is isomorphism decidable in polynomial time, but a generating set for the coset of isomorphisms $\mathrm{Iso}(G_0/G_1, H_0/H_1)$ can be found efficiently [4]. After this initial step, a positive solution to the following problem would show that GPI $\in$ P:

Problem 1. Given two groups G, H with characteristic elementary Abelian subgroups A and B, resp., compute $\mathrm{Iso}(G, H)$ from $\mathrm{Iso}(G/A, H/B)$ in time $\mathrm{poly}(|G|)$.

In fact, by developing a heuristic algorithm for Problem 1 in [12, Sect. 5], Cannon & Holt obtained a practical algorithm for GPI, but their algorithm uses

[1] The exact quotation from Babai's 1995 survey [2] is: "None of the problems mentioned in this section, with the possible exception of isomorphism of groups given by a Cayley table, is expected to have polynomial time solution."

[2] Miller [30] attributes this algorithm to Tarjan.

a backtrack search that does not have good worst-case guarantees.[3] Still, this is a very natural approach, and the algorithm in [4] solves the first step—testing isomorphisms for $G/\operatorname{Rad}(G)$—to this approach in the Cayley table model. In [4] they thus use this to solve GPI in the case that $\operatorname{Rad}(G)$ is trivial.

To the best of our knowledge, the only previous result about Problem 1 with a worst-case analysis in the Cayley table model is by Babai and the second author [5], who solved the case when $A \cong \mathbb{Z}_p^k$ and the Sylow p-subgroup[4] of G/A is trivial; that is, when $p \nmid |G/A|$. This was the key to the main result in [5], namely, a polynomial-time isomorphism test for groups with Abelian Sylow towers (see definition below, just before Corollary 1).

In this paper, we solve Problem 1 under certain conditions on the Sylow subgroups of G, more general than the aforementioned one for [5]. Furthermore, these conditions are very natural, as they are aligned with the celebrated tame-wild dichotomy in the representation theory of associative algebras [8,15].

For an algebra L over an infinite field, classifying its indecomposable representations up to isomorphism is a fundamental problem. Roughly speaking, the nicest possibility is when there are only finitely many indecomposables, in which case L is of *finite type*. Beyond this, some algebras have the property that their indecomposables come in finitely many one-parameter families in each fixed dimension d,[5] possibly with finitely many exceptions. While this can be much more complicated than finite type, it is still "classifiable;" such algebras are said to be of *tame type*.[6] Finally, some algebras L have the peculiar property that any representation of *any* algebra can be "embedded as" (or "simulated by") a representation of L; such algebras are called *wild*. Drozd's celebrated dichotomy theorem [16] says that every algebra over an infinite field is either tame or wild.

In the case of groups, there is an explicit description of the three cases (see [8, Theorem 4.4.4]): let p be the characteristic of the field $\mathbb{F}$. $\mathbb{F}G$ is of finite type iff $p = 0$, or $p > 0$ and the Sylow p-subgroup of G is cyclic. G is of tame type, but not finite, iff $p = 2$ and the Sylow 2-subgroup of G is dihedral, semi-dihedral, or generalized quaternion (see Sect. 2 for definitions). All other cases are wild.

[3] Due to different goals and settings, it is natural for Cannon & Holt and us to adopt different algorithmic ideas. That is, Cannon & Holt work with more succinct representations of groups, and their goal is to obtain algorithms fast in practice. We work with the "redundant" Cayley tables, and aim for worst-case analysis.

[4] Though Sylow p-subgroups of a group are not unique, all of them are isomorphic; hence we may refer to "the" Sylow p-subgroup.

[5] For readers not familiar with this concept, here is an example to illustrate intuitively what one-parameter families mean. For an algebraically closed field $\mathbb{F}$, the Jordan blocks form a one-parameter family with the eigenvalue $\lambda \in \mathbb{F}$ as the parameter. The indecomposable d-dimensional representations of $\mathbb{F}[x]$ are given exactly the $d \times d$ Jordan blocks. Defining tame and wild rigorously requires terminology that is unnecessary for this article; we refer to [8, Sect. 4.4] for a comprehensive introduction.

[6] Finite type can be considered as a special case of tame type, namely when the number of one-parameter families is 0. In the literature, some authors take the definition of "tame type" to explicitly exclude finite type. We do not adopt that convention here.

Suppose a group G has a normal subgroup A isomorphic to $\mathbb{Z}_p^d$, and let $Q = G/A$. G is called a *tame extension* of A by Q, if $\overline{\mathbb{F}}_p Q$ is of tame type.[7] We solve Problem 1 exactly for groups of this form. Note that the Sylow p-subgroup being cyclic already generalizes the condition for [5].

Theorem 1. *Suppose G, H come from the class of groups that have characteristic subgroups of the same type and isomorphic to the elementary Abelian subgroup $\mathbb{Z}_p^d$. There is a polynomial-time algorithm to compute the coset of isomorphisms* $\mathrm{Iso}(G,H)$ *from the coset of isomorphisms* $\mathrm{Iso}(G/\mathbb{Z}_p^d, H/\mathbb{Z}_p^d)$*, if G is a tame extension of $\mathbb{Z}_p^d$, namely if the Sylow p-subgroups of $G/\mathbb{Z}_p^d$ are cyclic, dihedral, semi-dihedral, or generalized quaternion.*

The condition on $G/\mathbb{Z}_p^d$ is satisfied by several well-known group classes:

- Groups with dihedral Sylow 2-subgroups are classified [7,20]: Let $\mathrm{O}(G)$ be the maximal normal odd-order subgroup. If G has a dihedral Sylow subgroup, $G/\mathrm{O}(G)$ must be isomorphic to one of: (i) a subgroup of $\mathrm{P\Gamma L}_2(\mathbb{F}_q)$ containing $\mathrm{PSL}_2(\mathbb{F}_q)$; (ii) the alternating group A_7; (iii) a Sylow 2-subgroup of G.
- The Sylow 2-subgroup of $\mathrm{SL}_2(\mathbb{F}_q)$ is generalized quaternion when q is odd [19, p. 42] (or see [14, Corollary 4.12]).
- If D is a division ring, then any Sylow subgroup of a finite subgroup of $D\backslash\{0\}$ is cyclic or generalized quaternion (see [14, Corollary 4.10]).
- The Sylow 2-subgroups of the following groups are semi-dihedral: $\mathrm{PSL}_3(\mathbb{F}_q)$ for $q \equiv 3 \pmod 4$, $\mathrm{PSU}_3(\mathbb{F}_q)$ for $q \equiv 1 \pmod 4$, the Mathieu group M_{11}, and $\mathrm{GL}_2(\mathbb{F}_q)$ for $q \equiv 3 \pmod 4$ (see, e.g., [1]).

Theorem 1 allows us to solve GPI in P for a class of groups that we now describe. Following [5], we say that a group G has a *Sylow tower* if there is a normal series $\mathrm{id} = G_\ell \lhd \cdots \lhd G_1 \lhd G_0 = G$ where each G_i/G_{i+1} is isomorphic to a Sylow subgroup of G. We say that G has an *elementary Abelian Sylow tower* if furthermore all its Sylow subgroups are elementary Abelian. The proof of the following corollary is straightforward.

Corollary 1. *The coset of isomorphisms between two groups G, H can be computed in polynomial time when (1)* $\mathrm{Rad}(G)$ *has an elementary Abelian Sylow tower, and (2) for any prime p dividing* $|\mathrm{Rad}(G)|$*, the Sylow p-subgroup of* $G/\mathrm{Rad}(G)$ *is cyclic, dihedral, semi-dihedral, or generalized quaternion.*

We now compare our result with the previous one [5]. Firstly, a critical difference is that in our setting we need to deal with both actions and cohomology classes (see Sect. 3). In the setting of [5], the Schur–Zassenhaus Theorem implies that the cohomology classes are always trivial, so this part does not appear in [5] at all. Secondly, to deal with actions (Problem 3), though we follow the algorithmic framework of [5], for the supporting algorithmic subroutines, we need to use some sophisticated algorithms in computational algebra (see Sect. 2), while in [5]

[7] $\overline{\mathbb{F}}_p$ is the algebraic closure of $\mathbb{F}_p$. Though it is not standard to apply "tame" to extensions, this is justified by the main mathematical results of this paper.

the corresponding subroutines are rather straightforward. Finally, we bound the running time of our algorithms by proving size bounds on representations and on group cohomology in the tame case, using an explicit description of representations from the literature, and applying previously known results on group cohomology. This was not needed in [5].

More broadly, to achieve Theorem 1, for the first time in the worst-case analysis of GpI, we step into the regime of modular representation theory—that is, when the characteristic of the underlying field divides the order of the group. This theory is much less well-understood than ordinary representation theory. As the reader may see later, to solve Problem 1 in general seems to require certain deep use of this theory. We hope this article serves as a first step in this direction.

Organization. In Sect. 2 we present some preliminaries. In Sect. 3 we show how the splitting strategy of [22] applies in this case, and in Sect. 4 we give an overview of the proofs; detailed proofs can be found in [23]. Finally, in Sect. 5 we discuss the general relationship between GpI and the tame-wild dichotomy.

2 Preliminaries

Notations and Definitions. For a prime p, $\mathbb{F}_p$ denotes the field of size p. The characteristic of a field $\mathbb{F}$ is denoted $\mathrm{char}(\mathbb{F})$. $\mathrm{M}(n,p)$ is the set of $n \times n$ matrices over $\mathbb{F}_p$, and $\mathrm{GL}(n,p)$ is the group of $n \times n$ invertible matrices of $\mathbb{F}_p$. For $n \in \mathbb{N}$, $[n] := \{1, \dots, n\}$. $\mathrm{Sym}(\Omega)$ denotes the symmetric group over a set Ω; when $\Omega = [n]$ we write S_n. A permutation group over Ω is a subgroup of $\mathrm{Sym}(\Omega)$.

$\mathbb{Z}_k$ denotes the cyclic group of order k. A group is elementary Abelian if it is isomorphic to $\mathbb{Z}_p^d$ for some prime p and some integer d. The dihedral groups (of order a power of 2) are $\mathrm{D}_{2^m} = \langle x, y \mid x^2 = y^{2^m} = 1, yx = xy^{-1} \rangle$. The semi-dihedral or quasi-dihedral groups are $\mathrm{SD}_{2^m} = \langle x, y \mid x^2 = y^{2^m} = 1, yx = xy^{2^{m-1}-1} \rangle$. The (generalized) quaternion groups are $\mathrm{GQ}_{2^m} = \langle x, y \mid x^2 = y^{2^{m-1}}, yx = xy^{-1} \rangle$. D_{2^m}, SD_{2^m}, and GQ_{2^m} are of order 2^{m+1}; D_{2^1} is the Klein four group. Since D_{2^m}, SD_{2^m}, or GQ_{2^m} are generated by 2 elements, we can test in P whether a given group is D_{2^m}, SD_{2^m}, or GQ_{2^m}.

General Group Theory. A p-group is a group of order p^k for some $k > 0$. A Sylow p-subgroup of G is a p-subgroup of G of maximal order; by the Sylow theorems, this order is the maximal order of p dividing $|G|$, and all Sylow p-subgroups are conjugate in G. Given the Cayley table of a group, a Sylow p-subgroup can be found in polynomial time.

A subgroup N of G is *characteristic* if N is sent to itself by every automorphism of G. A *characteristic subgroup functor* is a function $\mathcal{S}$ from finite groups to finite groups such that (1) $\mathcal{S}(G) \leq G$ for all G, and (2) any isomorphism $\varphi \colon G_1 \to G_2$ restricts to an isomorphism $\varphi|_{\mathcal{S}(G_1)} \colon \mathcal{S}(G_1) \to \mathcal{S}(G_2)$. In particular, it follows that $\mathcal{S}(G)$ is always characteristic in G. Examples of characteristic subgroup functors include most "natural" characteristic subgroups such as the center, the derived subgroup, and the terms of the derived, lower central, and upper central series. A characteristic subgroup functor is Abelian (resp. elementary Abelian), if $\mathcal{S}(G)$ is Abelian (resp., elementary Abelian) for all G. **Convention:** In this paper, whenever we say "characteristic subgroup" we mean the image of an implied characteristic subgroup functor.

Indecomposable Modules. As representations of a group Q over a field $\mathbb{F}$ are the same as modules over the group algebra $\mathbb{F}Q$, we shall use the terms module and representation interchangeably. For two representations θ and η, we use $\theta \cong \eta$ to denote that they are equivalent. Let M be a module of an algebra L. M is *indecomposable* if it cannot be written as a direct sum of two submodules. The decomposition of M into a direct sum of indecomposables is essentially unique:

Theorem 2 (Krull–Schmidt (see, e.g., [8, Theorem 1.4.6]**)).** *Let ϕ and ψ be two linear representations of a group Q. Suppose $\phi = \iota_1^{d_1} \oplus \cdots \oplus \iota_\ell^{d_\ell}$ and $\psi = \iota_1^{e_1} \oplus \cdots \oplus \iota_\ell^{e_\ell}$, where ι_i's are indecomposable and pairwise non-isomorphic, and all $d_i, e_i \geq 0$. Then $\phi \cong \psi$ iff $d_i = e_i$ for every $i \in [\ell]$.*

2-cohomology Classes. Let Q be a group, and A an Abelian group. An action θ of Q on A is a group homomorphism $Q \to \mathrm{Aut}(A)$. A *2-cocycle* w.r.t. the action θ is a function $f : Q \times Q \to A$ satisfying the 2-cocycle identity $f(p,q) + f(pq,r) = \theta_p(f(q,r)) + f(p,qr)$. The set of all 2-cocycles is an Abelian group under pointwise addition, denoted $Z^2(Q,A,\theta)$. Given a function $u : Q \to A$, the function $b_u(q,q') = u(q) + \theta_q(u(q')) - u(qq')$ is a *2-coboundary* $b_u : Q \times Q \to A$. The set of 2-coboundaries is a subgroup of $Z^2(Q,A,\theta)$, denoted $B^2(Q,A,\theta)$. The quotient group $H^2(Q,A,\theta) := Z^2(Q,A,\theta)/B^2(Q,A,\theta)$ is the group of *2-cohomology classes.* For f and f' in $Z^2(Q,A,\theta)$, if $f - f' \in B^2(Q,A,\theta)$ (representing the same cohomology class), they are called *cohomologous*, denoted $f \simeq f'$.

Preliminaries for Algorithms. As customary in permutation group algorithms [36], a permutation group is represented in algorithms by a set of generators. The automorphism group of a group G is represented as a permutation group in $\mathrm{Sym}(G)$. A coset of a permutation group is represented by a single coset representative together with a set of generators for the subgroup. A representation of Q is given by listing the images of $q \in Q$ explicitly. Two representations θ and η are *equal*, denoted $\theta = \eta$, if $\theta(q) = \eta(q)$ for every $q \in Q$; compare with $\theta \cong \eta$. A 2-cohomology class is represented by a 2-cocycle f, which in turn can be viewed as a matrix over $\mathbb{Z}_p$ of size $d \times |Q|^2$ when $A \cong \mathbb{Z}_p^d$.

Proposition 1 ([22])**.** *Given two 2-cocycles f and f' w.r.t. the action $\theta : Q \to A$ ($A = \mathbb{Z}_p^d$), whether $f \simeq f'$ can be decided in time* $\mathrm{poly}(|Q|, d, \log p)$.

Theorem 3 (Module isomorphism [9,13,25]**).** *Given two tuples of matrices $(A_1, \ldots, A_n)$, $(B_1, \ldots, B_n)$, $A_i, B_j \in M(d,p)$, there exists a deterministic* $\mathrm{poly}(d,n,\log p)$*-time algorithm that finds $C \in \mathrm{GL}(d,p)$ s.t. for every $i \in [n]$, $CA_i = B_iC$, if such C exists.*

Theorem 4 (Finding units in a matrix algebra [11]**).** *Given a linear basis of a matrix algebra L in* $\mathrm{M}(d,p)$*, a generating set of the unit group of L can be computed deterministically in time* $\mathrm{poly}(d,p)$.

Theorem 5 (Decomposing into indecomposables [13]**).** *Given a module M over an algebra L over a finite field $\mathbb{F}$, a direct sum decomposition of M can be computed in time polynomial in the input size and* $\mathrm{char}(\mathbb{F})$.

Theorem 6 (Parametrized setwise transporter problem [5]**).** *Given a set of generators of $P \leq \mathrm{S}_t$, and $S, T \subseteq [t]$ with $|S| = |T| = k$, $P_{S\to T} := \{\sigma \in P \mid S^\sigma = T\}$ can be computed in time* $\mathrm{poly}(t, 2^k)$.

3 The divide and conquer strategy for Problem 1

Now we briefly recall the divide and conquer strategy from [22], and how it applies to the particular case of Problem 1. Problem 1 requires us to compute the isomorphisms of G, H from the isomorphisms of $G/\mathbb{Z}_p^d, H/\mathbb{Z}_p^d$. It is then natural to examine how the quotient group $G/\mathbb{Z}_p^d$ and the characteristic subgroup $\mathbb{Z}_p^d$ are related by G; this is the starting point for the strategy from [22].

Given a group G and an Abelian characteristic subgroup A of G, let $Q := G/A$; we denote this situation $A \hookrightarrow G \twoheadrightarrow Q$. The *extension data* of $A \hookrightarrow G \twoheadrightarrow Q$ consists of two functions: the (conjugation) *action* $\theta : Q \times A \to A$ defined by $(q, a) \to qaq^{-1}$, and the *2-cocycle* $f_s : Q \times Q \to A$, depending on a transversal or *section* $s : Q \to G$—for each coset $q \in G/A$ assign $s(q) \in q$—and defined by $f_s(p, q) := s(p)s(q)s(pq)^{-1}$. $\mathrm{Aut}(A) \times \mathrm{Aut}(Q)$ acts naturally on θ and f_s.

In Problem 1, we are given two groups G and H, and their respective characteristic subgroups A and B (recall our convention about characteristic subgroup *functors* from Sect. 2). Note that if $G \cong H$, then $A \cong B$ and $G/A \cong H/B$. We first test whether $A \cong B$; this is easy because they are Abelian. Recall that we are given $\mathrm{Iso}(G/A, H/B)$; if it is empty then $G \ncong H$. Therefore, at this point we have either determined that $G \ncong H$, or we have $A \cong B$ (identified as A), and $G/A \cong H/B$ (identified as Q). This is the divide step of the strategy.

But these conditions are not sufficient to conclude $G \cong H$ (e. g., compare D_{4k} and $\mathbb{Z}_2 \times D_{2k}$ with k odd, both as $\mathbb{Z}_2 \hookrightarrow G \twoheadrightarrow D_{2k}$), so we are yet to conquer. Since every element of G has a unique expression as $as(q)$ for $a \in A, q \in Q$, $\mathrm{Iso}(G, H)$ embeds as a subgroup of $\mathrm{Aut}(A) \times \mathrm{Aut}(Q)$. When $A \cong \mathbb{Z}_p^d$, we have $\mathrm{Aut}(A) \cong \mathrm{GL}(d, p)$; $\mathrm{Aut}(Q)$ is given to us as part of $\mathrm{Iso}(G/A, H/B)$. By [22, Lemma II.2], $\mathrm{Iso}(G, H)$ consists exactly of those $(\alpha, \beta) \in \mathrm{Aut}(A) \times \mathrm{Aut}(Q)$ that make the two extension data the same.[8] Following [22], we refer to the problem of computing the coset in $\mathrm{Aut}(A) \times \mathrm{Aut}(Q)$ consisting of elements sending one group to the other as Extension Data Pseudo-congruence (or EDPC):

Problem 2. Let $A \cong \mathbb{Z}_p^d$. Given $\mathrm{Aut}(Q)$ and the extension data (θ, f) and (η, g) of $A \hookrightarrow G \twoheadrightarrow Q$ and $A \hookrightarrow H \twoheadrightarrow Q$, respectively, compute $\{(\alpha, \beta) \in \mathrm{Aut}(A) \times \mathrm{Aut}(Q) : \theta^{(\alpha,\beta)} = \eta, \text{ and } f^{(\alpha,\beta)} \simeq g\}$.

At first sight, EDPC asks for (α, β) that sends θ to η and f to g, *simultaneously*. However, note that $f \in H^2(Q, A, \theta)$: To define the space in which f lives

[8] Note that the condition that A, B be *characteristic* subgroups is crucial; if A and B are merely normal subgroups, then this does not hold in general. See [22] for details.

relies on θ in the first place. On the other hand, θ has no dependence on f. Therefore, EDPC reduces to solving the following two problems, *in order*. (We shall refer to Problem 3 as ACTION COMPATIBILITY or ACTCOMP, and Problem 4 as COHOMOLOGY CLASS ISOMORPHISM or CCISO.)

Problem 3. Suppose we are given a group Q by its Cayley table, $\text{Aut}(Q)$ by a set of generators, and two linear representations $\theta, \eta : Q \to \text{GL}(d, p)$ by listing images of Q explicitly. Compute a set of generators for the coset $\{(\alpha, \beta) \in \text{GL}(d, p) \times \text{Aut}(Q) \mid \theta^{(\alpha,\beta)} = \eta\}$, in time $\text{poly}(|Q|, p^d)$.

Problem 4. Suppose we are given a group Q by its Cayley table, a representation $\theta \colon Q \to \text{GL}(d, p)$ by listing the images of Q explicitly, and two 2-cocycles $f, g \colon Q \times Q \to \mathbb{Z}_p^d$ in $Z^2(Q, \mathbb{Z}_p^d, \theta)$. Furthermore we are given generators for $\{(\alpha, \beta) \in \text{GL}(d, p) \times \text{Aut}(Q) \mid \theta^{(\alpha,\beta)} = \theta\}$. Compute generators for the coset $\{(\alpha, \beta) \in \text{GL}(d, p) \times \text{Aut}(Q) \mid f^{(\alpha,\beta)} \simeq g \text{ and } \theta^{(\alpha,\beta)} = \theta\}$, in time $\text{poly}(|Q|, p^d)$.

4 Overview of Algorithms for ActComp and CCIso

In this section we give an overview of the algorithms for ACTCOMP and CCISO when $\overline{\mathbb{F}}_pQ$ is tame, thereby proving Theorem 1. See [23] for detailed proofs.

The algorithm for ACTCOMP goes as follows. Given representations $\theta, \eta : Q \to \text{GL}(d, p)$, decompose them into a direct sum of indecomposables (Theorem 5), and group them by isomorphism types (Theorem 3): $\theta = \iota_1^{d_1} \oplus \iota_2^{d_2} \oplus \cdots \oplus \iota_\ell^{d_\ell}$, and $\eta = \iota_1^{e_1} \oplus \iota_2^{e_2} \oplus \cdots \oplus \iota_\ell^{e_\ell}$. (Some d_i's and/or e_j's may be 0.) By Theorem 2, $\theta \cong \eta$ iff $d_i = e_i$ for all $i \in [\ell]$. To take into account the effect of $\text{Aut}(Q)$, consider the induced action of $\text{Aut}(Q)$ on the indecomposables of $\mathbb{F}_pQ$. Firstly, compute the closure of $I = \{\iota_1, \ldots, \iota_\ell\}$ under $\text{Aut}(Q)$, denoted as $\text{Clo}(I)$. Viewing $\text{Aut}(Q)$ as a permutation group on the domain $\text{Clo}(I)$, we need to compute the coset in $\text{Aut}(Q)$ that sends those indecomposables in θ of multiplicity m, to those indecomposables in η of multiplicity m, for every $m \in [d]$. For each $m \in [d]$, this is a setwise transporter problem, so applying Theorem 6 sequentially gives an efficient algorithm—provided that we can upper bound the number of indecomposables of dimension d, and thereby $|\text{Clo}(I)|$, by $\text{poly}(|Q|, p^d)$. We prove that for the tame type this holds. This does not follow directly from the definition of the tame–wild dichotomy, since that requires the underlying field to be infinite, whereas we care about representations over a *finite* field and need an upper bound on the *number* of indecomposables. We are nonetheless able to prove the upper bound by analyzing the explicit description of the indecomposable families for tame group algebras in the literature. This may be viewed as the first main technical contribution of this work. On the other hand, for the wild type this upper bound fails badly [33]. Finally, by Theorems 4 and 3 we can compute, for each $\beta \in \text{Aut}(Q)$ that make θ and η isomorphic, the coset $\alpha \in \text{GL}(d, p)$ that make $\theta^{(\alpha,\beta)} = \eta$.

We then give an algorithm for CCISO that takes the coset of action compatibilities as its input. As for ACTCOMP, the idea is to view the group of action compatibilities as a permutation group on $H^2(Q, \mathbb{Z}_p^d, \theta)$. Then given two 2-cocycles (representing two 2-cohomology classes), the problem becomes a point-wise transporter problem, a classical problem in permutation group algorithms that is polynomial-time solvable [36]. For this algorithm to be efficient for our purpose, we need to upper bound $|H^2(Q, \mathbb{Z}_p^d, \theta)|$ as $\mathrm{poly}(|Q|, p^d)$ when $\overline{\mathbb{F}}_p Q$ is tame. Using some standard cohomological yoga combined with known but deep results on group cohomology [24], we show that, amazingly, this is true. This is the second main technical contribution of this work. This finishes the overview.

5 Discussion

Generally speaking (if somewhat glibly), there are two overarching reasons an instance of an isomorphism problem can be easy (not just group isomorphism): 1) there are very few possible isomorphisms to check, or 2) there aren't very many isomorphism classes and/or they have an explicit classification. Although this is a coarse caricature of reality,[9] we believe it provides a useful viewpoint. The results of [26,35,38] use the classification of Abelian groups (2); the results of [3,4] roughly fall under (1): The number of isomorphisms is only $n^{O(\log\log n)}$, and then they use dynamic programming, an algorithm for code equivalence, and results on finite simple groups to reduce this to polynomial time; the results of [5,28,32] fall under (2) in the strong sense that they rely on the fact that the number of irreducible representations of a group G in characteristic p that doesn't divide G is *finite*, and all other representations are direct sums of these; and the results of [29] use an essentially *finite* classification of type (2) to reduce to (1) (see [10]). In this paper, we show that when (2) holds—of which tameness is a general interpretation—isomorphism can be tested in P.

Because of the universal property of wildness, an explicit classification is believed to be impossible for wild problems. This does not rule out structural information, nor does it rule out efficient algorithms to decide when two points are equivalent under a wild equivalence relation [9,13,25]. However, the wild problems that arise in GpI are often "wilder than wild" [6](analogous to a problem being NP-hard but not in NP), and seem to pose a core difficulty for GpI.

The reasons (1) and (2)—or rather, their absence—also partially explain the widely held belief that nilpotent groups of class 2—those G for which G modulo its center is Abelian—are the hardest cases of group isomorphism, despite the lack of a formal reduction. Option (1) is ruled out, because even for p-groups of class 2 (nilpotent groups of class 2 and order a power of the prime p) in which every element is of order p, there are roughly $n^{O(\log n)}$ possible isomorphisms to check.[10] Option (2) is also ruled out, because the p-groups of class 2 form a wild classification problem [37], and in fact, one that is "strictly wilder" than classifying the representations of finite-dimensional algebras [6].

Furthermore, the only algorithms [10,18,29] with worst-case guarantees for GpI on nontrivial classes of p-groups also rely in a key way on tameness. Garzon

[9] For example, we recognize that this may not apply to certain algorithms for GraphI.

[10] This is essentially because $\mathrm{Aut}(\mathbb{Z}_p^k) \cong \mathrm{GL}(k, \mathbb{F}_p)$, which is of size $\sim p^{k^2} = n^{\Theta(\log n)}$.

& Zalcstein show that GpI for so-called "P_3 groups" can be solved in polynomial time; their proof shows that P_3 groups are central products of certain 2-groups, and the centrally indecomposable non-Abelian P_3 groups of a given order fall into finitely many one-parameter families, i.e., the classification of such groups is tame.[11] The polynomial-time algorithm for quotients of generalized Heisenberg groups [29] was recently generalized to groups of genus 2 [10], and the authors make it clear that this algorithm depends in a crucial way on tameness.

These facts, the upper bounds in this paper, and the lower bound on the number of indecomposables in wild type, suggest that the border between tame and wild may be the key border to cross on the way to putting GpI into P.

Acknowledgment. We thank Gábor Ivanyos for pointing out to us reference [11]. J. A. Grochow is supported by an SFI Omidyar Fellowship, and Y. Qiao by Australian Research Council DECRA DE150100720 during this work.

References

1. Alperin, J.L., Brauer, R., Gorenstein, D.: Finite groups with quasi-dihedral and wreathed Sylow 2-subgroups. Trans. Amer. Math. Soc. **151**, 1–261 (1970)
2. Babai, L.: Automorphism groups, isomorphism, reconstruction. In: Graham, R.L., Grötschel, M., Lovász, L. (eds.) Handbook of Combinatorics (vol. 2), pp. 1447–1540. MIT Press, Cambridge (1995)
3. Babai, L., Codenotti, P., Grochow, J.A., Qiao, Y: Code equivalence and group isomorphism. In: Proceedings of 22nd SODA, pp. 1395–1408 (2011)
4. Babai, L., Codenotti, P., Qiao, Y.: Polynomial-time isomorphism test for groups with no Abelian normal subgroups. In: Czumaj, A., Mehlhorn, K., Pitts, A., Wattenhofer, R. (eds.) ICALP 2012, Part I. LNCS, vol. 7391, pp. 51–62. Springer, Heidelberg (2012)
5. Babai, L., Qiao, Y.: Polynomial-time isomorphism test for groups with Abelian Sylow towers. In: Dürr, C., Wilke, T. (eds.) 29th International Symposium on Theoretical Aspects of Computer Science, STACS 2012. LIPIcs, vol. 14, pp. 453–464. Schloss Dagstuhl - Leibniz-Zentrum fuer Informatik (2012). doi:10.4230/LIPIcs.STACS.2012.453
6. Belitskii, G.R., Sergeichuk, V.V.: Complexity of matrix problems. Linear Algebra Appl **361**, 203–222 (2003). Ninth Conference of the International Linear Algebra Society (Haifa, 2001)
7. Bender, H.: Finite groups with dihedral Sylow 2-subgroups. J. Algebra **70**(1), 216–228 (1981)
8. Benson, D.J.: Representations and Cohomology: Volume 1, Basic Representation Theory of Finite Groups and Associative Algebras. Cambridge University Press, Cambridge (1998). Cambridge Studies in Advanced Mathematics
9. Brooksbank, P.A., Luks, E.M.: Testing isomorphism of modules. J. Algebra **320**(11), 4020–4029 (2008)

[11] Formally, this has not yet been shown to be tame in the precise sense. But in the general sense of having only finitely many one-parameter families of indecomposables of each size, this is without a doubt a tame classification.

10. Brooksbank, P.A., Maglione, J, Wilson, J.B.: A fast isomorphism test for groups of genus 2. arXiv:1508.03033 [math.GR] (2015)
11. Brooksbank, P.A., O'Brien, E.A.: Constructing the group preserving a system of forms. Int. J. Algebra Comput. **18**(02), 227–241 (2008)
12. Cannon, J.J., Holt, D.F.: Automorphism group computation and isomorphism testing in finite groups. J. Symb. Comput. **35**, 241–267 (2003)
13. Chistov, A.L., Ivanyos, G., Karpinski, M.: Polynomial time algorithms for modules over finite dimensional algebras. In: ISSAC, pp. 68–74 (1997)
14. Conrad, K: Generalized quaternions (2013). http://www.math.uconn.edu/~kconrad/blurbs/grouptheory/genquat.pdf
15. Curtis, C.W., Reiner, I.: Representation Theory of Finite Groups and Associative Algebras. Interscience Publishers, New York (1966). AMS Chelsea Publishing Series
16. Drozd, J.A.: Tame and wild matrix problems. In: Dlab, V., Gabriel, P. (eds.) Representation Theory II. Lecture Notes in Mathematics, pp. 242–258. Springer, Heidelberg (1980)
17. Felsch, V., Neubüser, J.: On a programme for the determination of the automorphism group of a finite group. In: Leech, P.J. (ed.) Computational Problems in Abstract Algebra (Proceedings of a Conference on Computational Problems in Algebra, Oxford, 1967), pp. 59–60. Pergamon Press, Oxford (1970)
18. Garzon, M., Zalcstein, Y.: On isomorphism testing of a class of 2-nilpotent groups. J. Comput. Syst. Sci. **42**(2), 237–248 (1991)
19. Gorenstein, D.: Finite groups, 2nd edn. Chelsea Publishing Co., New York (1980)
20. Gorenstein, D, Walter, J.H.: The characterization of finite groups with dihedral Sylow 2-subgroups. I-III. J. Algebra, 2:85–151, 218–270, 354–393 (1965)
21. Grochow, J.A.: Matrix isomorphism of matrix Lie algebras. In: IEEE Conference on Computational Complexity, pp. 203–213 (2012). Also available as arXiv:1112.2012 and ECCC TR11-168
22. Grochow, J.A., Qiao, Y.: Algorithms for group isomorphism via group extensions and cohomology. In: IEEE Conference on Computational Complexity (CCC14), pp. 110–119 (2014). Also available as arXiv:1309.1776 [cs.DS] and ECCC Technical Report TR13-123
23. Joshua A. Grochow and Youming Qiao. Polynomial-time isomorphism test of groups that are tame extensions. arXiv:1507.01917 [cs.DS] (2015)
24. Guralnick, R., Kantor, W.M., Kassabov, M., Lubotzky, A.: Presentations of finite simple groups: profinite and cohomological approaches. Groups Geom. Dyn. **1**(4), 469–523 (2007)
25. Ivanyos, G., Karpinski, M., Saxena, N.: Deterministic polynomial time algorithms for matrix completion problems. SIAM J. Comput. **39**(8), 3736–3751 (2010)
26. Kavitha, T.: Linear time algorithms for Abelian group isomorphism and related problems. J. Comput. Syst. Sci. **73**(6), 986–996 (2007)
27. Köbler, J., Schöning, U., Torán, J.: The Graph Isomorphism Problem: Its Structural Complexity. Birkhauser Verlag, Basel (1993)
28. Le Gall, F.: Efficient isomorphism testing for a class of group extensions. In: Proceedings of 26th STACS, pp. 625–636 (2009)
29. Lewis, M.L., Wilson, J.B.: Isomorphism in expanding families of indistinguishable groups. Groups Complex. Cryptol. **4**(1), 73–110 (2012)
30. Miller, G.L.: On the $n^{\log n}$ isomorphism technique (a preliminary report). In: Proceedings of 10th ACM STOC, pp. 51–58. ACM Press, New York (1978)
31. Mulmuley, K.: On P vs. NP and geometric complexity theory. J. ACM **58**(2), 5 (2011)

32. Qiao, Y, Sarma, J.M.N., Tang, B.: On isomorphism testing of groups with normal Hall subgroups. In: Proceedings of 28th STACS, pp. 567–578 (2011)
33. Rickard, J.: Answer to: the number of indecomposable modules of finite groups over finite fields of a fixed dimension. http://mathoverflow.net/a/194773/8012
34. Rosenbaum, D: Bidirectional collision detection and faster algorithms for isomorphism problems. arXiv:1304.3935 [cs.DS] (2013)
35. Savage, C.: An $O(n^2)$ algorithm for Abelian group isomorphism. Technical report, North Carolina State University (1980)
36. Seress, Á.: Permutation Group Algorithms. Cambridge University Press, Cambridge (2003)
37. Sergeĭčuk, V.V.: The classification of metabelian p-groups. In: Matrix Problems (Russian), pp. 150–161. Akad. Nauk Ukrain. SSR Inst. Mat., Kiev (1977)
38. Vikas, N.: An $O(n)$ algorithm for Abelian p-group isomorphism and an $O(n \log n)$ algorithm for Abelian group isomorphism. J. Comput. Syst. Sci. **53**(1), 1–9 (1996)

Quantum Algorithm for Triangle Finding in Sparse Graphs

François Le Gall and Shogo Nakajima(✉)

Department of Computer Science,
Graduate School of Information Science and Technology,
The University of Tokyo, Hongō, Japan
nakajimashogo@is.s.u-tokyo.ac.jp

Abstract. This paper presents a quantum algorithm for triangle finding over sparse graphs that improves over the previous best quantum algorithm for this task by Buhrman et al. [SIAM Journal on Computing, 2005]. Our algorithm is based on the recent $\tilde{O}(n^{5/4})$-query algorithm given by Le Gall [FOCS 2014] for triangle finding over dense graphs (here n denotes the number of vertices in the graph). We show in particular that triangle finding can be solved with $O(n^{5/4-\epsilon})$ queries for some constant $\epsilon > 0$ whenever the graph has at most $O(n^{2-c})$ edges for some constant $c > 0$.

1 Introduction

Background. Triangle finding asks to decide if a given undirected graph $G = (V, E)$ contains a cycle of length three, i.e., whether there exist three vertices $u_1, u_2, u_3 \in V$ such that $\{u_1, u_2\} \in E$, $\{u_1, u_3\} \in E$ and $\{u_2, u_3\} \in E$. This problem has received recently a lot of attention, for the following reasons.

First, several new applications of triangle finding have been discovered recently. In particular, Vassilevska Williams and Williams have shown a surprising reduction from Boolean matrix multiplication to triangle finding [17], which indicates that efficient algorithms for triangle finding may be used to design efficient algorithms for matrix multiplication, and thus also for a vast class of problems related to matrix multiplication. Relations between variants of the standard triangle finding problem (such as triangle finding over weighted graphs) and well-studied algorithmic problems (such as 3SUM) have also been shown in the past few years (see for instance [16,18]).

Second, triangle finding is one of the most elementary graph theoretical problems whose complexity is unsettled. In the time complexity setting, the best classical algorithm uses a reduction to matrix multiplication [11] and solves triangle finding in time $O(n^{2.38})$, where n denotes the number of vertices in G. In the time complexity setting again, Grover search [10] immediately gives, when applied to triangle finding as a search over the set of triples of vertices of the graph, a quantum algorithm with time complexity $\tilde{O}(n^{3/2})$, which is still the best known upper bound for the quantum time complexity of this problem.[1] In the

[1] In this paper the notation $\tilde{O}(\cdot)$ removes polylog$n$ factors.

K. Elbassioni and K. Makino (Eds.): ISAAC 2015, LNCS 9472, pp. 590–600, 2015.
DOI: 10.1007/978-3-662-48971-0_50

query complexity setting, where an oracle to the adjacency matrix of the graph is given and only the number of calls to this oracle is counted, a surge of activity has lead to quantum algorithms with better complexity. Magniez, Santha and Szegedy [15] first presented a quantum algorithm that solves triangle finding with $\tilde{O}(n^{1.3})$ queries. This complexity was later improved to $O(n^{1.296...})$ by Belovs [4], then to $O(n^{1.285...})$ by Lee, Magniez and Santha [13] and Jeffery, Kothari and Magniez [12], and further improved recently to $\tilde{O}(n^{5/4})$ by Le Gall [8]. The main open problem now is to understand whether this $\tilde{O}(n^{5/4})$-query upper bound is tight or not. The best known lower bound on the quantum query complexity of triangle finding is the straightforward $\Omega(n)$ lower bound.

Another reason why triangle finding has received much attention from the quantum computing community is that work on the quantum complexity of triangle finding has been central to the development of algorithmic techniques. Indeed, all the improvement mentioned in the previous paragraph have been obtained by introducing either new quantum techniques or new paradigms for the design of quantum algorithms: applications of quantum walks to graph-theoretic problems [15], introduction of the concept of learning graphs [4] and improvements to this technique [13], introduction of quantum walks with quantum data structures [12], association of combinatorial arguments with quantum walks [8].

Triangle Finding in Sparse Graphs. The problem we will consider in this paper is triangle finding over sparse graphs (the graphs considered are, as usual, undirected and unweighted). If we denote m the number of edge of the graph (i.e., $m = |E|$), the goal is to design algorithms with complexity expressed as a function of m and n. Ideally, we would like to show that if $m = n^{2-c}$ for any constant $c > 0$ then triangle finding can be solved significantly faster than in the dense case (i.e., $m \approx n^2$). Besides its theoretical interest, this problem is of practical importance since in many applications the graphs considered are sparse.

Classically, Alon, Yuster and Zwick [1] constructed an algorithm exploiting the sparsity of the graph and working in time $O(m^{1.41})$, which gives better complexity than the $O(n^{2.38})$-time complexity mentioned above when $m \leq n^{1.68}$. Understanding whether an improvement over the dense case is also possible for larger values m is a longstanding open problem. Note in the classical query complexity setting it is easy to show that the complexity of triangle finding is $\Theta(n^2)$, independently of the value of m.

In the quantum setting, using amplitude amplification, Buhrman et al. [6] showed how to construct a quantum algorithm for triangle finding with time and query complexity $O(n+\sqrt{nm})$. This upper bound is tight when $m \leq n$ since the $\Omega(n)$-query lower bound for the quantum query complexity of triangle finding already mentioned also holds when m is a constant. Childs and Kothari [7] more recently developed an algorithm, based on quantum walks, that detects the existence of subgraphs in a given graph. Their algorithm works for any constant-size subgraph. For detecting the existence of a triangle, however, the upper bound they obtain is $\tilde{O}(n^{2/3}\sqrt{m})$ queries for $m \geq n$, which is worse that the bound obtained in [6]. Buhrman et al.'s result in particular gives an improvement over the $\tilde{O}(n^{5/4})$-query quantum algorithm whenever $m \leq n^{3/2}$.

A natural question is whether a similar improvement can be obtained for larger values of m. For instance, can we obtain query complexity $\tilde{O}(n^{5/4-\epsilon})$ for some constant $\epsilon > 0$ when $m \approx n^{1.99}$? A positive answer would show that even a little amount of sparsity can be exploited in the quantum query setting, which is not known to be true in the classical setting as mentioned in the previous paragraph.

Our Results. In this paper we answer positively to the above question. Our main result is as follows.

Theorem 1. *There exists a quantum algorithm that solves, with high probability, the triangle finding problem over graphs of n vertices and m edges with query complexity*

$$\begin{cases} O(n+\sqrt{nm}) & \text{if } 0 \leq m \leq n^{7/6}, \\ \tilde{O}(nm^{1/14}) & \text{if } n^{7/6} \leq m \leq n^{7/5}, \\ \tilde{O}(n^{1/6}m^{2/3}) & \text{if } n^{7/5} \leq m \leq n^{3/2}, \\ \tilde{O}(n^{23/30}m^{4/15}) & \text{if } n^{3/2} \leq m \leq n^{13/8}, \\ \tilde{O}(n^{59/60}m^{2/15}) & \text{if } n^{13/8} \leq m \leq n^2. \end{cases}$$

For the dense case (i.e., $m \approx n^2$) we recover the same complexity $\tilde{O}(n^{5/4})$ as in [8]. Whenever $m = n^{2-c}$ for some constant $c > 0$ (in particular, for $m \approx n^{1.99}$), we indeed obtain query complexity $\tilde{O}(n^{5/4-\epsilon})$ for some constant $\epsilon > 0$ depending on c. The query complexity of our algorithm is better than the query complexity of Buhrman et al.'s algorithm [6] whenever $m \gtrsim n^{7/6}$. When $m \lesssim n^{7/6}$ we obtain the same complexity $O(n + \sqrt{nm})$ as in [6].

2 Preliminaries

2.1 Query Complexity for Graph-Theoretic Problems

In this paper we adopt the standard model of quantum query complexity for graph-theoretic problems. The presentation given below will follow the description of this notions given in [8].

For any finite set T and any $r \in \{1, \ldots, |T|\}$ we denote $\mathcal{S}(T, r)$ the set of all subsets of r elements of T. We use the notation $\mathcal{E}(T)$ to represent $\mathcal{S}(T, 2)$, i.e., the set of unordered pairs of elements in T.

Let $G = (V, E)$ be an undirected and unweighted graph, where V represents the set of vertices and $E \subseteq \mathcal{E}(V)$ represents the set of edges. We write $n = |V|$. In the query complexity setting, we assume that V is known, and that E can be accessed through a quantum unitary operation $\mathcal{O}_G$ defined as follows. For any pair $\{u, v\} \in \mathcal{E}(V)$, any bit $b \in \{0, 1\}$, and any binary string $z \in \{0, 1\}^*$, the operation $\mathcal{O}_G$ maps the basis state $|\{u, v\}\rangle|b\rangle|z\rangle$ to the state

$$\mathcal{O}_G|\{u, v\}\rangle|b\rangle|z\rangle = \begin{cases} |\{u, v\}\rangle|b \oplus 1\rangle|z\rangle & \text{if } \{u, v\} \in E, \\ |\{u, v\}\rangle|b\rangle|z\rangle & \text{if } \{u, v\} \notin E, \end{cases}$$

where $\oplus$ denotes the bit parity (i.e., the logical XOR). We say that a quantum algorithm computing some property of G uses k queries if the operation $\mathcal{O}_G$, given as an oracle, is called k times by the algorithm. We also assume that we know the number of edges of the input graph (i.e., we know $m = |E|$). All the results in this paper can be easily generalized to the case where m is unknown.

Quantum Enumeration. Let $f_G\colon \{1,\ldots,N\} \to \{0,1\}$ be a Boolean function depending on the input graph G, and let us write $M = f^{-1}(1)$. Assume that for any $x \in \{1,\ldots,N\}$ the value $f_G(x)$ can be computed using at most t queries to $\mathcal{O}_G$. Grover search enables us to find an element x such that $f_G(x) = 1$, if such an element exists, using $\tilde{O}(\sqrt{N/M} \times t)$ queries to $\mathcal{O}_G$. A folklore observation is that we can then repeat this procedure to find all the elements $x \in \{1,\ldots,N\}$ such that $f_G(x) = 1$ with $\tilde{O}\left(\left(\sqrt{\frac{N}{M}} + \sqrt{\frac{N}{M-1}} + \cdots + \sqrt{\frac{N}{1}}\right) \times t\right) = \tilde{O}\left(\sqrt{N \times M} \times t\right)$ queries. We call this procedure *quantum enumeration.*

Quantum Walk Over Johnson Graphs. Let T be a finite set and r be a positive integer such that $r \leq |T|$. Let $f_G\colon \mathcal{S}(T,r) \to \{0,1\}$ be a Boolean function depending on the input graph G. We say that a set $A \in \mathcal{S}(T,r)$ is marked if $f_G(A) = 1$. Let us consider the following problem. The goal is to find a marked set, if such a set exists, or otherwise report that there is no marked set. We are interested in the number of calls to $\mathcal{O}_G$ to solve this problem. The quantum walk search approach developed by Ambainis [2] solves this problem using a quantum walk over a Johnson graph.

The Johnson graph $J(T,r)$ is the undirected graph with vertex set $\mathcal{S}(T,r)$ where two vertices $R_1, R_2 \in \mathcal{S}(T,r)$ are connected if and only if $|R_1 \cap R_2| = r-1$. In a quantum walk over a Johnson graph $J(T,r)$, the state of the walk corresponds to a node of the Johnson (i.e., to an element $A \in \mathcal{S}(T,r)$). A data structure $D(A)$, which in general depends on G, is associated to each state A. There are three costs to consider: the set up cost S representing the number of queries to $\mathcal{O}_G$ needed to construct the data structure of the initial state of the walk, the update cost U representing the number of queries to $\mathcal{O}_G$ needed to update the data structure when one step of the quantum walk is performed (i.e., updating $D(A)$ to $D(A')$ for some $A' \in \mathcal{S}(T,r)$ such that $|A \cap A'| = r-1$), and the checking cost C representing the number of queries to $\mathcal{O}_G$ needed to check if the current state A is marked (i.e., checking whether $f_G(A) = 1$). Let $\varepsilon > 0$ be such that, for all input graphs G for which at least one marked set exists, the fraction of marked states is at least ε. Ambainis [2] (see also [14]) has shown that the quantum walk search approach outlined above finds with high probability a marked set if such set exists (or otherwise report that there is no marked set) and has query complexity $\tilde{O}\left(\mathsf{S} + \frac{1}{\sqrt{\varepsilon}}\left(\sqrt{r} \times \mathsf{U} + \mathsf{C}\right)\right)$.

2.2 Quantum Algorithm for Dense Triangle Finding

In this subsection we outline the $\tilde{O}(n^{5/4})$-query quantum algorithm, on which our algorithm is mainly based, for triangle finding over a dense graph by Le Gall [8]. We actually present a version of this algorithm that solves the following slightly more general version of triangle finding, since this will be more convenient when describing our algorithms for sparse graphs in the next section: given two (non necessarily disjoint) sets $V_1, V_2 \subseteq V$, find a triangle $\{v_1, v_2, v_3\}$ of G such that $v_1 \in V_1$ and $v_2, v_3 \in V_2$, if such a triangle exists. Note that the original triangle finding problem is the special case $V_1 = V_2 = V$.

Let V_1 be any subset of V. For any sets $X \subseteq V_1$ and $Y \subseteq V$, we define the set $\Delta_G(X,Y) = \mathcal{E}(Y) \setminus \bigcup_{u \in X} \mathcal{E}(N_G(u))$, where $N_G(u)$ denotes the set of neighbors of u. For any vertex $w \in V$, we define the set $\Delta_G(X,Y,w) = \Big\{\{u,v\} \in \Delta_G(X,Y) \mid \{u,w\} \in E \text{ and } \{v,w\} \in E\Big\}$. An important concept used in [8] is the notion of *k-good sets*.

Definition 1. *Let k be any constant such that $0 \leq k \leq 1$, and V_1 be any subset of V. A set $X \subseteq V_1$ is k-good for (G, V_1) if the inequality $\sum_{w \in V_1} |\Delta_G(X,Y,w)| \leq |Y|^2 |V_1|^{1-k}$ holds for all $Y \subseteq V$.*

Lemma 1 *([8]). Let k be any constant such that $0 \leq k \leq 1$. Suppose that X is a set obtained by taking uniformly at random, with replacement, $\lceil 3|V_1|^k \log n \rceil$ elements from V_1. Then X is k-good for (G, V_1) with probability at least $1 - 1/n$.*

Lemma 1 was proved in [8] only for the case $V_1 = V$, but the generalization is straightforward.

Let a, b and k be three constants such that $0 < b < a < 1$ and $0 < k < 1$. The values of these constants will be set later. The quantum algorithm in [8] works as follows.

The algorithm first takes a set $X \subseteq V_1$ obtained by choosing uniformly at random $\lceil 3|V_1|^k \log n \rceil$ elements from V_1, and checks if there exists a triangle of G with a vertex in X and two vertices in V_2. This can be done using Grover search with

$$O\left(\sqrt{|X| \times |\mathcal{E}(V_2)|}\right) = \tilde{O}\left(|V_1|^{k/2} |V_2|\right) \tag{1}$$

queries. If no triangle has been reported, we know that any triangle of G with one vertex in V_1 and two vertices in V_2 must have an edge in $\Delta_G(X, V_2)$.

Now, in order to find a triangle with an edge in $\Delta_G(X, V_2)$, if such a triangle exists, the idea is to search for a set $A \in \mathcal{S}(V_2, \lceil |V_2|^a \rceil)$ such that $\Delta_G(X, A)$ contains an edge of a triangle. To find such a set A, the algorithm performs a quantum walk over the Johnson graph $J(V_2, \lceil |V_2|^a \rceil)$. The states of this walk correspond to the elements in $\mathcal{S}(V_2, \lceil |V_2|^a \rceil)$. The state corresponding to a set $A \in \mathcal{S}(V_2, \lceil |V_2|^a \rceil)$ is marked if $\Delta_G(X, A)$ contains an edge of a triangle of G. In case the set of marked states is not empty, the fraction of marked states is $\varepsilon = \Omega\left(|V_2|^{2(a-1)}\right)$. The data structure of the walk stores the set $\Delta_G(X, A)$. Concretely, this is done by storing the couple $(v, N_G(v) \cap X)$ for each $v \in A$, since this information is enough to construct $\Delta_G(X, A)$ without using any additional query. The setup cost is $\mathsf{S} = |A| \times |X| = \tilde{O}(|V_2|^a |V_1|^k)$ queries. The update cost is $\mathsf{U} = 2|X| = \tilde{O}(|V_1|^k)$ queries. The query complexity of the quantum walk is

$$\tilde{O}\left(\mathsf{S} + \sqrt{1/\varepsilon}\left(|V_2|^{a/2} \times \mathsf{U} + \mathsf{C}\right)\right), \tag{2}$$

where C is the cost of checking if a state is marked.

The checking procedure is done as follows: check if there exists a vertex $w \in V_1$ such that $\Delta_G(X, A)$ contains a pair $\{v_1, v_2\}$ for which $\{v_1, v_2, w\}$ is a triangle of G. For any $w \in V_1$, let $Q(w)$ denote the query complexity of checking if there exists a pair $\{v_1, v_2\} \in \Delta_G(X, A)$ such that $\{v_1, v_2, w\}$ is a triangle of G. Using Ambainis' variable cost search [3] this checking procedure can be implemented using $\mathsf{C} = \sqrt{\sum_{w \in V_1} Q(w)^2}$ queries. It thus remains to give an upper bound on $Q(w)$. Let us fix $w \in V_1$. First, a tight estimator of the size of $\Delta_G(X, A, w)$ is computed: the algorithm computes an integer $\delta(X, A, w)$ such that $|\delta(X, A, w) - |\Delta_G(X, A, w)|| \leq \frac{1}{10} \times |\Delta_G(X, A, w)|$, which can be done in $\tilde{O}(|V_1|^k)$ queries using (classical) sampling. The algorithm then performs a quantum walk over the Johnson graph $J(A, \lceil |V_2|^b \rceil)$. The states of this walk correspond to the elements in $\mathcal{S}(A, \lceil |V_2|^b \rceil)$. We now define the set of marked states of the walk. The state corresponding to a set $B \in \mathcal{S}(A, \lceil |V_2|^b \rceil)$ is marked if B satisfies the following two conditions:

1. There exists a pair $\{v_1, v_2\} \in \Delta_G(X, B, w)$ such that $\{v_1, v_2\} \in E$ (i.e., such that $\{v_1, v_2, w\}$ is a triangle of G);
2. $|\Delta_G(X, B, w)| \leq 10 \times |V_2|^{2(b-a)} \times \delta(X, A, w)$.

The fraction of marked states is $\varepsilon' = \Omega\left(|V_2|^{2(b-a)}\right)$. The data structure of the walk will store $\Delta_G(X, B, w)$. Concretely, this is done by storing the couple (v, e_v) for each $v \in B$, where $e_v = 1$ if $\{v, w\} \in E$ and $e_v = 0$ if $\{v, w\} \notin E$. The setup cost is $\mathsf{S}' = \lceil |V_2|^b \rceil$ queries since it is sufficient to check if $\{v, w\}$ is an edge for all $v \in B$. The update cost is $\mathsf{U}' = 2$ queries. The checking cost is $\mathsf{C}'_w = O\left(\sqrt{|\Delta_G(X, B, w)|}\right) = O\left(\frac{|V_2|^b}{|V_2|^a}\sqrt{\delta(X, A, w)}\right) = O\left(\frac{|V_2|^b}{|V_2|^a}\sqrt{|\Delta(X, A, w)|}\right)$. We thus obtain the bound $Q(w) = \tilde{O}\left(|V_1|^k + \mathsf{S}' + \sqrt{1/\varepsilon'}\left(|V_2|^{b/2} \times \mathsf{U}' + \mathsf{C}'_w\right)\right)$, and conclude that

$$\mathsf{C} = \tilde{O}\left(\sqrt{|V_1|}\left(|V_1|^k + \mathsf{S}' + \frac{|V_2|^{b/2} \times \mathsf{U}'}{\sqrt{\varepsilon'}}\right) + \frac{|V_2|^{b-a}}{\sqrt{\varepsilon'}} \times \sqrt{\sum_{w \in V_1} |\Delta(X, A, w)|}\right).$$

The final key observation is that, since the set X is k-good for (G, V_1) with high probability, as guaranteed by Lemma 1, the term $\sum_{w \in V} |\Delta(X, A, w)|$ in the above expression can be replaced by $O(|V_2|^{2a}|V_1|^{1-k})$, which enables us to express C as a function of a, b and k, and then the complexity of the second part of the algorithm (Expression (2)) as a function of a, b and k. The complexity of the whole algorithm (the maximum of Expressions (1) and (2)) can thus be written as a function of a, b and k as well.

For the original triangle finding problem (i.e., for the case $V_1 = V_2 = V$), taking $a = \frac{3}{4}$ and $b = k = \frac{1}{2}$ gives query complexity $\tilde{O}(n^{5/4})$.

3 Quantum Algorithm for Sparse Triangle Finding

In this section we describe our quantum algorithm for triangle finding in sparse graphs and prove Theorem 1.

Let d be a real number such that $0 \leq d \leq 1$. The value of this parameter will be set later. We define the two subsets of V, $\mathcal{V}_h^d = \{v \in V \mid \deg(v) \geq \frac{9}{10} \times n^d\}$ and $\mathcal{V}_l^d = \{v \in V \mid \deg(v) \leq \frac{11}{10} \times n^d\}$. A crucial observation is that $|\mathcal{V}_h^d| = O(m/n^d)$, since the graph G has m edges. The following proposition shows how to efficiently classify all the vertices of V into vertices in $\mathcal{V}_h^d$ and vertices in $\mathcal{V}_l^d$.

Proposition 1. *There exists a quantum algorithm using $Q_1 = \tilde{O}(n^{1-d}\sqrt{m})$ queries that partitions the set V into two sets V_h^d and V_l^d such that, with high probability, $V_h^d \subseteq \mathcal{V}_h^d$ and $V_l^d \subseteq \mathcal{V}_l^d$.*

Proof. Let v be any vertex in V. Using quantum counting [5] we can compute, using $\tilde{O}\left(\sqrt{\frac{n}{n^d}}\right)$ queries, a value $a(v)$ such that $|a(v) - \deg(v)| \leq n^d/100$ with probability at least $1 - 1/\mathrm{poly}(n)$. We use $a(v)$ to classify v as follows: we decide " v is in $\mathcal{V}_h^d$ " if $a(v) \geq n^d$, and decide " v is in $\mathcal{V}_l^d$ " if $a(v) < n^d$. This decision is correct with probability at least $1 - 1/\mathrm{poly}(n)$.

We can thus apply quantum enumeration as described in Sect. 2.1 to obtain a set $V_h^d \subseteq V$ of vertices such that, with high probability, all the vertices in V_h^d are in $\mathcal{V}_h^d$ and all the vertices in $V \setminus V_h^d$ are in $\mathcal{V}_l^d$. We then take $V_l^d = V \setminus V_h^d$. The overall complexity of this approach is $\tilde{O}\left(\sqrt{n \times \frac{m}{n^d}} \times \sqrt{\frac{n}{n^d}}\right) = \tilde{O}(n^{1-d}\sqrt{m})$ queries, since $|\mathcal{V}_h^d| = O(m/n^d)$. □

In the remaining of the section we assume that the algorithm of Proposition 1 outputs a correct classification (i.e., $V_h^d \subseteq \mathcal{V}_h^d$ and $V_l^d \subseteq \mathcal{V}_l^d$), which happens with high probability. In particular we assume that $|V_h^d| = O(m/n^d)$. We will say that a vertex $v \in V$ is d-high if $v \in V_h^d$, and say it is d-low if $v \in V_l^d$. Once the vertices have been classified, checking if G has a triangle can be divided into four subproblems: checking if G has a triangle with three d-low vertices, checking if G has a triangle with two d-low vertices and one d-high vertex, checking if G has a triangle with one d-low-degree vertex and two d-high vertices, and checking if G has a triangle with three high-degree triangles. We now present six procedures to handle these cases (for some cases we present more than one procedure to allow us to choose which procedure to use according to the value of m).

Proposition 2. *Let a_1, k_1 and b_1 be any constants such that $0 < a_1, k_1 < 1$ and $0 < b_1 < a_1$. There exists a quantum algorithm that finds a triangle of G consisting of three d-low vertices, if such a triangle exists, with high probability using $Q_2 = \tilde{O}(n + n^{k_1/2}m^{1/2} + n^{a_1+d/2+k_1-1/2} + n^{1/2+d/2+k_1-a_1/2} + n^{3/2+k_1/2-a_1} + n^{1+b_1+d/2-a_1} + n^{3/2-b_1/2} + n^{3/2-k_1/2})$ queries.*

The proof of Proposition 2 will use the following key lemma. The proof of this lemma can be found in the full version of the paper [9].

Lemma 2. *Let k be any constant such that $0 < k < 1$. Suppose that X is a set of size $|X| = \lceil 3n^k \log n \rceil$ obtained by taking uniformly at random vertices* *from* V_l^d. *Then, with probability at least* $1 - \frac{1}{n \exp(\frac{231}{10} n^{d+k-1} \log n)}$, *the inequality* $|N_G(v) \cap X| < \frac{33}{10} n^{d+k-1} \log n + 2 \log n$ *holds for all vertices* $v \in V_l^d$.

Proof (Proof of Proposition 2). We adapt the algorithm for the dense case presented in Sect. 2.2. We take $V_1 = V_2 = V_l^d$, and $X \subseteq V_1$ of size $|X| = \lceil 3|V_1|^{k_1} \log n \rceil$.

We replace the first step of the algorithm, which checks if there exists a triangle of G with a vertex in X and two vertices in V_2, by the following procedure based on [6]. We take a random edge $\{u, v\} \in \mathcal{E}(V_2) \cap E$ and then try to find a vertex w from X such that $\{u, v, w\}$ is a triangle of G. Note that this can be implemented using two Grover searches in $\tilde{O}(\sqrt{|\mathcal{E}(V_2)|/|\mathcal{E}(V_2) \cap E|} + \sqrt{|X|})$ queries, and that in the worst case (i.e., when there is only one triangle) the success probability of this approach is $\Theta(1/|\mathcal{E}(V_2) \cap E|)$. Using amplitude amplification we can then check with high probability the existence of such a triangle with total query complexity

$$\tilde{O}\left(\sqrt{|\mathcal{E}(V_2) \cap E|} \times (\sqrt{|\mathcal{E}(V_2)|/|\mathcal{E}(V_2) \cap E|} + \sqrt{|X|})\right) = \tilde{O}(n + \sqrt{n^{k_1} m}). \quad (3)$$

We now show how to adapt the second step of the algorithm presented in Sect. 2.2 to exploit the sparsity of the graph. First, as observed in [8], the cost of estimating the size of $\Delta_G(X, A, w)$ can be reduced to $\tilde{O}(\sqrt{n^{k_1}})$ queries by using quantum counting instead of random sampling (quantum counting was not used in [8] since it did not result in any speed-up for the dense case, but for the sparse case this is necessary). We now describe our main ideas to exploit the sparsity of the graph, and show how to reduce the cost of two quantum walks.

First, we describe how to reduce the setup cost S and the update cost U as follows. By Lemma 2, we know that $|N_G(v) \cap X| < t$ for all $v \in V_2$, where $t = \frac{33}{10} n^{d_1+k_1-1} \log n + 2 \log n$. Therefore we can use quantum enumeration to find all vertices in $N_G(v) \cap X$ with $\tilde{O}\left(\sqrt{\frac{|X|}{t}} + \cdots + \sqrt{\frac{|X|}{1}}\right) = \tilde{O}(\sqrt{|X|t})$ queries. The setup cost S is thus $\tilde{O}\left(|A| \times \sqrt{|X|t}\right) = \tilde{O}(n^{a_1+k_1+d/2-1/2})$ queries, and the update cost $\mathsf{U} = \tilde{O}(\sqrt{|X|t}) = \tilde{O}(n^{k_1+d/2-1/2})$ queries.

Next, we describe how to reduce the setup cost S'. This set up requires to obtain the couple (v, e_v) for each $v \in B$, where where w is a fixed vertex in V_1, $e_v = 1$ if $\{v, w\} \in E$ and $e_v = 0$ if $\{v, w\} \notin E$. Let $\mu(w) = n^d \times \frac{|B|}{|V|} = n^{d+b_1-1}$ be the average of $|N_G(w) \cap B|$ over all B. We use quantum enumeration to find at most $10 \times \mu(w)$ vertices in $N_G(w) \cap B$ from B. Thus the cost of this procedure is $\mathsf{S}' = \tilde{O}\left(\sqrt{\frac{|B|}{|\mu(w)|}} + \cdots \sqrt{\frac{|B|}{1}}\right) = \tilde{O}(\sqrt{|B| \times |\mu(w)|}) = \tilde{O}(n^{b_1+d/2-1/2})$ queries. Note that this procedure will not correctly prepare the database for all B's (since $|N_G(w) \cap B|$ may exceeds $10 \times \mu(w)$ for some B's); it will prepare correctly the database only for a large fraction of the B's. This is nevertheless not a problem since the initial state of the quantum walk is a uniform superposition of all the B's: this procedure will thus prepare a state close enough to the ideal state, which will modify only in a negligible way the final success probability of the whole walk.

We also modify the definition of a marked state for the second walk (we add one condition). Namely, the state corresponding to a set $B \in \mathcal{S}(A, \lceil |V_2|^{b_1} \rceil)$ will be marked if B satisfies the following three conditions:

1. there exist two vertices $v_1, v_2 \in B$ such that $\{v_1, v_2\} \in E$ (i.e., such that $\{v_1, v_2, w\}$ is a triangle of G);
2. $|\Delta_G(X, B, w)| \leq 10 \times |V_2|^{2(b_1 - a_1)} \times \delta(X, A, w)$;
3. $|N_G(w) \cap B| \leq 10 \times \mu(w)$.

It is easy to show that adding the third condition does not change significantly the fraction of marked states.

The checking procedure of the second walk (and thus its cost C'_w) is the same as in the dense case.

By evaluating the performance of the walks as done for the dense case in Sect. 2.2, but replacing Expression (1) by Expression (3) and replacing in the evaluation of Expression (2) the quantities S, U, S' and the cost of estimating $|\Delta_G(X, A, w)|$ by the expressions we just derived, we obtain the claimed query complexity for the whole algorithm. For instance, the checking cost of the first walk is

$$\begin{aligned}\mathsf{C} &= \tilde{O}\left(\sqrt{|V_1|}\left(|V_1|^{k_1/2} + \mathsf{S}' + \frac{|V_2|^{b_1/2} \times \mathsf{U}'}{\sqrt{\varepsilon'}}\right) + \frac{|V_2|^{b_1 - a_1}}{\sqrt{\varepsilon'}} \times \sqrt{\sum_{w \in V_1} |\Delta(X, A, w)|}\right) \\ &= \tilde{O}\left(n^{1/2 + k_1/2} + n^{b_1 + d/2} + n^{1/2 + a_1 - b_1/2} + n^{1/2 + a_1 - k_1/2}\right)\end{aligned}$$

queries. □

The proofs of the following five propositions can be found in [9].

Proposition 3. *Let a_2, k_2 and b_2 be any constants such that $0 < a_2 < 1$, $1 < n^{k_2} < |V_h^d|$ and $0 < b_2 < a_2$. A triangle of G consisting of two d-low vertices and one d-high vertex can be detected with high probability using $Q_3 = \tilde{O}(n + n^{k_2/2} m^{1/2} + n^{a_2 + d + k_2} m^{-1/2} + n^{1 + d + k_2 - a_2/2} m^{-1/2} + n^{1 + k_2/2 - a_2 - d/2} m^{1/2} + n^{1 + b_2 - a_2 - d/2} m^{1/2} + n^{1 - b_2/2 - d/2} m^{1/2} + n^{1 - d/2 - k_2/2} m^{1/2})$ queries.*

Proposition 4. *Let a_3, k_3 and b_3 be constants such that $1 < n^{a_3} < |V_h^d|$, $0 < k_3 < 1$ and $0 < b_3 < a_3$. A triangle of G consisting of two d-high vertices and one d-low vertex can be detected with high probability using $Q_4 = \tilde{O}(n + n^{k_3/2} m^{1/2} + n^{a_3 + k_3} + n^{k_3 - a_3/2 - d} m + n^{1/2 + k_3/2 - a_3 - d} m + n^{b_3 - a_3 - d/2} m + n^{1/2 - b_3/2 - d} m + n^{1/2 - d - k_3/2} m)$ queries.*

Proposition 5. *A triangle of G consisting of three d-high vertices can be detected with high probability using $Q_5 = \tilde{O}((m/n^d)^{5/4})$ queries.*

Proposition 6. *Let b_4 be any constant such that $0 < b_4 < 1$. A triangle of G consisting of three d-low vertices can be detected with high probability using $Q_6 = \tilde{O}(n^{b_4 + d/2} + n^{3/2 - b_4/2} + n^{1/2 + d})$ queries.*

Proposition 7. *A triangle consisting of at least one d-high vertex can be detected with high probability using $Q_7 = O(n + n^{-d/2}m)$ queries.*

We are now ready to prove Theorem 1.

Proof (Proof of Theorem 1). From Propositions 1, 2, 3, 4, 5, 6 and 7 and the discussion before Proposition prop:B2, the query complexity of our whole algorithm is min $\left[(Q_1+Q_6+Q_7), (Q_1+Q_3+Q_4+Q_5+Q_6), (Q_1+Q_2+Q_3+Q_4+Q_5)\right]$. We write $m = n^\ell$, for $0 \leq \ell \leq 2$, and optimize below the parameters.

If $\frac{7}{6} \leq \ell \leq \frac{7}{5}$, the query complexity is upper bounded by $Q_1 + Q_6 + Q_7 = \tilde{O}(n^{1+\ell/2-d} + n^{3/2-b_4/2} + n^{b_4+d/2})$, which is optimized by taking $b_4 = 1 - \frac{\ell}{7}$ and $d = \frac{3\ell}{7}$, giving the upper bound $\tilde{O}(n^{1+\ell/14})$.

If $\frac{7}{5} \leq \ell \leq \frac{3}{2}$, the query complexity is upper bounded by $Q_1 + Q_6 + Q_7 = \tilde{O}(n^{3/2-b_4/2} + n^{1/2+d} + n^{\ell-d/2})$, which is optimized by taking $b_4 = \frac{8}{3} - \frac{4\ell}{3}$ and $d = \frac{2\ell}{3} - \frac{1}{3}$, giving the upper bound $\tilde{O}(n^{1/6+2\ell/3})$.

If $\frac{3}{2} \leq \ell \leq \frac{13}{8}$, the query complexity is upper bounded by $Q_1+Q_3+Q_4+Q_5+Q_6 = \tilde{O}(n^{3/2-b_4/2}+n^{1/2+d}+n^{a_2+d+k_2-\ell/2}+n^{1+b_2+\ell/2-a_2-d/2}+n^{1+\ell/2-b_2/2-d/2}+n^{1+\ell/2-d/2-k_2/2}+n^{b_3+\ell-a_3-d/2}+n^{1/2+\ell-b_3/2-d}+n^{1/2+\ell-d-k_3/2})$, which is optimized by taking $a_2 = \frac{3}{10} + \frac{3\ell}{10}$, $a_3 = \frac{23\ell}{15} - \frac{59}{30}$, $b_2 = k_2 = \frac{1}{5} + \frac{\ell}{5}$, $b_3 = k_3 = \frac{14\ell}{15} - \frac{16}{15}$, $b_4 = \frac{22}{15} - \frac{8\ell}{15}$ and $d = \frac{4}{15} + \frac{4\ell}{15}$, giving the upper bound $\tilde{O}(n^{23/30+4\ell/15})$.

If $\frac{13}{8} \leq \ell \leq 2$, the query complexity is upper bounded by $Q_1+Q_2+Q_3+Q_4+Q_5 = \tilde{O}(n^{a_1+d/2+k_1-1/2}+n^{1+b_1+d/2-a_1}+n^{3/2-b_1/2}+n^{3/2-k_1/2}+n^{a_2+d+k_2-\ell/2}+n^{1+b_2+\ell/2-a_2-d/2}+n^{1+\ell/2-b_2/2-d/2}+n^{1+\ell/2-d/2-k_2/2}+n^{b_3+\ell-a_3-d/2}+n^{1/2+\ell-b_3/2-d}+n^{1/2+\ell-d-k_3/2})$, which is optimized by taking $a_1 = \frac{3}{4}$, $a_2 = \frac{19}{20} - \frac{\ell}{10}$, $a_3 = \frac{3\ell}{5} - \frac{9}{20}$, $b_1 = k_1 = \frac{31}{30} - \frac{4\ell}{15}$, $b_2 = k_2 = \frac{19}{30} - \frac{\ell}{15}$, $b_3 = k_3 = \frac{7}{30} + \frac{2\ell}{15}$ and $d = \frac{4\ell}{5} - \frac{3}{5}$, giving the upper bound $\tilde{O}(n^{59/60+2\ell/15})$.

For the case $\ell \leq \frac{7}{6}$ we can obtain a better upper bound by using the $O(n + \sqrt{nm})$-query algorithm by Buhrman et al. [6]. This upper bound actually corresponds to a degenerate case appearing in our approach: the case $d = 0$. Indeed, observe that without loss of generality we can assume that $\deg(v) \geq 1$ for all vertices $v \in V$ (for instance by adding dummy vertices to the graph). In this case we have $\mathcal{V}_h^d = V$ for $d = 0$, which means that we do not need to apply the algorithm of Proposition 1 in order to obtain a classification: we simply output $V_h^0 = V$ and $V_l^0 = \emptyset$ (i.e., all the vertices of the graph are 0-high). The only type of triangles we need to consider is triangles with three 0-high vertices, which can be found with complexity $O(n + \sqrt{nm})$ by Proposition 7 (in this case the algorithm of Proposition 7 is exactly the same as the algorithm in [6]). □

Acknowkedgments. The authors are grateful to Mathieu Laurière, Keiji Matsumoto, Harumichi Nishimura and Seiichiro Tani for helpful comments. This work is supported by the Grant-in-Aid for Young Scientists (B) No. 24700005, the Grant-in-Aid for Scientific Research (A) No. 24240001 of the JSPS and the Grant-in-Aid for Scientific Research on Innovative Areas No. 24106009 of the MEXT.

References

1. Alon, N., Yuster, R., Zwick, U.: Finding and counting given length cycles. Algorithmica **17**, 354–364 (1997)
2. Ambainis, A.: Quantum walk algorithm for element distinctness. SIAM J. Comput. **37**(1), 210–239 (2007)
3. Ambainis, A.: Quantum search with variable times. Theory Comput. Syst. **47**(3), 786–807 (2010)
4. Belovs, A.: Span programs for functions with constant-sized 1-certificates: extended abstract. In: Proceedings of the 44th Symposium on Theory of Computing, pp. 77–84 (2012)
5. Brassard, G., Høyer, P., Tapp, A.: Quantum counting. In: Larsen, K.G., Skyum, S., Winskel, G. (eds.) ICALP 1998. LNCS, vol. 1443, pp. 820–831. Springer, Heidelberg (1998)
6. Buhrman, H., Dürr, C., Heiligman, M., Høyer, P., Magniez, F., Santha, M., de Wolf, R.: Quantum algorithms for element distinctness. SIAM J. Comput. **34**(6), 1324–1330 (2005)
7. Childs, A.M., Kothari, R.: Quantum query complexity of minor-closed graph properties. SIAM J. Comput. **41**(6), 1426–1450 (2012)
8. Le Gall, F.: Improved quantum algorithm for triangle finding via combinatorial arguments. In: Proceedings of the 55th IEEE Annual Symposium on Foundations of Computer Science, pp. 216–225 (2014)
9. Le Gall, F., Nakajima, S.: Quantum algorithm for triangle finding in sparse graphs. arXiv:1507.06878
10. Grover, L.K.: A fast quantum mechanical algorithm for database search. In: Proceedings of the 28th Symposium on the Theory of Computing, pp. 212–219 (1996)
11. Itai, A., Rodeh, M.: Finding a minimum circuit in a graph. SIAM J. Comput. **7**(4), 413–423 (1978)
12. Jeffery, S., Kothari, R., Magniez, F.: Nested quantum walks with quantum data structures. In: Proceedings of the 24th Annual ACM-SIAM Symposium on Discrete Algorithms, pp. 1474–1485 (2013)
13. Lee, T., Magniez, F., Santha, M.: Improved quantum query algorithms for triangle finding and associativity testing. In: Proceedings of the 24th Annual ACM-SIAM Symposium on Discrete Algorithms, pp. 1486–1502 (2013)
14. Magniez, F., Nayak, A., Roland, J., Santha, M.: Search via quantum walk. SIAM J. Comput. **40**(1), 142–164 (2011)
15. Magniez, F., Santha, M., Szegedy, M.: Quantum algorithms for the triangle problem. SIAM J. Comput. **37**(2), 413–424 (2007)
16. Patrascu, M.: Towards polynomial lower bounds for dynamic problems. In: Proceedings of the 42nd Symposium on Theory of Computing, pp. 603–610, (2010)
17. Williams, V.V., Williams, R.: Subcubic equivalences between path, matrix and triangle problems. In: Proceedings of the 51th Symposium on Foundations of Computer Science, pp. 645–654 (2010)
18. Williams, V.V., Williams, R.: Finding, minimizing, and counting weighted subgraphs. SIAM J. Comput. **42**(3), 831–854 (2013)

Graph Drawing and Planar Graphs

On Hardness of the Joint Crossing Number

Petr Hliněný[1] (✉) and Gelasio Salazar[2]

[1] Faculty of Informatics, Masaryk University Brno, Brno, Czech Republic
hlineny@fi.muni.cz
[2] Instituto de Fisica, Universidad Autonoma de San Luis Potosi, San Luis Potosi, Mexico
gsalazar@ifisica.uaslp.mx

Abstract. The Joint Crossing Number problem asks for a simultaneous embedding of two disjoint graphs into one surface such that the number of edge crossings (between the two graphs) is minimized. It was introduced by Negami in 2001 in connection with diagonal flips in triangulations of surfaces, and subsequently investigated in a general form for small-genus surfaces. We prove that all of the commonly considered variants of this problem are NP-hard already in the orientable surface of genus 6, by a reduction from a special variant of the anchored crossing number problem of Cabello and Mohar.

1 Introduction

Motivated by his investigation on diagonal flips in triangulations of surfaces [5], Negami introduced in [6] the concept of *joint crossing numbers*. The general setup consists of two graphs embeddable on the same surface, and the problem is to find a simultaneous embedding into this surface, so that the number of edge crossings is minimized (Since both graphs are embedded, every crossing must involve an edge from each of the graphs.).

In Negami's original definition, the embedded graphs were allowed to share vertices and edges (this is the *diagonal crossing number*). In the subsequent papers on joint crossing numbers, the attention has been restricted to the case in which the corresponding graphs are disjoint. This mainstream case is the one we focus on in this work, and we restrict the attention to orientable surfaces.

Within this case (the graphs G_1, G_2 to be jointly embedded in the same surface Σ are disjoint), three variants proposed by Negami have been studied. In the first one, the aim is to minimize the number of crossings in *any* embedding of the disjoint union G_1+G_2 of G_1 and G_2; this is simply the *joint crossing number*. In the second variant, the *joint homeomorphic crossing number*, embeddings of G_1 and G_2 are already given, and one must embed G_1+G_2 so that the restriction of this embedding to each G_i is homeomorphic to the prescribed embedding of G_i. In the third, and most restricted variant, the *joint orientation-preserving*

P. Hliněný—Supported by the research centre Institute for Theoretical Computer Science (CE-ITI); Czech Science foundation project No. P202/12/G061.
G. Salazar—Supported by the CONACYT Grant 222667.

K. Elbassioni and K. Makino (Eds.): ISAAC 2015, LNCS 9472, pp. 603–613, 2015.
DOI: 10.1007/978-3-662-48971-0_51

homeomorphic crossing number, in addition, the restrictions of the embedding of $G_1 + G_2$ to each G_i must be orientation-preserving homeomorphic to the prescribed embedding of G_i (See the next section for more rigorous definitions.).

Relatively little is known on either of these variants. In [6], Negami bounded the homeomorphic crossing number in terms of the Betti numbers of the graphs and the genus of Σ. In [1], Archdeacon and Bonnington calculated the exact homeomorphic crossing number of two graphs embedded in the projective plane, and also gave lower and upper bounds, within a constant factor of each other, for the case in which the host surface is the torus (Negami also obtained some nontrivial bounds for toroidal joint embeddings in [6]). Richter and Salazar investigated in [7] the case in which both graphs are densely embedded.

The associated algorithmic problems are the following:

JOINT CROSSING NUMBER
Input: Graphs G_1, G_2 embeddable in a given surface Σ, and an integer k.
Question: Is the joint crossing number of G_1 and G_2 in Σ at most k?

JOINT HOMEOMORPHIC CROSSING NUMBER
Input: Embeddings of each of two disjoint graphs G_1, G_2 in a surface Σ, and k.
Question: Is the joint homeomorphic crossing number of G_1 and G_2 at most k?

JOINT OP-HOMEOMORPHIC CROSSING NUMBER
Input: Embeddings of each of two disjoint graphs G_1, G_2 in a surface Σ, and k.
Question: Is the joint orientation-preserving homeomorphic crossing number of G_1 and G_2 in Σ at most k?

It follows from [1, Theorem 2.2] that the last two problem variants are easy in the projective plane (it suffices to calculate the dual widths of the embeddings). The aforementioned results also suggest (although this is an open problem) that optimal solutions in the case of Σ being the torus, can always be obtained in a particularly nice way: embed all the vertices of one of the graphs, say G_1, in the same face of the other graph G_2 (and then route the excessive edges of G_1 across G_2). This nice property ceases to be true for the homeomorphic variant already in the double torus [7], and our results imply that the property fails really badly for all higher genus surfaces and all problem variants.

In his comprehensive survey [8] of the many different variants of crossing number definitions, Marcus Schaefer marks the complexity of all the aforementioned variants of the joint crossing number as open. These problems are all easily seen to be in NP, so the open problem is their hardness. Our main result in this paper settles this question.

Theorem 1.1. JOINT CROSSING NUMBER, JOINT HOMEOMORPHIC CROSSING NUMBER, *and* JOINT OP-HOMEOMORPHIC CROSSING NUMBER *are NP-hard problems in any orientable surface of genus* 6 *or higher. This remains true even if the inputs are restricted to simple* 3*-connected graphs.*

The proof of this theorem is via a chain of reductions from a special variant of the anchored crossing number problem of Cabello and Mohar [3].

The rest of this paper is organized as follows. In Sect. 2 we give rigorous definitions of the variants of the joint embedding problem we analyze, and review some basic concepts. To work out the reduction from Cabello-Mohar's anchored crossing number, we devise joint embeddings in which certain vertices of one of the graphs are required to lie in prescribed faces of the other embedding. These *face-anchored joint embeddings* are developed in Sect. 3. An additional fine-tuning of the construction is given in Sect. 4. The reduction to the anchored crossing number is laid out in Sect. 5. Finally, in Sect. 6 we present some concluding remarks, among which we give back a slight strengthening of the main result of aforementioned [3].

2 Basic Concepts

We follow the standard notation of letting $G_1 + G_2$ denote the graph obtained as the disjoint union of two graphs G_1, G_2. A *toroidal grid* of size $p \times q$ is the Cartesian product of a p-cycle with a q-cycle; this is a 4-regular graph consisting of an edge disjoint union of q copies of a p-cycle and p copies of a q-cycle. For each integer $h \geq 0$, we let $\mathcal{S}_h$ denote the orientable surface of genus h.

We recall that in a *drawing* of a graph G in a surface Σ, vertices are mapped to points and edges are mapped to simple curves (arcs) such that the endpoints of an arc are the vertices of the corresponding edge; no arc contains a point that represents a non-incident vertex. For simplicity, we often make no distinction between the topological objects of a drawing (points and arcs) and their corresponding graph theoretical objects (vertices and edges). A *crossing* in a drawing is an intersection point of two edges in a point other than a common endvertex. An *embedding* of a graph in a surface is a drawing with no edge crossings.

Let G_1, G_2 be disjoint graphs, both of which embed in the same orientable surface Σ. A drawing G^0 of the graph $G_1 + G_2$ in Σ is called a *joint embedding of* (G_1, G_2) if the restriction of G^0 to G_i, for each $i = 1, 2$, is an embedding. Furthermore, if prescribed embeddings G_1^0, G_2^0 of G_1, G_2 are given and the restriction of G^0 to G_i is homeomorphic (respectively, orientation-preserving homeomorphic) to G_i^0, $i = 1, 2$, then G^0 is a *joint homeomorphic embedding of* (G_1^0, G_2^0) (respectively, *joint orientation-preserving homeomorphic embedding of* (G_1^0, G_2^0)) in Σ. Loosely speaking, in the joint homeomorphic variant(s), one is only allowed to "deform" the prescribed embeddings of G_1, G_2 across the host surface.

Note that in any joint embedding of (G_1, G_2), crossings may arise only between an edge of G_1 and an edge of G_2. The *joint crossing number* of (G_1, G_2) in Σ is the minimum number of crossings over all joint embeddings of (G_1, G_2) in Σ. The *joint homeomorphic crossing number* and *joint orientation-preserving homeomorphic crossing number* are defined analogously.

In order to resolve the ordinary and homeomorphic variants of joint crossing number problems at once, we introduce the following. An instance (G_1, G_2) of the joint crossing number problem in Σ is called *orientation-preserving homeo-invariant* if the input graphs G_1, G_2 are given together with embeddings G_1', G_2' in Σ, and the following holds: there exists a joint embedding G^0 of (G_1, G_2), achieving the joint crossing number, such that the subembedding of G^0 restricted to G_i is orientation-preserving homeomorphic to G'_i, for $i = 1, 2$.

Note the important difference—while in the joint orientation-preserving homeomorphic crossing number problem we require the considered joint embeddings to respect the given homeomorphism classes of G_1^0, G_2^0 (a *restriction*), for an orientation-preserving homeo-invariant instance we admit all joint embeddings, but we know that some of the optimal solutions will respect the homeomorphism classes of G'_1, G'_2 (a *promise*). We call OP-Homeo-Invariant Joint Crossing Number problem the ordinary Joint Crossing Number problem with inputs restricted only to orientation-preserving homeo-invariant instances.

The following is a useful artifice in crossing numbers research. In a *weighted* graph, each edge is assigned a positive number (the *weight, or thickness* of the edge). Now the *weighted joint crossing number* is defined as the ordinary joint crossing number, but a crossing between edges e_1 and e_2, say of weights t_1 and t_2, contributes $t_1 t_2$ to the weighted joint crossing number. The weighted variants of the joint homeomorphic crossing number and of the joint orientation-preserving homeomorphic crossing numbers are defined analogously. The following reduction is easily proved using folklore tricks for transforming weighted graphs into ordinary ones; the fact that we are not in the plane does not play a role here.

Proposition 2.1 (folklore). *There is a polynomial-time reduction from the weighted joint crossing number problem, with edge weights encoded in unary, to the unweighted joint crossing number problem. Moreover, this reduction can preserve* 3*-connectivity and simplicity of the graphs.*

3 Face-Anchored Joint Embeddings

For the purpose of intermediate reduction we introduce the following variant of the concept of joint embedding of (G_1, G_2). Assume that $C_1, \dots, C_k$ are cycles of the graph G_1 such that there exists an embedding of G_1 in Σ in which each of $C_1, \dots, C_k$ is a facial cycle. Let $a_1, \dots, a_k \in V(G_2)$. A joint embedding G^0 of (G_1, G_2) in Σ is called *face-anchored with respect to* $\{(C_i, a_i) : i = 1, \dots, k\}$, if the restriction of G^0 to G_1 contains a face α_i bounded by C_i such that the vertex a_i of G_2 is drawn inside α_i, for all $i = 1, \dots, k$. The pairs (C_i, a_i) are the *face anchors* of this joint embedding problem, where each α_i bounded by C_i is an *anchor face* and each a_i is an *anchor vertex*.

We will consider face-anchored joint embeddings and their crossing number only in the case of Σ being the sphere $\mathcal{S}_0$ and k being a constant, and then we specifically speak about *face-anchored joint planar embeddings*, and call the corresponding algorithmic problem k-FA Joint Planar Crossing Number. If inputs of this problem are restricted only to instances which are orientation-preserving homeo-invariant (cf. Section 2), then we speak about the OP-Homeo-Invariant k-FA Joint Planar Crossing Number problem.

Theorem 3.1. *For every integer $h \geq 1$, there is a polynomial-time reduction from the* OP-Homeo-Invariant h-FA Joint Planar Crossing Number

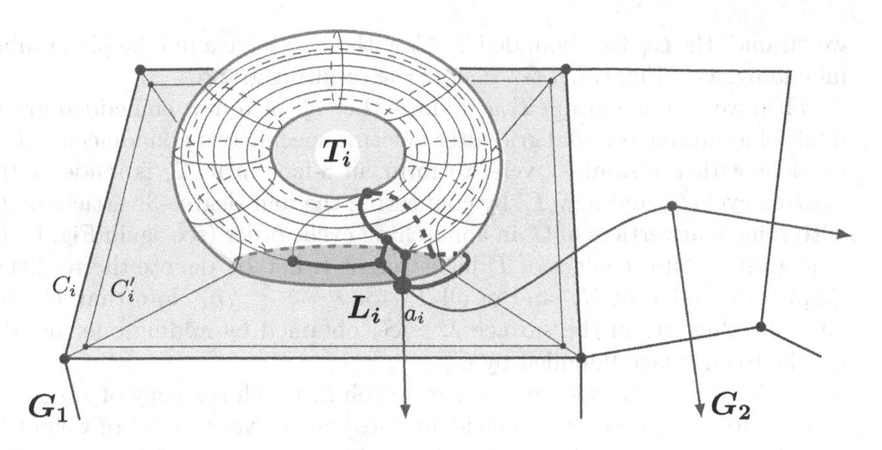

Fig. 1. A schematic detail of replacing one face anchor with a toroidal gadget, as used in the proof of Theorem 3.1 (the torus attaches to the light-gray face via the gray hole) (Color figure online).

problem to the OP-Homeo-Invariant Joint Crossing Number *problem in the surface* $\mathcal{S}_h$. *This reduction preserves connectivity of the involved graphs, and the target problem can then be restricted to simple 3-connected graphs.*

Proof. By Proposition 2.1, we may consider the source crossing problem as *unweighted* and to reduce to the *weighted* joint crossing number problem in $\mathcal{S}_h$, as long as the weights are polynomial in the input size.

Consider an unweighted input (G_1, G_2) of the OP-Homeo-Invariant h-FA Joint Planar Crossing Number problem, given along with the h face anchors $\{(C_i, a_i) : i = 1, \ldots, h\}$, and with planar embeddings G_1', G_2' of G_1, G_2 witnessing the homeo-invariant property. To prove the theorem it suffices to construct (in polynomial time) a pair (H_1', H_2') of $\mathcal{S}_h$-embedded graphs such that, denoting by H_1, H_2 the corresponding abstract graphs, the following holds:

- if s is the (unknown) face-anchored joint planar crossing number of (G_1, G_2), then the joint orientation-preserving homeomorphic (weighted) crossing number of (H_1', H_2') is at most $f(s)$ (for a suitable function f); and
- if the joint crossing number of (H_1, H_2) is at most $f(s)$ for some integer s, then the face-anchored joint planar crossing number of (G_1, G_2) is at most s.

We may assume that each cycle C_i is of length at least 4 (otherwise, we just subdivide it). Our construction of (H_1', H_2') can be shortly outlined as follows.

(i) We assign to every edge of $G_1 + G_2$ the same suitable weight p ("medium thick"). The purpose is that already a change in one crossing between G_1 and G_2 would cause a difference of p^2 in the target problem, a value larger than all future required crossings between "light" edges of weight 1 and other edges of weight up to p.

(ii) For each $i = 1, \ldots, h$, we create a disjoint copy C_i' of weight 1 of the anchor cycle C_i, and connect each vertex of C_i' with its master copy in C_i. Informally, we "frame" the G'_1-face bounded by C_i with C'_i to force a unique plane subembedding, as in Fig. 1. Let G_1^+ denote the resulting graph.

Then we create a graph T_i as follows. Let T_i^0 be a new embedded graph made of a suitable toroidal grid after deleting specific two nonadjacent edges incident with the same 4-cycle, to form an 8-face in it. T_i is made of the existing cycle C'_i and new T_i^0 by connecting the four degree-3 vertices of T_i^0 with some four vertices of C'_i in a matching cyclic order (see again Fig. 1, the blue graph). All the edges of T_i have weight 1. Let H_1 denote the resulting graph—the union of G_1^+ and of all T_i, for $i = 1, \dots, h$. Note that H_1 has an embedding H'_1 in the surface $\Sigma \simeq \mathcal{S}_h$ obtained by adding one toroidal handle to each face bounded by C'_i.

(iii) For each $i = 1, \dots, h$, we create a new graph L_i which is a copy of $K_{3,3}$ with seven of its edges (except two incident ones) made "very thick" of weight t_i. Let H_2 denote the graph made of G_2 and all L_i after identifying one vertex of L_i with a_i, for $i = 1, \dots, h$. Then H_2 has an embedding H'_2 in Σ, such that G'_2 is a subembedding of H'_2. See the red graph in Fig. 1.

The informal purpose of such construction is two-fold; first, the nonplanar graph L_i must "use" some of the handles of Σ, and second, the thick edges of L_i cannot cross any edge of G'_1 which now have weight p. Consequently, each L_i is "confined" to one of the G'_1-faces α_j bounded by C'_j. Moreover, it will be shown that no two $L_i, L_{i'}$ for $i \neq i'$ are confined to the same face α_j.

(iv) Additional detailed arguments ensure that (iii) actually confines L_i, and hence also the anchor vertex a_i, to α_i for $i = 1, \dots, h$. Briefly explaining this argument: for a sufficiently large integer t we choose $t_i := (h+1-i) \cdot t$, and we choose the grid in each T_j gadget such that the least number of edges of T_j that have to be crossed by a noncontractible loop on the toroidal handle of T_j equals $g_j := 5 + j$. It is then an easy exercise in calculus to argue that the joint crossing number is minimized only if "t_i is matched with g_i".

In other words, informally, an optimal joint embedding solution of (H_1, H_2) must "contain" a feasible solution of (G_1, G_2), and an optimal orientation-preserving homeomorphic solution of (G'_1, G'_2) "generates" a good orientation-preserving homeomorphic solution of (H'_1, H'_2). Further technical details clarifying the stated proof outline, and making the target graphs 3-connected, are left for the full preprint [4] due to space restrictions. □

4 Multiplying Face Anchors

Recall that our ultimate goal is to find a reduction from a special variant of the anchored crossing number problem [3], described in Sect. 5. This can already be achieved with Theorem 3.1, but such an approach would require an unbounded number of face anchors (and hence unbounded genus in the Joint crossing number problem). We thus present the following construction which "multiplies" the number of available face anchors, albeit in a special position.

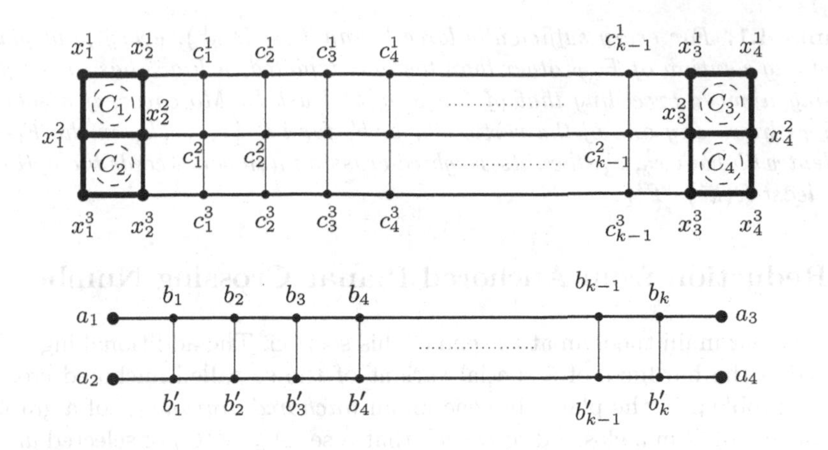

Fig. 2. The graphs F_1 (top) and F_2 (bottom) of the face-anchored joint planar embedding problem $\mathcal{F}_{k,T}$; the precise weights of the edges are specified in (F1)–(F6) below.

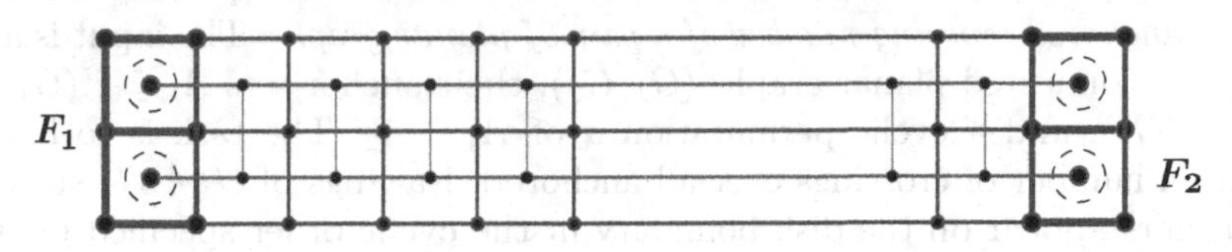

Fig. 3. Supposed crossing-optimal face-anchored joint planar embedding of $\mathcal{F}_{k,T}$.

Let F_1 be the graph of the $3 \times (k+3)$ plane grid, and F_2 be obtained from the $2 \times (k+2)$ plane grid by removing the two side edges (making a "ladder"), with notation as in Fig. 2. Let C_1 denote the cycle $(x_1^1, x_2^1, x_2^2, x_1^2)$ of F_1, and C_2, C_3, C_4 the cycles $(x_1^2, x_2^2, x_2^3, x_1^3)$, $(x_3^1, x_4^1, x_4^2, x_3^2)$, $(x_3^2, x_4^2, x_4^3, x_3^3)$. The weights of the edges of F_1 are as follows (where T is a large integer):

(F1) weight T^3 for the six edges $x_2^1x_2^2, x_2^2x_2^3, x_3^1x_3^2, x_3^2x_3^3, x_4^1x_4^2, x_4^2x_4^3$ and
(F2) weight T^4 for the remaining eight edges induced on the vertex set $\{x_j^i : i \in \{1,2,3\}, j \in \{1,2,3,4\}\}$ (yes, this part is intentionally not symmetric),
(F3) weight T^2 for every "horizontal" edge on the shortest paths from x_2^i to x_3^i, for $i = 1, 2, 3$,
(F4) weight jT for the "vertical" edges $c_j^1c_j^2$ and $c_{k-j}^2c_{k-j}^3$, for $j = 1, 2, \ldots, k-1$.

The weights of the edges of F_2 are as follows:

(F5) weight t_{j-1} for the "horizontal" edges $b_{j-1}'b_j'$ and $b_{k+2-j}b_{k+1-j}$, for $j = 1, 2, \ldots, k+1$ where $b_0 = a_1, b_0' = a_2, b_{k+1} = a_3, b_{k+1}' = a_4$, and t_j is defined by $t_0 = k^3$ and $t_j = t_{j-1} + j$,
(F6) weight $k+1$ for all the "vertical" edges b_jb_j', for $j = 1, 2, \ldots, k$.

Finally, we shortly denote by $\mathcal{F}_{k,T}$ the joint planar embedding instance of (F_1, F_2) with the set of four face anchors $\{(C_i, a_i) : i = 1, 2, 3, 4\}$. The details of the following claim are left for the full preprint [4] due to space restrictions:

Lemma 4.1. *For every sufficiently large k and $T = \Omega(k^6)$, every joint planar embedding solution of $\mathcal{F}_{k,T}$ other than the one depicted in Fig. 3 has its weighted crossing number exceeding that of Fig. 3 by at least T. Moreover, if a solution of $\mathcal{F}_{k,T}$ draws any one of the vertices b_i or b'_i for $i \in \{1, \ldots, k\}$ in the F_1-face incident with both x_2^1, x_3^1, then its weighted crossing number exceeds the optimum by at least $\Omega(k^3) \cdot T^2$.*

5 Reduction from Anchored Planar Crossing Number

We prove our main theorem at the end of this section. The additional ingredient we need is the hardness of a special variant of the so-called anchored crossing number problem in the plane. In general, an *anchored drawing* [3] of a graph G is a drawing of G in a closed disc D such that a set $A \subseteq V(G)$ of selected *anchor* vertices are placed in specific points of the boundary of D and the rest of the drawing lies in the interior of D.

We shall use the following very restrictive version of the problem which we call the *anchored crossing number of a pair of planar graphs*: The input is a pair of disjoint connected planar graphs (G_1, G_2), their anchor sets $A_1 \subseteq V(G_1)$ and $A_2 \subseteq V(G_2)$, and a cyclic permutation σ of $A_1 \cup A_2$. The task is to find the minimum number of crossings over all anchored drawings of $G_1 + G_2$ such that the anchors appear on the disk boundary in the cyclic order specified by σ. As before, the problem is considered in the edge weighted form.

Theorem 5.1 (Cabello and Mohar, [3]). The anchored weighted crossing number problem of the pair of planar graphs (G_1, G_2), with anchor sets (A_1, A_2) and permutation σ, is NP-hard even under the following assumptions:

(A1) each of the graphs G_1,G_2 itself has a unique anchored embedding, and
(A2) there is a partition $A_2 = A_2^1 \cup A_2^2 \cup A_2^3 \cup A_2^4$ such that, for $i = 1, 2, 3, 4$, the set A_2^i is consecutive in σ restricted to A_2, and the set of edges incident with A_2^i forms a minimum weight cut in G_2 separating A_2^i from $A_2 \setminus A_2^i$.

Figure 4 illustrates the hardness construction used in [3], and the conditions (A1) and (A2) of Theorem 5.1, which are not explicitly stated in [3] but can easily be verified there.

Notice, moreover, in Fig. 4 that the graph G_2 also has some "diagonal" minimum weight cuts which use the dashed red edges of weight only $w - 1$. Hence, for example, every minimum weight cut of G_2 between $A_2^1 \cup A_2^2$ and $A_2^3 \cup A_2^4$ has to use some of the dashed red edges and so cannot have all its edges incident to A_2. Consequently, the partition of A_2 into the four sets in (A2) is not just an artifact of the visual shape of G_2 in Fig. 4 but necessity.

We now establish the final key reduction required to prove Theorem 1.1.

Theorem 5.2. *There is a polynomial reduction from the special anchored crossing number problem given in Theorem 5.1 to the* OP-Homeo-Invariant 6-FA Joint Planar Crossing Number *problem.*

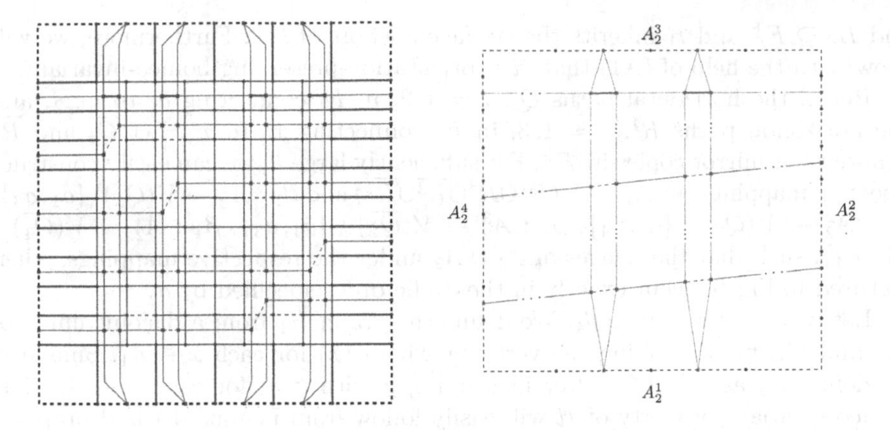

Fig. 4. An example of the construction of a hard anchored crossing number instance (G_1, G_2) taken from [3]: G_1 is blue and G_2 is red (detailed alone on the right). The solid thin red edges all have weight w (where w is a large integer) and the middle dashed red edges have weight $w - 1$ (Color figure online).

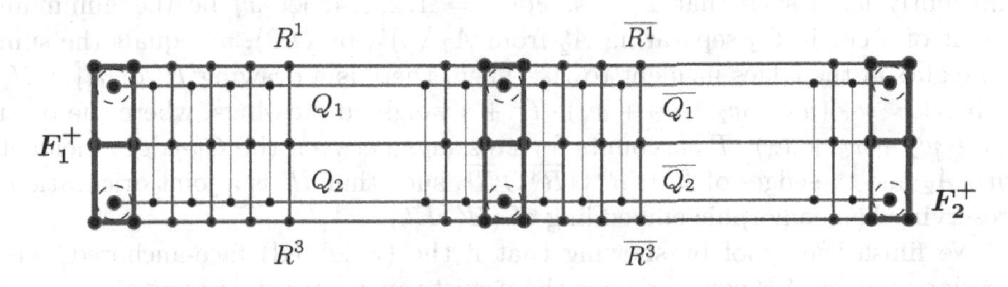

Fig. 5. A "join" of the instance $\mathcal{F}_{k,T}$ from Fig. 3 and of its horizontal mirror copy, giving a planar joint embedding instance $\mathcal{F}^+$ with 6 face anchors.

Proof. We will use the instance $\mathcal{F}_{k,T}$ (where sufficiently large k, T will be specified later) of joint planar embedding of (F_1, F_2) with the face anchors $\{(C_i, a_i) : i = 1, 2, 3, 4\}$, from Sect. 4 in the following way. The graph F_1 is joined with its mirror copy such that the anchor faces C_3, C_4 get identified with $\bar{C}_3, \bar{C}_4$ of the copy in a "horizontal mirror" way, resulting in the graph F_1^+. Similarly, F_2^+ results by joining F_2 with its mirror copy and identifying a_3, a_4 with the copies $\bar{a}_3, \bar{a}_4$, respectively. The resulting instance of joint planar embedding of (F_1^+, F_2^+) with the six face anchors $\{(C_1, a_1), (C_2, a_2), (C_3 = \bar{C}_3, a_3 = \bar{a}_3), (C_4 = \bar{C}_4, a_4 = \bar{a}_4), (\bar{C}_1, \bar{a}_1), (\bar{C}_2, \bar{a}_2)\}$, as depicted in Fig. 5, will be shortly denoted by $\mathcal{F}^+$.

Let $\mathrm{cr}(\mathcal{F}^+)$ shortly denote the (optimum) weighted crossing number of this instance $\mathcal{F}^+$, which equals twice the value by Lemma 4.1.

Consider an instance of the anchored crossing number problem, i.e., a pair of weighted planar graphs (G_1, G_2) with anchor sets (A_1, A_2) and permutation σ satisfying (A1) and (A2) for the partition $A_2 = A_2^1 \cup A_2^2 \cup A_2^3 \cup A_2^4$ in a suitable cyclic order. Our aim is to construct from it an instance $\mathcal{H}$ of face-anchored joint planar crossing number, formed by a pair of graphs (H_1, H_2), such that $H_1 \supseteq F_1^+$

and $H_2 \supseteq F_2^+$ and $\mathcal{H}$ inherits the six face anchors of $\mathcal{F}^+$. Furthermore, we will show with the help of (A1) that $\mathcal{H}$ is orientation-preserving homeo-invariant.

Recall the horizontal paths Q_i, $i = 1, 2$, in F_2 connecting a_i to a_{i+2}, and the horizontal paths R^j, $j = 1, 3$, in F_1 connecting x_2^j to x_3^j. Let $\overline{Q_i}$ and $\overline{R^j}$ denote their mirror copies in $\mathcal{F}^+$. For sufficiently large k, we can easily construct injective mappings $\alpha : A_1 \to V(R^1 \cup R^3 \cup \overline{R^3} \cup \overline{R^1})$ and $\beta_1 : A_2^1 \to V(Q_1) \setminus \{a_1, a_3\}$, $\beta_2 : A_2^2 \to V(Q_2) \setminus \{a_2, a_4\}$, $\beta_3 : A_2^3 \to V(\overline{Q_2}) \setminus \{\bar{a}_1, a_3\}$, $\beta_4 : A_2^4 \to V(\overline{Q_1}) \setminus \{\bar{a}_2, a_4\}$, such that the images of $A_1 \cup A_2$ under the respective mappings, when pictured in Fig. 5, occur exactly in the cyclic order specified by σ.

Let $\beta = \beta_1 \cup \beta_2 \cup \beta_3 \cup \beta_4$. We define the graph H_1 from a disjoint union of F_1^+ and G_1, by identifying the vertex x with $\alpha(x)$ for each $x \in A_1$. Similarly, we define H_2 as $F_2^+ \cup G_2$ after identifying y with $\beta(y)$ for each $y \in A_2$. The homeo-invariant property of $\mathcal{H}$ will easily follow from Lemma 4.1 and property (A1) for the following pair of embeddings (H_1', H_2'): for $i = 1, 2$, H_i' is the unique plane embedding of H_i such that the restriction of H_i' to F_i^+ is as in Fig. 5.

Let the weighted anchored crossing number of (G_1, G_2) with anchor sets (A_1, A_2) and cyclic permutation σ equal s. We assume that $T = \Omega(k^6)$ is chosen sufficiently large such that $T > s$. For $i = 1, 2, 3, 4$, let w_i be the minimum weight of a cut in G_2 separating A_2^i from $A_2 \setminus A_2^i$; by (A2), w_i equals the sum of weights of the edges incident to A_2^i. Then, there is a drawing H' of $H_1 + H_2$ with $\mathrm{cr}(\mathcal{F}^+) + (w_1 + w_2 + w_3 + w_4) \cdot T^2 + s$ weighted crossings, where the term $(w_1 + w_2 + w_3 + w_4) \cdot T^2$ accounts for crossings between the G_2-edges incident with A_2 and the edges of $R^1 \cup R^3 \cup \overline{R^3} \cup \overline{R^1}$, such that H' is a joint orientation-preserving homeomorphic embedding of (H_1', H_2').

We finish the proof by showing that if the (weighted) face-anchored joint crossing number of $\mathcal{H}$ equals r, then there exists an anchored drawing of (G_1, G_2) respecting (A_1, A_2) and σ, with at most $r' := r - (w_1 + w_2 + w_3 + w_4) \cdot T^2 - \mathrm{cr}(\mathcal{F}^+)$ crossings. This will then automatically imply that the aforementioned drawing H' which is joint orientation-preserving homeomorphic to (H_1', H_2'), is also an optimal solution of $\mathcal{H}$. The details are again left for the full preprint [4]. □

Proof. (of Theorem 1.1). Theorem 1.1 for the JOINT CROSSING NUMBER problem and genus 6 follows imediately by the chain of reductions from Theorem 3.1, and Theorems 5.1 and 5.2. For genus greater than 6, it suffices to add dummy face anchors in the reduction of Theorem 3.1. Finally, for hardness of the HOMEOMORPHIC and OP-HOMEOMORPHIC variants, we can simply use the same reductions—by the orientation-preserving homeo-invariant promise, the (hard) instances produced by the chain of reductions have the same solution value in all the three problem variants. □

6 Conclusions

The following is another immediate consequence of Theorems 3.1 and 5.1:

Theorem 6.1. *The h-FA* JOINT PLANAR CROSSING NUMBER *problem is NP-hard for every $h \geq 6$.*

There is yet another interesting consequence. The main result of aforementioned [3] is that Crossing Number is NP-hard even on *almost-planar graphs*, i.e. those which can be made planar by removing one edge. Their hardness reduction, derived from hard anchored crossing number instances as shown in Fig. 4, essentially uses an unbounded number of vertices of arbitrarily high degrees. Elaborating on the reduction of our proof of Theorem 5.2, while using a special gadget derived from $\mathcal{F}^+$ turned inside out, we can give back the following strengthening (we omit the proof due to space constraints):

Theorem 6.2 (slight improvement upon [3]). The Crossing Number problem remains NP-hard even if the input is restricted to almost-planar graphs having a bounded number, namely at most 16, vertices of degree greater than 3.

Note that, on the other hand, Cabello and Mohar [2] prove that Crossing Number is solvable in linear time if the input is an almost-planar graph with all vertices except for the two of the planarizing edge having degree at most 3.

Another natural extension of our results would be to prove Theorem 1.1 for non-orientable surfaces. This is not inherently difficult—it suffices to replace the toroidal gadgets T_i (cf. Fig. 1) with suitable projective grids, and to use a crosscap instead of each toroidal handle. However, a formal statement would require us to repeat most of the arguments of the proof of Theorem 3.1, and hence we refrain from giving the full statement in this short paper.

A question worth further investigation is how small the genus in Theorem 1.1 and the number of face anchors in Theorem 6.1 can be for the statements to hold.

References

1. Archdeacon, D., Bonnington, C.P.: Two maps on one surface. J. Graph Theory **36**(4), 198–216 (2001)
2. Cabello, S., Mohar, B.: Crossing number and weighted crossing number of near-planar graphs. Algorithmica **60**(3), 484–504 (2011)
3. Cabello, S., Mohar, B.: Adding one edge to planar graphs makes crossing number and 1-planarity hard. SIAM J. Comput. **42**(5), 1803–1829 (2013)
4. Hliněný, P., Salazar, G.: On hardness of the joint crossing number (2015). Full version arXiv:1509.01787
5. Negami, S.: Diagonal flips in triangulations on closed surfaces, estimating upper bounds. Yokohama Math. J. **45**(2), 113–124 (1998)
6. Negami, S.: Crossing numbers of graph embedding pairs on closed surfaces. J. Graph Theory **36**(1), 8–23 (2001)
7. Bruce Richter, R., Salazar, G.: Two maps with large representativity on one surface. J. Graph Theory **50**(3), 234–245 (2005)
8. Schaefer, M.: The graph crossing number and its variants: a survey. Electronic Journal of Combinatorics, #DS21, May 15, 2014

An $O(n^\epsilon)$ Space and Polynomial Time Algorithm for Reachability in Directed Layered Planar Graphs

Diptarka Chakraborty(✉) and Raghunath Tewari

Department of Computer Science and Engineering,
Indian Institute of Technology Kanpur, Kanpur, India
{diptarka,rtewari}@cse.iitk.ac.in

Abstract. Given a graph G and two vertices s and t in it, *graph reachability* is the problem of checking whether there exists a path from s to t in G. We show that reachability in directed layered planar graphs can be decided in polynomial time and $O(n^\epsilon)$ space, for any $\epsilon > 0$. The previous best known space bound for this problem with polynomial time was approximately $O(\sqrt{n})$ space [1].

Deciding graph reachability in SC is an important open question in complexity theory and in this paper we make progress towards resolving this question.

1 Introduction

Given a graph and two vertices s and t in it, the problem of determining whether there is a path from s to t in the graph is known as the graph reachability problem. Graph reachability problem is an important question in complexity theory. Particularly in the domain of space bounded computations, the reachability problem in various classes of graphs characterize the complexity of different complexity classes. The reachability problem in directed and undirected graphs, is complete for the classes non-deterministic log-space (NL) and deterministic log-space (L) respectively [2,3]. The latter follows due to a famous result by Reingold who showed that undirected reachability is in L [3]. Various other restrictions of reachability have been studied in the context of understanding the complexity of other space bounded classes (see [4–6]). Wigderson gave a fairly comprehensive survey that discusses the complexity of reachability in various computational models [7].

The time complexity of directed reachability is fairly well understood. Standard graph traversal algorithms such as DFS and BFS solve this problem in linear time. We also have a $O(\log^2 n)$ space algorithm due to Savitch [8], however it requires $O(n^{\log n})$ time. The question, whether there exists a single algorithm that decides reachability in polynomial time and polylogarithmic space is unresolved. In his survey, Wigderson asked whether it is possible to design a polynomial time algorithm that uses only $O(n^\epsilon)$ space, for some constant $\epsilon < 1$ [7]. This question is also still open. In 1992, Barnes, Buss, Ruzzo and Schieber made

K. Elbassioni and K. Makino (Eds.): ISAAC 2015, LNCS 9472, pp. 614–624, 2015.
DOI: 10.1007/978-3-662-48971-0_52

some progress on this problem and gave an algorithm for directed reachability that requires polynomial time and $O(n/2^{\sqrt{\log n}})$ space [9].

Planar graphs are a natural topological restriction of general graphs consisting of graphs that can be embedded on the surface of a plane such that no two edges cross. *Grid graphs* are a subclass of planar graphs, where the vertices are placed at the lattice points of a two dimensional grid and edges occur between a vertex and its immediate adjacent horizontal or vertical neighbor.

Asano and Doerr provided a polynomial time algorithm to compute the *shortest path* (hence can decide reachability) in grid graphs which uses $O(n^{1/2+\epsilon})$ space, for any small constant $\epsilon > 0$ [10]. Imai et al. extended this to give a similar bound for reachability in planar graphs [1]. Their approach was to use a space efficient method to design a separator for the planar graph and use divide and conquer strategy. Note that although it is known that reachability in grid graphs reduces to planar reachability in log space, however since this class (polynomial time and $O(n^{1/2+\epsilon})$ space) is not closed under log space reductions, planar reachability does not follow from grid graph reachability. Subsequently the result of Imai et al. was extended to the class of *high-genus* and *H-minor-free* graphs [11]. Recently Asano et al. gave a $\tilde{O}(\sqrt{n})$ space and polynomial time algorithm for reachability in planar graphs, thus improving upon the previous space bound [12]. More details on known results can be found in a recent survey article [13].

In another line of work, Kannan et al. gave a $O(n^\epsilon)$ space and polynomial time algorithm for solving reachability problem in *unique path graphs* [14]. Unique path graphs are a generalization of *strongly unambiguous* graphs and reachability problem in strongly unambiguous graphs is known to be in SC (polynomial time and polylogarithmic space) [15,16]. Reachability in strongly unambiguous graphs can also be decided by a $O(\log^2 n/\log\log n)$ space algorithm, however this algorithm requires super polynomial time [17]. SC also contains the class *randomized log space* or RL [18]. We refer the readers to a recent survey by Allender [19] to further understand the results on the complexity of reachability problem in UL and on certain special subclasses of directed graphs.

Our Contribution

We show that reachability in directed layered planar graphs can be decided in polynomial time and $O(n^\epsilon)$ space for any constant $\epsilon > 0$. A layered planar graph is a planar graph where the vertex set is partitioned into layers (say L_0 to L_m) and every edge occurs between layers L_i and L_{i+1} only. Our result significantly improves upon the previous space bound due to [1,12] for layered planar graphs.

Theorem 1. *For every $\epsilon > 0$, there is a polynomial time and $O(n^\epsilon)$ space algorithm that decides reachability in directed layered planar graphs.*

Reachability in layered grid graphs (denoted as LGGR) is in UL which is a subclass of NL [20]. Subsequently this result was extended to the class of all planar graphs [21]. Allender et al. also gave some hardness results for the reachability problem in certain subclasses of layered grid graphs. Specifically they showed that, 1LGGR is hard for NC^1 and 11LGGR is hard for TC^0 [20]. Both these problems are however known to be contained in L though.

As a consequence of our result, it is easy to achieve the same time-space upper-bound for the reachability problem in *upward planar graphs*. We say that a graph is upward planar if it admits an upward planar drawing, i.e., a planar drawing where the curve representing each edge should have the property that every horizontal line intersects it in at most one point. In the domain of graph drawing, it is an important topic to study the upward planar drawing of planar DAGs [22,23]. It is NP-complete to determine whether a planar DAG with multiple sources and sinks has an upward planar drawing [24]. However, given an upward planar drawing of a planar DAG, the reachability problem can easily be reduced to reachability in a layered planar graph using only logarithmic amount of space and thus admits the same time-space upper bound as of layered planar graphs.

Firstly we argue that its enough to consider layered grid graphs (a subclass of general grid graphs). We divide a given layered grid graph into a courser grid structure along k horizontal and k vertical lines (see Fig. 1). We then design a modified DFS strategy that makes queries to the smaller graphs defined by these gridlines (we assume a solution in the smaller graphs by recursion) and visits every reachable vertex from a given start vertex. The modified DFS stores the highest visited vertex in each vertical line and the left most visited vertex in each horizontal line. We use this information to avoid visiting a vertex multiple number of times in our algorithm. We choose the number of horizontal and vertical lines to divide the graph appropriately to ensure that the algorithm runs in the required time and space bound.

The rest of the paper is organized as follows. In Sect. 2, we give some basic definitions and notations that we use in this paper. We also state certain earlier results that we use in this paper. In Sect. 3, we give a proof of Theorem 1.

2 Preliminaries

We will use the standard notations of graphs without defining them explicitly and follow the standard model of computation to discuss the complexity measures of the stated algorithms. In particular, we consider the computational model in which an input appears on a read-only tape and the output is produced on a write-only tape and we only consider an internal read-write tape in the measure of space complexity. Throughout this paper, by log we mean logarithm to the base 2. We denote the set $\{1, 2, \cdots, n\}$ by $[n]$. Given a graph G, let $V(G)$ and $E(G)$ denote the set of vertices and the set of edges of G respectively.

Definition 1 (Layered Planar Graph). *A planar graph $G = (V, E)$ is referred as* layered planar *if it is possible to represent V as a union of disjoint partitions, $V = V_1 \cup V_2 \cup \cdots \cup V_k$, for some $k > 0$, and for any two consecutive partitions V_i and V_{i+1}, there is a planar embedding of edges from the vertices of*

V_i to that of V_{i+1} and there is no edge between two vertices of non-consecutive partitions.

Now let us define the notion of layered grid graph and also note that grid graphs are by definition planar.

Definition 2 (Layered Grid Graph). *A directed graph G is said to be a $n \times n$* grid graph *if it can be drawn on a square grid of size $n \times n$ and two vertices are neighbors if their L_1-distance is one. In a grid graph a edge can have four possible directions, i.e., north, south, east and west, but if we are allowed to have only two directions north and east, then we call it a* layered grid graph.

We also use the following result of Allender et al. to simplify our proof [20].

Proposition 1 ([20]). *Reachability problem in directed layered planar graphs is log-space reducible to the reachability problem in layered grid graphs.*

2.1 Class nSC and its properties

$\mathsf{TISP}(t(n), s(n))$ denotes the class of languages decided by a deterministic Turing machine that runs in $O(t(n))$ time and $O(s(n))$ space. Then, $\mathsf{SC} = \mathsf{TISP}(n^{O(1)}, (\log n)^{O(1)})$. Expanding the class SC, we define the complexity class nSC (short for `near-SC`) in the following definition.

Definition 3 (Complexity Class `near-SC` or nSC). *For a fixed $\epsilon > 0$, we define $\mathsf{nSC}_{\epsilon} := \mathsf{TISP}(n^{O(1)}, n^{\epsilon})$. The complexity class nSC is defined as*

$$\mathsf{nSC} := \bigcap_{\epsilon>0} \mathsf{nSC}_{\epsilon}.$$

We next show that nSC is closed under log-space reductions. This is an important property of the class nSC and will be used to prove Theorem 1. Although the proof is quite standard, but for the sake of completeness we provide it here.

Theorem 2. *If $A \leq_l B$ and $B \in \mathsf{nSC}$, then $A \in \mathsf{nSC}$.*

Proof. Let us consider that a log-space computable function f be the reduction from A to B. It is clear that for any $x \in A$ such that $|x| = n$, $|f(x)| \leq n^c$, for some constant $c > 0$. We can think that after applying the reduction, $f(x)$ appears in a separate write-once output tape and then we can solve $f(x)$, which is an instance of the language B and now the input length is at most n^c. Now take any $\epsilon > 0$ and consider $\epsilon' = \frac{\epsilon}{c} > 0$. $B \in \mathsf{nSC}$ implies that $B \in \mathsf{nSC}_{\epsilon'}$ and as a consequence, $A \in \mathsf{nSC}_{\epsilon}$. This completes the proof.

3 Reachability in Layered Planar Graphs

In this section we prove Theorem 1. We show that the reachability problem in layered grid graphs (denoted as LGGR) is in nSC (Theorem 3). Then by applying Proposition 1 and Theorem 2 we have the proof of Theorem 1.

Theorem 3. LGGR $\in$ nSC.

To establish Theorem 3 we define an auxiliary graph in Sect. 3.1 and give the required algorithm in Sect. 3.2.

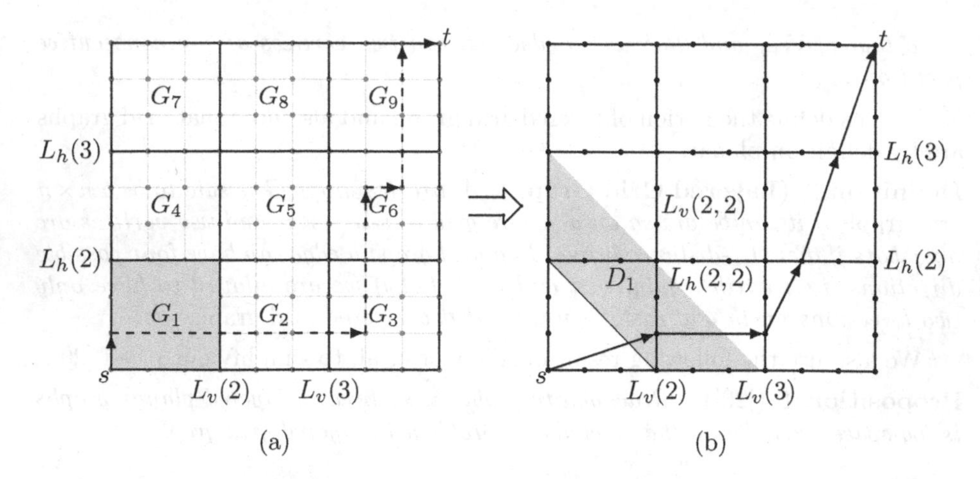

Fig. 1. (a) An example of layered grid graph G and its decomposition into blocks (b) Corresponding auxiliary graph H

3.1 The Auxiliary Graph H

Let G be a $n \times n$ layered grid graph. We denote the vertices in G as (i, j), where $0 \leq i, j \leq n$ and without loss of generality, we can assume that $s = (0, 0)$ and $t = (n, n)$. Let k be a parameter that determines the number of pieces in which we divide G. We will fix the value of k later to optimize the time and space bounds. Assume without loss of generality that k divides n. Given G we construct an auxiliary graph H as described below.

Divide G into k^2 many *blocks* (will be defined shortly) of dimension $n/k \times n/k$. More formally, the vertex set of H is

$$V(H) := \{(i, j) \mid i \text{ or } j \text{ is a non-negative multiple of } n/k.\}$$

Note that $V(H) \subseteq V(G)$. We consider k^2 many blocks $G_1, G_2, \cdots, G_{k^2}$, where a vertex $(i, j) \in V(G_l)$ if and only if $i'\frac{n}{k} \leq i \leq (i'+1)\frac{n}{k}$ and $j'\frac{n}{k} \leq j \leq (j'+1)\frac{n}{k}$, for some integer $i' \geq 0$ and $j' \geq 0$ and the vertices for which any of the four inequalities becomes equality, will be referred as *boundary vertices*. Moreover, we have $l = i' \cdot k + j' + 1$. $E(G_l)$ is the set of edges in G induced by the vertex set $V(G_l)$.

For every $i \in [k+1]$, let $L_h(i)$ and $L_v(i)$ denote the set of vertices, $L_h(i) := \{(i', j') | j' = (i-1)\frac{n}{k}\}$ and $L_v(i) := \{(i', j') | i' = (i-1)\frac{n}{k}\}$. When it is clear from the context, we will also use $L_h(i)$ and $L_v(i)$ to refer to the corresponding gridline in H. Observe that H has $k+1$ vertical gridlines and $k+1$ horizontal gridlines.

For every pair of vertices $u, v \in V(G_l) \cap V(H)$ for some l, add the edge (u, v) to $E(H)$ if and only if there is a path from u to v in G_l, unless $u, v \in L_v(i)$ or $u, v \in L_h(i)$ for some i. Also for every pair of vertices $u, v \in V(G_l)$ for some l, such that $u = (i_1, j_1)$ and $v = (i_2, j_2)$, where $i_1 = i_2 = i'\frac{n}{k}$ for some i' and

$j_1 = j'\frac{n}{k}$, $j_2 = (j'+1)\frac{n}{k}$ for some j', or $j_1 = j_2 = j'\frac{n}{k}$ for some j' and $i_1 = i'\frac{n}{k}$, $i_2 = (i'+1)\frac{n}{k}$ for some i', we add an edge between u and v in the set $E(H)$ if and only if there is a path from u to v in G_l and we call such vertices as *corner vertices*.

Before proceeding further, let us introduce a few more notations that will be used later. For $j \in [k]$, let $L_h(i,j)$ denote the set of vertices in $L_h(i)$ in between $L_v(j)$ and $L_v(j+1)$. Similarly we also define $L_v(i,j)$ (see Fig. 1). For two vertices $x, y \in L_v(i)$, we say $x \prec y$ if x is *below* y in $L_v(i)$. For two vertices $x, y \in L_h(i)$, we say $x \prec y$ if x is *right of* y in $L_h(i)$. Note that we consider these two type of orderings to ensure that for any $x, y \in V(H)$ reachable from s in H, if $x \prec y$, then x will be traversed by our algorithm before y.

Lemma 1. *There is a path from s to t in G if and only if there is path from s to t in the auxiliary graph H.*

Proof. As every edge (a,b) in H corresponds to a path from a to b in G, so if-part is trivial to see. Now for the only-if-part, consider a path P from s to t in G. P can be decomposed as $P_1P_2\cdots P_r$, such that P_i is a path from x_i to x_{i+1}, where x_i is the first vertex on P that belongs to $V(G_l)$ and x_{i+1} be the last vertex on P that also belongs to $V(G_l)$, for some l and in a layered grid graph, for such x_i and x_{i+1}, we have only following two possibilities:

1. x_i and x_{i+1} belong to different horizontal or vertical gridlines; or
2. x_i and x_{i+1} are two corner vertices.

Now by the construction H, for every i, there must be an edge (x_i, x_{i+1}) in H for both the above cases and hence there is a path from s to t in H as well. □

Now we consider the case when two vertices $x, y \in V(H)$ belong to the same vertical or horizontal gridlines.

Claim 1. *Let x and y be two vertices contained in either $L_v(i)$ or $L_h(i)$ for some i. Then deciding reachability between x and y in G can be done in log space.*

Proof. Let us consider that $x, y \in L_v(i)$, for some i. As the graph G under consideration is a layered grid graph, if there is a path between x and y, then it must pass through all the vertices in $L_v(i)$ that lies in between x and y. Hence just by exploring the path starting from x through $L_v(i)$, we can check the reachability and it is easy to see that this can be done in log space, because the only thing we need to remember is the current vertex in the path. Same argument will also work when $x, y \in L_h(i)$, for some i and this completes the proof. □

Now we argue on the upper bound of the length of any path in the auxiliary graph H. The idea is to partition the set $V(H)$ into $2k+1$ partitions in such a way that any two consecutive vertices on a path in H lie on two different partitions.

Lemma 2. *Any path between s and t in H is of length $2k$.*

Proof. Let us first define the sets $D_0, D_1, \cdots, D_{2k}$ (e.g., shaded region in Fig. 1(b) denotes D_1), where

$$D_l := \{(i,j) | (i'-1)\frac{n}{k} \leq i < i'\frac{n}{k}, (j'-1)\frac{n}{k} \leq j < j'\frac{n}{k} \text{ and } i'+j' = l+1\}.$$

Now consider $D'_l := D_l \cap V(H)$ for $0 \leq l \leq 2k$. Clearly, $D'_0, D'_1, \cdots, D'_{2k}$ induce a partition on $V(H)$. Now let us take any path $s = x_1 x_2 \cdots x_r = t$, from s to t in H, denoted as P. Observe that by the construction of H, for any two consecutive vertices x_i and x_{i+1} for some i, if $x_i \in D'_l$ for some l, then $x_{i+1} \in D'_{l+1}$ and $s \in D'_0$, $t \in D'_{2k}$. As a consequence, $r = 2k+1$ and hence length of the path P is $2k$.

3.2 Description of the Algorithm

We next give a modified version of DFS that starting at a given vertex, visits the set of vertices reachable from that vertex in the graph H. At every vertex, the traversal visits the set of outgoing edges from that vertex in counter-clockwise order.

In our algorithm we maintain two arrays of size $k+1$ each, say A_v and A_h, one for vertical and the other for horizontal gridlines respectively. For every $i \in [k+1]$, $A_v(i)$ is the *topmost* visited vertex in $L_v(i)$ and analogously $A_h(i)$ is the *leftmost* visited vertex in $L_h(i)$. This choice is guided by the choice of traversal of our algorithm. More precisely, we cycle through the outgoing edges of a vertex in counter-clockwise order.

We perform a standard DFS-like procedure, using the tape space to simulate a stack, say S. S keeps track of the path taken to the current vertex from the starting vertex. By Lemma 2, the maximum length of a path in H is at most $2k$. Whenever we visit a vertex in a vertical gridline (say $L_v(i)$), we check whether the vertex is lower than the i-th entry of A_v. If so, we return to the parent vertex and continue with its next child. Otherwise, we update the i-th entry of A_v to be the current vertex and proceed forward. Similarly when visit a horizontal gridline (say $L_h(i)$), we check whether the current vertex is to the right of the i-th entry of A_h. If so, we return to the parent vertex and continue with its next child. Otherwise, we update the i-th entry of A_h to be the current vertex and proceed. The reason for doing this is to avoid revisiting the subtree rooted at the node of an already visited vertex.

Lemma 3. *Let G_l be some block and let x and y be two vertices on the boundary of G_l such that there is a path from x to y in G. Let x' and y' be two other boundary vertices in G_l such that (i) there is a path from x' to y' in G and (ii) x' lies on one segment of the boundary of G_l between vertices x and y and y' lies on the other segment of the boundary. Then there is a path in G from x to y' and from x' to y. Hence, if (x,y) and (x',y') are present in $E(H)$ then so are (x,y') and (x',y).*

Proof. Since G is a layered grid graph hence the paths x to y and x' to y' must lie inside G_l. Also because of planarity, the paths must intersect at some vertex in G_l. Now using this point of intersection, we can easily show the existence of paths from x to y' and from x' to y. □

Lemma 4 will prove the correctness of our algorithm.

Lemma 4. *Let u and v be two vertices in H. Then starting at u our algorithm visits v if and only if v is reachable from u in H.*

Proof. It is easy to see that every vertex visited by the algorithm is reachable from u since the algorithm proceeds along the edges of H.

By induction on the shortest path length to a vertex, we will show that if a vertex is reachable from u then the algorithm visits that vertex. Let $B_d(u)$ be the set of vertices reachable from u that are at a distance d from u. Assume that the algorithm visits every vertex in $B_{d-1}(u)$. Let x be a vertex in $B_d(u)$. Without loss of generality assume that x is in $L_v(i,j)$ for some i and j. A similar argument can be given if x belongs to a horizontal gridline. Further, let x lie on the right boundary of a block G_l. Let $W_x = \{w \in B_{d-1}(u) | (w,x) \in E(H)\}$. Note that by the definition of H, all vertices in W_x lie on the bottom boundary or on the left boundary of G_l.

Suppose the algorithm does not visit x. Since x is reachable from u via a path of length d, therefore W_x is non empty. Let w be the first vertex added to W_x by the algorithm. Then w is either in $L_h(j)$, or in $L_v(i-1)$. Without loss of generality assume w is in $L_h(j)$. Let z be the value in $A_v(i)$ at this stage of the algorithm (that is when w is the current vertex). Since x is not visited hence $x \prec z$. Also this implies that z was visited by the algorithm at an earlier stage of the algorithm. Let w' be the ancestor of z in the DFS tree such that w' is in $L_h(j)$. There must exist such a vertex because z is above the j-th horizontal gridline, that is $L_h(j)$.

Suppose if w' lies to the left of w then by the description of the algorithm, w is visited before w'. Hence x is visited before z. On the other hand, suppose if w' lies to the right of w. Clearly w' cannot lie to the right of vertical gridline $L_v(i)$ since z is reachable from w' and z is in $L_v(i)$. Let w'' be the vertex in $L_h(j+1)$ such that w'' lies in the tree path between w' and z (See Fig. 2). Observe that all four vertices lie on the boundary of G_l. Now by applying Lemma 3 to the four vertices w, x, w' and w'' we conclude that there exists a path from w' to x as well. Since $x \prec z$, x must have been visited before z from the vertex w'. In both cases, we see that z cannot be $A_v(i)$ when w is the current vertex. Since z was an arbitrary vertex such that $x \prec z$, the lemma follows. □

We next show Lemma 5 which will help us to achieve a polynomial bound on the running time of the algorithm.

Lemma 5. *Every vertex in the graph H is added to the set S at most once in the algorithm.*

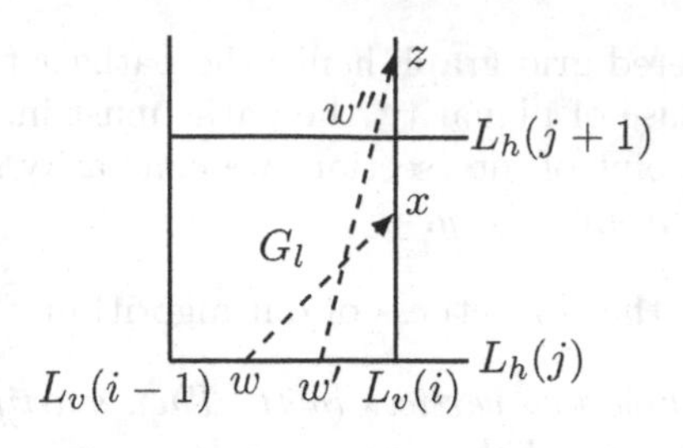

Fig. 2. Crossing between two paths

Proof. Observe that a vertex u in $L_v(i)$ is added to S only if $A_v(i) \prec u$, and once u is added, $A_v(i)$ is set to u. Also during subsequent stages of the algorithm, if $A_v(i)$ is set to v, then $u \prec v$. Hence $u \prec A_v(i)$. Therefore, u cannot be added to S again.

We give a similar argument if u is in $L_h(i)$. Suppose if u is in $L_v(i)$ for some i and $L_h(j)$ for some j, then we add u only once to S. However we update both $A_v(i)$ and $A_h(j)$. □

Our algorithm does not explicitly compute and store the graph H. Whenever it is queried for an edge (x, y) in H, it recursively runs a reachability query in the corresponding sub grid graph of G such that x is in the bottom left corner and y is in the top right corner of that sub grid graph and produces an answer. The base case is when a query is made to a grid graph of size $k \times k$. For the base case, we run a standard DFS procedure on the $k \times k$ size graph.

In the algorithm, until S is non-empty, in every iteration either an element is added or an element is removed from S. Hence by Lemma 5, the loop that check whether S is non-empty, iterates at most $4nk$ times. Inside that loop, there is another loop which cycles through all the neighbors of a vertex and hence iterates for at most $2n/k$ times where each iteration makes a constant number of calls to check the presence of an edge in an $n/k \times n/k$ sized grid. Let $\mathcal{T}(n)$ and $\mathcal{S}(n)$ be the time and space required to decide reachability in a layered grid graph of size $n \times n$ respectively. Then,

$$\mathcal{T}(n) = \begin{cases} 8n^2(\mathcal{T}(n/k) + O(1)) & \text{if } n > k \\ O(k^2) & \text{otherwise.} \end{cases}$$

Hence, $\mathcal{T}(n) = O\left(n^{3\frac{\log n}{\log k}}\right)$.

Since we do not store any query made to the smaller grids, therefore the space required to check the presence of an edge in H can be reused. A_v and A_h are arrays of size $k+1$ each. By Lemma 2, the number of elements in S at any stage of the algorithm is bounded by $2k$. Therefore,

$$\mathcal{S}(n) = \begin{cases} \mathcal{S}(n/k) + O(k \log n) & \text{if } n > k \\ O(k^2) & \text{otherwise.} \end{cases}$$

Hence, $\mathcal{S}(n) = O\left(\frac{k}{\log k} \log^2 n + k^2\right)$.

Now given any constant $\epsilon > 0$, if we set $k = n^{\epsilon/2}$, then we get $\mathcal{T}(n) = O(n^{6/\epsilon})$ and $\mathcal{S}(n) = O(n^\epsilon)$. This proves Theorem 3.

Acknowledgement. We thank N. V. Vinodchandran for his helpful suggestions and comments. The first author would like to acknowledge the support of Research-I Foundation and ACM-India/IARCS Travel Grants.

References

1. Imai, T., Nakagawa, K., Pavan, A., Vinodchandran, N., Watanabe, O.: An $O(n^{1/2+\epsilon})$-space and polynomial-time algorithm for directed planar reachability. In: 2013 IEEE Conference on Computational Complexity (CCC), pp. 277–286 (2013)
2. Lewis, H.R., Papadimitriou, C.H.: Symmetric space-bounded computation. Theor. Comput. Sci. **19**, 161–187 (1982)
3. Reingold, O.: Undirected connectivity in log-space. J. ACM **55**(4), 1–24 (2008)
4. Reingold, O., Trevisan, L., Vadhan, S.: Pseudorandom walks on regular digraphs and the RL vs. L problem. In: Proceedings of the Thirty-Eighth Annual ACM Symposium on Theory of Computing, STOC 2006, pp. 457–466. ACM, New York (2006)
5. Chung, K.M., Reingold, O., Vadhan, S.: S-t connectivity on digraphs with a known stationary distribution. ACM Trans. Algorithms **7**(3), 30:1–30:21 (2011)
6. Lange, K.J.: An unambiguous class possessing a complete set. In: Proceedings of the 14th Annual Symposium on Theoretical Aspects of Computer Science, STACS 1997, pp. 339–350 (1997)
7. Wigderson, A.: The complexity of graph connectivity. In: Havel, Ivan M., Koubek, Václav (eds.) MFCS 1992. LNCS, vol. 629, pp. 112–132. Springer, Heidelberg (1992)
8. Savitch, W.J.: Relationships between nondeterministic and deterministic tape complexities. J. Comput. Syst. Sci. **4**, 177–192 (1970)
9. Barnes, G., Buss, J.F., Ruzzo, W.L., Schieber, B.: A sublinear space, polynomial time algorithm for directed s-t connectivity. In: Proceedings of the Seventh Annual Conference on Structure in Complexity Theory, pp. 27–33 (1992)
10. Asano, T., Doerr, B.: Memory-constrained algorithms for shortest path problem. In: CCCG (2011)
11. Chakraborty, D., Pavan, A., Tewari, R., Vinodchandran, N.V., Yang, L.: New time-space upperbounds for directed reachability in high-genus and H-minor-free graphs. In: 34th International Conference on Foundation of Software Technology and Theoretical Computer Science, FSTTCS 2014, December 15–17, 2014, New Delhi, pp. 585–595 (2014)

12. Asano, T., Kirkpatrick, D.G., Nakagawa, K., Watanabe, O.: $\tilde{O}(\sqrt{n})$-space and polynomial-time algorithm for planar directed graph reachability. In: Proceedings of the 39th International Symposium on Mathematical Foundations of Computer Science, MFCS 2014, Part II, Budapest, Hungary, August 25–29, 2014, pp. 45–56 (2014)
13. Vinodchandran, N.V.: Space complexity of the directed reachability problem over surface-embedded graphs. Technical report TR14-008, I (2014)
14. Kannan, S., Khanna, S., Roy, S.: STCON in directed unique-path graphs. In: Hariharan, R., Mukund, M., Vinay, V. (eds.) IARCS Annual Conference on Foundations of Software Technology and Theoretical Computer Science. Leibniz International Proceedings in Informatics (LIPIcs), vol. 2, pp. 256–267. Schloss Dagstuhl-Leibniz-Zentrum fuer Informatik, Dagstuhl, Germany (2008)
15. Buntrock, G., Jenner, B., Lange, K.J., Rossmanith, P.: Unambiguity and fewness for logarithmic space. In: Budach, L. (ed.) Fundamentals of Computation Theory. Lecture Notes in Computer Science, vol. 529, pp. 168–179. Springer, Heidelberg (1991)
16. Cook, S.: Deterministic CFL's are accepted simultaneously in polynomial time and log squared space. In: Proceedings of the Eleventh Annual ACM Symposium on Theory of Computing, pp. 338–345. ACM (1979)
17. Allender, E., Lange, K.: RUSPACE(log n) $\subseteq$ DSPACE ($\log^2$ n / log log n). Theor. Comput. Syst. **31**(5), 539–550 (1998)
18. Nisan, N.: RL $\subseteq$ SC. In: Proceedings of the Twenty Fourth Annual ACM Symposium on Theory of Computing, pp. 619–623 (1995)
19. Allender, E.: Reachability problems: an update. In: Cooper, S.B., Löwe, B., Sorbi, A. (eds.) CiE 2007. LNCS, vol. 4497, pp. 25–27. Springer, Heidelberg (2007)
20. Allender, E., Barrington, D.A.M., Chakraborty, T., Datta, S., Roy, S.: Planar and grid graph reachability problems. Theor. Comput. Syst. **45**(4), 675–723 (2009)
21. Bourke, C., Tewari, R., Vinodchandran, N.V.: Directed planar reachability is in unambiguous log-space. ACM Trans. Comput. Theor. **1**(1), 1–17 (2009)
22. Di Battista, G., Tamassia, R.: Upward drawings of acyclic digraphs. In: Göttler, H., Schneider, H.-J. (eds.) WG 1987. LNCS, vol. 314. Springer, Heidelberg (1988)
23. Battista, G.D., Liu, W., Rival, I.: Bipartite graphs, upward drawings, and planarity. Inf. Process. Lett. **36**(6), 317–322 (1990)
24. Garg, A., Tamassia, R.: Upward planarity testing. Order **12**(2), 109–133 (1995). http://dx.doi.org/10.1007/BF01108622

Constant Query Time $(1+\epsilon)$-Approximate Distance Oracle for Planar Graphs

Qian-Ping Gu(✉) and Gengchun Xu

School of Computing Science, Simon Fraser University,
Burnaby, BC V5A1S6, Canada
{qgu,gxa2}@sfu.ca

Abstract. We give a $(1+\epsilon)$-approximate distance oracle with $O(1)$ query time for an undirected planar graph G with n vertices and non-negative edge length. For $\epsilon > 0$ and any two vertices u and v in G, our oracle gives a distance $\tilde{d}(u,v)$ with stretch $(1+\epsilon)$ in $O(1)$ time. The oracle has size $O(n \log n(\log n/\epsilon + f(\epsilon)))$ and pre-processing time $O(n \log n(\log^3 n/\epsilon^2 + f(\epsilon)))$, where $f(\epsilon) = 2^{O(1/\epsilon)}$. This is the first $(1+\epsilon)$-approximate distance oracle with $O(1)$ query time independent of ϵ and the size and pre-processing time nearly linear in n, and improves the query time $O(1/\epsilon)$ of previous $(1+\epsilon)$-approximate distance oracle with size nearly linear in n.

Keywords: Distance oracle · Planar graphs · Approximate algorithms · Graph decomposition

1 Introduction

Finding a distance between two vertices in a graph is a fundamental computational problem and has a wide range of applications. For this problem, there is a rich literature of algorithms. This problem can be solved by a single source shortest path algorithm such as the Dijkstra and Bellman-Ford algorithms. In many applications, it is required to compute the shortest path distance in an extreme short time. One approach to meet such a requirement is to use distance oracles.

A distance oracle is a data structure which keeps the pre-computed distance information and provides a distance between any given pair of vertices very efficiently. There are two phases in the distance oracle approach. The first phase is to compute the data structure for a given graph G and the second is to provide an answer for a query on the distance between a pair of vertices in G. The efficiency of distance oracles is mainly measured by the time to answer a query (*query time*), the memory space required for the data structure (*oracle size*) and the time to create the data structure (*pre-processing time*). Typically, there is a trade-off between the query time and the oracle size. A simple approach to compute a distance oracle for graph G of n vertices is to solve the all pairs shortest paths problem in G and keep the shortest distances in an $n \times n$

K. Elbassioni and K. Makino (Eds.): ISAAC 2015, LNCS 9472, pp. 625–636, 2015.
DOI: 10.1007/978-3-662-48971-0_53

distance array. This gives an oracle with $O(1)$ query time and $O(n^2)$ size. A large number of papers have been published for distance oracles with better measures on the product of query time and oracle size, see Sommer's paper for a survey [14].

Planar graphs are an important model for many networks such as the road networks. Distance oracles for planar graphs have been extensively studied. Djidjev proves that for any size $S \in [n, n^2]$, there is an exact distance oracle with query time $O(n^2/S)$ for weighted planar graphs [6]. There are several exact distance oracles with size $O(S)$ and more efficient query time for different ranges of S, for example, an oracle by Wulff-Nilsen [17] with $O(1)$ query time and $O(n^2(\log\log n)^4/\log n)$ size for weighted directed planar graphs and an oracle by Mozes and Sommer [13] with query time $O((n/\sqrt{S})\log^{2.5} n)$ and size $S \in [n\log\log n, n^2]$ for weighted directed planar graph. Readers may refer to Sommer's survey paper [14] for more details.

Approximate distance oracles have been developed to achieve very fast query time and near linear size for planar graphs. For vertices u and v in graph G, let $d_G(u,v)$ denote the distance between u and v. An oracle is called an α-approximate oracle or with *stretch* α for $\alpha \geq 1$ if it provides a distance $\tilde{d}(u,v)$ with $d_G(u,v) \leq \tilde{d}(u,v) \leq \alpha d_G(u,v)$ for u and v in G. An oracle is called one with an *additive stretch* $\beta \geq 0$ if it provides a distance $\tilde{d}(u,v)$ with $d_G(u,v) \leq \tilde{d}(u,v) \leq d_G(u,v) + \beta$. For $\epsilon > 0$, Thorup gives a $(1+\epsilon)$-approximate distance oracle with $O(1/\epsilon)$ (resp. $O(1/\epsilon + \log\log \Delta)$, where Δ is the longest finite distance between any pair of vertices in G) query time and $O(n\log n/\epsilon)$ size for undirected (resp. directed) planar G with non-negative edge length [15]. Klein further simplifies Thorup's oracles for undirected planar graphs [9]. Kawarabayashi et al. give a $(1+\epsilon)$-approximate distance oracle with $O((1/\epsilon)\log^2(1/\epsilon)\log\log(1/\epsilon)\log^* n)$ query time and $O(n\log n\log\log(1/\epsilon)\log^* n)$ size for undirected planar graphs with non-negative edge length [8]. The query times of the oracles above are fast but still at least $O(1/\epsilon)$.

Distance oracles with constant query time are of both theoretical and practical importance [4,5]. Our main result is an $O(1)$ query time $(1+\epsilon)$-approximate distance oracle for undirected planar graphs with non-negative edge length.

Theorem 1. *Let G be an undirected planar graph with n vertices and non-negative edge length and let $\epsilon > 0$. There is a $(1+\epsilon)$-approximate distance oracle for G with $O(1)$ query time, $O(n\log n(\log n/\epsilon + f(\epsilon)))$ size and $O(n\log n(\log^3 n/\epsilon^2 + f(\epsilon)))$ pre-processing time, where $f(\epsilon) = 2^{O(1/\epsilon)}$.*

The oracle in Theorem 1 has a constant query time independent of ϵ and size nearly linear in the graph size. This improves the query times of the previous works [8,15] that are (nearly) linear in $1/\epsilon$. Wulff-Nilsen gives an $O(1)$ time exact distance oracle for G but the oracle has size $O(n^2(\log\log n)^4/\log n)$ which is much larger than a function nearly linear in n.

The result in Theorem 1 can be generalized to an oracle described in the next theorem.

Theorem 2. *Let G be an undirected planar graph with n vertices and non-negative edge length, $\epsilon > 0$ and $1 \leq \eta \leq 1/\epsilon$. There is a $(1+\epsilon)$-approximate* *distance oracle for G with $O(\eta)$ query time, $O(n \log n(\log n/\epsilon + f(\eta\epsilon)))$ size and $O(n \log n(\log^3 n/\epsilon^2 + f(\eta\epsilon)))$ pre-processing time, where $f(\eta\epsilon) = 2^{O(1/(\eta\epsilon))}$.*

Our results build on some techniques used in the previous approximate distance oracles for planar graphs. Thorup [15] gives a $(1+\epsilon)$-approximate distance oracle for planar graph G with $O(1/\epsilon)$ query time. Informally, some techniques used in the oracle are as follows: Decompose G into a balanced recursive subdivision; G is decomposed into subgraphs of balanced sizes by shortest paths and each subgraph is decomposed recursively until every subgraph is reduced to a pre-defined size. A path Q separates vertices u and v in G if a shortest path between u and v intersects Q. For each subgraph X of G, let $\mathcal{P}(X)$ be the set of shortest paths used to decompose X. For each path $Q \in \mathcal{P}(X)$ and each vertex u in X, a set $P_Q(u)$ of $O(1/\epsilon)$ vertices called *portals* on Q is selected. For vertices u and v separated by some path Q in X, $\min_{p \in P_Q(u), q \in P_Q(v), Q \in \mathcal{P}(X)} d_G(u,p) + d_G(p,q) + d_G(q,v)$ is used to approximate $d_G(u,v)$. The oracle keeps the distances $d_G(u,p)$ and $d_G(p,v)$.

The portal set $P_Q(u)$ above is vertex dependent. For a path Q in G of length $d(Q)$, there is a set P_Q of $O(1/\epsilon)$ portals such that for any vertices u and v separated by Q, $\min_{p \in P_Q} d_G(u,p) + d_G(p,v) \leq d_G(u,v) + \epsilon d(Q)$ [11]. Based on this and a scaling technique, Kawarabayashi et al. [8] give another $(1+\epsilon)$-approximate distance oracle: Create subgraphs of G such that the vertices in each subgraph satisfy certain distance property (scaling). Each subgraph H of G is decomposed by shortest paths into a ρ-division of H which consists of $O(|V(H)|/\rho)$ subgraphs of H, each has size $O(\rho)$. For each subgraph X of H, let $\mathcal{B}(X)$ be the set of shortest paths used to separate X from the rest of H. For each path $Q \in \mathcal{B}(X)$, a portal set P_Q is selected. For vertices u and v separated by some path $Q \in \mathcal{B}(X)$, $\min_{p \in P_Q, Q \in \mathcal{B}(X)} d_H(u,p) + d_H(p,v)$ is used to approximate $d_G(u,v)$. This oracle does not keep the distances $d_H(u,p)$ and $d_H(p,v)$ but uses the distance oracle in [13] to get the distances. By choosing an appropriate value ρ, the oracle has a better product of query time and oracle size than that of Thorup's oracle.

We also use the scaling technique to create subgraphs of G. We decompose each subgraph H of G into a balanced recursive subdivision as in Thorup's oracle. For each subgraph X of H and each shortest path Q used to decompose X, we choose one set P_Q of $O(1/\epsilon)$ portals on Q for all vertices in X. A new ingredient in our oracle is to use a more time efficient data structure to approximate $d_G(u,v)$ instead of $\min_{p \in P_Q, Q \in \mathcal{P}(X)} d_H(u,p) + d_H(p,v)$. Using an approach in [16], we show that the vertices in $V(X)$ can be partitioned into $s = f(\epsilon)$ classes $A_1, ..., A_s$ such that for every classes A_i and A_j, there is a key portal $p_{ij} \in P_Q$ and for any $u \in A_i$ and $v \in A_j$, if u and v are separated by Q then $d_H(u,p_{ij}) + d_H(p_{ij},v) \leq (1+\epsilon)d_G(u,v)$ and $d_H(u,p_{ij}) + d_H(p_{ij},v)$ can be computed in $O(1)$ time. This gives a $(1+\epsilon)$-approximate distance oracle with $O(1)$ query time.

The rest of the paper is organized as follows. In the next section, we give preliminaries of the paper and review the techniques on which our oracles build. In Sect. 3, we present distance oracles with additive stretch. In Sect. 4, we give the $(1+\epsilon)$-approximate distance oracles which use the additive stretch oracles as subroutines. The final section concludes the paper.

2 Preliminaries

An undirected graph G consists of a set $V(G)$ of vertices and a set $E(G)$ of edges. For a subset $A \subseteq E(G)$, we denote by $V(A)$ the set of vertices incident to at least one edge of A. For $A \subseteq E(G)$ and $W \subseteq V(G)$, we denote by $G[A]$ and $G[W]$ the subgraphs of G induced by A and W, respectively. A graph H is a subgraph of G if $V(H) \subseteq V(G)$ and $E(H) \subseteq E(G)$. For $A \subseteq E(G)$ and $B = E(G) \setminus A$, the vertex set $S = V(A) \cap V(B)$ is a *separator* of G that decomposes G into subgraphs $G[A]$ and $G[B]$.

A path between vertices u and v in G is a sequence of edges $e_1, .., e_k$, where $e_i = \{v_{i-1}, v_i\}$ for $1 \leq i \leq k$, $u = v_0$, $v = v_k$, and the vertices $v_0, ..., v_k$ are distinct. Let $l(e)$ be the length of edge e in G. The length of path $Q = e_1, ..., e_k$ is $d(Q) = \sum_{1 \leq i \leq k} l(e_i)$. A path Q is a shortest path between vertices u and v if $d(Q)$ is the minimum among those of all paths between u and v. The distance between vertices u and v in G, denoted by $d_G(u,v)$, is the length of a shortest path between u and v. For each vertex u in G, the *eccentricity* of u is $\lambda(u) = \max_{v \in V(G)} d_G(u,v)$. The *radius* of G is $r(G) = \min_{u \in V(G)} \lambda(u)$. The diameter of G is $d(G) = \max_{u \in V(G)} \lambda(u)$.

A graph is planar if it has a planar embedding (a draw on a sphere without edge crossing). In the rest of this paper, graphs are undirected planar graphs with non-negative edge length unless otherwise stated.

A separator S of G decomposes G into at least two connected subgraphs. A set $\mathcal{P}$ of shortest paths in graph G is a *shortest path separator* of G if $S = \cup_{Q \in \mathcal{P}} V(Q)$ is a separator of G. A *recursive subdivision* of G is a structure that G is decomposed into subgraphs by a separator and each subgraph is decomposed recursively until each subgraph is reduced to a pre-defined size. In this paper, we use only shortest path separators.

A recursive subdivision of G can be viewed as a rooted tree T_G with G the root node. Each node in T_G with node degree one is called a *leaf node*, otherwise an *internal node*. Each internal node X of T_G is decomposed by a shortest path separator $\mathcal{P}(X)$ into subgraphs $X_1, \ldots, X_c$, $c \geq 2$. Each X_i, $i = 1, \ldots, c$ is a child node of X in T_G. For each internal node X, let $\mathcal{B}(X)$ be the set of boundary paths separating X from the rest of G. Based on a result by Lipton and Tarjan [12] on the balanced separators of G, Thorup [15] shows the following result.

Lemma 1. *[15] For a graph G, a $\frac{1}{2}$-balanced recursive subdivision T_G of G can be computed in $O(n \log n)$ time such that for each internal node X of T_G, $|V(X_i)| \leq |V(X)|/2$, $1 \leq i \leq c$, $|\mathcal{P}(X)| = O(1)$ and $|\mathcal{B}(X)| = O(1)$.*

The recursive subdivision of G in Lemma 1 will be used in our oracles. Note that since the size of a subgraph is reduced by at least 1/2, the depth of T_G is bounded by $\log(n)$.

Let Q be a shortest path in G and $\epsilon > 0$. Thorup shows that for every vertex u in G, there is subset $P_Q(u) \subseteq V(Q)$ of $O(1/\epsilon)$ vertices such that for any vertices u and v separated by Q (a shortest path between u and v intersects Q) then

$$d_G(u,v) \leq \min_{p \in P_Q(u), q \in P_Q(v)} d_G(v,p) + d_G(p,q) + d_G(q,v) \leq (1+\epsilon) d_G(u,v).$$

The vertices of $P_Q(u)$ are called *portals* on Q for u. For every subgraph X in a $\frac{1}{2}$-balanced recursive subdivision of G and every shortest path $Q \in \mathcal{B}(X) \cup \mathcal{P}(X)$, by keeping the distance from each vertex u in X to every portal in $P_Q(u)$ explicitly, Thorup shows the following result.

Lemma 2. *[15] For graph G and $\epsilon > 0$, there is a $(1+\epsilon)$-approximate distance oracle with $(1/\epsilon)$ query time, $O(n \log n/\epsilon)$ size and $O(n \log^3 n/\epsilon^2)$ pre-processing time. Especially for $\epsilon = 1$, there is a 2-approximate distance oracle for G with $O(1)$ query time, $O(n \log n)$ size and $O(n \log^3 n)$ pre-processing time.*

Our oracles will use this oracle for $\epsilon = 1$ (any constant works) to get a rough estimation of $d_G(u,v)$.

To reduce the query time to a constant independent of ϵ, we will use a portal set P_Q independent of vertex u. For vertices u and v separated by a path Q, $d_G(u,v) = \min_{p \in V(Q)} d_G(u,p) + d_G(p,v)$. For a $P_Q \subseteq V(Q)$, $\min_{p \in P_Q} d_G(u,p) + d_G(p,v)$ approximates $d_G(u,v)$. The following result will be used.

Lemma 3. *[11] For a path Q in G and $\epsilon > 0$, a set P_Q of $O(1/\epsilon)$ vertices in $V(Q)$ can be selected in $O(|V(Q)|)$ time such that for every vertices u and v separated by Q, $d_G(u,v) \leq \min_{p \in P_Q} d_G(u,p) + d_G(p,v) \leq d_G(u,v) + \epsilon d(Q)$.*

The set P_Q in Lemma 3 is called an *ϵ-portal set*. To apply the ϵ-portal set to our oracle, we further need to guarantee $d_G(u,v) = \Omega(d(Q))$ for vertices u and v in question. We will use the *sparse neighborhood covers* introduced in [1–3] of G to achieve this goal.

Lemma 4. *[3] For G and $\gamma \geq 1$, connected subgraphs $G(\gamma,1), \ldots, G(\gamma,n_\gamma)$ of G with the following properties can be computed in $O(n \log n)$ time:*

1. *For each vertex u in G, there is at least one $G(\gamma,i)$ that contains u and every v with $d_G(u,v) \leq \gamma$.*
2. *Each vertex u in G is contained in at most 18 subgraphs.*
3. *Each subgraph $G(\gamma,i)$ has radius $r(G(\gamma,i)) \leq 24\gamma - 8$.*

3 Oracle with Additive Stretch

We first give a distance oracle which for any vertices u and v in G and any $\epsilon_0 > 0$ returns $\tilde{d}(u,v)$ with $d_G(u,v) \leq \tilde{d}(u,v) \leq d_G(u,v) + 7\epsilon_0 d(G)$. Based on the scaling technique in [8] and Lemma 4, this oracle will be extended to an oracle stated in Theorem 1 for G in the next section.

We start with a basic data structure which keeps the following information:

- A $\frac{1}{2}$-balanced recursive subdivision T_G of G in Lemma 1, each leaf node in T_G has size $O(2^{(1/\epsilon_0)})$.
- For every leaf node X of T_G and all $u, v \in V(X)$, $\tilde{d}(u, v)$ with

$$d_G(u, v) \leq \tilde{d}(u, v) \leq d_G(u, v) + 3\epsilon_0 d(G).$$

- For each internal node X of T_G, an ϵ_0-portal set P_Q for every shortest path $Q \in \mathcal{P}(X) \cup \mathcal{B}(X)$. For every P_Q, every $u \in V(X)$ and every portal $p \in P_Q$, distance $\hat{d}(u, p)$ with

$$d_G(u, p) \leq \hat{d}(u, p) \leq d_G(u, p) + \epsilon_0 d(Q).$$

The data structure above gives a distance oracle with $3\epsilon_0 d(G)$ additive stretch and $O(1/\epsilon_0)$ query time: Given vertices u and v in G, let X_u and X_v be the leaf nodes of T_G containing u and v, respectively. If $X_u = X_v$ then $\tilde{d}(u, v)$ can be found in $O(1)$ time. Otherwise, let X be the nearest common ancestor of X_u and X_v. It is shown in [7] that X can be found in $O(1)$ time. Let $q = \arg_{p \in P_Q, Q \in \mathcal{B}(X) \cup \mathcal{P}(X)} \min\{d_G(u, p) + d_G(p, v)\}$. From $\hat{d}(u, p) \leq d_G(u, p) + \epsilon_0 d(Q)$, $\hat{d}(p, v) \leq d_G(p, v) + \epsilon_0 d(Q)$, Lemma 3 and $d(Q) \leq d(G)$,

$$\begin{aligned} d_G(u, v) \leq & \min_{p \in P_Q, Q \in \mathcal{B}(X) \cup \mathcal{P}(X)} \hat{d}(u, p) + \hat{d}(p, v) \leq \hat{d}(u, q) + \hat{d}(q, v) \\ & \leq d_G(u, q) + d_G(q, v) + 2\epsilon_0 d(Q) \leq d_G(u, v) + 3\epsilon_0 d(G). \end{aligned}$$

This distance can be computed in $O(1/\epsilon_0)$ time because $|P_Q| = O(1/\epsilon_0)$ and $|\mathcal{B}(X) \cup \mathcal{P}(X)| = O(1)$.

Now we reduce the query time for internal nodes in the above oracle to a constant independent of ϵ_0. For $z > 0$, let $f(z) = 2^{O(1/z)}$. Based on an approach in [16], we show that for each internal node X and each path $Q \in \mathcal{B}(X) \cup \mathcal{P}(X)$, the vertices in $V(X)$ can be partitioned into $f(\epsilon_0)$ classes such that for any two classes A_i and A_j, there is a key portal $p_{ij} \in P_Q$ and for every $u \in A_i$ and every $v \in A_j$ separated by Q, $\hat{d}(u, p_{ij}) + \hat{d}(p_{ij}, v) \leq d_G(u, v) + 7\epsilon_0 d(G)$. By keeping the classes and key portals, the query time is reduced to $O(1)$. We first define the classes.

Definition 1. Let Q be a shortest path in G, $P_Q = \{p_1 ..., p_l\}$ be an ϵ_0-portal set on Q and $r(G) \leq D \leq d(G)$. The vertices of G are partitioned into classes based on $\hat{d}(u, p_i), p_i \in P_Q$ as follows. For each vertex u, a vector $\boldsymbol{\Gamma}_u = (a_1, ..., a_l)$ is defined such that for $1 \leq i \leq l$, $a_i = \lceil \hat{d}(u, p_i)/(\epsilon_0 D) \rceil$. Vertices u and v are in the same class if and only if $\boldsymbol{\Gamma}_u = \boldsymbol{\Gamma}_v$.

We show some properties of the classes defined above in the next two lemmas.

Lemma 5. *Let Q be a shortest path in G and P_Q an ϵ_0-portal set on Q. Let $r(G) \leq D \leq d(G)$ and let A_i and A_j be any two classes of vertices in G defined in Definition 1. There is a key portal $p_{ij} \in P_Q$ such that for any vertices $u \in A_i$ and $v \in A_j$ separated by Q, $d_G(u, v) \leq \hat{d}(u, p_{ij}) + \hat{d}(p_{ij}, v) \leq d_G(u, v) + 7\epsilon_0 d(G)$.*

Proof. We choose arbitrarily a vertex $x \in A_i$ and a vertex $y \in A_j$ separated by Q. Let $p_{ij} = \arg_{p_i \in P_Q} \min\{\hat{d}(x, p_i) + \hat{d}(p_i, y)\}$ be the key portal. For any $u \in A_i$ and $v \in A_j$ separated by Q, let $q = \arg_{p_i \in P_Q} \min\{d_G(u, p_i) + d_G(p_i, v)\}$ and let $p = \arg_{p_i \in P_Q} \min\{\hat{d}(u, p_i) + \hat{d}(p_i, v)\}$. Then $\hat{d}(u,p) + \hat{d}(p,v) \leq \hat{d}(u,q) + \hat{d}(q,v) \leq d_G(u,q) + d_G(q,v) + 2\epsilon_0 d(Q) \leq d_G(u,v) + 3\epsilon_0 d(G)$ because $\hat{d}(u,q) \leq d_G(u,q) + \epsilon_0 d(Q)$, $\hat{d}(q,v) \leq d_G(q,v) + \epsilon_0 d(Q)$, P_Q is an ϵ_0-portal set and $d(Q) \leq d(G)$. Since $u, x \in A_i$, $v, y \in A_j$ and and $D \leq d(G)$,

$$\begin{aligned}\hat{d}(u, p_{ij}) + \hat{d}(p_{ij}, v) &\leq \hat{d}(x, p_{ij}) + \hat{d}(p_{ij}, y) + 2\epsilon_0 d(G)\\ &\leq \hat{d}(x,p) + \hat{d}(p,y) + 2\epsilon_0 d(G) \leq \hat{d}(u,p) + \hat{d}(p,v) + 4\epsilon_0 d(G).\end{aligned}$$

Therefore,

$$d_G(u,v) \leq \hat{d}(u,p_{ij}) + \hat{d}(p_{ij},v) \leq \hat{d}(u,p) + \hat{d}(p,v) + 4\epsilon_0 d(G) \leq d_G(u,v) + 7\epsilon_0 d(G).$$

This completes the proof of the lemma. □

Lemma 6. *The total number of classes by Definition 1 is $f(\epsilon_0)$.*

Proof. Each element a_i in a vector $\boldsymbol{\Gamma}_u = (a_1, .., a_l)$ has $O(1/\epsilon_0)$ different values. Since P_Q is an ϵ_0-portal set, $l = O(1/\epsilon_0)$ and there are $(1/\epsilon_0)^{O(1/\epsilon_0)}$ different classes. As shown in [16], the upper bound on the number of classes can be reduced to $2^{O(1/\epsilon_0)}$. For each vector $\boldsymbol{\Gamma}_u = (a_1, .., a_l)$, let $\boldsymbol{\Gamma}^*_u = (a_1, (a_2 - a_1), (a_3 - a_2), .., (a_l - a_{l-1}))$. Then $\boldsymbol{\Gamma}_u = \boldsymbol{\Gamma}_v$ if and only if $\boldsymbol{\Gamma}^*_u = \boldsymbol{\Gamma}^*_v$. Since $\sum_{2 \leq i \leq l} |a_i - a_{i-1}| = O(1/\epsilon_0)$, there are $2^{O(1/\epsilon_0)}$ different vectors of $(a_1, |a_2 - a_1|, |a_3 - a_2|, .., |a_l - a_{l-1}|)$. The i'th element of $(a_1, (a_2 - a_1), (a_3 - a_2), .., (a_l - a_{l-1}))$ is $|a_i - a_{i-1}|$ or $-|a_i - a_{i-1}|$. Therefore, there are $2^{O(1/\epsilon_0)}$ different vectors $(a_1, (a_2 - a_1), (a_3 - a_2), .., (a_l - a_{l-1}))$. □

Now we are ready to show a data structure DS_0 for our oracle with $7\epsilon_0 d(G)$ additive stretch. DS_0 contains the basic data structure given above and the following additional information:

- For each internal node X of T_G and each shortest path $Q \in \mathcal{B}(X) \cup \mathcal{P}(X)$, let $A^Q_1, ..., A^Q_s$ be the classes of vertices in $V(X)$ defined in Definition 1. For each vertex $u \in V(X)$, we give an index $I^Q_X(u)$ with $I^Q_X(u) = i$ if $u \in A^Q_i$; and an $s \times s$ array C_Q with $C_Q[i,j]$ containing the key portal p^Q_{ij} for classes A^Q_i and A^Q_j.

For data structure DS_0 we have the following three lemmas.

Lemma 7. *For graph G and $\epsilon_0 > 0$, data structure* DS_0 *can be computed in $O(n(\log^3 n/\epsilon_0^2 + f(\epsilon_0)))$ time.*

Due to the space limit, we only give an outline of the proof. Let T_G be the recursive subdivision of G in DS_0 and $b = 2^{(1/\epsilon_0)}$. It takes $O(n \log n)$ time to compute T_G (Lemma 1). For each leaf node X and all u and v in X, we find

$d_X(u,v)$ in $O(b^2)$ time and $\tilde{d}(u,v) = \min\{d_X(u,v), \min_{p \in P_Q, Q \in \mathcal{B}(X)} \hat{d}(u,p) + \hat{d}(p,v)\}$ in $O(b^2\epsilon_0)$ time. Therefore, it takes $O(nb^2/\epsilon_0) = O(nf(\epsilon_0))$ time to compute the distances $\tilde{d}(u,v)$ kept in all leaf nodes of T_G.

For each internal node X and every path $Q \in \mathcal{P}(X)$, we compute an ϵ_0-portal set P_Q, an auxiliary set A_Q which is an $(\epsilon_0/\log n)$-portal set on Q, and the distance $\hat{d}(u,p)$ for u in X and $p \in P_Q \cup A_Q$. For the root node, the computation is performed on G. For an internal node $X \neq G$, the computation is done using X and the distances $\hat{d}(u,p)$ for u in X and p in $A_{Q'}$, where Q' is a path in $\mathcal{B}(X)$. Since T_G has depth $O(\log n)$, it takes $O(n \log^3 n/\epsilon_0^2)$ time to compute all portal sets and distances $\hat{d}(u,p)$.

The value D for computing the classes can be found in $O(n)$ time. Since there are $O(n)$ internal nodes, by Lemma 6, it takes $(n)f(\epsilon_0)(1/\epsilon_0) = O(nf(\epsilon_0))$ time to compute all classes and key portals. Therefore, DS_0 can be computed in $O(n(\log^3 n/\epsilon_0^2 + f(\epsilon_0)))$ time. □

Lemma 8. *For graph G and $\epsilon_0 > 0$, the space requirement for data structure* DS_0 *is* $O(n(\log n/\epsilon_0 + f(\epsilon_0)))$.

Proof. Let T_G be the recursive subdivision of G in DS_0 and $b = 2^{(1/\epsilon_0)}$. Then T_G has $O(n)$ leaf nodes and $O(n)$ internal nodes. Each leaf node requires $O(b^2)$ space to keep the distances $\tilde{d}(u,v)$ for u, v in the node. Therefore, the space for all leaf nodes is $O(nb^2) = O(nf(\epsilon_0))$. By Lemma 2, the sum of $|V(X)|$ for all nodes X in T_G is $O(n \log n)$. From $|\mathcal{B}(X) \cup \mathcal{P}(X)| = O(1)$ for every X and $|P_Q| = O(1/\epsilon_0)$ for each $Q \in \mathcal{B}(X) \cup \mathcal{P}(X)$, the total space for keeping the distances $\hat{d}(u,v)$ between vertices and portals is $O(n \log n/\epsilon_0)$. By Lemma 6, the space for the classes $A_1^Q, .., A_s^Q$ in each internal node X is $f(\epsilon_0)$ for every $Q \in \mathcal{B}(X) \cup \mathcal{P}(X)$. Since there are $O(n)$ internal nodes, the total space for the classes in all nodes is $O(nf(\epsilon_0)) = O(nf(\epsilon_0))$. Therefore the space requirement for the oracle is $O(n(\log n/\epsilon_0 + f(\epsilon_0)))$. □

Lemma 9. *For graph G and $\epsilon_0 > 0$, $\tilde{d}(u,v)$ with $d_G(u,v) \leq \tilde{d}(u,v) \leq d_G(u,v) + 7\epsilon_0 d(G)$ can be computed in $O(1)$ time for any u and v in G using data structure* DS_0.

Proof. Let T_G be the recursive subdivision of G in DS_0. Let X_u and X_v be the leaf nodes of T_G that contains vertices u and v in G, respectively. The nearest common ancestor X of X_u and X_v can be found in $O(1)$ time [7]. If $X_u = X_v$ then $\tilde{d}(u,v)$ can be found in $O(1)$ time. Otherwise, for each path $Q \in \mathcal{B}(X) \cup \mathcal{P}(X)$, assume that $u \in A_i^Q$ and $v \in A_j^Q$, and let p_{ij}^Q be the key portal for A_i^Q and A_j^Q. By Lemma 5, $\tilde{d}(u,v) = \min_{p_{ij}^Q, Q \in \mathcal{B}(X) \cup \mathcal{P}(X)} \hat{d}(u, p_{ij}^Q) + \hat{d}(p_{ij}^Q, v) \leq d_G(u,v) + 7\epsilon_0 d(G)$. Since $|\mathcal{B}(X) \cup \mathcal{P}(X)| = O(1)$ and the key portal p_{ij}^Q can be found in $O(1)$ time for each path $Q \in \mathcal{B}(X) \cup \mathcal{P}(X)$, $\tilde{d}(u,v)$ can be computed in $O(1)$ time. □

From Lemmas 7, 8 and 9, we have the following result.

Theorem 3. *For graph G and $\epsilon_0 > 0$, there is an oracle which gives a distance $\tilde{d}(u,v)$ with $d_G(u,v) \leq \tilde{d}(u,v) \leq d_G(u,v) + 7\epsilon_0 d(G)$ for any vertices u and v in G with $O(1)$ query time, $O(n(\log n/\epsilon_0 + f(\epsilon_0)))$ size and $O(n(\log^3 n/\epsilon_0^2 + f(\epsilon_0)))$ pre-processing time.*

We can make the oracle in Theorem 3 a more generalized one: For integer η satisfying $1 \leq \eta \leq 1/\epsilon_0$, we partition each path $Q \in \mathcal{B}(X) \cup \mathcal{P}(X)$ into η segments $Q_1, .., Q_\eta$, compute the classes $A_1^{Q_l}, .., A_s^{Q_l}$ of vertices in $V(X)$ for each segment Q_l, $1 \leq l \leq \eta$, and key portal $p_{ij}^{Q_l}$, and use

$$\tilde{d}(u,v) = \min_{p_{ij}^{Q^l}, 1 \leq l \leq \eta, Q \in \mathcal{B}(X) \cup \mathcal{P}(X)} \hat{d}(u, p_{ij}^{Q_l}) + \hat{d}(p_{ij}^{Q_l}, v)$$

to approximate $d_G(u,v)$. By this generalization, we get the following result.

Theorem 4. *For graph G, $\epsilon_0 > 0$ and $1 \leq \eta \leq 1/\epsilon_0$, there is an oracle which gives a distance $\tilde{d}(u,v)$ with $d_G(u,v) \leq \tilde{d}(u,v) \leq d_G(u,v) + 7\epsilon_0 d(G)$ for any vertices u and v in G with $O(\eta)$ query time, $O(n(\log n/\epsilon_0 + f(\eta\epsilon_0)))$ size and $O(n(\log^3 n/\epsilon_0^2 + f(\eta\epsilon_0)))$ pre-processing time.*

4 Oracle with $(1+\epsilon)$ Stretch

For $\epsilon > 0$, by choosing an $\epsilon_0 = \frac{\epsilon}{7c}$ where $c > 0$ is a constant, the oracle in Theorem 3 gives a $(1+\epsilon)$-approximate distance oracle for graph G with $d_G(u,v) \geq d(G)/c$ for every u and v in G. For graph G with $d_G(u,v)$ much smaller than $d(G)$ for some u and v, we use a scaling approach as described in [8] to get a $(1+\epsilon)$-approximate distance oracle. The idea is that we compute a set of oracles as described in Theorem 3, each for a computed subgraph H of G. Given u and v, we can find in $O(1)$ time a constant number of subgraphs (and the corresponding oracles) such that the minimum value returned by these oracles is a $(1+\epsilon)$-approximation of $d_G(u,v)$. Therefore a $(1+\epsilon)$-approximate distance for any u, v can be computed in constant time.

Let l_m be the smallest positive edge length in G. We assume $l_m \geq 1$ and the case where $l_m < 1$ can be easily solved in a similar way by normalizing the length of each edge e of G to $l(e)/l_m$. For each scale $\gamma \in \{2^i | 0 \leq i \leq \log d(G)\}$, we compute a sparse cover $\mathcal{C}_\gamma = \{G(\gamma, j), j = 1, ..., n_\gamma\}$ of G as in Lemma 4. For each γ, we contract every edge $e = \{u,v\}$ of length $l(e) < \gamma/n^2$ in G and then compute $\mathcal{C}_\gamma$. Each edge of G appears in subgraphs $G(\gamma, j)$ for $O(\log n)$ different scales [8]. The data structure DS_1 for our $(1+\epsilon)$-approximate distance oracle keeps the following information:

- A 2-approximate distance oracle DS_T of G in Lemma 2.
- Subgraphs $G(\gamma, j)$ and for each subgraph $G(\gamma, j)$, an oracle $\mathrm{DS}_0(\gamma, j)$ in Theorem 3 with $\epsilon_0 = \epsilon/c'$, $c' > 0$ is a constant to be specified below.

Lemma 10. *For graph G and $\epsilon > 0$, $\tilde{d}(u,v)$ with $d_G(u,v) \leq \tilde{d}(u,v) \leq (1+\epsilon)d_G(u,v)$ can be computed in $O(1)$ time for any u and v in G using data structure* DS_1.

Proof. We assume $\epsilon > 5/n$, otherwise a naive exact distance oracle with $O(1)$ query time and $O(n^2)$ space can be used for the lemma. Given vertices u and v in G, oracle DS_T gives $\tilde{d}_T(u,v)$ with $d_G(u,v) \leq \tilde{d}_T(u,v) \leq 2d_G(u,v)$ in $O(1)$ time (Lemma 2). If $\tilde{d}_T(u,v) = 0$ then 0 is returned as $d_G(u,v)$. Otherwise, given $\tilde{d}_T(u,v)$, a scale γ with $\gamma/2 < \tilde{d}_T(u,v) \leq \gamma$ can be found by computing $\log(\tilde{d}_T(u,v))$[1]. By Lemma 4, there is a $G(\gamma,j)$ that contains u and every w with $d_G(u,w) \leq d_G(u,v) \leq \gamma$ and $d(G(\gamma,j)) = O(\gamma) = O(d_G(u,v))$. It's easy to see that $\mathrm{DS}_0(\gamma,j)$ returns a minimum distance among all the oracles at this scale containing u,v. Besides there exists a constant $c_1 > 0$ such that $d(G(\gamma,j)) \leq c_1 d_G(u,v)$ for any u,v and γ selected according to u,v. By oracle $\mathrm{DS}_0(\gamma,j)$, we get a distance $\tilde{d}_0(u,v)$ with $d_{G(\gamma,j)}(u,v) \leq \tilde{d}_0(u,v) \leq d_{G(\gamma,j)}(u,v) + 7\epsilon_0 d(G(\gamma,j))$. Since $G(\gamma,j)$ is a subgraph obtained from G with every edge e contracted for $l(e) < \gamma/n^2$, $d_{G(\gamma,j)}(u,v) \leq d_G(u,v)$. Let L be the largest sum of the lengths of the contracted edges in any path in G. Then $d_G(u,v) \leq d_{G(\gamma,j)}(u,v) + L$ and $L < \gamma/n \leq \frac{4}{5}\epsilon d_G(u,v)$. From $\gamma \leq 2\tilde{d}_T(u,v) \leq 4d_G(u,v)$ and $\epsilon > 5/n$. Let $\tilde{d}(u,v) = \tilde{d}_0(u,v) + \gamma/n$. Then

$$\begin{aligned} d_G(u,v) \leq \tilde{d}(u,v) &\leq d_{G(\gamma,j)}(u,v) + 7\epsilon_0 d(G(\gamma,j)) + \gamma/n \\ &\leq d_G(u,v) + 7c_1\frac{\epsilon}{c'}d_G(u,v) + \frac{4}{5}\epsilon d_G(u,v). \end{aligned}$$

By choosing $c' = 35c_1$, we have $d_G(u,v) \leq \tilde{d}_G(u,v) \leq (1+\epsilon)d_G(u,v)$. By Lemma 2, it takes $O(1)$ time to compute $\tilde{d}_T(u,v)$. The time for finding the right scale r is $O(1)$. From Lemma 4, there are $O(1)$ graphs $G(\gamma,j)$ containing u and v. From this and Theorem 3, it takes $O(1)$ time to compute $\tilde{d}(u,v)$. □

Lemma 11. *Data structure* DS_1 *requires* $O(n\log n(\log n/\epsilon + f(\epsilon)))$ *space and can be computed in* $O(n\log n(\log^3 n/\epsilon^2 + f(\epsilon)))$ *time.*

Proof. DS_T requires space $O(n\log n)$. Each $\mathrm{DS}_0(\gamma,j)$ requires space $O(n_{\gamma j} \log n_{\gamma j}/\epsilon + n_{\gamma j}f(\epsilon))$, where $n_{\gamma j} = |V(G(\gamma,j))|$. It is shown in [8] that each edge of G appears in $O(\log n)$ graphs of $G(\gamma,j)$. From this, $\sum_{\gamma,j} n_{\gamma j} = O(n\log n)$ and DS_1 requires space $O(n\log n(\log n/\epsilon + f(\epsilon)))$.

It takes $O(n\log^2 n)$ time to compute the sparse neighborhood covers. The value D for computing the classes can be computed in $O(n_{\gamma j})$ time. The time for computing $\mathrm{DS}_0(\gamma,j)$ for each $G(\gamma,j)$ is $O(n_{\gamma j}(\log^3 n_{\gamma j}/\epsilon^2 + f(\epsilon)))$. Therefore, DS_1 can be computed in $O(n\log n(\log^3 n/\epsilon^2 + f(\epsilon)))$ time. □

From Lemmas 10 and 11, we get Theorem 1 which is restated below.

Theorem 5. *For* $\epsilon > 0$, *there is a* $(1+\epsilon)$*-approximate distance oracle for* G *with* $O(1)$ *query time,* $O(n\log n(\log n/\epsilon + f(\epsilon)))$ *size and* $O(n\log n(\log^3 n/\epsilon^2 + f(\epsilon)))$ *pre-processing time.*

Using the oracle in Theorem 4 instead of DS_0, we get Theorem 2.

[1] We assume that $d(G)/l_m$ can be stored in one computer word of $O(1)$ bits.

5 Concluding Remarks

It is open whether there is a $(1+\epsilon)$-approximate distance oracle with $O(1)$ query time and size nearly linear in n for weighted directed planar graphs. For undirected planar graphs, it is interesting to reduce oracle size and pre-processing time (the function $f(\epsilon)$) for the oracles in this paper. Experimental studies for fast query time distance oracles are worth investigations.

References

1. Awerbuch, B., Berger, B., Cowen, L., Peleg, D.: Near-linear time construction of sparse neighborhood covers. SIAM J. Comput. **28**(1), 263–277 (1998)
2. Awerbuch, B., Peleg, D.: Sparse partitions. In: Proceedings of the 31st IEEE Annual Symposium on Foundations of Computer Science (FOCS 1990), pp. 503–513 (1990)
3. Busch, C., LaFortune, R., Tirthapura, S.: Improved sparse covers for graphs excluding a fixed minor. In: Proceedings of the 26th Annual ACM symposium on Principles of Distributed Computing (PODC 2007), pp. 61–70 (2007)
4. Chechik, S.: Approximate distance oracle with constant query time. In: Proceedings of the 46th Annual ACM Symposium on Theory of Computing (STOC 2014), pp. 654–663 (2014)
5. Delling, N., Goldberg, A., Pajor, T., Werneck, R.: Robust exact distance queries on massive networks. In: Microsoft Research Technique report, MSR-TR-2014-12 (2014)
6. Djidjev, H.: Efficient algorithms for shortest path problems on planar graphs. In: Proceedings of the 22nd International Workshop on Graph-Theoretical Concepts in Computer Science (WG 1996), pp. 151–165 (1996)
7. Harel, D., Tarjan, R.: Fast algorithms for finding nearest common ancestor. SIAM J. Comput. **13**, 338–355 (1984)
8. Kawarabayashi, K., Sommer, C., Thorup, M.: More compact oracles for approximate distances in undirected planar graphs. In: Proceedings of the 24th Annual ACM-SIAM Symposium on Discrete Algorithms (SODA 2013), pp. 550–563 (2013)
9. Klein, P.N.: Preprocessing an undirected planar network to enable fast approximate distance queries. In: Proceedings of the Thirteenth Annual ACM-SIAM Symposium on Discrete Algorithms. Society for Industrial and Applied Mathematics (2002)
10. Klein, P.N., Mozes, S. Sommer, C.: Structured recursive separator decompositions for planar graphs in linear time. In: Proceedings of the 45th Annual ACM symposium on Theory of computing (STOC 2013), pp. 505–514 (2013)
11. Klein, P.N., Subramanian, S.: A fully dynamic approximation scheme for shortest paths in planar graphs. Algorithmica **22**(3), 235–249 (1998)
12. Lipton, R.J., Tarjan, R.E.: A separator theorem for planar graphs. SIAM J. Appl. Math. **36**(2), 177–189 (1979)
13. Mozes, S., Sommer, C.: Exact distance oracles for planar graphs. In: Proceedings of the 23rd Annual ACM-SIAM Symposium on Discrete Algorithms (SODA 2012), pp. 209–222 (2012)
14. Sommer, C.: Shortest-path queries in static networks. ACM Comput. Surv. **46**(4), 1–31 (2012). (Article No. 45)
15. Thorup, M.: Compact oracles for reachability and approximate distances in planar digraphs. J. ACM (JACM) **51**(6), 993–1024 (2004)

16. Weimann, O., Yuster, R.: Approximating the diameter of planar graphs in near linear time. In: Fomin, F.V., Freivalds, R.U., Kwiatkowska, M., Peleg, D. (eds.) ICALP 2013, Part I. LNCS, vol. 7965, pp. 828–839. Springer, Heidelberg (2013)
17. Wulff-Nilsen, C.: Algorithms for planar graphs and graphs in metric spaces. Ph.D. thesis, University of Copenhagen (2010)

Partitioning Graph Drawings and Triangulated Simple Polygons into Greedily Routable Regions

Martin Nöllenburg[1], Roman Prutkin[2](✉), and Ignaz Rutter[2]

[1] Algorithms and Complexity Group, TU Wien, Vienna, Austria
noellenburg@ac.tuwien.ac.at
[2] Institute of Theoretical Informatics,
Karlsruhe Institute of Technology, Karlsruhe, Germany
{roman.prutkin,rutter}@kit.edu

Abstract. A greedily routable region (GRR) is a closed subset of $\mathbb{R}^2$, in which each destination point can be reached from each starting point by choosing the direction with maximum reduction of the distance to the destination in each point of the path. Recently, Tan and Kermarrec proposed a geographic routing protocol for dense wireless sensor networks based on decomposing the network area into a small number of interior-disjoint GRRs. They showed that minimum decomposition is NP-hard for polygons with holes.

We consider minimum GRR decomposition for plane straight-line drawings of graphs. Here, GRRs coincide with self-approaching drawings of trees, a drawing style which has become a popular research topic in graph drawing. We show that minimum decomposition is still NP-hard for graphs with cycles, but can be solved optimally for trees in polynomial time. Additionally, we give a 2-approximation for simple polygons, if a given triangulation has to be respected.

1 Introduction

Geographic or geometric routing is a routing approach for wireless sensor networks that became popular recently. It uses geographic coordinates of sensor nodes to route messages between them. One simple routing strategy is greedy routing. Upon receipt of a message, a node tries to forward it to a neighbor node that is closer to the destination than itself. However, delivery cannot be guaranteed, since a message may get stuck in a local minimum or *void*. More advanced geometric routing protocols employ strategies like face routing [2] and related techniques based on planar graphs to get out of local minima; see [5,15] for an overview.

An alternative approach is to decompose the network into components, such that in each of them greedy routing is likely to perform well [9,18,20]. A global data structure of preferably small size is used to store interconnectivity between components. One such network decomposition approach has been recently proposed by Tan and Kermarrec [19]. They model the network as a polygonal region with obstacles or holes inside it and try to partition this region into a minimum

K. Elbassioni and K. Makino (Eds.): ISAAC 2015, LNCS 9472, pp. 637–649, 2015.
DOI: 10.1007/978-3-662-48971-0_54

number of polygons, in which greedy routing works between any pair of points. They call such components *greedily routable regions (GRRs)*. For intercomponent routing, region adjacencies are stored in a graph. The protocol is able to guarantee finding paths of bounded stretch. The size of the routing state depends on the number of GRRs. The authors prove that partitioning a polygon with holes into a minimum number of regions is NP-hard and propose a simple heuristic. Its solution may strongly deviate from the optimum even for very simple polygons; see Fig. 1b.

In this paper, we approach the problem of finding minimum or approximately minimum GRR decompositions by first considering the special case of partitioning drawings of graphs, which can be interpreted as very thin polygons. We notice that in this scenario, GRRs coincide with *increasing-chord* drawings of trees as studied by Alamdari et al. [1].

A *self-approaching* curve is a curve, where for any point t' on the curve, the distance to t' decreases continuously while traversing the curve from the start to t' [11]. An *increasing-chord* curve is a curve that is self-approaching in both directions. The name is motivated by the equivalent characterization as those curves, where for any four points a, b, c, d in this order along the curve, it is $\text{dist}(b, c) \leq \text{dist}(a, d)$.

A graph drawing is self-approaching or increasing-chord if each pair of vertices is joined by a self-approaching or increasing-chord path, respectively. The study of self-approaching and increasing-chord graph drawings was initiated by Alamdari et al. [1]. They studied the problem of recognizing whether a given graph drawing is self-approaching and gave a complete characterization of trees admitting self-approaching drawings. In our own previous work [17], we studied self-approaching and increasing-chord drawings of triangulations and 3-connected planar graphs. Furthermore, the problem of connecting given points to an increasing-chord drawing has been investigated [1,8].

Contributions. First, we show that partitioning a plane graph drawing into a minimum number of increasing-chord components is NP-hard. This strengthens the result of Tan and Kermarrec [19] for polygons with holes. Next, we consider plane drawings of trees. We show how to model the decomposition problem using Minimum Multicut, which provides a polynomial-time 2-approximation. We then solve the partitioning problem for trees optimally in polynomial time using dynamic programming. Finally, we use the insights gained for decomposing graphs and apply them to the problem of minimally decomposing simple triangulated polygons into GRRs. We provide a polynomial-time 2-approximation for decompositions that are formed along chords of the triangulation.

2 Preliminaries

Greedily Routable Regions were introduced by Tan and Kermarrec [19] as follows.

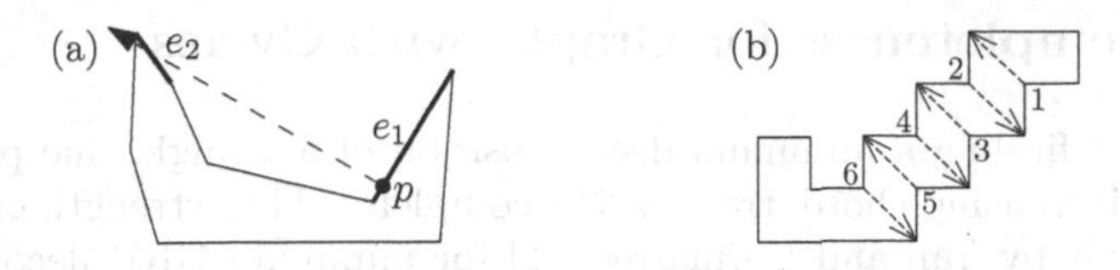

Fig. 1. (a) Normal ray $\text{ray}_{e_1}(p)$ and a pair of conflicting edges e_1, e_2. (b) The heuristic in [19] splits a non-greedy region by a bisector at a maximum reflex angle. If the splits are chosen in order of their index, seven regions are created, although two is minimum (split only at 6).

Definition 1 ([19]). *A polygon $\mathcal{P}$ is a* greedily routable region (GRR), *if for any two points $s, t \in \mathcal{P}$, $s \neq t$, point s can always move along a straight-line segment within $\mathcal{P}$ to some point s' such that $|s't| < |st|$.*

In the full version of our paper [16] we show that this definition is equivalent to the one used in the abstract.

A *decomposition* of a polygon $\mathcal{P}$ is a partition of $\mathcal{P}$ into simple polygons P_i, $i = 1, \ldots, k$, such that $\bigcup_{i=1}^{k} P_i = \mathcal{P}$ and no P_i, P_j with $i \neq j$ share an interior point. A decomposition of $\mathcal{P}$ is a *GRR decomposition* if each component P_i is a GRR. Using the concept of a *conflict relationship* between polygon edges (see Fig. 1a), Tan and Kermarrec give a convenient characterization of GRRs.

Definition 2 (Normal ray). *Let $\mathcal{P}$ be a polygon, $e = uv$ a boundary edge and p a point on uv. Let $\text{ray}_{uv}(p)$ denote a ray with origin in p orthogonal to uv, such that all points on this ray sufficiently close to p are not in the interior of $\mathcal{P}$. (If p is an acute reflex angle, no such ray exists.)*

Definition 3 (Conflicting edges). *Let e_1 and e_2 be two edges of a polygon $\mathcal{P}$. If for some point p in the interior of e_1, $\text{ray}_{e_1}(p)$ intersects e_2, then e_1* conflicts *with e_2.*

A polygon is a GRR if and only if it has no pair of conflicting edges [19].

Now consider a straight-line plane drawing Γ of a graph $G = (V, E)$. We identify the edges of G with the corresponding line segments of Γ and the vertices of G with the corresponding points. Planar straight-line drawings can be considered as infinitely thin polygons. The routing happens along the edges of Γ, and we call Γ a GRR if for any two points $s \neq t$ on Γ there exists a point s' on the same edge as s, such that $|s't| < |st|$.

Assume $\text{ray}_{e_1}(s)$ for an interior point s on an edge e_1 of Γ crosses another edge e_2 in point t. Then, any movement along e_1 starting from s increases the distance to t. We call such edges *conflicting*. It is easy to see that Γ is a GRR if it contains no pair of conflicting edges. Obviously, such a drawing Γ contains no cycles. In fact, a straight-line drawing of a tree is increasing-chord if and only if it has no conflicting edges [1], which implies the following lemma.

Lemma 1. *The following two properties are equivalent for a straight-line drawing Γ to be a GRR. 1) Γ is connected and has no conflicting edges; 2) Γ is an increasing-chord drawing of a tree.*

3 NP-Completeness for Graphs with Cycles

We show that finding a minimum decomposition of a straight-line plane drawing Γ into increasing-chord trees is NP-complete. This strengthens the NP-hardness result by Tan and Kermarrec [19] for minimum GRR decompositions of polygons with holes.

First, we prove NP-hardness. Both our proof and the proof in [19] are reductions from the NP-complete problem Planar 3SAT [14]. Recall that a Boolean 3SAT formula φ is called *planar*, if the corresponding variable clause graph G_φ having a vertex for each variable and for each clause and an edge for each occurrence of a variable (or its negation) in a clause is a planar graph. In fact, G_φ can be drawn in the plane such that all variable vertices are aligned on a vertical line and all clause vertices lie either to the left or to the right of this line and connect to the variables via E- or ∃-shapes [13]. The variable gadgets in [19] are cycles formed by T-shaped polygons which can be made arbitrarily thin. Thus, in the case of straight-line plane drawings we can use very similar variable gadgets (see Fig. 2). The clause gadgets in [19], however, are squares, at which three variable cycles meet. This construction cannot be adapted for straight-line plane drawings and we have to construct a significantly different clause gadget; see Fig. 3.

Consider a variable gadget consisting of k T-shapes as shown in Fig. 2. On each T-shape we place one black and one white point as shown in the figure. It is easy to verify that neither two black points nor two white points can be in one increasing-chord component. Thus, a minimum GRR decomposition of a variable gadget contains at least k components. If it contains exactly k components, then each component must contain one black and one white point, and there are exactly two possibilities. Each black point has exactly two white points it can share a GRR with, and once one pairing is picked, it fixes all the remaining pairings. The corresponding possibilities are shown in Fig. 2a and 2b and will be used to encode the values *true* and *false*, respectively.

To pass the truth assignment of a variable to a clause it is part of, we use *arm* gadgets. In total twelve variations of the arm gadget will be used, depending on the position of the literal in the clause, the position of the clause, and whether the literal is negated or not. Since in G_φ each clause c connects to three variables, we denote these variables or literals as the *upper*, *middle*, and *lower* variables of c. Similarly, an arm of c is called an *upper*, *middle*, or *lower* arm if it belongs to a literal of the same type in c. An arm is called a *right* (resp. *left*) arm if it belongs to a clause that lies to the right (resp. left) of the vertical variable line. Finally, an arm of c is *positive* if the corresponding literal is positive in c and it is *negative* otherwise.

The basic principle of operation of any arm gadget is the same; as an example consider the right upper positive arm in Fig. 2. The full version [16] covers the remaining arm types. Note that each arm can be arbitrarily extended both horizontally and vertically to reach the required point of its clause gadget. We select again black and white points (also called *distinguished* points) on the line

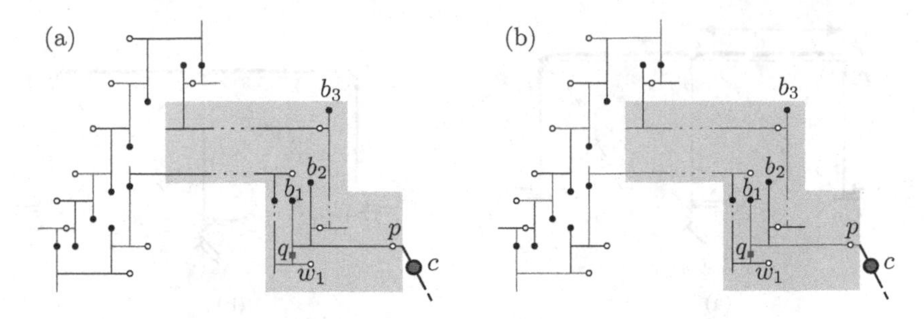

Fig. 2. Variable gadget with a right upper positive arm. (a) *true* and (b) *false* variable state (Color figure online).

segments of the arm gadget. We shall prove the following property which is crucial for our construction.

Property 1. 1. Consider a drawing Γ_i of a variable gadget together with all of its arms. Then, neither two black nor two white points on Γ_i can be in one GRR. In a minimum GRR decomposition of Γ_i, each component has one black and one white point, and exactly two such pairings of points are possible, one for each truth assignment.

2. Consider two such drawings Γ_i, Γ_j for two different variables. Then, no distinguished point of Γ_i can be in the same GRR as a distinguished point of Γ_j.

Part 1 of Property 1 extends the same property that we already showed for variable gadgets without arms to the case including all arms. It is easy to verify that it holds in all our constructions of the arm gadgets. Part 2 follows from the way the arms are connected by a clause, i.e., in Fig. 3 no pair of points from p_i, p_j, p_k can be in same GRR.

For each arm gadget we select a special *red* point q; see Fig. 2. Point q is neither white nor black. The clause gadget (green in Fig. 3, partly drawn in Fig. 2) is connected to the arm by a horizontal segment with a distinguished point p on its end, which is either black or white depending on the arm type. Each clause has a distinguished point c chosen as shown in Fig. 3. If points q and p are in the same GRR, then this GRR obviously cannot contain any point of a green clause segment (see Fig. 2b). We prove in the full version [16] that points p and q are in the same GRR exactly for the decomposition corresponding to the variable state that does *not* satisfy the clause.

Lemma 2. *In a minimum GRR decomposition, the distinguished point c of a clause gadget can share a GRR with a black or white point of an arm gadget if and only if the corresponding literal is in the* true *state.*

The remaining ingredient is to show the following property.

Property 2. If a variable assignment satisfies a clause, then its entire clause gadget can be contained in a GRR of the corresponding arm.

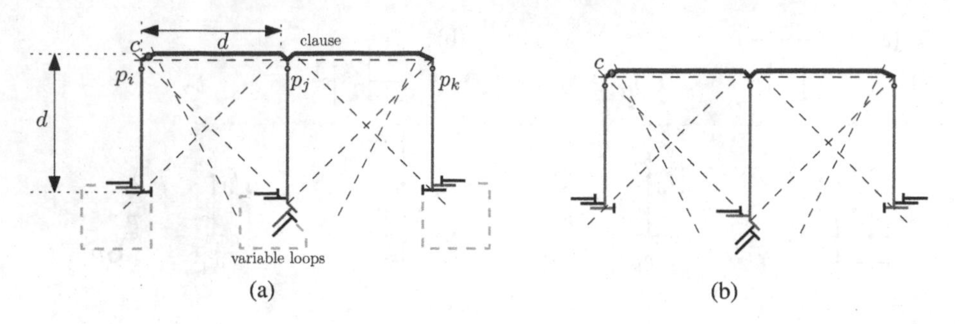

Fig. 3. Clause gadget rotated by 90°. (a) *true* and (b) *false* clause state (Color figure online).

In Fig. 3a, each variable is in a state that satisfies the clause. The lengths of the blue and green segments are chosen such that each blue component can be merged with the clause gadget (green) into a single GRR as indicated by the dashed lines. Finally, we can prove the NP-hardness result by showing that any satisfying truth assignment for a formula φ yields a GRR decomposition into a fixed number k of GRRs, where k is the total number of black points in our construction. Likewise, we can show that any decomposition into k GRRs necessarily satisfies each clause in φ. The full proof is in [16].

Theorem 1. *For $k \in \mathbb{N}_0$, deciding whether a plane straight-line drawing can be partitioned into k increasing-chord components is* NP*-complete.*

4 Trees

We consider greedy tree decompositions, or GTDs. For trees, greedy regions correspond to increasing-chord drawings. In the following, we consider a plane straight-line drawing of a tree $T = (V, E)$, $|V| = n$. We identify the tree with its drawing, the vertices with the corresponding points and the edges with the corresponding line segments. We want to partition it into a minimum number of increasing-chord subdrawings.

In such a partition, each pair of components shares at most one point. We make a restriction by only allowing contacts of the following type.

Definition 4 (Proper contacts). *Two drawings of trees with the only common point p have a* proper contact *if p is a leaf in at least one of them.*

This definition forbids GRRs to have contacts as in Fig. 4a. First, assume T is split only at its vertices. In the full version [16] we show how to drop this restriction and adapt our algorithms to compute minimum or approximately minimum GRR decompositions of plane straight-line tree drawings which allow splitting tree edges at interior points.

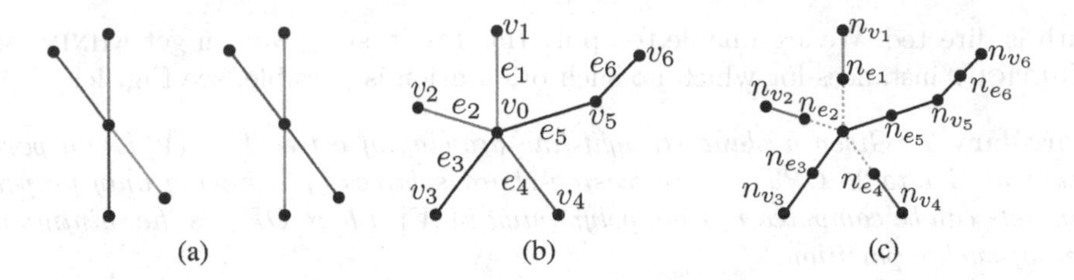

Fig. 4. (a) Orange and blue GRRs have non-proper contacts. (b) Tree drawing decomposed in GRRs. Edge pairs $\{e_1, e_2\}$, ..., $\{e_4, e_5\}$, $\{e_5, e_1\}$ as well as $\{e_1, e_6\}$, $\{e_4, e_6\}$ are conflicting. (c) Minimum Multicut instance constructed according to the proof of Proposition 3. No edge orientation respecting all paths between the terminals exists. Dashed edges form a solution.

4.1 2-Approximation Using Multicut

We show how to partition the edges of T into a minimum number of increasing-chord components with proper contacts using Minimum Multicut on trees. Given an edge-weighted graph $G = (V, E)$ and a set of terminal pairs $\{(s_1, t_1), \ldots, (s_k, t_k)\}$, an edge set $S \subseteq E$ is a *multicut* if removing S from G disconnects each pair s_i, t_i, $i = 1, \ldots, k$. A multicut is minimum if the total weight of its edges is minimum.

For the complexity of Minimum Multicut on special graph types, see the survey by Costa et al. [7]. Computing Minimum Multicut is NP-hard even for unweighted binary trees [3], but has a polynomial-time 2-approximation [10].

Consider a plane straight-line drawing of a tree $T = (V, E)$. We construct a tree T_M as follows. Tree T_M has a vertex n_v for each vertex $v \in V$ and a vertex n_e for each edge $e \in E$. For each $e = uv \in E$, edges $n_u n_e$ and $n_e n_v$ are in T_M.

The set of terminals is defined as $\{(n_{e_1}, n_{e_2}) \mid \text{edges } e_1, e_2 \in E \text{ are conflicting}\}$.

Lemma 3. *Let the set of edges E' of T_M form a* Minimum Multicut. *Assume removing E' from T_M disconnects T_M into connected components $C_1^M, \ldots, C_k^M$. Then, components $C_i = \{e \in E \mid n_e \in C_i^M\}$ form a minimum GRR decomposition of T.*

Proof. Consider a multicut E' of T_M, $|E'| = k - 1$. Consider a component C_i^M. Then, the edges in C_i are conflict-free and form a connected subtree T_i of T. Thus, T_i is a GRR by Lemma 1.

Next, consider a GRR decomposition of T into k subtrees $T_i = (V_i, E_i)$ with proper contacts. Assume T_i, T_j touch at vertex $v \in V$. Let edge $e = uv$ be in T_i, and let v be a leaf in T_i. Then, we add edge $n_e n_v$ of T_M to set S; see Fig. 4b and 4c. It is $|S| = k - 1$. After removing S from T_M, no connected component contains vertices n_{e_1}, n_{e_2} for a pair of conflicting edges e_1, e_2. Thus, S is a multicut. □

Note that Minimum Multicut is polynomial in directed trees [6], i.e., trees whose edges can be directed such that for each terminal pair (s_i, t_i), the s_i-t_i

path is directed. We are unable to apply this result, since we can get MINIMUM MULTICUT instances for which no such orientation is possible, see Fig. 4c.

Corollary 1. *Given a plane straight-line drawing of a tree $T = (V, E)$, a partition of E into $2 \cdot OPT - 1$ increasing-chord subtrees of T having only proper contacts can be computed in time polynomial in $|V|$, where OPT is the minimum size of such a partition.*

4.2 Optimal Solution

In the following we show how to compute an optimal solution for this problem in polynomial time via a dynamic program.

Theorem 2. *Given a plane straight-line drawing of a tree $T = (V, E)$, a partition of E into minimum number of increasing-chord subtrees of T (minimum GTD) having only proper contacts can be computed in time $O(|V|^7)$.*

As it is the case with minimum partitions of simple hole-free polygons into convex [4] or star-shaped [12] components, our algorithm is based on dynamic programming. Assume T is rooted at vertex r with degree 1. For each vertex u with parent π_u, let T_u be the subtree of u together with edge $\pi_u u$. A minimum partition is constructed from the solutions of subinstances as follows. We shall store minimum partitions of T_u for various possibilities of the greedy component containing u. We shall call this component the *root* component. For subtrees T_{u_1}, ..., T_{u_d} whose only common vertex is u, a minimum partition of $T' = \bigcup_i T_{u_i}$ is formed by choosing partitions of T_{u_i} and possibly merging some of the components containing u_i, i.e., the root components of T_{u_i}.

Given a tree root, the number of different subtrees containing it might be exponential, e.g., it is $\Theta(2^n)$ in a star. The key observation for our algorithm is that we do not need to store a partition for each possible root component. We require the following notation.

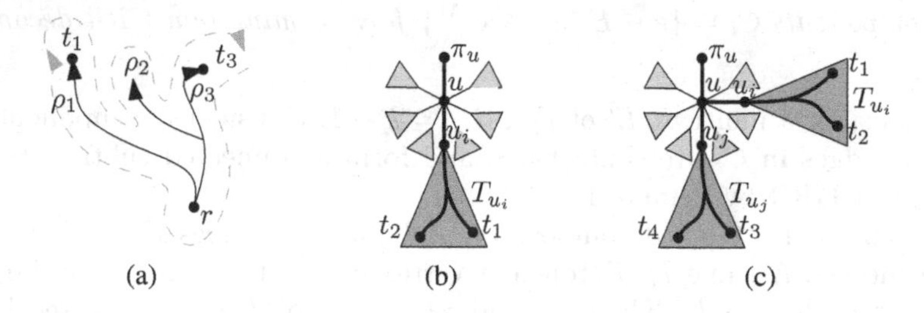

Fig. 5. (a) Path ρ_2 is clockwise between paths ρ_1 and ρ_3. (b), (c) Proof of Lemma 5.

Definition 5 (Path clockwise between). *Consider directed non-crossing paths ρ_1, ρ_2, ρ_3 with common origin r, endpoints t_1, t_2, t_3 and, possibly, common prefixes. Let V_i be vertices of ρ_i, $i = 1, 2, 3$, and let T be the tree formed by the union of ρ_1, ρ_2 and ρ_3. We say that ρ_2 is* clockwise between *ρ_1 and ρ_3, if the clockwise traversal of the outer face of T visits t_1, t_2, t_3 in this order; see Fig. 5a.*

Note that in Definition 5 the three paths may (partially) coincide. Lemma 4 shows that to decide whether a union of two subtrees is increasing-chord, it is sufficient to consider only the two pairs of "outermost" root-leaf paths of each subtree. This result is crucial for limiting the number of representative decompositions that need to be considered during our dynamic programming approach. The statement of the lemma is illustrated in Fig. 6a. The proof is found in the full version [16].

Lemma 4. *Let T_1, T_2 be trees drawn with increasing chords having the only common vertex r. Let all tree edges be directed away from r. Let paths ρ_1, ρ_2 in T_1 and ρ_3, ρ_4 in T_2 be paths from r to a leaf, such that:*

- *every directed path from r in T_1 is clockwise between ρ_1 and ρ_2;*
- *every directed path from r in T_2 is clockwise between ρ_3 and ρ_4;*
- *for $i = 1, \ldots, 4$, path ρ_i is clockwise between ρ_{i-1} and ρ_{i+1} (mod 4).*

If the union of paths $\rho_1, \ldots, \rho_4$ is drawn with increasing chords, then so is the union of trees T_1 and T_2.

We now describe our dynamic program in detail. For a root component R of T_u, let the *leftmost path* (or, respectively, the *rightmost path*) be the simple path in R starting at π_u which always chooses the next counterclockwise (clockwise) edge. Let $\Delta = 0, \ldots, 3$. For each pair of vertices t_1, t_2 in T_u, cell $\tau_\Delta[u, t_1, t_2]$ of a table τ_Δ stores the size of a minimum GRR decomposition of T_u, in which the root component has the π_u-t_1 path and the π_u-t_2 path as its leftmost and rightmost path, respectively, and such that u has degree $\Delta + 1$ in the root component. Furthermore, define $\tau[u, t_1, t_2] = \min_{\Delta=0,\ldots,3} \tau_\Delta[u, t_1, t_2]$ and $\tau[u] = \min_{t_1,t_2} \tau[u, t_1, t_2]$.

Clearly, for each leaf u, it is $\tau_0[u, u, u] = 1$, and $\tau_\Delta[u, t_1, t_2] = \infty$ for all other values of Δ, t_1, t_2. Let v be the only neighbor of the tree root r. Then, $\tau[v]$ is the size of a minimum GRR decomposition of T.

For ease of presentation, we use the following notations. Vertex u is not a leaf and has children $u_1, \ldots, u_d$. Vertices t_1, t_2 are in the subtree of u_i, vertices t_3, t_4 are in the subtree of u_j, vertices t_5, t_6 are in the subtree of u_k and vertex t_7 in the subtree of u_ℓ. For $q = 1, \ldots, 7$, let ρ_q denote the u-t_q path. It is always assumed that the t_q are such that paths $\rho_1, \ldots, \rho_7$ are in this clockwise order.

Lemma 5. *We have the recurrences*

1. $\tau_1[u,t_1,t_2] = \begin{cases} \tau[u_i,t_1,t_2] + \sum_{m\neq i}\tau[u_m] & \rho_1\cup\rho_2\cup\pi_u u \text{ increasing-chord} \\ \infty & \text{otherwise} \end{cases}$
2. $\tau_2[u,t_1,t_4] = \min_{t_2,t_3}\{\tau[u_i,t_1,t_2] + \tau[u_j,t_3,t_4] - 1 + \sum_{m\neq i,j}\tau[u_m]\}$
3. $\tau_3[u,t_1,t_5] = \min_{t_3}\{\tau[u_i,t_1,t_1] + \tau[u_j,t_3,t_3] + \tau[u_k,t_5,t_5] - 2 + \sum_{m\neq i,j,k}\tau[u_m]\}$
4. $\tau_0[u,u,u] = \mathbf{min}\{\sum_i \tau[u_i] + 1,$
5. $\quad \min_{t_1,t_2,t_3,t_4}\{\tau[u_i,t_1,t_2] + \tau[u_j,t_3,t_4] + \sum_{m\neq i,j}\tau[u_m]\},$
6. $\quad \min_{t_1,t_2,t_3,t_4,t_5,t_6}\{\tau[u_i,t_1,t_2] + \tau[u_j,t_3,t_4] + \tau[u_k,t_5,t_6] - 1 + \sum_{m\neq i,j,k}\tau[u_m]\},$
7. $\quad \min_{t_1,t_3,t_5,t_7}\{\tau[u_i,t_1,t_1] + \tau[u_j,t_3,t_3] + \tau[u_k,t_5,t_5] + \tau[u_\ell,t_7,t_7] - 2 + \sum_{m\neq i,j,k,\ell}\tau[u_m]\ \},$
8. and $\tau_0[u,\cdot,\cdot] = \infty$ otherwise.

where the minimizations in lines 2, 3, 5, 6, and 7 consider only vertices such that $\rho_1\cup\cdots\cup\rho_4\cup\pi_u u$, $\rho_1\cup\rho_3\cup\rho_5\cup\pi_u u$, $\rho_1\cup\cdots\cup\rho_4$, $\rho_1\cup\cdots\cup\rho_6$, *and* $\rho_1\cup\rho_3\cup\rho_5\cup\rho_7$ *are increasing-chord, respectively.*

Proof. Assume u is not a leaf. Each increasing-chord tree drawing either has maximum degree at most three or is a subdivision of $K_{1,4}$ (star with four leaves) [1]. This fact limits the number of possibilities of how the root component of T_u can be created from the root components of T_{u_i}, $i = 1,\ldots,d$.

Consider a GRR decomposition of T_u of size x with root component R which has π_u-t_1 and π_u-t_2 paths as its leftmost and rightmost path, respectively. If u has degree 2 in R, then it induces GRR decompositions of all T_{u_m} with $m = 1,\ldots,d$ with sizes x_m. Note that $x = \sum_{m=1}^{d} x_m$. By definition, $\tau[u_m] \leq x_m$. Moreover, the root component of T_{u_i} is $R' = R - \pi_u u$ and has ρ_1 and ρ_2 as its leftmost and rightmost path; see Fig. 5b. Therefore, $\tau[u_i,t_1,t_2] \leq x_i$. It follows that the right-hand side of recurrence 1 is at most x. Thus, the right-hand side of recurrence 1 is bounded by its left-hand side.

Conversely, assume that the right-hand side of recurrence 1 sums to $x < \infty$. Then, there exist GRR decompositions of T_{u_m} of size $\tau[u_m]$ for $m \neq i$, and a GRR decomposition of T_{u_i} of size $\tau[u_i,t_1,t_2]$ whose root component R' has ρ_1 and ρ_2 as leftmost and rightmost path, respectively. By Lemma 4 and the assumption that $\rho_1\cup\rho_2\cup\pi_u u$ is increasing-chord, it follows that $R' + \pi_u u$ is increasing-chord. Together, we obtain a GRR decomposition of T_u of size x, whose root component has its leftmost and rightmost paths ending at t_1 and t_2, respectively. Thus, the left-hand side of recurrence 1 is bounded by its right-hand side.

This finishes the proof for recurrence 1. The remaining recurrences can be proved analogously. The notable differences are as follows. In recurrence 2, the root component of the decomposition of T_u is obtained by merging root components of decompositions of T_{u_i} and T_{u_j} as well as edge $\pi_u u$; see Fig. 5c. Hence, the number of components in the whole decomposition decreases by 1. In recurrence 3, we merge three root components with edge $\pi_u u$. Then, u has degree 4, and, thus, each of the three root components must be a path. In recurrence 4, the root component of a decomposition of T_u is the single edge $\pi_u u$. It may, however, be beneficial to merge up to four root components of children of u, provided their boundary paths are such that their union is a GRR; see Lemma 4. □

Theorem 2 follows by applying Lemma 5 to T bottom-up. For the running time, the limiting factor is considering all possible choices of $t_1, \ldots, t_6$ in recurrence 5 to compute $\tau_0[u, u, u]$. In the full version [16] we show how to compute minimum partitions where we also allow splits in the interior of edges.

Theorem 3. *An optimal partition of a plane straight-line tree drawing in GRRs with proper contacts allowing edge splits can be computed in $O(|V|^{14})$ time.*

5 Triangulations

We now consider partitioning a hole-free polygon P with a fixed triangulation into a minimum number of GRRs by cutting it along chords of P contained in the triangulation. For such decompositions we only consider GRRs each consisting of a group of triangles of the triangulation that together form a simple polygon without articulation points.

We reduce the problem to MINIMUM MULTICUT on trees and use it to give a polynomial-time $(2 - 1/\mathrm{OPT})$-approximation, where OPT is the number of GRRs in an optimal partition. Recall that a polygon is a GRR if and only if it has no conflict edges [19]. Let Δ_{uvw} be the triangle defined by three non-collinear points u, v, w.

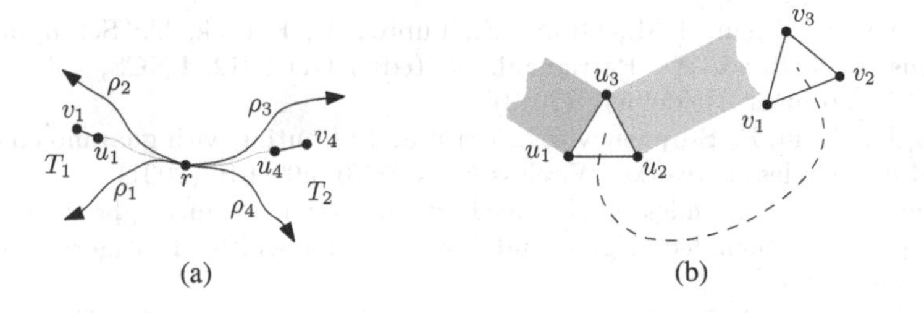

Fig. 6. (a) Statement of Lemma 4. (b) Conflicting triangles (Color figure online).

From now on, let triangles $\tau_1, \ldots, \tau_n$ form a triangulation of a simple hole-free polygon P, and let T be its corresponding dual binary tree. For simplicity we use τ_i to refer both to a triangle in P and its dual node in T.

Definition 6 (Projection of an edge). *For three non-collinear points u_1, u_2, u_3, let $proj_{u_1}(u_2u_3)$ denote the set of points covered by shifting u_2u_3 orthogonally to itself and away from u_1 (blue in Fig. 6b).*

Definition 7 (Conflicting triangles). *Let $\tau_i = \Delta_{u_1u_2u_3}$ and $\tau_j = \Delta_{v_1v_2v_3}$ be two triangles such that the two edges dual to u_1u_2 and v_1v_2 are on the τ_i-τ_j path in T. We call τ_i, τ_j conflicting, if $proj_{u_1}(u_2u_3) \cup proj_{u_2}(u_1u_3)$ contains an interior point of τ_j.*

Lemma 6. *Let $T' \subset T$ be a subtree of T and let P' be the corresponding simple polygon dual to T'. Then P' is a GRR if and only if no two triangles τ, τ' in P' are conflicting.*

By Lemma 6, the decompositions of P in k GRRs correspond bijectively to the multicuts E' of T with $|E'| = k - 1$ where the terminal pairs are the pairs of conflicting triangles; see [16] for the proof. We now use the 2-approximation for MINIMUM MULTICUT on trees [10] to give a $(2 - 1/\mathrm{OPT})$-approximation for the minimum GRR decomposition of P. Let E' be a 2-approximation of MINIMUM MULTICUT in T with respect to the pairs of conflicting triangles. By the above observation the minimum multicut for T has size $\mathrm{OPT} - 1$, hence $|E'| \leq 2\,\mathrm{OPT} - 2$, which in turn yields a decomposition into $2\,\mathrm{OPT} - 1$ regions. Thus the approximation guarantee is $2 - 1/\mathrm{OPT}$.

Theorem 4. *There is a polynomial-time* $(2 - 1/\mathrm{OPT})$*-approximation for minimum GRR decomposition of triangulated simple polygons.*

Acknowledgements. The second author thanks Jie Gao for pointing him to the topic of GRR decompositions.

References

1. Alamdari, S., Chan, T.M., Grant, E., Lubiw, A., Pathak, V.: Self-approaching graphs. In: Didimo, W., Patrignani, M. (eds.) GD 2012. LNCS, vol. 7704, pp. 260–271. Springer, Heidelberg (2013)
2. Bose, P., Morin, P., Stojmenović, I., Urrutia, J.: Routing with guaranteed delivery in ad hoc wireless networks. Wireless Netw. **7**(6), 609–616 (2001)
3. Calinescu, G., Fernandes, C.G., Reed, B.: Multicuts in unweighted graphs and digraphs with bounded degree and bounded tree-width. J. Algorithms **48**(2), 333–359 (2003)
4. Chazelle, B., Dobkin, D.: Optimal convex decompositions. In: Computational Geometry, pp. 63–133 (1985)
5. Chen, D., Varshney, P.K.: A survey of void handling techniques for geographic routing in wireless networks. Commun. Surv. Tutor. **9**(1), 50–67 (2007)
6. Costa, M., Létocart, L., Roupin, F.: A greedy algorithm for multicut and integral multiflow in rooted trees. Oper. Res. Lett. **31**(1), 21–27 (2003)
7. Costa, M.C., Létocart, L., Roupin, F.: Minimal multicut and maximal integer multiflow: A survey. Eur. J. Oper. Res. **162**(1), 55–69 (2005)
8. Dehkordi, H.R., Frati, F., Gudmundsson, J.: Increasing-chord graphs on point sets. J. Graph Algorithms Appl. (2015, to appear). doi:10.7155/jgaa.00348
9. Fang, Q., Gao, J., Guibas, L., de Silva, V., Zhang, L.: Glider: gradient landmark-based distributed routing for sensor networks. In: INFOCOM 2005, pp. 339–350. IEEE (2005)
10. Garg, N., Vazirani, V., Yannakakis, M.: Primal-dual approximation algorithms for integral flow and multicut in trees. Algorithmica **18**(1), 3–20 (1997)
11. Icking, C., Klein, R., Langetepe, E.: Self-approaching curves. Math. Proc. Camb. Phil. Soc. **125**, 441–453 (1999)

12. Keil, J.M.: Decomposing a polygon into simpler components. SIAM J. Comput. **14**(4), 799–817 (1985)
13. Knuth, D.E., Raghunathan, A.: The problem of compatible representatives. SIAM J. Discrete Math. **5**(3), 422–427 (1992)
14. Lichtenstein, D.: Planar formulae and their uses. SIAM J. Comput. **11**(2), 329–343 (1982)
15. Mauve, M., Widmer, J., Hartenstein, H.: A survey on position-based routing in mobile ad hoc networks. IEEE Netw. **15**(6), 30–39 (2001)
16. Nöllenburg, M., Prutkin, R., Rutter, I.: Partitioning graph drawings and triangulated simple polygons into greedily routable regions (2015). CoRR arXiv:1509.05635
17. Nöllenburg, M., Prutkin, R., Rutter, I.: On self-approaching and increasing-chord drawings of 3-connected planar graphs. In: Duncan, C., Symvonis, A. (eds.) GD 2014. LNCS, vol. 8871, pp. 476–487. Springer, Heidelberg (2014)
18. Tan, G., Bertier, M., Kermarrec, A.M.: Convex partition of sensor networks and its use in virtual coordinate geographic routing. In: INFOCOM 2009, pp. 1746–1754. IEEE (2009)
19. Tan, G., Kermarrec, A.M.: Greedy geographic routing in large-scale sensor networks: a minimum network decomposition approach. IEEE/ACM Trans. Netw. **20**(3), 864–877 (2012)
20. Zhu, X., Sarkar, R., Gao, J.: Shape segmentation and applications in sensor networks. In: INFOCOM 2007, pp. 1838–1846. IEEE (2007)

Computational Complexity II

A New Approximate Min-Max Theorem with Applications in Cryptography

Maciej Skórski(✉)

Cryptology and Data Security Group, University of Warsaw, Warsaw, Poland
maciej.skorski@mimuw.edu.pl

Abstract. We propose a novel proof technique that can be applied to attack a broad class of problems in computational complexity, when switching the order of universal and existential quantifiers is helpful. Our approach combines the standard min-max theorem and convex approximation techniques, offering quantitative improvements over the standard way of using min-max theorems as well as more concise and elegant proofs.

Keywords: Min-max theorems · Convex approximation · Cryptography

1 Introduction

1.1 The Min-Max Theorem

The celebrate von Neumann min-max theorem [12] states that every finite, two-player, zero-sum game has an equilibrium in mixed strategies. That is, the maximum value of the minimum expected gain for one player is equal to the minimum value of the maximum expected loss for the other. Any zero-sum game can be represented as a payoff matrix

$$A = [A(x,y)]_{x\in X, y\in Y}$$

where $A(x,y)$ is the payoff in case when the X-player chooses strategy $x \in X$ and the Y-player chooses strategy $y \in Y$, understood as a gain for the X-player and a loss of the Y-player. The basic moves $x \in X, y \in Y$ are called *pure strategies* (think of one of 3 options in the rock-paper-scissors game). We allow the players to use randomized strategies, which are called *mixed strategies* (think of picking a random answer in the rock-paper-scissors game) represented formally as distributions $p_X(\cdot), p_Y(\cdot)$ over X and Y respectively, and analyze the expected payoff

$$\mathbb{E}_{y\sim p_Y, x\sim p_X} A(x,y) = \sum_y \sum_x p_Y(y) p_X(x) A(x,y).$$

The full version of this paper is available online at http://arxiv.org/abs/1506.06633.
M. Skórski—This work was partly supported by the WELCOME/2010-4/2 grant founded within the framework of the EU Innovative Economy Operational Programme.

K. Elbassioni and K. Makino (Eds.): ISAAC 2015, LNCS 9472, pp. 653–663, 2015.
DOI: 10.1007/978-3-662-48971-0_55

If the player X goes first, she can guarantee her gain to be at least

$$\mathsf{MaxGain}(X) = \max_{p_X} \min_{y} \mathbb{E}_{x \sim p_X} A(x,y),$$

and when the player Y goes first he guarantees his lost to be at most

$$\mathsf{MinLoss}(Y) = \min_{p_Y} \max_{x} \mathbb{E}_{y \sim p_Y} A(x,y),$$

where in both equations we used the fact that the second player always achieves the best response with some pure strategy. The min-max theorem guarantees that we have an equilibrium between the players.

Theorem (Min-Max Theorem [12]). With the notation as above (and players using mixed strategies), we have

$$\mathsf{MaxGain}(X) = \mathsf{MinLoss}(Y).$$

Many more general versions of the min-max theorem exist. All of them assure the equality

$$\sup_{x \in X} \inf_{y \in Y} f(x,y) = \inf_{y \in Y} \sup_{x \in X} f(x,y)$$

under certain conditions imposed on the sets X, Y (for example both convex and compact subsets of a locally convex topological space) and the function f (for example continuity, convexity in y and concavity in x). The proofs typically use fixed point theorems. Min-Max theorems have a lot of applications in game theory, statistical decision theory, economy and theoretical computer science. In this paper we focus on applications in cryptography, and the simplest version will be enough for our discussion.

1.2 Switching the Order of Quantifiers by the Min-Max Theorem

The min-max theorem may be used to change the order of quantifiers (minimization corresponds to the existential quantifier and maximization corresponds to the universal quantifier). A very good example is the classical hardcore lemma due to Impagliazzo [8]. The lemma stated informally says that if for every algorithm A there exists a large set of inputs on which A fails to compute a fixed function f, then in fact there exists a large set of inputs on which every algorithm fails to compute f with probability close to $\frac{1}{2}$. This particular lemma falls into a broad class of results in complexity theory which can be proven using the min-max theorem. We explain this technique before giving more examples.

THE GENERAL FRAMEWORK. Let $\mathcal{A}$ be a class of test functions (for example poly-size circuits) over a set of possible inputs I and $\mathcal{C}$ be a class of distributions over I satisfying certain desired properties (for example samplability, high density, high entropy etc.), and v be a payoff function quantifies how well A performs on the input X (for example, unpredictability or distinguishing advantage). Suppose that we want to prove the existence of a distribution with certain properties for which every algorithm has bad (or alternatively good) performance.

Dream Statement. There is a distribution over inputs (with some certain properties) such that every algorithm performs badly/well.

$$\exists X \in \mathcal{C}\ \forall \mathrm{A} \in \mathcal{A} \quad v(\mathrm{A}, X) \leqslant c \tag{1}$$

In many cases, it is much easier to prove a weaker version, which gives the existence of a distribution with desired properties but only for a chosen algorithm.

Weak Statement. For every algorithm there is a distribution over inputs (with some certain properties) such that it performs badly/well.

$$\forall \mathrm{A} \in \mathcal{A}\ \exists X \in \mathcal{C}: \quad v(\mathrm{A}, X) \leqslant c \tag{2}$$

Note that this condition is considerably weaker. Indeed, we will see that in many applications proving the existence of a suitable distribution X for a fixed algorithm A is actually trivial. But the big question is whether Eq. (2) implies Eq. (1)

Does the Weak Statement imply the Dream Statement? Suppose that Eq. (2) holds. Can we conclude that Eq. (1) also holds, with possibly somewhat weaker class $\mathcal{A}$ and a weaker parameter c?

Note that we allow for some loss in quality (a weaker class of algorithms or a weaker payoff). Indeed, if both sets $\mathcal{C}$ and $\mathcal{A}$ are convex the answer is trivially "yes", by the min-max theorem. However, in most applications the set $\mathcal{A}$ consists of efficient algorithms (circuits of a bounded size) and is not convex, because taking a mixed strategy corresponds to combining many algorithms by (possibly) inefficient sampling. For the same reason, the set $\mathcal{C}$ might not be convex. However, we might "embed" non-convex sets $\mathcal{A}$ and $\mathcal{C}$ into "almost" convex hulls of $\mathcal{A}', \mathcal{C}'$ which are (hopefully) still sufficiently good for our purpose, by taking moderately long mixed strategies, instead of arbitrarily long. Indeed, let

$$\forall \mathrm{A} \in \mathcal{A}\ \forall X \in \mathcal{C}\ \exists \mathrm{A}' \in \operatorname{conv} \mathcal{A}'\ \exists X' \in \operatorname{conv} \mathcal{C}': \quad |v(A, X) - v(A', X')| \leqslant \delta \tag{3}$$

where the conv operator denotes the convex hull. We get the following

Approximate Min-Max Theorem. If the condition (3) holds, then the Weak Statement implies the Dream Statement is true with $\mathcal{A}$ and $\mathcal{C}$ replaced by $\mathcal{A}'$ and $\mathcal{C}'$.

1.3 Our Contribution

Summary. This framework is well known (cf. [1,7,14,17,18] to mention only some papers closely related to our cryptographic applications). What we offer, is a *novel approximation technique*. Previous works used to find A′ and X' in convex hulls by a trivial Chernoff approximation argument. We observe that much better results are obtained with a carefully chosen *convex approximation technique*. Indeed, it turns out that in many cases the quantity $|v(A, X) - v(A', X')|$ can be upper bounded by the *Hölder Inequality* which involves moments of A and X.

These moments may be better estimated based on properties of the sets $\mathcal{A}$ and $\mathcal{C}$ which leads to quantitative improvements. We stress that the key component is *the right choice of Hölder conjugates*, that is the exponents for the corresponding L_p, L_q spaces.

ADVANTAGES AND APPLICATIONS OF OUR FRAMEWORK. Using our technique we prove a whole bunch of results, reproving what is already known in a more clear and concise way, improving quantitative bounds, or obtaining new results. Details are given in Sect. 2.

1.4 Related Works

The work of [18] provides a tool to derive good bounds for certain sets $\mathcal{C}$, in the uniform settings. We stress that we consider only non-uniform adversaries here. In fact, our results can be probably made uniform by the use of constructive versions of auxiliary results on convex approximations we have applied (for example [4]). Anyway, uniform settings are not important for most of our applications like leakage-resilient crypto. While [18] gives hard bounds, we provide *a framework equipped with a different technique of handling* $\mathcal{C}$. Our technique can exploit *moment conditions*, which is impossible in [18]. We stress that the crucial component of our technique is the

2 Applications

We briefly recall some basic notation and conventions. We say that two distributions X_1, X_2 are (s, ϵ)-indistinguishable if for every A of size s we have $|\mathbb{E}\mathrm{A}(X_1) - \mathbb{E}\mathrm{A}(X_2)| \leqslant \epsilon$.

2.1 Impagliazzo Hardcore Lemma

IMPAGLIAZZO HARDCORE LEMMA. Suppose that are given a function $f : \{0,1\}^n \to \{0,1\}$ that is mildly hard to predict by a class of circuits; for every circuit A from our class, $\mathrm{A}(x)$ and $f(x)$ agree on at most, say, a 0.99 fraction of inputs x. This might happen when there is a set of noticeable size on which f is extremely hard to predict, meaning that there is (almost) no advantage over a random guess. This set could be as big as a $0.02 = 2(1 - 0.99)$ fraction of input. Indeed, if f cannot be guessed better than with probability $\frac{1}{2}$ on this set, then the probability that D agrees with f is at most $0.02 \cdot \frac{1}{2} + 0.98 \cdot 1 = 0.99$.

Quite surprisingly, this intuitive characterization is true. The first such result was proved by Impagliazzo [8], with a sub-optimal hardcore density. An improved version with the optimal density of the hardcore set was found by Holenstein [7]. Below we present the best possible result due to Klivans and Servedio, the lower bound was given in [11].

Theorem 1 (Optimal Unpredictability Hardcore Lemma [10]**).** *Let* $f : \{0,1\}^n \rightarrow \{0,1\}$ *be* ϵ*-unpredictable by circuits of size* s *under a distribution* V*, that is*

$$\Pr_{x\leftarrow V}[\mathrm{A}(x) = f(x)] \leqslant 1 - \frac{\epsilon}{2}, \quad \textit{for every } \mathrm{A} \textit{ of size at most } s. \tag{4}$$

Then for any $\delta \in (0,1)$ *there exists a event* E *of probability* ϵ *such that* f *is* $1-\delta$ *unpredictable under* $V|E$ *by circuits of size* $s' = \Omega\, s\delta^2/\log(1/\epsilon)$*, that is*

$$\Pr_{x\leftarrow V|E}[\mathrm{A}(x) = f(x)] \leqslant \frac{1+\delta}{2}, \quad \textit{for every } \mathrm{A} \textit{ of size at most } s'. \tag{5}$$

Our Contribution. We reprove Theorem 1 using the framework discussed in Sect. 1.3. Our approach has the following advantages over the related works:

(a) It is derived from the *standard min-max theorem.* Previous proofs which achieve optimal parameters require involved iterative arguments [10,18].
(b) It is *modular and much simpler* than all alternative proofs. Indeed, the argument of Holenstein is non-optimal and involved. Also the argument given by Vadhan and Zheng depends on a non-trivial trick attributed to Nissan and Levy (which improves the hardcore density from $\frac{\epsilon}{2}$ to ϵ) and the machinery is much heavier. Our approach does not require this trick and follows the most intuitive strategy: show that there is a hardcore for every fixed adversary and then switch the order of quantifiers.
(c) We have identified *the reason for non-optimality in previous proofs.* Some authors even suggested that it might be impossible to get the tight parameters using the standard min-max theorem [18]. We show that this is not true. The problem is not with the standard min-max theorem but with an inadequate approximation argument in previous works, which do uniform approximation [7].

A comparison is given below in Table 1.

A sketch of proof. Assume without losing generality that $f : \{0,1\}^n \rightarrow \{-1,1\}$[1]. Define the payoff v as the unpredictability of f by A under X

$$v(A,X) \stackrel{\text{def}}{=} \Pr_{x\leftarrow X}[f(x) = A(x)] = \frac{1+\mathbb{E}_{x\leftarrow X}\mathrm{A}(x)\cdot f(x)}{2},$$

Table 1. Hardcore lemmas obtained by different techniques.

Author	Technique	Hardcore Density	Complexity Loss
[8]	boosting (constructive approx.)	$\Pr[E] = \frac{\epsilon}{2}$	$O(\delta^{-2}\cdot \mathrm{poly}(1/\epsilon))$
[7]	standard min-max + Hardcore Optimization	$\Pr[E] = \epsilon$	$O(n\delta^{-2})$
[11]	complicated boosting (constructive approx.)	$\Pr[E] = \epsilon$	$O(\log(1/\epsilon)\delta^{-2})$
[18]	complicated boosting (constructive approx.)	$\Pr[E] = \epsilon$	$O(\log(1/\epsilon)\delta^{-2})$
this paper	simple min-max + L_p-approx	$\Pr[E] = \epsilon$	$O(\log(1/\epsilon)\delta^{-2})$

[1] We consider $\{-1,1\}$ outputs for technical convenience. Equivalently we could state the problem for $\{0,1\}$.

and note that this definition makes sense also for circuits with real outputs. Let the property set $\mathcal{C}$ consists of conditional distributions of the form $X = V|E$ where $\Pr[E] \geqslant \epsilon$ and E may vary[2] ; note that $\mathcal{C}$ is convex. Define $\mathcal{A}$ as the set of real-valued[3] circuits of size s, and let $\mathcal{A}'$ be the set of circuits of size $s' = \frac{s}{\delta^{-2}\log(1/\epsilon)}$. It is not hard to see that the assumption (4) implies

Proposition 1 (Weak Statement). *For every* $\mathrm{A} \in \mathcal{A}$ *we have* $v(X, A) \leqslant 0$ *for some* $X \in \mathcal{C}$.

Now we analyze what happens when we replace $\mathcal{A}$ by $\mathrm{conv}(\mathcal{A}')$. We claim that

Proposition 2 (Approximation Step). *For every* $\mathrm{A}' \in \mathrm{conv}(\mathcal{A}')$ *we have* $v(X, \mathrm{A}') \leqslant \delta$ *for some* $X \in \mathcal{C}$.

To prove this, we show that the Hölder Inequality implies for A, A' and $X \in \mathcal{C}$

$$|v(X, \mathrm{A}) - v(X, \mathrm{A}')| \leqslant \frac{1}{2}\left(\mathbb{E}_{x \leftarrow V}\left(\frac{\mathbf{P}_{V|E}(x)}{\mathbf{P}_V(x)}\right)^q\right)^{\frac{1}{q}} \cdot \left(\mathbb{E}_{x \leftarrow V}\left|\mathrm{A}(x) - \mathrm{A}'(x)\right|^p\right)^{\frac{1}{p}}$$

for any $p, q \geqslant 1$, $\frac{1}{p} + \frac{1}{q} = 1$. Now we can argue that

(a) $\left(\mathbb{E}_{x \leftarrow V}\left(\frac{\mathbf{P}_{V|E}(x)}{\mathbf{P}_V(x)}\right)^q\right)^{\frac{1}{q}} \leqslant \epsilon^{-\frac{1}{p}}$ (by the extreme points technique).

(b) $\left(\mathbb{E}_{x \leftarrow V}\left|\mathrm{A}(x) - \mathrm{A}'(x)\right|^p\right)^{\frac{1}{p}} = O\left(\sqrt{\frac{p}{\ell}}\right)$ for some A which is of complexity ℓ relative to $\mathcal{A}'$[4] (by standard facts on convex-approximation [3]).

Setting $\ell = \delta^{-2}\log(1/\epsilon)$ (so that $A \in \mathcal{A}$), taking $X = V|E$ which corresponds to A' according to Proposition 1, setting $p = 2\log(1/\epsilon)$ and putting this all together we get Proposition 5. This implies the following statement

Proposition 3 (Strong Statement). *For some* $X \in \mathcal{C}$ *we have* $v(X, A) \leqslant \delta$ *for every* $\mathrm{A} \in \mathcal{A}'$.

which proves Theorem 1 ($|v(X, A)| \leqslant \delta$ follows by considering $\mathcal{A}'$ closed under complements).

2.2 A (new) Optimal Hardcore Lemma for Metric Pseudoentropy and Applications to Transformations

Pseudoentropy notions extend classical information-theoretic entropy notions into computational settings. The following most widely used entropy notions capture what it means to be "computationally close" to a high entropy distribution.

---

[2] We can think of measures M such that $M(\cdot) \leqslant \mathbf{P}_V(\cdot)$ and $\sum_x M(x) \geqslant \epsilon$. Every $X \in \mathcal{C}$ can be written as $\mathbf{P}_X(\cdot) = M(\cdot)/\sum_x M(x)$ for one of these measures M.

[3] Following related works [5,14] we use circuits with real outputs for technical reasons.

[4] That is, A is a convex combination of ℓ members of A'.

Definition 1 (HILL Pseudoentropy [6]**).** *Let Y be a distribution with the following property: there exists Y' of min-entropy at least k such that for every* A *of size at most s we have $|\mathbb{E}\mathrm{A}(Y) - \mathbb{E}Y')| \leqslant \epsilon$. Then we say that X has k bits of HILL entropy of quality (s, ϵ) and denote by $\mathbf{H}^{\mathrm{HILL}}_{s,\epsilon}(Y) \geqslant k$.*

Definition 2 (Metric Pseudoentropy [1]**).** *Let Y be a distribution with the following property: for every* A *of size at most s there exists Y' of min-entropy at least k such that we have $|\mathbb{E}\mathrm{A}(Y) - \mathbb{E}Y')| \leqslant \epsilon$. Then we say that X has k bits of metric entropy of quality (s, ϵ) and denote by $\mathbf{H}^{\mathrm{Metric}}_{s,\epsilon}(Y) \geqslant k$.*

Pseudoentropy is an important research area, with applications in deterministic encryption, memory delegation [2], pseudorandom generators [6,18]. Metric Pseudoentropy is much easier to deal with, and fortunately can be converted into HILL entropy with some loss in quality parameters (s, ϵ).

OUR CONTRIBUTION. The following results shows that any distribution with metric pseudoentropy of "moderate" quality has a kernel of HILL entropy with "strong" quality. We also conclude the optimal Metric-HILL transformation.

Theorem 2 (A HILL-pseudoentropy hardcore for metric pseudoentropy). *Suppose that $\mathbf{H}^{\mathrm{Metric}}_{s,\epsilon}(Y) \geqslant n - \Delta$, for some $Y \in \{0,1\}^n$. Then there is an event E, of probability $1 - \epsilon$ such that $\mathbf{H}^{\mathrm{HILL}}_{s',\delta}(Y|E) \geqslant n - \Delta$ with $s' = \Omega(s\delta^2/(\Delta + 1))$ for every δ. In particular, $\mathbf{H}^{\mathrm{HILL}}_{s',\epsilon+\delta}(Y) \geqslant n - \Delta$.*

One possible application of this fact is amplifying hardness of pseudoentropy with poor quality. Imagine that we have many independent samples $X_1, X_2, \ldots, X_n$ from a distribution with a substantial entropy amount ($\Delta \ll n$) but of weak advantage $\epsilon = 0.99$. We can use the result above to show that pseudoentropy in $X_1, X_2, \ldots, X_n$ is roughly $(1-0.99)(n-\Delta)$ with good quality (see [16] for more details). Below we briefly compare this result with related works.

(a) Our result is *far stronger than the classical result due to Barak et al.* [1] about the transformation. Not only we replace the factor n by Δ, but also show the existence of a hardcore in the intermediate step.
(b) This result *unifies and improves our recent results* [15,16]. The corollary $\mathbf{H}^{\mathrm{HILL}}_{s',\epsilon+\delta}(Y) \geqslant n - \Delta$ was the same (and optimal) but the hardcore E was found with worse complexity $s' = \Omega(s \cdot \delta^2/n)$.
(c) Our result *explains the nature of the Metric-HILL transformation.* The HILL pseudoentropy hardcore is an intermediate step in going from Metric pseudoentropy to HILL pseudoentropy.

Our result is illustrated in Fig. 1. The parameters are optimal (see [16]).

A SKETCH OF PROOF. Let $\mathcal{A}$ be the set of real-valued circuits of size s and let $\mathcal{A}'$ be the set of circuits of size $s' = s\delta^2/(\Delta+1)$. Let $\mathcal{C}$ consists of the conditional distributions X of the form $X'|E$, where $\Pr[E] \geqslant 1 - \epsilon$ and $\mathbf{H}_{\infty}(X') \geqslant n - \Delta$; this set is convex. The payoff is defined as $v(X, A) \stackrel{def}{=} \mathbb{E}\mathrm{A}(Y) - \mathbb{E}\mathrm{A}(X)$. It is easy to see[5] that we have

[5] This is trivial for boolean A and somewhat more tricky for real-valued A. A short proof is given implicitly in [5].

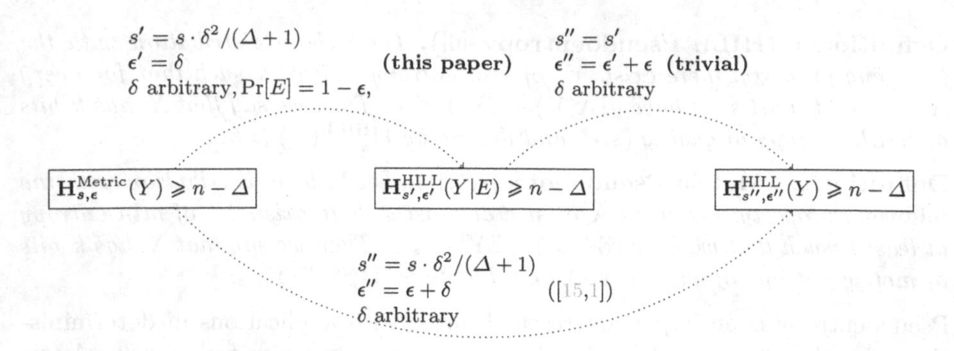

Fig. 1. The Metric-to-HILL pseudoentropy transformation.

Proposition 4 (Weak Statement). $\forall \mathrm{A} \in \mathcal{A}\ \exists X \in \mathcal{C}\quad v(X, A) \leqslant 0.$

Now we analyze what happens when $\mathcal{A}$ is replaced by $\mathrm{conv}(\mathcal{A}')$.

Proposition 5 (Approximation Step). *For every* $\mathrm{A}' \in \mathrm{conv}(\mathcal{A}')$ *we have* $v(X, \mathrm{A}') \leqslant \delta$ *for some* $X \in \mathcal{C}$.

To prove this, by the Hölder Inequality for any A, A' and $X \in \mathcal{C}$ we show

$$|v(X, \mathrm{A}) - v(X, \mathrm{A}')| \leqslant \left(\mathbb{E}_{x\leftarrow U}\left(2^n \mathbf{P}_{Y|E}(x)\right)^q\right)^{\frac{1}{q}} \cdot \left(\mathbb{E}_{x\leftarrow U}\left|\mathrm{A}(x) - \mathrm{A}'(x)\right|^p\right)^{\frac{1}{p}}$$

for any $p, q \geqslant 1$, $\frac{1}{p} + \frac{1}{q} = 1$ and the uniform distribution U. Now we argue that

(a) $\left(\mathbb{E}_{x\leftarrow U}\left(2^n \mathbf{P}_{Y|E}(x)\right)^q\right)^{\frac{1}{q}} \leqslant 2^{\frac{\Delta}{p}}$ (by the extreme points technique).
(b) $\left(\mathbb{E}_{x\leftarrow V}\left|\mathrm{A}(x) - \mathrm{A}'(x)\right|^p\right)^{\frac{1}{p}} = O\left(\sqrt{\frac{p}{\ell}}\right)$ for some A which is of complexity ℓ relative to $\mathcal{A}'$ (by standard facts on convex-approximation [3]).

Setting $\ell = \delta^{-2}(\Delta+1)$ (so that $A \in \mathcal{A}$), taking $X = X'|E$ which corresponds to A' according to Proposition 1, setting $p = \Delta + 1$ and putting this all together we get Proposition 5. This implies the following statement

Proposition 6 (Strong Statement). $\exists X \in \mathcal{C}\ \forall \mathrm{A} \in \mathcal{A}'\quad v(X, A) \leqslant \delta.$

This directly implies Theorem 2 (as before, we consider A' closed under complements). More details can be found in ??.

2.3 A (fixed) Construction of a Simulator for Auxiliary Inputs

In [9] there is a theorem, which says that any short information Z about X can be efficienly simulated from X, Below we state the corrected version [13].

Theorem 3 (Simulating auxiliary inputs, flaws fixed). *For any random variable* $X \in \{0,1\}^n$, *any correlated* $Z \in \{0,1\}^\lambda$ *and every choice of parameters* (ϵ, s) *there is a randomized function* $\mathsf{Sim} : \{0,1\}^n \to \{0,1\}^\lambda$ *of complexity* $O\left(s \cdot 2^{4\lambda}\epsilon^{-4}\right)$ *such that* Z *and* $\mathsf{Sim}(X)$ *are* (ϵ, s)*-indistinguishable given* X.

This result is the key component in the simplified analysis of the EUROCRYPT'09 stream cipher construction. Using Theorem 3, as described in [9], one proves the resilience of the cipher (assuming bounded leakage in every round) and if the underlying weak PRF is (s,ϵ)-secure against two queries on random inputs. The cipher security (s',ϵ') is related to (s,ϵ) by a polynomial loss in ϵ.

OUR CONTRIBUTION. We describe a flaw in the proof and improve the corrected bound by a significant super polynomial factor. Below we briefly describe the significance of our result

(a) *Discovered flaws* in the recent (TCC'14) analysis of the EUROCRYPT'09 stream cipher. The alternative bounds seem correct but are much weaker. In particular, we get *no meaningful security with the AES* used as a weak PRF in this construction[6]. This raises the problem of *whether the cipher built on AES is secure or not.* We would need a simulator with a loss of only $O(\epsilon^{-2})$ not ϵ^{-4} in complexity.
(b) A *simpler* construction based on the min-max theorem. Based on the framework in Sect. 1.3 we give an alternative proof achieving the simulator complexity of $O\left(s \cdot 2^{2\lambda}\epsilon^{-4}\right))$. The gain of $2^{2\lambda}$ over the original approach, which is a power of ϵ for recommended values of parameters [9], comes from the use of convex approximation techniques. Our proof is considerably simpler and quantitatively better than in [9] (in particular we don't need to use the min-max theorem twice depending on what is the value of the game). Also, it is much simpler than the alternative approach of Vadhan and Zheng [18], yet yields comparable results for small leakages (see Table 2).
(c) A *clear bound on the security level*, in terms of the time-success ratio. We derive a clear formula which shows what fraction of the security of the original weak PRF is transformed into security of the stream cipher. This analysis shows that we are *far from good and provable secure* leakage-resilient stream ciphers as we lose over $\frac{5}{6}$ of original security. For more details, see Table 2.

In Table 2 we compare the strength of the simulator theorems in terms of implied security for this construction. To our knowledge, this is the first analysis of the time-success ratio for this technique. For more details we refer to ??.

MORE ON THE FLAWS. In the claimed better bound $O\left(s \cdot 2^{3\lambda}\epsilon^{-2}\right)$ there is a mistake on page 18 (eprint version), when the authors enforce a signed measure to be a probability measure by a mass shifting argument. The number M defined there is in fact a function of x and is hard to compute, whereas the original proof assumes that this is a constant independent of x. In the alternative bound $O\left(s \cdot 2^{3\lambda}\epsilon^{-2}\right)$ a fixable flaw is a missing factor of 2^{λ} in the complexity (page 16 in the eprint version), which is because what is constructed in the proof is only a probability mass function, not yet a sampler [13].

[6] The final bounds on the cipher security depends on the simulator complexity and are given by $\epsilon' = O\left(\sqrt{2^{\lambda}\epsilon}\right)$ and $s' = s \cdot 2^{-4\lambda}\epsilon'^4$. We can't prove then even very weak security $\epsilon' = 2^{-32}$ having $\lambda = 10$ bits of leakage!.

Table 2. Security of the EUROCRYPT'09 stream cipher instantiated with a wPRF having 2^k keys and λ bits of leakage, obtained from different simulator results. Every attacker of size s succeeds with prob. at most $s/2^{k'}$.

Author	Technique	Simulator Complexity	Implied Security
[9]	Standard Min-Max + L_∞-approx	$s_h = s \cdot 2^{4\lambda}\epsilon^{-4}$	$k' = \frac{k}{6} - \frac{5}{6}\lambda$
[18]	Complicated Boosting	$s_h = s \cdot 2^{\lambda}\epsilon^{-2} + 2^{\lambda}\epsilon^{-4}$	$k' = \frac{k}{6} - \frac{1}{3}\lambda$
this paper	Standard Min-Max + L_p-approx	$s_h = s \cdot 2^{2\lambda}\epsilon^{-4}$	$k' = \frac{k}{6} - \frac{1}{2}\lambda$

A SKETCH OF THE PROOF. Let $\mathcal{A}$ be the set of real-valued circuits of size s and let $\mathcal{A}'$ be the set of circuits of size $s' = s \cdot 2^{-2\lambda}\epsilon^2$. Let $\mathcal{C}'$ consists of the distributions of the form $X, h(X)$, where h is computable in size $s \cdot 2^{\lambda}$; this set is *not* convex. Let $\mathcal{C}$ be the set of all circuits of size $s \cdot 2^{2\lambda}\epsilon^{-2}$. The payoff is defined as $v(h, A) \overset{def}{=} \mathbb{E}\mathrm{A}(X, h(X)) - \mathbb{E}\mathrm{A}(X, Z)$. It is easy to see that we have

Proposition 7 (Weak Statement). $\forall \mathrm{A} \in \mathcal{A}\ \exists h' \in \mathcal{C}' \quad v(h', A) \leqslant 0.$

Indeed, consider h^+_{A} which for every x outputs this value z for which $\mathrm{A}(x,z) = \max \mathrm{A}(x,\cdot)$ and h^-_{A} which for every x outputs this value z for which $\mathrm{A}(x,z) = \min \mathrm{A}(x,\cdot)$. Both are of complexity $O\left(2^{\lambda}\right)$. Since we have $\mathbb{E}\mathrm{A}(X, h^-(X)) \leqslant \mathbb{E}\mathrm{A}(X,Z)$ and $\mathbb{E}\mathrm{A}(X,Z) \leqslant \mathbb{E}\mathrm{A}(X, h^+(X))$, setting h' to be a distribution over h^+ and h^- that is $\Pr[h'(x) = z] = \theta \cdot \Pr[h^-(x) = z] + (1-\theta) \cdot \Pr[h^+(x) = z]$, we get $v(h', \mathrm{A}) = 0$ with some θ. In the next step we replace $\mathcal{A}'$ by $\mathrm{conv}(\mathcal{A}')$.

Proposition 8 (Approximation 1). $\forall A \in \mathrm{conv}\mathcal{A}'\ \exists h' \in \mathcal{C}' : v(h', \mathrm{A}') \leqslant \epsilon.$

This follows from the standard Chernoff Bound approximation argument[7] as

$$|v(h', A) - v(h', A')| = |\mathbb{E}\mathrm{A}(X, h'(X)) - \mathbb{E}\mathrm{A}'(X, h'(X))| \leqslant \sup_{x,z} |\mathrm{A}(x,z) - \mathrm{A}'(x,z)|.$$

Now we replace $\mathcal{C}'$ by conv $\mathcal{C}'$. Here a more delicate approximation is required.

Proposition 9 (Approximation 2). *For every* A *and every* $h' \in \mathrm{conv}\ \mathcal{C}'$ *there exists* $h \in \mathcal{C}$ *such that* $v(h, \mathrm{A}) \leqslant v(h', \mathrm{A}) + \epsilon$.

This follows because by the Hölder Inequality applied to $p = q = 2$ we obtain

$$|\mathbb{E}\mathrm{A}(X, h'(X)) - \mathbb{E}\mathrm{A}(X, h(X))| \leqslant 2^{\frac{\lambda}{2}} \cdot \left(\mathop{\mathbb{E}}_{x \sim X} \sum_{z} \left| \mathbf{P}_{x,h(x)}(x,z) - \mathbf{P}_{x,h'(x)(x,z)} \right|^2 \right)^{\frac{1}{2}},$$

and by the standard results on convex approximation [4] the second factor is at most $\ell^{-\frac{1}{2}}$ for some h of complexity ℓ with respect to $\mathcal{C}'$. We put $\ell = 2^{\lambda}\epsilon^{-2}$. From the proven propositions we obtain the final result.

Proposition 10 (Strong Statement). $\exists h \in \mathcal{C}\ \forall \mathrm{A} \in \mathcal{A}' \quad v(h, \mathrm{A}) \leqslant 2\epsilon.$

[7] A can be viewed as a distribution on $\mathcal{A}'$ we simply pick ℓ independent samples $\{\mathrm{A}_i\}_i$ and try to find an approximator of the form $\mathrm{A}' = \frac{1}{\ell}\sum_{i=1}^{\ell} \mathrm{A}_i$. It deviates by more than ϵ at (x,z) with probability $\exp(-2\ell\epsilon^2)$. We combine this with the union bound.

2.4 More Applications

For more applications we refer interested readers to the full version. They include the optimal Dense Model Theorem, a better auxiliary input simulator for bounded-variance adversaries (new), and a proof that every high-conditional entropy source can be efficiently simulated (new, extending [17]).

References

1. Barak, B., Shaltiel, R., Wigderson, A.: Computational analogues of entropy. In: Arora, S., Jansen, K., Rolim, J.D.P., Sahai, A. (eds.) RANDOM 2003 and APPROX 2003. LNCS, vol. 2764, pp. 200–215. Springer, Heidelberg (2003)
2. Chung, K.-M., Kalai, Y.T., Liu, F.-H., Raz, R.: Memory delegation. In: Rogaway, P. (ed.) CRYPTO 2011. LNCS, vol. 6841, pp. 151–168. Springer, Heidelberg (2011)
3. Docampo, D., Hush, D.R., Abdallah, C.T.: Constructive function approximation: theory and practice. In: Intelligent Methods in Signal Processing and Communications. Birkhauser Boston Inc. (1997)
4. Donahue, M.J., Darken, C., Gurvits, L., Sontag, E.: Rates of convex approximation in non-hilbert spaces. Constructive Approximation **13**, 187–220 (1997)
5. Fuller, B., O'Neill, A., Reyzin, L.: A unified approach to deterministic encryption (...). TCC 2012 (2012)
6. Hastad, J., Impagliazzo, R., Levin, L.A., Luby, M.: A pseudorandom generator from any one-way function. SIAM J. Comput. **28**, 1364–1396 (1999)
7. Holenstein, T.: Key agreement from weak bit agreement. In: STOC 2005 (2005)
8. Impagliazzo, R.: Hard-core distributions for somewhat hard problems. In: FOCS 36 (1995)
9. Jetchev, D., Pietrzak, K.: How to fake auxiliary input. In: Lindell, Y. (ed.) TCC 2014. LNCS, vol. 8349, pp. 566–590. Springer, Heidelberg (2014)
10. Klivans, A.R., Servedio, R.A.: Boosting and hard-core set construction. Mach. Learn. **51**(3), 217–238 (2003)
11. Lu, C.-J., Tsai, S.-C., Wu, H.-L.: On the complexity of hard-core set constructions. In: Arge, L., Cachin, C., Jurdziński, T., Tarlecki, A. (eds.) ICALP 2007. LNCS, vol. 4596, pp. 183–194. Springer, Heidelberg (2007)
12. Neumann, J.: Zur theorie der gesellschaftsspiele. Math. Ann. **100**, 295–320 (1928)
13. Pietrzak, K.: Private communication, may (2015)
14. Reingold, O., Trevisan, L., Tulsiani, M., Vadhan, S.: Dense subsets of pseudorandom sets. FOCS 2008 (2008)
15. Skorski, M.: Metric pseudoentropy: Characterizations, transformations and applications. In: ICITS (2015)
16. Skorski, M.: Nonuniform indistinguishability and unpredictability hardcore lemmas: new proofs and applications to pseudoentropy. In: ICITS (2015)
17. Trevisan, L., Tulsiani, M., Vadhan, S.: Regularity, boosting, and efficiently simulating every high-entropy distribution. CCC 2009 (2008)
18. Vadhan, S., Zheng, C.J.: A uniform min-max theorem with applications in cryptography. In: Canetti, R., Garay, J.A. (eds.) CRYPTO 2013, Part I. LNCS, vol. 8042, pp. 93–110. Springer, Heidelberg (2013)

Give Me Another One!

Mike Behrisch[1], Miki Hermann[2(✉)], Stefan Mengel[2], and Gernot Salzer[1]

[1] Technische Universität Wien, Vienna, Austria
{behrisch,salzer}@logic.at

[2] LIX (UMR CNRS 7161), École Polytechnique, Palaiseau, France
{hermann,mengel}@lix.polytechnique.fr

Abstract. We investigate the complexity of an optimization problem in Boolean propositional logic related to information theory: Given a conjunctive formula over a set of relations, find a satisfying assignment with minimal Hamming distance to a given assignment that satisfies the formula (NearestOtherSolution, NOSol).

We present a complete classification with respect to the relations admitted in the formula. We give polynomial-time algorithms for several classes of constraint languages. For all other cases we prove hardness or completeness regarding poly-APX, NPO, or equivalence to a well-known hard optimization problem.

1 Introduction

We investigate the solution spaces of Boolean constraint satisfaction problems built from atomic constraints by means of conjunction and variable identification. We study a minimization problem in connection with Hamming distance: Given an instance of a constraint satisfaction problem in the form of a generalized conjunctive formula over a set of atomic constraints, the problem asks to find a satisfying assignment with minimal Hamming distance to a given assignment that satisfies the formula (NearestOtherSolution, NOSol).

As it is common, we analyze the complexity of our optimization problem through a parameter, representing the atomic constraints allowed to be used in the constraint satisfaction problem. We give a complete classification of the complexity of approximation with respect to this parameterization. It turns out that our problems can either be solved in polynomial time, or they are complete for a well-known optimization class, or else they are equivalent to well-known hard optimization problems.

Our study can be understood as a continuation of the minimization problems investigated by Khanna et al. in [11], especially that of MinOnes. The MinOnes

Mike Behrisch and Gernot Salzer—Supported by Austrian Science Fund (FWF) grant I836-N23.

Miki Hermann—Supported by ANR-11-ISO2-003-01 Blanc International grant ALCOCLAN.

Stefan Mengel—Supported by QUALCOMM grant administered by École Polytechnique.

K. Elbassioni and K. Makino (Eds.): ISAAC 2015, LNCS 9472, pp. 664–676, 2015.
DOI: 10.1007/978-3-662-48971-0_56

Table 1. Boolean co-clones with bases.

iS_0^k	$\{\text{or}^k\}$	iL	$\{\text{even}^4\}$	iN	$\{\text{dup}^3\}$
iS_1^k	$\{\text{nand}^k\}$	iL_2	$\{\text{even}^4, \neg x, x\}$	iN_2	$\{\text{nae}^3\}$
iS_{00}^k	$\{\text{or}^k, x \to y, \neg x, x\}$	iV	$\{x \vee y \vee \neg z\}$	iI	$\{\text{even}^4, x \to y\}$
iS_{10}^k	$\{\text{nand}^k, \neg x, x, x \to y\}$	iV_2	$\{x \vee y \vee \neg z, \neg x, x\}$	iI_0	$\{\text{even}^4, x \to y, \neg x\}$
iD_1	$\{x \oplus y, x\}$	iE	$\{\neg x \vee \neg y \vee z\}$	iI_1	$\{\text{even}^4, x \to y, x\}$
iD_2	$\{x \oplus y, x \to y\}$	iE_2	$\{\neg x \vee \neg y \vee z, \neg x, x\}$	iM_2	$\{x \to y, \neg x, x\}$

optimization problem asks for a solution of a constraint satisfaction problem with the minimal Hamming weight, i.e., minimal Hamming distance to the 0-vector. Our work generalizes this by allowing the given vector to be any, potentially also non-0-vector. Moreover, our work can also be seen as a generalization of questions in coding theory.

It turns out that our problem NOSol lacks compatibility with existential quantification, which makes classical clone theory inapplicable. Therefore, we have to resort to weak co-clones requiring only closure under conjunction and equality. To dispose of the latter we apply the theory developed in [14], as well as minimal weak bases of Boolean co-clones from [12].

2 Preliminaries

An n-ary *Boolean relation* R is a subset of $\{0,1\}^n$; its elements $(b_1, \ldots, b_n)$ are also written as $b_1 \cdots b_n$. Let V be a set of variables. An *atomic constraint*, or an *atom*, is an expression $R(\boldsymbol{x})$, where R is an n-ary relation and $\boldsymbol{x}$ is an n-tuple of variables from V. Let $\mathcal{L}$ be the collection of all non-empty finite sets of Boolean relations, also called *constraint languages*. For $\Gamma \in \mathcal{L}$, a Γ*-formula* is a finite conjunction of atoms $R_1(\boldsymbol{x_1}) \wedge \cdots \wedge R_k(\boldsymbol{x_k})$, where the R_i are relations from Γ and the $\boldsymbol{x_i}$ are variable tuples of suitable arity.

An *assignment* is a mapping $m\colon V \to \{0,1\}$ assigning a Boolean value $m(x)$ to each variable $x \in V$. If we arrange the variables in some arbitrary but fixed order, say as a tuple $(x_1, \ldots, x_n)$, then the assignments can be identified with vectors from $\{0,1\}^n$. The i-th component of a vector m is denoted by $m[i]$ and corresponds to the value of the i-th variable, i.e., $m[i] = m(x_i)$. The *Hamming weight* $\text{hw}(m) = |\{i \mid m[i] = 1\}|$ of m is the number of 1s in the vector m. The *Hamming distance* $\text{hd}(m, m') = |\{i \mid m[i] \neq m'[i]\}|$ of m and m' is the number of coordinates on which the vectors disagree. The *complement* $\overline{m}$ of a vector m is its pointwise complement, $\overline{m}[i] = 1 - m[i]$.

An assignment m satisfies the constraint $R(x_1, \ldots, x_n)$ if $(m(x_1), \ldots, m(x_n)) \in R$ holds. It satisfies the formula φ if it satisfies all of its atoms; m is said to be a *model* or *solution* of φ in this case. We use $[\varphi]$ to denote the set of models of φ. Note that $[\varphi]$ represents a Boolean relation. In sets of relations represented this way we usually omit the brackets. A *literal* is a variable v, or its negation $\neg v$. Assignments m are extended to literals by defining $m(\neg v) = 1 - m(v)$.

We shall need the following Boolean functions and relations later: By $x \oplus y$ we denote addition modulo 2 and $x \equiv y$ means $x \oplus y \oplus 1$. Further, we let $\mathrm{nae}^3 := \{0,1\}^3 \setminus \{000, 111\}$, $\mathrm{dup}^3 := \{0,1\}^3 \setminus \{010, 101\}$ and $\mathrm{even}^4 := \{(a_1, a_2, a_3, a_4) \in \{0,1\}^4 \mid \oplus_{i=1}^4 a_i = 0\}$, as well as $S_0 := [x_1 \wedge x_4 \equiv x_2 \wedge x_3]$, $S_1 := [S_0(x_1, x_2, x_3, x_1)]$ and $S_2 := [x_1 \vee x_2 \rightarrow x_3]$. Moreover, for $k \geq 1$ we define $\mathrm{or}^k := \{0,1\}^k \setminus \{0 \cdots 0\}$ and $\mathrm{nand}^k := \{0,1\}^k \setminus \{1 \cdots 1\}$.

Throughout the text we refer to different types of Boolean constraint relations following Schaefer's terminology [13] (see also [4,6]). A Boolean relation R is (1) *1-valid* if $1 \cdots 1 \in R$ and it is *0-valid* if $0 \cdots 0 \in R$, (2) *Horn* (*dual Horn*) if R can be represented by a formula in conjunctive normal form (CNF) having at most one unnegated (negated) variable in each clause, (3) *monotone* if it is both Horn and dual Horn, (4) *bijunctive* if it can be represented by a CNF having at most two variables in each clause, (5) *affine* if it can be represented by an affine system of equations $Ax = b$ over $\mathbb{Z}_2$, (6) *complementive* if for each $m \in R$ also $\overline{m} \in R$. A set Γ of Boolean relations is called 0-valid (1-valid, Horn, dual Horn, monotone, affine, bijunctive, complementive) if *every* $R \in \Gamma$ satisfies that property.

A formula constructed from atoms by conjunction, variable identification, and existential quantification is called a *primitive positive formula* (*pp-formula*). We denote by $\langle \Gamma \rangle$ the set of all relations that can be expressed using relations from $\Gamma \cup \{=\}$, conjunction, variable identification, and existential quantification. The set $\langle \Gamma \rangle$ is called the *co-clone* generated by Γ. A *base* of a co-clone $\mathcal{B}$ is a set of relations Γ, such that $\langle \Gamma \rangle = \mathcal{B}$. All co-clones, ordered by set inclusion, form a lattice. Together with their respective bases, which were studied in [5], some of them are listed in Table 1. In particular the sets of relations being 0-valid, 1-valid, complementive, Horn, dual Horn, affine, bijunctive, 2affine (both bijunctive and affine), and monotone each form a co-clone denoted by iI_0, iI_1, iN_2, iE_2, iV_2, iL_2, iD_2, iD_1, and iM_2, respectively.

We will also use a weaker closure than $\langle \Gamma \rangle$, called *conjunctive closure* and denoted by $\langle \Gamma \rangle_\wedge$, where the constraint language Γ is closed under conjunctive definitions, but not under existential quantification or addition of explicit equality constraints.

Minimal weak bases of co-clones are bases with certain additional properties. Since we rely on only some of them, we shall not define this term but refer the reader to [12,14].

Theorem 1. *If Γ is a minimal weak base of a co-clone, then $\Gamma \subseteq \langle \Gamma' \rangle_\wedge$ for any base Γ'.*

Lagerkvist computed weak bases for all Boolean co-clones in [12]. From there we infer that each co-clone $\mathcal{B} \in \{\mathrm{iE}, \mathrm{iE}_0, \mathrm{iE}_1, \mathrm{iE}_2, \mathrm{iN}, \mathrm{iN}_2, \mathrm{iI}\}$ has a singleton minimal weak base $\{R_\mathcal{B}\}$, in which $R_{\mathrm{iE}} := (S_1 \times \{0,1\}) \cap (\{0,1\} \times S_2)$, $R_{\mathrm{iE}_0} := R_{\mathrm{iE}} \times \{0\}$, $R_{\mathrm{iE}_1} := S_1 \times \{1\}$, $R_{\mathrm{iE}_2} := S_1 \times \{0\} \times \{1\}$, $R_{\mathrm{iN}} := \mathrm{even}^4 \cap S_0$, $R_{\mathrm{iN}_2} := [R_{\mathrm{iN}}(x_1, \ldots, x_4) \wedge \bigwedge_{i=1}^4 x_{i+4} = \neg x_i]$ and $R_{\mathrm{iI}} := [S_1(x_1, x_2, x_3) \wedge S_1(\neg x_4, \neg x_2, \neg x_3)]$.

We assume that the reader has a basic knowledge of approximation algorithms and complexity theory, see e.g. [1,6]. For reductions among decision problems we use polynomial-time many-one reduction denoted by $\leq_{\mathrm{m}}$. Many-one equivalence between decision problems is written as $\equiv_{\mathrm{m}}$. For reductions among optimization problems we employ approximation preserving reductions (AP-reductions), represented by $\leq_{\mathrm{AP}}$. AP-equivalence of optimization problems is stated as $\equiv_{\mathrm{AP}}$. Besides, the following approximation complexity classes in the hierarchy PO $\subseteq$ APX $\subseteq$ poly-APX $\subseteq$ NPO occur.

We also need a slightly non-standard variation of AP-reductions between optimization problems $\mathcal{P}_1$, $\mathcal{P}_2$: Viz., $\mathcal{P}_1$ AP-*Turing-reduces* to $\mathcal{P}_2$ if there is a polynomial-time oracle algorithm $\mathbb{A}$ and a constant $\alpha \geq 1$ such that for all $r > 1$ on any input x for $\mathcal{P}_1$ we have

- if all oracle calls within $\mathbb{A}$ upon inputs for $\mathcal{P}_2$ are answered with feasible solutions for $\mathcal{P}_2$, then $\mathbb{A}$ outputs a feasible solution for $\mathcal{P}_1$ on input x, and
- if for every call in $\mathbb{A}$ the oracle answers with an r-approximate solution, then $\mathbb{A}$ computes a $(1 + (r - 1)\alpha + o(1))$-approximate solution for $\mathcal{P}_1$ on input x.

It is straightforward to check that AP-Turing-reductions are transitive. Moreover, if $\mathcal{P}_1$ AP-Turing-reduces to $\mathcal{P}_2$ with constant α and $\mathcal{P}_2$ has an $f(n)$-approximation algorithm, then there is an $\alpha f(n)$-approximation algorithm for $\mathcal{P}_1$.

To relate our problem to well-known optimization problems we make the following convention: For optimization problems $\mathcal{P}$ and $\mathcal{Q}$ we say that $\mathcal{Q}$ is $\mathcal{P}$-*hard* if $\mathcal{P} \leq_{\mathrm{AP}} \mathcal{Q}$, i.e. if $\mathcal{P}$ reduces to it. Moreover, $\mathcal{Q}$ is called $\mathcal{P}$-*complete* if $\mathcal{P} \equiv_{\mathrm{AP}} \mathcal{Q}$. We use these notions in particular with respect to the following problems from [11], taking parameters $\Gamma \in \mathcal{L}$.

Problem MinOnes(Γ). Given a conjunctive formula φ over relations from Γ, a solution is any assignment m satisfying φ. The goal is to minimize the Hamming weight $\mathrm{hw}(m)$.

Problem WeightedMinOnes(Γ). Given a conjunctive formula φ over relations from Γ and a weight function $w\colon V \to \mathbb{N}$ on the variables V of φ, a solution is again any assignment m satisfying φ. The objective is to minimize the value $\sum_{x:m(x)=1} w(x)$.

We now define some well-studied problems to which we will relate our problems. Note that these problems do not depend on any parameter.

Problem MinDistance. Given a matrix $A \in \mathbb{Z}_2^{k \times l}$ any non-zero vector $x \in \mathbb{Z}_2^l$ with $Ax = 0$ is considered a solution. The aim is to minimize the Hamming weight $\mathrm{hw}(x)$.

Problem MinHornDeletion. For a conjunctive formula φ over relations from the constraint language $\{[x \vee y \vee \neg z], [x], [\neg x]\}$, an assignment m satisfying φ is feasible. The objective is given by the minimum number of unsatisfied conjuncts of φ.

MinDistance and MinHornDeletion are NP-hard to approximate within $2^{\Omega(\log^{1-\varepsilon}(n))}$ for all $\varepsilon > 0$ [8,11]. Thus, unless P = NP, both are inequivalent to any problem $\mathcal{P} \in$ APX.

We also use the classic satisfiability problem SAT(Γ), asking for a conjunctive formula φ over a $\Gamma \in \mathcal{L}$, if φ is satisfiable. Schaefer presented in [13] a complete classification of complexity for SAT(Γ). His dichotomy theorem proves that

$\mathsf{SAT}(\Gamma)$ is in P if Γ is 0-valid ($\Gamma \subseteq \mathrm{iI}_0$), 1-valid ($\Gamma \subseteq \mathrm{iI}_1$), Horn ($\Gamma \subseteq \mathrm{iE}_2$), dual Horn ($\Gamma \subseteq \mathrm{iV}_2$), bijunctive ($\Gamma \subseteq \mathrm{iD}_2$), or affine ($\Gamma \subseteq \mathrm{iL}_2$); otherwise it is NP-complete. Moreover, we need the decision problem $\mathsf{AnotherSAT}(\Gamma)$, asking for a conjunctive formula φ over Γ and a model m, if there is another model $m' \neq m$ for φ. In [10] Juban completely classified the complexity of AnotherSAT. His dichotomy result shows $\mathsf{AnotherSAT}(\Gamma)$ to be polynomial-time decidable if Γ is both 0- and 1-valid ($\Gamma \subseteq \mathrm{iI}$), complementive ($\Gamma \subseteq \mathrm{iN}_2$), Horn ($\Gamma \subseteq \mathrm{iE}_2$), dual Horn ($\Gamma \subseteq \mathrm{iV}_2$), bijunctive ($\Gamma \subseteq \mathrm{iD}_2$), or affine ($\Gamma \subseteq \mathrm{iL}_2$); or else to be NP-complete.

3 Results

Here we present the formal definition of our considered problem, with parameter $\Gamma \in \mathcal{L}$, and our results; the proofs follow in subsequent sections.

Problem $\mathsf{NearestOtherSolution}(\Gamma)$, $\mathsf{NOSol}(\Gamma)$
Input: A conjunctive formula φ over relations from Γ and an assignment m satisfying φ.
Solution: Another assignment m' satisfying φ.
Objective: Minimum Hamming distance $\mathrm{hd}(m, m')$.

Theorem 2. *For every $\Gamma \in \mathcal{L}$ the optimization problem* $\mathsf{NOSol}(\Gamma)$ *is*

(i) in PO *if*
 (a) Γ is bijunctive ($\Gamma \subseteq \mathrm{iD}_2$) *or*
 (b) $\Gamma \subseteq \langle x_1 \vee \cdots \vee x_k, x \to y, \neg x, x\rangle$ for some $k \in \mathbb{N}$, $k \geq 2$ ($\Gamma \subseteq \mathrm{iS}_{00}^k$) *or*
 (c) $\Gamma \subseteq \langle \neg x_1 \vee \cdots \vee \neg x_k, x \to y, \neg x, x\rangle$ for some $k \in \mathbb{N}$, $k \geq 2$ ($\Gamma \subseteq \mathrm{iS}_{10}^k$)*;*
(ii) MinDistance*-complete if Γ is exactly affine* ($\mathrm{iL} \subseteq \langle\Gamma\rangle \subseteq \mathrm{iL}_2$)*;*
(iii) MinHornDeletion*-complete under* AP*-Turing-reductions if Γ is*
 (a) exactly Horn ($\mathrm{iE} \subseteq \langle\Gamma\rangle \subseteq \mathrm{iE}_2$) *or*
 (b) exactly dual Horn ($\mathrm{iV} \subseteq \langle\Gamma\rangle \subseteq \mathrm{iV}_2$)*;*
(iv) in poly-APX *if Γ is*
 (a) exactly both 0-valid and 1-valid ($\langle\Gamma\rangle = \mathrm{iI}$) *or*
 (b) exactly complementive ($\mathrm{iN} \subseteq \langle\Gamma\rangle \subseteq \mathrm{iN}_2$)*,*
 where $\mathsf{NOSol}(\Gamma)$ *is n-approximable but not ($n^{1-\varepsilon}$)-approximable unless* P = NP*;*
(v) and NPO*-complete otherwise* ($\mathrm{iI}_0 \subseteq \langle\Gamma\rangle$ *or* $\mathrm{iI}_1 \subseteq \langle\Gamma\rangle$)*.*

The optimization problem can be transformed into a decision problem as usual. We add a bound $k \in \mathbb{N}$ to the input and ask if $\mathrm{hd}(m, m') \leq k$. This way we obtain the corresponding decision problem $\mathsf{NOSol}^{\mathrm{d}}$. Its complexity follows immediately from the theorems above. All cases in PO become polynomial-time decidable, whereas the other cases, which are APX-hard, become NP-complete. This way we obtain a dichotomy theorem classifying the decision problems as polynomial or NP-complete for all finite sets of relations Γ.

4 Duality and Inapplicability of Clone Closure

The problem NOSol is not compatible with existential quantification as the following shows:

Example 3. Consider the relation $R = \{00000, 01111, 10101\}$ and let (φ_R, m) be an instance of NOSol with $\varphi_R = R(x_1, \dots, x_5)$ and $m = 10101$. Both $m_1 = 00000$ and $m_2 = 01111$ are feasible solutions of φ_R and $\mathrm{hd}(m, m_1) = \mathrm{hd}(m, m_2) = 3$. Hence m_2 is an optimal solution of (φ_R, m). Let $m' = 1010$, $m'_1 = 0000$, and $m'_2 = 0111$ be new tuples, constructed from m, m_1, and m_2 respectively, by truncating the last coordinate. Hence, they are the solutions of $(\exists x_5\, \varphi_R, m')$. However, note that $\mathrm{hd}(m', m'_1) = 2$ and $\mathrm{hd}(m', m'_2) = 3$. The tuple m'_2 is not an optimal solution of $(\exists x_5\, \varphi_R, m')$.

Because of this incompatibility, we cannot prove an AP-equivalence result between any two NOSol problems parametrized by constraint languages generating the same co-clone. Yet, similar results hold for the conjunctive closure.

Proposition 4. *Let Γ and Γ' be constraint languages. If $\Gamma' \subseteq \langle \Gamma \rangle_\wedge$ holds then we have the reductions $\mathsf{NOSol}^{\mathrm{d}}(\Gamma') \leq_{\mathrm{m}} \mathsf{NOSol}^{\mathrm{d}}(\Gamma)$ and $\mathsf{NOSol}(\Gamma') \leq_{\mathrm{AP}} \mathsf{NOSol}(\Gamma)$.*

Proof. For similarity it suffices to show that $\mathsf{NOSol}(\Gamma') \leq_{\mathrm{AP}} \mathsf{NOSol}(\Gamma)$ if $\Gamma' \subseteq \langle \Gamma \rangle_\wedge$.

Let a formula φ with a model m be an instance of $\mathsf{NOSol}(\Gamma')$. As $\Gamma' \subseteq \langle \Gamma \rangle_\wedge$, every constraint $R(x_1, \dots, x_k)$ of φ can be written as a conjunction of constraints upon relations from Γ. Substitute the latter into φ, obtaining φ'. Now (φ', m) is an instance of $\mathsf{NOSol}(\Gamma)$, where φ' is only polynomially larger than φ. For φ and φ' have the same variables and hence the same models, also the nearest other models of φ and φ' are the same. □

For a relation $R \subseteq \{0,1\}^n$, its *dual* relation is $\mathrm{dual}(R) = \{\overline{m} \mid m \in R\}$, i.e., the relation containing the complements of tuples from R. We naturally extend this to sets of relations Γ by putting $\mathrm{dual}(\Gamma) = \{\mathrm{dual}(R) \mid R \in \Gamma\}$. Since taking complements is involutive, duality is a symmetric relation. By inspecting the bases of co-clones in Table 1, we deduce that many co-clones are duals of each other, e.g. iE_2 and iV_2.

We now show that it suffices to consider one half of Post's lattice of co-clones.

Lemma 5. *For every Boolean constraint language Γ we have the mutual reductions $\mathsf{NOSol}^{\mathrm{d}}(\Gamma) \equiv_{\mathrm{m}} \mathsf{NOSol}^{\mathrm{d}}(\mathrm{dual}(\Gamma))$ and $\mathsf{NOSol}(\Gamma) \equiv_{\mathrm{AP}} \mathsf{NOSol}(\mathrm{dual}(\Gamma))$.*

Proof. For a Γ-formula φ and an assignment m to φ we construct a $\mathrm{dual}(\Gamma)$-formula φ' by substitution of every atom $R(\boldsymbol{x})$ by $\mathrm{dual}(R)(\boldsymbol{x})$. Then m satisfies φ if and only if $\overline{m}$ satisfies φ', $\overline{m}$ being the complement of m. Moreover, $\mathrm{hd}(m, m') = \mathrm{hd}(\overline{m}, \overline{m'})$. □

5 Finding Another Solution Closest to the Given One

5.1 Polynomial-Time Cases

Since we cannot take advantage of the clone closure, we must proceed differently. We use the following result based on a previous theorem of Baker and Pixley [2].

Proposition 6 (Jeavons et al. [9]). *Every bijunctive constraint $R(x_1, \ldots, x_n)$ is equivalent to $\bigwedge_{1 \leq i \leq j} R_{ij}(x_i, x_j)$, where R_{ij} is the projection of R to the coordinates i and j.*

Proposition 7. *If Γ is bijunctive ($\Gamma \subseteq \mathrm{iD}_2$) then $\mathsf{NOSol}(\Gamma)$ is in* PO.

Proof. According to Proposition 6 we may assume that the formula φ is a conjunction of atoms $R(x, y)$ or a unary constraint $R(x, x)$ in the form $[x]$ or $[\neg x]$. Unary constraints can be eliminated and their value propagated into the other clauses, since they fix the value for a given variable.

For each variable x we construct a model m_x of φ with $m_x(x) \neq m(x)$ such that $\mathrm{hd}(m_x, m)$ is minimal among all models with this property. Initially we set $m_x(x)$ to $1 - m(x)$ and $m_x(y) := m(y)$ for all variables $y \neq x$ and mark x as flipped. If m_x satisfies all atoms we are done. Otherwise let $R(u, v)$ be an atom falsified by m_x. If u and v are marked as flipped, the construction fails, a model m_x with the property $m_x(x) \neq m(x)$ does not exist. Otherwise the uniquely determined variable v in $R(u, v)$ is not marked as flipped. Set $m_x(v) := 1 - m(v)$, mark v as flipped, and repeat the process.

If m_x does not exist for any variable x, then m is the sole model of φ and the problem is not solvable. Otherwise choose one of the variables x for which $\mathrm{hd}(m_x, m)$ is minimal and return m_x as second solution m'. □

Proposition 8. *If $\Gamma \subseteq \mathrm{iS}_{00}^k$ or $\Gamma \subseteq \mathrm{iS}_{10}^k$ for some $k \geq 2$ then $\mathsf{NOSol}(\Gamma)$ is in* PO.

Proof. We perform the proof only for iS_{00}^k. Lemma 5 implies the same result for iS_{10}^k.

The co-clone iS_{00}^k is generated by $\Gamma' := \{\mathrm{or}^k, [x \to y], [x], [\neg x]\}$. According to [7], this set Γ' is also a so-called *plain basis* of iS_{00}^k, i.e. we may assume that our inputs (φ, m) contain conjunctive formulas φ over these relations and equality, without existential quantification.

Note that $x \vee y$ is a polymorphism of Γ, i.e., for any two solutions m_1, m_2 of φ we have that the assignment $m_1 \vee m_2$ which is defined by $(m_1 \vee m_2)(x) = m_1(x) \vee m_2(x)$ for every x is also a solution of φ. It follows that we get the optimal solution m' for the instance φ and m by either flipping some values 1 of m to 0 or flipping some values 0 of m to 1 but not both. To see this, assume the optimal solution m' flips both ones and zeros, then $m' \vee m$ is a solution of φ that is closer to m than m' which is a contradiction.

The main idea is to compute for each variable x of φ the distance of the solution m_x, which is minimal among the solutions of φ which differ from m

on the variable x, and flip only ones or only zeros. Then the algorithm chooses one m_x closest to m as m' and returns it. Since m and m' differ in at least one variable, this yields the correct result.

We describe the computation of m_x. If $m(x) = 0$, we flip x to 1 and propagate iteratively along equalities $x = z$ and $x \to y$-constraints, i.e., if $x \to y$ is a constraint of φ and $m(y) = 0$, we flip y to 1 and propagate. This process terminates after at most n flips, as we only flip from 0 to 1 and no variable is flipped more than once. If the resulting assignment satisfies φ, this is our m_x. Otherwise, there is no satisfying assignment which we get by flipping x and only flipping 0 to 1 and thus no candidate m_x with the desired properties. If $m(x) = 1$, we flip x to 0 and propagate backward along equalities $x = z$ and binary implications, i.e., if $y \to x$ is a constraint of φ and $m(y) = 1$, we flip y to 0 and iterate. Again, if the result satisfies φ, this is our m_x; else, there is no candidate m_x for this variable. Finally, return the candidate m_x being closest to m if it exists, otherwise there is no feasible solution. □

5.2 Hard Cases

Lemma 9. *Let Γ be a constraint language. If $\mathrm{iI}_1 \subseteq \langle\Gamma\rangle$ or $\mathrm{iI}_0 \subseteq \langle\Gamma\rangle$ holds then finding a feasible solution for* $\mathsf{NOSol}(\Gamma)$ *is* NPO*-hard. Otherwise,* $\mathsf{NOSol}(\Gamma) \in$ poly-APX.

Proof. Finding a feasible solution to $\mathsf{NOSol}(\Gamma)$ is exactly the problem $\mathsf{AnotherSAT}(\Gamma)$ which is NP-hard if and only if $\mathrm{iI}_1 \subseteq \langle\Gamma\rangle$ or $\mathrm{iI}_0 \subseteq \langle\Gamma\rangle$ according to Juban [10]. If $\mathsf{AnotherSAT}(\Gamma)$ is polynomial-time decidable, we can always find a feasible solution for $\mathsf{NOSol}(\Gamma)$ if it exists. Obviously, every feasible solution is an n-approximation of the optimal solution, where n is the number of variables of the input. □

Tightness Results. It will be convenient to consider the following decision problem.

Problem: $\mathsf{AnotherSAT}_{<n}(\Gamma)$
Input: A conjunctive formula φ over relations from Γ and an assignment m satisfying φ.
Question: Is there another satisfying assignment m' of φ, different from m, such that $\mathrm{hd}(m, m') < n$, where n is the number of variables of φ?

Note that $\mathsf{AnotherSAT}_{<n}(\Gamma)$ is not compatible with existential quantification. Let $\varphi(y, x_1, \ldots, x_n)$ with the model m be an instance of $\mathsf{AnotherSAT}_{<n}(\Gamma)$ and m' its solution satisfying $\mathrm{hd}(m, m') < n + 1$. Let m_1 and m_1' be the corresponding vectors to m and m', respectively, with the first coordinate truncated. When we existentially quantify the variable y in φ, producing $\varphi_1(x_1, \ldots, x_n) = \exists y\, \varphi(y, x_1, \ldots, x_n)$, then both m_1 and m_1' are solutions of φ', but we cannot guarantee $\mathrm{hd}(m_1, m_1') < n$. Hence we need the equivalent of Proposition 4 for this problem, whose proof is analogous.

Proposition 10. $\mathsf{AnotherSAT}_{<n}(\Gamma') \leq_{\mathrm{m}} \mathsf{AnotherSAT}_{<n}(\Gamma)$ *for* $\Gamma, \Gamma' \in \mathcal{L}$, $\Gamma' \subseteq \langle \Gamma \rangle_{\wedge}$.

Proposition 11. *If* $\Gamma \in \mathcal{L}$ *with* $\langle \Gamma \rangle = \mathrm{iI}$ *or* $\mathrm{iN} \subseteq \langle \Gamma \rangle \subseteq \mathrm{iN}_2$, *then* $\mathsf{AnotherSAT}_{<n}(\Gamma)$ *is* NP*-complete.*

Proof. Containment in NP is clear, so it only remains to show hardness. Since the considered problem is not compatible with existential quantification, we cannot use clone theory and therefore we will consider the three co-clones iN_2, iN and iI individually, making use of minimal weak bases.

Case $\langle \Gamma \rangle = \mathrm{iN}$*:* We show a reduction from $\mathsf{AnotherSAT}(R)$ where $R = \{000, 101, 110\}$ which is NP-hard by [10]. Since R is 0-valid, $\mathsf{AnotherSAT}(R)$ is still NP-complete if we restrict it to instances $(\varphi, \mathbf{0})$, where φ is a conjunctive formula over R and $\mathbf{0}$ is the constant 0-assignment. Thus we can perform a reduction from this restricted problem.

By Theorem 1 and Proposition 10 we may assume that Γ contains the minimal weak base relation R_{iN}. Given a formula φ over R, we construct another formula φ' over R_{iN} by replacing every constraint $R(x_i, x_j, x_k)$ with $R_{\mathrm{iN}}(x_i, x_j, x_k, w)$, where w is a new global variable. Moreover, set m to the constant 0-assignment. This construction is a many-one reduction from the restricted version of $\mathsf{AnotherSAT}(R)$ to $\mathsf{AnotherSAT}_{<n}(\Gamma)$.

To see this, observe that the tuples in R_{iN} that have a 0 in the last coordinate are exactly those in $R \times \{0\}$. Thus any solution of φ can be extended to a solution of φ' by assigning 0 to w. Assume that φ' has a solution m which is not constant $\mathbf{0}$ or constant $\mathbf{1}$. Because R_{iN} is complementive, we may assume that $m(w) = 0$. But then m restricted to the variables of φ is not the constant 0-assignment and satisfies all constraints of φ. This completes the proof of the first case.

Case $\langle \Gamma \rangle = \mathrm{iN}_2$*:* We show a reduction from $\mathsf{AnotherSAT}_{<n}(R_{\mathrm{iN}})$ which is NP-hard by the previous case. Reasoning as before, we may assume that Γ contains $R_{\mathrm{iN}_2} = \{m\overline{m} \mid m \in R_{\mathrm{iN}}\}$. Given an R_{iN}-formula φ over the variables $x_1, \ldots, x_n$, we construct an R_{iN_2}-formula over the variables $x_1, \ldots, x_n, x'_1, \ldots, x'_n$ by replacing $R_{\mathrm{iN}}(x_i, x_j, x_k, x_\ell)$ with $R_{\mathrm{iN}_2}(x_i, x_j, x_k, x_\ell, x'_i, x'_j, x'_k, x'_\ell)$. Moreover, we define an assignment m' to φ' by setting $m'(x_i) := m(x_i)$ and $m'(x'_i) := \overline{m}(x_i)$. It is easy to see that this construction is a reduction from $\mathsf{AnotherSAT}_{<n}(R_{\mathrm{iN}})$ to $\mathsf{AnotherSAT}_{<n}(\Gamma)$.

Case $\langle \Gamma \rangle = \mathrm{iI}$*:* Note that by restricting the first argument of the minimal weak base relation R_{iI} to 0, we get the relation $\{0\} \times R$ with $R := \{000, 011, 101\}$. By [10] we have that $\mathsf{AnotherSAT}(R)$ is NP-complete. Now we proceed similarly to the first case, observing that the only solution m such that $m(w) = 1$ is the constant 1-assignment. □

Proposition 12. *For* $\Gamma \in \mathcal{L}$ *such that* $\langle \Gamma \rangle = \mathrm{iI}$ *or* $\mathrm{iN} \subseteq \langle \Gamma \rangle \subseteq \mathrm{iN}_2$ *and any* $\varepsilon > 0$ *there is no polynomial-time* $n^{1-\varepsilon}$*-approximation algorithm for* $\mathsf{NOSol}(\Gamma)$*, unless* P = NP*.*

Proof. Assume that there is a constant $\varepsilon > 0$ with a polynomial-time $n^{1-\varepsilon}$-approximation algorithm for $\mathsf{NOSol}(\Gamma)$. We will show how to use this algorithm to solve $\mathsf{AnotherSAT}_{<n}(\Gamma)$ in polynomial time. Proposition 11 completes the proof.

Let (φ, m) be an instance of $\mathsf{AnotherSAT}_{<n}(\Gamma)$ with n variables. If $n = 1$, then we reject the instance. Otherwise, we construct a new formula φ' and a new assignment m' as follows. Let k be the smallest integer greater than $1/\varepsilon$. Choose a variable x of φ and introduce $n^k - n$ new variables x^i for $i = 1, \ldots, n^k - n$. For every $i \in \{1, \ldots, n^k - n\}$ and every constraint $R(y_1, \ldots, y_\ell)$ in φ, such that $x \in \{y_1, \ldots, y_\ell\}$, construct a new constraint $R(z_1^i, \ldots, z_\ell^i)$ by $z_j^i = x^i$ if $y_j = x$ and $z_j^i = y_j$ otherwise; add all the newly constructed constraints to φ in order to get φ'. Moreover, we extend m to an assignment of φ' by setting $m'(x^i) = m(x)$. Now run the $n^{1-\varepsilon}$-approximation algorithm for $\mathsf{NOSol}(\Gamma)$ on (φ', m'). If the answer is $\overline{m'}$ then reject, otherwise accept.

We claim that the algorithm described above is a correct polynomial-time algorithm for the decision problem $\mathsf{AnotherSAT}_{<n}(\Gamma)$ when Γ is complementive. Polynomial runtime is clear. It remains to show its correctness. If the only solutions to φ are m and $\overline{m}$, then, as $n > 1$, the approximation algorithm must answer $\overline{m'}$ and the output is correct. Assume that there is a satisfying assignment m_s different from m and $\overline{m}$. The relation Γ is complementive, hence we may assume that $m_s(x) = m(x)$. It follows that φ' has a satisfying assignment m_s' for which $\mathrm{hd}(m_s', m') < n$ holds. But then the approximation algorithm must find a satisfying assignment m'' for φ' with $\mathrm{hd}(m', m'') < n \cdot (n^k)^{1-\varepsilon} = n^{k(1-\varepsilon)+1}$. Since the inequality $k > 1/\varepsilon$ holds, it follows that $\mathrm{hd}(m', m'') < n^k$. Consequently, m'' is not the complement of m' and the output of our algorithm is again correct.

When Γ is not complementive but both 0-valid and 1-valid ($\langle\Gamma\rangle = \mathrm{iI}$), we perform the expansion algorithm described above for each variable of the formula φ and reject if the result is the complement for each run. The runtime remains polynomial. □

MinDistance-Equivalent Cases. In this section we show that affine co-clones give rise to problems equivalent to $\mathsf{MinDistance}$. The upper bound is easy.

Lemma 13. *For affine $\Gamma \in \mathcal{L}$ ($\Gamma \subseteq \mathrm{iL}_2$) the problem* $\mathsf{NOSol}(\Gamma)$ *reduces to* $\mathsf{MinDistance}$.

Proof. Let the formula φ and the model m be an instance of $\mathsf{NOSol}(\Gamma)$ over the variables $x_1, \ldots, x_n$. Clearly, φ can be written as $A\boldsymbol{x} = \boldsymbol{b}$ and m is a solution of this affine system. As any solution of $A\boldsymbol{x} = \boldsymbol{b}$ can be written as $m' = m + m_0$ where m_0 is a solution of $A\boldsymbol{x} = \boldsymbol{0}$, the problem becomes equivalent to computing the solutions of this homogeneous system of small weight. But this is exactly the $\mathsf{MinDistance}$ problem. □

The following lemma can be easily proved, since the equivalence relation $[x \equiv y]$ is the solution set of the linear equation $x + y = 0$. The relation $[x]$ is represented by the equation $x = 1$ whereas the relation $[\neg x]$ is represented by $x = 0$.

Lemma 14. $\mathsf{NOSol}(\{\mathrm{even}^4\}) \equiv_{\mathrm{AP}} \mathsf{NOSol}(\{\mathrm{even}^4, [x], [\neg x]\})$.

Corollary 15. *For $\Gamma \in \mathcal{L}$ with* $\mathrm{iL} \subseteq \langle\Gamma\rangle \subseteq \mathrm{iL}_2$ *we have* MinDistance $\leq_{\mathrm{AP}}$ NOSol(Γ).

Proof. We show an AP-reduction to NOSol($\{\mathrm{even}^4, [x], [\neg x]\}$). Since every system of linear equations can be written as a conjunction over relations in iL_2, the claim follows. □

MinHornDeletion-Equivalent Cases. As in Proposition 11 the need to use conjunctive closure instead of $\langle\ \rangle$ causes a case distinction in the proof of the following result.

Lemma 16. *If Γ is proper Horn* ($\mathrm{iE} \subseteq \langle\Gamma\rangle \subseteq \mathrm{iE}_2$) *then one of the following relations is in* $\langle\Gamma\rangle_\wedge$: $[x \to y]$, $[x \to y] \times \{0\}$, $[x \to y] \times \{1\}$, *or* $[x \to y] \times \{01\}$.

Proof. Supposing that $\langle\Gamma\rangle = \mathrm{iE}$, we get from Theorem 1 that R_{iE} belongs to $\langle\Gamma\rangle_\wedge$. Observe that $R_{\mathrm{iE}}(x_1, x_1, x_1, x_4) = [x_1 \to x_4]$ and thus $[x \to y] \in \langle R_{\mathrm{iE}}\rangle_\wedge \subseteq \langle\Gamma\rangle_\wedge$ which concludes this case. The case $\langle\Gamma\rangle = \mathrm{iE}_0$ leads to $[x \to y] \times \{0\} \in \langle\Gamma\rangle_\wedge$ in a completely analogous manner. The cases $\langle\Gamma\rangle = \mathrm{iE}_1$ and $\langle\Gamma\rangle = \mathrm{iE}_2$ lead to $[x \to y] \times \{1\} \in \langle\Gamma\rangle_\wedge$ and $[x \to y] \times \{01\} \in \langle\Gamma\rangle_\wedge$, respectively, by observing that $(x_1 \equiv x_1 \wedge x_3) = x_1 \to x_3$. □

Lemma 17. *If a constraint language $\Gamma \in \mathcal{L}$ is proper Horn* ($\mathrm{iE} \subseteq \langle\Gamma\rangle \subseteq \mathrm{iE}_2$), *then* NOSol($\Gamma$) *is* MinHornDeletion*-hard.*

Proof. Reduction from MinOnes($\Gamma \cup \{[x]\}$) which is MinHornDeletion-hard by [11]. Consider first the case in which $[x \to y] \in \langle\Gamma\rangle_\wedge$. By Proposition 4 we may assume that $[x \to y] \in \Gamma$. Let φ be a $\Gamma \cup \{[x]\}$-formula. We construct φ' as follows. Replace each atomic formula $R(y_1, \ldots, y_k)$ in φ, where $R \in \Gamma$, by its conjunctive normal form decomposition, which yields a formula φ''. Since $R \in \Gamma \subseteq \mathrm{iE}_2$ holds, each clause occurring in this decomposition contains at most one unnegated variable. Those that contain negated variables are 0-valid, and so is their conjunction. The remaining ones, which are not 0-valid, are just single variables (literals). Next, replace all literals y from φ'' by $x \to y$, where x is a global new variable. Finally, add $v \to x$ for all variables v of φ to get φ'.

Observe that φ' is 0-valid. Moreover, the other solutions of φ' are exactly the solutions of φ extended by the assignment $x := 1$, because whenever one of the variables v takes the value 1, the clause $v \to x$ forces x to 1 which in turn enforces the unary clauses y of φ by the implications $x \to y$. It follows that $\mathrm{OPT}(\varphi) + 1 = \mathrm{OPT}(\varphi', \mathbf{0})$.

Moreover, for every r-approximate solution m' of φ' we first check whether $m = \mathbf{0}$ is a solution of φ. In case it is, $\mathrm{OPT}(\varphi) = 0$ and we trivially have $\mathrm{hw}(m) \leq 2r\mathrm{OPT}(\varphi)$. Otherwise, $\mathrm{OPT}(\varphi) \geq 1$ and we get a solution m of φ by restriction to the variables of φ with the weight $\mathrm{hw}(m) = \mathrm{hd}(\mathbf{0}, m') - 1 \leq r(\mathrm{OPT}(\varphi', \mathbf{0})) - 1 \leq r(\mathrm{OPT}(\varphi) + 1) - 1 \leq 2r\mathrm{OPT}(\varphi)$. In any case, we have thus $\mathrm{hw}(m) \leq 2r\mathrm{OPT}(\varphi)$ which shows that the construction is an AP-reduction with $\alpha = 2$.

For the other cases of Lemma 16 we argue similarly. The only difference is the introduction of some new variables, forced to constant values by the respective relation from Lemma 16. It is easy to see that these constants do not change the rest of the analysis. □

The proof of the following corollary requires a reduction to a similar problem, namely NearestSolution (NSol), which differs from NOSol in the point that the input assignment m does not need to satisfy the input formula φ; if it does, then m is the optimal solution (see [3] for details).

Corollary 18. *If* $\Gamma \in \mathcal{L}$ *is proper Horn* ($\mathrm{iE} \subseteq \langle\Gamma\rangle \subseteq \mathrm{iE}_2$) *or proper dual-Horn* ($\mathrm{iV} \subseteq \langle\Gamma\rangle \subseteq \mathrm{iV}_2$) *then* NOSol($\Gamma$) *is* MinHornDeletion*-complete under* AP*-Turing-reductions.*

Proof. Hardness follows from Lemma 17 and duality. Moreover, NOSol(Γ) can be AP-Turing-reduced to NSol($\Gamma \cup \{[x], [\neg x]\}$) as follows: Given a Γ-formula φ and a model m, we construct for every variable x of φ a formula $\varphi_x = \varphi \wedge (x = \overline{m}(x))$. Then for every x we run an oracle algorithm for NSol($\Gamma \cup \{[x], [\neg x]\}$) on (φ_x, m) and output one result of these oracle calls that is closest to m.

We claim that this algorithm is indeed an AP-Turing reduction. To see this observe first that the algorithm always computes a feasible solution, unless only m satisfies φ. Moreover, we have $\mathrm{OPT}(\varphi, m) = \min_x(\mathrm{OPT}(\varphi_x, m))$. Let $A(\varphi, m)$ be the answer of the algorithm on (φ, m) and let $B(\varphi_x, m)$ be the answers to the oracle calls. Consider a variable x^* such that $\mathrm{OPT}(\varphi, m) = \min_x(\mathrm{OPT}(\varphi_x, m)) = \mathrm{OPT}(\varphi_{x^*}, m)$, and assume that $B(\varphi_{x^*}, m)$ is an r-approximate solution of (φ_{x^*}, m). Then we get

$$\frac{\mathrm{hd}(m, A(\varphi, m))}{\mathrm{OPT}(\varphi, m)} = \frac{\min_y(\mathrm{hd}(m, B(\varphi_y, m))}{\mathrm{OPT}(\varphi_{x^*}, m)} \leq \frac{\mathrm{hd}(m, B(\varphi_{x^*}, m))}{\mathrm{OPT}(\varphi_{x^*}, m)} \leq r.$$

Thus the algorithm is indeed an AP-Turing-reduction from NOSol(Γ) to NSol($\Gamma \cup \{[x], [\neg x]\}$). Note that NSol($\Gamma \cup \{[x], [\neg x]\}$) reduces to MinHornDeletion (see [3]). Duality completes the proof. □

6 Concluding Remarks

The studied problem is in PO for bijunctive constraints. If the constraints are implication hitting set bounded by k for some $k \geq 2$, the problem NOSol still remains in PO. The situation is more complicated for Horn constraints and dual Horn constraints, where the task becomes equivalent to MinHornDeletion. The next complexity stage of the solution structure is characterized by affine constraints, where we can apply standard linear algebra techniques to prove equivalence with the MinDistance-problem. The penultimate stage of solution structure complexity is represented by constraints, for which the existence of a solution is guaranteed by their definition, but we do not have any other exploitable information. We need a guarantee of at least two solutions. The existence of a second solution is guaranteed by iN_2 being complementive. Our problem belongs to the class poly-APX for these constraints. We can even exactly pinpoint the polynomial (n, i.e. arity of the formula) for which we can get a polynomial-time approximation. This complexity result indicates that we cannot get a suitable approximation for these types of the considered optimization problem. All other cases cannot be approximated in polynomial time at all.

References

1. Ausiello, G., Crescenzi, P., Gambosi, G., Kann, V., Marchetti-Spaccamela, A., Protasi, M.: Complexity and Approximation: Combinatorial Optimization Problems and Their Approximability Properties. Springer-Verlag, Heidelberg (1999)
2. Baker, K.A., Pixley, A.F.: Polynomial interpolation and the Chinese Remainder Theorem for algebraic systems. Math. Z. **143**(2), 165–174 (1975)
3. Behrisch, M., Hermann, M., Mengel, S., Salzer, G.: Minimal distance of propositional models (2015). CoRR abs/1502.06761
4. Böhler, E., Creignou, N., Reith, S., Vollmer, H.: Playing with Boolean blocks, Part II: Constraint satisfaction problems. SIGACT News **35**(1), 22–35 (2004)
5. Böhler, E., Reith, S., Schnoor, H., Vollmer, H.: Bases for Boolean co-clones. Inf. Process. Lett. **96**(2), 59–66 (2005)
6. Creignou, N., Khanna, S., Sudan, M.: Complexity classifications of Boolean constraint satisfaction problems. In: SIAM Monographs on Discrete Mathematics and Applications, vol. 7. SIAM, Philadelphia (PA) (2001)
7. Creignou, N., Kolaitis, P.G., Zanuttini, B.: Structure identification of Boolean relations and plain bases for co-clones. J. Comput. Syst. Sci. **74**(7), 1103–1115 (2008)
8. Dumer, I., Micciancio, D., Sudan, M.: Hardness of approximating the minimum distance of a linear code. IEEE Trans. Inf. Theory **49**(1), 22–37 (2003)
9. Jeavons, P., Cohen, D., Gyssens, M.: Closure properties of constraints. J. Assoc. Comput. Mach. **44**(4), 527–548 (1997)
10. Juban, L.: Dichotomy theorem for the generalized unique satisfiability problem. In: Ciobanu, G., Păun, G. (eds.) FCT 1999. LNCS, vol. 1684, pp. 327–337. Springer, Heidelberg (1999)
11. Khanna, S., Sudan, M., Trevisan, L., Williamson, D.P.: The approximability of constraint satisfaction problems. SIAM J. Comput. **30**(6), 1863–1920 (2000)
12. Lagerkvist, V.: Weak bases of boolean co-clones. Inf. Process. Lette. **114**(9), 462–468 (2014)
13. Schaefer, T.J.: The complexity of satisfiability problems. In: Proceedings 10th Symposium on Theory of Computing (STOC 1978), San Diego (California, USA), pp. 216–226 (1978)
14. Schnoor, H., Schnoor, I.: Partial polymorphisms and constraint satisfaction problems. In: Creignou, N., Kolaitis, P.G., Vollmer, H. (eds.) Complexity of Constraints. LNCS, vol. 5250, pp. 229–254. Springer, Heidelberg (2008)

On the Complexity of Computing Prime Tables

Martín Farach-Colton and Meng-Tsung Tsai(✉)

Rutgers University, New Brunswick, NJ 08901, USA
{farach,mtsung.tsai}@cs.rutgers.edu

Abstract. Many large arithmetic computations rely on tables of all primes less than n. For example, the fastest algorithms for computing $n!$ takes time $\mathcal{O}(\mathrm{M}(n\log n)+\mathrm{P}(n))$, where $\mathrm{M}(n)$ is the time to multiply two n-bit numbers, and $\mathrm{P}(n)$ is the time to compute a prime table up to n. The fastest algorithm to compute $\binom{n}{n/2}$ also uses a prime table. We show that it takes time $\mathcal{O}(\mathrm{M}(n)+\mathrm{P}(n))$.

In various models, the best bound on $\mathrm{P}(n)$ is greater than $\mathrm{M}(n\log n)$, given advances in the complexity of multiplication [8,13]. In this paper, we give two algorithms to computing prime tables and analyze their complexity on a multitape Turing machine, one of the standard models for analyzing such algorithms. These two algorithms run in time $\mathcal{O}(\mathrm{M}(n\log n))$ and $\mathcal{O}(n\log^2 n/\log\log n)$, respectively. We achieve our results by speeding up Atkin's sieve.

Given that the current best bound on $\mathrm{M}(n)$ is $n\log n 2^{\mathcal{O}(\log^* n)}$, the second algorithm is faster and improves on the previous best algorithm by a factor of $\log^2\log n$. Our fast prime-table algorithms speed up both the computation of $n!$ and $\binom{n}{n/2}$.
Finally, we show that computing the factorial takes $\Omega(\mathrm{M}(n\log^{4/7-\varepsilon} n))$ for any constant $\varepsilon > 0$ assuming only multiplication is allowed.

Keywords: Prime tables · Factorial · Multiplication · Lower bound

1 Introduction

Let $\mathrm{P}(n)$ be the time to compute prime table T_n, that is, a table of all primes from 2 to n. The best bound for $\mathrm{P}(n)$ on a log-RAM[1] is $\mathcal{O}(n/\log\log n)$, using the Sieve of Atkin, and $\mathcal{O}(n\log^2 n\log\log n)$ on the multitape Turing machine (TM), a standard model for analyzing prime table computation, factorial computation, and other large arithmetic computations [13,24,25]. This TM algorithm is due to Schönhage et al. [24] and is based on the Sieve of Eratosthenes.

M. Farach-Colton and M.-T. Tsai—Work supported by CNS-1408782 and IIS-1247750.

[1] In the standard RAM model, words have $O(\log n)$ bits on which arithmetic operations can be performed in constant time, say, so that array access takes constant time. We follow the convention of previous papers [14] on this topic by emphasizing this point in the name *log-RAM*.

K. Elbassioni and K. Makino (Eds.): ISAAC 2015, LNCS 9472, pp. 677–688, 2015.
DOI: 10.1007/978-3-662-48971-0_57

The main result of this paper is two algorithms that improve the time to compute T_n on a TM. One runs in $\mathcal{O}(n \log^2 n / \log\log n)$ and thus speeds up Schönhage's algorithm by a factor of $\log^2 \log n$.

The other has a running time that depends on the time to multiply large numbers. Let $\mathrm{M}(a, b)$ be the time to multiply an a-bit number with a b-bit number, and let $\mathrm{M}(a) = \mathrm{M}(a, a)$. We make the standard assumption [18] that $f(n) = \mathrm{M}(n)/n$ is a monotone non-decreasing function. Then we give a prime-table algorithm that runs in time $\mathcal{O}(\mathrm{M}(n \log n))$ on a TM. Fürer's algorithm [13] gives the best bound for $\mathrm{M}(n)$ on a TM, which is $n \log n 2^{\mathcal{O}(\log^* n)}$, a bound that was later achieved by a different method by De et al. [8], so our second algorithm is currently slower than the first algorithm.

Prime tables are used to speed up many types of computation. For example, the fastest algorithms for computing $n!$ depend on prime tables [6,24,27]. Schönhage's algorithm [24] is fastest and takes time $\mathcal{O}(\mathrm{M}(n \log n) + \mathrm{P}(n))$.

The number of bits in $n!$ is $\Theta(n \log n)$, and Borwein [6] conjectured that computing $n!$ takes $\Theta(\mathrm{M}(n \log n))$ time. On the log-RAM, Fürer [14] showed that $\mathrm{M}(n) = \mathcal{O}(n)$. So on the log-RAM, the upper bound of Borwein's conjecture seems to be true, since $\mathrm{M}(n \log n)$ dominates $\mathcal{O}(n/\log\log n)$ for now.

On a TM, there is a simple lower bound of $\Omega(n \log n)$ to compute $n!$, since that is the number of TM characters needed to represent the output. This contrasts with the $\mathcal{O}(n)$-word output on the log-RAM. On the other hand, no $\mathcal{O}(\mathrm{M}(n \log n))$-time algorithm was known in this model, since before our improved prime-table algorithms, $\mathrm{P}(n)$ dominated $\mathrm{M}(n \log n)^2$. Using our $\mathcal{O}(\mathrm{M}(n \log n))$-time prime-table algorithm, the time to compute $n!$ is improved to $\mathcal{O}(\mathrm{M}(n \log n))$. If Borwein's conjecture turns out to be true, this algorithm will turn out to be optimal for computing $n!$.

Another use of prime tables is in the computation of binomial coefficients. The exact complexity of computing binomial coefficients hadn't been analyzed, but here we show that a popular algorithm based on [19] takes time $\mathcal{O}(\mathrm{M}(n) + \mathrm{P}(n))$. Thus our faster algorithm also improves this running time by $\log^2 \log n$.

Finally, we consider lower bounds for computing $n!$. Although we do not produce a general lower bound for computing $n!$ on a TM[3], we do show a lower bound for algorithms on the following restricted model. We do not restrict which operation can be used but we assume that the factorial $n!$ is output by a multiplication. We assume that a multiplication can only operate on two integers, each of which can be an integer of $o(n \log n)$ bits or a product computed by a multiplication. We note that all known algorithms adhere to this restriction [5,6,24,26,27], under which we show a lower bound

$$\Omega\left(\max_t \left\{ \mathrm{M}_{t^{1/2-\varepsilon}}\left(\frac{1}{t} n \log n \right), \frac{t}{w} n \log n \right\} \right) \text{ for } t \in [1, n], \tag{1}$$

where w denotes the word size in the model. Given an upper bound and a lower bound for $\mathrm{M}(n)$, we can simplify the lower bound in Eq. (1).

[2] We note that before Fürer's algorithm, the opposite was true. This is because before Fürer's algorithm, the best bound on $\mathrm{M}(n)$ was $\mathcal{O}(n \log n \log\log n)$ [25].

[3] And indeed, such a result would be a much more significant than any upper bound!.

On the Turing Machine, we know that $\mathrm{M}(n)$ has a simple linear lower bound $\Omega(n)$ and an upper bound $n \log n 2^{\mathcal{O}(\log^* n)}$ due to Fürer [13] and De et al. [8]. In that case, we have a lower bound in the multiplication model of

$$\Omega(\mathrm{M}(n \log^{4/7-\varepsilon} n)) \text{ for any constant } \varepsilon > 0. \quad (2)$$

On the log-RAM, we know that $\mathrm{M}(n)$ has a lower bound of $\Omega(n/\log n)$ because operations on $\mathcal{O}(\log n)$ bit words take at least constant time. The upper bound for $\mathrm{M}(n)$, also due to Fürer [14], is $\mathcal{O}(n)$. In that case, under the multiplication restriction, we have the same lower bound as Eq. (2). They coincide because both models have a $\log^{1+\varepsilon} n$ gap between the lower and upper bounds of $\mathrm{M}(n)$.

Our Techniques. To get the claimed $\log^2 \log n$ speedup, we replace the sieve of Eratosthenes with a careful multitape Turing machine implementation of Atkin's sieve. To show an upper bound of $\mathrm{P}(n)$ in terms of $\mathrm{M}(n)$, we prove an upper bound of the number of (integer) lattice points on the ellipses and truncated hyperbola specified by Atkin's conditions. Such bound establish that a majority of multiplications operate on short operands, thereby making them faster.

Organization. In Sect. 2, we present the related work for computing prime tables. We propose two algorithms in Sect. 3. Last, in Sect. 4, we show a lower bound of computing factorials. The related work, upper bounds for factorials and binomials, and inequalities of $\mathrm{M}(n)$ can be found in [12, App. A-C].

2 Background and Related Work

In this section, we present the relevant background on computing prime tables and defer those for factorials and binomial coefficients to [12, App. A].

The Sieve of Eratosthenes is the standard algorithm used in RAM model. It creates a bit table where each prime is marked with a 1 and each composite is marked with a 0. The multiples of each prime found so far are set to 0, each in $\mathcal{O}(1)$ time, and thus the whole algorithm takes time $\sum_{p \leq n} n/p = \mathcal{O}(n \log \log n)$. However, on a TM, each multiple of a prime cannot be marked in $\mathcal{O}(1)$ time. Instead, marking all the multiples of a single prime takes $\mathcal{O}(n)$ time, since the entire table must be traversed. Because there are $\mathcal{O}(\sqrt{n}/\log n)$ such primes, this approach takes $\mathcal{O}(n^{3/2}/\log n)$ time.

Schönhage et al. give an algorithm to compute a prime table from 2 to n in $\mathcal{O}(n \log^2 n \log \log n)$ time [24]. His algorithm, for each prime $p \leq \sqrt{n}$, generates a sorted list[4] of the multiples of p, and then merges the $\mathcal{O}(\sqrt{n}/\log n)$ lists so generated. The total number of integers on these lists is $\mathcal{O}(n \log \log n)$, each integer needs to be merged $\mathcal{O}(\log n)$ times, and each integer has $\mathcal{O}(\log n)$ bits. Therefore, Schönhage's algorithm has running time $\mathcal{O}(n \log^2 n \log \log n)$.

[4] It is not the case that each list occupy a tape; otherwise, $\omega(1)$ tapes are required. To merge these lists, put half of the lists on a tape, half on the other, merge them pairwise, output the sorted lists on another two tapes and recurse. In this way, 4 tapes are enough.

Alternatively, one can use the AKS primality test [1] on each integer in the range from 2 to n. The fastest known variant of the AKS primality test is due to Lenstra and Pomerance and takes $\tilde{\mathcal{O}}(\log^6 n)$ time per test on a TM. If Agrawal's conjecture [1] is true, it takes $\tilde{\mathcal{O}}(\log^3 n)$ time. Whether the conjecture is true or not, it would still take $\Omega(n \log^3 n)$ time to compute a prime table. One can use the base-2 Fermat test, $2^p \equiv 2 \pmod p$, to screen out a majority of composite numbers. This would take $\mathcal{O}(n \log n \mathrm{M}(\log n))$, which is dominated by the AKS phase. All prime numbers and $o(n/\log n)$ composite numbers can pass the base-2 Fermat test [16]. Therefore, it reduces the complexity by a $\log n$ factor. In this case, it would take a finer analysis of AKS and settling Agrawal's conjecture to determine the exact complexity of this algorithm. It would likely take $\tilde{\mathcal{O}}(n \log^2 n) = \mathcal{O}(n \log^2 n \log^k \log n)$ for some $k > 0$, and this would improve on Schönhage's algorithm if $k < 1$.

We show how to implement the Sieve of Atkin to achieve a running time $\min\{\mathcal{O}(n \log^2 n / \log\log n), \mathcal{O}(\mathrm{M}(n \log n))\}$ on the Turing Machine in Sect. 3.

3 Fast Algorithms for Atkin's Sieve

In this section, we give two algorithms for implementing Atkin's Sieve on a TM. The first runs in time $\mathcal{O}(n \log^2 n / \log\log n)$. The second runs in time $\mathcal{O}(\mathrm{M}(n \log n))$. Given the state of the art in multiplication, the first is faster. We present both, in case a faster multiplication algorithm is discovered.

3.1 Atkin's Sieve in $\mathcal{O}(n \log^2 n / \log\log n)$

We define notions before proceeding to the proof. A *squarefree* integer denotes an integer that has no divisor that is a square number other than 1. Let $\mathrm{N}_{f(x,y)}(k) = 0$ if there are even number of integer pairs (x, y) that have $x > 0, y > 0$ and $f(x, y) = k$; or 1, otherwise. Similarly, let $\mathrm{N}'_{f(x,y)}(k) = 0$ if there are even number of integer pairs (x, y) that have $x > y > 0$ and $f(x, y) = k$; or 1, otherwise. The key distinction is that the latter requires that $x > y$. In [2], Atkin and Bernstein classify potential primes into three categories and perform a unique primality test for each category based on N and N′, as stated in Theorem 1.

Theorem 1 ([2, Theorems 6.1–6.3]). *For every squarefree integer $k \in 1+4\mathbb{N}$, k is prime iff $\mathrm{N}_{x^2+4y^2}(k) = 1$; for every squarefree integer $k \in 1+6\mathbb{N}$, k is prime iff $\mathrm{N}_{x^2+3y^2}(k) = 1$; for every squarefree integer $k \in 11+12\mathbb{N}$, k is prime iff $\mathrm{N}'_{3x^2-y^2}(k) = 1$.*

We show how to compute $\mathrm{N}_{x^2+4y^2}(k)$ for all $k \in [1, n]$ in $\mathcal{O}(n \log^2 n / \log\log n)$ time. First, for each $x \in [1, n^{1/2}]$, one can enumerate a short list of $x^2 + 4 \cdot 1^2, x^2 + 4 \cdot 2^2, \ldots, x^2 + 4 \cdot (n^{1/2})^2$ in $\mathcal{O}(\sqrt{n} \log n)$ time because the differences between two consecutive terms form an arithmetic progression. Furthermore, each short list is already sorted. Then, we merge short lists pairwisely until a single sorted list is obtained; therefore, the running time is $\mathcal{O}(n \log^2 n)$ because there are $\mathcal{O}(n)$

integers, each of which has $\mathcal{O}(\log n)$ bits and is encountered $\mathcal{O}(\log n)$ times in the merge process. By removing the duplicates in the single sorted list, which takes $\mathcal{O}(n \log n)$ time, $\mathrm{N}_{x^2+4y^2}(k)$ is obtained for all $k \in [1, n]$.

To speed up this process by a factor of $\log\log n$, noted in [2], Atkin and Bernstein show that the integers on these short lists are seldom coprime to the first $\log^{1/2} n$ primes. There are $\mathcal{O}(n/\log\log n)$ such integers in total. One can speed up this process by screening out the integers on these short lists that are not coprime to the first $\log^{1/2} n$ primes. This filter step can be completed in $\mathcal{O}(n \log^{1/2} n\mathrm{M}(\log n))$ time and the reduced short lists can be merged in the desired time. The same technique can be applied to $\mathrm{N}_{x^2+3y^2}(k)$ and $\mathrm{N}'_{3x^2-y^2}(k)$ for all $k \in [1, n]$.

Lemma 2. *Computing* $\mathrm{N}_{x^2+4y^2}(k)$, $\mathrm{N}_{x^2+3y^2}(k)$ *and* $\mathrm{N}'_{3x^2-y^2}(k)$ *for all* k *in* $[1, n]$ *takes* $\mathcal{O}(n \log^2 n/\log\log n)$ *time on the Turing Machine.*

We computed the Atkin conditions but now we need to get rid of all non-squarefree numbers. Therefore, we show that generating all non-squarefree numbers requires $\mathcal{O}(n \log n)$ time in Lemma 3. Merging these three lists followed by screening out the list of non-squarefree numbers gives a prime table, as summarized in Theorem 4.

Lemma 3. *Generating a sorted list of all non-squarefree integers in the range* $[1, n]$ *takes* $\mathcal{O}(n \log n)$ *time on the Turing Machine.*

Proof. We first generate the sorted list L_1 of all non-squarefree integers that has a divisor p^2 for some prime $p < \log n$. We initialize an array of n bits as zeros, for each prime $p < \log n$, we sequentially scan the entire array to mark all mp^2 for integer m by counting down a counter from p^2 to 0. Note that it requires amortized $\mathcal{O}(1)$ time to decrease down the counter by 1 due to the frequency division principle [4]. Since there are $\mathcal{O}(\log n/\log\log n)$ such primes, the running time of this step is $\mathcal{O}(n \log n/\log\log n)$. We then convert the array into the sorted list L_1 as required, which takes $\mathcal{O}(n \log n)$ time.

Next, we generate a sorted list L_2 of all non-squarefree integers that has a divisor p^2 for some prime $p \geq \log n$. We generate a sorted short list for each such prime p, containing all the integers $mp^2 < n$ for some integer m. Then, we merge these sorted short lists. Note that there are $\sum_{p \geq \log n} n/p^2 = \mathcal{O}(n/\log n)$ integers on these short lists, each integer has $\mathcal{O}(\log n)$ bits, and each integer is encountered $\mathcal{O}(\log n)$ times in the merging process. The running time is thus $\mathcal{O}(n \log n)$. We are done by merging L_1 and L_2. □

Theorem 4. *The prime table* T_n *from* 2 *to* n *can be computed on the Turing Machine in time*

$$\mathrm{P}(n) = \mathcal{O}(n \log^2 n/\log\log n).$$

3.2 Atkin's Sieve in $\mathcal{O}(\mathrm{M}(n \log n))$

We show that sieve of Atkin can be realized in $\mathcal{O}(\mathrm{M}(n \log n))$ time on the Turing Machine. We apply multiplication to the computation of $N_{f(x,y)}(k)$ and

$N'_{f(x,y)}(k)$ for all $k \in [1, n]$. The balance of the work will take $\mathcal{O}(n \log n)$, and will thus be dominated by the multiplication.

An important aspect of the multiplication will be the number of bits needed in the multiplicands. For this, we need Lemma 5, stating an upper bound of the number of (integer) lattice points on the ellipses specified by the first two Atkin conditions and on the truncated hyperbola $3x^2 - y^2 = k$ for $x > y > 0$.

Lemma 5. *The number of integer pairs (x, y) that satisfy $x^2 + 4y^2 = k$ for any positive integer k coprime to 6 is bounded by $k^{\mathcal{O}(1/\log\log k)}$. The same bound holds for $x^2 + 3y^2 = k$ and $3x^2 - y^2 = k, x > y > 0$.*

Proof. Observe that every pair (x, y) that satisfies $x^2 + 4y^2 = k$ induces a unique pair $(x' = x, y' = 2y)$ that satisfies $x'^2 + y'^2 = k$. Therefore, the number of pairs (x, y) that satisfies the latter equation is no less than that of the former. It is known that, for any odd integer k, there are

$$\mathcal{O}\left(\sum_{d|k} (-1)^{(d-1)/2}\right) \tag{3}$$

integer pairs (x', y') that satisfy $x'^2 + y'^2 = k$ [15]. Since the number of divisors of an integer k is no more than $\mathcal{O}\left(k^{1/\log\log k}\right)$ due to Wigert [9], an upper bound for (3) is $\mathcal{O}(k^{1/\log\log k})$. Similarly, it is known that for any odd integer k there are

$$\mathcal{O}\left(\sum_{d|k} \left(\frac{-3}{d}\right)\right) \tag{4}$$

integer pairs (x, y) that satisfy $x^2 + 3y^2 = k$ [17], where $\left(\frac{a}{b}\right)$ denotes the Jacobi symbol. Because each Jacobi symbol has value no more than 1, an upper bound for (4) is $\mathcal{O}(k^{1/\log\log k})$ as desired.

We argue that, for any integer k coprime to 6, the number of integer pairs (x, y) that satisfy equation $3x^2 - y^2 = k, x > y > 0$ has the same bound. We first give a proof for the case that x, y, k are mutually relatively primes and then relax the restriction.

Let $k = p_1^{r_1} p_2^{r_2} \cdots p_t^{r_t}$ where the p_i's are distinct primes more than 3 and the r_i's are positive integers. Observe that every integer pair (x, y) that satisfy $3x^2 - y^2 = k, x > y > 0$ has the property that $x, y < k^{1/2}$. Therefore, every integer pair (x, y) that satisfy $3x^2 - y^2 = k, x > y > 0$ induces a unique pair $(x' \equiv x \bmod k, y' \equiv x \bmod k)$ that satisfies $3x'^2 - y'^2 \equiv 0 \pmod{k}$ as well as induces a pair $(x' \equiv x \bmod p_i^{r_i}, y' \equiv y \bmod p_i^{r_i})$ that satisfies $3x'^2 - y'^2 \equiv 0 \pmod{p_i^{r_i}}$.

We claim that any integer pair (x, y) that satisfies $3x^2 - y^2 \equiv 0$ (mod k) has a unique product $(yx^{-1} \bmod k)$, where the inverse x^{-1} exists since x and k are relatively prime. We give a proof by contradiction. Suppose (x_1, y_1) and (x_2, y_2) yield the same product $(yx^{-1} \bmod k)$, then $y_1 x_2 \equiv y_2 x_1 \pmod{k}$ or, equivalently,

$y_1x_2 = y_2x_1$ due to $x_1, y_1, x_2, y_2 < k^{1/2}$. Since x_1 and y_1 are relatively prime, and x_2 and y_2 are relatively prime, then $x_1 = x_2$, $y_1 = y_2$, a contradiction.

We show that the number of distinct products $(yx^{-1} \bmod k)$ is at most 2^t. Since $(x' \equiv x \bmod p_i^{r_i}, y' \equiv y \mod p_i^{r_i})$ satisfies $3x'^2 - y'^2 \equiv 0 \ (\text{mod} p_i^{r_i})$, $(a_i \equiv y'x'^{-1} \bmod p_i^{r_i})$ is a square root of 3 modulo $p_i^{r_i}$. There are at most two distinct square roots of 3 for each modulo $p_i^{r_i}$, $p_i > 3$ [20, Theorem 5.2]. By the Chinese Remainder Theorem, $(a_1, a_2, \ldots, a_t)$ is in a one-to-one correspondence to $(yx^{-1} \bmod k)$. Hence, there are at most 2^t distinct products $(yx^{-1} \bmod k)$ as desired.

Consequently, the number of integer pairs (x, y) that satisfy $3x^2 - y^2 = k, x > y > 0$ for any integer k coprime to 6 is bounded by

$$\mathcal{O}\left(k^{1/\log\log k}\right) \text{ for } x, y, k \text{ are relatively primes.}$$

For the case that two of x, y, k have common divisor $d > 1$, then the third one also has the divisor d. Then, one can divide x, y, k by the common divisor d, thus reducing to a case of x, y, k' being mutually relatively prime for $k' < k$. There are $\mathcal{O}(k^{1/\log\log k})$ such smaller k' and each smaller k' contributes $\mathcal{O}(k^{1/\log\log k})$ pairs (x, y) at most. We are done. □

Lemma 6. *Given a function $f(x, y) = ax^2 + by^2$ for $a > 0, b > 0$, $\mathrm{N}_{f(x,y)}(k)$ for all $k \in [1, n]$ can be computed in $\mathcal{O}(\mathrm{M}(n \log n))$ time.*

Proof. Any positive integer pair (x, y) that satisfies $f(x, y) = k$ has the property that $ax^2, by^2 < k$. We claim that a long multiplication on a pair of $\mathcal{O}(n \log n)$-bit integers suffices to compute $N_{f(x,y)}(k)$ for all $k \in [1, n]$.

For $i \in [1, n]$, let $\alpha_i = 1$ if some $ax^2 = i$, or otherwise $\alpha_i = 0$. Similarly, for $j \in [1, n]$, let $\beta_j = 1$ if some $by^2 = j$, or otherwise $\beta_j = 0$. Then, the following product of polynomials

$$\sum_{i=[1,n]} \alpha_i z^i \sum_{j \in [1,n]} \beta_j z^j$$

has the property that the coefficient of z^k modulo 2 is equal to $\mathrm{N}_{f(x,y)}(k)$. One can use a multiplication to replace the product of polynomials by replacing z with an integer base B. To avoid carry issue, we choose $B = \Theta(\log n)$ because the coefficient of z^k is at least bounded by $\mathcal{O}(n^2)$. Thus, the running time is $\mathcal{O}(\mathrm{M}(n \log n))$. □

Corollary 7. *Given functions $f(x, y) = x^2 + 4y^2, g(x, y) = x^2 + 3y^2$, $\mathrm{N}_{f(x,y)}(k)$ and $\mathrm{N}_{g(x,y)}(k)$ for all $k \in [1, n]$ can be computed in $\mathcal{O}(\mathrm{M}(n \log n / \log\log n))$ time.*

Proof. We use the algorithm stated in Lemma 6 but, due to Lemma 5, we can choose B to be $\Theta(\log n / \log\log n)$ rather than $\Theta(\log n)$. One needs to avoid the computation of $\mathrm{N}_{f(x,y)}(k)$ for k not coprime to 6 because $\mathrm{N}_{f(x,y)}(k)$ might require more than $\Theta(\log n / \log\log n)$ bits for such k. We avoid the computation of $\mathrm{N}_{f(x,y)}(k)$ for such k by classifying x^2, $4y^2$, $3y^2$ into groups according to their residue modulo 6. Then, multiplying these groups in pairs only if their sum is coprime to 6, which amplifies the complexity by a constant factor. □

Lemma 8. *Given a function $f(x, y) = 3x^2 - y^2$, $\mathrm{N}'_{f(x,y)}(k)$ for all $k \in [1, n]$ can be computed in $\mathcal{O}(\mathrm{M}(n \log n))$ time.*

Proof. Any positive integer pair (x, y) that satisfies $f(x, y) = k$ and $x > y$ has the property that $x, y < k^{1/2}$. We claim that $\log n$ multiplications suffice to compute $N'_{f(x,y)}(k)$ for all $k \in [1, n]$.

We relax the condition $x > y$ by divide and conquer and then process each subproblem as Lemma 6. We reduce the range of pairs (x, y), $0 < y < x < n^{1/2}$ to following three cases, let $h = n^{1/2}/2$: (1) $x \in [h, n^{1/2}]$ and $y \in [0, h)$, (2) $0 < y < x < h$, (3) $h \leq y < x < n^{1/2}$.

Note that case (1) can be computed by the product of n-term polynomial as what was done in Corollary 7 due to Lemma 5. Therefore, case (1) can be done in $\mathcal{O}(\mathrm{M}(n \log n / \log \log n))$ time. Besides, the number of pairs (x, y) in cases (2) and (3) is half of that in the original case. To match the claimed complexity, we recurse for $\log \log n$ levels, with a running time of $\mathcal{O}(\mathrm{M}(n \log n))$ and generate $\mathcal{O}(\log n)$ lists of pairs (x, y) sorted in ascending $f(x, y)$ and we use the first algorithm in Lemma 3 to merge them into a sorted list L_1 in $\mathcal{O}(n \log n)$ time. Note that, by the first algorithm, any pair of duplicated integers is discarded, since we only care about parity. After the recursion, the number of unprocessed pairs (x, y) is $\mathcal{O}(n/\log n)$. We merge the unprocessed pairs (x, y) into a single sorted list L_2 in ascending $f(x, y)$ by the second algorithm used in Lemma 3, which takes $\mathcal{O}(n \log n)$ time. Finally, we are done by merging L_1 and L_2. □

Combining Lemmas 3, 8 and Corollary 7, we can realize the sieve of Atkin with a few of long multiplications and some minor procedures doable in $\mathcal{O}(n \log n)$ time. As a result, we have Theorem 9.

Theorem 9. *The prime table T_n from 2 to n can be computed on the Turing Machine in time*

$$\mathrm{P}(n) = \mathcal{O}(\mathrm{M}(n \log n)).$$

4 Lower Bound

We present a lower bound for computing the factorial $n!$. We do not restrict which operation can be used but we assume that the factorial $n!$ is output by a multiplication. We assume that a multiplication can only operate on two integers, each of which can be an integer of $o(n \log n)$ bits or a product computed by a multiplication. Under this assumption, we show that computing the factorial $n!$ has a lower bound $\Omega(\mathrm{M}(n \log^{4/7-\varepsilon} n))$ for any constant $\varepsilon > 0$.

To show the claimed lower bound, we need some lemmas for $\mathrm{M}(n)$ and $\mathrm{M}_k(n)$, where $\mathrm{M}_k(n)$ denotes the optimal time to multiply k pairs of two n-bit integers. There is a subtle difference between $\mathrm{M}_k(n)$ and $k\mathrm{M}(n)$. $\mathrm{M}_k(n)$ denotes the optimal time to multiply k pairs of integers, possibly in parallel, because all these integers are given at the beginning; however, $k\mathrm{M}(n)$ denotes the optimal time to multiply k pairs of integers serially, one after another. Hence, $\mathrm{M}_k(n) \leq k\mathrm{M}(n)$. To proceed to the lower bound proof, we need three additional inequalities of

M(n) and M$_k$(n), which are stated and shown in [12, App. C]. [12, Lemma C.1] and [12, Lemma C.3] are simple facts about the Turing Machine model. [12, Lemma C.2] is based on the property of progression-free set [3,7,10,11,21–23].

Since we restrict that the factorial $n!$ is output by a multiplication, there must be a multiplication $a_1 \times b_1 = a_0 = n!$ in every algorithm. Besides, we restrict that only the integers of $o(n \log n)$ bits and intermediate products can be multiplied. Therefore, a_1, b_1 are small integers or the computed intermediate products. Let $|x|$ denote the number of bits in x.

if $|a_i| > |a_0|/2$, then a_i has more than $o(n \log n)$ bits. Therefore, a_i is also an intermediate product and assert the existence of a multiplication $a_{i+1} \times b_{i+1} = a_i$. We can repeat this until some $|a_i| \leq |a_0|/2$. We define t to be the step where it stops. Therefore, there must be t multiplications, $a_i \times b_i = a_{i-1}$ for all $i \in [1, t]$, in any algorithm that can compute the factorial. In other words, we have a lower bound of

$$\sum_{i \in [1,t]} \mathrm{M}(|a_i|, |b_i|). \tag{5}$$

W.l.o.g., let $|a_i| \geq |b_i|$ and therefore $|a_i| \geq |a_0|/4$ for all $i \in [1, t]$.

Let us simplify Eq. (5) by observing the distribution of b_i's. Consider that

$$a_t \prod_{i \in [1,t]} b_i = a_0 \text{ and } \sum_{i \in [1,t]} |b_i| \geq |a_0| - |a_t|,$$

then $\mu = (|b_1| + |b_2| + \cdots + |b_t|)/t \geq |a_0|/(2t)$. Furthermore, for any $\gamma \in [1, t]$, if there is no b_i more than $\gamma\mu$, then there are t/γ b_i's more than $\mu/2$, which is an extension of Markov's inequality. We are ready to show the lower bound in Lemma 10.

Lemma 10. *Computing the factorial $n!$ has a lower bound*

$$\Omega\left(\mathrm{M}_{t^{1/2-\varepsilon}}\left(\frac{1}{t} n \log n\right)\right)$$

where t is a parameter to be determined later.

Proof. By applying the extended Markov inequality to Eq. (5), one has the lower bound

$$\sum_{i \in [1,t]} \mathrm{M}(|a_i|, |b_i|) \geq \max_{\gamma \in [1,t]} \min \left\{ \mathrm{M}(|a_0|/4, \gamma\mu), \frac{t}{2\gamma} \mathrm{M}(|a_0|/4, \mu/2) \right\},$$

which is, by [12, Lemma C.1], more than

$$\max_{\gamma \in [1,t]} \min \left\{ \mathrm{M}\left(\frac{\gamma}{t} n \log n\right), \frac{t}{2\gamma} \mathrm{M}\left(\frac{1}{2t} n \log n\right) \right\}.$$

We convert the two terms to the same form and compare. We apply Lemma [12, Lemma C.2] for the first term and the mentioned $\mathrm{M}_k(n) \leq k\mathrm{M}(n)$ bound for the second term, thus obtaining

$$\max_{\gamma \in [1,t]} \min \left\{ \mathrm{M}_{(2\gamma)^{1-\varepsilon}}\left(\frac{1}{2t} n \log n\right), \mathrm{M}_{\frac{t}{2\gamma}}\left(\frac{1}{2t} n \log n\right) \right\}$$

for any constant $\varepsilon > 0$. Observe that $\mathrm{M}_k(a) \leq \mathrm{M}_\ell(a)$ if $k \leq \ell$. As a result, we have the following lower bound, by choosing $\gamma = t^{1/2+\varepsilon}/2$ for any constant $\varepsilon > 0$,

$$\Omega\left(\mathrm{M}_{t^{1/2-\varepsilon}}\left(\frac{1}{t} n \log n\right)\right).$$

□

Observe that Lemma 10 yields a good lower bound only if t is small. Our strategy is to find another lower bound which is good when t is large. Then, we can trade off between these lower bounds. We finalize the proof for the claimed lower bound in Theorem 11.

Theorem 11. *On a TM, computing the factorial $n!$ has a lower bound*

$$\Omega(n \log^{4/7-\varepsilon} n) \text{ for any constant } \varepsilon > 0.$$

Proof. By Lemma [12, Lemma C.1], one has

$$\sum_{i \in [1,t]} \mathrm{M}(|a_i|, |b_i|) \geq \sum_{i \in [1,t]} (|a_i| + |b_i|) \geq t \frac{|a_0|}{4}.$$

Combining the above lower bound and the lower bound shown in Lemma 10, we obtain

$$\Omega\left(\min_t \max\left\{\mathrm{M}_{t^{1/2-\varepsilon}}\left(\frac{1}{t} n \log n\right), tn \log n\right\}\right). \tag{6}$$

Again, we convert the two terms to the same form and compare. We apply Lemma [12, Lemma C.3] for the first term and apply the current upper bound of $\mathrm{M}(n) \leq n \log n 2^{\mathcal{O}(\log^* n)}$ for the second term. Then, the lower bound becomes

$$\Omega\left(\min_t \max \mathrm{M}\left(\frac{n \log n}{t^{3/4+\varepsilon}}\right), \mathrm{M}\left(\frac{tn}{2^{\mathcal{O}(\log^* n)}}\right)\right). \tag{7}$$

The optimal bound appears at $t = \log^{4/7-\varepsilon} n$ for any constant $\varepsilon > 0$ as desired. □

Corollary 12. *On a log-RAM, computing the factorial $n!$ has a lower bound*

$$\Omega(n \log^{4/7-\varepsilon} n) \text{ for any constant } \varepsilon > 0.$$

Proof. We replace the lower bound of $\mathrm{M}(n)$ in Eq. 6 with $\Omega(n/\log n)$ and replace the upper bound of $\mathrm{M}(n)$ in Eq. 7 with $\mathcal{O}(n)$. By similar analysis, we are done. □

5 Conclusion

The prime number few.

—Stephen Pinker

References

1. Agrawal, M., Kayal, N., Saxena, N.: Primes is in P. Ann. of Math. **2**, 781–793 (2002)
2. Atkin, A.O.L., Bernstein, D.J.: Prime sieves using binary quadratic forms. Math. Comput. **73**(246), 1023–1030 (2004)
3. Behrend, F.A.: On sets of integers which contain no three terms in arithmetical progression. Proc. Natl. Acad. Sci. USA **32**(12), 331–332 (1946)
4. Berkovich, S., Lapir, G.M., Mack, M.: A bit-counting algorithm using the frequency division principle. Softw. Pract. Exper. **30**(14), 1531–1540 (2000)
5. Boiten, E.A.: Factorisation of the factorial-an example of inverting the flow of computation. Periodica Polytechnica SER. EL. ENG. **35**(2), 77–99 (1991)
6. Borwein, P.B.: On the complexity of calculating factorials. J. Algorithms **6**(3), 376–380 (1985)
7. Bourgain, J.: Roth's theorems in progressions revisited. Tech. rep. (2007)
8. De, A., Kurur, P.P., Saha, C., Saptharishi, R.: Fast integer multiplication using modular arithmetic. In: 40th Annual ACM Symposium on Theory of Computing (STOC), pp. 499–506 (2008)
9. Dickson, L.E.: History of the Theory of Numbers: Divisibility and Primality. Dover Publications, New York (2005)
10. Elkin, M.: An improved construction of progression-free sets. In: 21st Annual ACM-SIAM Symposium on Discrete Algorithms (SODA), pp. 886–905 (2010)
11. Erdös, P., Turán, P.: On some sequences of integers. J. Lond. Math. Soc. **11**, 261–264 (1936)
12. Farach-Colton, M., Tsai, M.: On the complexity of computing prime tables. CoRR abs/1504.05240 (2015)
13. Fürer, M.: Faster integer multiplication. In: 39th Annual ACM Symposium on Theory of Computing (STOC), pp. 57–66 (2007)
14. Fürer, M.: How fast can we multiply large integers on an actual computer? In: Pardo, A., Viola, A. (eds.) LATIN 2014. LNCS, vol. 8392, pp. 660–670. Springer, Heidelberg (2014)
15. Grosswald, E.: Representations of Integers as Sums of Squares. Springer, New York (1985)
16. Guy, R.: Unsolved Problems in Number Theory. Springer, New York (2004)
17. Uspensky, J.V., Heaslet, M.A.: Elementary Number Theory. McGraw-Hill, New York (1939)
18. Knuth, D.E.: The Art of Computer Programming: Seminumerical Algorithms. Addison-Wesley, Reading (1997)
19. Kummer, E.: Über die ergänzungssätze zu den allgemeinen reciprocitätsgesetzen. Journal für die reine und angewandte Mathematik **44**, 93–146 (1852)
20. LeVeque, W.J.: Topics in Number Theory. Dover, New York (2002)
21. Moser, L.: On non-averaging sets of integers. Can. J. Math. **5**, 245–253 (1953)
22. Roth, K.F.: On certain sets of integers. J. Lond. Math. Soc. **s1–28**(1), 104–109 (1953)
23. Salem, R., Spencer, D.C.: On sets of integers which contain no three terms in arithmetical progression. Proc. Natl. Acad. Sci. USA **28**(12), 561–563 (1942)
24. Schönhage, A., Grotefeld, A., Vetter, E.: Fast algorithms: A Multitape Turing Machine Implementation. Wissenschaftsverlag, B.I, Mannheim (1994)

25. Schönhage, A., Strassen, V.: Schnelle multiplikation grosser zahlen. Computing **7**, 281–292 (1971)
26. Ugur, A., Thompson, H.: The p-sized partitioning algorithm for fast computation of factorials of numbers. J. Supercomput. **38**(1), 73–82 (2006)
27. Vardi, I.: Computational Recreations in Mathematica. Addison-Wesley, New York (1991)

Game Values and Computational Complexity: An Analysis via Black-White Combinatorial Games

Stephen A. Fenner[1](✉), Daniel Grier[2], Jochen Messner[3], Luke Schaeffer[2], and Thomas Thierauf[4]

[1] University of South Carolina, Columbia, SC, USA
fenner@cse.sc.edu
[2] Massachusetts Institute of Technology, Cambridge, MA, USA
{grierd,lrs}@mit.edu
[3] Ulm, Germany
jochen_messner@web.de
[4] Aalen University, Aalen, Germany
thomas.thierauf@htw-aalen.de

Abstract. A black-white combinatorial game is a two-person game in which the pieces are colored either black or white. The players alternate moving or taking elements of a specific color designated to them before the game begins. A player loses the game if there is no legal move available for his color on his turn.

We first show that some black-white versions of combinatorial games can only assume combinatorial game values that are numbers, which indicates that the game has many nice properties making it easier to solve. Indeed, numeric games have only previously been shown to be hard for NP. We exhibit a language of natural numeric games (specifically, black-white poset games) that is PSPACE-complete, closing the gap in complexity for the first time between these numeric games and the large collection of combinatorial games that are known to be PSPACE-complete.

In this vein, we also show that the game of Col played on general graphs is also PSPACE-complete despite the fact that it can only assume two very simple game values. This is interesting because its natural black-white variant is numeric but only complete for $\mathsf{P}^{\mathsf{NP}[\log]}$. Finally, we show that the problem of determining the winner of black-white Graph Nim is in P using a flow-based technique.

Keywords: Combinatorial games · Computational complexity · Graph Nim · Poset games · Black-white games · Numeric games · Col

Extended abstract. A full version of this paper is [8]. The first author was supported by NSF grant CCF-0915948. The second author was supported by the Barry M. Goldwater Scholarship and by the NSF Graduate Research Fellowship under Grant No. 1122374. The third and fifth authors were supported by DFG grant TH472/4.

K. Elbassioni and K. Makino (Eds.): ISAAC 2015, LNCS 9472, pp. 689–699, 2015.
DOI: 10.1007/978-3-662-48971-0_58

1 Introduction

This extended abstract considers perfect information, two-player combinatorial games. In particular, we investigate whether the value[1] of these games influences the computational complexity of deciding which player should win under optimal play. We consider games that follow the normal gameplay convention: the players alternate moves according to the rules of the game until no move is possible for some player; that player then loses the game.

A combinatorial game is *impartial* if the allowed moves depend only on the position of the game and not on which of the two players is currently moving. Examples of impartial games are NIM, poset games, and GEOGRAPHY, all of which have well-understood complexity [2,10,13]. In contrast, *black-white* games have no options common to both players at any position.[2] Examples include games such as chess, checkers, and go. We explore simple black-white variants of well-known games in the full paper [8].

There is a general theory of combinatorial games developed by Conway [5] and Berlekamp et al. [1] that has served as one of the major tools in the area. It can be thought of as a sort of generalization or analogy to the famous Sprague-Grundy theorem for impartial games [11,14], which neatly distills the properties of NIM that allow it to be solved in polynomial time. In particular, this general theory distinguishes a class of combinatorial games that correspond directly to real numbers and hence share arithmetic operations and order properties with the real numbers. We call these *numeric games*, and we review their properties in [8]. We encourage readers who are unfamiliar with game values to read the full paper [8], which gives a more thorough investigation of their properties. Numeric games are special in that it is never beneficial for either player to make a move: if a player can win by moving, (s)he can also win by skipping a turn. Notice that this is not a property universally held by all black-white games (e.g. AMAZONS, black-white NODE KAYLES, HEX).

The standard decision problem associated with a game is to determine whether a given player has a winning strategy. As pointed out in [6], most two-player games with bounded length are either PSPACE-complete or in P. Any two-player game can be artificially converted into a numeric game (without altering its complexity) by forcing alternation between the players. These "one-sided" games are numeric for a vacuous reason: at any position, only one of the two players has options. In fact such games can only take on values 1, 0, or −1. See [8] for details. Until now, however, except for such one-sided games, there were no classes of numeric games known (to the authors of this paper) to be PSPACE-complete. Furthermore, the few results that are known about numeric games only show NP-hardness (see Blue-Red HACKENBUSH [1]). In this paper, we present a natural class of two-player bounded-length numeric games that is PSPACE-complete, namely, black-white

[1] Informally, the value of a game indicates which player will win the game and by how much. A more precise definition, along with much additional background information and full proofs, is given in the full paper [8].

[2] These are sometimes called red-blue games in the literature.

poset games. Since natural numeric games have previously only been shown to be as hard as NP, there existed hope prior to this result that a game with a more restricted set of game theoretic values may be easier to play. By presenting a natural PSPACE-complete numeric game, we provide evidence that no such connection exists.

Despite the fact that numeric games have relatively simple game values, they can still assume game values that are arbitrarily large. Perhaps then it is not merely the nature of the values that affects the complexity of a game but also the number of game values that it can assume. To this end, we investigate the game of Col [1] played on general graphs. We prove that although this class of games can only assume two possible game values (albeit not both numeric), it is still PSPACE-complete.[3] Informally, this has the following consequence for playing many games of Col side-by-side for which we know the corresponding game values. We can perform an extremely simple computation to decide which game to play in, but to decide which move to make in that game, we would need to solve a PSPACE-complete problem!

Interestingly, if you take the game of Col and convert it to a black-white version of the game in the natural way, then the game *does* become simpler to solve. In particular, we show the game becomes $\mathsf{P}^{\mathsf{NP}[\log]}$-complete. We conclude the paper with a flow-based technique for solving black-white GRAPH NIM.

1.1 Black-White Poset Games

Games on partially ordered sets, called *poset games*, are a class of two-player impartial games that have been widely studied. Given a partially ordered set, a player's turn consists of choosing one element e from the set. This element e is then removed along with all elements in the set that are greater than e. Well-studied subfamilies of poset games are NIM, CHOMP, DIVISORS, and HACKENDOT. In the black-white version of the game each element of the set has a color, black or white, and players are only allowed to choose elements of their own color (but choosing an element still removes everything above it, regardless of color).

Grier [10] showed that (impartial) poset games are PSPACE-complete. His proof is a reduction from NODE KAYLES, showed PSPACE-complete by T.J. Schaefer, who also implicitly showed that black-white NODE KAYLES is PSPACE-complete [13]. To show that black-white poset games are also PSPACE-complete, an obvious approach is to adapt Grier's reduction to use the black-white version of NODE KAYLES. However, Grier's construction crucially relies on the fact that both players can remove the same elements, and there is no obvious way to circumvent this restriction. In Sect. 2, we introduce novel techniques to show that black-white poset games are PSPACE-complete.

Proposition 1, below, asserts that black-white poset games are all numbers in the sense of Conway [5] or Berlekamp et al. [1]. See [8] for general background and for definitions specific to poset games.

[3] To clear possible confusion, Col was mistakenly referenced as being proven PSPACE-complete in [4].

Proposition 1. *All black-white poset games are numbers.*

We prove Proposition 1 in [8]. It is important because it helps establish our assertion that there is a natural numeric game that is PSPACE-complete.

1.2 Generalized Col

The game of Col [1] is a two-player combinatorial strategy game played on a simple planar graph. During the game, the players alternate coloring vertices of the graph. One player colors vertices white and the other player colors vertices black. A player is not allowed to color a vertex neighboring a vertex of the same color. The first player unable to color a vertex loses.

A well-known theorem about Col is that the value of any game is either x or $x + *$ where x is a number. We remove the restriction that Col games be played on planar graphs and consider only those games in which no vertex is already colored. We prove that deciding whether an initially uncolored graph is a win for the first player is a PSPACE-complete problem. Furthermore, it is easy to adapt the theorem about Col to show that the versions of Col we consider only assume the two very simply game values 0 and $*$.

1.3 Nim on Graphs

The game of NimG simultaneously generalizes the well-known game of Nim and Geography. A graph G is given where each vertex contains a positive number of sticks, and a token rests on a designated start vertex. In the "move-remove" variant we consider[4]—due to Stockman et al. [15], which we call Vertex NimG—each move consists of moving the token along an edge to a vertex v then removing at least one stick from v. We will here consider the game on directed graphs and treat undirected graphs as a special case.

Geography and Nim are both special cases of Vertex NimG. Lichtenstein and Sipser [12] showed that Geography is PSPACE-complete for bipartite graphs,[5] hence NimG is PSPACE-hard, even in the bipartite case, i.e., the black-white version. Burke and George [3] consider another variant called Neighboring NimG, which corresponds to NimG on graphs where every vertex has a self-loop. They show that Neighboring NimG is PSPACE-hard already for *undirected* graphs with ≤ 2 sticks per vertex. In contrast, Geography on undirected graphs is in P [7].

All the considered extensions of Nim on graphs are in PSPACE when the number of sticks is polynomially bounded. However, it is an open problem whether the winner can be determined in PSPACE when we allow exponentially many sticks, i.e., where the numbers of sticks are given in binary. (Clearly, EXP is an upper bound for the general case.)

---

[4] Other variants are possible; see Fukuyama [9] for example, where sticks are placed on edges.

[5] Bipartite Geography is one-sided as described above, hence vacuously numeric.

Analogously with GEOGRAPHY, the black-white version of NIMG is equivalent to the game on bipartite graphs. Since the black-white version of GEOGRAPHY remains PSPACE-complete, this holds for black-white NIMG too. As our one "easiness" result, we show that the black-white version of NIMG on *undirected* graphs is contained in P, even for an exponential number of sticks.[6]

2 PSPACE-Completeness of Black-White Poset Games

Here we show that deciding the winner of a black-white poset game is PSPACE-complete. By standard methods the problem can be solved in polynomial space, so we will focus on the other half of this claim:

Theorem 1. *Black-white poset games are* PSPACE-*hard.*

The proof is by a reduction from true quantified Boolean formulas (TQBF), a PSPACE-complete problem. We give the details of the reduction, and prove its correctness in the following subsections.

2.1 Construction

Suppose we are given a fully-quantified boolean formula ϕ of the form

$$\exists x_1 \forall x_2 \exists x_3 \cdots \exists x_{2n-1} \forall x_{2n} \exists x_{2n+1} f(x_1, x_2, \ldots, x_{2n+1})$$

where $f = c_1 \wedge c_2 \wedge \cdots \wedge c_m$ is in conjunctive normal form, with clauses $c_1, \ldots, c_m$. We define a game (not a poset game) based on this formula, called the *TQBF game*, where players take turns assigning the variables either 0 or 1 in turn. That is, White chooses an assignment for x_1, Black chooses an assignment for x_2, and so on. When all the variables are assigned, the game ends and White wins if f is true under that assignment, otherwise Black wins.

We define our black-white poset game G based on ϕ as follows, where $(X, \leq)$ is the poset.

- The poset is divided into sections. There is a section (called a *stack*) for each variable, a section for the clauses (the *clause section*), and a section for fine-tuning the balance of the game (*balance section*).
- The ith stack consists of a set of incomparable *waiting nodes* W_i above (i.e., greater than) a set of incomparable *choice nodes* C_i. We also have a pair of *anti-cheat nodes*, α_i and β_i, on all stacks except the last stack. For odd i, the choice nodes are white, the waiting nodes are black, and the anti-cheat nodes are black. The colors are reversed for even i.
- The set of choice nodes C_i, consists of eight nodes corresponding to all configurations of three bits (i.e., $000, 001, \ldots, 111$), which we call the *left bit*, *assignment bit* and *right bit* respectively.

[6] If the number of sticks is polynomially bounded, then undirected black-white NIMG trivially reduces to undirected GEOGRAPHY and so is clearly in P.

- The number of waiting nodes is defined to be

$$|W_i| = (2n + 2 - i)M$$

where M is the number of non-waiting nodes in the entire game. We will use the fact that $|W_i| \geq |W_{i+1}| + M$ later in the proof.
- The anti-cheat node α_i is above nodes in C_i with right bit 0 and nodes in C_{i+1} with left bit 0. Similarly, β_i is above nodes in C_i with right bit 1 and nodes in C_{i+1} with left bit 1.
- The *clause section* contains a black *clause node* b_j for each clause c_j, in addition to a black *dummy node*. The clause nodes and dummy node are all above a single white *interrupt node*. The clause node b_j is above a choice node z in C_i if the assignment bit of z is 1 and x_i appears positively in c_j, or if the assignment bit of z is 0 and x_i appears negatively in c_j.
- The balance section or *balance game* is incomparable with the rest of the nodes. The game consists of eight black nodes below a white node, which is designed to have game-theoretic value $-7\frac{1}{2}$. All nodes in this section are called *balance nodes.*

The basic idea is that players take turns taking choice nodes, and the assignment bits of the nodes they choose constitute an assignment of the variables, $x_1, \ldots, x_{2n+1}$. The assignment destroys satisfied clause nodes, and it turns out that Black can win if there remains at least one clause node. The waiting nodes and anti-cheat nodes exist to ensure players take nodes in the correct order. The interrupt node and dummy node control how much of an advantage a clause node is worth (after the initial assignment), and the balance node ensures the clause node advantage can decide whether White or Black wins the game (Fig. 1).

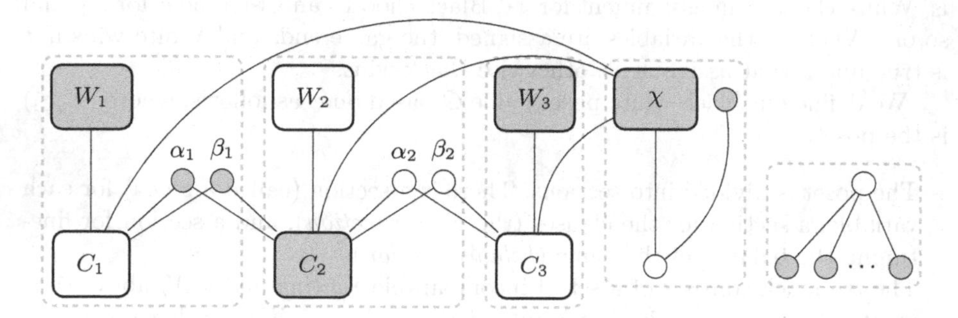

Fig. 1. An example game with three variables ($n = 1$). Circles represent individual nodes, blobs represent sets of nodes, and χ is the set of clause nodes. An edge indicates that some node in the lower node set is less than some node in the upper node set. The dotted lines divide the nodes into sections (stacks, clause section and balance section).

It is not hard to see that the number of nodes is polynomial in m and n, so the poset can be efficiently constructed from an instance of TQBF.

2.2 Strategy

We claim that White can force a win if and only if the formula is true. To show this, we need to give a strategy for White when the formula is true, and prove that it guarantees a win. We also need to show Black has a winning strategy when the formula is false.

Suppose that White and Black have an informal agreement to simulate the TQBF game in G by playing as follows. Suppose White's first move in the TQBF game is to assign x_1 to a_1. The corresponding move in G is to take a choice node in C_1, such that the assignment bit is a_1 and the other bits are arbitrary. Similarly, if Black's reply in TQBF is to assign x_2 to a_2 then he should take a choice node in C_2 with assignment bit a_2 and arbitrary right bit. White's first move destroyed either α_1 or β_1, so Black should choose the left bit of his reply to preserve the remaining black anti-cheat node in stack 1. Then White takes a node in C_3 such that the assignment bit reflects her assignment of x_3 in the TQBF game, the left bit preserves her anti-cheat node in the previous stack, and the right bit is arbitrary. This continues until White makes the final move in the TQBF game, corresponding to taking a choice node in C_{2n+1}. At this point the TQBF game ends, but there are still nodes in G; we assume the players continue under optimal play.

Assuming both players stick to the agreement, we claim (and will eventually prove) that the winner of the TQBF game is also the winner of G (under optimal play) and therefore deciding the winner of G tells us whether ϕ is true. This is complicated by the fact that players may *cheat* by taking the wrong nodes. Our goal is to show that the winner of the TQBF can also win G, even if the other player cheats.

A detailed discussion of strategies, with proofs, is in [8]. From this analysis we can prove our main result.

Theorem 2. *White has a winning strategy for G if and only if ϕ is true.*

The proof is in [8].

3 Generalized Col is PSPACE-Complete

Let Col be the language of Col games on uncolored general graphs where the first player has a winning strategy. Assume that the graphs are represented in some explicit manner, such as an adjacency matrix. We will show that Col is PSPACE-complete by giving a reduction from a game played on propositional formulas known to be PSPACE-complete [13]. The game, G_{pos}(POS CNF), is played on a positive CNF formula. The players take turns choosing a variable that appears in the formula. Player 1 sets variables to true, and Player 2 sets variables to false. Once all the variables have been chosen, Player 1 wins if the formula evaluates to true, and Player 2 wins if the formula evaluates to false.

Theorem 3. Col *is* PSPACE-*complete.*

Most of the rest of this section (Sect. 3) is dedicated to proving Theorem 3. The diagrams in the proof use the interpretation of Col in which the players remove vertices from the graph, tinting their neighbors so as to reserve them for the other player. Figure 2 shows this simple coloring scheme.

●	-	Only available to Black.
○	-	Only available to White.
•	-	Available to both players.

Fig. 2. Coloring scheme for Col graph.

Let G be the graph for some Col game which may already be partially colored. Assuming that vertices x and y in G are not already colored, we will let $G^{b(x)w(y)}$ denote the graph G where x has been chosen by Black and y has been chosen by White. Other game states are defined in an analogous fashion.

3.1 Preliminaries

We will first show that a slight variation of $\mathrm{G_{pos}}$(POS CNF) is also PSPACE-complete. Let G^*_{pos}(POS CNF) be identical to $\mathrm{G_{pos}}$(POS CNF) except that Player 1 sets variables to false in an attempt to make the formula false and Player 2 sets variables to be true with the goal opposite to that of Player 1. We will show that this game is also PSPACE-complete. Let X be the set of variables in the $\mathrm{G_{pos}}$(POS CNF) game and let $c_1, c_2, \ldots, c_m$ be the clauses. Let the G^*_{pos}(POS CNF) game be played with variables $X \cup \{u\}$ and formula $(c_1 \vee u) \wedge (c_2 \vee u) \wedge \ldots \wedge (c_m \vee u)$. Notice that if Player 1 does not make u false, then Player 2 will make u true and win the game. It is now easy to see that Player 1 wins the $\mathrm{G_{pos}}$(POS CNF) game iff Player 2 wins the G^*_{pos}(POS CNF) game.

3.2 Main Construction

Let $X = \{x_1, x_2, \ldots, x_n\}$ be the set of variables for the G^*_{pos}(POS CNF) game played on a CNF formula φ with clauses $C = \{c_1, c_2, \ldots, c_m\}$. We will construct a Col graph $G = (V, E)$ such that Player 1 wins the G^*_{pos}(POS CNF) game on φ iff the second player wins the Col game on G. The elements of G are as follows:

- $V = X \cup Y \cup C \cup \{z\}$.
- X is the set of variables in the G^*_{pos}(POS CNF) game.
- $Y = \{y_1, y_2, \ldots, y_n\}$ is a copy of the set of variables in the G^*_{pos}(POS CNF) game such that x_i refers to the same variable as y_i for $1 \leq i \leq n$.
- C is the set of clauses in the G^*_{pos}(POS CNF) game.
- $E = A \cup B \cup C \cup D$.
- $A = \{(y_i, x_i) \mid 1 \leq i \leq n\}$.
- $B = \{(x_i, c_j) \mid$ variable x_i appears in clause $c_j\}$.

– $C = \{(c_i, c_j) \mid 1 \leq i < j \leq m\}$.
– $D = \{(z, c_j) \mid 1 \leq j \leq m\}$.

An example of this construction on the formula is given in Fig. 3.

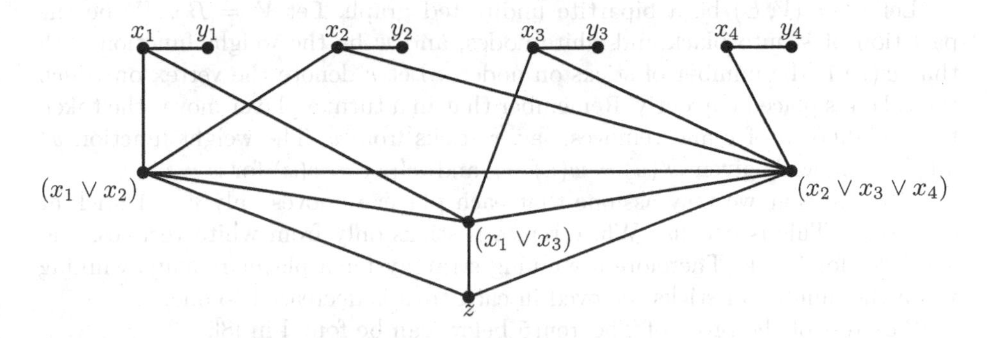

Fig. 3. Example of Col construction on $(x_1 \vee x_2) \wedge (x_1 \vee x_3) \wedge (x_2 \vee x_3 \vee x_4)$.

A complete analysis and proof of Theorem 3 are given in [8].

The complexity of Col does, in fact, stem from the vertices available to both players. First notice that the game of Col can also be thought of in the following manner. If Black chooses a vertex, delete that vertex and tint all neighboring vertices white, so that they are now only available to White. Similarly, if White chooses a vertex, delete that vertex and tint all neighboring vertices black, preserving them for Black. If a node is tinted both white and black, then it is available to neither player and can be deleted for clarity. For the purposes of displaying Col graphs in this paper, this interpretation will be used. Furthermore, this interpretation begets a natural black-white version of Col. That is, given a general graph where all nodes are initially tinted black or white, decide which player has the winning strategy.

Theorem 4. *Black-white* Col *is* $\mathsf{P}^{\mathsf{NP}[log]}$*-complete.*

See [8] for a proof of Theorem 4.

4 Black-White NimG on Undirected Graphs is in P

We consider the following problem associated with the game of NimG.

Undirected black-white NimG
Input: An undirected bipartite graph $G = (V, E)$, a weight function $w : V \longrightarrow \mathbb{N}$, and a node $v \in V$.
Question: Is v a winning position in the NimG game (G, w)?

If the vertex weights were polynomially bounded, then this game trivially reduces to undirected GEOGRAPHY and hence is in P. Here, the weights are given in *binary* and so may be exponential in the size of the input. We show the problem is still in P by providing a polynomial-time reduction to the maximum flow problem.

Let $G = (V, E)$ be a bipartite undirected graph. Let $V = B \cup W$ be the partition of V into black and white nodes, and w be the weight function such that $w(u)$ is the number of sticks on node u. Let v denote the vertex on which the token is placed currently. Remember that in a turn, a player moves the token to a neighbor u of v and removes, say, r sticks from u. The weight function w' after the move is given $w'(u) = w(u) - r$ and $w'(x) = w(x)$ for $x \neq u$.

Observe that we may assume that each player removes only $r = 1$ stick at each turn. This is because White removes sticks only from white vertices, and similarly for Black. Therefore a winning strategy for a player remains winning when the number of sticks removed in each turn is decreased to one.

The rest of the proof of Theorem 5 below can be found in [8].

Theorem 5. *Black-white* NIMG *on undirected graphs is in* P.

5 Conclusions and Open Problems

We have shown that it is PSPACE-hard to determine the winner of a black-white poset game, thus establishing a PSPACE-complete numeric game. We also show that COL played on uncolored general graphs is PSPACE-complete, which is the first game known to the authors that can only assume two very simple game theoretic values and still be PSPACE-complete. These two results cast doubt on the possibility that there is some connection between the range of values that a family of games can assume and the complexity of deciding the winner of a game in that family. An interesting open question is to definitively prove that no such connection exists. For instance, given (reasonable) game values x and y, is it possible to construct a PSPACE-complete game whose value always simplifies to either x or y? More concretely, now that we have a PSPACE-completeness result for a numeric game, can we hope to use it as a template for other numeric games with longstanding open complexity (e.g. Red-Blue HACKENBUSH)?

For NIMG, we have considered the black-white version on undirected graphs and have shown that it is decidable in P who wins even when one allows an exponential number of sticks. This is somewhat surprising given that winning gameplay may require an exponential number of moves. For all the other versions of NIMG that have been considered in the literature the exact complexity of the binary encoded versions is open. Some of these games are known to be PSPACE-hard, and yet we still do not know about membership in PSPACE.

Acknowledgments. We would like to thank an anonymous referee for pointing out that one-sided games, including directed bipartite GEOGRAPHY, are vacuously numeric.

References

1. Berlekamp, E.R., Conway, J.H., Guy, R.: Winning Ways for your Mathematical Plays. Academic Press, New York (1982)
2. Bouton, C.L.: Nim, a game with a complete mathematical theory. Ann. Math. **3**(1/4), 35–39 (1901)
3. Burke, K., George, O.: A PSPACE-complete graph Nim (2011). http://arxiv.org/abs/1101.1507v2
4. Cincotti, A.: Three-player Col played on trees is NP-complete. In: International MultiConference of Engineers and Computer Scientists 2009, pp. 445–447 (2009). Newswood Limited
5. Conway, J.H.: On Numbers and Games. Academic Press, New York (1976)
6. Demaine, E.D., Hearn, R.A.: Constraint logic: a uniform framework for modeling computation as games. In: 23rd Annual IEEE Conference on Computational Complexity, pp. 149–162. IEEE (2008)
7. Faenkel, A.S., Scheinerman, E.R., Ullman, D.: Undirected edge geography. Theoret. Comput. Sci. **112**, 371–381 (1993)
8. Fenner, S.A., Grier, D., Meßner, J., Schaeffer, L., Thierauf, T.: Game values and computational complexity: an analysis via black-white combinatorial games. Technical Report TR15-021, Electronic Colloquium on Computational Complexity, February 2015
9. Fukuyama, M.: A Nim game played on graphs. Theoret. Comput. Sci. **304**, 387–399 (2003)
10. Grier, D.: Deciding the winner of an arbitrary finite poset game is PSPACE-complete. In: Fomin, F.V., Freivalds, R.U., Kwiatkowska, M., Peleg, D. (eds.) ICALP 2013, Part I. LNCS, vol. 7965, pp. 497–503. Springer, Heidelberg (2013)
11. Grundy, P.M.: Mathematics and games. Eureka **2**, 6–8 (1939)
12. Lichtenstein, D., Sipser, M.: GO is polynomial-space hard. J. ACM **27**(2), 393–401 (1980)
13. Schaefer, T.J.: On the complexity of some two-person perfect-information games. J. Comput. Syst. Sci. **16**(2), 185–225 (1978)
14. Sprague, R.P.: Über mathematische Kampfspiele. Tohoku Math. J. **41**, 438–444 (1935–1936)
15. Stockman, G., Frieze, A., Vera, J.: The game of Nim on graphs: NimG (2004). http://www.aladdin.cs.cmu.edu/reu/mini_probes/2004/nim_graph.html

Online and Streaming Algorithms

Run Generation Revisited: What Goes Up May or May Not Come Down

Michael A. Bender[1], Samuel McCauley[1], Andrew McGregor[2], Shikha Singh[1], and Hoa T. Vu[2](✉)

[1] Stony Brook University, Stony Brook, NY 11794-2424, USA
{bender,smccauley,shiksingh}@cs.stonybrook.edu
[2] University of Massachusetts, Amherst, MA 01003, USA
{mcgregor,hvu}@cs.umass.edu

Abstract. We revisit the classic problem of run generation. Run generation is the first phase of external-memory sorting, where the objective is to scan through the data, reorder elements using a small buffer of size M, and output ***runs*** (contiguously sorted chunks of elements) that are as long as possible.

We develop algorithms for minimizing the total number of runs (or equivalently, maximizing the average run length) when the runs are allowed to be sorted or reverse sorted. We study the problem in the online setting, both with and without resource augmentation, and in the offline setting.

First, we analyze alternating-up-down replacement selection (runs alternate between sorted and reverse sorted), which was studied by Knuth as far back as 1963. We show that this simple policy is asymptotically optimal.

Next, we give online algorithms having smaller competitive ratios with resource augmentation. We demonstrate that performance can also be improved with a small amount of foresight. Lastly, we present algorithms tailored for "nearly sorted" inputs which are guaranteed to have sufficiently long optimal runs.

1 Introduction

External-memory sorting algorithms are tailored for data sets too large to fit in main memory. Generally, these algorithms begin their sort by bringing chunks of data into main memory, sorting within memory, and writing back out to disk in sorted sequences, called ***runs*** [7,11].

We revisit the classic problem of how to maximize the length of these runs, the ***run-generation problem***. The run-generation problem has been studied in its various guises for over 50 years [5–7,10,12,13,15].

The most well-known external-memory sorting algorithm is multi-way merge sort [1,9]. The multi-way merge sort is formalized in the ***disk-access machine***[1]

[1] The external-memory (or I/O) model applies to any two levels of the memory hierarchy.

K. Elbassioni and K. Makino (Eds.): ISAAC 2015, LNCS 9472, pp. 703–714, 2015.
DOI: 10.1007/978-3-662-48971-0_59

(***DAM***) model of Aggarwal and Vitter [1]. If M is the size of RAM and data is transferred between main memory and disk in blocks of size B, then an M/B-way merge sort has a complexity of $O\big((N/B)\log_{M/B}(N/B)\big)$ I/Os, where N is the number of elements to be sorted. This is the best possible [1].

A top-down description of multi-way merge sort follows. Divide the input into M/B subproblems, recursively sort each subproblem, and merge them together in one final scan through the input. The base case is reached when each subproblem has size $O(M)$, and therefore fits into RAM.

A bottom-up description of the algorithm starts with the base case, which is the run-generation phase. Naïvely, we can always generate runs of length M: ingest M elements into memory, sort them, write them to disk, and then repeat.

The point of run generation is to produce runs *longer* than M. After all, with typical values of N and M, we rarely need more than one or two passes over the data after the initial run-generation phase. Longer runs can mean fewer passes over the data or less memory consumption during the merge phase of the sort. Because there are few scans to begin with, even if we only do one fewer scan, the cost of a merge sort is decreased by a significant percentage.

Replacement Selection. The classic algorithm for run generation is called ***replacement selection*** [8,11]. Starting from an initially full internal-memory (or buffer), replacement selection proceeds as follows:

1. Pick the smallest element[2] at least as large as each element in the current run.
2. If no such element exists, then the run ends.
3. ***Eject*** that element, and ***ingest*** the next, so that the memory stays full.

Replacement selection can deal with input elements one at a time, even though the DAM model transfers input between RAM and disk B elements at a time. To see why, consider two additional blocks in memory, an "input block," which stores elements recently read from disk, and an "output block," which stores elements that have already been placed in a run. The algorithm can then ingest from the input block and eject to the output block one element at a time, while the blocks can be filled/emptied in chunks of size B.

Properties of Replacement Selection. It has been known for decades that when the input appears in random order, then the expected length of a run is $2M$ [7]. In [11], Knuth gives memorable intuition about this result, conceptualizing the buffer as a snowplow traveling along a circular track.

Replacement selection performs particularly well on nearly-sorted data and outputs runs much longer than M. For example, when each element in the input appears at a distance at most M from its actual rank, a single run is generated.

[2] Data structures such as heaps can identify the smallest elements in memory. But from the perspective of minimizing I/Os, this does not matter—computation is free in the DAM model.

On the other hand, replacement selection performs poorly on reverse-sorted data. It produces runs of length M, which is the worst possible.

Up-Down Replacement Selection. From the perspective of the sorting algorithm, it matters little, or not at all, whether the initially generated runs are sorted or reverse sorted.

This observation has motivated researchers to think about run generation where, each time a new run begins, the replacement-selection algorithm has a choice about whether to generate an ***up run*** or a ***down run***.

Knuth [10] analyzes the performance of replacement selection that alternates deterministically between generating up runs and down runs. He shows that for randomly generated data, this alternative policy performs *worse*, generating runs of expected length $3M/2$, instead of $2M$.

Martinez-Palau et al. [15] revive this idea in an experimental study. Their two-way-replacement-selection algorithms heuristically choose between whether the run generation should go up or down. Their experiments find that two-way replacement selection (1) is slightly worse than replacement selection for random input (in accordance with Knuth [10]) and (2) produces significantly longer runs on inputs that have mixed up-down runs and reverse-sorted inputs.

Our Contributions. The results in this paper complement these prior works. In contrast to Knuth's negative result for random inputs [10], we show that strict up-down alternation is the best possible for worst-case inputs. Moreover, we give better competitive ratios with resource augmentation, which helps explain why heuristically choosing between up and down runs based on the elements currently in memory may yield better solutions.

Up-down run generation boils down to figuring out, each time a run ends, whether the next run should be an up run or a down run. The objective is to minimize the number of runs in the output.[3] We establish the following:

1. *Analysis of Alternating-Up-Down Replacement Selection.* We prove that alternating-up-down replacement selection is 2-competitive. Furthermore, we show that this is the best possible for deterministic online algorithms.
2. *Resource Augmentation with Extra Buffer.* We analyze the effect of augmenting the buffer available to an online algorithm on its performance. We show that with a constant-factor-larger buffer, it is possible to perform better than twice optimal. Specifically, we design a deterministic online algorithm that, given a buffer of size $4M$, matches or beats any optimal algorithm with a buffer of size M. We also design a randomized online algorithm which is 7/4-competitive using a $2M$-size buffer.
3. *Resource Augmentation with Extra Visibility.* We show that performance factors can also be improved, without augmenting the buffer, if an algorithm has

[3] Note that for a given input, minimizing the number of runs is equivalent to maximizing the average length of runs.

limited foreknowledge of the input. In particular, we propose a deterministic algorithm which attains a competitive ratio of 3/2, using its regular buffer of size M, with a *lookahead* of $3M$ elements of the input (at each step).
4. *Better Bounds for Nearly Sorted Data.* We give algorithms that perform well on inputs that have some inherent sortedness. These results are reminiscent of previous literature studying sorting on inputs with "bounded disorder" [3] and adaptive sorting algorithms [4,14,16].
5. *PTAS for the Offline Problem.* We give a polynomial-time approximation scheme for the offline run-generation which guarantees a $(1+\epsilon)$-approximation[4] with a running time of $O\left(\left(\frac{1+\sqrt{5}}{2}\right)^{1/\epsilon} N \log N\right)$.

2 Up-Down Run Generation

In this section, we formalize the up-down run generation problem.

An instance of the up-down run generation problem is a stream I of N elements. The elements of I are presented to the algorithm one by one, in order. They can be stored in the memory of size M available to the algorithm, which we henceforth refer to as the ***buffer***. Each element occupies one slot of the buffer. In general, the model allows duplicate elements, although some results, particularly those in Sects. 5 and 7, do require uniqueness.

An algorithm A ***reads*** an element of I when A transfers the element from the input sequence to the buffer. An algorithm A ***writes*** an element when A ejects the element from its buffer and appends it to the ***output sequence*** S.

Every time an element is written, its slot in the buffer becomes free. Unless stated otherwise, the next element from the input takes up the freed slot. Thus, the buffer is always full, except when the end of the input is reached and there are fewer than M unwritten elements.

An algorithm can decide which element to eject from its buffer based on (a) the current contents of the buffer and (b) the last element written. The algorithm may also use $o(M)$ additional words to maintain its internal state (for example, it can store the direction of the current run). However, the algorithm cannot arbitrarily access S or I—it can only append elements to S, and access the next in-order element of I. We say the algorithm is at ***time step*** t if it has written exactly t elements.

A ***run*** is a sequence of sorted or reverse-sorted elements. The cost of the algorithm is the smallest number of runs that partition its output. Specifically, the number of runs in an output S, denoted $R(S)$, is the smallest number of mutually disjoint sequences $S_1, S_2, \ldots, S_{R(S)}$ such that each S_i is a run and $S = S_1 \circ \cdots \circ S_{R(S)}$ where $\circ$ indicates concatenation.

We let $\mathrm{OPT}(I)$ be the minimum number of runs of any possible output sequence on input I, i.e., the number of runs generated by the optimal offline algorithm. If I is clear from context, we denote this as OPT. Our goal is to give

[4] Due to space constraints, we defer some proofs to the full-version [2].

algorithms that perform well compared to OPT for every I. We call an online algorithm ***β-competitive*** if on any input, its output S satisfies $R(S) \leq \beta\text{OPT}$.

At any time step, an algorithm's ***unwritten-element sequence*** consists of the contents of the buffer, concatenated with the remaining (not yet ingested) input elements. For the sake of this definition, we assume that the elements in the buffer are stored in their arrival order (their order in the input sequence I).

Time step t is a ***decision point*** or ***decision time step*** for an algorithm A if $t = 0$ or if A finished writing a run at t. At a decision point, A needs to decide whether the next run will be increasing or decreasing.

Notation. We use $(x \nearrow y)$ to denote the increasing sequence $x, x+1, x+2, \ldots, y$ and $(x \searrow y)$ to denote the decreasing sequence $x, x-1, x-2, \ldots, y$.

Let $A = a_1, a_2, \ldots, a_k$. We use $A \oplus x$ to denote the sequence $a_1 + x, a_2 + x, \ldots a_k + x$. Similarly, we use $A \otimes x$ to denote the sequence $a_1 x, a_2 x, \ldots, a_k x$.

Let A and B be sequences. We say that A ***covers*** B if or all e, $e \in B \Rightarrow e \in A$. A ***subsequence*** of a sequence $A = a_1, \ldots, a_k$ is a sequence $B = a_{n_1}, a_{n_2}, \ldots, a_{n_\ell}$ where $1 \leq n_1 < n_2 < \ldots < n_\ell \leq k$.

3 Structural Properties

In this section, we identify structural properties about run generation and present the tools used to analyze our algorithms. We show that it is never a good idea to end a run early or to "skip over" an element (keeping it in the buffer even when it could have been added to the current run).

First, we show that adding elements to an input sequence never decreases the number of runs. Note that if S' is a subsequence of S, then $R(S') \leq R(S)$.

Lemma 1. *If I' is a subsequence of I, then* $\text{OPT}(I') \leq \text{OPT}(I)$.

A ***maximal increasing run*** is a run generated using the following rules (a ***maximal decreasing run*** is defined similarly):

1. Start with the smallest element in the buffer and always write the smallest element that is larger than the last element written.
2. End the run only when no element in the buffer can continue the run, i.e., all elements in the buffer are smaller than the last element written.

Lemma 2. *At any decision time step, a maximal increasing (decreasing) run r covers every other (non-maximal) increasing (decreasing) run r'.*

A ***proper algorithm*** is an algorithm that always writes maximal runs. We say an output is proper if it is generated by a proper algorithm. We show that there always exists an optimal proper algorithm.

Lemma 3. *For any input I, there exists a proper output S such that $R(S) =$* $\text{OPT}(I)$.

We use the following property of proper algorithms throughout the paper.

Property 1. Any proper algorithm satisfies the following two properties:

1. At each decision point, the elements of the buffer must have arrived while the previous run was being written.
2. A new element cannot be included in the current run if the last element output is larger (smaller) and the current run is increasing (decreasing).

The following observations and lemmas are used in the analysis of our algorithms.

Observation 1. *Consider algorithms A_1 and A_2 on input I. Suppose that at time step t_1 algorithm A_1 has written out all the elements that algorithm A_2 already wrote out by some previous time step t_2. Then, the unwritten-element sequence of algorithm A_1 at time step t_1 forms a subsequence of the unwritten-element sequence of algorithm A_2 at time step t_2.*

Lemma 4. *Consider a proper algorithm A. At some decision time step, A can write k runs $p_1 \circ \cdots \circ p_k$ or ℓ runs $q_1 \circ \cdots \circ q_\ell$ such that $|p_1 \circ \cdots \circ p_k| \geq |q_1 \circ \cdots \circ q_\ell|$.*

Then $p_1 \circ \cdots \circ p_k \circ p_{k+1}$, where p_{k+1} is either an up or down run, covers $q_1 \circ \cdots \circ q_\ell$; the unwritten-element sequence after A writes $p_1 \circ \cdots \circ p_{k+1}$ is a subsequence of the unwritten-element sequence after A writes $q_1 \circ \cdots \circ q_\ell$.

Proof. Since $|p_1 \circ \cdots \circ p_k| \geq |q_1 \circ \cdots \circ q_\ell|$, the set of elements that are in $q_1 \circ \cdots \circ q_\ell$ but not in $p_1 \circ \cdots \circ p_k$ have to be in the buffer when p_k ends. By Property 1, p_{k+1} will write all such elements.

The next theorem serves as a template for analyzing our algorithms. It lets us compare the output of our algorithm against that of the optimal in small *partitions*. We show that if in every partition i, an algorithm writes x_i runs that cover the first y_i runs of an optimal output (on the current unwritten-element sequence), and $x_i/y_i \leq \beta$, then the algorithm outputs no more than $\beta \cdot \text{OPT}$ runs.

Theorem 1. *Let A be an algorithm with output S. Partition S into k contiguous subsequences $S_1, S_2 \ldots S_k$. Let x_i be the number of runs in S_i. For $1 < i \leq k$, let I_i be the unwritten-element sequence after A writes S_{i-1}. Let $I_1 = I$, $I_{k+1} = \emptyset$ and $\alpha, \beta \geq 1$. For each I_i, let S'_i be the output of an optimal algorithm on I_i.*

If for all $i \leq k$, S_i covers the first y_i runs of S'_i, and $x_i/y_i \leq \beta$, then $R(S) \leq \beta \cdot \text{OPT}$. Similarly, if for all $i \leq k$, S_i covers the first y_i runs of S'_i, and $\mathbb{E}[x_i]/y_i \leq \alpha$, then $\mathbb{E}[R(S)] \leq \alpha \cdot \text{OPT}$.

Proof. Consider I'_i, the unwritten element sequence at the end of the first y runs of S'_{i-1} (we let $I'_1 = I$). We show that $\text{OPT}(I_i) \leq \text{OPT} - \sum_{j=1}^{i-1} y_i$ for all $1 \leq i \leq k$ using induction. Note that $\text{OPT}(I_1) = \text{OPT}$. Assume $\text{OPT}(I_i) \leq \text{OPT} - \sum_{j=1}^{i-1} y_i$. Since S_{i+1} covers the first y runs of S'_{i+1}, by Observation 1, I_{i+1} is a subsequence of I'_{i+1}. Then by Lemma 1, $\text{OPT}(I_{i+1}) \leq \text{OPT}(I'_{i+1})$.

For $i > 1$, $\text{OPT}(I'_{i+1}) = \text{OPT}(I_i) - y_i \leq \text{OPT} - \sum_{j=1}^{i} y_i$. Therefore, $\text{OPT}(I_{i+1}) \leq \text{OPT} - \sum_{j=1}^{i} y_i$. When $i = k$, we have $\text{OPT}(I_{k+1}) \leq \text{OPT} - \sum_{j=1}^{k} y_i$. But since I_{k+1} contains no elements, $\text{OPT}(I_{k+1}) = 0$, and we have $\sum_{j=1}^{k} y_i \leq \text{OPT}$. Since $R(S) = \sum_{j=1}^{k} x_i$, and $\sum_{i=1}^{k} x_i \leq \beta \sum_{i=1}^{k} y_i$, we have the following:

$$R(S) = \frac{\sum_{i=1}^{k} x_i}{\text{OPT}} \text{OPT} \leq \frac{\sum_{i=1}^{k} x_i}{\sum_{i=1}^{k} y_i} \text{OPT} \leq \beta \cdot \text{OPT}.$$

We also have the same in expectation, that is,

$$\mathbb{E}[R(S)] = \mathbb{E}[\sum_{i=1}^{k} x_i] \leq \alpha \sum_{i=1}^{k} y_i \leq \alpha \cdot \text{OPT}.$$

4 Up-Down Replacement Selection

We analyze *alternating-up-down replacement selection*, which deterministically alternates between writing (maximal) up and down runs. Knuth [10] showed that when the input elements arrive in a random order, alternating-up-down replacement selection performs worse than replacement selection (all up runs). We show that for deterministic online algorithms, alternating-up-down replacement selection is 2-competitive and optimal for *any* (adversarial) input.

Lemma 5. *Consider two inputs I_1 and I_2, where I_2 is a subsequence of I_1. Let S_1 and S_2 be proper outputs of I_1 and I_2 such that S_1 and S_2 have initial runs r_1 and r_2 respectively and r_1 and r_2 have the same direction. Let the unwritten-element sequence after r_1 and r_2 be I'_1 and I'_2 respectively. Then I'_2 is a subsequence of I'_1.*

Theorem 2. *Alternating up-down replacement selection is 2-competitive.*

Proof. We show that we can apply Theorem 1 to this algorithm with $\beta = 2$. In any partition that is not the last one of the output, the alternating algorithm writes a maximal up run r_u and then writes a maximal down run r_d. We must show that $r_u \circ r_d$ covers any run r_O written by a proper optimal algorithm on I_r, the unwritten element sequence at the beginning of the partition.

If r_O is an up run, then $r_O = r_u$ and thus is covered by $r_u \circ r_d$. If r_O is a down run, consider I', the unwritten-element sequence after r_u is written; I' is a subsequence of I_r. By Lemma 5 (with $I_1 = I_r$ and $I_2 = I'$), $r_u \circ r_d$ covers r_O.

In the last partition, the algorithm can write at most two runs while any optimal output must contain at least one run. Hence $x_i/y_i \leq 2$ in all partitions as required.

Theorem 3. *Let A be any online deterministic algorithm with output S_I on input I. Then there are arbitrarily long inputs I such that $R(S_I) \geq 2\text{OPT}(I)$.*

Furthermore, we show that no randomized algorithm can achieve a competitive ratio better than 3/2.

Theorem 4. *Let A be any online, randomized algorithm. Then there are arbitrarily long inputs such that* $\mathbb{E}[R(S_I)] \geq (3/2)\mathrm{OPT}(I)$.

5 Run Generation with Resource Augmentation

In this section, we consider two kinds of resource augmentation to circumvent the impossibility result on the performance of deterministic online algorithms.

- ***Extra buffer***: the algorithm's buffer is a constant factor larger (than the optimal).
- ***Extra visibility***: the algorithm has prescience—it can *see* a small number of incoming elements, but must read and write using the usual M-size buffer.

In this section, we assume that the input elements are unique, as duplicates nullify the power provided by augmentation. For example, c-visibility does not help if an input element is repeated c times.

We begin by analyzing the ***greedy algorithm*** for run generation. Greedy is a proper algorithm which looks into the future at each decision point, determines the length of the next up and down run and writes the longer run.

Greedy is not an online algorithm. However, it is central to our resource augmentation results. The idea of resource augmentation, in part, is that the algorithm can use the extra buffer or visibility to determine, at each decision point, which direction (up or down) leads to the longer next run.

We next look at some guarantees on the length of a run chosen by greedy (the ***greedy run***) and also on the run not chosen by greedy (the ***non-greedy run***).

Greedy Is Good but Not Great. First, we show that greedy is not optimal.

Lemma 6. *The greedy algorithm can be a factor of 3/2 away from optimal.*

Next, we show that all the runs written by the greedy algorithm (except the last two) are guaranteed to have length at least $5M/4$. In contrast, up-down replacement selection can have have runs of length M in the worst case.

Theorem 5. *Each greedy run except the last two has length* $\geq M + \lceil \lfloor M/2 \rfloor /2 \rceil$.

We bound how far into the future an algorithm must see to be able to determine which direction greedy would pick at a particular decision point. Intuitively, an algorithm should never have to choose between a very long up-run and a very long down-run. We formalize this idea in the following lemma.

Lemma 7. *Given an input I with no duplicate elements, let the two possible initial increasing and decreasing runs be r_1 and r_2. Then $|r_1| < 3M$ or $|r_2| < 3M$. This bound is tight; there is an input with $|r_1| = 3M$ and $|r_2| = 3M - 1$.*

Online Algorithms with Resource Augmentation. We present several online algorithms which use resource augmentation (buffer or visibility) to determine an up-down replacement selection strategy, beating the competitive ratio of 2.

Matching OPT Using $4M$-Size Buffer. We present an algorithm with $4M$-size buffer that writes no more runs than an optimal algorithm with an M-size buffer. Later on, we prove that $(4M-2)$-size is necessary even to be 3/2-competitive; thus this augmentation result is optimal up to a constant.

Consider the following deterministic algorithm with a $4M$-size buffer. The algorithm reads elements until its buffer is full. It then uses the contents of its buffer to determine, for an algorithm with buffer size M, if the maximal up run or the maximal down run would be longer. If the maximal up run is longer, the algorithm uses its full buffer (of size $4M$) to write a maximal up run; otherwise it writes a maximal down run.

Theorem 6. *Let A be the algorithm with a $4M$-size buffer described above. On any input, A writes no more runs than an optimal algorithm with M-size buffer.*

Proof Sketch. At each decision point, A determines the direction that a greedy algorithm on the same unwritten-element sequence (but with a buffer of size M) would pick. It is able to do so using its $4M$-size buffer because, by Lemma 7, we know the length of the non-greedy run is bounded by $3M$. Note that it does not need to write any elements during this step. In each partition, A writes a maximal run r in the greedy direction and thus covers the greedy run by Lemma 2. Furthermore, r covers the non-greedy run as well since all of the elements of this run must already be in A's initial buffer and hence get written out. An optimal algorithm (with M-size buffer), on the unwritten-element-sequence, has to choose between the greedy and the non-greedy run. Since A covers both the choices in one run, by Theorem 1, it is able to match or beat OPT. □

A natural question is whether resource augmentation boosts performance automatically, without using the greedy-run-simulation technique. The following lemma shows that this is not the case.

Lemma 8. *There exist inputs on which alternating up-down replacement selection with $4M$-size buffer does no better than it would with M-size buffer, that is, it produces twice the optimal number of runs.*

3/2-Competitive Using $4M$-Visibility. When we say that an algorithm has X-visibility ($X \geq M$) or $(X - M)$-lookahead, it means that the algorithm has knowledge of the next X elements of its unwritten-element-sequence.

The algorithm is only allowed to use the usual M-size buffer for reading and writing. Furthermore, the algorithm must read elements sequentially from input I, even if it sees future elements it would like to read or rearrange instead.

We present a deterministic algorithm which uses $4M$-visibility to achieve a competitive ratio of 3/2. At each decision point, similar to the algorithm in Theorem 6, we determine the direction leading to the longer (greedy) run using the $3M$-lookahead. However, unlike Theorem 6, an M-size buffer is too small to output this run. Instead, we show that it is possible to cover two runs of the optimal algorithm by writing three maximal runs—a greedy run, followed by two additional runs in the same direction and the opposite direction respectively.

Theorem 7. *Let* OPT *be the optimal number of runs given an M-size buffer on an input I with no duplicates. Then there exists an online algorithm A with an M-size buffer and $4M$-visibility such that A always outputs S satisfying $R(S) \leq (3/2)\mathrm{OPT}$.*

7/4-Competitive Using 2*M*-Size Buffer. A $2M$-size buffer is insufficient to determine the direction leading to the longer (greedy) run. Instead, suppose an algorithm picks a direction randomly, and writes a maximal run using a M-size buffer. It then uses the additional M buffer slots to simulate the opposite run.

With probability 1/2, the algorithm picks the greedy direction and can cover the first two runs of optimal (on the unwritten-element sequence) with three runs (as in Theorem 7). With probability 1/2, the algorithm picks the wrong direction. Consequently, writing four (alternating) runs cover two runs of the optimal. In expectation, this achieves a competitive ratio of $1/2(3/2) + 1/2(4/2) = 7/4$.

Theorem 8. *Let* OPT *be the optimal number of runs on input I given an M-size buffer, where I has no duplicate elements. Then there exists an online algorithm A with a $2M$-size buffer such that A always outputs S satisfying $\mathbb{E}[R(S)] \leq (7/4)\mathrm{OPT}$ and $R(S) \leq 2\mathrm{OPT}$.*

Lower Bound for Resource Augmentation. With less than $(4M - 2)$-augmentation, no deterministic online algorithm can be 3/2-competitive on all inputs. Thus, Theorems 6 and 7 are nearly tight.

Theorem 9. *With buffer size less than $(4M - 2)$, for any deterministic online algorithms A, there exists an input I such that if S is the output of A on I, then $R(S) \geq (3/2)\mathrm{OPT}$.*

6 Offline Algorithms for Run Generation

We give offline algorithms for run generation. The offline problem is the following—given the entire input, compute (using a polynomial-computation-time algorithm) the optimal strategy which, when executed by a run generation algorithm (with a buffer of size M), produces the minimum number of runs.

For any ϵ, a $(1 + \epsilon)$-approximation can be achieved by a brute force search on partitions of the output containing small number of runs. We improve the running time of this simple PTAS by pruning the suboptimal paths in this search.

Theorem 10. *There exists an offline algorithm A with output S such that $R(S) \leq (1+\epsilon)\mathrm{OPT}$. The running time of A is $O(\varphi^{1/\epsilon} N \log N)$ where $\varphi = (1+\sqrt{5})/2$.*

7 Run Generation on Nearly Sorted Input

We show that up-down replacement selection performs better on inputs with inherent sortedness. In particular, we say that an input is *c-**nearly-sorted*** if there exists a proper optimal algorithm which outputs runs of length at least cM.

Theorem 11. *There exists a randomized online algorithm A using M space in addition to its buffer such that, on any 3-nearly-sorted input I that has no duplicates, A is a $3/2$-approximation in expectation. Furthermore, A is at worst a 2-approximation regardless of its random choices.*

Theorem 12. *The greedy offline algorithm, i.e., picking the longer run at each decision point, is optimal on a 5-nearly-sorted input that contain no duplicates. The running time of the algorithm is $O(N)$.*

8 Conclusion and Open Problems

In this paper, we present an in-depth analysis of algorithms for run generation. We establish that considering both up and down runs can substantially reduce the number of runs in an external sort. The notion of up-down replacement selection has received relatively little attention since Knuth's negative result [10], until the experimental work of Martinez-Palau et al. [15].

The results in our paper complement the findings of Knuth [10] and Martinez-Palau et al. [15]. In particular, strict up-down alternation being the best possible strategy explains why heuristics for up-down run-generation lead to better performance. Moreover, our constant-factor competitive ratios with resource augmentation may guide followup heuristics and practical speed-ups.

We conclude with open problems. Can randomization help circumvent the lower bound of 2 on the competitive ratio of online algorithms? No randomized online algorithm can have a competitive ratio better than 3/2, but there is still a gap. What is the performance of the greedy offline algorithm compared to optimal? We show that greedy can be as bad as 3/2 times optimal. Is there a matching upper bound? Can we design a polynomial, exact algorithm for the offline run-generation problem? We find it intriguing that our attempts at an exact dynamic program all require maintaining too many buffer states to run in polynomial time.

Acknowledgments. We gratefully acknowledge Goetz Graefe and Harumi Kuno for introducing us to this problem and for their advice. This research was supported by NSF grants CCF 1114809, CCF 1217708, IIS 1247726, IIS 1251137, CNS 1408695, CCF 1439084, CCF 0953754, IIS 1251110, CCF 1320719, and by Google Research and Sandia National Laboratories.

References

1. Aggarwal, A., Vitter, J.S.: The input/output complexity of sorting and related problems. Commun. ACM **31**(9), 1116–1127 (1988)
2. Bender, M.A., McCauley, S., McGregor, A., Singh, S., Vu, H.T.: Run generation revisited: What goes up may or may not come down. arXiv preprint arXiv:1504.06501 (2015)
3. Chandramouli, B., Goldstein, J.: Patience is a virtue: revisiting merge and sort on modern processors. In: Proceedings International Conference on Management of Data, pp. 731–742 (2014)
4. Estivill-Castro, V., Wood, D.: A survey of adaptive sorting algorithms. ACM Comput. Surv. **24**(4), 441–476 (1992)
5. Frazer, W., Wong, C.: Sorting by natural selection. Commun. ACM **15**(10), 910–913 (1972)
6. Friend, E.H.: Sorting on electronic computer systems. J. ACM **3**(3), 134–168 (1956)
7. Gassner, B.J.: Sorting by replacement selecting. Commun. ACM **10**(2), 89–93 (1967)
8. Goetz, M.A.: Internal and tape sorting using the replacement-selection technique. Commun. ACM **6**(5), 201–206 (1963)
9. Graefe, G.: Implementing sorting in database systems. ACM Comput. Surv. **38**(3), 10 (2006)
10. Knuth, D.E.: Length of strings for a merge sort. Commun. ACM **6**(11), 685–688 (1963)
11. Knuth, D.E.: The Art of Computer Programming: Sorting and Searching. Adison-Wesley, Reading (1998)
12. Lin, Y.C.: Perfectly overlapped generation of long runs for sorting large files. J. Parallel Distrib. Comput. **19**(2), 136–142 (1993)
13. Lin, Y.C., Lai, H.Y.: Perfectly overlapped generation of long runs on a transputer array for sorting. Microprocess. Microsyst. **20**(9), 529–539 (1997)
14. Mallows, C.L.: Patience sorting. Bulletin Inst. Math. Appl. **5**(4), 375–376 (1963)
15. Martinez-Palau, X., Dominguez-Sal, D., Larriba-Pey, J.L.: Two-way replacement selection. Proc. VLDB Endow. **3**, 871–881 (2010)
16. Wikipedia: Timsort (2004). http://en.wikipedia.org/wiki/Timsort

Streaming Verification in Data Analysis

Samira Daruki[1(✉)], Justin Thaler[2], and Suresh Venkatasubramanian[1]

[1] School of Computing, University of Utah, Salt Lake City, USA
[2] Yahoo Labs, New York, USA
daruki@cs.utah.edu

Abstract. Streaming interactive proofs (SIPs) are a framework to reason about outsourced computation, where a data owner (the verifier) outsources a computation to the cloud (the prover), but wishes to verify the correctness of the solution provided by the cloud service. In this paper we present streaming interactive proofs for problems in data analysis. We present protocols for clustering and shape fitting problems, as well as an improved protocol for rectangular matrix multiplication. The latter can in turn be used to verify k *eigenvectors* of a (streamed) $n \times n$ matrix.

In general our solutions use polylogarithmic rounds of communication and polylogarithmic total communication and verifier space. For special cases (when optimality certificates can be verified easily), we present constant round protocols with similar costs. For rectangular matrix multiplication and eigenvector verification, our protocols work in the more restricted annotated data streaming model, and use sublinear (but not polylogarithmic) communication.

1 Introduction

Third party "cloud" services (from companies like Amazon, Google and Microsoft) perform intensive computational tasks on large data. Computing effort is split between a computationally weak "client" who owns the data and wishes to solve a desired task, and a "server" consisting of a cluster of compute nodes that performs the computation.

In this setting, how does a client verify that a computation has been performed correctly? The client here will have limited (streaming) access to the data, as well as limited ability to talk to the server. This motivates the study of interactive verification with extremely limited *sublinear space* (or streaming) verifiers. Such *streaming interactive proofs* (SIPs) have been developed for classic problems in streaming like frequency moment estimation and related graph problems.

The full version of this paper is available at http://arxiv.org/abs/1509.05514. This research was supported in part by the National Science Foundation under grants BIGDATA-1251049, CPS-1035565 and CCF-1115677.

K. Elbassioni and K. Makino (Eds.): ISAAC 2015, LNCS 9472, pp. 715–726, 2015.
DOI: 10.1007/978-3-662-48971-0_60

Our Contributions. We initiate a study of streaming interactive proofs for problems in data analysis. In what follows, we will refer to both SIPs and *annotated streaming protocols* which are a variant of SIPs (see Sect. 2 for more details).

Matrix Analysis. We present an annotated data streaming protocol (Sect. 3) for rectangular matrix multiplication over any field $\mathbb{F}$. Specifically, given input matrices $A \in \mathbb{F}^{k \times n}$ and $B \in \mathbb{F}^{n \times k'}$, our protocol computes their product using communication cost $k \cdot k' \cdot h \log |\mathbb{F}|$ and space cost $v \log |\mathbb{F}|$, for any desired pair of positive integers h, v satisfying $h \cdot v \geq n$. This improves on prior work [10] by a factor of k in space cost, and we prove that this tradeoff is optimal up to a factor of $\tilde{O}(\min(k, k'))$. The rectangular matrix multiplication protocol can in turn be used to verify k (approximate) eigenvectors of an $n \times n$ integer matrix A.[1]

Shape Analysis. We present a number of protocols for shape fitting and clustering problems. (i) We give *3-message* SIPs to verify a minimum enclosing ball (MEB) and the width of a point set *exactly* with $O(\log^c n)$ space and communication costs. Note that the MEB cannot be approximated to better than a constant factor by a streaming algorithm with space even polynomial in the dimension [1]. We show that the streaming hardness of the MEB problem holds even when the points are chosen from a discrete cube: this is important because our interactive proofs require discrete input (Sect. 4). (ii) We present $O(\log^c n)$ round protocols with $O(\log^c n)$ communication and verifier space for verifying optimal k-centers and k-slabs in Euclidean space (Sect. 5). (iii) We also show a 3-message protocol for verifying a 2-approximation to the k-center in a metric space by adapting the Gonzalez 2-approximation for k-center (Sect. 4).

Technical Overview. In our annotated data streaming protocol for matrix multiplication, we first observe that multiplying a $k \times n$ matrix A with an $n \times k'$ matrix B is equivalent to performing k' *matrix-vector* multiplications, one for each column of B. Rather than naively implement k' matrix-vector verification protocols [10], we exploit the fact that the k' matrix-vector multiplications are not independent, because the matrix A is held fixed in all of them.

For the k-center and k-slab problems, we must verify feasibility and optimality of a claimed solution. We verify feasibility by reducing to an instance of a range counting problem, for which a 2-message SIP exists [7]. For optimality, the prover must convince the verifier that *no* other feasible solution has lower cost. When $k = 1$, we show that there is a *sparse* witness of optimality, which the verifier can check directly using 3 messages, by reduction to range counting. For general k we instead observe that the certificate can be expressed as a sum over all solutions of potentially lower cost. Choosing a cost-based ordering of solutions converts this into a partial sum over a prefix of the ordered set of solutions. Our main tool is a way to verify such a sum in general, using $O(\log^c n)$ many messages.

[1] We cannot in general verify that the provided vectors are exact eigenvectors due to precision issues. Section 3 has details.

We note that while *core sets* are a natural witness for a property of a point set, they cannot always be computed by a streaming algorithm, nor is it clear that a claim of being a core set is easily verified. For the problems considered here, these issues preclude the use of a "simple" core set, requiring a more complex interactive protocol.

Prior Work on Streaming Verification. Chakrabarti et al. [5,6] introduced the notion of *annotations* in data streams, whereby an all-powerful prover could provide annotations to a verifier in order to complete a stream computation. Cormode et al. [11] introduced the SIP model which extends the annotated data streaming model to allow for multiple rounds of interaction between the prover and verifier. They introduced a streaming variant of the classical sum-check protocol [20], and used it to give logarithmic cost protocols for a variety of well-studied streaming problems. In subsequent works, protocols were developed in both models for graph problems and matrix-vector operations [10], *sparse streams* [4], and were implemented [9]. Most recently, Chakrabarti et al. [7] developed streaming interactive proofs of logarithmic cost that worked in $O(1)$ rounds, making use of an interactive protocol for the INDEX problem. Lower bounds on the cost of SIPs and their variants have also been studied [3,4,7,18]. These results make use of *Arthur-Merlin communication complexity* and related notions.

2 Preliminaries

Models. We work in the *streaming interactive proof* (SIP) model first proposed by Cormode et al. [11]. In this model there are two players, the prover P and verifier V. The input consists of a *stream* τ of n items from some universe. Let f be a function mapping a stream τ to any finite set $\mathcal{S}$. A k-message SIP for f works as follows. First, V and P read the input stream. During this phase, V computes some small secret state, which depends on τ and V's private randomness. Second, V and P then exchange k messages, after which V outputs a value in $\mathcal{S} \cup \{\perp\}$, where $\perp$ indicates that V is not convinced by P.

Any SIP for f must satisfy soundness and completeness. Completeness requires that there exists some prover strategy that causes the verifier to output $f(\tau)$ with probability $1 - \varepsilon_c$ for $\varepsilon_c \leq 1/3$. Soundness requires that for all prover strategies, the verifier outputs a value in $\{f(\tau), \perp\}$ with probability $1 - \varepsilon_s$ for some $\varepsilon_s \leq 1/3$. The values ε_c and ε_s are referred to as the completeness and soundness errors.[2] The *annotated data streaming model* of Chakrabarti et al. [5] essentially corresponds to one-message SIPs.[3]

Costs. In a SIP, the goal is to ensure that V uses sublinear space and that the protocol uses sublinear communication (number of bits exchanged between V

[2] All of our protocols achieve perfect completeness and soundness error $1/\text{poly}(n)$.

[3] While the original model allowed P to interleave information with the stream, most known annotated streaming protocols do not do so, and are thus 1-message SIPs.

and P) *after* stream observation. In our protocols, both V and P can execute the protocol in time quasilinear in the size of the input stream.

Input Model. We assume that each element of the input stream is a tuple (i, δ), where each i lies in a data universe $\mathcal{U}$ of size u, and $\delta \in \{+1, -1\}$. Negative values of δ model deletions. The data stream implicitly defines a frequency vector $\mathbf{a} = (a_1, \ldots, a_u)$, where a_i is the sum of all δ values associated with i in the stream.

Discretization. The protocols we employ make extensive use of finite field arithmetic. In order to apply these techniques to geometric problems, we assume that all input points are drawn from the discretized grid $\mathcal{U} = [m]^d$. The costs of our protocols will depend only logarithmically on m.

Protocols from Prior Work. We make use of three tools in our algorithms: Reed-Solomon fingerprints for testing vector equality, a two-message SIP of Chakrabarti et al. [7] for the PointQuery problem, and the streaming sum-check protocol of Cormode et al. [11]. We summarize the main properties of these protocols here: for more details, the reader is referred to the original papers, and the full version of this paper for Theorems 1 and 3.

Theorem 1 (Reed-Solomon Fingerprinting). *Suppose the input stream τ specifies two vectors $\mathbf{a}, \mathbf{a}' \in \mathbb{Z}^u$, guaranteed to satisfy $|\mathbf{a}_i|, |\mathbf{a}'_i| \leq u$ at the end of τ. There is a streaming algorithm using $O(\log u)$ space that satisfies the following properties: (i) If $\mathbf{a} = \mathbf{a}'$, then the algorithm outputs 1 with probability 1. (ii) If $\mathbf{a} \neq \mathbf{a}'$, then the algorithm outputs 0 with probability at least $1 - 1/u^2$.*

The PointQuery and RangeCount Protocols. An instance of the PointQuery problem consists of a stream of updates as described above followed by a query $q \in [u]$. The goal is to compute the coordinate $\mathbf{a}_q$. For RangeCount problem, let $(\mathcal{U}, \mathcal{R})$ be a range space and the input consist of a stream τ of elements (with size n) from the data universe $\mathcal{U}$ (with size u), followed by a range $R \in \mathcal{R}$. The goal is to verify a claim by P that $|R \cap \tau| = k$.

Theorem 2 (Chakrabarti et al. [7]). *Suppose the input to PointQuery satisfies $|\mathbf{a}_i| \leq \Delta$ at the end of the stream, for some known Δ. Then there is a two-message SIP for PointQuery on an input stream with length n, with space and communication each bounded by $O(\log u \cdot \log(\Delta + \log u))$. For RangeCount, there is a two-message SIP for RangeCount with space and communication cost bounded by $O(\log(|\mathcal{R}|) \cdot \log(n \cdot |\mathcal{R}|))$. In particular, for range spaces of bounded shatter dimension ρ, $\log|\mathcal{R}| = \rho \log n = O(\log n)$.*

Theorem 3 (Streaming Sum Check Protocol [11]). *Let g be a v-variate polynomial over $\mathbb{F}$, which may depend on the input stream τ. Denote the degree of g in variable i by $deg_i(g)$. Assume V can evaluate g at any point $\mathbf{r} \in \mathbb{F}$ with a streaming pass over τ, using $O(v \cdot \log|\mathbb{F}|)$ bits of space. There is an SIP for computing the function $F(\tau) = \sum_{\sigma \in \mathbb{F}^v} g(\sigma)$ that uses $O(v)$ messages and $O(\sum_{i=1}^{v} \deg_i(g) \cdot \log|\mathbb{F}|)$ communication, as well as $O(v \cdot \log|\mathbb{F}|)$ space.*

The GKR Protocol. Interactive proofs can be designed by algebrizing a circuit computing a function. One of the most powerful protocols of this form is due to Goldwasser et al. [14], and known as the GKR protocol. This was adapted to the streaming setting by Cormode et al. [11], yielding the following result.

Lemma 1 ([11,14]). *Let $\mathbb{F}$ be a finite field, and let $f \colon \mathbb{F}^u \to \mathbb{F}$ be a function of the entries of the frequency vector of a data stream (viewing the entries as elements of $\mathbb{F}$). Suppose that f can be computed by an $O(\log(S) \cdot \log(|\mathbb{F}|))$-space uniform arithmetic circuit $\mathcal{C}$ (over $\mathbb{F}$) of fan-in 2, size S, and depth d, with the inputs of $\mathcal{C}$ being the entries of the frequency vector. Then, assuming that $|\mathbb{F}| = \Omega(d \cdot \log S)$, f possesses an SIP requiring $O(d \cdot \log S)$ rounds. The total space cost is $O(\log u \cdot \log |\mathbb{F}|)$ and the total communication cost is $O(d \cdot \log(S) \cdot \log |\mathbb{F}|)$.*

3 Rectangular Matrix Multiplication and Eigenstructure

Eigenpair (eigenvalues and eigenvectors) computation is a key subroutine in data analysis. Eigenvalues of a streamed $n \times n$ matrix can be computed approximately [2], but there are no streaming algorithms to compute the *eigenvectors* of a matrix because of the output size.

Verifying the eigenstructure of a symmetric matrix A is more difficult than merely verifying that a claimed $(\lambda, \mathbf{v})$ is an eigenpair. This is because the prover must convince the verifier not only that each $(\lambda_i, \mathbf{v}_i)$ satisfies $A\mathbf{v} = \lambda\mathbf{v}$, but that the collection of eigenvectors together are orthogonal. Thus, the prover must prove that $VV^\top = D$ where V is the collection of eigenvectors and D is some diagonal matrix. Note however that this matrix multiplication check is *rectangular*: if we wish to verify that a collection of k eigenvectors are orthogonal, we must multiply a $k \times n$ matrix V by an $n \times k$ matrix $V^\top$.

We present an annotation protocol called MatrixMultiplication to verify such a *rectangular* matrix multiplication. Our protocol builds on the optimal annotations protocols for inner product and matrix-vector multiplication from [6] and [10]. We prove that our MatrixMultiplication protocol obtains tradeoffs between communication and space usage that are optimal up to a factor of $\tilde{O}(\min(k, k'))$.

Theorem 4. *Let A be a $k \times n$ matrix and B an $n \times k'$ matrix, both with entries in a finite field $\mathbb{F}$ of size $6n^3 \leq |\mathbb{F}| \leq 6n^4$. Let (h, v) be any pair of positive integers such that $h \cdot v \geq n$. There is a annotated data streaming protocol for computing the product matrix $C = A \cdot B$ with communication cost $O(k \cdot k' \cdot h \cdot \log n)$ bits and space cost $O(v \cdot \log n)$ bits. Moreover, any (online) annotated data streaming protocol for the problem requires the product of the space and communication costs to be at least $\Omega((k + k') \cdot n)$.*

Proof. We first recall the inner product protocol of Chakrabarti et al. [6]. Given input vectors $a, b \in \mathbb{F}^n$, the verifier in this protocol treats the n entries of a and b as a grid $[h] \times [v]$, and considers the unique bivariate polynomials $\widetilde{a}(X, Y)$ and $\widetilde{b}(X, Y)$ over $\mathbb{F}$ of degree at most h in X and v in Y satisfying $\widetilde{a}(x, y) = a(x, y)$ and $\widetilde{b}(x, y) = b(x, y)$ for all $(x, y) \in [h] \times [v]$. The verifier picks a random $r \in \mathbb{F}$, and evaluates $\tilde{a}(r, y)$ and $\tilde{b}(r, y)$ for all $y \in [v]$. As observed in [6], the verifier can compute $\tilde{a}(r, y)$ for any $y \in [v]$ in space $O(\log |\mathbb{F}|)$, with a single streaming pass over the input. Hence, the verifier's total space usage is $O(v \cdot \log |\mathbb{F}|)$. The prover then sends a univariate polynomial $s(X)$ of degree at most h, claimed to equal $g(X) = \sum_{y \in [v]} \tilde{a}(X, y) \cdot \tilde{b}(X, y)$. The verifier accepts $\sum_{x \in [h]} s(X)$ as the correct answer if and only if $s(r) = \sum_{y \in [v]} \tilde{a}(r, y) \cdot \tilde{b}(r, y)$.

Returning the matrix multiplication, let us denote the rows of A by $\mathbf{a}_1, \ldots, \mathbf{a}_k$ and the columns of B by $\mathbf{b}_1, \ldots, \mathbf{b}_k$. Notice that each entry C_{ij} of C is the inner product of $\mathbf{a}_i$ and $\mathbf{b}_j$.

The Prover's Computation. In our matrix multiplication protocol, the prover simply runs the above inner product protocol $k \cdot k'$ times, one for each entry C_{ij} of C. This requires sending $k \cdot k'$ polynomials, $s_{ij}(X)$: $(i, j) \in [k] \times [k']$, each of degree at most h. Hence, the total communication cost is $O(k \cdot k' \cdot h \cdot \log n)$.

The Verifier's Computation While Observing Entries of A. The verifier picks a random α and computes, for each $y \in [v]$, the quantity $s_y := \sum_i \tilde{a}_i(r, y)\alpha^i$. Using standard techniques [6], the verifier can compute each s_y with a single streaming pass over the entries of A, in $O(\log n)$ space. Hence, the verifier can compute all of the s_y values in total space $O(v \cdot \log n)$.

The Verifier's Computation While Observing Entries of B. For each $y \in [v]$, the verifier computes the quantity $s'_y := \sum_{j \in k'} \tilde{b}_j(r, y)\alpha^{k \cdot j}$. The reason that we define s'_y in this way is because it ensures that $s_y \cdot s'_y = \sum_{(i,j) \in [k] \times [k']} \tilde{a}_i(r, y) \cdot \tilde{b}_j(r, y)\alpha^{k \cdot j + i}$, which is just a fingerprint of the set of values $\{\tilde{a}_i(r, y) \cdot \tilde{b}_j(r, y)\}$ as (i, j) ranges over $[k] \times [k']$.

To check that all s_{ij} polynomials are as claimed, the verifier does the following. As the verifier reads the s_{ij} polynomials, she computes a fingerprint of the $s_{i,j}(r)$ values, i.e., the verifier computes $\sum_{i,j} s_{i,j}(r) \cdot \alpha^{j \cdot k + i}$. The verifier checks whether this equals $\sum_y (s_y \cdot s'_y)$. If so, the verifier is convinced that $A_{ij} = \sum_{x \in [h]} s_{ij}(x)$ for all $(i, j) \in [k] \times [k']$. If not, the verifier rejects.

Proof of Completeness. If the $s_{i,j}$ polynomials are as claimed, then:

$$\sum_{i,j \in [k] \times [k']} g_{i,j}(r) \cdot \alpha^{j \cdot k + i} = \sum_{i,j \in [k] \times [k']} \sum_{y \in [v]} \tilde{a}_i(r, y) \cdot \tilde{b}_j(r, y)\alpha^{j \cdot k + i}$$

$$= \sum_{y \in [v]} \sum_{i,j \in [k] \times [k']} \tilde{a}_i(r, y) \cdot \tilde{b}_j(r, y)\alpha^{j \cdot k + i} = \sum_{y \in [v]} s_y \cdot s'_y.$$

Proof of Soundness. If any of the $s_{i,j}$ polynomials are *not* as claimed (i.e., if $s_{ij}(X) \neq g_{ij}(X)$ as formal polynomials), then with probability at least $1 - h/|\mathbb{F}|$ over the random choice of $r \in \mathbb{F}$, it will hold that $s_{i,j}(r) \neq g_{ij}(r)$. In this event the verifier will wind up comparing the fingerprints of two different vectors, namely the $k \cdot k'$-dimensional vector whose (i, j)'th entry is $s_{i,j}(r)$, and the $k \cdot k'$-dimensional vector whose (i, j)'th entry is $\sum_{y \in [v]} \tilde{a}_i(r, y) \cdot \tilde{b}_j(r, y)$. These fingerprints will disagree with probability at least $1 - k \cdot k'/|\mathbb{F}|$. Hence, the probability that the prover convinces the verifier to accept is at most $h/|\mathbb{F}| + k \cdot k'/|\mathbb{F}|$. If $|\mathbb{F}| \geq 100 \cdot h \cdot k \cdot k'$, the soundness error will be bounded by $1/50$.

Lower Bound. Cormode et al. [10] proved a lower bound on the cost of (online) annotated data streaming protocols for *matrix-vector* multiplication (i.e., for multiplying a $k \times n$ matrix A by an $n \times 1$ matrix B). Specifically, their argument implies that if A is $k \times n$, then any protocol for multiplying A by a vector must have the product of the space and communication costs be at least $\Omega(k \cdot n)$. The claimed lower bound follows if $k > k'$ (the case of $k < k'$ is analogous).

On V's and P's Runtimes. Using Fast Fourier Transform techniques (cf. [9, Sect. 2]), the prover in the protocol of Theorem 4 can run in $O(k \cdot k' \cdot n \log n)$ total time, assuming the total number of updates to the input matrices A, B is $O(k \cdot k' \cdot n \log n)$. The verifier can run in time $O(\log n)$ per stream update. **The Eigenpair Verification Protocol.** We now show how to use Theorem 4 to verify that a claimed set of k eigenvalues and eigenvectors are indeed (approximate) eigenpairs of a given symmetric integer input matrix A. The protocol is cleanest to present assuming the entries of all of the claimed eigenvectors are integers, in which case the protocol can verify that the vectors are *exact* eigenvectors. We explain how to handle the general case at the end of the section.

The Case Where All Claimed Eigenvectors Have Integer Entries. The eigenpair verification protocol invokes MatrixMultiplication twice. In the first invocation, MatrixMultiplication is used to simultaneously verify that all claimed eigenpairs are indeed eigenpairs. Specifically, the MatrixMultiplication protocol is used to compute $C = A \cdot V$, where V is the matrix whose ith column equals the ith claimed eigenvector $\mathbf{v}_i$. The verifier use fingerprints to check that $C = V \cdot D$, where D is the diagonal matrix with entries corresponding to the claimed eigenvalues. In the second invocation, MatrixMultiplication is used to check that the claimed eigenvectors are orthogonal, by verifying that $V^\top V = D'$ for some diagonal matrix D' provided by the prover. Note that in both invocations of the MatrixMultiplication protocol, the verifier does not have the space to explicitly store the matrix V. Fortunately, storing V is not necessary, as within both invocations of the MatrixMultiplication protocol, V is treated as part of the input stream, and the MatrixMultiplication protocol does not require the verifier to store the input.

The General Case. We now sketch how to handle the general case when the entries of the claimed eigenvalues are not integers. The protocol guarantees in this general case that, for any desired error parameter ε, each claimed eigenpair $(\lambda_i, \mathbf{v}_i)$ satisfies $\|A\mathbf{v}_i - \lambda_i \mathbf{v}_i\|_2 \leq \varepsilon$. We follow the protocol of Cormode et al. [10]. Specifically, we reduce to the integer case by requiring the prover to round the entries of all claimed eigenvectors and eigenvalues to an integer multiple of ε' for some sufficiently small value ε' in such a way that the resulting eigenvectors are exactly orthogonal. It can be shown that there is some $\varepsilon' = 1/\text{poly}(n, \varepsilon^{-1})$ such that the rounding changes each entry of $A\mathbf{v}_i$ by at most ε/n^2. This ensures that the matrix V/ε' has integer entries all bounded in absolute value by $\text{poly}(n/\varepsilon)$. Hence each entry of V/ε' can be identified with an element of a finite field of size $\text{poly}(n, \varepsilon^{-1})$, and we can apply the integer matrix multiplication protocol to compute $A \cdot (V/\varepsilon')$ and $(V/\varepsilon')^\top (V/\varepsilon')$. The verifier checks that the latter result is a diagonal matrix, guaranteeing that the claimed eigenvectors are orthogonal. Given the former result, the prover can now convince the verifier that each entry of the former matrix is close enough to $(V/\varepsilon')\cdot D$ to ensure that $\|A\mathbf{v}_i - \lambda_i \mathbf{v}_i\|_2 \leq \varepsilon$.

Theorem 5. *Let A be a symmetric $n \times n$ integer matrix with entries bounded in absolute value by poly(n). Let k be an integer, let h and v be positive integers satisfying $h \cdot v \geq n$ and let $\varepsilon > 0$ be an error parameter. Then there is an annotated data streaming protocol for verifying that a collection of k eigenpairs $(\lambda_i, \mathbf{v}_i)$ are orthogonal, and each satisfies $\|A\mathbf{v}_i - \lambda_i \mathbf{v}_i\|_2 \leq \varepsilon$. The total communication cost is $O(k^2 \cdot h \cdot \log(n/\varepsilon))$ and the verifier's space cost is $O(v \cdot \log(n/\varepsilon))$.*

4 Shape Analysis in a Few Rounds

In this section, we give 3-message SIPs of polylogarithmic cost for finding an MEB and computing the width of a point set. The key here is to identify a sparse dual witness that proves optimality (or near-optimality) of the claimed (primal) solution and then check feasibility of both primal and dual solutions. We show how the verifier can perform both feasibility checks via a careful reduction to an instance of the RangeCount problem.

Verifying Minimum Enclosing Balls. Consider the Euclidean k-center problem with $k = 1$, otherwise known as the MEB: given a set of n points $P \subset \mathcal{U}$ in which $\mathcal{U} = [m]^d$, find a ball B^* of minimum radius that encloses all of them. No streaming algorithm that uses poly(d) space can approximate the MEB of a set of points to better than a factor of $\sqrt{2}$ by a coreset-based construction and $\frac{1+\sqrt{2}}{2}$ in general [1]. Also, the best streaming *multiplicative* $(1+\epsilon)$-approximation for the MEB uses $O((1/\epsilon)^{\frac{d}{2}})$ space [8].

The Protocol. The prover reads the input and sends the (claimed) minimum enclosing ball B. Our protocol reduces checking feasibility and optimality of B to carefully constructed instances of the RangeCount problem.

Checking Feasibility. We consider a new range space, in which the range set $\mathcal{B}$ is defined to consist of all balls with radius $j : j \in \{0, 1, \ldots, m^d\}$ and with centers in $[m]^d$. Notice that $|\mathcal{B}| = O(m^{2d})$. Using the protocol for RangeCount (Theorem 2), we can verify that the claimed solution B does in fact cover all points (because this will hold if and only if the range count of B equals the cardinality of the input point set $|P| = n$).

Checking Optimality. We will make use of the following well known fact about minimal enclosing balls, which was used as the main idea for developing an approximation algorithm for furthest neighbour problem, by Goel et al. [13]:

Lemma 2. *Let B^* be the minimal enclosing ball of a set of points P in $\mathbb{R}^d$. Then there exist at most $d+2$ points of P that lie on the boundary ∂B^* of B^* and contain the center of B^* in their convex hull.*

Putting it all Together. The complete 3-message MEB protocol works as follows.

1. V processes the data stream for RangeCount (with respect to $\mathcal{B}$ and P).
2. P computes the MEB B^* of P, then rounds the center c of the MEB to the nearest grid vertex. Denote this vertex by c^*. P sends c^* to V, as well as the radius r of B^*, and a subset of points $T \in P$ in which $\text{MEB}(T) = \text{MEB}(P)$. (Note that based on Lemma 2, $|T| \leq d+2$ suffices).
3. V first computes the center c of the MEB for the subset T and checks if c^* is actually the rounded value of c. Then V runs a RangeCount protocol with P to verify that the ball of radius $r+1$ and center c^* contains all of the input points. It then runs multiple copies of PointQuery to verify that the subset $|T| \leq d+2$ points provided by P are actually in the input set P.

Theorem 6. *There exists a 3-message SIP for the Minimum Enclosing Ball (MEB) problem with communication and space cost bounded by $O(d^2 \cdot \log^2 m)$.*

On V's and P's Runtimes. Assuming the distance function D under which the instance of MEB is defined satisfies mild "efficient-computability" properties, both V and P can be made to run in total time $\text{polylog}(m^d)$ per stream update in the protocol of Theorem 6. Specifically, it is enough that for any point $\mathbf{x} \in P$, there is a De-Morgan formula of size $\text{polylog}(m^d)$ that takes as input the binary representation of a ball $B \in \mathcal{B}$ and outputs 1 if and only if $\mathbf{x} \in B$. Under the same assumption on D, the prover P can be made to run in time $T + n \cdot \text{polylog}(m^d)$, where T is the time required to find the MEB of the input point set P. For details, see the full description of the PointQuery protocol of [7].

Streaming lower bounds on the grid. We note that restricting the points to a grid does not make the MEB problem easier for a streaming algorithm. We can show that lower bound for streaming MEB due to Agarwal and Sharathkumar [1] can be modified to work even if the points lie on a grid. The details of the proof can be found in the full version of this paper.

Verifying the Width of a Point Set. Let the width of a point set be the minimum distance between two parallel hyperplanes that enclose it. Like the MEB problem, the width of a point set can be approximated by a streaming algorithm using $O(1/\epsilon^{O(d)})$ space [8], without access to a prover. We present a similar protocol for verifying the width of a point set as well: details are in the full version of this paper.

Theorem 7. *Given a stream of n input points from $\mathcal{U} = [m]^d$, there is a three-message SIP for verifying the width of the input with space and communication cost bounded by $O(d^4 \cdot \log^2 m)$.*

Verifying Approximate Metric k-Centers. Using the same ideas as for the MEB, we can verify a 2-*approximation* to the metric k-center problem via the use of the $k+1$ points generated as witnesses by Gonzalez' 2-approximation algorithm [15]. More details are provided in the full version of the paper.

Theorem 8. *Let (X, d) be a metric space in which $|X| = m$. Given an input point set $|P| = n$ from (X, d), there is a streaming interactive protocol for verifying k-center clustering on P with space and communication costs bounded by $O(k + \log(|\mathcal{R}|) \cdot \log(n \cdot |\mathcal{R}|))$, in which $|\mathcal{R}| \leq m^{k+2}$.*

5 SIPs for General Clustering Problems

We present SIPs for two general clustering problems: k-center, and k-slab. Given a set of n points in $[m]^d$, a k-center is a set of k centers that minimize the maximum point-center distance. In the k-slab problem, the goal is to find k *hyperplanes* so as to minimize the maximum point-hyperplane distance.

k-Slabs. We first consider the k-slab problem. Even when $k = 2$ (and $d = 3$), this problem appears to be difficult to solve efficiently without access to a prover: in fact, it was shown that this problem does *not* admit a core set for arbitrary inputs [17]. Later, Edwards et al. [12] showed that if the input points are from $\mathcal{U} = [m]^d$ (as in our case), then there exists a coreset with size at most $(\frac{\log m}{\epsilon})^{f(d,k)}$ (exponential in dimension d), which provides a $(1 + \epsilon)$-approximation to k-slab problem. However, the k-slab problem does *not* admit a streaming algorithm to the best of our knowledge. As before, we can think of a "cluster" as described not by a single hyperplane, but as the region between two parallel hyperplanes that contain all the points in that cluster. The *width* of the cluster is the distance between the two hyperplanes. We now think of the k-slab objective as minimizing the maximum width of a cluster, a quantity we call the *width* of the k-slab.

Defining the Relevant Range Space. Each slab can be described by $d + 1$ points in $\mathcal{U} = [m]^d$ and a width parameter. A k-slab is a collection of k of such slabs. Let $\Re$ be the range space consisting set of all k-slabs. This range space has size $|\Re| = m^{kd^2+2kd}$. For any k-slab $\sigma \in \Re$, let $w(\sigma)$ denote its width. We assume a canonical ordering of the ranges $\sigma_1, \sigma_2, \ldots$, in increasing order of width (with an arbitrary ordering among ranges having the same width), as well as an effective enumeration procedure that given an index i returns the ith range in the canonical order. We assume the existence of a function $\mathcal{M} : \mathbb{R} \to \{-1, \ldots, |\Re| - 1\}$ that maps a width w to the smallest index i such that $w(\sigma_i) = w$, and to the null value -1 otherwise. The verifier can compute $\mathcal{M}$ by explicit enumeration using space for one range.

Stream Observation Phase of the SIP. Let $\tau = (p_1, p_2, \ldots, p_n)$ be the stream of input points. As the verifier sees the data points, it generates a *derived stream* τ'. For each point p_i in the actual input stream τ, V inserts into τ' all k-slabs $\sigma \in \Re$ which contain the point p_i. Since τ' is a deterministic function of τ and

P (who sees τ) can also produce τ' with no communication from V to P.[4] The frequency f_σ of the range $\sigma \in \tau'$ is the number of points that σ contains.

Proving Feasibility. After τ has passed, P supplies a candidate k-slab σ^* and claims that this has optimal width $w^* = w(\sigma^*)$. By applying the RangeCount protocol from Theorem 2 to the derived stream τ', V can check that $f^*_\sigma = n$ and is therefore feasible. This feasibility check requires only 3 messages.

Optimality. Proving optimality is more involved. Due to space constraints, we describe in the full version of paper how to use the GKR protocol to achieve this.

Protocol Costs. The total communication cost of the protocol $O(\log n \cdot \log(|\mathfrak{R}|) \cdot \log |\mathbb{F}|) = O(k \cdot d^2 \cdot \log m \cdot \log^2 n)$ bits. The total space cost is $O(\log(|\mathfrak{R}|) \cdot \log(|\mathbb{F}|)) = O(k \cdot d^2 \cdot \log m \cdot \log n)$ bits. The total number of rounds required is $O(\log n \cdot \log(|\mathfrak{R}|)) = O(k \cdot d^2 \cdot \log m \cdot \log n)$.

Theorem 9. *Given a stream of n points, there is a streaming interactive proof for computing the optimal k-slab, with space and communication bounded by* $O(k \cdot d^2 \cdot \log m \cdot \log^2 n)$. *The total number of rounds is* $O(k \cdot d^2 \cdot \log m \cdot \log n)$.

We can avoid the GKR protocol and reduce the number of rounds in Theorem 9 by a factor of $\log(n)$ using a technique introduced by Gur and Raz [16], and applied by Klauck and Prakash [19] to obtain an $O(\log |\mathfrak{R}|)$-round SIP for computing distinct items in a data stream by sacrificing perfect completeness and increasing the communication by a polylogarithmic amount.

k-Center. To verify solutions for Euclidean k-center, the relevant range space consists of unions of k balls of radius r, for all centers and radii in the grid. The size of this range space is m^{2kd}. We omit further details.

Theorem 10. *Given a stream of n input points, there is an SIP for computing the optimal k-center with space and communication bounded by* $O(k \cdot d \cdot \log m \cdot \log^2 n)$. *The total number of rounds is* $O(k \cdot d \cdot \log m \cdot \log n)$.

References

1. Agarwal, P.K., Sharathkumar, R.: Streaming algorithms for extent problems in high dimensions. In: Proceedings of the Twenty-First Annual ACM-SIAM symposium on Discrete Algorithms (SODA), pp. 1481–1489. Society for Industrial and Applied Mathematics (2010)
2. Andoni, A., et al.: Eigenvalues of a matrix in the streaming model. In: Proceedings of the Twenty-Fourth Annual ACM-SIAM Symposium on Discrete Algorithms (SODA), pp. 1729–1737. SIAM (2013)

[4] The running time increase for the mapping function and the derived stream can be avoided (as in Sect. 4) by observing that the frequency vector $f_{\mathbf{a}}$ is not arbitrary, since it tracks membership in ranges.

3. Babai, L., Frankl, P., Simon, J.: Complexity classes in communication complexity theory. In: 27th Annual Symposium on Foundations of Computer Science (FOCS), pp. 337–347. IEEE (1986)
4. Chakrabarti, A., Cormode, G., Goyal, N., Thaler, J.: Annotations for sparse data streams. In: Proceedings of the Twenty-Fifth Annual ACM-SIAM Symposium on Discrete Algorithms (SODA), pp. 687–706. SIAM (2014)
5. Chakrabarti, A., Cormode, G., McGregor, A.: Annotations in data streams. In: Albers, S., Marchetti-Spaccamela, A., Matias, Y., Nikoletseas, S., Thomas, W. (eds.) ICALP 2009, Part I. LNCS, vol. 5555, pp. 222–234. Springer, Heidelberg (2009)
6. Chakrabarti, A., Cormode, G., McGregor, A., Thaler, J.: Annotations in data streams. ACM Trans. Algorithms (TALG) **11**(1), 7 (2014)
7. Chakrabarti, A., Cormode, G., McGregor, A., Thaler, J., Venkatasubramanian, S.: On interactivity in Arthur-Merlin communication and stream computation. In: Electronic Colloquium on Computational Complexity (ECCC), vol. 20, p. 180 (2013)
8. Chan, T.M.: Faster core-set constructions and data-stream algorithms in fixed dimensions. Comput. Geom. **35**(1), 20–35 (2006)
9. Cormode, G., Mitzenmacher, M., Thaler, J.: Practical verified computation with streaming interactive proofs. In: Proceedings of the 3rd Innovations in Theoretical Computer Science Conference (ITCS), pp. 90–112. ACM (2012)
10. Cormode, G., Mitzenmacher, M., Thaler, J.: Streaming graph computations with a helpful advisor. Algorithmica **65**(2), 409–442 (2013)
11. Cormode, G., Thaler, J., Yi, K.: Verifying computations with streaming interactive proofs. Proc. VLDB Endowment **5**(1), 25–36 (2011)
12. Edwards, M., Varadarajan, K.R.: No coreset, no cry: II. In: Sarukkai, S., Sen, S. (eds.) FSTTCS 2005. LNCS, vol. 3821, pp. 107–115. Springer, Heidelberg (2005)
13. Goel, A., Indyk, P., Varadarajan, K.R.: Reductions among high dimensional proximity problems. In: Proceedings of the Twelfth Annual ACM-SIAM symposium on Discrete Algorithms (SODA), vol. 1, pp. 769–778. Citeseer (2001)
14. Goldwasser, S., Kalai, Y.T., Rothblum, G.N.: Delegating computation: interactive proofs for muggles. In: Proceedings of the Fortieth Annual ACM Symposium on Theory of Computing (STOC), pp. 113–122. ACM (2008)
15. Gonzalez, T.F.: Clustering to minimize the maximum intercluster distance. Theoret. Comput. Sci. **38**, 293–306 (1985)
16. Gur, T., Raz, R.: Arthur-Merlin streaming complexity. Inf. Comput. **243**, 145–165 (2015). 40th International Colloquium on Automata, Languages and Programming (ICALP 2013)
17. Har-Peled, S.: No, coreset, no cry. In: Lodaya, K., Mahajan, M. (eds.) FSTTCS 2004. LNCS, vol. 3328, pp. 324–335. Springer, Heidelberg (2004)
18. Klauck, H.: On Arthur-Merlin games in communication complexity. In: 26th Annual Conference on Computational Complexity (CCC), pp. 189–199. IEEE (2011)
19. Klauck, H., Prakash, V.: An improved interactive streaming algorithm for the distinct elements problem. In: Esparza, J., Fraigniaud, P., Husfeldt, T., Koutsoupias, E. (eds.) ICALP 2014. LNCS, vol. 8572, pp. 919–930. Springer, Heidelberg (2014)
20. Lund, C., Fortnow, L., Karloff, H.J., Nisan, N.: Algebraic methods for interactive proof systems. J. ACM **39**(4), 859–868 (1992)

All-Around Near-Optimal Solutions for the Online Bin Packing Problem

Shahin Kamali[1](✉) and Alejandro López-Ortiz[2]

[1] Massachusetts Institute of Technology, Cambridge, MA 02139, USA
skamali@mit.edu
[2] University of Waterloo, Waterloo, ON N2L 3G1, Canada
alopez-o@uwaterloo.ca

Abstract. In this paper we present algorithms with optimal average-case and close-to-best known worst-case performance for the classic online bin packing problem. It has long been observed that known bin packing algorithms with optimal average-case performance are not optimal in the worst-case. In particular First Fit and Best Fit have optimal asymptotic average-case ratio of 1 but a worst-case competitive ratio of 1.7. The competitive ratio can be improved to 1.691 using the Harmonic algorithm. Further variations of this algorithm can push down the competitive ratio to 1.588. However, these algorithms have poor performance on average; in particular, Harmonic algorithm has average-case ratio of 1.27. In this paper, first we introduce a simple algorithm which we term Harmonic Match. This algorithm performs as well as Best Fit on average, i.e., it has an average-case ratio of 1. Moreover, the competitive ratio of the algorithm is as good as Harmonic, i.e., it converges to 1.691 which is an improvement over Best Fit and First Fit. We also introduce a different algorithm, termed as Refined Harmonic Match, which achieves an improved competitive ratio of 1.636 while maintaining the good average-case performance of Harmonic Match and Best Fit. Our experimental evaluations show that our proposed algorithms have comparable average-case performance with Best Fit and First Fit, and this holds also for sequences that follow distributions other than the uniform distribution.

1 Introduction

An instance of the online bin packing problem is defined by a sequence $\sigma = \langle \sigma_1, \ldots, \sigma_n \rangle$ of *items* each having a *size* in the range $(0, 1]$. Items arrive one by one, and an algorithm should take an irrecoverable decision by placing each item into a bin without any knowledge about the forthcoming items. The goal is to pack items into a minimum number of bins of uniform capacity. Next Fit (NF) algorithm keeps one open bin. If an item does not fit in the open bin, it gets closed and a new bin is opened. First Fit algorithm (FF) maintains bins in the order they are opened and places each item in the first bin with enough space. If such a bin does not exist, a new bin is opened. Best Fit (BF) performs similarly to FF, except that it maintains bins in the decreasing order of their *levels*, where

K. Elbassioni and K. Makino (Eds.): ISAAC 2015, LNCS 9472, pp. 727–739, 2015.
DOI: 10.1007/978-3-662-48971-0_61

the level of a bin is the total size of items in it. An alternative approach is to partition items into a fixed number of classes and pack items of each class apart from other classes. An example is the Harmonic (HA) algorithm which defines K intervals $(1/2, 1], (1/3, 1/2], \ldots, (1/K, 1/(K-1)]$, and $(0, 1/K]$. Items with sizes in the same interval are treated separately using the Next Fit strategy.

Bin packing algorithms are usually compared through their average-case and worst-case performance. Under average-case analysis, it is assumed that item sizes are generated independently at random and follow a fixed distribution that is typically the uniform distribution over the interval $[0, 1)$. With this assumption, one can define the *asymptotic average-case performance ratio*, or simply *average ratio*, of an online algorithm $\mathbb{A}$ as $\lim_{n\to\infty} E\left[\frac{A(\sigma_{(n)})}{\text{OPT}(\sigma_{(n)})}\right]$, where $\sigma_{(n)}$ is a randomly generated sequence of length n and $A(\sigma)$ denotes the number of bins used by $\mathbb{A}$ for packing σ (the same notation is used for OPT). Next Fit has average ratio of 4/3 [5] while First Fit and Best Fit both have optimal average ratio of 1 [2]. To compare algorithms with average ratio of 1, a more precise measure of *expected waste* is defined as $E[A(\sigma_{(n)}) - s(\sigma_{(n)})]$, where $s(\sigma_{(n)})$ denotes the total size of items in $\sigma_{(n)}$. First Fit and Best Fit have expected waste of $\Theta(n^{2/3})$ and $\Theta(\sqrt{n}\lg^{3/4} n)$, respectively [16,21]. All online algorithms have expected waste of size $\Omega(\sqrt{n}\lg^{1/2} n)$ [21].

There are algorithms which are based on matching a "large" item with a "small". Throughout the paper, we call an item large if it is larger than 1/2 and small otherwise. Interval First Fit (IFF) algorithm [9] divides the unit interval into K intervals of equal length, namely $I_t = (\frac{t-1}{K}, \frac{t}{K}]$ for $t = 1, 2, \ldots, K$, where $K = 2j+1$ is an odd integer. The algorithm defines $j+1$ classes so that intervals I_τ and $I_{K-\tau}$ form class τ $(1 \le \tau \le j)$ and interval I_K forms class $j+1$. Items in each class are packed separately using a strategy similar to First Fit. Algorithm Online Match (OM) [7] also has a parameter K and declares two items as being *companions* if their sum is in the range $[1 - \frac{1}{K}, 1]$. A new bin is opened for each large item. For placing a small item x, the algorithm checks whether there is an open bin β with a large companion of x; in case there is, it places x in β and closes β. Otherwise, it packs x using the NF strategy in a separate list of bins. Matching Best Fit (MBF) algorithm is similar to Best Fit except that it closes a bin as soon as it receives the first small item. There is an online algorithm with expected waste of size $\Theta(\sqrt{n}\lg^{1/2} n)$ [22] which matches the lower bound of [21]. The above matching algorithms have promising average-case performance; however, they perform poorly in the worst case (see Table 1).

Competitive analysis is the standard worst-case measure for comparing online algorithms. Throughout the paper, by 'competitive ratio' of an online algorithm $\mathbb{A}$, we mean '*asymptotic* competitive ratio' of $\mathbb{A}$, which is defined as $inf\{r \ge 1 : \text{for some } N > 0, A(\sigma)/\text{OPT}(\sigma) \le r \text{ for all } \sigma \text{ with } \text{OPT}(\sigma) \ge N\}$. Next Fit has a competitive ratio of 2 while First Fit and Best Fit have the same ratio of 1.7 [11]. For large values of K, the competitive ratio of HA approaches to $T_\infty = \sum_{i=1}^{\infty} \frac{1}{t_i - 1}$, where $t_1 = 2$ and $t_{i+1} = t_i(t_i - 1) + 1, i \ge 1$. Members of a general framework of Super Harmonic algorithms [20] have even better competitive ratios. Similar to HA, these algorithms classify items by their sizes and pack items of the same class together. To improve over HA, a fraction of opened bins include items from different classes. These bins are opened with items of small sizes in the hopes of subsequently adding items of larger sizes. At the time of opening a bin, it is pre-determined how many items from each class should be placed in the bin, and it is guaranteed that the reserved spot is enough for any member of the class. Hence, the expected total size of items in the bin is less than 1, and the expected waste is linear to the number of opened bins. This implies that the average ratio of Super Harmonic algorithms is strictly larger than 1. Regarding the lower bound for competitive ratio of online algorithms, Balogh et al. [1] proved that no online algorithm can have a competitive ratio better than 1.54037. Table 1 includes a summary of the performance of bin packing algorithms.

In their survey of bin packing, Coffman et al. [4] state that *'All algorithms that do better than* First Fit *in the worst-case seem to do much worse in the average-case.'* In this paper, however, we show that this is not necessarily true and introduce an algorithm whose competitive ratio, average ratio, and expected wasted space are all at or near the top of each class. This also addresses a conjecture by Gu et al. [10] stated as *'Harmonic is better than* First Fit *in the worst-case performance, and* First Fit *is better than Harmonic in the average-case performance. Maybe there exists an on-line algorithm with the advantages of both First Fit and Harmonic.'*

Table 1. Average ratio, expected waste (under continuous uniform distribution), and competitive ratios for bin packing algorithms. Results in bold are our contributions.

Algorithm	Average ratio	Expected waste	Competitive ratio
Next Fit (NF)	$1.\bar{3}$ [5]	$\Omega(n)$	2
Best Fit (BF)	1 [2]	$\Theta(\sqrt{n}\lg^{3/4} n)$ [16,21]	1.7 [11]
First Fit (FF)	1 [16]	$\Theta(n^{2/3})$ [6,21]	1.7 [11]
Harmonic (HA)	1.2899 [15]	$\Omega(n)$	$\rightarrow T_\infty \approx 1.691$ [14]
Refined First Fit (RFF)	> 1	$\Omega(n)$	$1.\bar{6}\bar{6}$ [23]
Refined Harmonic (RH)	1.2824 [10]	$\Omega(n)$	1.636 [10,14]
Modified Harmonic (MH)	1.189 [17]	$\Omega(n)$	1.615 [18]
Harmonic++	> 1	$\Omega(n)$	1.588 [20]
Harmonic Match HM	**1**	$\boldsymbol{\Theta(\sqrt{n}\lg^{3/4} n)}$	$\boldsymbol{\rightarrow T_\infty \approx 1.691}$
Refined Harmonic Match (RHM)	**1**	$\boldsymbol{\Theta(\sqrt{n}\lg^{3/4} n)}$	**1.636**

1.1 Contribution

We introduce an algorithm called Harmonic Match (HM) which has a competitive ratio similar to Harmonic, i.e., approaches $T_\infty \approx 1.691$ for large values of K, where K is a parameter of the algorithm. For sequences generated uniformly and independently at random, Harmonic Match has an optimal average ratio of 1 and expected waste of $\Theta(\sqrt{n} \lg^{3/4} n)$ which is as good as Best Fit and better than First Fit. The idea behind Harmonic Match can be used in a general way to improve Super Harmonic algorithms. We illustrate this for the simplest member of this family, namely the Refined Harmonic algorithm of Lee and Lee [14]. We introduce a new algorithm called Refined Harmonic Match (RHM), which has a competitive ratio of at most 1.636. At the same time, the average ratio and expected waste of RHM are as good as those of Best Fit.

Harmonic Match and Refined Harmonic Match are easy-to-implement, and their running time is as good as Best Fit. This makes them useful in practical scenarios in which the worst-case scenarios might indeed happen. One example is the denial of service attacks in cloud [13] in which an adversary sends items (jobs or 'tenants) that form a worst-case sequence. In these cases, the advantage of RHM over Best Fit is significant from the perspective of cloud service providers. Although the analysis techniques used in this paper are straightforward, we use them to prove an important result that shows the average performance does not need to be compromised for better competitive ratios. For the bulk of this paper, we assume item sizes are distributed uniformly and independently in the interval $(0, 1]$. However, for a better picture on the average-case performance, we test them on sequences that follow other distributions. The results of our experiments suggest that Harmonic Match and Refined Harmonic Match have comparable performance with Best Fit and First Fit. At the same time, they have a considerable advantage over other members of the Harmonic family of algorithms. Due to space restrictions, many proofs have been removed. They will appear in the long version of the paper.

2 Harmonic Match Algorithm

Similarly to Harmonic algorithm, Harmonic Match has a parameter K and divides items into K classes based on their sizes. We use HM_K to refer to Harmonic Match with parameter K. The algorithm defines K pairs of intervals as follows. The i-th pair ($1 \leq i \leq K-1$) contains intervals $(\frac{1}{i+2}, \frac{1}{i+1}]$ and $(\frac{i}{i+1}, \frac{i+1}{i+2}]$. The K-th pair includes intervals $(0, \frac{1}{K+1}]$ and $(\frac{K}{K+1}, 1]$. An item x belongs to class i if the size of x lies in any of the two intervals associated with the i-th pair. Note that the intervals in HM_K are the same as Harmonic with parameter $K+1$ except that the interval $(\frac{1}{2}, 1]$ in the Harmonic algorithm is further divided into $K+1$ more intervals in Harmonic Match. This division enables "matching" large items with proportionally smaller items. The pair of intervals which form a class have the same length. This is essential for a good average-case performance for our uniform distribution on $(0, 1]$. The algorithm applies a strategy similar to Best Fit to place items inside each class. The Harmonic-type classification of items allows improvement on the competitive ratio.

The packing maintained by Harmonic Match includes two types of bins: the "mature" bins which are almost full and "normal" bins which become mature by receiving more items. For placing an item x, HM detects the class that x belongs to and applies the following strategy to place x. If x is a large item ($x > 1/2$), the algorithm opens a new bin and declares it as a normal bin. If x is small ($x \leq 1/2$), the algorithm applies the Best Fit (BF) strategy to place x in a mature bin. If there is no mature bin with enough space, the BF strategy is applied one more time to place x in a normal bin that contains the largest "companion" of x. A companion of x is a large item of the same class that fits with x in the same bin. In case x is placed in a bin (i.e., there is a normal bin with a companion of x) the selected bin is declared as a mature bin. Otherwise, the algorithm applies the Next Fit (NF) strategy to place x in a single normal bin maintained for that class; such a bin includes small items of the same class. If the bin maintained by NF does not have enough space, it is declared as a mature bin and a new NF-bin is opened.

Harmonic Match treats items of the same class in a similar way that Online Match does except that there is no restriction on the sum of the sizes of two companion items. To facilitate our analysis, we introduce the Relaxed Online Match (ROM) algorithm as a subroutine of HM. To place a large item, ROM opens a new bin. To place a small item x, it applies the Best Fit strategy to place x in an open bin with a single large item and closes the bin. If such a bin does not exists, ROM places x using the Next Fit strategy (and opens a new bin if necessary). Using ROM, we can describe the Harmonic Match algorithm in the following way. To place a small item, HM_K applies the Best Fit strategy to place it in a mature bin. Large items and the small items which do not fit in mature bins are treated using the ROM strategy along with other items of their classes. The bins which are closed by the ROM strategy are declared as mature bins.

2.1 Worst-Case Analysis

To analyze Harmonic Match, we observe that the classic Harmonic algorithm is *monotone* in the sense that removing an item does not increase the number of bins it opens.

Lemma 1. *Removing any item from an input sequence σ does not increase the number of bins used by the Harmonic algorithm for packing σ.*

Using the above lemma, we show that the number of bins used by HM_K for any sequence is no larger than that of Harmonic with parameter $K+1$ (HA_{K+1}). Informally speaking, the small items which are placed with large items in HM_K can be thought as being "removed" from the packing of Harmonic.

Lemma 2. *The number of bins used by Harmonic Match with parameter K (HM_K) to pack any sequence σ is no larger than that of Harmonic with parameter K (HA_{K+1}).*

Proof. We say a small item is *red* if it is placed in a bin with a large item in the packing of HM_K, and call it *white* otherwise. Consider a subsequence σ^- of σ in which red items are removed. We show $\text{HM}_K(\sigma) = \text{HA}_{K+1}(\sigma^-)$. Let σ_i denote the sequence formed by items of class i in HM_K $(1 \leq i \leq K)$. The number of bins opened by HM_K for σ_i is $l_i + \text{NF}(W_i)$ where l_i is the number of large items of σ_i and W_i is the sequence formed by the white items in σ_i. Let σ_i^- be a subsequence of σ_i in which red items are removed. Since small and large items are treated separately by HA_{K+1}, the number of bins used by HA_{K+1} for σ_i^- is also $l_i + \text{NF}(W_i)$, and we have $\text{HM}_K(\sigma_i) = \text{HA}_{K+1}(\sigma_i^-)$. Taking the sum over all classes, we get $\text{HM}_K(\sigma) = \text{HA}_{K+1}(\sigma^-)$. Since HA is monotone by Lemma 1, we have $\text{HA}_{K+1}(\sigma^-) \leq \text{HA}_{K+1}(\sigma)$, and $\text{HM}_K(\sigma) \leq \text{HA}_{K+1}(\sigma)$. □

For large values of K, the competitive ratio of Harmonic Match approaches $T_\infty \approx 1.691$. Indeed, the above upper bound is tight and we get the following result.

Theorem 1. *The competitive ratio of* HM_K *is equal to that of* HA_{K+1}*, i.e., it converges to* $T_\infty \approx 1.691$ *for large values of* K.

2.2 Average-Case Analysis

We study the average-case performance of the HM algorithm, assuming item sizes are distributed uniformly in the interval $(0, 1]$. Like most related work, we make use of the results related to the *up-right matching* problem. An instance of this problem includes n points generated uniformly and independently at random in a unit-square in the plane. Each point receives a $\oplus$ or $\ominus$ label with equal probability. The goal is to find a maximum matching of $\oplus$ points with $\ominus$ points so that in each pair of matched points the $\oplus$ point appears above and to the right of the $\ominus$ point. Let U_n denote the number of unmatched points in an optimal up-right matching of n points. For the expected size of U_n, it is known that $E[U_n] = \Theta(\sqrt{n} \lg^{3/4} n)$ [8,16,19,21]. Given an instance of bin packing defined by a sequence σ, one can make an instance of up-right matching as follows [12]. Each item x of size $s(x)$ in σ is plotted as a point in the unit square. The vertical coordinate of the point corresponds to the index of x in σ (scaled to fit in the square). If x is smaller than $1/2$, the point is labelled as $\oplus$ and its horizontal coordinate will be $1 - 2s(x)$ where $s(x)$ is the size of x; otherwise, the point will be $\ominus$ and its horizontal coordinate will be $2s(x) - 1$. A solution to the up-right matching instance gives a packing of σ in which the items associated with a pair of matched points are placed in the same bin. Note that the sum of the sizes of these two items is no more than the bin capacity. Also, in such a solution, each bin contains at most two items.

For our purposes, we study σ_t as a subsequence of σ which only includes items which belong to the same class in the HM algorithm. The items in σ_t are generated uniformly at random from $(\frac{1}{t+1}, \frac{1}{t}] \cup (\frac{t-1}{t}, \frac{t}{t+1}]$ where t is a positive integer. Since the two intervals have the same length, as we will describe, the items can be plotted in a similar manner on the unit square. Any bin packing algorithm which closes a bin after placing a small item can be used for the up-right matching problem. Each edge in the matching instance corresponds to a bin which includes one small and one large item. Recall that the algorithm

Matching Best Fit (MBF) is similar to Best Fit except that it closes a bin as soon as it receives an item with size smaller than or equal to $1/2$. So, MBF can be applied for the up-right matching problem. Indeed, it creates an optimal up-right matching, i.e., if we apply MBF on a sequence σ_t which is randomly generated from $(0,1]$, the number of unmatched points will be $\Theta(\sqrt{n_t}\lg^{3/4} n_t)$, where n_t is the length of σ_t [21]. We show the same result holds for the bin packing sequences in which items are taken uniformly at random from $(\frac{1}{t+1}, \frac{1}{t}] \cup (\frac{t-1}{t}, \frac{t}{t+1}]$.

Lemma 3. *For a sequence σ_t of length n_t in which item sizes are selected uniformly at random from $(\frac{1}{t+1}, \frac{1}{t}] \cup (\frac{t-1}{t}, \frac{t}{t+1}]$, we have $E[\text{MBF}(\sigma_t)] = n_t/2 + \Theta(\sqrt{n_t}\lg^{3/4} n_t)$.*

Proof. Define an instance of up-right matching as follows. Let x, with size $s(x)$, be the i-th item of σ_t $(1 \leq i \leq n_t)$. If x is small, plot a point with $\oplus$ label at position $(1-(s(x) \times t(t+1) - t), i/n_t)$; otherwise, plot a point with $\ominus$ label at position $(s(x)\times t(t+1)-(t^2-1), i/n_t)$. This way, the points will be bounded in a unit square. Since item sizes are generated uniformly at random from the two intervals and the sizes of the intervals are the same, the point locations and labels are assigned uniformly and independently at random. Hence, the number of unmatched points in the up-right matching solution by MBF is expected to be $\Theta(\sqrt{n_t}\lg^{3/4} n_t)$. The unmatched points are associated with the items in σ_t which are packed as a single item in their bins by MBF. Let sg denote the number of such items. We have $E[sg] = \Theta(\sqrt{n_t}\lg^{3/4} n_t)$. Except these sg items, other items are packed with exactly one other item in the same bin. So we have $\text{MBF}(\sigma_t) - sg = n_t/2$ which implies $E[\text{MBF}(\sigma_t)] = n_t/2 + E[sg] = n_t/2 + \Theta(\sqrt{n_t}\lg^{3/4} n_t)$. □

Recall that ROM is a subroutine of HM. The main difference between ROM and MBF is in placing small items without companions. For those, ROM applies the NF strategy while MBF opens a new bin for each item. Clearly, ROM has an advantage.

Lemma 4. *For any instance σ of the bin packing problem, the number of bins used by* ROM*to pack σ is no more than that of* MBF.

To prove the main result, we also need to show that MBF is monotone:

Lemma 5. *Removing an item does not increase the number of bins used by* MBF.

Provided with the above lemmas, we prove the main result of this section.

Theorem 2. *For packing a sequence σ of length n in which item sizes are selected uniformly at random from $(0,1]$, the expected wasted space of* HM *is $\Theta(\sqrt{n}\lg^{3/4} n)$.*

Proof. Let σ^- be a copy of σ in which the items which are placed in mature bins are removed. Let $\sigma_1^-, \ldots, \sigma_K^-$ be the subsequences of σ^- formed by items belonging to different classes of HM. We have:

$$\text{HM}(\sigma) = \sum_{t=1}^{K} \text{ROM}(\sigma_t^-) \leq \sum_{t=1}^{K} \text{MBF}(\sigma_t^-) \leq \sum_{t=1}^{K} \text{MBF}(\sigma_t)$$

The inequalities come from Lemmas 4 and 5, respectively. By Lemma 3, we have:

$$E[\text{HM}(\sigma)] \leq \sum_{t=1}^{K} \left(n_t/2 + \Theta(\sqrt{n_t}\lg^{3/4} n_t)\right) = \frac{n}{2} + \Theta(\sqrt{n}\lg^{3/4} n)$$

The last equation holds since K is a constant. The expected value of $s(\sigma)$, the total size of items in σ, is $n/2$. Consequently, for the expected waste of HM, we have the following equality which completes the proof:

$$E[\text{HM}(\sigma) - s(\sigma)] = n/2 + \Theta(\sqrt{n}\lg^{3/4} n) - n/2 = \Theta(\sqrt{n}\lg^{3/4} n)$$

□

3 Refined Harmonic Match

In this section, we introduce a slightly more complicated algorithm, called Refined Harmonic Match (RHM), which has a better competitive ratio than BF and HM while performing as well as them on average. Similar to HM, RHM classifies items based on their sizes. The classes defined for RHM are the same as those of HM_K with $K = 19$. The items which belong to class $t \geq 2$ are treated using the HM strategy. Namely, a set of mature bins are maintained. If an item fits in mature bins, it is placed there using the BF strategy; otherwise, it is placed together with similar items of its class using the ROM strategy. At the same time, the bins closed by the ROM strategy are declared as being mature. The only difference between HM and RHM in packing items of class 1, i.e., items in the range $(1/3, 2/3]$. RHM divides these items into four groups $a = (1/3, 37/96]$, $b = (37/96, 1/2]$, $c = (1/2, 59/96]$, and $d = (59/96, 2/3]$. To handle the sequences which result in the lower bound of T_∞ for competitive ratios of HA and HM, RHM designates a fraction of bins opened by items of type a to host the future c items. Note that the total size of a c item and an a item is no more than 1.

In what follows, we introduce an online algorithm called Refined Relaxed Online Match (RRM) as a subroutine of RHM that is specifically used for placing items of class 1. At each step of the algorithm, when two items of class 1 are placed in the same bin, that bin is declared to be mature and will be added to the set of mature bins maintained by the HM algorithm that packs items of other classes. RRM uses the following strategy to place an item x of class 1 ($x \in (1/3, 2/3]$). If x is a d-item, RRM opens a new bin for x. If x is a c item, the algorithms checks whether there are bins with an a item designated to be paired with a c item. In case there are, x is placed in a bin with an a item using the BF strategy; otherwise, a new bin is opened for x. For a and b items (small items of class 1), RRM uses the BF strategy to select a bin with enough space which includes a single large item (if there is such a bin). This is particularly important to guarantee a good average-case behavior. If x is a b item, the algorithms checks the bin with the highest level in which x fits; if such a bin includes a c or a b item, x is placed there. Otherwise (when there is no selected bin or when it has an a item), a new bin is opened for x. If x is an a item, the algorithm uses the BF strategy to place it into a bin with a d or c item. If no suitable bin exist, x is placed in a bin with a single a item (there is at most one such bin). If there is no such bin, a new bin is opened for x.

When a new bin is opened for an a-item, the bin will be marked to either include a c item or another a item in the future. We define A-bins as those which include two a items or a single a item designated to be paired with another a item, and define C-bins as those which include either a c item together with an a or a b item or a single a item designated to be paired with a c item in the future. RHM tries to maintain the number of A-bins as close to three times the number of C-bins as possible. Namely, when a bin is opened for an a item, if the number of A-bins is less than 3 times of C-bins, the bin is declared as an A-bin to host another a item later; otherwise, the open bin is declared as a C-bin to host a c item. This way, the number of A-bins is close to (but no more than) 3 times that of C-bins.

3.1 Worst-Case Analysis

In this section, we prove an upper bound of 1.636 for the competitive ratio of RHM. Since RHM applies HM for placing items of class $t \geq 2$, by Lemma 2, the number of bins opened by RHM for these items is no more than that of Harmonic. An analysis of the number of bins opened by the Harmonic algorithm gives the following lemma.

Lemma 6. *For the number of bins used by* RHM *to pack a sequence σ we have*

$$\text{RHM}(\sigma) \leq \text{RRM}(\sigma_{cl_1}) + n_X + \sum_{t=2}^{18} \left\lfloor \frac{n_t}{t+1} \right\rfloor + 20W'/19 + 20$$

in which σ_{cl_1} is the subsequence formed by items of class 1, n_X is the number of large items in classes other than class 1, n_t is the number of small items in class t, and W' is the total size of small items in class 19 (the last class).

Using the above lemma, we prove the following theorem.

Theorem 3. *The competitive ratio of* RHM *is at most* $373/228 < 1.636$.

To prove the theorem, in the packing of RRM for items of class 1, we define a_1-bins as those which only include one a-item designated to be paired with a c-item. We consider the following two cases and prove the theorem for each case separately.

- Case 1: There is at least one a_1-bin in the final packing.
- Case 2: There is no a_1-bin in the final packing.

Let $n_\tau(\tau \in \{a, b, c, d\})$ denote the number of items of class q in the input sequence. In both cases, we formulate the number of bins opened by RRM as a function of the number of items in each group (i.e., as a function of $n_a, n_b, n_c,$ and n_d). By definition of RRM, no c-bin and a_1-bin can exist at the same time. So, in Case 1, there is no c-bin in the packing. We can bound the number of C-bins by proving the inequality $3N_C \leq N_A + 3$ where N_C and N_A respectively denote the number of C-bins and A-bins. Using the definition of A-bins and C-bins, we show the number of bins opened by RRM is at most $n_d + 4n_a/7 + 4n_b/7 + 1$. Plugging this to Lemma 6 and applying a straightforward weighting function similar to that of Lee and Lee [14] completes the proof. In Case 2, we note that $N_A \leq 3N_C$ and use it to show the number of bins opened by RRM is at most $n_d + n_c + n_b/2 + 3n_a/7 + 2$. Applying another weighting function completes the proof. The details will appear in the long version of the paper.

3.2 Average-Case Analysis

We show that the average-case performance of RHM is as good as BF and HM. Except the following lemma, other aspects of the proof are similar to those in Sect. 2.2.

Lemma 7. *For any instance σ of the bin packing problem in which items are in the range $(1/3, 2/3]$, the number of bins used by* RRM *to pack σ is no more than that of Matching Best Fit (*MBF*).*

The key observation in the proof is that RRM uses the BF strategy to place a small item x in a bin which includes a large item. Note that small items are a and b items in the RRM algorithm. Only if such a bin does not exist, RRM deviates from the BF strategy (this is the main difference between RRM and Refined Harmonic of [14]). Given Lemma 7, a similar argument as the proof of Theorem 2 results in the following theorem.

Theorem 4. *For a sequence σ of length n in which item sizes are selected uniformly at random from $(0, 1]$, the expected wasted space of* RHM *is $\Theta(\sqrt{n} \lg^{3/4} n)$.*

4 Experimental Evaluation

The results of the previous sections indicate that HM and RHM have similar average-case performance as BF if item sizes are taken uniformly at random from the range $(0, 1]$. In this section, we expand the range of distributions beyond this distribution to further observe the performance of these algorithms. For that, we considered uniform distribution with different ranges for items sizes (ranges $(0, 1/2]$ and $(0, 1/10]$), as well as Normal and Weibull distributions with different parameters. We also considered uniform instances in which items are sorted in decreasing order of their sizes. The details about these distributions can be found in the long version of the paper. For all distributions, we computed the average number of bins used by different algorithms for packing 1000 sequences of length 100,000. For algorithms that classify items by their sizes, the number of classes K is set to 20.

We compute the *experimental average ratio* of an algorithm as the ratio between the observed expected number of bins used by the algorithm and that of Opt. We estimate the number of bins opened by Opt to be the total size of items. Figure 1 shows the bar chart for experimental average ratio of different online algorithms. It can be seen that Hm and Rhm, along with Bf and Ff, have a significant advantage over other algorithms.

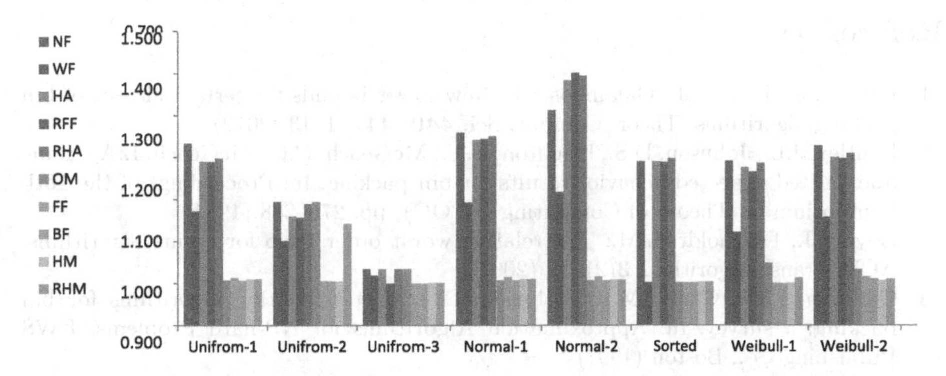

Fig. 1. The bar chart for the experimental average ratios of online bin packing algorithms. To make the results more visible, the vertical scale is changed to start at 0.9.

A difference between the packings of Hm and Rhm occurs when a number of small items of the first class (items of type a in Rhm) appear before any large item of the same class (an item of type c). In these cases, Rhm reserves some bins for subsequent large items (by declaring the bins to be C-bins). For symmetric distributions, where items of sizes x and $1-x$ appear with the same probability, it is unlikely that many small items appear before the next large item. Consequently, the average number of bins used by Hm and Rhm are the same. On the other hand, for asymmetric sequences where small items are more likely to appear, e.g., Uniform-2 with item sizes in the range $(0, 1/2]$, Hm has a visible advantage over Rhm. In these sequences, there is no reason to reserve bins for the large items since they are unlikely to appear.

5 Remarks

Hm and Rhm can be seen as variants of Harmonic and Refined Harmonic algorithms in which small and large items are carefully matched in order to improve the average-case performance. We believe that the same approach can be applied to improve the average performance of other Super Harmonic algorithms and in particular that of Harmonic++. Given the complicated nature of these algorithms, modifying them involves a detailed analysis which we leave as a future work.

It is possible to study the performance of bin packing algorithms using the relative worst order analysis [3]. Under this measure, when all items are larger than $\frac{1}{K+1}$, Harmonic with parameter K is strictly better than FF and BF by a factor of 6/5 [3]. Applying Lemma 2, when all items are larger than $\frac{1}{K+2}$, Harmonic Match with parameter K is strictly better than FF and BF. This provides another theoretical evidence for the advantage of Harmonic Match over BF and FF.

References

1. Balogh, J., Békési, J., Galambos, G.: New lower bounds for certain classes of bin packing algorithms. Theor. Comput. Sci. **440–441**, 1–13 (2012)
2. Bentley, J.L., Johnson, D.S., Leighton, F.T., McGeoch, C.C., McGeoch, L.A.: Some unexpected expected behavior results for bin packing. In: Proceedings of the 16th Symposium on Theory of Computing (STOC), pp. 279–288 (1984)
3. Boyar, J., Favrholdt, L.M.: The relative worst order ratio for online algorithms. ACM Trans. Algorithms **3**(2), 22 (2007)
4. Coffman, E.G., Garey, M.R., Johnson, D.S.: Approximation algorithms for bin packing: a survey. In: Approximation Algorithms for NP-hard Problems. PWS Publishing Co., Boston (1997)
5. Coffman, E.G., Hofri, M., So, K., Yao, A.C.C.: A stochastic model of bin packing. Inf. Control **44**, 105–115 (1980)
6. Coffman, E.G., Johnson, D.S., Shor, P.W., Weber, R.R.: Bin packing with discrete item sizes, part II: tight bounds on First Fit. Random Struct. Algorithms **10**(1–2), 69–101 (1997)
7. Coffman Jr., E.G., Lueker, G.S.: Probabilistic Analysis of Packing and Partitioning Algorithms. Wiley-Interscience Series in Discrete Mathematics and Optimization. Wiley, New York (1991)
8. Coffman, E.G., Shor, P.W.: A simple proof of the $O(\sqrt{n \log^{3/4} n})$ up-right matching bound. SIAM J. Discrete Math. **4**, 48–57 (1991)
9. Csirik, J., Galambos, G.: An $O(n)$ bin-packing algorithm for uniformly distributed data. Computing **36**(4), 313–319 (1986)
10. Gu, X., Chen, G., Xu, Y.: Deep performance analysis of refined harmonic bin packing algorithm. J. Comput. Sci. Technol. **17**, 213–218 (2002)
11. Johnson, D.S.: Near-optimal bin packing algorithms. Ph.D. thesis, MIT (1973)
12. Karp, R.M., Luby, M., Marchetti-Spaccamela, A.: Probabilistic analysis of multidimensional binpacking problems. In: Proceedings of the 16th Symposium on Theory of Computing (STOC), pp. 289–298 (1984)
13. Kousiouris, G.: Minimizing the effect of dos attacks on elastic cloud-based applications. In: Proceedings of International Conference on Cloud Computing and Services Science, pp. 622–628 (2014)
14. Lee, C.C., Lee, D.T.: A simple online bin packing algorithm. J. ACM **32**, 562–572 (1985)
15. Lee, C.C., Lee, D.T.: Robust online bin packing algorithms. Northwestern University, Technical report (1987)

16. Leighton, F.T., Shor, P.: Tight bounds for minimax grid matching with applications to the average case analysis of algorithms. Combinatorica **9**, 161–187 (1989)
17. Ramanan, P., Tsuga, K.: Average-case analysis of the modified harmonic algorithm. Algorithmica **4**, 519–533 (1989)
18. Ramanan, P.V., Brown, D.J., Lee, C.C., Lee, D.T.: On-line bin packing in linear time. J. Algorithms **10**, 305–326 (1989)
19. Rhee, W.T., Talagrand, M.: Exact bounds for the stochastic upward matching problem. Trans. AMS **307**(1), 109–125 (1988)
20. Seiden, S.S.: On the online bin packing problem. J. ACM **49**, 640–671 (2002)
21. Shor, P.W.: The average-case analysis of some online algorithms for bin packing. Combinatorica **6**, 179–200 (1986)
22. Shor, P.W.: How to pack better than Best-Fit: Tight bounds for average-case on-line bin packing. In: Proceedings of the 32nd Symposium on Foundations of Computer Science (FOCS), pp. 752–759 (1991)
23. Yao, A.C.C.: New algorithms for bin packing. J. ACM **27**, 207–227 (1980)

Serving Online Requests with Mobile Servers

Abdolhamid Ghodselahi(✉) and Fabian Kuhn

Department of Computer Science, University of Freiburg,
79110 Freiburg, Germany
{hghods,kuhn}@cs.uni-freiburg.de

Abstract. We study an online problem in which mobile servers have to be moved in order to efficiently serve at set of online requests. More formally, there is a set of n nodes and a set of k mobile servers that are placed at some of the nodes. Each node can potentially host several servers and the servers can be moved between the nodes. There are requests $1, 2, \ldots$ that are adversarially issued at nodes one at a time, where a request issued at time t needs to be served at all times $t' \geq t$. The cost for serving the requests is a function of the number of servers and requests at the different nodes. The requirements on how to serve the requests are governed by two parameters $\alpha \geq 1$ and $\beta \geq 0$. An algorithm needs to guarantee that at all times, the total service cost remains within a multiplicative factor α and an additive term β of the current optimal service cost.

We consider online algorithms for two different minimization objectives. We first consider the natural problem of minimizing the total number of server movements. We show that in this case for every k, the competitive ratio of every deterministic online algorithm needs to be at least $\Omega(n)$. Given this negative result, we then extend the minimization objective to also include the current service cost. We give almost tight bounds on the competitive ratio of the online problem where one needs to minimize the sum of the total number of movements and the current service cost. In particular, we show that at the cost of an additional additive term which is roughly linear in k, it is possible to achieve a multiplicative competitive ratio of $1 + \varepsilon$ for every constant $\varepsilon > 0$.

Keywords: Movement minimization · Competitive analysis · General cost function

1 Introduction

Consider of a company with several project teams which are located at different places. Moving a whole team to a new location is expensive, however depending on where new customers arrive, it might still be desirable to do. The cost for serving the customers at a certain location clearly depends (in a possibly non-linear way) on the number of project teams and on the number of customers

A full version of this paper is available at http://arxiv.org/abs/1404.5510 [12].

K. Elbassioni and K. Makino (Eds.): ISAAC 2015, LNCS 9472, pp. 740–751, 2015.
DOI: 10.1007/978-3-662-48971-0_62

at the location. Alternatively think of a distributed service that is offered on a large network such as the Internet. To offer the service, a provider might have a budget to place k servers in the network. The best placement of servers depends on the distribution of the users of the distributed service. As the set of users might grow (or even change arbitrarily) over time, from time to time, we might have to move some of the servers, even though migrating a whole server might be a relatively costly thing to do. These scenarios could be generally seen as a problem where servers are relatively large entities such that while they can be moved, doing this is a relatively costly operation, irrespective of, e.g., between which nodes a movement occurs. The above scenarios are applications of the abstract problem studied in this paper. The problem studied in this paper can be formally modeled as follows.

Assume two parameters α and β are given such that $\alpha \geq 1$ and max $\{\alpha - 1, \beta\} \geq 1$. There is a set V of n nodes and there are k mobile servers, where each server has to be placed at one of the nodes. Further, there are requests that arrive at the nodes in an online fashion and which need to be "permanently" served, i.e. an issued request at time t has to be served at all times $t, t+1, \ldots$. We assume that any node can potentially host an arbitrary number of servers. Formally, the cost for serving the requests at each node v, which is called *service cost* of node v, is given by a general cost function that depends on v, on the number of requests at node v, as well as on the number of servers placed at v. Generally, the more requests there are at some node, the more it costs to serve these requests. Further, if we place more servers at a given node, the cost for serving the requests at this node becomes smaller (formally defined in Sect. 2.2).[1] The requests arrive one by one and the task of an algorithm is to plan the movements of the k servers in a way to keep a *feasible configuration* of the k servers at all times. A configuration of servers is called feasible whenever the *total service cost*, that is the summation of service costs for all nodes, is upper bounded by $\alpha S_t^* + \beta$, where, S_t^* is the optimal total service cost at time t.

We consider two different objective functions. We first study a natural variant of the problem where the goal is to minimize the total number of movements. For this setting, we show that any deterministic online algorithm has a competitive ratio of at least $\Omega(n)$, independent of the value of k.

Given this negative result, we then consider an objective function where the cost at time t is the sum of the total number of movements up to time t and the total service cost at time t (shown by $Cost_t^{\mathcal{A}}$ for a given algorithm $\mathcal{A}$). We study a simple online greedy algorithm which a) only moves when it needs to move because the configuration is not feasible any more and b) always moves a server which improves the service cost as much as possible. We show that the total number of movements up to a time t of this online greedy algorithm can be upper bounded as a function of the optimal service cost S_t^* at time t. Most significantly, we show that even for $\alpha = 1$, for any $\varepsilon > 0$, as long as $\beta = \Omega(k + k/\varepsilon)$, at all

[1] The most basic cost function would incur a service cost of x whenever x requests are at a node with no server and a service cost of 0 for all requests at nodes with at least one server.

times t, the cost $Cost_t^{\mathcal{A}}$ of the greedy algorithm can be upper bounded by the cost $Cost_t^{\mathcal{O}}$ of an optimal algorithm as $Cost_t^{\mathcal{A}} \leq (1+\varepsilon)Cost_t^{\mathcal{O}} + O(\beta + k\log k)$. We also show that this result is essentially tight. In particular, an additive term which is at least linear in k is unavoidable (even for much larger multiplicative competitive ratio).

1.1 Related Work

When only considering the movement cost, the problem studied in the present paper falls into a class of movement problems introduced in [7]. In this version, the most similar of the classic problems is the k-server problem [18] or more specifically the paging problem [21] (equivalent to the k-server problem with uniform distances). In the k-server problem, every new request has to be served by moving some server to the location of the request and the only cost considered is the total movement cost. For general metric spaces, the best competitive ratios known are $2k-1$ [16] and $\tilde{O}(\log^2 k \log^3 n)$ [4]. The authors of [4] use a problem called the allocation problem (AP) to solve the k-server problem. The AP and also the results on the AP have some resemblances to the model and results in the present paper when considering the objective function based on service and movement costs. However, like k-server, in the AP the requests are served only once they arrive at the requested points while in our model the requests are permanently served and servers are not necessarily moved to the requested points.

When considering the variant of our problem where the service cost is included in the objective function, the problem can be seen as an online version of the mobile facility location problem (MFLP) with uniform distances. MFLP in general metrics was introduced in [7,11] as a movement problem. It can be seen as a generalization of the standard k-median and facility location problems [11]. The k-median and facility location problems have been widely studied in both operations research and computer science [3,5,6,8,13,15]. In [1,11], MFLP is modeled in such a way that the algorithm moves each facility and client to a point where in the final configuration, each client is at a node with some facility. The goal is to minimize the total movement cost of facilities and clients. The movement cost between the clients and the final configuration points could be interpreted as a service cost somewhat similar to what we use in this paper. Note that since in our case, requests need to be permanently served, we cannot model the service cost as a movement cost.

Classically, the cost of serving a request in the facility location problem is given by the distance from the request to the facility to which it is assigned. In a uniform metric, this corresponds to the most basic cost function that can be studied in our framework (service cost is equal to the number of requests at nodes with no servers). As described, we significantly generalize this basic service cost model. In the context of facility location, a similar approach was used in [14]. More concretely, in [14], it is assumed that the cost of a facility increases as a function of the requests it needs to serve.

There exist various natural models in which the locations of requests are not known in advance, and a solution must be built or maintained gradually over time without any knowledge about future requests like online facility location problem. The first algorithm for online facility location was introduced in [19]. For a broad discussion of models and results on online facility location problem, we refer to the survey in [10].

Finally, the problem studied in this paper has some resemblance to learning problems [2,17,20]. Somewhat similarly to expert learning algorithms where in essence, one converges to the "right set of experts", our algorithm has to converge to the "right set of nodes" to place its servers. However, in our case, the cost will usually be dominated by the total movement cost, i.e., the total cost for replacing the servers. In learning, switching to a different set of experts is usually not considered a (main) cost.

2 Problem Statement

We are given a set V of n nodes and there is a set of k servers. Further, there are requests $1, 2, \ldots$ that adversarially arrive one at a time. Moreover two parameters α and β are given such that

$$\alpha \geq 1 \quad \text{and} \quad \max\{\alpha - 1, \beta\} \geq 1. \tag{1}$$

We assume that at time $t \geq 1$, request t arrives at node $v(t) \in V$. For a node $v \in V$, let $r_{v,t}$ be the number of requests at node v after t requests have arrived, i.e., $r_{v,t} := |\{i \leq t : v(i) = v\}|$. In order to keep the *total service cost* small, an algorithm can move the servers between the nodes (if necessary, for answering one new request, we allow an algorithm to also move more than one server). However throughout the execution, each of the k servers is always placed at one of the nodes $v \in V$. We define a *configuration* of servers by integers $f_v \in \mathbb{N}_0$ for each $v \in V$ such that $\sum_{v \in V} f_v = k$. We describe such a configuration by a set of pairs as $F := \{(v, f_v) : v \in V\}$. The initial configuration is denoted by F_0.

Service Cost. We implicitly assume that if a node v has some servers, all requests at v are served by these servers. This also implies that the "assignment" of requests to servers can change over time and the service cost is not cumulative. Depending on the number of servers and the number of requests at a node $v \in V$, an algorithm has to pay some service cost to serve the requests located at v. This service cost of node v is defined by a *service cost function* σ_v such that $\sigma_v(x, y) \geq 0$ is the cost for serving y requests if there are x servers at node v. For convenience, for $t \geq 1$, we also define $\sigma_{v,t}(x) := \sigma_v(x, r_{v,t})$ to be the service cost with x servers at node v at time t. For some configuration F, we denote the total service cost at time t by $S_t(F) := \sum_{v \in V} \sigma_{v,t}(f_v) = \sum_{v \in V} \sigma_v(f_v, r_{v,t})$.

Feasible Configuration. We define a configuration F to be feasible at time t iff

$$S_t(F) < \alpha \cdot S_t^* + \beta \tag{2}$$

where S_t^* is the *optimal total service cost* at time t, i.e. $S_t^* := \min_F S_t(F)$. Note that S_t^* is not necessarily the same as the total service cost $S_t^{\mathcal{O}}$ of an optimal algorithm $\mathcal{O}$ at time t. We say that a configuration F^* is an *optimal configuration* at time t if $S_t(F^*) = S_t^*$.

Feasible Solution. For a given algorithm $\mathcal{A}$, we denote the solution at time t by $\mathcal{F}_t^{\mathcal{A}} := \{F^{\mathcal{A}}(i) : i \in [0,t]\}$, where $F^{\mathcal{A}}(t)$ is the configuration after reacting to the arrival of request t and where $F^{\mathcal{A}}(0) = F_0$. Note that for two integers $a \leq b$, $[a,b] := \{a, \ldots, b\}$ denotes the set of all integers between a and b. Further, for an integer $a \geq 1$, we use $[a]$ as a short form to denote $[a] := [1,a]$. The service cost of an algorithm $\mathcal{A}$ at time t is denoted by $S_t^{\mathcal{A}} := S_t(F^{\mathcal{A}}(t))$.

Movement Cost. We define the movement cost $M_t^{\mathcal{A}}$ of given algorithm $\mathcal{A}$ to be the total number of server movements by time t. Generally, for two feasible configurations, $F = \{(v, f_v) : v \in V\}$ and $F' = \{(v, f'_v) : v \in V\}$, we define the distance $\chi(F, F')$ between the two configurations as follows:

$$\chi(F, F') := \sum_{v \in V} \max\{0, f_v - f'_v\} = \frac{1}{2} \cdot \sum_{v \in V} |f_v - f'_v| . \tag{3}$$

The distance $\chi(F, F')$ is equal to the number of movements that are needed to get from configuration F to configuration F' (or vice versa). Based on the definition of χ, we can express the movement cost of an algorithm $\mathcal{A}$ with solution $\mathcal{F}_t^{\mathcal{A}} = \{F^{\mathcal{A}}(i) : i \in [0,t]\}$ as $M_t^{\mathcal{A}} = \sum_{i=1}^{t} \chi\left(F^{\mathcal{A}}(i-1), F^{\mathcal{A}}(i)\right)$.

2.1 Objective Functions

As described in Sect. 1, we consider two different objective functions.

Minimizing the Movement Cost. The goal is to keep the number of movements as small as possible. In other words, the cost $Cost_t^{\mathcal{A}}$ of an algorithm $\mathcal{A}$ is defined as $Cost_t^{\mathcal{A}} := M_t^{\mathcal{A}}$.

Minimizing the Combined Cost. The goal here is to minimize the overall cost of an algorithm $\mathcal{A}$, that is, we aim to keep $Cost_t^{\mathcal{A}} := S_t^{\mathcal{A}} + M_t^{\mathcal{A}}$ as small as possible.

2.2 Service Cost Function Properties

The service cost function σ has to satisfy a number of natural properties. First of all, for every $v \in V$, $\sigma_v(x, y)$ has to be monotonically decreasing in the number of servers x that are placed at node v and monotonically increasing in the number of requests y at v.

$$\forall v \in V \, \forall x, y \in \mathbb{N}_0 : \sigma_v(x, y) \geq \sigma_v(x+1, y) \tag{4}$$
$$\forall v \in V \, \forall x, y \in \mathbb{N}_0 : \sigma_v(x, y) \leq \sigma_v(x, y+1) \tag{5}$$

Further, the effect of adding additional servers to a node v should become smaller with the number of servers (convex property in x) and it should not decrease if the number of requests gets larger. Therefore, for all $v \in V$ and all $x, y \in \mathbb{N}_0$, we have

$$\sigma_v(x, y) - \sigma_v(x+1, y) \geq \sigma_v(x+1, y) - \sigma_v(x+2, y) \tag{6}$$
$$\sigma_v(x, y) - \sigma_v(x+1, y) \leq \sigma_v(x, y+1) - \sigma_v(x+1, y+1) \tag{7}$$

In the following, whenever clear from the context, we omit the superscript $\mathcal{A}$ in the algorithm-dependent quantities defined above.

3 Contributions

The following theorem provides a lower bound for any deterministic online algorithm that solves the problem of minimizing the total number of movements as described in Sect. 2.1. We remark that this lower bound as well as the lower bound in Theorem 3 even holds for the simple (and natural) scenario, where the service cost at a node with at least 1 server is 0 and the service cost at a node with 0 servers is equal to the number of requests at that node.

Theorem 1 (Lower Bound). *Assume that we are given parameters α and β which satisfy* (1) *and assume that the objective is to minimize the number of movements. Then, for any online algorithm $\mathcal{A}$, there exist an execution and a time $t > 0$ such that the competitive ratio between the number of movements by $\mathcal{A}$ and the number of movements of an optimal offline algorithm is at least $\Omega(n)$. More precisely for all $M_t^{\mathcal{O}} > 0$ and all $\Upsilon > 0$, there is an execution such that $M_t^{\mathcal{A}} \geq \frac{n}{2} \cdot M_t^{\mathcal{O}} + \Upsilon$.*

Given the large lower bound of Theorem 1, we adapt the objective function to also include the service cost. The following Theorems 2 and 3 upper and lower bound the achievable competitive ratio in this case. In Sect. 5.1, we describe a simple, deterministic online algorithm $\mathcal{A}$ with the following properties. For two given parameters α and β, $\mathcal{A}$ guarantees that at all times $t \geq 0$, (2) is met. Algorithm $\mathcal{A}$ guarantees (2) while keeping the total movement cost small. More precisely, we prove the following main theorem.

Theorem 2 (Upper Bound). *There is a deterministic algorithm $\mathcal{A}$ such that for all times $t \geq 0$, the following statements hold.*

– *If $\alpha = 1$ and $\beta = \Omega\left(k + \frac{k}{\varepsilon}\right)$ for an abitrary $\varepsilon > 0$,*

$$Cost_t^{\mathcal{A}} \leq (1+\varepsilon) Cost_t^{\mathcal{O}} + O(\beta + k \log k).$$

- *If* $\alpha = 1$ *and* $\beta = \Omega\left(\frac{k \cdot \log k}{\log\log k}\right)$*, for every* $\varepsilon \geq \log\log k / \log^{1-\delta} k$ *and any constant* $0 < \delta \leq 1$*,*

$$Cost_t^{\mathcal{A}} \leq (1+\varepsilon)\, Cost_t^{\mathcal{O}} + O(\beta)\,.$$

We also prove an almost matching lower bound. The total cost of both online and optimal offline algorithms are bounded by functions of the optimal service cost.

Theorem 3 (Lower Bound). *Given* $\alpha \geq 1$ *and* β *satisfying* (1)*, consider any deterministic online algorithm* $\mathcal{A}$ *and assume that* $\mathcal{O}$ *is an optimal offline algorithm. Then, when considering the combined objective function, there exist an execution and a time* $t > 0$ *such that the total costs of* $\mathcal{A}$ *and* $\mathcal{O}$ *can be bounded as follows.*

- *For* $\alpha = 1$ *and* $\beta = \Omega(k/\varepsilon)$ *for any* $\varepsilon > 0$*, it holds that*

$$Cost_t^{\mathcal{A}} \geq \left(1 + \varepsilon\left(1 - \frac{1+\varepsilon}{k}\right)\right) Cost_t^{\mathcal{O}} + \Omega(\beta + k\log k).$$

- *For* $\alpha = 1$ *and* $\beta = \Omega\left(\frac{k \cdot \log k}{\log\log k}\right)$ *for every* $\varepsilon \geq \log\log k / \log^{1-\delta} k$ *and any constant* $0 < \delta \leq 1$ *we obtain*

$$Cost_t^{\mathcal{A}} \geq \left(1 + \varepsilon\left(1 - \frac{1+\varepsilon}{k}\right)\right) Cost_t^{\mathcal{O}} + \Omega\left(\frac{k \cdot \log k}{\log\log k}\right).$$

Choosing $\alpha > 1$: The results of the above theorems all hold for $\alpha = 1$, i.e., an algorithm is always forced to move to a configuration which is optimal up to the additive term β. Even if α is chosen to be larger than 1, as long as we want to guarantee a reasonably small multiplicative competitive ratio (of order $o(k)$), an additive term of order $\Omega(k)$ is unavoidable. In fact, in order to reduce the additive term to $O(k)$, α has to be chosen to be of order k^{δ} for some constant $\delta > 0$. Note that in this case, the multiplicative competitive ratio grows to at least $\alpha \gg 1$. However, it might still be desirable to choose $\alpha > 1$. In that case, it can be shown that the movement cost $M_t^{\mathcal{A}}$ of our simple greedy algorithm $\mathcal{A}$ only grows logarithmically with the optimal service cost S_t^* (where the basis of the logarithm is α). As an application, this for example allows to be $(1+\varepsilon)$-competitive for any constant $\varepsilon > 0$ against an objective function of the form $\gamma \cdot S_t^{\mathcal{A}} + M_t^{\mathcal{A}}$ even if γ is chosen of order $k^{-O(1)}$.

4 Minimizing the Number of Movements

In this section we sketch the proof of Theorem 1. For a formal proof, we refer to [12]. For a given sequence of requests Σ, we first fix $\mathcal{A}$ as any deterministic online algorithm and $\mathcal{O}$ as an optimal offline algorithm. We distinguish two cases and we construct different executions for the two cases. We define *iterations* as subsequences of requests such that $\mathcal{A}$ needs to move at least once per iteration.

The number of movements by $\mathcal{A}$ is therefore at least the number of iterations of the execution.

Case $k \leq \lfloor n/2 \rfloor$. To start, we place a large number of requests on any $k-1$ nodes that initially have servers. We choose this number large enough so that no algorithm can ever move any of these $k-1$ servers. This essentially reduces the problem to $k=1$ and $n-k+1$ nodes. To bound the number of movements by $\mathcal{O}$, we then consider intervals of $n-k$ iterations such that $\mathcal{A}$ is forced to move in each iteration. During each interval, the requests are distributed in such a way that at the beginning of the i-th iteration of the interval there are at least $n-k-i+1$ nodes such that if $\mathcal{O}$ places a server on one of them, then (2) remains satisfied within the whole interval. Hence, $\mathcal{O}$ moves at most once per each interval.

Case $k > \lfloor n/2 \rfloor$. In this case there is some resemblance between the constructed execution and lower bound constructions for the paging problem. For simplicity assume that there are $n = k+1$ nodes (we let requests arrive at only $k+1$ nodes). At the beginning of each iteration we locate a sufficiently large number of requests on the node without any server of $\mathcal{A}$ such that (2) is violated. Thus, while in each iteration, $\mathcal{A}$ has to move at least one server to keep (2) satisfied, $\mathcal{O}$ only needs to move a server once in at least k iterations.

5 Minimizing Movements and Service Cost

We will now extend the objective function used in Sect. 4 by also including the service cost. We will see that this allows us to be able to compete against an optimal offline algorithm $\mathcal{O}$. In the rest of this section, first we devise a simple and natural online greedy algorithm. We then sketch the analysis of the algorithm in Sect. 5.2 and provide an almost tight lower bound in Sect. 5.3. For formal proofs, we refer to the full version of the paper [12].

5.1 Algorithm Description

The goal of our algorithm is two-fold. On the one hand, we have to guarantee that the service cost of the algorithm is always within some fixed bounds of the optimal service cost. On the other hand, we want to achieve this while keeping the overall movement cost low. Specifically, as we are given α and β in which (1) holds, we guarantee that at all times (2) remains satisfied. Condition (2) is maintained in the most straightforward greedy manner. Whenever after a new request arrives, (2) is not satisfied, the algorithm greedily moves servers until (2) holds again. Hence, as long as (2) does not hold, the algorithm moves a server that reduces the total service cost as much as possible. The algorithm stops moving any server as soon as the validity of (2) is restored.

Whenever the algorithm moves a server, it does a best possible move, i.e., a move that achieves the best possible service cost improvement. Thus, the algorithm always moves a server from a node where removing a server is as cheap as possible to a node where adding a server reduces the cost as much as possible. Therefore, for each movement m, we have

$$v_m^{src} \in \arg\min_{v \in V} \{\sigma_{v,\tau_m}(f_{v,m-1} - 1) - \sigma_{v,\tau_m}(f_{v,m-1})\} \text{ and } \quad (8)$$

$$v_m^{dst} \in \arg\max_{v \in V} \{\sigma_{v,\tau_m}(f_{v,m-1}) - \sigma_{v,\tau_m}(f_{v,m-1} + 1)\}, \quad (9)$$

where $\arg\min_v$ and $\arg\max_v$ denote the sets of nodes minimizing and maximizing the respective terms.

5.2 Analysis Overview

While the algorithm itself is quite simple, its analysis turns out relatively technical. We thus first describe the key steps of the analysis by discussing a simple case. We assume that the service cost at any node is equal to 0 if there is at least one server at the node and the service cost is equal to the number of requests at the node, otherwise. Further, we assume that we run the algorithm of 5.1 with parameters $\alpha = 1$ and $\beta = 0$, i.e. after each request arrives, the algorithm moves to a configuration with optimal service cost. Note that these parameter settings violate Condition (1) and we will therefore get a weaker bound than the one promised by Theorem 2.

First, note that in the described simple scenario, the algorithm clearly never puts more than one server to the same node. Further, whenever the algorithm moves a server from a node u to a node v, the overall service cost has to strictly decrease and thus, the number of requests at node v is larger than the number of requests at node u. Consider some point in time t and let

$$r_{\min}(t) := \min_{v \in V : f_{v,t} = 1} r_{v,t}$$

be the minimum number of requests among the nodes v with a server at time t. Hence, whenever at a time t, the algorithm moves a server from a node u to a node v, node u has at least $r_{\min}(t)$ requests and consequently, node v has at least $r_{\min(t)} + 1$ requests. Further, if at some later time $t' > t$, the server at node v is moved to some other node w, because the algorithm always removes a server from a node with as few requests as possible, we have $r_{\min}(t') \geq r_{\min}(t) + 1$. Consequently, if in some time interval $[t_1, t_2]$, there is some server that is moved more than once, we know that $r_{\min}(t_1) < r_{\min}(t_2)$. In our analysis, we partition time into phases, where the first phase starts at time 0 and where phases are maximal time intervals in which each server is moved at most once (cf. Definition 5.1 in [12]).

The above argument implies that after each phase $r_{\min}$ increases by at least one and therefore at any time t in phase p, we have $r_{\min}(t) \geq p - 1$ and at the end of phase p, we have $r_{\min}(t) \geq p$. In [12], the more general form of this statement appears in Lemma 5.1. There, γ_p is defined to be the smallest service cost improvement of any movement in phase p ($\gamma_p = 1$ in the simple case considered here), and Lemma 5.1 shows that $r_{\min}$ grows by at least γ_p in phase p. Assume that at some time t in phase p, a server is moved from a node u to a node v. Because node u already had its server at the end of phase $p-1$, we have $r_{u,t} = r_{\min}(t) \geq p-1$. Consequently, at the end of phase p, there is at least one node (the source of the last movement) that has no server and at least $p-1$ requests. The corresponding (more technical) statement in the general analysis appears in Lemma 5.3 in [12].

We will bound the total cost of the online algorithm and an optimal offline algorithm from above and below, respectively, as a function of the optimal service cost. Hence, the ratio between these two total costs provides the desired competitive factor. Our algorithm guarantees that at all times, the service cost is within fixed bounds of the optimal service cost (in the simple case here, the service cost is always equal to the optimal service cost). Knowing that there are nodes with many requests and no servers, therefore allows to lower bound the optimal service cost. In the general case, this is done by Lemmas 5.6 and 5.7 in [12]. In the simple case, considered here, as at the end of phase p, there are k nodes with at least p requests (the nodes that have servers) and there is at least one additional node with at least $p-1$ requests, we know that at the end of phase p, the optimal service cost is at least $p-1$. Consequently, the online algorithm (in the simple case) pays exactly the optimal service cost (as mentioned before, in the general case, the service cost is within fixed bounds of the optimal service cost) and at most $(p-1)k$ as movement cost. Hence, the total cost paid by online algorithm is at most a factor $k+1$ times the optimal service cost since the optimal service cost is at least $p-1$. By choosing α which is slighly larger than 1 and a larger β ($\beta \geq k$), the algorithm becomes more lazy and one can show that the difference between the number of movements of $\mathcal{A}$ and the optimal service cost becomes significantly smaller. Also note that by construction, the service cost of $\mathcal{A}$ is always at most $\alpha S_t^* + \beta \leq \alpha S_t^{\mathcal{O}} + \beta$.

When analyzing our algorithm, we mostly ignore to take into account the movement cost of an optimal offline algorithm. We only exploit the fact that by the time $\mathcal{A}$ decides to move a server for the first time, any other algorithm must also move at least one server and therefore the optimal offline cost becomes at least 1.

5.3 Lower Bound

The aim of this section is to prove our lower bound theorem stated in Sect. 3. As discussed in Sect. 3, the lower bound even holds for a natural special case where each node $v \in V$ can only have either 0 or 1 servers.

Assume that we are given parameters $\alpha \geq 1$ and β such that (1) holds and an algorithm $\mathcal{A}$ which guarantees that (2) remains satisfied at all times t. In the following, let $\mathcal{O}$ be any optimal offline algorithm. Given $\mathcal{A}$, we construct an execution in which $\mathcal{A}$ has to perform a large number of movements while the optimal service cost does not grow too much. Analogously to the analysis of the upper bound, we divide time into phases such that in each phase, $\mathcal{A}$ has to move $\Omega(k)$ servers and the optimal service cost grows as slowly as possible. For p phases, we define a sequence of integers $k/3 \geq n_1 \geq n_2 \geq \ldots n_p \geq 1$ and values $\Gamma_1 < \Gamma_2 < \cdots < \Gamma_p$. In the following, let v be a free node if v does not have a server. Roughly, at the beginning of a phase i, we choose a set N_i of n_i (ideally) free nodes and make sure that all these nodes have Γ_i requests. Note that constructing an execution means to determine where to add the request in each iteration. The value Γ_i is chosen large enough such that throughout Phase i a service cost of $n_i\Gamma_i$ is sufficiently large to force an algorithm to move. Hence, whenever there are n_i free nodes with Γ_i requests, $\mathcal{A}$ has to move at least one server to one of these nodes. For each such movement, we pick another free node that currently has less than Γ_i requests and make sure it has Γ_i requests. We proceed until there are k nodes with Γ_i requests at which point the main part of the phase ends. Except for the nodes in N_i, each of the k nodes with Γ_i requests leads to a movement of $\mathcal{A}$ and therefore, $\mathcal{A}$ has to move at least $k - n_i = \Omega(k)$ servers in Phase i. At the end of Phase i, we can guarantee that there are exactly k nodes with Γ_i requests, n_i nodes with Γ_{i-1} requests, $n_{i-1} - n_i$ nodes with Γ_{i-2} requests, etc. Assuming that for all v, $\sigma_v(x, y) = (1 - x)y$, we can then compute the optimal service cost after Phase p as $n_p\Gamma_{p-1} + \sum_{i=3}^{p}(n_{i-1} - n_i)\Gamma_{i-2}$. The service cost paid by $\mathcal{A}$ at time t can not be smaller than S_t^*. By contrast, the optimal offline algorithm moves at most $n_{i-1} - n_i + 1$ times in each Phase $i > 1$ (in the first phase it moves just once) and at most n_p at the end of the last phase to locate its servers in the optimal configuration. Therefore by the end of Phase p, $\mathcal{O}$ has to pay at most $O(n_1 + p)$ as the total movement cost. If we choose $p \geq k$, the total movement cost paid by $\mathcal{O}$ is $O(p)$ by end of Phase p, while the online algorithm has to pay $\Theta(pk)$ in total by this time. The service cost of $\mathcal{O}$ equals the optimal service cost at the end of Phase p. By choosing the values n_i appropriately, we obtain the claimed bounds.

6 Future Work

A possible way to extend the work of this paper could be to study an online version of MFLP [11] (OMFLP). In [4], it is shown that by exploiting the randomized low-stretch hierarchical tree decomposition of [9], it is possible to obtain a polylogarithmic competitive ratio for the k-server problem. Combined with the general cost functions studied in the present paper, a similar approach could work for OMFLP. On each level of the hierarchical decomposition, the cost of each subtree can potentially be modeled using a cost function similar to what we use in the present paper. Note that the lower bound of Theorem 3 already applies to OMFLP, even for a uniform underlying metric.

References

1. Ahmadian, S., Friggstad, Z., Swamy, C.: Local-search based approximation algorithms for mobile facility location problems. In: Proceedings of the 24th Symposium on Discrete Algorithms (SODA), pp. 1607–1621 (2013)
2. Arora, S., Hazan, E., Kale, S.: The multiplicative weights update method: a meta-algorithm and applications. Theory Comput. 8(1), 121–164 (2012)

3. Arya, V., Garg, N., Khandekar, R., Meyerson, A., Munagala, K., Pandit, V.: Local search heuristics for k-median and facility location problems. J. Comput. **33**(3), 544–562 (2004)
4. Bansal, N., Buchbinder, N., Madry, A., Naor, J.S.: A polylogarithmic-competitive algorithm for the k-server problem. In: Proceedings of the 52nd Symposium on Foundations of Computer Science (FOCS), pp. 267–276 (2011)
5. Byrka, J., Aardal, K.: An optimal bifactor approximation algorithm for the metric uncapacitated facility location problem. J. Comput. **39**(6), 2212–2231 (2010)
6. Charikar, M., Chekuri, C., Feder, T., Motwani, R.: Incremental clustering and dynamic information retrieval. In: Proceedings of the 29th Symposium on Theory of Computing (STOC), pp. 626–635 (1997)
7. Demaine, E.D., Hajiaghayi, M., Mahini, H., Sayedi-Roshkhar, A.S., Oveisgharan, S., Zadimoghaddam, M.: Minimizing movement. Tran. Algorithms (TALG) **5**(3), 30 (2009)
8. Drezner, Z., Hamacher, H.W.: Facility Location: Applications and Theory. Springer Science & Business Media, Heidelberg (2004)
9. Fakcharoenphol, J., Rao, S., Talwar, K.: A tight bound on approximating arbitrary metrics by tree metrics. In: Proceedings of the 35th Symposium on Theory of Computing (STOC), pp. 448–455 (2003)
10. Fotakis, D.: Online and incremental algorithms for facility location. SIGACT News **42**(1), 97–131 (2011)
11. Friggstad, Z., Salavatipour, M.R.: Minimizing movement in mobile facility location problems. Trans. Algorithms (TALG) **7**(3), 28 (2011)
12. Ghodselahi, A., Kuhn, F.: Serving online demands with movable centers. arXiv preprint arXiv:1404.5510 (2014)
13. Guha, S., Khuller, S.: Greedy strikes back: improved facility location algorithms. In: Proceedings of the 9th Symposium on Discrete Algorithms (SODA), pp. 649–657 (1998)
14. Hajiaghayi, M.T., Mahdian, M., Mirrokni, V.S.: The facility location problem with general cost functions. Networks **42**(1), 42–47 (2003)
15. Jain, K., Mahdian, M., Markakis, E., Saberi, A., Vazirani, V.V.: Greedy facility location algorithms analyzed using dual fitting with factor-revealing LP. J. ACM **50**(6), 795–824 (2003)
16. Koutsoupias, E., Papadimitriou, C.H.: On the k-server conjecture. J. ACM **42**(5), 971–983 (1995)
17. Littlestone, N., Warmuth, M.K.: The weighted majority algorithm. Inf. Comput. **108**(2), 212–261 (1994)
18. Manasse, M.S., McGeoch, L.A., Sleator, D.D.: Competitive algorithms for server problems. J. Algorithms **11**(2), 208–230 (1990)
19. Meyerson, A.: Online facility location. In: Proceedings of the 42nd Symposium on Foundations of Computer Science (FOCS), p. 426 (2001)
20. Shalev-Shwartz, S.: Online learning and online convex optimization. Found. Trends Mach. Learn. **4**(2), 107–194 (2011)
21. Sleator, D.D., Tarjan, R.E.: Amortized efficiency of list update and paging rules. Commun. ACM **28**(2), 202–208 (1985)

String and DNA Algorithms

An In-place Framework for Exact and Approximate Shortest Unique Substring Queries

Wing-Kai Hon[1], Sharma V. Thankachan[2], and Bojian Xu[3](✉)

[1] Department of Computer Science, National Tsing Hua University, Hsinchu, Taiwan
wkhon@cs.nthu.edu.tw

[2] School of Computational Science and Engineering, Georgia Institute of Technology, Atlanta, USA
sthankac@cc.gatech.edu

[3] Department of Computer Science, Eastern Washington University, Cheney, USA
bojianxu@ewu.edu

Abstract. We revisit the exact shortest unique substring (SUS) finding problem, and propose its approximate version where mismatches are allowed, due to its applications in subfields such as computational biology. We design a generic in-place framework that fits to solve both the exact and approximate k-mismatch SUS finding, using the minimum $2n$ memory words plus n bytes space, where n is the input string size. By using the in-place framework, we can find the exact and approximate k-mismatch SUS for every string position using a total of $O(n)$ and $O(n^2)$ time, respectively, regardless of the value of k. Our framework does not involve any compressed or succinct data structures and thus is practical and easy to implement.

Keywords: String pattern matching · Shortest unique substring · In-place algorithms

1 Introduction

We consider a **string** $S[1..n]$, where each character $S[i]$ is drawn from an alphabet $\Sigma = \{1, 2, \ldots, \sigma\}$. We say the character $S[i]$ **occupies** the string position i. A **substring** $S[i..j]$ of S represents $S[i]S[i+1]\ldots S[j]$ if $1 \le i \le j \le n$, and is an empty string if $i > j$. We call i the **start position** and j the **ending position** of $S[i..j]$. We say the substring $S[i..j]$ **covers** the kth position of S, if $i \le k \le j$. String $S[i'..j']$ is a **proper substring** of another string $S[i..j]$ if

Authors are listed in alphabetical order. Due to the page limit, a run example for Sect. 5 and all missing proofs and pseudocode can be found in the arXiv version of this paper.

K. Elbassioni and K. Makino (Eds.): ISAAC 2015, LNCS 9472, pp. 755–767, 2015.
DOI: 10.1007/978-3-662-48971-0_63

$i \leq i' \leq j' \leq j$ and $j' - i' < j - i$. The **length** of a non-empty substring $S[i..j]$, denoted as $|S[i..j]|$, is $j - i + 1$. We define the length of an empty string as zero.

The **Hamming distance** of two non-empty strings A and B of equal length, denoted as $H(A, B)$, is defined as the number of string positions where the characters differ. A substring $S[i..j]$ is **k-mismatch unique**, for some $k \geq 0$, if there does not exist another substring $S[i'..j']$, such that $i' \neq i$, $j - i = j' - i'$, and $H(S[i..j], S[i'..j']) \leq k$. A substring is a **$k$-mismatch repeat** if it is not k-mismatch unique.

Definition 1 (k-mismatch SUS). *For a particular string position p in S and an integer k, $0 \leq k \leq n-1$, the* k-mismatch shortest unique substring (SUS) covering position p, *denoted as SUS_p^k, is a k-mismatch unique substring $S[i..j]$, such that (1) $i \leq p \leq j$, and (2) there does not exist another k-mismatch unique substring $S[i'..j']$, such that $i' \leq p \leq j'$ and $j' - i' < j - i$.*

We call 0-mismatch SUS as **exact SUS**, and the case $k > 0$ as **approximate SUS**. For any k and p, SUS_p^k must exist, because at least S itself can be SUS_p^k, if none of its proper substrings is SUS_p^k. On the other hand, there might be multiple choices for SUS_p^k. For example, if $S = \mathtt{abcbb}$, SUS_2^0 can be either $S[1,2] = \mathtt{ab}$ or $S[2,3] = \mathtt{bc}$, and SUS_2^1 can be either $S[1..3] = \mathtt{abc}$ or $S[2..4] = \mathtt{bcb}$. Note that in Definition 1, we require $k < n$, because finding SUS_p^n is trivial: $SUS_p^n \equiv S$ for any string position p.

Problem (k-mismatch SUS Finding). Given the string S, the value of $k \geq 0$, and two empty integer arrays A and B, we want to work in the place of S, A, and B, such that, in the end of computation: (1) S does not change. (2) Each $(A[i], B[i])$ pair saves the start and ending positions of the rightmost[1] SUS_i^k, i.e., $S[A[i]..B[i]] = SUS_i^k$, using a total of $O(n)$ time for $k = 0$ and $O(n^2)$ time for any $k \geq 1$.

1.1 Prior Work and Our Contribution

Exact SUS finding was proposed and studied recently by Pei et al. [7], due to its application in locating snippets in document search, event analysis, and bioinformatics, such as finding the distinctness between closely related organisms [3], polymerase chain reaction (PCR) primer design in molecular biology, genome mapability [2], and next-generation short reads sequencing [1]. The algorithm in [7] can find all exact SUS in $O(n^2)$ time using a suffix tree of $O(n)$ space. Following their proposal, there has been a sequence of improvements [5,8] for exact SUS finding, reducing the time cost from $O(n^2)$ to $O(n)$ and alleviating the underlying data structure from suffix tree to suffix array of $O(n)$ space. Hu et al. [4] proposed an RMQ (range minimum query) technique based indexing structure, which can be constructed in $O(n)$ time and space, such that any future exact SUS covering any interval of string positions can be answered in $O(1)$ time. In this work, we make the following contributions:

[1] It is our arbitrary decision to resolve the ties by picking the rightmost choice. Our solution can also be easily modified to find the leftmost choice.

(1) We revisit the exact SUS finding problem and also propose its approximate version where mismatches are allowed, which significantly increases the difficulty as well as the usage of SUS finding in subfields such as bioinformatics, where approximate string matching is unavoidable due to genetic mutation and errors in biological experiments.
(2) We propose a generic in-place algorithmic framework that fits to solve both the exact and approximate k-mismatch SUS finding, using $2n$ words plus n bytes space. It is worth mentioning that $2n$ words plus n bytes is the minimum memory space needed to save those n calculated SUSes: (1) It needs 2 words to save each SUS by saving its start and ending positions (or one endpoint and its length) and there are n SUSes. (2) It needs another n bytes to save the original string S in order to output the actual content of any SUS of interest from queries. Note that all prior work [4,5,7,8] use $O(n)$ space but there is big leading constant hidden within the big-oh notation (see the experimental study in [5]).
(3) After the suffix array is constructed, all the computation in our solution happens in the place of two integer arrays, using non-trivial techniques. It is worth noting that our solution does not involve any compressed or succinct data structures, making our solution practical and easy to implement. Our preliminary experimental study shows that our solution for exact SUS finding is even faster than the fastest one among [5,7,8][2], in addition to a lot more space saving than them, enabling our solution to handle larger data sets. Due to page limit, we will deliver the details of our experimental study in the journal version of this paper.

2 Preparation

A **prefix** of S is a substring $S[1..i]$, $1 \leq i \leq n$. A **proper prefix** $S[1..i]$ is a prefix of S where $i < n$. A **suffix** of S is a substring $S[i..n]$, denoted as S_i, $1 \leq i \leq n$. S_i is a **proper suffix** of S, if $i > 1$.

For two strings A and B, we write $\mathbf{A = B}$ (and say A is **equal** to B), if $|A| = |B|$ and $H(A, B) = 0$. We say A is lexicographically smaller than B, denoted as $\mathbf{A < B}$, if (1) A is a proper prefix of B, or (2) $A[1] < B[1]$, or (3) there exists an integer $k > 1$ such that $A[i] = B[i]$ for all $1 \leq i \leq k-1$ but $A[k] < B[k]$.

The **suffix array** $SA[1..n]$ of S is a permutation of $\{1, 2, \ldots, n\}$, such that for any i and j, $1 \leq i < j \leq n$, we have $S[SA[i]..n] < S[SA[j]..n]$. That is, $SA[i]$ is the start position of the ith smallest suffix in the lexicographic order. The **rank array** $RA[1..n]$ is the inverse of the suffix array, i.e., $RA[i] = j$ iff $SA[j] = i$. The **k-mismatch longest common prefix (LCP)** between two strings A and B, $k \geq 0$, denoted as $LCP^k(A, B)$, is the LCP of A and B within Hamming distance k. For example, if $A =$ `abc` and $B =$ `acb`, then: $LCP^0(A, B)$

[2] Note that the work of [4] studies a different problem and its computation is of the query-answer model, and thus is not comparable with [5,7,8] and ours.

is $A[1] = B[1] = \texttt{a}$ and $|LCP^0(A, B)| = 1$; $LCP^1(A, B)$ is $A[1..2] = \texttt{ab}$ and $B[1, 2] = \texttt{ac}$ and $|LCP^1(A, B)| = 2$.

Definition 2 (k-mismatch LSUS). *For a particular string position p in S and an integer k, $0 \leq k \leq n-1$, the* k-mismatch left-bounded shortest unique substring (LSUS) starting at position p, *denoted as $LSUS_p^k$, is a k-mismatch unique substring $S[p..j]$, such that either $p = j$ or any proper prefix of $S[p..j]$ is not k-mismatch unique.*

We call 0-mismatch LSUS as **exact LSUS**, and the case $k > 0$ as **approximate LSUS**. Observe that for any k, $LSUS_1^k = SUS_1^k$ always exists, because at least S itself can be $LSUS_1^k$. However, for any $k \geq 0$ and $p \geq 2$, $LSUS_p^k$ may not exist. For example, if $S = \texttt{dabcabc}$, none of $LSUS_i^0$ and $LSUS_j^1$ exists, for all $i \geq 5$, $j \geq 4$. It follows that some string positions may not be covered by any k-mismatch LSUS. For example, for the same string $S = \texttt{dabcabc}$, positions 6 and 7 are not covered by any exact or 1-mismatch LSUS. On the other hand, if any $LSUS_p^k$ does exist, there must be only one choice for $LSUS_p^k$, because $LSUS_p^k$ has its start position fixed on p and need to be as short as possible. Note that in Definition 2, we require $k < n$, because finding $LSUS_p^n$ is trivial as $LSUS_1^n \equiv S$ and $LSUS_p^n$ does not exist for all $p > 1$.

Definition 3 (k-mismatch SLS). *For a string position p in S and an integer k, $0 \leq k \leq n-1$, we use SLS_p^k to denote the* shortest k-mismatch LSUS covering position p.

We call 0-mismatch SLS as **exact SLS**, and the case $k > 0$ as **approximate SLS**. SLS_p^k may not exist, since position p may not be covered by any k-mismatch LSUS at all. For example, if $S = \texttt{dabcabc}$, then none of SLS_p^0 and SLS_p^1 exists, for all $p \geq 6$. On the other hand, if SLS_p^k exists, there might be multiple choices for SLS_p^k. For example, if $S = \texttt{abcbac}$, SLS_2^0 can be either $LSUS_1^0 = S[1..2]$ or $LSUS_2^0 = S[2..3]$, and SLS_3^1 can be any one of $LSUS_1^1 = S[1..3]$, $LSUS_2^1 = S[2..4]$, and $LSUS_3^1 = S[3..5]$. Note that in Definition 3, we require $k < n$, because finding SLS_p^n is trivial as $SLS_p^n \equiv S$ for all p.

Lemma 1. *For any k and p: (1) $LSUS_1^k$ always exists. (2) If $LSUS_p^k$ exists, then $LSUS_i^k$ exists, for all $i \leq p$. (3) If $LSUS_p^k$ does not exist, then none of $LSUS_i^k$ exists, for all $i \geq p$.*

Lemma 2. *For any k and p, $|LSUS_p^k| \geq |LSUS_{p-1}^k| - 1$, if $LSUS_p^k$ exists.*

Lemma 3. *For any k and p, SUS_p^k is either SLS_p^k or $S[i..p]$, for some i, $i + |LSUS_i^k| - 1 < p$. That is, SUS_p^k is either the shortest k-mismatch LSUS that covers position p, or a right extension (through position p) of a k-mismatch LSUS.*

For example, let $S = \texttt{dabcabc}$, then: (1) SUS_3^0 can be either $S[3..5] = LSUS_3^0$, or $S[1..3]$, which is a right extension of $LSUS_1^0 = S[1]$. (2) $SUS_5^0 = S[4..5] = LSUS_4^0$. (3) $SUS_6^0 = S[4..6]$, which is a right extension of $LSUS_4^0 = S[4..5]$. (4) $SUS_4^1 = S[3..5] = LSUS_3^1$. (5) $SUS_6^1 = S[3..6]$, which is a right extension of $LSUS_3^1$.

The next lemma further says that if SUS_p^k is an extension of an k-mismatch LSUS, SUS_p^k can be quickly obtained from SUS_{p-1}^k.

Lemma 4. *For any k and p, if $SUS_p^k = S[i..p]$ and $i + |LSUS_i^k| - 1 < p$, i.e., SUS_p^k is a right extension (through position p) of $LSUS_i^k$, then the following must be true: (1) $p > 2$; (2) the rightmost character of SUS_{p-1}^k is $S[p-1]$; (3) $SUS_p^k = SUS_{p-1}^k S[p]$, the substring SUS_{p-1}^k appended by the character $S[p]$.*

3 The High-Level Picture

In this section, we present an overview of our in-place framework for finding both the exact and approximate SUS. The framework is composed of three stages, where all computation happens in the place of three arrays, S, A, and B, each of size n. Arrays A and B are of integers, whereas array S always saves the input string. The following table summarizes the roles of A and B at different stages by showing their content at the end of each stage.

Stages	$A[i]$	$B[i]$
1	Used as temporary workspace during stage 1, but the content is useless for stages 2 and 3.	Ending position of $LSUS_i^k$, if $LSUS_i^k$ exists; otherwise, NIL.
2	The largest j, such that $LSUS_j^k$ is an SLS_i^k, if SLS_i^k exists; otherwise, NIL.	Ending position of $LSUS_i^k$, if $LSUS_i^k$ exists; otherwise, NIL.
3	Start position of the rightmost SUS_i^k	Ending position of the rightmost SUS_i^k

Stage 1 (Sect. 4). We take the array S that saves the input string as input to compute $LSUS_i^k$ for all i, in the place of A and B. At the end of the stage, each $B[i]$ saves the ending position of $LSUS_i^k$, if $LSUS_i^k$ exists. Since each existing $LSUS_i^k$ has its start position fixed at i, at the end of stage 1, each existing $LSUS_i^k = S\big[i..B[i]\big]$. For those non-existing k-mismatch LSUSes, we assign NIL to the corresponding B array elements. The time cost of this stage is $O(n)$ for exact LSUS finding ($k = 0$), and is $O(n^2)$ for approximate LSUS finding, for any $k \geq 1$.

Stage 2 (Sect. 5). Given the array B (i.e., the k-mismatch LSUS array of S) from stage 1, we compute the rightmost SLS_i^k, the rightmost shortest LSUS covering position i, for all i, in the place of A and B. At the end of stage 2, each $A[i]$ saves the largest j, such that $LSUS_j^k$ is an SLS_i^k, i.e., the rightmost $SLS_i^k = S\big[A[i]..B[A[i]]\big]$, if SLS_i^k exists; otherwise, we assign $A[i] = $ NIL. Array

B does not change during stage 2. The time cost of this stage is $O(n)$, for any $k \geq 0$.

Stage 3 (Sect. 6). Given A and B from stage 2, we compute SUS_i^k, for all i, in the place of A and B. At the end of stage 3, each $(A[i], B[i])$ pair saves the start and ending positions of the rightmost SUS_i^k, i.e., $SUS_i^k = S[A[i]..B[i]]$. The time cost of this stage is $O(n)$, for any $k \geq 0$.

The arXiv version of this paper has the pseudocode of the in-place procedures that we will describe in Sects. 4.1, 4.2, 5, and 6.

4 Finding k-mismatch LSUS

The goal of this section is that, given the input string S and two integer arrays A and B, we want to work in the place of A and B, such that $B[i]$ saves the ending position of $LSUS_i^k$ for all existing $LSUS_i^k$; otherwise, $B[i]$ is assigned `NIL`. We take different approaches in finding the exact LSUS ($k = 0$) and approximate LSUS ($k \geq 1$).

4.1 Finding Exact LSUS ($k = 0$)

Lemma 5 (Lemma 7.1 in [6]). *Given a string S of size n, drawn from an alphabet of size σ, we can construct the suffix array SA of S in $O(n)$ time, using $n + \sigma$ words plus n bytes, where the space of n bytes saves S, the space of n words saves SA, and the extra space of σ words is used as the workspace for the run of the SA construction algorithm.*

Given the input string S, we first use the $O(n)$-time suffix array construction algorithm from [6] to create the SA of S, where the array A is used to save the SA and the array B is used as the workspace. Note that $\sigma \leq n$ is always true, because otherwise we will prune from the alphabet those characters that do not appear in the string. After SA (saved in A) is constructed, we can easily spend another $O(n)$ time to create the rank array RA of S (saved in B): $RA[SA[i]] \leftarrow i$ (i.e., $B[A[i]] \leftarrow i$), for all i. Next, we use and work in the place of A (i.e., SA) and B (i.e., RA) to compute the ending position of each existing $LSUS_i^0$ and save the result in $B[i]$, using another $O(n)$ time.

Definition 4.

$$x_i = \begin{cases} \left| \, LCP^0\left(S[i..n], S\left[SA\left[RA[i]-1\right]..n\right]\right) \right|, & \textit{if } RA[i] > 1 \\ 0, & \textit{otherwise} \end{cases}$$

$$y_i = \begin{cases} \left| \, LCP^0\left(S[i..n], S\left[SA\left[RA[i]+1\right]..n\right]\right) \right|, & \textit{if } RA[i] < n \\ 0, & \textit{otherwise} \end{cases}$$

That is, x_i (y_i, resp.) is the length of the LCP of $S[i..n]$ and its lexicographically preceding (succeeding, resp.) suffix, if the preceding (succeeding, resp.) suffix exists.

Fact 1. *For every string position* i, $1 \leq i \leq n$:

$$LSUS_i^0 = \begin{cases} S\left[i..i + \max\{x_i, y_i\}\right], & \text{if } i + \max\{x_i, y_i\} \leq n \\ \text{not existing}, & \text{otherwise.} \end{cases}$$

First, observe that in the sequence of x_i's, if $x_i > 0$, then $x_{i+1} \geq x_i - 1$ must be true, because at least $S[SA[RA[i] - 1] + 1..n]$ can be the lexicographically preceding suffix of $S[i + 1..n]$, and they share the leading $x_i - 1$ characters. That means, when we compute x_{i+1}, we can skip over the comparisons of the first $x_i - 1$ pair of characters between $S[i + 1..n]$ and its lexicographically preceding suffix. It follows that, given the SA and RA of S and using the above observation, we can compute the sequence of x_i's in $O(n)$ time. Using the similar observation, we can compute the sequence of y_i's in $O(n)$ time, provided that S and its SA and RA are given.

Second, since we can compute the sequences of x_i's and y_i's in parallel (i.e., compute the sequence of (x_i, y_i) pairs), we can use Fact 1 to compute the sequence of $LSUS_i^0$ in $O(n)$ time. Further, since $RA[i]$ is used only for retrieving the lexicographically preceding and succeeding suffixes of $S[i..n]$ when we compute the pair (x_i, y_i), we can save each computed $LSUS_i^0$ (indeed, $i+\max\{x_i, y_i\}$, the ending position of $LSUS_i^0$) in the place of $RA[i]$ (i.e., $B[i]$). In the case $i + \max\{x_i, y_i\} > n$, meaning $LSUS_i^0$ does not exist,we will assign NIL to $RA[j]$ (i.e., $B[j]$) for all $j \geq i$ (Lemma 1). The overall time cost for computing the sequence of $LSUS_i^0$ is thus $O(n)$, yielding the following lemma.

Lemma 6. *Given the character array* S *of size* n *that saves the input string, and the integer arrays* A *and* B*, each of size* n*, we can work in the place of* S*,* A*, and* B*, using* $O(n)$ *time, such that at the end of the computation,* S *does not change,* $B[i]$ *saves the ending of position of* $LSUS_i^0$*, if* $LSUS_i^0$ *exists (otherwise,* $B[i] =$ NIL*).*

4.2 Finding Approximate LSUS ($k \geq 1$)

Definition 5. *For a particular string position* p *in* S *and an integer* k, $0 \leq k \leq n - 1$, *the* k-mismatch left-bounded longest repeat (LLR) *starting at position* p*, denoted as* LLR_p^k*, is a* k*-mismatch repeat* $S[p..j]$*, such that either* $j = n$ *or* $S[p..j + 1]$ *is* k*-mismatch unique.*

Fact 2. *(1) If* $|LLR_p^k| < n - p + 1$*, i.e., the ending position of* LLR_p^k *is less than* n*, then* $LSUS_p^k = S\left[p..p + |LLR_p^k|\right]$*, the substring of* LLR_p^k *appended by the character following* LLR_p^k*. (2) Otherwise,* $LSUS_p^k$ *does not exist.*

Our high-level strategy for finding $LSUS_i^k$ for all i is as follows. We first find LLR_i^k for all i. Then we use Fact 2 to find each $LSUS_i^k$ from LLR_i^k: If LLR_i^k does not end on position n, we will extend it for one more character on its right side and make the extension to be $LSUS_i^k$; otherwise, LLR_i^k does not exist. Next, we explain how to find LLR_i^k, for all i.

Clearly, $|LLR_i^k| = \max\{|LCP^k(S_i, S_j)|, j \neq i\}$, for all i. The way we calculate $|LLR_i^k|$ for all i is simply to let every pair of two distinct suffixes to be compared with each other. In order to do so, we work over $n-1$ phases, named as $\mathcal{P}_1$ through $\mathcal{P}_{n-1}$. On a particular phase $\mathcal{P}_\delta$, we compare suffixes S_i and $S_{i-\delta}$ for all $i = n, n-1, \ldots, \delta+1$. Obviously, over these $n-1$ phases, every pair of distinct suffixes have been compared with each other exactly once. Over these $n-1$ phases, we simply record in $B[i]$, which is initialized to be 0, the length of the longest k-mismatch LCP that each suffix S_i has seen when compared with any other suffixes. Next, we explain the details of a particular phase $\mathcal{P}_\delta$.

On a particular phase $\mathcal{P}_\delta$, $1 \le \delta \le n-1$, we compare suffixes S_i and $S_{i-\delta}$ for all $i = n, n-1, \ldots, \delta+1$. When we compare S_i and $S_{i-\delta}$, we save in $A[1..k+1]$, which is initialized to be empty at the beginning of each phase, the leftmost mismatched $k+1$ positions in S_i. We will see later how to update $A[1..k+1]$ efficiently over the progress of a particular phase and use it to update the B array.

We treat $A[1..k+1]$ as a circular array, i.e., $i-1 = k+1$ when $i = 1$, and $i+1 = 1$ when $i = k+1$. Let `size`, which is initialized to be 0 at the beginning of each phase, denote the number of mismatched positions being saved in $A[1..k+1]$ so far in $\mathcal{P}_\delta$. We can describe the work of phase $\mathcal{P}_\delta$, inductively, as follows.

1. We compare S_n and $S_{n-\delta}$ by only comparing $S[n]$ and $S[n-\delta]$, since $S_n = S[n]$.
   (a) If $S[n] \neq S[n-\delta]$: Save n in any position in $A[1..k+1]$; $\texttt{size} \leftarrow 1$.
   (b) $B[n] \leftarrow \max\{B[n], 1\}$; $B[n-\delta] \leftarrow \max\{B[n-\delta], 1\}$.
2. Suppose we have finished the comparison between the suffixes S_{i+1} and $S_{i+1-\delta}$, for some i, $\delta+1 \le i \le n-1$. The leftmost $k+1$ mismatched positions (if existing) between them have been saved in the circular array $A[1..k+1]$. Let $A[\texttt{cursor}]$ be the element that is saving the first mismatched position (if existing) between the two suffixes.
3. Next, we compare the suffixes S_i and $S_{i-\delta}$ by only comparing $S[i]$ and $S[i-\delta]$, since S_{i+1} and $S_{i+1-\delta}$ have been compared. Remind that $\texttt{cursor}-1$ below is in its cyclic manner.
   (a) If $S[i] \neq S[i-\delta]$: $\texttt{cursor} \leftarrow \texttt{cursor}-1$; Save i in $A[\texttt{cursor}]$ and overwrite the old content in $A[\texttt{cursor}]$ if there is; $\texttt{size} \leftarrow \min\{\texttt{size}+1, k+1\}$.
   (b) If $\texttt{size} < k+1$: $B[i] \leftarrow \max\{B[i], n-i+1\}$; $B[i-\delta] \leftarrow \max\{B[i-\delta], n-i+1\}$.
   (c) Else: $B[i] \leftarrow \max\{B[i], A[\texttt{cursor}-1]-i\}$; $B[i-\delta] \leftarrow \max\{B[i-\delta], A[\texttt{cursor}-1]-i\}$. Note that $A[\texttt{cursor}-1]$ is saving the $(k+1)$th mismatched position between S_i and $S_{i-\delta}$.

After the computation of all LLR_i^k is finished, using the above $n-1$ phases, each $B[i]$ is saving $|LLR_i^k|$. Next, we can use Fact 2 to convert each LLR_i^k to $LSUS_i^k$ by simply checking each $B[i]$: If $i + B[i] - 1 < n$, i.e., LLR_i^k does not end on position n, then we assigne $B[i] = i + B[i]$, the ending position of $LSUS_i^k$; otherwise, we assign $B[i] = \texttt{NIL}$, meaning $LSUS_i^k$ does not exist.

The computation of all LLR_i^k takes $n-1$ phases and each phase clearly has no more than n comparisons, giving a total of $O(n^2)$ time cost. The procedure of converting each LLR_i^k to $LSUS_i^k$ spends another $O(n)$ time. Altogether, we get an $O(n^2)$-time in-place procedure for finding approximate LSUS, for any $k \geq 1$.

Lemma 7. *Given the character array S of size n that saves the input string, the integer arrays A and B, each of size n, and the value of integer $k \geq 1$, we can work in the place of S, A, and B, using $O(n^2)$ time, such that at the end of the computation, S does not change, $B[i]$ saves the ending of position of $LSUS_i^k$, i.e., $LSUS_i^k = S[i..B[i]]$, if $LSUS_i^k$ exists; otherwise, $B[i] = \texttt{NIL}$.*

5 Finding k-mismatch SLS

Now we are given the array B, where each $B[i]$ saves the ending position of $LSUS_i^k$ if $LSUS_i^k$ exists and NIL otherwise. In this section, we want to work in the place of A and B, such that in the end of computation: $A[i]$ saves j, such that $LSUS_j^k$ is the rightmost SLS_i^k, if such j exists; otherwise, $A[i] = \texttt{NIL}$. That means, in the end of this section, the rightmost $SLS_i^k = S\big[A[i]..B[A[i]]\big]$, if SLS_i^k exists; otherwise, $A[i] = B[i] = \texttt{NIL}$.

Recall that some k-mismatch LSUS may not exist and some positions may not be covered by any k-mismatch LSUS (see the examples after Definition 2). Further, due to Lemmas 1 and 2, we know such positions that are not covered by any k-mismatch LSUS must comprise a continuous chunk on the right end of string S.

Definition 6. *Let $LSUS_r^k$, $1 \leq r \leq n$, be the rightmost existing k-mismatch LSUS of the input string S. Let z, $1 \leq z \leq n$, be the rightmost string position that is covered by any k-mismatch LSUS of the string S.*

Again, due to Lemmas 1 and 2, it is trivial to find the values of r and z in $O(n)$ time: scan array B (i.e. LSUS array) from right to left, and stop when seeing the first non-NIL B array element, which is exactly $B[r]$, then $z = B[r]$. If $z < n$, we can then simply set $A[i] = \texttt{NIL}$ for all $i > z$. Recall that $B[i] = \texttt{NIL}$ already for all $i > z$ from stage 1. In the rest of this section, we only need to work with the two subarrays $A[1..z]$ and $B[1..z]$, making $A[i]$ to be the start position of the rightmost SLS_i^k, for all $i \leq z$.

Let $B[1..z]$ and an integer r, $1 \leq r \leq z$, be the input, where (1) $B[1..r]$ is of monotonically nondecreasing integers (Lemma 2), with $i \leq B[i]$, (2) $B[r+1..z]$ are all NIL, if $r < z$, and (3) $B[r] = z$.

We can use each $B[i]$, $i \leq r$, as a compact representation of the interval $I_i = (i, B[i])$. Let $\mathcal{I} = \{\ I_i \mid i \in [1..r]\ \}$, and $\ell_i = |B[i] - i + 1|$ be the length of I_i. Let $A[1..z]$ be an output array such that $A[j] = i$, where I_i is the rightmost shortest interval in $\mathcal{I}$ that covers j.

Definition 7. *For an interval I_i, we define the* effective covering region *with respect to the previous intervals $\mathcal{I}_{<i} = \{ I_k \mid k < i \}$ to be $[t_i, B[i]]$ where*

$$t_i = \max \Big\{ i, \ \max \{B[k] + 1 \mid I_k \text{ is shorter than } I_i, k < i\} \Big\}.$$

We call t_i the starting point of the effective covering region of I_i.

The effective covering region of I_i is exactly those regions that would set I_i as the answer, provided that all the intervals $I_{<i}$ before I_i are present, and all the intervals $I_{>i} = \{ I_k \mid k > i \}$ are absent.

We next define t_i^{-1} as a list[3], such that $j \in t_i^{-1}$ if and only if $t_j = i$. Observe that since $t_i \geq i$ by definition, any value j in t_i^{-1} must have $j \leq i$, and the effective region of I_j must cover i.

Lemma 8. *For $i = 1, 2, \ldots, z$: $A[i] = \max \bigcup_{k=1}^{i} t_k^{-1} = \max \{ A[i-1], \max t_i^{-1} \}$.*

Lemma 9. *Suppose that all t_i, $1 \leq i \leq r$, can be generated incrementally in $O(n)$ time. Then, we can obtain all* max t_i^{-1}*, $1 \leq i \leq z$, in $O(n)$ time.*

Indeed, we may scan t_i from right to left, i.e., $i = r, r-1, \ldots, 1$, and update max t_i^{-1} as we proceed. Firstly, if $t_i > i$, we set t_i^{-1} = undefined. Else, let $j = t_i$ (whose value is at least i), and we check if t_j^{-1} is defined: If not, simply set $t_j^{-1} = i$; otherwise, no update is needed.

The advantage of the 'right-to-left' approach is that we can construct t_i^{-1} in-place, by re-using the memory space of t_i. To see why it is so, by the time we need to update a certain entry $j = t_i$ at step i, the information t_j has been used (and will never be used), so that we can safely overwrite the original entry, storing t_j, to store t_j^{-1} instead. This gives the following corollary.

Corollary 1. *Suppose that all t_i's are generated, and are stored in a certain array $A[1..z]$. Then, we can obtain* max t_i^{-1} *for all i's, in-place, by storing the results in the same array $A[1..z]$; the time cost is $O(n)$.*

Our goal is to make our algorithm in-place. Suppose that we can have in-place incremental generation of t_i. Then, by the above lemma, we may store max t_i^{-1} temporarily at $A[i]$; afterwards, by the second equality of Lemma 8, we can compute the correct output A by a simple scan of A from left to right.

Thus, to make the whole process in-place, it remains to show how t_i can be computed in $O(n)$ time, in-place. For this, we define $\texttt{pred}[i]$ to be the largest j (if it exists) such that $j < i$ and length of I_j is shorter than I_i. It is easy to check that if $\texttt{pred}[i] = j$ is defined, then $t_i = \max \{ B[j] + 1, i \}$ (and $t_i = i$ otherwise).[4] Moreover, $\texttt{pred}[i]$ for all i's can be computed incrementally, with

[3] In actual run, t_i^{-1} saves the largest number in that list, as we will see more clearly later.

[4] For each $j' < j$, if $I_{j'}$ covers i, I_j would also cover i; in such a case, $B[j]+1 \geq B[j']+1$. For each $j' \in [\texttt{pred}[i], i-1]$, $I_{j'}$ is longer than I_i.

a way analogous to the construction of the failure function in KMP algorithm: we check $\texttt{pred}[i-1], \texttt{pred}[\texttt{pred}[i-1]], \texttt{pred}[\texttt{pred}[\texttt{pred}[i-1]]]$, and so on, until we obtain j in the process such that I_j is shorter than I_i, and set $\texttt{pred}[i] := j$.[5] If such j does not exist, we set $\texttt{pred}[i] = \texttt{NIL}$. The running time is bounded by $O(n)$.

This gives the following $O(n)$-time in-place algorithm (where B is read-only):

1. Compute $\texttt{pred}[i]$, $i = 1, 2, \ldots, r$, and store this in $A[i]$. Note that this step requires the length information of the intervals of I_i, which can be obtained in $O(1)$ time, on the fly, from $B[i]$.
2. Scan $A[1..r]$ (i.e., pred) incrementally, and obtain t_i from the above discussion. Save the value of t_i in $A[i]$. Note that this step requires the access to the original B.
3. Scan $A[1..r]$ (i.e., t_i) from right to left, and obtain max t_i^{-1} decrementally (stored in $A[i]$) by Corollary 1.
4. Scan $A[1..z]$ (i.e., max t_i^{-1}) incrementally ($i = 1, 2, \ldots, z$), and obtain the desired $A[i]$ by the second equality in Lemma 8.

Lemma 10. *Given the integer array A and B, each of size n, where each $B[i]$ saves the ending position of $LSUS_i^k$, if $LSUS_i^k$ exists and* NIL *otherwise, we can work in the place of array A and B, using $O(n)$ time, such that, in the end of computation, array B does not change, and $A[i]$ saves j, where $LSUS_j^k$ is the rightmost SLS_i^k, if such j exists; otherwise, $A[i] = \texttt{NIL}$. That is, $SLS_i^k = S[A[i]..B[A[i]]]$, if SLS_i^k exists; otherwise, $A[i] = B[i] = \texttt{NIL}$.*

6 Finding k-mismatch SUS

Now we have array A, where $A[i] = j$, such that $LSUS_j^k$ is the rightmost SLS_i^k, if position i is covered by any k-mismatch LSUS; otherwise, $A[i] = \texttt{NIL}$. Note that $A[i] = j$ is recording the start position of the rightmost SLS_i^k already, because $LSUS_j^k$ starts on position j. We also have array B, where $B[i] = i + |LSUS_i^k| - 1$, the ending position of $LSUS_i^k$, if $LSUS_i^k$ exists; otherwise, $B[i] = \texttt{NIL}$.

Step I. We want to transform A and B, such that each $(A[i], B[i])$ pair saves the start and ending positions of SLS_i^k, if SLS_i^k exists; otherwise, $(A[i], B[i]) = (\texttt{NIL}, \texttt{NIL})$. Since each $A[i]$ is already recording the start position of SLS_i^k already, as we have explained at the beginning of this section, we only need to make changes to array B. We first set $B[i] = \texttt{NIL}$ for all $i > z$ (Definition 6). Then, we scan array B from right to left, starting from position z through 1, and set each $B[i] = B[A[i]]$, the ending position of the rightmost SLS_i^k. Because the leftmost position that any existing $LSUS_i^k$ can cover is position i, we know $A[i] \leq i$ and we no longer need $B[i]$ (i.e., the information of $LSUS_i^k$) after SLS_i is computed. Therefore, it is safe to record SLS_i^k by overwriting $B[i]$ by $B[A[i]]$ (i.e., the ending position of SLS_i^k), in this right-to-left scan.

[5] Intuitively, pred defines the shortcuts so that we can skip some intervals in $I_{<i}$ to compute t_i.

Step II. We use arrays A and B to calculate SUS_i^k for each i and save the result in the place of A and B, i.e., each $(A[i], B[i])$ pair saves the start and ending position of SUS_i^k. Because of Lemmas 3 and 4, we can use arrays A and B to compute each SUS_i^k inductively, as follows:

1. $SUS_1^k = LSUS_1^k = SLS_1^k = S[A[1]..B[1]]$.
2. For $i = 2, 3, \ldots, n$, we compute SUS_i^k:
   (a) If $(A[i], B[i]) = (\texttt{NIL}, \texttt{NIL})$, meaning SLS_i^k does not exist, we set SUS_i^k to be SUS_{i-1}^k appended by the character $S[i]$, i.e., $SUS_i^k = S[A[i-1]..B[i-1]+1]$, and save SUS_i^k by setting $(A[i], B[i]) = (A[i-1], B[i-1]+1)$;
   (b.) Else, if SUS_{i-1}^k ends at position $i-1$ and $SUS_{i-1}^k S[i] = S[A[i-1]..B[i-1]+1]$ is shorter than $SLS_i^k = S[A[i]..B[i]]$, we set $(A[i], B[i]) = (A[i-1], B[i-1]+1)$;
   (c) Else, $SUS_i^k = SLS_i^k$ and thus we leave $A[i]$ and $B[i]$ unchanged.

Lemma 11. *Given arrays A and B:*

- $A[i] = j$, *such that* $LSUS_j^k$ *is the rightmost* SLS_i^k, *if* SLS_i^k *exists; otherwise,* $A[i] = \texttt{NIL}$.
- $B[i] = i + |LSUS_i^k| - 1$, *the ending position of* $LSUS_i^k$, *if* $LSUS_i^k$ *exists; otherwise,* $B[i] = \texttt{NIL}$.

we can work in the place of A and B, using $O(n)$ time, such that, in the end of computation, each $(A[i], B[i])$ saves the start and ending positions of SUS_i^k, i.e., $SUS_i^k = S[A[i]..B[i]]$, $i = 1, 2, \ldots, n$.

By concatenating the claims in Lemmas 6, 7, 10, and 11, we get the final result.

Theorem 1. *Given an array S of size n that saves the input string, two integer arrays A and B, each of size n, and the value of integer $k \geq 0$, we can work in the place of arrays S, A, and B, using a total of $O(n)$ time for $k = 0$ and $O(n^2)$ time for any $k \geq 1$, such that in the end of computation, S does not change, each $(A[i], B[i])$ pair represents the start and ending positions of the rightmost SUS_i^k, i.e., $SUS_i^k = S[A[i]..B[i]]$.*

References

1. Adaş, B., Bayraktar, E., Faro, S., Moustafa, I.E., Külekci, M.O.: Nucleotide sequence alignment and compression via shortest unique substring. In: Ortuño, F., Rojas, I. (eds.) IWBBIO 2015, Part II. LNCS, vol. 9044, pp. 363–374. Springer, Heidelberg (2015)
2. Derrien, T., Estell, J., Marco Sola, S., Knowles, D.G., Raineri, E., Guig, R., Ribeca, P.: Fast computation and applications of genome mappability. PLoS ONE **7**(1), e30377 (2012)
3. Haubold, B., Pierstorff, N., Möller, F., Wiehe, T.: Genome comparison without alignment using shortest unique substrings. BMC Bioinform. **6**, 123 (2005)

4. Hu, X., Pei, J., Tao, Y.: Shortest Unique Queries on Strings. In: Moura, E., Crochemore, M. (eds.) SPIRE 2014. LNCS, vol. 8799, pp. 161–172. Springer, Heidelberg (2014)
5. İleri, A.M., Külekci, M.O., Xu, B.: A simple yet time-optimal and linear-space algorithm for shortest unique substring queries. Theor. Comput. Sci. **562**, 621–633 (2015). Also in CPM 2014
6. Nong, G.: Practical linear-time $O(1)$-workspace suffix sorting for constant alphabets. ACM Trans. Inf. Syst. (TOIS) **31**(3), 15:1–15:15 (2013)
7. Pei, J., Wu, W.C.H., Yeh, M.Y.: On shortest unique substring queries. In: Proceedings of IEEE International Conference on Data Engineering (ICDE), pp. 937–948 (2013)
8. Tsuruta, K., Inenaga, S., Bannai, H., Takeda, M.: Shortest unique substrings queries in optimal time. In: Geffert, V., Preneel, B., Rovan, B., Štuller, J., Tjoa, A.M. (eds.) SOFSEM 2014. LNCS, vol. 8327, pp. 503–513. Springer, Heidelberg (2014)

Inferring Strings from Full Abelian Periods

Makoto Nishida[1], Tomohiro I.[2(✉)], Shunsuke Inenaga[1], Hideo Bannai[1], and Masayuki Takeda[1]

[1] Department of Informatics, Kyushu University, Fukuoka, Japan
{inenaga,bannai,takeda}@inf.kyushu-u.ac.jp

[2] Department of Computer Science, TU Dortmund, Dortmund, Germany
tomohiro.i@cs.tu-dortmund.de

Abstract. Strings u, v are said to be *Abelian equivalent* if u is a permutation of the characters appearing in v. A string w is said to have a *full Abelian period* p if $w = w_1 \cdots w_k$, where all w_i's are of length p each and are all Abelian equivalent. This paper studies reverse-engineering problems on full Abelian periods. Given a positive integer n and a set D of divisors of n, we show how to compute in $O(n)$ time the lexicographically smallest string of length n which has all elements of D as its full Abelian periods and has the minimum number of full Abelian periods not in D. Moreover, we give an algorithm to enumerate all such strings in amortized constant time per output after $O(n)$-time preprocessing. Also, we show how to enumerate the strings which have all elements of D as its full Abelian periods in amortized constant time per output after $O(n)$-time preprocessing.

1 Introduction

A positive integer p is said to be a period of a string w of length n if $w[i] = w[i+p]$ for all $1 \leq i \leq n - p$. Periodicity of strings is one of the most classical topics in combinatorics on words, and has been extensively studied in the literature, from both the combinatorics point of view (e.g., see [8,13]) and the algorithmics point of view (e.g., see [10,12]).

Among a number of extensions and generalizations of periods of strings, this paper deals with *Abelian periodicity* of strings. Strings u, v are said to be *Abelian equivalent* if u is a permutation of the characters appearing in v. For instance, aabbc and bacba are Abelian equivalent. A string w is said to have a *full Abelian period* p, if $w = w_1 \cdots w_k$, where $k = \frac{n}{p}$ and all w_i's are Abelian equivalent. A string w is said to have an *Abelian period* p if $w = yz$, where y has a full Abelian period p, and z is a string shorter than p s.t. the number of occurrences of each character c in z is no more than that in the prefix $y[1..p]$ of length p of y. A string w is said to have a *weak Abelian period* p if $w = xy$, where y has an Abelian period p, and x is a string shorter than p s.t. the number of occurrences of each character in x is no more than that in the prefix $y[1..p]$ of length p of y. The prefix x of w is called the head w.r.t. the weak Abelian period p.

K. Elbassioni and K. Makino (Eds.): ISAAC 2015, LNCS 9472, pp. 768–779, 2015.
DOI: 10.1007/978-3-662-48971-0_64

Fici et al. [6] proposed an $O(n \log\log n)$-time algorithm to compute all full Abelian periods and an $O(n^2)$-time algorithm to compute all Abelian periods for a given string of length n. Recently, Kociumaka et al. [11] showed an optimal $O(n)$-time algorithm to compute all full Abelian periods, and an improved $O(n(\log\log n + \log \sigma))$-time algorithm to compute all Abelian periods, where σ is the alphabet size. Fici et al. [7] presented an $O(n^2\sigma)$-time algorithm to compute all weak Abelian periods, and later Crochemore et al. [5] gave an improved $O(n^2)$-time solution to the problem.

This paper studies "reverse-engineering" problems on full Abelian periods. Namely, given a positive integer n and a set D of divisors of n, the task is to find a string on alphabet Σ of length n which has all elements of D as its full Abelian periods. Let $\mathcal{S}(n, D, \Sigma)$ denote the set of possible solutions to the problem. It is trivial if we do not restrict the number of full Abelian periods the string contains, since string c^n with any character c is a solution to any given set D. On the other hand, it is not always possible to construct a string such that D is *exactly* the set of its full Abelian periods. Hence, we will identify the smallest superset of D (which will be denoted by Q_D) that represents the set of inevitable elements to have all elements of D as full Abelian periods and consider the problem of computing a string having the smallest superset as the set of full Abelian periods. Let $\mathcal{S}'(n, D, \Sigma)$ denote the set of possible solutions to this problem. Firstly, we present an $O(n)$-time algorithm to compute the lexicographically smallest string in $\mathcal{S}'(n, D, \Sigma)$ (see Problem 1 and Theorem 1). Next, we show how to enumerate $\mathcal{S}(n, D, \Sigma)$ in amortized constant delay after $O(n)$-time preprocessing (see Problem 2 and Theorem 2). Finally, we show how to enumerate $\mathcal{S}'(n, D, \Sigma)$ in amortized constant delay after $O(n)$-time preprocessing (see Problem 3 and Theorem 3).

These results can be obtained by discovering and/or utilizing combinatorial properties on full Abelian periods. Particularly, it is non-trivial to achieve amortized constant-time delay enumerations since we are not even allowed to output strings naively, which takes $\Theta(n)$ time per output. In order to overcome this difficulty, we show that our search tree, which is well-controllable by multiset permutations, can represent output strings effectively while designing an efficient pruning method for $\mathcal{S}'(n, D, \Sigma)$ enumeration.

Related Work. Reverse-engineering strings from given string data structures and from regularities on strings is a well-studied class of problems, see [3,9,14,16] for some of the recent developments.

Blanchet-Sadri et al. [2] described an algorithm which, given two positive integers p, q, computes a string of length $2\mathrm{lcm}(p, q) - 2$ or $2\mathrm{lcm}(p, q) - 3$ having both p, q as its weak Abelian periods and containing $\gcd(p, q) + 1$ distinct characters, where $\mathrm{lcm}(p, q)$ and $\gcd(p, q)$ denote the least common multiple and the greatest common divisor of p, q, respectively. They also extend their algorithm to compute a string containing don't-care symbols (or holes). Unfortunately, the running times of their algorithms are not analyzed in [2]. Our results in this paper are different from theirs at least in that (1) we deal with full Abelian periods, (2) our input D can contain more than two positive integers, and (3) we give an enumeration algorithm to output all solutions.

2 Preliminaries

Let Σ be an ordered *alphabet* of size σ, and c_i be the ith lexicographically smallest character in Σ for each $1 \le i \le \sigma$. We assume that for each $c_i \in \Sigma$, its rank i in Σ is already known and can be computed in constant time. An element of Σ^* is called a *string*. The length of a string w is denoted by $|w|$. The empty string ε is the string of length 0, namely, $|\varepsilon| = 0$. For a string $w = xyz$, strings x, y, and z are called a *prefix*, *substring*, and *suffix* of w, respectively. The ith character of a string w of length n is denoted by $w[i]$ for $1 \le i \le n$. For $1 \le i \le j \le n$, let $w[i..j] = w[i] \cdots w[j]$, i.e., $w[i..j]$ is the substring of w starting at position i and ending at position j in w.

Two strings u, v are said to be *Abelian equivalent* iff u is a permutation of v, which we denote by $u \simeq v$. Note that $\simeq$ is an equivalence relation. A Parikh vector [15] of a string $u \in \Sigma^*$, denoted $\mathbf{P}_u$, is an array of length σ such that for any $1 \le i \le \sigma$, $\mathbf{P}_u[i]$ stores the number of occurrences of character c_i in u. We define $\mathbf{P}_u = \mathbf{P}_v$ iff $\mathbf{P}_u[i] = \mathbf{P}_v[i]$ for all $1 \le i \le \sigma$. Clearly, two strings $u \simeq v$ are Abelian equivalent iff $\mathbf{P}_u = \mathbf{P}_v$. For any two Parikh vectors $\mathbf{P}$ and $\mathbf{Q}$, let $\oplus$ and $\ominus$ be the operators such that $\mathbf{P} \oplus \mathbf{Q} = \langle \mathbf{P}[1] + \mathbf{Q}[1], \ldots, \mathbf{P}[\sigma] + \mathbf{Q}[\sigma] \rangle$ and $\mathbf{P} \ominus \mathbf{Q} = \langle \mathbf{P}[1] - \mathbf{Q}[1], \ldots, \mathbf{P}[\sigma] - \mathbf{Q}[\sigma] \rangle$, respectively. For any Parikh vector $\mathbf{P}$ and rational r, let $r\mathbf{P} = \langle r\mathbf{P}[i], \ldots, r\mathbf{P}[\sigma] \rangle$.

Let w be any non-empty string of length n. A divisor p of n is said to be a *full Abelian period* of w if the substrings $w[(k-1)p + 1..kp]$ for all $1 \le k \le n/p$ are Abelian equivalent. Let $FAP(w)$ denote the set of full Abelian periods of string w. For instance, $FAP(\mathtt{abcabcacbbaccacbba}) = \{6, 9, 18\}$.

For any positive integer n, let $Div(n)$ denote the set of divisors of n. We will use the following result to analyze the efficiency of our algorithms.

Lemma 1 (E.g. see Theorem 13.12 of [1]). *Let $\delta > 0$ be any constant. Then, there exists an integer n_δ such that $|Div(n)| = n^{(1+\delta)\ln 2/\ln\ln n}$ for any $n \ge n_\delta$, where $\ln x$ denotes the natural logarithm of x.*

For any set S of positive integers, let $\gcd(S)$ denote the greatest common divisors of all elements in S. Let $N_n = \{1, \ldots, n\}$ denote the set of positive integers up to n. Kociumaka et al. [11] showed that N_n can be preprocessed in $O(n)$ time so that later, given two integers $x, y \in N_n$, $\gcd(\{x, y\})$ can be answered in $O(1)$ time. The next corollary is then immediate.

Corollary 1. *N_n can be preprocessed in $O(n)$ time so that later, given a subset $S \subseteq N_n$, $\gcd(S)$ can be answered in $O(|S|)$ time.*

We consider reverse-engineering problems on full Abelian periods. A simplest kind of such problems would be: Given a subset D of divisors of a positive integer n, compute a string of length n which has every element of D as its full Abelian period. However, this problem is uninteresting, since string c^n for any character $c \in \Sigma$ is a solution to the problem (note that $D \subseteq FAP(c^n) = Div(n)$ always holds). To make it more interesting and meaningful, we restrict the number of elements of $FAP(w)$ that do not belong to a given D, and consider the following problem.

Problem 1. Given a positive integer n and $D \subseteq Div(n)$, compute the lexicographically smallest *string w of length n s.t. $D \subseteq FAP(w)$ and $|FAP(w) - D|$ is smallest possible.*

We also deal with the two following enumerating problems:

Problem 2. Given a positive integer n, $D \subseteq Div(n)$, and an ordered alphabet Σ, compute the set $\mathcal{S}(n, D, \Sigma)$ of all strings w of length n over Σ s.t. $D \subseteq FAP(w)$.

Problem 3. Given a positive integer n, $D \subseteq Div(n)$, and an ordered alphabet Σ, compute the set $\mathcal{S}'(n, D, \Sigma)$ of all strings w of length n over Σ s.t. $D \subseteq FAP(w)$ and $|FAP(w) - D|$ is smallest possible.

3 Algorithms

In this section, we present our algorithms to solve Problems 1–3.

3.1 New Properties on Full Abelian Periods

We begin with some properties of full Abelian periods of strings, which are useful to solve the problems.

The next lemma shows a necessary-and-sufficient condition for $p \in Div(n)$ to be a full Abelian period of a string of length n.

Lemma 2. *Let n be a positive integer and let $p \in Div(n)$. For any string w of length n, $p \in FAP(w)$ iff $\mathbf{P}_{w[1..pi]} = \frac{pi}{n}\mathbf{P}_w$ for any $1 \le i \le \frac{n}{p}$.*

Proof. ($\Rightarrow$) If $p \in FAP(w)$, then $\mathbf{P}_{w[1..p]} = \mathbf{P}_{w[p+1..2p]} = \cdots = \mathbf{P}_{w[n-p+1..n]}$. Hence, $\mathbf{P}_{w[1..pi]} = \frac{pi}{n}\mathbf{P}_w$ for any $1 \le i \le \frac{n}{p}$.

($\Leftarrow$) If $\mathbf{P}_{w[1..pi]} = \frac{pi}{n}\mathbf{P}_w$ for any $1 \le i \le \frac{n}{p}$, then $\mathbf{P}_{w[p(i-1)+1..pi]} = \frac{pi}{n}\mathbf{P}_w \ominus \frac{p(i-1)}{n}\mathbf{P}_w = \frac{p}{n}\mathbf{P}_w$. Namely, $\mathbf{P}_{w[1..p]} = \mathbf{P}_{w[p+1..2p]} = \cdots = \mathbf{P}_{w[n-p+1..n]} = \frac{p}{n}\mathbf{P}_w$, and hence, p is a full Abelian period of w. □

Lemma 2 motivates us to consider the positions x where $\mathbf{P}_{w[1..x]} = \frac{x}{n}\mathbf{P}_w$.

Definition 1. *Let n be a positive integer and let $D \subseteq Div(n)$ be a subset of divisors of n. Let $S_D = \{kp \mid p \in D, 1 \le k \le \frac{n}{p}\}$ and $m = |S_D|$. The multiples factorization of interval $[1, n]$ w.r.t. D, denoted MF_D, is a sequence $g_1, \ldots, g_m$ of non-empty intervals such that $\bigcup_{i=1}^{m} g_i = [1, n]$ and for each $1 \le i \le m$, $g_i = [s_{i-1} + 1, s_i]$ where $s_0 = 0$ and s_i is the ith smallest element of S_D.*

For instance, if $n = 18$ and $D = \{6, 9\}$, then $S_D = \{6, 9, 12, 18\}$ and hence $MF_D = [1, 6], [7, 9], [10, 12], [13, 18]$.

The strings in $\mathcal{S}(n, D, \Sigma)$ can be characterized by S_D or MF_D as follows:

Corollary 2. *Let n be a positive integer and let $D \subseteq Div(n)$.*

- *For any string w, $w \in \mathcal{S}(n, D, \Sigma)$ iff $\mathbf{P}_{w[1..x]} = \frac{x}{n}\mathbf{P}_w$ for any $x \in S_D$. $\mathbf{P}_{w[1..x]} = \frac{x}{n}\mathbf{P}_w$.*
- *For any string w, $w \in \mathcal{S}(n, D, \Sigma)$ iff $\mathbf{P}_{w[s_{i-1}+1..s_i]} = \frac{|g_i|}{n}\mathbf{P}_w$ for any $g_i = [s_{i-1} + 1, s_i] \in MF_D$.*

For any positive integer n and $D \subseteq Div(n)$, let $Q_D = \{\ell p \in Div(n) \mid p \in D, 1 \leq \ell \leq \frac{n}{p}\}$, i.e., Q_D is the set of multiples of elements of D which divide n.

It is easy to see from Lemma 2 that if p is a full Abelian period of string w of length n then $\ell p \in Div(n)$ with $1 \leq \ell \leq \frac{n}{p}$ is also a full Abelian period of w. Then, we get the following lemma:

Lemma 3. *For any string w of length n and $D \subseteq FAP(w)$, $Q_D \subseteq FAP(w)$.*

By Lemma 3, if $\{w \in \Sigma^n \mid Q_D = FAP(w)\}$ is not empty, $\{w \in \Sigma^n \mid Q_D = FAP(w)\}$ is the solution to Problem 3. In Sect. 3.2, we will show that the lexicographically smallest string in $\{w \in \Sigma^n \mid Q_D = FAP(w)\}$ can be achieved by a binary string, implying that $\{w \in \Sigma^n \mid Q_D = FAP(w)\}$ is not empty if $|\Sigma| \geq 2$. Since there is no point in considering reverse-engineering problems on a unary alphabet, we assume $|\Sigma| \geq 2$.

Let us now consider the relationship between the number of different characters in strings that are solutions to the problems, and $\gcd(D)$. Ilie and Constantinescu [4] showed that if a string w of length n has *weak* Abelian periods p, q with $\gcd(p, q) = 1$ and satisfies $n \geq 2pq - 1$, then w is a unary string (see Sect. 1 for the definition of weak Abelian periods). Assume a string w has weak Abelian periods p, q such that $\gcd(p, q) \geq 2$ and $||x_p| - |x_q|| = y \gcd(p, q)$ for some non-negative integer y, where x_p, x_q denote the heads of the weak Abelian periods p, q, respectively. Blanchet-Sadri et al. [2] showed that if the above string w satisfies $n \geq 2\text{lcm}(p, q) - 1$, then w contains at most $\gcd(p, q)$ different characters, where $\text{lcm}(p, q)$ denotes the least common multiple of p, q. Since full Abelian periods are special cases of weak Abelian periods, these results can be applied to our problems when n is sufficiently long. However, in our case n can be as small as $\text{lcm}(D)$, where $\text{lcm}(D)$ is the least common multiple of all elements of D. Hence their results cannot be applied directly to our problems for a short string length n. In what follows, we give a lemma which is specialized for full Abelian periods and holds for any n divisible by $\text{lcm}(D)$.

Lemma 4. *For any positive integer n, let $D \subseteq Div(n)$. Then, any string w of length n satisfying $D \subseteq FAP(w)$ can contain at most* $\gcd(D)$ *different characters.*

Proof. Let Σ_w be the set of characters occurring in w and let $\sigma_w = |\Sigma_w|$. Let $d = \gcd(D)$ and $\mathbf{P}' = \frac{d}{n}\mathbf{P}_w$.

First, we show that every element of $\mathbf{P}'$ is a non-negative integer. Assume on the contrary that, for some $1 \leq j \leq \sigma$, $\mathbf{P}'[j] = \frac{x}{y}$ where both $x \geq 1$ and

$y \geq 2$ are integers, and the fraction $\frac{x}{y}$ is irreducible. By assumption, for any $p \in D$, $p \in FAP(w)$. It follows from Lemma 2 that $\frac{1}{p}\mathbf{P}_{w[1..p]}[j] = \frac{1}{n}\mathbf{P}_w[j]$, which yields $\frac{d}{p}\mathbf{P}_{w[1..p]}[j] = \frac{d}{n}\mathbf{P}_w[j] = \mathbf{P}'[j] = \frac{x}{y}$, and thus $\mathbf{P}_{w[1..p]}[j] = \frac{px}{dy}$. Since both $\mathbf{P}_{w[1..p]}[j]$ and $\frac{p}{d}$ are integers and since y is relatively prime to x, y must be a divisor of $\frac{p}{d}$. This implies that dy is a divisor of p. Since the above argument holds for any $p \in D$ and since $y \geq 2$, we have $\gcd(D) \geq dy > d$, a contradiction. Hence every element of $\mathbf{P}'$ is a non-negative integer.

Since the sum of entries of $\mathbf{P}'$ is d and each entry is a non-negative integer, there are at most d entries in $\mathbf{P}'$ that are positive integers. As $\mathbf{P}' = \frac{d}{n}\mathbf{P}_w$, w can contain at most d distinct characters. □

The following corollary is immediate from Lemma 4.

Corollary 3. *Let n be a positive integer and $D \subseteq Div(n)$. If $\gcd(D) = 1$, then $\{c_i^n \mid 1 \leq i \leq \sigma\}$ is the solution to Problems 2 and 3.*

Due to Corollaries 1 and 3, we can solve the problems in $O(n)$ time in the case where $\gcd(D) = 1$. In what follows, we consider the case where $\gcd(D) \geq 2$.

3.2 Inferring the Lexicographically Smallest String

In this subsection, we show how, given $D \subseteq Div(n)$, to compute the lexicographically smallest string w with $FAP(w) = Q_D$, which is the solution to Problem 1.

Lemma 5. *Assume $d = \gcd(D) \geq 2$. Let w be the string of length n such that $w = f_1 \cdots f_m$, where $f_0 = \varepsilon$, $m = |MF_D|$, and for each $1 \leq i \leq m$, $f_i = c_1^{|f_i| - |f_i|/d} c_2^{|f_i|/d}$ and $[|f_1 \cdots f_{i-1}| + 1, |f_1 \cdots f_i|]$ is the ith subinterval of MF_D. Then, this string w is the solution to Problem 1.*

Proof. Firstly, we show that w satisfies $FAP(w) = Q_D$. Let $\mathbf{B} = \langle d - 1, 1, 0, \ldots, 0 \rangle$, then $\mathbf{P}_w = \frac{n}{d}\mathbf{B}$. Notice that $\mathbf{P}_{w[1..x]} = \frac{x}{n}\mathbf{P}_w = \frac{x}{d}\mathbf{B}$ iff $x \in S_D$. Then by Lemma 2, $FAP(w) = Q_D$.

Next, we show that w is the lexicographically smallest string of length n such that $FAP(w) = Q_D$. In so doing, consider any string u of length n such that $\mathbf{P}_u = Q_D$. By the definition of MF_D, for any $g_i \in MF_D$ there exist $p_i, q_i \in D$ such that $g_i = [\ell p_i + 1..kq_i]$ for some $\ell \geq 0$ and $k \geq 1$. Let $h_i = u[g_{i-1}+1..g_i]$ for any $1 \leq i \leq m$. If every element of D belongs to $FAP(u)$, then using Corollary 2 we get $\mathbf{P}_{h_i} = \frac{|h_i|}{n}\mathbf{P}_u$, and hence $\frac{1}{|h_i|}\mathbf{P}_{h_i} = \frac{1}{n}\mathbf{P}_u$ for any $1 \leq i \leq m$. Thus, if $\frac{1}{|h_i|}\mathbf{P}_{h_i} \neq \frac{1}{n}\mathbf{P}_u$ for some $1 \leq i \leq m$, then some element of D does not belong to $FAP(u)$. Now, let us consider to edit the string $w = f_1 \cdots f_m$ to another string w' without changing the set of full Abelian periods, namely, $FAP(w') = FAP(w)$. By the above arguments, we cannot edit each factor f_i of w so that $\frac{1}{|f_i|}\mathbf{P}_{f_i} \neq \frac{1}{n}\mathbf{P}_w$, and hence the only allowed edit operations on w is swapping characters inside each f_i. However, since each $f_i = c_1^{|f_i| - \frac{|f_i|}{d}} c_2^{\frac{|f_i|}{d}}$ is the lexicographically smallest string satisfying $\frac{1}{|f_i|}\mathbf{P}_{f_i} = \frac{1}{n}\mathbf{P}_w$, swapping characters inside f_i only increases the lexicographical rank of the string. Hence, $w = f_1 \cdots f_m$ is the lexicographically smallest string such that $FAP(w) = Q_D$. □

In order to compute the string w of Lemma 5, we use the following lemma:

Lemma 6. *For any positive integer n and $D \subseteq Div(n)$, Q_D and MF_D can be computed in $O(n)$ time and space.*

Proof. We compute Q_D as follows: For every $1 \leq x \leq n$ with $x \in Div(n)$, scan D and check if there is an element in D that divides x. It takes $O(n+|D||Div(n)|) = O(n + |Div(n)|^2) = O(n + n^{o(1)}) = O(n)$ time thanks to Lemma 1.

To compute MF_D in $O(n)$ time, we use the following fact: An integer x is a multiple of $d \in D$ iff $\gcd(x, n)$ is a multiple of $d \in D$. That is, $x \in MF_D$ iff $\gcd(x, n) \in Q_D$. Hence, given Q_D, we can check if $x \in MF_D$ or not in constant time after $O(n)$-time preprocessing for gcd queries [11]. □

The next theorem is immediate from Lemmas 5 and 6.

Theorem 1. *Problem 1 can be solved in $O(n)$ time.*

3.3 Enumerating $\mathcal{S}(n, D, \Sigma)$

In this subsection, we show how to solve Problem 2. Assume $d = \gcd(D) \geq 2$, since otherwise, Corollary 3 gives an optimal solution. Furthermore, it suffices to consider an alphabet Σ of size $\sigma \leq d$ due to Lemma 4.

We consider the set $\mathbb{B}_d = \{\mathbf{P}_u \mid u \in \Sigma^d\}$ of Parikh vectors for strings on Σ of length d. Our algorithm enumerates, for every Parikh vector $\mathbf{B} \in \mathbb{B}_d$, the solutions w in $\mathcal{S}(n, D, \Sigma)$ with $\frac{d}{n}\mathbf{P}_w = \mathbf{B}$. Since $\frac{d}{n}\mathbf{P}_w \in \mathbb{B}_d$ holds for any solution w, all the solutions can be output without omission by this approach.

One of the components of our algorithm is to enumerate the permutations of a given multiset. We denote by $\mathcal{E}(\mathbf{P})$ an enumerator of the permutations of a multiset represented by a Parikh vector $\mathbf{P}$. Although there are several sophisticated algorithms to enumerate multiset permutations (e.g. see [17,18]), we employ a simple backtracking algorithm: $\mathcal{E}(\mathbf{P})$ chooses a character c in $\mathbf{P}$ and invokes $\mathcal{E}(\mathbf{P}')$ recursively, where $\mathbf{P}'$ is the Parikh vector obtained by decreasing the entry for c by one in $\mathbf{P}$. By using a linked list to represent the non-zero values in $\mathbf{P}$, each backtracking can be done in $O(1)$ time. In addition, we allow $\mathcal{E}(\mathbf{P})$ to dump all characters at once if $\mathbf{P}$ contains no more than one distinct characters, i.e., for any character c and any integer $k \geq 1$, $\mathcal{E}(\mathbf{P}_{c^k})$ discards c^k in constant time. By doing so, the multiset permutations can be enumerated in constant time per output.

One application of $\mathcal{E}$ is the enumeration of $\mathbb{B}_d$. Observe that there is a one-to-one correspondence between a binary string x of length $d + \sigma - 1$ containing d 0's and $\sigma - 1$ 1's and a Parikh vector $\mathbf{P}$ for a string of length d: We can define $\mathbf{P}$ such that, for any $1 \leq i \leq \sigma$, $\mathbf{P}[i]$ is the number of 0's between the ith 1 and the $(i + 1)$th 1 in $1x1$. Hence, $\mathcal{E}(\mathbf{P}_{0^d 1^{\sigma-1}})$ can enumerate $\mathbb{B}_d$ in $O(\sigma + d)$ time per output.

When $\mathbf{B}$ is fixed, we compute w with $\frac{d}{n}\mathbf{P}_w = \mathbf{B}$ in the common backtracking approach, i.e., we traverse strings by determining characters from left to right. In the end, the output strings are represented by the search tree.

Let $MF_D = g_1, \ldots, g_m$. We consider a backtracking program $\mathcal{P}(n, D, \mathbf{B})$ that is composed of a series of m multiset permutation enumerators $\mathcal{E}(\frac{|g_1|}{d}\mathbf{B}), \ldots, \mathcal{E}(\frac{|g_m|}{d}\mathbf{B})$. For any position $k \in g_i = [s_{i-1}+1, s_i]$, $w[k]$ is computed by the ith enumerator, i.e., $w[k]$ is the $(k - s_{i-1})$th character of outputs of the ith enumerator. Note that the output of $\mathcal{P}(n, D, \mathbf{B})$ is the strings w such that $\mathbf{P}_{w[s_{i-1}+1..s_i]} = \frac{|g_i|}{d}\mathbf{B} = \frac{|g_i|}{n}\mathbf{P}_w$ for any $g_i = [s_{i-1}+1, s_i] \in MF_D$. Hence the next lemma is immediate from Corollary 2.

Lemma 7. *The output of $\mathcal{P}(n, D, \mathbf{B})$ is $\{w \in \mathcal{S}(n, D, \Sigma) \mid \frac{d}{n}\mathbf{P}_w = \mathbf{B}\}$.*

Theorem 2. *Problem 2 can be solved in $O(n + |\mathcal{S}(n, D, \Sigma)|)$ time and $O(n)$ working space.*

Proof. Firstly, we compute Q_D in $O(n)$ time and space using Lemma 6. By Lemma 7, $\mathcal{P}(n, D, \mathbf{B})$ enumerates all solutions with $\frac{d}{n}\mathbf{P}_w = \mathbf{B}$. Then, $\mathcal{S}(n, D, \Sigma)$ can be enumerated by running $\mathcal{P}(n, D, \mathbf{B})$ for every $\mathbf{B} \in \mathbb{B}_d$. If $\mathbf{B}$ contains only a single character c, we just output c^n. If $\mathbf{B}$ contains more than two distinct characters, we take the search tree of $\mathcal{P}(n, D, \mathbf{B})$ as a representation of the output strings. While most internal nodes of the search tree are branching, some exceptions appear when a single kind of character remains in an enumerator. However, this enumerator discards all these same characters at once, and thus, its single child becomes either a leaf or a branching node. Therefore the size of the search tree is bounded by a constant factor of the number of leaves (output strings). Since each backtracking can be done in $O(1)$ time, $\mathcal{S}(n, D, \Sigma)$ can be enumerated in $O(1)$ time per output. The working space is $O(n)$ since only the information on the current node and its ancestors is needed for the traversal. □

3.4 Enumerating $\mathcal{S}'(n, D, \Sigma)$

In this subsection, we modify the algorithm of Sect. 3.3 to solve Problem 3.

Recall that we assume $d = \gcd(D) \geq 2$. Since c^n is not a solution, we only consider $\mathbf{B}$ containing more than two distinct characters.

The task is to prune unnecessary search space leading to strings with $Q_D \subset FAP(w)$. Namely, we must prevent any element in $Div(n) - Q_D$ from being a full Abelian period of the enumerated strings. The next observation claims that, in so doing, we do not have to care about all the elements in $Div(n) - Q_D$.

Observation 1. *Let w be a string of length n. For any $p, p' \in Div(n)$ such that p is divisible by p', p' is not a full Abelian period of w if p is not a full Abelian period of w.*

Let $E_D = \{p \in Div(n) - Q_D \mid$ no $p' \in Div(n) - Q_D$ exists of which p is a proper divisor$\}$. In what follows, we focus on how to prevent any element in E_D from being a full Abelian period.

It follows from Lemmas 2 and 7 that for any output string w of $\mathcal{P}(n, D, \mathbf{B})$ and $p \in Div(n)$, $p \in FAP(w)$ iff $\mathbf{P}_{w[1..hp]} = \frac{hp}{d}\mathbf{B} = \frac{hp}{n}\mathbf{P}_w$ for any $1 \leq h < n/p$. The point is that we can notice whether $p \in FAP(w)$ or not when $w[1..n-p]$ is determined. Note that for any output string w of $\mathcal{P}(n, D, \mathbf{B})$ and any position $j \in S_D$, $\mathbf{P}_{w[1..j]} = \frac{j}{d}\mathbf{B}$ holds (see Definition 1 for the definition of S_D). For any $p \in E_D$, let $C_p = \{kp \mid 1 \leq k < \frac{n}{p}\} - S_D$. By the definition of S_D, p and $n-p$ are always in C_p. Then, our pruning method is based on the next corollary:

Corollary 4. *For any output string w of $\mathcal{P}(n, D, \mathbf{B})$ and $p \in E_D$, $p \in FAP(w)$ iff $\mathbf{P}_{w[1..j]} = \frac{j}{d}\mathbf{B}$ for any $j \in C_p$.*

Based on Corollary 4, we can prune the traversal at $w[1..n-p]$ when $p \in E_D$ satisfies the right-hand condition of Corollary 4. However, a naive checking procedure would take $O(n)$ time whenever we come to a string of length $n-p$ for some $p \in E_D$. Our time bound is achieved by showing that

- a data structure can be maintained during the traversal so that we can check the condition in constant time, and
- the size of the pruned search tree is bounded by a constant factor of the number of outputs.

A Data Structure to Check the Condition in Constant Time. First, we show the following lemma:

Lemma 8. *Let n be a positive integer and let $D \subseteq Div(n)$. For any $p \in E_D$, $C_p = \{j \mid \gcd(j, n) = p\}$.*

Proof. By definition, any positive integer j is in C_p iff j is divisible by p, but not divisible by some $q \in Q_D$. We show that the right-hand condition is equivalent to $\gcd(j, n) = p$.

If $\gcd(j, n) = p$, it is clear that j is divisible by p. It also tells that j is not divisible by some $q \in Q_D$, since otherwise, there must be some $q \in Q_D$ that divides p, which contradicts that $p \in E_D$.

For the opposite direction, assume on the contrary that $\gcd(j, n) \neq p$, i.e., p is a proper divisor of $\gcd(j, n)$ as x is divisible by p. Since j is not divisible by some $q \in Q_D$, neither is $\gcd(j, n)$. However, this implies that $\gcd(x, n) \in E_D$, which contradicts that $p \in E_D$. □

Lemma 8 implies that each position j corresponds to at most one element p in E_D with $\gcd(j, n) = p$. Then, in the preprocessing phase, we do the following:

Lemma 9. *For any positive integer n and $D \subseteq Div(n)$, we can compute E_D and for every $j \in E_D$ with $j \neq \min C_p$ its predecessor $\max\{j' \mid j > j' \in C_p\}$ in $O(n)$ time and space.*

While traversing strings, we maintain a bit vector V of length n dynamically. When we come to a string $w[1..i]$, $V[j] = 1$ iff $j \leq i$, there is $p \in E_D$ with $j \in C_p$, and $\mathbf{P}_{w[1..j']} = \frac{j'}{d}\mathbf{B}$ for any $j' \in C_p$ with $j' \leq j$. Now consider $\mathcal{P}(n, D, \mathbf{B})$ extends $w[1..i]$ to $w[1..i]c$. Provided that we know if $\mathbf{P}_{w[1..i]c} = \frac{i+1}{d}\mathbf{B}$ or not, it is easy to update the bit vector in constant time: We only need to update $V[i+1]$ to be 1 iff $\mathbf{P}_{w[1..i]c} = \frac{i+1}{d}\mathbf{B}$, there is $p \in E_D$ with $i+1 \in C_p$, and $V[j'] = 1$, where j' is a predecessor of $i+1$ in C_p (the last condition is ignored if $i+1$ does not have a predecessor).

Using V we can check the right-hand condition of Corollary 4 in constant time: The condition holds iff $V[n-p]$ becomes 1 for some $p \in E_D$.

We have two issues left: How to update V when $\mathcal{P}(n, D, \mathbf{B})$ extends $w[1..i]$ to $w[1..i]c^k$ with $k > 1$; how to check if $\mathbf{P}_{w[1..i]c} = \frac{i+1}{d}\mathbf{B}$ or not in constant time.

Regarding the former issue, note that the situation occurs only when $|w[1..i]c^k| \in S_D$, and thus, $\mathbf{P}_{w[1..i]c^k} = \frac{i+k}{d}\mathbf{B}$. Since $\mathbf{B}$ contains at least two distinct characters, $\mathbf{P}_{w[1..i]c^{k'}} = \frac{i+k}{d}\mathbf{B} \ominus \mathbf{P}_{c^{k-k'}} \neq \frac{i+k'}{d}\mathbf{B}$ for any $k' < k$. Then, actually we do nothing on V in this case, i.e., leave $V[i+1..i+k]$ as 0's.

Regarding the latter issue: When we come to a string $w[1..i]$, we maintain the maximum value in $\{\mathbf{P}_{w[1..i]}[h]/\mathbf{B}[h] \mid c_h \text{ is a character in } \mathbf{B}\}$ and the number of distinct characters achieving this value. We can notice that $\mathbf{P}_{w[1..i]} = \frac{i}{d}\mathbf{B}$ when all characters in $\mathbf{B}$ achieve the maximum value. It is easy to maintain these values during the traversal as the values may only increase. We use $O(n)$ space to store the values for all ancestors.

The Size of the Pruned Search Tree. We estimate the size of the search tree of $\mathcal{P}'(n, D, \mathbf{B})$, which is an augmented version of $\mathcal{P}(n, D, \mathbf{B})$ with the pruning method based on Corollary 4. The nodes at which we pruned the traversal are said to be *bad*, and the other nodes are said to be *good*. We identify a node by its corresponding string. A good leaf corresponds to an output string.

The following two lemmas give properties on good/bad nodes.

Lemma 10. *Let u be a good node with $|u| < n$. At most one child of u is bad.*

Lemma 11. *Let u be a good node with $|u| < n$. At least one child of u is good.*

Lemma 11 implies that a good node contains at least one output in its subtree. Further, we can bound the size of the pruned search tree as follows:

Lemma 12. *The size of the search tree of $\mathcal{P}'(n, D, \mathbf{B})$ is bounded by a constant factor of the number of good leaves.*

Proof. We consider the tree obtained by deleting bad nodes from the search tree and show that three non-branching nodes in the tree cannot be in a row. For any non-branching node u, one of the following conditions holds for some character c; (1) its child is uc^k with $k \geq 1$ and $|uc^k| = S_D$, (2) there is a bad node uc (in the original search tree). Assume that the child of u is not a leaf.

- Case (1): The next character will be generated by a fresh enumerator having at least two distinct characters. Since $\mathbf{P}_{uc^k} = \frac{|uc^k|}{d}\mathbf{B}$, $\mathbf{P}_{uc^k c'} \neq \frac{|uc^k c'|}{d}\mathbf{B}$ for any character c', and hence, uc^k is a branching node.

– Case (2): Since $\mathbf{P}_{uc} = \frac{|uc|}{d}\mathbf{B}$, $\mathbf{P}_{uc'c''} \neq \frac{|uc'c''|}{d}\mathbf{B}$ for any characters c' and c''. Hence, the child of u cannot be a non-branching node of Case (2).

By the above argument, a path of non-branching nodes can be of length two, which is composed of nodes of Case (2) followed by Case (1), but cannot be longer. Therefore, the number of internal nodes is bounded by a constant factor of the number of good leaves. By Lemma 10, the number of bad nodes is at most the number of internal nodes, and hence, the statement holds. □

Complexities. Putting all together, we get the following result:

Theorem 3. *Problem 3 can be solved in* $O(n + |\mathcal{S}'(n, D, \Sigma)|)$ *time and* $O(n)$ *working space.*

Proof. We use Lemma 9 to preprocess the input in $O(n)$ time so that each backtracking can be conducted in $O(1)$ time. We enumerate $\mathcal{S}'(n, D, \Sigma)$ by running $\mathcal{P}'(n, D, \mathbf{B})$ for every $\mathbf{B} \in \mathbb{B}_d$ (excluding $\mathbf{B}$ consisting of a single kind of characters). By Lemma 12, the size of the search tree is bounded by the number of output strings. By taking the search trees as a representation of the output strings, we achieve the time complexity. We only need $O(n)$ working space. □

References

1. Apostol, T.M.: Introduction to Analytic Number Theory. Undergraduate Texts in Mathematics. Springer, Heidelberg (1976)
2. Blanchet-Sadri, F., Simmons, S., Tebbe, A., Veprauskas, A.: Abelian periods, partial words, and an extension of a theorem of Fine and Wilf. RAIRO - Theor. Inf. Appl. **47**(3), 215–234 (2013)
3. Cazaux, B., Rivals, E.: Reverse engineering of compact suffix trees and links: a novel algorithm. J. Discrete Algorithms **28**, 9–22 (2014)
4. Constantinescu, S., Ilie, L.: Fine and Wilf's theorem for Abelian periods. Bull. EATCS **89**, 167–170 (2006)
5. Crochemore, M., Iliopoulos, C.S., Kociumaka, T., Kubica, M., Pachocki, J., Radoszewski, J., Rytter, W., Tyczynski, W., Walen, T.: A note on efficient computation of all Abelian periods in a string. Inf. Process. Lett. **113**(3), 74–77 (2013)
6. Fici, G., Lecroq, T., Lefebvre, A., Élise Prieur-Gaston, Smyth, W.F.: Quasi-linear time computation of the Abelian periods of a word. In: PSC 2012, pp. 103–110 (2012)
7. Fici, G., Lecroq, T., Lefebvre, A., Prieur-Gaston, E.: Computing Abelian periods in words. In: PSC 2011, pp. 184–196 (2011)
8. Fine, N.J., Wilf, H.S.: Uniqueness theorems for periodic functions. Proc. Am. Math. Soc. **16**, 109–114 (1965)
9. Gawrychowski, P., Jeż, A., Jeż, Ł.: Validating the Knuth-Morris-Pratt failure function, fast and online. Theory Comput. Syst. **54**(2), 337–372 (2014)
10. Knuth, D.E., Morris, J.H., Pratt, V.R.: Fast pattern matching in strings. SIAM J. Comput. **6**(2), 323–350 (1977)
11. Kociumaka, T., Radoszewski, J., Rytter, W.: Fast algorithms for Abelian periods in words and greatest common divisor queries. In: STACS 2013, pp. 245–256 (2013)

12. Lifshits, Y.: Processing compressed texts: a tractability border. In: Ma, B., Zhang, K. (eds.) CPM 2007. LNCS, vol. 4580, pp. 228–240. Springer, Heidelberg (2007)
13. Lothaire, M.: Combinatorics on Words. Cambridge Mathematical Library, Cambridge (1997)
14. Nakashima, Y., Okabe, T., I, T., Inenaga, S., Bannai, H., Takeda, M.: Inferring strings from Lyndon factorization. In: Csuhaj-Varjú, E., Dietzfelbinger, M., Ésik, Z. (eds.) MFCS 2014, Part II. LNCS, vol. 8635, pp. 565–576. Springer, Heidelberg (2014)
15. Parikh, R.: On context-free languages. J. ACM **13**(4), 570–581 (1966)
16. Starikovskaya, T., Vildhøj, H.W.: A suffix tree or not a suffix tree? J. Discrete Algorithms **32**, 14–23 (2015)
17. Takaoka, T.: An $O(1)$ time algorithm for generating multiset permutations. In: Aggarwal, A.K., Pandu Rangan, C. (eds.) ISAAC 1999. LNCS, vol. 1741, pp. 237–246. Springer, Heidelberg (1999)
18. Williams, A.: Loopless generation of multiset permutations using a constant number of variables by prefix shifts. In: SODA 2009, pp. 987–996 (2009)

Toehold DNA Languages are Regular (Extended Abstract)

Sebastian Brandt, Nicolas Mattia, Jochen Seidel(✉), and Roger Wattenhofer

ETH Zurich, Zurich, Switzerland
{brandts,nmattia,seidelj,wattenhofer}@ethz.ch

Abstract. We explore a method of designing algorithms using two types of DNA strands, namely rule strands (rules) and input strands. Rules are fixed in advance, and their task is to bind with the input strands in order to produce an output. We present algorithms for divisibility and primality testing as well as for square root computation. We measure the complexity of our algorithms in terms of the necessary rule strands. Our three algorithms utilize a super-constant amount of complex rules.

Can one solve interesting problems using only few—or at least simple—rule strands? Our main result proves that restricting oneself to a constant number of rule strands is equivalent to deciding regular languages. More precisely, we show that an algorithm (possibly using infinitely many rule strands of arbitrary length) can merely decide regular languages if the structure of the rules themselves is simple, i.e., if the rule strands constitute a regular language.

1 Introduction

DNA is sometimes considered as an alternative to orthodox silicon-based technologies for computing. But how powerful can a DNA-based computer be? In this paper, we analyze the computational expressiveness of toehold DNA computing, c.f. [15,20]. In the toehold method, a DNA computation takes place in a *soup*, i.e., a container filled with solution, in which DNA strands are floating around and binding with each other. Designing a DNA algorithm is equivalent to designing a set of DNA strands (the *rule strands*) such that, when strands representing an input (*input strands*) are added, a desired output, e.g., an indicator for YES or NO, is produced. An "execution" of a DNA algorithm corresponds to multiple steps of *strand binding*, where initially only rule and input strands bind, incrementally forming larger and larger molecules that become available for binding. To ensure that bindings occur in the desired manner, special care must be taken when designing the DNA strands. The details are explained in Sect. 2, where we also define how DNA algorithms formally operate.

Using DNA programming techniques described in Sect. 3, DNA algorithms for square root computing and primality testing can be implemented. Our algorithms (presented in the full version) use a super-constant amount of rules.

All algorithms and proofs are presented in the full version of this paper, available at http://disco.ethz.ch/publications/ISAAC2015-dna.pdf.

K. Elbassioni and K. Makino (Eds.): ISAAC 2015, LNCS 9472, pp. 780–790, 2015.
DOI: 10.1007/978-3-662-48971-0_65

Moreover, the rules are rather complex in the sense that the description of each strand relies on the ability to count, and on the knowledge of an upper bound for the input. It is therefore natural to ask whether this complexity is necessary. Our main contribution is to answer this question affirmatively. Specifically, in Sect. 4 we show that if the rule strands describing some algorithm $\mathcal{A}$ form a regular language (i.e., are not complex), then $\mathcal{A}$ decides some regular language. We further establish that a constant number of rules is sufficient to express any DNA algorithm deciding a regular language.

Related Work. Utilizing DNA for the purpose of computation was first explored by Adleman [2], who solved a seven-city instance of the Hamiltonian path problem, and Lipton [10], who suggested a method to solve the SATISFIABILITY problem. After these first algorithmic usages of DNA, the idea to use the computational capacity to control devices on a molecular level emerged (see, e.g., [5]). These early techniques rely on enzymes to perform the desired task.

Our studies are motivated by the rise of the toehold exchange method [15], which is a way to perform DNA computations without relying on enzymes. The benefit of this method is that synthesizing the DNA strands needed for it (see, e.g., [19], in particular the supplementary material) is a simpler process than producing the building blocks for enzyme-based DNA computations. Strand displacement using toehold exchange is described thoroughly in [20] and simulated in [9]. The toehold technique enables the design of DNA circuits using only DNA strands, thus disposing of the necessity of other molecules, like enzymes. An essential building block in toehold-based computation is the *seesaw gate* [14], which allows the design of arbitrary DNA circuits. The seesaw gate was used to, e.g., compute approximate majority [6] with a protocol analyzed in [3]. Following this line of work, we utilize toeholds to initiate strand binding.

It is well understood that DNA molecules can form complex structures (see, e.g., [12,16] for an overview). For instance, Winfree [18] investigated how DNA strands can bind to form linear duplex strands, or duplex "tree" strands using junctions, which connect more than one strand in one point. He found that the linear strands correspond to a transition sequence in a finite automaton, whereas the trees correspond to a derivation tree of a context-free grammar. Note that these two structures do not return an output in the classical sense—the only "output" is a DNA molecule in which every base is bound. In contrast to that, in our work, we add input to the strand binding process, and are interested in an output, e.g., in the form of a YES or NO answer.

The techniques used in Adleman's construction inspired the study of so-called sticker systems [7], which are a generalized version of binding processes where the binding relation is not necessarily symmetric. Other studied language operations that are motivated by DNA interactions encompass the superposition [4], the PA-matching [8], and the hairpin [11,13] operator. This line of work examines the effect of these operations on a language's classification within the Chomsky hierarchy. In our studies, we also utilize methods from formal language theory in order to describe the strand binding process, which ultimately allows us to derive a lower bound on the complexity of DNA algorithms for non-regular languages.

2 Model

DNA Basics. DNA strands consist of nucleotides (or bases), linked together in a specific order. There are four types of nucleotides: Adenine, Cytosine, Guanine and Thymine (or simply A, C, G, and T). At an atomic level, DNA strands have two types of extremities: a so-called 5' end on one side, and a 3' end on the other side. The two extremities assign a *direction* to a DNA strand, and throughout this paper we orient nucleotide sequences of DNA strands from the 5' end towards the 3' end. When displayed, an arrow indicates the direction from the 5' end to the 3' end, as in Fig. 1a.

An A can *bind* (via hydrogen bonds) with a T on a different strand, and similarly a G can bind with a C. This is the process that gives DNA its stability, and its helicoidal structure. Pairs of nucleotides which are able to bind in this manner are called *Watson-Crick-complementary* (or simply *WK-complementary*). That is, A and T are WK-complementary, and G and C are WK-complementary. Sequences of nucleotides can bind as well, given that those sequences have opposite directions and are complementary. For instance, the strands ATCG and CGAT from Fig. 1a can bind completely, whereas the two strands ATCG and TAGC (reversed CGAT) cannot.

Notation. To ease readability, sequences of nucleotides are commonly grouped into so-called *domains*. A domain is represented by a single character displayed in `teletype` font. Two domains `g` and `h` are complementary to each other if the nucleotide sequence of `g` is the WK-complement of `h`'s sequence. Complementary domains will be represented as overlined, e.g., $\mathtt{g} = \bar{\mathtt{h}}$. Note that for a sequence `gh`, the WK-complementary $\overline{\mathtt{gh}}$ is $\bar{\mathtt{h}}\bar{\mathtt{g}}$. Please refer to Fig. 1b for an illustration.

For a set S of strands we denote by $\bar{S}$ the set containing all WK-complements of strands in S. For two strands σ, τ, the *(concatenated) strand* $\sigma\tau$ is the strand obtained from concatenating the nucleotide sequences of σ and τ. For two sets S, T of strands we write ST for the set $\{\sigma\tau : \sigma \in S, \tau \in T\}$. Similarly, when σ is a strand, we also write σS and $S\sigma$ for the sets $\{\sigma\}S$ and $S\{\sigma\}$, respectively. For positive integers i, we write S^i for the set SS^{i-1}, and by convention S^0 contains only the empty strand ε. We denote by S^* the set $\cup_{i \geq 0} S^i$. The notation for sets of strands naturally extends to sets of domains, which can be viewed as sets containing the corresponding strands.

$\overrightarrow{\text{ATCG}}$
$\overleftarrow{\text{TAGC}}$

(a)

$\overrightarrow{\mathtt{abc}} \equiv \overrightarrow{\text{ATCGATTCTC}}$
$\overleftarrow{\text{TAGCTAAGAG}} \equiv \overleftarrow{\bar{\mathtt{a}}\bar{\mathtt{b}}\bar{\mathtt{c}}} \equiv \overrightarrow{\bar{\mathtt{c}}\bar{\mathtt{b}}\bar{\mathtt{a}}} \equiv \overrightarrow{\overline{\mathtt{abc}}}$

(b)

Fig. 1. (a) Two complementary DNA strands. The arrowhead shows the 3' end. (b) The strand `abc` composed of the three domains $\mathtt{a} \equiv$ ATCG, $\mathtt{b} \equiv$ ATT, and $\mathtt{c} \equiv$ CTC binds with its complement $\overline{\mathtt{abc}}$.

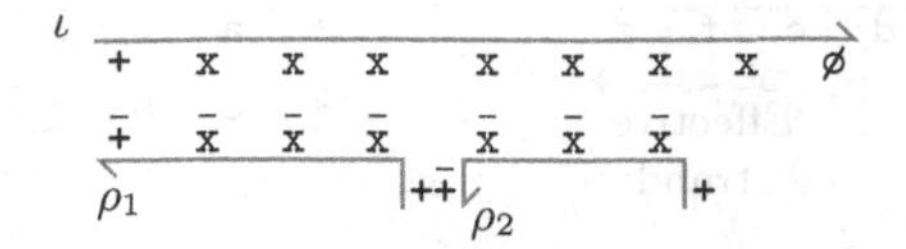

Fig. 2. Checking whether 7 is divisible by 3 using DNA strands. The upper strand ι represents the input in unary. Both strands ρ_1 and ρ_2 of the form $\rho = +\bar{x}^3\bar{+}$ bind to ι, first ρ_1 and then ρ_2. The unmatched domains +xø represent the remainder

Example. Consider, as an example, the question whether some integer d is divisible by another integer q (see Fig. 2). To answer this question, let +, ø, and x be domains, and denote by $\iota = +x^d ø$ the strand composed of d repetitions of x, delimited by + and ø. We refer to ι as the *input strand* for our question.

The idea is to successively let ι bind with multiple copies of the strand $\rho = +\bar{x}^q\bar{+}$. Note that ρ can only partially bind to ι. In particular, the first copy of ρ interacting with ι will bind with one + and q x domains. Since DNA is not completely rigid, the unmatched + domain of ρ becomes available for binding. Next, a second copy of ρ binds with the + part from the previous ρ-strand and the next q x domains of the input strand. Each step corresponds to subtracting q, and we end the process when the remainder of the division is left. An "output" can now be obtained by checking for all possible remainders of the division.

Rule Strands and DNA Algorithms. Let $\mathcal{U}$ be a universe of domains so that $\mathcal{U} \cap \bar{\mathcal{U}} = \emptyset$, i.e., if some domain x is in $\mathcal{U}$, then $\bar{x}$ is not. Let $\Sigma \dot{\cup} \Delta \dot{\cup} \Lambda$ be a partition of $\mathcal{U}$. We refer to Σ as the set of *input domains*, to $\Delta \cup \overline{\Delta}$ as the set of *delimiter domains* (*delimiters* for short), and to $\Psi = \Sigma \cup \overline{\Sigma} \cup \Lambda \cup \overline{\Lambda}$ as the set of *rule domains*. In the strand binding process, the delimiter domains will function as *toeholds* that initiate the binding. A strand $\rho \in d_1\Psi^*d_2$, where d_1 and d_2 are delimiters, is referred to as a *rule strand*. A collection $\mathcal{A}$ of rule strands is called a *DNA algorithm*. The input to a DNA algorithm is specified in the form of an *input strand*, which is a strand ι of the form $+\Sigma^*ø$, where + and ø are two fixed delimiters chosen from the set Δ.

A DNA algorithm is "executed" in a *soup*, i.e., a container filled with solution, in which the rule strands of some algorithm $\mathcal{A}$ are floating around. The execution is initiated by adding the input strand ι. We assume that all strands in the soup, i.e., the rule strands and the input strand, are present sufficiently many times.

All strands in the soup share the property of starting and ending with delimiters, and that delimiters appear only at the ends of a strand. When two strands σ and τ meet, they may bind and form a new strand which we call an *effective strand*, see Fig. 3. The binding occurs along some prefix of σ and a corresponding complementary suffix of τ (or the other way around, when the roles of σ and τ are switched). This means that two strands always bind (at least) at two complementary delimiters. The effective strand resulting of this binding is composed of the unmatched prefix of τ and the unmatched suffix of σ, in this order. Note that the new effective strand also has the form of a rule strand. Effective strands behave like any other strand, and from now on we will not distinguish between

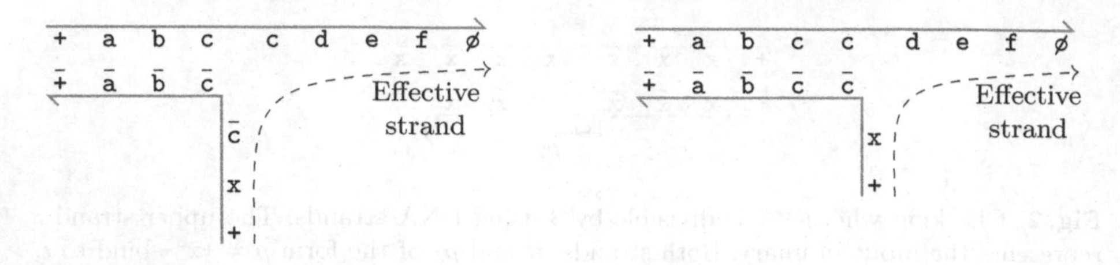

Fig. 3. Two examples of strand binding. On the left, not all possible domain bindings are involved, since one more c on the upper strand could bind with one more c̄ of the lower strand; the resulting effective strand is +xc̄cdefø. On the right, all domains have bound; the resulting effective strand is +xdefø;

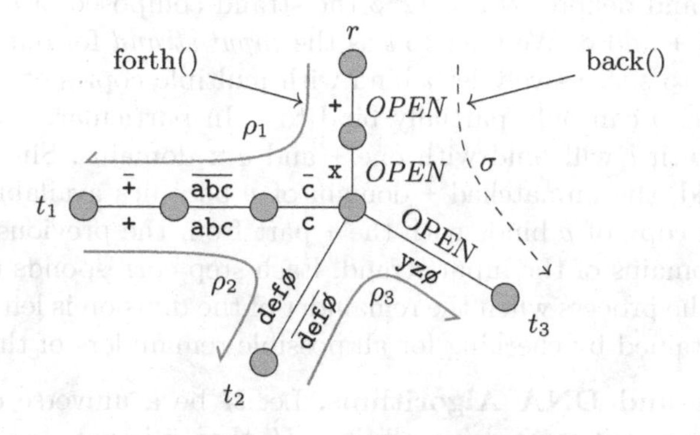

Fig. 4. An assembly T for σ = +xyzø with root r. The strands read in a pre-order traversal between the leaf pairs (r, t_1), (t_1, t_2), and (t_2, t_3) are ρ_1 = +xc̄c̄b̄ā+̄, ρ_2 = +abccdefø, and ρ_3 = ø̄f̄ēd̄yzø, respectively.

strands and effective strands. We will use the terms *rule strand* and *input strand* to stress when a strand is not an effective one.

The process of strands binding with each other represents the computation performed by $\mathcal{A}$. The resulting structure can be described as a tree, see Fig. 4.

Definition 1. *Fix a set S of strands. Let $T = (V, E, \text{forth}, \text{back})$ be a rooted ordered*[1] *tree, with nodes and edges in V and E, respectively, and two functions* forth, back, *each assigning a label from $(\mathcal{U} \cup \bar{\mathcal{U}})^* \dot{\cup} \{\text{OPEN}\}$ to every edge in E, where* OPEN *is a special value. Denote by r the root of t, and for convenience consider r to be a leaf. We say that T is an S*-assembly for σ *(or just* assembly for σ *when S is clear from the context) if T satisfies the following conditions.*

(i) For all $e \in E$, if $\text{back}(e) \neq \text{OPEN}$, then $\text{back}(e) = \overline{\text{forth}(e)}$.
(ii) There is a unique path p from r to its rightmost descendant, s.t. the concatenated forth *labels on p are σ, all edges e on p have the label $\text{back}(e) =$ OPEN, and no other edges in T have an* OPEN *label.*

[1] A tree is ordered if the children of every node are ordered, e.g., from left to right.

(iii) *Consider any two leafs* t_1, t_2 *with* $t_2 \neq r$, *so that* t_1 *is the last leaf visited before* t_2 *in a pre-order traversal of* T, *and denote by* $t_{1,2}$ *their nearest common ancestor. The word* g *obtained by concatenating the* back *labels on the path from* t_1 *to* $t_{1,2}$ *and the* forth *labels on the path from* $t_{1,2}$ *to* t_2 *is in* S.

In this work, we focus on decision problems, i.e., problems where the output is either YES or NO. Let ι be an input strand. We say that $\mathcal{A}$ *accepts* ι if there is a $(\mathcal{A} \cup \{\iota\})$-assembly for $+\emptyset$. Thus, the ability to produce the strand $+\emptyset$ corresponds to a YES output of $\mathcal{A}$, whereas the absence thereof corresponds to a NO output. We denote by $\mathcal{L}(\mathcal{A})$ the set of input strands accepted by $\mathcal{A}$, and say that $\mathcal{A}$ *decides* $\mathcal{L}(\mathcal{A})$. In practice, the strand $+\emptyset$ can be detected using fluorescence techniques [1].

Finite Automata. For a deterministic finite automaton (DFA) B we denote by $\mathcal{L}(B)$ the *regular language* accepted by B. Please refer to a standard textbook (e.g., [17]) for a thorough introduction to formal languages.

3 DNA Algorithms

When designing (DNA) algorithms it is convenient to use building blocks for solving reoccurring tasks. We will now introduce three such building blocks which we use in our algorithms, namely *gluing*, *substituting*, and *aggregating*. For that, in the remainder of this section, let $\mathtt{d}_i, 1 \leq i \leq 4$, be delimiters, and let p, x, s, y be arbitrary sequences of domains from Ψ^*.

Gluing. The gluing building block transforms the two strands $\sigma_1 = \mathtt{d}_1 p x \mathtt{d}_2$ and $\sigma_2 = \mathtt{d}_3 y s \mathtt{d}_4$ into the new strand $\mathtt{d}_1 p s \mathtt{d}_4$, formed of the prefix $\mathtt{d}_1 p$ of σ_1 and the suffix $s\mathtt{d}_4$ of σ_2. This is achieved by using the *gluing strand* $\overline{\mathtt{d}_2}\,\overline{x}\,\overline{y}\,\overline{\mathtt{d}_3}$. The strand binding is illustrated in Fig. 5a.

Substituting. The purpose of the substituting building block is to substitute a *whole* strand $\sigma_1 = \mathtt{d}_1 x \mathtt{d}_2$ with a strand $\sigma_2 = \mathtt{d}_3 y \mathtt{d}_4$. In general, this cannot be achieved with a single rule strand since more than two delimiter domains would be required to appear on it. Instead, in our algorithms, we introduce two new domains u and v for every substitution from x to y of the above form.

The replacement is now performed in three steps using three rule strands τ_1, τ_2, and τ_3 (illustrated in Fig. 5b) as follows. First, the strand $\tau_1 = \mathtt{d}_3 \mathtt{u} \overline{\mathtt{x}}\,\overline{\mathtt{d}_1}$ binds with σ_1 to form the intermediate strand $\iota_1 = \mathtt{d}_3 \mathtt{u} \mathtt{d}_2$. Next, the strand $\tau_2 = \overline{\mathtt{d}_2}\,\overline{\mathtt{u}} \mathtt{v} \mathtt{d}_4$ is used to form the intermediate strand $\iota_2 = \mathtt{d}_3 \mathtt{v} \mathtt{d}_4$ by binding with ι_1. In the last step, the strand $\tau_3 = \mathtt{d}_3 y \overline{\mathtt{v}}\,\overline{\mathtt{d}_3}$ binds with ι_2 to form the desired strand σ_2. As a short-hand for these three rule strands, we write the *substitution rule* $\sigma_1 \rightarrow \sigma_2$, e.g., $\mathtt{+abc}\emptyset \rightarrow \mathtt{+z}\emptyset$.

The above rule strands ensure that if only τ_1, or only τ_1 and τ_2 are applied, but not τ_3, then either u or v remain on the effective strand. Since for each substitution rule new domains u and v are introduced, they cannot be matched by any strand that is not from this substitution. Thus, we may assume that substitutions are either applied in full, or not at all. (The strands obtained by applying at most two of the three rules have no effect on the soup's output.)

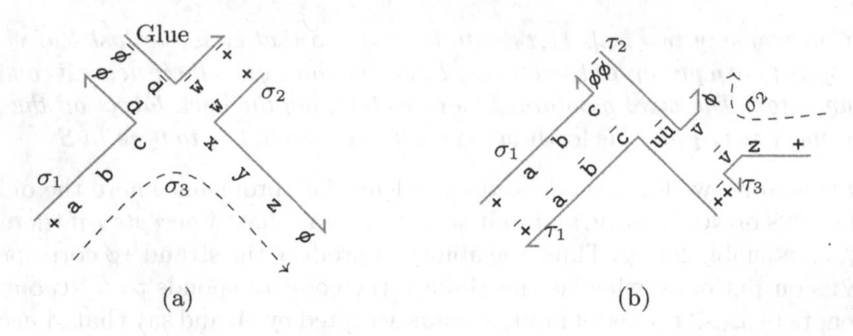

Fig. 5. (a) Gluing the two strands σ_1 and σ_2 with $\overline{\text{ødw+}}$ yields the effective strand σ_3. (b) Substituting the strand $\sigma_1 = \texttt{+abc}\emptyset$ by $\sigma_2 = \texttt{+z}\emptyset$ using the three strands τ_1, τ_2, and τ_3. The effective strand indicated by the dashed line is σ_2, as desired.

Aggregating. By combining the two building blocks, it is possible to *aggregate* two whole strands $\sigma_1 = \texttt{d}_1x\texttt{d}_2$ and $\sigma_2 = \texttt{d}_3y\texttt{d}_4$ into a strand $\sigma_3 = \texttt{d}_5z\texttt{d}_6$, i.e., σ_3 is only obtained if both σ_1 and σ_2 are present. For that, let u and v be new domains. We add the following strands: (1) the substitution $\sigma_1 \to \texttt{+u}z\texttt{d}_6$, (2) the substitution $\sigma_2 \to \texttt{d}_5\texttt{v}\emptyset$, and (3) the gluing strand $\overline{\emptyset\texttt{vu+}}$. We abbreviate this set of rules by writing $\sigma_1 \wedge \sigma_2 \to \sigma_3$, and note that also larger aggregations are possible by applying the principle inductively.

Synopsis. Using the above techniques, the idea for divisibility testing explained in Sect. 2 can be extended to obtain a DNA algorithm for primality testing. The details are deferred to the full version of this paper. In its unary version the algorithm consists of $O(n)$ rules. Note that with this many rules it is possible to describe a simple DNA algorithm that matches every strand ι_p, with p prime, to produce a YES-output. With binary inputs it is possible to devise a primality testing algorithm using $O(\sqrt{n}\log n)$ rules, i.e., less than the number of primes $p \leq n$, which is $\Theta(n/\log n)$. Regardless of the way the input is presented, enumerating the rules for our primality testing algorithm requires knowledge of $\sqrt{n}$. In the full version we also present an algorithm to actually compute $\sqrt{n}$ using DNA strands.

4 Regular DNA Algorithms

The rule sets corresponding to our algorithms form context-free or even context-sensitive languages, i.e., the rules for the two algorithms are "complex". One might hope that this complexity is not necessary. However, in this section we show that such complex rule sets cannot be avoided (for the aforementioned problems), and that in fact rule sets that form a regular language are not very powerful algorithms. Specifically, we are going to establish the following theorem.

Theorem 1. *Let $\mathcal{A}$ be a DNA algorithm. If $\mathcal{A}$, interpreted as a set of strands, is a regular language, then $\mathcal{L}(\mathcal{A})$ is regular.*

For the remainder of this section, fix some DNA algorithm $\mathcal{A}$ such that $\mathcal{A}$ is a regular language. It will be convenient to assume that $+\bar{+}$ and $\bar{\phi}\phi$ are both in $\mathcal{A}$. While this changes neither the language detected by $\mathcal{A}$ nor the fact that $\mathcal{A}$ is a regular language, the assumption allows us to consider only $(\mathcal{A} \cup \{\iota\})$-assemblies for $+\phi$ that begin and end with strands from $\mathcal{A}$. We denote by $B = (Q, \mathcal{U} \cup \bar{\mathcal{U}}, \delta, s, F)$ a DFA satisfying $\mathcal{L}(B) = \mathcal{A}$.

Consider any accepted input strand ι and an $(\mathcal{A} \cup \{\iota\})$-assembly T of $+\phi$. Note that forth(e) contains a delimiter if and only if one endpoint of e is a leaf or the root. By definition, in a pre-order traversal of T the paths between two successive leafs are labeled with strands $(\rho_1, \ldots, \rho_l)$ in $\mathcal{A}$. For each strand $\rho_i \in \mathcal{A}$ there is an accepting transition sequence in B, which consequently corresponds to the path in T from which ρ_i was obtained. Our life would be simple if $\rho_i \in \mathcal{A}$ for all i, since we would have to deal only with strands from a regular language. The main difficulty in our proof of Theorem 1 is to handle the case where $\rho_i = \iota$.

In our proof of Theorem 1 we use assemblies to investigate how exactly the strands from $\mathcal{A}$ and the input strands bind to form $+\phi$. To describe the language $\mathcal{L}(\mathcal{A})$, we consider all possible assemblies for $+\phi$ that can be formed with some input strand and strands from $\mathcal{A}$. More precisely, we ask the question: What are the possible input strands with which $+\phi$ can be assembled? A key ingredient to answering this question is the notion of a *junction*, which will allow us to answer the question in a recursive manner.

Definition 2. *Let Q be a set of states, let $v, w \in Q$, and let $J \subseteq Q \times Q$ with $(v, w) \notin J$. The triple (v, w, J) is called a* junction. *An* instance of the junction *(v, w, J) is a sequence $q = (q_1, \ldots, q_\ell)$ with entries in $Q \dot{\cup} \{\text{NULL}\}$, where* NULL *is a special value not contained in Q, satisfying*

(i) $q_1 = v$ and $q_\ell = w$,
(ii) for all i, if $q_i \neq \text{NULL}$, then either $(q_i, q_{i+1}) \in J \cup \{(v, w)\}$, or $q_{i+1} = \text{NULL}$ and $(q_i, q_{i+2}) \in J \cup \{(v, w)\}$, and
(iii) all entries in q are pairwise distinct, except possibly q_1 and q_ℓ (when $v = w$).

The basic idea behind our proof is to define a language $\mathcal{I}$ such that (1) $\mathcal{I}$ is regular, (2) every $\iota \in \mathcal{I}$ is accepted by $\mathcal{A}$, and (3) every ι accepted by $\mathcal{A}$ is in $\mathcal{I}$. The definition of $\mathcal{I}$ is encapsulated in the recursive *sealing operator* $\mathcal{X}$, which we will define shortly. Claim (1) will then be established by the fact that the recursion $\mathcal{X}$ is finite. Claims (2) and (3) will be confirmed by relating the recursion $\mathcal{X}$ with appropriate assemblies, for which we will identify nodes in an assembly with junctions.

For the definition of $\mathcal{X}$, we set the empty union to $\emptyset$ and the empty intersection to $+\Sigma^*\phi$, i.e., all valid input strands. Let $I(v, w, J)$ be the set of instances of the junction (v, w, J). The sealing operator is defined using two sub-operators C and D as follows.

$$\mathcal{X}(v, w, J) := \bigcup_{q \in I(v,w,J)} \left(\bigcap_{\substack{(q_i, q_{i+1}): \\ q_i \neq NULL \neq q_{i+1}}} C(q_i, q_{i+1}, J) \cap \bigcap_{i: q_i = NULL} D(q_{i-1}, q_{i+1}, J) \right)$$

For $C(x,y,J)$, we denote by $H_{x,y,J}$ the set of pairs $(z_1,z_2) \in J \setminus \{(x,y)\}$ for which there are two transition sequences $x \to z_1$, $z_2 \to y$ in B such that the corresponding words w_1 and w_2 satisfy $w_1 = \overline{w_2}$. Note that in particular $H_{x,y,J}$ is finite, since it is contained in J. Now, the sets C are defined as follows.

$$C(x,y,J) := \begin{cases} +\Sigma^*\emptyset, & \text{if } \exists (z_1,z_2) \in H_{x,y,J} \text{ s.t. } z_2 = s \text{ and } z_1 \in F \\ \bigcup_{(z_1,z_2)\in H_{x,y,J}} \mathcal{X}(z_1,z_2,J \setminus \{(x,y),(z_1,z_2)\}), & \text{otherwise.} \end{cases}$$

For $D(x,y,J)$, we construct a finite automaton $B_{K,x,y}$ with ε-transitions as follows: Let B_K be the automaton B supplemented with the transitions $\delta(k_1,\varepsilon) = k_2$ for all $(k_1,k_2) \in K$. Next, let B'_K be a copy of B_K, and denote by x' the copy of the state x in B'_K. The automaton $B_{K,x,y}$ is now obtained by taking B_K's starting state, B'_K's accepting states, and adding the transition $\delta(y,\varepsilon) = x'$. With this, the sets D are defined as follows.

$$D(x,y,J) := \bigcup_{K \subseteq J \setminus \{(x,y)\}} \left(\overline{\mathcal{L}(B_{K,x,y})} \cap \bigcap_{(z_1,z_2)\in K} \mathcal{X}(z_1,z_2,J \setminus \{(x,y),(z_1,z_2)\}) \right).$$

Basically, $\mathcal{X}$ assigns a language to any junction (v,w,J). The intricate choice of $\mathcal{X}$ provides that $\mathcal{X}(v,w,J)$ contains exactly the input strands that "seal" the junction (v,w,J) with a sub-tree T' of some assembly T, such that in T' only junctions (v',w',J') with $(v',w') \notin J$ and $J' = J \setminus \{(v',w')\}$ appear.

We set the language $\mathcal{I}$ to

$$\mathcal{I} := \bigcup_{w:\delta(w,\emptyset)\in F} \mathcal{X}(\delta(s,+),w,Q \times Q \setminus \{(\delta(s,+),w)\}),$$

and follow the plan to prove Theorem 1 as outlined above. The first step is to show that $\mathcal{I}$ is regular, which is asserted by the following lemma.

Lemma 1. *For any $v,w \in Q$ and $J \subseteq Q \times Q \setminus \{(v,w)\}$, the language $\mathcal{X}(v,w,J)$ is regular.*

Our proofs for Lemma 1 and all remaining lemmas and theorems are presented in the full version of this paper. In our effort to establish Theorem 1, the following lemma confirms the aforementioned claim (2).

Lemma 2. *If $\iota \in \mathcal{I}$, then ι is accepted by $\mathcal{A}$.*

To obtain the opposite direction of the statement in Lemma 2, consider any $(A \cup \{\iota\})$-assembly T for $+\emptyset$. We now describe a procedure that assigns an additional label to T, thus obtaining the *state-marked assembly* T'. Specifically, to each node u in T, we assign $\deg(u)+1$ labels $\text{qmark}(u,1),\ldots,\text{qmark}(u,\deg(u)+1)$ from $Q\dot\cup\{NULL\}$, where $\deg(u)$ denotes the number of u's children.

To obtain the labels, we traverse T "together with B", feeding to B the forth labels when traversing an edge to the ith child the first time (in forward direction), and set $\text{qmark}(u, i)$ to B's current state. On the way back to u's parent, we assign the label $\text{qmark}(u, \deg(u) + 1)$, respectively. At leaf nodes, the DFA B is reset to its starting state.

We will later use the sequence defined for a node u by the qmark labels to obtain junctions and junction instances corresponding to u. For that, it is convenient to define two short-hands for reading the labels qmark. Consider any node u and denote by $u_1, \dots, u_{\deg(u)}$ the children of u. The first short-hand, qpair, assigns to each edge (u, u_i) a pair of states as follows.

$$\text{qpair}(u, u_i) := \begin{cases} (\text{qmark}(u, i-1), \text{qmark}(u, i+1)), & \text{if } \text{qmark}(u, i) = \mathit{NULL} \\ (\text{qmark}(u, i), \text{qmark}(u, i+2)), & \text{if } \text{qmark}(u, i+1) = \mathit{NULL} \\ (\text{qmark}(u, i), \text{qmark}(u, i+1)), & \text{otherwise,} \end{cases}$$

where we set $\text{qmark}(u, 0) = \text{qmark}(u, \deg(u) + 2) = \mathit{NULL}$.

The second short-hand $\text{junct}(u)$ assigns a junction to every node u. If u is a leaf, then the assignment depends on $\text{qmark}(u', i + 1)$, where u' is the parent of u, and u is the ith child of u'. Specifically, $\text{junct}(u) = (\text{qmark}(u), x, \emptyset)$, where x is s if $\text{qmark}(u', i + 1) \in Q$ and NULL otherwise. If u is not a leaf, denote by v and w the first and the last non-NULL entry in the sequence of qmark labels for node u, respectively. Let further p be the path from T's root to u, and denote by $P \subseteq Q \times Q$ the set $\{\text{qpair}(e) : e \in p \text{ and qpair}(e) \text{ has no } \mathit{NULL} \text{ entries}\}$. The junction $\text{junct}(u)$ is now $(v, w, Q \times Q \setminus (P \cup \{(v, w)\}))$.

Lemma 3. *Let T be a state-marked assembly for some strand σ. There is a state-marked assembly T' for σ such that*

(i) all nodes v in T have $\deg(v) \neq 1$, except the root,
(ii) at every node v in T', the sequence $(\text{qmark}(v, i), \dots, \text{qmark}(v, j))$ is an instance of the junction $(\text{qmark}(v, i), \text{qmark}(v, j), J)$ for some J, where i and j are the first and last index for which $\text{qmark}(v, i)$ and $\text{qmark}(v, j)$ are not NULL, *and $i \in \{1, 2\}$, $j \in \{\deg(v), \deg(v) + 1\}$, and*
(iii) on every simple path P in T' from the root to a leaf, for every two edges e, e' on P, if all entries in $\text{qpair}(e)$ and $\text{qpair}(e')$ are from Q (i.e., not NULL*), then it holds that $\text{qpair}(e) \neq \text{qpair}(e')$.*

The above technical Lemma 3 essentially states the existence of a normal form for assemblies. This normal form is key in our proof for the next lemma, which establishes the missing part of Theorem 1.

Lemma 4. *If $\iota \in \mathcal{L}(\mathcal{A})$, then $\iota \in \mathcal{I}$.*

Theorem 1 is now established with help of Lemmas 1, 2 and 4. Since finite languages are regular, it follows from Theorem 1 that $\mathcal{L}(\mathcal{A})$ is regular if the size of $\mathcal{A}$ is finite. Lastly, in the full version we also establish the following.

Theorem 2. *Let $\mathcal{L}$ be a language over the alphabet Σ with $\varepsilon \notin \mathcal{L}$. If $\mathcal{L}$ is regular, then there is a constant size DNA algorithm that decides $+\mathcal{L}\emptyset$.*

References

1. Lakowicz, J.R.: DNA technology. In: Lakowicz, J.R. (ed.) Principles of Fluorescence Spectroscopy, pp. 705–740. Springer, New York (2006)
2. Adleman, L.M.: Molecular computation of solutions to combinatorial problems. Science **266**(5187), 1021–1024 (1994)
3. Angluin, D., Aspnes, J., Eisenstat, D.: A simple population protocol for fast robust approximate majority. Distrib. Comput. **21**(2), 87–102 (2008)
4. Bottoni, P., Labella, A., Manca, V., Mitrana, V.: Superposition based on Watson-Crick-like complementarity. Theory Comput. Syst. **39**(4), 503–524 (2006)
5. Breaker, R.R.: Engineered allosteric ribozymes as biosensor components. Curr. Opin. Biotechnol. **13**(1), 31–39 (2002)
6. Cardelli, L., Csikász-Nagy, A.: The cell cycle switch computes approximate majority. Scientific reports 2, September 2012
7. Kari, L., Păun, G., Rozenberg, G., Salomaa, A., Yu, S.: DNA computing, sticker systems, and universality. Acta Informatica **35**(5), 401–420 (1998)
8. Kobayashi, S., Mitrana, V., Păun, G., Rozenberg, G.: Formal properties of PA-matching. Theoret. Comput. Sci. **262**(1–2), 117–131 (2001)
9. Lakin, M.R., Phillips, A.: Modelling, simulating and verifying turing-powerful strand displacement systems. In: Cardelli, L., Shih, W. (eds.) DNA 17 2011. LNCS, vol. 6937, pp. 130–144. Springer, Heidelberg (2011)
10. Lipton, R.J.: DNA solution of hard computational problems. Science **268**(5210), 542–545 (1995)
11. Manea, F., Martín-Vide, C., Mitrana, V.: Hairpin lengthening: language theoretic and algorithmic results. J. Logic Comput. **25**(4), 987–1009 (2015). doi:10.1093/logcom/exs076. http://logcom.oxfordjournals.org/content/25/4/987.abstract
12. Patitz, M.: An introduction to tile-based self-assembly and a survey of recent results. Nat. Comput. **13**(2), 195–224 (2014)
13. Păun, G., Rozenberg, G., Yokomori, T.: Hairpin languages. Int. J. Found. Comput. Sci. **12**(06), 837–847 (2001)
14. Qian, L., Winfree, E.: Scaling up digital circuit computation with DNA strand displacement cascades. Science **332**(6034), 1196–1201 (2011)
15. Seelig, G., Soloveichik, D., Zhang, D.Y., Winfree, E.: Enzyme-free nucleic acid logic circuits. Science **314**(5805), 1585–1588 (2006)
16. Seeman, N.C.: An overview of structural DNA nanotechnology. Mol. Biotechnol. **37**(3), 246–257 (2007)
17. Sipser, M.: Introduction to the Theory of Computation. International Thomson Publishing, Boston (1996)
18. Winfree, E., Yang, X., Seeman, N.C.: Universal computation via self-assembly of DNA: some theory and experiments. In: DNA Based Computers II. DIMACS, vol. 44, pp. 191–213. American Mathematical Society (1996)
19. Zhang, D.Y., Turberfield, A.J., Yurke, B., Winfree, E.: Engineering entropy-driven reactions and networks catalyzed by DNA. Science **318**(5853), 1121–1125 (2007)
20. Zhang, D.Y., Winfree, E.: Control of DNA strand displacement kinetics using toehold exchange. J. Am. Chem. Soc. **131**(47), 17303–17314 (2009)

Author Index

Printed in the United States
By Bookmasters